Proceedings

AA002456

2008 Workshop on Power Electronics and Intelligent Transportation System

PEITS 2008

Proceedings

2008 Workshop on Power Electronics and Intelligent Transportation System

Guangzhou, China
4–5 August 2008

Edited by
Qi Luo
Weitao Zheng

Co-Sponsored by
2008 International Symposiums on Electronic Commerce and Security (ISECS 2008)
IEEE SMC
IEEE Computer Society
Institute of Electrical and Electronics Engineers, USA
Guangdong University of Business Studies, China
Wuhan University of Technology, China
Wuhan Institute of Technology, China
Intelligent Information Technology Application Research Institute
Engineering Technology Press, Hong Kong

Los Alamitos, California

Washington • Tokyo

Copyright © 2008 by The Institute of Electrical and Electronics Engineers, Inc.

All rights reserved.

Copyright and Reprint Permissions: Abstracting is permitted with credit to the source. Libraries may photocopy beyond the limits of US copyright law, for private use of patrons, those articles in this volume that carry a code at the bottom of the first page, provided that the per-copy fee indicated in the code is paid through the Copyright Clearance Center, 222 Rosewood Drive, Danvers, MA 01923.

Other copying, reprint, or republication requests should be addressed to: IEEE Copyrights Manager, IEEE Service Center, 445 Hoes Lane, P.O. Box 133, Piscataway, NJ 08855-1331.

The papers in this book comprise the proceedings of the meeting mentioned on the cover and title page. They reflect the authors' opinions and, in the interests of timely dissemination, are published as presented and without change. Their inclusion in this publication does not necessarily constitute endorsement by the editors, the IEEE Computer Society, or the Institute of Electrical and Electronics Engineers, Inc.

IEEE Computer Society Order Number P3342
BMS Part NumberCFP0875E-PRT
ISBN 978-0-7695-3342-1
Library of Congress Number 2008929487

Additional copies may be ordered from:

IEEE Computer Society	IEEE Service Center	IEEE Computer Society
Customer Service Center	445 Hoes Lane	Asia/Pacific Office
10662 Los Vaqueros Circle	P.O. Box 1331	Watanabe Bldg., 1-4-2
P.O. Box 3014	Piscataway, NJ 08855-1331	Minami-Aoyama
Los Alamitos, CA 90720-1314	Tel: + 1 732 981 0060	Minato-ku, Tokyo 107-0062
Tel: + 1 800 272 6657	Fax: + 1 732 981 9667	JAPAN
Fax: + 1 714 821 4641	http://shop.ieee.org/store/	Tel: + 81 3 3408 3118
http://computer.org/cspress	customer-service@ieee.org	Fax: + 81 3 3408 3553
csbooks@computer.org		tokyo.ofc@computer.org

Individual paper REPRINTS may be ordered at: <reprints@computer.org>

Editorial production by Lisa O'Conner
Cover art production by Joe Daigle/Studio Productions
Printed in the United States of America by The Printing House

   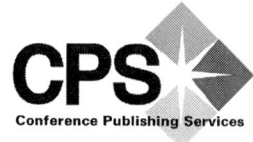

IEEE Computer Society
Conference Publishing Services (CPS)
http://www.computer.org/cps

2008 Workshop on Power Electronics and Intelligent Transportation System

PEITS 2008

Table of Contents

Message from the Workshop Chair..xiv
PEITS 2008 Organizing Committee...xv
PEITS 2008 Committee Members...xvi

Session 1: Wireless Communication

A Cross-Layer Fast Handover Mechanism for IEEE 802.16e Networks
with HMIPv6 Mobility 3
    Yilin Zhang, Fuqiang Liu, and Xinhong Wang

The Design of Manufacturing Execution System Based on RFID 8
    Jiwei Hua, Hexu Sun, Tao Liang, and Zhaoming Lei

Dynamic Channel Assignment in Wireless LANs 12
    Bo Wang, William Wu, and Yongqiang Liu

Routing Optimized Video Transmission over Mobile Ad Hoc Networks 18
    Yang He, Yuanxin Ouyang, Chao Li, and Zhang Xiong

High-Frequency Surface-Wave Radar Real Time Frequency Selecting Based
on Frequency Randomly Hopping Signal 23
    Xu Liu and Changjun Yu

A New Highway Incident Detection Method Based on Wireless Positioning
Technology 27
    Hang Wen

Session 2: Information Security

A Dynamic ID-Based User Authentication and Key Agreement Scheme
for Multi-Server Environment Using Bilinear Pairings 33
  Jianyan Geng and Lijiang Zhang

A Novel Deterministic Packet Marking Scheme for IP Traceback 38
  Zhaoyang Qu and Chunfeng Huang

A Fast Method of Local Network Topology 42
  Yubin Yao, Dan Wang, Chuanqi Wang, and Yong Li

Invulnerability Assessment for Mobile Ad Hoc Networks 48
  Cong Jin and Shu-Wei Jin

Discovery of Unknown Visual Information Based on Target Tracking 52
  Cong Jin and Shu-Wei Jin

A Novel Email Virus Propagation Model 56
  Cong Jin, Jun Liu, and Qing-Hua Deng

Scheme of Electronic Seal Based on Public Key Infrastructure 61
  Jiuhua Zhang

Session 3: Control System

CAN-Based Integrated Control Strategy for the Drivetrain of Commercial
Vehicles Equipped with Automated Manual Transmission 67
  Jingxing Tan, Xiaofeng Yin, and Liang Yin

Delimitation of Traffic Coordinated Control Sub-Area for the Road Network
Containing Freeway 72
  Nan Jiang, Xiaoguang Yang, and Shoupeng Tang

A Multi-Path Access Control Scheme in Core Packet Network 77
  Xuekang Sun and Wanyi Gu

Performance Optimization Based on Dynamic Bandwidth Allocation
for Networked Motion Control Systems 83
  Weiquan Zhao and Di Li

A Digital Optimal Control Method for DC-DC Converters Based on Simple
Model 88
  Zhaoxia Leng, Jian Liu, Qingfeng Liu, and Huamin Wang

Design and Practice of an Elevator Control System Based on PLC 94
  Xiaoling Yang, Qunxiong Zhu, and Hong Xu

Corrosion Diagnosis of Conductors and Down Lead Lines for Grounding
Grids Maintenance 100
  Yunfeng Ni, Jian Liu, Sen Wang, and Zhizhong Li

First-Order Optimality Conditions for Convex Semi-Infinite Inequality
Constrained Programming with Noncompact Sets 105
  Meixia Li and Jinchuan Zhou

Study on Synthesis Control of Power Quality for Electrified Railway 110

Xiangzheng Xu and Baichao Chen

Delay Synchronizing Rectification Control Strategy of Bi-Directional
Converter in a Novel Stand-Alone PV System 113

Xinhua Zhang, Zhenyue Huang, and Zhiling Liao

Control Strategy of Three-Phase AC/DC Voltage-Source Converters Based
on Storage Function 117

Jiuhe Wang, Peirong Xia, and Jinlong Zhang

Adaptive Optimal Iterative Learning Control for Local Ramp Metering 122

Shangtai Jin and Zhongsheng Hou

Simulation Study on a Single Neuron PID Control System of DC/DC
Converters 127

Xinggui Wang, Bing Xu, and Lei Ding

H Infinity Robust Controller for Electric Bicycles 131

HaoBin Zhou, Bo Long, and BingGang Cao

Vector Control System of Induction Motor Based on Fuzzy Control Method 136

HaoBin Zhou, Bo Long, and BingGang Cao

Session 4: Multimedia information Processing

Improving Image Encryption Using Multi-Chaotic Map 143

Yikui Zhai, Shuyong Lin, and Qiong Zhang

Multi-View Video Coding Using Color Correction 149

Lixian Shangguan and Jifeng Sun

Improved Voice Activity Detection Based on Iterative Spectral Subtraction
and Double Thresholds for CVR 153

Xiangbin Li, Guo Li, and Xueren Li

A Segmentation and Extraction Method for Crops of Image in Complex
Background 157

Hongjuan Kou, Shuguang Zhang, and Dening Zhang

Research of Online Signature Recognition Based on Energy Feature 161

Jinxu Guo, Jianbin Zheng, and Bo Lu

Session 5: Signal Processing Application in Communications

Optimum of Intersection Signal Timing Based on IGA 167

Rong-Hua Du

Spectrum Sensing Based on Cyclostationarity 171

Shiyu Xu, Zhijin Zhao, and Junna Shang

Traffic Signal Optimization with Vehicles Queue and the Number
of Pedestrians Non-Complying at Single Intersection 175

Zhi-Na Zhou, Zhong-Ke Shi, and Ying-Feng Li

Communication Reduction for Continuous Extreme Values Monitoring Over
Distributed Data Streams 181

Xiaoming Peng, Yanxiang He, and Li Tian

A New Approach to MANet Routing Based on Erasure Codes 188

Zheng Chen, Xiaojing Wang, Sheng Cao, and Dan Tang

Session 6: Neural Networks and Computational Intelligence

Analyzing QAR Data Using K-PCA 195

Xiao-Rong Feng and Xing-Jie Feng

An Improvement Route Generation Algorithm for Bus Network Design 199

Yikui Mo, Jun Deng, and Jingyuan Wang

Forecasting Railway Network Data Traffic: A Model and a Neural Network
Solution Algorithm 203

Teng Jing

Prediction of Logistics Amount Based on Neural Networks 209

Lianyu Wei and Zhaowei Liu

A Hybrid Neural Network-Based IE and IMM Architecture for Target
Tracking 214

Jian Rong, Xiu Wang, Xiaochum Zhong, and Haitao Zhang

A Forecasting Model Based Support Vector Machine and Particle Swarm
Optimization 218

Qi Wu, Hong-Sen Yan, and Hong-Bing Yang

A New Web Service Discovery Method Based on Semantic 223

Wuling Ren and Zhujun Xu

Research on Multi-Attribute Decision-Making Method Based on AHP
and Outranking Relation 227

Liguo Fan and Feng Zuo

Nearest Neighbor-Clustering Algorithm Based on Hierarchical Optimization
Strategy 233

Jie Wang and Guoqiang Jiang

Model and Algorithm for Initial Route Planning 237

Shiyan Xu and Yuyao He

Integration of an Improved Particle Swarm Algorithm and Fuzzy Neural
Network for Shanghai Stock Market Prediction 242

Fu-Yuan Huang

The Collaborative Filtering Recommendation Based on Similar-Priority
and Fuzzy Clustering 248

SongJie Gong

Support Vector Machine Based Approach for State Estimation of Iraqi Super
Grid Network 252

Afaneen A. Abod, Abdullah H. Abdullah, and Mohammed K. Abd

Gradual Approaching Method for Distribution Network Dynamic
Reconfiguration 257

HuPing Yang, YunYan Peng, and Ning Xiong

Application of Wavelet Packet Analysis and Improved LSSVM on Rotating
Machinery Fault Diagnosis 261

Lingling Zhao and Kuihe Yang

An Urban Map Matching Algorithm Using Rough Sensor Data 266

Shu Liu, Zongying Shi, Mingguo Zhao, Wenli Xu, and Kai Zhang

A Neighborhood-Based Trajectory Clustering Algorithm 272

Yunxin Tao and Dechang Pi

A Personal Query Framework Based on Personal and Role Ontology 276

Liying Fang, Jianzhuo Yan, Pu Wang, and Bin Shi

The Study of the Sentence Format in Nature Language Understanding 280

Jianming Miao and Quan Zhang

Queuing Analysis for Reconfigurable Computing 284

*Kourosh Hassanli, Ali Khayatzadeh Mahani, Hadishahriar Shahhoseini,
and Ebrahim Teimoury*

Session 7: Theory, Design and Implementation of Circuits & Systems

Extension and Application of Space Syntax — A Case Study of Urban Traffic
Network Optimizing in Beijing 291

Xinqi Zheng, Lu Zhao, Meichen Fu, and Shuqing Wang

The Parameter of Automatic Voltage Regulator's Effect on Steady State
Stability Limit of Turbine Generator 296

Yong-Gang Li, Jun-Jie Fu, Yan-Jun Zhao, and Qiang Liu

The Integer Ambiguity Transformation Study in Mixed Navigation Based
on Real-Time Kinematic 301

Yun Chen, Jijing Wang, and Renying Wang

The Research of Car Rear-End Warning Model Based on MAS and Behavior 305

Jun Liang, Xian-Yi Cheng, and Xiao-Bo Chen

A Voltage Fluctuation and Flicker Monitoring System Based on Wavelet
Transform 310

Zhenmei Li, Jin Shen, Peiyu Wei, and Tianze Li

Robust Points Tracking Method Using Euclidean Reconstruction
for Augmented Reality 315

Peng Chen, Rui Zhang, and Zhang Gao

Prediction and Analyze of PCB Common-Mode Radiation Based
on Current-Driven Mode 320

Zhiliang Zhang and Fengxun Gong

The Experimental Study About the Influence of Methanol/Diesel Fuel Mixture
on Diesel Engine Performance 324
    Chu Weitao

The Development of Multifunctional Intellectual Electrical Parameters Test
Instrument 328
    Zhifang Yang

A Linear Design of Temperature Detecting Circuits with Pt Resistance 332
    Zhifang Yang

Session 8: Power System

The Evaluation of Success Degree in Electric Power Engineering Project
Based on Principal Component Analysis and Fuzzy Neural Network 339
    Baoqian Duan, Yun Tang, Li Tian, and Qingchao Liu

A Canceling Noise Research for Underground Mine Powerline Carrier
Communication Based on Adaptive Theory 345
    Shaoliang Wei, Jialin Cao, Yimin Chen, Fengyu Cheng, Deming Nie, and Hengwen Li

Research on Circulation of Parallel Three-Phase Converters in MW Wind
Power System 349
    Chunxue Wen, Jianlin Li, Xiaoguang Zhu, and Honghua Xu

SOA Based Electric Power Real-Time Data Warehouse 355
    Cuiru Wang and Shuangxi Liu

Unified Control of Photovoltaic Grid-Connection and Power Quality
Managements 360
    Xiaogao Chen, Qing Fu, Shijie Yu, and Longhua Zhou

Research on Testing Technology of High-Power Photovoltaic Arrays 366
    Yinglei He, Qing Fu, Xiangfeng Li, and Xiaogao Chen

Direct Power Control Based on Three-Phase Shunt Hybrid Power Filter Case
Study 370
    Wenjin Dai and Ming Chen

A Novel Three-Phase Active Power Filter Based on Instantaneous Reactive
Power Theory 375
    Wenjin Dai and Yu Wang

Study and Design of Three-Phase Voltage-Source PWM Rectifier in Movable
Power Station 380
    Xinggui Wang and Xiaoying Li

Modeling and Simulation of Monitor-Control Network in Ship Power Station 384
    Dandan Chen, Li Xia, and Haifeng Wang

A New Harmonic Analysis Method for AC/DC/AC Converters 389
    Zhaorui Lv, Li Xia, and Zhengguo Wu

Simulation and Optimization of the Power Station Coal-Fired Logistics
System Based on Witness Simulation Software 394
Yabin Li and Tiejun Ci

Study on Voltage Stability of Distribution Networks 399
Xiaomeng Wu, Jian Liu, and Suli Yan

A New Modeling Method for the Switching Power Converter 404
Weiping Wang, Yujie Shi, and Dianbing Yang

Research on Wind Power Systems 408
Luo Qi

Simulation Study of Fuzzy Based Unified Power Flow Controller on Power
Flow Controlling 411
Lütfü Saribulut and Mehmet Tümay

Session 9: Intelligent Transportation System

Framework on Hierarchical Optimization of Traffic Count Location for City
Traffic System 419
Heng Wang, KePing Li, Jian Sun, and Ying Liu

Game Theory Choice Model for Public Transportation Priority — Take
Nanjing for Example 423
Chen Wang and Jun Chen

Research on the Establishment of Xi'an Advanced Public Transport System 430
Zhongxiang Feng, Jing Liu, Xianyuan Dong, and Kunpeng Gao

A Forecast Model of Urban Passenger Flow Containing New Railway Project 435
Hui Xie, Kefei Yan, and Ya Wen

Distributing Stations in a Given Urban Rail Transit Line 440
Jin-Li Wei and Meng-Meng Zhang

Intelligence Toll Management System of Highway Traffic 444
Bo Yan and Yehua Huang

Congestion Pricing and Sustainable Development of Urban Transportation
System 449
Jianhu Zheng

Solving Traveling Salesman Problems by Genetic Differential Evolution
with Local Search 454
Jian Li, Peng Chen, and Zhiming Liu

A Vehicle Scheduling Model and Efficient Algorithm for Single Bus Line 458
Jingqi Xu, Haode Liu, and Jing Teng

Stress Test of Road Network Capacity with Cube Software 462
Li Li, Yu Quan, Bian Yang, and Yi Ping

Traffic Congestion Information Promulgating System Based on GIS-T 467
Bing Zhang, Wei Deng, and Ling Mao

Optimization for Urban Traffic Assignment by Congestion Toll Levied
on Reginal Expressway 472
    Meng-Meng Zhang and Jinli Wei

Active Safety in Autonomous Land Vehicle 476
    Daxue Liu, Xiangjing An, Zhenping Sun, and Hangen He

Context Sensitive Design — Route Selection Design of Changsha-Jishou
Freeway 481
    Yingxue Zhang, Qisen Zhang, Chunhua Han, and Zhaohui Liu

Modeling of Urban Traffic System Based on Dynamic Stochastic Fluid Petri
Net 485
    Jingyu Li and Qiqiang Li

Study on Driving Behavior and Traffic Conflict at Highway Intersection 492
    Xing Ge, Jian Lu, Qiao-Jun Xiang, and Peng-Ying Wang

Study on Information Sharing Platform of Distributed Transportation
Monitoring Systems Based on Interoperability and Internet 496
    Jian Lee, Yuan-Hua Jia, and Gu-Chang Ao

Quantitative Measure for Transportation Network Efficiency 500
    Jin Qin, Feng Shi, Li-Xin Miao, and Qun Chen

Combined Land-Use and Transportation Demand Modeling Based
on Equitableness 505
    Ming Yang, Xiucheng Guo, and Lan Wu

Safety Evaluation for First Class Highway Design Based on Unascertained
Measure Model 511
    Sheng-Wen Tu, Xiu-Cheng Guo, and Jian-Ming He

Some Remarks on ON/OFF Network Traffic 515
    Chu Chen, Yong Xu, and Ling Zhang

A Target Classification Algorithm Based on Transportation Sensing Network 520
    Xun-Xue Cui, Guo-Xin Qiu, Jian-Qin Zeng, Li-Jun Xing, and Qi Liu

The Optimal Route Guidance Modeling for ITS Influenced by Dynamic
Factors 525
    Wensheng Zhang and Mingsheng Wang

Research on Intelligent Transportation System Technologies and Applications 529
    Luo Qi

Session 10: Other Topics

Study on the Turnover Intention of Knowledge Employees Influenced
by Person-Organization Fit (POF) 535
    Pei-Lan Guan and Xiao-Jun Wu

An Approach for Multi-Dimensional Separation Concerns at Architecture
Level 541
    Lin-Lin Zhang, Shi Ying, You-Cong Ni, Jing Wen, and Kai Zhao

GADAM: An Authorization Model Based on Attribute Delegation in Grid 546
  Rongbin Wang, Shuyu Chen, and Jing Li

Research on the Application of Probationary System in Tacit Knowledge
Sharing 551
  Zhihong Li and Jun Li

Research on Factors Influencing Knowledge Transfer and Managerial
Mechanisms in the Community of Practice 556
  Zhihong Li, Jun Li, and Minxia Li

BA Extended Model Based on the Competition Factors 561
  Shaohua Tao, Chaofeng Guo, Jingju Gao, and Song Hang

A Reflective NetGAP Logic Framework Design 565
  Songsen Yu, Yiju Zhan, Qing-Lin Cai, and Yong-Hua Wang

Research on the Real Option Decision Model about Venture Capital Exit 569
  Qing Yang, Ane Pan, and Jue Li

Study on Optimal Model and Algorithm of Sorting Sequence of Stage Plan
on Marshalling Station 574
  Lei Li and BingMou Cui

A New Class of Highly Fault Tolerant Erasure Code for the Disk Array 578
  Dan Tang, Xiaojing Wang, Sheng Cao, and Zheng Chen

A Hybrid Active Compensation Method for Current Balance Based on Y,d11
Connection Traction Transformer 582
  Wang Guo, Ren Enen, and Tian Mingxing

Research on Virtual Disassembly Simulation Based on Constraint Matrix 587
  Shi-Ting Li, Bo Zhu, and Qi Cai

The Study on E-Commerce Standardization in China 592
  Shaohua He and Fan Yang

The Application of the Analytical Hierarchy Process in Performance
Evaluation System in Commercial Bank's IT Department 601
  Jiang-Tao Wang and Hong Zhou

Transform Government Functions Effectively, and Innovate New Rural
Industrial Development Model — Based on the Theory and Empirical Study
of Deqing 608
  Li Bing Ning

A Dietary Investigation and Analysis on Students Majored in P.E. and Sports ... 613
  Yanxia Peng

Short Term Traffic Flow Prediction Based on Online Learning SVR 616
  Dehuai Zeng, Jianmin Xu, Jianwei Gu, Liyan Liu, and Gang Xu

Short Term Traffic Flow Prediction Using Hybrid ARIMA and ANN Models 621
  Dehuai Zeng, Jianmin Xu, Jianwei Gu, Liyan Liu, and Gang Xu

Author Index 627

Message from the PEITS 2008 Workshop Chair

PEITS 2008 that is in conjunction with International Symposium on Electronic Commerce and Security (ISECS 2008) will be held in Guangzhou, China, 4-5 August, 2008. We cordially invite you to attend the 2008 Workshop on Power Electronics and Intelligent Transportation System (PEITS 2008), which will be held in August 2008 in Guangzhou, China. Just like the name of the Workshop, the theme for this workshop is Advancing Power Electronics and Intelligent Transportation System.

Welcome to PEITS 2008. Welcome to Guangzhou, China, 4-5 August 2008. PEITS 2008 is a workshop of International Symposium on Electronic Commerce and Security (ISECS 2008) which is co-sponsored by Engineering Technology Press, Hong Kong, Guangdong University of Business Studies, China, Wuhan University of Technology, China. It is also co-sponsored IEEE SMC, IEEE Computer Society, and Intelligent Information Technology Application Research Institute. Much work went into preparing a program of high quality. We received about 375 submissions. Every paper was reviewed by 2-3 program committee members, about 124 were selected as regular papers, representing a 33 % acceptance rate for regular papers.

The purpose of PEITS 2008 is to bring together researchers and practitioners from academia, industry, and government to exchange their research ideas and results and to discuss the state of the art in the areas of the workshop. In addition, the participants of the workshop will have a chance to hear from renowned keynote speakers IEEE & IET Fellow Prof. Chin-Chen Chang from National Chung Hsing University, Taiwan, Associate Editor, *IEEE Transactions on Systems, Man & Cybernetics,* Part C, Prof. Ben K.M. Sim from Chinese University of Hong Kong, Hong Kong, and IEEE Fellow Prof. Ben M. Chen at the National University of Singapore, Singapore.

We thank Mr. Thomas Baldwin, the mangers of IEEE Computer Society Press, USA, who enthusiastically support our workshop. Thanks also go to Miss Lisa O'Conner for her wonderful editorial service to this proceeding.

We would like to thank the workshop chair, organization staff, and the members of the program committees for their hard work. Special thanks go to Prof. Weitao Zheng from Wuhan University of Technology, China.

We hope that PEITS 2008 will be successful and enjoyable to all participants. We look forward to seeing all of you next year at the PEITS 2009.

Weitao Zheng, *Wuhan University of Technology, China*

PEITS 2008
Organizing Committee

Honorary Chairs

Charles Rubenstein, *IEEE Technology Management Council*
Chin-Chen Chang, *National Chung Hsing University*
Jiaqing Wu, *Guangdong University of Business Studies, China*

Program Committee Chairs

Jun Ni, *University of Iowa, USA*
Qihai Zhou, *Southwestern University of Finance and Economics, China*
Yongjun Chen, *Guangdong University of Business Studies, China*

Workshop Chair

Weitao Zheng, *Wuhan University of Technology, China*

Finance Chairs

Qi Luo, *Wuhan Institute of Technology, China*
Fei Yu, *Peoples' Friendship University of Russia, Russia*

IEEE SMC Technical Committees

Feng Hsu, *NASA at Johnson Space Center, Texas form Technical Committee on Systems Safety and Security*
A G Hessami, *Atkinsglobal Advanced Technology Group, UK form Technical Committee on Systems Safety and Security*
Sun-Bae Cho, *Yonsei University, Korea*
Haibin Zhu, *Nipissing University, Canada*
Ben K.M. Sim, *Hong Kong Baptist University, Hongkong*

Publication Chair

Qi Luo, *Wuhan Institute of Technology, China*
Fei Yu, *Peoples' Friendship University of Russia, Russia*

Local Organizing Chairs

Feng Yang, *Guangdong University of Business Studies, China*

PEITS 2008
Committee Members

Prof. Uskov V., *Bradley University, USA*
Prof. Masoud M., *University of Canberra, Australia*
Prof. David A., *Monash University, Australia*
Prof. Aijun A., *York University, Canada*
Prof. Vo Ngoc A., *University of Melbourne, Australia*
Prof. Hiroki A., *Hokkaido University, Japan*
Prof. Michael B., *University of New South Wales, Australia*
Prof. Hideo B., *Kyushu University, Japan*
Prof. Michael B., *University of Konstanz, Germany*
Prof. Hendrik B., *Katholieke Universities Leuven, Belgium*
Prof. Francesco B., *ISTI-C.N.R., Italy*
Prof. Ulf B, *Technical University Berlin, Germany*
Prof. Rui C., *LIACC/FEUP University of Porto, Portugal*
Prof. Longbing C., *University of Technology, Sydney, Australia*
Prof. Tru Hoang C., *Ho Chi Minh City University of Technology, Vietnam*
Prof. Sanjay C., *University of Sydney, Australia*
Prof. Arbee C., *National Chengchi University, Taiwan*
Prof. Ming-Syan C., *National Taiwan University, Taiwan*
Prof. Chang- Tsun L., *University of Warwick, UK*
Prof. Mike B., *University of Adelaide, Australia*
Prof. Lang W., *University of Adelaide, Australia*
Prof. Angela W., *University of Adelaide, Australia*
Prof. Andrew K., *University of Oxford, UK*
Prof. Jia Z., *Northern Illinois University, USA*
Prof. Martin W., *University of Liverpool, UK*
Prof. Huang H., *Hunan Agricultural University, China*
Prof. Guangxue Y., *Jiaxing University, China*
Prof. Quanyuan W., *National University of Defense Technology, China*
Prof. Miaoliang Z., *Zhejiang University, China*
Prof. Howell Y., *MIT Lincoln Lab, USA*
Prof. Hiroshi I., *University of Tokyo, Japan*
Prof. Kazuo I., *Kyoto University of Tokyo*
Prof. Rocco S., *Columbia University, USA*
Prof. Baowen X., *Southeast University, China*
Prof. Qihai Z., *Southwestern University Of Finance and Economics, China*
Prof. Guiping L., *Hunan Agricultural University, China*
Prof. Renfa L., *Hunan University, China*
Prof. Honghua T., *Wuhan Institute of Technology, China*
Prof. Xiaoling.W., *Wuhan Institute of Technology, China*
Prof. Prof. Yongjun C., *Guangdong University of Business Studies, China*
Prof. Prof. Guangquan Z., *Suzhou University, China*
Prof. Chen X., *Hunan University, China*
Prof. Jun N., *University of Iowa, USA*
Prof. Jixin W., *Huazhong Normal University, China*

Prof. Yi Z., *Huazhong Normal University, China*
Prof.Qingtang, L., *Huazhong Normal University, China*
Prof. Zhixiang H., *Changsha University of Science and Technology, China*
Prof. Yi. L., *Nanjing Normal University, China*
Prof. Feiyue, W., *City University of Hong Kong, Hong Kong*
Prof. Cheng, J., *George Mason University, USA*
Prof. Zhihai,W., *Beijing Jiaotong University, China*
Prof. Ronghuai., *Beijing Normal University, China*
Prof. Anido L., *University of Vigo, Spain*
Prof. Albert, R., *University of Southern California, USA*
Prof. Edmond, P., *Nanyang Technological University, Singapore*
Prof. Rhalibi, A., *Liverpool John Moores University, UK*
Prof. Huawen W., *Wuhan University, China*
Prof. Xiaogong Y., *Wuhan University, China*
Prof. Shiying G., *Huazhong University of Science and Technology, China*
Prof. Tjeerd, P., *University of Twente, Faculty of Behavioral Sciences, Netherlands*

Session 1

Wireless Communication

A Cross-Layer Fast Handover Mechanism for IEEE 802.16e Networks with HMIPv6 Mobility

Yilin Zhang, Fuqiang Liu, Xinhong Wang
Broadband Wireless Communication and Multimedia Laboratory
Tongji University, Shanghai, 201804, China
zhangyl021@gmail.com, liufuqiang@mail.tongji.edu.cn

Abstract

IEEE 802.16e standard is considered to be one of the most promising solutions for next generation broadband wireless networks. Still, handover latency in IEEE 802.16e networks is an important issue that affects real-time applications. In this paper, a cross-layer fast hierarchical handover mechanism (CLFH) for IEEE 802.16e networks is introduced. The handover procedure of IEEE 802.16e is integrated with fast and hierarchical MIPv6 handover to achieve performance improvements. Moreover, the 802.16e network re-entry procedures are optimized to reduce the link layer handover delay. The proposed mechanism is discussed in detail and compared with a mechanism proposed in IETF draft. The simulation results illustrate that the proposed mechanism achieves better performance in the handover latency and the service disruption time.

1. Introduction

A variety of wireless access technologies have been developed for different requirements of mobile users. As the use of IP-based access network increasing, next generation network environment is naturally moving towards all-IP-based networks. The mobile WiMAX protocol, based on IEEE 802.16e standard, is considered to be one of the most convincible technologies for next generation broadband wireless networks due to its IP-based characteristics. The current standard specifies the physical layer and MAC layer protocol between base station (BS) and mobile stations (MS) for mobile wireless environment [1].

One of the research challenges for next generation all-IP-based wireless networks is the design of seamless mobility management techniques that take advantage of IP-based technologies to achieve global roaming among various access technologies [2]. In the current network, a typical handover operation takes several seconds, which is unacceptable for time-sensitive applications like VOIP and video streaming. The key to achieve seamless mobility management depends on finding an optimized handover mechanism that satisfies the limited handover delay tolerable for real-time application.

In 802.16e, handover procedure is an important part discussed in MAC layer. However, from the IP-based service point of view, the whole handover procedure shall involve not only the MAC layer (L2) but also IP layer (L3). The total handover duration becomes unacceptable for time-sensitive applications. Therefore, the use of link layer information to reduce the handover latency has gained attention recently. The Fast Mobile IPv6 (FMIPv6) protocol enables a MN to quickly detect at the L3 that it has moved to a new subnet by receiving L2 triggers, and the MN can gather anticipative information about the new access router (AR) when it is still connected to the previous subnet [3]. An IETF draft that implement the FMIPv6 Handover over IEEE 802.16e networks is presented in [4]. The major work of this mechanism is by setting up a tunnel between the IP layer of serving BS (sBS) and target BS (tBS) during handover so that data packets will be forwarded to the tBS even before the handover procedure is completed. The use of link layer information significantly reduces the service disruption time during handover operation. However, each time a mobile node (MN) moves between IP subnets, it needs to execute MIP registration process with home agent (HA) and corresponding node (CN) in L3 handover. WiMAX is a wireless metropolitan area network technology, which means the BS has a large coverage area, so the frequent MIP registration processes will increase the network load and affect the total handover latency.

Hierarchical Mobile IPv6 (HMIPv6) protocol is proposed by employing a hierarchical network structure to reduce handover latency [5]. Fast Handover for Hierarchical MIPv6 (F-HMIPv6) protocol takes the advantage of FMIPv6 and HMIPv6, which is developed on the foundation of FMIPv6 and introduces the Mobility Anchor Point (MAP) entity in HMIPv6 to provide a better solution for micro mobility [6]. When MN moves in the same MAP domain, it does not need to register to HA so that it can reduce signal payload and handover latency. But F-HMIPv6 does not consider macro mobility case. Fast Macro Mobility Handovers in HMIPv6, a draft in [7], provides a macro mobility solution, but it still suffers long registration delay for MN to bind with both the new MAP and HA. In this paper, we propose a new cross-layer fast handover mechanism for IEEE 802.16e networks based on HMIPv6, which is donated as CLFH. CLFH mechanism provides solution for both micro (intra-MAP) and macro (inter-MAP) mobility cases. The MIP registration procedure is executed in advance so as to speed up the total handover process, and the IEEE 802.16e network re-entry procedures are optimized to reduce the L2 handover delay.

The rest of this article is organized as follows. The network model and the proposed mechanism are described in Section 2. To illustrate the performance achievement, a mathematical analysis is conducted and simulation results are examined in Section 3. Finally, Section 4 concludes this paper.

2. Proposed mechanism

In this section, the network model of CLFH mechanism is presented and the handover procedure of CLFH mechanism is introduced in detail.

2.1. Network model

Figure 1 illustrates the network model of the proposed CLFH mechanism which is based on HMIPv6 system architecture.

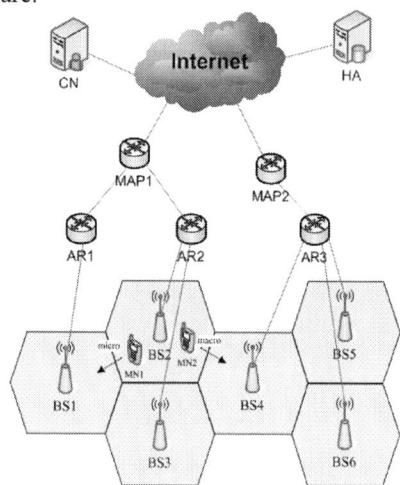

Figure 1 Network model of CLFH handover mechanism

The CLFH mechanism considers both intra-MAP and inter-MAP handover cases. In intra-MAP handover case, a MN moves between subnets in the same MAP domain, e.g., the handover of MN1 from BS2 to BS1 in Figure 1. In inter-MAP handover case, a MN moves between subnets of different MAP domains, e.g., the handover of MN2 from BS2 to BS4 in Figure 1.

Identical with HMIPv6 scheme, a MN has two CoA, on-link CoA (LCoA) and regional CoA (RCoA). When CN communicates with MN, it sends data packets to the RCoA of MN, and MAP then forwards the packets to the LCoA of MN. In intra-MAP handover case, MN changes its LCoA within the same MAP domain, new LCoA binding with MAP is required. And in inter-MAP handover case, MN needs to acquire a new RCoA as well as a new LCoA and perform registration process to both new MAP (nMAP) and HA.

2.2. CLFH handover mechanism for IEEE 802.16e networks

The detail message flow of the proposed mechanism in intra-MAP and inter-MAP handover cases are shown in Figure 2 and Figure 3 respectively. The same functionality of the fast handover procedure is kept, but the anchor point is moved from the pervious AR (pAR) to the MAP. For efficient handover, four L2 triggers in [4] are used in our mechanism: New Link Detected (NLD), Link Handover Impend (LHI), Link Switch (LSW), and Link Up (LUP). The handover procedure is divided into three stages: network topology acquisition, handover preparation and handover execution.

In network topology acquisition stage, it is the same in intra-MAP and inter-MAP handover cases. At first, as defined in 802.16e, the sBS broadcasts MOB_NBR-ADV message to the MS periodically. When MN moves to the boundary location of sBS, it starts scanning with the neighboring BSs and selects candidate BSs for handover based on the channel information. Once a new BS is detected through scanning, MN triggers a NLD to report that a new link is detected. After the channel scanning, the MS can understand the next AR is in the same or different MAP domains. In L3, the MN tries to get information about the new AR by exchange of Router Solicitation for Proxy (RtSolPr) and Proxy Router Advertisement (PrRtAdv) messages with pAR.

When a handover is needed, the MS sends sBS the MOB_MSHO-REQ message indicating one or more tBSs. In response, sBS sends the MOB_BSHO-RSP message to select one tBS. After that, L2 of the MN triggers a LHI to L3 to report that a L2 handover decision has been made and its execution is imminent. Then in intra-MAP handover case, as shown in Figure 2, the MN configures a new LCoA using the network prefix of the tBS and initiates handover by send Fast Local BU (FLBU) message to the MAP. After that, Handover Initiation (HI) and Handover Acknowledgement (HACK) messages occur between the MAP and new AR (nAR) to check the validity of new LCoA by DAD procedure and to establish a temporary tunnel. Once the tunnel is set up, MAP sends Fast Local Binding Acknowledgement (FLBACK) message to the MN, which means the new LCoA has been bound with RCoA, and the packets destined for the MN begin to forward to the nAR through tunnel. In inter-MAP handover case, as shown in Figure 3, the MN configures both new LCoA and new RCoA. After forming these addresses, the MN sends a FBU message to the pMAP to initiate handover. Then pMAP exchanges HI and HACK with nAR and nMAP to check the validity of new LCoA and new RCoA by DAD procedures and to establish a temporary tunnel between the pMAP and the nAR. After that, the pMAP will return a FBACK to the MN and forwards packets destined for the MN to nAR through tunnel. When the MN receives the FBACK message, the LSW trigger is generated to signal L2 that L3 handover preparation is completed.

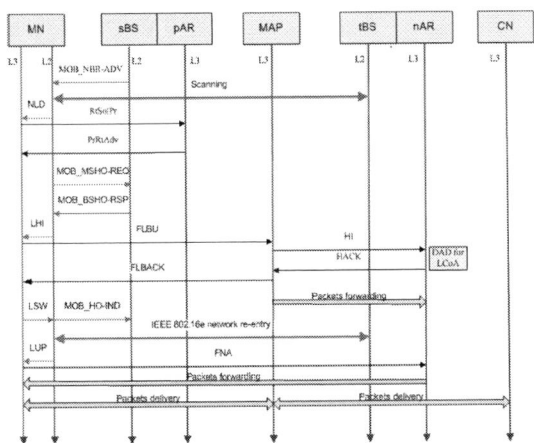

Figure 2 CLFH mechanism for 802.16e (intra-MAP)

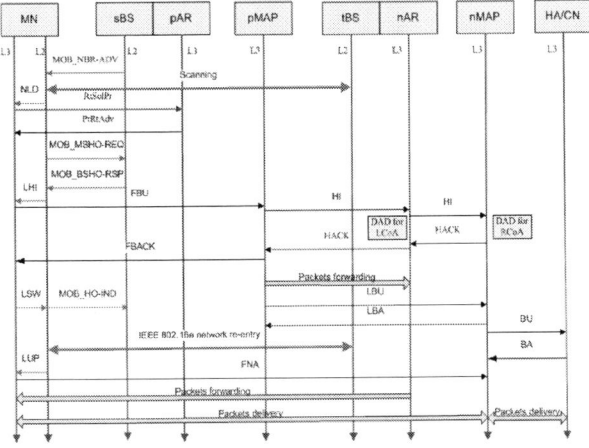

Figure 3 CLFH mechanism for 802.16e (inter-MAP)

Then handover execution stage starts. L2 of the MN sends the MOB_HO-IND message immediately, switches its link to the tBS, and starts network re-entry procedures. The 802.16e network re-entry procedures are optimized in CLFH mechanism, and the detail is discussed in Section 3. As soon as completing the network re-entry, the LUP trigger makes L3 of the MN send Fast Neighbor Advertisement (FNA) message to nAR. Then nAR delivers the buffered packets to the MN and no more registration process is needed in intra-MAP handover case,

as illustrated in Figure 2. However, in inter-MAP handover case, the MN must execute MIP registration process. In our mechanism, registration process is designed to execute simultaneous with the network re-entry procedures to reduce the total handover latency. During the network re-entry procedures, the pMAP exchanges LBU and LBA messages with nMAP to bind the MN's new LCoA with its new RCoA, and the nMAP registers the MN's new RCoA with HA and CN through BU and BA messages, as illustrated in Figure 3. When network re-entry procedures and MIP registration process are all finished, the whole handover operation is completed and the MN can normally communicate with CN.

3. Performance evaluation

In this section, we evaluate and compare the handover performance of the proposed mechanism with the draft in [4] by numerical analysis and simulation work.

3.1. Performance analysis

In this paper, we consider the handover latency as the time interval from the moment when the wireless link is going down to the moment when the MN can normally send and receive packets to/from CN. In our analysis, the wireless link is going down when the LHI trigger is sent. The MN can normally communicate with CN after the completion of whole handover process. We focus here on network layer handover. Therefore, we do not consider link layer detection delay. In order to analyze the performance of the proposed mechanism, we assume our analysis based on IEEE 802.16e OFDMA TDD system. We define some parameters as follows:

T_f: Frame duration of 802.16e system.

T_s: Delay between MN and AR.

T_{ld}: Delay of the wired link in one hop.

T_{a_b}: Delay between node a and node b.

N_{a_b}: Number of hops between node a and node b.

T_{DAD}: Average delay of the DAD procedure.

The mechanism in [4] is also designed for 802.16e networks, so we take it to compare with the proposed CLFH mechanism. The draft describes both the predictive and reactive modes. We only consider predictive mode for simplify. The total handover latency of the draft in [4] (referred to as T_{FMIPv6}) is calculated as:

$$T_{FMIPv6} = 2T_{pAR\_nAR} + T_{DAD} + T_{L2HO} + 2T_{MN\_HA}$$
$$+ 2T_{MN\_CN} + 3T_s$$
$$= 2(N_{pAR\_nAR} + N_{nAR\_HA} + N_{nAR\_CN}) \times T_{ld} + T_{DAD}$$
$$+ T_{L2HO} + 7T_s \qquad (1)$$

T_{L2HO} is L2 handover delay, the conventional L2 handover delay is derived below.

$$T_{L2HO} = T_{sync} + T_{cont\_rng} + T_{auth} + T_{reg} + T_f \quad (2)$$

T_{sync} is the average time required to synchronize with new downlink, which can be done within 2 frames. T_{cont_rng} is the average time required for contention based ranging process, which includes contention resolution processes such as CDMA ranging detection and random backoff. It typically needs minimum of 6 frames roundtrip delay plus a random handling duration. T_{auth} is the average time required for re-authorization, which needs 2 frames plus a handling duration of about 100ms. T_{reg} is the average time required for re-registration, which needs 2 frames plus a handling duration of about 10ms. And the MOB_HO-IND message takes 1 frame.

It takes a long time for network re-entry procedures in conventional L2 handover scheme. So it is proposed to combine the some of the network re-entry procedures with MIP handover to speed up the total handover time. We optimize L2 handover by the use of non-contention based ranging, pre-authentication and pre-registration. Optimized L2 handover delay can be represented by

$$T'_{L2HO} = T_{sync} + T_{rng} + T_{reg}/2 + T_f \quad (3)$$

where T_{rng} is the average time required for non-contention based ranging process, which needs 2 frames plus a random handling duration. The re-authentication procedure, which is the most time-consuming procedure in network re-entry procedures, is moved into the L3 handover preparation process and executes in advance so that the delay of re-authorization can be eliminated. In pre-registration, tBS gets registration information through the backbone, but tBS may send an unsolicited registration response message to MS, so the registration delay can be reduced to a half.

The total handover latency of intra-MAP and inter-MAP handover cases of the CLFH mechanism (referred to as T_{CLFH}^{intra} and T_{CLFH}^{inter}) are calculated as (4) and (5), respectively.

$$T_{CLFH}^{intra} = 2T_{MN\_MAP} + 2T_{AR\_MAP} + T_{DAD} + T'_{L2HO} + T_s \quad (4)$$

$$
\begin{aligned}
T_{CLFH}^{inter} &= \max(T'_{L2HO} + T_s, 2T_{pMAP\_nMAP} + 2T_{nMAP\_HA} + 2T_{nMAP\_CN}) \\
&\quad + 4T_{AR\_MAP} + 2T_{pMAP\_nMAP} + T_{DAD} \\
&= \max(T'_{L2HO} + T_s, 2(N_{pAR\_nAR} + N_{nAR\_HA} + N_{nAR\_CN} - 4) \times T_{ld}) \\
&\quad + 4T_{AR\_MAP} + 2(N_{pAR\_nAR} - 2) \times T_{ld} + T_{DAD} \quad (5)
\end{aligned}
$$

The service disruption time is defined as the period when the MN cannot send or receive packets. In our analysis, the MN disconnects from the sBS after sending MOB_HO-IND message. It cannot send or receive packets from the new link until it sends a FNA message to nAR after network re-entry procedures. According to the handover procedure of CLFH mechanism, both the intra-MAP and inter-MAP cases have the same disruption time.

The service disruption time of FMIPv6 and CLFH mechanisms are as follows:

$$D_{FMIPv6} = T_{L2HO} + T_s \quad (6)$$

$$D_{CLFH} = T'_{L2HO} + T_s \quad (7)$$

3.2. Simulation results

We present the effects of different parameters on the handover latency and the service disruption time based on previous analysis with simulation works. Here we set $T_s = T_f + 1ms$, $T_{MN\_MAP} = T_s + 2ms$. And we assume $N_{pAR\_nAR} = 5$, $N_{nAR\_HA} = 20$, $N_{nAR\_CN} = 25$. The DAD time is chosen from the uniform distribution: [500ms, 800ms].

Figure 4 shows a comparison of the handover latency between FMIPv6 and CLFH mechanisms increases with frame duration. The handover latency is greatly reduced in CLFH mechanism. The performance of intra-MAP handover case is the best because no more MIP registration procedure is needed in that case. In inter-MAP case the handover latency changes little as the frame duration increasing, since in that case the MIP registration process affects the total handover latency mostly.

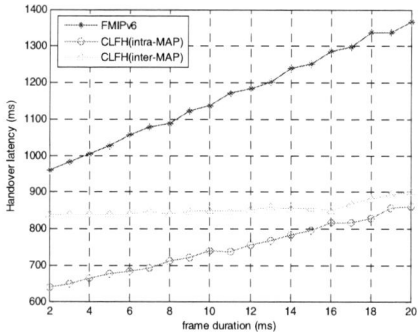

Figure 4 Handover latency v.s. frame duration

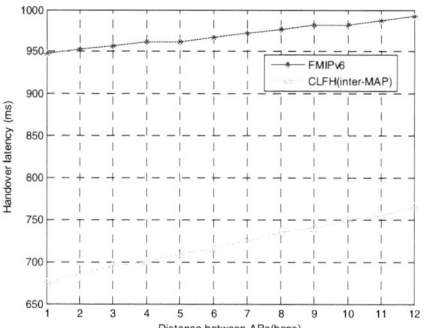

Figure 5 Handover latency v.s. distance between ARs

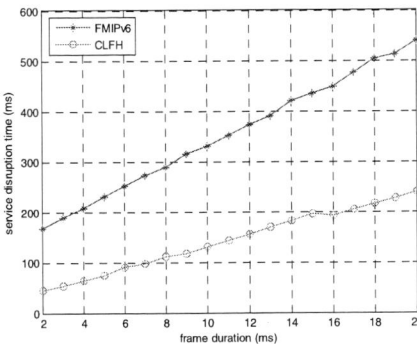

Figure 6 Disruption time v.s. frame duration

Figure 5 shows the effect of the distance between pAR and nAR to the handover latency. We assume T_f is 5 ms here. The handover latency increases with the number of hops between pAR and nAR, and the effect to CLFH mechanism is more significant. The intra-MAP case of CLFH mechanism is not considered here because the number of hops between pAR and nAR is fixed to 2 in our network model.

Figure 6 shows the effect of frame duration to service disruption time. It indicates that the disruption time increases as the frame duration increases. The CLFH mechanism takes the advantage of FMIPv6 to forward packets through tunnel even before the handover procedure is completed, so the disruption time mostly depends on the L2 network re-entry delay. The service disruption time decreases apparently in CLFH mechanism since we adopt optimized L2 handover scheme that further reduce the disruption time.

4. Conclusions

In this paper, we propose CLFH mechanism to provide solution for micro and macro mobility cases by combining the handover procedures of IEEE 802.16e with fast and hierarchical MIP handover mechanisms. By numerical analysis and simulation study, we show that the CLFH mechanism has better handover performance over the existing mechanism. Our research does not consider link layer detection, which also occupies a large amount of the global handover latency. There are more messages involved in inter-MAP handover case that may cause higher signaling cost. More precise study for these aspects is needed in future work.

Acknowledgement

This work was supported in part by Science and Technology Breakthrough Plan of Shanghai Science and Technology Committee (No.07dz15006_2) and National 863 Project of Study on Vehicle Communication Network based on Mobile Relay Technique (No. 2007AA01Z239).

References

[1] IEEE Std 802.16e (2005). "IEEE Standard for Local and Metropolitan Area Network-Part 16: Air Interface for Fixed and Mobile Broadband Wireless Access Systems".

[2] Ian F. Akyildiz, Jiang Xie (2004). "A survey of mobility management in next-generation all-IP-based wireless systems". *IEEE Wireless Communications*. Vol. 11, No. 4, pp.16-28.

[3] R. Koodli (2005). "Fast Handover for Mobile IPv6," IETF RFC-4068.

[4] H.-J. Jang (2008). "Mobile IPv6 Fast Handovers over IEEE 802.16e Networks," IETF Internet-draft, work in progress, draft-ietf-mipshop-fh80216e-07.txt.

[5] H. Soliman (2005). "Hierarchical Mobile IPv6 Mobility Management," IETF RFC-4140.

[6] H.Y. Jung (2005). "Fast Handover for Hierarchical MIPv6," IETF Internet-draft, draft-jung-mobopts-fhmipv6-00.txt.

[7] Y.S. Mun (2007). "Fast Macro Mobility Handovers in HMIPv6," IETF Internet-draft, work in progress, draft-mun-mipshop-fhmacro-02.txt.

The Design of Manufacturing Execution System Based On RFID

Jiwei Hua, Hexu Sun, Tao Liang, Zhaoming Lei
Department of Automation, Hebei University of Technology, Tianjin, 300130, China
E-mail: huajiwei@yeah.net

Abstract

From the supply chain, now RFID technology is gradually applied to the core of manufacturing process. By adopting the RFID technology in workshop layer, the exact real-time information that is obtained from RFID, can be seamlessly integrated into the manufacturing execution system. Therefore it can create additional value, and increase productivity for the enterprise. The technical advantages of applying RFID technology to the manufacturing execution system are analyzed detailedly by this paper. Taking the textile industry as object, the design of manufacturing execution system based on RFID is presented by this paper. And it provide perfection system solution for enterprise realizing real-time informatization.

1. Introduction

Manufacturing execution system (MES) is defined as a shop floor control system which includes either manual or automatic labor and production reporting as well as on-line inquiries and links to tasks that take place on the production floor. MES includes links to work orders, receipt of goods, shipping, quality control, maintenance, scheduling, and other related tasks. The MES (Manufacturing Execution System) is the connecting link in the enterprise information integration system, which is the bridge of information between the Manufacturing and management, so it is the key to the development of enterprise informationization. RFID(Radio Frequency Identification) is new automatic identification technology, which has many advantages, such as a long-distance contact, non-mechanical wear, programmable, a bigger storage and more flexible memory. RFID technology is very suitable for data acquisition and process control in the industrial manufacturing sites, because of its characteristics of waterproof, antimagnetic, high temperature resistance, etc. All of those characteristics of RFID technology can be relied on by manufacturing enterprise to obtain real-time information, which can be seamlessly integrated into the manufacturing execution system and decision of management can be transferred to the industrial manufacturing sites. The technical advantages of applying RFID technology to the manufacturing execution system are analyzed detailedly by this paper. Taking the textile industry as object, the

design of manufacturing execution system based on RFID is presented by this paper. And it provides perfection system solution for enterprise realizing real-time informatization.

2. The advantage of applying RFID technology in manufacturing execution system

Applying RFID technology to the manufacturing process, the real-time production data can be written into RFID tags, such as process data, raw material data, quality data, etc. Therefore, enterprise can master real-time workshop production data, such as Production Schedule, raw material quantity in-process, work-in-process storage. Through real-time information mining analysis, enterprise can realize lean production at the workshop floor. Those dynamic manufacturing information is integrated to the enterprise management system, what make management can dynamic analyze production plan implementation, respond quickly to the changing of the market, and adjust the plan of production and purchase. The production quality can be managed dynamic, provide basic information for production quality traceability.

Because of applying RFID technology in those notes of production, enterprise can obtain the real-time information of manufacturing process to improve the production efficiency and reduce the production cost, perfect production management. At the same time, the real-time information can be transferred to upstream and downstream enterprises. The more real-time information is shared by enterprises, the more utilization values of real-time information are improved. Just a lot of RFID technologies are used, therefore the enterprise management is more intelligent and automation.

3. System design

3.1. The analysis of application background

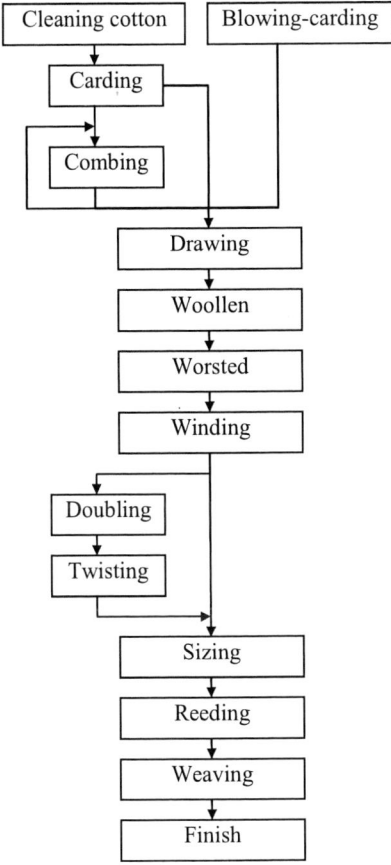

Figure 1. The production process of
cotton spinning industry

The example of a certain textile enterprise manufacturing execution system is given by this paper. According to the characteristic of manufacturing process, the general design scheme of manufacturing execution system based on RFID technology. Considering the manufacturing process, the textile, especially cotton spinning industry is hybrid industry system of process and discrete, and the manufacturing process of cotton yarn is closer to process industry. The production process of cotton spinning industry is shown in figure1. The process cleaning-cotton and carding can be incorporated into blowing-carding, because the equipments are different. The times of drawing is determined by production process. It is also the case with doubling and twisting. There are many kinds of equipment at one process. Therefore, the process should be determined by not only production variety but also production situation.

Considering the production process, the cotton spinning industry is semi-continuous process industry and it can be taken for continuous process industry as a whole. It is different between the cotton spinning industry and general process industry that there are different types of equipments at same process. Its characteristics are as follows:

(1) Multi-equipment serial;
(2) Multi-equipment parallel;
(3)Some productions have the characteristics of reverse process and skip process;
(4) Multiple specifications simultaneous on production line;
(5)Many production constraints (equipment productivity, exchange time, production process);
(6) Production process complex.

3.2. The general architecture of system

According to the analysis of enterprise status, there is many problems in the enterprises management, such as data resources wasting, data transmission inaccuracy and not in time. To counter these problems, manufacturing

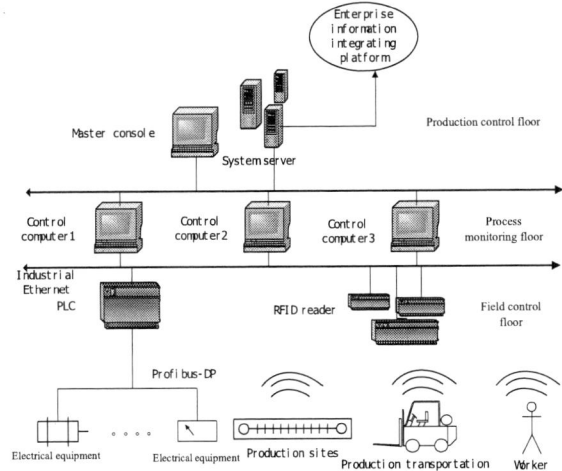

Figure 2 The general architecture of system

execution system is divided into three floors, which are production control floor, process monitoring floor, field control floor. In the field control floor, the PROFIBUS field bus is adopted to real-time monitor electrical equipment at production field. Through using RFID technology, the key data of production process can be real-time acquired, and also work-in-process storage, worker attendance. The real-time data of production site is monitored by process monitoring floor, and scheduling optimization of production. The production is integrated into management system by production control floor, therefore enterprise management can accurately obtain operation information of production site in time. The production system can be seamlessly integrated into the management system, what make production data be sharing, and many control systems can be comprehensively analyzed and uniform planned. The general architecture of system is shown in Figure 2.

3.3. System function design

Through the feasibility analysis for enterprise status, the manufacturing execution system is divided into six subsystems. That are as follows: workshop management, equipment management, monitoring management, quality management, cotton assorting management, production plan, all of those are shown in Figure 3. The concrete function of every subsystem is as follows:

(1) Workshop Management Subsystem:

Workshop management play an important role in production management as acceptance production order, organizing production, recording production data, controlling production, providing information feedback for higher authorities, guaranteeing production stationary. The concrete modules are as follows: production team management, reporting production data, plan declaration at bottom layer.

(2) Equipment Management Subsystem:

Equipment is the important port of enterprise fixed assets, and the base of production. The usage rate and intact rate of equipments are directly related to enterprise production efficiency and organizing production condition. All of equipments should be registered, and the key equipment must be established account. The every

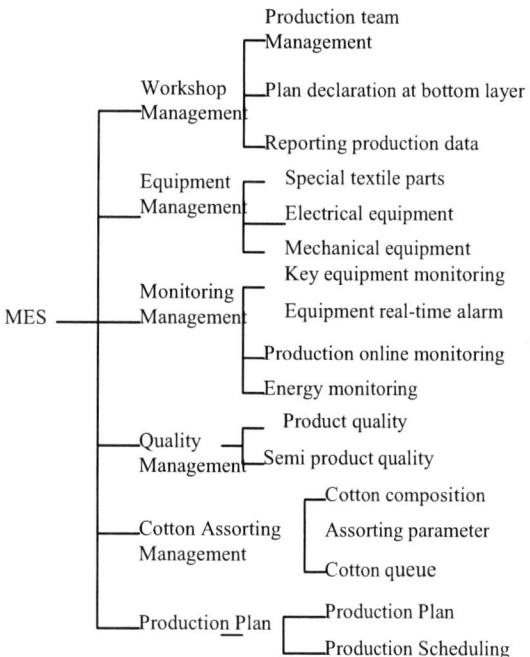

Figure 3 The architecture of system function

archive content of equipment consists: technical data, maintenance content, cycle and record. It is important that special textile parts should be administered specially, such as top-roller, ring, reeds, etc.

(3) Monitoring Management Subsystem

Monitoring management subsystem bases RFID and PROFBUS. The industrial configuration software FIX is used to monitor equipment in production sites, and whole production process. The production data and work-in-process storage of important process should be acquired real-time, which have great influence on production. The consumption of water and electricity can be monitored automatically.

(4) Quality Management Subsystem

Quality management includes two ports, which are product quality management and semi product quality management. From the raw material to production, the quality at every node of production process should be tracked by using RFID technology, and at same time establishing the feedback mechanism of quality backtracking. The production quality trend chart in the important node of production can be established, therefore enterprise can visually analyze quality statistics to find out the key of quality fluctuation and improve the quality of product.

(5) Cotton Assorting Management Subsystem

Cotton assorting management subsystem applies the method of classification-queue to assort cotton. The classification is that cotton is divided into one class, which suits some tex yarn. The queue is that the different batch cotton is arranged a team, the regions, property, type of which are close. After one batch cotton has been used up, another batch cotton of same team will be used to replace it.

Cotton assorted by computer base standard technology parameters of cotton yarn, which provides optional batch number of cotton ,according to technology parameters of present cotton and the condition of classification-queue tables. After the new batch cotton has been used, the technology parameters will be recalculated, and the price information of cotton will be transferred to the quotation system

(6) Production Plan Subsystem

In fact, the enterprise will guarantee equipment under the condition of full load operation. After obtaining new production plan, the enterprise must stop some production for vacating equipments and special textile parts to meet new order. Then enterprise will calculate the gross profit of a single production and make the comprehensive resources of production reach equilibrium with new plan, what base related data, such as equipment data, work center data, process design, special textile parts information, etc. At last enterprise make optimization adjustment of present plan and append new plan into present master plan for generating new master plan. According to the new master plan and fact production condition, the enterprise can arrange production.

4. The application of RFID integrated with MES

Enterprise middleware technology is extended to RFID field by RFID middleware, which can shield the diversity and complexity of RFID equipments, and provide great support for background system. Therefore more widely RFID equipments can be applied in enterprise. The RFID middleware framework of this system is shown in figure4.

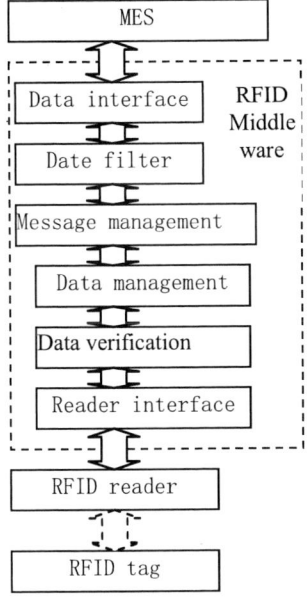

Figure 4 RFID integrated with MES

The RFID data from different types of reader must receive adaptability treatment by the reader interface, and then the RFID data will be unified form. After data verification, the RFID data will be packed accordance with protocol of user defining. The packed message will be cached in message server by message management. All of these messages will be classified base on their content, and same class messages should be arranged in the same message queue. After the repeat RFID data are filtered by data filter module, the filtered data will be transferred to center database. At the same time, the data access module of data interface can provide the interface of center database for enterprise application and remote application system.

5. Conclusion

Through integrating RFID technology into present manufacturing execution system, the more great information chain can be established. The accurate information can be transferred in time, therefore the productivity can be enhanced and the utilization ratio of assets can be improved. The higher level quality control and every kinds of on-line measurement can provide information base for monitoring management of production process and optimization adjustment of production plan. Relying on RFID technology, the enterprise can implement Lean Production and differential production. Considering its production capability, the enterprise should establish reasonable production plan and auxiliary plan to meeting market change. Avoiding waste of repeat build, the enterprise should establish suitable design of system to meet itself condition. It is more important that manager and engineer should realize fundamental change in the management idea through implementing manufacturing execution system.

References

[1] Radio frequency techno logy for automated manufacturing and logistics control [J]. The International Journal of Advanced Manufacturing Technology, 2003, 21: 7692774.

[2] A Real-Time Scheduling and Rescheduling System Based on RFID for Semiconductor Foundry Fabs. Journal of the Chinese Institute of Industrial Engineers, Vol. 24, No. 6, pp. 437-445 (2007)

[3] Dirk H, Mapping and localization with RFID technology, Proceedings of the 2004IEEE International Conference on Robotics & Automation, New Orieans, LA April,2004, pp. 1015-1020

[4] Kitayoshi H, Sawaya K. Long range passive RFID-tag for sensor networks. IEEE Vehicular Technology Conference.2005:2696-2700

[5] D. McFarlane. Networked RFID in Industrial Control: Current and Future. Emerging Solutions For Future Manufacturing Systems. Springer, 2005.

[6] Cheng Tao, Li Jin. Analysis and simulation of RFID anti-collision algorithms .IEEEProceeding Advanced Communi-cation Technology. Phoenix Park,Korea. 2007, :697-701 .

Dynamic Channel Assignment in Wireless LANs

Bo Wang[1], William Wu[2], Yongqiang Liu[3]

[1]*Institute of Computing Technology, Chinese Academy of Sciences*
[2]*Electrical Engineering Dept. Stanford University*
[3]*NEC Laboratories China*
bwang@jdl.ac.cn, willywu@stanford.edu, liuyongqiang@research.nec.com.cn

Abstract

A traffic-aware and client-aware algorithm for dynamically assigning channels to Access Points (APs) in Wireless Local Access Networks (WLANs) is proposed. Traffic loads and Received Signal Strength (RSS) values are used to characterize interference. The problem of selecting channels to minimize total interference then reduces to the Max k-Cut problem, which we efficiently solve with semidefinite programming (SDP) relaxation. Testbed experiments demonstrate that our traffic-aware algorithm can significantly improve the quality of a channel assignment in terms of total network throughput and channel utilization. Under a randomly generated continuous traffic distribution, throughput gains of 40% on average over static channel assignment are observed.

1. Introduction

The explosive popularity of Wireless Local Area Networks (WLANs) in recent years has led to a dramatic rise in the density of WiFi Access Points (APs) in wireless environments. The wireless spectrum is divided into "channels" – bands over which APs can operate. A natural problem that arises is how to assign channels to the APs so as to increase the total network throughput. A well-established heuristic approach is to minimize the interference between APs using the same channel. Due to the broadcast nature of wireless, as the AP density of WLANs continues to increase, this problem of interference is becoming more severe and important.

Currently, network administrators often manually decide on a static channel assignment for APs based on RF profiles [1]. However, traffic loads in a network tend to vary with time, and consequently, such assignments do not result in the best performance.

In this work, we propose a centralized dynamic channel assignment algorithm that efficiently adapts the channel assignments alongside variations in traffic. Our approach incorporates the observed traffic demands of wireless APs and clients, as well as their power-inferred locations, in order to increase network throughput as a whole. The key points and advantages of our algorithm may be summarized as follows:

(1) We adopt the heuristic of minimizing sum network interference to increase overall network throughput. Experiments in our paper confirm the validity of this.

(2) Our algorithm is "traffic-aware", constantly monitoring the traffic patterns of both APs and stations. This allows us more efficient channel usage by designing channel assignments that accommodate to the traffic distribution.

(3) Our algorithm is "client-aware". This means that in addition to APs, clients also take part in measuring and reporting received signal power and traffic, thereby giving us a better picture of interference in the network.

(4) We use actual measurements of signal power and traffic load to intuitively define and measure interference. This measurement-based approach allows us to avoid making simplistic modeling assumptions for the wireless channel and MAC protocol, as such theoretical models are known to be especially inaccurate indoors. It also allows us to assess interference at a higher resolution than otherwise.

(5) Characterizing interference in terms of signal power has the desirable property of *additivity*, which will more easily allow us to address simultaneous interferers. Namely, additivity allows us to reduce the minimization problem to Max k-Cut, which can then be relaxed into a semidefinite programming problem that can be efficiently solved.

Due to all of these design characteristics, the techniques proposed in this paper improve channel assignment for WLANs, exploiting opportunities both to make use of idle channels, and also to reuse active channels. Our experiments show that our channel assignment algorithm substantially improves network throughput by 40% on average when compared to a static assignment scheme.

Our paper is structured as follows: In Section 2, we give some background on WLANs. Section 3 presents an analysis and model for interference. Section 4 describes the reduction to Max k-Cut, and the SDP relaxation used to approximately solve it. Section 5 gives a high-level summary of our dynamic channel assignment algorithm. Implementation issues are discussed in Section 6, and experiments are presented in Section 7. Related works are discussed in Section 8. Lastly, we conclude in Section 9.

2. Background on WLANs

A WLAN consists of two kinds of participants: *APs* and *stations* (users). APs are like the service providers of Internet access for stations. When we wish to refer to either APs or users, without distinction, we may call them *nodes*. Each station *associates* itself with exactly one AP to receive service. We assume that stations are always associated with the AP from which received signal power is strongest. As the station surfs the Internet, data is then sent back and forth between the station and its AP. Stations associated with an AP are called that AP's *clients*. There are no direct client-to-client data flows, nor any direct AP-to-AP data flows.

We define a *cell* as an AP together with all stations that are associated with it. Each node in a cell must use the same *channel*, which is a frequency band over which communications take place. The 802.11 signal frequency range is divided into a number of these channels. For instance, 802.11b/g operates in the 2.4GHz unlicensed frequency band, and provides a set of 11 channels, among which there are only 3 non-overlapping channels. 802.11a operates in the 5.4GHz unlicensed frequency band, and provides 12 non-overlapping channels. In this paper, we always assume that channels are non-overlapping; however, our work can be extended to the case of partially overlapping channels easily [2].

Lastly, we address carrier-sensing. When a node wishes to transmit, it first measures the signal strength it is currently sensing on its channel. If this amount exceeds a certain Clear Channel Assessment (CCA) threshold (e.g., -85 dBm for NIC cards in our testbed), then to prevent signal collision, the CSMA (Carrier-Sense Multiple Access) protocol does not allow transmission to be initiated.

3. Interference Model

3.1. Motivations for Minimizing Interference

We first suggest some general rationale for wanting to minimize interference. Ultimately, we hope to increase throughput, which we assume has a relationship proportional with SINR, the signal-to-interference-and-noise ratio. This relationship can be seen in Table 1 [3], which describes the 802.11 a/g specification of SINR requirements for different transmission rates on wireless links.

Table 1 SINR vs. Data rate In 802.11 a/g

SINR (dB)	6	7.8	9	10.8	17	18.8	24	24.6
Data rate (Mbps)	6	9	12	18	24	36	48	54

We express SINR as

$$SINR = \frac{Intended\ Signal}{Background\ Noise + Interference\ Signal}.$$

The "Background Noise" is usually constant. "Intended Signal" refers to the amount of signal the receiver senses from the node that intended to communicate with it. Lastly, "Interference Signal" refers to the superposition of signals received from other nodes that are broadcasting on the same channel. In an environment with few mobile users (e.g., an office), the locations of senders and receivers are roughly fixed over some time scale. Thus we may consider the "Intended Signal" to also be fixed. Hence, only the "Interference Signal" may change with channel assignment, so we focus on reducing this quantity in order to increase SINR.

3.2. Motivation for Traffic-Awareness and Client-Awareness

We now look at a small example showing how client information can help us find opportunities for channel reuse. Consider Figure 1. There are 2 APs (AP1, AP2) and 4 stations (STA 1-4). STA1 and STA2 connect with AP1, STA3 and STA4 connect with AP2. The circles depict the regions that each AP covers. There is an "interference region" in which these circles overlap; to fully cover the network, such regions are always present. Assume all transmissions are on downlink: that is, APs send to stations. Now note that STA2 and STA3 are in the interference region. That means when STA2 is receiving from AP1, it will be subject to interference from AP2. Similarly, STA3 suffers interference from AP1 when receiving from AP2. On the other hand, pairs AP1-STA1 and AP2-STA4 are free from interference. Thus, we can deduce which nodes lie in the interference region only by comparing the relative interference powers received at the stations.

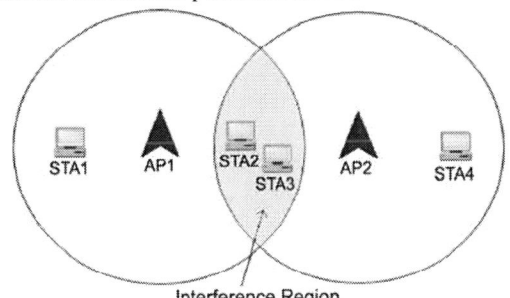

Interference Region

Figure 1 Interference Example.

In this scenario, the decision of how to assign channels to APs depends on the distribution of client traffic. When STA2 and STA3 are quiet with no throughput, and STA1 and STA4 are downloading, we could allocate the same channel to these two APs without interference. This is the notion of *channel reuse*. On the other hand, when STA1 and STA4 are quiet, but STA2 and STA3 are downloading, we must assign different channels to the APs to avoid interference. Thus, the interference in the network is a function of both the locations of nodes and the traffic distribution. The example also shows how it is useful to be aware of client traffic information in addition to AP traffic information. If we relied only on capturing congestion information at the AP, it would have been very difficult to distinguish between the two aforementioned scenarios in an algorithmic manner.

3.3. A Model of Interference

In the absence of interference, the link capacity would be the product of the maximum sending rate of the sender and the delivery ratio of the link. However, interference may reduce the maximum feasible sending rate of the sender. It may also reduce the probability of the receiver successfully receiving a packet, due to collisions. We now seek a model of interference in terms of measurable observables that accounts for both the sender's and receiver's perspective.

Before describing the model, we informally define the term "load". The "sending load" of a node is the fraction of time it's transmissions acquire the medium when it wishes to transmit to various recipients (in the case of a client, there is a single recipient – its AP; in the case of an AP, all of its clients could be recipients). Similarly, the "receiving load" is the fraction of time its receptions acquire the medium when the node wishes to receive from various transmitters. Intuitively, the load value reflects the *probability* of the transmission or reception of a node at any time point. In Section V, we discuss how load information can be collected in practice.

We first consider the effects of traffic distribution on interference. Given two nodes A and B, let $Interference_A^B$ denote the interference A experiences due to B over some predetermined time interval. Since interference only occurs when there are ongoing transmissions at B, $Interference_A^B$ must be proportional to the *sending load* of B, denoted by $Lsend_B$. Furthermore, B affects A only when A is *active* – that is, when A is either transmitting or receiving. If A desires to transmit, CSMA may force it to back off from acquiring the channel due to the transmissions of B. And if A is receiving while B is transmitting, then naturally the SINR is affected. Thus, if we denote the *receiving load* of A by $Lrecv_A$, then $Interference_A^B$ must also be proportional to ($Lsend_A$ + $Lrecv_A$), the fraction of time over which A is active.

We now consider the effect of received signal power. The shorter the distance between A and B, and the fewer the obstacles between them, the stronger the strength of the signal received by A from B, and thus the larger is $Interference_A^B$. This attenuated signal power can be quantified by RSS_A^B (Received Signal Strength), a quantity reported by the NIC.

Ignoring multipath fading effects on the interference, the above considerations naturally lead us to the following proposed interference metric:

$$Interference_A^B = RSS_A^B \bullet Lsend_B \bullet (Lsend_A + Lrecv_A). \quad (1)$$

We now define the interference *between A and B* as

$$I(A, B) = Interference_A^B + Interference_B^A.$$

This is intuitive because there are two directions of interference between two nodes, and we wish to account for both. Due to the broadcast nature of wireless, it is reasonable to assume that the power of interfering signals combines additively. (Note it is crucial here that we use units of mW and not dBm.) We can thus naturally extend our definition of interference between non-singleton sets of nodes as

$$I(\{A_1, ..., A_m\}, \{B_1, ..., B_n\}) = \sum_{1 \le j \le m, 1 < k \le n} I(A_j, B_k). \quad (2)$$

4. Relation With Max k-Cut

4.1. Reduction to MkC

In this part, we formally state the interference minimization problem and show how it reduces to MkC. Let cells be denoted by $C_1, C_2, ..., C_N$, where C_i contains AP_i and all stations associated with AP_i, and N is the number of APs. For a high node density scenario, due to CSMA, there can be only one node transmitting in the same cell at any given time. Also, all available channels are orthogonal. Thus, *interference can only occur between different cells using the same channel*. The total network interference can then be written as

$$I_{Total} := \sum_{1 \le i < j \le N} I(C_i, C_j) Ch(i, j),$$

where $I(C_i, C_j)$ is the interference that would be produced if cells i and j use the same channel, and $Ch(i, j) = 1$ if i and j use the same channel, or 0 otherwise. Suppose that there are k channels available. The problem is then to minimize I_{Total} over the possible assignments of cells to channels $[1:k]$.

We may think of this as a graph-theoretic problem as follows: First construct a complete, undirected graph $G = (V, E)$, where the vertex set $V = \{C_1, C_2, ..., C_N\}$, and the edge set $E = \{E_{ij}\}$ for $i, j \in [1:N]$ and $i \ne j$. To each edge, we assign the weight $W_{ij} := I(C_i, C_j)$. Then choosing an assignment to minimize I_{Total} is the Min k-Partition (MkP) problem, which is NP-hard. Since the sum of all edge weights in the graph is constant, we can instead formulate the equivalent problem of maximizing the sum weight of edges that go *across* partitions. That is,

$$\text{Maximize} \sum_{1 \le i < j \le N} W_{ij} D(i, j),$$

where $D(i, j) = 1 - Ch(i, j)$. This is the Max k-Cut (MkC) problem.

4.2. SDP Relaxation for MkC

We perform an SDP relaxation of MkC according to the technique suggested in [4]. We will use the following fact

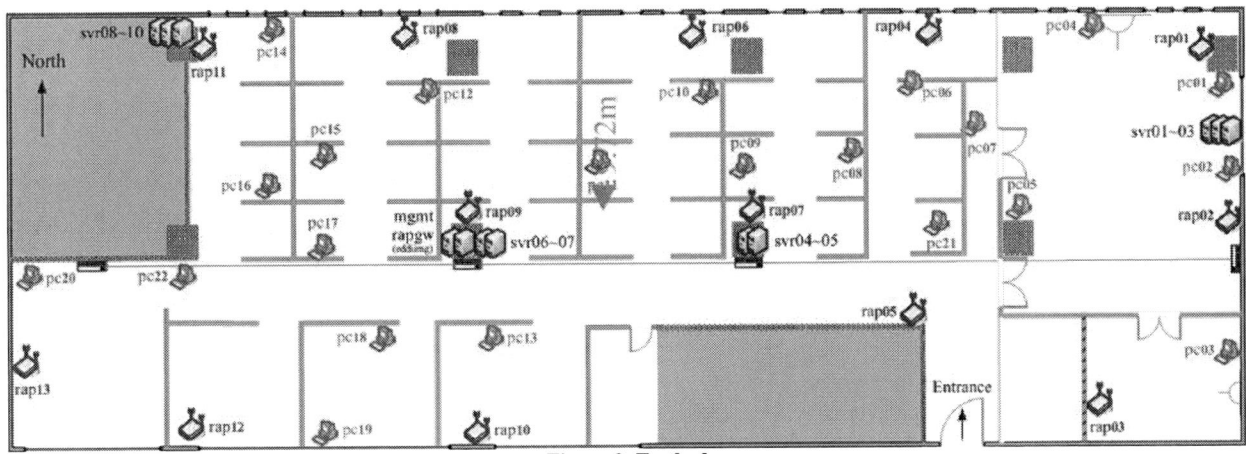

Figure 2 Testbed.

from linear algebra: for $n \geq k$, there exists a set of k vectors $R := \{r_1, \ldots, r_k\}$ such that $r_i^T r_j = -1/(k-1)$ for $i \neq j$, and $r_i^T r_j = 1$ otherwise. We will use these vectors to indicate whether cells i and j use different or identical channels, respectively. Then we formulate our optimization problem as:

$$\text{Maximize (over } x_i \in R \text{)}: \quad \frac{k-1}{k} \sum_{1 \leq i < j \leq N} W_{ij}(1 - x_i^T x_j),$$

where x_i is a variable drawn from R that represents the channel assigned to cell i. Defining $X_{ij} := x_i^T x_j$, and discarding constant terms in the objective, we have

$$\text{Minimize (over } X \succeq 0 \text{)}: \quad tr(WX)$$

$$\text{s.t. } X_{ii} = 1, \; X_{ij} \in \{\frac{-1}{k-1}, 1\}$$

where we wish to minimize over all symmetric positive definite matrices X satisfying the constraints. Relaxing the second constraint as $\frac{-1}{k-1} \leq X_{ij} \leq 1$ then produces an SDP. We can efficiently solve for X with an SDP solver [5], and then apply the polynomial time approximation heuristic of Frieze and Jerrum [6], which uses randomized hyperplane rounding to efficiently generate an approximate solution. To the scale of our testbed, in a few milliseconds the algorithm finds the optimal solution every time.

5. Summary of Dynamic Channel Assignment

Our approach is summarized as follows:

Toward the end of every T second time interval:
1. Report traffic and RSS information to the server.
2. If the network has "sufficiently changed",
   a. Compute co-channel interference between cells.
   b. Compute channel reassignments to minimize sum network interference.
   c. If the potential interference reduction is substantial, then reassign channels.

Note that several variables may be manipulated to adjust the frequency of channel reassignment. Firstly, the time interval T can be made larger to reduce overhead, or smaller to increase resolution in detecting network changes; in our experiment, we used 10 seconds. In the last step, if the network load is too heavy at the moment, or if the expected interference reduction is small, we can opt to not go ahead with channel reassignment. Generally, the frequency of reassignment should be adjusted according to the needs of the application at hand.

6. Implementation Issues

Below we list some issues to clarify our implementation.
- *AP-to-AP RSS Measurements*: To measure RSS values between APs, we conduct a round-robin initialization procedure [7] in which each AP broadcasts a beacon alone while all other APs are quiet and record their RSS values. These measurements are transmitted to the server. Since APs are immobile, we do not expect the network to deviate significantly from these initial measurements, and thus, the AP-to-AP RSS values are kept fixed.
- *Station-to-Station and AP-to-Station RSS Measurements*: Whenever a station has no load, it switches through the available channels and broadcasts probe beacons in each, while also recording RSS values of probe beacons emitted from other idle stations. APs also overhear these beacons and record their RSS values. In this way, station-affiliated RSS values are periodically measured and transmitted to the server, which then updates its RSS information.
- The *sending (receiving) load* of a node is the ratio of observed sent (received) throughput to the maximum possible throughput it can achieve, where the latter is based on its current data rate. In our model, if one wished to replace loads with *demands*, an exponentially weighted moving average may be used to predict demands, as suggested in [8].
- A cell's channel can be changed immediately after the AP broadcasts a special message indicating the change to its clients. This latency would be expected to be around 1-2

ms, which is small compared to the period of time that the cell stays on a channel [9].

7. Performance Evaluation

We set up a 500 M^2 wireless testbed on our office floor containing 13 APs, 22 wireless stations (PCs), and 10 wired servers, as shown in Fig. 2. Each machine is equipped with 802.11/a/b/g NICs using the MadWiFi driver [10]. We used 802.11a, which has more non-overlapping channels in practice. The cards have RTS/CTS disabled, and are set to maximum transmission power. The nodes are positioned sufficiently densely such that almost any two can sense each other.

The 22 wireless stations are uniformly associated with the 13 APs according to their natural spatial proximity; on average, each AP connects with 2 stations. The signal quality between each station to its associated AP is at least 27 dBm, meaning that if there is no interference, an AP or station can use its 54 Mbps physical data rate to achieve a UDP throughput of about 38 Mbps. In all of the following experiments, the traffic pattern is constant bit rate (CBR) UDP traffic, generated at wired servers, and transmitted to the stations through the APs.

Our Dynamic Channel Assignment (DCA) algorithm is compared against a Static Channel Assignment (SCA) that is described as follows: (1) Construct an interference graph by only using RSS information between APs, and (2) Run the MkC solver to compute a one-time channel assignment. SCA models an intelligent RF profiling-based static assignment that does not use traffic or client-side information, and is similar in design to [11]. Such static channel assignment methods are still most commonly used in existing WLANs today.

7.1. Simple Demonstration of DCA

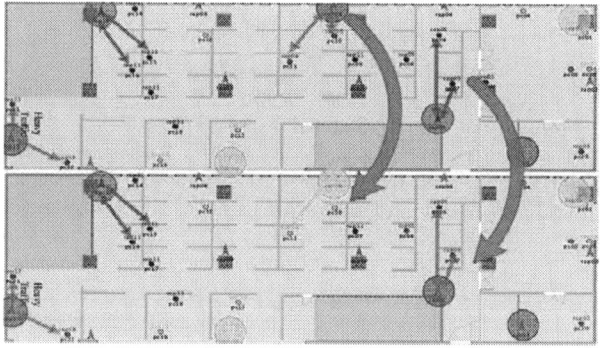

Figure 3 Demonstration of SCA vs. DCA.

We first show a simple experiment to visually illustrate the decisions made by our algorithm. In Fig. 3, we have 7 APs (1,3,5,6,10,11,13) working over 3 available channels, where each channel is indicated by a different color. The small arrows between AP and station indicate traffic flows. We see that there are 4 cells possessing traffic. In the top picture above, we see a channel assignment made by SCA. Since SCA does not consider traffic flows, only two out of three channels are used. On the other hand, the bottom picture in Fig. 3 shows

the channel reassignment made by DCA for the same traffic distribution. Large magenta arrows indicate the differences in these two channel assignments. Note that DCA utilizes all three channels. Furthermore, since the number of busy cells is larger than the number of available of channels, we must have at least two cells using the same channel. DCA assures that those two cells are furthest away from each other, thereby incurring little interference. In this simple scenario, throughput gains can be more than 50%.

7.2. Varying Number of Available Channels

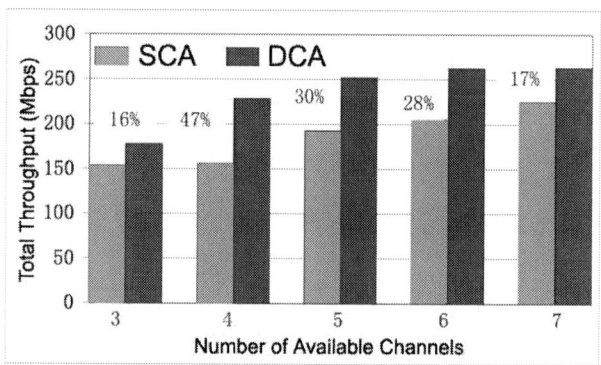

Figure 4 Throughput vs. Number of Available Channels. Bars marked with DCA's percentage improvement over SCA.

In our second experiment, we first specified an initial traffic distribution among 13 working APs, in which each one of the seven APs 1, 2, 3, 6, 10, 11, and 13 was given a heavy flow (> 38 Mbps demand) to send, while the flows through remaining APs are very light (< 1 Mbps demand). Let k denote the number of available channels. We use SCA first and monitor the average sum network throughput over 10 minutes. Afterwards, use DCA instead to dynamically change channel assignments, and monitor throughput for another 10 minutes. We repeat these two steps five times for each k ranging from 3 to 7. Lastly, we average the throughputs over all five trials for DCA and SCA. These averages are plotted vs. k in Fig. 4. Note that since the maximum UDP throughput is 38 Mbps, the sum network throughput can be at most $7 \times 38 = 266$ Mbps. We see that when k=5, DCA almost achieves this upper bound with two channels reused, allowing seven flows to be transmitted without interference. Total throughput under DCA continues to increase and dominate that of SCA as k increases. DCA always makes use of all available channels. However, SCA cannot guarantee this, and thus, on average, it still does not achieve 266 Mbps even when k=7.

7.3. Adaptation to Continuously Changing Traffic

In our third experiment, we set k=5, and change the traffic distribution every 10 minutes. At the start of each iteration, we randomly choose 6 out of the 13 APs to transmit a heavy flow to one of its stations, while the remaining APs engage in light flows. Our DCA will be triggered when there is a sufficiently large change in traffic distribution. The monitored throughput in Fig. 5 plots the continuous traffic throughput under DCA and SCA, and shows that DCA dominates SCA. The throughput gains are roughly 40% on average. This occurs because the traffic is often not uniform amongst channels, and thus DCA can seize opportunities to make use of idle channels, and also reuse channels, at all times. This experiment demonstrates the practical benefits of our algorithm by simulating the continuous traffic flows of the real world.

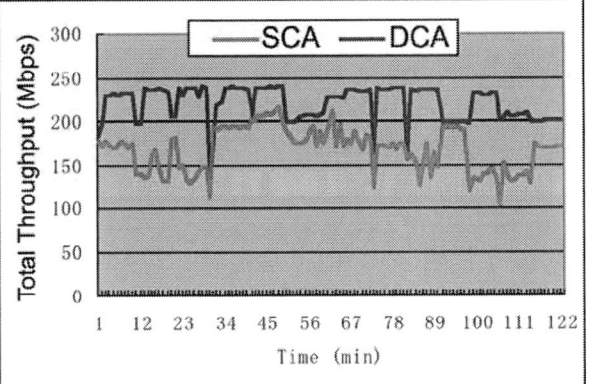

Figure 5 Throughput vs. Time Under Continuous Randomized Traffic.

8. Related Works

The channel assignment problem has been well studied for cellular networks [12], but the solutions there are not readily adapted to WLANs due to many differing characteristics between these networks. For instance, in cellular networks, the regularity of cells allowed for one-shot channel assignments that rarely change; however, the ever-changing nature of WLANs require dynamic solutions that vigilantly monitor interference in an efficient way. In [13], Mishra et al. used the number of clients which are able to associate with two APs to build the metric of interference between any two cells, and reduced the interference minimization problem to graph coloring. This work was extended in [14], where the authors argued that clients have a better view of interference, and hence channel assignment should be conducted from the client perspective. Also, they combined load balancing with channel assignment. However, inaccuracies in their approach arise because client activity is not homogeneous, and neither are the levels of interference they experience. Rozner et al. [8] accounts for traffic, but uses an ideal path-loss model and other assumptions to compute the interference range. Our work is the first to give an approach that accounts for all three of the following: (a) traffic-awareness, (b) client-awareness, and (c) uses a measurement-based interference metric.

9. Conclusions

In this work, we proposed an intuitive model of interference that makes use of both traffic and client awareness, and is measurement-based. We then proposed and implemented an efficient dynamic channel assignment algorithm over WLANs, reducing network interference, and increasing aggregate throughput use by 40% on average over a static channel assignment scheme in our testbed experiments.

For future work, we hope to conduct experiments over a larger testbed, and extend our algorithm to partially overlapped channels.

References

[1] Jim Geier, "Assigning 802.11b access point channels," Wi-Fi Planet.

[2] A. Mishra, V. Shrivastava, S. Banerjee, and W. Arbaugh: "Partially overlapped channels not considered harmful." SIGMETRICS/Performance 2006: 63-74.

[3] V. Mhatre, K. Papagiannaki, and F. Baccelli. "Interference Mitigation through Power Control in High Density 802.11 WLANs," In IEEE INFOCOM, 2007.

[4] M.X. Goemans and D.P. Williamson. "Improved Approximation Algorithms for Maximum Cut and Satisfiability Problems Using Semidefinite Programming," J.ACM, 42, 1115-1145, 1995.

[5] CSDP: C Library for Semidefinite Programming. https://projects.coin-or.org/Csdp/

[6] A. Frieze and M. Jerrum. "Improved approximation algorithms for MAX k-CUT and MAX BISECTION." Algoritmica, 18,1997.

[7] L. Qiu, Y. Zhang, F. Wang, M. Han, and R. Mahajan, "A General Model of Wireless Interference", Mobicom 2007

[8] E. Rozner, Y. Mehjta, A. Akella, L. Qiu, "Traffic-aware channel assignment in wireless LANs," in Mobicom 2006.

[9] I. Ramani and S. Savage, "Syncscan:practical fast handoff for 802.11 infrastructure networks," in Proceedings of IEEE Infocom, 2005.

[10] Madwifi: Multiband Atheros Driver For Wifi. http://madwifi.sourceforge.net.

[11] K. Ishii, T. Osawa, NEC Corporation(JP), "Wireless LAN for reestablishing wireless links between hosts according to monitor desired and undesired signals." US 6,434,132 B1.

[12] T. Rappaport, "Wireless Communications: Principles and Practice." Prentice Hall, 1996.

[13] A. Mishra, S. Banerjee, and W. Arbaugh, "Weighted coloring based channel assignment for wlans," Mobile Computer Communications Review (MC2R), vol. 9, no. 3, 2005.

[14] A. Mishra, V. Brik, S. Banerjee, A. Srinivasan, and W. Arbaugh, "A client-driven approach for channel management in wireless LANs," in IEEE Infocom, 2006.

Routing Optimized Video Transmission over Mobile Ad Hoc Networks

Yang He, Yuanxin Ouyang, Chao Li, and Zhang Xiong
School of Computer Science and Engineering, Beihang University
Beijing 100083, China
heyang@cse.buaa.edu.cn

Abstract

In mobile ad hoc networks, the communication node always produces a great deal of routing control packets while performing the route discovery and maintenance, which seriously impacts the real-time video transmission quality. Therefore, we propose a new Routing Control Packets Limited Forwarding (RCPLF) algorithm that can reduce the routing overhead of on-demand routing protocol. In our algorithm, the route discovery is limited within a reasonable region, so that the number of routing control packets can be reduced and the bandwidth occupied by them can be saved effectively. Thus RCPLF algorithm is beneficial to improve the real-time video transmission quality. The simulation experiments have been carried out to measure the performance. In addition, any routing protocol using flooding based route discovery can use this algorithm to optimize video transmission performance.

1. Introduction

The mobile ad hoc networks with characteristics of high flexibility and adaptability boom very broad requirements in military and civil applications. Thereinto, the real-time video communication over ad hoc networks is considered to be a more spectacular area. In mobile ad hoc networks, the real-time video communication is always impacted by frequent link interruptions, collisions, limited bandwidth and so on. Therefore, there are great challenges of supporting the real-time video transmission in these environments. Some existing researches focus on video coding technique. Multiple description coding (MDC) can improve the video communication fault tolerance [1-3]. Others pay more attention to the video transmission technique. Multicast routing for multiple destination video transmission can effectively save bandwidth and improve communication efficiency [4-6]. Multipath video transmission can balance network traffic and reduce network congestion [2, 7, 8].

To maintain the continuous real-time video communication in mobile ad hoc networks, the routing protocol need reestablish the routes frequently for adapting the dynamic change of network topology, which will bring a great deal of routing control packets and produce high routing overhead. Especially, in high mobility or heavy traffic environments, the number of routing control packets may be even more than the data packets. Moreover, routing control packets in the MAC layer are always transmitted with a higher priority, and in mobile ad hoc networks, these packets take very much bandwidth and bring severe influence on the video transmission, causing heavy delivery delay and loss.

In ad hoc networks, there have been a great deal of prior works in the area of reducing routing overhead by limiting routing control packets forwarding region [9-12]. Robert [10] added prior route information in the route discovery request packets to limit their forward region and reduce routing overhead, but the length of packets is increased. Xuhui [11] used two phases route discovery to localize the discovery region firstly and search for the route in this region secondly, which can reduce routing overhead, whereas increase the route reestablishment delay. Several other methods [9, 12] utilize particular facilities, such as the GPS device or directional antenna, to restrict the route discovery region or direction. However, there are rare researches discussing the influence of routing overhead on real-time video transmission.

In this paper, we investigate the influence of reducing routing overhead to real-time video transmission over wireless mobile ad hoc networks. The rest of this paper is structured as follows. First, in Section 2, we propose a novel routing overhead optimized algorithm which can significantly decrease the amount of routing control packets of the on-demand routing protocol. Then, we verify the effectiveness by applying it to DSR protocol [13] and AODV protocol [14], and study the impact of decreasing routing overhead on real-time video transmission quality in Section 3. Finally, we draw some conclusions in Section 4.

2. Routing Control Packets Limited Forwarding Algorithm

Generally, the on-demand routing protocol always implements an overall network flooding to find and establish a route. However, when a route is interrupted, the new route usually can be found around the prior routing area, so the transmission region of routing control packets will be limited around there. Based on this idea, we design a novel Routing Control Packets Limited Forwarding (RCPLF) algorithm. It consists of three components: the

978-0-7695-3342-1/08 $25.00 © 2008 IEEE
DOI 10.1109/PEITS.2008.112

maintenance of the Minimum Hop Count table, the routing control packets forwarding limitation strategy and correction strategy. For the first component, when a node overhears the data packet forwarding in the network, it will calculate the minimum hop count from itself to the destination node of the data packet and cache it in the local Minimum Hop Count Table. In fact, only the node around the routing path can overhear the packet and get the minimum hop count to the destination. For the second component, when some route is interrupted, the source node will initiate the route discovery procedure with the forwarding limitation strategy. For the third component, nodes around the interrupted link of the prior route may have big position changes that cause the route interruption, so the forwarding correction strategy is performed around the area of the interrupted link, where more nodes will forward the routing control packet, which leads to an improvement to the route reestablishment successful ratio.

2.1. Maintenance of the Minimum Hop Count Table

Each node in the network maintains a Minimum Hop Count Table (*MHCT*), in which each record stores the minimum hop count information from itself to some destination node. The structure of the table is very simple and only includes three fields: the destination node ID (*DN*), the minimum hop count (*MHC*) and the refresh time (*$T_{refresh}$*). Each node keeps overhearing the delivering data packets in the shared wireless channel. Whenever it receives one, it will refresh the *MHCT* according to the routing information in the packet header. More details are showed in the following:

(1) The data packet header has to contain the hop count information from the forwarding node to the destination node. It is only one byte size, and must be refreshed according to the local routing table whenever it is forwarded.

(2) A node, which overhears a data packet, must determine whether it is on the routing path of this packet. If the node is on the routing path, it should forward the data packet according to the routing information, calculate the minimum hop count to the destination node, and update it into the *MHCT*. However, if the node just overhears a data packet forwarded near the region, it will not forward the packet, but get the hop count value from the data packet header, increase the value by one, and update it into the *MHCT*. Note that a node maybe have overheard the same data packet and repeatedly calculated the hop count to the same destination node for many times, however, only the minimum hop count value will be recorded in the *MHCT*.

(3) While a node updates its *MHCT*, it should refresh the *$T_{refresh}$* field by the current time *T_{now}*. And when *T_{now}*-*$T_{refresh}$*>*$T_{MHC\text{-}timeout}$*, it means that this record is time-out and should be deleted from the *MHCT*.

2.2. Forwarding Limitation Strategy

When the route from the source node to the destination node is interrupted and needs to be reestablished, the following measures will be performed:

(1) The source node will get the minimum hop count to the destination, add it into the hop count limitation field (*H_{RREQ}*) of the Route Request (RREQ) packet, and broadcast the packet.

(2) When an intermediate node receives a non-repeated RREQ packet, it will extract the value of *H_{RREQ}* from the packet and lookup its *MHCT* to check whether there is a minimum hop count record to the destination node. If the record exists and the value of *MHC* is less than *H_{RREQ}*, the RREQ packet will be rebroadcast with *H_{RREQ}* set as *MHC*. Otherwise, it will not be forwarded.

(3) When the destination node receives a RREQ packet, it will send a Route Reply (RREP) packet back to the source node through the reversing routing path acquired from the RREQ Packet.

It is noteworthy that when the source node needs to establish the route to a destination node in the first time or the forwarding limitation strategy fails to discover a route, a whole network flooding process will be implemented for a larger scope of route discovery.

2.3. Forwarding Correction Strategy

When the node in the end of the prior interrupted link closer to the source node receives a non-repeated RREQ packet, it will update the *H_{RREQ}* of the RREQ packet with the minimum hop count value to the destination node according to its *MHCT*. Meanwhile, it will make a correction to the value of *H_{RREQ}* field, increase it by *H_{inc}*, and then forward the RREQ packet. The correction of *H_{RREQ}* will increase the number of nodes which satisfy the conditions of forwarding RREQ packets in the local region near the prior interrupted link, and improve the route reestablishment successful ratio.

2.4. Sample

Figure 1 displays the procedure of route discovery with RCPLF algorithm in a simple scenario. In Figure 1(a), the black bold solid line represents a route from the source node *s* to the destination node *d*, and the number in the node represents the minimum hop count to the destination node stored in its local Minimum Hop Count table. Because of the movements of the node *i*, the network topology changes and the route between the node *s* and node *d* is interrupted. Figure 1(b) and Figure 1(c) are two different scenarios which are caused by different positions the node *i* moves to. Figure 1(b) just uses the forwarding limitation strategy to reestablish the route. Figure 1(c) involves the forwarding correction strategy more. For the latter position, the node i adds the value of H_{RREQ} of the RREQ packet (H_{inc}=1) to reestablish the route. In all the Figures, the grey nodes represent the nodes which are involved in sending the RREQ packets, the arrows mean the RREQ packets, and the numbers beside the arrows represent the H_{RREQ} values of the RREQ packets. In Figure 1, it shows that the region in the network which is involved in forwarding RREQ packets has been greatly restrained with RCPLF algorithm, and the amount of routing control packets forwarding in the network has decreased significantly.

3. Simulation Results

Applying RCPLF algorithm to DSR and AODV protocols, we use the real-time video transmission as a traffic model and some simulation experiments to verify the optimization effect of the algorithm in the aspect of routing overhead. Conclusively, we analyze the influence of the network routing overhead on the real-time video transmission quality.

The simulation experiments are based on NS-2 platform extended by the Monarch research group in CMU [15], EvalVid video simulation framework [16, 17], and its extension on NS-2 platform [18]. 50 nodes are distributed in an 800m × 800m square area randomly. Each node moves following the random Waypoint model with a randomly chosen speed distributed between 0 and maximum speed. The distributed coordination function (DCF) of IEEE 802.11 for wireless LANs is used as the MAC layer. The radio model is based on the Lucent Technologies WaveLAN product, which is a shared-media radio with a transmission rate of 2Mbps, and a radio range of 250m. The standard QCIF sequence Foreman, which has 400 frames and is encoded to MPEG4 by FFMPEG [19] with the group of pictures (GOP) pattern IBBPBBPBB, is used as the video source. 5 pairs of sender and receiver are chosen and establish connections randomly in the first 100s in the simulation for video transmission. The frame rate is set to 25fps and average video bit rate is set to 65.8Kbps. Each video connection transmits the Foreman sequence 25 times repeatedly and the total number of sent video is 10000 frames. The simulation is 500s long, and results are averaged over 20 runs.

Some major preset parameters of RCPLF algorithm are customized as follows: $T_{MHC-timeout}$=5s, H_{inc}=1. We evaluate the performance of the video communication in different node maximum speed scenes using the following metrics: the routing overhead, the video frame delivery ratio, the Peak Signal-to-Noise Ratio (PSNR) of the video before and after delivery, and the video frame delivery delay. The results are shown in Figure 2-5.

Figure 2 shows the result of the routing overhead against the node mobility rate, measured in the maximum speed. As expected, after applying RCPLF algorithm, the

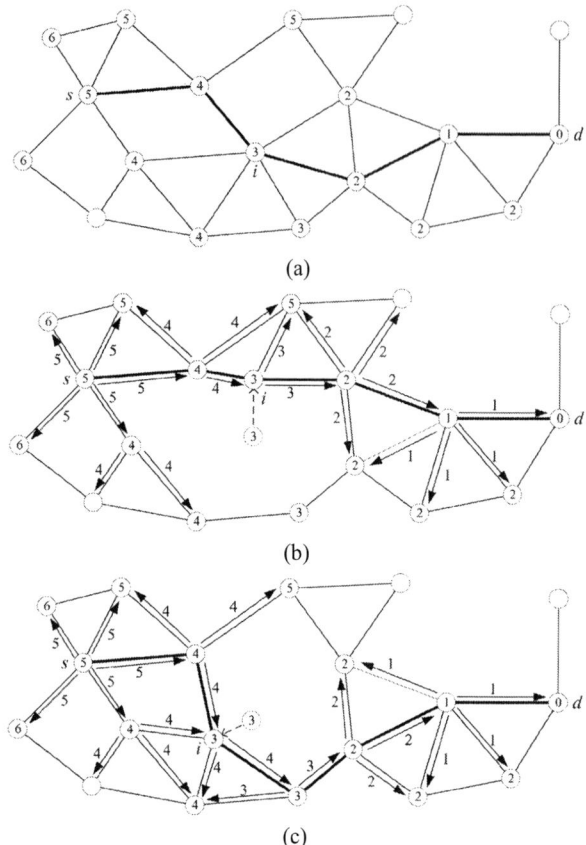

(a)

(b)

(c)

Figure 1 RCPLF illustration. (a) A route from the source node s to the destination node d is established. (b) The routing control packets forwarding limitation strategy is used for a route discovery. (c) Both the routing control packets forwarding limitation strategy and the correction strategy are used for the route discovery.

routing overhead of DSR and AODV has a remarkable reduction by about 40% to 60%, which indicates that RCPLF algorithm can decrease a large amount of routing control packets. Meanwhile, it does favor to the data packets transmission in a better network environment, such as less packet conflictions.

Figure 3 and Figure 4 show the comparison of four schemes on the aspects of the video frame delivery ratio and average PSNR when increasing maximum node speed. The average PSNR can objectively represent the quality of video received in the destination node. RCPLF improves the frame delivery ratio by about 3% to 10%, compared with the original DSR and AODV protocols, which brings an improvement to the video quality for about 1db to 2db on the aspect of the average PSNR. Although DSR has lower routing overhead than AODV, the frame delivery ratio and average PSNR are not as good as AODV. Analysis shows that DSR use a strategy called replying from cache, which admits the intermediate nodes generate RREP packets based on their routing cache instead of rebroadcast the RREQ packets. This strategy can prevent the RREQ packets from flooding throughout the network and bring a remarkable reduction of the routing overhead, whereas the source node will probably use a stale route to delivery video data, which causes the frame delivery ratio degradation of DSR. Our algorithm makes a reasonable restriction to the forwarding region of routing control packets. It can not only decrease the routing overhead, but also bring performance improvements on the video frame delivery ratio and quality.

For video transmission, the video frame delivery end-to-end delay is an important metric, which may have a great impact on the video service response latency, video stream continuity, and the video buffer size provided by the receiving terminal. Figure 5 shows the result. For DSR and AODV, RCPLF can decrease the video frame end-to-end delay by about 20%. These achievements are due to the deduction of routing overhead and a better network environment.

For a further analysis, we select a scene that the maximum node speed is 10m/s and compare the original DSR protocol with RCPLF optimizing version. One video bit stream between the same source node and the destination node is chosen, and the results are presented in Figure 6, where the x-axis represents the number of the video frames which are received or recovered in the destination node, the left side y-axis represents the video frame PSNR which is corresponding to the upper curve in the figure, and the right side y-axis which is corresponding to the lower curve represents how many packets are lost per frame. It shows that the video quality is fine (about 30db) when there are fewer packets lost, whereas more packets loss will lead to a bad video quality in a latter period of time. In addition, because the experiments mainly focus on the impact of the routing overhead on the quality of video transmission, we do not make priority management to video frames. The loss of the key frames is another reason which decreases video quality significantly. In Figure 6(a), when using DSR protocol, the video quality decreases heavily because of the huge loss of video frames. However, in Figure 6(b), the DSR protocol is optimized by RCPLF, so the decrease of routing overhead reduces the video frames loss which brings an improvement to the video transmission quality. The average PSNR of the video frames has increased from 25.08db to 27.52db, namely, the increase range is about 10%. In Figure 7, we select some representative received frames which are marked in Figure 6. It shows that the frame qualities achieve remarkable improvements.

4. Conclusions

In this paper, we propose a Routing Control Packets Limited Forwarding algorithm which can reduce the routing overhead in mobile ad hoc networks. We also investigate the impact of the control packets on video transmission quality. The simulation experiments show

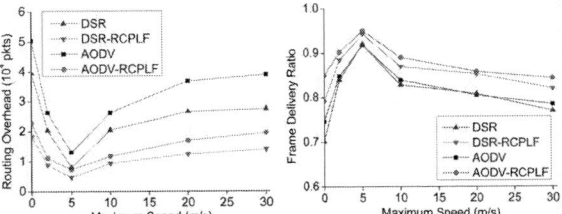

Figure 2 Routing overhead Figure 3 Frame delivery ratio

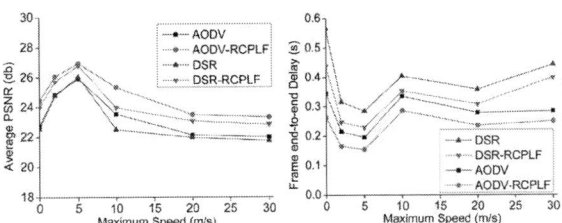

Figure 4 Average PSNR Figure 5 Frame end-to-end delay

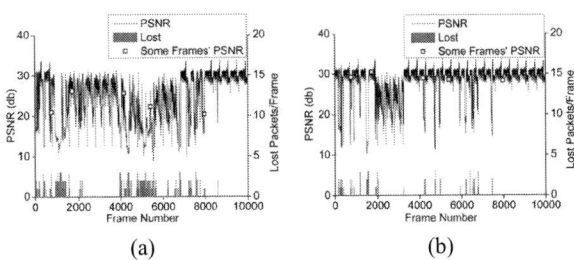

(a)   (b)

Figure 6 Video frames PSNR and lost packets per frame in the scene with 5 video bit stream connections and the maximum mobility speed of 10m/s. (a) Performance of the original DSR protocol (Average PSNR = 25.08db). (b) Performance of the DSR protocol optimized by RCPLF algorithm (Average PSNR = 27.52db).

Frame_757 Frame_1718 Frame_4168 Frame_5399 Frame_7962
(a)

Frame_757 Frame_1718 Frame_4168 Frame_5399 Frame_7962
(b)

Figure 7 Some selected frames. (a) Performance of the original DSR protocol. (b) Performance of the DSR protocol optimized by RCPLF algorithm.

that RCPLF algorithm can reduce routing overhead by about 40% to 60%. Moreover, RCPLF algorithm can improve the video transmission quality on the aspects of the video frame average PSNR and the video transmission delay efficiently.

Acknowledgements

We would like to thank the anonymous referees for their comments. This work is partly supported by the Science and Technology Foundation of China Aero under Grant No.05E551010, the Key Subject Foundation of Beijing under Grant No.XK100060423 and the Graduate Innovation and Practice Foundation of Beihang University under Grant No.2007.20.

References

[1] V. K. Goyal, "Multiple description coding: compression meets the network", *Signal Processing Magazine, IEEE,* vol. 18, pp. 74-93, 2001.

[2] N. Gogate, C. Doo-Man, S. S. Panwar, and A. Y. W. Yao Wang, "Supporting image and video applications in a multihop radio environment using path diversity and multiple description coding", *IEEE Transactions on Circuits and Systems for Video Technology,* vol. 12, pp. 777-792, 2002.

[3] Y. Wang, A. R. Reibman, and S. Lin, "Multiple Description Coding for Video Delivery", *Proceedings of the IEEE,* vol. 93, pp. 57-70, 2005.

[4] M. Shiwen, C. Xiaolin, Y. T. Hou, and H. D. A. S. H. D. Sherali, "Multiple description video multicast in wireless ad hoc networks", in *First International Conference on Broadband Networks,* 2004, pp. 671-680.

[5] C. Chow and H. Ishii, "Video Multicast over Mobile Ad Hoc Networks: Multiple-Tree Multicast Ad Hoc on-Demand Distance Vector Routing Protocol (MT-MAODV)", in *IEEE 18th International Symposium on Personal, Indoor and Mobile Radio Communications,* 2007, pp. 1-5.

[6] W. Wei and A. Zakhor, "Multiple Tree Video Multicast Over Wireless Ad Hoc Networks", *IEEE Transactions on Circuits and Systems for Video Technology,* vol. 17, pp. 2-15, 2007.

[7] S. Mao, S. S. Panwar, and Y. T. Hou, "On optimal partitioning of realtime traffic over multiple paths", in *24th Annual Joint Conference of the IEEE Computer and Communications Societies. Proceedings, IEEE INFOCOM,* 2005, pp. 2325-2336 vol. 4.

[8] M. Shiwen, L. Shunan, W. Yao, S. S. A. P. S. S. Panwar, and A. Y. L. Yihan Li, "Multipath video transport over ad hoc networks", *Wireless Communications, IEEE,* vol. 12, pp. 42-49, 2005.

[9] K. Young-Bae and H. V. Nitin, "Location-aided routing (LAR) in mobile ad hoc networks". vol. 6: Kluwer Academic Publishers, 2000, pp. 307-321.

[10] C. Robert, eda, R. D. Samir, and K. M. Mahesh, "Query localization techniques for on-demand routing protocols in ad hoc networks", *Wireless Networks* vol. 8, pp. 137-151, 2002.

[11] H. Xuhui, L. Yong, M. J. Lee, and T. N. A. S. T. N. Saadawi, "Route update and repair in wireless sensor networks", in *Consumer Communications and Networking Conference,* 2004, pp. 82-87.

[12] G. Hrishikesh, J. Tarun, C. Carlos De Morais, and P. A. Dharma, "DRP: An Efficient Directional Routing Protocol for Mobile Ad Hoc Networks", *IEEE Transactions on Parallel and Distributed Systems,* vol. 17, pp. 1438-1451, 2006.

[13] The Dynamic Source Routing Protocol for Mobile Ad Hoc Networks (DSR). IETF MANET Working Group, 2004. (Accessed at http://tools.ietf.org/html/draft-ietf-manet-dsr-10.)

[14] Ad hoc On-Demand Distance Vector (AODV) Routing. IETF MANET Working Group, 2003. (Accessed at http://tools.ietf.org/html/draft-ietf-manet-aodv-13.)

[15] Wireless and Mobility Extensions to ns-2. The Rice Monarch Project 2000. (Accessed at http://www.monarch.cs.rice.edu/cmu-ns.html.)

[16] EvalVid - A Video Quality Evaluation Tool-set. 2003. (Accessed at http://www.tkn.tu-berlin.de/research/evalvid/.)

[17] J. Klaue, B. Rathke, and A. Wolisz, "EvalVid - A Framework for Video Transmission and Quality Evaluation", in *the 13th International Conference on Modelling Techniques and Tools for Computer Performance Evaluation,* Urbana, Illinois, USA, 2003, pp. 255-272.

[18] K. Chih-Heng, L. Cheng-Han, S. Ce-Kuen, and H. Wen-Shyang, "A Novel Realistic Simulation Tool for Video Transmission over Wireless Network", in *Proceedings of the IEEE International Conference on Sensor Networks, Ubiquitous, and Trustworthy Computing,* 2006, pp. 275-283.

[19] FFmpeg 2007. (Accessed at http://ffmpeg.mplayerhq.hu/.)

High-frequency Surface-wave Radar Real Time Frequency Selecting based on Frequency Randomly Hopping Signal

Xu Liu[1], Changjun Yu[2]

1Research Institute of electronic Engineering,
Harbin Institute of technology, Harbin,150001, China
2Research Institute of electronic Engineering,
Harbin Institute of technology, Harbin,150001, China
liuxuyueli2006@163.com, yuchangjun@263.net

Abstract

Serious radio interferences are distributed in the operating frequency band of High-frequency surface-wave radar (HFSWR) which makes it difficult to find silent bandwidth especially at night to work, the performance of HFSWR is decreased greatly. Frequency randomly hopping signal which breakthrough the restriction of continual spectrum has become an effective anti-jamming method. Based on the study of HF spectrum forecasting measure, the frequency selecting methods of frequency hopping signal radar was studied in this paper. Forecast with deterministic model followed by state partition was proposed, which has higher reliability than traditional deterministic model and statistical model. The rule of most silent channel was used to select the optimum frequency band and genetic algorithm was used to optimize frequency combination. The noise floor and the signal side-lobe level are minimized effectively and it improves the detection capability of HFSWR in heavily congested electromagnetic environments.

1. Introduction

Serious radio interferences are distributed in the operating frequency band（3—15 MHz）of HFSWR[1], with high level noise and dense interferences, it is difficult to find a continuous clean bandwidth. For HF radar using phase coding signals and LFM signals, in order to avoid interferences, it have to reduce signal bandwidth, badly effect the detection capability of HFSWR.

Frequency randomly hopping signal can breakthrough the restriction of continual spectrum, as long as there are amount of discontinuous narrow bandwidth, when there summation can satisfy the requiring of signal bandwidth, can make radar avoid serious radio interferences[2]. So, in heavily congested electromagnetic environments, as anti-jamming efficient measure in frequency domain frequency randomly hopping signal is getting broad recognition.

Random nature of radio interferences leads to the non-uniformity of hop intervals, using general pulse compression technology process echo can not gain perfect pinnacle, around main-lobs always distribute many high side-lobs. In multi target condition, the side-lobs of strong object can submerge weak signals, which can lower the dynamic range of the system, increase the percent of false alarm[8], accordingly restrict the actual application of frequency randomly hopping signal. The side-lobs are in connection with frequency combination, so how to match the best combination by using frequency spectrum from inspect system is the key of its application.

This paper first studied HF spectrum forecasting measure, forecast with deterministic model followed by state partition was proposed, which has higher reliability than traditional deterministic model and statistical model. The rule of most silent channel was used to select the optimum frequency band and genetic algorithm was used to optimize frequency combination. The noise floor and the signal side-lobe level are minimized effectively and it improves the detection capability of HFSWR in heavily congested electromagnetic environments.

2. Signal Analyzing

Frequency randomly hopping signal in one modulation period can denote as [4]:

$$s(t) = \sum_{n=0}^{N-1} rect\left(\frac{t - T_{pw}/2 - nT_{pp}}{T_{pw}}\right) \exp\left(j2\pi\left(f_0 + \Delta B_n\right)t + \varphi_0\right)$$

(1)

Here N is pulse number in one modulation period, T_{pw} is the pulse wide, T_{pp} is the pulse period, f_0 is carrier frequency, φ_0 is initial phase, ΔB_n is the offset of the nth frequency relative to f_0.

Its time (range) auto correlation function is

$$R(\tau) = \int_0^{T_{wp}} s(t)s^*(t-\tau)\,dt, -T_{pw} \le \tau \le T_{pw} \quad (2)$$

From (2) we get

$$R(\tau) = \left(T_{pw} - |\tau|\right)\sum_{n=0}^{N-1} \exp\left(j2\pi(f_0 + \Delta B_n)\tau\right), \quad -T_{pw} \le \tau \le T_{pw}$$

（3）

It is obvious that frequency combination affect maximum range side-lobs.

3. Spectrum Forecasting

Selecting silent frequency is to ensure the radar system can work normally in the next period, so forecast veracity is the precondition of optimizing frequency combination.

This paper using actual spectrum measured in Wei-Hai, frequency range is 5-13Mhz, frequency resolution is 1.25Khz, accumulation periods is 108s,then we get 6400 frequency point , data of 27 batch. Use the former 4 batches to forecast every time .Comparing the result with actual data, forecasting with deterministic model followed by state partition has higher reliability.

3.1. Markov forecasting

Statistical model is used to forecast the state of channels, State partition is decide by threshold. Threshold=Noise floor + Constant (unit : dB). Noise floor is estimated by MM arithmetic[5]. Using Markov [6]to forecast, if the current channel state is S(n),then the next batch state is $S(n+1) = S(n) \cdot P$, P is the state transfer probability matrix ,it can be get from frequencies transfer probability matrix and rules estimation.

3.2. Four states to two states forecasting

To forecast with Markov-chain, the estimate of state transfer probability matrix needs more historical states. This paper proposed a simple way to forecast, it uses the current two patches, in order to minimize information losses, so we extend the state range to four, while the object state is still two, which is called four states to two states forecast.

Adding or decreasing an offset on the base of basic threshold, we get three thresholds. According to the amplitude of power spectrum, there are four states of the channels, which can be defined respectively as 0, 0.25, 0.75 and 1.

Suppose the current patch is n, its silent exponent is $I(n)$, the former patch is $n-1$, and its silent exponent is $I(n-1)$, the next patch state is $S(n+1)$. According to the different state combination of the current two patches:

1) If $I(n)+I(n-1)<1$, then $S(n+1)=0$;

2) If $I(n)+I(n-1)>1$, then $S(n+1)=1$;

3) While $I(n)+I(n-1)=1$, if $I(n)<0.5$, then $S(n+1)=0$, else $S(n+1)=1$.

3.3. Exponential smoothing forecasting then four states to two states forecasting

Combine exponent forecasting and four states to two states forecasting can improve reliability. Suppose the current patch is n, its silent exponent is $I(n)$, after exponent forecast, the patch is n+1, and its silent exponent is $I(n+1)$,the next patch is $S(n+1)$, according to the different state combination of the current patch and the forecasting patch ,the mapping relation is as bellow.

1) If $I(n)+I(n+1)<1$,then $S(n+1)=0$;

2) If $I(n)+I(n+1)>1$,then $S(n+1)=1$;

3) While $I(n)+I(n+1)=1$, if $I(n+1)<0.5$, then $S(n+1)=0$, else $S(n+1)=1$.

Using the actual data, take 6-26 patches as objective patch, we get their reliability probability of three measures in fig1. Threshold offset is 1dB in four state to two state forecast, and the slip coefficient is 0.5 in exponential smoothing forecasting.

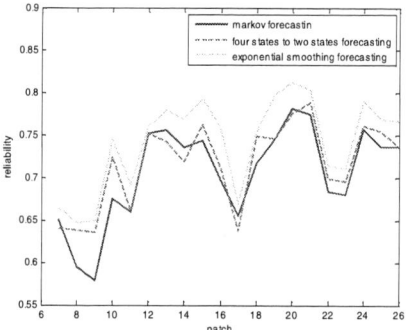

Figure 1 Compare of reliability with three forecasting methods

4. Study on frequency selecting methods

Before selecting frequency we need define child frequency segment in all frequency segments and its silent frequency, adopting smoothing slip-window method to traverse in all frequency segments, the width of slip-window is the max spectrum span 100Khz, the smallest silent bandwidth is 1.25Khz.

The rule of most silent channel: first select a acceptable threshold, then calculate the silent channel number, take the silent channel number as selecting rules.

The rule of max side-lob least: code and optimize silent frequency among every slip-window, take the max side-lob least of the optimized frequency combination as selecting slip-window rules.

After selected frequency rules, we need to optimized a number of frequency combination in selected silent frequency channel. GA has systematical and parallelism, it have better searching capability, if there are enough iteration generations, GA can finally find the optimized combination. So we select GA to code and optimize frequency combination, for the quantity sequence number, 0 is vacancy, 1 is available, then every frequency combination can be represented by a binary digit. The objective function is to minimize max side-lob, if amplitude is unitary, side-lobs are negative, so we take its absolute value as objective function. Hence, according to limitative GA, we can get optimized silent frequency combination on different selecting rules.

The rule of max side-lob least can get better frequency combination in selected silent channels, but in selected segment always exist radio interferences , average power are much higher. While the rule of most silent channel can ensure less interference, provide much room to select frequency combination, getting side-lob least frequency combination more easily.

Exponential smoothing forecasting then four states to two states forecasting on current working frequency can get the next working frequency spectrum and the state of silent channel, then applying different selecting rules, and by GA gain frequency combination of lesser side-lobs. Selecting frequency can by the rules of average power least, the rules of power threshold least, the rules of undulate combine. Average power less, max side-lobs lower and the stabilization of selected channel are always incompatible, which can be selected by different electromagnetic environments and selecting goals.

5. Test and evaluate selecting effect

A certain electromagnetic environments can get different frequency segments by different selecting rules, in order to evaluate selecting effect, we need to test the frequency combination taking the noise floor lower and the max signal side-lob lower as criterion.

Using actual spectrum data to select by the rule of most silent channel and the rule of max side-lob least separately, adding emulation target echo to process in velocity and range domain. Emulation parameters are as table 1.

Select frequency 100 times by the two methods separately and process, the statistic result is as fig2 and fig3. (Both GA are 10 times iterations)

The probability of signal side-lobs level gained by the rule of most silent channel higher than the rule of max side-lob least is 0.81, the probability of noise floor gained by the rule of most silent channel lower than the rule of max side-lob least is 0.72, which are consistent with expectation. The compare of max signal side-lobs level and noise floor with the two methods are given in table2 and table 3.

The probability of signal side-lobs level gained by the rule of most silent channel higher than the rule of max side-lob least is 0.81, the probability of noise floor gained by the rule of most silent channel lower than the rule of max side-lob least is 0.72, which are consistent with expectation. The compare of max signal side-lobs level and noise floor with the two methods are given in table2 and table 3.

From table2 and table3, we can see the frequency combination selected by the rule of most silent channel can gain lower noise floor and its signal side-lobs level is stable, which can improve the detection capability of HFSWR. And through a mass of emulation, the noise floor will be lowered in 1-3dBc after forecasting.

TABLE 1 Emulation parameters in processing

T_{pp}	3ms	T_{pw}	0.5ms	T_{wp}	450ms	P_c	0.6	P_m	0.01
N	15	M	128	R	150Km	V	36Km/h	f0	7Mhz
B_n	2.5Khz	GA	10	Span	100Khz	Coef	99	f_s	100Kz

Table 2 Statistic of noise floor by the different selecting methods

noise floor(dBc)	max	min	mean
rule of max side lob least	-67.3702	-81.3848	-75.3011
rule of most silent channel	-72.7986	-82.7144	-78.2164

TABLE3 Statistic of max signal side-lobs level by the different selecting methods

max signal side-lob(dB)	max	min	mean
rule of max side lob least	-2.6935	-19.5233	-14.2695
rule of most silent channel	-12.6050	-16.1017	-14.1506

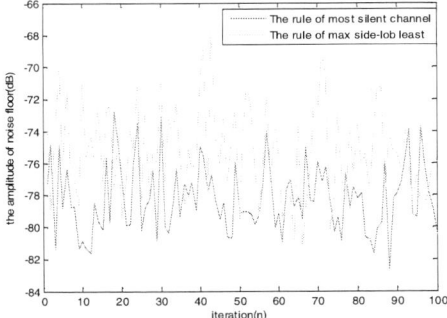

Fig.2 Chang of noise floor along with n by the two different selecting methods

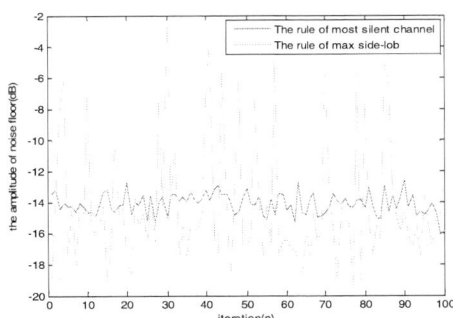

Fig3 Chang of max signal side-lob along with n by the two different selecting methods

5. Conclusion

This paper studied the frequency selecting methods of frequency hopping signal radar in HFSWR. Forecast with deterministic model followed by state partition was proposed, which has higher reliability than traditional deterministic model and statistical model. The rule of most silent channel was used to select the optimum frequency band and genetic algorithm was used to optimize frequency combination, which can make frequency hopping signal sending in the next working period has lower side-lob. It effectively minimized the noise floor and the signal side-lobe level and improved the adaptability of HFSWR in heavily congested electromagnetic environments.

References

[1] G.D. McNeal (1997). "Bandwidth occupancy profile of mid-Atlantic HF band Users". Radio Science. Vol. 32, No. 5, pp.2037-2045.

[2] S.D. Green, S.P. Kingsley (1997). "Improving the Range Time Side-lobes of Large Bandwidth Discontinuous Spectra HF Radar Waveforms". The Seventh International Conference on HF Radio Systems and Techniques, Nottingham, 1997: 246-250.

[3] Wei Yin-sheng, LIU Yong-tan(2002). "Random Interrupted HF radar waveform designing and processing. Electronic transaction". Vol. 3, No. 3, pp.437-440.

[4] Zhang Dong-po, Liu Xing-zhao (2004). "A Processing Method for Random Spectrum Discontinuous Radar Signal". Modern Radar. Vol. 8, No.8, pp. 42-44.

[5] V.M.Kutuzov. "Synthesis of Non-Regular Mutitone Signals and Algorithms of Their Processing". Proceedings of ICSP'96: 813-816

[6] D.J. Percival, M. Kraetzl, M.S. Britton. "Markov Model for HF Spectral Occupancy in Central Australia". IEE Conference of the 7th International Conference on HF Radio Systems and Techniques. Nottingham, 1997, 441:7-10

[7] D.J. Percival. "A Markov model for HF spectral occupancy in central Australia". IEE Conference Publication. 1997,411:14-18

[8] WEI Yin-sheng, LIU Yong-tan(2004). "Genetic Algorithm Realization of Optimum Frequency Coding of Signals with Discontinuous Spectra". Modern Radar. Vol. 1, No. 1, pp.14-16.

A New Highway Incident Detection Method Based on Wireless Positioning Technology

Hang Wen

Transportation College of Southeast University
zeal-hw@sohu.com

Abstract

An innovative motorist-initiated incident detection method using wireless positioning technology, such as Global Positioning System (GPS) and Cellular Phone Positioning System (CPS) is developed. Then, the basic detection algorithm is presented. The position information and operating characteristics of reporting vehicles are detected using wireless positioning system and is transferred by Global System for Mobile Communications (GSM) networks. According to the information, the Traffic Management Center (TMC) can position the incident accurately in short order. The validity of this incident detection method is proved by a simulation model and the detecting performances are evaluated under different traffic and incident conditions.

1. Introduction

With the development of highway networks in the world, the problem of congestion is becoming more and more serious. These congestions consist of recurrent congestion plus the additional (non-recurrent) congestion caused by accidents, breakdowns, and other random events, for example, inclement weather and debris. Evidently, quick and adequate detection of incidents and fast response of emergency services will reduce consequences of incidents considerably. Many counties have paid much attention to the management of highway traffic incidents. Since 1960s, many kinds of incident detection (ID) methods have appeared which can be sorted as manual detection methods and automatic detection methods generally.

The manual detection techniques are the earliest, most enforceable methods, and they are also most commonly used. Generally speaking, the main advantages of manual detection techniques are convenient, direct, economical and efficient. The Automatic Incident Detection (AID) methods can be classified into two kinds, the indirect method and the direct method. Most of AID methods belong to the former. Though the automatic incident detection techniques play more and more important roles in traffic management, the manual detection techniques are still the leading method in actual operation by experience. Even in America, the most popular methods to find traffic incidents are no doubt the manual methods. In China, where the automatic incident detection techniques are still under development, the use of automatic detection techniques is even infrequently. So, an innovative real-time incident detection method should be developed based on new technologies.

2. Methodologies

2.1. Process of incident detection

As shown in Figure 1, motorist-initiated highway incident detection methods based on wireless positioning techniques mainly use two ways to find incidents.

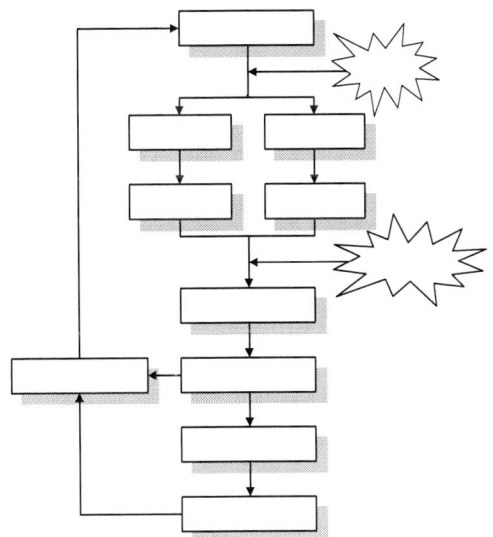

Figure 1. Process of motorist-initiated incident detection

On one hand, when the motorist has found his/her vehicle broken down or other vehicles met with an incident, he/she can observe for a while or immediately press the alarm button on in-car terminal. The in-car terminal can give an alarm to Traffic management Centre (TMC) via Global System for Mobile Communications (GSM) network.

On another hand, when vehicle encounters the shake, incline or collision, the in-car sensor can detect the situation of vehicle automatically, comparing detecting

978-0-7695-3342-1/08 $25.00 © 2008 IEEE
DOI 10.1109/PEITS.2008.61

data with thresholds (it can receive other in-car equipment's feedback information, for example, when airbag ejects, the in-car sensor will estimate the vehicle got involved in an incident) and estimate if the vehicle has been involved into an incident. If incident is affirmed, the in-car terminal will send a default alarm message to TMC via GSM.

It is important to send the information of incident position to TMC via wireless positioning and GSM techniques effectively and immediately. Now, there are two kinds of techniques available. One is the Global Positioning System (GPS); the other is Wireless Phone Positioning System (or Cellular Phone Positioning System, CPS). The key factors which influence the performance of certain positioning techniques include positioning accuracy, positioning frequency, reliability and network scope, etc.

2.1. Framework of Incident Detection System

As shown in Figure 2, the typical motorist-initiated incident detection system consists of three proportions which are in-car terminal, public GSM network and TMC.

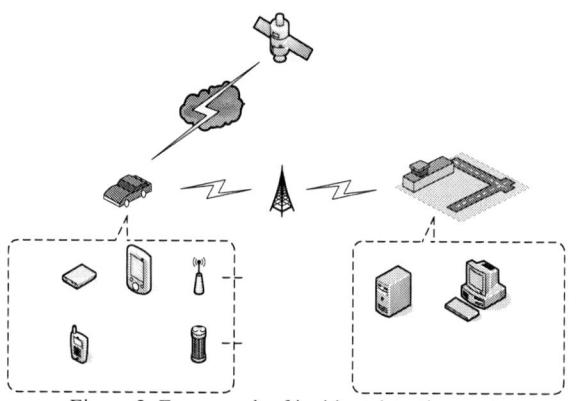

Figure 2. Framework of incident detection system

1) Traffic management Centre (TMC)

The TMC, being the core of the incident detection system, is made up of communication gateway and information processing terminal. The main task of TMC is to receive the report information from road users and to make proper response. For one thing, it receives alarm reports from the in-car terminals, and then passes them to the information processing terminal which can deal with the information based on Geographic Information System (GIS). And then, TMC responds to the verification information of traffic incident from information processing terminal and send them to the decision-maker of TMC.

2) In-car terminal

The in-car terminal consists of three core parts, GPS module, GSM module and control module as well as some optional parts such as airbag sensors, CPS module, etc. The GPS module is used to receive positioning signal from satellites. The GSM module is used to exchange information with TMC via GSM network. And the control module can detect the running characteristics of vehicles, such as driving direction, velocity and acceleration of vehicle and the location of vehicle. The airbag sensor can trigger the alarm button and send the alarm report automatically when the vehicle gets involved into an incident and the airbags eject. The CPS module is used to get the positioning information from GSM base stations.

The basic operation mode of the in-car terminal is given as follows: As soon as motorist presses alarm button, the in-car terminal receives GPS positioning signal which will be sent to TMC via GSM network after being pre-processed by the in-car control module. What's more, if the in-car terminal is equipped with airbag sensors, the vehicle could send alarm report automatically when it comes across a casualty, and if the in-car terminal has CPS module, the GPS positioning mode can be changed to GSM positioning mode automatically when the GPS signals are disturbed.

3) GSM network

GSM network, through which alarms including GPS or CPS positioning information can be accurately sent to TMC, is the data link between the in-car terminal and the TMC.

3. Algorithms

3.1. Incident detection algorithms

In an incident situation, a typical driver would reduce speed as the incident came into the driver's view because of safety or curiosity. The driver would maintain this reduced speed until he or she passes the incident location, whereupon the driver would accelerate back to or even exceed his upstream speed. This phenomenon is commonly called rubbernecking or gawking. While a vehicle get involved into an accident, it might gradually or suddenly decelerate to stop. According to these principles, TMC could position the incident spot based on some rules.

1) Case of single report

When TMC receives only one incident report, it could make conclusions as follows based on the velocity, acceleration and location of the vehicle giving the alarm:

➢ The vehicle giving alarm is right in or just passes the incident spot if it is in a low speed and being accelerated.

➢ The vehicle giving alarm is getting close to the incident spot if it keeps running in a low and uniform velocity.

➤ The vehicle giving alarm has passed by the incident spot if it keeps running in a high and uniform velocity.
➤ The vehicle giving alarm is getting close to the incident spot if it is in a high speed and being decelerated.
➤ The vehicle giving alarm is involved in an incident or waiting in a queue caused by congestion if it has stopped.

It is important to define the conceptions such as "high speed", "low speed", "uniform velocity", "accelerate" and "decelerate". That is to say, we should give some thresholds to both velocity and acceleration. To make proper thresholds, we should consider lots of factors such as road conditions, traffic conditions, characteristics of alarming vehicles, air conditions etc. In the meantime, the distance between alarm vehicle and incident spot should be estimated based on history or investigation data.

2) Case of multi-reports

The TMC can verify the incident information including position and severity of the incident more reliably if it receives reports from more than two vehicles, namely multi-reports. Of course, the precondition of using multi-reports is that the conclusions drawn from every report are coincident in logic. If the conclusions drawn from multi-reports are incompatible, the TMC had better to use the prepared algorithms, for example, Artificial Neural Network Model, to estimate the position and severity of an incident. Here we use "Cluster Computing Method" to define this procedure.

3.2. Incident verification method

Incident verification is to verify if an incident has happened, where it has happened, how severe it is and other incident information. There are several commonly used methods for incident verification including:
➤ Design certain incident detection algorithms with a check method that requires TMC to receive a certain number of reports before any action is initiated;
➤ Closed circuit TV cameras viewed by operators;
➤ Dispatch field unit (e.g. police or service patrols) to the incident site;
➤ Communication with aircraft operator by the police, the media, or an information service provider;
➤ Information from cellular calls.

4. Feasibility Evaluations

From literature reviewed, there is no precedent of traffic incident detection method based on wireless positioning technology that has been used on a large scale in the world. Consequently, in the absence of field data, a simulation test model was developed to measure the effectiveness of this incident detection method theoretically.

4.1. Modeling approach

A microscopic, time-stepping simulation model was used to generate data on the temporal and spatial sequence of individual driver arrivals upstream of an incident under different traffic flow rates and incident types. The data generated by the simulation model were then fed into a discrete probability model to represent the binary responses of drivers reporting or not reporting incidents they observed on the highway. In summary, to get the desired operational measures, two models were applied sequentially: the simulation model followed by the probability model.

4.2. Simulation results

The examination of the system performance was based on the analysis of the aggregate influence of the variation of input parameters of the simulation model and the probability model. The simulation input variables and the probability modeling parameters were varied independently and applied sequentially. The three major variables were the proportion of vehicles with in-car terminal, the severity of the incident, and false alarm reduction procedure.

1) Proportion of vehicles with in-car terminal (p_1)

Figure 3 shows the detection performance of the system detecting a disabled vehicle blocking one lane of a simulated three-lane directional freeway under variable traffic flow conditions.

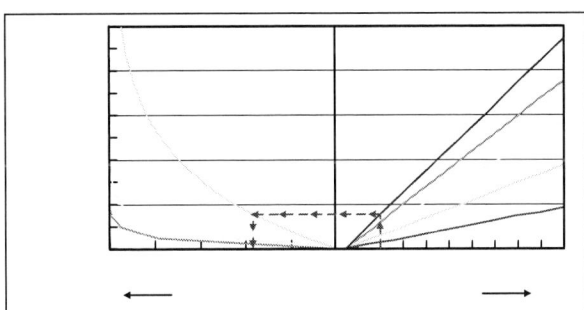

Figure 3. Effect of proportion of vehicles with in-car terminal

It is shown that, the probability of detection will be enhanced with the increase of traffic volume and proportion of vehicles with in-car terminal (p_1). Commonly, as the traffic volume keeps steady in short term, p_1 seems to be the key variable of incident detection performance. For example, when the traffic volume equals 2000 vehicle per hour per lane, if p_1 is 0.01, the detection rate will never exceed 49% within 2 minutes; and if p1 is 0.1, the detection rate will reach one hundred percent.

2) Severity of incident

Figure 4 shows how detection performance is affected by the severity of the incident. The data used in Figure 4 were generated by simulation model using 1000 vehicle/h/lane as traffic flow rates and 0.1 as proportion of vehicles with in-car terminal. The incident is divided into four types: shoulder incident, one-lane blocking incident, two-lane blocking incident and three-lane blocking incident.

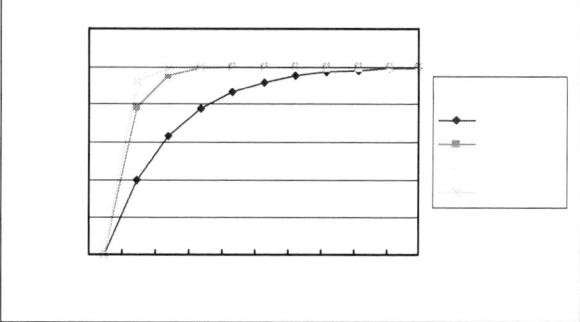

Figure 4. Effect of severity of incident

It is obviously that the detection performance is improved with the increase of severity of incident. It is because that motorist will be puzzled if the vehicle has met the incident or just break down when he/she find a vehicle stopped at shoulder. The simulation result also shows that the measure of mean time of detection (MTD) can't be used to evaluate the detection performance under different severity of incident:

3) Minimum number of reports (K)

A false alarm is defined as the report of an incident by a vehicle when no incident has occurred. In order to reduce false alarms, a highway agency can institute a procedure that would require a TMC to receive a certain number of reports before any action is initiated. Here, K is defined as a nonnegative integer specifying minimum number of reports about an incident that should be received by TMC before a verification action is initiated and five cases were considered, $K = 1, 2, 3, 4,$ and 5. Figure 5 shows how detection performance is affected by the minimum number of reports.

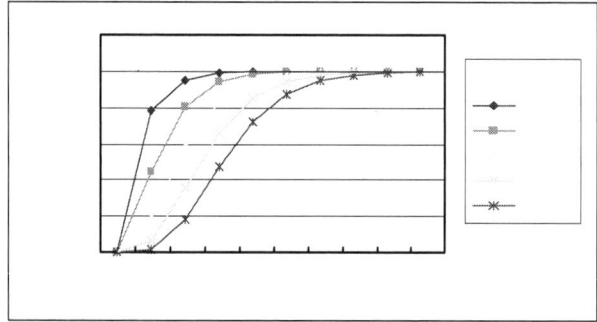

Figure 5. Effect of minimum number of reports

With the increase of K, detection performance declines evidently, especially in the initial several minutes. As K is designed to reduce the false alarm rate to concentrate the personnel and material resources to the verified incident, the TMC can set up a series of variable minimum number of reports according to different environments to equilibrate the detection performance and false alarm rate.

5. Conclusions

The motorist-initiated incident detection method based on wireless positioning technology is proved to be feasible and:

➢ be applicable in online environments, restricting computation times and type of input data.
➢ be calibrated with a relatively small amount of data;
➢ be cost-effective;
➢ be robust with respect to measurement errors and be relatively insensitive to parameter settings (for example traffic volume); and
➢ be generic and easily transferable to other locations.

The preconditions of using this ID method consist of increase of the proportion of vehicles with wireless positioning in-car terminal, improvement the propaganda, and properly designed incentive mechanism on potential reporting motorists.

6. References

[1] X.Y. Zhang, Q.D. Da, and G.W. Zhang. "Technology of highway automatic incident detection". System Engineering Theory and Application, vol. 6, pp. 118-124, 2006. (In Chinese)

[2] H.M. Wen. "Study on the key theories and technologies of incident management system on freeways". Changchun: Ph.D. thesis of Jilin University, January 2002. (In Chinese)

[3] D.H.Sun, W.X. Chen. "Expressway automatic incident detection method using mobile phone positioning", Journal of Highway and Transportation Research and Development, vol. 23, no. 2, pp. 133-136, 2006. (in Chinese)

[4] B.L. Smith, H. Zhang, M. Fontaine, and M. Green. Final report of ITS center project: Cellphone probes as an ATMS tool. http://cts.virginia.edu, 2003.

[5] R.N. Mussa, J.E. Upchurch, "Monitoring urban freeway incidents by wireless communications". Transportation Research Record 1748, 2001.

[6] D.H. Wang. Traffic Flow Theory. Beijing: China Communications Press. September 2002. (in Chinese)

Session 2

Information Security

A Dynamic ID-based User Authentication and Key Agreement Scheme for Multi-server Environment Using Bilinear Pairings

Jianyan Geng

Department of Mathematics, Wanjie Medical College, Zibo, 255213, P. R. China
gengjianyan@gmail.com

Lijiang Zhang

Key Lab of Cryptographic Technology and Information Security, Ministry of Education
Shandong University, Jinan, 250100, P. R. China
zhlijiang@gmail.com

Abstract

Remote user authentication is suited to solve the privacy and security problems for the single server architecture environment. With the rapid development of Internet service, more and more network architectures are used in multi-server environments. In this paper, a user authentication scheme for multi-server environment is presented. The proposed scheme is based on bilinear pairings. Moreover, it does not only implement mutual authentication and session key agreement but also achieve user's anonymity.

1 Introduction

With the rapid development of Internet service, remote user authentication scheme becomes an important issue for practical applications, especially in electronic commerce. In 1981, Lamport [9] proposed a remote password authentication scheme for single server environment and then a variety of schemes [11, 20, 5, 2, 16, 3, 12, 8] have been proposed to improve the security, the cost or the efficiency.

Recently, more and more network architectures are used in multi-server environments. In such multi-server network architecture, a network user must prove to the servers that he is a legitimate remote login user before he can get any service from the various servers. In conventional password authentication methods, each network user does not only need to log into various remote servers repetitively but also needs to remember various user IDs and passwords. Therefore, users often have to write down their numerous IDs and passwords. This is insecure because these IDs and passwords may easily leak out. Another problem in using traditional remote login methods to authenticate a user is that the

server often uses a verification table that stores each user's ID and password. Therefore, the server requires extra memory space to store the verification table and these schemes will be partially or totally broken if the verification tables are stolen by the adversary. Moreover, any users who want to can get service from multiple servers must repeat registration to each server and it make the scheme inefficient.

A secure and efficient remote user authentication scheme for multi-server environment usually meets the following requirements [7]:

1. **Single registration.** Users only must register at the registration center once and can use all the permitted services in eligible servers.

2. **Change password securely and freely.** It allows the cardholder to change his password freely after assuring the legality of card holder.

3. **Mutual authentication and key agreement.** It allows the users and servers to authenticate with each other and then negotiate a session key to protect the transmitting message.

4. **Low computation and communication cost.** Due to the power constraints and small flash memory of smart cards, they may not provide a powerful computation capability and high bandwidth.

5. **No verification table.** No verification or password table is stored in a server.

6. **Security**. The authentication scheme must be able to resist all kinds of attacks such that it can be applied in the real world.

In this paper, we propose a dynamic ID-based user authentication and key agreement scheme for multi-server

environment using bilinear pairings. We show that our proposed scheme can resist replay attack, forgery attack, server spoofing attack, stolen verifier attack and insider attack. Our scheme does not only implement mutual authentication and session key agreement but also achieve user's anonymity. Moreover, the proposed scheme achieve low-computation requirement for the smart card's applications.

The remainder of this paper is organized as follows: some related work is given in section 2 and definition and properties of bilinear pairings be described briefly in section 3. We present our novel authentication scheme in section 4 and analysis the performance and security of the proposed scheme in section 5. Finally, we make conclusion in section 6.

2 Related Work

Due to the widespread applications of Internet service, the study of remote user authentication scheme for multi-server environment has received considerable attention and some schemes have been proposed [10, 18, 13, 14, 7, 19]. These schemes can be divided into two types, namely hash-based authentication and public-key based authentication.

In 2000, Lee and Chang [10] proposed a user identification and key distribution scheme for multi-server environment that is based on the difficulty of factorization and hash function. Subsequently, Tsaur et al. [18] proposed a remote user authentication scheme based on RSA cryptsystem in 2001. At the same time, Li et al. [13] proposed a multi-server authentication scheme using neural networks. The major drawback of this approach is it spends too much time on training neural networks. In [14], Lin et al. proposed a remote authentication scheme based on the discrete logarithm problem for multi-server environments. However, in this scheme every user must have a large amount of memory to store the public parameters for authentication later that is impractical for smart-card based applications. In 2004, Juang [7] proposed an efficient multi-server user authentication and key agreement scheme based on hash function and symmetric-key cryptosystem. Recently, Liao and Wang [19] presented a secure and efficient authentication scheme with anonymity for multi-server environment that is based on hash function.

Recently, the bilinear pairings have been found as important applications in cryptography and allowed us to construct practical identity (ID) based cryptographic schemes [1, 15, 17]. In 2006, Das et al. [4] proposed a remote user authentication scheme using the properties of bilinear pairings for single server environment. In 2007, Yoon et al. [21] demonstrates that Das et al.'s scheme was still vulnerable to an impersonation attack and an off-line password guessing attack and presented an improved remote user authentication and key agreement scheme for single-server environ-

ment.

3 Bilinear Pairings

We briefly describe the basic definition and properties of bilinear pairings in this section.

Suppose G_1 is an additive cyclic group generated by P, whose order is a large prime q, and G_2 is a multiplicative cyclic group of the same order. A map $\hat{e} : G_1 \times G_1 \rightarrow G_2$ is called a bilinear mapping if it satisfies the following properties:

1. Bilinear: $\hat{e}(aP, bQ) = \hat{e}(P, Q)^{ab}$, for all $P, Q \in G_1$ and all $a, b \in Z_q^*$.

2. Non-degenerate: there exists $P, Q \in G_1$ such that $\hat{e}(P, Q) \neq 1$.

3. Computable: there is an efficient algorithm to compute $\hat{e}(P, Q)$ for all $P, Q \in G_1$.

Typically, the bilinear mapping can be derived from either the Weil or the Tate pairing on an elliptic curve over a finite field. A more comprehensive description about bilinear mapping and its applications can refer to [1, 6, 15, 17].

Now, we review some well-known problems to be used in the security analysis of our proposed scheme.

1. Elliptic Curve Discrete Logarithm Problem (ECDLP): Consider the equation $Q = kP$, where $P, Q \in G_1$. It is easy to calculate Q given k and P, but it is hard to determine k given P and Q.

2. Bilinear Computational Diffie-Hellman Problem (BCDHP): Given (P, aP, bP, cP) for $a, b, c \in Z_q^*$, it is hard to compute $v \in G_2$ such that $v = \hat{e}(P, P)^{abc}$.

4 The Proposed Scheme

In this section, we propose an efficient and secure authentication and key agreement scheme for multi-server environment that is based on bilinear pairings. The proposed scheme achieves user's anonymity by using dynamic ID. The new scheme is composed of four phases: initialization phase, registration phase, login & session key agreement phase and password change phase. Figure **??** shows the proposed authentication scheme. The scheme works as follows:

Initialization Phase: Let G_1 be an additive cyclic group of prime order q generated by P and G_2 a multiplicative cyclic group of the same order as G_1. Registration center RC selects a cryptographic hash function $H : \{0, 1\}^* \rightarrow G_1$ and a one way collision-resistance function $f : \{0,1\}^* \rightarrow \{0,1\}^*$. $\hat{e} : G_1 \times G_1 \rightarrow G_2$ is a bilinear map from additive group G_1 to multiplicative group G_2.

The RC selects a secret key s and computes its public key $Pub_{RC} = sP$. Suppose that $\{S_1, S_2, \cdots, S_m\}$ is a set of servers in a multi-server environment. RC selects a secret key x_j for the server S_j where $j = 1, 2, \cdots, m$ and computes the server's public key $Pub_j = x_j P$. Then, the RC publishes the system parameters $(G_1, G_2, q, P, \hat{e}(\cdot), Pub_{RC}, Pub_{j\{j=1,2,\cdots,m\}}, f(\cdot), H(\cdot))$ and keeps s secret. The RC sends $x_j, f(s)$ to server S_j via a secure channel.

Registration Phase: Whenever user U_i wants to access the resources of the service provider S_j, he has to submit his identity ID_i and $H(PW_i)$ to RC. Then, RC performs the following computations:

1. Compute $SID_i = H(ID_i, ID_{RC})$, $P_i = s \cdot SID_i$, $V_i = P_i + H(ID_i\|H(PW_i))$, $Ver_i = f(P_i)$ and $AID_i = \hat{e}(H(ID_{RC}), SID_i)^{f(s)}$ where $\|$ is a concatenation operation and ID_{RC} is the RC's identity.

2. Store $SID_i, V_i, Ver_i, AID_i, H(\cdot), f(\cdot)$ in a smart card and issue to the user U_i through a secure channel.

Login & Session Key Agreement Phase: When the user U_i wants to access the sources of the remote server S_j, he inserts the smart card and provides his identity ID_i^* and password PW_i^*. Then the smart card executes the following steps:

1. Compute $P_i^* = V_i - H(ID_i^*\|H(PW_i^*))$ and checks whether $f(P_i^*)$ and Ver_i is equal or not. If yes, the validity of the user can be assured and proceeds to the next step; otherwise, reject the login request.

2. Choose random $r_1 \in Z_q^*$ and computes $C_1 = r_1 P$, $CID_i = SID_i + r_1 \cdot Pub_j$, $h = f(N_i\|C_1)$ and $W = r_1^{-1}(P_i^* + h'P)$.

3. Send the login request message $< CID_i, C_1, N_i, W >$ to the server provider S_j.

Upon receiving the login request message $< CID_i, C_1, N_i, W >$, the server provider S_j performs the following steps:

1. Compute $SID_i^* = CID_i - x_j \cdot C_1$ and checks whether the equation $\hat{e}(W, C_1) \overset{?}{=} \hat{e}(SID_i, Pub_{RC}) \cdot \hat{e}(P, P)^h$ holds. If not, the server S_j rejects the login request and terminates this session.

2. Choose random $r_2 \in Z_q^*$ and computes $C_2 = r_2 P$.

3. Compute session key $sk = f(r_2 \cdot C_1)$. Then, compute $AID_i^* = \hat{e}(H(ID_{RC}), SID_i)^{f(s)}$ and $Ver = f(AID_i^*, C_1, N_i, sk)$.

4. Send back the message $< C_2, Ver >$ to the smart card.

Upon receiving the message $< C_2, Ver >$, the smart card performs the following steps:

1. Compute session key $sk^* = f(r_1 \cdot C_2)$ and checks whether $f(AID_i, C_1, N_i, sk^*)$ and Ver is equal or not. If yes, the validity of the server S_i is authenticated, otherwise, the session is interrupted.

2. Compute $Ver' = f(AID_i, C_2, N_i, sk^*)$ and sends back the message Ver' to the server S_j.

Upon receiving the message Ver', the server provider S_j computes $f(AID_i^*, C_2, N_i, sk)$ and compares it with Ver'. If it holds, the identity of U_i can be authenticated.

Password Change Phase: When the user U_i wants to change his password, he inserts his smart card to card reader and inputs ID_i^* and PW_i^*. Then the smart card computes $P_i^* = V_i - H(ID_i^*\|H(PW_i^*))$ and checks whether $f(P_i^*)$ and Ver_i is equal or not. If yes, the smart card allows the user to resubmit a new password PW_i^{new}, and then compute $V_i^{new} = P_i^* + H(ID_i^*\|H(PW_i^{new}))$. Thus, V_i stored in smart card can be replaced with V_i^{new}.

Equation $\hat{e}(W, C_1) = \hat{e}(SID_i, Pub_{RC}) \cdot \hat{e}(P, P)^h$ can be deduced as follows:

$$
\begin{aligned}
\hat{e}(W, C_1) &= \hat{e}(r_1^{-1}(P_i + hP), r_1 P) \\
&= \hat{e}(P_i + hP, P)^{r_1^{-1} \cdot r_1} \\
&= \hat{e}(P_i, P) \cdot \hat{e}(hP, P) \\
&= \hat{e}(s \cdot SID_i, P) \cdot \hat{e}(P, P)^h \\
&= \hat{e}(SID_i, sP) \cdot \hat{e}(P, P)^h \\
&= \hat{e}(SID_i, Pub_{RC}) \cdot \hat{e}(P, P)^h.
\end{aligned}
$$

5 Security and Performance Analysis

We analyze the security and performance of the proposed authentication scheme in this section. We show that our proposed scheme can resist replay attack, forgery attack, server spoofing attack, stolen verifier attack and insider attack. Moreover, the proposed scheme can achieve mutual authentication and session key agreement.

1. *Replay attack.* A replay attack is a form of network attack in which a valid data transmission is maliciously or fraudulently repeated or delayed. If the adversary replays an old message $< CID_i, C_1, N_i, W >$ to the server S_j, then the server sends message $< C_2^*, Ver^* >$ back to the adversary. However, the adversary can not obtain the corresponding session key $sk^* = f(r_1 \cdot C_2^*)$ and $Ver^* = f(AID_i, C_2^*, N_i, sk^*)$ without knowing r_1 and AID_i. On the other hand, if the the adversary replays an old verification message C_2, Ver to the user where Ver is associated with N_i^*, then the user computes $f(AID_i, C_1, N_i, sk^*)$ and checks whether is is equal to Ver. They are not equivalent obviously since N_i is not equal to N_i^*.

2. *Forgery attack and server spoofing attack.* When the adversary wants to masquerade the legal user U_i to pass the verification of the server S_j, the adversary must construct a valid login request message $< CID_i, C_1, N_i, W >$. Without knowing the ID_i^* and PW_i^*, the adversary cannot compute P_i^* to forge a valid $W = r_1^{-1}(P_i + hP)$. Similarly, the proposed scheme can protect the user from cheating by the masqueraded server provider since the adversary can not obtain AID_i^* and Ver without the knowledge of x_j and $f(s)$.

3. *Stolen verifier attack.* Since the scheme has no maintain a verification table or password table in the remote server, nobody obtain any verifiable information from the remote server to threaten the proposed scheme. So the scheme can resist the stolen verifier attack.

4. *Insider attack.* In many scenarios, the user uses a common password to access several systems for his convenience. If the user login request is password-based and the server maintains password or verifier table for login request verification, an insider of server could impersonate user's login by stealing password and gets access of the other systems. In the registration phase, since user U_i registers to RC with $ID_i, H(PW_i)$, the insider of RC can not obtain PW_i directly. When the user U_i login to server S_j, the insider can obtain $P_i = s \cdot SID_i$ and $f(s)$ instead of s. It is infeasible to get s from P_i under ECDLP assumption. Also, it is hard to get s from $f(s)$ since it is protected by one way function.

5. *User's anonymity* The user U_i will send the login message $< CID_i, C_1, N_i, W >$ to the server provider S_j. Hence, an attacker may incept and analyze the login message. However, it is infeasible to derive ID_i from the login message because H is a cryptographic hash function. Furthermore, the login message is dynamic in each login phase. Therefore, our scheme can protect user's anonymity.

6. *Password change securely.* when the smart card was stolen, unauthorized users cannot change the password of the card without knowing the U_i's password PW_i, because the smart card can verify P_i using the stored Ver_i in the password change phase. Therefore, the proposed scheme provides secure password change.

7. *Session key agreement.* In the login & session key agreement phase, user computes key $sk^* = f(r_1 \cdot C_2)$ and the remote server computes key $sk = f(r_2 \cdot C_1)$. Since $sk^* = f(r_1 \cdot C_2) = f((r_1 r_2) \cdot P) = f(r_2 \cdot C_1) = sk$, the user U_i and the server S_j get common session key sk. Moreover, an attacker who can get C_1, C_2 via public channel can not obtain sk under ECDLP assumption.

8. *Mutual authentication.* Mutual authentication means that both the user and remote system are authenticated to each other within the same protocol. Firstly, the server authenticates the user U_i by checking $\hat{e}(W, C_1) = \hat{e}(SID_i, Pub_{RC}) \cdot \hat{e}(P, P)^h$, because it is hard to compute W without knowing the knowledge of U_i^* and r_1. Moreover, the server also authenticates the user U_i by verifying $Ver^* = f(AID_i^*, C_2, N_i, sk)$. An attacker who does not know both AID_i and r_1 can not pass the authentication. Similarly, user authenticates the server S_j by verifying $Ver = f(AID_i, C_1, N_i, sk^*)$. An adversary can not calculate AID_i^* without knowing x_j and $f(s)$. Therefore, mutual authentication between user and server is achieved.

9. *Forward security.* The forward security is defined as the assurance that any previous session keys will not compromised if the master secret key s is leaked. In our scheme, any session keys are related to r_1, r_2, which is dynamic in different session. Hence, it is not helpful to deduce the session keys used in the past that the secret key s is leaked.

Since the computation ability of the smart card is limited, the authentication scheme must take efficiency into consideration. We only consider the computational cost of login & session key agreement phase. In our scheme, smart card needs five scalar multiplications of elliptic curve point and two hash to point operation; also, three scalar multiplications and four bilinear pairing operations are performed by the server. As the pairing operation is costly, so our scheme takes high computation cost. However, pairing operation is done by the server with large computation system, thereby the computation cost of the verification process is not a constraint.

We summarize the functionality of the proposed scheme and make comparison with some related schemes in table 1.

Table 1. The functionality comparison between our proposed scheme and the others

	Ours	Juang [7]	Tsuar [18]	Lin [14]
Single registration	Yes	Yes	Yes	No
Securely change password	Yes	No	No	No
Mutual authentication	Yes	Yes	No	No
Session key agreement	Yes	Yes	No	No
User's anonymity	Yes	No	No	No
No verification table	Yes	Yes	Yes	Yes
Forward security	Yes	Yes	No	No
Replay attack resistant	Yes	Yes	Yes	No

6 Conclusion

In this paper, we present a novel multi-server user authentication and key agreement scheme using bilinear pairings. We show that our proposed scheme can resist replay attack, forgery attack, server spoofing attack, stolen verifier attack and insider attack. Our scheme does not only implement mutual authentication and session key agreement but also achieve user's anonymity. Moreover, the proposed scheme achieve low-computation requirement for the smart card's applications.

Acknowledgement

The second author is supported by the National Science Fund for Distinguished Young Scholars (Grant No. 60525201) and by the National Grand Fundamental Research 973 Program of China (Grant No. 2007CB807902 and Grant No. 2007CB807903).

References

[1] D. Boneh and M. Franklin. Identity-based encryption from the weil pairing. *Advances in Cryptology*, LNCS 2139:213–229, 2001.

[2] C. K. Chan and L. M. Cheng. Cryptanalysis of a remote user authentication scheme using smart cards. *IEEE Trans. Consumer Electron*, 46(3):992–993, 2000.

[3] C. C. Chang and K. F. Hwang. Some forgery attacks on a remote user authentication scheme using smart cards. *Informatics*, 14(3):289–294, 2003.

[4] M. L. Das, A. Saxena, V. P. Gulati, and D. B. Phatak. A novel remote user authentication scheme using bilinear pairings. *Computers & Security*, 25(3):184–189, 2006.

[5] M. S. Hwang and L. H. Li. A new remote user authentication scheme using smart cards. *IEEE Trans. Consumer Electron*, 46(1):28–30, 2000.

[6] A. Joux. A one round protocol for tripartite diffie-hellman. *Proceedings of Algorithmic Number Theory Symposium*, LNCS 1838:385–394, 2000.

[7] W. S. Juang. Efficient multi-server password authenticated key agreement using smart cards. *IEEE Trans. Consum. Electron.*, 50(1):251–255, 2004.

[8] W. C. Ku and S. T. Chang. Impersonation attack on a dynamic id-based remote user authentiation scheme using smart cards. *IEICE Trans. Common.*, E88-B(5):2165–2167, 2005.

[9] L. Lamport. Password authentication with insecure communication. *Communication of the ACM*, 24(11):770–772, 1981.

[10] W. B. Lee and C. C. Chang. User identification and key distribution maintaining anonymity for distributed computer network. *Comput. Syst. Sci.*, 15(4):211–214, 2000.

[11] R. E. Lennon, S. M. Matyas, and C. H. Mayer. Cryptographic authentication of time-invariant quantities. *IEEE Trans. Common.*, 6:773–777, 1981.

[12] K. C. Leung, L. M. Cheng, A. S. Fong, and C. K. Chen. Crytpanalysis of a remote user authentication scheme using smart cards. *IEEE Trans. Consumer Electron*, 49(3):1243–1245, 2003.

[13] L. Li, I. Lin, and M. Hwang. A remote password authentication scheme for multi-server architecture using neural networks. *IEEE Trans. Neural Netw.*, 12(6):1498C1504, 2001.

[14] C. Lin, M. S. Hwang, and L. H. Li. A new remote user authentication scheme for multi-server architecture. *Future Gener. Comput. Syst.*, 1(19):13–22, 2003.

[15] K. G. Paterson. Id-based signature from pairings on elliptic curves. *Electron. Lett*, 38(18):1025–1026, 2002.

[16] J. J. Shen, C. W. Lin, and M. S. Hwang. A modified remote user authentication scheme using smart cards. *IEEE Trans. Consumer Electron*, 49(2):414–416, 2003.

[17] N. P. Smart. An identity based authentication key agreement protocol based on pairing. *Electron. Lett*, 38:630–632, 2002.

[18] W. J. Tsuar, C. C. Wu, and W. B. Lee. A flexible user authentication for multiserver internet services. *Networking-JCN2001*, LNCS2093:174C183, 2001.

[19] Y. P. L. S. S. Wang. A secure dynamic id based remote user authentication scheme for multi-server environment. *Comput. stand. interfaces*, doi:10.1016/j.csi.2007.10.007, 2007.

[20] S. M. Yen and K. H. Liao. Shared authentication token secure against replay and weak key attack. *Information Processing Letters*, pages 78–80, 1997.

[21] E. J. Yoon, W. S. Lee, and K. Y. Yoo. Secure remote user authentication scheme using bilinear pairings. *WISTP 2007*, LNCS4462:102–114, 2007.

A Novel Deterministic Packet Marking Scheme for IP Traceback

Qu Zhaoyang
School of Information Engineering,
Northeast Dianli University,
Jilin City, Jilin Province, China
E-mail: qzywww@mail.nedu.edu.cn

Huang Chunfeng
School of Information Engineering,
Northeast Dianli University,
Jilin City, Jilin Province, China
E-mail: hcf431@126.com

Abstract

This paper proposes a Novel Deterministic Packet Marking (NDPM) scheme for IP traceback. It mainly marks packets with IP addresses and the number of current Autonomous System (AS) through Border Gateway Protocol (BGP) routers. According to the marked information in the packets, victims can not only trace the attack source to the original AS, but also can filter the malicious packets, hence making it possible to alleviate the impact caused by attack traffic on the victims. It resolves the disadvantages of the traditional packet marking methods with heavy computational overhead and low speed of tracing back to the attack source. At the same time, it doesn't need all the routers' participation in marking in the attack path, which greatly reduces the overhead of routers.

Keywords: AS; Deterministic Packet Marking; IP Traceback; Computer Security

1. Introduction

Distributed denial-of-service (DDoS) attacks are the most common and effective network attack means in the Internet today. They are easy to implement, difficult to defense and trace, and pose an immense threat to the Internet. According to these features, researchers in this field have proposed many attack source tracing methods in recent years. The main methods [1] are: packet marking, logging, link testing, ICMP(Internet Control Message Protocol) traceback message, overlay network and so on.

One of the most effective defense mechanisms proposed was Path Identifier (Pi) [2]. As the Pi method uses the router's IP address to construct the path information of each packet, which was stored in each packet's identification field. However, because of the limitation of the identification field, only two bits of

resulted message digest of router's IP address are used, which results in the same path information representing different paths. To address this problem, Changhyun Beak et al. [3] proposed using Link-ID instead of IP addresses or routers to construct the path information of each packet. A Link-ID was the information of path between BGP routers in AS and each BGP router's connection to the outside of the AS. However, when the size of internal BGP Link-ID table is large, the hashing is necessary in the path Link-ID field, it can only provide a packet filtering rule for the victim, but can't trace the attack source to the original AS. Guang Jin, Jiangang Yang, Wei Wei and Yabo Dong have proposed Across-Domain Deterministic Packet Marking (ADDPM) [4] for IP traceback. In the scheme, it uses the 30-bit space in IP header reserved for fragmented traffic. Three deterministic markings are recorded into a packet at both the ingress router of source domain and the border router of destination domain respectively. The both routers' IP addresses and the source AS number are marked. With the marked information, the victim can trace to the remote attack origin by the markings. However, there are plenty of ingress routers in the Internet, it can't make sure that all the ingress routers are safe and the attackers can write forged information in the packets to mislead in filtering and tracing.

To resolve the above problem, this paper proposes a novel deterministic packet marking scheme for IP traceback.

2. The NDPM scheme

2.1. Design motivation of the NDPM algorithm

Our work is motivated by the following factors:
A. The number of BGP routers is far less than that of ingress routers in the Internet. Hence, making all the ingress routers safe is very difficult, if not impossible, while it can be realized to make sure that all the BGP routers are safe. If using a new

deterministic packet marking algorithm researches on these limited BGP routers, its efficiency will be greatly improved.

B. The AS number is 16 bits in length while IPv4 address is 32-bit length (IPv6 address length is 128 bits).Thus, encoding the AS number needs less header space than encoding IP address.

2.2. Basic idea of the NDPM algorithm

When a packet reaches the BGP router, it checks if the packet is en-router from other BGP router or other AS. If the check result is negative, firstly the router marks its information as the packet's initial router information and writes the current 16-bit AS number into the packet. Secondly, it concatenates the BGP router's information and current AS number. At last, it writes the hash result of the concatenation into the verification field. According to the marked information from the received packets, the victim firstly uses the verification field to decide whether the packet is a forged packet or not, and then filters the attack packets.

2.3. Coding scheme of the NDPM algorithm

The scheme stores the marking fields as shown in Figure 1.

Initial Router ID 10 bits	Verification 10 bits	ASID 16 bits

Figure 1. Marking fields of NDPM

Almost all the packet marking schemes utilize the fields in IP header which are reserved for fragmented traffic. The 16-bit identification field and the Reserve Flag (RF) bit are used by most previous schemes [5] [6]. Yet the 17 bits is not sufficient to convey too much address information in DPM-like methods. Recently we notice that the 13-bit offset field is also used more and more [7] [8] [9]. We agree that the occupation is feasible, for fragmented traffic is seldom and if the usage is redefined.

In order to make full use of available space in IP header, besides using 16-bit identification field, this scheme obtains more space through overloading the offset field and the TOS field. From Figure 1 we know, the coding scheme of the NDPM algorithm totally needs 36 bits for marking. The 16-bit source AS number is exactly put into the 16-bit identification field. And we place the 10-bit initial router ID (the 10-bit hash result of the initial BGP router's IP address) into the offset field and put the 10-bit hash result of the concatenation into the TOS field

and the last 2 bits of the offset field in the packet's IP header.

Figure 2 shows the concrete coding of the scheme in the IP header.

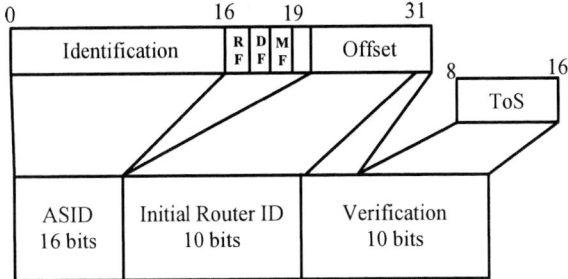

Figure 2. Coding scheme of NDPM

2.4. The NDPM algorithm realization

Figure 3 shows the marking process at BGP routers.

Main Process

/*At the BGP routers */
H = Hash(IP)   /*IP is the current BGP router IP address*/
ASN = The number of the Current AS

If (! the Packet is en-router from other BGP Router or other AS)
    { Write H into the Initial Router ID field
    Write ASN into the ASID field
    Concat(H, ASN)
    Write Hash(Concat(H, ASN)) into the Verification field
    }

Figure 3. Marking process at BGP routers

When the packet reaches the BGP Router, it checks out that the initial router information of the packet is not marked. It marks the 10-bit hash result of its IP address into the initial router ID field and 16-bit AS number into the ASID field in its forwarding packets, subsequently makes a concatenation with the initial router ID field and the ASID field, and then writes the hash result of the concatenation into the verification field.

When the packet arrives at the victim, it calculates the hash value of the concatenation with the initial router ID field and the ASID field. If the hash result is not equal to the value of the verification field, it shows that the packet is a forged packet, the victim drops it. If the result is equal, the victim filters the attack packets according to the marked information in the ASID field, the Initial Router ID field and the verification field. During the process of packet filtering, we introduce the notion of the threshold filtering to allow the victim to lower the false negative rate at the expense of raising the false positive rate (the rate at which attack packets are accepted). The intuition behind threshold filters is that it may be in the victim's best interest to accept a small number of attack packets if that allows it to accept a large number of legitimate users' packets. The threshold filter is simply a value $t_i \left(0 \le i < 2^{36} \right)$ chosen by the victim for each possible mark value i. Given the number of attack packets a_i with the mark i, and the number of users' packets u_i with the mark i, if the ratio of attack packets to the total packets is such that:

$$\frac{a_i}{a_i + u_i} > t_i \qquad (1)$$

Then the victim drops all packets with the mark i. For example, a threshold value of $t_3 = 0.25$ would allow a victim to admit all packets with a mark of 3, provided that attack traffic comprises less than 25% of all traffic with the mark three.

3. Theoretical analysis of the NDPM algorithm

(1) Computational load

Because the scheme doesn't need the recombination of the address fragments and attack path reconstruction, it can trace the attack source to the BGP router in the source AS. It only needs hash functions to process on the IP address, while marking the initial router information and the ending router information. The computational load is rather small, which can be negligible.

(2) Router overhead

In the traditional scheme for IP traceback, it needs all the routers in the attack path to participate in marking, and this scheme only needs BGP router's participation. According to the statistics, the average number of the router that a packet passes through from the source to the destination is approximately seventeen in the Internet, while the number of AS is approximately three [10]. Supposing that the packet passes through two BGP routers in each AS, the router overhead can be calculated by $1 - \dfrac{3 * 2}{17} = 0.6471$, which can reduce the router overhead more than sixty-four percent.

(3) Traceback time

In the marking process, the number of the source AS is marked in the ASID field of each packet. That is to say, each packet has the complete 16-bit source AS number, so it can trace the original AS of the attack packets without combination of fragments, hence making it possible to trace the DDoS source on a real-time basis.

4. Simulation environment and results

4.1. Simulation environment

In order to test the performance of the proposed scheme, we have done simulation experiments by using Network Simulator version 2 (NS2) [11]. Each node represents an AS. Based on this simulator, the victim is employed in node 9. At the same time, we employ three slave systems in node 1, 3 and 14, and install the packet marking algorithm on each BGP routers. BGP routers mark the packets and the victim filters the attack packets according to the marking information marked by BGP routers. Figure 4 shows the simulation environment in NS2.

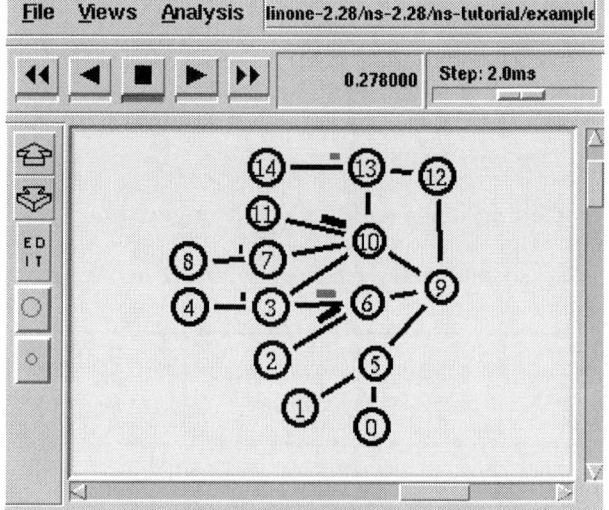

Figure 4. **Simulation environment**

4.2. Simulation results

As it is shown in Figure 5, in the course of simulation, the attacks occur three times. The first time happens at the fifteenth second, the second time takes place at the thirtieth second, and the third time at the forty-fifth second. Each of the attacks lasts five seconds. From the graph, we can see it clearly, when the defense mechanism is not employed, the victim is flooded with packets which are approximately 8000 during the attacks. With the proposed scheme, the number of packets is reduced between 2000 and 3000. So it can be acknowledged that the proposed deterministic packet marking scheme can defense the DDoS attacks and alleviate the impact caused by attack traffic on the victims more effectively.

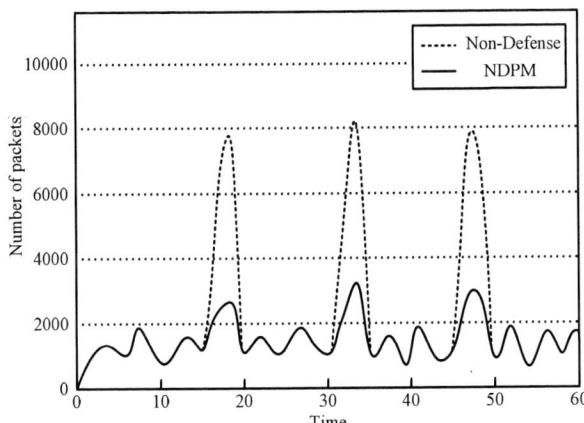

Figure 5. Simulation results from the experiments

5. Conclusion

As the proposed scheme obtains more space through overloading the offset field and TOS field, it resolves the shortage of storage space in the common path identification methods. At the same time, because each packet has the complete 16-bit source AS number, it can trace the DDoS attack source to the original AS in real-time, while the Link-ID method can not trace the DDoS attack source at all. In all, the proposed deterministic packet marking scheme is extremely light-weight, both on the routers for marking, and on the victims for decoding and filtering.

6. References

[1] C. Douligeris, and A. Mitrokotsa, "DDoS attacks and defense mechanism: classification and state-of-the-art", Computer Networks, vol. 44, pp.643-666, 2004.

[2] Abraham Yaar, Adrian Perrig, and Dawn Song, "Pi: A Path Identification Mechanism to Defend against DDoS Attacks", Proceedings of Symposium on Security and Privacy, Page(s): 93-107, 2003.

[3] Changhyun Beak, Junaid Ahsenali Chaudhry, Keonsoo Lee, Seungkyu Park, and Minkoo Kim, "A Novel Packet Marketing Method in DDoS Attack Detection", American Journal of Applied Sciences, 4(10):741-745, 2007.

[4] Guang Jin, Jiangang Yang, Wei Wei, and Yabo Dong, "Across-Domain Deterministic Packet Marking for IP Traceback", Second International Conference on Communications and Networking in China, 22-24 Aug. 2007, Page(s):382-386.

[5] S. Savage, D. Wetherall, A. Karlin, et al., "Practical network support for IP traceback", in Proc. ACMSIGCOMM, Stockholm, Sweden, 2000, pp.295-306.

[6] A. Belenky and N. Ansari, "IP traceback with deterministic packet marking", IEEE Communications Letters, vol.7, pp. 162-164, 2003.

[7] Z. Gao and N. Ansari, "A practical and robust inter-domain marking scheme for IP traceback", Computer Networks, vol. 51, pp.732-750, 2007.

[8] Y. Kim, J. Jo, F. Merat, et al., "Defeating distributed denial-of-Service attack with deterministic bit marking", in Proc. IEEE Globecom, Dec. 2003.

[9] C. Gong and K. Sarac, "Toward A More Practical Marking Scheme for IP Traceback", in Proc. BroadNETS, San Jose, CA, Oct. 2006.

[10] V. Paruchuri, A. Durresi, and L. Barolli, "FAST: Fast Autonomous System Traceback", 21st International Conference on Advanced Networking and Applications, 21-23 May 2007, Page(s):498-505.

[11] Jae Chung and Mark Claypool, "NS by Example", WPI WORCESTER POLYTECHNIC INSTITUTE Computer Science, 2002.

A Fast Method of Local Network Topology

Yubin Yao[1], Dan Wang[1], Chuanqi Wang[2] and Yong Li[2]

1. *Marine Engineering College, Dalian Maritime University, Dalian, China*
2. *Yantai Dongfang Electronic Information Industry Co. Ltd, Yantai, China*
yyb@newmail.dlmu.edu.cn

Abstract

Only few switches in a real-time power system are operated during a short interval between the executions of the network topology algorithm, and switch activities just have local affect on network topology. So a fast local network topology method using optimal search direction is presented in this paper. The proposed method has three main procedures: determining optimal search direction, substation configuration and network configuration. Firstly the proposed method determines generalized overhanging branches and finds the optimal search directions for them. Then the substation configuration reconfigures buses only at substation voltage levels affected by switch activities. At last the network configuration determines network connectivity for islands in optimal search direction which can greatly decrease search scope. The presented method is fast, effective and reliable, and experimental results illustrate the effectiveness of the method.

1. Introduction

In modern operation control centers, the power system network topology algorithm is basic and critical algorithm for a number of energy management system (EMS) and distribution management system (DMS) functions, including state estimation, operator's load flow, contingency analysis, and dispatcher training simulator (DTS). And the network topology program is executed periodically, so a fast algorithm of the network topology is of great importance.

Being algorithm for the power system network connectivity, the network topology algorithm can be accomplished by the search method [1-5] and matrix method [6]. The search method includes the depth-first search method [1-3] and the breadth-first method [4-5]. The search method is widely used in large power system because of its high speed.

In most cases, there are only a few switch status changes since the last execution of the network topology for the real-time network. So it is computationally desirable to obtain the current network topology by partially updating on the previous network topology, and many local network topology methods are presented [1-4].

There are two procedures in the network topology algorithm, one is grouping nodes into buses according to switch's status, and the other is determining network connectivity for islands. Because reconfiguring the buses is only carried out at the substation voltage levels affected by the switch activities and the number of the nodes in each substation voltage level is small, tracing through the entire substation voltage level is very fast. On the contrary tracing through the entire network to determine the post-switching network connectivity is very time-consuming.

Based on the investigation on the specific property of the electric network, a fast local network topology method using optimal search direction is presented in this paper. The proposed method has three main procedures: determining optimal search direction, substation configuration and network configuration. The procedure of the optimal search direction determination determines generalized overhanging branches and finds the optimal search directions for them. The substation configuration reconfigures buses only at substation voltage levels affected by switch activities. The network configuration determines network connectivity for islands using optimal search direction which can greatly decrease search scope.

To reduce search scope effectively is critical for the network configuration, so optimal search direction is defined in this paper. Firstly this paper defines generalized overhanging buses and generalized overhanging branches, then defines the direction which the generalized overhanging branches direct into the generalized overhanging bus as the optimal search direction. When the network topology algorithms detects that the change of switch status occurs, it firstly groups the nodes into buses based on the current switch status, if the open of the switch results in the split of the bus, network configuration is startup. If there is one but one of the new buses that has generalized overhanging branches from it, then determine the network connectivity for islands beginning form that bus; otherwise it searches buses in general way.

2. Brief Introduction to Network Topology

2.1. Terminology

To make the paper clear, terminology used in this paper is listed as follows:.
(a) Node: The physical jointed point of the equipments.
(b) Switch: An equipment which open and close the circuits, such as circuit break, disconnector.
(c) Bus: A group of nodes connected by closed switches.
(d) Branch: A two-end equipment with definite impedance, such as transmission line, two-winding transformer, series capacitor, series inductor. A three-winding transformer is regarded as three branches.
(e) Island: A part of power system which connected by branches. An island with at least one generation is called energized island; an island without any generation is called de-energized island.
(f) Overhanging bus: A bus which connects only one branch.
(g) Overhanging branch: A branch which connects to an overhanging bus.
(h) Generalized overhanging bus: A bus which connects only one branch after removing all overhanging branches and generalized overhanging branches. Generalized overhanging bus used in this paper includes overhanging bus.
(i) Generalized overhanging branch: A branch which connects to a generalized overhanging bus. Generalized overhanging branch used in this paper includes overhanging branch.

2.2. Data Structures

The topology processing input data tables usually given as a node-switch oriented model. Figure 1 shows the circuit diagram for a small example power system with three substations.
The mainly data needed for network topology are as follows:
(a) Switch array: Data for all switches, each switch has two entries, the second entry for the same switch with opposite-end. The fields of switch data include from-node, to-node, index to next switch with same from-node, status, type.
(b) Branch array: data for all branches, each branch has two entries, the second entry for the same branch with opposite-end. The fields of branch data include from-bus, to-bus, index to next branch with same from-bus, optimal search direction flag, and type.
(c) Node array: data for all nodes, the fields of node data include the bus index the node belongs to, index into the switch arrays corresponding to the first switch incident to the node, status.
(d) Bus array: data for all buses, the fields of bus data include the island index the bus belongs to, index into the branch arrays corresponding to the first branch incident to the bus, status.
(e) Island array: data for all islands, the fields of island data include the number of buses in each island, flag for being energized or de-energized.

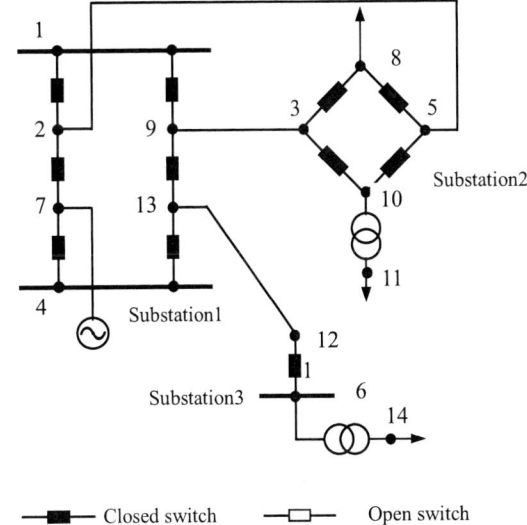

Figure 1. Node / switch model for a small example power system.

2.3. Network Topology Process

A conventional network topology algorithm consists of two main procedures: substation configuration and network configuration.

2.3.1. Substation configuration. This procedure determines the buses at each substation by analyzing the switch statuses. Each bus is constituted by a group of nodes connected by closed switches. The input data for this procedure are switch array and node array, the result of this procedure is mainly bus array.

2.3.2. Network configuration. This procedure determines the islands by analyzing network connectivity. The input data for this procedure are branch array and bus array, and establishes island index for each bus.

Although the two procedures employ different data and yield different result, they use the same method. The proposed method uses breadth-first search method.

The bus / branch model by network topology for Figure 1 is shown in Figure 2.

Because the procedure of reconfiguring the buses is only carried out at the substation voltage levels affected by the switch activities and the number of the nodes in each substation voltage level is small, tracing through the entire substation voltage level is very fast. On the contrary tracing through the entire network to determine the post-switching network connectivity is very time-consuming.

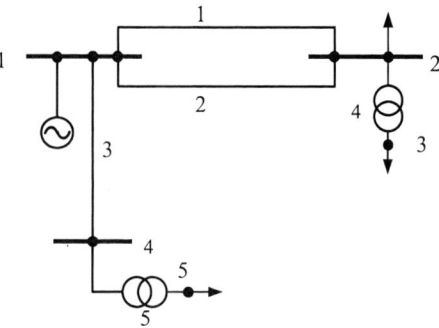

Figure 2. Bus / branch model for Figure 1.

3. Network Change Caused by Switch Activity

3.1. Close of the Switch

When a switch closes, one of the following conditions may occur:
- (a) If both ends of the switch belong to the same pre-switching bus, the close of the switch has no infection on the network topology.
- (b) If the ends of the switch belong to different pre-switching buses in the same island, the close of the switch just causes mergence of the two buses.
- (c) If the ends of the switch belong to different pre-switching buses in the different islands, the close of the switch causes both mergence of buses and mergence of islands.

3.2. Open of the Switch

When a switch opens, one of the following conditions may occur:
- (a) If both ends of the switch belong to the same post-switching bus, the open of the switch has no infection on the network topology.
- (b) If the ends of the switch belong to different post-switching buses in the same island, the open of the switch just causes split of the bus.
- (c) If the ends of the switch belong to different post-switching buses in the different islands, the open of the switch causes not only the split of the bus but also the split of the island.

4. The proposed Method of Local Network Topology

The proposed local network topology algorithm consists of three main procedures: optimal search direction determination, substation configuration and network configuration.

4.1. Optimal Search Direction Determination

The steps to determine optimal search direction are as follows:
- (a) Count the number of branches connected to the bus for every bus. The bus array built during the execution of the overall network topology is used in this step.
- (b) Designate the bus without branch connected to as isolated bus; designate the bus with only one branch connected to as generalized overhanging bus.
- (c) Designate the branch connected to a generalized overhanging bus as generalized overhanging branch, and designate the direction to generalized overhanging bus as optimal search direction.
- (d) Remove all generalized overhanging branches, go to step (a) to find new generalized overhanging branches. Repeat this procedure until no more new generalized overhanging branch being found.

During this procedure, the flag filed of branch array is set. Because one generalized overhanging branch has two entries in branch array, only the flag filed of the entry with to-bus being generalized overhanging bus is set to 1, that is, the direction from a bus to a generalized overhanging bus is defined as optimal search direction, otherwise the flag filed of the entry is set to 0.

4.2. Substation Configuration

If the network topology program detects a change of switch's status in power system, it starts substation configuration. According to the action of the switch, the substation configuration divided into two sub-procedures

The steps for the sub-procedure of the switch close are as follows:
- (a) Check the buses the switch connected to, if the ends of the switch belong to different buses, merge the two buses with the small bus number.
- (b) Check if the pre-switching buses the switch connected to belong to the same island. If not, merge the two islands with a small island number or energized one.
- (c) Change the node array, branch array and bus array

to reflect the change of buses.

The steps for the sub-procedure of the switch open are as follows:

    (a) Reconfigure the buses only at the substation voltage levels affected by the switch activities.

    (b) Check the post-switching buses the ends of the switch belong to, if the ends of the switch belong to different post-switching buses, that means the former buses is split.

    (c) If the split of the buses occurs, then check if the new buses both have branches connected to. If it is true, that means the split of island may occurs, then set the flag for star-up of network configuration.

    (d) Change the node array, branch array and bus array to reflect the change of buses.

4.3. Network Configuration

If the flag for star-up of network configuration is set, then network topology program is startup, the steps of the substation configuration are as follows:

    (a) Check if there is one but one of the affected buses that has generalized overhanging branches from it, then go to next step (b1) or (b2) according to the conditions.

    (b1) If one bus meets the condition in step (a), that means the split of the island occurs, then determine the network connectivity for islands beginning form that bus, the buses being reached remain in the former island, the buses not being reached form a new island.

    (b2) If no bus meets the condition in step (a), then determine the network connectivity for islands from one of the affected buses until reach the other one, which means that the island does not split. If the other affected bus is not reached before a complete island formed, which can assert that the island splits, and the buses being reached remain in the former island, the buses not being reached form a new island.

    (c) Change the bus array and island array to reflect the change of islands.

From the pre-switching network shown in Figure 2, it can find that the buses 3, 4, 5 are generalized overhanging buses; and the branches 3, 4, 5 are generalized overhanging branches.

Part fields of the branch array are listed in Table 1. Each branch has two entries in branch array. For example, line 3 has two entries in branch array, the from-bus and to-bus of one entry are bus 1 and bus 4 respectively, the from-bus and to-bus of the other entry are bus 4 and bus 1 respectively.

Table 1 The branch array for the system shown in figure 2

Branch Name	From-bus	To-bus	Flag
Branch 1	1	2	0
Branch 1	2	1	0
Branch 2	1	2	0
Branch 2	2	1	0
Branch 3	1	4	1
Branch 3	4	1	0
Branch 4	2	3	1
Branch 4	3	2	0
Branch 5	4	5	1
Branch 5	5	4	0

If switch 1 (the one connected between nodes 6 and 12) is open, the new network topology is shown in Figure 3.

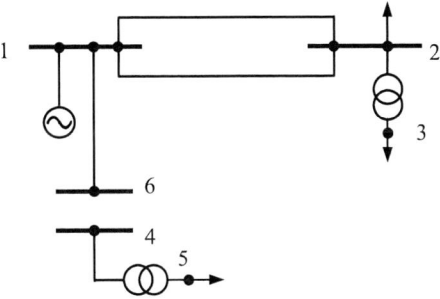

Figure 3. Bus / branch model for Figure 1 after opening a switch.

The substation configuration finds that the pre-switching bus 4 has split two buses, which one bus is keep the old bus number, the other is assigned a new bus number. And the flag for star-up of network configuration is set because that bus split has found and need to determine if the split of the island will occur.

The network configuration finds that new bus 4 has a generalized overhanging branch from it, while new bus 6 has none. Such the split of the island is assured, then determines the network connectivity for islands from bus 4. After this procedure, the post-switching islands are obtained. There are two islands, bus 4 and 5 are in island 1, and bus 1, 2, 3 and 6 are in island 2.

5. Case Study

In this section, a real network named Yantai distribution network is used to testify our conclusions. The bus / branch model of Yantai distribution network is shown in Figure 4.

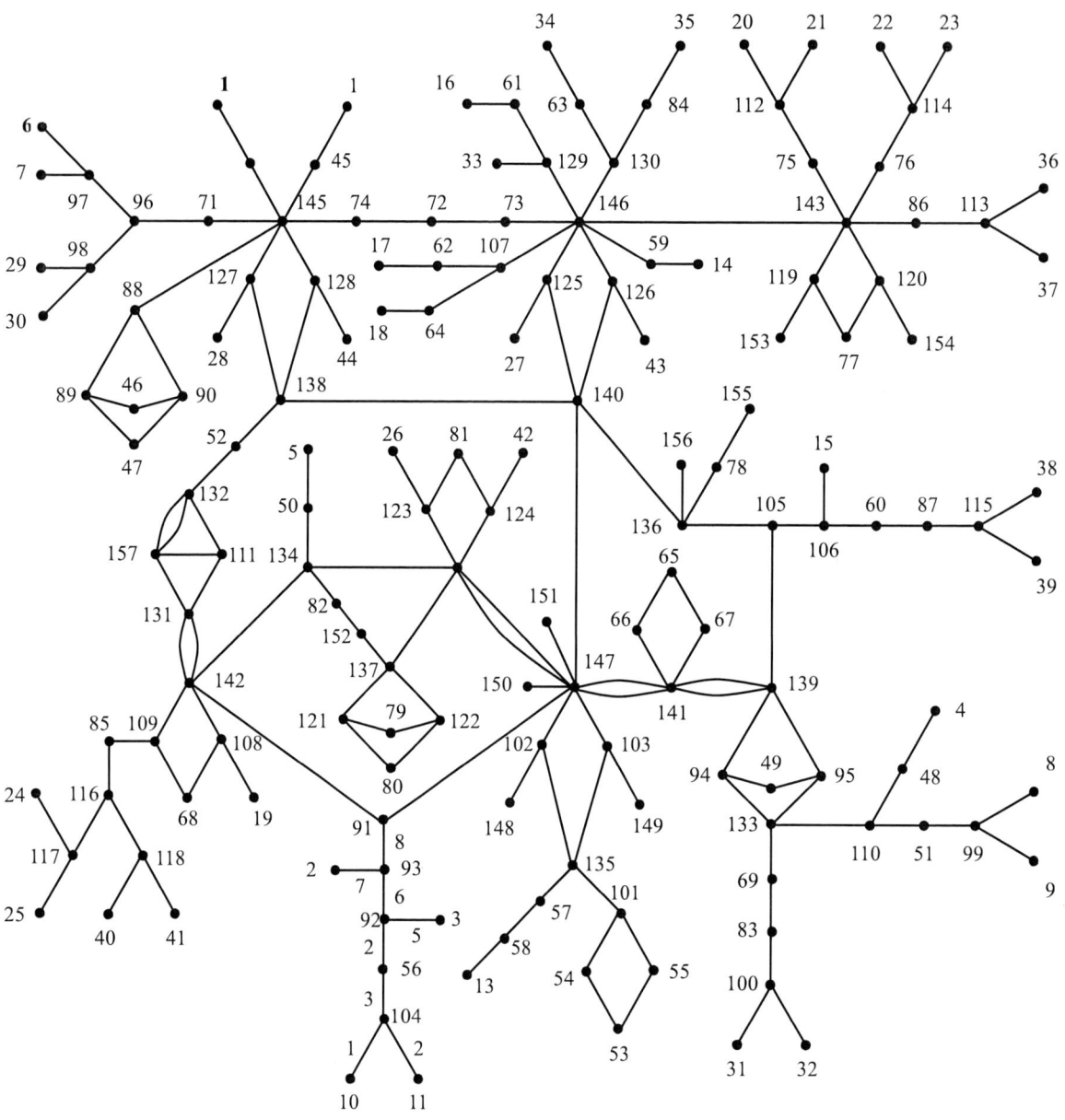

Figure 4. A real power system network.

During substation configuration, 157 buses and 182 branches including parallel branches are found in the energized island. During the procedure of the optimal search direction determination, 92 generalized overhanging branches are found. Over 50 percent of the branches in this network are generalized overhanging branches, and search path in the optimal search direction

is very short. So to determine network connectivity for islands using optimal search direction can greatly decrease the search scope.

Some generalized overhanging branches are listed in Table 2.

Table 2 Part of the branch array for the system shown in figure 4

Branch Name	From-bus	To-bus	Flag
Branch 1	10	104	0
Branch 1	104	10	1
Branch 2	11	104	0
Branch 2	104	11	1
Branch 3	56	104	1
Branch 3	104	56	0
Branch 4	56	92	0
Branch 4	92	56	1
Branch 5	3	92	0
Branch 5	92	3	1
Branch 6	92	93	0
Branch 6	93	92	1
Branch 7	2	93	0
Branch 7	93	2	1
Branch 8	91	93	1
Branch 8	93	91	0

6. Conclusions

The power system network topology algorithm is basic and critical algorithm in modern operation control centers, and is executed periodically, so a fast algorithm of the network topology is of great importance. The number of switches with status changed is very small during a short interval between the executions of the network topology algorithm, and switch activities have just local affect on the network topology. So it is feasible to use local network topology program to detect the switch activities and obtain the post-switching topology by modifying the former network topology results.

While reconfiguring the buses only at the substation voltage levels affected by switch activities is a fast procedure, reducing the searching scope in the process of determining network connectivity for islands is not easy task. Based on the investigation on the specific property of the electric network, this paper presents a fast local network topology method. The proposed method defines generalized overhanging branches and optimal search direction for the generalized overhanging branches. Determines network connectivity for islands in optimal search direction can greatly decrease the search scope. There has high percentage of generalized overhanging branches in real power system, so the proposed method has considerable practical value. The results of a practical example illustrate the effectiveness of the proposed method.

7. Acknowledgement

This work was supported in part by NSFC, P. R. China under Grant 60674037.

8. References

[1] Bose A, Clements K. "A Real-Time Modeling of Power Networks," Proceedings of the IEEE, vol. 75, pp 1607-1622, Dec. 1987.

[2] Prais M, BoseA. "A Topology Processor That Tracks Network Modifications Over Time," IEEE Trans on Power Systems, vol. 3, pp 992-998, Aug. 1988.

[3] YehsakulPD, DabbaghchiI. "A Topology-Based Algorithm for Tracking Network Connectivity," IEEE Transon Power Systems, vol.10, pp 339-345, 1995.

[4] Zhu Wendong, Liu Guangyi, Yu Erkeng, et al. "The Fast Calculation Method of Local Power Network Topology"(in Chinese), Power System Technology, vol. 20, no. 3, pp 30-33, Mar. 1996

[5] Yao Yubin, Jin Wenzhuan, Jin Li. "Fast Network Topology Method for a Distribution Network"(in Chinese), Relay, vol. 33, no. 19, pp 31-35, Oct. 2005.

[6] Wang Xiangzhong, Li Xiaolan. "Topology Identification of Power Network Based on Incidence Matrix"(in Chinese), Power System Technology, vol. 25, no. 2, pp 10-12, 16, Feb, 2001

Invulnerability Assessment for Mobile Ad Hoc Networks

Cong Jin[1], Shu-Wei Jin[2]

[1] *Department of Computer Science,*
Central China Normal University, Wuhan 430079, P.R.China
[2] *No.1 Middle School Attached to Central China Normal University,*
Wuhan 430223, P.R.China
E-mail: jincong@mail.ccnu.edu.cn

Abstract

Usually, a mobile Ad Hoc network is confronted with four different damages: random failure, mobile attack, energy limit and selective attack. These attacks will not only destroy network topology, but also will make the transferring capacity and bandwidth of network to be influenced. An assessment of mobile Ad Hoc network invulnerability is proposed based on the characteristic of mobile Ad Hoc network. A definition of invulnerability rate is proposed. The invulnerability rate of network is decided by three factors, i.e., connectivity, transferring capacity, and bandwidth of network. At last, we give two experiments to evaluate the performance of the proposed invulnerability assessment method. The results show that mobile Ad Hoc network has the very strong invulnerability.

1. Introduction

Wireless networks[1-5] have become increasingly popular in the past few decades, particularly within the 1990's when they are being adapted to enable mobility and wireless devices became popular. There are currently two kinds of mobile wireless networks. The first is known as infrastructured networks with fixed and wired gateways. Typical applications of this type of "one-hop" wireless network include wireless local area networks (WLANs). The second type of mobile wireless network is the infrastructureless mobile network, commonly known as the Ad Hoc network[1]. An Ad Hoc network is usually a self-organizing and self-configuring "multi-hop" network which does not require any fixed infrastructure. In an Ad Hoc network, all nodes are dynamic and arbitrarily located, and are required to relay packets for other nodes in order to deliver data across the network.

Ad Hoc networks are suited for use in situations where infrastructure is either not available, not trusted, or should not be relied on in times of emergency. A few examples include: military solders and equipments in the battlefield, sensor networks for various research purposes, emergency rescue after an earthquake or flood, and temporary offices such as campaign headquarters.

In Ad Hoc network, nodes are dynamic, arbitrarily located, and random movement. Because an Ad Hoc network hasn't any fixed infrastructure, it must deal with limitations such as high power consumption, low bandwidth, high error rates and arbitrary movements of nodes et al. When two mobile nodes are within the overlay range of communicating each other, they can communicate directly. However, the communicable overlay range of every mobile node is limit, and if the distance between two nodes exceeds the distance of direct communicating, other nodes are needed for communicating. Once a node occur fault, the rest nodes can still work normally by combination freely. Therefore, an Ad Hoc network must have invulnerability. In this paper, we will research invulnerability assessment of mobile Ad Hoc network.

2. Some definitions

Generally, the network invulnerability is thought that the ability of the network still working normally when occurring network fault. There is the relation closely between the network invulnerability and network topology. The network topology can determine that whether the rest part of network still maintain connective state when network node or network link occur fault[6-9]. Therefore, network topology influences directly invulnerability of the network. Usually, the network invulnerability is definite by cohesion and connectivity.

Cohesion For a connective network, let CH_{ij} is minimal link number need deleted for breaking path between nodes v_i and v_j, then the cohesion of network is given by
$$CH = \min_{i,j}(CH_{ij})$$

Connectivity For a connective network, let CN_{ij} is minimal node number need deleted for breaking path between nodes v_i and v_j, then the connectivity of network is given by
$$CN = \min_{i,j}(CN_{ij})$$

We notice that mobile Ad Hoc network topology changes randomly with time. For any a pair node, the link number and node number need deleted aren't all constants, so the cohesion and connectivity of Ad Hoc network can't be obtained quantificationally. It is more difficult than traditional network to assess the invulnerability of mobile Ad Hoc network.

However, in a small time compartment, mobile Ad Hoc network topology may be thanked to be invariant. So, when a link between two nodes is destroyed, if there is a backup link so that they is still connective, signal may be transferred by the backup link. However, a part bandwidth of the backup link will be taken up so that original transfer is influenced. Therefore, network bandwidth also influences invulnerability of Ad Hoc network.

Based on above analysis, some new invulnerability assessments definition of mobile Ad Hoc network is proposed in the paper.

Definition 2.1 Let v_i and v_j be two consecutive nodes, E_{ij} be a link between two nodes v_i and v_j, and T_{ij} be information capacity transferred from v_i to v_j. If E_{ij} is broken off, and a path, with node v_{a0}, v_{a1}, ... , v_{at}, is a backup path from v_i to v_j, where $v_{a0}= v_i$, $v_{at} = v_j$, and $B_{s\ s+1}$ (s=0, 1, ... , t-2, t-1) is bandwidth of the backup path, then TC_{ij} is said to invulnerability transfer capacity from v_i to v_j and have value

$$TC_{ij} = \min B_{s\ s+1} \quad (s = 0,1,...,t-2,t-1)$$

ICR_{ij} is said to invulnerability capacity rate from v_i to v_j and have value

$$ICR_{ij} = \frac{TC_{ij}}{T_{ij}}.$$

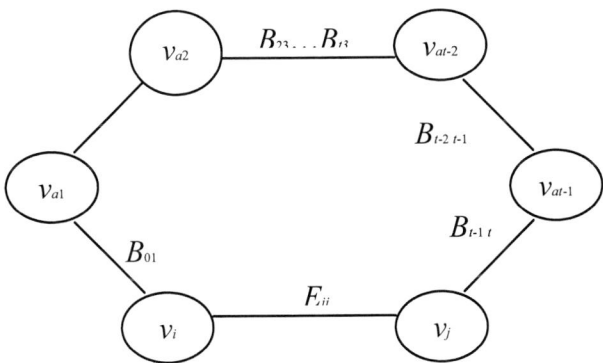

Figure 1. Calculation of invulnerability capacity rate

Definition 2.2 In mobile Ad Hoc network, let v_i and v_j be two nodes, and T_{ij} be information capacity transferred from v_i to v_j. Node v_{a0}, v_{a1}, ... , v_{at} are gone through one by one for transferring information from v_i to v_j, and node v_{b0}, v_{b1}, ... , v_{bk} are gone through one by one according to another path for transferring information from v_i to v_j. Besides $v_{a0}= v_i$, $v_{at} = v_j$, $v_{b0}= v_i$, and $v_{bk} = v_j$, the rest node all does not different from each other. According to Definition 2.1, $TC_{s\ s+1}$ (s=0, 1, ... , t-2, t-1) is invulnerability transfer capacity from v_i to consecutive node v_j, and $B_{s\ s+1}$ (s=0, 1, ... , k-2, k-1) is rest bandwidth from v_{bs} to $v_{b\ s+1}$. Then ICR_{ij} is said to invulnerability capacity rate from v_i to v_j and have value

$$ICR_{ij} = \min\{TC_{s\ s+1}/T_{ij}\} \quad (s = 0,1,...,t-2,t-1).$$

IBR_{ij} is said to invulnerability bandwidth rate from v_i to v_j and have value

$$IBR_{ij} = \min\{B_{s\ s+1}/T_{ij}\} \quad (s = 0,1,...,k-2,k-1).$$

Remark. If v_i and v_j are consecutive node, then

$$ICR_{ij} = IBR_{ij}.$$

Definition 2.3 In mobile Ad Hoc network, the average value of invulnerability capacity rate of any two nodes is called invulnerability capacity rate of mobile Ad Hoc network and is defined by *ICR*.

Definition 2.4 In mobile Ad Hoc network, the average value of invulnerability bandwidth rate of any two nodes is called invulnerability bandwidth rate of mobile Ad Hoc network and is defined by *IBR*.

The calculation of invulnerability capacity rate is shown in Figure 1.

We also notice that there is the relation closely between the invulnerability of mobile Ad Hoc network and the randomness of network topology, and there also is the relation closely between the randomness of network topology and time. In other words, the coordinates of the network node are function of time. Thus, to describe coordinates of the network node by probability theory are very suitable undoubtedly.

Definition 2.5 In mobile Ad Hoc network, if rest nodes are organized afresh after some nodes destroyed so that network topology is still connective, then the probability of this event is called invulnerability topology rate defined by *ITR*.

Definition 2.6 Invulnerability rate of mobile Ad Hoc network is defined by flowing formula

$$IR = \frac{1}{3}(ITR + ICR + IBR).$$

3. Mobile Ad Hoc network model for calculating ITR

3.1 Network model

For researching invulnerability rate of mobile Ad Hoc network quantificationally, a network model is given as follows.

In a given region, there is a mobile Ad Hoc network with n nodes, original coordinate of every node v_i ($i = 0,1,...,n$) is $(x_i(0),y_i(0))$, azimuth angle is φ_i, and travel speed is V_i. Communication region of every node v_i ($i = 0,1,...,n$) is a disk with radius R. This shows that two nodes can communicate directly if and only if their distance isn't larger than R.

In t time, coordinate of node v_i ($i = 0,1,...,n$) is

$$\begin{cases} x_i(t) = x_i(0) + V_i\, t\cos\varphi_i \\ y_i(t) = y_i(0) + V_i\, t\sin\varphi_i \end{cases}$$

then the distance of any two nodes v_i and v_j is

$$d_{ij}(t) = \sqrt{[x_i(t) - x_j(t)]^2 + [y_i(t) - y_j(t)]^2}.$$

Where,

$$\begin{cases} x_i(t) - x_j(t) = x_i(0) + V_i\, t \cos\varphi_i - x_j(0) + V_j\, t \cos\varphi_j \\ y_i(t) - y_j(t) = y_i(0) + V_i\, t \sin\varphi_i - y_j(0) + V_j\, t \sin\varphi_j \end{cases}$$

So, incidence matrices for describing direct connectivity can be obtained as follows:

$$H_{n \times n}(t) = \begin{pmatrix} h_{11}(t) & h_{12}(t) & \cdots & h_{1n}(t) \\ h_{21}(t) & h_{22}(t) & \cdots & h_{2n}(t) \\ \cdots & \cdots & \cdots & \cdots \\ h_{n1}(t) & h_{n2}(t) & \cdots & h_{nn}(t) \end{pmatrix}.$$

Where, $h_{ij}(t) = \begin{cases} 1, & d_{ij}(t) \leq R \\ 0, & other \end{cases}, (i, j = 0, 1, ..., n)$.

$h_{ij}(t) = 1$ shows that nodes v_i and v_j can be direct connectivity in t time, and $h_{ij}(t) = 0$ shows that nodes v_i and v_j can't be direct connectivity in t time.

3.2 Calculation method of ITR

After obtaining incidence matrices $H_{n \times n}(t)$, we give the algorithm of calculating ITR as follows:

***Step* 1** For giving node select probability p, $0 < p < 1$, let $m = \lfloor pn \rfloor$ be number of node selected. a_1, a_2, ... , a_m are random number generated randomly in interval $[0,1]$, and $\lfloor na_1 \rfloor, \lfloor na_2 \rfloor, ..., \lfloor na_m \rfloor$ are ordinal notation of selected nodes.

***Step* 2** Let all elements corresponding ordinal notation $\lfloor na_1 \rfloor, \lfloor na_2 \rfloor, ..., \lfloor na_m \rfloor$ about row and column in incidence matrices $H_{n \times n}(t)$ be zero, and a new incidence matrices $\overline{H}_{n \times n}(t)$ is obtained.

***Step* 3**

(1) Let all elements of diagonal line in incidence matrices $\overline{H}_{n \times n}(t)$ be zero. We select arbitrary non-zero element h_{ij} corresponding node v_i.

(2) We found out row notation of all elements with value 1 of jth column. For convenience and without loss of generality, suppose that these row notations are R_1, R_2, R_3, and node v_i is direct connectivity with node v_{iR_1}, v_{iR_2}, v_{iR_3} respectively. So, we can get a node set $\overline{v}_1 = \{ v_i, v_{iR_1}, v_{iR_2}, v_{iR_3} \}$.

(3) After let all elements of ith row are zero, we find out all elements with value 1 in R_1th, R_2th, and R_3th column respectively, and record their row notation. These nodes are direct connectivity with node v_{iR_1}, v_{iR_2}, v_{iR_3} respectively. After the same node deleted, we can get a node set $\overline{v}_2$.

(4) Repeat *Step* 2, up to no longer appear new element in the node set. So, we may get final node set $\overline{v}_{end}$. If

$|\overline{v}_{end}| = \lfloor \alpha(n - m) \rfloor$, where α is valid connectivity node rate of network, then we think that mobile Ad Hoc network is connectivity, else isn't connectivity.

***Step* 4** Select appropriate test times $N(t)$, repeat Step 2 to Step 4, record times $C(t)$ of network connectivity, and get invulnerability topology rate in t time as follows

$$ITR(t) = \frac{C(t)}{N(t)}.$$

4. Experimental results

4.1 Experimental parameters

To evaluate the performance of the proposed invulnerability assessment method of Ad Hoc network, a set of experiments is performed under the following conditions.

In initial time, suppose that network has $n = 150$ nodes distributed uniformly in the region with size 1000×1000 square meter, communication region of every node is a disk with radius $R = 120$, azimuth angle φ_i of every node is a random variable satisfying a uniform distribution on the interval $[0, 2\pi]$, and travel speed V_i is another random variable satisfying a uniform distribution on the interval $[0, 3]$.

The ratio number of node selected randomly is $p = 7\%$ and $p = 15\%$ respectively. In t time, select test times $N(t) = 500$, and $\alpha = 1$.

For network topology, add capacity and bandwidth for every link. The method is described as follows

If υ is a random number satisfying a uniform distribution on the interval $[5, 20]$, then $\lfloor \upsilon \rfloor$ is information capacity transferred between two consecutive nodes.

If τ is a random number satisfying a uniform distribution on the interval $[5, 15]$, then $\lfloor \tau \rfloor$ is bandwidth of a link between two consecutive nodes.

For convenience, suppose node will be reversing rotation automatically when a node arrive the region boundary.

4.2 Experimental results

We calculate a times invulnerability rate every 25 seconds. When $p = 7\%$, the invulnerability rate of mobile Ad Hoc network is shown in Figure 2, and average invulnerability rate is 0.9216. This shows that mobile Ad Hoc network has the very strong invulnerability.

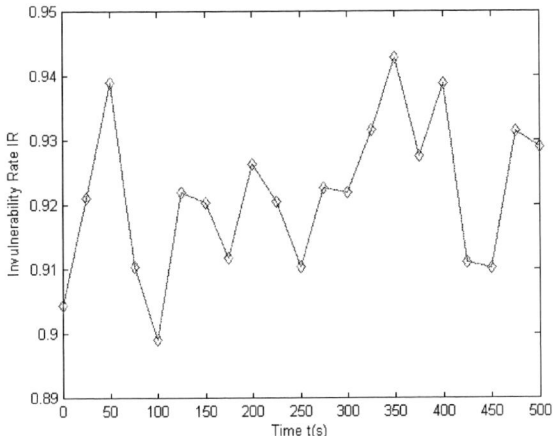

Figure 2. Invulnerability rate of mobile Ad Hoc network when $p = 7\%$

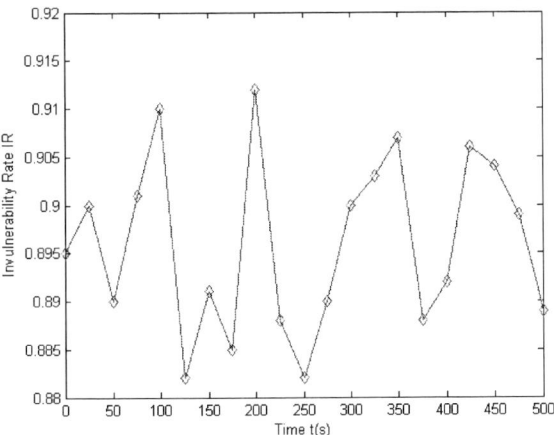

Figure 3. Invulnerability rate of mobile Ad Hoc network when $p = 15\%$

When $p = 15\%$, the invulnerability rate of mobile Ad Hoc network is shown in Figure 3, and average invulnerability rate is 0.8959. In this case, though $\lfloor pn \rfloor = 22$ nodes have been destroyed, the invulnerability rate of mobile Ad Hoc network still is above 89%. This shows that mobile Ad Hoc network still has the very strong invulnerability.

We notice that, if $\alpha < 1$, the invulnerability rate of mobile Ad Hoc network is more good than $\alpha = 1$.

5. Conclusion

The invulnerability of mobile Ad Hoc network is a challenging task due to mobility and the resulting inherent dynamic network topology. The nodes in Ad Hoc networks are usually restricted devices in respect to their energy sources, computational capabilities and communication range. For mobile Ad Hoc network, the current researches focus mostly on the routing protocols and MAC, and it is less about the invulnerability of mobile Ad Hoc network. In this paper, a new quantificational assessment method of the invulnerability of mobile Ad Hoc network is proposed. The new method can not only be used for assessment invulnerability of mobile Ad Hoc network, but also can be used for optimizing the network model, and we will further investigate this problem.

References

[1] Anthony H Dekker, Bernard D Colbert. Network robustness and graph topology. *Proceedings of the 27th Conference on Australasian Computer Science*, Dunedin, New Zealand, 2004, pp. 359-368.

[2] Petter Holme, Beom Jun Kim. Attack vulnerability of complex networks. *Physical Review E*, 2002, 65(5): pp. 05610.

[3] Paolo Crucitti, Vito Latora. Error and attack tolerance of complex networks. *Physics A*, 2004, 340, pp. 388-39.

[4] Wilkov R. Analysis and design of reliable computer networks. *IEEE Transactions on Communication*, 1972, 20(3): pp. 660-678.

[5] Page L B, Perry J E. Reliability polynomials and link importance in networks. *IEEE Transactions on Reliability*, 1994, 43(1): pp. 51-58.

[6] Sanso B, Soumis F.Communication and transportation network reliability using routing models. *IEEE Transactions on Reliability*, 1991, 40(1): pp. 29-38.

[7] Bin Wang. Towards cost-effective provisioning and survivability in ultra high speed Networks. http://computing.fnal.gov/cd/DOE/ultranet/talk-bin-wang.htm.

[8] Wing O, Demetrion P. Analysis of probabilistic networks. *IEEE Transactions on Communication Technology*, 1964, 12(3): pp. 38-40.

[9] Newport K T. Design of survivable communication networks under performance constraints. *IEEE Transactions on Reliability*, 1991, 40(4): pp. 433-440.

Discovery of Unknown Visual Information Based on Target Tracking

Cong Jin[1], Shu-Wei Jin [2]

[1] *Department of Computer Science,*
Central China Normal University, Wuhan 430079, P.R.China
[2] *No.1 Middle School Attached to Central China Normal University,*
Wuhan 430223, P.R.China
E-mail:jincong@mail.ccnu.edu.cn

Abstract

To discover important but previously unknown information in video sequences, a discovery method of unknown visual information based on target tracking is proposed in this paper. It provides a worthy approach for support decision. The proposed discovery method analyzes the video using target tracking. This method, adopting detection and correspondence of feature points, target identification and tracking, and target occlusion techniques etc, analyzes video sequences. In the video, using above techniques, we not only study the spatio-temporal relationships of moving target to obtain important information accurately, but also still study the target model problem. This discovery method can be applied in actual domain.

1. Introduction

The rapid progress of data collection tools, advanced database system technologies, and the World Wide Web (WWW) technologies has precipitated the explosive growth of vast amounts of data in various forms. Hence, there is an increasing need for discovering important and previously unknown knowledge from the complex types of data.

To discover information, the analysis of the video can provide a worthy approach for various managements. For example, issues associated with extracting traffic movement and recognizing accident information from real time video sequences is discussed in the literature [1]. What is missing in these efforts is to model and index the data for on-line analysis, storage or later pattern analysis. In order to identify and track the temporal and relative spatial positions of moving targets in video, it is necessary to have target-based representation of video data. Our discovery method of unknown visual information focuses on obtaining targets in the frame and their traces across the frames. For this purpose, attention has been devoted to find motion targets for tracking. So, it is used that detecting video frames to find feature points corresponds to a target that is meaningful to human viewers.

In this paper, a discovery method of unknown visual information based on target tracking is proposed. Based on the trajectory information of targets, some previously unknown or non-intuitive knowledge can be figured out and be used to support decision making.

The organization of this paper is as follows. In the next Section, we describe the discovery framework of unknown visual information. The knowledge discovery process and the core technique of video data that includes detection and correspondence of the feature points, target identification and tracking, and target occlusion etc are introduced in Section 3. Along with the discussion, a real-life example traffic video sequence is used in Section 4. Section 5 presents the conclusions.

2. Method description

In this paper, a discovery method of unknown visual information is proposed. For better comprehending our scheme, the framework of this scheme is firstly given. The scheme can be represented in the framework shown in the block diagram of Figure 1.

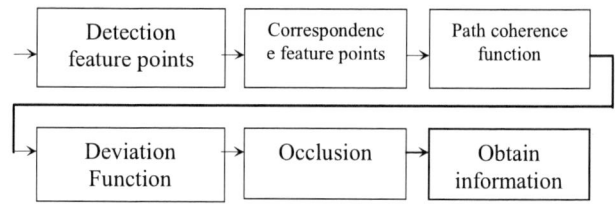

Figure 1 Building blocks for video data mining

A detailed discussion of the individual blocks will be deferred in next section.

3. Discovery information from video data

Video analysis can discover and provide useful information. To the best of our knowledge, the current multimedia based transportation applications and research work either do not connect to databases or have limited capabilities to index and store the collected data (such as videos) in their databases. Therefore, those applications cannot provide organized, unsupervised, conveniently accessible, and easy-to-use video information to the planners. In order to discover and provide some important but previously unknown knowledge from the video to the planners, video information discovery technique needs to be employed. In this paper, the proposed video information discovery technique includes detection and correspondence of feature points, target identification and tracking, and target occlusion, etc.

Tracking of motion target in a sequence of frames is a very important technique. If there is only one target in the image sequence, the task can often be solved easily, but if there are many targets moving simultaneously and independently, more complex approaches are needed to incorporate individual target motion-based constraints. In this situation, motion constraints described earlier should be examined. Consequently, it is possible to formulate the notion of path coherence which implies that the motion of a target at any point in an image sequence will not change abruptly[2].

Motion detection based on correspondence of the feature points work for inter-frame time intervals that cannot be considered small enough. Detection of corresponding target point in subsequent images is a fundamental part of this method.

The first step of the method is to find significant points in all images of the sequence-points least similar to their surrounding representing target corners, borders, or any other characteristic features in an image that can be tracked over time. Point detection is a matching procedure, which looks for correspondences between these points.

3.1. Detection of feature points

Corners play a significant role in detection of feature points. Corners in images can be located using local detectors. The Achard-Rouquet[3] detector is based on the face that the angle between two gradient vectors is important around corners. This detector is used extensively, and the performance of the detector is satisfied.

The corner strength is hence calculated using cross product of the neighbourhood gradients.

$$k_{ij} = \sum_{u,v \in V_{i,j}} \left\| \overrightarrow{grad}(f(i,j)) \wedge \overrightarrow{grad}(f(u,v)) \right\|^2 \quad (1)$$

$$= \sum_{u,v \in V_{i,j}} \left\| \overrightarrow{grad}(f(i,j)) \right\|^2 \cdot \left\| \overrightarrow{grad}(f(u,v)) \right\|^2$$

$$\cdot \sin^2[\overrightarrow{grad}(f(i,j)), \overrightarrow{grad}(f(u,v))]$$

where $f(i,j)$ represents the luminance of the image at the coordinate (i,j), and $V_{i,j}$ represents the neighbourhood of the pixel (i,j) and denotes the cross product. Given:

$$X = \delta f / \delta x, \ Y = \delta f / \delta y \quad (2)$$

And developing the equation (1), a normalized version of the detector is obtained

$$k = \frac{f_x^2 <f_y^2> + f_y^2 <f_x^2> - 2 f_x f_y <f_x f_y>}{<f_x^2> + <f_y^2>}. \quad (3)$$

Where $< >$ denotes the mean applied on the neighbourhood $V_{i,j}$.

The set of the feature points is obtained using a threshold of the value k and selecting local maximal. The Achard-Rouquet detector applied on *Lena* is shown on Figure 2.

Figure 2 Achard-Rouquet detector response

3.2. Correspondence of Feature Points

Assuming that feature points have been located in all images of a sequence, a correspondence between points in consecutive images is sought. Many approaches may be applied to seek an optimal correspondence, and we use method of the literature [4] in this paper, because this method is more simple.

Let $A_1 = \{x_m\}$ be the set of all feature points in the first image, and $A_2 = \{y_n\}$ the feature points in the second image. Let c_{mn} be a vector connecting points x_m and y_n (c_{mn} is thus a velocity vector; $y_n = x_m + c_{mn}$). Let the probability of correspondence of two points x_m and y_n be P_{mn}. Two points x_m and y_n can be considered potentially corresponding if their distance satisfies the assumption of maximum velocity,

$$|x_m - y_n| \leq c_{max} \quad (4)$$

Where c_{max} is the maximum distance a point may move in the time interval between two consecutive images. Two correspondences of points $x_m y_n$ and $x_k y_l$ are termed consistent if

$$|c_{mn} - c_{kl}| \leq c_{dif} \quad (5)$$

Where c_{dif} is a preset constant derived from prior knowledge. Clearly, consistency of corresponding point pairs increases the probability that a correspondence pair is correct.

3.3. Path Coherence Function

The path coherence function ϕ will be normalized, i.e., $\phi(\cdot) \in [0,1]$, and it represents a measure of agreement between the derived target trajectory and the motion constraints.

Let the trajectory T_i of a target i be represented by a sequence of points in the projection plane,

$$T_i = (X_i^1, X_i^2, ..., X_i^n) \quad (6)$$

Where X_i^k represents a (three dimensional) trajectory points in image k of the sequence, shown in Figure 3.

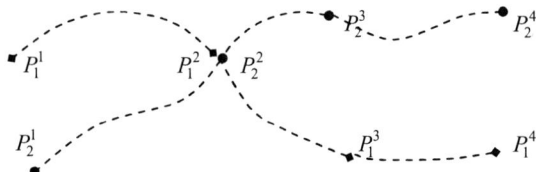

Figure 3 The trajectory of two targets moving simultaneously and independently

Let x_i^k be the projection image coordinates associated with the point X_i^k. Then the trajectory can be expressed in vector form:

$$T_i = (x_i^1, x_i^2, ..., x_i^n) \qquad (7)$$

3.4. Deviation Function

Deviation in the path can be used to measure the path coherence. Let d_i^k be the deviation in the path of the point i in the image k:

$$d_i^k = \phi(\ \overline{x_i^{k-1}x_i^k},\ \overline{x_i^k x_i^{k+1}}\ ) \qquad (8)$$

or $\qquad d_i^k = \phi(\ x_i^{k-1}, x_i^k, x_i^{k+1}) \qquad (9)$

Where $\overline{x_i^{k-1}x_i^k}$ represents the motion vector from point x_i^{k-1} and ϕ is the path coherence function. The deviation D_i of the entire trajectory of the target i is then $D_i = \sum_{k=2}^{n-1} d_i^k$.

Similarly, for m trajectories of m targets in the image sequence, the overall trajectory deviation D can be determined as $D = \sum_{i=1}^{m} D_i$.

With the overall trajectory deviation defined in this way, the multiple target trajectory tracking can be solved by minimizing the overall trajectory deviation D.

By above analysis and describe, the path coherence function ϕ can be expressed as follows:

$$\phi(P_i^{k-1}, P_i^k, P_i^{k+1})) = w_1(1 - \cos\theta) + w_2(1 - 2\frac{\sqrt{s_k s_{k+1}}}{s_k + s_{k+1}}$$

$$= w_1(1 - \frac{\left|\overline{x_i^{k-1}x_i^k} \cdot \overline{x_i^k x_i^{k+1}}\right|}{\left\|\overline{x_i^{k-1}x_i^k}\right\| \cdot \left\|\overline{x_i^k x_i^{k+1}}\right\|}) + w_2(1 - 2\frac{\sqrt{\left\|\overline{x_i^{k-1}x_i^k}\right\| \cdot \left\|\overline{x_i^k x_i^{k+1}}\right\|}}{\left\|\overline{x_i^{k-1}x_i^k}\right\| \cdot \left\|\overline{x_i^k x_i^{k+1}}\right\|}) \qquad (10)$$

Where the angle θ and distances s_k, s_{k+1} are given by Figure 4. The weights w_1, w_2 reflect the importance of direction coherence and velocity coherence.

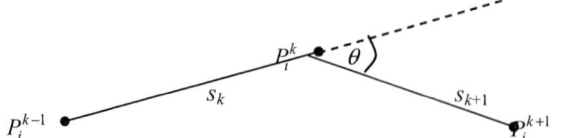

Figure 4 Path coherence function - definition of the angle θ and distances s_k, s_{k+1}

3.5. Occlusion

Spatio-temporal approaches to analysis of image sequences with multiple independently moving targets represent to motion analysis. When simultaneously tracking several targets with independent motion, target occlusion is almost guaranteed to occur. Consequently, some targets may partially or completely disappear in some image frames which can result in errors in target trajectory. If minimization of the overall trajectory deviation D is performed using the given path coherence function, it is assumed that the same number of targets is detected in each image of the sequence and that the detected target points consistently represent the same targets in every image. Clearly, this is not the case if occlusion occurs.

To overcome the occlusion problem, additional local trajectory constraints must be considered and trajectories must be allowed to be incomplete if necessary. Incompleteness may reflect occlusion, appearance or disappearance of a target, or missing target points due to changed target aspect resulting from motion or simply due to poor target detection. Thus, additional motion assumptions that were not reflected in the definition of the path coherence function ϕ, e.g., maximum velocity, must be incorporated. An algorithm was presented in the literature [5], which finds the maximum set of complete or partially complete trajectories and minimizes the sum of local smoothness deviations for all identified trajectories. Local smoothness deviation is constrained not to exceed a preset maximum ϕ_{max} and the displacement between any two successive trajectory points X_i^k, X_i^{k+1} must be less than a preset threshold d_{max}. To deal efficiently with incomplete trajectories, we use phantom points as substitutes for the missing trajectory points. These hypothetical points allow each potential trajectory to be treated as complete and permit consistent application of the optimization function.

4. Experiment

In this paper, the discovery method of unknown visual information is applied to a real-life video sequence. The video sequence consists of 10 frames from a traffic turn a corner place. For simplicity and real-time processing purpose, the video frames only were of size 256 rows $\times$ 256 columns, 8 bit grayscale images, shown in Figure 5.

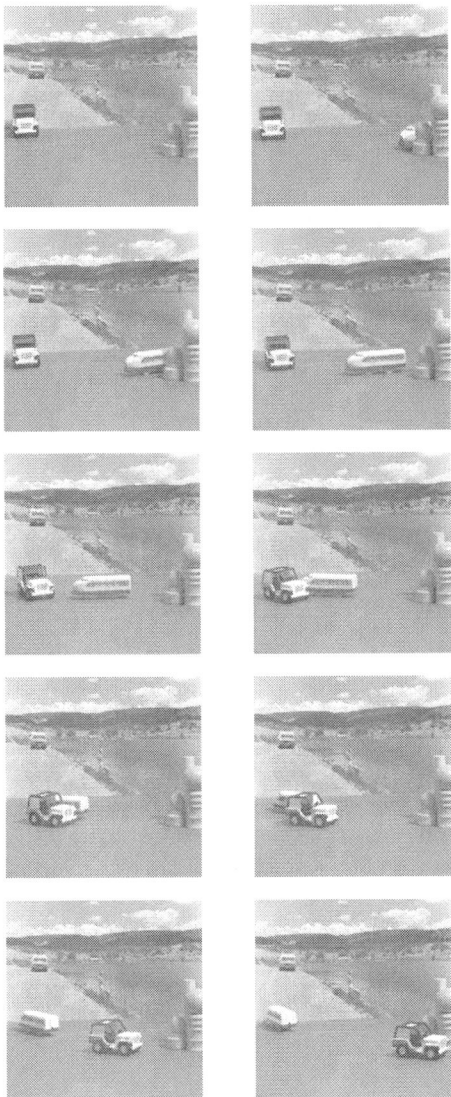

Figure 5 The original video sequences

As only the vehicles are important for our experiment, we focus on vehicle identification and tracking. After the feature points of the video frames are detected, the work of corresponding feature points is realized. According to our method, the identification and tracking of the vehicle can complete.

From the experiment, we can find that there are two vehicles with independent motion and a vehicle with static state in this video sequences. According to spatio-temporal approaches to analysis of image sequences, making use of the trajectories of the feature points, the two moving vehicles ever drive right against the face. In other words, two car's directions aren't the same.

The experiment results demonstrate the knowledge discovery process (i.e., the vehicle tracking) from the image sequences using the proposed discovery method of unknown visual information. The discovered information can be applied to traffic applications so that spatio-temporal queries can be answered, such as the estimate of traffic flow, and detection of the vehicle direction etc.

5. Conclusion

In this paper, we present work on extracting and modeling the spatio-temporal relationships of moving targets from video sequences. From the experiment results, it can be seen that our discovery method of unknown visual information can model not only the number of moving targets, but also the judge the direction.

Much work needs to be done to automatically collect, index, store and analyze spatio-temporal information from real-time video streams. The paper is only an initial method of such systems. We intend to extend the present work in the following directions:

(1) Integrate the methods of information indexing, analyzing and retrieving into our method.

(2) Extend our method from static background to complex background;

(3) Extend our method from daytime operations to night time domain.

References

[1] Z.Y.Fu, W.M.Hu, and T.N.Tan. "Similarity based vehicle trajectory clustering and anomaly detection", *IEEE International Conference on Image Processing*, 2005, pp. 602-605

[2] S.C.Chen, M.L.Shyu, and C.C.Zhang. "An intelligent framework for spatio-temporal vehicle tracking". *The 4th International IEEE Conference on Intelligent Transportation Systems*, Oakland, CA, USA. 2001, pp. 213-218

[3] R.Jain, R.Kasturi, and B.G.Schunck. *Machine Vision*. McGraw-Hill, New York. 1995

[4] J.Devars, C.Achard-Rouquet, and E.Bigorgne. "Un detecteur de point carateristiques sur des images multispectrales". *Extension versun detecteur subpixellique*. In GRETSI 99, 1999, pp.627-630

[5] I.K.Sethi, and R.Jain. "Finding trajectories of feature points in a monocular image sequence". *IEEE Transactions on Pattern Analysis and Machine Intelligence*. 1987, pp. 56-73

A Novel Email Virus Propagation Model

Cong Jin, Jun Liu, Qing-Hua Deng
Department of Computer Science,
Central China Normal University, Wuhan 430079, P.R.China
E-mail: jincong@mail.ccnu.edu.cn

Abstract

In this paper, two new parameters, User Vigilance *and* Removing Time, *are incorporated into the classical email virus propagation model SEIR. New model extended the classical virus propagation model SEIR. The experimental results showed that these parameters greatly influence the virus propagation. Meanwhile, the simulations demonstrate that improved model can be used for calculating the costs of virus outbreak.*

1. Introduction

Currently, email virus constitutes one of the major Internet security problems. Usually a virus email has an attachment file that contains copy of the virus. The virus hides the attachment file's executable property by forging it to be any type of files, like image, word document, etc. When an email user clicks on this attachment, the virus program will be activated and infect the local computer.

For years, the email virus propagation models based on epidemiological theories of human epidemic disease have been researched. Anderson and May[1] investigated spreading characters of various infectious disease, and added Exposed state to Susceptible-Infected-Removed model. Then SEIR virus model appeared, and typical states of SEIR model in Table 1.

Table 1 Typical states of SEIR model

S	Susceptible
E	Exposed to infection
I	Infected
R	Removed

Although the SEIR email virus propagation model achieved a better performance, the SEIR model has three shortcomings as follows:

(1) In SEIR, time is divided into some discrete steps to describe the model[2,3]. Transitions between individuals in each state of the models are described by simple probabilities in every time steps. The simplified email virus propagation model can't reflect the actual situation of virus diffusing.

(2) Virus propagation is simulated by same virus model from email virus breakout to immunization. Thus, such model can't reflect the actual situation accurately.

(3) Traditional SEIR model neglected difference of the email users. Quite a lot Internet users less understand virus hided in email attachment. Email users usually give an appropriate trust to emails from their friends. Email with virus may be opened without suspiciously, and not be scanned by anti-virus software. The situation is called that users have little vigilance.

In this paper, two new parameters, *i.e. Removing Time* and *User Vigilance*, are incorporated to SEIR model for improving SEIR model. Two new parameters of the email virus propagation model play an important role on improving the model performance.

2. Proposed email virus propagation model

The general process of the email virus infection is described as follows.

First, the virus is released into the wild by its creator. The virus is spreads freely, infecting user's machines in the network. In the beginning of the virus spreading, the serious virus is not noticed or alerted. Meanwhile, anti-virus techniques are not developed. So email users haven't abilities to remove the virus. After the virus has spread for some time, anti-virus company works to isolate the virus and generates an anti-virus technique used to detect the presence of the virus. This process can keep on some time. The time T of anti-virus technique used is called as *Removing Time*. So, new email virus spread model contains two phases.

Time $t < T$: Virus spreads freely.

Time $t \geq T$: Anti-virus technique presents, and the most users start to remove virus.

A. *t < T Phase*

Before the virus can spread unchallenged, the user state only has three cases: *Susceptible* (*S*), *Exposed* (*E*) and *Infected* (*I*), no remove states. In this situation, the infected users become more and more because of no appearing anti-virus software. Figure 1 represents this state.

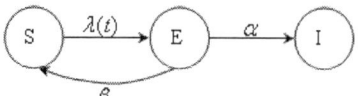

Figure 1 *t < T* **virus model**

In following discussion, the meanings of the some signs are as follows:

$P(S \rightarrow E) = \dfrac{rI(t)}{N}$: Rate of a susceptible user becoming *Exposed* state; N is total number of users.

r : Clustering coefficient.

$P(S \rightarrow I) = \alpha$: Rate of a *Exposed* susceptible user becoming *Infected* state.

$P(E \rightarrow S) = \beta$: Rate of an *Exposed* user becoming *Susceptible* (such as, by deleting email with virus).

Therefore, the state of Figure 1 can be described by the following equations[3]:

$$\begin{cases} \dfrac{dS(t)}{dt} = -\lambda(t)S(t) + \beta E(t) \dots \dots \dots \dots (1) \\ \dfrac{dE(t)}{dt} = \lambda(t)S(t) - (\beta + \alpha)E(t) \dots \dots \dots (2) \\ \dfrac{dI(t)}{dt} = \alpha E(t) \dots \dots \dots \dots \dots \dots (3) \\ \dfrac{dR(t)}{dt} = 0 \dots \dots \dots \dots \dots \dots \dots (4) \end{cases}$$

Where, $S(t)$, $E(t)$, $I(t)$, and $R(t)$ are number of users with *Susceptible*, *Exposed*, *Infected* and *Removed* state respectively at every time step. Thus $\dfrac{ds(t)}{dt}, \dfrac{dE(t)}{dt}, \dfrac{dI(t)}{dt}, \dfrac{dR(t)}{dt}$ are the increasing rates of *Susceptible*, *Exposed*, *Infected* and *Removed* users respectively at every time step. $\lambda(t)$ is a virus propagation function and vary with the time. In the beginning, the value of $\lambda(t)$ is small, then it increases mildly with virus spreads, and more infected users appear. In Equation (1), $\lambda(t)S(t)$ is number of susceptible users changing into exposed at time t. Consider probability of some user may discard suspicious email with virus attachment, $\beta E(t)$ means the number of increasing susceptible users at time t. Therefore $\beta E(t) - \lambda(t)S(t)$ is changer rate of susceptible users. Furthermore, owing to no presence of anti-virus software, so the value of $\dfrac{dR(t)}{dt}$ keeps 0.

It should be noted that $\lambda(t)$ doesn't reveal the factor of network congestion when many computers are infected.

B. $t \geq T$ Phase

When $t \geq T$, anti-virus technique has developed to cleanup or isolate virus. In fact, not all users install or update the anti-virus software for isolating the virus. If a user adopts the anti-virus software with a high probability for detecting and removing the virus, he can obtain a safer environment when connecting to the Internet. This user is called with a high vigilance. However, many Internet users haven't much understanding about the importance of anti-virus software. So, they don't install or update the anti-virus software in time on their computers. These users have great threat to other users. The portion of users called low vigilance will continue to infect computers from their email address book. Therefore, the virus spreading parameter *User Vigilance* should models by *User Vigilance* δ defined as follows:

$$User\ Vigilance\ \delta = \frac{The\ number\ of\ installing\ anti-virus\ software}{The\ total\ number\ N\ of\ users}$$

User Vigilance δ indicates user rate of installing anti-virus software. $\delta \in [0, 1]$. 0 indicates all users don't install or update anti-virus software, and 1 indicates all users install or update anti-virus software.

For high vigilance δ users, the anti-virus technique is distributed, and email virus is cleaned up. Therefore, email virus spreading and cleanup can be modeled as shown in Figure 2. *Susceptible*, *Exposed*, and *Infected* states directly become *Removed* state with a high rate. The model reflects that high vigilance users cause email virus accelerates to die.

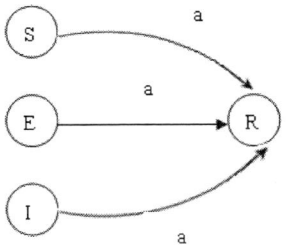

Figure 2 High vigilance users virus model

For low vigilance $1 - \delta$ users, the virus continues to spread and infect others users because of no installing anti-virus software in time. So, the virus propagation model may be simplified into anti-virus technique ($t<T$). This phase considers infected users becoming *Removed* state at quite a small rate k ($k \ll a$). Similarly, users are thought to be one of four states: *Susceptible*, *Exposed*, *Infected*, and *Removed*. Figure 3 represents this state.

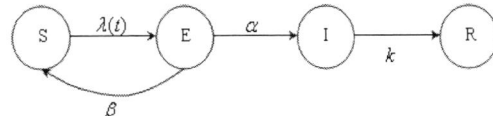

Figure 3 Low vigilance users virus model

Based on the above analytic, for $t \geq T$ phase, virus propagation model can be described by the following equations:

$$\begin{cases} \dfrac{dS(t)}{dt} = (1-\delta)(-\lambda(t)S(t) + \beta E(t)) - \delta a S(t) \quad\text{............................(5)} \\[2mm] \dfrac{dE(t)}{dt} = (1-\delta)(-\lambda(t)S(t) - (\alpha+\beta)E(t)) - \delta a E(t) \quad\text{..................(6)} \\[2mm] \dfrac{dI(t)}{dt} = (1-\delta)(\alpha E(t) - kI(t)) - \delta a I(t) \quad\text{......................................(7)} \\[2mm] \dfrac{dR(t)}{dt} = (1-\delta)kI(t) + \delta a(S(t) + E(t) + I(t)) \quad\text{............................(8)} \end{cases}$$

In Equation (8), $\delta a(S(t) + E(t) + I(t))$ is total number of changing into *Removed* users from high vigilance users. Note that *Removed* rate a is usually large quantity. $(1-\delta)kI(t)$ and $\delta a(S(t) + E(t) + I(t))$ is interpreted as all removed users at time t. Equation (7) reveals that the population of infected users at time t will reduce at big rate, since most high vigilance users installed anti-virus software. From these equations, user vigilance play an important role for controlling the virus propagation.

3. Experiment results

Email virus propagation is affected by many parameters in the email virus model. The influence of some parameters $\lambda(t)$, β, γ, for virus propagation behavior, has already been researched. In this paper, we only concern another two key factors, *i.e.*, *Removing Time* t and *User Vigilance* δ.

In simulation experiment, let email viruses be only transferred by users email address books. Thus email address relationship between users' address books forms a logical network for email viruses. We let the email network has 10000 email users, *i.e.*, N=10000, user clustering coefficient r =10, and initial infected users are 10 ($I(0) = 0$). Other parameters β, α, a are set as 0.0088, 0.0022, and 0.2, respectively.

A. Initial Results

The Figure 4(a) and (b) provide a simple comparison between proposed email virus model after incorporating new parameters *User Vigilance* and *Removing Time* and traditional SEIR.

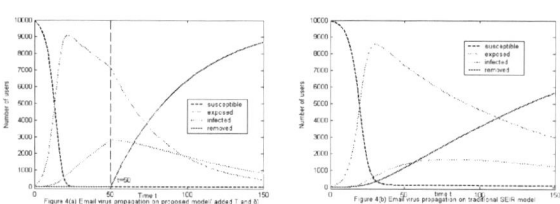

(a) Proposed Virus Model (b) Traditional SEIR model
Figure 4 Comparison between proposed virus model and traditional SEIR

Figure 4(a) assumes that anti-virus technique presents at time 50, *i.e.* T=50. It shows that the number of infected

users increases quickly and accumulates a high value at time 50 before anti-virus software appearing, while removed users keep 0. In other words, email virus would infect freely all email users without anti-virus software. After the anti-virus software is available (i.e. $t > T$), the number of infected users drops and removed users increases when quite a lot users install or update new anti-virus software. But the speed of infected users going up and removed users going down are determined by one vital parameter, *i.e.*, *User Vigilance*, and which will be deeply discussed in Section 3.2.

Figure 4(b) reveals that removed users immediately appear when virus attempts to spread, and the size of infected users is smaller than Figure 4(a). This is because the traditional SEIR virus model assumed that, as long as email virus has break out, email users have strategy to control the virus spreading further. However, this assumption is not consistent with objective fact.

B. Effect of User Vigilance δ

In proposed virus model, *User Vigilance* δ is a vital parameter and is related to how many users install the new anti-virus software. To perform effects of *User Vigilance*, δ is set as three different value (0.1, 0.2, 0.5), and let T=50 (This means that anti-virus technique is developed at time 50). The numerical curve of $I(t)(R(t), E(t), S(t))$ with different δ will be discussed in Figure 5.

Figure 5(a) clearly shows the outbreak size, *i.e.*, number of infected users, for varying *User Vigilance* after anti-virus technique takes action. The general trend of curve is that $I(t)$ goes down gradually at time t>50. The higher the *User Vigilance* is, the faster speed infected users decrease. This effect is probably interpreted as email users with higher vigilance accelerate the virus fading away. That is to say, the duration of outbreak is more short for a big δ, and it results in a weak cost.

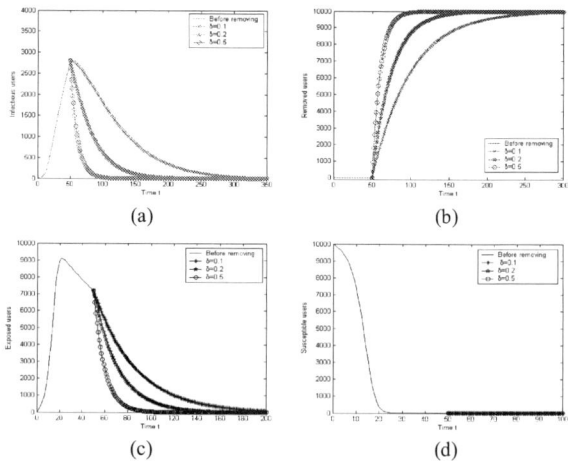

(a) (b)

(c) (d)
Figure 5 Effects of *User Vigilance* δ

Figure 5(b) illustrates that number of removed users keep zero (t<50) and then increase gradually, later tend to stable state in general. Solid line marked with asterisk raises more slowly than other two lines. Thus, increasing δ means increased efficiency of cleanup (It indicates that more users installed or updated the new anti-virus software for centralized immunization). As showed in Table 2, when δ =0.1, the email virus is fully removed at time 310, but δ reaches 0.5, removing virus wins a success just at 105. This result would help us further understanding about why *User Vigilance* is great important to remove email virus.

Table 2 Some discrete data about virus spreading

User Vigilance δ	0.5	0.2	0.1
Time of Fully Removed t	105	182	310
Outbreak Duration $\triangle t$	55	132	260

Figure 5(c) and (d) give the change of exposed users, *i.e.*, susceptible users, with varied δ . According to experiment Figure 5(d), susceptible user already disappeared before anti-virus technique appearing, so the value of $S(t)$ obviously continues to keep zero. Figure 5(c) shows that the number of susceptible users drops dramatically. This is because these users become removed at a high rate.

C. Effect of Removing Time T

A question addressed by proposed model is "Whether the *User Vigilance* is the most important parameter". Another parameter *Removing Time T* is simulated in Figure 6. Figure 6 shows the costs from varying T =5, 10, 20, and 50 respectively. The general trend is that number of infected users decrease gradually as anti-virus software is available (t>T).

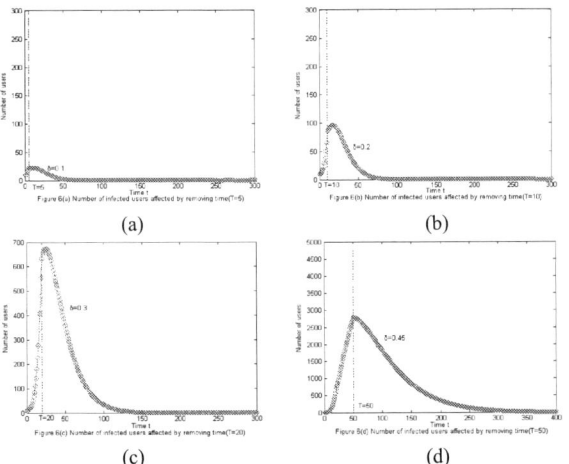

(a) (b)

(c) (d)

Figure 6 Number of infected users affected by *Removing Time T*

From Figure 6, the effects of *Removing Time* mainly embody follows.

(1) For a small T, the maximum of infected users are not significant, while larger T results in increasing outbreak (Table 3). That's to say, generating anti-virus technique quickly can greatly reduce costs, but it will produce opposite situation if the new anti-virus software presents lately.

Table 3 Outbreak degree of virus with different *Removing Time T*

Removing Time T	5	10	20	50
Time t	9	18	30	62
Maximum rate of infected users	< 0.23%	< 0.96%	6.74%	28.02%

In particular, in Figure 6(a), (b), (c), we notice that the number of infected users don't decay right away when t>T , but to experience a slow growth. Figure 6(d) should have a similar effect, however this kind of phenomenon don't appear obviously because the proportion of vertical axis is big. Before anti-virus technique is distributed, the virus spreads unhindered quickly. The size of infected users may accumulate a high degree during virus spreading freely. So anti-virus software will take some time to make the infected users became small. This fact gives a well understanding why the maximum of infected users don't appear at time T.

(2) For a large T, the outbreak duration will last a long time in despite of email users with high *User Vigilance* δ . Conversely (*i.e.*, a small T), the system may be not suffer the great losses even though δ is not quite large. This phenomenon is not surprising because the network may accumulate more infected users for a large T. It indicates that anti-virus technique present lately, and email virus already spreads a long time. Therefore, the time that infected users are made immune ($I \rightarrow R$) can be long.

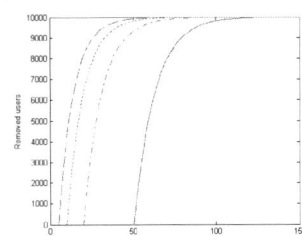

Figure 7 Number of removed users affected by removing time *T*

In summary, simulations of the model reveal that the time of anti-virus technique appearing, *i.e.*, *Removing Time T*, a vital factor to control virus email propagation. If anti-virus software is used before the large-scale outbreak of email virus, it is easy to defeat virus propagation; otherwise, it is hard to defeat despite of higher *User*

Vigilance. Based on this conclusion, one feasible measure may suggest that mail server enhances a mechanism through increasing probability of filter spam, prolonging average time of sending or receiving emails before anti-virus technique is available. The much time of researching anti-virus software will win for workers of network information security.

4. Conclusions

An email virus model based on the epidemiological is proposed in this paper. The research extends previous work by incorporating two new parameters into SEIR. Analysis and simulation show that new parameters are vital factors in controlling virus spreading. Meanwhile, the more promptly anti-virus industry develops corresponding anti-virus technique, the less outbreak email viruses result in. However, even if people have ability to remove this email virus, system will undergo a great loss if most email users have much low vigilance.

Finally, the paper also gives insight into the relative importance of the research of new anti-virus technique for the entire anti-virus industry.

References

[1] R.M.Anderson, R.M.May. Infectious diseases of humans: dynamics and control. Oxford: Oxford University Press, 1991, pp. 304-318

[2] R.P.Satorras, A.Vespignani. Epidemic dynamics in finite size scale-free networks. Physical Review E, 65: 035108, 2002

[3] B.K.Mishra, D.Saini. "Mathematical models on computer viruses". Applied Mathematics and Computation, 2007, 187(2): 929-936

Scheme of Electronic Seal Based on Public Key Infrastructure

Zhang Jiuhua

Department of Physics and Electron Communication,
Leshan Teachers College, Leshan, SiChuan, 614000, China
lstcwd@163.com

Abstract

This paper improves the general electronic seal scheme based on PKI platform. The original scheme has low efficiency and complicated process when seal's authenticity is verified. In this new scheme, before the seal is used for the time a random code will be created by CA. The random code and the electronic seal image will be printed together. User can easily complete the seal security anti-counterfeiting verification under security strength with no high requirement, through contacting CA for checking the random.

1. Introduction

With the rapid development of information and internet technology in the world, electronic commerce has gradually been new mode of commerce activity.

Compared with traditional model, electronic commerce has many good qualities such as low cost, high efficiency, convenient management and so on. The electronic contract is an important part of electronic commerce, and digital signature is one of the important technology instrumentalities for guaranteeing the security of electronic commerce. The traditional digital signature technology can solve the security problems of electronic document such as legitimacy, authenticity, integrality traceability and so on. But many users still hope to see company seal or individual handwritten image in electronic documents; electronic seal technique solves the problem that the traditional digital signature technique can not do it. At present, almost electronic seal system scheme are based on the existing digital signature schemes whit PKI platform and digital watermark technology. In this system scheme, it is complicated to verify the paper document printed from the electronic document with electronic seal. Firstly, the paper document needs to be converted into digital data, and then the data need to be accounted by computer to complete the process of verification. In this paper, a new scheme improves the problem.

2. PKI technology description

Internet Security Glossary defines public-key infrastructure (PKI) as the set of hardware, software, people, policies, and procedures needed to create, manage, store, distribute, and revoke digital certificates based on asymmetric cryptography. The principal objective for developing PKI is to enable secure, convenient, and efficient acquisition of public keys. PKI is one of the important techniques for solving network security problem.

The typical PKI application system includes security certificate authority (CA), digital certificate repository, key pair recovery system, certificate revocation system, and application program interface (API). The fowling contents are their deifications:

(1) Certificate authority:

The issuer of certificates and certificate revocation lists (CRLs). It may also support a variety of administrative functions, although these are often delegated to one or more Registration Authorities.

(2) Digital certificate repository:

A generic term used to denote any method for storing certificates and CRLs so that they can be retrieved by End Entities which is essential elements of PKI system.

(3) Key pair recovery:

Key pairs can be used to support digital signature creation and verification, encryption and decryption. It is important to provide a mechanism to recover the necessary decryption keys when normal access to the keying material is no longer possible, otherwise it will not be possible to recover the encrypted data. Loss of access to the decryption key can result from forgotten passwords, corrupted disk drives, damaged to hardware tokens, and so on. Key pair recovery allows end entities to restore their key pair from an authorized key backup facility.

(4) Certificate revocation:

An authorized person advises a CA of an abnormal situation requiring certificate revocation. Reasons for revocation include private key compromise, change in

978-0-7695-3342-1/08 $25.00 © 2008 IEEE
DOI 10.1109/PEITS.2008.126

affiliation, and name change as well as various identity certificates in life.

(5) Application program interface:

The value of the PKI enables users to conveniently use of encryption, digital signatures and other security services. So a complete PKI must provide a good application interface system to enable various applications can interact with PKI through a safe, consistent and credible manner and ensure integrity and convenience of the security network environment

3. Electronic seal description

Electronic seal, in a large sense, refers to all the contents that exist in electronic forms, attach to and logically associate with the electronic document, can be used to identify the electronic document user's authentication, assure the integrality and validity of the electronic document, and show the signer's agreement on the facts recorded in the electronic document.

According to these characteristics, people put forth some demands to the E-seal system. A security electronic seal system should achieve the following goals:

(1) Authentication:

Before the transferring of the E-document and seal files, the users should first affirm the authentication of both sides through special technology, in order to assure that both users' identity cannot be fake or counterfeit.

(2) Confidentiality:

Hide the sensitive electronic document content and store the seal files with keys. Even if the data is intercepted and captured, the hacker cannot get to know the content.

(3) Integrity:

Demand the receiver should be able to test the integrity and authenticity of the received E-document, and judge whether the stamped E-document has been interpolated.

(4) Non-Repudiation:

Once the transferring of the E-document and seal files is finished, through specific control through specific control and technology, the sender cannot deny the data he has sent, and the receiver cannot deny the data he has received.

4. Electronic seal system structure

Seal certificates of the electronic seal system come from the CA server. Through CA servers, the electronic seal system finishes works of generating certificate, publishing certificate and revoking certificate. At the same time, seal server is responsible for database maintenance and management. And the seal database includes seal image, seal certificate key and records of authorization usage seal.

Because seal key is stored in seal server, so seal server should be set in a high security environment including physical isolation and network security and so on. Meanwhile, there should be a whole log file to record the seal usage situation. CA servers and seal servers manage seals, secret keys, certificates and accept the application of usage seal by web servers.

5. Implementation scheme based on PKI

In this electronic seal system implementation scheme based on PKI, the digest of original document is embedded by a watermark in seal image after encrypted by client private key. So, it is an important step to embed and extract watermark in the scheme. This scheme mainly includes two function modules: attaching electronic module and seal and verification module.

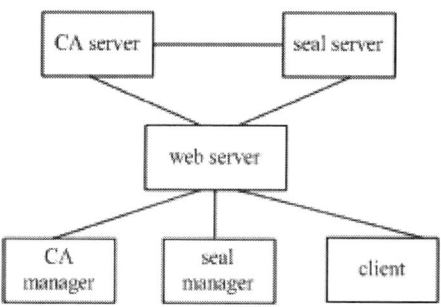

Fig. 1 Electronic seal system structure

5.1. Attaching electronic seal module

In this module, user's major works to be complete are application for usage seal, making digest of the original documents, digital signature, and embedding watermark, attaching electronic seal and sent attached to another user who will verify the electronic seal.

The whole process of attaching electronic seal is shown in detail in Fig.2.

(1) User makes use of the certificate to apply for usage seal image toward CA. And CA will generate a random code for the application at the time that will be used with electronic seal together.

(2) Digest of the original document is obtained by Hash algorithm. The digest is embedded as a mark in seal image after encrypted by client private key.

(3) Seal image embedded watermark and random code together also may be regarded as electronic seal which will be stamped together on original document.

5.2. Verification module

In this module, many operation processes are contrary with the former processes in first module. Main processes are: receiving document, separation electronic seal, verification signature, calculation digest and so on. Details are shown in Fig.3.

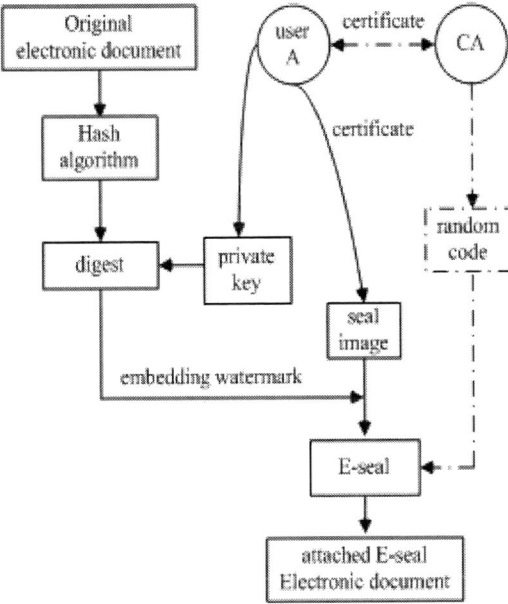

Fig. 2 Attaching Electronic Seal Process

(1) Seal image embedded watermark and random code are separated respectively from electronic document stamped electronic seal.

(2) After extracting watermark from the seal image, user makes use of the public key obtained from CA to decrypt it and can get digest of the original document.

(3) User can get another digest, comes from the document separated from received document attached electronic seal, by Hash calculation.

(4) Through comparing the two digest, user can get a verification conclusion.

In Fig.2 and Fig.3, the dashed line fractions are optional according to different security levels. Under lower security level situation, user can easily only take use of the random code through making the relationship with CA to verify the seal and document and need not to the more complicated operations: extraction watermark,

decryption, Hash calculation and so on. Otherwise, random code verification can operation together them to get the higher security.

6. Conclusion

The design of the electronic seal system based on PKI is comprehensive usage of the digital signature and the watermark technique. The random code mechanism, provided in this implication scheme, increases the verification flexibility and improves the capability of different operation environment. At he same time, the random sequence code authentication function can increase the safety of the system to defend the false

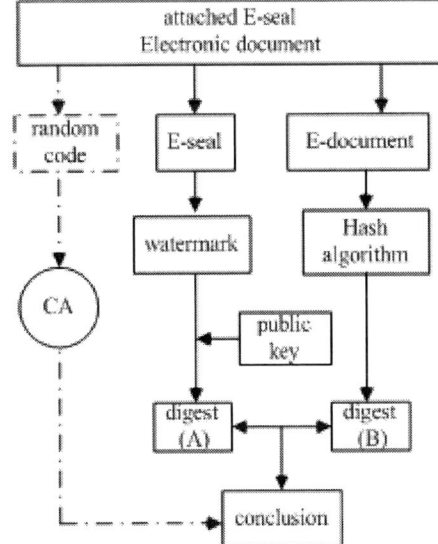

Fig. 3 Electronic seal system structure

capability.

Acknowledgement

It is a project supported by Education Department of SiChuan Province(2006B074).

References

[1] HOUSLEY R, POLK W, SOLO D, Internet X..509 Public Key Infrastructure Certificate and CRL Profile [S].RFC2459, IFTF, January1999.

[2] WANG Fei, TANG Guang-ming and SUN Yi-feng, electronic Seal System Based on Fragile Watermark and Digital Signature, Computer Application & Research, 2004(4)

[3] LI Xin, SUN Yufang, An Electric Seal System Based on Public Key Infrastructure, Computer Science, 2004Vol.31NO.2

[4] GUO Zhengrong, ZHOU Cheng, Implementation of Electronic Seal System Based on Public Key Infrastructure, Computer Science, 2006Vol.33No.9

[5] ZHANG Qiuyu, YU Dongmei, Electronic seal system based on digital watermark and implementation, Computer Engineering and Design,2007Vol.28NO.14

[6] Liu Feng , Han Zhong-liang, Research of the PKI-based Electronic Seal in Network Transfer Computer Applications, 2004,VoI 23, NO3

Session 3

Control System

2008 Workshop on Power Electronics and Intelligent Transportation System

CAN-based Integrated Control Strategy for the Drivetrain of Commercial Vehicles Equipped with Automated Manual Transmission

Jingxing Tan, Xiaofeng Yin, and Liang Yin
School of Transportation and Automotive Engineering, Xihua University
xiaofengyin@vip.sina.com

Abstract

In traditional control strategy for the drivetrain of commercial vehicles equipped with Automated Manual Transmission (AMT), no information exchanges between the Transmission Control Unit (TCU) and the Engine Control Unit (ECU). Such independent operations of these two parts of drivetrain inevitably reduce operational safety as well as starting and shifting performance of vehicles. To overcome the shortcoming of independent operations, an integrated control strategy based on Controller Area Network (CAN) for the drivetrain of vehicles during the starting and shifting processes of AMT was put forward. In this paper, we summarize the communication protocol and characteristics of CAN bus at first, and then define the communication information specification between TCU and ECU, which are classified into instructive and shared information according to their different functionalities. Based on that, a CAN-based integrated control strategy is put forward, which is proven to be practicable through the tests of an AMT prototype vehicle.

1. Introduction

To get better operational safety as well as starting and shifting performance than those of the traditional independent control strategy, AMT and engine are needed to be controlled in parallel during the starting and shifting processes of commercial vehicles equipped with AMT. In this paper, a CAN-based integrated control strategy for the starting and shifting processes of AMT is put forward, in which, CAN bus is used as communication infrastructure to exchange control instructions and shared messages between TCU and ECU.

The rest of the paper is organized as follows. Section 2 summarizes the communication protocol and characteristics of CAN bus. Section 3 introduces a CAN-based architecture for drivetrain integrated control system. Section 4 analyzes the common information that can be shared by TCU and ECU or should be passed from one to another, and then classifies them into shared information as well as instructive information, finally defines these information in accordance with CAN protocol specification. Section 5 discusses the proposed CAN-based control strategy for the starting and shifting process of AMT. Section 6 evaluates the proposed strategy through starting and shifting tests of an AMT prototype vehicle. The paper concludes with Section 7.

2. Controller area network

The CAN specification version 2.0 [1], which has being adopted widely in today's automotive industry, was constituted by Bosch in September 1991. In this specification, Part A defines the standard message format that has 11-bit identifier; while Part B defines the format for both standard and extended messages, and the latter has 29-bit identifier. In November 1993, ISO released the international standard for CAN, ISO11898, which paved the way for CAN's application. In addition, there also exist other protocols based on CAN, such as SAE J1939 [2], CANOpen [3], etc.

As a serial communication bus efficiently supporting distributed real-time control with very high level of security, CAN's application ranges from high speed network to low cost multiplex wiring [4]. The CAN communication interface usually integrates the functionalities of CAN physical and data link layers such that it could be able to deal with frame forming, which includes bit stuffing, data block coding, CRC checking, priority arbitration, etc. It is a multi-master bus and can use twisted-pair, coaxial cable or optical fiber as communication media with bit rate up to 1Mbps [5].

One advantage of CAN protocol is that it abolishes the traditional node address coding while adopts the communication data block coding, which makes the number of nodes in a network free of limitation and meanwhile can let various nodes receive the same data at

---

* The work reported in this paper was supported in part by MOP of China, Scientific Research Fund of Sichuan Provincial Education Department, and Key Scientific Research Fund of Xihua University.

978-0-7695-3342-1/08 $25.00 © 2008 IEEE
DOI 10.1109/PEITS.2008.20

the same time. This characteristic of CAN is very useful in the distributed real-time automotive embedded control systems. Besides, the 8-byte maximum data length can satisfy the general purpose for the transmission of control instructions, work status and measured data in vehicle network [6].

3. Architecture for drivetrain integrated control system

On the basis of traditional manual mechanical transmission, AMT adds electronic control unit, i.e., TCU, to realize the automation of the starting, shifting, and other related processes of vehicles [7]. In traditional AMT system, TCU takes care of all controls of engine fuel-supply, clutch disengagement and engagement, and gear-selecting and shifting during the starting and shifting processes, according to, (1)

the information needed by TCU includes position of selecting lever, position of accelerate pedal, throttle

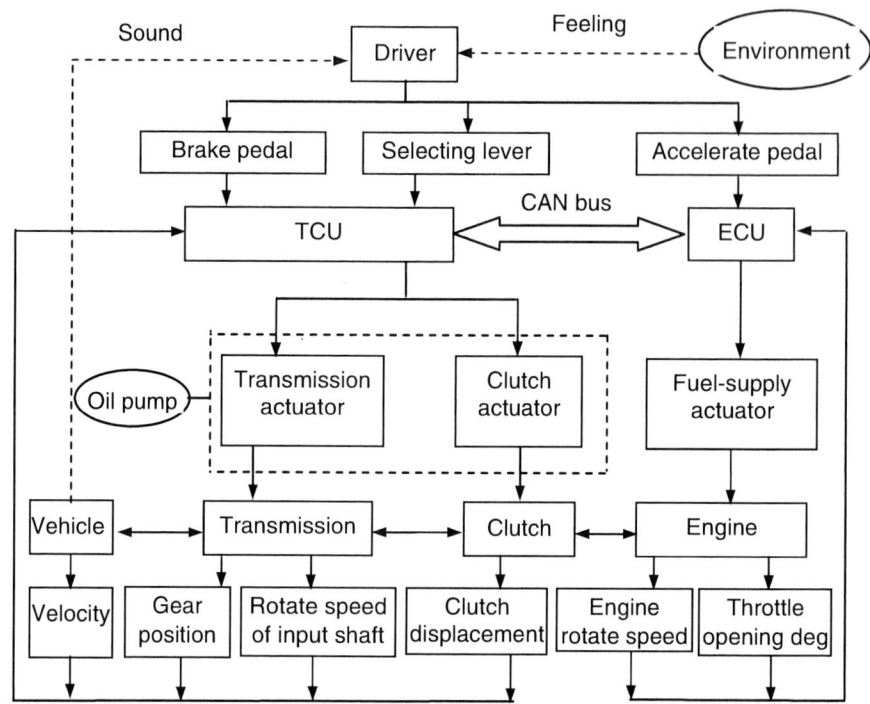

Figure 1. Architecture for drivetrain integrated control system

the pre-set control rules, such as gear-shifting schedule, clutch engagement rule, fuel-supply rule, etc., (2) the driver's intention expressed by selecting lever, accelerate pedal, and brake pedal, etc., as well as (3) vehicle work status information including engine rotated speed, rotate speed of input shaft of transmission, vehicle velocity, throttle opening degree, and gear position, etc.

However, according to the traditional control strategy, both TCU and ECU take care of the fuel-supply control of engine simultaneously and independently, during the starting and shifting processes of vehicle, which may cause collision and thus inevitably reduces operational safety and starting and shifting performance. Therefore, we need to treat the drivetrain as a whole during the starting and shifting processes of vehicle. That means we need to coordinate the control of transmission and engine in these processes. In this paper, we proposed a CAN-based integrated control architecture for the drivetrain of vehicles equipped with AMT, as shown in Figure 1, TCU and ECU exchange information via CAN bus.

4. Definition of communication information

During the starting and shifting processes of vehicle,

opening degree, engine rotate speed, rotate speed of the input shaft, vehicle velocity, clutch displacement, gear position, braking signal, hydraulic oil temperature, etc. And meanwhile, ECU needs information, including position of accelerate pedal, throttle opening degree, engine rotate speed, rotate speed of the input shaft, vehicle velocity, clutch displacement, gear position, temperature, pressure, air flux, etc., to perform engine control. According to the functionalities served in the control systems, these information can be classified into 2 types, one is shared information that is used to reduce the number of sensors and provide redundant information for fault diagnosis, and the other is instructive information used to set the control target for related controlled object in the integrated control system.

The position of accelerate pedal, throttle opening degree, engine rotate speed, rotate speed of the input shaft, vehicle velocity, gear position, and clutch displacement are treated as shared information during the processes of starting and shifting, Regarding the instructive information, we only consider the target rotate speed of engine that is used to regulate the engine rotate speed in the processes of starting and shifting. The priority of the instructive information should be higher than that of the shared information. Regarding the shared information, we can specify their priorities according to their degree of importance.

In this paper, the extended format is used to define the communication information in the integrated drivetrain control according to the above principles. The definition of communication information is shown as Table 1, in which the instructive information only occurs in the starting and shifting process, while others may always occur during the normal operational period of ECU and TCU.

Table 1. Definition of communication information

Type	Source	Destination	Period (ms)	Data length (Bytes)	Byte 1 Byte 2	Byte 3	Byte 4	Byte 5	Byte 6-8
						Data definition			
Instructive	TCU	ECU	10	2	Target rotate speed of engine (rpm)	Reserved	Reserved	Reserved	Reserved
Shared	ECU	TCU	50	4	Engine rotate speed (rpm)	Throttle opening degree (%)	Accelerate pedal pos. (%)	Reserved	Reserved
Shared	TCU	ECU	100	5	Rotate speed of input shaft (rpm)	Velocity (km·h⁻¹)	Clutch displ. (%)	Gear pos.	Reserved

5. CAN-based integrated control strategy

5.1. Starting process

The position of accelerate pedal reflects the driver's demand on the output power of engine during the starting process. To overcome the road resistance, considering the velocity change in the short period of starting is very small, we use the engine rotate speed corresponding to the maximum torque under various throttle opening degree, on the engine torque characteristic diagram, as the target of engine speed control during the starting process. This choice can satisfy the driver's starting intention and meanwhile prevent engine from flameout and rolling, which can then make engine operate stably, reduce the noise, and get a smooth starting.

To control the engine rotate speed during the starting process, the current information of the position of accelerate pedal is sent from ECU to TCU as shared information at first, when TCU receives this message, it will calculate the target engine rotate speed according to the curve shown in Figure 2, and then the calculated result will be sent back to ECU as instructive information. ECU will use this information as target rotate speed in its rotate speed control algorithm, and try to maintain the actual engine rotate speed at this target value through regulating the amount of fuel supplied. In parallel with the engine rotate speed control, TCU controls the clutch engagement speed according to the deviation between the actual and target rotate speed of engine, and sends shared information, including the rotate speed of the input shaft, vehicle velocity, clutch displacement, etc., to ECU periodically, which will be used by ECU to determine if it should terminate the engine rotate speed control. Once the

starting process is completed, TCU will stop sending the instructive information to ECU.

5.2. Shifting process

The gear shifting process consists of six or seven actions, which includes clutch disengagement, fuel-supply reducing, gear shifting-out, gear position selecting (this

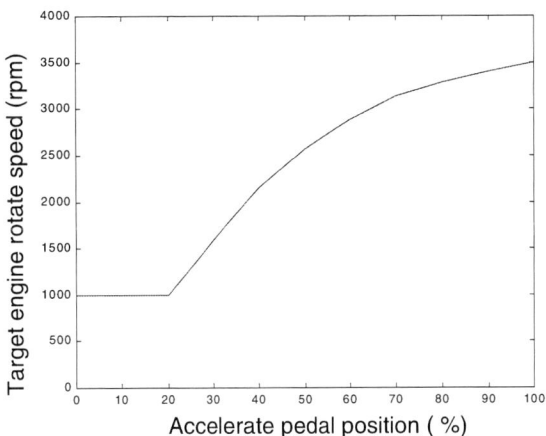

Figure 2. Target engine rotate speed

action occurs in some gear shifting process), gear shifting-in, clutch engagement, and fuel-supply resuming. In this action sequence, engine is under idle load from the thorough disengagement to the beginning of engagement of clutch. To avoid rapid increasing of engine rotate speed because of this idle operation, we need to control the engine rotate speed during the shifting process of AMT. During this period, if engine rotate speed is too high, fuel consumption will increase and engine noise will rise; on the contrary, if engine rotate speed is too low, shift jerk will occur because the vehicle inertia will inversely drag the engine when clutch resumes engagement. So how to choose an appropriate target engine rotate speed is very important for shifting control.

Since the shifting process is quite short and vehicle inertia is very large, we can assume that the velocity of vehicle will keep invariable during the period of automatic shifting (*Assumption 1*). Using the rotate speed of the input shaft, at the moment when gearbox is shifted into the new gear position, as the target engine rotate speed, the above side effects can be overcome.

The schematic diagram of how to determine the target engine rotate speed during the shifting process is shown as Figure 3. Where, t_0 is the time at the beginning of clutch disengagement, t_1 is the time at the moment of gear shifting into a new gear position, ω_e is the engine rotate speed, ω_i is the rotate speed of the input shaft of gearbox.

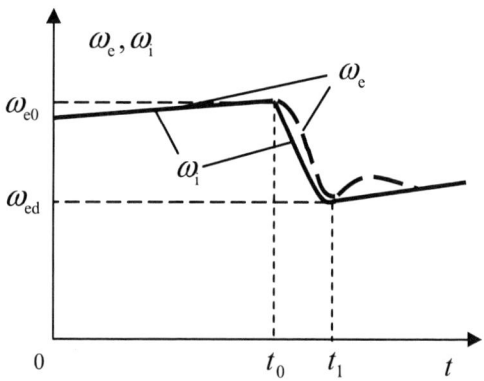

Figure 3.   Schematic diagram of the target engine rotate speed determination

Assuming i_{gk} and $i_{g(k+1)}$ are the current gear ratio and the target gear ratio respectively, i_0 is the gear ratio of main reducer, r is the radius of wheel, and ω_{e0} is the engine rotate speed at the time of t_0, the velocity of vehicle at the beginning of gear shifting can be described as formula (1).

$$v_{a(k)} = 0.377 \frac{r\omega_{e0}}{i_{gk}i_0} \tag{1}$$

Considering the rotate speed of the input shaft of gearbox at the time t_1 as the target engine rotate

speed ω_{ed}, the velocity of vehicle at the time t_1 can be described as formula (2).

$$v_{a(k+1)} = 0.377 \frac{r\omega_{ed}}{i_{g(k+1)}i_0} \tag{2}$$

From *Assumption 1*, we have $v_{a(k)} = v_{a(k+1)}$. According to formula (1) and (2), the target engine rotate speed can be educed as follows.

$$\omega_{ed} = \frac{i_{g(k+1)}}{i_{gk}} \omega_{e0} \tag{3}$$

To realize the integrated control for drivetrain during the shifting process, ECU sends the information of engine rotate speed at the moment of clutch disengagement as shared information at first, once TCU receives this message, it will calculate the target engine rotate speed using formula (3), according to the gear ratios of the current and the target gear positions. Because the clutch disengagement process is quit short and the response of actuator has time delay, TCU starts to send the calculated target engine rotate speed as instructive information back to ECU periodically when the clutch is disengaged about 80%. ECU will then use this information as control target in its rotate speed control algorithm, and try to maintain the actual engine rotate speed at this target value through regulating the amount of fuel supplied. In parallel with the engine rotate speed control, TCU completes the gear shifting-out, gear selecting, and gear shifting-in control, and when the actual engine rotate speed is close to its target, it initiates the engagement control of clutch, which controls the clutch engagement displacement and speed according to the deviation between the actual and target rotate speed of engine, and sends shared information, including rotate speed of the input shaft, vehicle velocity, clutch displacement, etc., to ECU periodically, which will be used to determine if ECU should terminate the engine rotate speed control. Once the shifting process is completed, TCU will stop sending the instructive information to ECU.

6. Experimental evaluation

To evaluate the proposed integrated control strategy, we fulfilled CAN-based starting and shifting control experiments using a prototype AMT test vehicle.

Figure 4 shows the result of the starting process test when the accelerate pedal is pressed down to 50% (for this test vehicle, the position of accelerate pedal equals to the throttle opening degree) of its full displacement. As shown in Figure 4, the engine rotate speed is controlled within the scope of 2000 – 2500 rpm, and its undulation is also very small, the time used for clutch engagement is very short (only about 2.2s). These result in an acceptable performance with small friction work and smooth starting.

Figure 5 shows the test result of a dynamic gear shifting process from gear position 1 to gear position 2 with full throttle opening degree. As shown in Figure 5, the test vehicle starts gear shifting when the engine rotate speed is about 5500 rpm, ECU and TCU cooperate to control engine, clutch, and transmission in the whole shifting process, which results in a quite short shifting time (only about 0.9s). This shifting process reduces the power-interrupted time of drivetrain that almost doesn't affect the dynamic performance of vehicle.

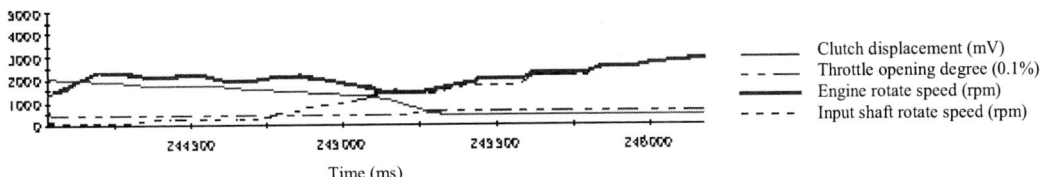

Figure 4. Starting process test

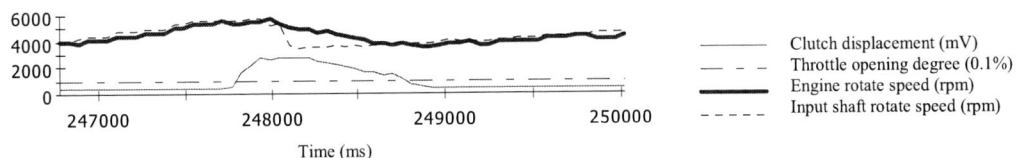

Figure 5. Shifting process test

7. Conclusion

For the purpose of getting better operational safety and starting and shifting of commercial vehicle equipped with AMT, as well as reducing the wiring harness and information redundancy, a CAN-based integrated control strategy for the drivetrain is proposed in this paper. The proposed strategy uses CAN as the communication bus for the integrated control of vehicle drivetrain during the starting and shifting processes. We introduced an architecture for drivetrain integrated control system, defined the communication information specification between ECU and TCU based on the information analysis of both, and discussed the strategy in details. Experimental evaluation shows that the proposed strategy is practicable.

References

[1] Robert Bosch GmbH, *CAN Specification (Version 2.0)*, 1991.
[2] SAE International, *SAE J1939 – Recommended Practice for a Serial Control and Communications Vehicle Network*, 2000.

[3] CiA, *CAN Application Layer for Industrial Applications*, 2005
[4] Y. Rao, *The Principle and Application Technology of Field Bus CAN (Second Edition)*, Beijing Aeronautic and Aerospace University Press, 2007.

[5] C. J. Menon, and S. Shimura, "Future Trends in Networking", *SAE Technical Paper 2003-01-3738*, 2003.
[6] S. M. Mahmud, and S. Alles, "In-Vehicle Network Architecture for the Next-Generation Vehicles", *SAE Technical Paper 2005-01-1531*, 2005.
[7] X.Yin, *Study on the Architecture and Supporting Software Development of Vehicle Power-train Automatic Manipulating System [PhD Dissertation]*, Jilin University, Changchun, 2002.

Delimitation of Traffic Coordinated Control Sub-area for the Road Network Containing Freeway

Nan Jiang, Xiaoguang Yang, Shoupeng Tang
School of Transportation Engineering,
Tongji University, ShangHai, 201804, China
nancyjiang1226@hotmail.com, yangxg@mail.tongji.edu.cn, tangshoupeng123@163.com

Abstract

A practical approach is presented for determining boundary of urban traffic signal coordinated control sub-area when freeway exists. It is based on the analysis of quantitative measure of interconnection between intersection and freeway ramps. According to these quantitative indexes, those intersection(s) and ramp(s) that relate with each other closely should be delimited in the same sub-area and be coordinately controlled. The proposed approach lays a foundation for research on coordinated control of freeway ramps and adjacent intersections which is an essential element in improving efficiency and mitigating congestion of freeway network.

1. Introduction

Practical and robust traffic control strategies are essential for relieving traffic congestion. Specially, coordinating traffic signals of several closely related intersections is a good method to improve traffic efficiency and reduce vehicular delays. So, how to reasonably delimit urban traffic signal coordinated control sub-area is very important since it is the precondition of employing coordinated control strategy.

Some researches have proposed the method of defining scope of coordinated control sub-area [1], but they focused on only ground road network without considering freeways. Besides, while there has been substantial amount of research conducted in the area of optimal corridor control [2][3], the main purpose of which is coordinately control freeway ramps and their adjacent intersections, most of them assume the scope of coordinated control area is predefined.

As a matter of fact, distribution of traffic flow, available routes and other traffic characteristics of one area's road network change a lot when freeways become part of this net. Accordingly, the scope of coordinated control sub-area also changes as Figure 1 shows. So, this study presents a method of determining coordinated control sub-area when freeway exists basing on analyzing connection relationship between ramps and ground intersections.

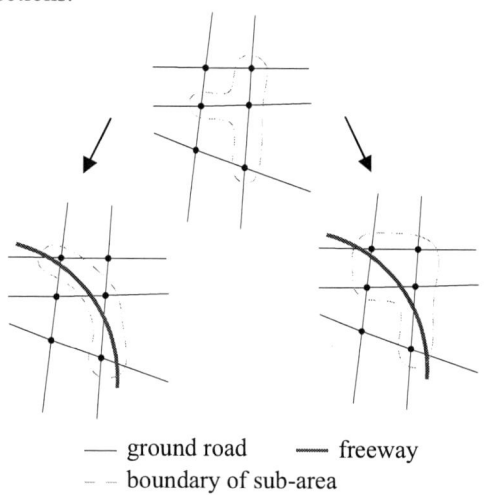

— ground road   — freeway
- - boundary of sub-area

Figure 1 Possible two changes of coordinated control sub-area's scope after freeway enters

2. Indexes necessary for doing delimitation

For the need of doing delimitation, some indexes should be obtained firstly.

2.1. Congestion index of freeway ramp

Because of the uneven distribution of traffic demand and difference in capacity, the level of congestion is different among different intersections and ramps. Congestion of a whole road network often begins from several key intersections and ramps. If no proper measure is taken, congestion of these key nodes will finally develop into a vast congestion over the whole net. So, delimitation of coordinated control sub-area needs to find out key nodes firstly, and then find out those connected nodes and routes which cause the traffic jam accordingly.

Congestion index which represents nodes' congestion level is proposed as the criterion for choosing key nodes in this study. Since there has been substantial amount of research conducted in the area of ground intersection's congestion index [1][4], this study focuses on only ramp's congestion index.

The congestions index used in this study quantifies the congestion level of a ramp by combining measurements from detectors in both ramp and its nearby mainline as follows:

$$C^t = \partial \frac{(P_r^t + V_r^t)}{(1 + V_r^t)} \cdot \beta \frac{(P_f^t + V_f^t)}{(1 + V_f^t)}$$

Where:

C^t = congestion index of a freeway connecting area during t on a 0 to 1.0 scale.

V_r^t = number of vehicles passed detector in ramp during t.

V_f^t = number of vehicles passed detector in on-ramp downstream mainline or off-ramp upstream mainline during t.

P_r^t (P_f^t) = 1.0 if detector in ramp (mainline) is occupied by a vehicle at the end of t, 0.0 else.

∂ , β = weights for detector in ramp and mainline respectively, $\partial + \beta = 1$.

C^t can be gotten by using only volume and presence detection commonly available from loop detectors [5][6]. Further, by adjusting the value of ∂ and β for certain area, the congestion index can reflect certain special traffic states of freeway. For example, we can increase the value of β if one on-ramp's nearby downstream mainline is detected as bottleneck area.

2.2. Connection index between ramps

Referring to some coordinated ramp metering algorithms, a coordinated ramp metering zones are typically three to six miles in length. So, this study assumes that only ramps far from the key ramp not more than 6 miles are related with this key ramp, i.e., only connection indexes between them and the key ramp will be calculated.

Further, since one on-ramp's meter rate greatly depends on that of its nearby on-ramps, and one off-ramp's traffic state is also easily affected by that of its nearby off-ramps, this study assumes that the connection index between the key on-ramp (off-ramp) and other on-ramps (off-ramps) in the same metering zone is 1.0.

For an on-ramp and an off-ramp, the connection level between them is closely related with the travel time and traffic flow's distribution between them. So, this study proposes following formula to calculate the connection index between an on-ramp and an off-ramp:

$$I_{on,off} = \varepsilon \frac{r_{on,off}}{t}$$

Where:

ε = parameter to reflect practical condition of freeway.

$r_{on,off}$ = the rate of vehicles flow from the on-ramp to the off-ramp with respect to all vehicles enter freeway from the on-ramp. This rate can be gotten by referring former traffic distribution data.

t = travel time between the on-ramp and the off-ramp.

2.3. Connection index between on-ramp and intersection

To measure the connection between one freeway on-ramp and its correlated intersection, this study quotes the form of reasonable connection index in America's *Traffic control system Handbook* [7].

$$I_{g,on} = \frac{0.5}{1+t} \left[\frac{n q_{max}}{\sum_{i=1}^{n} q_i} - 1 \right]$$

Where:

$I_{g,on}$ = connection index between one freeway on-ramp and its correlated ground intersection.

t = travel time between intersection and converging point of on-ramp and mainline.

n = the number of traffic flows that run from intersection to on-ramp. For a common four-leg intersection, n = 3.

q_{max} = maximum of q_i .

$\sum_{i=1}^{n} q_i$ = sum of volume of all traffic flows arriving at on-ramp, i.e., the traffic demand of on-ramp.

2.4. Connection index between off-ramp and intersection

The congestion of an off-ramp shows its correlated intersection's insufficiency in discharging traffic since all traffic on off-ramp will enter that intersection. At this time, the signal of ground intersection needs to be adjusted and coordinated with off-ramp control policy. So, the connection index between off-ramp and intersection proposed by this study can be expressed by congestion level of off-ramp as follows:

$$I_{off,g} = \frac{V_{off}^t}{Ca_{off}}$$

Where:

V_{off}^t = detected traffic volume of off-ramp during t.

Ca_{off} = capacity of off-ramp, vpt.

3. Steps of delimitation

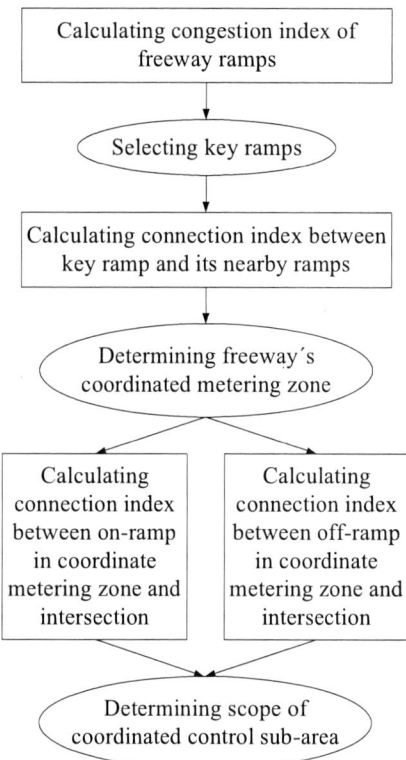

Figure 2 The steps of delimitation

Based on these necessary indexes, this study proposes the process of delimiting coordinated control sub-area considering the existence of freeway which can be divided into several steps as Figure 2 shows.

3.1. Selecting key ramps

After getting one ramp's congestion index C^t, whether it is a key ramp or not can be judged according to following criterion:

$$if \quad C^t \geq C_d, \quad this\ area\ is\ a\ key\ node$$

$$C^t < C_d, \quad this\ area\ is\ not\ a\ key\ node$$

Where:

C_d = the desired level of ramp congestion index to achieve freeway maximum throughput.

If one ramp is a key node, coordinating control strategies for it and other ramps and its adjacent intersections should be taken so that to decrease the congestion level of this key ramp. Otherwise, those nodes' present control policy could be maintained.

3.2. Determining freeway's coordinated metering zone

After selecting out key ramps, connection indexes between it and other ramps can be calculated according to the method introduced in 2.2. Then, by comparing those indexes with their corresponding threshold value, it is easy to find out which ramps are closely related and should be coordinately metered. Putting these ramps in a common zone, then, scope of this coordinated metering zone is determined.

3.3. Determining scope of coordinated control sub-area

After determining the scope of coordinated metering zone, which ground intersections that directly connected with those ramps should be added in this zone, similarly, can be determined by comparing $I_{g,on}$ and $I_{off,g}$ with their corresponding critical value $I_{g,on}^{cr}$ and $I_{off,g}^{cr}$, i.e., when the value of connection index between an intersection and a ramp is bigger than its critical value, these two nodes should be coordinated.

So, by adding those ground intersections that closely related with ramps into the freeway's coordinated metering zone, the boundary of that zone is expanded and form a new scope covers ramps, freeway mainline between those ramps and ground intersections.

For further intersections that not directly connected with ramps in coordinated metering zone, whether they should be included in this new scope can be decided according to some former researches on delimitation of ground intersections' coordinated control sub-area's [1]. So, this study does not consider these intersections and regards the new scope mentioned above is just the boundary of coordinated control sub-area considering the existence of freeway that we want to get.

4. Case study

Figure 3 is a simplified expression of road network including one freeway. The steps of delimitation proposed above will be further illustrated base on it.

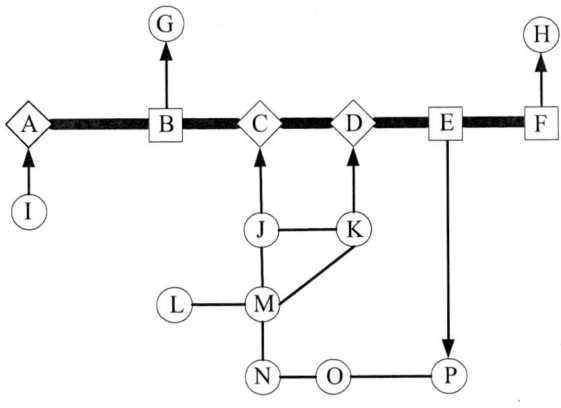

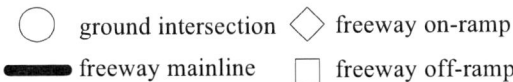

Figure 3 Simplified structure of road network

Step 1: Congestion index of each ramp is calculated and C is judged as a key node.

Step 2: All ramps that far from C not more than 6 miles are found out. They are off-ramp B, E and on-ramp D.

Step 3: The connection indexes between C and B, D, E are calculated and shown as the number on arc CB, CD and CE. Assuming 0.7 is the value of threshold connection index between different ramps, it is easy to find out that B and D should be coordinated with C. So, a coordinated metering zone of freeway can be decided as Figure 4 shows.

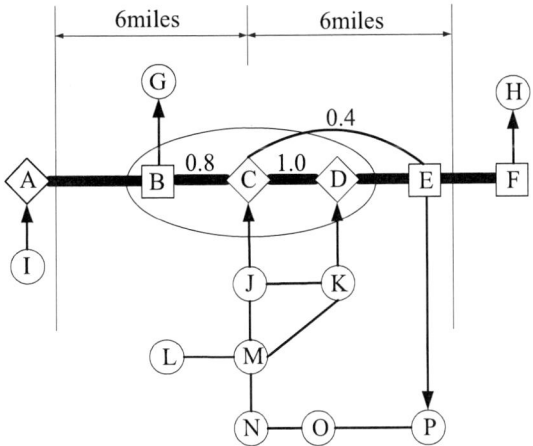

Figure 4 Coordinated metering zone of freeway

Step 4: Ramps in coordinated metering zone are ramp B, C and D. The ground intersections that directly connected with them are intersection G, J and K respectively. The connection indexes between those ramps and intersections are calculated so that to decide whether G, J and K should be added into the coordinated metering zone determined in last step.

Connection index between on-ramp C, D and intersections J, K are determined using formula introduced in 2.4 and shown as the number on arrow JC, KD in Figure 5. Connection index between off-ramp B and intersection G is calculated according to the method proposed in 2.5 and also shown as the number on arrow BG. Assuming 0.6 is the value of threshold connection index between on-ramp and correlated intersection, and 0.5 is that between off-ramp and connected intersection, it is not difficult to see that intersection J and K should be coordinated with ramp C and D respectively while G shouldn't. So, the scope of coordinated metering zone get in last step is expanded to contain intersection J and K. This new scope is the boundary of coordinated control sub-area, which is shown in Figure 5.

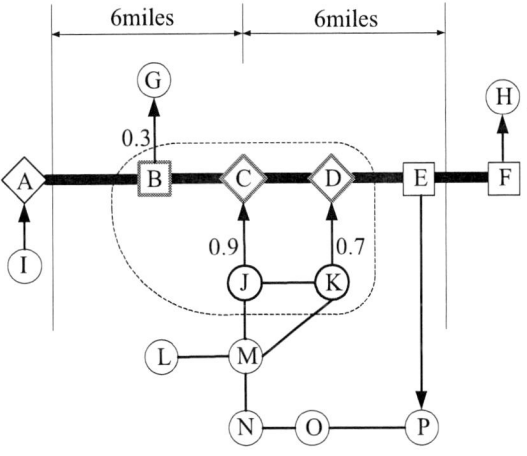

Figure 5 Boundary of coordinated control sub-area

5. Conclusion and further work

Basing on analyzing connection relationship between different types of nodes in road network, this study presents a method of delimiting traffic signal coordinated control sub-area for the road network including freeways. The calculation of ramp's congestion index, connection index between two ramps, on-ramp and intersection, and off-ramp and intersection are proposed. The steps of delimitation are explained in detail and illustrated by case study. Since required data are easy to be obtained, the proposed method is practical and feasible. However, the determination of those indexes' threshold values still needs further research work.

Acknowledgement

This research was partly supported by the National 863 Project in China, No.2006AA11Z205.

References

[1] LIN Yu, JIANG Nan, and XIAO Yonglai, "Algorithm of Delimiting Urban Traffic Signal Coordinate Control Subarea", 14th World Congress on Intelligent Transport Systems, Beijing, 2007, 4059.

[2] H. M. Zhang, W. W. Recker, "On Optional Freeway Ramp Control Policies for Congested Traffic Corridors", Transportation Research Part B 33, 1999, pp.417-436.

[3] van den Berg, M., B. De Schutter, J. Hellendoorn, "Integrated Model Predictive Control for Mixed Urban and Freeway Networks", The 83rd Annual Meeting of TRB, 2004, 001327.

[4] Hanbali, R.M., Fornal, C.J., "Methodology for Evaluating Effectiveness of Traffic-responsive Systems on Intersection Congestion and Traffic Safety", Transportation Research Record, Transportation Research Board, Washington, D. C., 1997, pp. 137-149.

[5] Kwon, E., and Stephanedes, Y., "Development of an Adaptive Control Strategy in a Live Intersection Laboratory", Transportation Research Record, Transportation Research Board, Washington, D. C., 1998, pp. 123-129.

[6] Kwon, E., Kim, S. and Kwon, T., "Pseudo Real-time Evaluation of Adaptive Traffic Control Strategies Using Hardware-in-Loop Simulation", The 27th Annual Conference of the IEEE Industrial Electronics Society. Denver, Colorado, 2001, PP. 1910-1914.

[7] Federal Highway Administration, Traffic Control Systems Handbook, U.S. DOT, 1975.

A Multi-path Access Control Scheme in Core Packet Network

SUN Xuekang , Gu Wanyi
School of Telecommunication Education,
Beijing University of Posts and Telecommunications,
Beijing 100088, China
sunxuekang6395@126.com

Abstract

The current best-effort Internet is adverse to deliver real media information since it is provided with an uncertain delay property in the traditional routing, flow control and retransmission. The research on the access control mechanism is more significant along with the development of the media applications. In this paper, a multi-path access control scheme on cost in core packet network is proposed to make it highly scalable and robust. In which, the scheduling and routing model in core network is presented, and the link cost of core network is calculated, then the optimized route is selected in term of the minimum cost. Simulation results show that the cost provided by this access control scheme would be nearly accordant with the performance of services in the network。

1. Introduction

Stateless characteristic is very important in traditional Internet network. As a result of this stateless architecture, the Internet is both highly scalable and robust. However, as Internet becomes a global commercial infrastructure that is expected to support a plethora of new applications such as IP telephony, interactive TV and e-commerce, the best effort service will no longer be sufficient. In consequence, there is an urgent need to have stateless characteristic in order to increase the efficiency of network resources, add the throughput of services and decrease the transport delays.

In the architecture of supplying the Differentiated Services (DiffServ), the two-level structure is introduced in, that is, the core network and access network. Core network is composed of stateless core nodes. Access network includes many edge nodes, in each of which the state information is inserted into the packet header.

2. Overview of access control scheme

This work is supported by National High-Tech Research and Development plan of China (grant No. 2006AA01Z246; 2007AA10Z235); Education & Research Fund of the Scholars returned from oversea provided by National Education Committee.

Access control scheme of core network includes the packet scheduling and routing and so on. The packet scheduling determines the transmitting sequence of packets, while the routing determines the transmitting direction of packets.

There are two kinds of routing algorithms, the centralized and distributed algorithms [1][2].The parameters used in the centralized algorithm, such as network throughput and delay, are constant for a long time. Therefore when a network node or link is in trouble, or when a new node applies to join the network, it is difficult to access due to the topology change. The used parameters are changed with dynamic network states in a distributed algorithm such as the distance vector routing algorithm and the link state routing algorithm, so the selected routing must be updated in time.

When the distance vector routing algorithm is used in a packet network, every router must preserve a vector list expressed as the output link and the used time, or the distance (the distance is measured with the number of hops and delay, or the queue length in fact). On above information, an optimal path is selected in term of the pass minimum entity to get to the destination. Because we needn't consider the link rate of every hop, this algorithm is simple. But the router must transmit part or full of the route list to its neighbor. As a result, when the topology is changed, there are problems of the infinitude computation and lower convergent rate. To overcome above disfigurements, the link state routing algorithm is proposed.

The basic idea of the link state routing algorithm is that every link has a suit of measure states in term of its QoS. The network maintains momentarily a topology map in witch the shortest path can be calculated by link state information. It is obvious that the topology map must be updated in time. Thus the router only needs to transport the link state information to the other routers in the same area. This algorithm has the rapid convergent velocity, provides with the function of discriminating interior and exterior routing as well as supports the capability of the multi-path transportation. Open shortest path first (OSPF) protocol is one of type applications.

The simplest scheduling algorithm is FIFO, but it doesn't support DiffServ. Algorithms such as RED, WFQ and DRR are proposed later. However RED algorithm is sensitive to network states and performance parameters. DRR is not suitable for the service delay demand. WFQ can fairly realize the dynamic scheduling, but it cannot strictly guarantee to higher quality services and provide with larger overhand. Then several improved algorithms are appeared such as W2FQ and SFQ which need smaller calculating amount in the highly fair condition.

On above researches, a multi-path access control scheme on cost in the stateless core network is proposed in this paper. The purpose is that the stateless idea and token parameter associated with the user fare are introduced into the algorithm to make the user fare associate with the obtained quality of service and keep the stateless core network scalable and robust.

3. A Multi-path Access Control scheme in core network

3.1. Stateless core network

The QoS of communication network, including throughput, jitter and latency, is becoming more and more important along with the fast development of different services. Therefore, the core network safety is important. Now, it's considered that stateless mesh configuration is suitable to core network. Comparing with Ring Network, it still suffers from its complex construction and high expense though it can guarantee its higher reliability and stronger survivability. However, the mesh network used the recovering technique needs free throughput less than ring network (it needs 50% reserved bandwidth), so the efficiency of mesh network is improved [3][4].

3.2. Token cost evaluation

During the transition of a packet from one edge node to another, the certain cost associated with the priority of service is expended. Services of various classes are entered different queues in a node. Meanwhile, the link cost is related to the needed bandwidth and the current performance of the link. The number of token is described as the used cost. In a two-level network, each edge node reserves an indication label that describes the amount of expended cost per rate between two edge nodes. The procedure of transmission will be carried out as in Fig.1:

It is obvious that a packet is initially transmitted from a source to its neighboring edge node, and then a label which includes the amount of current possessed tokens and the multiple N of used token cost per rate Nci is inserted into the packet header (e.g. the amount of the

used tokens per N kbps is NNci). Nci is decided by the history information stored in the indication label, the class of service and priority of service. The labeled packet is transmitted to its neighboring core node. Based on the configured strategy and current network states, the expended tokens of a node or link can be calculated in core node, and then is deducted from the left tokens of the packet before the packet is transferred to next node.

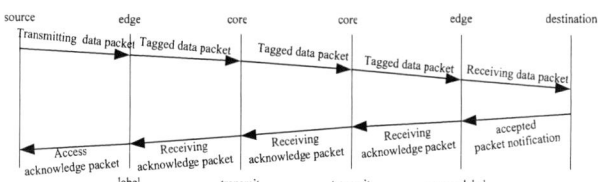

Figure.1 Procedure of transmission

When the packet arrives at the edge node, it is transmitted to the destination after its label is eliminated. If it is successfully received, the destination will generate an accepted packet. At first, it is transported to the connected edge node. Then a receiving acknowledge packet is transferred to the edge node connected with the source. Finally, a access acknowledge packet is delivered to the source. Based on the received acknowledge packet which contains the total left tokens during the transmission, the edge node can decide the actual cost. If it's necessary, the parameters could be updated on the basis of the comparison with original information in the cost indication label.

It is emphasized that the cost could be precisely computed in post-payment service. But in pre-payment service, the packet token cost given by the edge node could be less than the required tokens. Therefore, when the packet arrives at the node, if the left tokens are equals to 0 or less than the demand of this node or the service, the network will provide BE service for it. As a result, the QoS would be degraded. If Nci connected with some edge nodes is absent in the token indication label, the edge node should get this information by sending the detection packet to the network.

3.3. A Multi-path access control scheme

3.3.1 MRED-tdiffser scheduling

In this section we present our stateless scheduling algorithm, called MRED-tdiffser, which approximates the behavior of WFQ [5] [6][7]. It eliminates the correlation among flows in WFQ and introduces a token cost associated with the user fare to distinguish QoS. In a core node, the scheduling strategy of the multi-queue RED is used to adapt to the delay demands of media services.

We first consider a router without buffer and with total output token capacity M, where the arrival token rate of flow i is v_{itoken}. If a lot of flows get to the router at same time, its aggregate token rate $v_{token} = \sum_{i=1}^{n} v_{itoken}$. In term of token allocation principle, the larger the token rate of a flow, the bigger its bandwidth. The largest sharable token rate of flow i is given by

$$K_i(t) = \frac{v_{itoken}(t)M}{\sum_{i=1}^{n} v_{itoken}(t)} = \frac{B_{itoken}(t)M}{\sum_{i=1}^{n} B_{itoken}(t)} = \eta B_{itoken}(t) \quad (1)$$

Where η is represented the link state. Thus the accepted token rate of flow i is defined as

$$K_{itoken}(t) = \min\{v_{itoken}(t), K_i(t)\}$$

Then dropping probability is:

$$p_i = \max(0,1 - \frac{K_i(t)}{v_{itoken}(t)}) = \max(0,1 - \frac{\eta B_{itoken}(t)}{v_{itoken}(t)}) \quad (2)$$

We notice that the above algorithm is defined for a system without buffer, in which the arrival token rates are known exactly. Our task now is to extend this approach to the situation in real routers where there is substantial buffering, and the arrival token rates are not known. It needs to estimate token rate $v_{itoken}(t)$ and sharable token rate $K_i(t)$.

(1) Token rate

Token rate v_{itoken} is estimated at each node. We use exponential averaging to estimate the token rate of a flow.

$$v_{itoken}^{\ n} = (1 - e^{-\frac{T_i^k}{K}}) \frac{N \cdot N_{Ci}^{\ k}}{T_i^k} + e^{-\frac{T_i^k}{T_i^k}} v_{itoken}^{\ n-1} \quad (3)$$

where $v_{tokeni}^{\ n}$ and $v_{tokeni}^{\ n-1}$ respectively denote the token rate of flow i at time t_n and t_{n-1}; T_i^k is equal to the arrival differential time of the kth and $(k-1)^{th}$ packet of flow i; K is a constant; N_{Ci} is denoted as the token cost per rate of flow i. Usually the token cost of a node is N times N_{Ci}. N is a real number.

(2) Link largest sharable token rate

To give intuition, we consider again the above flow model and assume system operation in term of the dropping probability in Eq.(2). The current accepted token rate is an estimation function of the current accepted aggregate token rate without buffer, which we denote by $\hat{K}$ (t). Then we have

$$F[(\hat{K}(t)] = \sum_{i=1}^{n} \min(v_{itoken}(t), \hat{K}_i(t))$$

Note that along with the increase of network load, $F[\hat{K}(t)]$ shows a piecewise-linear, increasing

characteristic. If the link is congested, $F[\hat{K}(t)]$=M. The bandwidth of per flow is according to its token. To avoid keeping per flow state，we instead to compute $\hat{K}_i(t)$ by using aggregate measurement of $F[\hat{K}(t)]$ and v_{token}. Thus when the proportion of arrival token rates is changed, $\hat{K}_i(t)$ is updated. Now we approximate $F[\hat{K}(t)]$ by a linear function that intersects the origin. It's estimated as

$$\hat{K}_i^{\ n}(t) = \hat{K}_i^{\ n-1}(t) \frac{M}{v_{token}(t)}$$

In term of token allocation principle, we must have

$$B_{token\ i}^{\ n}(t) = B_{token\ i}^{\ n-1}(t) \frac{M}{v_{token}(t)} \quad (4)$$

Where $B_{tokeni}^{\ n}$ 和 $B_{tokeni}^{\ n-1}$ respectively denote the bandwidth of flow i at time t_n and t_{n-1}. M is total token capability of a link.

3.3.2 Multi-path routing

The multi-path routing function includes two sides. One is the collection of the momentarily updated network state information. The other is the calculation of the usable path. This algorithm uses the feedback mechanism and introduces the token cost associated with the user fare. Without congestion, the function of dynamic cost C_{token} that can reflect the link load is defined as

$$C_{token}(t) = \frac{a}{1 - v_{token}(t)/M}.$$

It is obvious that the cost function is instantaneously changed with arrival aggregate token rate. When the link utilization approaches 1, an incrementally small change of arrival aggregate token rate can result in an arbitrary large cost change. To solve this problem, we use the iterative formula to compute the link cost. At the same time, we use dropping probability p to correct it.

$$C_{token}(t_n) = \frac{a}{1 - p} + C_{token}(t_{n-1}) \frac{v_{token}(t_{n-1})}{M} \quad (5)$$

where $C_{token}(t_n)$ and $C_{token}(t_{n-1})$ respectively denote the link cost at the time t_n and t_{n-1}; a is the initial cost of the link; $v_{token}(t_{n-1})$ is the aggregate token rate at time t_{n-1}.

The optimized route is selected in term of the minimum cost. Thus the token rate of the network achieves dynamic load-balancing by the token cost control of each hop. Generally, the token cost of the link connected two core nodes is equals to the product of the link cost and the carried traffic rate. Note that the token cost among different core nodes is distinct. Therefore, the total token cost of the link which includes several core nodes is equal to the sum of each section token cost.

On the condition of larger network loads, it is appeared that some of the users want to occupy the same link. To avoid this phenomenon, even if the arrival aggregate token rate is close to or overrun the total link token capability M, the average dropping probability of links can be used to control the direction of token flows in order to token load-balancing.

3.4. Procedure of packet scheduling

The scheduling content of stateless core nodes includes the detection of flow rate and the token rate, the allocation of bandwidth, the calculation of dropping probability as well as the update of label. Fig. 2 depicts the scheduling procedure of the stateless core node. The arrival packets from many edge nodes at first are received by parameter statistic module in which NN_{ci} is calculated in term of parameter N and class of service showed in received packet. The results are delivered to the token rate estimation module, the token cost modulation and dropping probability calculation module. Based on the class of service, the arrival packet flows from parameter statistic module are queued in M-RED queue module.

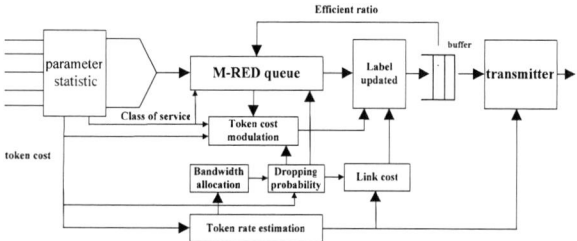

Figure 2 core node's function configuration

The token rate estimation module takes the responsibility for the calculation of token rate, and then transmits it to the bandwidth allocation and transmission module. The function of the bandwidth allocation module is the allocation bandwidth based on token demand. Then its result is transmitted to the dropping probability module for the control of M-RED queues and link cost module. In M-RED queue module, the order is based on the dropping probability of the flow, then packets are transmitted to label module by using the method of multi-queue round. In label module, token is updated (e.g. deduce node and link scheduling token cost), and then updated packets is restored in a buffer. Based on the efficient ratio of the buffer, the length of queues can be regulated. When the capacity of services is high to some extent, the congestion can be declined by dropping packets. As this operation will result in falling quality of service, token cost can be adjusted with its class of service in token modulation

module. Based on token rate, the data packets are transported in the transmission module.

4. The analysis of simulation result
4.1. Simulation model

Fig.3 gives am example of network topology which is composed of four core nodes to illuminate the process of routing. s_1 and s_2 are sources . d is a destination. e_1, e_2 and e_d separately denote edge nodes. c_1, c_2, c_3 and c_4 are core nodes that form the mesh core network. The capability of each link is supposed to be 10Mbit/s.The delay is 0.01ms between source or destination and the edge node, 0.1ms between core node and edge node, 0.1ms among the cores.

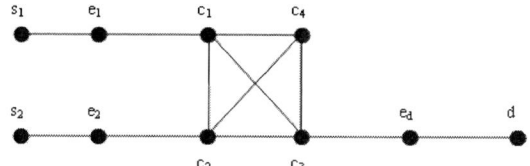

Figure.3 Network topology

In term of the characteristic of the two-level network, we use two kinds of packet scheduling modes, RED in edge node, MRED-tdiffserv (see Eq.(2)~Eq.(4)) in cores.

For the convenience, s_1 and s_2 use the mode of CBR to control the token cost of flows. It is supposed that the token cost of 1Mb/s expends 1 token, the speed of s_1 and s_2 is respectively 10Mb/s and the packet length is 50 bytes. By using Eq.(5), the token cost of links is obtained in different times. The optimized route is selected in term of the cost minimum.

Note that the initial token cost (a) of the link is showed in Table.1. It is obvious that there are three initial paths such as c_1-c_3、c_1-c_4-c_3 and c_1-c_2-c_3 in core network when packets are transmitted from s_1 to the destination d (see Fig.1). Since the least cost value of the link is equal to 11, c_1-c_3 is selected as an actual path in this algorithm. The cost of the link can be approximately obtained by Eq.(5) at other times(see Table.2).

Table.1The initial cost of Links

link	c_1-c_2	c_1-c_4	c_2-c_3	c_3-c_4	c_1-c_3	c_2-c_4
a	10	10	10	10	11	11

Table2. Link State on Cost

time (s) / flow	0	2	4	6	8	10
s_1	c_1—c_3	c_1—c_4—c_3	c_1—c_3	c_1—c_4—c_3	c_1—c_3	c_1—c_4—c_3
s_2	c_2—c_3	c_2—c_3	c_2—c_1—c_3	c_2—c_3	c_2—c_3	c_2—c_3

As said in Table.1, the cost of links at different times is varied with network states. Fig.4 shows the token cost rate of s_1 and s_2 flow (flow1 and flow2). Therefore their time average value is nearly equal. Since we suppose the velocity of s_1 or s_2 flows is respectively 10Mbit/s and the capability of the link connected with c_3 and e_d is also 10Mbit/s, the drop pocket phenomenon takes place in this link. Figure.5 shows the instantaneous allocated bandwidth of flows. And the average bandwidths of flow1and flow2 within 10 Sec. are separately 4.7Mbit/s and 5.1Mbit/s. As we can see in the two figures, the bandwidths of the different accepted flows would be nearly accordant with their token cost (as mentioned above, flow1 and flow2 are nearly related to 1：1).

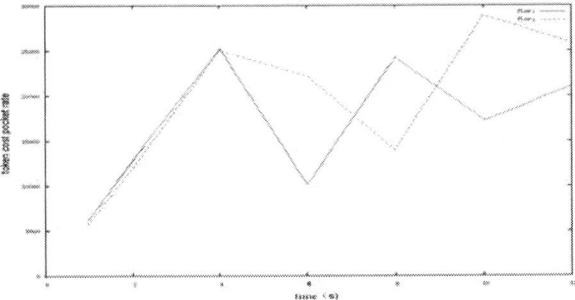

Figure.4 The token cost of s_1 and s_2 flow

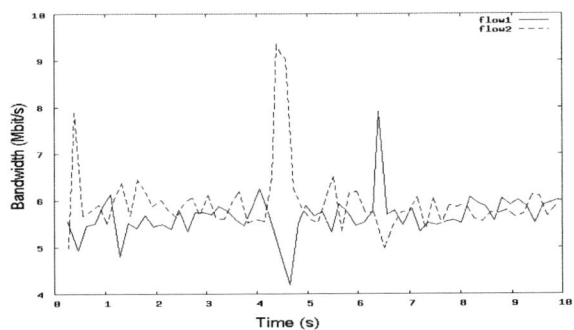

Figure.5 The instantaneous allocation bandwidth

It is obvious that the drop pocket phenomenon takes place since there are two flows through c_1—c_3 at the time T=4.When the initial token cost (a) of the link is 11 (see Eq.1), the token cost will increase about 22 (actually larger than 22). So far token flows will escape from this link in next time.

4.2. Performance evaluation

In traditional Internet, there are all kinds of scheduling algorithms to make network have different performances [8][9][10]. To exactly evaluate the proposed algorithm on cost in stateless core network, it is compared with FIFO-Droptail, DRR and SFQ (They are used in the same network configuration and routing algorithm as above simulation). The results compared with MRED-tdiffserv

are given in Table 3. It denoted that s_1 and s_2 flow are 5Mb/s. Simulation results show that this algorithm is not adverse to the delay property. It is because of the multi-queue order and queue-length control that the average time of packet queue is decreased. Along with the increase of the core network's scope, we illustrate that the multi-path access control scheme is used in all kind of simple networks given in Fig.6 。The performance of stateless core network is depicted in Table.4 (The simulation condition is same as above).

Table 3. Performance Compare

algorithm index	Droptail-FIFO	DRR	SFQ	MRED-tdiffserv
average delay (s)	0. 069885	0. 083497	0. 069897	0. 042638
lost packet rate (%)	0. 0083	0. 0042	0. 0096	0. 01002

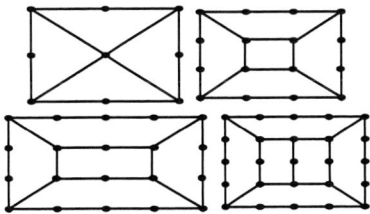

Figure 6 Multi-node Core network

On the scheduling and control of the data link, the complexity results from the maintenance of consistent and dynamic states in a distributed environment. As the router in core network uses the stateless core scheduling algorithm, it doesn't handle the lot of conventional requests related with services. Thus the complexity of this algorithm on time and space doesn't change with the number of flows and the type of services. Therefore it is easiest to reduce the complexity of the scheduling algorithm as there is a natural tradeoff between the complexity and flexibility of the scheduler. Though we used CBR flows to simulate the access control scheme, the result still has universal.

5. Conclusions

The token cost flow is introduced in the proposed multi-path access control scheme while it makes the scheduling stateless. It can preferably solve the problem of congestion under the network source competition to achieve the load-balancing and large throughput of the network. Note that this scheme can automatically illustrate the corresponding relationship of the distribution of token flows and user fare.

Table 4 Performance of the Multi-node Network

node index	4	9	16	20	25
average delay (s)	0. 04264	0.04268	0.04292	0.04308	0.04318
lost packet rate (%)	0. 0100	0.0125	0.0145	0.01711	0.01871

References

[1]A.Patooghy,A.Sarbazi-Azad.Performance Comparison of Partially Adaptive Routing Algorithms. Proceeding of the 20th international Conference on Advanced Information Networking and Applications (AINA'06)。

[2] Piet Van Mieghem and Fernando A,Kuipers. Concepts of Exact QoS Routing Algorithms. IEEE/ACM Transactions on Networking. Vol.12,No.5,October 2004.

[3] Stoica I.Stateless core: a scalable approach for quality of service in the internet [D]. Carnegie Mellon University, Department of Electrical and Computer Engineering, December 15, 2000

[4] Shareq Rahman, Chita R. Das. Parallel Simulation of Mesh Touting Algorithms. IEEE Proceedings of the 16th ICDCS.

[5] Anirudha Sahoo and D.Manjunath. Revisiting WFQ: Minimum Packet Lengths Tighten Delay and Fairness Bounds. IEEE Communications Letters,VOL.11.No.4 April.2007

[6] Y.-S. Yen, W. Chen, J.-C. Zhhuang and H.-C. Chao, Sliding weighted fair queueing scheme for real-time applications. IEE Proc.-Commun., Vol. 152, No. 3, June 2005

[7] Peng Wang, David L.Mills. A Probabilistic Approach for Achieving Fair Bandwidth Allocations in CSFQ. Proceedings of the 2005 Fourth IEEE International Symposium on Network Computing and Applications (NCA'05).

[8] Eitan Altman，Tania Jimenez, Simulation analysis of RED with short lived TCP connections. Available online at www.sciencedirect.com Computer Networks 44(2004) p631~641

[9]D. E. Wrage and J. Liebeherr. A near-optimal packet scheduler for Qos network. In Proceeding of INFOCOM'97，pages 576-583. Kobe. Janpen.1997

[10]D.C.Stephens, J.C.R.Bennet, and H.Zhang. Implementing scheduling algorithms in high speed networks. IEEE Journal of Selected Area on Communication: Special Issue on Next-generation IP switch and Router 17(6):1145-1158, June 1999.

Performance Optimization Based On Dynamic Bandwidth Allocation for Networked Motion Control Systems

Zhao Weiquan[1,2], Li Di [1]

1 School of Mechanical Engineering,
South China University of Technology,Guangzhou, 510640, China
2 School of Software,
Dongguan University of Technology,Dongguan ,523808, China
zhaowq@dgut.edu.cn, zhao.weiquan@mail.scut.edu.cn

Abstract

Limited network bandwidth and static bandwidth allocation method are the key factors to affect the performance of Network Motion Control Systems. To solve inefficiency of traditional static scheduling strategy, a dynamic bandwidth allocation method based on current QoC is proposed. Implementation algorithm and step of the dynamic bandwidth allocation are presented.

1. Introduction

Networked Motion Control Systems is a type of special Networked Control Systems, which is composed of controllers and multi-axes motor drivers and can transfer motion control instruction and receive motion control state synchronously and real-timely. NMCS has been applied to large-scale distributed system(automatic production line in factories and plants, pocess control, etc.) and small-sized equipment system(advanced automobile and airplane, robots, types of production equipments, etc.). For the control demand and from the communication perspective, network must provide deterministic real-time communication, so the communication rate is restricted. Though there are some prospective communication bus for motion control(such as sercos, synqnet, Ethernet Powerlink etc.), relative high cost and technique protection restricts its wide use. So nowadays most practical communication networks are bandwidth limited. On the other hand, in the real industrial evironment, for flexible and agility manufacture, control system structrue and wokload may be changed frequently. As a consequence of bandwidth limitation and workload variation, the network resources available for transmitting messages among axes control loops are prone to be nondeterministic, which cause unpredictable network delay, data dropout and jitter. So system QoC(Qualty of Control) may be deteriorated and even resulting in certain axes control loops to be unstable. Therefore, the overall performance of NMCS that consists of several axes loops

depends on not only the design of control algorithms, but also how to allocate the bandwidth resources, i.e. the scheduling of the shared network resource. This is especially true when the available bandwidth is time varying and sometimes scarce.

The reduction of the sampling interval can improves the control loop's performance, but a shorter sampling interval requires more network bandwidth to transmit more sensor data or control data, which increases the network traffic load. This may even affect the system stability and performance of the control loop if the maximum available network bandwidth is exceeded. Therefore, a co-design method of control and communication system must be applied in analying and designing an NMCS. The main evaluation measures of the network QoS are time delay and packet loss statistics, network efficiency, and network utilization. These measures can be used to determine the capability of the network medium and to provide information to specify control parameters such as the sampling period. The network efficiency is the ratio of the total transmitting time to the time used to send messages, including queuing time, block time, and so on [1]. There are several control specifications can be used to evaluate the control QoC such as phase margin, rising time, steady-state error, integral of the absolute value of the error (IAE), integral of the time multiplied by the absolute value of the error (ITAE), and so on. In this paper, based on current control performance QoC, a co-design method including network bandwidth scheduling and adaptive control adjustment for NMCS that optimizes overall control performance and network-bandwidth usage is presented.

The remainder of this paper is organized as follows. Section 2 describes the basic idea of dynamic bandwidth allocation and the relationship between sampling period and bandwidth. Section 3 proposes the implementation algorithm of dynamic bandwidth allocation. Section 4 concludes and points out the future research topics.

2. Basic Idea of Dynamic Bandwidth Allocation

Conventional bandwidth allocation and scheduling techniques focused on static strategies that would guarantee average control performance at the expense of permanently occupying the available bandwidth. From the control perspective, the static bandwidth allocation method is an "open-loop" solution because once established at system set-up, the static scheduling can not be adjusted at run time. However, due to network bandwidth limitation in some cases, not all control loops can simultaneously gain enough bandwidth allocation to provide the best possible control performance. Static techniques may not be efficient when changing conditions occurs at the control-application or network conditions, because pre-assigned bandwidth resources could be underutilized. Ideally, these underutilized network resources could be made available to other applications to provide more functionality. The basic idea is that making scheduling decisions to allocate limited bandwidth based on the performance information of each axes, called "closed loop" technique. A network cotroller can been designed that can adaptively adjusts its control algorithm(controller parameters) according to the QoS changes like varying time delays and packet losses. This can be regarded as an adaptive control problem which can modify its parameters on-line according to the changing network QoS parameters. The resulting control algorithm would explicitly depend on the current measurements of network QoS parameters. Elementary idea is proposed that designing network middleware to acquire network QoS, then based on QoS deciding sampling period and acquiring optimal control performance by adjusting controller plus[2]. For acquiring QoS needing artificial adusting, the method is time-consuming and influencing the actual effect.

The sampling period hi can be obtained from the bandwidth utilization bi to be assigned to the ith control loop according to the following equation: $h_i = \frac{c_i}{b_i}$,

$$0 \leq b_i \leq 1 \ (1).$$

Where the operation time ci indicates the time in the best case required to finish a control operation for control loop i, which only includes the time for data processing such as sensor sampling, controller calculating, and actuator actuating, and the time for transmitting the data packets from the sensor node to the controller node and from the controller node to the actuator node. The available bandwidth bi is the parameter that denotes the portion of the network bandwidth assigned to control loop i. The network utilization is defined as the ratio of the total time used to transmit data and the total running time, which the sum of the ration of message transmission times and message periods of all devices. If the network

utilization approaches zero, there is network bandwidth available for other functionalities or control purposes. If the network utilization approaches one, the network becomes saturated, and it is difficult to increase the sampling rates of control devices or add more nodes. Then network bandwidth reallocation to adjust the traffic load or network redesign is needed.

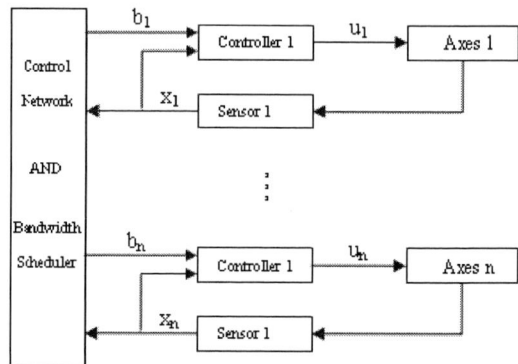

Figure 1. NMCS architecture with bandwidth scheduler

If assuming that the periodic sensor data from the sensor node and the control data from the controller node are packetized to an identical bit length L_i. If the data rate of the network medium is R, the best-case one-way data transmission time is: $T_{i,t} = \frac{L_i}{R}$ (2). Let $T_{i,p}$ be the time needed for data processing such as sensor sampling (for A/D conversion), actuator actuating (for D/A conversion), and controller calculating the control data for control loop i in the best case. The operation time c_i can be expressed as: $c_i = 2T_{i,t} + T_{i,p}$ (3). For each known control loop i, $T_{i,p}$ and $T_{i,t}$ in the best case can be measured and computed. Thus from (3) we can assume that c_i is a constant, then smaller h_i indicates bigger b_i from (1). So we can know that the rationale behind (1) is: a higher sampling frequency requires more bandwidth allocation to transmit more data by network. Based on the above discussion, there are several special cases to be considered:

1) When transmission time is equal to sampling period

   $h_i = c_i$, i.e., for $b_i = \dfrac{c_i}{h_i}$, so $b_i = 1$, that is to say the

   100% of the network bandwidth is used by control loop i, and no other control loops are allowed to share the network bandwidth. This is the case of NMCS with only one axes control loop;

2) When sampling period is equal to the maximum allowable delay bound (MADB) , i.e., $h_i = D_i$, where D_i is the maximum allowable delay bound (MADB) of the control loop i, then the minimum network bandwidth utilization of control loop i can be denotes as: $(b_i)_{\min} = c_i / D_i$ (4);

3) Supposing there are N control loops. While assuring all the other control loops are assumed to acquire their

minimum bandwidth, the most available bandwidth could be assigned to control loop i is:

$$(b_i)_{max} = 1 - \sum_{j \neq i}^{N} (b_j)_{min} \quad (5).$$

In some cases, due to network bandwidth limitation, not all systems can simultaneously obtain enough bandwidth allocation to transfer data and execute at their highest sampling frequency. Investigating how to perform the optimal bandwidth allocation to obtain the optimal control performance for each closed-loop system thus to optimize the overall system QoC is our main consideration. According to the conclusion from the document [3], the rationale of dynamic bandwidth allocation is described as[4]: when a controlled plant is in equilibrium, the assigned execution rate (or sampling period) may not be required. That is, the assigned bandwidth may be wasted, and it can be reduced for the sake of saving bandwidth usage and enhancing the bandwidth utilization of other control loops. On the other hand, when a controlled plant is affected by a perturbation, increasing its assigned bandwidth by adding the underutilized bandwidths of other control loops in equilibrium may accelerate system recovery from the perturbation and improve its system performance. When the overall network bandwidth resource is limited, this network-bandwidth allocation method is particularly useful.

Document [5] showed that the feedback-control loop performance directly depends on the loop delay, which is defined as the interval between the instant when the sensor node samples data and the instant when the actuator actuates the control command. The MADB of control loop i can be obtained from a conventional stability criterion and performance analysis[6]. In order to guarantee system stability and control performance, two control measures can be used to determine the maximum allowable loop delay: phase margin ϕ and the closed-loop bandwidth ω_{bw}. To guarantee acceptable control performance such as response rapidness and smoothness, the "rule of thumb" for selecting the sampling frequencies mentioned can be used to estimate the MADB D_i for control loop i, and D_i can be estimated by $D_i \leq T_{i,bw} / 20$ (6), where $T_{i,bw} = 2\pi / \omega_{i,bw}$ (7) and $\omega_{i,bw}$ is the closed-loop bandwidth of control loop i. Alternatively, the MADB D_i could also be estimated by $D_i \leq t_{i,r} / 4$ (8), where $t_{i,r}$ is the rise time of closed-loop system i. Let e_i denote the error of control loop i, the distance of the state from its equilibrium that can be expressed: $e_i = \|x_i(k)\|$ (9).

We can define a performance criterion for control loop i to relates control performance (such as the error) with available bandwidth utilization as $e_i = f(b_i)$ (10). Generally speaking, the less the bandwidth allocation is, the worse the control performance (i.e., the larger the error). Thus for the convenience of study, rerefering to the method in document [3], an approximated linear relation can be denoted as: $e_i = f(b_i) \approx \frac{\beta_i}{b_i}$ (11).

Where the parameter β_i is specific to each control loop and can be determined prior to the implementation of the NMCS by evaluating the control performance of each control loop for a broad range of sampling rates or bandwidth allocation or the urgency dgree of the control loop.

3. Dynamic Bandwidth Allocaiton Algorithm

Let the vector $x_i(t) = [x_i^1(t), ..., x_i^n(t)]^T$ is the state of the system of each ith axes(i = 1, . . . , n), at time t, where $u_i(t)$ is the control input to the axes, the linear time invariant vector dierential equations can be described as:

$$\begin{cases} \dot{x}_i(t) = A_i x_i(t) + B_i u_i(t) \quad (12) \\ y_i(t) = C_i x_i(t) + D_i u_i(t) \quad (13) \end{cases} t \in \Re^+$$

Where Ai, Bi, Ci, Di are coefficient vector. If the input ui(t) is given by a discrete state feedback control law discribed by the discretization of (12) and (13), the new system dynamics are specified by the state-space representation given in (14), where Li(hi) is the control law: $x_{i,n+1} = \Phi_i(h_i) x_{i,n} + \Gamma_i(h_i) L_i(h_i) x_{i,n}$ (14)

The dynamic bandwidth allocation problem is how to dynamically assign an avaliable bandwidth bi to each control loop according to the control performance and network bandwidth availability such that the overall QoC of the NMCS can be optimized. Obviously to realize dynamic bandwidth allocation, controller need to support variable sampling period, that is to say period should satisfy $h_i \in [h_i^{min}, ..., h_i^{max}]$. At the system level, each control loop ci can be characterized by its bandwidth bi, its performance criterion pi, and its axes error ei, represented by $c_i = \{b_i, p_i, e_i\}$. With this information, for a given set of n control closed loops, c1, . . . , cn, the problem is to determine the bandwidth bi(i = 1, . . . , n), such that all control operations are schedulable and the overall axes control system performance is maximized through adjusting sampling period.

The basic constraint of bandwidth allocation is $\sum_{i=1}^{N} b_i \leq 1$, i.e., the total bandwidth utilization must not exceed the whole network capacity. Then the current additionally available bandwidth utilization is $b_a = 1 - \sum_{i=1}^{N} b_i$ (15). If $\sum_{i=1}^{N} b_i \geq 1$, then the NMCSs are not schedulable with the current choice of network medium. Choosing another network or reducing the number of the control loops is needed.

If using conventional method each control loop is allocated with a fixed bandwidth, there may be a waste of the network bandwidth since each control loop has its own control and traffic requirement and some control loops may not necessarily need a fixed bandwidth when they are in equilibrium. In order to provide services to the maximal number of control loops with their QoC requirements and to achieve high utilization of the bandwidth resources, the bandwidth allocated to each control loop needs to be minimized. So we defined the following cost function to be minimized $f(p_i(b_i, e_i))_i = a_{i,1} e_i + a_{i,2} b_i$ (16), where $a_{i,1}$ and $a_{i,2}$ are the weighting coefficients for control loop i. The optimization object of bandwidth allocation is to find a suitable bandwidth utilization bi that can minimize the network bandwidth usage and maximize the system performance (i.e. minimize the error) as well. When considering all the control loops, the optimization function $f(p_i(b_i, e_i))$ for the whole system can be denoted as following:

$$f(p_i(b_i, e_i)) = \sum_{i=1}^{N} f(p_i(b_i, e_i))_i = \sum_{i=1}^{N} (a_{i,1} e_i + a_{i,2} b_i)$$ (17).

Because all the control loops are assumed to be independent, if each $f(p_i(b_i, e_i))_i$ of the ith control loop is minimal, then $f(p_i(b_i, e_i))$ is also the minimal. so the optimal bandwidth allocation for each control loop can achieve the QoC optimization of the overall system with the upper cost function. There are also three notable special cases to be considered:

1) When the ith control loop is in equilibrium, i.e. $e_i = 0$, the optimal bandwidth allocation of control loop i is: $(b_i)_{opt} = (b_i)_{min} = c_i / D_i$ (18), where $(b_i)_{opt}$ is the optimal bandwidth utilization for control loop i. Substituting $D_i \leq T_{i,bw} / 20$ and $D_i \leq t_{i,r} / 4$ into (18) obtaining following: $(b_i)_{opt} = 20 c_i / T_{i,bw}$; $(b_i)_{min} = 4 c_i / t_{i,r}$, the corresponding optimal sampling period can be denoted as: $(h_i)_{opt} = T_{i,bw} / 20$ (19) or $(h_i)_{opt} = t_{i,r} / 4$ (20).

2) When control loop i is affected by perturbation, i.e. $e_i \neq 0$, substituting $e_i = f(b_i) \approx \frac{\beta_i}{b_i}$ (11) and

$$f(p_i(b_i, e_i))_i = a_{i,1} e_i + a_{i,2} b_i$$ (16)

differentiating $f(p_i(b_i, e_i))_i$ with respect to bi, optimal bandwidth is: $(b_i)_{opt} = \sqrt{\dfrac{(1 + a_{i,1}) \beta_i}{a_{i,2}}}$ .The corresponding sampling period can be denoted as:

$$(h_i)_{opt} = \frac{c_i}{\sqrt{\dfrac{(1 + a_{i,1}) \beta_i}{a_{i,2}}}} .$$

3) To the control loop i with the highest processing demands, i.e., $h_i \leq h_j, j = 1, ..., N, j \neq i$, by assigning all the additionally available bandwidth allocation to the loop, further performance improvement can be obtained, i.e. $(b_i)_{opt} = \sqrt{\dfrac{(1 + a_{i,1}) \beta_i}{a_{i,2}}} + b_a$, where b_a is the current additionally available bandwidth utilization, so optimal bandwidth allocation can be further denoted

$$(b_i)_{opt} = \sqrt{\frac{(1 + a_{i,1}) \beta_i}{a_{i,2}}} + 1 - \sum_{i}^{n} b_i = 1 - \sum_{j \neq i}^{N} b_j .$$ The

corresponding sampling period is $(h_i)_{opt} = \dfrac{c_i}{1 - \sum_{j \neq i}^{N} b_j}$.

The concret steps of the optimal algorithm can be described as following:

1) Decides each axes control loop basic control parameters: the number of control loop N, the weighting coefficients of each control loop $a_{i,1}$ and $a_{i,2}$, control loop specific coefficients β_i, closed-loop bandwidth ω_{bw} and rise time $t_{i,r}$;

2) Decides network basic parameters: bit length Li of data package, data rate of the network medium R, best case data processing time $T_{i,p}$;

3) Calculate the best-case one-way data transmission time $T_{i,t}$, estimate MADB D_i, acqure bandwidth range $[(b_i)_{min}, (b_i)_{max}]$;

4) Acoording to $\sum_{i=1}^{N} b_i \leq 1$, judge network schedulability, if not satisfy then goto step 2 to reassign network for satisfying schedulable condition.

5) Based on the formula $(b_i)_{opt} = \sqrt{\dfrac{(1 + a_{i,1}) \beta_i}{a_{i,2}}}$, decides optimal bandwidth allocation for each control loop,

further acoording to $(h_i)_{opt} = \dfrac{c_i}{\sqrt{\dfrac{(1+a_{i,1})\beta_i}{a_{i,2}}}}$, decides

optimal sampling period.

6) Based on whether the error is beyond the bound, calculates optimal sampling period. If the error is within the bound $(h_i)_{opt} = T_{i,bw}/20$ or $(h_i)_{opt} = t_{i,r}/4$; if the error is beyond the bound $(h_i)_{opt} = \dfrac{c_i}{\sqrt{\dfrac{(1+a_{i,1})\beta_i}{a_{i,2}}}}$.

4. Conclusion

The optimal scheduling algorithm can be implemented as a part of the control algorithm in each axes control loop. To each control loop, the controllers with three different sampling periods are designed prior to the system implementation. During the system run-time, each control loop keeps monitoring the system error to check if it is zero (or within a preset threshold, i.e. deadband method), then the decision of which controller(i.e. control algorithm) should be used is made based on this system error information. If system error is zero (or within a given threshold), then the smallest network bandwidth utilization (largest sampling period) is assigned to this control loop. When the system error is large (experiencing perturbation), there are two cases. If the current control loop has the highest processing demands, then the optimal bandwidth utilization is assigned as

$$(b_i)_{opt} = \sqrt{\frac{(1+a_{i,1})\beta_i}{a_{i,2}}} + 1 - \sum_{i}^{n} b_i = 1 - \sum_{j \neq i}^{N} b_j$$,that is to

say, all the currently available bandwidth should be assigned to this control loop to ensure its best QoC. Otherwise, the optimal bandwidth utilization is assigned to optimize the overall QoC as the following formula:

$$(b_i)_{opt} = \sqrt{\frac{(1+a_{i,1})\beta_i}{a_{i,2}}}$$. The presented scheduling

algorithm requires controllers capable of running with different sampling periods. To a given system, sampling period can be assigned according to following three metods: $(h_i)_{opt} = T_{i,bw}/20$ or $(h_i)_{opt} = t_{i,r}/4$;

$$(h_i)_{opt} = \frac{c_i}{\sqrt{\dfrac{(1+a_{i,1})\beta_i}{a_{i,2}}}} ; \quad (h_i)_{opt} = \frac{c_i}{1 - \sum_{j \neq i}^{N} b_j}.$$

For keeping system stability and meeting performance requirements, accordingly adapting the gains or some control parameters is necessary. These control algorithm

can be designed prior to the system running. The presented optimal bandwidth allocation algorithm is feasible for it need not too much computation resources. On the other hand, the algorithm can be extendded to multi-loop NMCS simultaneously experience perturbations by introducing prioritization mechanism. Each control loop can be assigned a priority number according to their processing demands and QoC specifications. The decision of assigning additional bandwidth can be made based on the priorities of these control loops.

Acknowledgement

This work is supported by National Natural Science Foundation of China (NSFC 50505075), Natural Science Foundation of Guangdong province (05103543) and Program for New Century Excellent Talents in University .

References

[1] F.-L. Lian, J. R. Moyne, and D. M. Tilbury, "Performance evaluation of control networks: Ethernet, ControlNet and DeviceNet," IEEE Control Systems Magazine, vol. 21, no. 1, pp. 66–83, February 2001.

[2] Tipsuwan Y., Chow M. -Y., Network-based controller adaptation for network QoS negotiation and deterioration, The 27th Annual Conference of IEEE Industrial Electronics Society, Denver: IEEE Press, 2001, 1794-1799.

[3] M. Velasco, P. Martí and M. Frigola, Bandwidth management for distributed control of highly articulated robots, in Proc. of IEEE International Conference on Robotics and Automation, Barcelona, Spain, pp. 266–271, April 2005.

[4] Y. Halevi and A. Ray, Integrated communication and control systems: Part I—analysis, Journal of Dynamic Systems, Measurement and Control, vol. 110, no. 4, pp. 367–373, December 1988.

[5] P. Marti, G. Fohler, K. Ramamritham, and J.M. Fuertes, Improving Quality-of-Control using Flexible Timing Constraints: Metric and Scheduling Issues, Proc. of the 23rd IEEE Real-Time System Symposium, Austin, TX, USA, December 2002.

[6] Kim D S, Lee Y S, KwonW H, et al. Maximum allowable delay bounds of networked control systems [J]. Control Engineering Practice, 2003, 11 (11): 1301~1313.

A Digital Optimal Control Method for DC-DC Converters Based on Simple Model

LENG Zhaoxia[1], LIU Jian[2], LIU Qingfeng[1], WANG Huamin[1]

1. Xi'an University of Technology, Shaanxi, Xi'an, 710048, CHINA
2. Xi'an University of Science & Technology, Shaanxi, Xi'an, 710054, CHINA
Liuqingfeng_lzx@yahoo.com.cn

Abstract

A novel optimal control strategy based on simple model is presented, which is easy to be realized adopting digital way for DC-DC converter. The discrete model of converter is founded based on state transfer matrix and power-exponent functions are linearized to obtain simple discrete model. The object function for optimization control is designed based on the simple model of converter and the optimal control is obtained. The control error of simple model is modified according to the output error of actual system and model for favorable adaptability. In this paper, the simple discrete model of Boost converter is founded and optimal control is obtained, digital PI is adopted to adjust control error. Simulation and experiment results validate that the output voltage of converter can be controlled accurately and quickly on various work conditions.

1. Introduction

DC-DC converters are applied in many electric power supply systems for small size, low weight, high efficiency etc[1]. In last years, many productions on converter control have been produced, for example: some novel control strategies of converters based on analog circuit are presented in literature [2-4], some non-linear control methods are introduced in PWM converters in literature [5-6], such as fuzzy control and sliding mode control etc. The dynamic response characteristic of PWM converters can be improved by adopting Fuzzy control and sliding mode control, but the realization of control strategies based on analog circuit is comparatively complex, and the control performance of analog control circuit is easily affected by the temperature and aging characteristic of circuit elements, environment factor etc, it is inconvenient to adjust the parameters of analog control circuit.

Comparing with analog control system, digital control system has the following advantages: short design cycle, facility and variety, easy modularizations manage etc[7-8]. With gradually development of chip technology, digital control would be rapidly developed in DC-DC power converters.

The accuracy of mathematics model would decide the control performance of controller designed based on the mathematics model of DC-DC converter, some research work about converter modeling are carried on in literature [9-11]. Precise mathematics model is useful to describe various operation states of converter, but the calculation of model is excessively complex and it is difficult to be realized based on digital control, moreover the accuracy of model and control effect would be affected by the uncertainty parasitic parameters of converter.

Based on above-mentioned reasons, a digital optimal control strategy based on simple model of DC-DC converter is presented in this paper. The ideal optimization control is deduced according to the simple model of converter and the control strategy is modified according to the error between the output of actual system and the output of ideal model. The digital realization of control method is easy for simple mathematics model and simple control strategy, these influence aroused by the uncertainty parasitic parameters of converter and the error introduced in modeling process can be compensated by modifying ideal optimization control. Simulation and experiment results validate that the output voltage of converter can be controlled accurately and quickly on various work conditions.

2. Basic principle

The basic principle of control strategy is showed in figure 1, the main function of digital controller is marked with broken line frame. The optimization value of duty ratio is deduced based on optimization control theory and simple model of converter, and the optimization duty ratio is directly applied to simple model, so the output voltage of model can be obtained. According to the error between the output voltage of model and the output voltage of actual system, the optimization duty ratio is modified, and the modified control value will be applied to actual converter system.

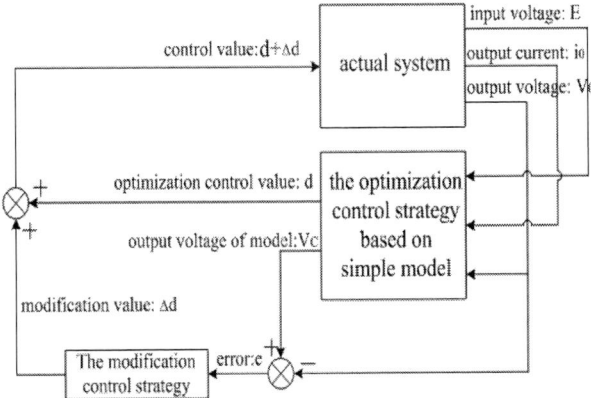

Figure 1 The sketch of control strategy

First, the simple discrete model of DC-DC converter is founded. Then the object function of optimization control is designed based on the simple model, expressions 1 shows the object function.

$$Min \quad f = (x_{1k} - V)^2 + (x_{2k} - I)^2 \quad (1)$$

here, x_{1k} and x_{2k} is respectively the output voltage and inductance current of converter at the k th sampling time (in this paper, the two values in expression 1 are deduced based on cycle state of discrete model before sampling time k), V is the reference value of output voltage of converter, I is the reference value of inductor current. I is determined by the power balance between the input and output of converter, which needs to know the value of load resistance. In fact, the output voltage and current of converter can be sensed to calculate the load resistance. For example, I of Boost converter is showed in expression 2[12].

$$I = \frac{V^2}{RE} \quad (2)$$

here, R is load resistance, E is input voltage.

d_k is the duty ratio of the k th switching cycle, the optimization control d_k can be obtained by minimizing the object function(expression 1), then d_k is directly applied to simple model. Modification control Δd is determined according to the output error of simple model and actual system, various digital controllers can be designed to realize modification control Δd, simple digital PI is adopted in this paper. The output voltage of converter can rapidly arrive at the reference voltage on various work conditions when the modified control $d_k + \Delta d_k$ is applied to actual system. Input voltage, output voltage and output current of actual system need to be sampled for realizing control on various work conditions, the control of input perturbation is fulfilled according to the sampling value of input voltage and the

control of load perturbation is fulfilled according to the sampling values of output voltage and current.

In this paper, the digital optimal control strategy of Boost converter based on simple model is investigated

3. The realization of control strategy for CCM Boost Converter

3.1. The simple model of CCM Boost converter

The circuit topology of Boost converter is showed in figure 2. E is input voltage, i_L is inductor current, V_C is capacitance voltage(that is the output voltage of simple model), R is load.

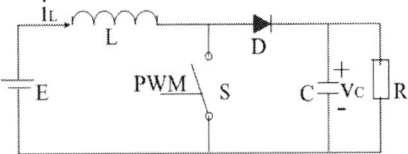

Figure 2 The circuit topology of Boost converter

Expression 3 describes the two linear operation states of converter which operates in CCM(continuous inductor current mode):

$$\begin{cases} \dot{x} = A_1 x + B_1 E & (S \; turns \; on) \\ \dot{x} = A_2 x + B_2 E & (S \; turns \; off) \end{cases} \quad (3)$$

here, state variable $x = [V_C \; i_L]^T$,

$$A_1 = \begin{bmatrix} -1/RC & 0 \\ 0 & 0 \end{bmatrix} \quad , \quad A_2 = \begin{bmatrix} -1/RC & 1/C \\ -1/L & 0 \end{bmatrix} \quad ,$$

$$B_1 = \begin{bmatrix} 0 \\ 1/L \end{bmatrix}, B_2 = B_1, E \text{ is input voltage.}$$

Expression 3 can be transformed into expression 4 by adopting state transfer matrix.

$$x_{(k+1)T} = G_2 G_1 x_{kT} + G_2 H_1 + H_2 \quad (4)$$

$$G_1 = e^{A_1 d_k T} ,$$

$$H_1 = \int_{kT}^{(k+d_k)T} e^{A_1((k+d_k)T-\tau)} B_1 E d\tau$$

$$G_2 = e^{A_2(1-d_k)T} ,$$

$$H_2 = \int_{(k+d_k)T}^{(k+1)T} e^{A_2((k+1)T-\tau)} B_2 E d\tau$$

here, T is switching cycle.

The model described in expression 4 is complex and the calculation of power-exponent function is fussy, so it is inconvenient for digital realization that the control strategy is designed based on the complex model. If DC-DC converter operates in high frequency, the discrete simple model of CCM Boost converter can be deduced with the linearization of the power-exponent function in expression 4, the result is showed in expression 5.

$$x_{(k+1)T} = Ax_{kT} + Bx_{kT}d_k + FE \qquad (5)$$

here:

$$A = \begin{bmatrix} 1 - T/RC & T/C \\ -T/L & 1 \end{bmatrix}, \quad B = \begin{bmatrix} 0 & -T/C \\ T/L & 0 \end{bmatrix},$$

$$F = \begin{bmatrix} 0 \\ T/L \end{bmatrix}.$$

3.2. The realization of control strategy

Substituting the output voltage and inductor current of model deduced according to expression 5 into expression 1, the control value d_k is solved by minimizing expression 1, d_k is showed in expression 6.

$$d_k = \frac{-\alpha_1\beta_1 - \alpha_2\beta_2}{\alpha_1^2 + \alpha_2^2} \qquad (6)$$

here: $\alpha_1 = -Tx_{2k}/C$, $\alpha_2 = Tx_{1k}/L$,

$\beta_1 = (1 - T/RC)x_{1k} + (T/C)x_{2k} - V$,

$\beta_2 = (-T/L)x_{1k} + x_{2k} + (T/L)E - I$.

The restriction condition of duty ratio is: $0.1 \le d_k \le 0.9$.

In the kth switching cycle, d_k is directly applied to simple model, then the output voltage and inductor current of simple model of converter at the $k+1$th sampling time can be deduced based on expression 5.

Although it is convenient for digital realization that the control strategy of output voltage is designed on simple model of Boost converter, actual circuit has a few parasitic parameters, such as inductor equivalent resistance R_L and capacitance equivalent series impedance ESR, different control error would occur in various application field when the control strategy designed based on simple model is directly applied to actual system. In this paper, modification control Δd is adopted to compensate the control error, so the advantage of designing control strategy based on simple model is retained and the shortage is compensated.

So, the control value $d_k + \Delta d_k$ is applied to actual Boost converter in the kth switching cycle, Δd_k is produced by PI controller according to e_k which is the error between the output voltage of actual system and the output voltage of simple model at the kth sampling time.

The expression of digital PI controller is described in

expression 7:

$$\Delta d_k = k_p * [(1 + T/T_i) * e_k - e_{k-1}] + \Delta d_{k-1} \quad (7)$$

here, k_p is proportion coefficient and T_i is integral time constant.

For realizing the control strategy above-mentioned, the output voltage of actual system is sampled to obtain voltage error e_k; For realizing the control of input perturbation, the input voltage of converter is sampled to update input voltage E in simple model; For realizing the control of load perturbation, the output current is sampled and the load resistance of simple model is updated according to the relation between output voltage and output current..

3.3. The stability analysis of control strategy

The discrete expression of converter can be obtained by substituting expression 6 into expression 5:

$$x_{(k+1)T} = f(x_{kT}) \qquad (8)$$

The stability condition of discrete system is: if the modules of eigenvalues of Jacobi matrix are less than 1 at steady operation point, the discrete system is stability. So, the stability condition of the control strategy based on simple model of Boost converter is:

$$L < 2C$$

the detailed process of deducing is described in appendix.

4. The simulation and experiment results

The parameters of Boost converter are as follows: $E = 5\text{V}$, $L = 100\mu\text{H}$, $R_L = 0.36\Omega$, $C = 100\mu\text{F}$, $ESR = 0.013\Omega$, $R = 10\Omega$, $f = 100\text{kHz}$, the reference output voltage $V = 12\text{V}$. The simulation results of Boost converter adopting the control strategy directly based on simple model are showed in figure 3, (a) is start-up process; (b) is the adjusting process of load change and input change, the sequence are: ① R changes from 10Ω to 20Ω, ② E changes from 5 V to 8 V, ③ E changes from 8 V to 5 V, ④ R changes from 20Ω to 10Ω. The corresponding simulation results of Boost converter adopting the control strategy with modification control are showed in figure 4.

Contrasting the results of figure 3 with the results of figure 4, if the control strategy directly based on simple model is adopted, the adjusting process of converter is rapid, but it is impossible for converter to realize exact control of output voltage because of parasitic parameters, the steady error of output voltage is always existent. If the control strategy with modification control is adopted, the output voltage of converter can be exactly controlled in various dynamic processes and the adjusting speed is relatively rapid. The digital control strategy presented in this paper is effective according to the results showed in figure 3 and figure 4.

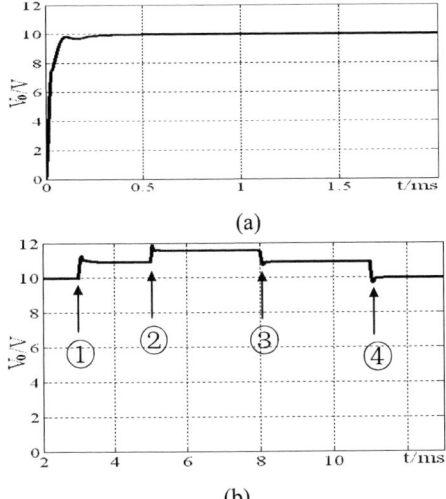

(a)

(b)

Figure 3 The simulation result without modification control

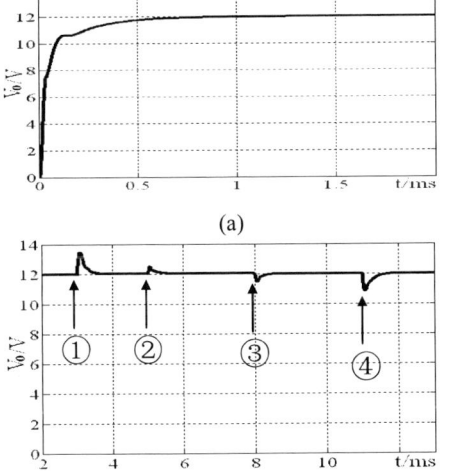

(a)

(b)

Figure 4 The simulation result with modification control

The amplitude of impulse voltage in the output is one of main parameters describing transient characteristics of closed-loop switching converter. The control idea presented in this paper is minimizing the error between output voltage and reference voltage, so the obvious impulse voltage wouldn't occur in the start-up process of converter adopting the control strategy presented in this paper.

The experiment results are showed in figure 5, the switching frequency $f = 100$kHz, the parameters and operation conditions of converter is accordant with the status in simulation. The input voltage, output voltage and output current of converter are sampled by adopting Hall

sensor, DSP2812 is adopted to complete A/D conversion and realize control of converter, the output signal of DSP is separated from the driving circuit of switching by adopting optocoupler. In figure 5, (a) shows the start-up process; (b) shows the adjusting process of load changing from $10\,\Omega$ to $20\,\Omega$; (c) shows the adjusting process of input changing from $5\,\text{V}$ to $8\,\text{V}$; (d) shows the adjusting process of input changing from $8\,\text{V}$ to $5\,\text{V}$; (e) shows the adjusting process of load changing from $20\,\Omega$ to $10\,\Omega$.

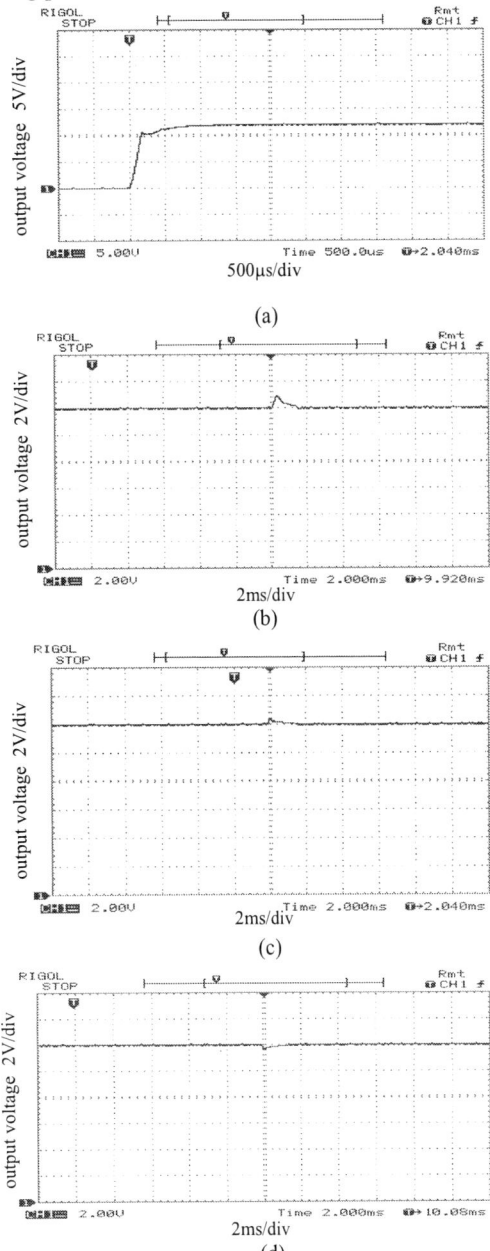

(a)

(b)

(c)

(d)

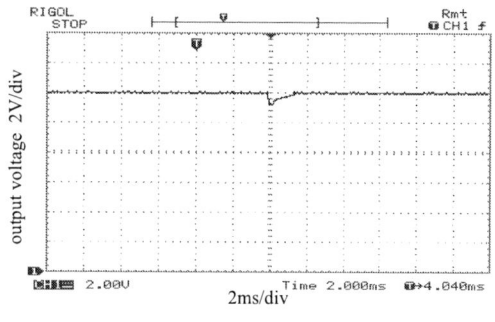

(e)

Figure 5 The experiment result

5. Conclusion

In this paper, a novel optimal control of DC-DC converter based on simple model is presented. Adopting this control strategy, a rapid and no-impulse voltage control of output voltage can be realized in the start-up process of converter, the output voltage of converter can be exactly and rapidly controlled at various operation condition. For simple expression of control strategy, it is easy to be realized by adopting digital controller. Simulation and experiment results are given as a verification of the control strategy.

References

[1] Wu Zhong, Li Hong, Zuo Peng, Liu Wei-zhi. "Cascade control of DC/DC boost converters". *Proceedings of the CSEE*, 2002, 22(1):110~115

[2] Xu Feng, Xu Dian-guo, Liu Yu-xiu. "A novel nonlinear control method for PWM DC-DC converter with optimizing transient response". *Proceedings of the CSEE*, 2003, 23(12):133~139

[3] Wang Feng-yan, Xu Jian-ping. "Modeling and analysis of V²C controlled buck converte". *Proceedings of the CSEE*, 2006, 26(2):121~126

[4] Lin Wei-ming, Huang Shi-peng, Zhang Guan-sheng, Chen Wei. "Research on nonlinear control strategy for PWM DC-DC converters". *Proceedings of the CSEE*, 2001, 21(3):19~22

[5] Wang Wei, Yi Jian-qiang, Zheng Yao-lin, Zhao Dong-bin. "Fuzzy neural network control method for DC-DC converter". *JOURNAL OF SYSTEM SIMULATION*, 2004,16(11):2567~2574

[6] Ahmed M, Kuisma M, Tolsa K, Silventoinen P. "Implementing sliding mode control for buck converter". *2003 IEEE 34th Annual Power Electronics Specialist Conference*, June 2003, Vol. 2:634~637

[7] Guo L, Hung Y J, Nelms R M. "Digital implementation of sliding mode fuzzy controllers for boost converters". *Conference Proceedings of IEEE Applied Power Electronics Conference and Exposition* , March 2006:1424~1429

[8] Guo Tang-shi, Chen Jin-yun, Lin Long-feng, Yin Hua-jie. "New digital control for low-power switching converter". *Power Electronics*, 2003,37(1):63~65

[9] Reatti A, Kazimierczuk M K. "Small-signal model of PWM converters for discontinuous conduction mode and its application for Boost converter". *IEEE Transactions on Circuits and Systems-I*, 2003, 50(1):65~73

[10] Zhu M, He Y, Luo F L. "A novel modeling method for multi-state boost converters". *IEEE International Conference on Industrial Technology, 2005. ICIT 2005,14-17 Dec.* 2005:1~6

[11] Zhang Bo, Qu Ying. "Study on the experiment and the precise discrete model of bifurcation and chaos for bUCK DC/DC converter". *the Proceeding of CSEE*, 2003, 23(12):99~103

[12] Zhang W F, Feng G, Liu Y F. "A direct duty cycle calculation algorithm for digital power factor correction(PFC) implementation". *2004 35th Annual IEEE Power Electronics Specialists Conference.* 2004:2326~2332

Appendix

$$
\left.\frac{\partial x_{(k+1)T}}{\partial x_{kT}}\right|_{x=X} = \left.\frac{\partial f}{\partial x_{kT}}\right|_{x=X} = \left[\begin{array}{cc} 1 - \dfrac{T}{RC} - \dfrac{T}{C} x_{2k}\dfrac{\partial d_k}{\partial x_{1k}} & \dfrac{T}{C} - \dfrac{T}{C}d_k - \dfrac{T}{C} x_{2k}\dfrac{\partial d_k}{\partial x_{2k}} \\[4mm] -\dfrac{T}{L} + \dfrac{T}{L}d_k + \dfrac{T}{L} x_{1k}\dfrac{\partial d_k}{\partial x_{1k}} & 1 + \dfrac{T}{L} x_{1k}\dfrac{\partial d_k}{\partial x_{2k}} \end{array}\right]_{x=X},
$$

here: $X = [V \quad I]^T$ is steady operation point.

$$d_k\big|_{x=X} = \frac{\dfrac{T^2}{C^2}x_{2k}^2 + \dfrac{T}{C}x_{2k}[(1-\dfrac{T}{RC})x_{1k}-V] + \dfrac{T^2}{L^2}x_{1k}^2 - \dfrac{T}{L}x_{1k}(x_{2k}+\dfrac{T}{L}E-I)}{\dfrac{T^2}{C^2}x_{2k}^2 + \dfrac{T^2}{L^2}x_{1k}^2}\Bigg|_{x=X} \,,$$

$$= 1 - \frac{E}{V}$$

$$\frac{\partial d_k}{\partial x_{1k}}\bigg|_{x=X} = \frac{\dfrac{T}{C}x_{2k}(1-\dfrac{T}{RC})-\dfrac{T}{L}(x_{2k}+\dfrac{T}{L}E-I)}{\dfrac{T^2}{C^2}x_{2k}^2+\dfrac{T^2}{L^2}x_{1k}^2} - 2\dfrac{T^2}{L^2}x_{1k}\frac{\dfrac{T}{C}x_{2k}[(1-\dfrac{T}{RC})x_{1k}-V]-\dfrac{T}{L}x_{1k}(x_{2k}+\dfrac{T}{L}E-I)}{(\dfrac{T^2}{C^2}x_{2k}^2+\dfrac{T^2}{L^2}x_{1k}^2)^2}\Bigg|_{x=X}$$

$$= \frac{\dfrac{T}{C}I - \dfrac{T^2}{RC^2}I + \dfrac{T^2}{L^2}E}{\dfrac{T^2}{C^2}I^2 + \dfrac{T^2}{L^2}V^2}$$

$$\frac{\partial d_k}{\partial x_{2k}}\bigg|_{x=X} = \frac{\dfrac{T}{C}[(1-\dfrac{T}{RC})x_{1k}-V]-\dfrac{T}{L}x_{1k}}{\dfrac{T^2}{C^2}x_{2k}^2+\dfrac{T^2}{L^2}x_{1k}^2} - 2\dfrac{T^2}{C^2}x_{2k}\frac{\dfrac{T}{C}x_{2k}[(1-\dfrac{T}{RC})x_{1k}-V]-\dfrac{T}{L}x_{1k}(x_{2k}+\dfrac{T}{L}E-I)}{(\dfrac{T^2}{C^2}x_{2k}^2+\dfrac{T^2}{L^2}x_{1k}^2)^2}\Bigg|_{x=X} \,,$$

$$= \frac{\dfrac{T^2}{RC^2}V - \dfrac{T}{L}V}{\dfrac{T^2}{C^2}I^2 + \dfrac{T^2}{L^2}V^2}$$

Jacobi matrix: $\dfrac{\partial f}{\partial x_{kT}}\bigg|_{x=X} = \begin{bmatrix} \dfrac{C^2V^2 - \dfrac{2TC}{R}V^2}{L^2I^2+C^2V^2} & \dfrac{CTEV+CLVI}{L^2I^2+C^2V^2} \\[4mm] \dfrac{LCIV - \dfrac{2TL}{R}VI}{L^2I^2+C^2V^2} & \dfrac{L^2I^2 + \dfrac{TL}{R}V^2}{L^2I^2+C^2V^2} \end{bmatrix} = \begin{bmatrix} \rho_1 & \rho_2 \\ \rho_3 & \rho_4 \end{bmatrix}.$

eigenvalues are: $\lambda = \dfrac{(\rho_1+\rho_4)\pm\sqrt{(\rho_1+\rho_4)^2-4(\rho_1\rho_4-\rho_2\rho_3)}}{2} = \dfrac{(\rho_1+\rho_4)\pm\sqrt{(\rho_1+\rho_4)^2}}{2}$

so the stability condition is: $(\rho_1+\rho_4) = (1 + \dfrac{\dfrac{TL}{R}V^2 - \dfrac{2TC}{R}V^2}{L^2I^2+C^2V^2}) < 1 \Rightarrow L < 2C$

Design and Practice of an Elevator Control System Based on PLC

Xiaoling Yang[1,2], Qunxiong Zhu[1], Hong Xu[1]

[1] College of Information Science &Technology,
Beijing University of Chemical Technology, Beijing 100029, China
[2] Automation College of Beijing Union University,Beijing,100101, China
yxl_lmy@ sina.com, zhuqx@mail.buct.edu.cn,

Abstract

This paper describes the development of 2 nine-storey elevators control system for a residential building. The control system adopts PLC as controller, and uses a parallel connection dispatching rule based on "minimum waiting time" to run 2 elevators in parallel mode. The paper gives the basic structure, control principle and realization method of the PLC control system in detail. It also presents the ladder diagram of the key aspects of the system. The system has simple peripheral circuit and the operation result showed that it enhanced the reliability and performance of the elevators.

1. Introduction

With the development of architecture technology, the building is taller and taller and elevators become important vertical transportation vehicles in high-rise buildings. They are responsible to transport passengers, living, working or visiting in the building, comfortable and efficiently to their destinations. So the elevator control system is essential in the smooth and safe operation of each elevator. It tells the elevator in what order to stop at floors, when to open or close the door and if there is a safety-critical issue.

The traditional electrical control system of elevators is a relay-controlled system. It has the disadvantages such as complicated circuits, high fault ratio and poor dependability; and greatly affects the elevator's running quality. Therefore, entrusted by an enterprise, we have improved electrical control system of a relay-controlled elevator in a residential building by using PLC. The result showed that the reformed system is reliable in operation and easy for maintenance.

This paper introduces the basic structure, control principle and realization method of the elevator PLC control system in detail.

2. System structure

The purpose of the elevator control system is to manage movement of an elevator in response to user's requests. It is mainly composed of 2 parts:

2.1. Electric power driving system

The electric power driving system includes: the elevator car, the traction motor, door motor, brake mechanism and relevant switch circuits.

Here we adopted a new type of LC series AC contactors to replace the old ones, and used PLC's contacts to substitute the plenty of intermediate relays. The circuits of traction motor are reserved. Thus the original control cabinet's disadvantages, such as big volume and high noise are overcome efficiently.

2.2. Signal control system

The elevator's control signals are mostly realized by PLC. The input signals are: operation modes, operation control signals, car-calls, hall-calls, safety/protect signals, door open/close signal and leveling signal, etc. All control functions of the elevator system are realized by PLC program, such as registration, display and elimination of hall-calls or car-calls, position judgment of elevator car, choose layer and direction selection of the elevator, etc. The PLC signal control system diagram of elevator is showed in Figure 1.

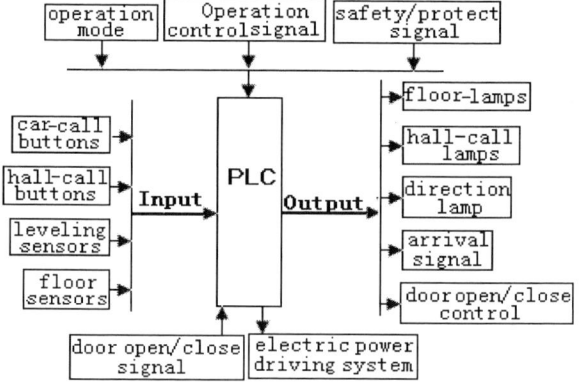

Figure 1 PLC signal control system diagram

2.3. Requirements

The goal of the development of the control system is to control 2 elevators in a 9-storey residential building.

For each elevator, there is a sensor located at every floor. We can use these sensors to locate the current position of the elevator car. The elevator car door can be opened and closed by a door motor. There are 2 sensors on the door that can inform the control system about the door's position. There is another sensor on the door can detect objects when the door is closing. The elevator car's up or down movement is controlled by a traction motor.

Every floor, except the first and the top floor, has a pair of direction lamps indicating that the elevator is moving up or down.

Every floor, has a seven segment LED to display the current location of the elevator car.

The first step for the development of the elevator control is to define the basic requirements. Informally, the elevators behavior is defined as follows.

(1) Running with a single elevator

Generally, an elevator has three operation states: normal mode, fire-protection mode and maintenance mode. The maintenance mode has the highest priority. Only the maintenance mode is canceled can the other operation modes be implemented. The next is fire-protection mode, the elevator must return to the bottom floor or base station immediately when the fire switch acts. The elevator should turn to normal operation mode when the fire switch is reset. Under normal operation mode, the control system's basic task is to command each elevator to move up or down, to stop or start and to open and close the door. But is has some constraints as follows:

Each elevator has a set of 9 buttons on the car control panel, one for each floor. These buttons illuminate when they are pressed and cause the elevator to visit the corresponding floor. The illumination is canceled when the corresponding floor is visited by the elevator.

Each floor, except the first and the top floor, has two buttons on the floor control panel, one to request an up-elevator, one to request a down-elevator. These buttons illuminate when they are pressed. The illumination is canceled when an elevator visits the floor, then moves in the desired direction.

The buttons on the car control panel or the floor control panel are used to control the elevator's motion.

The elevator cannot pass a floor if a passenger wants to get off there.

The elevator cannot stop at a floor unless someone wants to get off there.

The elevator cannot change direction until it has served all onboard passengers traveling in the current direction, and a hall call cannot be served by a car going in the reverse direction.

If an elevator has no requests, it remains at its current floor with its doors closed.

(2) Parallel running with two elevators

In this situation, there are two elevators to serve the building simultaneously. It runs at 7am to 9am and 5pm to 7pm every day.

When an elevator reaches a level, it will test if the stop is required or not. It will stop at this level when the stop is required.

At the same time, to balance the number of stops, the operation of two elevators will follow a certain dispatching principle.

An elevator doesn't stop at a floor if another car is already stopping, or has been stopped there.

The normal operation of elevators is implemented by cooperation of its electric power driving system and logic control system.

3. Software design

Due to the random nature of call time, call locations and the destination of passengers, the elevator control system is a typical real-time, random logic control system. Here we adopted collective selective control method with siemens PLC S7-200 CPU226 and its extension modules. There are 46 input points and 46 output points in the system. The I/O points are showed in Table1 and Table 2.

Table 1 Input points

description	address
1-8 floor up hall-call	I0.0-I0.7
2-9 floor down hall-call	I1.0-I1.7
1-9 floor car-call	I2.0-I2.7, I3.0
1-9 arrival sensor	I3.1-I3.7, I4.0-I4.1
door open button	I4.2
door close button	I4.3
door close location switch	I4.4
door open location switch	I4.5
up leveling sensor	I4.6
down leveling sensor	I4.7
fire switch	I5.0
driver operation switch	I5.1
touch panel switch of car door	I5.2
overload	I5.3
Forced speed changing switch	I5.4
full load	I5.5

Table 2   Output points

description	address
1-8 floor up hall-call lamp	Q0.0-Q0.7
2-9 floor down hall-call lamp	Q1.0-Q1.7
1-9 floor car-call lamp	Q2.0-Q2.7, Q3.0
up moving lamp	Q3.1
down moving lamp	Q3.2
Seven segment LED display of	Q3.3-Q3.7
elevator's position	Q4.0-4.1
door opening	Q4.2
door closing	Q4.3
up moving	Q4.4
down moving	Q4.5
full load lamp	Q4.6
high speed operation	Q4.7
low speed operation	Q5.0
acceleration of starting	Q5.1
deceleration of braking	Q5.2-Q5.4
alarm beeper	Q5.5

About software designing, we adopt the modularized method to write ladder diagram programs. The information transmission between modules is achieved by intermediate register bit of PLC.

The whole program is mainly composed of 10 modules: hall-call registration and display module, car-call registration and display module, the signal combination module, the hall-call cancel module, the elevator-location display module, the floor selection module, the moving direction control module, the door open/close module, the maintenance operation module and the dispatching module under parallel running mode.

The design of the typical modules is described as follows:

3.1. Hall-call registration and display

There are two kinds of calls in an elevator: hall-call and car-call. When someone presses a button on the floor control panel, the signal will be registered and the corresponding lamp will illuminate. This is called hall-call registration.

When a passenger presses a button in the elevator car, the signal will be registered and with the corresponding lamp illuminated. This is called car-call registration.

Figure2 shows the ladder diagram of up hall-calls registration and display. The self-lock principle is used to guarantee the calls' continuous display.

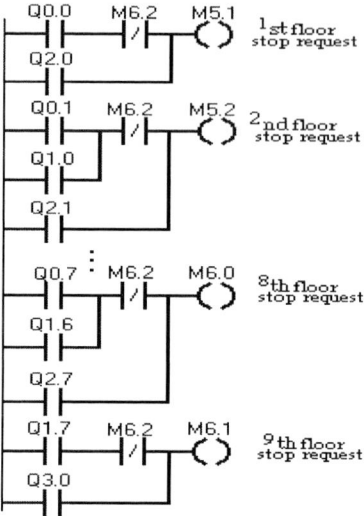

Figure 2 up hall-call registration and display

3.2. The collective selection of the calls

Here the collective selection control rules are used. As showed in Figure3, M5.1-M5.7, M6.0 and M6.1 are auxiliary relays in PLC. They denote the stopping request signal of 1_{st} to 9_{th} floor respectively. The auxiliary relay M6.2 denotes the elevator driver's operation signal. When there is a call in a certain floor, the stopping signal of corresponding floor will output. When the elevator is operated by the driver, the hall-calls will not be served. And the elevator cannot pass a floor at which a passenger wishes to alight.

3.3. The cancellation of the calls

The program of this module can make the elevator response the hall-calls which have the same direction as the car's current direction, and when a hall-call is served, its registration will be canceled. The ladder diagram of up hall-calls' cancellation is showed in Figure4.

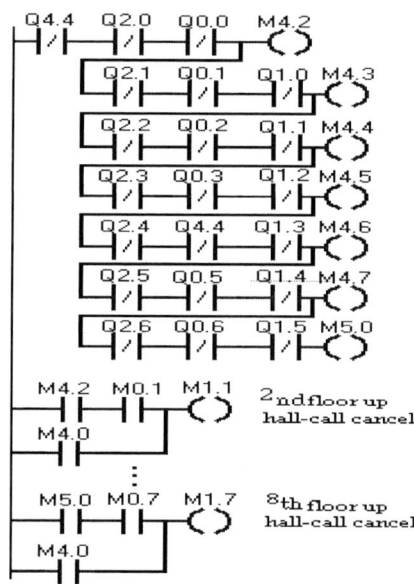

Figure 3 The combination of the calls

Figure 4 The cancellation of up calls

In Figure4, the auxiliary relay M4.0 is the up moving flag of the elevator. When the current direction of the elevator is up, M4.0's contacts are closed; on the contrary, when the current direction of the elevator is down, M4.0's contacts are opened. M0.1 to M0.7 denotes the car-calls' stopping request signal of floor 2 to floor 8 respectively.

This program has two functions:

(1) Make the elevator response the normal down hall-calls when it is moving down, and when a down hall-call is served, its registration is canceled.

(2) When the elevator is moving up, the corresponding floor's down hall-call it passing by is not served and the registration is remained.

The cancellation of down hall-calls is reversed with up hall-calls.

3.4. Elevator's direction

The elevator may be moving up or down, depending on the combination of hall-calls and car-calls. The following ladder diagram in Fig.5 illustrates that the elevator will move up.

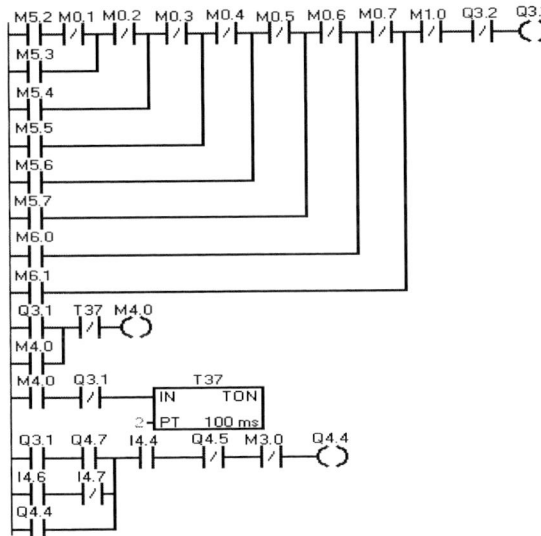

Figure 5 Up moving of the elevator

Figure5 shows that when the calls corresponding floor is higher than the elevator's current location, the elevator will go up. Here the auxiliary relay M4.0 is used as the up-moving flag. When the elevator is moving up, the up-moving lamp is illuminated, so the M4.0 is connected on. When the elevator arrives the top floor, the up-moving lamp is off and the timer starts. After 0.2s, the M4.0 is disconnected, the up-moving display is off. Here we used M4.0 to replace Q3.1 which can ensure the cancellation's reliability.

3.5. Elevator's floor-stopping

Figure6 shows the ladder diagram of the elevator's floor-stopping function.

As showed in Figure6, M6.4 is the flag of floor-stopping signal. M6.6 is the floor-stopping signal sent by the driver. M7.0 is the fire signal sent by the fire switch. And M6.7 is the forced speed changing signal. When either of these contacts act, the system should send out the floor-stopping signal.

4. Minimum waiting time algorithm

In traffic of elevator systems, there are two types of control task usually. The one is the basic control function to command each elevator to move up or down, to stop or

start and to open and close the door. The other is the control of a group of elevators.

The main requirements of a group control system in serving both, car and hall calls, should be: to provide even service to every floor in a building; to minimize the time spent by passengers waiting for service; to minimize the time spent by passengers to move from one floor to another; to serve as many passengers as possible in a given time[1].

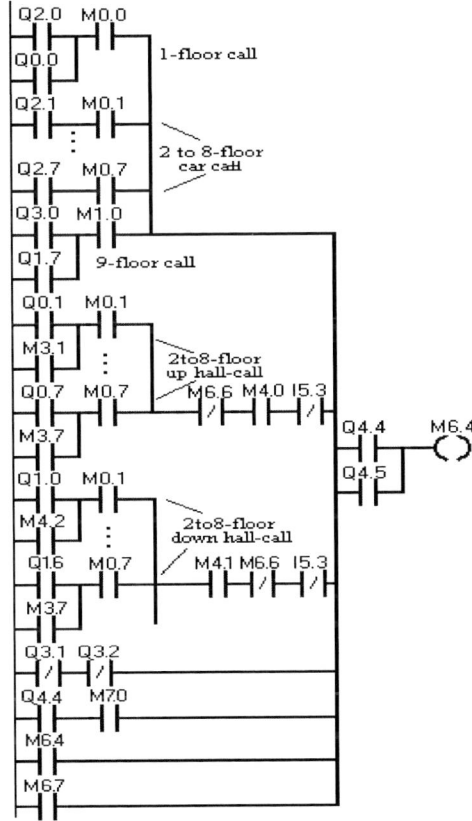

Figure 6 The elevator's floor-stopping

There are many dispatching algorithms for elevator's group control. Such as Nearest-neighbor Algorithm[2], which the elevator always serve the closet request next; Zoning Algorithm[3] which by analyzing the traffic of elevator system with unequal floor and population demand to dispatch the elevator; and Odd-even rule, which an elevator only serves the odd floor and the other only serves the even floor.

The Nearest-neighbor Algorithm minimizes the length of the elevator's empty move to the next request. it usually has very small average waiting times, but individual waiting times can become quite large[2]. The Zoning Algorithm usually used in buildings which has heavy traffic situations, such as the office building at lunch time.

Compared to the office building and shopping mall, the traffic flow of residential buildings is relatively low and even in every floor. Secondly, people usually think of elevators as purely functional objects and the experience of riding an elevator is time waited for most of them. Furthermore, there exist immense problems when attempting to satisfy all requirements.

Considering all of the reasons above, we adopted the "minimum waiting time" algorithm to realize the 2 elevators' parallel running[4].

4.1. Evaluation function

The goal of the "minimum waiting time" algorithm is to predict the each elevator's response time according to all calls, and select the elevator which has the shortest response time to serve.

When there is a call, the system calculates out the function values of each elevator according the evaluation function showed in (1) and (2):

$$J(*)=Min[J(1),J(2),...,J(n)] \qquad (1)$$
$$J(i)=T_r(i)+KT_d(i)+KT_o(i) \quad i=1,2,...,n \qquad (2)$$

$J(i)$ is the evaluation index of each elevator; $T_r(i)$ denotes the time of the elevator directly moving to the destination corresponding the latest call from its current floor; $T_o(i)$ denotes the additional acceleration and deceleration time of a floor-stop of the elevator; $T_d(i)$ denotes the average time of the passenger boarding and alighting the elevator; and K is the sum of hall-calls and car-calls. But when a hall-call and a car call corresponds the same floor, the K is only calculated one time.

4.2. Calculation of minimum waiting time

In equation (2), K is a certain value, T_o and T_d can be obtained by means of statistics. $T_r = T*L$, where T denotes the average time of the elevator passing by one floor; L denotes the desired floors of the elevator from current floor to the hall-call floor.

In order to calculate the L value, we defined the 2 elevators are A and B respectively; Y_A,Y_B denotes the current floor of elevator A and B respectively. H is the corresponding key value when a hall-call button is pressed, and H=floor number of the hall-call.

We defined 4 tables for the PLC realization: up hall-call registration table, down hall-call registration table, car-call registration table of A and car-call registration table of B. When a certain call button is pressed, its floor value is recorded in corresponding table.

Here we take elevator A as an example. First, define the variable M_A, M_B and M_W. Where M_A, M_B denotes the extreme value of car-calls with same direction of A or B's movement respectively.

When elevator A is up-moving, set M_A is equal to the maximum value in car-call registration table A; when elevator A is down-moving, set M_A is equal to the minimum value in car-call registration table A.

M_W denotes the extreme value of hall-calls with same direction of A's movement.

When elevator A is up-moving and up-hall-call value$\geq Y_A$, set $M_W=0$; otherwise, set M_W is equal to the minimum value in up-hall-call registration table A. When elevator A is down-moving and up-hall-call value$\leq Y_A$, set $M_W=0$; otherwise, set M_W is equal to the maximum value in down-hall-call registration table A .

Thus, we can determine the L value according to Y_A, H, M_A and M_W. There are 3 situations:

(1) When the hall-call's direction is opposite to elevator A's movement:

$$L=|Y_A-M_A|+|M_A-H| \qquad (3)$$

(2) When the hall-call's direction is same as elevator A's movement and it is in the front of elevator A:

$$L=|Y_A-H| \qquad (4)$$

(3) When the hall-call's direction is same as the elevator A's movement and it is in the back of elevator A:

$$L=|Y_A-M_A |+|M_A-M_W|+|H-M_W| \qquad (5)$$

So the i-th floor's minimum waiting time can be calculated by (6) as follows:

$$Time(i)=TL(i)+KT_d(i)+KT_o(i) \quad i=1,2,...,n \qquad (6)$$

When the calls change during the operation of elevators, the system calculates the minimum waiting time of each elevator. Then it allocates the current call to the elevator which has small value. When the each elevator has the same value, then the current call is prior to elevator A.

When an elevator is wrong or not in service, the system can exit the dispatching algorithm and turns to a single elevator running mode.

4.3. Algorithm realization

Compared with single elevator running mode, the parallel running mode is mainly different at the processing method about hall-calls. The former uses collective selective control method, and the latter uses dispatch rule combined with collective selective control method.

Here the system is to control a 9-storey building, so we choose two Siemens S7-200 PLCs(CPU226) and its Extensive Modules to control the single elevator respectively. And by using PPI Protocol to realize the communication between 2 PLCs.

The PPI Protocol adopts master-slave communication mode, so we defined elevator A as the master and elevator B as the slave. By communication program, the 2 PLCs can exchange the massage such as the current position, hall-calls or car-calls and moving direction. Then by using "minimum waiting time" algorithm, the system realizes the optimal operation of 2 elevators.

Figure7 shows the ladder program of the car-calls extreme value calculation of elevator A.

In Figure7, VB121~VB130 is the register address of elevator A's car-call corresponding to each floor, Q3.1 is the up-moving lamp of elevator *A*, and the car-calls extreme value is saved in VB120.

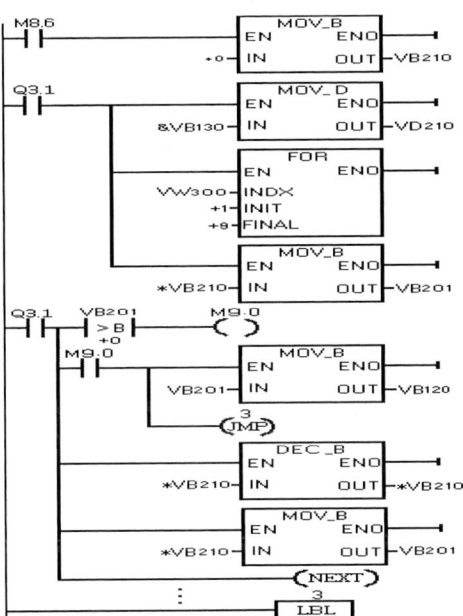

Figure 7 The car-calls extreme value calculation of elevator A

5. Conclusions

In this paper, we have improved an old elevator control system by using PLC, and realized the group control of 2 elevators. The new control system has been operated for 1 year, and its operation scenarios are as follows:

(1) Down–Peak

This traffic condition concerns people out of the building in the morning between 7am to 9am.

(2) Up–Peak

This condition concerns people entering the building between 5pm to 7pm.

(3) Other

It covers the day from 6:00 to 0:00 except the two situations above. And in this situation, there is only one elevator running.

The results are expressed via an average waiting time and maximum waiting time(both given in seconds) are collected in Tables 3 and 4.

Table 3
Average and maximum waiting time(before reformed)

	Average	Maximum
Down–peak	63.20	240.33
Up–peak	52.78	235.26
Other	43.25	215.43

Table 4
Average and maximum waiting time(reformed)

	Average	Maximum
Down–peak	30.12	203.33
Up–peak	27.81	195.20
Other	37.32	186.43

Due to the nonparallel running before the reform, so the average waiting time and maximum waiting time of down–peak and the up–peak are very longer than the reformed. The practice results have showed the better performance of the improved control system.

References

[1] Ricardo Gudwin, Fernando Gomide, Marcio (1998). "A Fuzzy Elevator Group Controller With Linear Context Adaptation". *IEEE World Congress on Computational Intelligence*. Vol. 12, No. 5, pp.481-486.

[2] Philipp Friese, Jorg Rambau (2006). "Online-optimization of multi-elevator transport systems with reoptimization algorithms based on set-partitioning models". *Discrete Applied Mathematics* .No. 154, pp.1908-1931.

[3] Zheng Yanjun, Zhang Huiqiao, Ye Qingtai, Zhu Changming. (2001). "The Research on Elevator Dynamic Zoning Algorithm and It's Genetic Evolution". *Computer Engineering and Applications*, No. 22, pp.58-61.

[4] Xiaodong Zhu, Qingshan Zeng (2006). "A Elevator Group Control Algorithm for Minimum Waiting Time Based On PLC". *Journal of Hoisting and Conveying Machiner*, No. 6, pp.38-40

Corrosion Diagnosis of Conductors and Down Lead Lines
For Grounding Grids Maintenance

Ni Yunfeng[1] Liu Jian[1,2]
1. *School of Electrical Engineering, Xi'an University of Science & Technology, Xi'an,710054*
nnyfeng@163.com

Wang Sen[2] Li Zhizhong[2]
2. *Shaanxi Electric Power Research Institute, Xi'an,710054*
Wangsen01@yahoo.com.cn

Abstract

To improve the efficiency of the corrosion diagnosis, based on the hierarchical model, an actual grounding grid can be divided into six levels, Such as Actual grids, Quasi- Circuit-Domain, Circuit-Domain, Circuit-Network, Touchable Network and Intrinsic Network. Based on above model, a novel node partition based approach is put forward. In which, a grounding grid is divided into several small-scale sub-grids. Taking the down lead line resistances into consideration, a set of linear equations for the conductance of the intrinsic branches in each sub-grid is deduced and solved, respectively, without the nonlinear iteration. Corrosion diagnosis of the down lead lines is also carried out for each sub-grid, respectively. A dynamic planning based optimization procedure to reduce the work in the maintenance is proposed, with which, the best test scheme can be worked out. An experimental grounding grid with sixty branches is used as an example. The results of corrosion diagnosis show that the proposed approaches are feasible with the maintenance works remarkably reduced.

1. Introduction

The grounding grid is an important power apparatus to ensure the stable operation of electric power system and safety of operators. It usually can be damaged by the corrosion after years of operating.

Recently, many works have been done on grounding grid corrosion diagnosis [1]. The method of grounding grid corrosion diagnosis can be classified into two kinds such as electro-magnetic approaches and circuit theory approaches.

In reference [2], electro-magnetic approaches to analysis the performance of grounding grids and to find the break positions in grounding grids were given. Unfortunately, electro-magnetic approaches cannot give a satisfactory diagnosis results for the branches corroded but not broken. In Ref.[3] an electric circuit method based on Tellegon theory for grounding grids diagnosis was presented. The approaches based Artificial and Tabu search corrosion diagnosis was described in Ref.[4].

A successful approach of corrosion diagnosis for grounding grids was put forward in Ref.[5]. Shifting the positions of the current exciter within the touchable nodes and measuring the voltages as many as possible for each excitation, the number of equations was enlarged remarkably. Thus, the accuracy was improved remarkably.

Ref.[6] illustrated a method with Topology–Splitting and Hierarchical Simplification for evaluating the testability of grounding grids, based on which, we can simplify an actual grid into an intrinsic grid, in which the whole nodes are touchable and the whole branches are unique certain

Although many achievements have been made, the following problems are still needed to investigate.

The effect of down lead line resistances to voltage measured is ignored and the approach of down lead line corrosion diagnosis is not appeared above references. In additional, during the test for corrosion diagnosis, the maintenance work is larger, how to improve the efficiency of maintenance work is to be considered, optimization of test scheme is discussed and a new method of producing the best test scheme is presented in paper.

2. Hierarchical modeling of grounding grids

A grounding grid can be converted into six levels: such as the Actual Grid (AG), the Quasi-Circuit Domain Grid (QCD), the Circuit Domain grid (CD), the Circuit Network (CN), the Touchable Grid (TG) and the Intrinsic Grid (IG).

AG is the actual topology of a resistance grid. Taking the touchable nodes as splitting points, an AG can be split into many sub-grids. If a sub-grid is not able to be further split, it is called a Quasi-Circuit-Domain (QCD).

As for a QCD, we can merge the series or parallel branches as much as possible but keep all touchable nodes. We call the result of the above process a Circuit-Domain (CD). Based on circuit theory, we can eliminate all un-touchable nodes in a CD and model it by equivalent branches between each two touchable nodes. We call the result of above process a Circuit Network (CN). A CN is the grid formed by all of the CDs.

A TG is the grid formed by all of the CNs. We can form an IG by merging the parallel branches in the TG as much as possible but keep all touchable nodes. The branches in the IG are called intrinsic branches. The QCDs, CDs and CNs are called units of the levels of AG, CN and TG, respectively.

The resistance of branches in AG can be classified into two kinds, such as clear-branch and undetermined-branch., according to the principle proposed in [5], with which the nodes of intrinsic grid are all touchable nodes, the IBs are all determined as soon as the test is much sufficient.

For example, an actual grounding grid is shown in Fig.1. There are eighteen touchable nodes in the grounding grid, which is indicated by solid circle in Fig.1. According to the hierarchical simplification mode, it can be decomposed into six levels and its AG and IG shown in Fig.1 (a) and (b), respectively.

According to the method of testability evaluation proposed in Ref.[6], the clear branches are shown by hollow rectangles and the uncertain branches are shown by shadow rectangles.

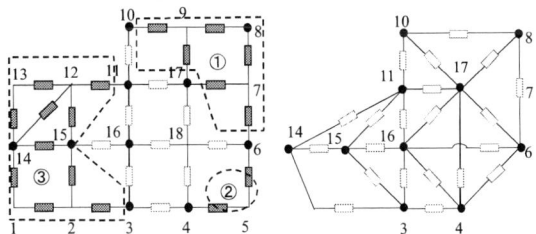

(a) Actual grid   (b) Intrinsic Grid
Fig.1 Decomposition of the grounding grid

3. Novel approach of corrosion diagnosis

As for the i-th touchable node in a region, according to KCL, we have

$$I - E = \sum_{(j \in \beta) \cap (j \notin \alpha)} V_{j,i} G_{i,j} \qquad (1)$$

Where, $V_{i,j}$ is the voltage between node i and j, $G_{i,j}$ is the admittance of the intrinsic branch between node i and j. I_i is the injection current of node i. β is the set of the nodes adjacent with node i, and are have obtained, we call the node i as a common node.

Where, $E = \sum_{k \in \alpha} V_{k,i} G_{i,k}$ α is the set of intrinsic branches with the resistances obtained

Similarly, choosing the different node as the common node and completing a test as the test for node-i, a linear equation group which includes all the intrinsic branch resistance can be established. We have

$$I - E = VG \qquad (2)$$

By solving the equations, the resistance of IB can be worked out. We call above test as the test for node-i.

In the course of testing, to avoid the influences of the resistance of down lead line, we should select the test

scheme in which the DC exciter source has no common down lead line with the instrument of voltage measured, so we must put the exciter outside i-th partition. The formula (1) is become Homogeneous Linear Equation, namely: $\qquad VG = 0 \qquad (3)$

There is only zero solution for this equation group.

3.1 Conductance Ratios of intrinsic branches in a sub-grid

As for node-i block, Selecting one of branches randomly, assuming the branch is $G_{i,h}$, if the row element of the formula (3) is divided by $G_{i,h}$, we have:

$$VK = W \qquad (4)$$

Where, $K_{j,h} = G_{i,i}/G_{i,h}$, W is a constant vector.

The variable of the liner equation group is K which described the ratio of the intrinsic branch between the each branch and $b_{i,h}$ in the region of node-i.

Adopting the algorithm of Gauss elimination to solve the equation group, each element of **K** can be worked out.

Similarly, we can choice the different node as the common node and complete the test as the test for node-i, a linear equation group, which includes all the ratio of intrinsic branch between each branch and base conductance $G_{i,h}$ can be established by solving the equation group, all branch ratio based on the base conductance $G_{i,h}$ would be worked out.

Based on above results of solution, by taking one pole of the exciter on the common node which connected with the based conductance, and putting the other pole outside the i-th partition and testing. We may supply an equation which includes two variables such as the base conductance and RDLL.

The RDLL is determined by the method illustrated in section 3, so the equation which supplements can be solved, the base conductance would be worked out. Further, we may calculate all resistance of IB.

3.2 Resistances of Down Lead Lines

A partition of node-O described in Fig.2, there are four touchable nodes, O, X, Y and Z of grounding grids, respectively.

Assuming $b_{i,j}$ represents the branch between the node-i and node-j, thus, $b_{O,X}, b_{O,Y}$ and $b_{O,Z}$ are intrinsic branches of the common node-i respectively, $b_{O',O}, b_{X',X}, b_{Y',Y}$ and $b_{Z',Z}$ are branches of the each down lead line, respectively.

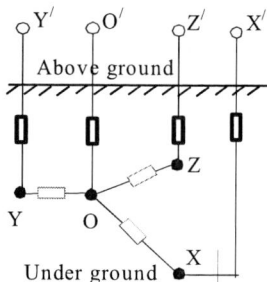

Fig.2 The IBs and its down lead lines in region node-O

3.2.1 Resistances of the down lead lines of the non-common touchable nodes

The method to solve the resistance of down lead line is described as follows. In Fig.2, as for to obtain $G_{X',X}$, we need to supply an independent test for it by method in section 3.1, The conductance of $b_{X',X}$ are included in the supplement equation and the rests of down lead line resistances are not.

The independent equation can be described as below:

$$V_{X',O'} - I/G_{X',X} + V_{Y',O'}K_{Y,X} + V_{Z',O'}K_{Z,X} = 0 \quad (5)$$

Where, $V_{A',B'}$ represents the measuring voltage between the touchable node A' and touchable node B' above ground. The DC source value is I.

$$K_{Y,X} = G_{O,Y}/G_{O,X}, \quad K_{Z,X} = G_{O,Z}/G_{O,X}.$$

Such ratios can be worked out by solving two equations which described ratio based on taking the $b_{O,X}$ as the base conductance. So the conductance of down dead line $b_{X',X}$ would be represented by $G_{X',X}$, and we have

$$G_{X',X} = I/(V_{X',O'} + V_{Y',O'}K_{Y,X} + V_{Z',O'}K_{Z,X}) \quad (6)$$

Further more, we have: $R_{X',X} = 1/G_{X',X}$
(7)

Where, $R_{X',X}$ is the RDLL $b_{X',X}$. Similarly, we can obtain the resistance of other non-common node down lead line

3.2.2 Resistances of the down lead lines of the non common touchable nodes

By taking one pole of the DC exciter to the common touchable node and the other of exciter pole to connect with random touchable node outside the partition of common node and testing.

An equation which includes RDLL can be established, we have:

$$V^{(1)}_{X',O'} + V^{(1)}_{Y',O'}K_{Y,X} + V^{(1)}_{Z',O'}K_{Z,X} =$$
$$I/G_{O',O}(1 + K_{Y,X} + K_{Z,X}) + I(G_{O,X})^{-1}$$
(8)

Where, $K_{Y,X}$ and $K_{Z,X}$ have been obtained in section 3.1

Obviously, there are two variables $G_{O',O}$ and $G_{O,X}$ in (10), and two variables both located to right of equation.

Due to having two variables in (8), we can add to establish a new equation by same method.

Having observed the new added equation and (8), we may find that they are relative between two equations. A conclusion can be drawn that the RDLL of a common touchable node can not be solved. So a new approach which called it Method of Exchanging Region (MEG) is described as follows.

Selecting an adjacent node of the common node-O as a new common node and changing the common node-O as a non-common node, repeating test as section 3.2.1, the resistance of the common node-O can be solved.

Based on the results of solution resistance of common node-O, replacing it to (8), the intrinsic branch resistance denoted by $G_{O,X}$ can be worked out.

3.2.3 Actual grounding grid branch resistance evaluation

Resistances of all intrinsic branches in the region of node-i would be worked out by adopting the approach described in section 3.

Based on above results, adopting the method proposed in reference [6], resistances of clear-branch in grounding grids can be traced to calculate step by step.

4. Test scheme optimization

As for an actual grounding grid, adopting the method proposed in section 3 and executing a test, the resistance of all intrinsic branches would be obtained

4.1 Maintenance work evaluation

The test scheme corresponding to the solving resistance of IB in region of common node can be divided four steps to carry out.

Test scheme analyzed has shown that maintenance work would be different greatly with the different test scheme. The different test scheme has the different sequence of selecting the common node during its testing.

In order to express the optimization of a test scheme, the concept of maintenance work would be defined as below.

$$F = \sum_{i \in \lambda} f_i$$

(9)

Where, λ is the set of common nodes chosen, taking touchable nodes i as the common node, fi is maintenance work, we have: $f_i = S_i + T_i$ (10)

Where, S_i is the times of exciter needed moving and Ti is the times of voltage measured in test face to node-i .

Supposed M intrinsic branches connected with the common node-i, in which N intrinsic branches resistance are un-certain. Solving the whole IBs in the partition of node-i, four steps can accomplish the testing whose maintenance work needs to describe with four steps.

According to the certain branch whether is included in the partition or not, the method to solve the intrinsic branches resistance can be classified into two kinds.

(i) Assuming that the partition of node-i has M intrinsic branches with N of them are unknown, the exciter need to be shifted for N times, and in each exciting position, the M voltages are measured. We can establish linear equation group which includes N equations. To establish the equation group, N times excitation and N×M times voltage measured are needed to be done.

Now, we can supply a dependent test and establish a dependent equation which includes the RDLL. Do above test, only one times excitation and 1×M times voltage measured are needed. Owning to M intrinsic branches in this partition, M+1 RDLLs are needed to solve. Thus, its maintenance work is (M+1)× (M+1).

(ii) As all the intrinsic branches are uncertain, work can be divided four steps to be calculated.

Step1: for solving the ratio of intrinsic branch.

According to the section 3.2, to solve the ratio of IBs in the partition of node-i, the maintenance work is:

$$f_i = N (M + 1) \qquad (11)$$

Step2: maintenance work for the RDLLs of every non-common node in the partition of node-i.

$$f_i = (M - 1)(M + 1) \qquad (12)$$

Step3: maintenance work for solving the RDLLs of common nodes in the partition of node-i.

According to the new Method of Exchanging Region proposed in section 3.3, it can be solved.

Assuming j is the new replacing common-node chosen in the partition of node-i, there is K intrinsic branch connected with it.

Case 1: if the intrinsic branches connected with node-j are uncertain, taking node-j as the common node, firstly, the ratio of intrinsic branches can be solved by independent linear equations. its maintenance work is:

$$f_i = K (K + 1) \qquad (13)$$

Case 2: in the partition of node-j, the number of intrinsic branch connected with node-j is K, if the intrinsic branches K_1 connected with node-j are unknown. Based on the method illustrated in section 3.2, we need supply a dependent test and establish a dependent equation including the resistance of common node-i down lead line.

To establish the equation, its maintenance work can be expression by follows:

$$f_i = K + 1 \qquad (14)$$

Step4: maintenance work for the intrinsic branch resistance in the partition of node-i

If the RDLLs of all nodes in the partition of node-i is known, by doing a test as facing common node, a equation including the resistance of common node-i down lead line and $G_{O,X}$, can be established, to establish the

equation, its maintenance work is:

$$f_i = M + 1 \qquad (15)$$

Thus, the total work is sum adding from (11) to (15). The optimization of test scheme can be described as

$$\text{Object function} \quad Min \quad F = \sum_{i \in \lambda} f_i \qquad (16)$$

Constraint condition: the resistance of all intrinsic branches in λ should be worked out.

To optimize the test scheme, we should select appropriate nodes with a suitable sequence and carry out the corresponding test for the selected nodes in turn.

4.2 Dynamic programming approach

In the proposed dynamic programming based approach, the selected touchable nodes are used as the vertexes of a graph besides a starting vertex labeled as 0-th vertex.

Taking the starting point and the chosen touchable node as the node, using the sides to describe the order of priority of the two chosen touchable nodes then the weight coefficient of the sides is the maintenance work of the reachable node which is chosen later. The node of the most outside layer is called leaf node; the group of sides which connect the leaf node and the side of the starting points is called the route of this leaf node, and the sum of the weight coefficient of the sides included in the route is the weight coefficient of this route; if the nodes in the route of a certain leaf node are the same, and different only in order, they are called repeated route; if the testing plan corresponding a route of a leaf node can solve all the problems in intrinsic branches, then this leaf node is end leaf node.

The optimization of the testing plan can be made by consulting the dynamic scheme.

i). Calculating the measuring connection f test for each of the touchable nodes respectively, and choosing the smallest node corresponding f as the node of the first layer and establishing the side between the starting points, and the weight coefficient.

ii). Choosing a current non end leaf node in the rest of the nodes, calculating the measuring connection f of each touchable node except the nodes in the route of the leaf node, and choosing the smallest node corresponding f as the first outside layer of the leaf node and establish its side between the leaf nodes. Its weight coefficient is the corresponding maintenance work, and the newly chosen nodes become new leaf nodes instead of the original ones.

iii). Deciding whether there is repeated routes. If one of the routes which repeat each other is kept and the others are deleted, then go back to step iii, Otherwise, go on to step iv.

iv). Deciding whether all the current leaf nodes are end leaf nodes. If the answer is yes, go on to step v. If no, go back to step ii.

v). If the testing plan corresponding to the route of

some end leaf nodes with the smallest weight coefficient is the best optimizing testing plan, its weight coefficient is the corresponding smallest maintenance work.

5. Experiment

An experimental grounding grid with sixty branches shown in Fig.4 is used as an example. The touchable nodes are Node 0, 3, 5, 8, 10, 12, 13, 15, 17, 18, 20, 21, 22, 23, 24, 25, 27, 29, 30, 32, 33 and 34, which are illustrated by solid circles in the Fig.4. The resistances of the labeled branches are shown in Table 1. The resistances of other branches are all 0.1Ω.

Table 1 Resistance of abnormal branches of the grounding grid (Ω)

Type of Branch	/	//	///		O	×
Resistance	0.19	0.36	0.47	0.57	0.78	1.35

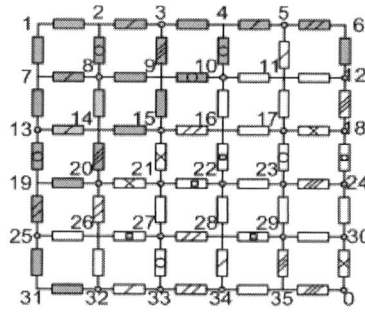

Fig.4 An experiment grounding grid

By the hierarchical model described in section 2, the intrinsic network of the experiment grounding grid is shown in Fig.5.

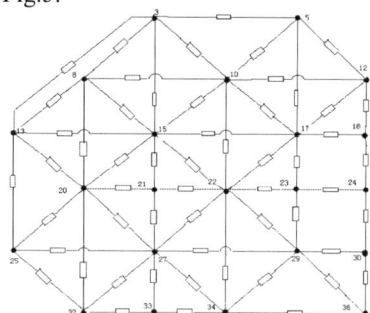

Fig.5 The IG and its intrinsic branch

The constant DC source with amplitude 30A is used as an exciter, with which the test scheme of complete measurement is done in the experiment. The total maintenance work is 364 by adopting the method produce the best test scheme.

Having obtained the intrinsic branch resistance, we can calculate resistances of clear-branch in actual grounding grids. The result shows that the error of average relative is 0.8692%, and the maximal relative error is 4.6572%. At the same time, the resistance of down lead line can be diagnosed also.

6. Conclusion

We may draw the following conclusions:
i). With the proposed model, a large scale grounding grid can be divided into several small scale sub-grids to solve.

ii). With the proposed approach, the resistances of the intrinsic branches and the down lead lines can be worked out by a linear method without iteration.

iii). The proposed dynamic planning based procedure can obtain the optimal test scheme with the minimum maintenance work.

iv).An experimental grounding grid with sixty branches is used as example, the results of corrosion diagnosis show that the proposed approach is feasible

Reference

[1] Zhang Liping, Yuan Jiansheng and Li Zhongxin, "Numerical calculation of substation grounding systems"[J]. Proceedings of the CSEE, 1999, 19(5), pp. 76-79 (in chinese).

[2] Zhang Bo, Cui Xiang and Zhao Zhibin et al, "Analysis of grounding grids at large scale substations in frequency domain"[J],Proceedings of the CSEE, 2002, 22(9), pp.59-63.

[3] Hu Jun, Zeng Rong and He Jinliang ,"Novel method of corrosion diagnosis for grounding grid"[C]. Proceedings of the 2000 International Conference on Power System Technology, Perth, Australia, 2000(3), pp.1365-1370.

[4] Xiao Xinhua,Liu Hua and Chen Xianlu, "Analysis of theory and method about the corrosion as well as the broken point of the grounding grid"[J],Journal of Chongqing University,2001,24(3),pp.72-75 (in chinese).

[5] Liu Jian,Wang Jianxin andWang Sen,"An improved algorithm of corrosion diagnosis for grounding grid &its evaluation"[J], Proceedings of the CSEE,2005,25 (3),pp.71-77 (in chinese).

[6] Wang Shuqi, Liu Jian and Ni Yunfeng, "Hybrid Approach Based on Rule and Numerical Methodology to Evaluate Testability of Branches of Grounding Grid"[J], Shaanxi Electric Power,2008,(1),pp. 15-18 (in chinese).

First-order Optimality Conditions for Convex Semi-infinite Inequality Constrained Programming with Noncompact Sets *

Meixia Li
School of Mathematics and Information Science
Weifang University,
Weifang, Shandong, 261061, China
limeixia001@163.com

Jinchuan Zhou
Department of Mathematics, School of Science
Beijing Jiaotong University
Beijing, 100044, China
jinchuanzhou@yahoo.com.cn

Abstract

In this paper, we study the optimal value function on a noncompact set and obtain the expressions of the subderivative and subdifferential of the optimal value function. Based on some characterizations of subderivative and subdifferential of the objective function and constraint functions, we develop the first-order necessary and sufficient optimality conditions for convex semi-infinite inequality constrained programming in which the index sets are not necessarily compact.

1. Introduction

Consider the following convex semi-infinite inequality constrained programming (CSICP):

$$\min\{\psi^0(x)|\psi^j(x) \leq 0, j \in J\},$$

where

$$\psi^j(x) = \sup\{\phi^j(x,y)|y \in Y_j\}, \quad Y_j \subseteq R^{m_j}(j \in \bar{J}).$$

$$J = \{1, 2, \cdots, q\}, \bar{J} = \{0, 1, 2, \cdots, q\}.$$

The functions $\phi^j(x,y) : R^n \times R^{m_j} \to R$ is convex in $x \in R^n$ for each $y \in Y_j$, and Y_j is a nonempty set in R^{m_j}. Compared with the standard semi-infinite programming, the index set Y_j is not necessarily compact here.

In the recent years, (generalized) semi-infinite programming problem and its broad range of applications have been an active study area in mathematical programming. For example, Vázquez and Rückamann [9] proposed two extensions of the Kuhn-Tucker constraint qualification for the class of generalized semi-infinite programming problems.

Cánovas et al.[1] presented the metric regularity of the mapping defined by semi-infinite constraint systems of inequalities and equalities. Using the variational analysis technique, in terms of the epi-coderivative, Zheng and Yang [11] provided Lagrange multiplier rules for a class of semi-infinite optimization problems where all functions are lower semicontinuous or locally Lipschitz. Under the assumptions that the index set $Y(x)$, depending on the variable x, is uniformly compact, several authors have developed the first-order or second-order optimality conditions for generalized semi-infinite programming problems; see [8] and references therein. Using a unified framework by writing the corresponding constraints in a form of cone inclusions, Shapiro [7] developed the weak and strong duality properties of convex semi-infinite programming. For a thorough study of this subject, refer to the survey paper [3] and book [2] which contains several works on (generalized) semi-infinite programming problem and other related topics such as semidefinite programming, optimal control, wavelets, and others.

However, most of these results require a common essential assumption that Y_j is compact, which is quite strong and does not often occur in practice. Therefore, it is necessary and interesting to study the generalized semi-infinite programming without this compactness restriction. Based on the above motivation, Wang et al.[10] developed the first-order optimality conditions for generalized semi-infinite min-max programming problems over the relative interior of the effective domain of ψ^0. In this paper, by using the characterizations of the objective function, we develop the first-order optimality conditions for the convex semi-infinite inequality constrained programming(CSICP), which generalizes the corresponding results given in [4], since no compactness assumption on the index sets is imposed.

The rest of this paper is organized as follows. Section 2 contains some basic notation and preliminary results. Section 3 is devoted to the optimality conditions for (CSICP).

---

*This research was supported by the National Natural Science Foundation of China (10571106).

978-0-7695-3342-1/08 $25.00 © 2008 IEEE
DOI 10.1109/PEITS.2008.52

2. Preliminaries

We first recall the related concepts and conclusions which are the main tools for our theoretical analysis.

Let A be a nonempty set in R^n. The closure, convex hull, interior, and relative interior of A are denoted by cl$A$, co$A$, int$A$, and ri$A$, respectively.

The feasible direction cone $F_A(\hat{x})$ of A at $\hat{x} \in A$ is defined as

$$F_A(\hat{x}) := \{v \in R^n | \exists \hat{\lambda} > 0, \hat{x} + \lambda v \in A, \forall \lambda \in (0, \hat{\lambda})\}.$$

The cone of tangents of A at $\hat{x} \in A$, denoted by $T_A(\hat{x})$, is defined as

$$T_A(\hat{x}) := \{v \in R^n | \exists \lambda^n \searrow 0, v^n \to v, \hat{x} + \lambda^n v^n \in A\}.$$

Let $f : R^n \to \bar{R} := [-\infty, +\infty]$ be an extended real valued function. Then f is said to be proper if the effective domain dom$f := \{x \in R^n | f(x) < +\infty\} \neq \emptyset$ and the set $\{x \in R^n | f(x) = -\infty\} = \emptyset$. The directional derivative of the function f at x in the direction $w \in R^n$, denoted by $f'(x; w)$, is given by the following limit if it exists

$$f'(x; w) := \lim_{\lambda \downarrow 0} \frac{f(x + \lambda w) - f(x)}{\lambda}.$$

If f is proper convex and $x \in \text{dom} f$, then $f'(x; w)$ exists for any $w \in R^n$, where the derivative value is either finite or infinite.

The subderivative function $df(\hat{x})(\hat{w}) : R^n \to \bar{R}$ is defined as

$$df(\hat{x})(\hat{w}) := \liminf_{\substack{w \to \hat{w} \\ \lambda \downarrow 0}} \frac{f(\hat{x} + \lambda w) - f(\hat{x})}{\lambda},$$

and the subdifferential of f at $\hat{x}$ is defined as

$$\partial f(\hat{x}) = \{v \in R^n | \langle v, w \rangle \leq df(\hat{x})(w), \forall w \in R^n\}. \quad (2.1)$$

3. Main Results

We begin by considering the sup-type function defined as follows:

$$\psi(x) = \sup\{\phi(x, y) | y \in Y\}, \quad (3.1)$$

where $\phi : R^n \times R^m \to R$ is a real-valued function and Y is a nonempty set in R^n. Let $\nabla_x f(\hat{x}, \hat{y})$ denote the gradient of f with respect to x evaluated at $(\hat{x}, \hat{y})$. To address the subdifferentiability properties of the sup-type function, the following assumption is needed.

Assumption 3.1 (i) *The function $\phi(x, y)$ is continuous and is convex in $x \in R^n$ for each $y \in Y$.*
(ii) *The gradient $\nabla_x \phi(\cdot, \cdot)$ is continuous on $R^n \times Y$ and $\sup\{\|\nabla_x \phi(x, y)\| | \ y \in \theta(x, \delta)\} < +\infty$ for some $\delta < \psi(x)$, where $\theta(x, \delta) = \{y \in Y | \ \phi(x, y) \geq \delta\}$.*
(iii) *The function $\nabla_x \phi(\cdot, y)$ is equi-continuous at each $x \in R^n$ with respect to $y \in Y$, i.e., for any $x \in R^n$ and $\varepsilon > 0$, there exists a neighborhood $N(x, r)$ with $r > 0$ such that $\|\nabla_x \phi(x_1, y) - \nabla_x \phi(x_2, y)\| < \varepsilon$, whenever $x_1, x_2 \in N(x, r)$ and $y \in Y$.*

Let's define $\hat{Y}(x) = \{y \in Y | \ \phi(x, y) = \psi(x)\}$ and $\theta(x, \delta) = \{y \in Y | \phi(x, y) \geq \delta\}$.

The following lemma, due to Wang et al.[10], gives a characterization of the directional derivative of ψ.

Lemma 3.2 [[10],Theorem 2.1] *Suppose that ϕ satisfies Assumption 3.1 and $\hat{x} \in \text{dom} \psi$. Then the following results hold:*
(i) *if $w \in F_{\text{dom} \psi}(\hat{x})$, then*

$$\psi'(\hat{x}; w) = \lim_{\delta \nearrow \psi(\hat{x})} \sup\{\langle \nabla_x \phi(\hat{x}, y), w \rangle | \ y \in \theta(\hat{x}, \delta)\};$$

(ii) *if $w \notin T_{\text{dom} \psi}(\hat{x})$, then*

$$\psi'(\hat{x}; w) = +\infty.$$

The following lemma, credited to Rockafellar [6], gives the relationship between the directional derivative and the subderivative.

Lemma 3.3 [[6], Proposition 8.21] *For a convex function $f : R^n \to \bar{R}$ and any point $\hat{x}$ where f is finite, the limit*

$$h(w) := \lim_{\tau \downarrow 0} \frac{f(\hat{x} + \tau w) - f(\hat{x})}{\tau}$$

exists for all $w \in R^n$ and

$$df(\hat{x})(w) = (\text{cl} h)(w), \quad \forall w \in R^n. \quad (3.2)$$

In view of Lemmas 3.2 and 3.3, we can prove the one of main results in this section, which gives a structural expression of the subderivative.

Theorem 3.4 *Suppose that ϕ satisfies Assumption 3.1 and $\hat{x} \in \text{dom} \psi$. Then the following results hold:*
(i) *if $w \in T_{\text{dom} \psi}(\hat{x})$, then*

$$d\psi(\hat{x})(w) = \lim_{\delta \nearrow \psi(\hat{x})} \sup\{\langle \nabla_x \phi(\hat{x}, y), w \rangle | \ y \in \theta(\hat{x}, \delta)\};$$

(ii) *if $w \notin T_{\text{dom} \psi}(\hat{x})$, then*

$$d\psi(\hat{x})(w) = +\infty.$$

Proof. We first show that $d\psi(\hat{x})(w) = +\infty$ for all $w \notin T_{\text{dom}\psi}(\hat{x})$. Indeed, if $w \notin T_{\text{dom}\psi}(\hat{x})$, for any sequences $\{\lambda_i\}$ converging to 0 and $\{w_i\}$ converging to w, we must have $\hat{x} + \lambda_i w_i \notin \text{dom}\psi$ for all i sufficiently large. Hence, by the definition of subderivative of ψ at $\hat{x}$, we get $d\psi(\hat{x})(w) = +\infty$. Lemma 3.3 implies that for all $w \in R^n$,

$$d\psi(\hat{x})(w) = \text{cl}\psi'(\hat{x}; w) = \liminf_{w' \to w} \psi'(\hat{x}; w'). \quad (3.3)$$

Note that $T_{\text{dom}\psi}(\hat{x}) = \text{cl}F_{\text{dom}\psi}(\hat{x})$ and that $\psi'(\hat{x}, w) = +\infty$ for all $w \notin T_{\text{dom}\psi}(\hat{x})$ by Lemma 3.2. Therefore, for any $w \in T_{\text{dom}\psi}(\hat{x})$, we have

$$\begin{aligned}
&\text{cl}\psi'(\hat{x}; w) \\
&= \liminf_{w' \to w} \psi'(\hat{x}; w') \\
&= \liminf_{\substack{w' \to w \\ w' \in F_{\text{dom}\psi}(\hat{x})}} \psi'(\hat{x}; w') \\
&= \liminf_{\substack{w' \to w \\ w' \in F_{\text{dom}\psi}(\hat{x})}} \pi(w), \quad (3.4)
\end{aligned}$$

where $\pi(w) = \lim_{\delta \nearrow \psi(\hat{x})} \sup\{\langle \nabla_x \phi(\hat{x}, y), w \rangle \mid y \in \theta(\hat{x}, \delta)\}$. It is easy to see that π is convex. We now show that π is lower semi-continuous. This is equivalent to showing that the level set $\text{lev}_{\leq \alpha} \pi = \{w \mid \pi(w) \leq \alpha\}$ is closed for every $\alpha \in R$; that is, for any given sequence $\{w_n\}$ in $\text{lev}_{\leq \alpha} \pi$ with limit $\hat{w}$, we need to show that $\hat{w} \in \text{lev}_{\leq \alpha} \pi$. Since $w_n \in \text{lev}_{\leq \alpha} \pi$ for all n, we have

$$\lim_{\delta \nearrow \psi(\hat{x})} \sup\{\langle \nabla_x \phi(\hat{x}, y), w_n \rangle \mid y \in \theta(\hat{x}, \delta)\} \leq \alpha. \quad (3.5)$$

Noting that $\theta(\hat{x}, \delta)$ is monotonically contractive when δ is monotonically increasing, Assumption 3.1 implies that there exist $M > 0$ and $\delta_0 < \psi(\hat{x})$ such that

$$\sup\{\|\nabla_x \phi(\hat{x}, y)\| \mid y \in \theta(\hat{x}, \delta)\} \leq M, \quad \forall \delta \in (\delta_0, \psi(\hat{x})).$$

Consequently,

$$\begin{aligned}
&\sup\{\langle \nabla_x \phi(\hat{x}, y), \hat{w} \rangle \mid y \in \theta(\hat{x}, \delta)\} \\
&= \sup\{\langle \nabla_x \phi(\hat{x}, y), \hat{w} - w_n + w_n \rangle \mid y \in \theta(\hat{x}, \delta)\} \\
&\leq \sup\{\langle \nabla_x \phi(\hat{x}, y), \hat{w} - w_n \rangle \mid y \in \theta(\hat{x}, \delta)\} \\
&\quad + \sup\{\langle \nabla_x \phi(\hat{x}, y), w_n \rangle \mid y \in \theta(\hat{x}, \delta)\} \\
&\leq \sup\{\langle \nabla_x \phi(\hat{x}, y), w_n \rangle \mid y \in \theta(\hat{x}, \delta)\} + M\|\hat{w} - w_n\|.
\end{aligned}$$

Taking the limit in the above inequality as δ approaches $\psi(\hat{x})$, it follows from (3.5) that

$$\pi(\hat{w}) \leq M\|\hat{w} - w_n\| + \alpha,$$

from which and the fact $w_n \to \hat{w}$ we get $\pi(\hat{w}) \leq \alpha$, that is, $\hat{w} \in \text{lev}_{\leq \alpha} \pi$. The lower semicontinuity of π and (3.4)

imply that

$$\begin{aligned}
d\psi(\hat{x})(w) &= \liminf_{\substack{w' \to w \\ w' \in F_{\text{dom}\psi}(\hat{x})}} \pi(w') \\
&\geq \liminf_{w' \to w} \pi(w') \\
&= \pi(w). \quad (3.6)
\end{aligned}$$

To obtain the reverse inequality, choose some $u \in \text{ri}F_{\text{dom}\psi}(\hat{x}) = \text{ri}T_{\text{dom}\psi}(\hat{x})$. If $\pi(w) = +\infty$, then $d\psi(\hat{x})(w) = \pi(w) = +\infty$. If $\pi(w) < +\infty$, using the convexity of π and the fact that $\lambda u + (1 - \lambda)w \in \text{ri}F_{\text{dom}\psi}(\hat{x})$ for each $\lambda \in (0, 1)$ (see [5], Theorem 6.1), we have

$$\begin{aligned}
d\psi(\hat{x})(w) &= \liminf_{\substack{w' \to w \\ w' \in F_{\text{dom}\psi}(\hat{x})}} \pi(w') \\
&\leq \liminf_{\lambda \searrow 0} \pi(\lambda u + (1 - \lambda)w) \\
&\leq \pi(w).
\end{aligned}$$

The above inequality, together with (3.6), implies that $d\psi(\hat{x})(w) = \pi(w)$ and the result then follows.

Theorem 3.5 below shows that the subdifferential $\partial\psi(\hat{x})$ defined in (2.1) coincides with the usual subdifferential from convex analysis (see [5], Section 23).

Theorem 3.5 *Suppose that ϕ satisfies Assumption* 3.1. *If $\hat{x} \in \text{dom}\psi$, then*

$$\begin{aligned}
\partial\psi(\hat{x}) &= \{v \in R^n \mid \langle v, w \rangle \leq d\psi(\hat{x})(w), w \in R^n\} \\
&= \{v \in R^n \mid \langle v, w \rangle \leq \psi'(\hat{x}; w), w \in R^n\}.
\end{aligned}$$

Proof. Let's denote

$$A = \{v \in R^n \mid \langle v, w \rangle \leq d\psi(\hat{x})(w), w \in R^n\}$$

and

$$B = \{v \in R^n \mid \langle v, w \rangle \leq \psi'(\hat{x}; w), w \in R^n\}.$$

Lemma 3.3 implies that $d\psi(\hat{x})(w) = \liminf_{w' \to w} \psi'(\hat{x}; w')$ for all $w \in R^n$, which in turn implies that $d\psi(\hat{x})(w) \leq \psi'(\hat{x}; w)$ for all $w \in R^n$. Hence, $A \subseteq B$.

To show the converse, let $v \in B$. For any fixed $w \in R^n$, we have

$$\begin{aligned}
\langle v, w \rangle &= \lim_{w' \to w} \langle v, w' \rangle \\
&\leq \liminf_{w' \to w} \psi'(\hat{x}, w') \\
&= d\psi(\hat{x})(w).
\end{aligned}$$

This means that $v \in A$. Hence, $B \subseteq A$ and the result follows.

In the following, we study the first-order optimality conditions for the convex semi-infinite inequality constrained programming(CSICP) by using the characterizations of the objective function and constraint functions.

Let the feasible region of (CSICP) denote by X_I, that is,

$$X_I = \{x \in R^n | \psi(x) \le 0\},$$

where $\psi(x) = \max_{j \in J} \psi^j(x)$. Given any $\hat{x} \in X_I$, consider the following unconstrained problem:

(P) $$\min_{x \in R^n} F(x),$$

where $F(x) = \max\{\psi^0(x) - \psi^0(\hat{x}), \psi(x)\}$. In what follows, we shall study the relationship between the local minimum of (P) and (CSICP). Without loss of generality, we can assume that the set $X_I \cap \text{dom}(\psi^0)$ is nonempty. In addition, the following assumption is needed.

Assumption 3.6 *The set $\hat{Y}^j(x)$ is nonempty for all $j \in \bar{J}$ and $x \in X_I \cap \text{dom}(\psi^0)$, where $\hat{Y}^j(x) = \{y \in Y_j | \phi^j(x, y) = \psi^j(x)\}$.*

Theorem 3.7 *Considering the problems (P) and (CSICP), the following statements are true.*

1. *If $\hat{x}$ is a local minimum of (CSICP), then $\hat{x}$ is a local minimum of (P).*

2. *Suppose that the Assumptions 3.1 and 3.6 hold. If $\hat{x}$ is a local minimum of (P) and $0 \notin \partial\psi(x)$ for any x with $\psi(x) = 0$, then $\hat{x}$ is a local minimum of (CSICP).*

Proof. The feasibility of $\hat{x}$ means that $\psi(\hat{x}) \le 0$, and hence,

$$F(\hat{x}) = 0. \tag{3.7}$$

Furthermore, the local optimality of $\hat{x}$ implies the existence of $\rho > 0$ such that $\psi^0(x) \ge \psi^0(\hat{x})$ for all $x \in B(\hat{x}, \rho) \cap X_I$. For any given $x \in B(\hat{x}, \rho)$, consider the following two cases.

Case 1. If $x \in X_I$, then $\psi(x) \le 0$, and hence,

$$\begin{aligned} F(x) &= \max\{\psi^0(x) - \psi^0(\hat{x}), \psi(x)\} \\ &= \psi^0(x) - \psi^0(\hat{x}) \\ &\ge 0. \end{aligned} \tag{3.8}$$

Case 2. If $x \notin X_I$, then $\psi(x) > 0$, and hence,

$$\begin{aligned} F(x) &= \max\{\psi^0(x) - \psi^0(\hat{x}), \psi(x)\} \\ &\ge \psi(x) \\ &> 0. \end{aligned} \tag{3.9}$$

Putting the facts (3.7)-(3.9) together, we obtain the local optimality of $\hat{x}$ for (P).

To address the second part, we assume, without loss of generality, that $X_I \subseteq \text{dom}(\psi^0)$. Since $\hat{x}$ is a local minimum of (P), there exists $\rho > 0$ such that

$$F(x) \ge F(\hat{x}) = 0 \quad \text{for all } x \in B(\hat{x}, \rho). \tag{3.10}$$

To complete the proof, we need to show the existence of $\hat{\rho} > 0$ such that $\psi^0(x) \ge \psi^0(\hat{x})$ for all $x \in B(\hat{x}, \hat{\rho}) \cap X_I$. For any given $x \in B(\hat{x}, \hat{\rho}) \cap X_I$, the following two cases are considered here. First, if $\psi(x) < 0$, then the nonnegativity of $F(x) = \max\{\psi^0(x) - \psi^0(\hat{x}), \psi(x)\}$ implies that $\psi^0(x) \ge \psi^0(\hat{x})$. Second, if $\psi(x) = 0$, then $0 \notin \partial\psi(x)$ by hypothesis. According to the relationship between the subdifferential and the directional derivative established in Theorem 3.5, there exists a vector $\omega \in R^n$ such that $\psi'(x, \omega) < 0$, that is,

$$\psi'(x, \omega) = \lim_{\lambda \downarrow 0} \frac{\psi(x + \lambda w) - \psi(x)}{\lambda} = \lim_{\lambda \downarrow 0} \frac{\psi(x + \lambda w)}{\lambda} < 0,$$

where the directional derivative is well-defined due to the convexity of the function ψ^j. Therefore, there exists a sequence $x_i = x + \lambda_i w \in B(\hat{x}, \hat{\rho})$ converging to x such that $\psi(x_i) < 0$, which in turn implies $x_i \in X_I \subseteq \text{dom}(\psi^0)$. The feasibility of x_i and the nonnegativity of $F(x_i)$ imply that $\psi^0(x_i) \ge \psi^0(\hat{x})$. Taking the limit in both sides and using the continuity of ψ^0, we obtain $\psi^0(x) \ge \psi^0(\hat{x})$. This completes the proof.

An interesting results is that some assumptions presented in the foregoing theorem can be dropped when $\hat{x}$ is a strict local minimum of (P). This is formalized by the following theorem.

Theorem 3.8 *If $\hat{x} \in X_I$, then $\hat{x}$ is a strict local minimum of (CSICP) if and only if $\hat{x}$ is a strict local minimum of (P).*

Proof. The necessity can be proved by following almost the same argument as in Theorem 3.7. We now show the sufficiency. The strict local optimality of $\hat{x}$ implies the existence of $\rho > 0$ such that

$$F(x) > F(\hat{x}) \quad \text{for all } x \in B(\hat{x}, \rho) \text{ with } x \ne \hat{x}. \tag{3.11}$$

Noting that $F(\hat{x}) = 0$ and using (3.11), we obtain that $F(x) = \max\{\psi^0(x) - \psi^0(\hat{x}), \psi(x)\} > 0$, which, together with the fact that $\psi(x) \le 0$ for any $x \in X_I$, implies that $\psi^0(x) > \psi^0(\hat{x})$. This completes the proof.

By closely following the proof of Theorem 3.4, we can obtain the following results, which characterizes the subderivative $dF(\hat{x})$.

Theorem 3.9 *Suppose that each ϕ^j for $j \in \bar{J}$ satisfies Assumption 3.1 and $\hat{x} \in X_I \cap \text{dom}\psi^0$. Then the following results hold:*
(i) if $w \in T_{\text{dom}F}(\hat{x})$, then

$$dF(\hat{x})(w) = \max\{d\psi^j(\hat{x}; w) | j \in \{0\} \cup \hat{J}(\hat{x})\};$$

(ii) if $w \notin T_{\text{dom}F}(\hat{x})$, then

$$dF(\hat{x})(w) = +\infty,$$

where for any $j \in \bar{J}$,

$$d\psi^j(\hat{x})(w) = \lim_{\delta \nearrow \psi^j(\hat{x})} \sup\{\langle \nabla_x \phi^j(\hat{x}, y), w \rangle | \, y \in \theta^j(\hat{x}, \delta_j)\},$$

$$\theta^j(\hat{x}, \delta_j) = \{y \in Y_j | \, \phi^j(\hat{x}, y) \geq \delta_j\}$$

and

$$\hat{J}(\hat{x}) = \{j \in J | \, \psi^j(\hat{x}) = F(\hat{x})\}.$$

With these preparations, the optimality condition for (CSICP) can be stated.

Theorem 3.10 *Suppose that each ϕ^j for $j \in J$ satisfies Assumption 3.1. If $\hat{x}$ is a local minimum of (CSICP), then*

$$dF(\hat{x})(w) \geq 0.$$

Furthermore, if the above inequality is strict except at $w = 0$, that is,

$$dF(\hat{x})(w) > 0 \quad \text{for all } w \in R^n \text{ with } w \neq 0, \quad (3.12)$$

then $\hat{x}$ is a strict local minimum of (CSICP).

Proof. In view of Theorem 3.7, the local minimum of (CSICP) is also the local minimum of (P). The first result follows immediately from the definition of the subderivative $dF(\hat{x})(w)$.

To show the second result, suppose, on the contrary, that $\hat{x}$ is not a strict local minimizer. In other words, there must exist a sequence $\{x_k\}_{k=1}^{\infty} \subseteq X_I \cap \operatorname{dom}\psi^0$ such that $x_k \to \hat{x}$ as $k \to \infty$, and

$$\psi^0(x_k) - \psi^0(\hat{x}) \leq 0 \quad \text{for all } k \in N. \quad (3.13)$$

For any given $j \in \hat{J}(\hat{x})$, noting that $\psi^j(\hat{x}) = F(\hat{x}) = 0$, we have

$$\psi^j(x_k) - \psi^j(\hat{x}) \leq 0. \quad (3.14)$$

Let $h_k = \frac{x_k - \hat{x}}{\|x_k - \hat{x}\|}$. Clearly, $\{h_k\}$ is bounded, and hence, by passing to a subsequence if necessary, we can assume that $\{h_k\}$ is convergent and the limit point is h_0 with $\|h_0\| = 1$. Note that

$$x_k = \hat{x} + \|x_k - \hat{x}\| \frac{x_k - \hat{x}}{\|x_k - \hat{x}\|} \in X_I \cap \operatorname{dom}\psi^0. \quad (3.15)$$

However, by the definition of F, we have $X_I \cap \operatorname{dom}\psi^0 \subseteq \operatorname{dom}(F)$. Thus, it follows this fact and (3.15) that $h_0 \in T_{\operatorname{dom}(F)}(\hat{x})$. Combining (3.13) and (3.14) yields that for all $j \in \{0\} \cup \hat{J}(\hat{x})$,

$$\frac{\psi^j(\hat{x} + \|x_k - \hat{x}\| \frac{x_k - \hat{x}}{\|x_k - \hat{x}\|}) - \psi^j(\hat{x})}{\|x_k - \hat{x}\|} \leq 0,$$

which further implies that

$$
\begin{aligned}
& d\psi^j(\hat{x})(h_0) \\
= & \liminf_{\substack{u \to h_0 \\ \lambda \downarrow 0}} \frac{\psi^j(\hat{x} + \lambda u) - \psi^j(\hat{x})}{\lambda} \\
\leq & \liminf_{k \to \infty} \frac{\psi^j(\hat{x} + \|x_k - \hat{x}\| \frac{x_k - \hat{x}}{\|x_k - \hat{x}\|}) - \psi^j(\hat{x})}{\|x_k - \hat{x}\|} \\
\leq & 0.
\end{aligned}
$$

According to Theorem 3.9, we have

$$dF(\hat{x})(h_0) = \max\{d\psi^j(\hat{x})(h_0)|j \in \{0\} \cup \hat{J}(\hat{x})\} \leq 0,$$

which contradicts (3.12). This completes the proof.

References

[1] M.J.Cánovas, A.L.Dontchev, M.A.López and J.Parra, *Metric regularity of semi-infinite constraint systems*, Math. Programming, **104** (2005), 329–346.

[2] M.A.Goberna and M.A.López, Semi-infinite Programming-Recent Advances, Kluwer Academic Publishers,Boston, 2001.

[3] R.Hettich and K.O.Kortanek, *Semi-infinite programming: theory, methods, and applications.* SIAM Review, **35** (1993), 380–429.

[4] E.Polak, Optimization: Algorithms and Consistent Approximation, Springer-Verlag, New York, 1997.

[5] R.T.Rockafellar, Convex Analysis, Princeton University Press, Princeton, 1970.

[6] R.T.Rockafellar and R.J.Wets, Variational Analysis, Springer, New York, 1998.

[7] A.Shapiro, *On duality theory of convex semi-infinite programming*, Optimization, **54** (2005), 535–543.

[8] O.Stein, *First-order optimality conditions for degenerate index sets in generalized semi-infinite optimization*, Math. Oper. Res., **26** (2001), 565–582.

[9] F.G.Vázquez and J.-J.Rückmann, *Extensions of the Kuhn-Tucker constraint qualification to generalized semi-infinite programming*, SIAM J. Optim., **15** (2005), 926–937.

[10] C.Y.Wang, X.Q.Yang and X.M.Yang, *Optimal value functions of generalized semi-infinite min-max programming on a noncompact set*, Sci. China Ser. A, **48** (2005), 261–276.

[11] X.Y.Zheng and X.Q.Yang, *Lagrange multipliers in nonsmooth semi-infinite optimization problems*, Math. Oper. Res., **32** (2007), 168–181.

Study on Synthesis Control of Power Quality for Electrified Railway

Xiangzheng Xu[1,2]
1.Wuhan University, Wuhan, China
2. East China Jiao Tong University, Chian
ecjtuxxz@sina.com

Baichao Chen
Dept. of Electrical Engineering Wuhan
University, Wuhan, China
xzxu@4y.com.cn

Abstract

This paper analyzes necessity of synthesis control of power quality aiming at characteristics of traction power supply system for electrified railway. In order to achieve real-time detection and compensation for power quality of electric railway traction power system, this paper provides a detection method of selective harmonic current, so vertiginous harmonic current of traction load is detected. Synchronously vertiginous passive power is gained basing on hilbert digital phase-shifting filter. And imbalance load current is balanced through regulating magnetron static VAR compensator (SVC) device. The kind new injection-type hybrid active power filter (HAPF) combined with magnetron SVC make to compensate reactive power effectively, eliminate harmonic drastically and balance load imbalance current availably, and improve reliability and security of electrified railway greatly.

1. Introduction

With rapidly economic development, these problems such as population growth continuously, environmental pollution, energy shortages, inadequate transportation capacity, etc, are becoming increasingly serious. And it has constrained the sustainable development of society. Electrified railway has big transport capacity, high-volume, speediness, energy-saving, environment-friendly, and many other advantages, which will replace the traditional internal combustion engine traction. Electrified railway system as a special power user, its traction load has the four features (non-linear, non-sinusoidal and non-symmetry, non-continuity), and produce reactive power, harmonics and negative sequence current and otherwise electricity hazards [1]. These electricity hazards will badly affect the power grid and various electrical equipments, and not only endanger other user safety of the public power grid, but also endanger reliable operation of electrified railway. As a result, the comprehensive management of power quality for traction power supply system, to improve the traction power

substation reliability and safe operation of the electrified railway is of great significance.

In order to achieve real-time detection and compensation of electrified railway power quality, the inspection method of designated degree harmonic current is adopted and the rapidly changing specific degree harmonic current composition of traction load is detected. The rapidly changing active power and reactive power ingredients are detected based on hilbert digital phase-shifting filter. Through adjusting magnetron SVC device the imbalance load is balanced. This new injection-type hybrid active power filter combined with magnetron SVC make to compensate effectively passive power, eliminate drastically harmonic and improve load imbalance current of electrified railway.

2. Harmonic current detection method

Harmonic current detection method is the means which all harmonic current will be the sum total harmonic current i_{ah} after parallel computing every harmonic current, namely, it is compensated current command signal. PWM pulse signal is produced from current controller, when instruction signal of compensating current is made, and to control the inverter produce a harmonic current with equal magnitude and opposite polarity of actual harmonic current. The current in the power grids contains only fundamental component. The compensated current is compared with the actual compensation current to form a closed-loop control of the current track.

As to detecting harmonic current of three-phase circuit, the three-phase signal should be changed into mutually vertical α and β signal of two-phase [2]. And then its rotation transformation can be taken. Method mentioned above is applied in the single phase system. Auxiliary current with delayed 90° to fundamental current is formed. Take detection of the 5th harmonic current for example, detecting principle is showed in Figure 1.

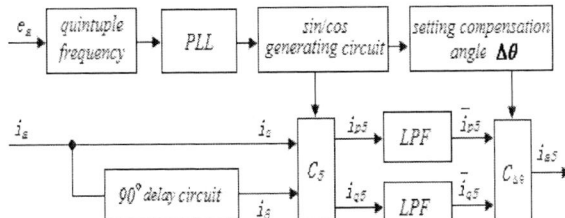

Figure 1. **Detecting principle of single-phase 5th harmonic**

3. Reactive power detection method

Based on hilbert digital phase-shifting filter the reactive power measurement principle diagram is shown in Figure 2.

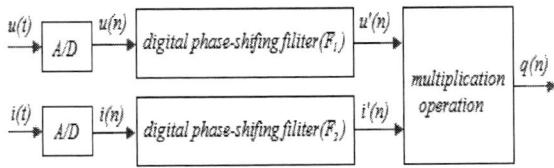

Figure 2. **Reactive power detection principle diagram**

Firstly, a certain phase voltage and current signals of the power grids are implemented A / D conversion at the same sampling rate, and discrete digital voltage signal $u(n)$ and digital current signal $i(n)$ are gained. Secondly, a pair of above digital signals is carried out phase shifter processing by F_1 and F_2 filter respectively. And a pair of complex digital signal $u'(n)$ and $i'(n)$ are made, the phase angle is 90° between the fundamental wave of the voltage and current and harmonic components. Then the two complex signals are multiplied to make instantaneous reactive power $q(n)$.Finally, the reactive power Q is made after $q(n)$ is filtered.

4. Balancing asymmetric load

Any non-grounding Y connected load can be expressed as Δ connected load through Y-Δ transformation. Ideal load compensator is substituted using an arbitrary three-phase passive admittance. When the passive admittance is paralleled with the load, it is equal to a real symmetric load toward power supply. The passive admittance is combined with power factor correction to form a ideal compensation network with Δ connected fashion, which each branch of triangle has three parallel compensated susceptance [3].

$$B_r^{ab} = -B_1^{ab} + (G_1^{ca} - G_1^{bc})/\sqrt{3}$$
$$B_r^{bc} = -B_1^{bc} + (G_1^{ab} - G_1^{ca})/\sqrt{3} \qquad (1)$$
$$B_r^{ca} = -B_1^{ca} + (G_1^{bc} - G_1^{ab})/\sqrt{3}$$

A cycle of P and Q are easily gained using instantaneous reactive power theory [4]. Virtual value of lines voltage can be obtained through the voltage sensor. Therefore, the compensated susceptance can be calculated, and mutative load can be fleetly conducted real-time tracking, and the compensation device capacity is rapidly adjusted, three-phase imbalance voltage and current are improved by compensated voltage and current. Because of the special nature of electrified railway, traction arm is usually a phase and b phase. There is load between a phase and c phase or b phase and c phase, thus, there is nothing between a phase and b phase. $P_{ab} = Q_{ab} = 0$. Q is considered the inductive reactive power. The compensation capacity of the device is set as follows:

$$Q_r^{ab} = -(\frac{U_{ab}^2}{U_{ca}^2}P_{ca} - \frac{U_{ab}^2}{U_{bc}^2}P_{bc})/\sqrt{3}$$

$$Q_r^{bc} = -Q_{bc} + \frac{U_{bc}^2}{U_{ca}^2}P_{ca}/\sqrt{3} \qquad (2)$$

$$Q_r^{ca} = -Q_{ca} - \frac{U_{ca}^2}{U_{bc}^2}P_{bc}/\sqrt{3}$$

From Equation (1) and (2), it can be found that the compensation capacity of the device is impacted by active power and reactive power between any two phases, that is, the compensation capacity may be a positive or a negative. If a group magnetron reactor and fixed capacitors (magnetron SVC) are shunted between every two-phase (a phase and b phase, b phase and c phase, c phase and a phase), the compensated capacity is dynamically adjusted with mutative load within the limits of requisite capacity. Each of the compensation devices can be produced capacitive reactive power or inductive reactive power, thereby, the reactive power of power network is compensated, and the power factor is increased, synchronously the asymmetry load of the three-phase is greatly improved.

5. Hybrid active power filter

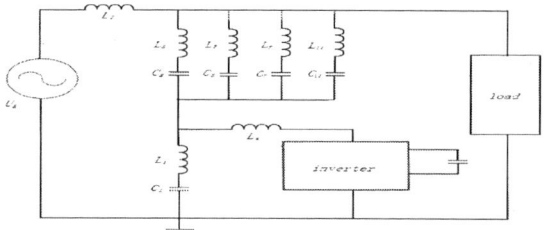

Figure 3. **Hybrid active power filter principle diagram**

In Figure 3, branch L_1-C_1 produces series resonance in the fundamental wave frequency. In the same way, branch L_3-C_3, branch L_5-C_5, branch L_7-C_7 and branch L_{11}-C_{11} produce series resonance respectively in 3[th]and 5[th]and 7th and 11th harmonic frequencies [5]. The hybrid active power filter make use of resonant characteristics of LC circuit to structure injection loop, thus inverter almost does not bear the fundamental voltage, and the capacity of the inverter is greatly reduced. Fundamental wave branch produces resonance in the fundamental frequency, the branch is similar to a short circuit, and LC filter can compensate reactive power. The APF and fundamental resonant branch are in series, the series branch is equal to impedance in harmonic frequencies, which is in series with power supply. At this point, load harmonic current is injected passive power filter circuit. Filtering effect of passive filter is greatly improved.

6. Comprehensive compensation scheme

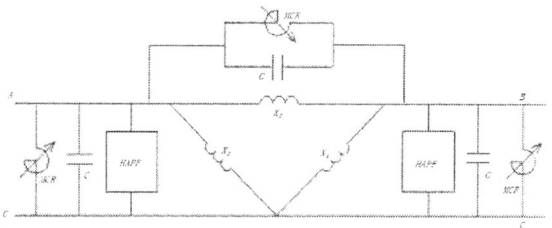

Figure 4. **The functional block diagram of Comprehensive compensation**

The kind new injection-type hybrid active power filter (HAPF) combined with magnetron SVC make to compensate reactive power effectively, eliminate harmonic drastically and improve load imbalance current. Principle Circuit is shown on Figure 4. Magnetron SVC can compensate reactive power which is composed of magnetron reactor (MCR) and fixed capacitor (FC). Take into account each group reactor capacity configuration of magnetron SVC is different, and single-phase MCR is now widely used in electrified railway reactive power compensation system because of its technology reliability, accordingly three single-phase MCR are adopted in device of magnetron SVC compensation device.

Magnetron SVC depends on the control angle regulation of the magnetron reactor, reactor capacity is smoothly adjusted through changing control angle of thyristor. According to survey results the compensated capacity of compensation device is real-time calculated by different configurations of magnetron SVC between a phase and b phase or b phase and c phase or c phase and a phase. Changing trigger angle of thyristor, when reactive power is compensated, load imbalance current is improved synchronously. Load harmonic of electrified railway is suppressed by hybrid active power filter. Capability and cost of HAPF are reduced consumedly, and the reliability of the system is improved. So from the reliability, economic, technical point of view the comprehensive scheme is the best compensated scheme.

7. Conclusions

Comprehensive compensation scheme make use of the advantages of MCR and HAPF. Reactor capacity is smoothly adjusted through changing control angle of thyristor. When reactive power is compensated, synchronously load imbalance current is eliminated. The voltage of APF bearing is reduced through resonant principle, the capacity and cost of equipment are reduced, and work efficiency and reliability are improved distinctly. Active power filter (APF) is used to remedy filtering effects of passive power filter (PPF), and absorb the remaining harmonic, and prevent PPF from overload. Practice has proved that the scheme can not only suppress harmonics, but also compensate reactive power and negative sequence current. Compensation effect is prominent, and operation is safe. The traction power substation reliability and safe operation of electrified railway are improved greatly.

References

[1] Chao Zhang, "An algorithm for random harmonic current detection of active power filter", *Electric machines and control,* Vol.6, 2003, pp.252-255.
[2] Han Xue-jun, Wang Jing-hao, "Research on integrated compensation of three phase unbalance load", *Power system technology,* Vol.30, 2006, pp.289-291.
[3] Hong Zheng, and Zhen-qing Yi, "Harmonic analysis and reckoning of traction transformer station for electrified railways", *Journal of jiangsu university (Natural Science Edition)* , Vol.24, 2006, pp.76-70.
[4] Yue Wang, and Jun Yang, and Zhao-an Wang, "Study on hybrid power filter used in electrified railway system", *Proceedings of the CSEE,* Vol.23, 2003, pp.23-27.
[5] Zhang Ya di, "Study of compensating three phase imbalance of electrified railway power based on the mcr with two-stage iron core," *Electric drive,* Vol.37, 2007, pp.52-55.

2008 Workshop on Power Electronics and Intelligent Transportation System

Delay Synchronizing Rectification Control Strategy of Bi-directional Converter in a Novel Stand-alone PV System

Xinhua Zhang, Zhenyue Huang, Zhiling Liao

School of Electrical and Information Engineering, Jiangsu University

Zhenjiang, 212013, Jiangsu Province, China

E-mail: zxinh@ujs.edu.cn, jsdxlzl@ujs.edu.cn

Abstract

A novel stand-alone photovoltaic (PV) system is proposed, which exhibits the advantages of better protection and more efficient control on charge/discharge of the battery. Furthermore, it can realize energy management of the system. The key point of energy management is how to control the bi-directional converter effectively. Considering traditional soft start strategy would be ineffective for bi-directional converter in the double sources application for the possibility of damaging the power device, a new soft start control strategy named delay synchronizing rectification control strategy is proposed. A 500W prototype converter is built to verify the theoretical analysis and the control strategy.

1. Introduction

With the rapid depletion of the conventional fossil fuels, energy crisis and environmental pollution become more and more serious. The solar energy photovoltaic (PV) system will play an important role in alleviating energy crisis, reducing environmental pollution and greenhouse effect [1-4]. The PV power system is classified by the stand-alone PV system and the grid-connected PV system. In traditional stand-alone PV system the battery is directly connected to the DC bus and charge/discharge of

the battery is under no control, which may result in large charge/discharge current and shortening battery life [2, 5].

In this paper a novel stand-alone PV system is proposed, as shown in Fig.1. The uni-directional Buck DC-DC converter converts the wide-range voltage of solar cell (150-350VDC) into the steady voltage of 100VDC at DC bus. A 48V battery is in paralleled with the DC Bus via a bi-directional DC-DC converter, which can transmit energy in both directions. Buck-Boost bi-directional converter is selected in the system, for the merits of simple configuration, easy control and rapid response [5-7].

Soft start method can reduce input inrushing current of DC-DC converter at startup, which is widely used in uni-directional DC-DC converters. In the proposed stand-alone PV system in Fig.1, the bi-directional converter is at double sources application—DC bus at high voltage port (V_H) and battery at low voltage port (V_L), where traditional soft start method fails to serve. This paper will analyze the drawback of traditional soft start in double sources application and proposes a novel soft start strategy named delay synchronizing rectification control strategy. A 500W prototype converter is built to verify the control strategy.

2. The drawback of traditional soft start in double sources application

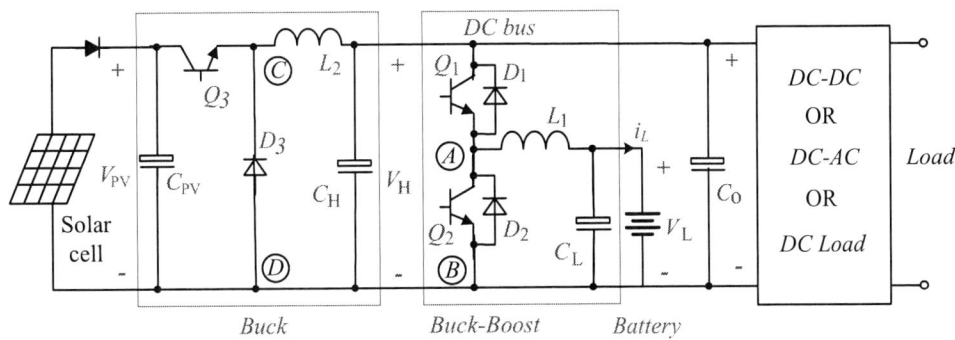

Fig.1 the proposed novel stand-alone PV system

978-0-7695-3342-1/08 $25.00 © 2008 IEEE

DOI 10.1109/PEITS.2008.57

113

Fig.2 shows the block diagram of Buck-Boost bi-directional DC-DC converter of double sources in the stand-alone PV system. V_H is at high-voltage port and V_L is at low-voltage port. Switches Q1 and Q2 conduct complementally. The bi-directional converter can work in Buck or Boost mode according to the operation state of the system. In Buck mode, Q1 is controlled to regulate the output voltage, therefore Q1 is defined as active switch, and Q2 is defined as passive switch. Likewise in Boost mode, Q2 is defined as active switch and Q1 is defined as passive switch. Passive switch is working under synchronizing rectification.

Traditional soft start method makes the active switch's duty cycle increase gradually, so output voltage increases slowly. At double sources application, if the bi-directional converter works in Buck mode, as shown in Fig.2(a), V_H supply energy to V_L. In traditional soft start, the duty cycle of Q1 is small and that of Q2 is large at startup. V_H-V_L is applied on inductor with short time per switch period while $-V_L$ with long time. The inductor current inversely increases rapidly, as shown in Fig.3(a). As a result, the inductor is saturated and devices are damaged by large current. Likewise, if the bi-directional converter works in Boost mode, as shown in Fig.2(b), V_L supply energy to V_H. In traditional soft start, the duty cycle of Q2 is small and that of Q1 is large at startup. V_L is applied on inductor with short time per switch period while $-(V_H-V_L)$ with long time. The inductor current inversely increases rapidly, as shown in Fig.3(b). As a result, the inductor is saturated and devices are damaged by large current. Therefore, traditional soft start strategy cannot be used in double sources application. Another new soft start control strategy should be employed.

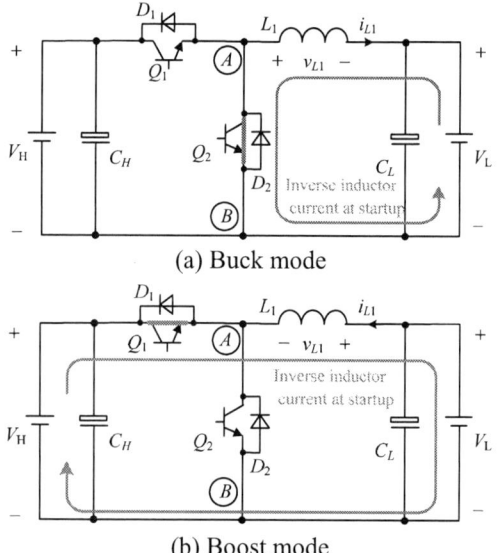

(a) Buck mode

(b) Boost mode

Fig.2 Buck-Boost bi-directional DC-DC converter of double sources in the stand-alone PV system

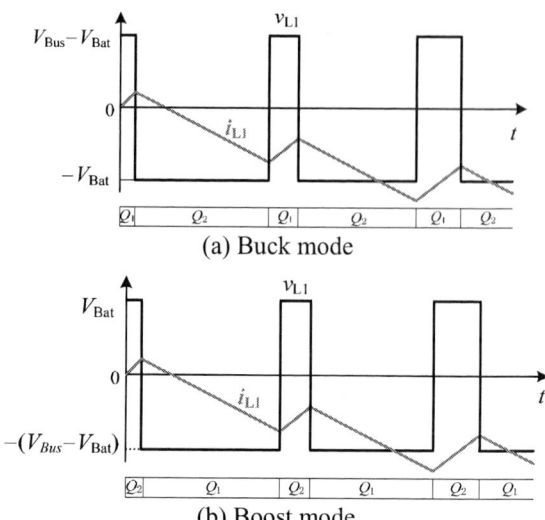

(a) Buck mode

(b) Boost mode

Fig.3 Inverse inductor current in traditional soft start

3. Delay synchronizing rectification control strategy

As analysis above, complemented operation of active switch and passive switch in Buck or Boost mode will produce some problems. If passive switch is shut and let inductor current i_{Lf} freewheel through its body diode during soft start, there will be no path for inverse current. However, in steady state, the efficiency of the converter will be reduced. Therefore, after steady state, passive switch is driven for synchronizing rectification, which is defined as delay synchronizing rectification control method in this paper. The control circuit block diagram is shown in Fig.4. After startup, V_{DD} charges capacitor C and the capacitor voltage V_C increases gradually. It is logically "OR" with Enable Signal V_E and its inverted one. Then Q_{E1} and Q_{E2} are obtained. Q_{E1} "AND" with PWM_1 and give out drive signal $Q_{1\text{-div}}$. Q_{E2} "AND" with PWM_2 and give out drive signal $Q_{2\text{-div}}$. Fig.5(a) shows the illustrative graphs of delay synchronizing rectification control method in Buck mode. At this time, Enable Signal $V_E=0$, PWM_1 is chosen and Buck mode is implemented. The time of passive drive delay t_d is determined by the time coefficient RC $= \tau$. After startup, when Vc increases to the threshold voltage of "OR" gate (half of V_{DD}), $Q_{2\text{-div}}$ is released to implement synchronizing rectification. Fig.5(b) shows the illustrative graphs of delay synchronizing rectification control method in Boost mode. At this time, Enable Signal $V_E =1$, PWM_2 is chosen and Boost mode is implemented. The time of passive drive delay t_d is also determined by the time coefficient RC$= \tau$. After startup, when Vc increases to the threshold

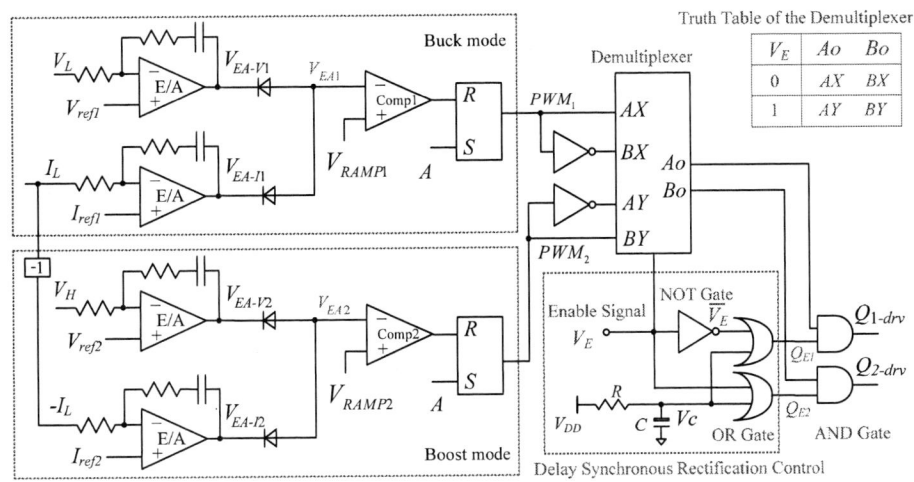

Fig.4 Control circuit of bi-directional converter

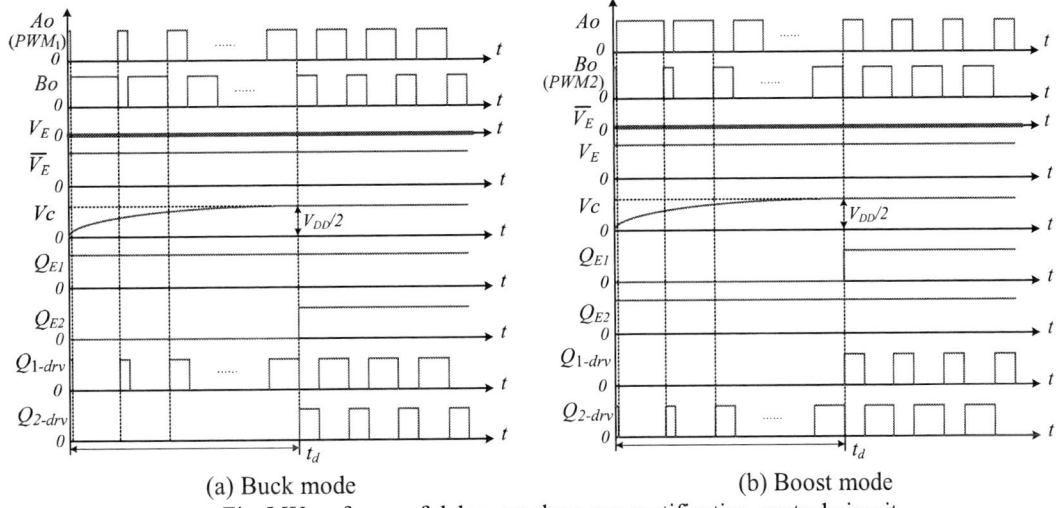

(a) Buck mode   (b) Boost mode

Fig.5 Waveforms of delay synchronous rectification control circuit

voltage of "OR" gate (half of V_{DD}), Q_{1-div} is released to implement synchronizing rectification.

When the bi-directional DC-DC converter works in fault mode during over voltage or over current, it is immediately shut down and the capacitor rapidly discharges. In this way, delay synchronizing rectification control method could be implemented each time during bi-directional DC-DC converter soft start.

4. Experimental results

A 500W prototype converter is built in our lab, whose parameter is as follows: VH=90~110VDC, VL=44~56VDC, P＝500W, fs＝100kHz. Fig.6(a) and Fig.6(b) show the experimental waveforms of delay synchronizing rectification in buck mode and boost mode, respectively. After the duty cycle of the active switch increases from zero to steady-state value, the passive switch is driven and then operates complementarily to active switch. From the experimental results it can be seen that the delay synchronizing rectification control strategy works well both in buck mode and in boost mode.

5. Conclusions

This paper analyzes the drawbacks of traditional soft start in double sources application of bi-directional converter in PV power system, and then proposes a novel soft start control strategy named delay synchronizing rectification control strategy. During soft start, the passive switch is shut and let the inductor current freewheel through its body diode, there will be no path for inverse current. After steady state, passive switch is driven for synchronizing rectification. A 500W prototype converter is built to verify the theoretical analysis and the effectiveness of the control strategy

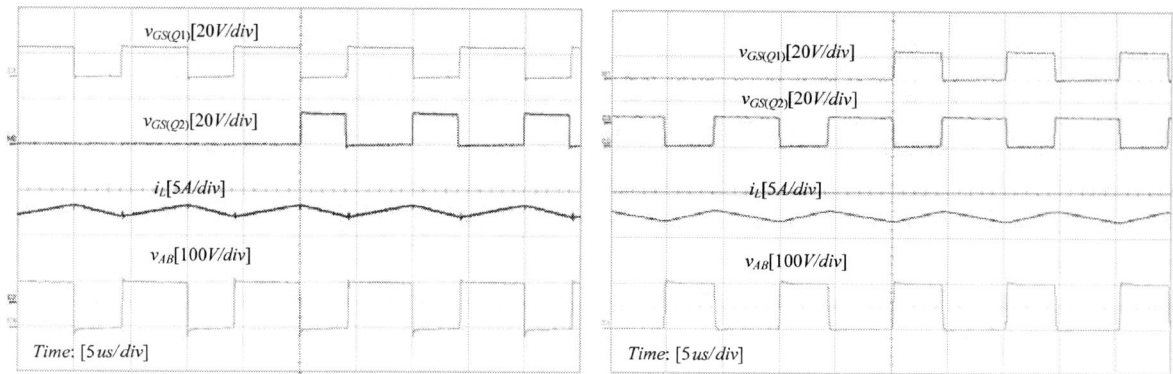

(a) Delay synchronous rectification control of buck   (b) Delay synchronous rectification control of boost

Fig.6 Experimental waveforms

References

[1] Park J H, Ahn J Y, Cho B H and Yu G J. "Dual-Module-Based Maximum Power Point Tracking Control of Photovoltaic Systems," IEEE Trans. Ind. Electron., Vol53, No.4, 2006, pp.1036-1047.

[2] Pacheco V M, Freitas L C, Vieira J B, Coelho E A A and Farias V J. "Stand-alone photovoltaic energy storage system with maximum power point tracking," In Proc. IEEE APEC, 2003, pp.97-102.

[3] Esram T, Kimball J W, Krein P T, Chapman P L and Midya P. "Dynamic Maximum Power Point Tracking of Photovoltaic Arrays Using Ripple Correlation Control," IEEE Trans. Power Electron., Vol21, No.5, 2006, pp.1282-1291.

[4] Koizumi H and Kurokawa K. "A Novel Maximum Power Point Tracking Method for PV Module Integrated Converter," In Proc. IEEE PESC, 2005, pp.2081-2086.

[5] Duryea S, Islam S and Lawrance W. "A battery management system for stand-alone photovoltaic energy systems," IEEE Industry Applications Magazine, Vol7, No.3, pp.67-72.

[6] D.M. Sable, F.C. Lee and B.H. Cho, "A zero-voltage-switching bidirectional battery charger/discharger for the NASA EOS satellite," in Proc. IEEE APEC, 1992, pp.614-621.

[7] J.N. Marie-Francoise, H. Gualous and A. Berthon, "DC to DC converter with neural network control for on-board electrical energy management," in Proc. IEEE IPEMC, 2004, Vol. 2, pp. 521-525.

2008 Workshop on Power Electronics and Intelligent Transportation System

Control Strategy of Three-Phase AC/DC Voltage-Source Converters Based on Storage function

Wang Jiuhe, Xia Peirong, Zhang Jinlong

School of Automation ,Beijing Information Science and Technology University, Beijing, China

wjhyhrwm@163.com

Abstract

In order to improve the properties of AC/DC converters, scholars in different countries have began to apply nonlinear control theory based on storage function to AC/DC converters. For this purpose, this paper introduces the study status of three-phase AC/DC Voltage-Source Converters based on storage function, which includes Lyapunov control theory, passivity control theory using Euler-Lagrange(EL) model and port controlled Hamiltonian with dissipation (PCHD)model, And proposes a new design method of passivity controller based on EL model of AC/DC converter. Meanwhile this paper analyzes nonlinear control strategies above.

1. Introduction

Three-phase AC/DC voltage-source converter has many advantages, such as sinusoidal current control and unity power factor (UPF) on the AC side, and DC output voltage control on the DC side, and reversible power flow between both sides etc.. Therefore, "green transformation" of electric energy is realized by AC/DC converter. The nonlinear control strategy of AC/DC converter based on storage function has many advantages such as fast response and good stable performances, high power factor and low input current total harmonic disturbance(THD), and simpler structure etc., which are interested by scholars in different countries[1]. In order to promote above strategies, the paper introduces the control principle of Lyapunov control and passivity control theory using EL model and PCHD model, and proposes a new design method of passivity controller based on EL model.

The paper is organized as follow. In Section 2 Math model in synchronous rotating *dq* coordinate is given. Section 3 introduces Lyapunov control strategy of PWM rectifier and passivity control strategy of PWM rectifier in Section 4. Finally, Section 5 analyzes above control strategies. Section 6 states our conclusions and points to further research.

2. Math model of AC/DC converter

The power circuit of Three-phase AC/DC voltage-source converter is shown in Fig.1. In order to set up Math model, it is assumed that the AC voltage is a balanced three-phase supply, the filter reactor is linear, IGBT is ideal switch and lossless. u_u, u_v and u_w are the

phase voltages of three-phase balanced voltage source and i_u, i_v, and i_w are phase currents in Fig.1. S_u, S_v, and S_w are the switching functions of AC/DC converter. S_j （ *j* =u、v、w） is defined as:

$$S_j = \begin{cases} 1, & S_j \text{ closed} \\ 0, & \bar{S}_j \text{ closed} \end{cases} \quad (1)$$

u_{DC} is the DC output voltage, R and L mean resistance and inductance of filter reactor, respectively, C is the DC side capacitance, R_L is the DC side load, u_{ru}, u_{rv} and u_{rw} are the input voltages of rectifier, and i_L is load current.

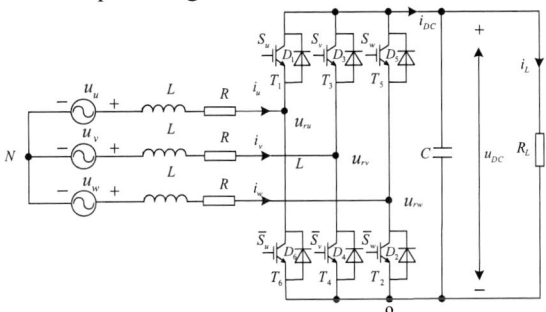

Fig.1 Power circuit of Three-phase AC/DC voltage-source converter

From Fig.1, the math model of three-phase AC/DC voltage-source converter can be written as:

$$\begin{cases} L\dfrac{di_u}{dt} = u_u - Ri_u - \left(\dfrac{2S_u - S_v - S_w}{3}\right)u_{DC} \\ L\dfrac{di_v}{dt} = u_v - Ri_v - \left(\dfrac{2S_v - S_w - S_u}{3}\right)u_{DC} \\ L\dfrac{di_w}{dt} = u_w - Ri_w - \left(\dfrac{2S_w - S_u - S_v}{3}\right)u_{DC} \\ C\dfrac{du_{DC}}{dt} = i_u S_u + i_v S_v + i_w S_w - i_L \end{cases} \quad (2)$$

According to (2), Math model in synchronous rotating *dq* coordinate is given as follows:

$$\begin{cases} L\dfrac{di_d}{dt} = \omega L i_q - Ri_d - S_d u_{DC} + u_d \\ L\dfrac{di_q}{dt} = -\omega L i_d - Ri_q - S_q u_{DC} + u_q \\ C\dfrac{du_{DC}}{dt} = \dfrac{3}{2}(S_d i_d + S_q i_q) - i_L \end{cases} \quad (3)$$

where ω is angular frequency of AC voltage.

3. Lyapunov control strategy

3.1. Determining the equilibrium point

978-0-7695-3342-1/08 $25.00 © 2008 IEEE
DOI 10.1109/PEITS.2008.59

For balanced three-phase supply, $u_d = U_m, u_q = 0$, U_m is the amplitude of the AC phase voltage. In order to realize unity power factor(UPF) and DC output voltage control, let that the desired equilibrium values are $u_{DC}^* = u_{DCR}$, $i_d^* = I_m$ and $i_q^* = 0$, $S_d = S_{do}$, and $S_q = S_{qo}$. S_{do} and S_{qo} represent the steady-state values of the switching functions S_d and S_q, respectively. The amplitude I_m of the AC phase current, S_{do} and S_{qo} in steady operation can be written as:

$$I_m = \frac{1}{2}\left[\frac{U_m}{R} - \sqrt{\left(\frac{U_m}{R}\right)^2 - \frac{8u_{DCR}^2}{3RR_L}}\right] \quad (4)$$

and

$$\begin{cases} S_{do} = \dfrac{U_m - RI_m}{u_{DCR}} \\[3mm] S_{qo} = -\dfrac{\omega LI_m}{u_{DCR}} \end{cases} \quad (5)$$

3.2. Controller Design

Based on[2], Let us define a positive definite Lyapunov function candidate as:

$$V(x) = \frac{1}{2}Lx_1^2 + \frac{1}{2}Lx_2^2 + \frac{1}{3}Cx_3^2 \quad (6)$$

where $x_1 = i_d - I_m, x_2 = i_q, x_3 = u_{DC} - u_{DCR}$.

To guarantee that $\dot{V}(x)$ is negative definite, controller expressed by S_d and S_q is given by

$$\begin{cases} S_d = S_{do} + \Delta S_d \\ S_q = S_{qo} + \Delta S_q \end{cases} \quad (7)$$

where

$$\Delta S_d = \begin{cases} -(S_{do} + S_{dom}) & \alpha(u_{DCR}x_1 - I_mx_3) < -(S_{do} + S_{dom}) \\ \gamma(u_{DCR}x_1 - I_mx_3) & -(S_{do} + S_{dom}) \le \alpha(u_{DCR}x_1 - I_mx_3) \le S_{dom} - S_{do} \\ S_{dom} - S_{do} & (u_{DCR}x_1 - I_mx_3) > S_{dom} - S_{do} \end{cases},$$

$$\Delta S_q = \begin{cases} -(S_{qom} + S_{qo}) & \beta x_2 < -(S_{qom} + S_{qo}) \\ \beta x_2 & S_{qom} - S_{qo} \le \beta x_2 \le -(S_{qom} + S_{qo}) \\ S_{qom} - S_{qo} & \beta x_2 > S_{qom} - S_{qo} \end{cases},$$

$$S_{dom} = \sqrt{\frac{1}{3} - S_{qom}^2},$$

S_{qom} represents the possible maximum steady-state value that S_q can take for maximum i_L (assuming that u_{DC} is kept constant). α and β are arbitrary real constants.

4. passivity control strategy
4.1. passivity control strategy based on EL model

4.1.1. EL model of AC/DC converter.
We can transform (3) into (8) which expressed by matrix [3,4]as follow:

$$\begin{bmatrix} L & 0 & 0 \\ 0 & L & 0 \\ 0 & 0 & \frac{2}{3}C \end{bmatrix}\begin{bmatrix} \frac{di_d}{dt} \\ \frac{di_q}{dt} \\ \frac{du_{DC}}{dt} \end{bmatrix} + \begin{bmatrix} 0 & -\omega L & S_d \\ \omega L & 0 & S_q \\ -S_d & -S_q & 0 \end{bmatrix}\begin{bmatrix} i_d \\ i_q \\ u_{DC} \end{bmatrix}$$

$$+ \begin{bmatrix} R & 0 & 0 \\ 0 & R & 0 \\ 0 & 0 & \frac{2}{3R_L} \end{bmatrix}\begin{bmatrix} i_d \\ i_q \\ u_{DC} \end{bmatrix} = \begin{bmatrix} u_d \\ u_q \\ 0 \end{bmatrix} \quad (8)$$

From(8), EL model of rectifier is obtained[5] as follow:

$$M\dot{x} + Jx + \mathfrak{R}x = u \quad (9)$$

With positive definite diagonal matrix M, a skew symmetric matrix J, i.e. $J = -J^T$, and a symmetric positive semi-definite matrix $\mathfrak{R} \ge 0$. The matrices are expressed by

$$M = \begin{bmatrix} L & 0 & 0 \\ 0 & L & 0 \\ 0 & 0 & \frac{2}{3}C \end{bmatrix}, x = \begin{bmatrix} i_d \\ i_q \\ u_{DC} \end{bmatrix} = \begin{bmatrix} x_1 \\ x_2 \\ x_3 \end{bmatrix}, J = \begin{bmatrix} 0 & -\omega L & S_d \\ \omega L & 0 & S_q \\ -S_d & -S_q & 0 \end{bmatrix},$$

$$\mathfrak{R} = \begin{bmatrix} R & 0 & 0 \\ 0 & R & 0 \\ 0 & 0 & \frac{2}{3R_L} \end{bmatrix}, u = \begin{bmatrix} u_d \\ u_q \\ 0 \end{bmatrix}.$$

4.1.2. Controller Design Ⅰ.
Determining the equilibrium point see 3.1. Controller design is as follow. For the error state vector $x_e = x - x^*$, its dynamic equation based on (9) is then expressed as:

$$M\dot{x}_e + Jx_e + \mathfrak{R}x_e = u - M\dot{x}^* - Jx^* - \mathfrak{R}x^* \quad (10)$$

Let V denote the error storage function in the following form:

$$V = \frac{1}{2}x_e^T Mx_e \quad (11)$$

The desired asymptotic behavior of the error dynamics (10) can be achieved by performing a damping injection. To carry out this job, consider the following desired error dissipation matrix:

$$\mathfrak{R}_d x_e = (\mathfrak{R} + R_a)x_e \quad (12)$$

where R_a is positive definite diagonal matrix.

After a straightforward calculation using (10) and (12), the error dynamics with desired damping becomes

$$M\dot{x}_e + Jx_e + \mathfrak{R}x_e = u - (M\dot{x}^* + Jx^* + \mathfrak{R}x^* - R_ax_e) \quad (13)$$

From(13), in order to $M\dot{x}_e + Jx_e + \mathfrak{R}x_e = 0$ u is given by

$$u = M\dot{x}^* + Jx^* + \mathfrak{R}x^* - R_ax_e \quad (14)$$

then $\dot{V} = -x_e^T(\mathfrak{R} + R_a)x_e < 0$.

In view of $M\dot{x}^* = \mathbf{0}$, passivity controller obtained from(14) is written as:

$$\begin{cases} S_d = \dfrac{U_m + R_{a1}i_d - I_m(R_{a1} + R)}{u_{DCR}} \\[3mm] S_q = \dfrac{u_{lq} + R_{a2}i_q - \omega L I_m}{u_{DCR}} \end{cases} \quad (15)$$

4.1.3. Controller Design Ⅱ.
In order to improve the properties, a new method designing controller is introduced below. (13) can be transformed as

$$M\dot{x}_e + \mathfrak{R}_d x_e = u - (M\dot{x}^* + J(x^* + x_e) + \mathfrak{R}x^* - R_a x_e) \quad (16)$$

From(16), u is given by

$$u = M\dot{x}^* + Jx + \mathfrak{R}x^* - R_a x_e \quad (17)$$

Based on $M\dot{x}^* = \mathbf{0}$, passivity controller obtained from(17) is written as:

$$\begin{cases} S_d = \dfrac{u_d + \omega L i_q - R I_m + R_{a1}(i_d - I_m)}{u_{DC}} \\[3mm] S_q = \dfrac{-\omega L i_d + R_{a2}i_q}{u_{DC}} \end{cases} \quad (18)$$

4.2. passivity control strategy based on PCHD model

4.2.1. PCHD model of AC/DC converter

Let $x_1 = Q = C u_{DC}$, and $x_2 = \psi_1 = L i_d$ and $x_3 = \psi_2 = L i_q$. By orthogonal transform (2) can be expressed as:

$$\begin{cases} \dfrac{dx_1}{dt} = S_d \dfrac{x_2}{L} + S_q \dfrac{x_3}{L} - \dfrac{x_1}{R_L C} \\[3mm] \dfrac{dx_2}{dt} = u_d - R\dfrac{x_2}{L} + \omega x_3 - S_d \dfrac{x_1}{C} \\[3mm] \dfrac{dx_3}{dt} = u_q - R\dfrac{x_3}{L} - \omega x_2 - S_q \dfrac{x_1}{C} \end{cases} \quad (19)$$

Let $H(x)$ denotes the total energy of AC/DC converter, $H(x) = x_1^2/2C + x_2^2/2L + x_3^2/2L$. Then (19) can be expressed as:

$$\begin{bmatrix} \dot{x}_1 \\ \dot{x}_2 \\ \dot{x}_3 \end{bmatrix} = \left(\begin{bmatrix} 0 & S_d & S_q \\ -S_d & 0 & \omega L \\ -S_q & -\omega L & 0 \end{bmatrix} - \begin{bmatrix} \frac{1}{R_L} & 0 & 0 \\ 0 & R & 0 \\ 0 & 0 & R \end{bmatrix} \right) \begin{bmatrix} \frac{\partial H}{\partial x_1} \\ \frac{\partial H}{\partial x_2} \\ \frac{\partial H}{\partial x_3} \end{bmatrix} + \begin{bmatrix} 0 & 0 & 0 \\ 1 & 0 & 0 \\ 0 & 1 & 0 \end{bmatrix} \begin{bmatrix} u_d \\ u_q \\ 0 \end{bmatrix}$$

$$(20)$$

(20) is simplified as

$$\dot{x} = (J(x) - \mathfrak{R}(x))\frac{\partial H(x)}{\partial x} + gu \quad (21)$$

where

$$J = \begin{bmatrix} 0 & S_d & S_q \\ -S_d & 0 & \omega L \\ -S_q & -\omega L & 0 \end{bmatrix}, \mathfrak{R} = \begin{bmatrix} 1/R_L & 0 & 0 \\ 0 & R & 0 \\ 0 & 0 & R \end{bmatrix}, g = \begin{bmatrix} 0 & 0 & 0 \\ 1 & 0 & 0 \\ 0 & 1 & 0 \end{bmatrix}, u = \begin{bmatrix} u_d \\ u_q \\ 0 \end{bmatrix},$$

$$J = -J^T, \mathfrak{R} = \mathfrak{R}^T.$$

(21) is PCHD model of AC/DC converter.

4.2.2. Design procedure of controller

We derive a PDE parameterized by the chosen matrices whose solutions characterize all the energy functions that can be assigned. Finally, from this family of solutions we choose one that satisfies the minimum requirement and compute the control. More precisely, the final objective is to find a static state-feedback control $u = \beta(x)$ such that the closed-loop dynamics is a PCH system with dissipation of the form[6]

$$\dot{x} = [J_d(x) - \mathfrak{R}_d(x)]\frac{\partial H_d(x)}{\partial x} \quad (22)$$

where the new energy function $H_d(x)$ has a strict local minimum at the desired equilibrium x^*, and $J_d(x) = -J_d^T(x)$ and $\mathfrak{R}_d(x) = \mathfrak{R}_d^T(x) \geq 0$ are some desired interconnection and damping matrices, respectively.

Given $J(x,u), \mathfrak{R}(x), H(x), g(x,u)$ and the desired equilibrium to be stabilized $x^* \in R^n$. Assume we can find the functions $\beta(x), J_a(x), \mathfrak{R}_a(x)$ and a vector function $K(x)$ satisfying

$$[J(x,\beta(x)) + J_a(x) - (\mathfrak{R}(x) + \mathfrak{R}_a(x))]K(x)$$
$$= -[J_a(x) - \mathfrak{R}_a(x)]\frac{\partial H}{\partial x}(x) + g(x,\beta(x)) \quad (23)$$

and such that
(i) (Structure preservation)

$$J_d(x) := J(x,\beta(x)) + J_a(x) = -[J(x,\beta(x)) + J_a(x)]^T$$
$$\mathfrak{R}_d(x) := \mathfrak{R}(x) + \mathfrak{R}_a(x) = [\mathfrak{R}(x) + \mathfrak{R}_a(x)]^T \geq 0$$

(ii) (Integrability) $K(x)$ is the gradient of a scalar function. That is,

$$\frac{\partial K}{\partial x}(x) = \left[\frac{\partial K}{\partial x}(x)\right]^T$$

(iii) (Equilibrium assignment) $K(x)$, at x^*, verifies

$$K(x^*) = -\frac{\partial H}{\partial x}(x^*)$$

(iv) (Lyapunov stability) The Jacobian of $K(x)$, at x^*, satisfies the bound

$$\frac{\partial K}{\partial x}(x^*) > -\frac{\partial^2 H}{\partial x^2}(x^*)$$

Under these conditions, the closed-loop system $u = \beta(x)$ will be a PCH system with dissipation of the form (22), where

$$H_d(x) \triangleq H(x) + H_a(x)$$

and

$$\frac{\partial H_a}{\partial x}(x) = K(x)$$

Furthermore, x^* will be a (locally) stable equilibrium of the closed-loop. It will be asymptotically stable if, in addition, the largest invariant set under the closed-loop dynamics contained in

$$\left\{ x \ \in R^n \left| \left[\frac{\partial H}{\partial x}(x) \right]^T \ \mathscr{R}_d(x)\frac{\partial H}{\partial x}(x) = 0 \right. \right\}$$

equals $\{x^*\}$. An estimate of its domain of attraction is given by the largest bounded level set $\{ x \ \in R^n | H_d(x) \le c \}$.

4.2.3. Controller Design

Determining the equilibrium points sees 3.1. Controller design is as follow. From (21) gu is constant, then S_d and S_q are seen as control variable. According to (21) S_d and S_q are achieved to realize $x^* = \arg\min H_d(x)$ under conditions (i), (ii), (iii) and (iv). The matching objective is achieved if and only

$$(J_d - \mathscr{R}_d)\frac{\partial H_a}{\partial x} + (J_a - \mathscr{R}_a)\frac{\partial H}{\partial x} - gu = 0 \qquad (24)$$

where $J_d = J + J_a, \mathscr{R}_d = \mathscr{R} + \mathscr{R}_a, H_d = H + H_a$.
Defining

$$K(x) = [k_1(x), k_2(x), k_3(x)]^T = \left[\frac{\partial H_a}{\partial x_1}, \frac{\partial H_a}{\partial x_2}, \frac{\partial H_a}{\partial x_3} \right]^T \text{ and fixing}$$

the interconnection and damping matrices $J_d = J$ and $\mathscr{R}_d = \mathscr{R}$, that is $J_a = 0$ and $\mathscr{R}_a = 0$. Equation (24) simplifies to

$$\left[\begin{bmatrix} 0 & S_d & S_q \\ -S_d & 0 & \omega L \\ -S_q & -\omega L & 0 \end{bmatrix} - \begin{bmatrix} \frac{1}{R_L} & 0 & 0 \\ 0 & R & 0 \\ 0 & 0 & R \end{bmatrix} \right] \begin{bmatrix} k_1 \\ k_2 \\ k_3 \end{bmatrix} - \begin{bmatrix} 0 \\ u_d \\ u_q \end{bmatrix} = 0 \quad (25)$$

then (25) can be expressed as:

$$\begin{cases} -k_1/R_L + S_d k_2 + S_q k_3 = 0 \\ -S_d k_1 - R k_2 + \omega L k_3 - u_d = 0 \\ -S_q k_1 - \omega L k_2 - R k_3 - u_q = 0 \end{cases} \qquad (26)$$

From (26), we obtain the equation expressing S_d and S_q as follows:

$$\begin{cases} S_d = \dfrac{-R k_2 + \omega L k_3 - u_d}{k_1} \\ S_q = \dfrac{-\omega L k_2 - R k_3 - u_q}{k_1} \end{cases} \qquad (27)$$

Now, we substitute (27) into the first equation in (26) to get the k_1 expression as:

$$k_1 = -\sqrt{R_L(R k_2^2 + u_d k_2 + R k_3^2 + u_q k_3)} \qquad (28)$$

In order to simply the design of controller, one can take $k_1 = k_1(x_1)$, $k_2 = k_2(x_1)$ and $k_3 = k_3(x_1)$, and consequently, using the integrability condition(ii) as follows:

$$\frac{\partial k_1}{\partial x_2} = \frac{\partial k_2}{\partial k_1}, \frac{\partial k_1}{\partial x_3} = \frac{\partial k_3}{\partial k_1}, \frac{\partial k_2}{\partial x_3} = \frac{\partial k_3}{\partial k_2} \qquad (29)$$

One gets that $k_2 = A_1$ and $k_3 = A_2$ are constants from (29).

In order to obtain the extremum of H_d at $x^* = [x_1^*, x_2^*, x_3^*]^T$, the equilibrium condition $\frac{\partial H_d}{\partial x} = 0$ is

$$\begin{cases} \dfrac{x_1^*}{C} + k_1 = 0 \\ \dfrac{x_2^*}{L} + A_1 = 0 \\ \dfrac{x_3^*}{L} + A_2 = 0 \end{cases} \qquad (30)$$

For balanced three-phase supply $u_d = \sqrt{\frac{3}{2}}U_m$ and $u_q = 0$,

We obtain $x_1^* = C u_{DCR}$, $x_2^* = \sqrt{\frac{3}{2}}L I_m$ and $x_3^* = 0$.

From(28) and (30), k_1, k_2 and k_3 are expressed as:

$$\begin{cases} k_1 = -\sqrt{\dfrac{3}{2}R_L(I_m U_m - RI_m^2)} \\ k_2 = A_1 = -\dfrac{x_2^*}{L} = -\sqrt{\dfrac{3}{2}}I_m \\ k_3 = A_2 = -\dfrac{x_3^*}{L_2} = 0 \end{cases} \qquad (31)$$

(31) satisfies condition (iii) based on power conservation $\frac{3}{2}(I_m U_m - RI_m^2) = \frac{u_{DCR}}{R_L}$ at x^*. From (31) H_a is

$$H_a = -x_1\sqrt{\frac{3}{2}R_L(I_m U_m - RI_m^2)} - \frac{x_2^*}{L}x_2 \qquad (32)$$

and H_d is

$$H_d = \frac{x_1^2}{2C} + \frac{x_2^2}{2L} + \frac{x_3^2}{2L} - x_1\sqrt{\frac{3}{2}R_L(I_m U_m - RI_m^2)} - \frac{x_2^*}{L}x_2 \quad (33)$$

In order to guarantee that H_d has a minimum at x^*, the Hessian of H_d has to obey

$$\left. \frac{\partial^2 H_d}{\partial x^2} \right|_{x=x^*} > 0$$

From(33)

$$\left. \frac{\partial^2 H_d}{\partial x^2} \right|_{x=x^*} = \begin{bmatrix} \dfrac{1}{C} & 0 & 0 \\ 0 & \dfrac{1}{L} & 0 \\ 0 & 0 & \dfrac{1}{L} \end{bmatrix} \qquad (34)$$

which is always positive definite, so the minimum condition is satisfied. Substituting (31) into (27), the control laws can be expressed in terms of the output voltage u_{DC} and i_L:

$$\begin{cases} S_d = \dfrac{RI_m - U_m}{-\sqrt{u_{DC}(I_m U_m - RI_m^2)/i_L}} \\ S_q = \dfrac{\omega L I_m}{-\sqrt{u_{DC}(I_m U_m - RI_m^2)/i_L}} \end{cases} \qquad (35)$$

5. Analysis of control strategies based on storage function

5.1. Analysis of Lyapunov control strategy

From (4), (5) and (7), the controller is complex. Although controller (7) can guarantee that $\dot{V}(x)$ is negative definite and that system is stable, it cannot transform (3) into decoupling. So the dynamic state performances of AC/DC converters based on Lyapunov control strategy is no good.

5.2. Analysis of passivity control strategy on EL model

5.2.1. Controller Design I.
Compared with (7),passivity controller(15) is simple. But substituting (15) into (3), then (3) is not decoupling. When damp R_a is injected the system is a couple one. R_a only accelerate the rate of convergence. When system operates in ideal stable state, i.e. $i_d = I_m$, $i_q = 0$ and $u_{DC} = u_{DCR}$, equations in (3) hold. In practice $i_d \neq I_m$ and $i_q \neq 0$ caused by various perturbation influence the dynamic and static state performances of AC/DC converters.

5.2.2. Controller Design II.
In order to analyze the performances of AC/DC converter using Controller design II, we substitute (19) into(3) and (3) can be written as:

$$
\begin{cases}
\dfrac{L}{R+R_{a1}}\dfrac{di_d}{dt} + i_d = I_m \\[2mm]
\dfrac{L}{R+R_{a2}}\dfrac{di_q}{dt} + i_q = 0 \\[2mm]
Cu_{DC}\dfrac{du_{DC}}{dt} = \dfrac{3}{2}U_m i_d - \dfrac{3}{2}RI_m i_d + \dfrac{3}{2}R_{a1} i_d (i_d - I_m) \\[2mm]
\qquad\qquad + \dfrac{3}{2}R_{a2} i_q^2 - \dfrac{u_{DC}^2}{R_L}
\end{cases}
\tag{36}
$$

When R_{a1} and R_{a2} are chosen big, i_d fast steady at I_m and i_q at 0, then (36) is changed as

$$
\begin{cases}
\dfrac{L}{R+R_{a1}}\dfrac{di_d}{dt} + i_d = I_m \\[2mm]
\dfrac{L}{R+R_{a2}}\dfrac{di_q}{dt} + i_q = 0 \\[2mm]
Cu_{DC}\dfrac{du_{DC}}{dt} = \dfrac{3}{2}U_m I_m - \dfrac{3}{2}RI_m^2 - \dfrac{u_{DC}^2}{R_L}
\end{cases}
\tag{37}
$$

The third equation in (37) satisfies power conservation, and u_{DC} steady at u_{DCR}. (37) shows that AC/DC converter can be decoupled by control law (19) and control law (19) can improve the dynamic and static state performances of AC/DC converters.

5.3. Analysis of passivity control strategy based on PCHD model

Passivity control controller based on PCHD model is (35). Compared with (7), (35) is simple. But the controller not transform (3) into decoupling, and the dynamic state performances of AC/DC converter is no good. The advantage of (35) is to need not AC-current sensor, which decreases maintain work and increases reliability for AC/DC converter.

6.Conclusions

AC/DC converter using control strategy based on storage function has good dynamic and static state performances. According to treatise and analysis from above, the controller (19) can transform (3) into decoupling and AC/DC converter has most dynamic and static state performances compared with other control strategies. Although controller (35) not transform (3) into decoupling, it can omit AC-current sensor. So controllers (19) and (35) have application perspective and are researched further.

References

[1] Li Zhengxi, Wang Jiuhe, Li Huade. "Review on nonlinear control strategies of three phase boost type PWM rectifiers",*Electric Drive*, 2006, vol.36, No.1, pp 9-13.

[2] Hasan Kömürcügil, Osman Kükrer. "Lyapunov-based control for three-phase PWM AC/DC voltage-source converters" , *IEEE Trans. Power Electron*, 1998, vol.13, No.5, pp801-813.

[3] Qiao Shutong, Jiang Jianguo. "Output Error Passivity Control of Three-Phase Boost-Type PWM Rectifiers", *TRANSACTIONS OF CHINA ELECTROTECHNICAL SOCIETY*, 2007, vol.22,No2, pp68-73.

[4] Yang Wei, Wang Jiuhe, Song Xiaohui. "Passivity based control strategy of three phase PWM voltage rectifiers", *Journal of Liaoning Technical University*, 2007, vol.26, Suppl. I , pp 173-175.

[5] Tzann-Shin Lee. "Lagrangian modeling and passivity-based control of three-phase AC/DC voltage-source converters", *Industrial Electronics, IEEE Transactions on*,2004, vol.51, No4, pp892-902.

[6] R. Ortega, A. van der Schaft. "Interconnection and damping assignment passivity-based control of port-controlled Hamiltonian systems" , *Automatica*, 2002, vol.38, pp 585–596.

2008 Workshop on Power Electronics and Intelligent Transportation System

Adaptive Optimal Iterative Learning Control for Local Ramp Metering

Shangtai Jin Zhongsheng Hou

Advanced Control Systems Lab, Beijing Jiaotong University, Beijing, China
E-mail: jst1101@163.com, zhshhou@bjtu.edu.cn

Abstract

In this work, a novel adaptive optimal iterative learning control algorithm (AOILC) is applied to address the traffic density control via ramp metering in a macroscopic level freeway environment. The traffic density control problem is formulated into an output tracking problem and the initial traffic density is variable with iteration change. Rigorous analyses and intensive simulations show the effectiveness of the algorithm.

1. Introduction

Iterative learning control (ILC) has been intensively studied over the past two decades [1-4]. ILC was originally proposed in the robotics community as an intelligent teaching mechanism for robot manipulators [5]. The basic idea of ILC is to improve the control signal for the present operation cycle by feeding back the control error in the previous cycle. Nowadays, ILC has become one of the most active research areas in control theory and applications. It is one of the most effective methodologies for repeatable control environment which deals with repeated tracking control tasks for deterministic systems. Specifically, ILC improves the transient response and tracking performance in time domain when the system executes the same motion under essentially the same initial conditions.

The formulation of ILC design problem could be divided into three categories, the first is the contraction mapping based ILC control design methods. So far, most of the ILC publications belong to this category [6, 7]. This category is belong to the almost a model-free method. The second one is the energy function based ILC, which the dynamics in state space has been incorporated in ILC design [8, 9]. The third one is the optimization based ILC control design method, in which explicit optimization objective is introduced into the ILC control design, and the monotonic error sequence could be achieved [10-13].

The optimization based ILC control design method is also called model-based ILC methods. The model based ILC algorithms proposed were based on the known linear time-invariant model of the controlled plant. So the algorithms become hyper-sensitive to model uncertainties and it cannot be used directly in the practice.

There are certain additional traits and requirements found in model-based optimal ILC approaches. First, dynamics of almost all practical processes are intrinsically nonlinear, and the nonlinearities become exposed when the processes are operated over a wide range of conditions. For this reason, it is desirable to derive an optimal ILC control algorithms that can accommodate nonlinear system models. Secondly, to model the practical plants is not an easy thing, and sometimes it is impossible in the view of cost or accuracy. Hence, it is desirable to design directly the input-output based ILC control law for the ILC control task. So we can enjoy not only extra good properties of the optimal ILC, but also the little requirements on the system dynamic model of the prototype of the ILC as well.

The objective of this paper is to provide a more general and comprehensive framework for quadratic criterion based ILC that is capable of addressing all the issues that mentioned to be important for the process control applications. We first introduce a dynamic linearization method that can transform general discrete-time nonlinear model into a time-varying linearized model using a concept of pseudo-partial- derivative (PPD). Based on this discrete-time linearized model, the adaptive optimal ILC for discrete-time nonlinear systems is designed, and the mathematical properties are also discussed.

The rest of the paper is organized as follows. In section 2, the problem formulation and the model transformation are presented. In section 3, the adaptive optimal ILC is designed, and its properties are also discussed. Simulation results are provided in Section 4. Section 5 concludes the paper.

2. Problem formulation and model transformation

2.1 Problem formulation

The system to be controlled is described by the following SISO discrete-time nonlinear equation

$$y(t+1) = f(y(t), \cdots, y(t-n_y), u(t), \cdots, u(t-n_u)), \quad (1)$$

where $y(t), u(t)$ are the output and input at time t, $t \in \{0, 1, \cdots, T-1\}$, n_y, n_u are the unknown orders, and $f(\cdot)$ is an unknown nonlinear function.

978-0-7695-3342-1/08 $25.00 © 2008 IEEE
DOI 10.1109/PEITS.2008.80

The control task is the perfect tracking in a finite interval under a repeatable control environment. The perfect tracking task implies that a trajectory must be strictly followed from very beginning of the execution. The repeatable control environment implies (1) identical target trajectory and (2) same initial condition for all trials. In more details, the control objective for ILC is to design a sequence of appropriate control inputs $u_k(t)$ such that the system output $y_k(t)$ approaches the target trajectory $y_d(t)$, $t \in \{0,1,\cdots,T-1\}$, and $T > 0$ is a finite number. The subscript k denotes the k-th repeated control operation period and is called k-th learning "iteration". $y_d(t)$ is invariant in iteration domain.

2.2 Model transformation

The system dynamics in the k-th iteration
$$y_k(t+1) = f(y_k(t),\cdots,y_k(t-n_y),u_k(t),\cdots,u_k(t-n_u)). \quad (2)$$
The system dynamics in the (k-1)-th iteration
$$y_{k-1}(t+1) = f(y_{k-1}(t),\cdots,y_{k-1}(t-n_y),u_{k-1}(t), \\ \cdots,u_{k-1}(t-n_u)). \quad (3)$$
We first make two assumptions with regards to the system.

A1: The partial derivative of $f(\cdot)$ with respect to system output $y_k(t),\cdots,y_k(t-L_y+1)$ and control input $u_k(t),\cdots,u_k(t-L_u+1)$ is continuous.

A2: The system (2) is generalized Lipschitz, that is, satisfying
$$|\Delta y_k(t+1)| \le D\|\mathbf{H}_k(t)\| \quad \forall t \ and \ \|\mathbf{H}_k(t)\| \ne 0, \quad (4)$$
where D is a constant, $\Delta y_k(t+1) = y_k(t+1) - y_{k-1}(t+1)$, $\mathbf{H}_k(t) = \begin{bmatrix} \Delta y_k(t) & \cdots & \Delta y_k(t-L_y+1) & \Delta u_k(t) & \cdots & \Delta u_k(t-L_u+1) \end{bmatrix}$, $\Delta u_k(t) = u_k(t) - u_{k-1}(t)$, L_y and L_u are positive integer.

Remark 1: These assumptions of the system are reasonable and acceptable from a practical point of view. Assumption (A1) is a typical condition for many control laws which a general nonlinear system should satisfy. Assumption (A2) poses a limitation on the rate of change of the system output permissible before the control law to be formulated is applicable.

Theorem 1: For the nonlinear systems (2) and (3), when Assumptions (A1) and (A2) hold, then for the given L_y and L_u, there must exist $\boldsymbol{\theta}_k(t)$, called pseudo-partial-derivative (PPD), and $\|\boldsymbol{\theta}_k(t)\| \le D$ such that if $\|\mathbf{H}_k(t)\| \ne 0$, the system may be described as
$$\Delta y_k(t) = \boldsymbol{\theta}_k(t)^T \mathbf{H}_k(t), \quad (5)$$
where $\boldsymbol{\theta}_k(t) = \begin{bmatrix} \theta_{k,1}(t) & \cdots & \theta_{k,L_y}(t) & \theta_{k,L_y+1}(t) & \cdots & \theta_{k,L_y+L_u}(t) \end{bmatrix}^T$.

Proof:

Subtracting (3) from (2), and using the differential mean theorem, yields
$$\Delta y_k(t+1) = f(y_k(t),\cdots,y_k(t-n_y),u_k(t),\cdots,u_k(t-n_u)) \\ - f(y_{k-1}(t),\cdots,y_{k-1}(t-n_y),u_{k-1}(t),\cdots,u_{k-1}(t-n_u)) \\ = \frac{\partial f^*}{\partial y_k(t)}\Delta y_k(t)+\cdots+\frac{\partial f^*}{\partial y_k(t-L_y+1)}\Delta y_k(t-L_y+1)+\cdots \\ + \frac{\partial f^*}{\partial u_k(t)}\Delta u_k(t)+\cdots+\frac{\partial f^*}{\partial u_k(t-L_u+1)}\Delta u_k(t-L_u+1)+\psi_k(t), \quad (6)$$
where $\partial f^* / \partial u_k(t)$ represents the partial derivative value of f at some point in the interval $[u_k(t),u_{k-1}(t)]$, $\partial f^* / \partial y_k(t)$ represents the partial derivative value of f at some point in the interval $[y_k(t),y_{k-1}(t)]$ and
$$\psi_k(t) = f(y_{k-1}(t),\cdots y_{k-1}(t-L_y+1),y_k(t-L_y),\cdots,y_k(t-n_y), \\ u_{k-1}(t),\cdots,u_{k-1}(t-L+1),u_k(t-L)\cdots,u_k(t-n_u)) \quad (7) \\ - f(y_{k-1}(t),\cdots,y_{k-1}(t-n_y),u_{k-1}(t),\cdots,u_{k-1}(t-n_u)).$$

Considering the following equation with a variables $\boldsymbol{\eta}_k(t)$
$$\psi_k(t) = \boldsymbol{\eta}_k^T(t)\mathbf{H}_k(t), \quad (8)$$
Equation (8) must have at least a solution $\boldsymbol{\eta}_k^*(t)$ since condition $\|\mathbf{H}_k(t)\| \ne 0$ holds. In fact, it must have infinite solutions.

Let
$$\boldsymbol{\theta}_k(t) = \begin{bmatrix} \dfrac{\partial f^*}{\partial y_k(t)} & \cdots & \dfrac{\partial f^*}{\partial y_k(t-L_y+1)} & \dfrac{\partial f^*}{\partial u_k(t)} & \cdots & \dfrac{\partial f^*}{\partial u_k(t-L_u+1)} \end{bmatrix}^T + \boldsymbol{\eta}_k^*(t), \quad (9)$$
Replacing $\psi_k(t)$ of (7) by (9), then we have
$$\Delta y_k(t) = \boldsymbol{\theta}_k(t)^T \mathbf{H}_k(t).$$
The boundedness of PPD is the straightforward result from the assumption (A2) and (5).

Remark 2: Theorem 1 is an extension of the results in reference [14-16]. This theorem shows that $\boldsymbol{\theta}_k(t)$ is a differential signal in some sense along the iteration axis and bounded for any iteration number k. Furthermore, PPD $\boldsymbol{\theta}_k(t)$ is a slowly time-varying parameter along the iteration axis and its relation with $\mathbf{H}_k(t)$ may be ignored when $\|\mathbf{H}_k(t)\|$ is not too large.

Rewriting (5) in a compact form, we obtain
$$\mathbf{A}_k\mathbf{y}_k = \mathbf{A}_k\mathbf{y}_{k-1} + \mathbf{B}_k\Delta\mathbf{u}_k, \quad k = 0,1,2,\cdots \quad (10)$$
where $\Delta\mathbf{u}_k = [\Delta u_k(0),\cdots,\Delta u_k(T-1)]^T$, $\mathbf{y}_k = [y_k(1),\cdots,y_k(T)]^T$, $\mathbf{y}_{k-1} = [y_{k-1}(1),\cdots,y_{k-1}(T)]^T$,

$$\mathbf{A}_k = \begin{bmatrix} 1 & & & & \\ -\theta_{k,1}(1) & 1 & & & \\ & & \ddots & & \\ -\theta_{k,L_y}(L_y+1) & \cdots & -\theta_{k,1}(L_y+1) & 1 & \\ & \ddots & & & \ddots \\ & & -\theta_{k,L_y}(T-1) & \cdots & -\theta_{k,1}(T-1) & 1 \end{bmatrix}_{T \times T},$$

$$\mathbf{B}_k = \begin{bmatrix} \theta_{k,L_y+1}(0) & & & \\ & \ddots & & \\ \theta_{k,L_y+L_u}(L-1) & \cdots & \theta_{k,L_y+1}(L-1) & \\ & \ddots & \cdots & \ddots \\ & & \theta_{k,L_y+L_u}(T-1) & \cdots & \theta_{k,L_y+1}(T-1) \end{bmatrix}_{T \times T}.$$

3. Norm-optimal iterative learning control

3.1 Optimal learning control law algorithm

Let $y_d(t)$ be the desired output signal, $e_k(t) = y_d(t) - y(t)$ be the output error, and define the optimization design problem for the ILC controller as follows

$$\min_{\mathbf{u}_{k+1}} J_{k+1}(\mathbf{u}_{k+1}) = \min_{\bar{u}_{k+1}(t)} \left(\|\mathbf{A}_{k+1}\mathbf{e}_{k+1}\|^2 + \lambda \|\mathbf{u}_{k+1} - \mathbf{u}_k\|^2 \right), \quad (11)$$

subject to

$$\mathbf{A}_k \mathbf{e}_{k+1} = \mathbf{A}_k \mathbf{e}_k - \mathbf{B}_{k+1} \Delta \mathbf{u}_{k+1} \quad (12)$$

where $\lambda > 0$.

To simplify notions, we will hereafter use $\mathbf{e}_k = \mathbf{y}_d - \mathbf{y}_k$ to denote $\mathbf{e}_k(t) = \mathbf{y}_d(t) - \mathbf{y}(t)$.

Inserting (12) into (11), and differentiating the objective function, and setting the differential equal to zero yields the following norm-optimal ILC control law

$$\mathbf{u}_{k+1} = \mathbf{u}_k + (\lambda I + \mathbf{B}_{k+1}^T \mathbf{B}_{k+1})^{-1} \mathbf{B}_{k+1}^T \mathbf{A}_{k+1} \mathbf{e}_k. \quad (13)$$

Remark 4: when $L_y = 0, L_u = 1$, the norm optimal ILC control law becomes

$$u_{k+1}(t) = u_k(t) + \theta_{k+1}(t)(y_d(t) - y_k(t))/(\lambda + \|\theta_{k+1}(t)\|^2), \quad (14)$$
$$\forall t \in [0, T-1], k = 1, 2, \cdots.$$

this is the same as reference [17].

Since the PPD $\theta_k(t)$ is unknown, a new parameter estimation criterion function is used for the derivation of the estimator.

$$J(\theta_k(t)^T) = \left| \Delta y_{k-1}(t+1) - \theta_k(t)^T \mathbf{H}_{k-1}(t) \right|^2 + \mu \left\| \theta_k(t)^T - \hat{\theta}_{k-1}(t)^T \right\|^2, \quad (15)$$

where $\mu > 0$.

By using (3), the minimization of above criterion function gives estimation algorithm

$$\hat{\theta}_k(t) = \hat{\theta}_{k-1}(t) + \frac{\mathbf{H}_{k-1}(t)}{\mu + \|\mathbf{H}_{k-1}(t)\|^2} \times (\Delta y_{k-1}(t+1) - \hat{\theta}_{k-1}^T(t)\mathbf{H}_{k-1}(t)), \quad (16)$$

and control law algorithm becomes

$$\mathbf{u}_{k+1} = \mathbf{u}_k + (\lambda I + \hat{\mathbf{B}}_{k+1}^T \hat{\mathbf{B}}_{k+1})^{-1} \hat{\mathbf{B}}_{k+1}^T \hat{\mathbf{A}}_{k+1} \mathbf{e}_k. \quad (17)$$

In order to make parameter estimation algorithm (16) have stronger ability to track time-varying parameter, a reset measurement of estimation algorithm should be taken

$$\hat{\theta}_k(t) = \hat{\theta}_0(t), \quad \text{if } \left\| \hat{\theta}_k(t) \right\| \leq \varepsilon \text{ or } \|\mathbf{H}_{k-1}(t)\| \leq \varepsilon \quad (18)$$

where ε is a some small positive constant, $\hat{\theta}_0(t)$ is the initial estimation value of $\hat{\theta}_k(t)$.

In order to obtain the convergence and stability for the controller, another assumption about the system should be made.

A3: Estimation error of $\theta_k(t)$ is sufficiently small, such that $\left\| \hat{\mathbf{A}}_{k+1}^{-1} \hat{\mathbf{B}}_{k+1} \hat{\mathbf{B}}_{k+1}^T \hat{\mathbf{A}}_{k+1} - \mathbf{A}_{k+1}^{-1} \mathbf{B}_{k+1} \hat{\mathbf{B}}_{k+1}^T \hat{\mathbf{A}}_{k+1} \right\| < \sigma_{\min k+1}$,

where $\sigma_{\min k+1}$ is the smallest eigenvalue of $\hat{\mathbf{B}}_{k+1} \hat{\mathbf{B}}_{k+1}^T$

Theorem 2 Assume A1-A3 hold, Suppose that the plant described by (2) is controlled by (17) and the estimate $\hat{\theta}_k(t)$ is identified using (16) and (18), then

1) Estimate $\hat{\theta}_k(t)$ is bounded for all $t \in \{0, 1, \cdots T-1\}$ and $k \in \{0, 1, \cdots\}$.

2) $\|\mathbf{e}_{k+1}\| \leq \rho \|\mathbf{e}_k\|$, where $0 < \rho < 1$, $\lim_{k \to \infty} \|\mathbf{e}_k\| = 0$.

Proof:

1) Introduce the following notation:

$$\tilde{\theta}_k(t) = \hat{\theta}_k(t) - \theta_k(t). \quad (19)$$

From (5) we have

$$\Delta y_{k-1}(t+1) = \theta_k(t)^T \mathbf{H}_k(t). \quad (20)$$

When $\|\mathbf{H}_{k-1}(t)\| \leq \varepsilon$, from (18) we can see that $\hat{\theta}_k(t)$ is bounded. When $\|\mathbf{H}_{k-1}(t)\| \leq \varepsilon$ subtracting $\theta_k(t)$ from both sides of (18) yields

$$\tilde{\theta}_k(t) = \mathbf{H}_{k-1}(t)(\Delta y_{k-1}(t) - \mathbf{H}_{k-1}(t)^T \tilde{\theta}_{k-1}(t))/(\mu + \|\mathbf{H}_{k-1}(t)\|^2) \quad (21)$$
$$+ \tilde{\theta}_{k-1}(t) + \theta_{k-1}(t) - \theta_k(t).$$

Substituting (20) into (21) gives

$$\tilde{\theta}_k(t) = (I - \mathbf{H}_{k-1}(t)\mathbf{H}_{k-1}(t)^T)\tilde{\theta}_{k-1}(t)/(\mu + \|\mathbf{H}_{k-1}(t)\|^2) \quad (22)$$
$$+ \theta_{k-1}(t) - \theta_k(t).$$

Let $\Xi(t) = (I - \mathbf{H}_{k-1}(t)\mathbf{H}_{k-1}(t)^T)\tilde{\theta}_{k-1}(t)/(\mu + \|\mathbf{H}_{k-1}(t)\|^2)$, computing $\|\Xi(t)\|^2$ yields

$$\|\Xi(t)\|^2 = \|\tilde{\theta}_{k-1}(t)\|^2 + \left(-2 + \frac{\|\mathbf{H}_{k-1}(t)\|^2}{\mu + \|\mathbf{H}_{k-1}(t)\|^2} \right) \frac{\|\mathbf{H}_{k-1}(t)^T \tilde{\theta}_{k-1}(t)\|^2}{\mu + \|\mathbf{H}_{k-1}(t)\|^2}. \quad (23)$$

Since $\theta_k(t)$ is an iteration dependent column vector and $\mu > 0$, we have

$$\|\Xi(t)\|^2 < \|\tilde{\theta}_{k-1}(t)\|^2. \quad (24)$$

This implies that there exist positive constant $d < 1$ such that following inequality holds

$$\left\| \tilde{\theta}_k(t) \right\| \leq d^k \|\tilde{\theta}_0(t)\| + 2D(1 - d^k)/(1 - d). \quad (25)$$

In view of (25), $\tilde{\boldsymbol{\theta}}_k(t)$ is bounded, $\boldsymbol{\theta}_k(t)$ is bounded, so all of the $\hat{\boldsymbol{\theta}}_k(t)$ is bounded.

Substituting (17) into (12)

$$\begin{aligned}
\mathbf{e}_{k+1} &= \mathbf{e}_k - \mathbf{A}_{k+1}^{-1}\mathbf{B}_{k+1}\Delta\mathbf{u}_{k+1} \\
&= \left(I - \mathbf{A}_{k+1}^{-1}\mathbf{B}_{k+1}\left(\lambda I + \hat{\mathbf{B}}_{k+1}^{T}\hat{\mathbf{B}}_{k+1}\right)^{-1}\hat{\mathbf{B}}_{k+1}^{T}\hat{\mathbf{A}}_{k+1}\right)\mathbf{e}_k
\end{aligned} \quad (26)$$

Taking norms on both sides of (26) we have

$$\begin{aligned}
\|\mathbf{e}_{k+1}\| &= \left\|\left(I - \mathbf{A}_{k+1}^{-1}\mathbf{B}_{k+1}\left(\lambda I + \hat{\mathbf{B}}_{k+1}^{T}\hat{\mathbf{B}}_{k+1}\right)^{-1}\hat{\mathbf{B}}_{k+1}^{T}\hat{\mathbf{A}}_{k+1}\right)\mathbf{e}_k\right\| \\
&\leq \frac{\lambda + \left\|\hat{\mathbf{A}}_{k+1}^{-1}\hat{\mathbf{B}}_{k+1}\hat{\mathbf{B}}_{k+1}^{T}\hat{\mathbf{A}}_{k+1} - \mathbf{A}_{k+1}^{-1}\mathbf{B}_{k+1}\hat{\mathbf{B}}_{k+1}^{T}\hat{\mathbf{A}}_{k+1}\right\|}{\lambda + \sigma_{\min k+1}}\|\mathbf{e}_k\|.
\end{aligned} \quad (27)$$

From assumption 3 and (27) we can conclude
$\|\mathbf{e}_{k+1}\| \leq \rho\|\mathbf{e}_k\|$, and $\lim_{k\to\infty}\|\mathbf{e}_k\| = 0$.

4. Simulation

In order to verify the effectiveness of the AOILC approach, we simulate a freeway traffic flow process in the presence of a large exogenous disturbance (modeled by an exiting flow in an off-ramp during a period). The learning process is iterated for 50 cycles.

The system to be controlled is the same as reference [18], the macroscopic traffic flow model is

$$\rho_i(t+1) = \rho_i(t) + (q_{i-1}(t) - q_i(t) + r_i(t) - s_i(t))T/L_i, \quad (28)$$

$$q_i(t) = \rho_i(t)v_i(t), \quad (29)$$

$$\begin{aligned}
v_i(t+1) &= v_i(t) + (V(\rho_i(t)) - v_i(t))T/\tau \\
&+ v_i(t)(v_{i-1}(t) - v_i(t))T/L_i \\
&- \nu T(\rho_{i+1}(t) - \rho_i(t))/(\tau L_i(\rho_i(t) + \kappa)),
\end{aligned} \quad (30)$$

$$V(\rho_i(t)) = v_{free}(1 - (\rho_i(t)/\rho_{jam})^l)^m. \quad (31)$$

The definition of parameters in model and the setting of parameters are similar to reference [19, 20].

Boundary conditions are summarized as follows
$\rho_0(t) = q_0(t)/v_1(t), v_0(t) = v_1(t), \rho_{N+1}(t) = \rho_N(t), v_{N+1}(t) = v_N(t).$

Consider a long segment of freeway that is subdivided into 5 sections. The length of each section is 0.5 km. The desired density is $30 veh/km$. The initial traffic volume entering section 1 is 1500 vehicles per hour. The initial density, mean speed, and other parameters used in this model are listed in Table 1. There exist an on-ramp with known traffic demands in section 3 and an off-ramp with unknown exiting traffic flow in section 4. The parameters used in this model are listed in Table 1. In order to illustrate the proposed AOILC algorithm can overcome the limitations on identical initial condition of traditional ILC, a random varying along iteration axis $\rho_i(0)$ is used. In the simulation we choose $\lambda = 0.001, \mu = 1$. Fig.1 shows the tracking error of the traffic flow density in section 3 and Fig.2 shows density tracking performance

in section 3 with random initial conditions, where the learning error norm is a usual vector norm.

For practical urban freeway traffic control system, form fig.1 and fig.2 we can see that the proposed AOILC can overcome the limitations of traditional ILC with respect to initial condition, achieve the perfect tracking except initial point.

Table1: Parameters associated with the traffic model

v_{free}	ρ_{jam}	l	m	κ	τ
80	80	1.8	1.7	13	0.01
T	ν	$q_0(k)$	$r_i(0)$	α	
0.00417	35	1500	0	0.95	

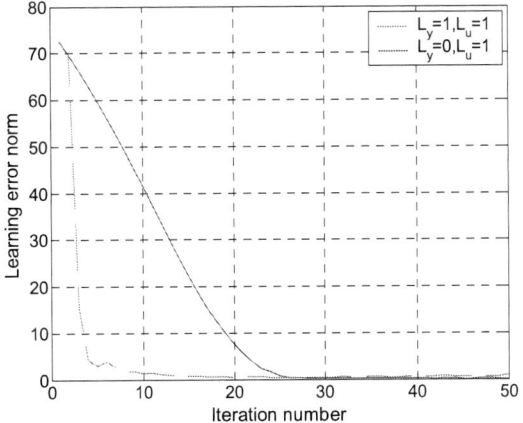

Fig.1 The norm of tracking error of traffic flow density in section 3 with random initial conditions

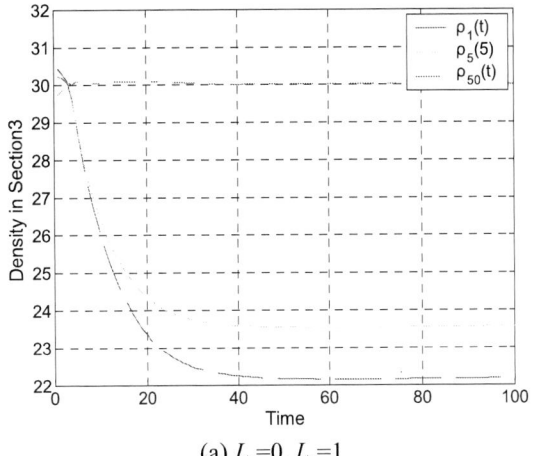

(a) $L_y=0, L_u=1$

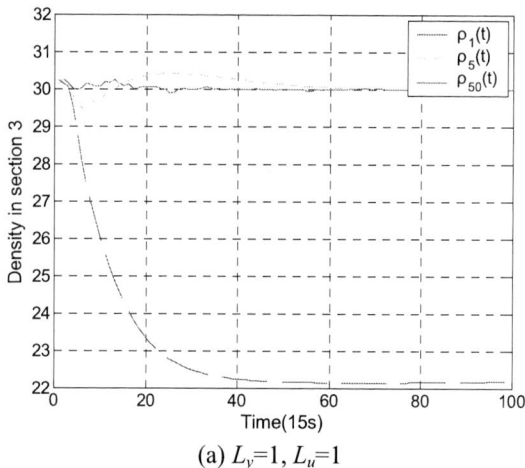

(a) $L_y=1$, $L_u=1$

Fig.2 Traffic flow density tracking performance in section 3 with random initial conditions

5. Conclusion

The AOILC scheme based on the discrete-time dynamic linearized model is proposed in this paper, and the convergence over the entire finite time interval can be guaranteed by theoretical analysis when the initial conditions are randomly varying along the iteration axis. Moreover, this paper covers the results of Owens (1996), which is the special case of the AOILC in LTI system. The main features of the AOILC scheme is that the controller design only depends on the I/O data of the dynamic system. The simulation results show effectiveness of proposed algorithm.

6. References

[1] K. L. Moore, Iterative Learning Control for Deterministic Systems, Springer-Verlag, Advances in Industrial Control Series, 1993.
[2] M.X. Sun, and B.J. Huang, Iterative Learning control, National Defence Industry Press, Beijing, 1998. (In Chinese)
[3] Y.Q. Chen, and C.Y. Wen, Iterative Learning Control: Convergence, Robustness and Application, Lecture Notes in Control and Information Sciences 248, Springer, 1999.
[4] J.X. Xu, and Y. Tan, Linear and Nonlinear Iterative Learning Control, Lecture Notes in Control and Information Sciences, 291, Springer, 2003.
[5] S. Arimoto, S. Kawamura, and F. Miyazaki, Bettering operation of robots by learning. Journal of Robotic Systems, 1984, 1(1), 123-140.
[6] C.J. Chen, A discrete iterative learning control for a class of nonlinear time-varying systems, IEEE Transactions on Automatic Control, 1998, 43(5),748-752.
[7] T.Y. Kuc, J. S. Lee, and K. Nam, An iterative learning control theory for a class of nonlinear dynamic systems, Automatica, 1992, 28(6), 1215–1221.
[8] C. Ham, Z.H. Qu, and J.H. Kaloust, Nonlinear learning control for a class of nonlinear systems based on Lyapunov's

direct method, In Proceeding of 1995 IEEE American Control Conference, 3024-3028.
[9] J. X. Xu, and Y. Tan, A composite energy function based learning control approach for nonlinear systems with time varying parametric uncertainties, IEEE Transactions on Automatic Control, 2002, 47 (11), 1940-1945.
[10] N. Amann, D.H. Owens, and E. Rogers, Iterative learning control for discrete-time systems with exponential rate of convergence. IEE Proc.-Control Theory and Applications, 1996, 143(2), 217-224.
[11] J.H. Lee, K.S. Lee, and W.C. Kim, Model-based iterative learning control with a quadratic criterion for time-varying linear systems, Automatica, 2000, 36(5), 641-657.
[12] J.J. Hatonen, D.H. Owens, and K.L. Moore, An algebraic approach to iterative learning control. International Journal of Control, 2004, 77(1), 45-54.
[13] J.J. Hatonen, Issues of algebra and optimality in iterative learning control. PhD dissertation, University of Oulu, Finland, 2004.
[14] Z.S. Hou, The parameter identification, adaptive control and model-free learning adaptive control for nonlinear systems, PhD thesis, Northeastern University, China,1994. (in Chinese)
[15] Z.S. Hou, and W.H. Huang, Model-free learning adaptive control of a class of SISO nonlinear systems. Proc of Amer. Control Conf., Albuquerque, New Mexico, 1997, 343-344.
[16] Z.S. Hou , Non-Parametric Model and its Adaptive Control. Science press, Beijing, 1999. (in Chinese)
[17] Chi R.H., Hou Z.H., Dual-stage optimal iterative learning control for nonlinear non-affine discrete-time systems[J], Acta Automatica Sinica, 2007, 33(10), 1061-1065
[18] M. Parageorgiou, J.M. Blosseville, and H. Hadj-Salem, Modelling and real time control on traffic flow on the southern part of Bpulevard Peripherique in Paris. Part I: Modelling; Part II: Coordinated on-ramp metering, Transportation Research, 1990, A24, 345-370.
[19] Z.S. Hou, J.X. Xu, and H.W. Zhong, Freeway Traffic Control Using Iterative Learning Control-Based Ramp Metering and Speed Signalling, IEEE Transaction on vehicular technology, 2007, 56 (2), 466-477.
[20] Z.S. Hou, J.X. Xu, and J.W. Yan, An Iterative Learning Approach for Density Control of Freeway Traffic Flow via Ramp Metering, Transportation Research Part C, 2008, 16, 71-97

Simulation Study on A Single Neuron PID Control System of DC/DC Converters

Xinggui Wang[1], Bing Xu [2], Lei Ding [3]

[1] *Department of Electrical and information Engineering,*
Lanzhou University of Science and Technology, Lanzhou, Gansu, 730050, China
[2] *Department of Electrical and information Engineering,*
Lanzhou University of Science and Technology, Lanzhou, Gansu, 730050, China
[3] *Department of Economics and Management,*
liaoning shihua univercity, Fushun Liaoning, 113001, China
wangxg82@tom.com, jinduoxia33@sohu.com, xiaodingding1204@sohu.com

Abstract

In order to improve the traditional PID control system charged with the parameters perturbed sensitive defects, and use excellent knowledge extraction of neural network and learning ability, people start to put the neural network control strategy into the DC / DC converters, also to design a single neuron PID controller. Furthermore, we just make the Sepic converter as an example to research simulation of its performance. As a result, the simulation indicates that the system has a good dynamic performance and robustness.

1. Introduction

DC / DC converters usually use PWM to achieve output voltage stability. PWM control system is essentially a discrete nonlinear system, therefore it is difficult to structure the precision transfer function. At present, the most extensive practical application of the control strategy is PID control. Its parameter must be selected based on the precise object model in practical situations; moreover, its parameters often determine the adjustment through repeated testing method [1]. Once the circuit parameters have been identified, PID controller parameters would be defined, so the system can only make effects to a certain scope of the disturbance to show good performance[2].

Neural networks which got strong non-linear approach and adaptive capacity have been widely used to solve many control problems, and the application of the DC / DC converters has been well researched [3]. However, the more the number of neural network layers are, the more complex the network structure can be, and the more complicated the neural network training process will be. As a result, the huge time for learning of the value will be not conducive to real-time control, also the system will reduce the reaction time[4]. Therefore, it is very important to do the research of reducing the complexity of the neural network controller.

A single neuron as the basic unit of neural networks has self-learning and adaptive capacity; on the other hand, the structure is simple, and the calculation is easy. Furthermore, the traditional PID regulator can be also simple in structure and adjusted easily and so on. The constitution of single neuron PID controller with self-learning and adaptive capacity is not only simple structure, clear learning algorithm physical meaning, small amount of calculation, but also adapting to DC / DC converter load of rapid fluctuations, which has strong robustness.

2. The design that based on a single neuron PID controller

2.1. The structure of controller

When single neuron PID controller detected the disturbance of the DC / DC converters, it should be able to predict the corresponding changes in output, and give out appropriate control signals which pulse width adjustable and in order to stabilize the output voltage. Based on the single neuron PID controller of the DC / DC converter system as shown in Figure 1. The output of Neural network controller adjust the PWM pulse width which can control the DC / DC converter for direct.

978-0-7695-3342-1/08 $25.00 © 2008 IEEE
DOI 10.1109/PEITS.2008.94

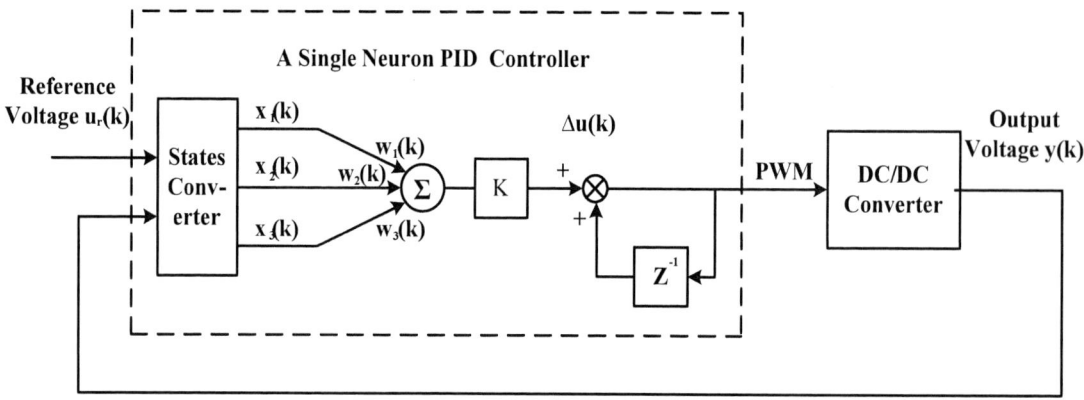

Figure 1 Based on the single neuron PID controller structure

2.2. The realization of the single neurons PID controller

The state input of single neuron PID is:

$$x_1(k) = e(k) ; \tag{1}$$

$$x_2(k) = e(k) - e(k-1) ; \tag{2}$$

$$x_3(k) = e(k) - 2e(k-1) + e(k-2) ; \tag{3}$$

This method has obviously physical meaning: $x_1(k)$ is the integration points, and it can eliminate the steady-state error, improve the control accuracy; $x_2(k)$ is the proportion points, and it can quickly reduce the tracking error; $x_3(k)$ is differentiation points, and it can improve the response rate and reduce the overshoot [5].

$$\Delta u(k) = K \sum_{i=1}^{3} x_i(k) w_i(k) ; \tag{4}$$

Among them, $\Delta u(k)$ is the output increase of the moment k neurons; K is the proportion of neurons. $w_i(k)$ is coefficient for the importation of state and the corresponding connection weights, it reflects the controlled and targeted response to the dynamic nature of the process, Single neuron controller through its own learning strategy kept the weighted coefficient of adjustment to achieve adaptive, self-organization functions, then achieve rapid elimination of bias and get to the steady-state results[6]. The controller in the use of Hebb rules and a combination of supervised learning algorithm. In the second method of borrowing the idea of performance indicators to adjust the weights in order to achieve control the bias of the output deviation. Presume performance indicators is :

$$J(k) = \frac{1}{2}[y(k+1) - u_r(k+1)]^2 ; \tag{5}$$

We suppose that the right to the adjustment factor walk along the negative gradient to search, then we obtain

$$\Delta w_i(k+1) = w_i(k+1) - w_i(k) = \eta_i e(k+1) \frac{\partial y(k+1)}{\partial u(k)} \frac{\partial u(k)}{\partial w_i(k)} ; \tag{6}$$

and η_i is study rate (i=P, I, D).

Bring (4) into (6), then we get (7)

$$\Delta w_i(k+1) = \eta_i K e(k+1) e(k) \frac{\partial y(k+1)}{\partial u(k)} ; \tag{7}$$

To ensure the convergence and robustness of this single neurons control of learning algorithm, we will let the learning algorithm be standardization and then we can see that

$$\left. \begin{array}{l} u(k) = u(k+1) + K \sum_{i=1}^{3} \overline{w_i(k)} x_i(k) \\[2mm] \overline{w_i(k)} = \dfrac{w_i(k)}{\sum_{i=1}^{3} |w_i(k)|} \\[2mm] w_1(k+1) = w_1(k) + \eta_I K e(k+1) e(k) \dfrac{\partial y(k+1)}{\partial u(k)} \\[2mm] w_2(k+1) = w_2(k) + \eta_P K e(k+1) e(k) \dfrac{\partial y(k+1)}{\partial u(k)} \\[2mm] w_3(k+1) = w_3(k) + \eta_D K e(k+1) e(k) \dfrac{\partial y(k+1)}{\partial u(k)} \end{array} \right\} \tag{8}$$

3. Simulation results and analysis

In this paper, we set Sepic-DC / DC converters for example, and use the designed single neuron PID controller to control it. When we set different reference voltage of single neuron PID controller, we study the dynamic nature of the controlled circuit. The structure of the circuit is shown below in Figure 2.

128

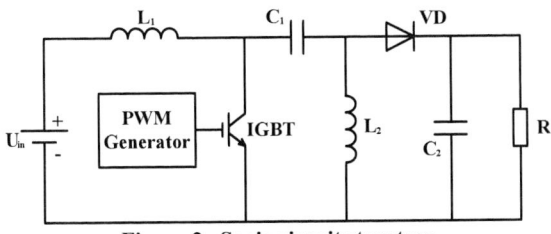

Figure 2 Sepic circuit structure

In this article we use Simulink of Matlab 6.5 simulation software to simulate, and use the module of SimPowerSystems structures to build Sepic circuit model. We use Simulink library modules to do single neuron PID algorithm sub-module, we control the system by adjusting the pulse width of PWM wave[7] .It is shown the Sepic-DC / DC converter simulation model of the structure controlled by single neuron PID controller infigure3.

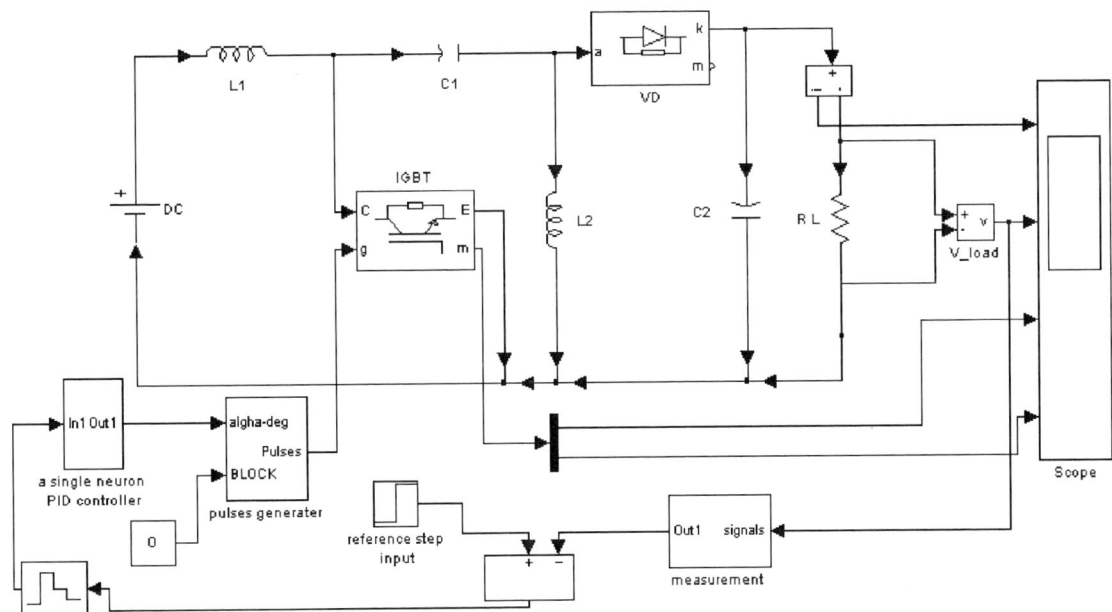

Fig.3. simulation structure model of the Sepic model of DC/DC converter which controlled by single neuron PID

The simulation uses ode23tb algorithm, and the PWM frequency is 10 KHz, simulation time is 0. 1s, the estimated initial input voltage is 200 V, component parameters: $L_1 = 0.6mH$, $C_1 = 0.82\mu F$, $L_2 = 0.05mH$, $C_2 = 100\mu F$. In the single neuron PID algorithm, input is the step response signal ur(k), that is the given voltage value. When the simulation time is 0.025 s, the value of ur(k) is 500 V and 100 V, the output voltage corresponding diagrams were in Figure 4 and 5 respectively. The controller control the system by controlling the pulse width of PWM wave to make output voltage achieve given value .At the same time the values of the three PID parameters K_P , K_I, K_D are also changed, in order to adapt to the changes in the input parameters. So the controller based on the single neuron PID algorithm can change with the system input, and timely adjust the size of the pulse width of PWM wave to control the output voltage, let the DC / DC converters always work in the vicinity of the voltage setting. When changing the given voltage value, we will achieve the purpose of

regulating the output voltage, and the system has strong dynamic characteristics.

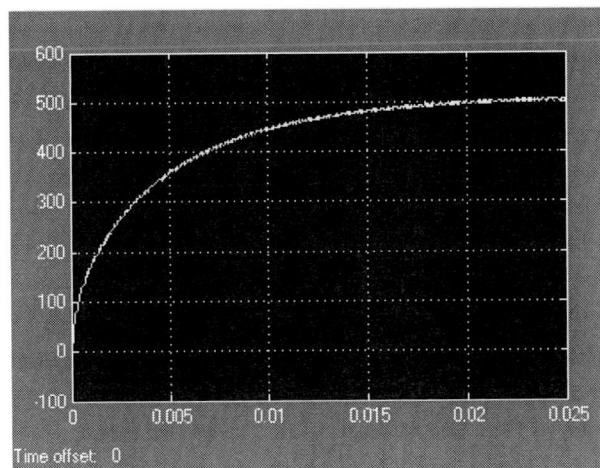

Figure 4 The response plans which ur(K)=500v

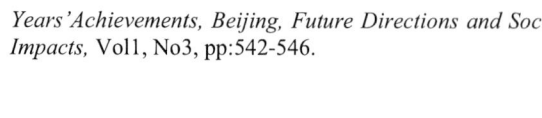

Years' Achievements, Beijing, Future Directions and Social Impacts, Vol1, No3, pp:542-546.

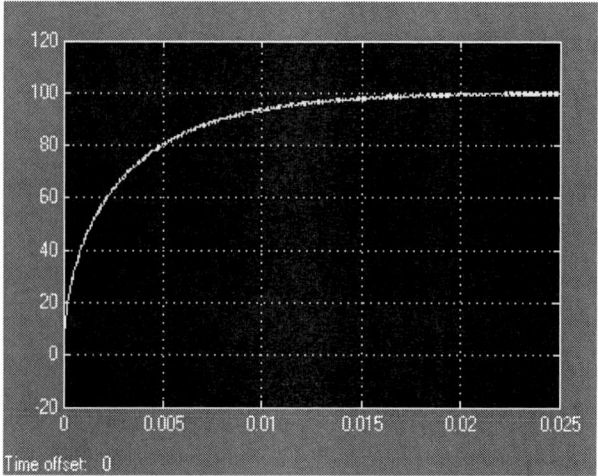

Figure 5 The response plans which ur(K)=100v

4. Conclusions

This article integrated neural network with the traditional PID control, and designed a single neuron PID controller, even if the system is superior to the traditional dynamic of the PID control, and the structure is easier than the multi-layer neural network controller and easier to fulfillment . The simulation results show that when single neurons Sepic PID controller was applied to the converter, it will achieve good control results, In addition, the controller also can be applied to other types of DC / DC converters, it has widely application prospects .

References

[1] Yan xiaozhao, Zhang xingguo(2006). *"Application of Increasing PID Controlling Method in Temperature Controlling System". Journal of Nantong University (Natural Science)*, Vol. 5, No. 4, pp.48-51 (in Chinese).

[2] Huang huanpao, Han jingqing(2002). *" Nonlinear PID controller and its applications in power plants ". IEEEPower System Technology*, No.3,pp:1513-1517.

[3] Sun xiao, Xu dehong, et al . "Neural-network-controlled Single-phase UPS Inverters with Improved Transient Response and Adaptability to Various Loads, Power Electronics and Drive Systems(1999)". *Proceedings of the IEEE International Conferenc.(PEDS'99).*No.2,pp:865-870.

[4] Babuska R ,et al(2002) . "Fuzzy self-t uning PI control of pH in fermentation" *.Engineering Applications of Artificial Intelligence*, Vol15,pp :3-1 5.

[5] Shi jun, Fang youtong, Ye yunyue,et al. (2003), "Vector control system based on single neuron PI for linear induction motor". *Transactions of China Electrotechnical Society* ,Vol18,No4,pp:94~98,(in Chinese).

[6] Tao yonghua. *New type PID control and appliance*, China Machine Press, Beijing

[7] Zhang qian, Wang yan, Wang lan(2006). "The Simulation Method of Intelligent Control Based on Dynamic Models Library and Its Application". *Proceedings of 2006 International Conference on Artificial Intelligence—50*

H∞ Robust Controller for Electric Bicycles

Zhou HaoBin[1, 2], Long Bo[2], Cao BingGang[2]

1   *Department of Material Science and Engineering*

Xi'an Shiyou University，Xian 710065, China

2   *School of Mechatronic Engineering*

Xian JiaoTong University, Xian 710049, China

hbzhou@xsyu.edu.cn

Abstract

The performance of electric bicycles is largely determined by their electric controllers, which is regarded as a core component in the whole system. We use the robust control theory to design a H_∞ robust controller for the current loop in the closed speed-current dual-loop system. Furthermore, we use the H_∞ robust controller to design an energy recovery controlling system based on the driving and energy regenerating circuit topology for widely used electric bicycles. Experiments and simulations show that the H_∞ robust controller out-performs the traditional PID controller in many respects including stability, error, response speed and driving distance per battery charge.

Key Words: Electric Bicycle, Energy Recovery, H_∞ robust control

1 Introduction

Similar to electric vehicles, which are getting more popular in the world due to environmental and economical concerns[3][11], electric bicycles have been widely used as replacements for motorcycles for short distanced urban commute in many regions , such as California China and other south-east Asia countries[1][4]. However, a major disadvantage of electric bicycles produced by using traditional technologies is that the driving distance per battery charge is short. And the batteries must be recharged frequently. The goal of this study is to improve the driving distance per batteries charge. In this paper, we focus on designing new control systems, specially the brake control systems, to achieve the goal, because the control systems are core parts acting as the "brains" in electric bicycles, and their properties crucially determine the performance the whole systems.

A major cause for the short driving distance per battery charge is that electric bicycles must be accelerating and braking frequently in typical city traffic environments. Large amount of energy is wasted due to the mechanical friction in the braking process. It seems natural to design new braking systems which recover the lost kinetic energy. In fact, this idea can be realized by using permanent magnet brushless DC motors (BLDCM),

which can work in both electric driving mode as well as electric generating mode[2][9][15]. Our control systems are designed to regenerating electric energy from braking kinetic energy based on using the permanent magnet BLDCM in electric bicycles. The new systems maintain the same wire connections in the permanent magnet BLDCM in working conditions. The control systems switch the motors between driving mode and electric regenerating mode by reversing the electric current directions in motors depending on the driving environments. When in the electric energy re-generating mode, the motors convert the braking kinetic energy into electric energy, and then re-charge back to the batteries, therefore to improve the driving distance per battery charge. In addition, the energy recovery systems reduce the noise pollution caused by mechanical braking process.

In designing electric energy regenerating braking systems, the control systems are the key components. Most of the current control systems use the PID controllers which are based on the feedback control concept[14]. This is due to the PID controller's simple design and easy to use properties. However, the PID control method has its intrinsic drawbacks. When it is used for nonlinear, time-varying, coupled processes with parameter and structural uncertainties, the PID controller cannot achieve satisfactory results, because it is not designed to control complex processes and does not have the ability to control certain components in the systems that are not considered in the mathematical model. On the other hand, riding electric bicycles must constantly face uncertain driving environments, and people have increased and different expectations on riding comfort. All these demand more complex control systems. Any by using robust control theory [10][16], we can achieve this.

Robust control theory is developed to apply the new developments in the modern control theory, which is often mathematical oriented, to engineering applications, so that it is more suitable and convenient to use for practical problems. The key idea is to re-formulate the characteristics of the systems in the frequency domain. It also takes into account of the uncertainty and error of the mathematical models in the design [13]. By assuming that there exists error in the frequencies and parameters between the mathematical models and the practical problems, the controllers are designed based on analytical strategies to guarantee achieving satisfactory performance for all objects being controlled within certain error ranges.

In this paper, we use H_∞ robust control theory, which has been widely used for many applications[5][6][7][8][12], to design new driving and energy regenerating brake control systems for electric bicycles. We also compare our designs with the commonly used PID based driving and braking control system. And our results demonstrate that the new design has better energy recovery property.

The paper is arranged as following. In the next two sections, we present the design for driving control system and braking control system respectively. The comparison between the new controller and the traditional PID controller is given in Section 4. And brief conclusions are given in Section 5.

2 Driving Control System Design

2.1 The Main Driving Control Circuit

A three-phase permanent magnet BLDCM for electric bicycles consists of a symmetric three-phase winding stator and a permanent magnet rotor. Currently, the commonly used BLDCM are usually two phase conducting. The switching on-or-off of the three-phase stator inverter is determined by the

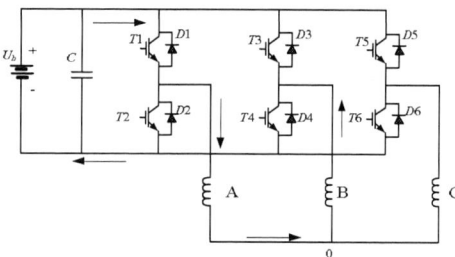

Figure 1   Three-phase full-bridge circuit and its driving current diagram.

output signals from the three position sensors of the rotor. Essentially, a BLDCM can also be viewed as a DC motor whose mechanical phase changing device is replaced by an electric phase changing device. A complete three-phase full-bridge circuit and its driving current diagram is shown in Figure 1 (the AB phase is used here as an example).

The Mathematical Model for the Driving Circuit

To simplify the analysis, we use a bi-polar three-phase permanent magnet BLDCM as an example and apply trapezoidal waves as its anti-potential. Without loss of generality, we make the following assumptions:

(1) The windings of the stator use three-phase star connection without a centerline.

(2) The windings are evenly distributed in the smooth inner surface of the stator. Alveolar effects are not considered.

(3) We ignore the energy loss caused by magnetic saturation, hysteretic and eddy current.

(4) No damping from the windings of the rotor and the permanent magnet.

Since the two-phase conducting brushless motor is equivalent to a DC motor, we can establish an equivalent circuit model with the state equations for two different cases: (a)   T1-conduction   ($0 \le t \le dT$)   and (b)T1-cutting-off.

Using the method of average and asymptotic analysis, we can separate the steady state and transient state to achieve the following linearized model for small signals.

$$
\dot{x} = \begin{bmatrix} \dfrac{-2(r_{ba}+2r)+r_d-(1-D)(r_{ba}+2r+r_d)}{2L} & \dfrac{-K_e}{L} \\ \dfrac{K_t}{J} & 0 \end{bmatrix} x + \begin{bmatrix} \dfrac{r_{ba}+2r+r_d}{2L} & \dfrac{-K_e}{L} \\ \dfrac{K_t}{J} & 0 \end{bmatrix} \cdot X \cdot d + \begin{bmatrix} \dfrac{v_b}{L} \\ 0 \end{bmatrix} \cdot d \quad (1)
$$

$$
y = \begin{bmatrix} 1-D & 0 \end{bmatrix} x + \begin{bmatrix} -1 & 0 \end{bmatrix} \cdot X \cdot d
$$

3 Design of Electric Energy Recovery Controller

3.1 The Electric Energy Regenerating Brake Control System

When the motor is in the braking mode, the brake control system achieves the goal of energy recovery by using the Boost chopper modulation of the three-phase inverter PWM controller to convert the kinetic braking energy into electricity and then charge back batteries, which consequently improves its driving distance. Figure 2 is the current flow diagram for the energy regenerating brake control system.

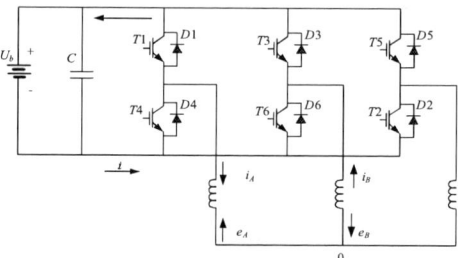

Figure 2   The current flow diagram for the energy Regenerative braking control system.

3.2 The Mathematical Model of Energy Regenerating brake controller

The brake system is similar to the BOOST converter. The equivalent circuit and its derivation are similar too. We can obtain a linearized small-signal model for the steady state and transient state by using asymptotic analysis for the brake system which is stated as following,

$$\dot{x} = (A_{ON} \cdot d + A_{OFF} \cdot (1-d)) \cdot x + (B_{ON} \cdot d + B_{OFF} \cdot (1-d)) \cdot w$$

$$= \begin{bmatrix} \dfrac{(r_m+r_s)d+(r_m+r_d+\frac{r_b r_c}{r_b+r_c})(1-d)}{L} & \dfrac{r_b(1-d)}{L(r_b+r_c)} \\ \dfrac{r_b(1-d)}{C_0(r_b+r_c)} & \dfrac{1}{C_0(r_b+r_c)} \end{bmatrix} x + \begin{bmatrix} \dfrac{1}{L} & \dfrac{r_c(1-d)}{L(r_b+r_c)} \\ 0 & \dfrac{1}{C_0(r_b+r_c)} \end{bmatrix} w \tag{2}$$

$$i_b = (C_{ON} \cdot d + C_{OFF} \cdot (1-d)) \cdot x + (D_{ON} \cdot d + D_{OFF} \cdot (1-d)) \cdot w$$

$$= \begin{bmatrix} \dfrac{r_c(1-d)}{r_b+r_c} & \dfrac{1}{r_b+r_c} \end{bmatrix} x + \begin{bmatrix} 0 & -\dfrac{1}{r_b+r_c} \end{bmatrix} w$$

4. The Design of H∞ Robust Controller for the Driving System

The standard MATLAB/SIMULINK toolbox is used to model and analyze the BLDCM system. We establish a mathematical model the brushless motor, and simulate the trapezoidal wave by using piecewise linearization for the inverse of the electric potential modulated in the system. And we design a closed dual-loop control system for the speed-current in the brushless motor as shown in Figure 3.

4.1 The Comparison Between H∞ Robust Controller and PID Controller.

We use dual-loop in our H∞ robust controller, where the inner current loop is controlled by the method of H∞ robust control. The goal is to find a controller C(s), so that the H∞ norm of the closed loop is ensured to be smaller than a given small integer Y. In this case, the controller can be expressed as:

$$C(s) = \begin{bmatrix} A_f & -Z \cdot L \\ F & 0 \end{bmatrix} \tag{3}$$

where

$$\left. \begin{array}{l} A_f = A + \gamma^{-2} B_1 B_1^T X + B_2 F + ZLC_2 \\ F = -B_2^T X, L = -YC_1^T, Z = (I - \gamma^{-2} YX)^{-1} \end{array} \right\} \tag{4}$$

And X and Y are the solutions of the following two algebraic Riccati equations:

$$\left. \begin{array}{l} A^T X + XA + X(\gamma^{-2} B_1 B_1^T - B_2 B_2^T)X + C_1 C_1^T = 0 \\ AY + YA^T + Y(\gamma^{-2} C_1^T C_1 - C_2^T C_2)Y + B_1^T B_1 = 0 \end{array} \right\} \tag{5}$$

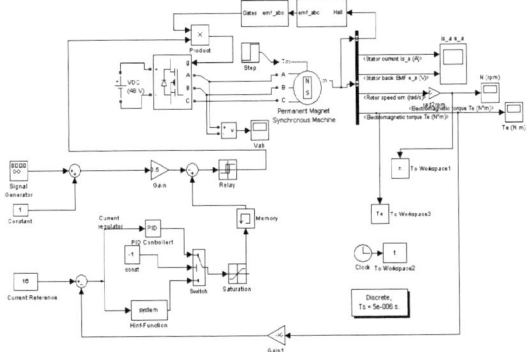

Figure 3　The design for the closed dual-loop control system for the speed-current in the brushless motor.

The conditions for the existence of such a H∞ controller are:

(1) D11 is small enough, and satisfies D11 < Y;

(2) The solution X in the Riccati equation is a positive definite matrix;

(3) The solution Y in the Riccati equation is a positive definite matrix;

(4) Eigenvalues of XY are smaller than Y^2.

By introducing weight functions [17]

$$W_1(s) = \frac{500(0.005s+1)^2}{(0.2s+1)^2}, V(s) = 0.25s^2 \times 10^{-4} \tag{6}$$

We can use the "*hinf()*" function in the robust control toolbox in MATLAB to design the H∞ controller which satisfies the general requirements on the mixed stability and quality stability. This controller is given as:

$$G_c(s) = \frac{960s^2 + 81350s + 50079}{s^3 + 487s^2 + 4791s + 11915} \tag{7}$$

where we take the control period $T_S = 0.001s$. Using a bilinear transform, we obtain the discrete controller as:

$$K_b(s) = \frac{1.386 \cdot 10^{-8} \cdot s^3 + 1.471 \cdot s^2 + 1379 \cdot s + 330.5}{s^3 + 1.165 \cdot 10^4 \cdot s^2 + 241.7 \cdot s + 1.252} \tag{8}$$

Similarly, by introducing the weight functions:

$$W_1(s) = \frac{0.4 \cdot s + 130}{s + 0.01}$$

$$W_2(s) = \frac{s + 50}{200}$$

$$W_3(s) = 20,$$

And using the robust control toolbox in MATLAB, we can find the H∞ robust controller for the braking system as:

$$K_b(s) = \frac{1.386 \cdot 10^{-8} \cdot s^3 + 1.471 \cdot s^2 + 1379 \cdot s + 330.5}{s^3 + 1.165 \cdot 10^4 \cdot s^2 + 241.7 \cdot s + 1.252} \tag{9}$$

If we take the sampling period as T=0.002s, and use a bilinear transformation, we obtain the discrete controller for the driving system as:

$$K_d(z) = \frac{2.252 \cdot 10^{-4} \cdot z^3 - 7.189 \cdot 10^{-6} z^2 - 2.251 \cdot 10^{-4} z + 7.29 \cdot 10^{-6}}{z^3 - 1.158 \cdot z^2 - 0.6839 \cdot z + 0.84194} \tag{10}$$

In Figures 4-6, we show the simulated responding curves of the driving and braking systems for the PID controller and H∞ robust controller respectively. The results indicate that the H∞ robust controller have better response time and stability than these of the PID controller.

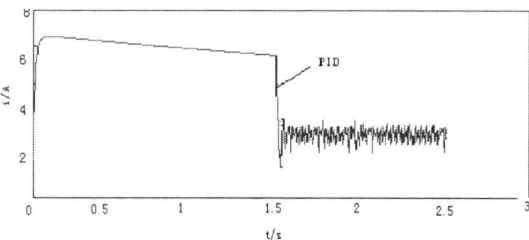

Figure 4 driving current and velocity waveforms for the PID controller

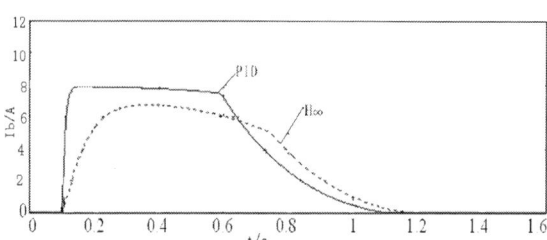

Figure 5 Driving current and velocity waveforms for the H∞ controller

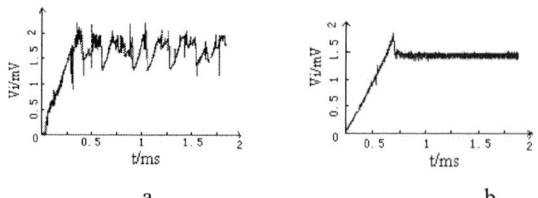

Fig7 different control algorithm of out put current waveform comparison while driving
a PID control algorithm
b. H∞control algorithm

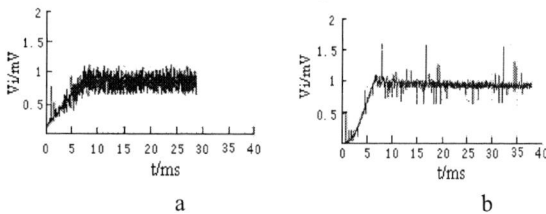

Fig8 different control algorithm of out put current waveform comparison while energy regenerating
a.PID control algorithm for regenerating
b. H∞control algorithm for regenerating

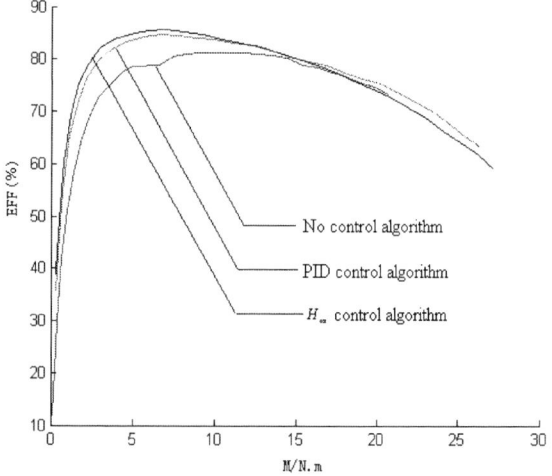

Fig 9 Comparison diagram of Three test controller efficiency

Figure 6 Braking current response waveform for the PID and H∞ controller

4.2 The Experimental Results for Energy Recovery

The H_∞ robust controller as described in the previous sections is designed and applied on BLDCM. Fig 7 is the comparison current diagram for different control arithmetic when the motor is in computation state. Fig 8 is the regenerating current waveform diagram comparison for different controlling arithmetic. Fig 9 is the efficiency comparison diagram for three scenarios: no controller, PID controller and H_∞ robust controller. As shown in the figures, experiments and test results show that the new controller has high efficiency than the traditional PID controller, even both controller's maximum efficiency surpass 80% as shown in Fig9. The test results also show that the system runs steady, and has excellent reliability and strong ability of banishing disturbance.

As we mentioned earlier, the ultimate goal to design the control system in electric bicycles is to improve the driving distance per battery charge by energy recovering through braking process. Our experiments measure the impacts on mileage for the newly designed control system. In the experiments, we use the same type of bicycles for all tests, which means their shapes, batteries, weights of the bicycles and the riders are the same. We compare the following three different control systems: EB0 (general control system without energy recovery ability), EB1 (PID controller for energy recovery), EB2 (H∞ robust controller for energy recovery). Figure 10 shows the mileage curves for the 48v electric bicycle (the motor is produced by Shanghai 21 Institute, the An Na Da 48v, 500W, permanent magnet BLDCM) by using 300m fixed track braking tests.

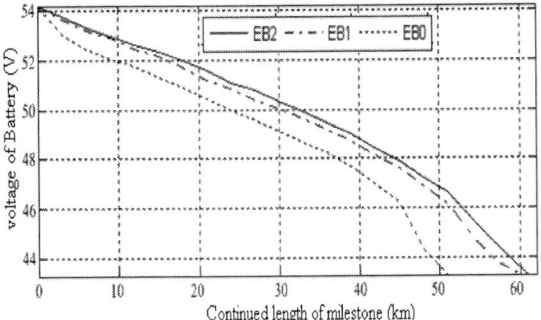

Figure 10 48V electric bicycle trip continued sentinel braking distance of 300m

To confirm performance of our newly designed robust controller, we carried out the driving distance experiments by using 36V and 48V (the rated voltage of batteries used on electric bicycles) batteries in typical city traffic environments. The results show that both of the energy regenerating control systems, which are designed on the basis of PID controller and H_∞ robust controller in the paper, can significantly extend the driving range. Comparing to the electric bicycle of which the control system is not energy regenerative, the control system with H_∞ robust controller can increase the driving range by 12.5%~26.0% and 10.5%~19.6%, for 36V and 48V batteries respectively; while the control system with PID controller can increase the driving range by 10%~22.4% and 8.7%~17.6%, for 36V and 48V batteries respectively. That means the control system with H_∞ robust controller can increase more than PID controller by 1.8%~2.6%.

5 Conclusions

1) We apply robust control theory and H∞ robust control method to the driving and braking control systems for electric bicycles. We design the H∞ robust controller for the inner loop in the speed-current closed dual-loop system. Numerical simulation results show that new controller is better than the traditional PID controller in many respects including stability, error and response speed. We have also performed the comparisons on the real products, and the experiments shows that the electric bicycles can consistently improve the driving distance from 1.8% to 2.6% depending on the road conditions and different batteries over the same bicycles equipped with PID controller.

2) The energy recovery system using H∞ robust controller is more robust and has shown better properties for the uncertainties in mathematical models, parameter measurements and resistance to the environmental disturbance than the PID controller. It provides more comfortable rides.

References

[1]Van Amburg B. *Emerging markets and players in the electric two-wheel industry in California and Asia[J]*. EVS14，1997.

[2]Bai Z. F., *Research and Development on Ultracapacitor-Battery Composite Power System of Electric Vehicle*. Xi'an, 2005.

[3]Chan C. C.，Wong Y. S.. The state of the art of electric vehicles technology [A]. *The 4th International Conference on Power Electronics and Motion Control[C]*. 2004. 46~57.

[4] Cui W. A., *Electric Bicycles [M]*, Mechanical Industrial Press, Beijing, 2001.5~7，28，45~49，192~200. (in Chinese)

[5] Doyle J. C.，Gbver K.，Hargoneger P. K., State-space solution to standard H_2 and H_∞ contrl problem. *IEEE Trans on Automatic Control[J]*. 1989，34(8)：831~847.

[6]Iwasak. All controllers for the general H_∞ control problem： LMI existence conditions and state space formula. *Automatica*[J]. 1994，30：1307~1317

[7]Kang W. Nonlinear H_∞ Control and Its Application to Rigid Spacecraft. *IEEE Trans on Automatic Control[J]*. 1995，40(7)：1281~1285.

[8]Kuraoka，Ohka H.，Ohba N.，Hosoe M.，Zhang S F. Application of H_∞ design to automotive fuel control. *IEEE Control Systems Magazine[J]*. 1990，10(3)：102~106.

[9]Li J. C., The Driving and Energy Recovery Braking System for Electric.*Vehicles [D]. Xian JiaoTong University*, Xian 2003.

[10]Mei S. W., Shen T. L. and Liu K. Z., *Modern Robust Control Theory and its Applications [M]*. Tsinghua University Press, Beijing, 2003. (in Chinese)

[11]Morita K.，et al. Advanced Clean Energy Vehicle Project in Japan. .*Proceedings of the 16th International Electric Vehicle Symposium[J]*，Beijing.1999，CD.

[12]Naim W R，Ben-Yaakov G. H_∞ control applied to boost power converters. *IEEE Trans on Power Electronics[J]*. 1997，12(4)：677~683.

[13]Shen T. L., *H∞ Control Theory and Applications [M]*.*Tsinghua University Press*, Beijing, 1996. (in Chinese)

[14]Tao Y. H., *New PID Control Systems and Applications [M]*, Mechanical Industrial Press, Beijing, 2005.17~20. (in Chinese)

[15]Zhang S., *The Principle of Brushless DC motor and Applications [M]*.Mechanical Industrial Press, Beijing, 1996.

[16]Zhou K. M., Doyle J. C., and Glover K., *Robust and Optimal Control [M]*. The National Defense Industrial.Press, 2002.

[17]Wu Xiudong, Xie Xueshu, Weight Selections in H_∞ Control, *Journal of Tsinghua University [J]*. 1997，37(1)：27~30. (in Chinese)

Vector Control System of Induction Motor Based on Fuzzy Control Method

Zhou HaoBin[1, 2], Long Bo[2], Cao BingGang[2]

1 Department of Material Science and Engineering
Xi'an Shiyou University，Xi'an 710065, China
2 School of Mechatronic Engineering
Xi'an JiaoTong University, Xi'an 710049, China
hbzhou@xsyu.edu.cn

Abstract

Aiming at non-linear model of multivariable induction motor, traditional PI method is hard to achieve satisfactory control purpose. Fuzzy control can carry the real-time control on system without the accurate mathematical model of induction motor. This paper elaborates induction motor vector control system and the basic principle of the fuzzy PI, making use of fuzzy reasoning, automatically adjusting the controller's parameters. Simulation results prove that fuzzy PI controller is superior to traditional PI controller in the aspect of response speed、steady state accuracy and disturbance attenuation.

Key words: fuzzy control   vector control adjustment parameter

1. Introduction

PI control algorithm is simple, realized easily, used in traditional induction motor speed regulating system usually. With the demand of save energy performance and technical and economic target of speed regulating system, PI control have some limitation. For complex model of induction motor, the parameter of the controller is hard auto adjust to adapt outside conditional variety, making the control of motor inaccuracy. To have better adaptability, realize the automatic adjustment of the controller parameter, can adopt the method of the fuzzy control theories[1]. The fuzzy control can realize the automatic adjustment of the control parameter, with strong adaptability and good speed governing. This paper apply vector control method on induction motor, use the decoupling of rotor flux linkage and torque, practice a continuous control, comparatively width speed governing scope. Electric current loop use space vector pulse width modulation, and design a kind of controller of self- adaptation fuzzy control, proceed simulation study through Matlab/Simulink, carry on contrast with traditional PI controller.

2 Vector control system

2.1 Induction motor model

Dynamic model of induction motor is a high、nonlinear、close coupled and multivariable system. Take forward of voltage、current、flux linkage of each winding as motor routine and right-hand screw rule. At this time, mathematical model of induction motor is constitute by voltage equation 、 flux linkage equation 、 torque equation[2] of two-phase arbitrary revolution coordinate as follows:

（1）   voltage equation

$$
\begin{bmatrix} u_{sd} \\ u_{sq} \\ u_{rd} \\ u_{rq} \end{bmatrix} = \begin{bmatrix} R_s + L_s p & -\omega_e L_s & L_m p & -\omega_e L_m \\ \omega_e L_s & R_s + L_s p & \omega_e L_m & L_m p \\ L_m p & -\omega_s L_m & R_r + L_r p & -\omega_s L_r \\ \omega_s L_m & L_m p & \omega_s L_r & R_r + L_r p \end{bmatrix} \begin{bmatrix} i_{sd} \\ i_{sq} \\ i_{rd} \\ i_{rq} \end{bmatrix}
$$

(1)

（2）   flux linkage equation

$$
\begin{cases} \psi_{sd} = L_s i_{sd} + L_m i_{rd} \\ \psi_{sq} = L_s i_{sq} + L_m i_{rq} \\ \psi_{rd} = L_m i_{sd} + L_r i_{rd} \\ \psi_{rq} = L_m i_{sq} + L_r i_{rq} \end{cases}
$$

(2)

（3）   electromagnetic torque equation

$$
T_e = n_p L_m \left(i_{sq} i_{rd} - i_{sd} i_{rq} \right)
$$

(3)

2.2 Vector principle

The basic principle of vector control is to simulate torque control rule of direct current motor in general three-phase AC motor, stator current vector is decomposed to field current component that generate magnetic flux and torque current component that generate torque, make them orthogonal and arm s length sale, and adjust independent. So, torque control of AC motor is similar to DC motor in principle. The key of vector control is to control amplitude and spatial location (frequency and phase).

In antecedent dynamic model analysis, as to synchronous revolution coordinate, $\omega_{dqs} = \omega_1$, $\omega_{dqr} = \omega_1 - \omega = \omega_s$, the inner part of rat cage motor is short-circuit, $u_{rd} = u_{rq} = 0$ state equations is achieved:

$$\frac{d\omega}{dt} = \frac{p_n^2 L_m}{JL_r}(i_{sq}\Psi_{rd} - i_{sd}\Psi_{rq}) - \frac{p_n}{J}T_L$$

$$\frac{d\Psi_{rd}}{dt} = -\frac{1}{T_r}\Psi_{rd} + (\omega_1 - \omega)\Psi_{rq} + \frac{L_m}{T_r}i_{sd}$$

$$\frac{d\Psi_{rq}}{dt} = -\frac{1}{T_r}\Psi_{rq} + (\omega_1 - \omega)\Psi_{rd} + \frac{L_m}{T_r}i_{sq}$$

$$\frac{di_{sd}}{dt} = \frac{L_m}{\sigma L_s L_r T_r}\Psi_{rd} + \frac{L_m}{\sigma L_s L_r}\omega\Psi_{rq} - \frac{R_s L_r^2 + R_r L_m^2}{\sigma L_s L_r^2}i_{sd} + \omega_1 i_{sq} + \frac{u_{sd}}{\sigma L_s}$$

$$\frac{di_{sq}}{dt} = \frac{L_m}{\sigma L_s L_r T_r}\Psi_{rq} - \frac{L_m}{\sigma L_s L_r}\omega\Psi_{rd} - \frac{R_s L_r^2 + R_r L_m^2}{\sigma L_s L_r^2}i_{sq} - \omega_1 i_{sd} + \frac{u_{sd}}{\sigma L_s}$$

(4)

In the equation: σ — motor leakage coefficient, $\sigma = 1 - L_m^2 / L_s L_r$, T_r — rotor electromagnetic time constant. $T_r = L_r / R_r$

If daxis is taken as direction of whole flux linkage vector, anti-clockwise 90 is q axis, it against to vector, named T axis, this two phase synchronous revolution coordinate is based on field orientation coordinate. M axis is rotor field orientation, T axis is torque control component. Like DC motor, it control individual field current and torque current so that control torque. So, $\Psi_{rd} = \Psi_{rm} = \Psi_r$, $\Psi_{rq} = \Psi_{rt} = 0$, $\dot{\Psi}_{rt} = 0$, state equations substituted is given:

$$\omega_{sl} = L_m i_{st} / T_r \psi_r \qquad (5)$$

$$i_{sm} = (T_r p + 1)\psi_r / L_m \qquad (6)$$

It can be seen, rotor field is generated by state current having no relation to torque component, speak from this meaning, field excitation component and torque component of state current is decoupling.

The above formula still expresses that both are first-order inertia link, time constan is rotor flux field time constant, when field excitation i_{sm} happen to change suddenly, the change of Ψ_r is objected by field excitation inertia, seeing from this, inertial function of DC motor field-winding is same to the Ψ_r.

Mathematical model is described as following structural style.

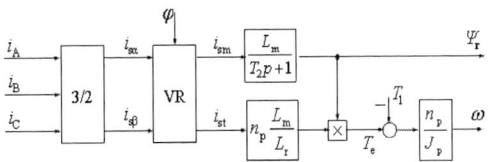

Figure 1 mathematical model of vector transformation and current decoupling of induction motor

3 Fuzzy PI controller design

Under the circumstance of nonlinear and parameter time-varying object controlled, accurate model can not be given. Although traditional PID control is simple, controlling result better, it can not solve the circumstance that model parameter occur variety, moreover, in practical spot, because of being subjected to parameter setting, traditional PID parameter setted is not good, its performance is not quite good, be badly to the adaptability. So, according to complex circumstance appllied and the control request of high-performance, having an urgent request to PID parameter self-setted, fuzzy control to PID setted is a good solution.

3.1 Fuzzy PI principle

Fuzzy control is computer intelligent control based on fuzzy set theory 、 fuzzy language 、 fuzzy logical-inference, its basic idea is presented by famous professor named L.A.Zadeh in California university in America, who achieved the great success in fuzzy control theory and application study through development of many years.

3.2 Fuzzy PI parameter setted

Fuzzy self adapting controller has many kind of structural style at the present time, but their operating principle is same to each other. Fuzzy self adapting controller take error e and error alteration ec as an input, amend parameter at a real time using fuzzy control regulation, these constitute fuzzy self adapting controller.

PID parameter fuzzy self adapting is to find fuzzy relationship of three parameter、 e and ec, check e and ec continuously in service, according to fuzzy control principle amending three parameter in real time, meeting the different need of controlled parameter in different time, making object controlled having good dynamic and static character.

The core of PID controller design that parameter fuzzy self-adapting is to create suitable fuzzy regulation, see figure 1、figure 2、figure 3 individually. At the present, the building of fuzzy regulation is achieved by generalizing experience of operation crews, analysing closed loop response of system, and using method of trial and error through many simulation test.

Establish fuzzy controller inferenced by Mamdani of 1 two-input and three-output using figure windows editor, supposed that input and output is (-3,3),corresponding language value is negative big(NB),negative middle(NM), negative small(NS), zero(Z), positive small(PS), positive middle(PM), positive big (PB), all the degree of membership function of input and output variance are trimf, see figure 2:

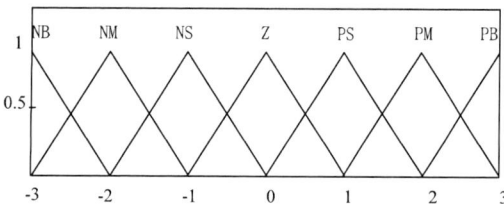

Figure 2　degree of membership function of e、ec、Kp、Ki、Kd

（1）fuzzy regulation table of Kp

Table 1　fuzzy regulation table of Kp

kp\ec e	NB	NM	NS	Z	PS	PM	PB
NB	PB	PB	PM	PM	PS	Z	Z
NM	PB	PB	PM	PS	PS	Z	NS
PS	PM	PM	PM	PS	Z	NS	NS
Z	PM	PM	PS	Z	NS	NM	NM
PS	PS	PS	Z	NS	NS	NM	NM
PM	PS	Z	NS	NM	NM	NM	NB
PB	Z	Z	NM	NM	NM	NB	NB

（2）fuzzy regulation table of Ki

Table 2　fuzzy regulation table of Ki

ki\ec e	NB	NM	NS	Z	PS	PM	PB
NB	NB	NB	NM	NM	NS	Z	Z
NM	NB	NB	NM	NS	NS	Z	Z
PS	NB	NM	NS	NS	Z	PS	PS
Z	NM	NM	NS	Z	PS	PM	PM
PS	NM	NS	Z	PS	PS	PM	PB
PM	Z	Z	PS	PS	PM	PB	PB
PB	Z	Z	PS	PM	PM	PB	PB

（3）fuzzy regulation table of Kd

Table 3　fuzzy regulation table of Kd

kd\ec e	NB	NM	NS	Z	PS	PM	PB
NB	PS	NS	NB	NB	NB	NM	PS
NM	PS	NS	NB	NM	NM	NS	Z
PS	Z	NS	NM	NM	NS	NS	Z
Z	Z	NS	NS	NS	NS	NS	Z
PS	Z	Z	Z	Z	Z	Z	Z
PM	PB	PS	PS	PS	PS	PS	PB
PB	PB	PM	PM	PM	PS	PS	PB

4 Simulation result

Simulation curve diagram is from figure 4 to figure 6. Figure 3 and figure 4 express individually that speed response curve of fuzzy PI and PI with a speed code valuation equal to 120rpm/min, rotational speed is zero at initial state, when time is equal to 0.5 second. Figure 5 and figure 6 express individually that torque response curve of fuzzy PI and PI with a torque code valuation equal to 150N.m, rotational speed is zero at initial state, when time is equal to 1 second.

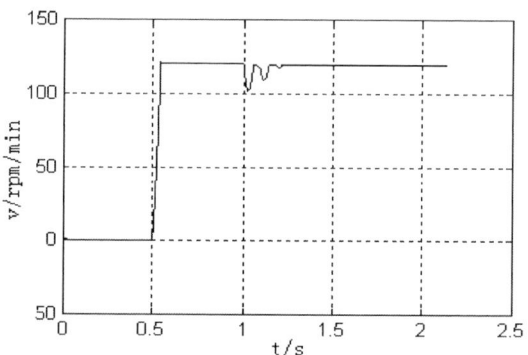

Figure 3 speed response curve of PI algorithm controller

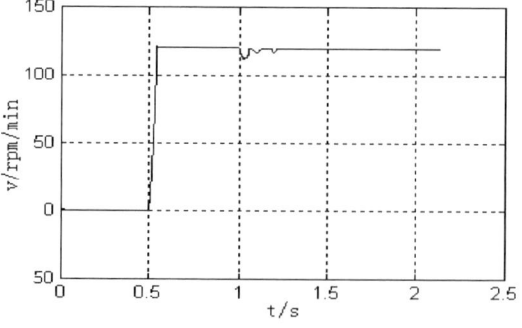

Figure 4 speed response curve of fuzzy PI algorithm controller

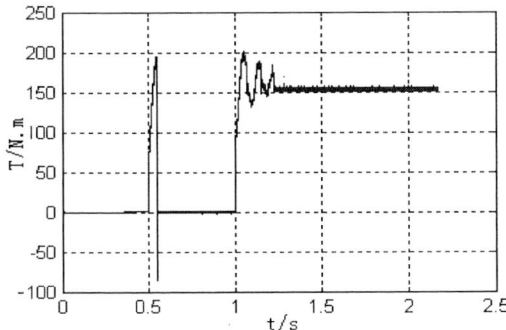

Figure 5 torque response curve of PI algorithm controller

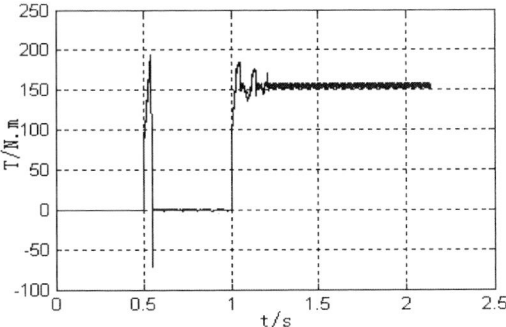

Figure 6 torque response curve of fuzzy PI algorithm controller

See from simulation result, with initial code valuation, fuzzy PI controller is better than traditional PI controller in speed response and overshoot suppression. When there is some load torque, the former is very excellent in against-interference ability and robust is strong. Under the circumstance of adding code valuation, fuzzy PI controller is superior to traditional PI controller in speed response and steady-state error, which meet the design request.

5 Conclusion

Aiming at the characteristic of induction motor with higher order、nonlinear、closed linkage, adopt vector control system according to rotor field orientation control method, integrate fuzzy and PI these two algorithm, design a kind of self-adapting fuzzy controller, proceed simulation test using fuzzy tool in Matlab, compare to traditional PI controller, test express that fuzzy PI controller is superior to traditional PI controller in speed response、steady-state accuracy and against-inference.

References

[1] Yi jigai, Hou yuanbin. *Intelligent Control Technology(Fifth Edition)[M]*.beijing: Beijing Industrial University Press,2004

[2]Chen boshi. *AC speed system[M]*. Beijing: Machinery Industry Press,2005.

[3] L.A.Zadeh. *Fuzzy Algorithms [J]*. Information and Control，1968，12

[4]Wen xin, Zhou lu, *MATLAB toolbox Fuzzy Logic Analysis and Application[M]*.beijing: Beijing Science Press，2001.

[5]A.Visioli. *Tuning of PI controllers with fuzzy logic[J]*.IEEE Proc-Control Theory，2001，148（1）：69～81.

[6]L.A.Zadeh.*Fuzzy Algorithms [J]*.Information and Control，1968（12）：28～30.

[7]Edson Bim . *Fuzzy Optimization for Rotor Constant Identification of an Indirect FOC Induction Motor Drive [J]*. IEEE Transactions on Industrial Electronics，2001，48（6）：1293～1295.

[8]Lin,C.T.Neural Fuzzy Control Systems with Structure and Parameter Learning,*World Scientific*.Singapore，1994.

[9]Karr C L，etal.Fuzzy control of PH using genetic algorithms.*IEEE Trans.on Fuzzy Systems*，1993，1（1）46～53.

[10] Driankov ， D.H.Hellendoorn ， M.Reinfrank.An Introduction to Fuzzy Control ， *Springervrelag* ， Berlin.1993.

Session 4

Multimedia information Processing

2008 Workshop on Power Electronics and Intelligent Transportation System

Improving Image Encryption Using Multi-chaotic Map

Yikui Zhai
Wuyi University
School of Information
Guangdong,China 529020
samuelzhaistu@126.com

Shuyong Lin
Shantou University Medical College
The First Affiliated Hopspital
Guangdong, China 515063
kola-lin@163.com

Qiong Zhang
Shantou University
Signal Processing Lab
Guangdong, China 515063
qzhang@stu.edu.cn

Abstract

Chen G et al. presented a 3D chaotic cat map based symmetric image encryption method and Kai W et al. broke it only with the knowledge of symbolic dynamics and some specially designed plain-images. In this letter we point out why the original scheme was vulnerable to the proposed attacks. Some essential weaknesses in the original scheme are discussed here. Based on the analysis, an improved encryption scheme with multi-chaotic map is presented to achieve higher security. In our proposed method, the 3D cat map with random scan processing is developed to increase the confusion process; the combinations of coupled map lattice (CML) model, tent map and logistic chaotic map are employed in the diffusion process to improve the initial sensitivity and security of the encryption system. In addition, entropy is used to evaluate the performance of the proposed encryption system. Experimental results show that the presented method can resist various kinds of attacks and provide an effective way for high security image encryption.

1. Introduction

With the rapid development of computer network technology, multimedia communications bring us much convenience, but they also give attacker or illegal user an opportunity, thus the security of communication becomes more and more important. For digital image, there are two major protection methods. One is information hiding which includes watermarking, anonymity, steganography and cover channel. The other is encryption which includes conventional encryption and chaotic encryption[12].

There is close relationship between chaos and cryptography. As it is known, chaotic systems have many good features, such as sensitivity to initial conditions and control parameters, ergodicity, and mixing property, which can be connected with the "confusion" and "diffusion" properties in cryptography. Thus, good cryptographic system

can be regarded as chaotic or pseudo-random system from the point of algorithm [2], since ideal cryptographic performance ensured by good cryptosystem is just like chaos generated from complex dynamical systems. In [9], Marco Gotz had shown some conventional steam ciphers which could exhibit chaotic behaviors. Due to the limitation of conventional cryptology and the distinct advantage of chaotic cryptography, chaotic cryptography was made a better candidate than many traditional ciphers for multimedia data encryption.

Basically, there are two major approaches to design digital image chaotic ciphers: in the first approach, chaotic system is used to generate pseudo-random key-stream to mask plaintext directly [6], which corresponds to stream ciphers; in the second approach, Plaintext and/or the secret key is used as the initial conditions and/or control parameters, iterating/counter-iterating chaotic systems multiple times to obtain cipher-text [1] [5], which corresponds to block ciphers. Recently, the two-dimensional chaotic map has been generalized to three-dimensional for designing a secure symmetric image encryption scheme, aiming to increase the security of the cryptosystem, such as 3D Baker map based image encryption scheme proposed by Mao YB [8] and 3D Cat map based encryption scheme by Chen G [3]. Nevertheless, their cryptosystems have many drawbacks. For example, there are still several weaknesses of 3D Cat map based image cryptosystem in [3] and it can be successfully cryptanalyzed only by rebuilding a valid equivalent 3D Cat matrix, estimating the equivalent initial conditions used in the diffusion process and restoring spatial permutation processes assisted with symbolic dynamics and specially designed plain-images [11]. In this letter, we study why the new chaotic cipher in [3] is vulnerable and propose a solution to improve its security. The securities of the improved scheme are studied theoretically and experimentally here. There are three major parts in the algorithm: spatial confusion process, diffusion process and key generator. Random-scan preprocessing and group-key are also included. Random-scan not only enlarges the space of secret

978-0-7695-3342-1/08 $25.00 © 2008 IEEE
DOI 10.1109/PEITS.2008.10

key, but also confuses the position of the first pixel, thus the attackers can hardly get the ciphertext of the first pixel or track the 3D cat cap map permutation process, which made it more difficult for attackers to cryptanalyze the diffusion key. Moreover, group-key ensures different keys in different rounds, which takes a good balance of the computation and security. The rest of this essay is organized as follows. Section 2 points out the weaknesses in [3]. Section 3 describes the proposed improved encryption scheme briefly, including spatial permutation process, diffusion process and the design of key generator. Section 4 is the experimental results and the security analysis. Finally, Section 5 gives out the conclusion of this essay.

2. Weaknesses in the original system

Some essential defects were found in the original scheme, which make the attacks available. If these defects are avoided, the attacks proposed in [11] will be infeasible, and the chaotic cryptosystem will be stronger from the cryptographic viewpoint.

Firstly, there are many equivalent control parameters. For example, any matrix A' making $A' \equiv A(mod M)$ and $|det(A')| = 1$ can also recover the shuffled-image cubes encrypted by the 3D Cat matrix A. And After n times of iterations, the pixel at position $(0,0,0)$ remains unchanged, that is,if $(x_0, y_0, z_0) = (0, 0, 0)$, after n times of iterations, $(x_0^n, y_0^n, z_0^n) = (0, 0, 0)$. So in a normal scan mode, the first pixel's position can not be permuted by chaotic cat map. This is actually a weakness of the permutation process based on the cat map. And it does some helps to the attackers although the permutation process is further strengthened by a diffusion process. Secondly, the diffusion process is too simple for the intruder to break the cryptosystem. Though logistic map is used, the process can not achieve high security. The intruder can search the equivalent initial value by exhausting and reorienting the range. So it is sufficient for the intruder to break the diffusion process only with the control parameters no matter the binary sequence is known or not. Note that $I(0) = S$ is only used to initialize $C(0)$ and useless for decipher procedure. Moreover, a thorough chosen-plain-text attack called the key recovery attack has been proposed to describe how to obtain the initial condition of Logistic maps assisted with symbolic dynamics [11]. Lastly, the key scheming is not secure enough. It seems unnecessary to use Chen's circuit in key scheming to generate the control parameters from the 128-bit binary sequence, and any chaotic map can be used instead. Moreover, in [3], the binary sequence is divided into eight segments,each two of which used to generate the parameters of the system respectively. So when the input key strings has only one difference in the last bit, some of the initial parameters in the system maybe the same in the encryption

process. This apparently reduces the security of the key scheming. Therefore, in the following, we will propose a improved scheme to avoid the above weaknesses and give a detail security analysis on it.

3. The improved scheme

3.1. 3D Cat map confusion process

2D cat map was extended to 3D cat in [3], as follows

$$\begin{bmatrix} x_{n+1} \\ y_{n+1} \\ z_{n+1} \end{bmatrix} = A \begin{bmatrix} x_n \\ y_n \\ z_n \end{bmatrix} mod 1 \qquad (1)$$

Where $A=[A_1, A_2, A_3]^T$, $A_1=[1 + a_x a_z b_y, a_z, a_y + a_x a_z + a_x a_y a_z b_y]$; $A_2 = [b_z + a_x b_y + a_x a_z b_y b_z, a_z b_z + 1, a_y a_z + a_x a_y a_z b_y b_z + a_x a_z b_z + a_x a_y b_y + a_x]$; $A_3 = [a_x b_x b_y + b_y, b_x, a_x a_y b_x b_y + a_x b_x + a_y b_y + 1]$.

Specially, by setting $a_x = a_y = a_z = b_x = b_y = b_z = 1$, one can obtain $det|A| = 1$, and the three eigenvalues of A ($\sigma_1 = 7.1842 > 1$, $\sigma_2 = 0.2430 > 1$, $\sigma_3 = 0.5728 > 1$) are much larger than those of 2D cat map. It represents that the corresponding largest Lyapunov characteristic exponents is $\lambda_{max} = ln 7.1842 > 0$. Therefore the 3D cat map is still chaotic, but more complex. While the determinant of the transformed matrix A remains 1, that is 3D cat map is also a 1-1 map, area-preserving and invertible. However, digital encryption is a transformation in finite set, chaotic map used for image encryption should be discretized and kept useful properties unchanged, such as the mixing property, the sensitivity to initial conditions and parameters.

3D cat map $C(i, j, k)$ can be discretized by

$$\begin{bmatrix} x_{n+1} \\ y_{n+1} \\ z_{n+1} \end{bmatrix} = A \begin{bmatrix} x_n \\ y_n \\ z_n \end{bmatrix} mod M \qquad (2)$$

Let $C_d(i, j, k), C(i, j, k)$ denote the discrete and continuous maps respectively. Then the above discretization satisfies the following asymptotic property:

$$\lim_{N \to \infty} max_{0 \leq i, j, k \leq N} |C(i/N, j/N, k/N) - C_d(i, j, k)| = 0 \quad (3)$$

Since a, b of A are positive integers, one can easily know $|det(A)| = 1$, which denotes the 3D cat map is a 1-1 map and its chaotic property such as the mixing property, the sensitivity to initial conditions and parameters kept unchanged.

In order to avoid the weaknesses of 3D cat map in position $(0, 0, 0)$ mention above, a simple method is proposed here to change the position of the pixels at the corners. That is, to change the normal scan order into a random one. Before iterations of chaotic map, a random-couple (r_x, r_y, r_z) is generated, which represents the position of a randomly selected pixel in the cubic image. Then, the whole cubic image shifts in x, y, z directions by r_x, r_y, r_z, respectively. That is, the left-bottom pixel shifts from $(0,0,0)$ to (r_x, r_y, r_z), which is shown on the right side of Figure1. Where the left of Figure1 is the normal scan mode. From right side of Figure1, we can see that the image is shifted, and the three parts (I, II and $\wedge$) outside are mapped into the corresponding parts of the original image. The random shift process changes the normal scan mode into a random one, so it is called a random-scan mode.

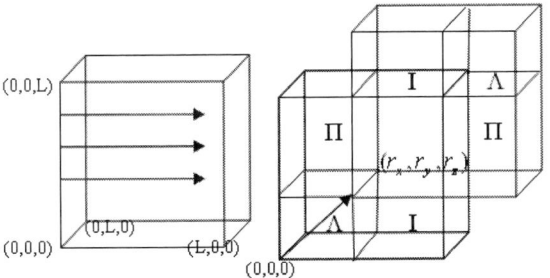

Figure 1. Comparison of normal scan and random scan mode

The 3D cat map combines with random scan mode is

$$
\begin{bmatrix} x_{n+1} \\ y_{n+1} \\ z_{n+1} \end{bmatrix} = A \begin{bmatrix} x_n + r_x \\ y_n + r_y \\ z_n + r_z \end{bmatrix} mod M \tag{4}
$$

From above, the modified map is still invertible, thus it can realize the invertible confusion process. There are two merits of the modified chaotic confusion process: firstly, the random couple can be generated by the key generator, which enlarges the key space M^3 times, where M denotes the height of the 3D cubic image; Secondly, it permutes the position of the first pixel, which will make it difficult for the attackers to obtain the plaintext of the first pixel and cryptanalyze the diffusion key by chosen-plain-text cryptanalytic attack. So it greatly increases the security of the cryptosystem.

3.2. Designation of diffusion process

It is great significant of the diffusion process in the image encryption. On one hand, it makes the discretized chaotic map non-invertible; On the other hand, it makes the statistical property of plain-text significantly changed by diffusing small changes of every pixel into the whole cipher-text, thus statistical properties become nonsense for attackers. For a security encryption scheme, diffusion mechanism is of great importance. Without it, attackers will be able to cryptan-

alyze cipher with useful information obtained by comparing plain-text and cipher-text. For the purpose of diffusion, 'XOR' and 'MOD' operation will be applied to each pixel between every two adjacent rounds of the 3D cat map in the new approach. To avoid the second weakness mentioned in Section 2, we aim to improve the diffusion process to obtain higher security here. Further details are given below. The diffusion process is realized by diffusion function $D_e(M_i, K_{di})$, which defined as

$$
\begin{cases}
d_{-1} = & K_{di}^1 \\
d_{-2} = & K_{di}^2 \\
m_{-1} = & K_{di}^3 \\
d_k = & m_k \oplus d_{k-1} \oplus g[y(m_{k-1} + d_{k-1}, L)] \\
& \oplus g[p(m_{k-1} + d_{k-2}, L)]
\end{cases} \tag{5}
$$

Where d_{-1}, d_{-2} and m_{-1} is used as the initial condition of diffusion equation above; m_k stands for the k_{th} pixel value of the image after confusion (M_i). d_k is the k_{th} pixel value after diffusion. L is the range of pixel value. $g(.)$denotes the quantization process given as

$$
g(X, L) = 2^L X \tag{6}
$$

Where $X = 0.x_0 x_1 \cdots x_L$, and x_i is a binary fraction. $p(x)$ defined as tent map

$$
p(x_{n+1}) = \begin{cases} x_n, & 0 \le x_n \le 1/2 \\ 1 - x_n, & 1/2 \le x_n \le 1 \end{cases} \tag{7}
$$

And $y(.)$denotes the chaotic CML map[4], defined as

$$
\begin{cases}
y_{n+1}(1) &= (1-\varepsilon)f[y_n(1) + \varepsilon f[y_n(2)] \\
y_{n+1}(i) &= (1-\varepsilon)f[y_n(i) + \varepsilon f[y_n(i+1)]
\end{cases} \tag{8}
$$

Where i ($i = 1, 2, \cdots, L$) stands for lattice point, and y_n satisfies $-1 \le y_n \le 1$. n is discrete-time step. $f(x)$stands for logistic map, which is

$$
f(x) = 4x(1 - x) \tag{9}
$$

ε is the coupling parameter, where $\varepsilon = 0.99$. The border satisfies $y_n(L + 1) = y_n(1)$, where L is the size of the lattice.Thus the CML mode is L dimensional spatial temporal chaotic dynamic system.

The corresponding de-diffusion process function is

$$
\begin{cases}
d_{-1} = & K_{di}^1 \\
d_{-2} = & K_{di}^2 \\
m_{-1} = & K_{di}^3 \\
m_k = & m_{k-1} \oplus d_{k-1} \oplus g[y(m_{k-1} + d_{k-1}, L)] \\
& \oplus g[p(m_{k-1} + d_{k-2}, L)]
\end{cases} \tag{10}
$$

CML is not only chaotic in both spatial and temporal direction, but also sensitive to initial values and boundary conditions.The application of spatial temporal CML chaotic map in the diffusion process can achieve much higher complexity and larger key space than logistic map and is more suitable for image encryption. With the combination of multi-chaotic map, the encryption system can reach higher security than the previous one[3].

3.3. Key generator

According to the basic criteria of cipher, a cryptosystem should be of high sensitivity to secret key, for instance, the cipher should be closely correlated with secret key. Here we use a good key generation mechanism to ensure the key sensitivity and achieve large enough key space of the cryptosystem . Furthermore, the input key sequence is designed to have changeable length ($\geq 128bit$). In the proposed method, there are n/n_0 groups of different keys. Different groups employ different keys and the same group employs the same key to iterate n_0 times, in order to take the good balance of the key space and complexity. The key generator consists of three parts: the initial values of the confusion process are $a_x, a_y, a_z, b_x, b_y, b_z$; the initial values of random scan preprocessing are r_x, r_y, r_z; the initial values of diffusion process are $d_t^{-2}, d_t^{-1}, m_t^{-1}$. Let $X_t^i (i = 1, 2, , 12)$ denote the above corresponding initial values respectively. And here the key generation function is defined as

$$X_t^i = f^T \left\{ \left(\frac{t^2 \sum_{j=1}^{12} X_{t-1}^j}{\sum_{j=1}^{12} jt} \right) mod1 \right\}, i = \{1, 2, \cdots, 12\} \quad (11)$$

Where $t \in \{1, 2, \ldots, n/n_0\}$,it means iteration time and each iteration takes the same key.

3.4. Image encryption scheme based on the improved method

The complete image encryption scheme consists of five steps of operations.

Step1. Use the key generator to generate all the initial values.

Step2. Pile up the 2D image to 3D. Suppose that the image to be encrypted is of W-pixels wide and H-pixels high. The pile up equation is

$$W \times H = N_1^3 + N_2^3 + \cdots + N_k^3 + N_R \quad (12)$$

Where $N_1, N_2, \ldots, N_k$ is the side length of each cube, and N_R is the remainder.

Step3. Use the random scan mode as the preprocessing of confusion and perform the 3D cat map based confusion process.

Step4. Perform the diffusion process.

Step5. Transform the 3D cubes back to a 2D image.

Note that, in generally both step 3 and step 4 run N times repeatedly to increase the system's security. The more times it runs the higher security it will get. However, the cost of computation will be higher and the encryption time will be longer as well.

To the end, the decryption process is similar to the encryption process, only with reverse operational sequences of Step 3 and Step 4, and has essentially the same algorithmic complexity and time consumption.

4. Experimental results and performance analysis

4.1. Results and Key sensitivity analysis

In this experiment, a 256 grey-scale image "Lena" of size 256×256 is used, as shown in Figure2.(a). When spreading up to 3D, pixels are randomly selected from the original image to form length= 8 cubes as the space of the confusion process. And The pixel to be encrypted is randomly selected to form the 3D cube in this cryptosystem. From the experiments, we can see that randomly selecting pixels during the encryption can not only avoid the block effect but also improve the short reappearance cycle length. With the $Key1 =' 123456789*ab2cdef'$, the encrypted image and the decrypted image are shown as Figure2.(b) and Figure2.(c) respectively. If there is only one bit deference in the key, say, $Key2 =' 123456789 * ab2cdee'$, from Figure2.(d), the encrypted image can't be decrypted correctly or even do the least help to the attackers.

 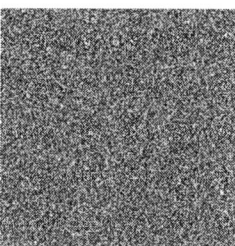

(a) original image   (b) en-image with *key1*

 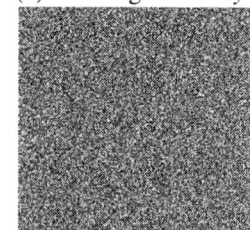

(c) de-image with *key1*   (d) de-image with *key2*

Figure 2. Results and Key sensitivity analysis

As a good encryption system should also have high sensitivity to key and large enough key space to resist any brute-force attack. In this paper, 9 parameters are used in confusion process, which space size is L^9, and 3 parameters are in diffusion process, the space size of which is 2^{3L}. The bigger size the image has, the wider range of the cube length can choose. As to a 256×256 sized image, the value of L can range from 8 to 32, thus the space size together can range from 2^{51} to 2^{141}. As group-key is used in this algorithm, different keys are used in different iteration groups, so the total key space is from 2^{51t} to 2^{141t}. Even the lower limit can ensure this encryption system to resist any brute-force attack. To test the key sensitivity of 3D cat map, the following method is adopted [7]. From different encryption keys($K - \triangle k$, K and $K + \triangle k$), the cipher-text is calculated respectively as

$$
\begin{aligned}
Y &= [C(X, K)]^n \\
Y_1 &= [C(X, K - |\triangle k|)]^n \\
Y_2 &= [C(X, K + |\triangle k|)]^n
\end{aligned} \quad (13)
$$

Where $[C()]^n$ denotes the n_{th} time iterations of chaotic map, and there is one bit difference among different keys. Thus the changed ratio of cipher-text is calculated as

$$
Cdr = \frac{Diff(Y, Y_1) + Diff(Y, Y_2)}{2N^2} \times 100\% \quad (14)
$$

Table 1. Changed ratio of cipher-text(Cdr) of varying in Cube length

Test	Cipher round						
	1	2	3	4	5	6	7
Cdr%(L=8)	99.92	99.85	99.73	97.83	99.93	99.91	99.93
Cdr%(L=16)	99.56	96.65	98.62	95.51	95.71	96.63	97.52
Cdr%(L=32)	95.61	95.72	96.56	99.61	97.83	95.69	97.65

Which N is the size of the plain-image, and $Diff()$ calculates the numbers of different pixels of the cipher-image. A and B are the same size images, and $A(i, j, k)$ is the pixel value in position (i, j, k). where $key1 = $ '123456789 * ab2cdef'; $key2 = $ '123456789 * ab2cdee'; $key3 = $ '123456789 * ab2cdeg'; that is one bit difference among keys. To change the value of space size $L = 8, 16, 32$ form Table 1 all the Cdr values of different L achieve above 99.50%. It proves that the cryptosystem has high key sensitivity. Here,when $L = 8$, the whole encryption system's best key sensitivity is achieved. So we use $L = 8$ in this paper to analysis the encryption performance.

4.2. Correlation analysis

1200 pairs of adjacent pixels from plain-image and cipher-image from different directions are randomly se-lected respectively, and the correlation coefficients are calculated. The results are shown in Table 2. It shows that,by comparing with the adjacent pixel correlation coefficients of plain-image, those of the cipher-image are indeed very small, which indicate that the encryption system has high de-correlation performance.

Table 2. Coefficients of the adjacent pixels of the Plain-image and Cipher-image

correlation analysis	Plain image	Cipher image
Vertical	0.9691	0.0051
Horizontal	0.9379	0.0042
Diagonal	0.922	0.0009

4.3. *NPCR* and *UACI* analysis

Number of pixels change rate (*NPCR*) means the number of pixels changed in cipher-text when only one pixel value changes in plaintext[3]. The larger the *NPCR*, the higher sensitivity the system has. Here the pixel value in the position of $(1, 3)$ of the plain-image is changed from 133 to 134. Table 3 shows that even in the first encryption the system can reach the good encryption performance and also keep stable at all iterations.

Unified average changing intensity (*UACI*) means the changing intensity of the corresponding pixels of the plain-image and cipher-image[3]. The larger the *UACI*, the more resistant to the differential attack the system will get. Table 3 shows that during the second encryption the system has already reached the satisfying encryption performance and kept stable at all iterations.

Table 3. NPCR and UACI test

Test	Iteration times n						
	1	2	3	4	5	6	7
NPCR(%)	99.53	99.61	99.56	99.67	99.61	99.55	99.64
UACI(%)	19.61	33.39	33.41	33.55	33.63	33.58	33.64

4.4. Entropy and Computation analysis

To study the robustness of the encryption system further, the concept of entropy from the area of informatics has been introduced here[10]. The entropy is defined as:

$$
H(R) = \sum_{i=0}^{255} f_k \lg(f_k) \quad (15)
$$

Where f_k is the frequency of the k_{th} pixel value. From the definition of the entropy, the smoother the picture is, the smaller the entropy is. And the minimum entropy is zero while the picture has only one pixel value. In contrary, the more disordered the picture is, the bigger the entropy is. And the maximum entropy of 8 is achieved while the probability of every pixel value is the same in the picture.

 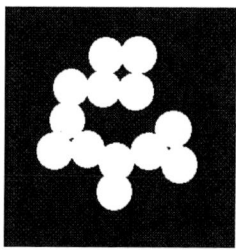

(a)cameraman   (b)circle

Figure 3. Images using in Entropy analysis

Figure 3 are plain-images of descending complexity used in this test. Table 4 shows the entropy of images before and after encryption. From Table 4, the entropy of all the encrypted images is more than 7.996, which means a high confusion is achieved in the encrypted system, thus the encryption system has a robust performance varying the different complexity of the original image.

Table 4. Entropy comparison of plain-image and cipher-image

	Cameraman	Circle
Before Encryption	7.0097	0.7954
After Encryption	7.9981	7.9962

The proposed algorithm can get different encryption speed via the selection of different computational complexity and key space according to different security level, thus it can take the balance in the need of complexity and security. Simulations show that, on a 1GHz CPU, 512M memory Computer, for an image of size 256×256, the encryption process is done in 0.5s. For a 512×512 size image, just 1s is needed. Therefore the proposed algorithm is fit for the different security level instantaneous network image encryption.

5. Conclusions

In this paper, some essential shortcomings of the previous proposed 3D chaotic Cat map algorithm [3] are studied and a new 3D chaotic cat map algorithm to improve its security by avoiding these weaknesses is proposed. In the new algorithm, Random-scan preprocessing mode to avoid the tracking of corner positions encryption, CML model and tent map in the diffusion process to increase the security and sensitivity, and group-key in different iterations to enlarge key space are introduced. Simulation results show that, our scheme is immune to the attacks presented by [11]. It has large enough key space which can resist differential attack, chosen-plaintext attack and statistical attack, and also achieves relatively high encryption security and fast encryption speed. Although the security of the original scheme has been improved, the linearity of the 3D cat map cannot be changed even with the random-scan preprocessing. It is a remained weakness in our improved scheme. So we believe that the security usage of the linearity chaotic map in the image encryption needs further study to solve this problem in the future.

References

[1] E. Alvarez and A. Fernamdez. New approach to chaotic encryption. *Physics Letters A*, 263(4-6):373–375, 1999.

[2] R. Brown and L. O. Chua. Clarifying chaos: Examples and counter examples. *Int.J.Bifurcation and Chaos*, 6(2):219–249, 1996.

[3] G. R. Chen, Y. B. Mao, and C. K. Chui. A symmetric image encryption scheme based on 3d chaotic cat maps. *Chaos, Solitons and Fractals*, 12:749–761, 2004.

[4] Z. R. Cherati and M. R. J. Motlagh. Control of spatiotemporal chaos in coupled map lattice by discrete-time variable structure control. *Physics Letters A*, 370(3-4):302–305, 2007.

[5] J. Fridrich. Image encryption based on chaotic maps. In *Proc. IEEE Int. Conference on systems, Man and Cybernetics*, volume 2, pages 1105–1110, 1997.

[6] T. Kohda and A. Tsuneda. Stream cipher systems based on chaotic binary sequences. *SCIS96-11C*, 1 1996.

[7] S. G. Lian, J. S. Sun, and Z. Q. Wang. A block cipher based on a suitable use of the chaotic standard map chaos. *Solitons and Fractals*, 26:117–129, 2005.

[8] Y. B. Mao, G. R. Chen, and S. G. Lian. A novel fast image encryption scheme based on the 3d chaotic baker map. *Int J Bifurcat Chaos*, 14(10):3613–3624, 2004.

[9] G. Marco, K. Kristina, and S. Wolfgang. Discrete-time chaotic encryption systems-partI : Statistical design approach. *IEEE.Trans.Circuits and Systems-I*, 44(10):963–970, 1997.

[10] E. Ott. *Chaos in dynamical systems*. Cambridge University Press, 2nd edition, 6 2005.

[11] K. Wang and W. J. Pei. On the security of 3d cat map based symmetric image encryption scheme. *Physics Letters A*, 343:432–439, 2004.

[12] L. H. Zhang, X. F. Liao, and X. B. Wang. An image encryption approach based on chaotic chaos. *Chaos, Solitons and Fractals*, 24(2):759–765, 2005.

Multi-view Video Coding Using Color Correction

Lixian Shangguan, Jifeng Sun

South China University of Technology, Guangzhou, China,510640

lxshangguan@qq.com, ecjfsun@scut.edu.cn

Abstract

The color variations between multi-view video sequences may degrade the inter-view prediction and result in low coding efficiency. In this paper we propose an efficient multi-view video coding scheme using dominant basic color mapping based color correction. The experimental coding results show that color correction has the potential to make multi-view video coding more efficient.

1. Introduction

In recent years, with the development of computer vision, computer graphics, multimedia and related field technologies, multi-view video has gained more and more interests and expected to be a next generation visual application[1]. The multi-view video is captured by many cameras simultaneously from different directions of the same scene. Because of the increased number of cameras, the multi-view video data become extremely large. Since the serious limitation of bandwidth and storage capacity in the system, we need to compress the multi-view sequence efficiently without sacrificing its visual quality significantly. [2]

Currently, multi-view video coding (MVC) is also investigated by the Joint Video Team (JVT). Now, the JVT is developing a Joint Multi-view Video Model (JMVM), which is based on the video coding standard H.264/AVC. [3] The prediction structure, which is proposed by HHI and called "Hierarchical B Pictures", has been adopted by the current JMVM [4]. The JMVM uses adaptive motion-compensated prediction and disparity compensated prediction to exploit the statistical dependencies within temporally successive pictures and neighboring views to reduce the temporal and inter-view redundancy.

Due to the dissimilar radiometric characteristics of different cameras and the variation of lighting conditions, significant luminance and chrominance discrepancies among different camera views often exist in multi-view video sequences. [5] These variations may reduce the number of matching blocks between current frames and reference frames and degrade coding efficiency. The color correction processing should be considered in JMVM to make inter-view prediction more efficient.

In this paper an efficient multi-view video coding scheme, which using dominant basic color mapping based color correction, is proposed. We present the experimental coding results showing that color correction has the potential to make multi-view video coding more efficient.

The rest of the paper is organized as follows. Section 2 describes the color correction algorithm for multi-view video. The experiment results are presented in section 3.

2. Algorithm for color correction

2.1 Related works

Since Reinhard reported a simple but very successful technique that transfers color characteristics from a source to a target image [6], color transfer research has gained more interests. In our work, we aim to make color correction on the current image, that means to transfer the color of reference image to the current image in multi-view video coding. The process should be automatic and the corrected image should be natural and similar to the reference image. To satisfy these two conditions, we transfer colors automatically in accordance with characteristics of human color perception.

Berlin and Kay [7] and other studies showed that human color perception can group similar colors into the eleven basic color categories (BCCs). In this paper, we adopt this theory and assume that we can categorize the color space into eleven perception-based color categories.

We can divide the algorithm into the following steps. Firstly, we categorize the color space into eleven basic color categories. Before categorization we should transfer the colors of image to CIELAB color space, in which the distance between points is directly proportional to perceived color difference. Secondly, the dominant basic colors (DBCs) of the image would be extracted based on their distribution probability. Finally, corrected image can be obtained via mapping dominant basic colors. The Fig.1 shows the flow chart of the algorithm.

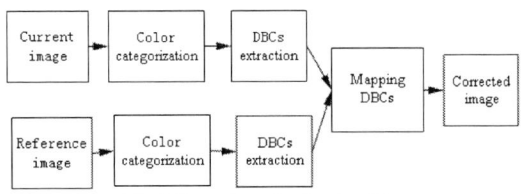

Fig.1 Flow chart for the algorithm

2.2 Color categorization method

In the algorithm, we performed the pixel clustering via probabilistic categorization, which is usually used in fuzzy color segment, rather than binary categorization. That means we estimate the probability iPxy $\in [0, 1]$ rather than iPxy $\in \{0, 1\}$ that pixel I(x, y) belongs to ith BCCs in the image. Probabilistic categorization can well help us avoid pseudo-contours, which appear if the input image contains some color gradation regions that step over the boundaries between two different BCCs and, therefore, are divided into different categories. This leads to a problem that similar pixel color values might be shifted in different color directions because we later transfer each pixel color according to its category. Therefore, unexpected pseudo-contours might be created.

In this section, each pixel is categorized to basic color categories. We denote this categorization as the "color categorization" of a pixel. The categorization method consists of two steps: initial color categorization and fuzzy color categorization.

In the initial color categorization, we cluster the space into eleven BCCs using k-means algorithm. An 11-D vector is used to denote the probabilities of each pixel belonging to each BCCs . In this step, each pixel completely belongs to one of the 11 BCCs. For example, if the color of a pixel is categorized in the third BCC, then the vector becomes p=(0,0,1,0,0,0,0,0,0,0,0).[8]

In order to avoid pseudo-contours, a novel probabilistic color categorization in the fuzzy color categorization step is performed. The calculation of one iteration step for an element of p is shown by equation (1). The iteration of diffusion continues until there are no more changes while processing, or until it reaches a predefined iteration number.[8]

$$p^i_{xy} = \frac{\sum\limits_{x',y' \in \delta} p^i_{x'y'} w((I(x,y),I(x',y')))}{\sum\limits_{x',y' \in \delta} w((I(x,y),I(x',y')))} \quad (1)$$

$$w(A,B) = 1 - \frac{1}{1 + e^{-0.5(\text{Dist}(A,B)-T)}}$$

Where p^i_{xy} is the probability that I(x,y) belongs to the category i, I(x,y) is the CIELAB value of the pixel (x,y)in the input image, Dist(A,B) is the Euclidean distance between A and B, T is the threshold value, δ is the set of eight neighborhoods of the pixel(x,y), w(A,B) is the weighting function that goes to 1 when color difference between A and B becomes small, and goes to 0 if it becomes large.

2.3 Extracting dominant basic colors

After the above operations, we can describe each pixel in the image with the 11 BCCs and the corresponding probabilities. We compute the mean μ_i

and standard deviation σ_i as:

$$\mu_i = \sum_{x,y} p^i_{xy} I(x,y) / \sum_{x,y} p^i_{xy}$$

$$\sigma_i = \sqrt{\sum_{x,y} p^i_{xy}(I(x,y) - \mu_i)^2 / \sum_{x,y} p^i_{xy}} \quad (2)$$

In order to extract dominant basic colors, we should obtain the percentage of the ith BCC occupying in the whole 11 BCCs .We can compute them by

$$p_i = \sum_{x,y} p^i_{xy} / (\sum_i \sum_{x,y} p^i_{xy}) \quad (3)$$

Then we sort the $\{p_i\}$ in descending order. If the accumulative percentage $\sum\limits_{i=1}^{M} p_i) < 0.9$, the corresponding colors are regarded as dominant basic colors.[9] M is the dominant basic colors number. The dominant basic color is represented as

$$DBC = \{c_i, p_i, \mu_i, \sigma_i\} \quad (4)$$

Where c_i denotes it is ith dominant basic color; p_i identifies its percentage value, μ_i and σ_i are the mean and color standard variation in CIELAB color space.

2.4 Correcting colors

The corrected color value in each category of pixel (x,y) can be computed by equation (5),via mapping their corresponding dominant basic colors.

$$I^{out}_i(x,y) = \frac{\sigma^{ref}_i}{\sigma^{in}_i}(I^{in}(x,y) - \mu^{in}_i) + \mu^{ref}_i \quad (5)$$

Where $I^{in}_i(x,y)$ is the color value of pixel (x,y), μ^{in}_i and σ^{in}_i are the mean and standard deviation of ith category in input image, μ^{ref}_i and σ^{ref}_i are the mean and standard deviation of ith category in reference image.

The final corrected color value of the pixel $I^{out}(x,y)$ is computed by linearly interpolating the transferred color values in each category

$$I^{out}(x,y) = \frac{\sum\limits_{i=1}^{M} p_i I^{out}_i(x,y)}{\sum\limits_{i=1}^{M} p_i} \quad (6)$$

Where p_i is the percentage of the ith BCC occupying in the whole 11 BCCs, $I^{out}(x,y)$ is the CIELAB color value of pixel (x, y) with three color channels of L(x, y), a(x, y)and b(x, y).

3. Experiments

3.1 Experimental scheme

In order to make multi-view video coding more efficient and provide users with more immersive impression, the experiment scheme is proposed and shown in Fig.2. We applied color correction to other view's decoded frames before the prediction scheme to use color-corrected frames for inter-view prediction. Therefore current frame would become more similar to the reference frame, and this color-coherence would make multi-view video easier and more efficient to be encoded.

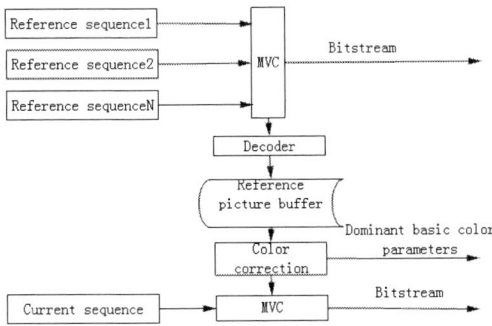

Fig. 2 MVC scheme based on color correction

We implemented our approach on JMVM6.0. JMVM has recently been used as reference software by the 3DAV ad-hoc group at MPEG/JVT meetings. The basic coding structure of JMVM is H.264/AVC, and its current reference structure is shown in Fig 3. This scheme uses hierarchical B picture coding in temporal direction. In order to use adjacent views as reference frames, S1, S3, and S5 sequences are coded after encoding of surrounding views. Color correction process can be applied to those views using adjacent views as reference frames.

The main coding parameters in experiments are listed in table 1. The other coding parameters configurations are well-established and used as default. We need do some changes to implement the experiments. Firstly, the color correction algorithm codes should be embedded in the JMVM. The 11 basic color categories information should be included in the frame header. Secondly, the presented scheme requires some high level syntax specification to signal that the encoder uses color correction frame as reference frame. And then the decoder will utilize the color correction frames as reference to decode the original frame accordingly.

Table 1 Coding parameters in experiments

Basis QP	26,28,30,32
Entropy coding	CABAC
Prediction structure	Hierarchical B pictures
GOPSize	8
NumberReferenceFrames	2

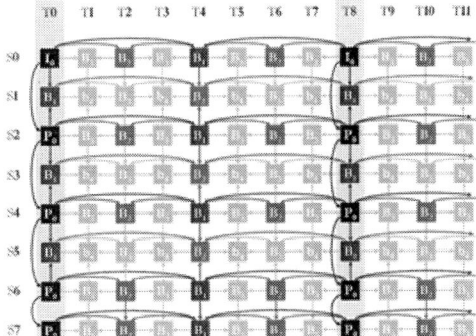

Fig.3 Inter-view-temporal prediction structure, using hierarchical B pictures.

3.2 Experimental results

Because the "Flamenco1" and "Golf2" test sequences, supported by KDDI Lab, have significant color variations between different views, they are very suitable for our experiments. All of the flowing experiments are based on them.

We choose two frames from different view of the two test sequences to verify the performance of color correction. As shown in Fig.4 and Fig.5, there is high similarity between the corrected images and reference images and this would result in more matching blocks and higher coding efficiency.

(a) input image   (b) reference image   (c) corrected image
Fig.4 Correction result for one frame of "Golf2"

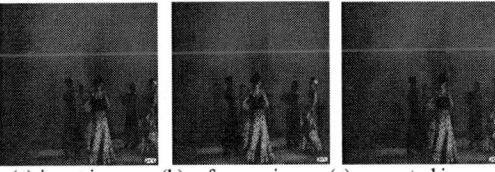

(a) input image   (b) reference image (c) corrected image
Fig.5 Correction result for one frame of "Flamenco1"

The major goal of the experiments is to compare the efficiency of multi-view video coding with color correction and without color correction. In order to make experiments more convenient, we implement the approach only on the former three views video sequences. The coding performance is compared only on view1 and view2, not including view0, because the color correction doesn't influence the coding gain of view0.

Some parameters mentioned above have to be transmitted from the encoder to the decoder. However, because their bit streams are much smaller than that for the encoded pictures, so we don't consider them in the experiments. The comparison curves of coding performances are shown in Fig.6. The coding performances of the proposed method are better than that without color correction. PSNR values of "Golf2" are improved about 1dB, but PSNR values of "Flamenco1" are close to the reference model. The reason is that there are less color variations between views and the quality of corrected images is not good enough to contribute to the coding efficiency.

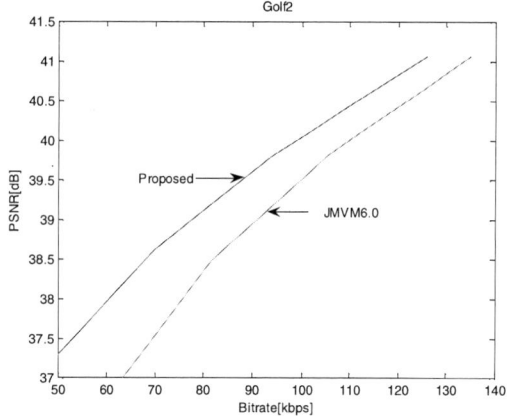

(a)Rate-distortion curves for Golf2

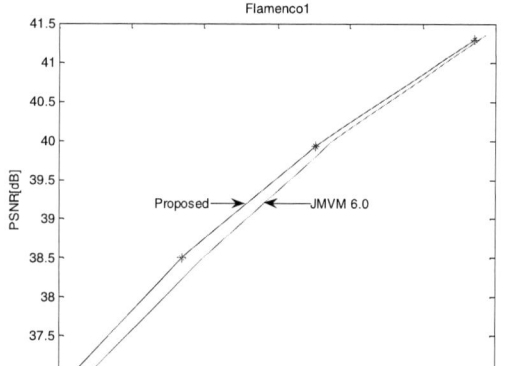

(b)Rate-distortion curves for Flamenco1
Fig.6 Results of multi-view video coding

4. Conclusion

In this paper, we have described an efficient color correction algorithm and multi-view video coding scheme using color correction. The color correction algorithm consists of color categorization, dominant basic color extraction and mapping. After the color correction, there would be more matching blocks between current frame and reference frame. As a result, the color correction with good quality can contribute to the coding efficiency about 1dB on average.

References

[1] "Applications and Requirements for 3DAV", document N5877, MPEG Trondheim Meeting, 2003
[2]Yo-Sung Ho and Kwan-Jung Oh "Overview of Multi-view Video Coding", Systems, Signals and Image

Processing, 2007, pp5-12
[3] A. Vetro, Y. Su, H. Kimata, and A. Smolic, ".Joint multi-view video model JMVM 6.0",. ITU-T and ISO/IEC Joint Video Team, Document JVT-Y207, Oct. 2007,Shenzhen.
[4] K. Mueller, P. Merkle, etal., "Multi-View Video Coding based on H.264/MPEG4-AVC Using Hierarchical B Pictures," PCS2006, Beijing, China, April 2006
[5] ISO/IEC JTC1/SC29/WG11,"Luminance and Chrominance Compensation for Multi-View Sequences Using Histogram Matching", Nice, France, 2005.
[6] Erik Reinhard, Michael Ashikhmin, Bruce Gooch, and Peter Shirley. "Color transfer between images", IEEE CG&A, 21:34－41, 2001.
[7] B. Berlin and P. Kay, "Basic Color Terms: Their Universality and Evolution", Berkley, CA: University California Press, 1969
[8] Y. Chang, S. Saito, and M. Nakajima, "Example-Based Color Transformation of Image and Video Using Basic Color Categories" IEEE Trans. Image Process. 16, 329 ,2007
[9] Feng Shao, Mei Yu," Dominant color extraction based color correction for multi-view images", Chinese Optics Letters, August-10,-2007

2008 Workshop on Power Electronics and Intelligent Transportation System

Improved Voice Activity Detection Based on Iterative Spectral Subtraction and Double Thresholds for CVR

Xiangbin Li[1], Guo Li[1], Xueren Li[2]

1 AAC Department. Engineering College,
Air Force Engineering University, Xi' an, Shaanxi, 710038, China
2 SR Department. Air Force Engineering University, Xi' an, Shaanxi, 710051, China
xianlxb@163.com

Abstract

Cockpit voice recorder (CVR) in aircraft black box records many cockpit voices, such as speaker voices, noises and background sounds with special meanings. Cockpit voices' complexity exacerbates analysis difficulty through traditional differentiating and hearing methods, so that fresh cockpit voices are not captured easily from non-stationary sounds. In this paper, by analyzing firstly thoroughly characteristics of cockpit voices, we develop an improved voice activity detection scheme based on iterative spectral subtraction and double thresholds Finally, to demonstrate the effectiveness of the proposed scheme, we make simulations with a section of speech (SNR=8) from standard voice bank and a section of true cockpit voices and compare the probabilities P_{cs} and P_{cn} of the three algorithms, where P_{cs} denotes probability of correctly detecting speech frames P_{cn} probability of correctly detecting noise frames. Simulations results are presented to demonstrate the effectiveness of the improved algorithm.

1. Introduction

Cockpit voice recorder (CVR) in aircraft black box records many cockpit voices, such as speaker voices, noises and background sounds with special meanings. They are complex non-stationary signals with characteristics of mutation, instantaneousness and singularity. In some special cases, speaker voices combined in background sounds play an important role in air accident investigation (AAI). However, when the aircraft is flying, especially when an accident is happening, the record condition of CVR is becoming worse. Therefore, fast extraction of speaker voices from cockpit voices is an important work in AAI. The conventional measure of AAI is based on "differentiating and hearing" method and simple audio processing, so more exact speaker voices can't be obtained easily.

Voice activity detection (VAD) is just detecting the beginning and ending of a section of speech signal, and achieves the goal of distinguishing speaker voice from background sounds. VAD in AAI requires proper detection and low computing cost with the purpose of gaining more time for AAI. The traditional double thresholds VAD based on short time energy and zero crossing rates acquires wonderful performance results in case of high signal to noise ratio(SNR), but has total failure when speaker voices are submerged in strong background sounds or low SNR, for example, when the accident happens.

In this paper, we develop an improved scheme based on double thresholds VAD with spectral subtraction. The paper is organized as follows: Section 2 analyzes characteristics of cockpit voices. Section 3 and 4 describe respectively the scheme of traditional VAD based on double thresholds and basic spectral subtraction. In section 5 and 6 present the improved scheme, and simulation results are also given in this section. Finally, some conclusions are drawn in section 7.

2. Voice characteristics of CVR

The frequency scope of cockpit voice recorded by CVR is very wide, about 150Hz to 6800 Hz, which brings some difficulty to sound separation. With the aim of facilitating sound separation, the cockpit voice is classified to three kinds: aviation noises, speaker voices and background sounds [1][2].

Aviation noises include additive noises and non-additive noises. Additive noises contain periodic noises, impulse, broad band noises, and speech interference. Non-additive noises are mainly sound residue and circuit noises. Non-additive noises can be transformed to additive ones by means of a particular transformation. However, more specifically, aviation noises include engine sound, exterior air current noise while flying, skating noise while takeoff and landing, circuitry noise in electrical equipments and circuits, motor noise droved by power when manipulating aircraft and so on.

978-0-7695-3342-1/08 $25.00 © 2008 IEEE
DOI 10.1109/PEITS.2008.84

153

Table 1 Characteristics of sounds of CVR

Sound	Characteristics in time domain	Characteristics in frequency domain	Short time energy	Zero crossing rate
surd	less obvious	Obvious, energy mainly locates in high frequency band	weak	high
sonant	obvious and periodic	has formant, energy mainly locates in low frequency band	strong	low
silence	less obvious	less obvious	weak	high
impulse noise	transient		strong	high

Background sounds mainly consist of various sounds except cockpit voices and aviation noises. Different background sound implies that special event has happened [3].Cockpit voices involve conversations between pilot and co-flyers, communication from control tower and speech for navigation and identification. Voice signals are a time-varying and non-stationary random process, but its characteristic keep unchangeable in a short time 10-30 ms because of relatively stability of vocal cords sound channel. Chinese language includes surd and sonant. Table 1 shows some characteristics of sounds of CVR.

3. Basic VAD algorithm based on double thresholds

3.1. Basic conception

(1) Short Time Energy (STE): The sound intensity of a speech series $x(n)$ is described by short time energy, which is defined as follows:

$$E_n = \sum_{m=-\infty}^{m} \left[x(m)w(n-m) \right]^2$$
$$= \sum_{m=n}^{n+N-1} \left[x(m)w(n-m) \right]^2 \quad (1)$$

Especially, STE of surd is very weak and STE of sonant is quite strong.

(2) Zero Crossing Rate (ZCR): The ZCR of a speech series $x(n)$ is defined as follows:

$$z_n = \sum_{m=-\infty}^{m} \left| \text{sgn}\left[x(n)\right] - \text{sgn}\left[x(n-1)\right] \right| w(n-m)$$
$$= \left| \text{sgn}\left[x(n)\right] - \text{sgn}\left[x(n-1)\right] \right| w(n) \quad (2)$$

where $\text{sgn}\left[x\right]$ is a sign function and $w(n)$ is a window function, which are defined as follows:

$$\text{sgn}\left[x\right] = \begin{cases} 1 & (x \geq 0) \\ -1 & (x < 0) \end{cases} \quad (3)$$

$$w(n) = \begin{cases} 1/2N & (0 \leq n \leq N-1) \\ 0 & (\text{others}) \end{cases} \quad (4)$$

3.2. VAD based on double thresholds

Generally, we use ZCR to detect sonant and STE to surd in practical applications [4]. The whole VAD process is divided to four sections: silence section (status=0), transition section (status=1), speech section (status=2) and end section. At the beginning of VAD, we set two thresholds for STE and ZCR each other, for example, high threshold T_{amp1} and T_{zcr1}, low threshold T_{amp2} and T_{zcr2}. Besides, we define a variable *count* as a speech counter, *silence* as silence counter, *minlen* as a minimum time threshold. Figure 1 shows flow of VAD based on double thresholds.

Many practices prove that this method can separate speeches from background noises effectively and efficiently in high SNR according to table 1. However, the aviation condition with low SNR and awful environment for record causes the method loses its own performance, because the speeches are submerged in strong aviation background noises. Therefore, former noise reduction and speech enhancement are becoming extremely important.

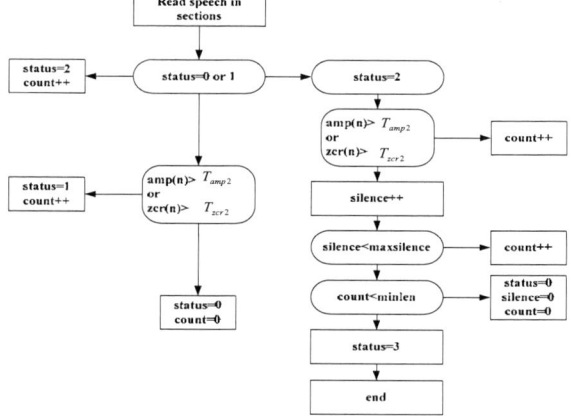

Figure 1 Flow chart of VAD based on double thresholds

4. Scheme of basic spectral subtraction

The basic spectral subtraction (SS) method is described briefly in this section. Assume that a noisy speech signal is expressed as

$$y(i) = s(i) + d(i), \quad 0 \leq i \leq N-1 \quad (5)$$

where $s(i)$ and $d(i)$ are a frame of clean speech and noise, respectively. Considering human's ear be no-sensitive to phase distortion, phase signal of noise is implemented when phase is restored [5][6]. In the frequency domain, equation (5) is expressed as

$$|Y_k|^2 = |S_k|^2 + |N_k|^2 + S_k N_k^* + S_k^* N_k \qquad (6)$$

where Y_k, S_k and N_k are discrete-time Fourier transforms (DFT) of $y(i)$, $s(i)$ and $d(i)$, respectively. Because $s(i)$ and $d(i)$ are independent and N_k is gauss distribution, equation (6) is expressed as (7) in frequency domain.

$$\left|\hat{S}_k\right|^2 = \left[\left[|Y_k|^2 - E\left[|N_k|^2\right]\right]\right]^{1/2}$$
$$= \left[|Y_k|^2 - \lambda_n(k)\right]^{1/2} \qquad (7)$$

For a frame of speech signal, we have $|Y_k|^2 = |S_k|^2 + \lambda_n(k)$, so estimate of original speech is expressed as

$$\left|\hat{S}_k\right|^2 = \left[\left[|Y_k|^2 - E\left[|N_k|^2\right]\right]\right]^{1/2}$$
$$= \left[|Y_k|^2 - \lambda_n(k)\right]^{1/2} \qquad (8)$$

where $\lambda_n(k)$ is statistical mean of unvoiced speech, $\left|\hat{S}_k\right|^2$ is amplitude of enhanced speech.

However, basic SS can generate much musical noises in residual noises. Some modified SS are proposed to reduce effect. Weighting factor α and power coefficient β are introduced into SS, so equation (8) is modified as

$$\left|\hat{S}_k\right| = \left[|Y_k|^\alpha - \beta\lambda_n^\alpha\left[|N_k|^2\right]\right]^{1/\alpha} \qquad (9)$$

Modified SS is degraded to basic SS when α =2 and β =1[5]. Other modified SS is showed in relative references [6][7]. Better enhancement performance can be gained by adjusting two parameters suitably, but voice distortion becomes severer as the degree of noise reduction is larger.

5. Proposed VAD based on improved spectral subtraction

5.1. Improved spectral subtraction

In this paper, we propose iterative spectral subtraction to formerly reducing noise and enhancing speech. This method uses basic SS or modified SS for appropriate times. The former enhanced speech becomes latter input signal, so music noise is seen as input noise to reduce again.

5.2. Proposed VAD based on improved spectral subtraction

On the basis of section above, we can firstly apply spectral subtraction for noisy sound of CVR to reducing noise and enhance speech, and then enhanced signal is filtered by a preceding filter, finally cockpit voice is extracted by means of double thresholds VAD. Figure 2 shows the flow chart of proposed VAD. The preceding filter is a high-pass filter, such as $1 - 0.9375z^{-1}$, which can filter low-frequency interference, especially interference of frequency 50Hz or 60Hz, and advance spectrum of high frequency which is useful for cockpit voice.

6. Experiment and simulation

6.1. Evaluation standard of VAD

For a wonderful VAD, two requirements must be taken into considered comprehensively: to detect more speech sections and more unvoiced speech sections. However, when VAD tries to detect more speech frames, it misjudges silence as speech or otherwise. The latter is ever worse than the former for accident investigation. Therefore, two evaluation standards are compared to weighing quantificationally the performance of VAD: probability of correctly detecting speech frame P_{cs} and probability of correctly detecting noise frame P_{cn}, which are expressed as

$$P_{cs} = \frac{N_1}{N_1^{hand}}, \quad P_{cn} = \frac{N_0}{N_0^{hand}} \qquad (10)$$

where N_1^{hand} and N_0^{hand} are relatively the overall number of hand-labeling speech frames and noise frames by hand-labeling, N_1 and N_0 are relatively number of being detected correctly by VAD.

6.2. Experiment results

In this paper, a section of speech in car (SNR =8) from standard voice bank Aurora2 and a section of true cockpit sound are used, simulation experiments based on traditional double thresholds VAD only and the proposed VAD are carried out. Figure 3 and table 2 compare the performance of various methods in different environment. Due to former spectral subtraction, the SNR increases, the curves of STE and ZCR become smoother and proper probability increases.

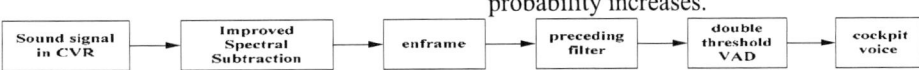

Figure 2   Flow chart of proposed VAD

Table 2 Experiment results

Method / Evaluation standard / Test samples	Traditional double thresholds VAD			Proposed VAD (Iterative number=1)			Proposed VAD (Iterative number=2)		
	SNR of original signal	P_{cs} (%)	P_{cn} (%)	SNR of enhanced signal	P_{cs} (%)	P_{cn} (%)	SNR of enhanced signal	P_{cs} (%)	P_{cn} (%)
In car(SNR=8)	8	85.4	86.2	10	91.2	93.1	14	95.2	96.5
True cockpit sound(SNR=1)	1	65.4	66.6	9	85.4	86.6	11	92.1	93.3

7. Conclusion

According to the characteristics of sound signal recorded in CVR, the objective of this paper proposes iterative spectral subtraction to improve the performance of traditional double thresholds VAD. Basic spectral subtraction has its own flaws: Because the current noise frame is replaced by the statistical mean of the whole noise, there are much musical noises in residual noises. To reduce the effect, we introduce iterative spectral subtraction. It can reduce efficiently and effectively music signal by adjusting appropriately the iterative number. The proposed algorithm can increase the SNR and reduce computational complexity for runtime requirement. Experiment results demonstrate that spectral subtraction which is used to noise reduction and speech enhancement to enhance SNR can provide feasible condition for traditional double thresholds VAD. Proposed VAD can also gain wonderful performance even if in the low SNR.

[1] James R. Cash. Group Chairman's Report of Investigation Sound Spectrum Study of Cockpit Voice Recorder, American Airlines Flight 587, DCA02MA001, Belle Harbor, NY, Nov. 12 ,2002.
[2] Ronald Stearman. Signal Analysis of Cockpit Voice Recorder Data, Report No. ASE46Q-FP1 [R]. The University of Texas, 2003.August. 16.
[3] YANG Lin, Cockpit Voice Recorder (CVR) and Laboratory Processing Methods [J]. China Civil Aviation, 2003, 29(12):21-22.
[4] An Improved Voice Activity Detection Using Higher Order Statistics [J]. IEEE Trans. on Acoustic, Speech, Signal Process, VOL.13.NO.5, SEPTEMBER 2005:965-974
[5] Sim B L, Tong Y C, Chang J S, and Tan C T. A parametric formulation of the generalized spectral subtraction method. IEEE Trans. on. Speech and Audio Processing, 1993, 6(4):325-337
[6] Boll S F. Suppression of acoustic noise in speech using spectral subtraction. IEEE Trans. on Acoustic, Speech, Signal Process, 1979, ASSP-27(2):113-120
[7] Y. Ephraim and D.Malah, "Speech enhancement using a minimum mean-square log-spectral amplitude estimator," IEEE Trans. Acoustic, Speech, Signal Processing, vol. 33, no. 2, pp 443–445, 1985.

Reference

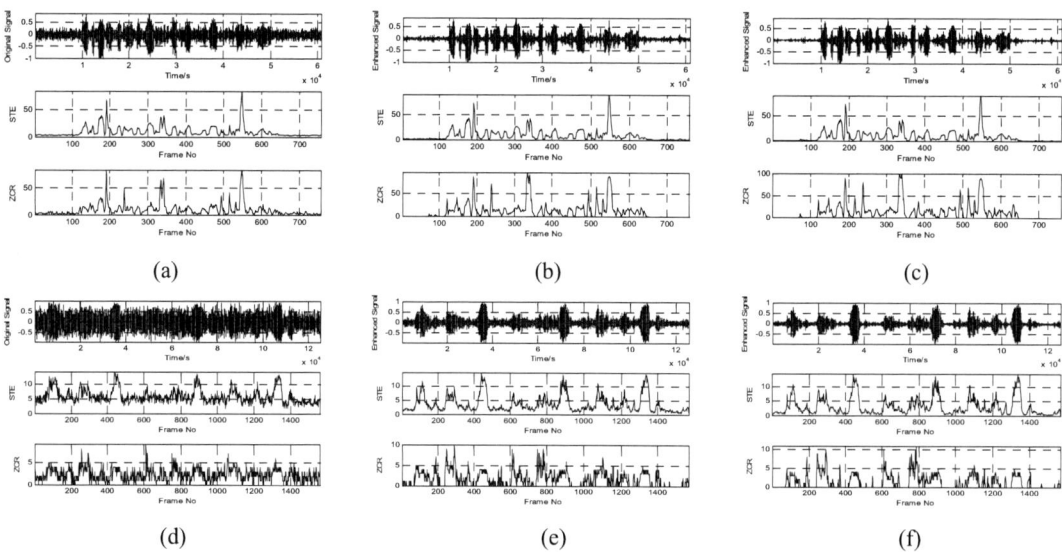

Figure 3 Performance of various methods in different environment. (a)(b)(c) a section of speech in car (SNR =8). (a) Traditional double thresholds VAD. (b) Proposed VAD (Iterative number=1). (c) Proposed VAD (Iterative number=2). (d)(e)(f) A section of true cockpit sound. (d) Traditional double thresholds VAD. (e) Proposed VAD (Iterative number=1). (f) Proposed VAD (Iterative number=2)

2008 Workshop on Power Electronics and Intelligent Transportation System

A Segmentation and Extraction Method for Crops of Image in Complex Background

Hongjuan Kou, Shuguang Zhang, Dening Zhang

Department of Mechanical and Electronic Engineering, Agricultural University of Hebei，
Baoding,Hebei,071001,China
khj828@sina.com, zsg59@163.com, zhangdn001@163.com

Abstract

In the traditional Fuzzy C-Means clustering algorithm (FCM), determination of initial cluster centers and the number of clusters is hard, and interference exists during image segmentation for space information of image isn't considered. In order to overcome the shortcomings above, initial cluster centers and the number of clusters were ascertained by ant colony algorithm, then combine the improved FCM algorithm that considering space information to reduce the image noise. The segmentation and extraction of crops in the image under the complex background was completed. The image that the interference still existed after segmentation would be disposed by the method of mathematical morphology filter. The background was removed completely and the segmentation effect was improved.

1. Introduction

Image segmentation is one of the fundamental problems for image processing and machine vision, the key points are: the image is divided into a number of aggregations of non-overlapped regional. These regional is either meaningful to the current task, or helpful to show the corresponding relationship between these regional and actual objects or certain parts of the objects[1]. Usually the purpose of image segmentation is carried out the image with analysis, identification, compression, encoding, the accuracy of the segmentation directly impacts on the effectiveness of follow-up tasks, so it is of great significance. In practical applications, with the impact of noise, inconspicuous regional borders between the background and objectives of some images add a certain ambiguity to the edge. How to filter out the background in the collected image information and to identify target accurately has become an important issue in the image fusion technology. Based on image fusion technology has played an important role in the agricultural sector day after day. In this paper, taking the images of greenhouse crops as an example, a method is developed that can accurately segment and extract crop images from the complex background.

Segmentation algorithm main include edge extraction methods and regional segmentation methods etc[2]. Fuzzy C-means (FCM) clustering segmentation algorithm has

been applied more widely. FCM algorithm divides the image pixel into several clusters through clustering, this clustering via iterative optimization objective function to realize the set partitioning. In recent years, some experts made some improvements on the FCM clustering algorithm, but there is still no effective and reasonable application of spatial information[3]. As image pixel has a high degree of spatial correlation[3]. If we do not consider this spatial information, image pixel noise may be misjudgement, and the number of clustering can not be pre-determined in the traditional FCM algorithm, and the cluster center is very difficult to determine.

This paper presents a new segmentation method, which has made some improvements to the FCM algorithm and combines the ant colony algorithm.　Mathematical morphology filtering method is applied to the image with small interference after segmentation.

2.Fuzzy C-Means (FCM) Clustering segmentation method

FCM clustering algorithm estimated by weighted similarity measurement that is based on each sample of clustering space and number of clustering centre, then iterative optimization with objective function, for confirming the best of clustering. Algorithm as follows:
Image $A = \{A_1, A_2, \cdots, A_N\}$ is the sample set, N is the number of pixels on sample set, and c $(2 \leq c \leq N)$ is the number of cluster. A function of FCM clustering algorithm as definition:

$$J_m(U,V:A) = \sum_{j=1}^{N} \sum_{i=1}^{c} (u_{ij})^m (d_{ij})^2 \qquad (1)$$

$$d_{ij}^{\;2} = \left\| a_j - v_i \right\|^2 \qquad (2)$$

Where, U is a collection of fuzzy membership, the following conditions are met

---

978-0-7695-3342-1/08 $25.00 © 2008 IEEE
DOI 10.1109/PEITS.2008.101

157

$$\left\{ \begin{array}{l} u_{ij} \in [0,1], \quad 1 \le i \le c, \quad 1 \le j \le N \\ \sum_{i=1}^{c} u_{ij} = 1, \quad 1 \le j \le N \\ \sum_{i=1}^{n} u_{ij} > 0, \quad 1 \le i \le c \end{array} \right. \tag{3}$$

$$u_{ij} = 1 \Big/ \left[\sum_{k=1}^{c} \left(\| a_j - v_i \| \Big/ \| a_j - v_k \| \right)^{2/(m-1)} \right] \tag{4}$$

$$v_i = \sum_{j=1}^{N} u_{ij}^{m} x_j \Big/ \sum_{j=1}^{N} u_{ij}^{m} \tag{5}$$

$m \in [1, \infty)$ is fuzzy weighted index; usually m = 2 is the ideal value; the V is the collection composed of c cluster centers $V = [v_i]$; d_{ij} is distance from the first j pixel to the i cluster center. The value of J is smaller, more compact clustering. The best classification results are when objective function of formula (1) achieves the minimum.

3. Improved algorithm---FCM Algorithm, based on Ant Clustering Algorithm

Height relevant of neighbourhood pixels does not considered by Traditional FCM Algorithm. Reference[4] defined a new space function, which fully used the effectively neighbourhood information.

$$h_{ij} = \sum_{k \in \Omega(a_j)} u_{ik} \tag{6}$$

Where the $\Omega(a_j)$ is the center windows of pixels a_j. When most neighbourhood pixels of one pixel are the same kind, the value of h_{ij} becomes higher. Then we update the membership with the space function.

$$u'_{ij} = u_{ij} h_{ij} \Big/ \sum_{k=1}^{c} u_{ij} h_{kj} \tag{7}$$

Space function heightens the original membership function, and reduces the impact of noise and pseudo-pixel in the process of image segmentation.

FCM is the iterative algorithm. The cluster center is required initialization before the cluster analysis. Good initialization algorithm will not only accelerate the convergence rate, but also get good segmentation result[5]. This paper makes use of Ant clustering algorithm to initialize cluster center.

Ant colony algorithm, known as ant algorithm, was developed as a bionic evolutionary algorithm[6] in 1992 by Italian scholars M. Dorigo, V.maniezzo and others who were inspired by route choice behaviour in the process of ants foraging. It shows that during ants foraging, they usually find the shortest path from the nest to food, and when the environment changes, they can still search out the best path. Ant achieve this function by leaving a kind of 'pheromone' in the course of the campaign. The later ants can feel the existence and intensity of the pheromone, so it caused that there are more ants chose the route. It becomes positive feedback process.

Ant cluster algorithm was proposed by P.S.Shelokar, it's a new ant algorithm for dealing with cluster[7]. In fact, the process of Ants foraging is a process of continuous clustering. The food is the centre of the clustering[8]. The approach as follows: chose pixel A_i in image $A_{m \times n}$ as an ant, each ant is a two-dimensional vector characterized pixel gray level and gradient:

$$A = \left\{ A \, \middle| \, A_i = (a_{i1}, a_{i2}), i = 1, 2, \cdots, N, N = m \times n \right\}$$

r is the radius of clustering, d_{ij} is the weighted Euclidean distance between the pixels A_i to the representatives A_j, $\tau_{ij}(t)$ is information concentration of the route of ant i to cluster center. P is the weighted factor, $\| \bullet \|_2$ is 2 - norm.

$$d_{ij} = \| P(A_i - A_j) \|_2 \tag{8}$$

$$\tau_{ij}(t) = \begin{cases} 1 & d_{ij} \le r \\ 0 & d_{ij} > r \end{cases} \tag{9}$$

The probability of samples A_i merger to A_j is $P_{ij}(t)$:

$$P_{ij}(t) = \frac{\tau_{ij}^{\alpha}(t) \eta_{ij}^{\beta}(t)}{\sum_{s=1}^{j} \tau_{sj}^{\alpha}(t) \eta_{sj}^{\beta}(t)} \tag{10}$$

There $S = \left\{ A_s \, \middle| \, d_{sj} \le r, s = 1, 2, \cdots, j, j+1, \cdots, N \right\}$, η_{ij} is a reciprocal of d_{ij}, α β are regulate the factor, they are accumulation information of pixel clustering process and the influence factor of the route that is selected by guiding function. We combine A_i to the area of A_j if $P_{ij}(t)$ was greater than P_0. Here set $C_j = \left\{ A_k \, \middle| \, d_{kj} \le r, k = 1, 2, \cdots, J \right\}$

Then we get the good clustering center,

$$O_j = \frac{1}{J} \sum_{k=1}^{J} A_k , \quad A_k \in C_j \tag{11}$$

and calculate the deviation error of j-cluster

$$D_j = \sum_{k=1}^{J} \left[\sum_{i=1}^{2} (x_{ki} - O_{ji})^2 \right]^{\frac{1}{2}} \tag{12}$$

and compute the overall error $\varepsilon = \sum_{j=1}^{k} D_j$, if $\varepsilon \le \varepsilon_0$ we can calculate the cluster center v_i and the number of cluster c.

We use c as the cluster number of FCM cluster algorithm, and the cluster center as FCM initialization cluster center, and then initialize each parameter.

4. Filtering method based on mathematical morphology

The segmentation results of using above method may have a small amount of interference, so mathematical morphology filtering methods is applied to filter the segmentation results. Mathematical morphology is a new method that applies to the field of image processing and pattern recognition. The basic idea is to measure and extract the corresponding shape in image using certain form of structure, in order to analyse and identify image. The verge information extraction of mathematical morphology is better than other algorithms. The basic operation of morphological filtering is corrosion, expansion, opening and closing set operations and deformation.

4.1 Expansion

Let A and B is the collection in z, Φ is the empty set, A is expanded by B, as $A \oplus B$, $\oplus$ is the expansion operator, defined as

$$A \oplus B = \left\{ x \mid \left[\left(\hat{b} \right)_x \cap A \right] \neq \Phi \right\} \quad (13)$$

B is called as structural elements. Above formula shows that the process of A expanded by B is the pool of location of the center pixel of B when there is at least one non-zero elements of intersection between the displacement of B and A.

4.2 The process of images corrosion

Let A and B is the collection in z, A is corrupted by B, as $A \Theta B$, Θ is the corrosion operator, defined as:

$$A \Theta B = \left\{ x \mid \left(B \right)_x \subseteq A \right\} \quad (14)$$

The result of A corrupted by B is x, A contains result of B displaced by x.

4.3 Open computing and closed computing

Open computing: first, operating image corrosion, then applying the expansion operation to the result of corrosion. It is defined as:

$$A \circ B = \left(A \Theta B \right) \oplus B \quad (15)$$

Opening algorithm can generally smooth image contours, weaken the narrow part and remove the fine outstanding.

Closing algorithm: First, operating image expansion, and then applying corrosion computing to the results of the expansion, defined as:

$$A \bullet B = \left(A \oplus B \right) \Theta B \quad (16)$$

Closing algorithm is also smoothing images contours, In contrast with the opening algorithm, it generally can integrate the narrow gap and the bends slender mouth, removed small hole, fill the gap profile. Open computing and closing computing at the closed meeting on the set and reverse the dual, namely:

$$\left(A \bullet B \right)^C = A^C \circ B^C \quad (17)$$

First, corrupting operation can be carried out for images, which will eliminate interference, and expansion of operations can be carried out later, making crop images restitute.

Figure 1. Crops of Image in Complex Background

Figure 2. Segmentation results of above-mentioned algorithm

5. Experimental results and analysis

Based on the environment of the greenhouse for growth of crops is special, and the background is complex. In order to verify the efficiency and practicability of the algorithm which is above-mentioned. This paper adopts a picture of greenhouse plants. Figure 2 is based on the algorithm.

The result shows that the method in this paper can separate crop plant from complex background images, with the ant clustering algorithm, we can confirm the number of clustering and clustering center, thereby this algorithm reduce the number of iterations in FCM algorithm. From the Figure 2 we can see that all the background of greenhouse environment is wiped off thoroughly.

6. Conclusion

This paper gives insight to the flaws that, in the complex background, it is difficult for crop segmentation and extraction, the cluster number of FCM clustering algorithm is difficult to determine. Proposed to improve FCM algorithm, received good result. The image with small interference after segmentation is filtered by the mathematical morphology filtering. So that the background is thoroughly removed. The results of the experiment show that the algorithm is effective and practical.

7. References.

[1] Aimin Wang, Lansun Shen. The review of image segmentation [J]. Measurement and Control Technology, 2002, 19(5), 1-6.

[2] PAL B, PAL S K. A review on image segmentation techniques [J].Pattern Recognition, 1993, 26(9): 1277-1294.

[3] DECORETX, DURAND F, SILLION FX,etal.Billboard clouds for extrememodel simplification[J].ACM Transactions on G raphics, 2003,22(3): 689-696.

[4] Chuang K S,Tzeng H L,Chen S,et al.Fuzzy c-means clustering with spatial information for image segmentation[J].Computerized Medical Imaging and Graphics,2006(30):9-15.

[5] HUANG Guo-rui, WANG Xu-fa, CAO Xian-bin. Ant colony optimization algorithm based on directional pheromone diffusion [J]. Chinese Journal of Electronics, 2006, 15(3):447-450.

[6] SHELOKAR PS, JAYARAMANVK, KULKAMI B D. An ant colony approach for clustering [J]. Analytica Chimica Acta, 2004, 509(2):187-195.

[7] Yang HF, Hou CZ. Image segmentation by ant colony algorithm based on two-dimensional gray [J]. Laser & Infrared, 2005, 35 (8):614-617.

[8] Licai Yang, Lina Zhao, Xiaoqing Wu. Medical image segmentation of fuzzy C-means clustering based on the ant colony algorithm [J].JOURNALOF SHANDONGUNIVERSITY (ENGINEERING SCIENCE), 2007, 37(3):51-5

Research of Online Signature Recognition Based on Energy Feature

Jinxu Guo, Jianbin Zheng, Bo Lu
School of Information Engineering
Wuhan University of Technology, Wuhan, 430070, P. R. China
guojx2000@126.com, zhengjb@whut.edu.cn

Abstract

The paper proposes an on-line signature recognition algorithm with signature energy as feature. The signature energy features at sharp trajectory change points are extracted by means of Daubechies wavelet decomposition of signature signal. Then, 15 points with most dominant energies are chosen. Finally, a new algorithm of classification is put forward, after dynamic time warping matching, with computation amount reduced greatly. Experiment findings show that false acceptance rate is 8 percent while false rejection rate is 0 percent.

1. Introduction

Signature recognition is one of the most important research areas in the field of identity recognition based on biometrics. The technology is also regarded as a front subject in many fields such as pattern recognition and signal processing. An important advantage of the signature recognition compared with other biometric attributes is its long tradition in many common commercial fields.

Generally, there are two main types of signature recognition, namely off-line and on-line signature recognition. Off-line signature recognition deals with the analysis of the signature image alone. The major drawback of this type of signature recognition is that the signature image alone constitutes a limited database for analysis, difficult to make an effective determination of the validity of the signature.

On-line signature recognition, in turn, consists of digitizing the signature as it is being produced. With this method, the information obtained will contain not only the signature image, but also time domain information, such as signing speed and acceleration. In addition, signing pressure can be recorded through handwriting pad. All this information can be combined to determine the validity of a signature much more effectively than what an off-line recognition system is capable of.

Owing to the signatures from the same subject with a certain degree of fluctuation, especially the nonlinearity of time axis, it is rather difficult to match one-by-one

between the signature signals, and large amounts of computation can be caused as well.

Plenty of methods have been proposed for on-line signature recognition. In particular, Darwish and Auda used a neural network as classifier to compare 210 signature features [1]. Nakanishi introduced wavelet to the signature verification [2], and decomposed the signature signal into 8 levels, but the result has not yet been reported so far.

Dynamic Time Warping (DTW) algorithm, having been widely used in speech recognition, provides for a nonlinear alignment of two signatures [3][4]. DTW also has the advantage of robustness. Even if the test sequence cannot be completely matched with the reference sequence, the algorithm can still match the test and the reference sequences well so long as the time order exists. It does this by evaluating various permitted pairings between the points of the two sequences, and by selecting the best alignment path based on some optimality criteria and search constrains. While matching the whole sequences between the test and reference signatures, DTW is time-consuming and much complicated.

Signature energy features at sharp trajectory change points, rather than the energy features at each different frequency channels, are extracted, after fast wavelet decomposition of signature signal. These features, if selected appropriately, are representative of the whole signature to a certain degree. In order to simplify DTW algorithm, in the paper, from another point of view, combining excellent performance of DTW with small amounts of signature energy features representing a signature. A new algorithm is as follows:

Firstly, the system realizes fast wavelet decomposition and reconstruction of the signature signal after preprocessing. Then, 15 most dominant signature energies of sharp trajectory change points in the signature are extracted, and kept its original order, prepared for DTW matching. Instead of the total points of both sequences, the amount of computation can be reduced greatly through DTW comparison between feature points of test and reference signatures. Finally, a new classification method is put forward.

2. Collection of database and preprocessing

2.1. Collection of database

The database consists of 1320 signatures obtained from 22 subjects signing their own signatures over a five-month period. Each of the subjects attended about three data-collection sessions per week during the five-month period. On the average, approximately 30 sitting signatures and 30 standing signatures were obtained from each subject. The signer subjects were chosen at random from a large group of volunteers in our laboratory. Six of the 22 subjects were women.

2.2. Preprocessing of signature signal

A microprocessor based system has been designed in our laboratory to collect on-line handwriting signature signal, with USB communication interface to computer. The embedded system real-time collects handwriting signature signals, including the position signals of X and Y. The space resolution is 4900×4900 and the sampling interval Δt is 10ms.

First, we must remove the gaps between strokes, and these gaps reflect the time that the handwriting pen does not contact the handwriting pad and the positions that strokes begin and end. Simultaneously, we must normalize the size, the length of the signature and so on, the effect after being processed as Figure 1.

3. Wavelet analysis and energy feature extraction

3.1. Wavelet decomposition and deconstruction

Wavelet multi-resolution analysis [5] was first put forward by Mallat in 1988. Using Mallat algorithm, the wavelet decomposition and the reconstruction method becomes extremely simple. Original signature signal S_0 can be decomposed as $S_0 = S_3 + d_3 + d_2 + d_1$, where S_i, d_i (i =1,2,3) are respectively low and high frequency component for a certain layer of signal decomposition. In the paper, db3 wavelet basis is used for the wavelet decomposition and reconstruction, and signal is decomposed into 3 levels. Figure 2 shows the wavelet decomposition and reconstruction wave of X-axis signal of a signature.

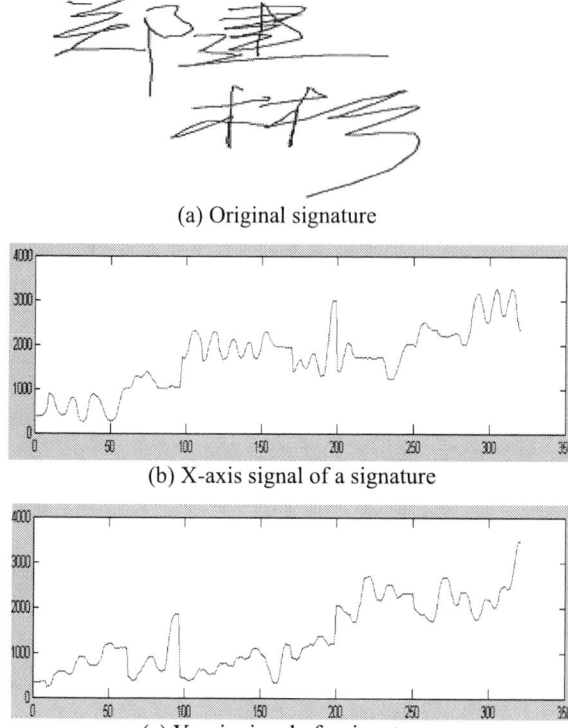

(a) Original signature

(b) X-axis signal of a signature

(c) Y-axis signal of a signature

Figure 1. Original signature and its acquisition signal

3.2. Signature energy feature extraction

Through large amounts of experimental study, signature energy, with different frequency component in the inflection point in the signature signal, is closely related with the writing characteristics at each sharp trajectory change point. d1, which is characteristics of individuality, reflects the tremor while writing, but it is unstable. d3 is stable, however, its individuality is not obvious. Taking all the factors into consideration, we add d1, d2 and d3 together, as Figur3 shows. Based on the wave of d1+d2+d3, above all, we should find out positions of the sharp change points, then compute signature energy [6] as in (1).

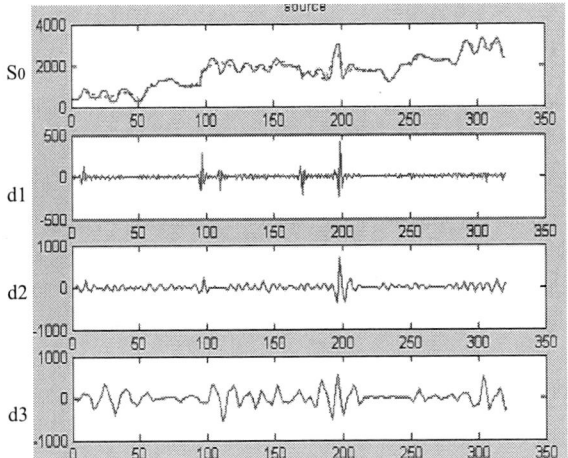

Figure 2. Wavelet decomposition and reconstruction of signature signal

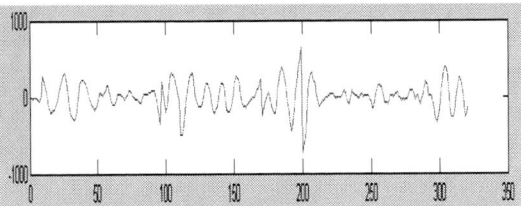

Figure 3. d1+d2+d3 wave

$$e_i = \frac{1}{n \times 2 + 1} \sum_{j=i-n}^{i+n} |A_j| \qquad (1)$$

e_i is the signature energy values, while A_j is the value at j in Figure 4. The window size is $n \times 2 + 1$, while the window center i is the sharp change point. We usually take $n = 2$. For example, all the signature energies at 15 sharp change points are curved as Figure 4. Owing to the wavelet decomposition, the signature energy feature has nothing to do with their initial position.

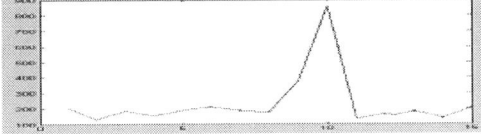

Figure 4. Signature energy curve

4. Energy feature matching based on DTW and result analysis

DTW algorighm tries to find the corresponding relations of each points between two sequences through recursive method, and can minimize the distance of the two sequences, and can calculate this distance. In this article, the goal of DTW algorithm is to get an optimal time-allignment matching path between signature energy features of reference pattern R and test pattern T. The brief realization process is as follows [3][4]:

Suppose P is a warping function mapping the feature points of test signal to these of reference signal:

$$P = p(1), p(2), ..., p(k), ..., p(K) \qquad (2)$$

where $p(k) = (i(k), j(k))$, i and j representing the respective feature point of reference signal R (total length is I) and test signal T (total length is J), it expresses the k th times matching between the i th signature energy feature point of R and the j th point of T. The limit condition of warping function is:

① Monotonicity: $i(k-1) \le i(k)$, $j(k-1) \le j(k)$

② Continuity: $i(k) - i(k-1) \le 1$, $j(k) - j(k-1) \le 1$

③ Boundary: $i(1) = 1$, $j(1) = 1$, $i(K) = I$, $j(K) = J$

④ Window: $|i(k) - j(k)| \le r$, where r is the permit width of the window. It cannot be discussed here for lack of space. We define:

$$d(p_k) = d((i(k), j(k))) = \|R_i - T_j\| \qquad (3)$$

The essence of DTW algorithm is to seek the matching path P which makes minimum the total distance from R to T, namely

$$D(P) = \underset{P}{Min} \sum_{k=1}^{K} d(p_k) \qquad (4)$$

From the DTW above, P is the most optimized matching path between R and T. In the paper, with search window width $r = 3$, 15 points with most dominant signature energies are extracted, in its original order, then the DTW matching method is employed. Figure 5 shows the signature energy curves of a subject prior to and after the DTW matching (X-axis signature signal).

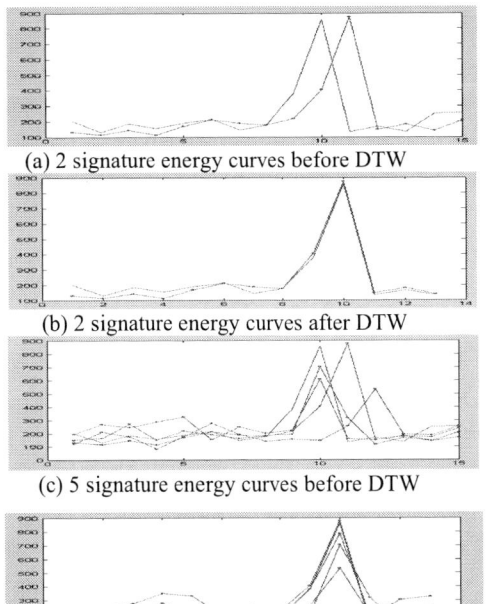

(a) 2 signature energy curves before DTW

(b) 2 signature energy curves after DTW

(c) 5 signature energy curves before DTW

163

(d) 5 signature energy curves after DTW

Figure 5. Comparison of signature energy feature curves of subject A

After DTW matching, we can take out the signature energies of the corresponding M (for example, $M = 13$) points to form respective feature vector E and e, and calculate its matching value according to (5):

$$ n_1 = \frac{1}{M} \cdot \sum_{i=1}^{M} Max(E_i, e_i) \Big/ Min(E_i, e_i) \qquad (5) $$

where E_i is the i th point signature energy value of reference signature, while e_i is the corresponding i th point signature energy value of test signature after DTW, with n_0 as threshold value. If $n_1 \le n_0$, the reference signature and test signature may be regarded as matched, namely, the two signatures may be regarded as from the same subject. Otherwise, if $n_1 > n_0$, the two signatures are not from the same subject. When $n_1 = 1$, the two signatures are completely matched. The bigger n_1 is, the worse the matching is.

The matching algorithm has the characteristics that threshold selection is only concerned with relative amplitude, having nothing to do with absolute amplitude of signature energy feature, so it is possible to set a common threshold value for any different subjects. It is not convenient for us to set different thresholds for different subjects, so this may provide a solution to the threshold problem bothering us for a long time. Meanwhile, if signature energy at any inflection point is great enough, n_1 is much bigger, then the two signatures cannot be regarded as from the same subject. Only if signature energies between reference signature and test sample at every inflection point are matched, may the two signatures be regarded as from the same subject. If threshold n_0 is properly set (for example, $n_0 = 1.9$), it is likely that false acceptance rate (FAR) is small while false rejection rate (FRR) is kept very small. Table 1 shows the experimental result.

Table1. Experimental result on the condition of $n_0 = 1.9$

Signer	Mean of n_1	Deviation of n_1	FAR	FRR
Same subject	1.51	0.16	8.0%	0%
Different subjects	2.95	0.91		

5. Conclusion

Signature forgery can be divided into random forgery and skilled forgery. Random forgeries usually occupy more than 90% of forgery signatures, the algorithm in the paper can eliminate random forger quickly through computer. As for skilled forgery, it is possible to be eliminated through next fine matching step, like stroke

matching between break points, or matching the whole sequences between the test and reference signatures through the same DTW algorithm.

Large amounts of experimental findings show that false acceptance rate is 8.0 percent while false rejection rate is 0 percent for random forgery and part of skilled forgery in the signature database in our laboratory. This fast algorithm can be used in the first stage of signature recognition in order to eliminate large amounts of forgeries. It is one kind of great potential identity recognition method.

References

[1] A.M.Darwish, and G.A.Auda, "A New Composite Feature Vector for Arabic Handwritten Signature Verification", *Proc. IEEE Int. Conf. On Acoustics*, 1994, Vol. 2, pp.613-616.
[2] I.Nakanishi, N.Nishiguchi, Y.Itoh and Y.Fukui, "On-line Signature Verification Method Utilizing Feature Extraction Based on DWT", *IEEE Trans. Inf. Theory*, 2003, Vol. 38, No. 2, pp.691-712.
[3] M.Parizeau, R.Plamondon, "A Comparative Analysis of Regional Correlation, Dynamic Time Warping, and Skeletal Tree Matching for Signature Verification", *IEEE Trans. On Pattern Analysis and Machine Intelligence*, 1990, Vol. 12, No. 7, pp.710-717.
[4] R.Martens, L.Claesen, "On-line Signature Verification by Dynamic Time Warping", *Proc. of the 13th Int. Conf. on Pattern Recognition*, 1996, pp.38-42.
[5] S.G.Mallat, "A Theory for Multiresolution Signal Decomposition: The Wavelet Representation", *IEEE Trans. On Pattern Analysis and Machine Intelligence*, July 1989, Vol. 11, No. 7, pp.674-693.
[6] M.C.Lee, and C.M.Pun, "Texture Classification Using Dominant Wavelet Packet Energy Features", *Proc. of the 4th IEEE Southwest Symposium on Image Analysis and Interpretation*, April 2000, pp.301-304.

Session 5

Signal Processing Application
in Communications

2008 Workshop on Power Electronics and Intelligent Transportation System

Optimum of Intersection Signal Timing based on IGA

DU Rong-hua

Changsha University of Science and Technology, Changsha, 410076, China

CSDRH@163.COM

Abstract

In this paper, the traditional research methods of the intersection signal timing and optimization are discussed and analyzed. An improved signal timing model, which considers delay time, stop frequency and traffic capacity as the performance index of signal timing optimization model, is presented. The performance index weighting coefficient will be adjusted to adapt different traffic conditions. A new algorithm, based on immunogenetics, is proposed to optimize traffic signal timing. The simulation results show that the new signal timing method reduces the average delay time and the average stop frequency and increases the traffic capacity of the intersection.

1. Introduction

Optimizing signal timing based on the traffic features of intersection is to seek signal timing scheme which enables user interest and management efficiency to achieve the optimum [1].Currently, a single intersection signal timing mainly adopts F-B method, which is proposed by British scholar F. Webster and B. Cobber based on statistical equilibrium theory in the 1950s [2]. Domestic and international scholars have improved FB [3]. FB assumes that the vehicle's delay time is cumulative, but actually arriving and departing of vehicles are discrete events, FB is still difficult to be truly effective way to resolve the traffic problems, we need to find other more reasonable timing method.

In this paper, multi-phase intersection signal control is researched; an improved method is presented, which considers the delay time, the stop frequency and traffic capacity as the performance index of signal timing optimization model. A new algorithm, based on immunogenetics, is proposed to optimize traffic signal timing. The simulation results show that the new method reduce the average delay time and the average stop frequency, increase the traffic capacity of the intersection.

2. Signal Timing Model

2.1 Cycle length and timing

According to the Webster delay formula, the literature [4] obtained the best cycle length C and the effective green time g_i:

$$C = \frac{1.5L + 5}{1 - Y} \qquad (1)$$

$$L = \sum_{i=1}^{n} (l + I - A) \qquad (2)$$

$$Y = \sum \max \left[y_i^1, y_i^2 \cdots \right] \qquad (3)$$

$$g_i = \frac{y_i}{Y}(C - L) \qquad (4)$$

where:

L: the total loss time of per cycle, s;

I: change interval, s;

Y: the sum of the maximum y_i of each phase, y_i represents traffic intensity of i phase, is ratio between traffic flow and saturated flow;

A: yellow interval.

Based on road safety and drivers patience for waiting, the cycle length of four-phase intersection: $60s < C < 180s$, the minimum effective green time of each phase of is 10 s, So each phase must meet the following constraints:

$$10 < t_i < C - 30 \qquad (5)$$

2.2 Average delay

As Webster delay formula only applies to a smaller saturation, so the average delay of vehicles uses that formula [5]:

$$d_i = \frac{(C - g_i)^2}{2C(1 - y_i)} + \frac{C - g_i}{2Cq_i} + \frac{q_i C^2}{2S_i g_i (S_i g_i - q_i C)} \qquad (6)$$

Where:

q_i: average approach flow rate at signal controlled intersections（vch/s）;

S_i: Saturation Flow Rate at signal controlled intersections（vch/s）;

$y_i = q_i / s_i$, so the total delay time of all vehicles entered the intersection in one cycle length is:

$$D = \sum_{i=1}^{n} d_i q_i C \qquad (7)$$

2.3 Traffic capacity and average stop frequency

978-0-7695-3342-1/08 $25.00 © 2008 IEEE
DOI 10.1109/PEITS.2008.16

167

In the signal control intersections, vehicles can pass through the stop lines in the effective green time, so the traffic capacity of phase i is [4]:

$$Q_i = S_i \times g_i / C \qquad (8)$$

Average stop frequency of phase i is [4]:

$$H_i = 0.9 \times \frac{1 - g_i}{1 - y_i} \qquad (9)$$

2.4 Signal timing optimization model

This paper not only choose delay time、stop frequency and traffic capacity as performance indexes of objective function, but also considers the actual situation of China's urban road traffic. The performance index weighting coefficient should be adjusted to adapt different traffic conditions. [6]

When time delay and stop frequency reduces, traffic capacity will increase, and thus the objective function adopts negative value of the traffic capacity.

$$\min_x f(x) = \sum (k_i^1 d_i q_i C + k_i^2 H_i - k_i^3 Q_i) \qquad (10)$$

$$s.t \begin{cases} g_i - a_i > 0 \\ \sum (g_i + l_i) \le b \\ 0.75 \le \dfrac{y_i C}{g_i} \le 0.95 \\ 1 \le i \le n \end{cases} \qquad (11)$$

Where:

a_i: The minimum effective green time

l_i: the loss time of phase i;

b: the maximum - period;

In the traffic normal period, the principle of the signal timing is to minimize the delay and parking of vehicles in intersections; in the traffic peak period, the focus is to improve the traffic capacity of the intersection, so the ratio of the weighted coefficient of the delay and stopping should increase when the traffic flow rate of intersection decrease, so [6]:

$$k_i^1 = S_i y_i (1.0 - Y) \qquad (12)$$

$$k_i^2 = S_i y_i (1.0 - Y) \qquad (13)$$

$$k_i^3 = \frac{3600 Y}{C} \qquad (14)$$

the saturation of intersection (saturation is the ratio of the traffic flow with the traffic capacity, the traffic capacity is equal to the product of the saturation traffic flow and the green radio) will be limited, 0.75 is to avoid to increase delays and parking of vehicles unnecessarily when the traffic capacity is far greater than the traffic demand, 0.95 is to avoid congestion caused by big saturation [7].

3. Algorithm Design

In this paper, the immune genetic algorithm, based on genetic algorithms, fully considers antibodies containing excellent genes as role in the evolution of the population, and expresses concern about good growth and low-affinity antibody, ensembles antibodies similar to antigen. Crossover operation of antibodies can accelerate the speed of convergence, and ensure the strongest adaptive antibody not to mutate in the latter part of the evolution, and maintain stability of superior antibody.

3.1 Immune genetic operator

(1) Selection operator

In the evolutionary process, the diversity of population will reduce premature phenomena. So the similar individual must be excluded as far as possible to spread to the entire solution space. Algorithm adjusts the choice probability by introduction of antibody concentration factor. Then it could maintain the diversity of the population, and ensure that the excellent antibody to enter the larger the probability of to the next generation [9, 10].

Affinity degree refers to match degree between antibody and antigen. In the evolutionary process of Population, the initial antibody cells will clone while encountering antigen, high affinity cells have a stronger ability to reproduce, rapidly improve Affinity degree of the survival cells.

Definition 1: Affinity degree between antigen and antibody k is:

$$A_k = \frac{1}{1 + t_k}, t_k = \frac{f(x_k) - f^{\min}(x)}{f^{\max}(x) - f(x_k) + \delta} \qquad (15)$$

Where, t_k shows combined degree between antigen and antibody k, $f^{\min}(x), f^{\max}(x)$ are minimum and maximum of objective function values, and δ is a little Positive Number, so dividend is not zero.

Definition 2: for a given number N of the population, concentration of antibody k is:

$$C_k = \frac{\sum\limits_{i=1}^{N} f(i)}{N} \qquad (16)$$

$$f(i) = \begin{cases} 1, A_i > \sigma \\ 0, A_i \le \sigma \end{cases} \qquad (17)$$

Where, C_k is concentration of antibody k, N is Total number of antibodies, $f(i)$ is concentration function of antibody, σ is concentration Threshold Value, also is concentration Inhibition radius, $0 < \sigma \le 1$.

Definition 3: based on Affinity degree and concentration of antibody, antibody selection probability p_k is:

$$p_k = \alpha \frac{A_k}{\sum\limits_{j=1}^{N} A_j} + (1-\alpha)\frac{1}{N} e^{-\frac{c_k}{\beta}} \qquad (18)$$

Where, α, β are regulation factor, $0 < \alpha, \beta < 1$, and A_k is affinity degree of antibody, the greater affinity degree, the greater choice probability; the greater the concentration of antibodies, the smaller choice probability.

(2) Crossover Operator

The real coding can express a greater scope of the solution space, which also be used in this algorithm, if the cross-computing uses linear reorganization, parents can generate slightly larger than in the definition of an arbitrary point line. To assume two cross-substituting father individual x_1 and x_2, the offspring $x_i^{'}$ is:

$$x_i^{'} = x_1 + r_i(x_2 - x_1) \qquad (19)$$

where, $r_i \in [-0.5, 1.5], i = 1,2$. If offspring exceeds the solution space of objective function, the value of the principle of offspring is: if $x_i^{'}(j) > b_j$, then $x_i^{'}(j) = 2b_j - x_i^{'}(j)$; if $x_i^{'}(j) < a_j$, then $x_i^{'}(j) = 2a_j - x_i^{'}(j)$.

(3) Mutation Operator

Mutation means imitating biological variation so as to increase the diversity of the population [11]. Mutation operator may introduce new entity which does not appear in the initial population, and may resume missing excellent genes after selection and crossover. As x is the father of the former generation, $x^{'}$ is the offspring, $x^{'} = x \pm 0.5L\lambda$, $\lambda = \sum\limits_{i=0}^{m} \frac{a(i)}{2^i}$, $a(i)$ is the random probability value between [0, 1]; L is range of the variables. To ensure that the optimal antibody is not destroyed, it is necessary to inhibit the process of mutation so that the mutation rate and the affinity of the antibody are inversely proportional.

3.2 Steps of IGA

Step 1: initialization:

（1）Initialization of parameters, the real coding is used for antibodies, coding length is l, and the solution space is S, the number of evolution $k = 0$, set convergence conditions.

（2）Initialization of the population, according to experience of experts in the field, randomly generated initial antibody N in the space of objective function, integrated to $X(0)$.

Step 2: create a good antibody memory pool, calculate affinity degree of each antibody of population $X(k)$ with antigen, and extract the best n_1 antibody from the entire population create a good memory pool X_M.

Step 3: judgment of Convergence, if meet the convergence conditions, then terminate the process of evolution, the best antibody is taken out from good memory pool X_M, which is the best global optimal solution, if the conditions are not satisfied with convergence, entered Step 4.

Step 4: Evolution groups, antibodies of $X(k)$ obtain choice probability in accordance with the choice operator, antibodies which choice probability is r will be composed of population X_s, and after crossover and mutation of X_s, and the population $X^{'}(k)$ will be received.

Step 5: Calculate affinity degree, Calculate antibody's affinity degree of population $X^{'}(k)$.

Step 6: Update groups, $X^{'}(k)$ compare with $X(k)$ in accordance with the size of affinity degree, replace low affinity antibodies of the father population, generate the next generation antibody $X(k+1)$, and update good memory pool.

Step 7: If $k < \lambda$, λ as the greatest number of evolution, to the steps three to judge convergence.

4. Simulating Test

The paper takes SongGuiYuan intersection of Changsha, shown in Figure 1, as an example to establish the four-phase traffic flow model. The four directions of the intersection set up a dedicated right turn lane, the minimum green time of four-phase is 15 s, the largest cycle length is 180 s, phase loss time is 5 s, the saturation flow is 2000 vch/h.

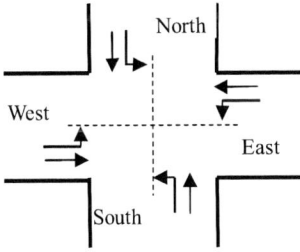

Fig 1 Distribution of Traffic Flow

Consider formula (10) as optimizing objective function to optimize signal timing. To test the effectiveness of this algorithm, we take $y_1 = 0.30$, $y_2 = 0.26$, $y_3 = 0.28$, $y_4 = 0.32$ and $y_1 = 0.42$, $y_2 = 0.38$, $y_3 = 0.40$, $y_4 = 0.45$ as case to study. Select 15 traffic flows at the intersection as the initial population to solve signal timing scheme of different traffic flow. Main parameters are: initial population size is 15, to develop an excellent memory for the number selection antibodies during creating good memory antibody pool $n_1 = 8$, the biggest evolution of the number $\lambda = 100$, concentration threshold $\sigma = 0.5$, and regulatory factors $\alpha, \beta = 0.5$. From the results shown in the table, we can conclude that the algorithm based on this paper is more reasonable than other algorithms, the new method reduces the average delay time and the average stop frequency , increases the traffic capacity of the intersection.

Table 1 Comparing Performance Index

Algorithm	C	g_1	g_2	g_3	g_4	d	q	h
IGA	126	35	26	25	20	35.7	1632	0.68
Webster	135	40	15	20	30	45.2	1450	0.85
Survey	143	35	15	20	35	48.6	1476	0.79
$y_1 = 0.30$		$y_2 = 0.26$		$y_3 = 0.28$		$y_4 = 0.32$		

Algorithm	C	g_1	g_2	g_3	g_4	d	q	h
IGA	175	45	30	32	48	55.2	1805	0.74
Webster	180	40	35	38	47	68.3	1621	0.88
Survey	173	38	35	30	50	79.2	1576	0.82
$y_1 = 0.42$		$y_2 = 0.38$		$y_3 = 0.40$		$y_4 = 0.45$		

where: Signal time, s; g_i : Effective green time; d : average vehicle delay, s; h : average stop frequency; q : Traffic capacity, vch/h

In order to verify that immune genetic algorithm (IGA) is more advanced than other general genetic algorithm (GA), this paper take traffic normal data shown in table 1 as an example to compare signal timing scheme and average delay of IGA and GA in different evolutionary algebra . In order to obtain optimal timing, IGA must evolutes 36 and SGA must evolutes 58, the results show that convergence rate of IGA speed up.

5. Conclusion

In this paper, the traditional research methods of the intersection signal timing and Optimization are analyzed. An improved method is presented, which considers delay time, stop frequency and traffic capacity as the performance index of signal timing optimization model, the performance index weighting coefficient will be adjusted to adapt different traffic conditions. A new algorithm based on immunogenetics is proposed to optimize traffic signal timing. The simulation results show that the new method reduces the average delay time and the average stop frequency and increases the traffic capacity of the intersection.

Acknowledgement

This work is supported by the Hunan Nature Science Foundation No.07JJ5068, and the Hunan Education Department Foundation No.07B003.

References

[1] YANG Jin-dong, YANG Dong-Yuan. Optimized Signal Time Model in Signaled Intersection. *Journal OF Tongji University*, 2001,29(7): 789～794 (in Chinese)

[2]WEBSTER F V·Traffic Signal Settings [J].*Road Research Laboratory Technical Paper*, 1958, 39(1):1-39·

[3]Yang Xiaoguang,Yang Peikun.Traffic Delays Simulation Algorithm in Stop lines of Intersection[J]·*Journal OF Tongji University*,1993,21(1):67-73·

[4] Yang Peikun. *Traffic management and control* [M].BeiJing：People's Communications Publishing House,1995

[5]Pu Qi,Tan Yonghao,Yiang Chao.Signal-planning Optimal Model for Intersection[J].*Journal of ShangHai TieDao University*,1999,20(4):31-34

[6] Gu Huaizhong, Wang Wei. A Global Optimization Simulated Annealing Algorithm for intersection Signal Timing [J]. *Journal OF SouthEast University*, 1998,28(3): 68-72 (in Chinese)

[7] YAN Yan-xia, LI Wen-quan. Ant Colony Optimization for Signalized intersection [J]. *Journal of Highway and Transport Research and Development*, 2006,23(11): 116-125 (in Chinese)

[8] Leandro Nunes de Castro, Jon Timmis. An Introduction to Artificial Immune Systems: *A New Computational Intelligence Paradigm* [M]. Springer-Verlag, 2002

[9] Xue Wen-tao. A Genetic algorithm based on immune learning mechanism and its application [J]. *Information and Control*,2008,37(1):9-17 (in Chinese)

[10] JIAO L C, WANG L. A novel genetic algorithm based on immunity [J].*IEEE Trans Systems, Man and Cybernetics*, 2000, 30(5):552-561.

[11] Hartmann S. A Competitive genetic algorithm for resource-constrained project scheduling [J].*Naval Research Logistics*, 1998, 45: 733-7

Spectrum Sensing Based on Cyclostationarity

Shiyu Xu, Zhijin Zhao, Junna Shang

School of Communication Engineering
Hangzhou Dianzi University
Hangzhou, China
xushiyu613@126.com , zhaozj03@hdu.edu.cn, shangjn@163.com

Abstract

Real time spectrum sensing with certain accuracy plays a key role in cognitive radio. Cyclostationary feature is used for spectrum sensing in this paper. Usually, cyclostationary feature detection requires high computation complexity, in this paper we analyze the performance of some frequencies and cycle frequencies for detection according to the licensed users' signal features, which reduce the complexity significantly. The best detection point is determined through simulation analysis on different detection points, and then we propose combination detection method using multiple detection points to obtain better performance. Results validate the effectiveness of the proposed method.

1. Introduction

Today, by unprecedented growth of wireless applications, the problem of spectrum scarce is becoming apparent. Most of the spectrum has been allocated to specific users, while other spectrum bands that haven't been assigned are overcrowded because of overuse. However, most of the allocated spectrum is idled in some times and locations. The Federal Communication Commission (FCC) research report [1] reveals that, seventy percent of the allocated spectrum is underutilized. So we need a technique to deal with the problem of spectrum underutilization, which makes the birth of cognitive radio. Cognitive radio [2][3]can sense external radio environment and learn from past experiences. It can also access to unused spectrum band dynamically without affecting the licensed users, in such a way to improve the spectrum efficiency. Sensing external radio environment quickly and accurately plays a key role in cognitive radio. Energy detection [4], pilot detection [4], and cyclostationary feature detection [4] are three commonly used spectrum sensing methods. Energy detection is easy to implement, but its performance degrades greatly under low signal-to-noise ratio (SNR) or with noise uncertainty. Pilot detection can detect signals with low SNR, but it needs the licensed user's prior knowledge and perfect synchronization, which is hard to realize in reality.

Cyclostationary feature detection can achieve high detection probability under low SNR, however, it usually needs high computation complexity. In [5], the author suggests to detect the cyclic features only in axis $f = 0$ and axis $\alpha = 0$, thus making the detection space from two dimensions to two one dimension and requiring less complexity.

Most of the papers [6][7] about spectrum sensing are based on given thresholds to analyze the detection performance, but in cognitive radio, we need to put the protection of licensed users in first and foremost, so we must ensure a predefined probability of detection (P_D). In reality, based on a given location and channel, the licensed users' signal parameters are known and the SNR is changing slowly, so we assume that we can obtain the licensed users' signal type and SNR before making detection. In this paper, using of the licensed users' prior knowledge, we only make detections in some specific frequencies and cycle frequencies, and combine multiple detection points to improve the performance further. And then given the P_D required by licensed users, the probability of false alarm (P_{FA}) under different SNRs is analyzed. Through the threshold adjustment, we reduce the P_{FA} to make better use of spectrum hole when the SNR is high and increase the P_{FA} to avoid interference to the licensed users when the SNR is low.

2. The principle of cyclostationarity

Modulated signals are in general coupled with cosine carrier, repeating spreading, over-sampling etc., resulting in built-in periodicity. When the signal's mean and auto-correlation exhibit periodicity, i.e., $m_x(t + T) = m_x(t)$, $R_x(t + T, u + T) = R_x(t, u)$, we call this signal a second-order cyclic statistics process[8]. The auto-correlation of signal $x(t)$ is defined as

978-0-7695-3342-1/08 $25.00 © 2008 IEEE
DOI 10.1109/PEITS.2008.41

$$R_X(t,\tau) = \lim_{N \to \infty} \frac{1}{2N+1} \sum_{n=N}^{N} x(t + \tau/2 + nT_0) x^*(t - \tau/2 + nT_0) \quad (1)$$

Since $R_X(t,\tau)$ is periodic with period T_0, it can be expressed as a Fourier series representation

$$R_X(t,\tau) = \sum_{m=-\infty}^{+\infty} R_X^{m/T_0}(\tau) e^{j2\pi mt/T_0} \quad (2)$$

$$R_X^{\alpha}(\tau) = \frac{1}{T_0} \int_{-\infty}^{\infty} R_X(t,\tau) e^{-j2\pi \alpha t} dt \quad (3)$$

where α is the second-order cycle frequency equals to m/T_0, $R_X^{\alpha}(\tau)$ is referred to as the cyclic autocorrelation function. The spectrum α coherence function (SCF) can be obtained from (3) as

$$S_X^{\alpha}(f) = \int_{-\infty}^{\infty} R_X^{\alpha}(\tau) e^{-j2\pi f\tau} d\tau = \frac{1}{T} X(f + \alpha/2) X^*(f - \alpha/2) \quad (4)$$

where $X(f)$ is the Fourier Transform of signal $x(t)$. From (4) we can find that $S_X^{\alpha}(f)$ is the correlation of the signal spectrum.

Different types of signals have different spectrum correlation features, in this paper we use BPSK signal with symbol period T_0 as an example for our research. From [9] we know that the SCF of BPSK signal have peaks at points $\alpha = \pm m/T_0, f = \pm f_c$ and $\alpha = \pm 2 f_c \pm m/T_0, f = 0 \quad (m = 0, \pm 1, \pm 2 \cdots)$. Fig. 1 demonstrates the SCF of a BPSK signal with symbol rate $R = 1/T_0 = 2 \times 10^6$, carrier frequency $f_c = 4 \times 10^6$ Hz. Fig. 2 depicts the contour figure of the SCF. The SCF of the Gaussian white noise is given in Fig. 3. These figures illustrate that the SCF of BPSK signal is different from the SCF of Gaussian white noise and cyclostationary features can be used for signal detection under low SNR environment.

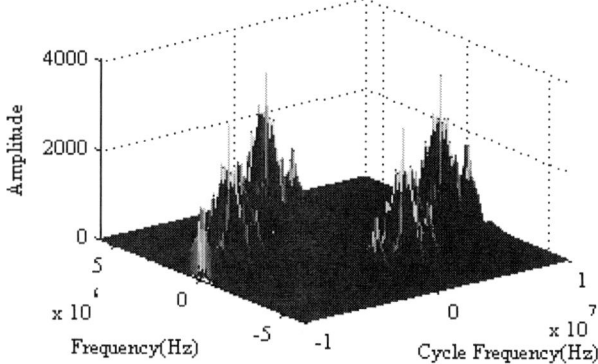

Figure 1. SCF of BPSK signal

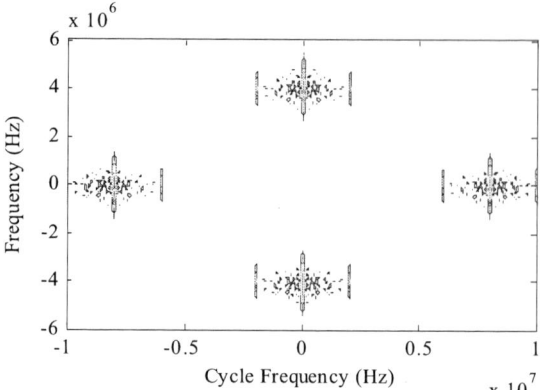

Figure 2. Contour figure of a BPSK signal

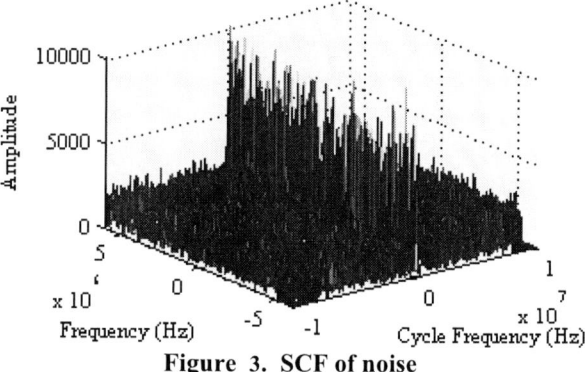

Figure 3. SCF of noise

3. Implementation

Usually, the cyclostationary detection requires high computation complexity and can't meet the real-time operation that required by spectrum sensing for cognitive radio. In order to reduce the complexity, in [5] the author suggests to make detection in axis $f = 0$ and axis $\alpha = 0$, and he also points out that when $f_c = f_s/4$, the detection point $f = 0$, $\alpha = 2 f_c$ performs best, here f_s denotes the sample rate. However, from [9] we can find that the points $\alpha = \pm m/T_0, f = \pm f_c$ also have peaks, which can be demonstrated as in Fig. 4 that depicts the SCF of a BPSK signal at cycle frequency domain under $f = f_c$. Obviously, [5] ignores the detection of these points.

In this paper, we assume that the characteristics of licensed users' signals are known and detect these signals under specific points. We use time-domain averaging and frequency-domain smoothing to obtain SCF as follows [9]

1) Determine points of cycle frequency and carrier frequency that we need to analyze.
2) Get M groups of data with length of N, compute FFT for each group of data

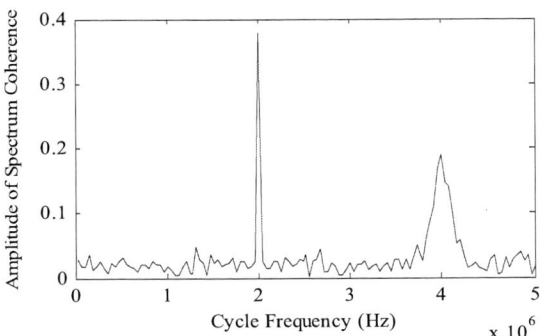

Figure 4. Cycle frequency domain profile of a BPSK signal

$$X_m(k) = \sum_{n=0}^{N-1} x_m(n)e^{-j2\pi nk/N} \qquad (5)$$

where $X_m(k)$ is the FFT of m-th data group.

3) Compute power spectrum density(PSD)

$$S_m(k) = \frac{1}{N} X_m(k) X_m^*(k) \qquad (6)$$

where $S_m(k)$ denotes the PSD of the signal from m-th group of data.

4) Compute SCF

$$S_m^{\alpha_i}(k) = \frac{1}{N} X_m(k + \alpha_i/2) X_m^*(k - \alpha_i/2) \qquad (7)$$

where $S_m^{\alpha_i}(k)$ is the SCF of i-th detection point.

5) Frequency-domain smoothing

$$S_m^{\alpha_i}(k)_{\Delta f} = \frac{1}{P} \sum_{p=-(P-1)/2}^{(P-1)/2} S_m^{\alpha_i}(k+p) \qquad (8)$$

where P is spectrum domain smoothing factor, $S_m^{\alpha_i}(k)_{\Delta f}$ is the SCF of i-th detection point from m-th group of data after spectrum domain smoothing.

6) Compute test statistics

$$I_m(\alpha_i) = \frac{S_m^{\alpha_i}(k)_{\Delta f}}{[S(k - \alpha_i/2)S(k + \alpha_i/2)]^{1/2}} \qquad (9)$$

$$I(\alpha_i) = \frac{1}{M} \sum_{m=1}^{M} I_m(\alpha_i) \qquad (10)$$

$$I(\alpha) = \sum_{i=1}^{D} w_i I(\alpha_i) \qquad (11)$$

where $I_m(\alpha_i)$ is the test statistics at i-th detection point that we obtain from m-th group of data, $I(\alpha_i)$ is the test statistics at i-th detection point and $I(\alpha)$ is the overall test statistics. D is the number of point in decision, w_i is the weight at i-th detection point.

7) Detection decision

If
$$I(\alpha) < \lambda \qquad \text{no signal}$$
$$I(a) > \lambda \qquad \text{signal exist}$$

where λ is the threshold determined by P_{FA}.

4. Simulation results

In this simulation, we choose BPSK signal with carrier frequency f_c=5 MHz, symbol rate R=2 Msps, and sample rate f_s =20 MHz. When $\alpha \neq 0$, the SCF of a BPSK will exhibit peaks at points $\alpha = \pm m/T_0, f = \pm f_c$ and $\alpha = \pm 2f_c \pm m/T_0$,$f = 0$ $(m = 0, \pm 1, \pm 2 \cdots)$, so we choose $\alpha = 2000000, f = f_c$, $\alpha = 8000000, f = 0$ and $\alpha = 10000000, f = 0$ for our simulation analysis. Fig. 5 is the detection performance of the BPSK signal at different detection points under $P_D = 0.99$, M=20, $P = 5$. It illustrates that the detection point $\alpha = 10000000, f = 0$ performs best, detection point $\alpha = 2000000, f = f_c$ takes the second place and the point $\alpha = 8000000, f = 0$ performs worst when only single point is used to detect. We use the best detection point $\alpha = 10000000, f = 0$ and the better detection point $\alpha = 2000000, f = f_c$ to detect at the same time to further improve the detection performance. Fig. 5 also shows that the performance of the combination detection method, the weight of the two detection points is chosen as 0.35 and 0.65 respectively. The performance is improved apparently. Of course, the combination detection method increases the computation complexity compare to single point detection, but from the section III we can find that the combination detection method only needs more correlation in step 4, so the complexity doesn't increase linearly as the detection point.

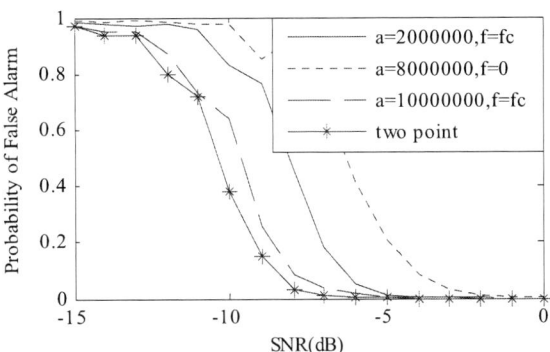

Figure 5. Performance under P_D =0.99

Fig. 6 is the detection performance of the BPSK signal on different detection points under $P_D = 0.95$, $M = 20$, $P = 5$. Comparing Fig. 5 with Fig. 6, we can find that when the required P_D decreases, the P_{FA} also decreases accordingly.

The performance using different number of data groups is studied as shown in Fig. 7, from which we can find that with the increasing of M the performance is also improved, but the computation complexity also increases linearly, so we need to make tradeoff between the complexity and the performance. Fig. 8 shows the receiver operating characteristic (ROC) of the performance under SNR=$-$12dB, it validates that the performance of this detection method is meaningful.

5. Conclusion

In this paper, spectrum sensing based on cyclostationarity in cognitive radio is considered. The second-order cyclic features built-in in modulated signals is used to detect the signals. Due to high complexity of cyclostationary feature detection, we choose to detect specific frequencies and cyclic frequencies based on the signal's feature to degrade complexity greatly. We compare the detection performance of different points to find the best detection points through simulation analysis and propose to combination detection method using multiple detection points to get better performance. Results validate the effectiveness of the proposed detection method.

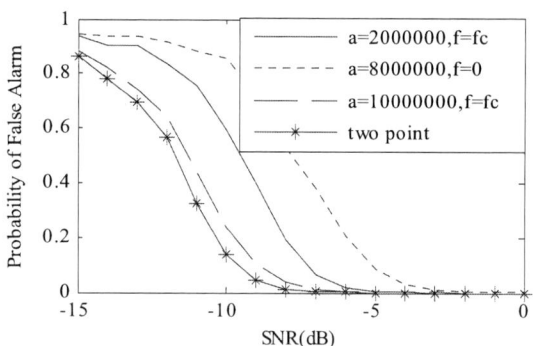

Figure 6. Performance under P_D =0.95

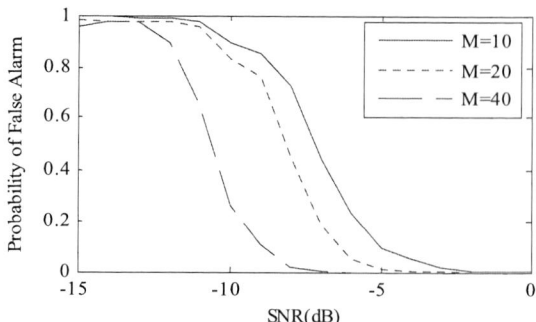

Figure 7. Performance with different groups

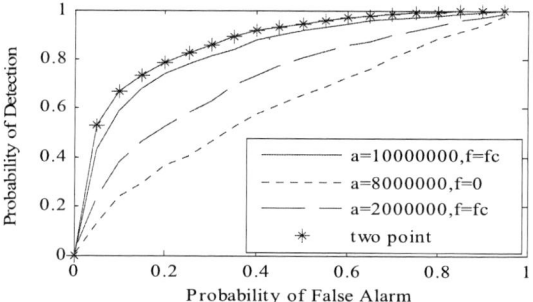

Figure 8. ROC of BPSK signal's detection

References

[1] Federal Communications Commission. Spectrum Policy Task Force Report. Washington: *ET Docket*, 2002:2-135.

[2] J Mitola, G.Q. Maguire, " Cognitive Radio:Making Software Rradios More Personal," *IEEE Personal Communications Magazine*, 6(4), pp.13-18, 1999.

[3] I .Akyildiz, Q Lee, M C.Vuran, "Next Generation/dynamic Spectrum Access/cognitive Radio Wireless Network: A Survey," *Computer Networks*, 50(13), pp.2127-2159, 2006.

[4] X Zhou, H Zhang, The Priciple and Application of Cognitive Radio. Beijing, *Beijing University of Posts and Telecommunications Press*, pp.16-25, 2007.

[5] Z Ye, J Grosspietsch,G Memik, "Spectrum Sensing Using Cyclostationary Spectrum Density for Cognitive Radios," *IEEE Workshop on Signal Processing Systems*, pp.1-6, 2007.

[6] K. Kim, I. Akbar, K.K. Bae, J.S. Um, "Cyclostationary Approaches to Signal Detection and Classificaton in CognitiveRadio," *IEEE International Symposium on DySPAN 2007*, pp.212-215, 2007.

[7] Y. Zhang, W.D. Yang, Y.M. Cai, "Cooperative Spectrum Sensing Technique," *WiCom2007*, pp.1167-1170, 2007.

[8] Z Huang, Y Zhou, W Jiang,The Process and Application of Cyclostationary Signals. *Beijing, Science Press*, pp.23-35, 2006.

[9] W Zhang, "Automatic Modulation Recognition of Digital Communication Signals," Doctor Thesis, *Graduate School of National University of Defense Technology*, 2006.

Traffic Signal Optimization with Vehicles Queue and the Number of Pedestrians Non-complying at Single Intersection

Zhi-Na ZHOU, Zhong-Ke SHI, Ying-Feng LI

College of Automation, Northwestern Polytechnical University
Email:Zhina_zhou@126.com

Abstract

In urban traffic signal optimization, the length of vehicle queue, waiting for green signal, is regarded as a performance index (PI). However, pedestrians are reluctant to waiting for green signal without person specially assigned to manage pedestrians in developing city, such as Xi'an. To the signal optimization problem of domestic urban traffic with pedestrians' noncompliance, a hybrid optimization algorithm is developed. The PI includes the length of vehicles queue and the number of pedestrians breaching signal. The number of pedestrians breach is estimated by a Monte Carlo method. The simulations are conducted in the condition of heavy traffic demands. The validation results indicate that the proposed method is able to improve the traffic condition for weak colony compared with the accustomed control method.

1. Introduction

Throughput is considered by accustomed optimization problem to traffic signal control [1, 2]. Much research has been conducted on vehicle delays at signalized intersections under a broad range of conditions, but the research on vehicle delay is much less considered with colony of pedestrian Signal noncompliance. However, in the large number of Chinese cities, traffic situations are significantly different. Researches show the proportion of walking in the total egression comes to 40% .At signalized intersection, the swarm of pedestrians is usually emerged. If there is no person specially assigned to manage pedestrians, pedestrians arriving during pedestrian non-green phases usually do not comply with traffic signals [4, 5].

Thus, the traditional method can't adapt the existing mixed traffic state in developing city, such as Xi'an. And pedestrian delay is regard as one of the key performance index to evaluate a signalized intersection's

Level-of-service [3]. Several models estimating pedestrian delays at signalized intersections have been developed [3-6] in recent years. But these models and methods are much less considered with colony of pedestrian Signal noncompliance. However, in the large number of Chinese cities, traffic situations are significantly different. At signalized intersection, the swarm of pedestrians is usually emerged. If there is no person specially assigned to manage pedestrians, pedestrians arriving during pedestrian non-green phases usually do not comply with traffic signals [5].

On the other hand, Genetic Algorithm (GA) has been widely used in optimization problem [1, 7]. Chen formulated a dynamic traffic flow model and developed an adaptive hybrid real-coded Genetic Algorithm (AHRGA) for optimizing the green times and cycle time [1, 2]. But the pedestrians are ignored in the traffic model.

In this paper, a method based on AHRGA and Monte Carlo simulation is developed to optimize the green times and cycle time. The presented method is fully considered that pedestrians swarm and follow the colony in breaching traffic signal.

2. Estimating the number of pedestrians breaching

Considering the traffic conditions in internal city, Li proposes a Monte Carlo model to estimate vehicle delay with swarming pedestrians Signal noncompliance at signalized intersections [8]. In this paper, the method is developed for estimating the number of pedestrians breaching traffic signal.

The model consists of a set of rules, and the algorithm of the model is shown in Fig. 1. The pedestrians are classified with arrivals during non-green signal and green signal. According to Fig. 1, the process is different to two kinds of arrivals.

978-0-7695-3342-1/08 $25.00 © 2008 IEEE
DOI 10.1109/PEITS.2008.47

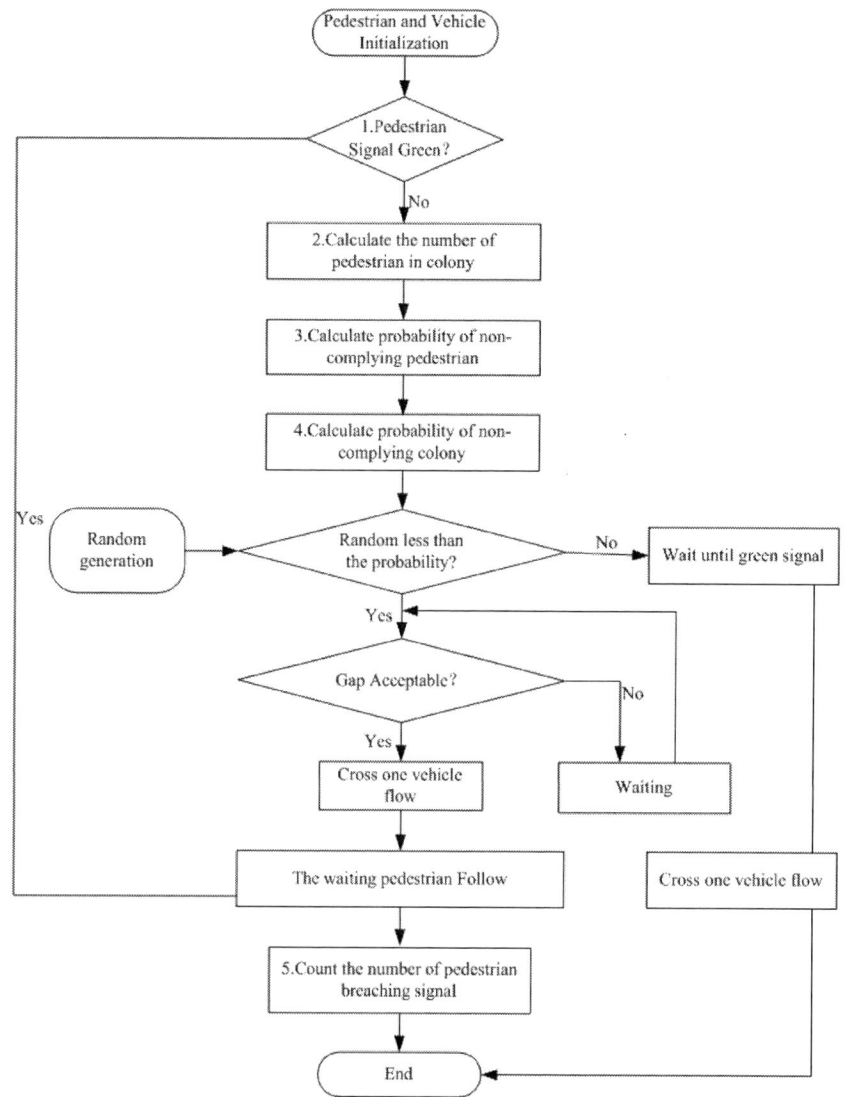

Fig. 1. Estimating the number of pedestrians breaching signal

2.1. Pedestrian waiting time for green signal

Assumed a cycle includes N phases, and pedestrian signal is green during the i-th phase ($i \in [1, N]$). In this paper, the time unit is second

Then,

$$t_{i,st} = \begin{cases} 0 & ,i = 1 \\ \sum_{j=1}^{i-1} t_j & ,i > 1 \end{cases} \quad (1)$$

Where t_j represents the signal time at j-th phase, j-th phase implement at $t_{i,st}$ during a cycle.

$$t_{ped,c} = \mod\left(t_{ped,a}^k, t_R + t_G\right) \quad (2)$$

Where k-th pedestrian arrive to the intersection

at $t_{ped,a}^k$, t_G is the green phase for pedestrians using the crosswalk, and t_R is the non-green phase.

If $t_{i,st} < t_{ped,c} < \left(t_{i,st} + t_i\right)$ is true, the pedestrians arrived at green signal. Otherwise, the pedestrians arrived at non-green signal.

When the k-th pedestrian arrived at non-green signal, the time for the k-th pedestrian waiting green signal is calculated, as follows:

$$t_{ped,w}^k = \begin{cases} t_{i,st} - t_{ped,c} & ,t_{ped,c} < t_{i,st} \\ t_R + t_G + t_{i,st} - t_{ped,c} & ,t_{ped,c} > \left(t_{i,st} + t_i\right) \end{cases} \quad (3)$$

2.2. The number of pedestrians in colony

According to the results from the field observations, the pedestrian is accustomed to follow the adjacent walker in breaching traffic signal. Then, it is necessary to identify whether colony for choosing uniform decision is gathered. A simplified algorithm is presented, as follows:

If $t_{ped,a}^k < t_{ped,a}^{k-1} + 2$ is true, the arrival choose to join in the colony, i.e. the arrival choose uniform decision with the colony.

If $t_{ped,a}^k < t_{ped,a}^{k-1} + 2$ is false, two kinds of cases are dealt with different methods. Firstly, the arrival makes alone a decision to breach traffic signal if the fore-colony had through a lane. Secondly, if the fore-colony waits for green signal, the arrival breach traffic signal according to $P_{ped,1}$ or chooses uniform decision with the colony. Thus, the fore-colony chooses to follow in cross if the later arrival breach traffic signal.

2.3. Probability of single pedestrian non-complying

From the large of field research, the estimation to the probability of single pedestrian non-complying is given, as follows:

$$p_{ped,w}^k = \tanh\left(at_{ped,w}^k\right) \tag{4}$$

Where $a = 0.01$ is constant, and indicates sensitivity to waiting time.

2.4. Probability of colony non-complying

According to Bayes's formula about conditional probability of independence event [9], the following formula can be established to estimate the probability of colony breaching during non-green signal to pedestrians:

$$p_{ped,n} = p_{ped,1} \sum_{i=0}^{n-1} \left(1 - p_{ped,1}\right)^i \tag{5}$$

where n is the number of pedestrians in colony, $p_{ped,1} \in [0,1]$ is the probability of single pedestrian choosing to breach traffic signal, $P_{ped,n}$ is the probability of colony breaching traffic signal and the number of pedestrians in the colony is n.

In general, the later arrivals might choose to follow the fore-walker in breach traffic signal at Signalized Intersections in Developing Cities. Then, the simulation algorithm and statistical module is conducted with the circumstance.

2.5. statistical modules

The statistical module obtains the number of pedestrians breaching signal. Thus, the variable is given, as follows:

$$N_{ped\_non\_signal} = f\left[q_v, q_{ped}, Signal\left(C, u_v, u_{ped}, ps\right)\right] \tag{6}$$

Where $N_{ped\_non\_signal}$ denotes the number of pedestrians breaching signal, q_v denotes the average arrival rate of vehicles, q_{ped} denotes the average arrival rate of pedestrians, $Signal\left(C, u_v, u_{ped}, ps\right)$ represents signal scheme and includes these parameters, as follows: cycle time, green ratio for vehicle, green ratio for pedestrian, phase sequence.

3. Signal optimization algorithm

3.1. Real-coded Genetic Algorithm (RGA)

RGA [1, 9] shows many advantages when to optimize the problem with continuous parameters. Steps of RGA used in most cases are summarized, as follows:

Step 1: According to (7), determine the genetic operators and the corresponding values

$$x_{i,j}^0 = c_j^d + r_i \cdot (c_j^u - c_j^d) \quad i \in [1, POPSIZE], j \in [1, n] \tag{7}$$

Where $r_i \in [0,1]$ is uniform distribution;

Step 2: Generation $k = 1$, randomly generate the initial population $p(0)$, and calculate the fitness of each individual;

Step 3: $k = k + 1$;

Step 4: According to the fitness, select new population $p'(k)$ from $p(k-1)$;

Step 5: Crossover the individuals in $p'(k)$, and generate the new population $p''(k)$, and calculate the fitness of each new individual;

Step 6: Mutate the individuals in $p''(k)$, and generate the next generation population $p(k)$;

Step 7: If the terminal condition is not satisfied, go to step 3, otherwise, the algorithm is finished.

3.2. Vary-Band stochastic search Algorithm

Vary-band stochastic search Algorithm [2] improves PI by searching the neighborhood of result from RGA, as follows:

Step 1: $X^0 = (x_1^0, x_2^0, \cdots, x_n^0)$ from RGA ,

$X^{best} = X^0$, initializes frequency of iterative operations and $t = 0$;

Step 2: According to (8) and (9), initializes search band

$$S^0 = ([c_1^{d,0}, c_1^{u,0}], [c_2^{d,0}, c_2^{u,0}], \cdots, [c_n^{d,0}, c_n^{u,0}])$$

Where $x_i^0 \in [c_i^d, c_i^u] \ \forall i = 1,2\cdots,n$;

$$c_i^{d,0} = \begin{cases} c_i^d & , x_i^0 \le c_i^d + 0.25(c_i^u - c_i^d) \\ c_i^d + 0.5(c_i^u - c_i^d) & , x_i^0 \ge c_i^d + 0.75(c_i^u - c_i^d) \\ c_i^d + 0.5(x_i^0 - c_i^d) & , else \end{cases} \quad (8)$$

$$c_i^{u,0} = \begin{cases} c_i^d + 0.5(c_i^u - c_i^d) & , x_i^0 \le c_i^d + 0.25(c_i^u - c_i^d) \\ c_i^u & , x_i^0 \ge c_i^d + 0.75(c_i^u - c_i^d) \\ x_i^0 + 0.5(c_i^u - x_i^0) & , else \end{cases} \quad (9)$$

Step 3: $X^t = X^{best}$;

Step 4: According to (10), X_j^{new} is given,

$$X_j^{new} = (x_{1,j}^{new}, x_{2,j}^{new}, \cdots, x_{n,j}^{new});$$

$$x_{i,j}^{new} = x_{i,j}^t + \delta_i \cdot (\sum_{m=1}^{R} r_m - R/2) \quad (10)$$

Where $i \in [1,n], j \in [1,N_r]$,

$$\delta_i = \frac{4 * \min\{c_i^{u,0} - x_{i,j}^t, x_{i,j}^t - c_i^{d,0}\}}{R},$$

And $r_m \in [0,1]$ is uniform distribution

Step 5: Calculate the fitness of each new individual;

Step 6: If $f(X_j^{new}) < f(X^{best})$, $X^{best} = X_j^{new}$ and go to step 7;

If $f(X_j^{new}) \ge f(X^{best})$ during continuous Ns_2 generation of iterative operations, go to step7; otherwise go to step 4;

Step 7: According to (11), reduce search band S^0;

$$\begin{cases} c_i^{d,0} = c_i^{d,0} + \xi \cdot (x_i^{best} - c_i^{d,0}) \\ c_i^{u,0} = c_i^{u,0} - \xi \cdot (c_i^{u,0} - x_i^{best}) \end{cases} \quad (11)$$

Where $i \in [1,n]$

Step 8: $t = t+1$, if $t > N_2$, X^{best} is never changed during continuous Ns_3 generation of iterative operations, the algorithm is finished, otherwise, go to step 3.

3.3. The objective function and constraint condition

The AHRGA requires parameters including the maximal number of generations, the number of individuals, the crossover probability, and the mutation probability for RGA. In this paper, the RGA used a maximum of 100 generations with a population size of 50; adaptive crossover probability is given, as follows:

$$P_c = \begin{cases} P_{c1} & , F < F_{avg} \\ (P_{c1} - P_{c2}) \cdot \exp(-\alpha_1 \cdot \frac{F - F_{avg}}{F_{max} - F}) + P_{c2} & , F \ge F_{avg} \end{cases}$$

$$P_m = \begin{cases} P_{m1} & , F < F_{avg} \\ (P_{m1} - P_{m2}) \cdot \exp(-\alpha_2 \cdot \frac{F - F_{avg}}{F_{max} - F}) + P_{m2} & , F \ge F_{avg} \end{cases}$$

where

$$P_{c1} = 0.99, \quad P_{c2} = 0.6, \quad \alpha_1 = 0.1,$$
$$P_{m1} = 0.1, \quad P_{m2} = 0.001, \quad \alpha_2 = 0.1,$$

F_{max} denotes the maximal value of individual's fitness;

F_{avg} denotes the average value of individual's fitness;

F denotes the value of individual's fitness.

The minimization of the PI, including length of vehicle queue and number of pedestrians breaching signal at the intersection for K periods, is adopted as the system objective. And two factors are given different weight to PI, as follows:

$$\min W_1 \sum_{k=1}^{K} \sum_{j=1}^{4} \sum_{l=1}^{3} \frac{\max\{0, Q_{jl}^t(k-1) + \sum_{i=1}^{l}(q_{jl}^i(k) - \mu_{jl}^i * \lambda_{jl}) * g^i(k)\}}{C(k)} \text{ s.t.}$$

$$+ W_2 \sum_{k=1}^{K} \sum_{j=1}^{4} \sum_{l=1}^{2} \frac{N_{ped\_non\_signal,jl}^k}{C(k)}\}$$

$$\begin{cases} \sum_{i=1}^{l} g^i(k) = C(k) \\ C_{min} \le C(k) \le C_{max} \\ g_{min}^i \le g^i(k) \le g_{max}^i \end{cases}$$

Where $i \in [1,I], k \in [1,K]$, λ denotes the saturation flow rate, green times g^i and cycle time $C(k)$ is control variables, C_{min} and C_{max} are the minimal and maximal cycle times respectively, g_{min} and g_{max} are the minimal and maximal green times respectively. μ_{jl}^i is a state variable indicating the state of the green signal, and can be expressed as follows:

$$\mu = \begin{cases} 1 & green\ signal \\ 0 & non\text{-}green\ signal \end{cases}$$

4. Simulation validating

Some parameters used in the simulation is assumed, as follows:

1) The average vehicle and pedestrian arrival rate, q_v and q_{ped}, complies with Poisson distribution;

2) The saturation flow volume for all lanes is 0.6 veh/sec;

3) $C_{min} = 60$, $C_{max} = 180$, the first and third phases: $g_{min} = 6$, $g_{max} = 90$, the second and fourth phases: $g_{min} = 6$, $g_{max} = 40$;

4) The study period is 3.

5) Phase state matrix is given, as follows:

μ [4][4][4]={

{{0,1,1,0},{0,0,1,1},{0,1,1,0},{0,0,1,1}},

{{1,0,1,0},{0,0,1,0},{1,0,1,0},{0,0,1,0}},

{{0,0,1,1},{0,1,1,0},{0,0,1,1},{0,1,1,0}},

{{0,0,1,0},{1,0,1,0},{0,0,1,0},{1,0,1,0}}};

6) The vehicles flow volume is given, as follows:

Heavy traffic demands:

{440,670,210}, {425,630,190},

{438,665,200}, {420,620,186};

7) The pedestrians flow volume is given, as follows:

Heavy traffic demands:

{3033, 2896}, {3128,2942},

{2746,2628}, {3213,3258};

8) Initial vehicles queue is given, as follows:

Heavy traffic demands:

{10,12,8},{10,12,8},{10,12,8},{10,12,8};

9) Length of vehicle queue and number of pedestrians breaching signal are given different weight to PI, as follows:

$W_1 = 1$, $W_2 = 0.5$

Table 1. The simulation result to signal scheme

	Cycle No.	g1	g2	g3	g4	C
considering pedestrians	1	44.6	8.4	30.7	23.7	107.4
	2	49.6	19.8	13.7	36.0	119.2
	3	44.6	31.0	7.7	25.1	108.4
ignoring pedestrians	1	27.9	27.9	50.8	39.0	145.5
	2	53.1	36.8	49.5	39.4	178.7
	3	57.4	26.2	52.4	34.4	170.4

Table 2. Comparison with the simulation results

considering pedestrians		ignoring pedestrians	
PI	Number of pedestrians breaching signal	PI	Number of pedestrians breaching signal
396.2	295.6	658.0	570.4
475.4	357.3	549.8	462.6
208.0	75.9	662.2	574.8

The result of the simulation considering pedestrians and ignoring pedestrians using AHRGA optimization method is shown respectively in Table 1 and Table 2. From Table 2, the conclusion can be drawn that the number of pedestrians breaching signal can be reduced by using the presented method in the condition of heavy traffic demands.

5. Conclusions

1. A Monte Carlo method is developed to estimate number of pedestrians breaching signal at single intersections in developing cities, such as Xi'an. The simulation results shows that the simulation model and the optimization method produce better effects than traditional method with heavy traffic volumes.

2. AHRGA is developed to optimize the green times and cycle time in single signalized intersections. And the PI includes the retained vehicles for waiting green signal and number of pedestrians signal noncompliance.

3. With heavy traffic volume, the method validity is certified by simulation.

Acknowledgements

This research was supported by National Science Foundation of P. R. China, No. 60134010 and cultivation foundation for magnitude item of education ministry. Their support is appreciated.

References

[1] X. F. Chen, Z. K. Shi, Real-coded Genetic Algorithm for Signal Timings Optimization of a Single Intersection, *Intersection. Proceedings of the First International Conference on Machine Learning and Cybernetics*, Beijing, China, 2002, Vol.4-5, 1245-1248.

[2] X. F. Chen, Research on Dynamic Optimization and Control Techniques for Urban Traffic Signal, *A Dissertation for the Degree of Philosophy Doctor*.

[3] R. Braun, M. Roddin, NCHRP report 189: quantifying the benefits of separating pedestrians and vehicles, *National Research Council*, Washington, DC, 1978.

[4] Q. Li, Z. Wang, J. Yang, J. Wang, Pedestrian delay estimation at signalized intersections in developing cities, *Transp. Res.A* 39 (2005) 61–73.

[5] J. Yang, Q. Li, Z. Wang, J. Wang, Estimating pedestrian delays at signalized intersections in developing cities by Monte Carlo method, *Mathematics and Computers in Simulation* 68 (2005) 329–337.

[6] M. Virkler, Pedestrian compliance effects on signal delay, *Transp. Res.* Rec. 1636 (1994) 88–91.

[7] Goldberg D. E., Genetic Algorithm in Search, Optimization, and Machine Learning. *Addison-Wisley Publishing Co., Reading*, Massachusetts, 1989.

[8] Li YingFeng and Shi ZhongKe, Vehicle delay model with colony of pedestrian Signal noncompliance at signalized intersections in developing cities, *The First International Conference of Transportation Engineering (ICTE)*, Chengdu, China, 2007, Vol 2, 1064-1069

[9] Eshelman L., Schaffer J. Real-coded genetic algorithms and interval-schema. In: *Whitley Ded. Foundations of Genetic Algorithms 2*. San Mateo: Morgan Kaufmann Publishers, 1992, 187~202.

2008 Workshop on Power Electronics and Intelligent Transportation System

Communication Reduction for Continuous Extreme Values Monitoring over Distributed Data Streams

Xiaoming Peng[1,2], Yanxiang He[1], Li Tian[2]

[1]*Computer School of Wuhan University, Wuhan, Hubei, 430072 P.R. China*
[2]*Department of Information & Command Automation, AFRA, Wuhan, Hubei, 430019, China*
Pengxm@speednet.net.cn

Abstract

We focus on aspects of physical distribution of streams, and address the problem of communication reduction for continuous extreme values monitoring over distributed data streams. We firstly develop an effective pruning technique to minimize the number of elements to be kept for extreme values queries. Then we consider the distributed environment, where remote nodes delay the data transmission as late as possible, and adopt the pruning strategy to filter local stream tuples, which is quite efficient in communication reduction. The method is extended to adaptively run in a degraded manner for resource limitation. Analytical analysis and experimental evidences show the efficiency of proposed approach on communication reduction.

1. Introduction

Recently, the problem of handling different types of continuous queries over data streams[1,2] becomes one of the hottest topics in database research field due to the development of sensor network and real-time monitoring technology. For most applications, the recent data are import than older ones, therefore, the sliding window model (i.e. considering "recent" portions of streams only)[1,2] has received considerable attention.

Data streams are distributed essentially [3], and sometimes we must consider the problem of communication reduction. A good case in point is sensor networks. The nodes in a wireless sensor network generate vast amounts of data that must be communicated to the network root (also known as the base station) using radio transmission. Nodes are battery powered, and radio usage dominates their energy consumption. In terms of power consumption, transmitting a single bit of data is equivalent to executing 800 instructions [4]. Therefore, compared with continuously streaming all data to the root and processing it there, we can greatly extend the lifetime of the network by developing query-specific plans that limit the amount of data transmitted by the nodes.

We address the problem of communication reduction for continuous extreme values monitoring (MAX or MIN) over distributed sliding window streams. The presentation of this paper bases on the MAX problem since the

discussion for MIN is identical. The reason why we focus on the MAX query is threefold. First, this query is a typical example of a more general exemplary aggregate. An exemplary aggregate [4] is one where the solution consists of one or more representative values from the network, as opposed to a summary, where the solution is computed over all the values. Second, in developing a solution to MAX, we hope to gain insight into optimizing this general class of monitoring queries. Finally, in practice, continuous MAX is useful and important for detecting abnormal or extreme behavior[13]. This type of behavior is not captured by summary aggregates such as mean or sum, establishing a fundamental difference between the applications of these two query types.

Without loss of generality, we consider the time-based sliding window. The problem considered in this paper is shown in Fig.1, which can be formulized as follows: a distributed data stream processing architecture comprises a collection of k remote sites N_i (i {1,..,k}) and a designated coordinator site C. Generally, direct communication between N_i are not allowed, instead, N_i exchanges messages only with C. Each N_i monitors a local stream S_i (i=1..k) and transmits tuples <t,v> to C continuously, indicating that at time t, the concerned object attribute value is v. Queries Q_j=CQ_MAX(W_j,S) are registered in C to continuously monitor the MAX values against sliding window W_j over the global stream

$$S = \bigcup_{i=1}^{k} S_i.$$

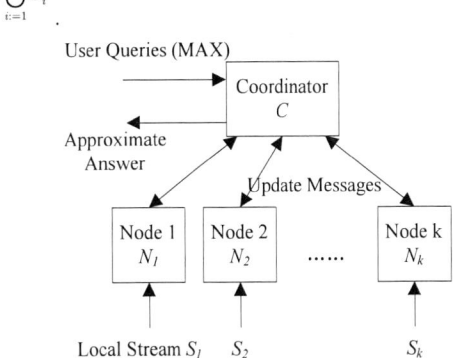

Figure 1. Problem formulation

It is obvious that enormous bandwidth consumption will be occurred if all update tuples are transmitted to C. On the other hand, the periodically sample method must consider the tradeoff between accuracy and communication. Moreover, a long period may lead to

978-0-7695-3342-1/08 $25.00 © 2008 IEEE
DOI 10.1109/PEITS.2008.77

delay or loss of event detecting. In this paper, we present an approach to reduce the communication for continuous MAX query with accuracy and real-time guarantee.

The rest of this paper is organized as follows. Section 2 reviews the related works. In section 3, after analyzing characters of sliding window and MAX query, a novel pruning technique is developed to minimize the number of elements to be kept for processing MAX query. In section 4, this pruning technique is employed in distributed environment for communication reduction. Section 5 experimentally verifies the efficiency of the proposed techniques. Lastly, section 6 concludes the paper and directions for future works.

2. Related works

Approaches used for communication reduction in distributed applications can be classified as general methods and special methods. General methods are universal and can be adopted for many queries, while special ones are designed for a given query, which can explore characters proper to this query and obtain better efficiency.

2.1. General Methods

Filter[5-8] is one of the effective general methods: Filters are installed in remote nodes, which define several local constraints. Only those data violated these constraints are needed to push to the coordinate node, and at the same time, filters are adjusted to fit the data changes. Based on the above main idea, Olston et al.[5] present a range caching method. In their case, the filter is represented as a range. The length of the range naturally provides a mechanism for controlling the trade-off between accuracy of the response to a query on the value, and the amount of communication between the root and the remote node. Reference [6] makes a further research on range caching method, where two operation called bound shrinking and bound growing are defined, which provide a mechanism to adaptively adjust the length of the range according to the data change. A.Jain[7] adopts Kalman filter[8] as a pruning and prediction approach for communication reduction.

Prediction models are also used as a general method for reducing communication [9, 10]. Every attribute measured by remote nodes is attached to a prediction model that will be used to predict future values of that attribute, and the same prediction model is shared with the coordinator. The coordinator employs these models to answer continuous queries, while the remote nodes check whether the prediction is close (with some precision) to the actual value or not. In this way, only significant values (when the prediction has a large deviation from the actual value) are needed to transmit to the coordinator, saving a great amount of communication while still guaranteeing

sufficient precision of query results. Three models named as static, linear-growth and velocity/acceleration are developed by Cormode[9], and reference [10] improved these three models from parameters' definition and the update policy.

2.2. Special Methods

In recent years, several methods are proposed specially for MAX queries [11-13]. Madden et al. [10] suggest strategies for running ad hoc queries. Even while operating in a "one-shot" setting they recognize the advantages of pushing threshold-based filters into the network. For example, while running a pipelined MAX query, they suggest aborting the query after some amount of time, determining the maximum value so far, and running a new query requiring nodes to only respond if they have an even larger value. The algorithm in [12] improves the range caching method[5] for MAX queries. When an ad hoc MAX query is received, the root sorts nodes by range upper bounds. The root queries each node in order one-by-one, while maintaining a running maximum value. If enough nodes have been searched such that all unsearched nodes have upper bounds below the running maximum, the running maximum is the solution. Silberstein et. al [13] identify and employ the use of constraint localization and dynamic routing as fundamental methods for reducing message traffic in sensor networks.

The approach proposed in this paper belongs to special methods, which can be used for extreme query only. We focus on the communication efficiency in exact continuous MAX queries, which is quite different from the existing researches for sensor networks and approximate ad hoc queries.

3. Data pruning strategy

A closer examination of sliding window model and MAX query allows us to abstract some new properties. Fig.2 shows an example of data points contained in a sliding window, where each 2D point contains two attributes: the arriving time (x axis) and the attribute value (y axis). Point in sliding window follows the first in first out rule, it is obvious that point c expires later than point a and b since it arrives later than them. On the other hand, the value of c is larger than that of *a* and *b*, it is immediate that in any sliding window containing c, neither a nor b will become the MAX point, so a and b are redundant and can be discarded safely. Inspired by the above observation, a novel pruning technique is proposed in this section.

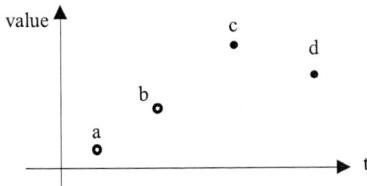

Figure 2. Character of extreme value on sliding window

Theorem 1. For point p_1 and p_2, if $p_1.t < p_2.t$ and $p_1.value \leq p_2.value$, then p_1 can be discarded with correctness guarantee for continuous MAX query processing in sliding windows.

Definition 1. [Key Point] For data set P contained in a sliding window and a point $p \in P$, p is called a *key point* if $\forall p' \in P, p'.t > p.t \Rightarrow p'.value < p.value$, . The subset of P containing all *key points* is called the *key points set*, and is denoted as KP.

Definition 2. [Neighbor Greater Than] For a key point $p \in P$, another key point $p' \in P$ is *neighbor greater than* p if $p'.value > p.value$ and $p'.t = max\{e.t|\ e \in KP$, $e.value > p.value \}$. In other words, p' is the latest one in all key points larger than p. We denote this relationship as $p' = NeighborGT(p)$. Conversely, p is *neighbor less than* p' and denoted as $p = NeighborLT(p')$.

We can see that KP is the remainder of P after adequate pruning according to theorem 1. Fig.3 shows an example of above definitions, where key points are presented by solid points and linked by a real line. For key point c, existence of any point with larger value (a or b) deprives the chance for c to become the MAX point. For all these points larger than c, the latest one (point b, which is *neighbor greater than* c) determines the time when c may become the MAX point: when point b is expired and removed from the sliding window, c has the chance to be the MAX one only if the newly arrived points after c are all less than it.

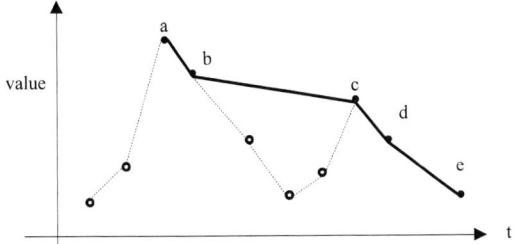

Figure 3. Illustration of Key Points on sliding window

The following conclusion can be deducted from theorem 1.

Corollary 1. KP contains the minimal number of points to be kept for exact continuous MAX query. Suppose that the distribution of value is independent, and restricted that data values are always distinct, then the average size of KP is $M=O(logN)$, where N is the cardinality of P.

Proof. Theorem 1 guarantees that all discarded points have no chance to become the MAX point at any time. For each *key point · p*, if it is not discarded until *NeighborGT(p)* expires (i.e. the newly arrived points after p are all less than it), then p will be the MAX point at that time. In other words, each key point is likely to be the MAX point in the future. So KP contains the minimal number of points to be kept for exact continuous MAX query.

According to definition 1, KP can be regarded as the maximum vectors problems in 2-dimensional space. From theorem 2 in [14], $M=O(logN)$ immediately follows.

4. Extreme values monitoring on distributed streams

In this section, a framework of Continuous Extreme Values Monitoring over Distributed Data Streams (*CEVMDS*) is proposed.

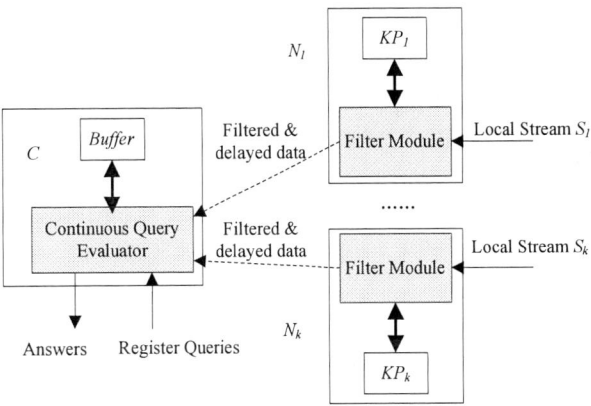

Figure 4. Framework of extreme values monitoring over distributed data streams

As shown in Fig.4, suppose that a continuous MAX query with sliding window width W is registered in C. Remote node N_i keeps the key points set KP_i against data of local stream S_i contained in W. Instead of sending the newly arriving point immediately to C, filter module of N_i uses it to prune the old KP_i dynamically according to theorem 1. Key points are cached in remote nodes as late as possible for effective pruning, and only filtered & delayed data are transmitted to C, saving a great number of communication. Main challenge of this *delay-transmission* strategy consists in how to guarantee the correctness and real-time of global result while local stream data are pruned and delayed.

4.1. Algorithms

One of the most important components in *CEVMDS* framework is filter module in remote node. The processing is described with pseudo code in Fig.5, where *Send(p)* represents the action of transmitting p to C, while parameter T_{exp} is defined for N_i to denote the time when next transmission will take place. Algorithm FM_Arrive describes the process when a new data arrives, while FM_Event is triggered when the current time t=T_{exp}.

Algorithm: *FM_Arrive*

Input: An incoming data p

Output: Change of KP_i ,and data transmission, if needed

1	p=the newly arriving data
2	$DP=\{p' \mid p' \in KP_i \quad and \quad p'.value \leq p.value \}$
3	$KP_i=KP_i-DP+\{p\}$
4	p'=the oldest key point in KP_i
5	$T_{exp}=p'.t+W$
6	If $KP_i= \{p\}$ then
7	Send(p)
8	End if

Algorithm: *FM_Event*

1	p= the oldest key point in KP_i
2	p'= NeighborLT(p)
3	$KP_i=KP_i-\{p\}$
4	Send(p')
5	$T_{exp}=p'.t+W$

Figure 5. Pseudo code description of Filter Algorithm

When a new data p arrives, N_i firstly modifies KP_i by adding p and deleting all points less than p (line 1-3 of FM_Arrive). Then parameter T_{exp} is updated as the expiration time of the oldest key point in KP_i (line 4,5). If the new KP_i contains only one data, indicating that p is larger than all the old key points, therefore p is sent to C in this occasion (line 6-8).

When $t=T_{exp}$, the oldest key point p is expired (i.e. outside the sliding window W), so it should be discarded from the system. *NeighborLT(p)* is transmitted, and its expiration time is used as the new value of parameter T_{exp} (line 4, 5).

Now we prove the correctness of above filter algorithm. According to the presentation in Fig.5, we can get:

Lemma 1. Each data point transmitted from N_i to C is the oldest valid *key point* of local stream S_i.

Theorem 2. Data points transmitted from remote nodes are adequate for coordinator node C to continuously monitor the exact MAX result.

Proof. We demonstrate the above conclusion from following two aspects: (1) the global MAX point must not be pruned in remote nodes; (2) the *delay-transmission* strategy adopted will not affect the real-time output of global result.

(1) Suppose that at time t, the global MAX point is p, that is for $\forall p' \in \left\{ p' \mid p' \in \bigcup_{i=1}^{k} S_i \quad and \quad t-W < p'.t \leq t \right\}$, $p'.value<p.value$ is hold. Without loss of generality,

suppose that $p \in S_i$, it is obvious that for $\forall p' \in S_i$ satisfying $t-W < p'.t \leq t$, $p'.value<p.value$ is also satisfied. In other words, no point arrive later than p in local stream S_i has a larger value than it. According to theorem 1 and definition 1, p is the oldest valid *key point* of S_i at time t. Lemma 1 shows that p must not be pruned.

(2) From algorithm description in Fig.5, we can know that for each cached point p, $\exists p' \in KP_i$ satisfies $p'.t<p.t$ and $p'.value>p.value$. Therefore p can not be the MAX point when p' is valid. The filter algorithm guarantees that p will be transmitted to C when *NeighborGT(p)* expires, in other words, when p is delayed and cached in N_i, *NeighborGT(p)* is always valid and has already been transmitted to C. So the coordinator C can generate the exact and real-time output although data are delayed and pruned.

Theorem 2 is proved by summing up the above analysis. And finally, the presented method trivially extends to count-based sliding windows.

4.2. Complexity Analysis

In both versions of the sliding window (i.e., count-based and time-based), the tuples are evicted in a *first-in-first-out* manner. Therefore, all the valid data points can be stored in a single list so that insertion and expiration access can be finished in constant time. According to theorem 1, points of KP_i list are sorted in a degressive order according to data value.

For a new arriving point, line 2 of FM_Arrive can be finished in $O(M)$ time by checking each point in KP_i, where M is the cardinality of KP_i. Line 4-8 can be processed in $O(1)$ time, so the complexity of operation for a new arriving point is $O(M)$. Parameter T_{exp} is defined, and FM_Event is triggered only when $t=T_{exp}$ follows, avoiding redundant processing for each point's expiration. From Fig.5 we can easily deduct that the performance of FM_Event is $O(1)$.

The storage cost of *CEVMDS* depends on the size of key points set. The KP_i list is sorted in a degressive order against data values, so DP in FM_Arrive algorithm (line 2) can be represented by a temp pointer; while parameters such as T_{exp}, p, p' can be stored in constant memory. Therefore the storage cost of *CEVMDS* is $O(M)$.

4.3. Extended Pruning Strategy

Considering the situation where computation and storage resource are limited in remote nodes so that the filter algorithm shown in Fig.5 can not be applied properly. The pruning strategy is extended in this section, which provides a schema to automatically run in a degraded manner for resource restriction. The extension aims at FM_Arrive because the proper execution of FM_Event is the footstone of correctness guarantee.

When CPU is busy (for example, the busy ratio excesses a predefined threshold), an extended pruning strategy which consumes less computation resource can be adopted (although the pruning efficiency may be degraded); when the CPU resource is adequate, the filter module then resume the primary pruning strategy. The following is a simple extended strategy for CPU restriction.

Strategy 1. When a new data point p arrives, the KP_i is not updated and p is directly sent to C.

If storage units are inadequate to store all the key points, strategy 1 can also be used to fit this restriction. However, we propose another strategy for this situation, which provides a better pruning efficiency than strategy 1.

Strategy 2. Suppose the oldest key point of KP_i is p, *NeighborLT(p)* can be sent to C before $t=T_{exp}$ satisfies if the storage resource is limited, then update T_{exp} and delete p from the system.

It is obvious that the coordinator C can receive more points when the above two strategies are adopted, so the correctness of global result is guaranteed. Research on developing other strategies which can not only suit for the resource restriction but also provide an effective pruning is remained as future steps.

5. Experimental evaluation

In this section we present the results of our experiments conducted on a Pentium 4 PC with a 2.8GHz processor, 1GB main memory, and Windows XP OS. All experiments were programmed in Borland Delphi 7.

5.1. Experimental Setting

All the experiments are evaluated based on the following three synthetic data set: (1)corr: data values tend to increase with time passing by. (2) indep: data values are independent. and (3) anti: data values tend to decrease against their generation time. All synthetic data range from 0 to 1, and these three kinds of data are illustrated in Fig.6. We generate 1,000,000 points for each type of dataset and only W of them are kept in the system. The new arriving data point replaces the oldest one of the existing W points, simulating the count-based sliding window environment..

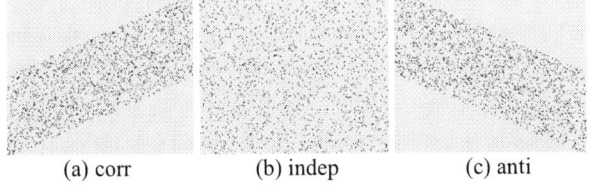

(a) corr   (b) indep   (c) anti

Figure 6. Synthetic data sets used during the experiments

5.2. Data Pruning Evaluation

Fig.7 indicates the relationship between cardinality of data sets and the number of their corresponding key points, where (a)-(c) illustrate the average size (AVGSize) and the max size (MAXSize) of KP with growth N (cardinality of data set) in these three types of synthetic data sets, while (d) shows the average percentage of key points. It can seen that the size of key points set is affected by data distribution. Generally speaking, with the same size of data set, key points of *inc* is the smallest, while the *dec* synthetic data set has most sky points. M shows an increase trend with the growth of N, however the increasing speed slows down. M is far smaller than N in all these three types of data sets. When $N>10,000$, the percentage of key points is no more than 1% for *dec* data set, while this proportion is less than 0.1% for *inc* and *indep* data sets. So the pruning technique proposed in this paper is quite effective for data reduction, therefore effective for communication reduction.

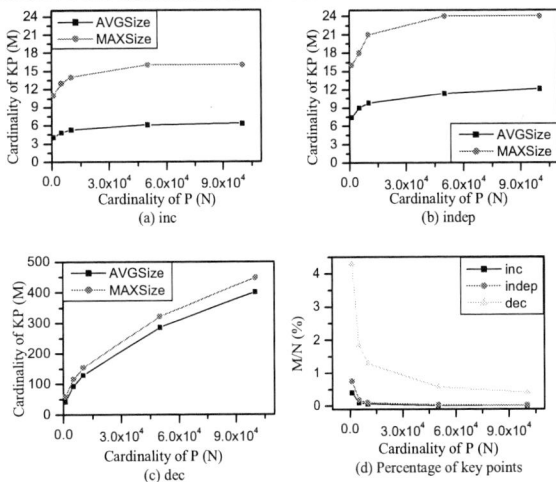

Figure 7. Efficiency of data pruning

5.3. Communication Reduction

The following two factors play important roles in reducing communication cost: on one hand, benefit from the pruning strategy adopted in remote nodes, a lot of redundant data points are discard, leading to the reduction of communication; on the other hand, the *delay-transmission* strategy requires extra information transmission to ensure the arriving order of data points. In specific, during the original scenario where all data points are pushed to coordinator C, the time when points are transmitted to C can be regarded as the points' arriving time, while under the *delay-transmission* strategy, the occasion of transmission is usually not equal to the time when points arrive the system, so the time information must be stored and transmitted for correctness guarantee. We can compute the total communication reduction efficiency by the following formula:

$$E = \frac{2M}{N} \times 100\% \qquad (1)$$

Where N is the number of original data points, and M is the size of key points set.

We investigated the communication reduction efficiency of the proposed approach in different data sets, and the result is shown in Fig.8.

We can see that the percentage of actual communication descend with the growth of sliding window's width. This is because the wider sliding window is, the longer data points are cached, increasing the possibility of points to be discarded.

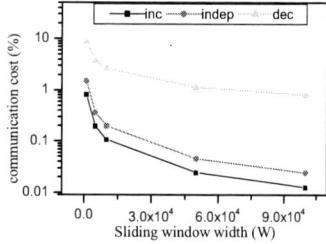

Figure 8. Efficiency of communication reduction

5.4. Degraded Processing

Suppose the degraded processing caused by CPU busy occupies 1/3 of the total processing time, we investigated the communication reduction efficiency of extended strategy 1, and the result is shown in Tab.1. Because strategy 1 directly send new arriving points to C without any pruning processing, the communication reduction efficiency lies on the percentage of CPU busy time. The actual communications are all tend to 33.3% in three types of data sets.

Table 1. Communication reduction with CPU limitation

W	inc(%)	indep(%)	dec(%)
1000	33.872	34.325	39.060
5000	33.462	33.573	35.810
10000	33.404	33.464	35.061
50000	33.350	33.364	34.095
100000	33.342	33.349	33.869

Take the *indep* data set as an example, and set W=100,000, we investigated the communication reduction efficiency when storage is limited. The result is shown in Fig.9, where x axis is the max number of key points can be stored in remote nodes, y axis shows the actual communication cost. It can be seen that the communication reduction efficiency benefits from the growth of max storage capability. However, when the storage capability excesses a certain threshold, the effect of capability growth on communication reduction is inconspicuous. This is because when the storage capability excesses the average size of key points set, the storage is not fully utilized. The optimization of storage is

relevant to the data distribution and the width of sliding window.

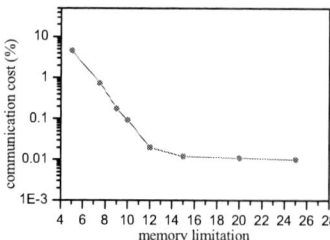

Figure 9. Communication reduction with storage limitation

6. Conclusions

In this paper, we address the communication reduction of continuous extreme query processing over distributed data streams, and propose a framework named *CEVMDS*, which adopts a *delay-transmission* strategy to pruning redundant data in remote nodes while guaranteeing the correctness of global result. The approach is extended to adaptively processing CPU/storage restriction. The proposed methods are experimentally evaluated to be effective and efficient.

In future steps, we plan to conduct a more extensive analysis on the cost-efficient processing of other queries in distributed environment.

Acknowledgement

This research has been supported by the national high-tech R&D program (863 programs) of China under Grant No.2007AA01Z138.

References

[1] B. Babcock, S. Babu, M. Datar, R. Motwani, and J. Widom, "Models and issues in data stream systems," in ACM SIGMOD-SIGACT-SIGART symposium on Principles of database systems (PODS). Madison, Wisconsin, 2002, pp. 1-16.

[2] Jin Cheqing, Qian Weining, Zhou Aoying. Analysis and Management of Streaming Data: A Survey [J]. Journal of Software (in Chinese). 2004,15(8): 1172~1181.

[3] G. Cormode and M. Garofalakis, "Streaming in a Connected World: Querying and Tracking Distributed Data Streams," in VLDB Tutorials. Seoul, Korea, 2006, pp. 1266.

[4] S. Madden, M. J. Franklin, J. Hellerstein, and W. Hong, "TAG: a Tiny AGgregation service for ad-hoc sensor networks," in Symposium on Operating Systems Design and Implementation (OSDI). Boston, 2002, pp. 131-146.

[5] C. Olston, B. T. Loo, and J. Widom, "Adaptive Precision Setting for Cached Approximate Values," in ACM SIGMOD international conference on Management of data. Santa Barbara, California, 2001, pp. 355-366.

[6] C. Olston, J. Jiang, and J. Widom, "Adaptive Filters for Continuous Queries over Distributed Data Streams," in

ACM SIGMOD international conference on Management of data. San Diego, California, 2003, pp. 563-574.

[7] A. Jain, E. Y. Chang, and Y. F. Wang, "Adaptive Stream Resource Management Using Kalman Filters," in ACM SIGMOD international conference on Management of data. Paris, France, 2004, pp. 11-22.

[8] R. E. KALMAN, "A new approach to linear filtering and prediction problems," Transactions of the ASME-Journal of Basic Engineering, vol. 82, pp. 35-45, 1960.

[9] G. Cormode and M. Garofalakis, "Sketching Streams Through the Net: Distributed Approximate Query Tracking," in International Conference on Very Large Data Bases (VLDB). Trondheim, Norway, 2005, pp. 13-24.

[10] L. Tian, A. Li, and P. Zou, "Research on Prediction Models over Distributed Data Streams," in WISE Workshop on Web-Based Massive Data Processing (WMDP 2006). Wuhan, China: LNCS 4526, 2006, pp. 25-36.

[11] [11] S. Madden, R. Szewczyk, M. J. Franklin, and D. Culler, "Supporting Aggregate Queries over AD-Hoc Wireless Sensor Networks," in IEEE Workshop on Mobile Computing Systems and Applications. Callicoon, New York, USA, 2002, pp. 49-58.

[12] Z. Liu, K. C. Sia, and J. Cho, "Cost-Efficient Processing of Min/Max Queries over Distributed Sensors with Uncertainty," in ACM symposium on Applied computing. Santa Fe, New Mexico, 2005.

[13] A. Silberstein, K. Munagala, and J. Yang, "Energy-Efficient Monitoring of Extreme Values in Sensor Networks," in ACM SIGMOD International Conference on Management of Data. Chicago, Illinois, 2006, pp. 169-180.

[14] J. L. Bentley, H. T. Kung, M. Schkolnick, and C. D. Thompson, "On the average number of maxima in a set of vectors and applications," Journal of the Assooanon for Computing Machinery, vol. 25, pp. 536-543, 1978.

A New Approach to MANet Routing based on Erasure Codes

Zheng Chen, Xiaojing Wang, Sheng Cao, Dan Tang

Chengdu Institute of Computer Applications, Chinese Academy of Sciences, Chengdu 610041

chenzheng.prc@gmail.com

Abstract

In this paper, we present a novel approach for mobile ad-hoc routing, which is called AOMDV-CB. The brand-new protocol is an extension of the well-known AOMDV (Ad-hoc On Demand Multi-path Distance Vector) protocol, holds a potential to drastically increase scalability in both the size and the diameter of the ad-hoc network with the advantage of erasure codes. Performance comparison of AOMDV-CB with AOMDV and AODV using ns-2 simulations is given in our paper, results shows that AOMDV-CB is able to achieve a remarkable improvement in the robust and efficiency of ad-hoc network.

1. Introduction

A mobile ad-hoc network (MANet) is a kind of wireless self-configuring network, which contains mobile routers connected by wireless links. The routers are free to move randomly and organize themselves arbitrarily; thus, the topology of the network may change rapidly and unpredictably.

MANet became a popular subject for research as laptops and 802.11/Wi-Fi wireless networking became widespread in the mid to late 1990s. Many of the academic papers evaluate protocols and abilities assuming varying degrees of mobility within a bounded space. Different protocols are then evaluated based on the packet deliver rate, the end-to-end delay, and other measures.

Routing for ad-hoc networks can be classified into two main types: table-driven routing protocols and on-demand routing protocols. Table-driven routing protocols attempt to maintain consistent information about the path from each node to every other node in the network. The Destination-Sequenced Distance-Vector Routing [1] (DSDV) protocol is a table driven algorithm that modifies the Bellman-Ford routing algorithm [2] to include timestamps that prevent loop-formation. Other well-known table-driven routing protocols include WRP [3] (Wireless Routing Protocol) CGSR [4] (Clusterhead Gateway Switch Routing) and TBRPF [5] (Topology Dissemination Based on Reverse-Path Forwarding).

On-demand routing protocols were designed with the aim of reducing control overhead, thus increasing bandwidth and conserving power at the mobile stations. These protocols limit the amount of bandwidth consumed by maintaining routes to only those destinations for which a source has data traffic. Therefore, the routing is source-initiated as opposed to table-driven routing protocols that are destination initiated. There are several recent examples of this approach such as DSR [6] (Dynamic Source Routing), AODV [7] (Ad Hoc On Demand Distance Vector Routing), LMR [8] (Lightweight Mobile Routing), TORA [9] (Temporally Ordered Routing Algorithm), ABR [10] (Associatively-Based Routing) and SSR [11] (Signal Stability Routing).

Though several performance studies of ad-hoc networks have shown that on-demand protocols incur lower routing overheads, however, they are not without performance problems. High route discovery latency together with frequent route discovery attempts in dynamic networks can affect the performance adversely. Multi-path protocols try to alleviate these problems by computing multiple paths in a single route discovery attempt. New route discovery is needed only when all paths fail. Multi-path protocols can also used to balance load by forwarding data packets on multiple paths. Some famous protocols have been extended into multi-path protocols, such as SMR [12] (Split Multi-path Routing), MP-DSR [13] (Multi-Path Dynamic Source Routing) and AOMDV [14] (Ad Hoc On Demand Distance Multi-path Vector Routing).

Multi-path protocols have magnificently improved the performance of ad-hoc networks, but with the increase of size and the diameter of the ad-hoc network, they showed their own shortcomings. The life time of a single path is decrease quickly, and the routing protocol has to change and rebuilt paths, it brings more overhead and reduces the reliability of the network. Secondly, although the existing multi-path protocols can generate a number of paths for point to point data transfer, but the paths the independent, non-organic links can be established. Based on the above considerations, we try to bring Erasure code technology into ad-hoc routing protocol, to make a number of paths into an organic one, to achieve further improvement of network performance.

The remainder of this paper is organized as follows. In section 2, we introduce the former AODV and AOMDV protocols briefly as it's the basement of our AOMDV-CB protocol. The details of AOMDV-CB will be presented in section 3. The methodologies of the performance evaluation, as well as the simulation environment are presented in section 4. The results of the quantitative comparison of the three protocols (AODV, AOMDV and

---

978-0-7695-3342-1/08 $25.00 © 2008 IEEE

DOI 10.1109/PEITS.2008.89

AOMDV-CB) are also discussed in that section. We conclude the paper in section 5.

2. Former works

2.1 AODV

The AODV protocol, which is brought out by Perkins and Royer at 1999, is designed for mobile nodes in ad-hoc network, where there often are changes in topology. The AODV protocol is based on on-demand route discovery. Because of that every node has different and limited local knowledge of the whole network. The fact that a node seeks information about the network, only when needed, is causing low overhead since nodes do not have to maintaining unnecessary route information.

To achieve route information, AODV uses three different kinds of messages, which are Route request (RREQ), Route Reply (RREP) and Route Error (RERR). AODV is using ring expansion when discovering new routes to limit flooding of the network and there by reducing overhead. The protocol is ideal for discovering neighbor nodes. If a node needs a route to a node in the other end of the network, the protocol will course a reasonable flooding of the network. Expansion ring search is a better strategy than doing a full scale search for the node. Likely some other node in the network has a valid route to the destination, and will send a RREP to source, and there by reducing overhead. By every RREQ a node sends, a sequence number is increased, this is used by the protocol to guarantee loop-freedom in paths found.

2.2 AOMDV

AOMDV, which is brought out by Marina and Das at 2001, offers a multipath, loop-free extension to AODV. It ensures that alternate paths at every node are disjoint, therefore achieves path disjointness without using source routing.

In AOMDV, each RREQ and respectively RREP defines an alternative path to the source or destination. Multiple paths are maintained in routing entries in each node. The routing entries contain a list of next-hops along with corresponding hop counts for each destination. To support multipath routing, route tables in AOMDV contain a list of paths for each destination. Two additional fields, hop-count and last-hop, are stored in the route entry to help address the problems of loop freedom, and path disjointness, respectively. To ensure loop-free paths AOMDV use the advertised hop-count value at node i for destination d. This value represents the maximum hop-count for destination d available at node i. Consequently, alternate paths at node i for destination d are accepted only with lower hop-count than the advertised hop count value. To ensure that paths in the route table are link-disjoint, a node discards a path advertisement that has either a common next hop or a common last hop as one already in the route table.

3. AOMDV-CB

AOMDV-CB is an extension of AOMDV protocol. We use erasure code technology in the protocol to increase the robust of the ad-hoc network. As we introduced, when the size and the diameter of the ad-hoc network increases, the packet drop ratio and delay increases rapidly. After analyzed the current ad-hoc multi-path protocols, we thought the problem maybe that the paths which the protocols achieved are the independent, so if one path is disjoint, the data on this path have to be dropped and retransfer. But if the protocol is erasure code based, the data could be recovered by the other packet on the other path without retransfer.

At first, we give the main idea of the design of the multi-path protocols based on erasure codes.

1. Maintain loop-free multi-path between each node.
2. Traffic between each node should be transferred on different path.
3. t data packets, which are transferred on different path, should be encoded into a group, padding with k parity packets that could tolerant k packet missing (erasure errors).
4. We thought array codes are feeds better needs of our AOMDV-CB theme because of its low complexity.

Based on the above several basic principles, we designed AOMDV-CB (Ad Hoc On Demand Distance Multi-path Vector Routing - Coding Based) protocol. In the following, we describe the detail of how to modify AOMDV into our protocol.

3.1 Route discovery and maintain

There is no significant difference between AOMDV and AOMDV-CB in the route discovery and maintain stage. We only need to make some small changes in several parameters to make sure that the protocol could provide more different paths between nodes.

Modify the parameters $aomdv\_\max\_paths\_$ and $aomdv\_prim\_alt\_path\_len\_diff\_$. If our scheme is based on erasure code with parameters $(n, n-k, k+1)$, then

$$aomdv\_\max\_paths\_$$
$$= aomdv\_\max\_paths\_ \times n/k$$
$$aomdv\_prim\_alt\_path\_len\_diff\_$$
$$= aomdv\_prim\_alt\_path\_len\_diff\_ \times n/k$$

3.2 Data transfer

In the RFC 3561, which is the IETF's specification of AODV, do not mention the detail of data transfer. This means AODV protocol weak up and achieves route information only when the higher layer protocol need to get a route to another node. Although no available specification about AOMDV, the description in the AOMDV papers are not discuss the detail of data transfer too. But in our protocol, because we want to encode each data packet, so we need to modify AOMDV to control the detail of data transfer. It's the biggest difference between AOMDV and AOMDV-CB. And the detail is illustrated as follow.

1. We make use of a special header carrying control information for data packets that can be included in any existing IP packet. The header must immediately follow the IP header in the packet.

2. The length of AOMDV-CB header is 32bits, it contains 4-octet, and the header has the following format:

Code Information 16bits	Packet ID 16bits

Where Code Information is a 16 bits field, the value of this field defines which erasure code this packet is encoded. We applied only one erasure codes in our implement, which is STAR code [15] with parameters $(10,7,4)$. If this field is all '1', means we use $(10,7,4)$ STAR code to encode the data packets, and if the value of this field is not all '1', means we use other erasure codes, which haven't complete in our implement yet but will be consider in the nearly future. Packet ID is also a 16bits field. The value of this field defines the ID of this Packet between nodes to nodes. According to this field, we could calculate the parity relationship of these data fields. For example, if we use $(10,7,4)$ STAR code, then the data packets which $Packet\ ID \bmod 10 = 0...6$ is the information packets and the data packets which PacketID mod $10 = 7...9$ are the parity packets.

3. Modify the data transfer method of the AOMDV. As we introduced, although the AOMDV could find and maintain more than one path, the data transfer in AOMDV is always choose the shortest path. It chooses another path only when the shortest path is disjoint. In our scheme, we want the data to be transferred in several different paths, so that it could show the advantage of our Code-Based scenario. In our implement, we could 3 different paths to transfer data packets.

4. Performance Evaluation

The simulation results for AOMDV-CB against AODV and AOMDV are presented. Ns-2.29 [16] was used for simulating the protocols. The AODV implementation is the original one in the ns-2 lib, and the AOMDV

implementation is based on the by implementation provided by akhayyat, which is available at [17].

To show the advantage of our protocol, which is suitable for large scale ad-hoc routing, we consider simulation over a 1200m×1200m containing 50 nodes, each node generate traffic at the rate of 2 packets per second and 8 packets per second, this constitutes a net load of 100 packets per second (moderate load) and 400 packets per second (high load) respectively. The MAC layer used is 802.11 with a 250m transmission range and a throughput of 2 Mbps. To represent mobility we choose the random way point mobility model with each node moving at a random speed between $(0, maxspeed]$, where maxspeed are 0.001m/s、5m/s、10m/s、20m/s、30m/s、40m/s and pause time is 0. The simulations were run for 500 simulated seconds.

We choose packet delivery ratio, average end-to-end delay of data packets, as the performance metrics of interest. Packet Delivery ratio is the fraction of CBR data packets received at the destination. Average data latency or the end-to-end delay includes the average of all possible delays for the data packets; from the time the data is transmitted to till it is received.

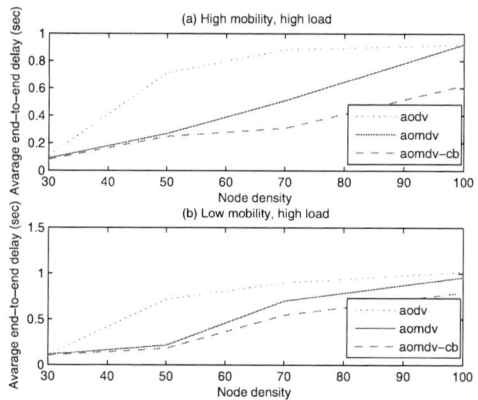

Figure.1. Data latency

Figure.1 is the compare of the average end-to-end delay of the data packets in the three protocols. We differentiate between two cases: high node mobility with high load (pause time 0ms, 20 connection, 4 pkt/sec, 512B) which is shown in figure.1.(a) and low node mobility with high load(pause time 120ms, 20 connection, 4 pkt/sec, 512B) which is shown in figure.1.(b). End-to-end delay for the AOMDV-CB protocol is higher than the other protocols because the node could decode a data packet when it received enough parity packets without waiting for the lagged packets. In low mobility and low node density scenarios end-to-end delay is approximately 0.8s, while in higher node density the end-to-end delay augments to1000ms. The reason that why the AOMDV-CB is not magnificent better than AOMDV when Node density ≤ 50, we thought is because the experimental field is too wide for the nodes that they could not establish enough path between each others.

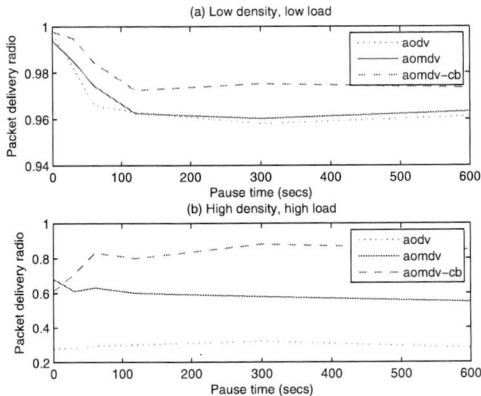

Figure.2. Packet delivery ratio

The data packet delivery ratio is also been presented for two different scenarios in figure.2. Low density with low load (30 nodes, 10 connection, 4 pkt/sec, 512 B) is shown in figure.2.(a) and High density with high load (100 nodes, 20 connection, 4 pkt/sec, 512 B) is shown in figure.2.(b). As we expect, in both of these two cases, data packet delivery ratio of AOMDV-CB is much better than the two traditional protocols.

5. Conclusions and Future Work

We bring the erasure correct code into the ad-hoc protocols, and modify the AOMDV with $[10,7,4]$ STAR code to establish a new protocol which is called AOMDV-CB. A lot of experimental comparisons have been given between AOMDV-CB, AOMDV and AODV. The results show that our new protocol has better performance on end-to-end delay and packet delivery ratio. While our scheme still has some shortage on routing overhead and other metrics which did not detailed discussed in this paper and it's our future work. Nevertheless, it's a new approach to the ad-hoc routing problem. Apply more erasure codes into our scheme and tests its performance is our future works.

6. References

[1]   C. E. Perkins, P. Bhagwat, "Highly Dynamic Destination-Sequenced Distance-Vector Routing (DSDV) for Mobile Computers," ACM SIGCOMM'94 Conference on Communications Architectures, Protocols and Applications, Oct. 1994, pp. 234-244

[2]   L. R. Ford Jr., D. R. Fulkerson, "Flows in Networks," Princeton Univ.Press, 1962.

[3]   S. Murthy, J. J. Garcia-Luna-Aceves, "A Routing Protocol for Packet Radio Networks," ACM International Conference on Mobile Computing and Networking, MOBICOM'95, November 1995, pp. 86-95

[4]   C.-C. Chiang, "Routing in Clustered Multi-hop, Mobile Wireless Networks with Fading Channel," Proc. IEEE SICON '97, Apr. 1997, pp. 197–211

[5]   R. Ogier, F. Templin, M. Lewis, "Topology Dissemination Based on Reverse-Path Forwarding (TBRPF)," RFC 3684, February 2004.

[6]   D. B. Johnson, D. A. Maltz, "Dynamic Source Routing in Ad-Hoc Wireless Networks," Mobile Computing (ed. T. Imielinski and H. Korth), Kluwer Academic Publishers, Dordrecht, The Netherlands, 1996, pp. 153–81

[7]   C. E. Perkins and E. M. Royer, "Ad-hoc On-Demand Distance Vector Routing," Proc. 2nd IEEE Wksp. Mobile Comp. Sys. and Apps., Feb. 1999, pp. 90–100

[8]   M. S. Corson and A. Ephremides, "A Distributed Routing Algorithm for Mobile Wireless Networks," ACM/Baltzer Wireless Networks J., vol. 1, no. 1, Feb. 1995, pp. 61–81.

[9]   V. D. Park and M. S. Corson, "A Highly Adaptive Distributed Routing Algorithm for Mobile Wireless Networks," Proc. INFOCOM '97, Apr. 1997.

[10]   C. K. Toh, "A Novel Distributed Routing Protocol to Support Ad-Hoc Mobile Computing," IEEE 15th Annual International Phoenix Conference on Computers and Communications, March 1996. pp 480-486

[11]   R. Dube et al., "Signal Stability based Adaptive Routing (SSA) for Ad-Hoc Mobile Networks," IEEE Pers. Commun., Feb. 1997, pp. 36–45.

[12]   S. J. Lee and M. Gerla, "Split multipath routing with maximally disjoint paths in ad hoc networks," ICC 2001. IEEE International Conference on Communications, Helsinki, 2001.

[13]   R. Leung, J. Liu, E. Poon, A. L. C. Chan, and B. Li, "MP-DSR: a QoS-aware multi-path dynamic source routing protocol for wireless ad-hoc networks," Local Computer Networks, 2001. Proceedings. LCN 2001. 26th Annual IEEE Conference on, Tampa, FL 2001.

[14]   M. K. Marina and S. R. Das, "Ad hoc on-demand multipath distance vector routing," Technical Report, Computer Science Department, Stony Brook University, 2003.

[15]   C. Huang, L. Xu, "STAR: An Effcient Coding Scheme for Correcting Triple Storage Node Failures," FAST-2005: 4th Usenix Conference on File and Storage Technologies, December, 2005.

[16]   ns   2:   Network   Simulator: (http://www.isi.edu/nsnam/ns/)

[17]   www.ccse.kfupm.edu.sa/~akhayyat/

Session 6

Neural Networks
and Computational Intelligence

2008 Workshop on Power Electronics and Intelligent Transportation System

Analyzing QAR Data Using K-PCA

Feng Xiao-rong

College of Computer Science & Technology,
Civil Aviation University of China, Tianjin
300300
fengxiaorong@163.com

Feng Xing-jie

College of Computer Science & Technology,
Civil Aviation University of China, Tianjin
300300
fxingjie@163.com

Abstract

This paper analyzes the drawbacks of traditional principal component analysis (PCA) firstly, and discusses the kernel principal component analysis (KPCA) as well as its drawbacks of high complexity secondly. Then it proposes the K-PCA method. Comparing with KPCA, the method proposed in this paper could achieve dimensionality reduction with faster speed. The results show that: the proposed method performs an experiment on QAR data has a good effect of dimensionality reduction and high correct classification rate.

1. Introduction

Quick Access Recorder (QAR) is widely adopted in flight-data recording system. Some developed countries in the world have put the flight data under the regular monitor and control, for example, in American , the accident and incident rate per 10 thousand hours has dropped from 3.6 (in 1950) to 0.15 after using QAR system . Most of aircrafts in China have been fixed with QAR system on the request of Civil Aviation Administration of China. QAR Data has become the important information for monitoring the quality of flights, detecting the working state of engines, diagnosing system failure and realizing three-dimensional analysis. QAR Data system is multi-variables in the order of time, so dimensionality reduction is necessary.

PCA [1] is the conventional method of dimensionality reduction. That is, mapping data to feature space, taking the processes of linear combination for original features, then sequencing variance of features and eliminating small ones. Traditional PCA is a kind of linear mapping and could not get prefect effect in normal conditions. It's extremely sensitive to outliers and missing values, which easily lead to incomplete or fault results. Experts have offered new thoughts to improve it, such as PCA based on fuzzy method, method of principal curves and surface proposed [2] by Hastie and Stuetzle.

With the development of research on Support Vector Machines (SVM) [3], the research on kernel has received more and more attention in recently years. KPCA method which proposed by Scholkopf [4-5] tries to map variable data into high-dimensional feature space and then carries out PCA. The complexity of computing kernel matrix of KPCA, O (pN2) is closely connected to the number and dimension of the samples, where p is the dimension of each sample, and N is the number of the samples. In order to increase computing efficiency, some researchers proposed a method named KPCA based on clustering, which regards the clustering centers as new samples and then use KPCA. By this way, the number of samples is reduced and complexity is also descended. However, this method also has drawbacks, that is, the number of clusters and the new samples both have some influence on KPCA. Aiming at the above problems, an improved PCA is proposed in this paper. It constructs covariance matrix by kernel function and then use PCA where the complexity is $O (p^2N)$. The result of experiments shows that the method proposed in this paper has better effect than KPCA, without referring to related parameters in advance. When the number of samples is too large, the correct classification rate is same to that of KPCA by adding numbers of principal components with less time consuming.

The remainder of this paper is organized as follows: Section2 introduces the PCA and KPCA methods, our proposed methods are described in detail in Section 3, this is followed by the experimental description and the corresponding results obtained in Section 4, and in Section 5 the related work conclusions and future work are presented.

2. PCA and KPCA methods

In this section, we briefly describe the PCA and KPCA methods, from which our proposed method is extended. For more details on these topics, please refer to [1, 4].

2.1 PCA.

PCA has been widely used for multivariate data analysis and dimensionality reduction. PCA method is to describe data in a coordinate system orthogonal transformation. The central idea of PCA is to reduce the dimensionality of a data set consisting of a large number of interrelated variables, while retaining the variances as much as possible in the data set .This is achieved by

---

978-0-7695-3342-1/08 $25.00 © 2008 IEEE

DOI 10.1109/PEITS.2008.8

195

transforming to a new set of variables, the principal components, which are uncorrelated and ordered so that the first few retain most of the variation in all of the original variables.

Let A is the original data set of size $n \times p$, where n is the number of items in the dataset and p is the number of variables for the dataset. Then PCA is performed by applying Singular Value Decomposition (SVD). That is, when a covariance matrix S is decomposed by SVD, $S = U\Lambda U^T$, a matrix U contains the variables' loadings for the principal components, and a matrix Λ has the corresponding variances along the diagonal. Then chooses the first k principal components, of which variances represent 95% of the total variance, and k<p. The new principal component can be described as:

$$\begin{cases} F_1 = a_{11}X_1 + a_{21}X_2 + \cdots + a_{p1}X_p \\ F_2 = a_{12}X_1 + a_{22}X_2 + \cdots + a_{p2}X_p \\ \cdots \\ F_k = a_{1k}X_1 + a_{2k}X_2 + \cdots + a_{pk}X_p \end{cases}$$

Proceeding in this way, $F_1, F_2, \cdots, F_k$, are called the first, second, $\cdots$, and k principal components respectively, and variances of which are decreased in turn.

2.2 KPCA

PCA method neglects smaller variance items, while preserves the larger variance, thereby reducing the data dimensions. But it is a linear transformation, only the linear correlation characteristics of the data can be extracted. While the non-linear problems are difficult to achieve the expected drop-dimensional effect .To solve this problem，Scholkopf proposed a new method which utilized the kernel trick and termed KPCA.

KPCA first mapping the data set into a high dimensional feature space F, then computing dot products in feature space by means of kernel functions in input space. In the end performs standard PCA in F.

Definition 1 A kernel is a function k, such that $K(x,z) = \langle \phi(x) \bullet \phi(z) \rangle$, for all x, z X, where ϕ is a mapping from X to a feature space F.

Definition 2 let the data set X is mapped into a feature space F by using an arbitrary nonlinear map.

$\phi : \Re^N \to F, x \mapsto X$. The covariance matrix in the feature space can be described as follows, assuming the data are centered:

$K_{ij} = K(\phi(x_i), \phi(x_j))(1 \le i, j \le n)$,an n×n Kernel Matrix, which is also called as Gram matrix.

Mercer's Conditions which the kernel function satisfies with has been presented in papers [4-5], i.e., the kernel matrix derived from the kernel function must be positive semidefinite.

The KPCA method steps: first compute the matrix K; second, compute its eigenvectors and normalize them in F; third, compute projections of a test point onto the eigenvectors.

3. OUR APPROACH (K-PCA)

3 .1 K-PCA

In this section, we firstly briefly describe the advantage and disadvantage of PCA as well as the KPCA method, and then propose our new approach in detail.

From section 2, we know that the central idea of PCA is to reduce the dimensionality of a data set consisting of a large number of interrelated variables, while retaining the variation as much as possible in the data set, by using an orthogonal transformation. While KPCA method is a nonlinear generalization of PCA in the sense .It is performing PCA in feature spaces of arbitrarily large dimensionality, and use the kernel function $K(x,z) = \langle \phi(x) \bullet \phi(z) \rangle$ recover the original PCA algorithm. Compared to PCA method, kernel PCA has the main advantage that no nonlinear optimization is involved. But the disadvantage of KPCA is high time consumption in computing Kernel matrix, in the case where we need to use a large number of observations.

Assume that we are given a set of n items, and each data item is an p dimensional column vector, i.e., $x_i \in X$ where $1 \le i \le n$. Assume that the data is mean centered, i.e., $\sum_{i-1}^{n} x_{ji} = 0$, for $1 \le j \le n$. The covariance matrix S can be computed as follows:

$$S = \frac{1}{n} \sum_{i=1}^{n} x_i \bullet x_i^T \text{ an p×p matrix.}$$

The PCA method diagonalizes the S to obtain the principal components, which can be achieved by solving the following eigenvalue problem:

$\lambda S = S\nu$

While the KPCA method is based on the Kernel matrix:

$K_{ij} = K(\phi(x_i), \phi(x_j))(1 \le i, j \le n)$,an n×n matrix. Then run traditional PCA in the feature spaces.

And n is far greater than p, so the KPCA has high time consumption , due to compute dot products in feature space by means of kernel functions in input space. In fact, the kernel method is a trick. It can be applied to any algorithm, as long as the algorithm has dot product [6, 7].

Definition 3 Assume that we are given a set of n items, and each data item is an p dimensional column vector, i.e., $x_i \in X$ where $1 \le i \le n$. Assume that the data is mean centered, i.e., $\sum_{i-1}^{n} x_{ji} = 0$, for $1 \le j \le n$. The covariance matrix C can be computed as follows:

$$C_{ij} = K(x_i, x_j)$$ an p×p matrix. Then diagonalizes the C to obtain the principal components. This method termed as K-PCA.

The Computational Complexity of K-PCA O (np2) is less than KPCA O (n2p) where n is more than p, and the results show that: the K-PCA performed an experiment on QAR data has a good effect of dimensionality reduction and high correct classification rate.

To perform K-PCA, the following steps have to be carried out: step 1: we compute the matrix $C_{ij} = K(x_i, x_j)$ .Step 2, we solve $\lambda C = Cv$ by diagonalizing C, and normalize the eigenvector expansion coefficients v^n by requiring $\lambda_n (v^n \bullet v^n) = 1$ .Step 3,to extract the principal components of a test point x, we then compute projections onto the eigenvectors by

$$F_i = \sum_{j=1}^{p} v_i^j X_p$$

3.2 The common Kernel function

1. Polynomial function of degree h
$$k(x_i, x_j) = (x_i \bullet x_j + c)^h, (h \ge 2, c \ge 0)$$
2. Gauss Radial Basis nuclear
$$k(x_i, x_j) = \exp\left(-\frac{\|x_i - x_j\|^2}{2\delta^2}\right)$$
3. Anh-function
$$k(x_i, x_j) = \tanh(-a(x_i \bullet x_j) - b)$$

4. Performance evaluation

4.1 Dataset

The experiments have been conducted on two different QAR data set A and B.

Table 1. Summary of data sets used in the experiments

	A	B
variables	120	100
items	39120	57740
labels	2	2

A has the same route but with different model, while B has the same route and model, but with different Pilots.

4.2 Method

For K-PCA, first need to construct kernel matrix C, as described in Section 3. Then choose Polynomial function

of degree h, and set c=2, h=2. In order to compute the classification accuracy of K-PCA, we choose 2, 4, 6,8,10 principal components, and performed 10 fold cross validation employing Support Vector Machine (SVM).

We Compared the performance of K-PCA with two other techniques, PCA and KPCA using the same kernel function with K-PCA.

4.3 Results

Figure 1 shows the time consumption in calculation nuclear matrix of KPCA and K-PCA.

As can be seen from figure 1, the Number of samples have obvious impact on KPCA, while small on K-PCA . With the number of samples increasing, the time-consuming of KPCA becomes more and more.

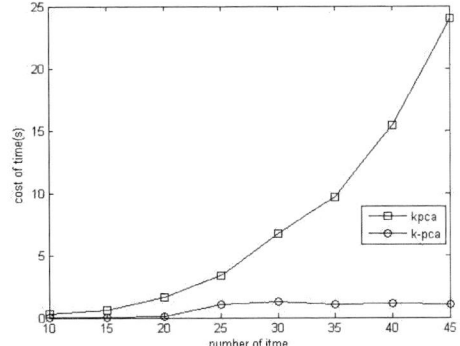

Figure1. Time of calculation nuclear matrix

Figure 2 illustrates the comparison of cumulative contribution rate on data set A, respectively using PCA, KPCA and K-PCA methods. From Figure 2, it can be observed that KPCA and K-PCA perform better than the PCA in Dimensionality reduction effect.

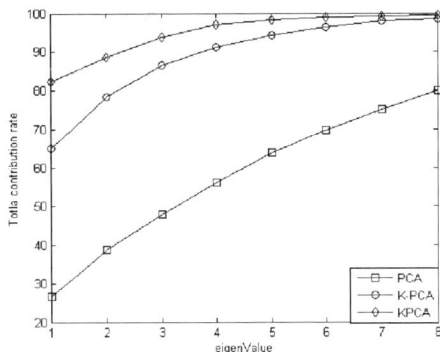

Figure 2.Comparison of cumulative contribution rate

Figure 3 represents the classification accuracies of the three techniques on the A data set. Respectively using only 8 features obtained by KPCA and K-PCA, the classification accuracy is over 90%, while the K-PCA has the low consumption of time than KPCA. Figure 4 shows the results of the classification accuracy for the B data set. Similarly as for the A data set, K-PCA performs as well as

KPCA method.

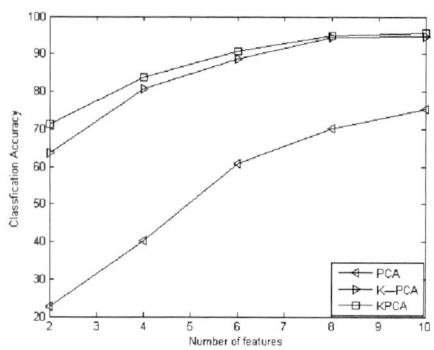

Figure 3.Correct classification rate of data A

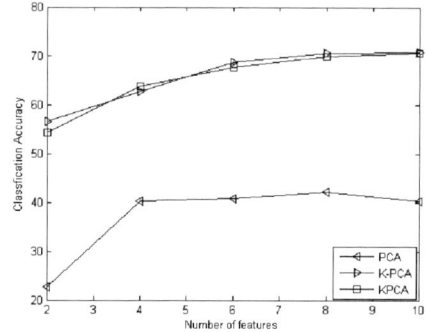

Figure 4.Correct classification rate of data B

5. Conclusions and future work

In this paper, we proposed a technique which based on the kernel function covariance matrix of principal component analysis. The results show that: the proposed method used for QAR data has a good effect of dimensionality reduction and high correct classification rate. We intend to extend this method into FOQA, Fuel savings and Flight Delays, comparing with other techniques such as GPCA [8] and K-LDA [9]

References

[1] I. T.Jolliffe. Principal Component Analysis.Springer,2002.

[2] Has tieT .Principal Curves and surfaces. Laboratory for computational statistics: [Technical Report 11].Stanford University Depts. of Statistics, 1984.

[3] Steve R.Gunn.Support Vector Machines for Classification and Regression. [Technical Report 11].Faculty of Engineering, Science and Mathematics School of Electronics and Computer Science.1998.

[4] B.Scholkopf, A.J.Smola, and K-R.Muller.Nonlinear component analysis as a kernel eigenvalue problem. *Neural Computation*, 10(5):1299-1319, 1998.

[5]K-R.Muller,S.MiKa, G.Ratsch, K.Tusda, and B.Sch0lkopf.An introduction to kernel-based learning algorithms.*IEEE Trans.Pattern Anal.* Machine Intell.12 (2):181-201, Match 2001.

[6]Park, C.H, Park, H.Nonliner feature extraction based on cancroids and kernel functions. *Pattern Recognition* 37,801-810,2004.

[7]Kim, S.W., Onmmen, B.J., On utilizing search methods to select subspace dimensions for kernel-based nonlinear subspaceclassifiers.*IEEETrans.Patt.*Anal.Mach.Intell.27.136-1 41, 2005.

[8] J. Ye, R. Jana Dan, and Q. Li. Gpca: an efficient dimension reduction scheme for image compression and retrieval. In KDD '04: *Proceedings of the tenth ACM SIGKDD international conference on Knowledge discovery and data mining*, pages 354–363, New York, NY, USA, 2004. ACM Press.

[9] H. Yoon, K. Yang, and C. Shahabi. Feature subset selection and feature ranking for multivariate time series. IEEE Trans.Knowledge Data Eng. - *Special Issue on Intelligent Data Preparation*, 17(9), September 2005.

An Improvement Route Generation Algorithm for Bus Network Design

Yikui MO[1], Jun DENG[2], and Jingyuan WANG[1]

[1]*College of Civil Engineering, Shenzhen University, Nanhai Road 3688, Shenzhen, P.R.C*
[2]*Urban Planning & Design Institute of Shenzhen,Zhenxing Road 3, Shenzhen, P.R.C*
E-mail： moyikui@126.com, dengj@upr.cn, wangjingyuan_01@sina.com

Abstract

In order to overcome the shortages of the conventional bus route generation algorithm which aims to maximize the number of direct travelers on the shortest path，this paper present an improvement route generation algorithm for the design of bus network, which aims to maximize the direct passenger-kilometers per unit length and minimize the average travel time of direct travelers on each route at the same time. At first, the paper analyzed the irrationality of the conventional bus network design model developed by others, revised the objective function appropriately, and built a new model of bus network design. Then a modified algorithm is developed particularly to solve this problem, this design algorithm is heavily guided by the demand matrix and allows the designer's knowledge to be implemented so as to reduce the search space. Finally, the applications of the new algorithm are illustrated with a numerical example. Numerical results indicate that the proposed algorithm would be efficient in practice.

1. Introduction

In recent years, the sensitive increase of congestion phenomena in the urban areas has produced important changes for the role reserved to the public transport. Critical phase for the planning of the public transport system is the bus network design problem. The bus network design problem has been studied by several authors in the past: Silman, Barzily, and Passy, (1974); Mandl, (1979); Hasselstrom, (1981); Ceder and Wilson, (1986); Wang Wei,(1992); Baaj and Mahmassani, (1995); Lin Boliang and Yang Fushe, (1999); and Wang Wei and Yang Xinmiao,(2002). In literature different approaches are proposed for the solution of the bus network design problem. But most of them are very difficult to implement. This paper proposes a new approach which is intended to be easier to implement and less demanding in terms of both data requirements and analytical sophistication than previous methods.

This paper is organized as follows: the next section introduces the basic idea of the proposed method for the bus network design problem and presents the mathematic model. In Section 3, a new solution algorithm for the bus network design problem is proposed. Computational results on a particular network are presented in Sections 4 and 5 contains conclusions.

2. Mathematic model

Among several kinds of methods proposed by different researchers in the past, the category of methods which aims to maximize the number of direct travelers on the shortest path is relatively simple and practical, such as the method proposed by Wang Wei et al. In this method, all routes are along the shortest paths between different O-D pairs, and which has the maximum direct travelers is selected firstly.

There are, however, some limitations for this method. At first, this method prefers to generate longer route than the shorter one since the accumulated direct travelers on the longer route is often more than the shorter one. Even the shorter route has higher direct traveler density, it will not be selected. It's unreasonable because the route which has higher direact traveler is more efficient and economical.

Secondly, the average travel time of passengers on the route along the shortest path is not always shorter than the others. Sometimes, the route on the second or third shortest path is the better one. This paper presents a model which includes two objective functions. The first objective is to maximize the efficiency of the transit network and the second is to minimize the total travel time. In order to keep it simple, and easy to implement, the method only takes the direct travelers into account. The direct passenger transport density is used to represent the efficiency of the transit network and the average detour coefficient of direct travelers is used to describe the travel time. The bus network design problem can be formulated as follows:

$$MaxF = \frac{\sum_{I \in R} \sum_{i \in N} \sum_{j \in N} q_{ij}^{I} \cdot l_{ij}^{I}}{\sum_{I \in R} l_{I}} \quad (1)$$

$$MinY = \frac{\sum_{I \in R} \sum_{i \in N} \sum_{j \in N} q_{ij}^{I} \cdot l_{ij}^{I}}{\sum_{I \in R} \sum_{i \in N} \sum_{j \in N} q_{ij}^{I} \cdot d_{ij}} \quad (2)$$

$$s.t.\begin{cases} l_{\min} \le l_I \le l_{\max} & (3) \\ l_I / w_I \le \eta & (4) \\ Q_I^{sum} \ge Q_{\min} & (5) \\ Q_I / \overline{q}_I \le \mu & (6) \\ N_a \le N_{Max} & (7) \\ Q_I \le Q_{IMax} & (8) \\ S_I \in S & (9) \end{cases}$$

Where F is the direct passenger-kilometers per unit length; Y is the average detour coefficient of direct travelers, which reflect how close the riding distance to space distance; q^I_{ij} is the number of direct travelers from bus stop i to j along the route I; l_I is the length of the route I; d_{ij} is the space distance between bus stop i and j; l^I_{ij} is the riding distance between bus stop i and j along the route I; R is the set of routes in the bus network; N is the set of bus stops in the network; l_{min} is the lower bound for l_I; l_{max} is the upper bound for l_I; w_I is the length of the shortest path through the road network; η is the upper bound of non-linear coefficient; Q_I^{sum} is the total direct passengers riding the route I; Q_{min} is the lower bound for Q_I^{sum}; Q_I is the hourly direct passenger flow per direction that pass through the maximum section of the route I; $\overline{q}_I$ is the average section passenger flow of the route I; μ is the maximum of section non-equilibrium factor of passenger flow; N_a is the number of routes on the road link a; N_{Max} is the upper bound for N_a; Q_{Imax} is the hourly direct passenger capacity per direction which can be calculated by the following equation:$Q_{IMax}=C_I\gamma\cdot\alpha$, where C_I is the hourly passenger capacity per direction, α is the transfer rate and γ is the level of service influence factor which is less than 1; S_I is the set of road links which has more than one bus route on it; S is the set of road links which is feasible for bus driving on.

There are seven constraints in this model, constraint (3) ensures that the length of route I is in a reasonable range. Constraint (4) is the non-linear coefficient restriction. Constraint (5) ensures there are enough passengers for route operating in a cost-efficient manner. Constraint (6) is the equilibrium distribution restriction of passenger flow. Constraint (7) ensures there are not too much routes on the same road link. Constraint (8) means that the passenger flow should not exceed the passenger capacity. Constraint (9) ensures that the road network used as the basis of bus network is feasible for bus driving on.

To demonstrate the rationality of the proposed method, this paper uses a simple numerical examples to compare the method with those discussed in the literature such as Mandl (1979) and Wang Wei (1992). As shown in the Figure 1,node s is the origin and t is the destination, i and j are middle nodes, the travel demand between the different pairs of nodes are shown in the figure. It is easy

to see that the path $s{\to}j{\to}t$ is more rational than $s{\to}i{\to}t$ to be selected as the bus route. The path $s{\to}j{\to}t$ has the higher direct passenger-kilometers per unit length and the lower average detour coefficient of direct travelers. According to the proposed method, the path $s{\to}j{\to}t$ will be selected as the bus route, where F_{sjt} is 277 *passenger-km/km* and Y_{sjt} is 1.03. But according to the method discussed by Mandl (1979) and Wang Wei (1992), the path $s{\to}i{\to}t$ will be selected as the bus route, where F_{sit} is 85 *passenger-km /km* and Y_{sit} is 1.04.

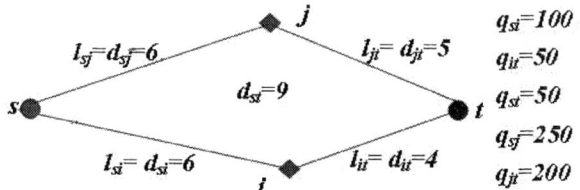

Figure 1. A simple example

3. The improvement algorithm

According to mathematic characteristics of the model discussed above, the input data needed for this method may be summarized as follows: a) Road network; b) Set of bus network nodes; c) Demand matrix; d) Space distance between every pairs of nodes; e) The maximum and minimum route length; f) The maximum non-linear coefficient; g) The hourly passenger capacity per direction; h) The transfer rate; i) The minimum passengers for route operating; j) The maximum of section non-equilibrium factor; k) The maximum routes on the same road link; l) Minimum percentage of total demand satisfied directly by set of routes. m) Minimum percentage of total demand satisfied by set of routes(via 0 or 1 transfers).

The solution algorithm for the bus network design problem can be arranged as follows:

Step 1: Identify the M node pairs with the highest demand; set $k = 1$.

Step 2: Generate all of routes that satisfy the constraint (3) to (6) between the kth node pair. Compute the value of Y of each route using the formula (2), then select the route which has the minimum value of Y as the recommended route between the kth node pair.

Step 3: If $k = M$, go to step 4, otherwise, set $k = k + 1$ and return to step 2.

Step 4: Compute the value of F of each recommended route using the formula (1), select the route which has the maximum value of F as the decided route. According to the number of routes on each road link, modify the impedance of road links using the following formula: $t_a=t_a^0\times K_a$, where t_a^0 is the initial impedance of link a, and K_a is the overlap coefficient which can be specified by the following Table 1.

Table 1. Recommended value of K_a

Number of routes on link a	0	1	2	3	4	5	6
K_a	1	1.25	1.56	1.95	2.44	3.05	3.81

Step 5: Compute the total demand satisfied directly using the routes generated so far. If that demand exceeds the minimum percent demand satisfied directly, go to step 6, otherwise, remove from the demand matrix all node pairs whose demand can be satisfied directly and return to step 1.

Step 6: Compute the total demand satisfied by the set of routes (via 0 or 1 transfers). If that demand exceeds the minimum percent total demand satisfied, terminate the algorithm process and output the bus network, otherwise, remove from the demand matrix all node pairs whose demand is satisfied with a maximum of two transfers and return to step 1.

4. Numerical example

In this section, to illustrate the applications of the model and algorithm proposed in this paper, we present a numerical example for the bus network design problem. Figure 2 shows a simple urban road network and traffic zones. The length of each road link are also shown in the figure. The travel demand matrix is shown in the following table 2.

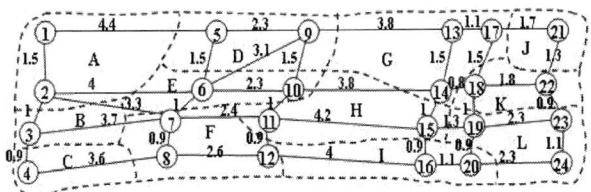

Figure 2. The test network and traffic zones

Besides the road network and demand matrix, the input data used in this example can be summarized as follows: l_{min}=5km, l_{max}=15km, η=1.1, Q_{min}=100, C_f=7920, α=0.6, γ=0.85, Q_{Imax}=4039, μ=1.5, Minimum percentage of total demand satisfied directly by set of routes is 0.6, Minimum percentage of total demand satisfied by set of routes(via 0 or 1 transfers) is 0.95.

There are five traffic zones that has the enough demand to be original and terminal stations，which are A,C,E,J and L. Except for those which don't satisfy the length constraint, there are eight possible O-D pairs to generate bus routes, which are (A,E), (A,J), (A,L), (C,E), (C,L), (J,C), (J,E) and (L,E).

The feasible routes that satisfy the constraint (3) and (4) between each O-D pair and the average detour coefficient of direct travelers of each route are shown in the following table 3.

Table 2. The travel demand matrix

O\D	A	B	C	D	E	F	G	H	I	J	K	L
A	68	168	418	33	289	19	93	118	49	289	26	636
B	168	18	103	34	107	28	27	68	47	90	13	201
C	426	93	68	52	334	28	61	80	34	418	52	576
D	34	33	43	15	33	18	23	13	50	42	11	110
E	292	93	351	32	32	34	39	48	67	318	52	570
F	23	34	19	17	33	16	23	23	18	25	21	35
G	90	30	56	23	34	23	13	15	25	58	12	125
H	125	66	84	13	61	24	16	14	34	66	17	107
I	51	42	33	43	66	20	21	33	12	67	23	94
J	298	90	409	43	314	23	56	67	64	27	33	668
K	32	15	51	12	48	23	13	18	24	33	13	51
L	632	193	584	107	584	33	118	101	86	666	60	28

Table 3. The feasible routes between each O-D pair

O-D	The feasible route	Value of Y	Value of F
A-E	①1-2-6 *	1.67	436
	②1-5-6	1.82	\
A-J	1-5-9-13-17-21	1.00	641
A-L	①1-2-7-11-12-16-20-24 *	1.20	1114
	②1-2-3-7-11-12-16-20-24	1.26	\
C-E	①4-8-7-6	1.33	\
	②4-3-2-6 *	1.15	1204
C-L	4-8-12-16-20-24	1.00	908
J-C	①21-17-13-9-6-7-8-4 *	1.22	1364
	②21-17-13-9-5-6-7-8-4	1.28	\
J-E	21-17-13-9-6	1.06	613
L-E	①24-23-22-18-14-10-6	1.11	\
	②24-23-19-18-14-10-6 *	1.07	805

It is known from the Table 3 that there are not too much feasible routes between each O-D pair, and the recommended route which we have marked with an asterisk has lower value of Y. The route (21-17-13-9-6-7-8-4) between J and C has the highest direct passenger-kilometers per unit length, so it is selected as the first route.

Remove the demand can be satisfied directly from the demand matrix and repeat the previous step until the total demand satisfied by the set of routes has exceeded the minimum percent. At last, we can obtain the optimal solution of the numerical example as follows: the route (21-17-13-9-6-7-8-4) between J and C, the route (1-2-7-11-12-16-20-24) between A and L, the route (4-3-2-6) between C and E, the route (24-23-19-18-14-10-6) between L and E.

5. Conclusion

In this paper, a new method for bus network design problem has been studied on a simulated network, which is intended to be easier to implement and less demanding in terms of both data requirements and analytical sophistication than previous methods. The proposed method can overcomes the shortages of the conventional method discussed by Mandl (1979),Wang Wei (1992), and et al. It is found that the model and the solution algorithm proposed in this paper is efficient in solving bus network design problem and will bring new light for the solution to it.

Acknowledgements

This work was supported by the Shenzhen University R/D Fund (No. 200817).

Reference

[1] Silman L. A., Barxily Z., and Passy U, "Planning the route system for urban buses", *Computers and Operations Research*, 1974, pp. 201-211.

[2] Mandl C. E. "Evaluation and optimization of urban public transportation networks", *European,* 1980, vol. 5, issue 6, pp.396-404

[3] Ceder, A., and N. H. M. Wilson, "Bus network design", *Transportation Research(Part B)*, 1986, 20B, pp.331-344.

[4] Hasselsttom D, "Public transportation planning-A mathematical programming approach", *Doctoral dissertation, Department of Business Administration,* University of Gothenburg, Sweden, 1981

[5] WANG Wei, "A simple and practical method for transit network optimization", *Computer and Communications*, 1992, pp. 57-59

[6] Baaj, M.H. and Mahmassani, H.S.. "Hybrid route generation heuristic algorithm for the design of transit networks", *Transportation Research* C 3, 1995, pp.31-50

[7] Lin Boliang,Yang Fushe and Li Peng, "Designing Optimal Bus Network for Minimizing Trip Times of Passenger Flows", *China Journal of Highway and Transport*, 112, 1999, pp.79-83

[8] Wang Wei, Yang Xinmiao and Chen Xuewu, *Urban Public Transportation System Planning and Management,* Science Publishing Company, Beijin, 2002

[9] Baaj M.H. & Mahmassani H.S, "TRUST: A Lisp program for the analysis of transit route configurations", *Transportation Research Record*, 1990, Vol.1283, pp. 125-135.

[10] Fusco G., Gori S., Petrelli M, "A heuristic transit network design algorithm for medium size towns", *Proceedings of 9th Euro Working Group on Transportation*, Bari, Italy, June 2002, pp. 652-656.

[11] Ceder A. & Israeli Y. "Design and evaluation of transit routes in urban networks", *Proceedings of the 3rd International Conference on Competition and Ownership in Surface Passenger Transport,* Ontario, Canada, 1993.

[12] Dhingra S.L., Muralidhar S. & Krishna Rao K.V. "Public transport routing and scheduling using genetic algorithms", *Paper presented at the CASPT 8th nternational Conference, June 2000 Berlin, Germany,* 2000.

Forecasting railway network data traffic: A model and a neural network solution algorithm

Teng Jing

School of Transportation Engineering, Tongji University, Shanghai, 200092, China
E-mail:tengjing@263.net

Abstract

Forecasting network data traffic is an important part of the function of planning and managing information systems. However, the contents of network data are so stochastic and complex that it is very difficult to establish stable functions to describe the mapping relationship between data flows and associated causal influences. In this paper, a multi-layer feed forward neural networks (NN) model is put forward to identify such relationship and the corresponding learning rule of NN, back-propagation (BP) algorithm, is given. In addition necessary estimation and validation processes are designed to ensure the successful implementation of the model proposed. The paper elucidates the application of NN model around the case of forecasting China railway Transportation Management Information Systems (TMIS) network traffic. The predictive results obtained demonstrate that the NN model and the solution algorithm are applicable for information planning on the TMIS network.

1. Introduction

In the telecommunications industry, planning network-loading capacity has become a focus, which usually includes two aspects of studies. One aspect aims at optimizing the costs of capacitated facilities of a network. For example, Magnanti develops modeling and solution approaches for loading facilities to satisfy the given demand at minimum cost[1]. Balakrishnan develops a decomposition algorithm for local access telecommunications network expansion planning based on a minimum cost optimization methodology[2]. Their studies are done based on known planning demand of network data traffic. The second aspect of past studies has been forecasting network data traffic, which is an important and practical work for planning future communication networks. This aspect can be divided

into two sub-areas of research, namely: forecasting market service demand for a network. For example, Wright presents a case study of the application of the Delphi Method for forecasting the market for broadband telecommunications in the year 2000[3]. He evaluates market demand from seven different viewpoints and the results were obtained for the market in the whole of North America with a focused case study of the Toronto urban core. The other sub-area is forecasting network data traffic through mining the mapping relationships between network service demand and network data traffic. However, this latter problem is very complex, because many network services work together simultaneously and almost all the services are real-time and stochastic. It is very difficult to find stable functions describing the above mapping relationships. In this paper, we try to apply neural networks to identify such mapping relationships and sequentially forecast the network data traffic based on future network service demand.

1.1. About TMIS

TMIS is the main component of Chinese railway information Engineering and was implemented progressively since 1992. The main sub-projects of TMIS are management information systems for: basic data, dispatching data, real-time vehicle-tracking data, freight transportation, passenger transportation, intermodal transportation and integrated transportation.

The main data saved, processed and transmitted in TMIS comes from the entire processes related with railway transportation management and operation. The information flow extracted from the data flow can be used for aiding different management decision-making levels. Currently, in the Chinese Ministry of Railways (MoR), there are three transportation management/operation levels, which are MoR, railway administration, and railway station and service section. The railway transportation equipment is almost all

978-0-7695-3342-1/08 $25.00 © 2008 IEEE
DOI 10.1109/PEITS.2008.23

distributed in the lower level, namely in the stations and service sections. Railway administration is usually in charge of a regional network. The transportation command center of MoR harmonizes railway administrations' operations and balances all the vehicle cars' distribution. Between each pair of adjacent longitudinal levels, there exist data flows in dual directions in TMIS. In addition, in order to realize trains formation, running and breakup processes, cargo loading and unloading processes, and passenger boarding and alighting processes, transverse data flows exist between the units or departments in the same level. Figure 1 simply shows the longitudinal and transverse data-flow processes.

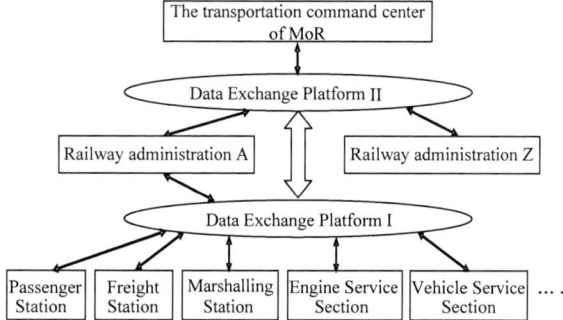

Figure 1. TMIS Data Flows

In Chinese "the Tenth Five-year plan" (From 2001 to 2005), the TMIS engineering projects have been completed. Now TMIS is being improved to support series of decision and operation systems, among which TDCS (Train Dispatching Command System) and CTC (Centralized Traffic Control System) are the most important.

1.2. Question specification

The rate of economic development and the growth in the regional economy translates directly in increases in demand for rail transport. With the expected rapid development of China, railway traffic will experience a significant increase. From the TMIS introduction we can see that the information service demand coming from transportation production is the key and direct element influencing network data traffic transmitted in TMIS. What is the relationship between the network data traffic and the service demand? How should the future TMIS network data traffic be estimated? Only by solving these questions, can the planner estimate the capacity of the TMIS communication network.

2. Methodology

In the entire data exchange platforms (Figure 1), the data traffic transmitted in platform I is the biggest,

because this level is directly in charge of dispatching trains and directing vehicle running, which needs real-time supervision. The information exchange relationship in platform I is shown in Tables 1.

Table 1. Origin-destination of TMIS data transmission on platform I

Station and section	Direction of data flow	Railway administration
Vehicle service section	↔	Vehicle dispatcher, Train operation dispatcher
Engine service section	↔	Engine dispatcher, Train operation dispatcher
Power service section	↔	Power dispatcher
Rail-building service section	↔	Train operation dispatcher, Station watcher
Marshalling yard	↔	Train operation dispatcher, Station watcher
Passenger department in station	↔	Passenger transportation dispatcher who is also in charge of baggage and package transportation, Passenger tickets-managing database
Freight department in station	↔	Freight transportation dispatcher who is also in charge of container transportation, Cargo tickets-managing database and container-managing database
Station watcher	↔	Train operation dispatcher
Vehicle/engine code scanner	→	Vehicle/engine code-identifying computer
Axle temperature detector	→	Axle temperature-managing computer
Container code detector	→	Container managing computer

The data exchange relationship in the other level is almost the same as in platform I, but the data contents are different. Data in platform II is mostly the statistical results and need not be transmitted in real-time. The data traffic transmitted in platform II is basically in proportion to the number of railway administration. However, the data traffic transmitted in platform I is not in proportion to the number of railway stations and railway service sections. In this paper we only solve the problem of data traffic transmitted in platform I. The problem related to platform II can be solved easily once the first question is solved.

In data exchange platform I, every minute there are so complex data transmitted. On the side of railway administration, dispatchers will communicate with stations or sections frequently. For example, the train dispatcher needs to issue dispatching orders to station staff and obtain information from the stations related to current vehicle status. If there is a marshalling station involved, the progress of trains being assembled must also be communicated. The freight transportation dispatcher needs to deal with freight cars distribution, knowing the progress of unloading and loading and freight flow status. The work progress of any production phase must be sent to the corresponding dispatchers and shared with relevant stations and sections at regular intervals. From the above, it can be seen that there exists many stochastic data exchanges in the transportation operation. In addition, the activities made by different people when dealing with the same work may be different. Hence, it is very difficult to define a function to describe the relationship between data traffic transmitted in real-time and the associated practical operation.

However, all TMIS functions are designed around the transportation production. Therefore, in a relative long period there may be some relationship between the sum data traffic transmitted and production scale or output of railway transportation. We hope to find the stable mapping relationship in data exchange platform I. If the stable mapping relationship is identified, we can forecast future network data traffic in the communication network according to the planning production scale and transportation output. Future communication network design capacity can be estimated based on the increasing rate of network data traffic.

It was decided to apply a multi-layer feed forward NN to form this forecasting model. NN has the ability to approximate a desired mapping relationship from training samples, in particular when there exist many complex influences and perhaps both linear and nonlinear relationships are present. Repeatedly training endows NN with the ability of "a black box"

3. Model Formulation

The model applied and described here is a two-layer fully connected feed forward NN, which has one input layer with multi-nodes representing independent variables; one hidden layer with sufficient multi-nodes; and one output layer with only one output node representing the dependent variable. Each node in the input layer receives a corresponding value of an input independent variable (x_i). The input independent variables are the factors influencing the network data traffic transmitted in the communication network. The

only dependent variable (y_h^2) output from the output layer represents the network data traffic transmitted.

Before NN application, the most important work is to endow NN with the identification function, which is the key to estimate the NN model. The procedure of NN learning such function is obtained through adjusting weights with sufficient samples. Therefore, a good learning rule is needed. Usually, the back-propagation (BP) learning rule is a valid method to calibrate the weights of a multi-layer feed forward NN. Here we assume that there are T samples namely T pairs of known input and output vectors $\{X(t), D(t)\}$ used to train NN. For a fully connected feed forward NN with K layer, besides inputting layer coded with 0, there are K layers coded with 1 to K. Then BP learning or training steps can be simply shown as in Figure 2.

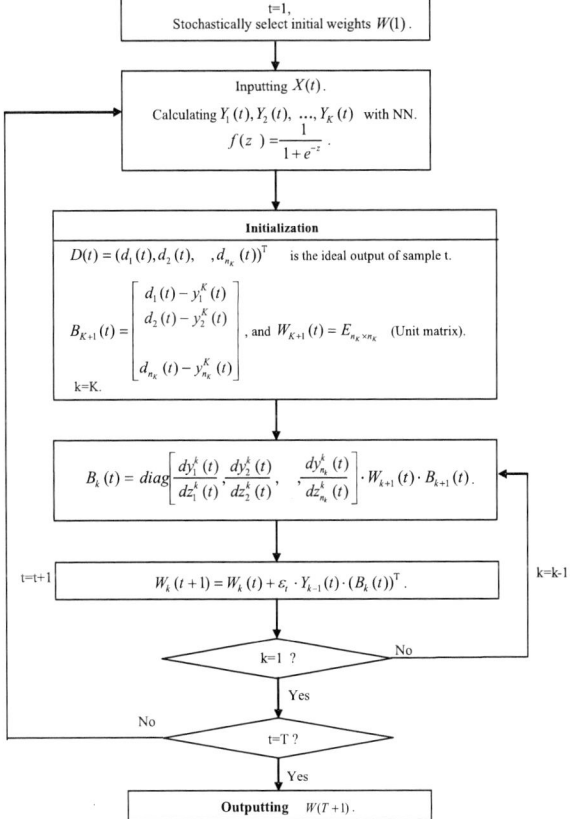

Figure 2. BP algorithm operation flow

Where:

t is the code of samples or code of learning steps, $t = 1, 2, \cdots, T$;

The weight matrix, input vector and output vector from 0th layer to 1st layer and from $(k\text{-}1)^{\text{th}}$ layer to k^{th} layer are expressed respectively as:

$$Z_1 = W_1^T X , \quad Y_1 = f(Z_1)$$
$$Z_k = W_k^T Y_{k-1}, \quad Y_k = f(Z_k), k = (2 \cdots K)$$

$W(1) = \{W_1(1), W_2(1), \cdots, W_K(1)\}$ are the initial weights stochastically selected;

$W(T+1)$ are the final weights outputted after being trained with T samples;

ε_t is a coefficient denoting learning efficiency of step t (here, ε_t is set as 1);

f, a Sigmond function, is selected as the activation function.

In each step (t), the process of weights adjusting begins from the last layer ($W_K(t)$) and ends at the first layer ($W_1(t)$). Therefore, the learning of network weights with all the training sets can be seen as a backward recursive procedure, which is also a characteristic of the BP algorithm.

4. NN implementation

Figure 3 shows the four steps used in the NN implementation. The main work is to decide the node number in the hidden-layer. In general practice, trial and error is usually applied[4]. Namely, we first set several candidate schemes (referred to as hidden-node schemes), each of which has a different number of nodes in the hidden layer.

Figure 3. The operation flow of NN

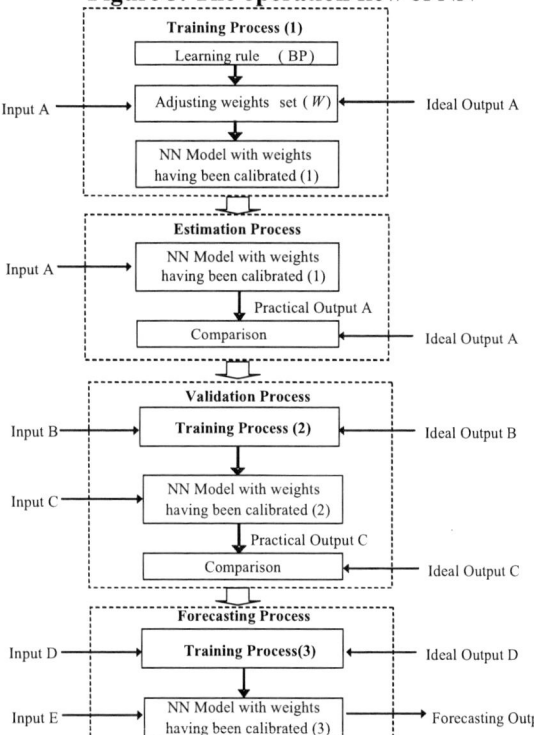

A series of steps will be performed to select the most adaptive one. All the hidden-node schemes must

be processed with the same sample set selected randomly from the full data set and compared under the same criteria.

4.1. Setting input and output valuables

The dependant variable output from the NN is the TMIS network data traffic, which can be directly measured from the ports of the servers working for the data exchange in platform I. The independent valuable input to NN are shown in Table 2. The influencing factors were selected according to the following two principles: the factors should be easily quantified and they should reflect the scale and output of railway transportation production.

Table 2. Independent valuables inputted to NN

Production scale-independent variables	Objects influenced
Number of marshalling stations	Train operation dispatching
Number of medium stations only set for trains operation	
Number of engine Service Sections	
Number of vehicle service sections	
Number of power service sections	
Number of rail-building service section	
Passenger turnover	
Freight turnover (Except containers)	
Number of freight train lines in train graph	
Number of passenger train lines in train graph	
Number of freight stations	Freight transportation dispatching
Freight bills	
Container turnover	
Number of passenger stations	Passenger transportation dispatching, passenger tickets
Passenger traffic	

In Table 2, the top eight independent variables mainly reflect the routine phase dispatching planning and statistical information. Whatever the station scale, the data size and format of the report handed up and the plan handed down are the same. The frequency of exchange between railway administrations and stations or sections in a given period is also the same for the same task. The other independent variables are selected to reflect the accumulation of the real-time data traffic.

In the Chinese railway system, the main transportation task is planned and assigned on a monthly basis. Six China administrations were selected and network data traffic from TMIS for recent four years were collected. The corresponding data listed in

Table 2 was also collected. Each data set consisted of 48 points (monthly for four years), giving a total of six data sets for each of the six administrations.

4.2. Training and Estimation Process

After setting the hidden-node schemes, the first step is the training process, whose goal is to minimize the total error for all examples in the training set (Figure 3). The learning algorithm, BP algorithm, needs to be used to adjust the weights set (W) recursively. Here we assume that there are P sample sets selected stochastically from the full data set to train the NN. In each sample set, there are N pairs of known input and output vectors. Each hidden-node scheme will be trained respectively and individually with the P sets. Following the weights calibration, each scheme has P weights sets ($W^1, W^2, \cdots, W^P$).

In the second step, the same P sample sets (Input A) used in the training process will be input into the NN again (Figure 3). This time the NN weight set has been calibrated and the error between the practical output and the actual output can be calculated. We will select the best-qualified scheme by comparing the errors of all the schemes.

For each scheme, the hidden-node scheme evaluation function can be expressed as equation 1, 2 and 3. The scheme with the smaller $G(W)$ and $G_{\max}(W)$ value will be selected.

$$G(W^i) = \left[\frac{1}{N} (\frac{D_i - Y_i}{D_i})^\mathrm{T} (\frac{D_i - Y_i}{D_i}) \right]^{1/2}, \quad (1)$$

$$G(W) = \frac{1}{P} \sum_{i=1}^{P} G(W^i), \quad (2)$$

$$G_{\max}(W) = \max(\left| \frac{d_{ij} - y_{ij}}{d_{ij}} \right|) \quad , \quad i \in (1 \cdots P) \quad \text{and}$$

$$j \in (1 \cdots N). \quad (3)$$

Here $D_i = (d_{i1}, d_{i2}, \cdots, d_{ij}, \cdots, d_{iN})$ is the actual output vector and $Y_i = (y_{i1}, y_{i2}, \cdots, y_{ij}, \cdots, y_{iN})$ is the practical output vector from the NN, corresponding to the sample set i;

N is the number of samples in each set;

P is the number of sample sets.

The estimation results are given in Table 3. The scheme with 30 nodes meets performs best at meeting the demand. It can be seen that in order to reach the stated level of precision enough number of samples are necessary to train NN. The hidden-node scheme with 30 nodes was used in the application described below.

Table 3. The estimation comparison of three hidden-node schemes with big sample set

Hidden-node Scheme	Set No.	$G(W)$ (%)		$G_{\max}(W)$ (%)	
		Practical	Criterion	Practical	Criterion
(a) 10 nodes	1	20.9	15	29.1	25
	2	18.5	15	25.7	25
	3	14.6	15	20.4	25
	Avg.	18.0	15	25.1	25
(b) 20 nodes	1	16.2	15	24.2	25
	2	12.8	15	21.9	25
	3	10.5	15	15.1	25
	Avg.	13.2	15	20.4	25
(c) 30 nodes	1	10.1	15	17.5	25
	2	13.7	15	19.6	25
	3	7.8	15	12.2	25
	Avg.	10.6	15	16.5	25

4.3. Validation process

In this step, the selected hidden-node scheme will be trained with new sample set (Input B in Figure 3) and a different sample set (Input C in Figure 3) will be selected to check the "best" hidden-layer scheme's validation. The purpose is to check the universality of the NN. The evaluation functions $G(W)$ and $G_{\max}(W)$ use equations 1, 2 and 3, except that the sample set changes and $P = 1$. If the values of $G(W)$ and $G_{\max}(W)$ accord with the criteria, the NN model with such hidden-node scheme will be applied.

The actual and the NN forecast values are compared in Figure 4. The error statistical data is given in Table 4. It was found that the forecast result can adequately satisfy the criteria. It was demonstrated that the NN trained with data from one set of regions, can be used to forecast traffic data for other regions.

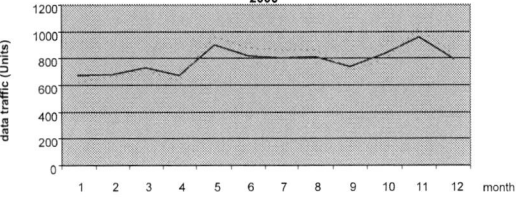

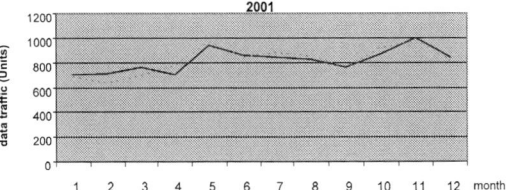

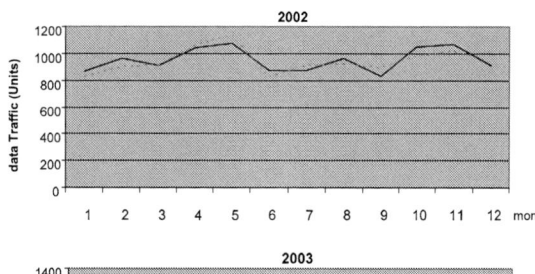

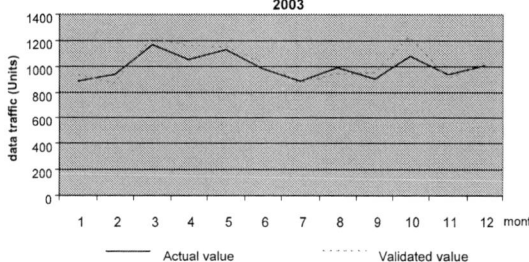

——— Actual value   ········ Validated value

Figure 4. Validation of the NN model using actual training data sets

Table 4. Error statistical analysis of Validation

Hidden-node Scheme	$G(W)$ (%)	$G_{\max}(W)$ (%)
30 nodes	4.98	13.4

Lastly, the validated NN can now be used to forecast future traffic data.

4.4. Forecasting process

The NN model was used to forecast the TMIS network data traffic of a railway administration in the west of China. The input values for the NN were defined according to the planning construction scale and railway traffic in the next five years. This data was available only on an annual basis. We estimated the average monthly data for the next five years, shown here as input E, in Figure 3.

The NN model was trained with the entire six sample sets together. The combination of these six sample sets can be seen as Input D in Figure 3. The average monthly network data traffics of the future five years have been forecasted as shown in Figure 5. After comparing the five average monthly network data traffics, the annual rate of increase in TMIS data traffic was estimated as shown in Figure 6.

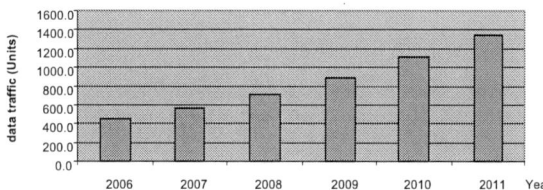

Figure 5. Average monthly network data traffic in the future five years

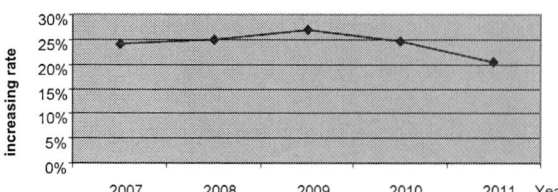

Figure 6. Yearly increasing rate curve of network data traffic

5. Conclusion

This paper offers an efficient method to forecast the network data traffic. The influences on network data traffic coming from transportation equipment scale and transportation production output have been considered. In this way, the planner can clearly grasp the requirement arising from a practical railway transportation production.

There are some issues which require further study and research. Although the NN model has basically satisfied the demand, the maximum absolute error, $G_{\max}(W)$, has reached near 17%. This figure should be reduced in the future. It is possible that other influencing factors not taken into account here also play a part.

6. Acknowledgement

The research was supported by Program for Young Excellent Talents in Tongji University.

7. References

[1] T. L. Magnanti, P. Mirchandani, R. Vachani, "Modeling and solving the two-facility capacitated network loading problem," *Operations Research*, vol.43, No.1, pp. 142-157, February 1995.

[2] A. Balakrishnan, T.L. Magnanti, R.T. Wong, "A decomposition algorithm for local access telecommunications network expansion planning," *Operations Research*, vol.43, No.1, 1995, pp. 58-76, February 1995.

[3] D. Wright., "Analysis of the market for access to broadband telecommunications in the year 2000," *Computers & Operations Research*, vol.25, No.1, pp. 127-138, January 1998.

[4] M. Mozolin, J.C. Thill, E. Lynn Usery, "Trip distribution forecasting with multilayer perceptron neural networks: A critical evaluation," *Transportation Research Part B*, vol.34, No.4, 53-73, January 2000.

Prediction of Logistics Amount Based on Neural Networks

Lianyu Wei, Zhaowei Liu

School of civil engineering , Hebei university of technology,Tianjin,300401,China
lzw-0606@163.com

Abstract

The logistics amount can be predicted in ANN (artificial neural networks) model. The influencing factors of the logistics amount and the basic thought of ANN prediction were analyzed. There are weakness existing in feedforward network and simple dynamic network prediction model. The methods together with dynamic feedback and causality was put forward and real time recurrent learning was used in order to make up the lack of the traditional method in the regional logistics system. The theory can be tested in the simulation example. The result get through the method is most close to the real situation and has high reliability.

Key works: neural networks, logistic amount, prediction, feedback system, traffic work

1. Introduction

Quantitative prediction methods which logistics forecasting commonly uses includes the analysis of the causal link of the regression model prediction and trend analysis of the time-series forecasting method extrapolation. There are some certain limitations and unsatisfactory points for the prediction methods mentioned above especially with logistics amount and the forecast impact on the logistics of the highly nonlinear factors unrecognized, predicting the outcome of a serious distortion requires the neural network methods logistics of the forecast, to make up the traditional method for the highly nonlinear problems.

Artificial neural network prediction is mainly for the logistics of a simple feed-forward model of causal relationship between forecasting method, using the simple model of the dynamic feedback of the current forecast only by the logistics of the past several years forecast, in theory there are a number of deficiencies, the two methods have certain limitations[1,2]. Combination of the two other systems in the prediction method has been applied in the forecast, but the logistics are not yet in the forecast system applications. Based on the purposes of the

forecasts for the regional logistics, we propose artificial neural network dynamic integrated forecasting methods and give specific methods to computation and simulation checking.

2. Analyze the main related causal factors of logistics

There are more relevant factors of regional logistics impact. Different logistics have a different major influencing factor, which could be divided into logistics issue, attracting of logistics and working capital to the total volume (the aggregate of non-issue and attracting).

Logistics issue (O) is the total of logistics amount which sent from the region to other field, the main influencing factors is the region's mineral resources, the level of economic development as well as the gap with outside the region. The related factors of issue are population, traffic mileage, the per-capita GNP, excess percentage of the added industrial value compared with the national average level, excess percentage of industrial investment compared with the national average level ,the percentage of mineral resources, export proportion of the vegetable and agriculture of the total output of the production, the proportion of industrial exportation of the total output of the production, and proportion of other agroforestry exportation. Here's exports outside the region, rather than abroad.

Logistics attracting volume (D) is the total of logistics amount which sent from the other field to region, the main affecting factors are the region's construction scale, the total population, the margin of the level of economic development between local areas and other field, the margin between the local and external consumption level and the degree of matching of related industries. The main influencing factors of logistics attract volume are the construction scale (proportion of construction of gross national product), the degree of the mineral information accord with local processing industry, the total of industrial output, population, the level of all the local related processing industries, the per capita consumption, the total of consuming goods of social retail sales and the ratio of agricultural production in the GNP.

The total Logistics turnover reflects temporary storage, processing volume which is needed in the process of the goods flow in a region, and is a decision-making basis as a regional logistics park and logistics centre construction, and mainly reflects the scale of the situation in the exchange of logistics between the regional economy and the other field economic. The main related factors of the total turnover are population, the level of the per capita consumption, and the added value percentage of industry augments profits compared with the national average level, and the added value percentage of industry profits compared with the national average level ,the total industrial output value, the proportion of vegetables agricultural "export" yield of the total output, the proportion of industrial "export" of the total output of the production, the proportion of other agroforestry "export" of the total output.

3. The basic idea of Artificial Neural Network Forecasting

At present, there are two kinds of methods for artificial neural network forecasting logistics.

(1) Feedforward prediction methods. A need to predict the logistics of $Y(t)$, all relevant factors by $X(t)=[x_1(t), x_2(t),..., x_m(t)]^T$ said its mathematical formula for

$$Y(t) = f[x_1(t), x_2(t),...,x_m(t)]^T, \quad M= 8 \text{ or } 9$$

The formula is a highly nonlinear model, the need for a feed-forward neural network model, BP algorithm to determine weight, by calculating to forecast the specific logistics[3,4], the specific calculation process is no longer discussed. The method does not consider the timing, it is simple and easy to model, but does not achieve the dynamic prediction.

(2) Simple time series prediction. One effective way is to try to find some experience of the time series observations and can be described in nonlinear multivariable function

$$y(t)= f[y(t-1),y(t-2),...,y(t-n)]$$

Among them, taking $y(t)(t=N,N-1,...,n)$ as a given time series sample value; f is a highly non-linear function, if f is a linear function, predict in the traditional AR model. The prediction needs a large number of nonlinear neural networks via samples training, and neural network can approximate arbitrary nonlinear continuous function, f is very flexible. When all the known value is the input for the neural network, the neural network is only step forecasts, in order to use neural networks for multi-step prediction, $y(t)$ predictive value is backed to the input, the corresponding data from other step shift, so predictable $y(t+1)$, and so on. Figure 1 is a block diagram of the training state. Since the methods have not been taken into account and when a large number of specific causal factors change, in particular the development of China's economy unconventional application, forecast is less effective.

Considering the advantages and disadvantages of the two methods, they will be integrated to consider two methods, namely, a comprehensive forecasting method.

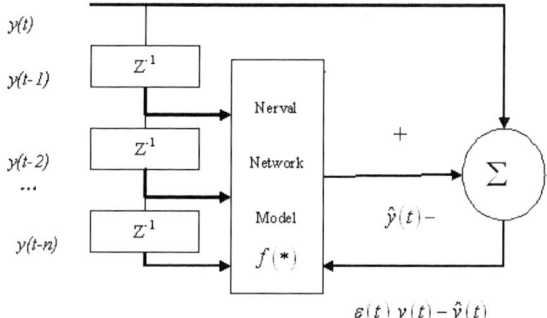

Figure 1. Neural network training logistics forecast state block diagram

4. Dynamic Integrated forecast of the logistics system

Considering X factors and taking into account factors affecting the dynamics of demand and historical time series data, they compose a comprehensive forecasting method [5]. The simplest form to forecast Y logistics demand is launching the various influencing factors and the logistics of the past state into space after the former gave to the network, the basic mathematical model structure

$$y(t+1)=f[x(t),x(t-1),...,x(t-p),$$
$$y(t-1), y(t-2)...,y(t-q)]$$

$x(t-j)$ is affecting factors vector, as $f(\bullet)$ can be any function of the nonlinear, so also called nonlinear model ARMA (NARMA), which is also known as the output delay NN model of the feedback network. Figure 2 shows the prediction system for the identification system diagram, switch S in the above situation is as serial and parallel model due to taking output of unknown system as the input of the network,

210

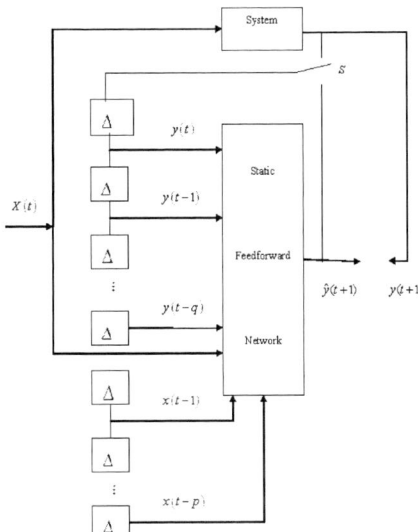

Figure 2. Logistics System delay unit forecast network diagram

the NN system after this study is still not fully represented, but the convergence of learning is better; switch S below is known as parallel model, the learning system can be fully representative system, but the convergence of learning cannot proved. Clearly the calculation of the model structure is still used BP algorithm, but it has historical data entered into the simulation system, in order to study and work offline process, they can only predict next year's data, but cannot they predict time-varying dynamic integrated forecast.

To solve this problem, we need to introduce feedback model [6], the network structure can be expressed in Figure 3, the algorithm called real-time recursive algorithm. In Figure 3, M(8 or 9) affecting logistics demand factors are plus input, using $x_i(t)$ as the i factors for the impact of logistics demand in t time (or n moment); system directly output variables is the logistics demand volume, the number is $c=1$; According output variables, taking into the needs of model, taking $N=6$ as calculating the number of units; including output modules and the number of hidden units, $N-c=N-1=5$ is the number of hidden units, $y(t)$ shows, including feedback, N-dimensional vector output, at time t, which c (1)-dimensional variables for the logistics needs. $X(t)$ and $y(t)$ together into M+N dimensional vector of input $u(t)$, A stands plus input which is a collection of various factors, B stands a joint input layer and a handling layer which are from the output feedback network in the process of input, a total of MN former connectivity and N^2 feedback link. There are N self feedback connections. W denotes $N(N+M)$ weights matrix. When we consider the threshold, we should give -1 to one of the m input.

The model used real-time algorithm is recursive algorithm, the specific steps for training

（1）$t=0$ the right can be given to the initial uniform distribution of random numbers, given the requirements of accuracy and other initial conditions. From $t=0$, t on each step, calculate the N output neurons in accordance with the following equation dynamic, which obtains the input $u_i(t)$.

Input value for Unit j in t time.

$$v_j(t) = \sum_i \omega_{ji}(t) u_i(t)$$

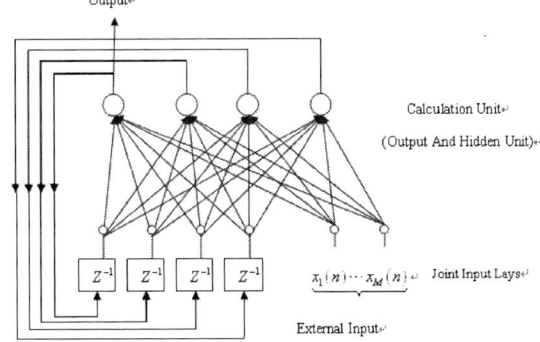

Figure 3.

The next moment the output j

$$y_j(t+1) = \varphi\left[v_j(t)\right]$$

(2) All j, k, l, calculated according to initial conditions $\prod_{kl}^j(t)$

$$\prod_{kl}^j(t+1) = \varphi'\left[v_j(t)\right] \bullet \left[\sum_{i \in B} \omega_{ji}(t) \prod_{kl}^j(t) + \delta_{kj} u_l(t)\right]$$

Among $\prod_{kl}^j(0) = 0$.

(3)Based on the calculation $\prod_{kl}^j(t)$ of the error signal $e_j(t) = d_j(t) - y_j(t)$, calculate the amendment of weights.

$$\Delta\omega_{kl}(t) = \eta \sum e_j(t) \prod_{kl}^j t$$

Among them, η as the learning step

(4)Revised weights $\omega_{kl}(t)$

$$\omega_{kl}(t+1) = \omega_{kl}(t) + \Delta\omega_{kl}(t)$$

In the above calculation, every sample modifies a weight, it uses the instantaneous error gradient, and this can be achieved when mixing forecast of the logistics under the dynamic state.

5. Calculation Example

Above forecast model and real-time recursive algorithm are applied to the logistics of a dynamic prediction [1], using Borland C programming for system simulation. Taking into the other training samples is a matrix of 50 9×5 (or 8×5), the number is very large; here only list a group of samples of the normalized data (Table 1, Table 2, Table 3).

Through computer simulation, finally receive dynamic weight between logistics issue, logistics attract volume and the total volume of the input layer and the hidden layer in working capital, here only excerpts forecast 2001 when the weights were shown in table 4, Table 5 and Table 6.

Table 1. The logistics of issuing samples

Year	1996	1997	1998	1999	2000
1	0.584	0.601	0.625	0.647	0.668
2	0.365	0.392	0.418	0.452	0.487
3	0.471	0.483	0.498	0.512	0.538
4	0.281	0.294	0.317	0.343	0.371
5	0.408	0.421	0.436	0.452	0.471
6	0.603	0.581	0.558	0.534	0.512
7	0.543	0.551	0.549	0.565	0.576
8	0.356	0.369	0.381	0.395	0.407
9	0.528	0.536	0.543	0.563	0.576
10	0.509	0.524	0.532	0.548	0.569

Table 2. The logistics of attracting samples

Year	1996	1997	1998	1999	2000
1	0.577	0.589	0.603	0.611	0.625
2	0.759	0.771	0.788	0.802	0.817
3	0.611	0.634	0.657	0.672	0.699
4	0.584	0.601	0.625	0.647	0.668
5	0.753	0.761	0.768	0.775	0.788
6	0.488	0.492	0.496	0.503	0.511
7	0.657	0.669	0.675	0.683	0.692
8	0.435	0.443	0.451	0.458	0.465
9	0.502	0.518	0.531	0.549	0.563

Table 3. Logistics total turnover samples

Year	1996	1997	1998	1999	2000
1	0.584	0.601	0.625	0.647	0.668
2	0.488	0.492	0.496	0.503	0.511
3	0.281	0.294	0.317	0.343	0.371
4	0.408	0.421	0.436	0.452	0.471
5	0.611	0.634	0.657	0.672	0.699
6	0.543	0.551	0.549	0.565	0.576
7	0.356	0.369	0.381	0.395	0.407
8	0.528	0.536	0.543	0.563	0.576
9	0.508	0.521	0.534	0.551	0.578

Further simulation was issued in 2001, the logistics issue $Y_1(2001)=0.501$, logistics attract volume $Y_2(2001)=0.623$, the total logistics $Y_3(2001)=0.531$. After normalization of the formula, $Y_1(2001)=70060\times10^4$ t, the logistics attract volume $Y_2(2001)=77380\times10^4$ t, the total logistics $Y_3(2001)=71860\times10^4$ t.

From the actual value, in 2001 the region issued amount is 0.521, attract capacity is 0.631; the total of logistics is 0.538 (all the data is after the normalization). The comparison between forecast and the actual value indicates that the application of neural networks for the logistics of the forecast has high accuracy.

According to the method, this paper forecasts all the logistics after 2002. Predictive value of logistics issue from 2002 to 2011 are 77770×10^4 t, 82120×10^4 t, 95260×10^4 t, 101190×10^4 t, 103580×10^4 t, 108400×10^4 t, 118050×10^4 t, 129440×10^4 t, 142380×10^4 t, 154670×10^4 t; predictive value of logistics attract are 79860×10^4 t, 86640×10^4 t, 91290×10^4 t, 95410×10^4 t, 95260×10^4 t, 98740×10^4 t, 107330×10^4 t, 117530×10^4 t, 126460×10^4 t, 137210×10^4 t; predictive value of Total turnover of the logistics are 76990×10^4 t, 85890×10^4 t, 90710×10^4 t, 101590×10^4 t, 111550×10^4 t, 124450×10^4 t, 136390×10^4 t, 142120×10^4 t, 153210×10^4 t, 163160×10^4 t. We can see that the demands for the logistics which region's economic development need show steady growth, which the increase of the logistics issue volume gradually overtop the logistics attracted volume, logistics turnover as economic development is increasing, and more than a logistics capacity and attract volume.

Table 4. Forecast the weight of the logistics issue in 2001 (6×15)

0.071	0.648	0.305	0.483	0.432	0.594	0.436	0.468	0.121	0.243	0.567	0.834	0.325	0.545	0.574
0.177	0.877	0.259	0.018	0.492	0.753	0.716	0.830	0.312	0.937	0.076	0.442	0.504	0.548	0.687
0.425	0.318	0.564	0.782	0.845	0.806	0.153	0.448	0.445	0.428	0.490	0.562	0.254	0.678	0.392
0.078	0.566	0.105	0.406	0.409	0.056	0.009	0.779	0.940	0.612	0.596	0.930	0.687	0.505	0.631
0.517	0.736	0.065	0.310	0.147	0.579	0.085	0.162	0.569	0.823	0.375	0.697	0.244	0.574	0.547
0.854	0.362	0.278	0.694	0.512	0.305	0.563	0.405	0.453	0.764	0.203	0.409	0.024	0.723	0.962

Table 5. Forecast the weight of the logistics attract in 2001 (6×14)

0.357	0.544	0.214	0.381	0.392	0.465	0.565	0.617	0.753	0.957	0.785	0.746	0.456	0.729
0.248	0.477	0.654	0.305	0.587	0.744	0.947	0.253	0.634	0.374	0.875	0.416	0.862	0.784
0.201	0.268	0.308	0.578	0.174	0.858	0.469	0.354	0.506	0.722	0.504	0.032	0.357	0.806
0.565	0.254	0.036	0.145	0.454	0.345	0.217	0.951	0.385	0.304	0.886	0.714	0.388	0.183
0.487	0.364	0.147	0.559	0.621	0.584	0.602	0.572	0.896	0.951	0.245	0.525	0.397	0.953
0.561	0.642	0.335	0.051	0.364	0.539	0.704	0.359	0.615	0.629	0.785	0.908	0.468	0.147

Table 6. Forecast the weight of the logistics turnover in 2001 (6×14)

0.334	0.543	0.123	0.418	0.823	0.364	0.841	0.395	0.294	0.412	0.546	0.362	0.358	0.632
0.564	0.202	0.503	0.532	0.947	0.603	0.125	0.276	0.903	0.225	0.287	0.305	0.216	0.904
0.641	0.142	0.482	0.697	0.365	0.076	0.357	0.482	0.644	0.451	0.284	0.254	0.651	0.574
0.576	0.746	0.589	0.243	0.604	0.806	0.268	0.307	0.456	0.554	0.396	0.321	0.254	0.643
0.242	0.412	0.315	0.061	0.552	0.247	0.729	0.261	0.276	0.354	0.451	0.159	0.305	0.071
0.507	0.694	0.645	0.489	0.848	0.724	0.668	0.854	0.346	0.029	0.354	0.052	0.506	0.301

6. Closing remarks

At present, the neural network system for the other prediction are more and dynamic feedback model for the forecast of logistics is not used. In this paper, feedback model and real-time feedback recursive model are used to calculate and predict logistics capacity and improve the net, BP algorithm and the shortcomings of logistics forecasting logistics are integrated dynamic non-linear forecasting models. Compared with the simulation results and the actual results, the model has a high reliability.

References

[1] Yan Pingfan, Zhang Changshui. Artificial neural network and simulation evolution calculation [M]. Beijing: Tsinghua Publishing House.2000.

[2] Draye J S. Dynamic recurrent NN:A dynamical analysis[J]. IEEE Trans SMC(B), 1996, 26:692-706.

[3] Fariborz Y, Partovi J B. Timing of monitoring and control of CPM project[J]. IEEE Transactions on Engineering Management, 1993, (1); 68-75.

[4] Singelmann H T. Computational capabilities of recurrent NARX NN[J]. IEEE Trans SMC(B), 1997, 27: 208-214.

[5] Williams R J. A learning algorithm for continually running fully recurrent NN[J]. Neural Computation, 1989, (1);270-280.

[6] Galicki M. Learning continuous trajectories in recurrent neural networks with time-dependent weights [J]. IEEE Trans NN，1999, 10: 741-755.

[7] Hebei Province People's Government. Hebei Economic Yearbook (2002 year)[M]. Beijing: China Statistics Press, 2002.

2008 Workshop on Power Electronics and Intelligent Transportation System

A hybrid neural network-based IE and IMM architecture for target tracking

Rong Jian and Wang Xiu
School of Physical Electronic
University of Electronic Science
and Technology of China, Chengdu,
610054

Zhong Xiaochun
Dept. of Applied Physics
Southwest Jiaotong University
Chengdu , 610031, China

Zhang Haitao
School of Physical Electronic
UESTC, Chengdu, 610054
tianmen822@yahoo.com.cn

Abstract

In order to enable a tracking system to work stably in the environment with fast maneuver and rapidly changing noise, a new hybrid architecture combining Interacting Multiple Model (IMM) and neural network-based Input estimate (IE) together is presented in this paper. In this architecture, IMM provides estimation of covariance of measurement noise to neural network-based IE, while IE enables the system to work effectively when the targets lead fast and complex maneuver, both of the outputs of IMM and NNIE will be fused in fusion module. In order to verify the effectiveness of this architecture, several simulations were leaded and the results prove it can work stably with rapidly changing noise and fast maneuver.

Keywords: IMM, Neural network-based IE, Target Tracking

1. Introduction

A good tracking algorithm or architecture is very important in the target tracking system including cooperative and uncooperative target tracking. In the real tracking scenario, the target can lead fast and complex maneuver, and also the process noise and measurement noise might change from time to time.

A lot of approaches have been developed to track targets, during all of the tracking approaches, Interacting Multiple Model (IMM) [1] and Input Estimate (IE) [2] have received great popularity because their practicality. But they still have some shortages to be conquered. For example, IMM can not work effectively when targets lead a fast and complex maneuver and IE works on the condition that the covariance of measurement noise and process noise are available. The previous limits of IMM and IE approaches call for some improvement. Luckily, in 1997 a neural network-based IMM [15] was introduced to solve the problem of tracking targets with fast and complex maneuver. But this approach also needs to know the covariance of measurement noise and process noise.
Noise identification has been a very active area in the past forty years, and it can be classified into four categories: Bayesian approach [3, 4, 5], maximum likelihood estimation [6, 7], covariance-matching techniques [8, 9], and correlation methods [11, 12]. But all existing methods are only valid for stationary noise or noise with slowly

varying statistics. Luckily, in 1994, X RONG LI, and YAAKOV BAR-SHALOM [14] introduced to use IMM approach to identify measurement and process noise which are valid for rapidly varying noise.

In that case, this paper presents a new architecture that combine IMM and neural network-based IE to enable the tracking system to work stably in the environment with fast maneuver and rapidly changing noise. IMM provides estimation of covariance of measurement noise to IE, while IE enables the system to work effectively when the targets lead fast and complex maneuver, both of the outputs of IMM and NNIE will be fused in a fusion module.

The remainder of this paper is organized as follows. Section II describes the details of this new architecture. In order to test the effectiveness of this architecture, in section III an illustrative example is outlined and the results are discussed. Finally, a conclusion of this work is given in section IV.

2. Description of new architecture

2.1. Analysis of IMM

Interacting Multiple Model (IMM) is one of the most popular algorithms in target tracking system. It can effectively track targets even when no movement model of the target is known in advance. Simple system model can be described as follows:

$$X_j(k)=F_j(k-1)X_j(k-1)+G_j(k-1)w_j(k-1) \quad \forall j \in M \quad (1)$$

$$z(k) = H_j(k)X_j(k) + v_j(k) \quad \forall j \in M \quad (2)$$

where M is the number of the models, $X_j(k)$ is the state at time k, $w_j(k)$ and $v_j(k)$ are covariance of process noise and measurement noise at time k, with covariances of $Q_j(k)$ and $R_j(k)$ respectively. One IMM system usually has several models, and every model has its' own probability of being effective at time k, denote as $\mu_j(k)$. The probability of jumping from one model to another model is governed by a first-order homogeneous Morkov Chain

$$P\{m_j(k+1) \mid m_i(k)\} = p_{ij} \quad \forall i, j \in M \quad (3)$$

where m(k) is the modal state at time k. The IMM algorithm is a recursive one. Each of cycle consists of four major steps: interaction (mixing), filtering, mode

978-0-7695-3342-1/08 $25.00 © 2008 IEEE
DOI 10.1109/PEITS.2008.28

214

update, and combination, as illustrated in Fig. 1 for a two-mode case. In each cycle, the initial condition for the filter matched to a certain mode is obtained by interacting (mixing) the state estimates of all filters at the previous time under the assumption that this particular mode is in effect at the current time. This is followed by a regular filtering (prediction and update) step, performed in parallel for each mode. Then a combination (weighted sum) of the updated state estimates for all filters yields the state estimate. The probability of a mode being in effect plays a key role as the weighting of the interaction (mixing) and the combination of states and covariances. Specifically, the mode probabilities are given by

$$\mu_j(k) = \frac{1}{c(k)} \sum_i p_{ij} \mu_i(k) N[v_j(k); 0, S_j(k)] \qquad (4)$$

where $c(k)$ is normalization constant, $v_j(k)$ and $S_j(k)$ are residual and its covariance, respectively at time k for the filter matched to mode j. More specific details of IMM can refer to [13].

In 1994, X RONG LI and YAAKOV BAR-SHALOM [14] introduced to use IMM to estimate the covariance of measurement noise even when it varies quickly. The covariance of the measurement noise can be attained with

$$R = \sum_i \mu_i(k) R_i(k) \qquad (5)$$

Unfortunately, tracking system using this algorithm might lose a target which leads a short-term maneuver and flies at a high speed. This shortage can be complemented when it cooperates with neural network-based which will be introduced next.

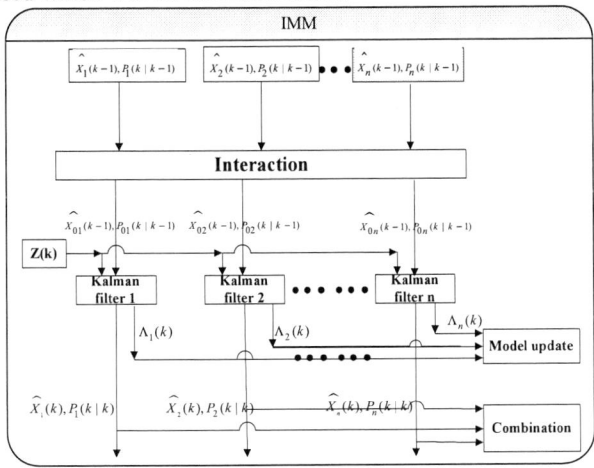

Figure 1. Major steps of cycle of IMM algorithm

2.2. Analysis of Neural network-based IE

Input Estimate is another famous approach in target tracking field, which uses a unique system model as follows:

$$X(k) = F(k)X(k-1) + Bu(k) + G(k-1)w(k) \qquad (6)$$

$$z(k) = H(k)X(k) + v(k) \qquad (7)$$

which have one more item $u(k)$ representing the acceleration of a target comparing with formula 1 and 2. A lot of methods have been developed to compute $u(k)$ online, neural network-based IE [15] is one of the effective methods which can adjust $u(k)$ online and are capable of reliably tracking even uncooperative targets executing fast and complex maneuvers in a number of applications. The basis of using neural network to abstract acceleration parameter is that when there is no maneuver, the mean of innovation sequence (a group of data consists of past innovation value) equals to zero, but when there is a maneuver, it no longer equals to zero.

Selection of features which input into neural network are very important to effectively abstract the acceleration, there are three basic entities that help acquire a good estimate of target maneuver: These are (1) intensity of acceleration, (2) direction of tangential velocity, and (3) initial velocity at the time of maneuver.

But this approach has a precondition that the system has a known covariance of measurement noise and process noise. But they are usually unavailable and vary along with time in the real tracking environment. If there is an approach which can offer the covariances of measurement noise and process noise, neural network-based IE will become an approach with high adaptability. So the idea of using IMM approach to estimate the covariance of measurement noise which can be adopted by neural network-based IE and also fusing the result from both of the methods is natural.

2.3. new architecture

Neural network-based IE can work in the tracking environment with fast and complex maneuver while IMM can not with the precondition that the covariance of measurement noise and process noise are available, so when they are not available, this approach might lose targets. Although there are a lot of methods which can estimate the covariance of measurement noise, most of them can only work correctly when the measurement noise changes smoothly or keeps constant. Luckily, IMM can offer the estimate of covariance of measurement noise even when the measurement noise is changing rapidly. In that case, we design a new architecture shown in figure 2. In this architecture, there are three modules: IMM module, neural network-based IE module and fusion module. 1) The IMM module takes the measurement value of sensor $z(k)$ as an input and outputs estimates of covariance of measurement noise $R(k)$, state estimate $\hat{x}(k)$ and the covariance matrix of the prediction error $P(k|k)$. 2) The Neural network-based IE takes $z(k)$ and $R(k)$ as

215

inputs, and outputs $\hat{x}(k)$ and $P(k\,|\,k)$. In this module, neural network has three inputs [15]:

v_1 : intensity of acceleration, defined as follows:

$$v_1(k) = \frac{r_x^2(k)}{S_{xx}(k)} + \frac{r_y^2(k)}{S_{yy}(k)} \qquad (8)$$

where $r(k) = [r_x r_y]$ is innovation sequence

$$r_k = (z_k - F_k \hat{x}_k^-) \qquad (9)$$

and $S_{xx}(k)$ and $S_{yy}(k)$ are the diagonal elements of covariance matrix of $r(k)$:

$$S(k) = HP(k\,|\,k-1)H^T + R \qquad (10)$$

v_2 : change in heading, defined as follows:

$$v_2(k) = \alpha_{LT}(k) - \alpha_{LT}(k-1) \qquad (11)$$

where $\alpha_{LT}(k)$ and $\alpha_{LT}(k-1)$ are heading estimate computed with past data points (i.e., N equals to three) at sampling instants k and (k-1) respectively,

$$\alpha_{LT}(k) = \mu[\sum_{i=1}^{N}(y_i - \overline{y})^2 / \sum_{i=1}^{N}(x_i - \overline{x})^2]^{1/2} \qquad (12)$$

where

$$\mu = \operatorname{sgn}\sum_{i=1}^{N}(y_i - \overline{y})(x_i - \overline{x}) \qquad (13)$$

$$\overline{x} = (1/N)\sum_{i=1}^{N}x_i \qquad (14)$$

$$\overline{y} = (1/N)\sum_{i=1}^{N}y_i \qquad (15)$$

v_3 : covariance of measurement noise $R(k)$.

3) The fusion module takes both of the outputs of IMM and neural network-based IE as inputs and outputs fusion estimate $\widehat{X}_f(k)$ using the rule of weighted averaging algorithm.

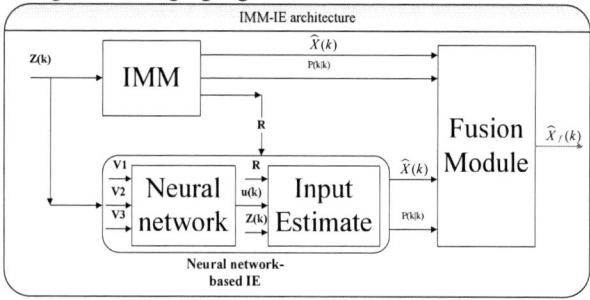

Figure 2. New architecture combining IMM with NN-IE

3. Simulation and performance evaluation of the architecture

In order to demonstrate the effectiveness of this architecture, a simulation is presented in this section. This simulation system has a sampling time of 1s. In IMM module, two pairs of models are used. Each pair has one CV (constant velocity) model and two current statistical models (CSM) with different measurement noise. Specific

parameters of CV and current statistical model are shown in table 1.

Table 1:

	CV I	CV II	CSM I	CSM II	CSM III	CSM IV
R	0.1	10	0.1	10	0.1	10
Q	$\begin{bmatrix} \frac{01}{3} & \frac{01}{2} \\ \frac{01}{2} & 1 \end{bmatrix}$	$\begin{bmatrix} \frac{01}{3} & \frac{01}{2} \\ \frac{01}{2} & 1 \end{bmatrix}$	------	------	------	------
α	---------	-----------	0.1	0.1	0.01	0.01

where Q , R and α are covariance of process noise, covariance of measurement noise and acceleration time constant respectively. We assume the maximum acceleration equals to $20\,m/s^2$. The covariance of process noise of CSM is decided by α and sampling time.

Neural network has three inputs and two outputs. The number of neutron in the hidden layer can only be decided after training. In order to have a good accuracy of estimating acceleration, we created 800 training data by simulation as following design:

Acceleration: range from -20 to 20, divided into 40 levels;

Covariance of measurement noise: range from 0.1 to 10, divided into 20 levels.

The training result shows 35 neurons are enough for the hidden layer.

Suppose a target is moving with the initial velocity of 10m/s, the covariance of measurement noise of sensor changes at instants 15 from 2 to 5 and then jumps from 5 to 8 at instants 18. The result of position error at x coordinates is shown in figure 3. From the result, we can tell when measurement noise changes, the estimate error of x coordinates offered by neural network-based IE reaches 15m while IMM have relatively stable result. The fusion of these two approaches is also good.

Another simulation was designed with the initial velocity of 20m/s and acceleration of $2\,m/s^2$. Then a maneuver happens at instants 10 with acceleration of $10\,m/s^2$, and it lasts for 3 sampling time, then the acceleration turns to be $5\,m/s^2$ and keep constant till end. The result of position error is given by figure 4, and it tells the truth that when fast maneuver happens, NN-IE is relatively stable than the IMM, and the fusion result is also acceptable.

These two results prove that this new architecture can work stably even when there are fast changing noise and fast maneuver.

216

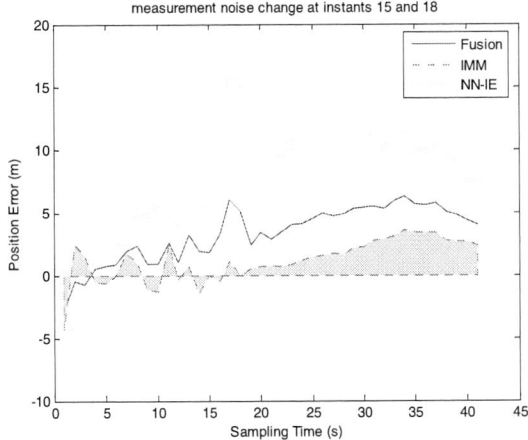

Figure 3. Comparison of position error of x coordinates offered by IMM, NN-IE and new architecture with fast changing measurement noise.

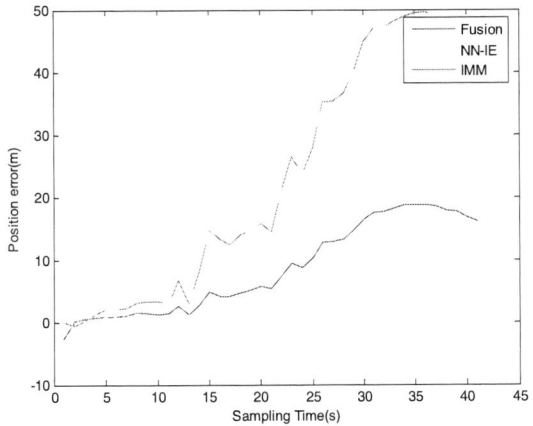

Figure 4.Comparison of position error of x coordinates offered by IMM, NN-IE and new architecture with fast maneuver.

4. Conclusion

The main purpose of this paper is to present a hybrid architecture to enable a tracking system to work effectively with fast changing noise and quick maneuver. The simulation results prove it can work stably.

5. Acknowledgement

This study is supported by the National Natural Science Foundation of China, NNSF number is 60572079.

6. References

[1] Blom H A P, "An efficient filter for abruptly changing systems", Proceedings of 23rd Conference on Decision and Control. IEEE, Las Vegas:,1984, pp.656-659

[2] Y.T.Chan, ATC.Hun, A.G.C. Hu, J.B. Plant, "A Kalman Filter Based Tracking Scheme with Input Estimation", Transaction on aerospace and electronic systems, pp. 237-244, 1979

[3] Alspach, D. L, "A parallel filtering algorithm for linear systems with unknown time varying noise statistics", IEEE Transactions on Automatic Control, AC-19, Oct. 1974, pp. 552-556.

[4] Alspach, D. L., and Abiri.A, "A Bayesian solution to the problem of the state estimation in an unknown noise environment", International Journal of Control, 19, 1974.

[5] Scharf, L. L., and Alspach, D. L, "Nonlinear state estimation in observation noise of unknown covariance", In Proceedings of 1976 Joint Automatic Control Conferences, Purdue University, West Lafayette, IN, July 1976, pp.637-641.

[6] Bohlin, T, " Four cases of identification of changing systems", In Mehra, R. K., and Lainiotis, D. G. (Eds.), System Identification: Advances and Case Studies. New York Academic Press, 1976.

[7] Smith, P. L., "Estimation of the covariance parameters of nonstationary time-discrete linear systems", In Proceedings of 2nd Symposium on Nonlinear Estimation Theory and Its Applications, San Diego, CA, Sept. 1971, pp.325-328.

[8] Hagar, H., and Bpley, B. D. "A sequential filter for estimating state noise covariances", In Proceedings of 4th Symposium on Nonlinear Estimation Theory and Its Applications, San Diego, CA, Sept. 1973, pp. 98-101.

[9] Meyers, K. A., and Tapley, B. D., "Adaptive sequential estimation with unknown noise statistics", IEEE Transactions on Automatic Control, AC-21, 4, Aug.1976, 52W23.

[10] Mehra, R. K. "On the identification of variances and adaptive Kalman filtering," IEEE Transactions on Automatic Control, AC-15, 2, Apr.1970, pp.175-184.

[11] Sinha, N. K., and Tom, A., "Adaptive state estimation for systems with unknown noise Covariances," International Journal of Systems Sciences, 18, 4(1977), pp.377-383.

[12] Hampton, R. L. T. (1973) On unknown state dependent noise, modeling errors and adaptive filtering," In Proceedings of 4th Symposium on Nonlinear Estimation Theory and Its Applications, San Diego, CA, Sept. 1973, pp.102-106..

[13] Henk A. P. Blom and Yaakov Bar-Shalom, "The Interacting Multiple Model Algorithm for Systems with Markovian Switching Coefficients", IEEE Transactions on automatic control. 33, No. 8, August 1988

[14] X Rong Li and Yaakov Bar-Shalow, "A Recursive Multiple Model Approach to Noise Identification", IEEE Transaction on aerospace and electronic systems" , July 1994, vol. 30, No. 3.

[15] Malur K. Sundareshan, Farid Amoozegar, "Neural network fusion capabilities for efficient implementation of tracking algorithms", Society of Photo-Optical Instrumentation Engineers, 1997, pp. 692–707.

A Forecasting Model Based Support Vector Machine and Particle Swarm Optimization

Qi WU, Hong-Sen YAN, Hong-Bing Yang

Key Laboratory of Measurement and Control of Complex Systems of Engineering, Ministry of Education, Southeast University, Nanjing, Jiangsu, 210096, China
wuqi7812@163.com

Abstract

In view of the bad forecasting results of the standard ε-support vector machine (SVM) for product sale series with the normal distribution noise, a SVM based on the Gaussian loss function named by $g-SVM$ is proposed. And then, a hybrid forecasting model for product sales and its parameter-choosing algorithm are presented. The results of its application to car sale forecasting indicate that the short-term forecasting method based on $g-SVM$ is effective and feasible.

1. Introduction

However, the standard SVM encounters some difficulty in real application. Some improved SVMs have been put forward to solve the concrete problem [1]. The standard SVM adopting ε-insensitive loss function has good generalization capability in some application [2-3]. But it is difficult to deal with the normal distribution noise parts of series. Therefore, the main contribution of this paper focuses on the establishment of a new SVM that can deals with the Gaussian noise parts of series.

For this end, combining the principal characteristics of product sale forecasting in manufacturing industries, a hybrid model is proposed, on the basis of SVM with Gaussian loss function called g-SVM and PSO with chaotic mapping (ECPSO).

The g-SVM is described in Section 2. Section 3 provides a new PSO called embedded chaotic PSO (ECPSO) to obtain the optimal parameters of g-SVM. And then gives the hybrid forecasting model based on the g-SVM and ECPSO. An application in car sale forecast is given, and then g-SVM is compared with the standard v-SVM and ARMA. Section 5 draws some conclusions.

2. The v-support vector machine with Gaussian loss function

For standard v-SVM, it is difficult to deal with the normal distribution noise of time series. To overcome the shortage of ε-insensitive loss of standard v-SVM, Gaussian function is selected as the loss function of v-SVM. Then, the optimal problem inhibiting the normal distribution noise of time series is described as follows:

$$\min_{w,b,\xi^{(*)},\varepsilon} \quad \tau\left(w,\xi^{(*)},\varepsilon\right)=\frac{1}{2}\|w\|^2+C\cdot\left(v\cdot\varepsilon+\frac{1}{2}\sum_{i=1}^{l}\left(\xi_i^2+\xi_i^{*2}\right)\right)$$

$$\text{s.t.} \begin{cases} \left(w\cdot x_i+b\right)-y_i \leq \varepsilon+\xi_i \\ y_i-\left(w\cdot x_i+b\right) \leq \varepsilon+\xi_i^* , \\ \xi^{(*)} \geq 0, \varepsilon \geq 0 \end{cases} \quad (1)$$

where $C>0$ is a penalty factor, $\xi_i^{(*)}$ ($i=1,...,l$) are slack variables and $v\in(0,1]$ is an adjustable regularization parameter, ε is also an adjustable tube' magnitude parameter. Parameter ε appears as the variable of optimal problem, its value is given by the final solution.

Problem (1) is a quadratic programming (QP) problem. The steps of its solution are described as follows:

Step 1: Suppose the training sample set $T=\left\{\left(x_1,y_1\right),\cdots,\left(x_i,y_i\right),\cdots,\left(x_l,y_l\right)\right\}$, where $x_i\in R^d$, $y_i\in R, i=1,\cdots,l$.

Step 2: Select the kernel function K, regularization parameter v and the penalty factor C. Construct the QP problem (6) of the g-SVM.

Step 3: By introducing Lagrangian multipliers, a dual problem can be defined as follows.

$$\max_{\alpha,\alpha^*} W\left(\alpha,\alpha^*\right) = \sum_{i=1}^{l}\left(\alpha_i^* - \alpha_i\right)y_i$$

$$-\frac{1}{2}\sum_{i,j=1}^{l}\left(\alpha_i^* - \alpha_i\right)\left(\alpha_j^* - \alpha_j\right)K\left(x_i, x_j\right) - \frac{1}{2C}\sum_{i=1}^{l}\left(\alpha_i^2 + \alpha_i^{*2}\right)$$

$$\text{s.t.}\begin{cases} \sum_{i=1}^{l}\left(\alpha_i^* - \alpha_i\right) = 0 \\ \sum_{i=1}^{l}\left(\alpha_i^* + \alpha_i\right) \le C \cdot v \\ \alpha_i, \alpha_i^* \in \left[0, C/l\right], i = 1, \cdots, l \end{cases} \tag{2}$$

The Lagrangian multipliers $\alpha_i^{(*)}$ can be determined by solving the problem (2).

Step 4: For a new input x, construct the following regression function

$$f(x) = \sum_{i=1}^{l}\left(\alpha_i^* - \alpha_i\right)K\left(x_i, x\right) + b \tag{3}$$

Parameter b can be computed by Eq. (4), select the two scalars α_j ($\alpha_j \in \left(0, l/C\right)$) and α_k^* ($\alpha_k^* \in \left(0, l/C\right)$), then we have

$$b = \frac{1}{2}\Big[y_j + y_k -$$

$$\left(\sum_{i=1}^{l}\left(\alpha_i^* - \alpha_i\right)K\left(x_i, x_j\right) + \sum_{i=1}^{l}\left(\alpha_i^* - \alpha_i\right)K\left(x_i, x_k'\right)\right)\Big] \tag{4}$$

Parameter ε can be given by Eq. (5) or Eq. (6).

$$\varepsilon = \sum_{i=1}^{l}\left(\alpha_i^* - \alpha_i\right)K\left(x_i, x_j\right) + b - y_j \tag{5}$$

or

$$\varepsilon = y_k - \sum_{i=1}^{l}\left(\alpha_i^* - \alpha_i\right)K\left(x_i, x_k\right) - b . \tag{6}$$

3. Embedded chaotic particle swarm optimization

There are not effective ways how to confirm the optimal parameters of the SVM model up to now. There exists crossover error in crossover validation method used commonly to determine penalty coefficient, controlling vector and kernel parameter. To overcome the shortage, a new PSO with chaotic mapping is proposed, namely embedded chaotic particle swarm optimization (ECPSO). ECPSO utilized

to optimize the parameters of g-SVM can increase the diversity of individual and searching ergodicity.

3.1. Standard particle swarm optimization

Similarly to evolutionary computation techniques, PSO uses a set of particles, representing potential solutions to the problem under consideration. The swarm consists of n particles. Each particle has a position $X_i = \left\{x_{i1}, x_{i2}, \cdots, x_{ij}, \cdots x_{im}\right\}$, a velocity $V_i = \left\{v_{i1}, v_{i2}, \cdots, v_{ij}, \cdots, v_{im}\right\}$, where $i = 1, 2, \cdots, n; j = 1, 2, \cdots, m$, and moves through a m-dimensional search space. According to the global variant of the PSO, each particle moves towards its best previous position and towards the best particle g in the swarm. Let us denote the best previously visited position of the i-th particle that gives the best fitness value as $P_i = \left\{p_{i1}, p_{i2}, \cdots, p_{ij}, \cdots, p_{im}\right\}$, and the best previously visited position of the swarm that gives best fitness as $pg = \left\{pg_1, pg_2, \cdots, pg_j, \cdots, pg_m\right\}$.

The change of position of each particle from one iteration to another can be computed according to the distance between the current position and its previous best position and the distance between the current position and the best position of swarm. Then the updating of velocity and particle position can be obtained by using the following equations:

$$v_{ij}^{k+1} = w \times v_{ij}^{k} + c_1 \times r_1 \times \left(p_{ij} - x_{ij}^{k}\right) + c_2 \times r_2 \times \left(pg_j - x_{ij}^{k}\right) \tag{7}$$

$$x_{ij}^{k+1} = x_{ij}^{k} + v_{ij}^{k+1} \tag{8}$$

where w is called inertia weight and is employed to control the impact of the previous history of velocities on the current one. Accordingly, the parameter w regulates the trade-off between the global and local exploration abilities of the swarm. A large inertia weight facilitates global exploration, while a small one tends to facilitate local exploration. A suitable value of the inertia weight w usually provides balance between global and local exploration abilities and consequently results in a reduction of the number of iterations required to locate the optimum solution. k denotes the iteration number, c_1 is the cognition learning factor, c_2 is the social learning factor, r_1 and r_2 are random numbers uniformly distributed in the range [0, 1].

3.2. The improvement of PSO

Similar to genetic algorithm (GA), PSO is a population based optimization tool. PSO is based on the metaphor of social interaction and communication such as bird flocking [4]. Original PSO is distinctly different from other evolutionary-type methods in a way that it does not use the filtering operation (such as crossover and mutation) and the members of the entire population are maintained through the search procedure so that information is socially shared among individuals to direct the search towards the best position in the search space. It has also fast converging characteristics and more global searching ability at the beginning of the run and a local searching near the end of the run. However, it has sometimes a slow fine-tuning ability of the solution quality. While solving problems with more local optima, it is more likely that PSO will explore local optima at the end of the run [5-6]. To avoid this problem, the improvement on PSO is made as follows: utilizing chaotic mapping operator [7] to generate child chaotic population of each initial particle and leading to expand the searching adjacent space of each initial particle.

A simplest chaotic mapping which was brought to the attention of scientists by May[8] that appears in nonlinear dynamics of biological population with chaotic behavior is logistic mapping, whose equation is the following:

$$X_{n+1} = \mu X_n \left(1 - X_n\right) \tag{9}$$

In Eq. (9), X_n is the n th chaotic number where n denotes the iteration number. Obviously, $X_n \in (0,1)$ under the conditions that the initial $X_0 \in (0,1)$ and that $X_0 \notin \{0.0, 0.25, 0.5, 0.75, 1.0\}$. $\mu = 4$ has been used in the experiments.

On the basis of analysis on the aforementioned chaotic system theory, the embedded chaotic particle swarm optimization (ECPSO) is proposed in this paper. The proposed ECPSO consists of two PSO dwelled in father process and child process respectively. The local optimal particle is obtained from child process, while the global optimal particle is obtained from father process. Child chaotic colony of each particle from child process consists of sequences generated from chaotic mapping. The local optimal particle obtained from child process substitutes the original particle from father process where it is necessary to make a random-based choice. By this way, it is intended to improve the global convergence and to prevent to stick on a local solution. In fact, however, these particles can't ensure the optimization's

ergodicity entirely in phase space, because they are random in traditional PSO.

Combining the product forecasting with g -SVM, a hybrid forecasting system shown in Fig. 1 can be described as follows: ECPSO optimizes the parameters of g -SVM by training samples. The optimal parameters are input into g -SVM, train the samples and obtain the support vectors, then forecast the sale series.

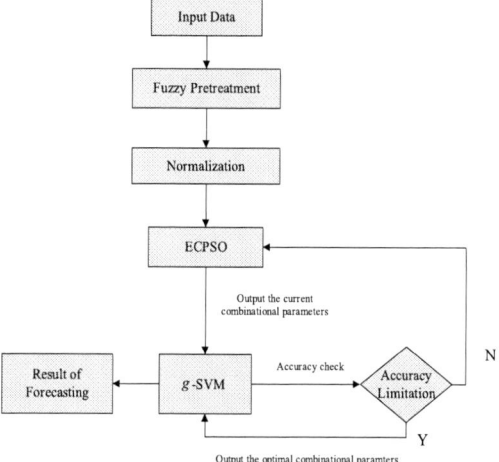

Figure 1. The hybrid model based on ECPSO and g -SVM

Many actual applications suggest that radial basis functions tend to perform well under general smoothness assumptions, so that they should be considered especially if no additional knowledge of the data is available. In this paper Gaussian radial basis function is used as the kernel function of g -SVM.

4. Application

In developing a SVM forecastor, the first important step is feature selection or feature extraction (new features are transformed from the original inputs). In the development of SVM, all available indicators can be used as the inputs, but irrelevant or correlated features could adversely impact the generalization performance due to the curse of dimensionality problem. Thus, it is critical to perform feature selection or feature extraction in SVM. In our experiments, car sale series are selected from past sale record in a typical company. The detailed characteristic data and sale series of these cars compose the corresponding training and testing sample sets. During the process of the car scale series forecasting, six influencing factors, viz., brand famous degree (BF), performance parameter (PP), form beauty (FB), sales experience (SE), oil price (OP) and dweller deposit(DD) are taken into account. All linguistic information of influencing factors is dealt with fuzzy comprehensive evaluation and forms numerical information.

The proposed g-SVM has been implemented in Matlab 7.1 programming language. The experiments are made on a 1.80GHz Core(TM)2 CPU personal computer (PC) with 1.0G memory under Microsoft Windows xp professional. The initial father process parameters of ECPSO are given as follows: number of particles: $n = 50$; particle dimension: $m = 6$; inertia weight: $w = 0.9$; positive acceleration constants: $c_1, c_2 = 2$; the maximal iterative number: $k_{max} = 100$; the fitness accuracy of the normalized samples is equal to 0.0002; The initial child process parameters of ECPSO are given as follows: inertia weight: $w = 0.9$; positive acceleration constants: $c_1, c_2 = 2$; the maximal iterative number: $k_{max} = 100$; the fitness accuracy of the normalized samples is equal to 0.0002. The performance comparison between the standard PSO and ECPSO is shown in Table 1. It is obvious that chaotic mapping can expand the searching adjacent space of each initial particle. Compared with PSO, ECPSO has better global searching capability and worse searching efficiency. From the view point of forecasting, researchers seek to good forecasting model by finding the optimal model parameters.

Table 1 The performance comparisons of PSO and ECPSO

Algorithm	Iterative number	MSE	
		Max	Min
PSO	8(Father process)	3.7e-4	2.8e-4
ECPSO	7(Child process)	3.0e-4	2.8e-4

The optimal combinational parameters are obtained by Algorithm ECPSO, viz., $C = 753$, $v = 0.86$ and $\sigma = 0.028$. Fig. 2 illuminates the sale series forecasting result given by ECPSO g-SVM.

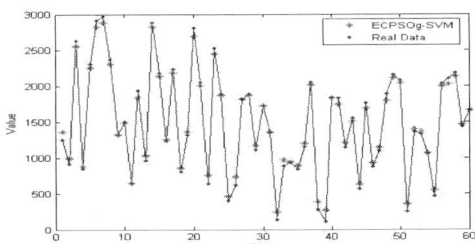

Figure 2. The forecasting results from ECPSO g-SVM model

To verify the forecasting capacity of ECPSO g-SVM based on ECPSO and v-support vector machine with Gaussian loss function, autoregressive moving average (ARMA), PSO v-SVM based on PSO and standard v-support vector machine and ECPSO v-SVM based on ECPSO and standard v-support vector machine are selected to deal with the above car sale series. Their results are shown in Table 2.

Table 2 The forecasting results from four models

No.	real data	Forecasting value			
		M 1	M 2	M 3	M 4
1	1665	1475	1666	1665	1665
2	1435	1566	1474	1439	1464
3	2183	1537	2147	2110	2137
4	2098	1513	2035	1999	2025
5	2027	1328	2007	1971	1997
6	461	1421	558	496	548
7	1060	1457	1065	1096	1065
8	1325	1464	1375	1330	1365
9	1363	1348	1338	1424	1398
10	252	1578	369	310	359
11	2078	1571	2052	2017	2043
12	2153	1523	2120	2083	2110

Note: M1: ARMA; M2: PSO v-SVM; M3: ECPSO v-SVM; M4: ECPSOg-SVM.

The indexes **MAE**, **MAPE** and **MSE** are used to evaluate the forecasting capacity of four models shown in Table 3. To represent the error trend well, the latest 12 months forecasting results are used to analyze the forecasting performance of the above models. It is obvious that the forecasting accuracy given by support vector machine excels the ones by autoregressive moving average (ARMA). The parameters optimized by ECPSO are more optimal to construct SVM model for the car sale series than ones by PSO.

The indexes **MAPE** and **MSE** provided by ECPSO g-SVM with Gaussian loss function are better than ones of ECPSO v-SVM with ε-insensitive loss function, while **MAE** of ECPSO g-SVM is worse than one of ECPSO v-SVM. Considering the non-linear influence from multidimensional data, it is found that support vector machine with Gaussian loss function can deal with some data with normal additive noise.

Table 3 Error statistic analysis on four different models

Model	MAE	MAPE	MSE
ARMA	518.7	0.7914	401030
PSO v-SVM	46.5	0.0468	3045
ECPSO v-SVM	42.7	0.0706	2933
ECPSO g-SVM	44.2	0.0668	2844

It is shown in Tables 2 and 3 that the selection of loss function is important for production sale forecasting. The optimal loss function in regression estimation is actually related to the noise in the data. For the normal additive noise, the Gaussian loss function is the best choice. For the noise with symmetric density, Huber's least modulus function performs best. Considering the enterprise sale environment, some errors exist inevitably in the process of data gather and estimation. Thus, the above forecasting results are satisfying. The results of application in car sale forecasting indicate that the forecasting method based on g-SVM is effective and feasible.

5. Conclusion

Product sale forecasting is the foundation of manufactory production planning and inventory controlling. Enterprise needs the prompt sale quantity urgently, arranges production planning and gets the optimal controlling of production cost in product life cycle time with the increase of product complexity and diversity.

Product sale series have the characteristic of multidimension, small sample and nolinearity. For some normal additive noise from sale data, it is difficult to deal with it by the standard support vector machine with ε-insensitive loss function. While the support vector machine with Gaussian loss function proposed in this paper can effectively inhibit the normal additive noise and obtain good generalization property.

Acknowledgements

This research is supported by the National Natural Science Foundation of China under grants 60574062 and by the National High Tech R&D Program of China under Grant 2007AA04Z112.

References

[1] J. Sun, G..S. Hong, M. Rahman, and Y.S. Wong, "The application of nonstandard support vector machine in tool condition monitoring system", Electronic Design, Test and Applications, 2004. DELTA 2004. Second IEEE International Workshop, 2004, pp. 295 – 300.

[2] H.S. Yan, and D. Xu, "An approach to estimating product design time based on fuzzy v-support vector machine", IEEE Transactions on Neural Networks, 2007, 18(3), pp. 721-731.

[3] H. Aytug, and S. Sayin, "Using support vector machines to learn the efficient set in multiple objective discrete optimization", European Journal of Operational Research, In Press, Corrected Proof, Available online, 2007.

[4] J. Kenedy, and R. Eberheart, "Particle swarm optimization", Proceedings of the IEEE International Conference on Neural Networks, 1995, pp. 1942-1948 .

[5] S.K.S. Fan, and E. Zahara, "A hybrid simplex search and particle swarm optimization for unconstrained optimization", European Journal of Operational Research, 2007, 181(2), 527–548.

[6] M.F. Tasgetiren, Y.C. Liang, M. Sevkli, and G. Gencyilmaz, "A particle swarm optimization algorithm for make span and total flow time minimization in the permutation flowshop sequencing problem", European Journal of Operational Research, 2007, 177(3), pp. 1930–1947.

[7] Y.F. Sun, and F.Q. Deng, "Chaotic parallel genetic algorithm with feedback mechanism and its application in complex constrained problem", Proceedings of IEEE Conference on Cybernetics and Intelligent Systems, Singapore, 2004, pp. 596-601.

[8] R. May, "Simple mathematical models with very complicated dynamics", Nature, 1976, 261, pp. 45-67.

A New Web Service Discovery Method Based on Semantic

Wuling Ren, Zhujun Xu
College of Computer Science and Information Engineering,
ZheJiang GongShang University, Hangzhou, ZheJiang, 310018, China
xuzhujun728@tom.com, rwl@zjgsu.edu.cn

Abstract

To discovery web services in Internet, many approaches have been proposed such as UDDI and DWS. The problem of those methods is that they are just a kind of simple syntax match based on keywords. They make the match inflexible, sometimes even unfaithful. This paper proposes a new algorithm to discover web services based on semantic. It uses tree-form data structure to describe the web services and give all the nodes a weight value by certain strategy, then compute the semantic similarity between the web services requested and the services registered. This algorithm makes services matching basing on semantic relationships at conceptual level. To validate the feasibility and effectiveness of the algorithm, we construct a self-developed prototype system named WSD. The experiments prove that this algorithm has high recall and precision than other methods.

1. Introduction

With the rapid development of Internet, web services technologies are becoming more and more important. Web services are the self-describing modular software application that can be defined, advertised, discovered, located and used across Internet by using a set of open regulations such as soap, xml [1]. The appearance of web services enable the interoperation of heterogeneous systems and the reuse of distributed system function in application development [2].

UDDI (Universal Description, Discovery and Integration) is the materialization of the SOA registry component [3], for publishing and discovering Web services. Providers may publish Web services along with their associated metadata in a UDDI registry. Though UDDI has many advantages which makes itself become the main registry mechanism, but it also has some defects that prevent its popularity:

(1) UDDI is based on XML and it can't provide semantic description.

(2) It can't describe the search target precisely.

(3) It can't measure the matching degree between the candidates and the search target.

The first two points influence the precision ratio, and the third point mainly influences the recall ratio.

Because UDDI method lacks semantic description, some new algorithms are proposed. The algorithm presented in document searches web services based on ontology property values, but it has to define the ontology property first. The method presented in document [4] does the services matching based on ontology of web services, but it has to define the hyponymy of ontology before matching.

The paper tries to solve these problems by computing the services' weight value, storing the web services in tree-form data structure and computing the semantic similarity to match web services. At last, the paper uses a self-developed prototype system to validating this method.

2. Web services description

The Web services could be described as: $WS=<CP, SP, I/O, Qos$ [5]$>$.

CP is the common properties, namely every web service is supposed to possess, just as service ID, service provider ID, service name, service version etc. SP is the special properties of services, it best represents the feature of the web service. I/O is the interface set of input and output of the service. Qos is the quality of web service. It could be described from many aspects, just like response time, service price, performance, reliability, security, credit worthiness, throughput, etc. The paper mainly tests Qos from response time, performance, and reliability.

The paper uses to describe the web service, and store the related information in tree-form data structure. To show the difference between web services on semantic and enhance the search efficiency, the paper stores SP information in the upper nodes, and gives them biggish values, stores CP information in the lower nodes, and gives them smaller values. If the upper nodes have been matched in searching, the lower nodes could be omitted. Therefore the efficiency is enhanced.

There are many methods to compute the weight value, the paper uses TF-IDF [6] function to calculate the weight:

$$weight = TF * IDF = TF * \lg[D / DF(W)]$$

TF is the word frequency of keyword W, D is the totality of the web services, DF(W) means how many times the keyword W appears in the service. The TF-IDF function doesn't take the quantity of web services into account. So the paper uses TFC weight-calculate method, which is very like TF-IDF function, but it adopts TLN as one consideration in calculating weight value:

$$weight(i) = \frac{[TF_{ij} * \log(D / DF_i)]}{\sqrt{\sum_{K=1}^{M} weight(i) = [TF_{ij} * \log(D / DF_i)]}}$$

If one web service has properties as$< W_1, W_2, ... W_n, >$, then the this web service could be expressed by a N-dimensional vector $T = <T_1, T_2, ... T_n>$. T_i is the weight value of W_i.

3. Web service matching

Web services matching demands high precision ratio, high recall ratio, high flexibility and high efficiency. The paper proposes a effective matching algorithm, its basic thoughts are: (1)According to the web service request, locate the knowledge domain; (2)Construct a tree to describe the registered web services, give every node a matching weight value by TFC weight-calculate method which is illuminated before; (3)Calculate the semantic similarity degree between the services. If the similarity degree is bigger than the matching weight value, return the current web service, or else return null to show no service is found.

3.1 The process of web services matching

At first, according to the query submission to locate the knowledge domain, assign every node of web service tree a weight value. Next, calculate the semantic similarity between services. At the same time, take two assistant steps to help the match:

(1) Pre-treat the information of web service request. It includes: segmenting the property words, extending the abbreviation, deleting the abandoned words, extracting the morpheme, normalizing similarity factors, etc. These treatments would enhance the recall ratio.

(2)Deal with the services matching result set. If the services' similarity degree is lower than the request, remove them from the returned set. It's related to the weight value of web service tree node. This step would enhance the precision ratio.

To show the web services discovery process clearly, the paper gives the flow chart as following, which describes a web service query from submission to return:

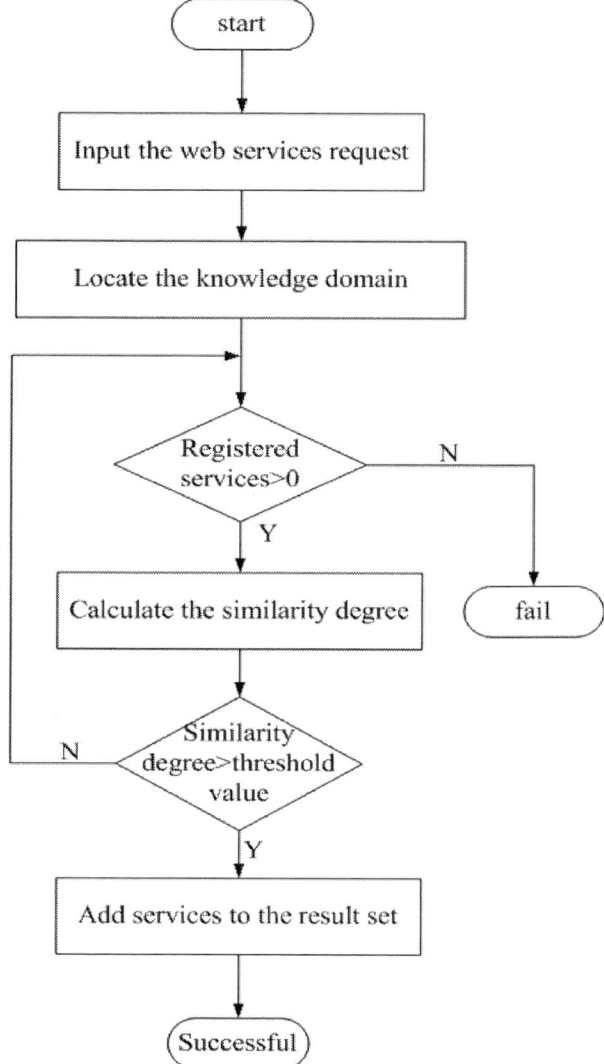

Figure 1 Flow chart for web services matching

3.2 Calculate semantic similarity degree

WordNet [7] is a large lexical database of English, developed under the direction of George A. Miller. Nouns, verbs, adjectives and adverbs are grouped into sets of cognitive synonyms (Synsets), each expressing a distinct concept. Synsets are interlinked by means of conceptual-semantic and lexical relations. At present, there many methods used to calculate the similarity by hyponymy , for example RES, LIN, JCN, etc.

The paper calculates the semantic similarity between web services based on WordNet and some other assistant techniques.

Suppose that one web service is described by m properties , expressed as $WS_m = (P_1, P_2, ... P_m)$, and another web service is $WS_n = (P_1, P_2, ... P_n')$, then the relationship of the two services could be expressed as following:

$$\begin{pmatrix} R_{11} & R_{12} & ... & R_{1m} \\ R_{21} & R_{22} & ... & R_{2m} \\ \vdots & \vdots & & \vdots \\ R_{n1} & R_{n2} & ... & R_{n3} \end{pmatrix}$$

R_{mn} means the degree of association, the concrete computing method is as following:

```
if(WS_m=WS_n'){
    R_mn=1;
else
{
    foreach(record of association table)
        if(P_1∈ WS_m && P_2∈ WS_n){
            i++;
            // v is a default parameter
            R_mn=(i-v)/(i+v);}
        else{
            R_mn=u; // u is a default parameter
            }
    }
}
```

The weight value of WS_m could be expressed as $T=(T_1, T_2, ...T_m)$, and weight value of WS_n could be expressed as $T=(T_1, T_2, ...T_n)$. The semantic similarity of WS_m and WS_m could be figured out as following:

```
S_i=S_i' = similarity_1= similarity_2=0;
for(i=1; i<=n; i++)
{
    for(i=1; i<=n; i++)
    {
        S_i=S_i+T_i'*R[i][j];
        // S_i is the association degree of WS_m
        S_i'=S_i'+T_i*R[i][j];
        // S_i' is the association degree of WS_n
    }
    similarity_1= similarity_1+S_i;
    similarity_2= similarity_2+S_i';
}
similarity=( similarity_1+ similarity_2)/2;
```

When similarity is calculated, we normalize the similarity factors, and compare it with the matching weight value. If the similarity degree bigger than the default threshold value, the target service is found.

4. Experiment and analysis

At present, there isn't a uniform measurement for the web service discovery. Because of different definitions and free comparison strategies, it's difficult to judge two discovery algorithms which is better. The common evaluating indicators are recall ratio and precision ratio. The paper uses them to validate the algorithm.

To validate the algorithm, a prototype system named WSD is constructed. It uses OWLS-TC V2 service test data [8]. The OWLS-TC's test data is collected from five different domains, and two hundred services are extracted from each domain as test data.

Argument UDDI Registry System (AURS) is famous in web services discovery field, so we compare AURS with WSD to estimate the WSD' performance.

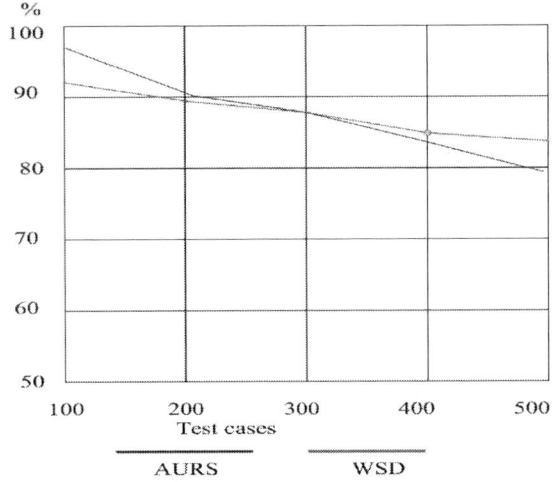

Figure 2 The precision ratio of AURS and WSD

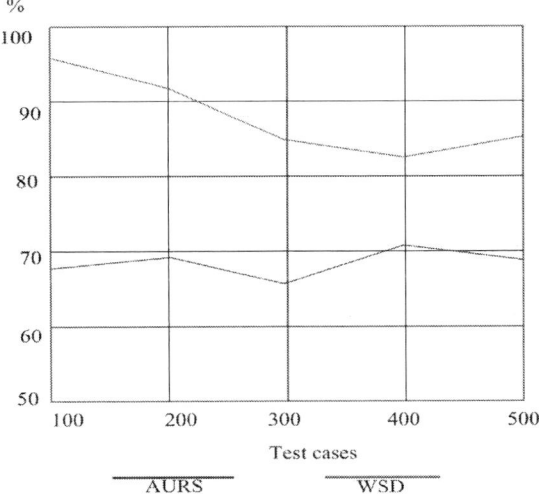

Figure 3 The recall ratio of AURS and WSD

Figure 2 and Figure 3 show the experimental result of AURS and WSD. The experimental result reveals that AURS has higher precision ratio than WSD when the test cases are less than 300. But with the increase of test cases and domains, AURS's precision ratio descends fast. That's because AURS is based on prolog inference rule, the services returned all match the services requests exactly, so scope of application is narrow. WSD's matching is based on web services' semantic similarity, so it's not confined to one given domain, and its recall ratio and precision ratio is high and stable comparing with AURS.

5. Conclusions

The paper extends the web service descriptive language OWL-S, and takes Qos of web service into account. Then store service's descriptive information into tree-form data structure., assign every tree node a matching weight value. Tree has layered structure, which helps enhance the search efficiency. When matching the web services, judge the matching degree by their semantic similarity. It helps enhance the recall ration and the precision ratio. The experimental result indicates that the WSD system has high performance comparing with Argument UDDI Registry System.

Acknowledgement

The authors thank editor-in-chief and anonymous reviewers for their comments and suggestions. The work is supported by the Science & Technology Plan of Zhejiang Province, China (No.2006C11239).

References

[1] D.Martin, M.Burstein, "Describing Web services using DAML-S and WSDL", *DAMLS Coalition Working Document*, Vol. 33, No. 10, pp.708-717.

[2] M.Dumas, J.Sullivan, M.Heravizadeh, D.Edmond and A.Hofstede, "Towards a semantic framework for service description". *W3C Recommendation*, 2004, pp.1003-1010.

[3] Dipanjan Chakraborty, Filip Perich, Sasikanth Avanchaet al.DReggie, "Semantic service discovery for M-Commerce applications", *Proceedings of the Workshop on Reliable and Secure Applications in Mobile Environment*, New Orleans, USA, 2001, pp. 28-31.

[4] Klein M., Bernstein, "A Searching services on the semantic Web using process ontologies", *Proceedings of the International Semantic Web Working Symposium (SWWS)*. Amsterdam, IOS Press, 2001, pp. 159~172.

[5] Deng ShuiGuang, Wu ZhaoHui, Kuang Li, Lin Chuan, Jin Yue-Ping, "Management of service flow in a flexible way", *Lecture Notes in Computer Science*, 2004, pp. 428-438.

[6] Paolucci.M, KawamuraT, SycaraK P, "Importing the Semantic Web in UDDI", *E-Business and the Semantic Web (WES 2002)*, 2006, pp.225-236.

[7] K.Sivashanmugam, K.Verma, A.Sheth and J.Miller, "Adding semantic to web services standards", *In Proceedings of the 1st International Conference on Web services (ICWS' 03)*, USA, 2007.

[8] Jian Wu and Zhaohui Wu, "Similarity-based Web service Matchmaking", *proceedings of the 2005 IEEE International Conference on Services Computing*, China, 2005.

2008 Workshop on Power Electronics and Intelligent Transportation System

Research on Multi-Attribute Decision-Making Method Based on AHP and Outranking Relation

Liguo Fan[1], Feng Zuo[2]

1. School of Business Administration, North China Electric Power Univ, Baoding 071003,China
2. JiLin Transmission and Transformation Engineering Co. Changchun 130021,China
fanliguo@sohu.com

Abstract

In this paper, we analyzed the disadvantage of traditional AHP, and presented a new decision-making (MADM) method based on outranking relation and AHP. The method firstly determined weight of each attribute by AHP, then used thresholds express uncertainty of the data according to decision-maker(DM)'s risk attitudes by outranking relation, also compared alternatives in pairs, calculated score of alternatives under single attribute, expressed both sides of alternative with die index of concordance and discordance, integrated the index of concordance an d discordance into the degree of outranking, finally obtained the final ranking of the alternatives using net credibility degree. The method used fully expert's opinion and the course of decision-making met men's real decision-making thinking custom. At the end, the method is explained by an example of decision-making of transmission network planning.

1. Introduction

Multi-attribute decision-making theory has widespread practical application in economic management and engineering project field. It is making quantitative and qualitative evaluation according to different attribute towards multiple competitive plans, then gets integrated evaluation and accomplishes decision. Multi-attribute decision-making method can make decision-maker select the most satisfied alternative based on a set of attributes in the circumstance of complexity of decision-making factor and uncertainty. It is an effective decision-making method.

In the past the decision problem mainly depend on the judgment and decision which was made by those planning experts who merely have some empirical knowledge. But this kind of subjective judgment either lacks of theory supporting or exists plenty of accidental factors. The incorrect evaluating for attributes or the disaccord judgment of experts makes that available information for decision-maker is inaccurate. Thus the result is not satisfied by using definite value to express the information. Obviously it is necessary for enhancing credibility degree of the result by considering variation of information in the course of decision-making. Meantime it demands that MADM can deal with imperfect preference information because MADM includes

decision-maker's preference information which is common imperfect. Aiming at above problem in MADM, outranking relation as a mode of treating uncertain information easily and practically is extensively applied lots of important project item.

In this paper, we propose a Multi-attribute decision method based on AHP and outranking relation and apples to a actual case of transmission network planning.

2. AHP

Hierarchy analysis is a practical multi-attribute decision method proposed by Saaty, a famous American professor of operation research of Pitsburg University[1]. The idea is decomposing complicated problems by building clear hierarchical structure firstly, scaling various attributes by using relative scale to compare in pairs and building judgment matrix layer-by-layer then solving local weight of judgment matrix secondly, calculating composite weight of each alternatives and sequences those alternatives lastly. The feature of the method is scaling judgment of human, disposing systematically various decision problem by quantitative and qualitative associated method and forming capacity of man-machine cooperation.

The basic step of AHP is:
1) Establishment of hierarchy structure,
2) Forming judgment matrix,
3) Verifying of consistency of judgment matrix,
4) Determining weight vector.

3. Outranking Relation
3.1. The principle of the method

Outranking relation is described: two alternatives are regarded as indifference if the attribute value of two alternatives is similar. Similar value in expressing is named "The indifference threshold". If the attribute value of one alternative is bigger than the attribute value of another alternative and if differential value is bigger than some value and we regard the former is superior than the latter, the value is commonly named "The strict preference threshold". we express the degree of one alternative being superior than another alternative by using fuzzy membership value under other circumstances.

978-0-7695-3342-1/08 $25.00 © 2008 IEEE
DOI 10.1109/PEITS.2008.42

227

3.2. ELECTRE III

ELECTRE[2,3] method is presented by Benoyown in 1969 and it is completed step-by-step. The key of ELECTRE method is using the concept of outranking relation[4] which demands decision-maker to regard alternative1 is superior than alternative2 according to some risk degree. Decision-maker can determine order of quality by a series of evaluating for outranking relation.

Outranking relation[5] is regarded the third method as treating uncertain factor except probability and fuzzy. Among ELECTRE III is the most representation method of applying outranking relation. It begins from decision-maker's point of view, establishes outranking relation according to decision-maker's risk attitude and considers the concept of imperfect compensation among attribute. The theory is easily accepted because it is embody of men's daily thought.

A MADM problem is usually formulated by a set of alternatives $A = \{a_1, a_2, \cdots a_n\}$, a set of estimate attribute $F = \{f_1, f_2, \cdots f_m\}$ and evaluation of estimate X_{ik} of alternatives under estimate attribute. Where X_{ik} presents evaluation of estimate of alternative a_i under estimate attribute f_k [6,8,10]. Without loss of generality, given that all the criteria values are to be maximized.

3.2.1. Determining threshold. For any ordered pair (a_i, a_k) of alternatives, the three thresholds are as follows:

The indifference threshold q_j: For the attribute f_j, a_i and a_k are indifferent if $a_{ij} - a_{kj} \leq q_j$ and $a_{kj} - a_{ij} \leq q_j$.

The strict preference threshold p_j : For the attribute f_j, a_i is strictly preferred to a_k if $a_{ij} - a_{kj} > p_j$.

The veto threshold v_j: For the attribute f_j, reject the hypothesis of outranking of a_i over a_k if $a_{kj} - a_{ij} \geq v_j$

It implies that: $v_j > p_j > q_j > 0$.

3.2.2. Calculating concordance. A concordance index $C(a_i, a_k)$ is computed for each ordered pair (a_i, a_k) of alternatives and defined by:

$$C(a_i, a_k) = \sum_{j=1}^{m} \omega_j c_j (a_i, a_k)$$

Where ω_j is the weight determining the relative importance of the attribute f_j and $c_j(a_i, a_k)$ is defined by:

$$c_j(a_i, a_k) = \begin{cases} 1, \text{当} X_{kj} - X_{ij} \leq q_j \\ 0, \text{当} X_{kj} - X_{ij} \geq p_j \\ (p_j + X_{ij} - X_{kj})/(p_j - q_j), otherwise \end{cases} \quad (1)$$

Where $c_j(a_i, a_k)$ shows the degree of concordance with the judgmental statement that a_i outranks a_k under the attribute f_j, the index of global concordance $C(a_i, a_k)$ represents the amount of evidence to support the concordance among all the attribute, under the hypothesis that a_i outranks a_k.

3.2.3. Calculating discordance.

$$d_j(a_i, a_k) = \begin{cases} 0, \text{当} X_{ij} - X_{kj} \leq p_j \\ 1, \text{当} X_{ij} - X_{kj} \geq v_j \\ (X_{ij} - X_{kj} - p_j)/(v_j - p_j), p_j < X_{ij} - X_{kj} \leq v_j \end{cases} \quad (2)$$

Where $d_j(a_i, a_k)$ shows the degree of discordance with the judgmental statement that a_i outranks a_k. Discordance includes non-compensable ingredient of attribute. It is unacceptable of using good data to substitute defective data.

3.2.4. Calculating the degree of outranking. The degree of outranking $S(a_i, a_k)$ of order alternatives couple (a_i, a_k) is defined by:

$$S(a_i, a_k) = \begin{cases} C(a_i, a_k), \text{当} J(i,k) \in \Phi \\ C(a_i, a_k) \prod_{j \in J(i,k)} \dfrac{1 - d_j(a_i, a_k)}{1 - C(a_i, a_k)}, otherwise \end{cases} \quad (3)$$

Where $J(i, k)$ is defined as the set of attribute for which $d_j(a_i, a_k) > C(a_i, a_k)$. If $J(i, k) = \Phi$, we have $d_j(a_i, a_k) \leq C(a_i, a_k)$ for all attribute, then, $S(a_i, a_k)$ is the same as $C(a_i, a_k)$. $S(a_i, a_k)$ shows the degree of credibility of outranking with the judgmental statement that a_i outranks a_k.

4. Decision Model Based on AHP and Outranking Relation

4.1. Establishment of hierarchy structure

Aiming at complicated decision-making problem we should define alternatives and determine target of the decision-making problem firstly. Secondly we should select target layer, attribute layer and alternatives layer of the problem. Finally we establish hierarchy structure of decision-making problem.

4.2. Determining of weight vector

We establish judgment matrix by scalarization of data according to upper hierarchy structure. The process of forming judgment matrix is the process of scalarize data. Scalarization is changing various basic data to canonical format which is compare right. We can use not only positive scalar in fuzzy comprehensive evaluation but also relative scalar in AHP to scalar data. Since the process of actual scoring indicates that judgment of comparison in pairs for object is easier and exacter than simultaneous judgment for multi-object comparison, the paper uses relative scalar method which mainly through experts' accurate comparison and scoring of data in pairs and thus to standardize and quantify them. The paper selects reciprocal 1~9 scale as scalar method of judgment matrix. What is called reciprocal scale points that if the result of comparison is a_{ij} about alternatives i relative to alternatives j, the result of comparison about alternatives j relative to alternatives i is a_{ji} which is reciprocal of a_{ij}.

4.3. Calculating concordance and discordance

Firstly we need to determine threshold. Attribute threshold commonly reflect imputation of comparing result among of alternatives. It is an important influencing factor to define comparing result of alternatives. Calculating concordance and discordance according to upper formula.

4.4. Calculating the degree of outranking

According to upper calculating result we calculate the degree of outranking by formula (3).

4.5. Calculating total score and Sequencing alternatives

Traditional ELECTRE III method needs to introduce other thresholds to sequence alternatives after determining the degree of outranking. This will undoubtedly increase the degree of difficulty for decision-maker using it. And traditional ELECTRE III method only gains partial sequence of alternatives. That is to say that the incomparable situation of certain two alternatives does exist. In order to avoid the defect, the paper adopts net credibility method to sequence alternatives based on reference [8]. The formula is followed:

$$\Phi(a_i) = \Phi^+(a_i) - \Phi^-(a_i) = \sum_{a_k=1}^{n} S(a_i, a_k) - \sum_{a_k=1}^{n} S(a_k, a_i),$$

$$a_i = 1, 2, \cdots, n \qquad (4)$$

where $\Phi(a_i)$ is net credibility. The size of $\Phi(a_i)$ determines order of quality. $\Phi^+(a_i)$ which describes degree of a_i outranking other alternatives is concordance. $\Phi^-(a_i)$ which describes degree of other alternatives outranking a_i is discordance.

5. Example

The paper researches a multi-attribute decision-making (MCDM) method in transmission network planning. A transmission network planning problem needs to consider some alternatives and relates to some attribute of sociality, reliability, economy, adaptability. It is good multi-attribute decision-making example.

5.1. Determining planning alternatives, defining planning object and establishing hierarchy structure

Because transmission network planning generally relates to important substations and lines between regions and planning scheme affects safety, reliability and economy of power system in life span of electric equipment, considering emphatically investment, operating cost, extension ability of planning scheme in the course of planning. Selecting four alternatives and twelve attribute and establishing hierarchy structure of AHP as fig.1 according to intention of decision-makers.

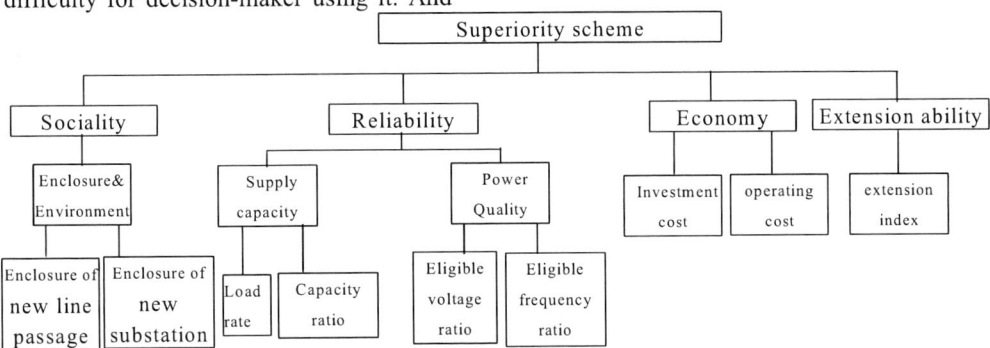

Fig. 1. AHP Hierarchy of the case

f_1 Enclosure and environment: enclosure of new line passages(mu). Representing floor area of line passage of planning network.

f_2 Enclosure and environment: enclosure of new substations (mu). Representing floor area of substations of planning network.

f_3 Supply availability: load rate(%). Representing the ratio of actual system load and system capacity.

f_4 Supply availability: capacity ratio. Representing the ratio of system capacity and system load.

f_5 Power quality: eligible voltage ratio(%).

f_6 Power quality: eligible frequency ratio (%).

f_7 Investment cost: capital cost(billion yuan). Representing elemental construction cost of planning network.

f_8 operating cost(billion yuan).Representing operating and management cost of planning network.

f_9 extension ability: extension index (%). Representing extension ability of planning network aiming at load growth, technical upgrade etc.

After proceeding to network analysis and approximate estimating of investment according to four alternatives, we get the performances of alternatives and change non-benefit index to benefit index by negative. The performances of alternatives are in Table 1.

Table1 Decision table of alternatives

a	f_1	f_2	f_3	f_4	f_5	f_6	f_7	f_8	f_9
a_1	-29940	-124	68	1.8	99.2	99.89	-19.03	-0.380	42
a_2	-22356	-143	70	1.7	99.1	99.22	-18.85	-0.377	68
a_3	-28933	-208	65	1.6	98.9	98.12	-16.69	-0.424	36
a_4	-30002	-232	62	1.9	99.4	99.96	-19.32	-0.336	73

5.2. Forming judgment matrix and determining weight

The paper selects reciprocal 1~9 scale as scalar method of judgment matrix. Decision-makers can determine structure of judgment matrix according to request of decision-making problem. The judgment matrix is as followed.

$$G = \begin{bmatrix} 1.00,1.00,0.25,0.50,0.50,0.50,0.20,0.25,0.50 \\ 1.00,1.00,0.25,0.50,0.50,0.50,0.20,0.25,0.50 \\ 4.00,4.00,1.00,2.00,2.00,2.00,0.80,1.00,2.00 \\ 2.00,2.00,0.50,1.00,1.00,1.00,0.40,0.50,1.00 \\ 2.00,2.00,0.50,1.00,1.00,1.00,0.40,0.50,1.00 \\ 2.00,2.00,0.50,1.00,1.00,1.00,0.40,0.50,1.00 \\ 5.00,5.00,1.25,2.50,2.50,2.50,1.00,1.25,2.50 \\ 4.00,4.00,1.00,2.00,2.00,2.00,0.80,1.00,2.00 \\ 2.00,2.00,0.50,1.00,1.00,1.00,0.40,0.50,1.00 \end{bmatrix}$$

Maximum eigenvalue of judgment matrix is 9.1618. Weight vector indicating attribute importance is as followed by normalized treating eigenvector.

[0.0370 ,0.0409 ,0.1449 ,0.0834 ,0.1404 , 0.1260 ,0.2780 ,0.0632 ,0.0862]

5.3. Defining the thresholds

The thresholds reflect comparison results of alternatives. This is the most influential factor for determining comparison result of alternatives. These values have relation with the importance of the criteria and the preferences of decision-makers. With the number of alternatives increasing resulting in enlarging range of attribute, the risk attitude of decision-makers alters correspondingly. For some attribute of sampling

uniformly such as f_3, f_4, f_5, f_6, f_8, Setting the indifference threshold and the strict preference threshold according to the number of alternatives: the indifference threshold=(maximum attribute value- minimum attribute value)/ the number of alternatives. Since the strict preference threshold reflects strict preference relation in one attribute of alternatives, the paper adopts the setting method which the strict preference threshold is three times [9] for the indifference threshold to assure rationality of setting. For some attribute of sampling non-uniformly such as f_1, f_2, f_9, Setting the thresholds to constant. According to the risk attitude of decision-makers the standard of setting the indifference threshold makes that it is indifferent for comparison of alternatives. For the attribute whose change is sensitive for decision-makers such as f_7, decision-makers generally think the indifference threshold doesn't overrun error range of the attribute value. The strict preference threshold is three times to the indifference threshold. For the attribute whose change is sensitive for decision-makers, if different value of two alternatives for one attribute overruns the value, decision-makers can't get order of quality of two alternatives. The value is defined as the veto threshold. It means that the attribute can't be compensated by other attribute. For f_1, f_2, f_3, f_4, f_5, f_6, f_8, f_9, the veto threshold is sampled as twenty times of different value of maximum attribute value and minimum attribute value. For f_7, the veto threshold of investment cost is sampled four billion yuan. The thresholds and weights are defined for the example, as in Table 2.

5.4. Calculating concordance and discordance

Calculating concordance according to formula(1). The result is in Table 3.

Table 2 Threshold value of alternatives

Thresholds	f_1	f_2	f_3	f_4	f_5	f_6	f_7	f_8	f_9
q_j	62	19	0.889	0.033	0.056	0.204	0.966	0.010	5
p_j	186	57	2.667	0.099	0.168	0.612	2.898	0.030	15
v_j	1240	380	17.78	0.66	1.12	4.08	4	0.2	100

Table 3 Concordance index of alternatives

a	a_1	a_2	a_3	a_4
a_1	1	0.7862	0.7652	0.6268
a_2	0.7354	1	0.8282	0.5870
a_3	0.3926	0.2780	1	0.5008
a_4	0.8142	0.7772	0.5733	1

Calculating concordance according to formula(2). The result is in Table 4~ Table 12.

Table 4 Discordance index for f_1

a	a_1	a_2	a_3	a_4
a_1	0	1	0.7789	0
a_2	0	0	0	0
a_3	0	1	0	0
a_4	0	1	0.8378	0

Table 5 Discordance index for f_2

a	a_1	a_2	a_3	a_4
a_1	0	0	0	0
a_2	0	0	0	0
a_3	0.0836	0.0248	0	0
a_4	0.1579	0.0991	0	0

Table 6 Discordance index for f_3

a	a_1	a_2	a_3	a_4
a_1	0	0	0	0
a_2	0	0	0	0
a_3	0.0220	0.1544	0	0
a_4	0.2205	0.3529	0.0220	0

Table 7 Discordance index for f_4

a	a_1	a_2	a_3	a_4
a_1	0	0	0	0.0018
a_2	0.0018	0	0	0.1800
a_3	0.1800	0.0018	0	0.3583
a_4	0	0	0	0

Table 8 Discordance index for f_5

a	a_1	a_2	a_3	a_4
a_1	0	0	0	0.0336
a_2	0	0	0	0.1387
a_3	0.1387	0.0336	0	0.3487
a_4	0	0	0	0

Table 9 Discordance index for f_6

a	a_1	a_2	a_3	a_4
a_1	0	0	0	0
a_2	0.0167	0	0	0.0369
a_3	0.3339	0.1407	0	0.3541
a_4	0	0	0	0

Table 10 Discordance index for f_7

a	a_1	a_2	a_3	a_4
a_1	0	0	0	0
a_2	0	0	0	0
a_3	0	0	0	0
a_4	0	0	0	0

Table 11 Discordance index for f_8

a	a_1	a_2	a_3	a_4
a_1	0	0	0	0.0824
a_2	0	0	0	0.0647
a_3	0.0824	0.1000	0	0.3412
a_4	0	0	0	0

Table 12 Discordance index for f_9

a	a_1	a_2	a_3	a_4
a_1	0	0.1294	0	0.1882
a_2	0	0	0	0
a_3	0	0.2000	0	0.2588
a_4	0	0	0	0

5.4. Calculating the degree of outranking

The degree of outranking is calculated by formula(3) according to concordance and discordance above. The result is in Table13.

Table 13 Outranking matrix

a	a_1	a_2	a_3	a_4
a_1	1	0	0.7206	0.6268
a_2	0.7354	1	0.8282	0.5870
a_3	0.3926	0	1	0.5008
a_4	0.8142	0	0.2179	1

5.5. Sequencing alternatives

The paper adopts net credibility method to sequence alternatives. The sequence result is calculated by formula (4) and is showed in Table 14.

Table 14 Net credibility degree of alternatives

Net credibility	Alternatives			
	a_1	a_2	a_3	a_4
$\Phi(a_i)$	-0.5948	2.1506	-0.8733	-0.6825

The order of quality of alternatives are made as a_2, a_1, a_4, a_3 by comparison of net credibility of alternatives.

6. Conclusion

The paper proposes a new multi-attribute decision-making (MADM) method based on AHP and outranking relation. The method is applied in transmission network planning decision-making successfully. The main work is as follows.

The method establishes complete judgment and decision-making structure based on AHP and outranking relation. Since ELECTRE III embodies analogue of thought process of man and AHP is availed widespread, The applicability of the method is available.

The paper adopts net credibility method to sequence alternatives. It simplifies traditional sequence method and improves efficiency and accuracy of calculating.

References

[1] Ge Shao-yun, Dong Zhi. Cabling reconstruction of urban electric network based on interval analytic hierarchy process. ELECTRIC POWER, 2004,37(10):34-37.

[2] Benayoun, R., Roy, B. and Sussman, N. (1996). Manual de Reference du Programme Electre. Notede Synthese et Formation, No. 25, Direction Scientifique SEMA, Paris, 1966.

[3] Hwang, C. L. and Yoon, K. (1981). Multiple Attribute Decision Making: Methods and Applications. Springer-Verlag, New York.

[4] Zeng Zong-yuan, Huang Cheng-long, Wang Jie-ren. Research of multinational investment of enterprise based on Outranking Methods.Science and management Association of 2004, 2004

[5] Nijkamp, P. (1974). A Multicriteria Analysis for Project Evaluation: Economic-Ecological Evaluation of Residential Environment. Papers of the Regional Science Association, Vol. 35, pp87-111.

[6] Roy, B. (1991). The Outranking Approach and the Foundations of the ELECTRE Method. Theoryand Decision, Vol. 31, pp 49-73.

[7] Rogers, M. and Bruen, M. (1998). A New System for Weighting Environmental Criteria for Use within ELECTRE III. European Journal of Operational Research. Vol. 107

[8] Hui-Fen Li,Jian-Jun Wang. An Improved Ranking Method for ELECTRE III. Wireless Communications, Networking and Mobile Computing, Sep2007

[9] Roy B,Present M. A programming method for determining which paris metro stations should be renovated[J]. European Journal of Operational Research，1986,24(2): 318-334.

[10] Stopford, J. M. and Wells, L. T. (1972). Managing the Multinational Enterprise, New York: Basic Books.

2008 Workshop on Power Electronics and Intelligent Transportation System

Nearest Neighbor-Clustering Algorithm Based on Hierarchical Optimization Strategy

Wang Jie, Jiang Guoqiang

College of Electrical Engineering, Zhengzhou University, Zhengzhou, 450001, China
wj@zzu.edu.cn, gqjiang567@163.com

Abstract

In order to overcome the shortcoming of nearest neighbor-clustering algorithm in the cluster center determined, the cluster width of the acquisition, and the hidden nodes learning. A FCM strategy is being proposed to determine the cluster center, introducing the target function and the LMS method to make the cluster width adjusted adaptively, and a pruning strategy is adopted to cut the redundant hidden nodes. The simulation results in the nearest neighbor-clustering based on hierarchical optimization strategy show that the algorithms are greatly improved in the learning accuracy and speed.

1. Introduction

RBF (RBF-Radial Basis Function) [1] ~ [6] is a neural network with a single hidden layer of the three layers. Because of RBF neural network with a simple network structure, rapid learning, and better promotion of capacity, it has been widely applied to function approximation, system identification, time series prediction, speech recognition, signal processing, real-time optimal control, combinatorial optimization, knowledge engineering, and many other areas. it need to determine the parameters of the base functions, including an implied value of the center, width, hidden layer to the output of the value and the right to connect nodes and other hidden layer in RBF neural networks. The system has no laws to follow on How to effectively determine the RBF neural network structure and network parameters. A nearest neighbor-clustering algorithm is adopted in the paper [1] for the RBF neural network learning, which is forming clusters adaptively on line. It does not need to determine the number of hidden nodes unit in advance, the RBF neural network can achieve good result by using the nearest neighbor-clustering algorithm and the algorithm is done on line learning. But it is inadequate in the cluster center determining, the acquisition of the cluster width, and the number of clusters studying. Therefore, in this paper, it presents a hierarchical optimization strategy to improve nearest neighbor-clustering algorithm and a

nonlinear function approaching is done by using the hierarchical optimization strategy, the simulation results show that the designed algorithm is simple, and having fast learning speed and high precision.

2. RBF neural network

RBF neural network has three layers, namely the input layer, hidden layer, the output layer, the input of the Vector right is 1, hidden units is mapped to the output layer by the Gaussian Function, as shown in (1).

$$R_i(X) = \exp(\frac{-\|x - c_i\|^2}{2\sigma^2}) \quad i = 1,2 \cdots N \quad (1)$$

$R_i(x)$: Hidden layer output of the number i ;

c_i : Cluster center of number i (center of Gaussian Function);

σ : Cluster width (width of Radial Basis Function);

N : The number of the clusters (the number of the hidden nodes).

The output of the RBF neural network is shown as (2).

$$f = \frac{\sum_{i=1}^{n} w_i \exp(-\frac{\|x^k - c_i\|^2}{\sigma^2})}{\sum_{i=1}^{n} \exp(-\frac{\|x^k - c_i\|^2}{\sigma^2})} \quad (2)$$

3. Nearest Neighbor-Clustering Algorithm
3.1. Nearest Neighbor-Clustering Algorithm steps

Nearest neighbor clustering algorithm [1] ~ [2] is one of the most simple, most basic cluster algorithms. It main idea is the cluster center is made by the first data input; next, the other data is examined by computing their distance d_{ik} to the existing clusters.

If $d_{ik \min} < \sigma$, (the σ is a predetermined value,) then the data is put into the exact cluster;

978-0-7695-3342-1/08 $25.00 © 2008 IEEE
DOI 10.1109/PEITS.2008.55

If $d_{ik\min} > \sigma$, a new cluster is set, the new cluster is made by the data input.

Specific algorithm as follows:

Step 1.for the first data x^1, making x^1 as the first cluster c_1, and setting the cluster width as σ.

Step 2.considering the data x^k, assumptions that it has existed n cluster centers, such as $c_1, c_2, c_3 \cdots c_n$, figuring the data x^k to the cluster's distance d_{ik}.

$$i = 1, 2, 3 \cdots n$$

If $d_{ik\min} < \sigma$, then making the data x^k as the cluster i.

If $d_{ik\min} > \sigma$, making the data x^k as the new cluster center, and $c_{n+1} = x^k$.

Step3.repeat steps 2, until all the dates are classified.

3.2. The shortcoming of Nearest Neighbor-Clustering Algorithm

(1)For the cluster center, the nearest neighbor clustering algorithm always make the first data input of the neural woks as the cluster center, this certainly can make the algorithm simple, but each cluster node may have many data such as $x^1, x^2, \cdots x^m$, the cluster center may not be excellent by choosing the first data, especially if $x^m > \cdots > x^2 > x^1$, $y^m > \cdots > y^2 > y^1$, which can not outstand the cluster center, in some papers, that is also proved the choice is unreasonable.

(2)It is only considering the similarity of the input dates, but not considering on the function output. The cluster formation do not fully reflect the relationship between input and output, it is also not using the output error function to learn the RBF neural network. It is fast in the speed, but it is at the expense of learning accuracy and precision, which is restricted in the occasions of needing high accuracy and precision.

(3)The cluster width σ can not change once determined, in fact the width σ should be changed ad aptly, because the width σ determines the number of the clusters, and it affects the speed and accuracy finally. The determining of σ needs experiments or expert giving, which is not conductive to the application of algorithm.

(4)The nearest neighbor clustering algorithm can adjust the hidden layer nodes adeptly, but it can only generate the hidden nodes, and this generation is increasing monotonous, for some large numbers of dates input, which will generate many redundant nodes without controlling the nodes, it will make the neural network too large and can not achieve good balance between fast learning speed and accuracy. Though it should control the

generation of the hidden nodes and make the nodes in a good level.

4. Hierarchical algorithm optimization strategy

Hierarchical optimization strategy: firstly, using the FCM (Fuzzy C-means) to determine the cluster center, secondly, using the target function and LMS algorithm make the cluster width adjusted adeptly, finally, introducing the pruning strategy to cut the redundant nodes.

4.1. Optimization the cluster center

Using the FCM to determine the cluster center, the object of the FCM is finding the $U \in [\mu_{ik}] \in M_f$ and $c = (c_1, c_2, \cdots, c_m)$, making the minimum of J.

$$J = \sum_{i=1}^{m} \sum_{k=1}^{n} (\mu_{ik})^2 \left\| x^k - c_i \right\|^2 \qquad (3)$$

$$(i = 1, 2, \cdots, m; k = 1, 2, \cdots n)$$

FCM algorithm as follows:

Step1.initial the parameter m, m is the number of cluster nodes given by the nearest neighbor clustering algorithm.

Step2.calculate μ_{ik}

$$\mu_{ik} = [\sum_{j=1}^{m} (\frac{x^k - c_i}{x^k - c_j})^2]^{-1} \qquad (4)$$

$$(i = 1, 2, \cdots, m; k = 1, 2, \cdots n)$$

Step3.update the c_i as follows:

$$c_i = \frac{\sum_{k=1}^{n} x^k (\mu_{ik})^2}{\sum_{k=1}^{n} (\mu_{ik})^2}, \quad (i = 1, 2, \cdots, m) \qquad (5)$$

Step4.if $\left\| U^k - U^{k-1} \right\| < \varepsilon$, then stop, (ε is a set threshold)

Else continue the step 2.

4.2. Adjust the cluster width σ adaptively

Choosing the target function (6) and the LMS algorithm adjust the cluster width σ.

$$E = \frac{1}{2}\sum_{i=1}^{k}(y - y_m)^2 \qquad (6)$$

$$\sigma_{k+1} = \sigma_k + \mu \frac{\partial E}{\partial \sigma_k} \qquad (7)$$

y : The target output;

y_m : The actual output;

μ : A set threshold.

4.3 Prune the redundant hidden nodes

The number of the hidden nodes determines the accuracy of the output, generally speaking, the more the hidden nodes, the higher learning accuracy of the system. At the same time, the learning time will be too long if the number of hidden nodes is too large in number, so it should prune some redundant nodes properly. Here define the nodes which are little affected to the output of the network as the redundant nodes, and remove the redundant nodes. Using the following pruning standard [5] to improve the hidden layer, and the specific standards as follows:

Step1.calculate all the hidden nodes output of the Radial Basis Function. $i = 1, 2, \cdots N$,

N : The number of the hidden nodes at present.

$$y_{k(i)} = w_i \exp(-\frac{\left\| x^k - c_i \right\|^2}{\sigma_k^2}) \qquad (8)$$

Step2.calculate the maximum absolute value of the hidden Radial Basis Function.

$$y_{k(i)\max} = \max(\left| y_{k(i)} \right|), i = 1, 2 \cdots N \qquad (9)$$

Step3.calculate each formal value of the Radial Basis Function output.

$$k_i = \left| \frac{y_{k(i)}}{y_{k(i)\max}} \right|, i = 1, 2 \cdots N \qquad (10)$$

Step4.if $k_i < \sigma_k$, then removes the node i , and prunes the w_i and the corresponding elements.

Step5.repeat the step1 to step 4 for all the dates input.

5. Experiment simulation and analysis

In this paper, introducing a non-linear function approximation problem to proof the superiority of the hierarchical optimization strategy.

The function is shown as （11）.

$$y = \sin(x) + \cos(x) + 0.1x^2 - 0.01x, x \in R \quad (11)$$

The number of the sample is 100, NUME=100, the sample is in [-5-5] at random.

Now, compare the algorithm HOSNNC (Nearest Neighbor-Clustering Algorithm Based on Hierarchical Optimization Strategy) and NNC (Nearest Neighbor-Clustering Algorithm) on three conditions.

(1)It only adjust the cluster by using the FCM, and name it algorithm NNC1, it only give the simulation graph NNC due to the space constraints.

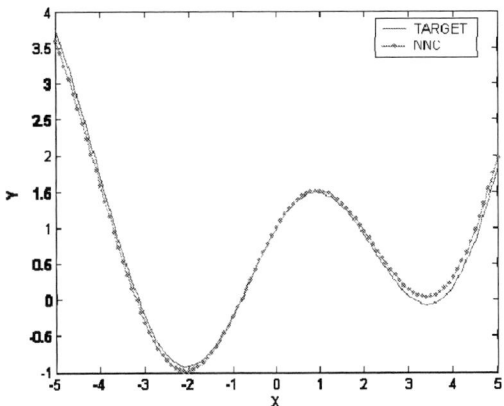

Figure 1 Function Approximation of NNC

Table 1 the comparison of NNC and NNC1

	Training time	Final hidden nodes	System error
NNC	300ms	41	0.180
NNC1	330ms	40	0.140

From the table 1, we can know by using the FCM in the algorithm NNC1, although the training time increased by 30 ms, but the system error had decreased by 20%.it can choose the algorithm NNC1 on occasions which need high accuracy but don't care learning time too much.

(2)Introducing the target function and LMS on the algorithm NNC1, adjust the cluster width σ adaptively, and name it algorithm NNC2.

Table 2 the comparison of NNC1 and NNC2

	Training time	Final hidden nodes	System error
NNC1	330ms	40	0.140
NNC2	398ms	35	0.080

From Table 2, when the cluster width σ is adjusted by the algorithm NNC2 ,the final number of cluster is declined by 5, the system error drops 42.9 percent, but only an increase of training time 68 ms.

(3)Finally, we introduce the pruning redundant nodes strategy on the basis of NNC2, namely the algorithm HOSNNC. The simulation map is as figure 2

235

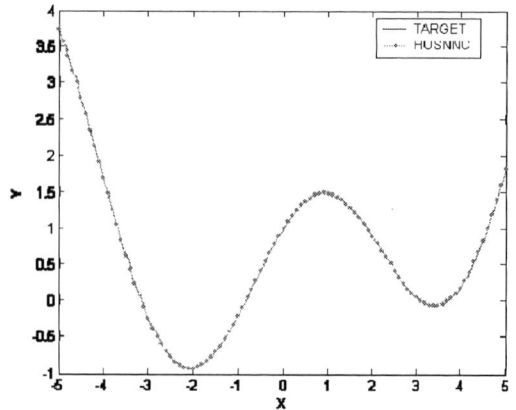

Figure 2 Function Approximation of HOSNNC

Table 3 the comparison of NNC2 and HOSNNC

	Training time	Final hidden nodes	System error
NNC2	398ms	35	0.080
HOSNNC	280ms	22	0.019

From the table 3, we know by using the pruning redundant nodes strategy in algorithm HOSNNC, the number of the cluster nodes has fallen to 22 from 35, which is nearly about 40%.it is known to all the number of the cluster nodes greatly affect the training time. Although increasing the steps of the computing, the decline in the number of cluster nodes largely offset the study time, it also reflects the superiority of algorithm by the decline in training time of 118 ms, and the system error also drops to 0.019 from 0.080.

It indicates that the training time has fallen 15%from table 1, table 3, figure 1 and figure3.the number of the cluster nodes has fallen to 22 from 40,and the system error has fallen to 0.019 from 0.180.it can make good balance between the training time and the learning accuracy by using the designed algorithm of HOSNNC .

6. Conclusions

This paper proposes a nearest neighbor clustering algorithm based on hierarchical optimization strategy. Introducing the FCM algorithm to determine the cluster center. Choosing the target function and the LMS algorithm adjust the cluster width σ adaptively, which improve the system's learning accuracy and promotion. Introducing pruning standard to improve the hidden layer, which can make the number of the hidden nodes in good level. The simulation results in the nearest neighbor-clustering based on hierarchical optimization strategy show the algorithms are greatly improved in the learning accuracy and speed. The designed algorithm has certain reference value and application value.

Acknowledgement

This work was supported by Fund for Outstanding Researchers of Henan Province, China, under Grant 074200510013

References

[1]Jian Zhou,Zhiyang Shui, "Improved Nearest Neighbor-Clustering Algorithm ",journal of Shandong institute of building material",1999 pp.122-124.
[2]Mingxing Zhu,Delong Zhang, "Study on the Algorithms of Selecting the Radial Basis Function Center", journal of Anhui university natural science edition",2000.pp.74-78.
[3]Qiurning Zhu, "A Global Learning Algorithm for a RBF networks", Neural Networks, 1999, PP.52-100.
[4]XU Lethal, "Rival Penalized Competitive Leaning for Clustering Analysis RBF Net and Curve Detections", IEEE Trans on NN, 1993, pp.636 -649.
[5]Yingwei L, Sundarajan N. and Saratchandran P.A, "Sequential Learning Scheme for Function Approximation Using Minimal Radial Basis Function Neural Networks", Neural Computation, 1997, pp.461-478.
[6]Euktai K, Park M, Sevnghwan J, etal, "A new approach to fuzzy modeling", IEEE Trans. On Fuzzy System,1997 PP.1029 -1039.

2008 Workshop on Power Electronics and Intelligent Transportation System

Model and Algorithm for Initial Route Planning

Shiyan Xu[1], Yuyao He [2]

[1] *School of Electronic and Control Engineering,*
Chang'an University, Xi'an, Shaanxi, 710064, China
2 College of Marine Engineering,
Northwestern Polytechnical University , Xi'an, Shaanxi,710072, China
shyxu@chd.edu.cn, yyhe@chd.edu.cn

Abstract

In this paper, we propose a user-optimal model, which takes into account the user equilibrium for initial route planning in the plain area. We consider a city with five highly compact central business districts (CBD). The road network of the city is relatively dense and is considered as a continuum. The traffic demand from any destination in the city to the CBD is assumed to be a function of both the destination location and the total travel cost to the CBD. Each commuter is assumed to patronize a CBD to minimize individual total cost. The total cost depends on the location, the traffic flow intensity and a cost associated with the congestion externality at the CBD. The model is firstly formulated as a calculus of variations problem. Then finite element method and evolutionary structural optimization (ESO) is used to solve and optimize the model. Finally we find the initial location of future artery through solving the continuum user equilibrium problem. A numerical example is given to demonstrate the effectiveness of the proposed methodology.

1. Introduction

Determining the best route through an area is one of the oldest spatial problems. The route scheme selection shall be based on technical standard, taking into considerations of such factor as cost, geology, vegetation, landuse, landcover and flood control, risk probabilities and coordination with local development planning program, etc.. The conventional route planning has solely been based on topographical considerations-gradient and curvature. During the last decade, a few attempts have been made to solve the route-planning problem, such as Remote Sensing technologies [1], GIS [2] and virtual reality systems [3]. Reference [4] presents a state-of-the-art review in three-dimensional (3D) analysis for transportation planning and highway design. All of these researches emphasize the technic and geology.

Road is built for commuters ultimately. But commuters' activity as an important condition of route planning is taken as powerful factor by few of researches. In this paper, traffic equilibrium and user-optimal pattern are used to plan initial route in the plain area. For emphasizing the effect of commuters who use the roads, Traffic equilibrium is imported to describe the commuters' behavior flexibly and effectively.

When modeling traffic equilibrium problems in a transportation system, there are two approaches classified in the literature: discrete modeling and continuum modeling. The discrete modeling approach is commonly adopted for the detailed planning and analysis of a transportation system. However, in the initial planning or regional study of transportation networks, for which the detailed information about road links either does not exist or is not readily available, the discrete modeling approach is not as effective. Moreover it's difficult for discrete modeling to plan the alignment and location of the road. In the continuum modeling approach, we are more interested in the general trend and pattern of the distribution and travel choice of users in the transportation system at the macroscopic rather than detailed level. Planners are more concerned about the philosophy and concept of the system design at this stage, such as the locations of an additional CBD, the expansion of the transportation system in different parts of the city and the locations of road, etc. [5]

A number of researches can be found in the literature about the continuous modeling of transportation network. There is promising formulation and solution algorithm for the continuum modeling of a problem with an arbitrary region configuration, in which the user equilibrium problem was formulated as a mathematical program [6]. The finite element method (FEM) [7] was then used to solve the program to determine the cost potential and flow intensity in the city. Based on this approach, reference [8] proposed a robust algorithm. Reference [9] developed continuum models for the determination of the market areas of competitive facilities. They made a substantial extension of the earlier continuous spatial modeling approach in spatial economics. Reference [10] set up a bi-

978-0-7695-3342-1/08 $25.00 © 2008 IEEE
DOI 10.1109/PEITS.2008.75

237

level model to find the optimal housing provision pattern that maximizes the total utility of the system. Although all of these works has formulated and analyzed the traffic equilibrium problem in considerable depth for continuous approaches, they don't consider the problem of route selection and alignment. However, in the initial phase of design, planners are more concerned about land-use problem. They want to know where should be road, residential area and CBD etc.

In light of these methodological advances, we propose a user-optimal model and use the continuous equilibrium approach to solve the location and alignment problem of city artery. A city with five CBD is considered. In the city, the commuters' origins are dependent on the total travel cost to the CBD. The users' route choices are assumed to follow the user equilibrium principle. That is, all of the users are free to make their choice of CBD and their optimal routes within the continuum to minimize their total facility and travel cost. Evolutionary structural optimization (ESO) method developed by reference [11], is reformed to be used to optimize distribution of the city. Through slowly removing inefficient material from the city, the initial location of artery which may provides a farther consultation for the planner is gotten.

The paper is organized as follow: In the next section, the model city with five CBD is described. The notation to be used and the specific equations of the problem are defined. In section3, the problem formulation is described, and Newtonian algorithm is proposed to solve the resultant problem. In section 4, the ESO method is used to optimize the structure of city to get the location of artery. A numerical example that demonstrates the effectiveness of the model is given in section 5.

2. Definitions and notation

Consider a city with five CBD as shown in Fig. 1, in which the transportation network is approximated as a continuum [12]. It is assumed that the CBD is sufficiently compact compared with the suburb of the city. The homes of commuters are widely dispersed over the whole city. The commuters will travel from their home locations through the continuum to patronize any one of the CBD. Denote the region of the city as Ω, the boundary of the city as Γ, and the location of the CBD as On, n=0, 1,···4. It is assumed that the CBD is of finite size embraced by a clockwise boundary segment Γcn, n=0, 1,··· 4. Denote Ω_n as the area of CBDn. The distribution of the travelers over the city Ω is assumed to be continuous and represented by a nonnegative, heterogeneous density function $q(x, y)$ where q is the total demand per unit area from the home location $(x, y) \in \Omega$. To consider the elasticity of travel demand, $q(x, y)$ is assumed to be a function of the minimum travel cost

$$q(x, y) = D[u(x, y), x, y] \qquad (1)$$

where $u(x, y)$ and $q(x, y)\mathrm{d}x\mathrm{d}y$, with $\mathrm{d}x$ a small distance around (x, y) measured in the x-direction, are the minimum travel cost and the travel demand that are generated from the location $(x, y) \in \Omega$ to travel to the CBD. The function D(•) is assumed to be monotonic decreasing to reflect the elastic nature of travel demand with respect to the total cost, and its inverse function exists.

The flow state at different locations in the city region is represented by a flow vector, $\mathbf{f}(x, y) = (f_x(x, y), f_y(x, y))$, where $f_x(x, y)$ and $f_y(x, y)$ are the flow flux in the x and y directions (measured in trips per unit time per unit width) respectively, which are the magnitude of the flow vector projected onto the x and y directions. The flow vector represents the movement direction at location $(x, y) \in \Omega$ when heading to CBDn in the two-dimensional plane. The flow intensity is defined as

$$\left| \mathbf{f}(x, y) \right| = \sqrt{f_x(x, y)^2 + f_y(x, y)^2} \qquad (2)$$

which is the norm of the flow vector at (x, y). This flow intensity can be interpreted as the traffic flow entering into a small sub-area embracing the point (x, y) through a unit width on the boundary (perpendicular to the flow direction). It is reasonable to assume that the higher the flow intensity, the more congested is within the sub-area due to greater interactions between traffic factors.

The local travel cost in the continuous domain of the city is assumed to be dependent on the local flow intensity and road configuration but not on direction (the isotopic case),

$$c(x, y, \mathbf{f}) = a(x, y) + b(x, y)\left| \mathbf{f}(x, y) \right|^{\gamma(x, y)} \qquad (3)$$

where $c(x, y, \mathbf{f})$ is the cost per unit distance of travel at co-ordinate $(x, y) \in \Omega$, and $a(x, y)$, $b(x, y)$ and $\gamma(x, y)$ are strictly positive scalar functions of the cost-flow relationship reflecting the local characteristics of the streets.

Suppose that a commuter makes a choice of the CBD that maximizes his or her satisfaction or minimize his or her generalized cost. Transportation cost and externality cost are considered to be the two major factors that make impacts on commuter spatial choice behavior. Thus, we use a deterministic, generalized cost function to represent the cost for commuters to patronize each facility,

$$G[H(x, y), O_n] = C[H(x, y), O_n] + S_n(Q_n) \qquad (4)$$

where $G[H(x, y), O_n]$ is the generalized cost for commuters whose homes are located at point $H(x, y)$ to patronize location of CBD O_n (n=0,1, $\cdots$ 4), $C[H(x,y),O_n]$ is the minimum transportation cost traveling from home location $H(x,y)$ to O_n, $S_n(Q_n)$ is the market externality function for CBDn, which is expressed in the same unit as transportation cost. The market externality function indicates the degree of market share of the CBD. These externality function can be decreasing to represent positive externality (efficiency), or increasing for negative externality (congestion), or convex for the cases of positive externality when market share is small and negative externality when market share is large.

Inside the domain of the city Ω, the flow vector and trip demand must satisfy the flow conservation condition as

$$\nabla \mathbf{f}(x,y) + q(x,y) = 0, \qquad \in \Omega \qquad (5)$$

Assuming that there is no traffic flow on the boundary of the city, we have

$$\mathbf{f} = \mathbf{0}, \qquad \text{on } \Gamma \qquad (6)$$

The commuters that will be attracted to a CBD becomes

$$Q_n = \iint_{\Omega_n} q(x,y)\mathbf{d}\Omega, \qquad n = 0,1,\cdots 4 \qquad (7)$$

From flow conservation principle at CBDn, we have

$$\oint_{\Gamma_{cn}} \mathbf{f} \cdot \mathbf{nd}\Gamma + Q_n = 0, \qquad n = \mathbf{0,1},\cdots 4 \qquad (8)$$

3. Formulation of the problem and solution algorithm

Given the fixed location and characteristics of each competitive CBD over the space, a spatial user-optimal equilibrium is an allocation of all commuter demand among the CBD such that each commuter who is located at H(x, y) is attracted by the CBD O_n so that the following two conditions are satisfied and the shortest path is used. One condition is that commuter patronizes the CBD with the lowest generalized cost in a deterministic manner. The other condition is that the elasticity of customer demand u(x, y) is assumed to be a function of the minimum generalized cost (transportation and externality cost) for commuters whose homes are located at point H(x, y) to patronize any of the CBD.

According Wardrop's user equilibrium condition, the travel time of any unused paths is greater than or equal to that of the used paths. We formulate the problem of user equilibrium for the case of elastic demand as the following mathematical model.

minimise $z(\mathbf{f}) = \sum_{n=0}^{1} \int_{0}^{Q_n} S_n(\zeta)\mathbf{d}\zeta + \iint_{\Omega}[a|\mathbf{f}| + \frac{b}{\gamma+1}|\mathbf{f}|^{\gamma+1}]\mathbf{d}\Omega$

$$(9a)$$

subject to

$$\nabla \mathbf{f}(x,y) + q(x,y) = 0, \qquad \forall (x,y) \in \Omega \qquad (9b)$$

$$\mathbf{f} = \mathbf{0}, \qquad \forall (x,y) \in \Gamma \qquad (9c)$$

$$u(x,y) = 0, \qquad \forall (x,y) \in \Gamma_c \qquad (9d)$$

$$\oint_{\Gamma_{cn}} \mathbf{f} \cdot \mathbf{nd}\Gamma + Q_n = 0, \qquad n = \mathbf{0,1},\cdots 4 \qquad (9e)$$

In this study, the whole city region is first discretized into a set of finite element and following modified Lagrangian function is used:

$$\overline{\Pi} = \sum_{i=1}^{N_E} \iint_{\Omega_i} (a|\mathbf{f}| + \frac{b}{\gamma+1}|\mathbf{f}|^{\gamma+1} + v(\nabla \mathbf{f} + q))\mathbf{d}\Omega + \sum_{n=0}^{4} \int_{0}^{Q_n} S_n(\zeta)\mathbf{d}\zeta +$$

$$\sum_{i=1}^{N_E} \int_{\partial\Omega_i \cap \Gamma} \mathbf{w}_\Gamma \cdot \mathbf{fd}\Gamma + \sum_{i=1}^{N_E} \int_{\partial\Omega_i \cap \Gamma_c} \sigma_{\Gamma_c} u\mathbf{d}\Gamma + \sum_{n=0}^{4} \pi_n (\oint_{\Gamma_{cn}} \mathbf{f} \cdot \mathbf{nd}\Gamma + Q_n)$$

$$(10)$$

where N_E is the number of elements in the mesh, Ω_i and $\partial\Omega_i$ are the domain and the boundary of the ith element respectively, and $v(x,y)$, $\mathbf{w}_\Gamma(x,y) = (w_{\Gamma x}(x,y) + w_{\Gamma y}(x,y))$, σ_{Γ_c} and π_n are the Lagrange multipliers associated with the constraints (9b), (9c), (9d) and (9e), respectively. Let $\delta\mathbf{f} = (\delta f_x, \delta f_y)$ be arbitrary function which vanish on the boundaries of the domain, i.e.

$$\delta\mathbf{f} = 0, \qquad \text{on } \Gamma \qquad (11)$$

We apply the shape function of the three-node linear triangle finite element to interpolate the variation of the flow vectors and cost potential functions. Let M_N is the number of nodes in the mesh, M_B is the number of nodes along Γ and M_C is the number of nodes along Γ_c. (x_i,y_i), (x_j,y_j) and (x_k,y_k) are the nodal coordinates of the triangular element. The Lagrangian in Eq. (10) can now be re-written as $\overline{\Pi}(\mathbf{\Psi})$, where

$$\mathbf{\Psi} = \text{Col}(\overline{\mathbf{f}}, \overline{\mathbf{u}}, \overline{\mathbf{w}}, \overline{\sigma}) \quad ,$$

$$\overline{\mathbf{f}} = \text{Col}(f_{xi}, f_{yi}, i = 1,2,\cdots, M_N) \quad ,$$

$$\overline{\mathbf{u}} = \text{Col}(u_i, i = 1,2,\cdots, M_N) \quad ,$$

$$\overline{\mathbf{w}} = \text{Col}(w_{xi}, w_{yi}, i = 1,2,\cdots, M_B) \quad ,$$

$$\overline{\sigma} = \text{Col}(\sigma_i, i = 1,2,\cdots, M_c) .$$

Now, let $\mathbf{\Psi}^0$ be an approximate solution of the problem. Expanding $\overline{\Pi}(\mathbf{\Psi})$ by Taylor's series at $\mathbf{\Psi}^0$ and neglecting the higher order term, we have

$$\overline{\Pi}(\mathbf{\Psi}) \cong \overline{\Pi}(\mathbf{\Psi}^0) + R^T(\mathbf{\Psi} - \mathbf{\Psi}^0) + \frac{1}{2}(\mathbf{\Psi} - \mathbf{\Psi}^0)^T H(\mathbf{\Psi} - \mathbf{\Psi}^0)$$

$$(12)$$

where $R = \nabla\overline{\Pi}(\mathbf{\Psi}^0)$ is the residual vector and $H = \nabla^2\overline{\Pi}(\mathbf{\Psi}^0)$ is the Hessian matrix of the Lagrangian.

For a stationary point, the derivatives with respect to all of the variables vanish (i.e. $\nabla \overline{\Pi}(\mathbf{\Psi}) = 0$). After rearranging, we can show that

$$\mathbf{\Psi} \cong \mathbf{\Psi}^0 - H^{-1}R \qquad (13)$$

This is also known as the Newtonian method for optimization problems. The residual vector and Hessian matrix can be evaluated in a similar way as that was reported in reference [8].

4. Structural optimization of the city

By slowly removing inefficient material from a structure, the shape of the structure evolves towards an optimum. This is the concept of ESO. A reliable sign of potential structural failure is excessive stress, on the other hand a reliable sign of inefficient material use is low stress. Ideally the stress on every part of the structure should be the same.

In this study, we use a rejection criterion based on local flow intensity level, where lowly flow stressed part is assumed to be under-utilized and will be removed. By gradually removing part with lower flow intensity, the flow intensity level in the new designs becomes more and more uniform. The detailed procedure is described as follows:

Step 1: Assume the traffic demand all over the city and evaluate the flow intensity of every finite element f_i, then find the maximum flow intensity of element f_{max}.

Step 2: If $(f_i / f_{max}) \leq RR_n$, an element displacement rate, then remove the element i and singular elements. If there is no more element removed, set $RR_n = RR_n + ER$, where ER is an evolution rate.

Step 3: load the traffic demand all over the city again and evaluate f_i and f_{max}, if all of element flow intensity $f_i \leq f_{lim}$, where f_{lim} is limitative flow intensity of element, go to Step 2.

Step 4: otherwise stop and the elements remained is solution.

5. Numerical example

Consider the modeled city. The finite element mesh shown in Figure 1 is used for the analysis. We take the similar demand and cost-flow relationship function as reference [9].

$$q = 100e^{-0.5u}, \quad \text{veh/h/km}^2$$

where u is measured in hours. The cost-flow relationship is specified as

$$c = 0.01 + 0.1 \times 10^{-4} |\mathbf{f}|^{1.2}, \quad \text{h/km}$$

throughout the city. The quadratic convex externality cost function for the CBD is given as

$$S_n(x) = 1.8 - 0.2 \times 10^{-3} x + 0.1 \times 10^{-7} x^2$$

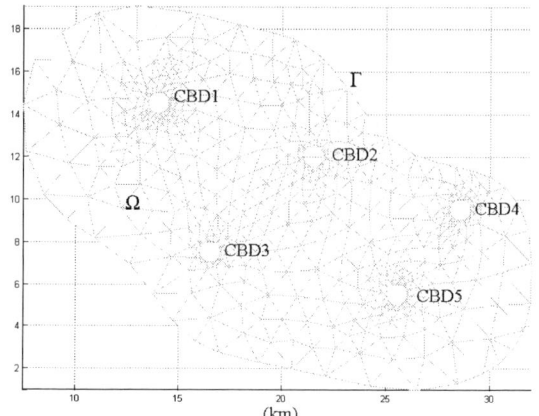

Figure 1 The modeled city

$\mathbf{\Psi}^0$, the initial guess solution of the problem, is arbitrarily chosen as an unit vector $\{\mathbf{1}\}$ with all elements being equal to 1, and the acceptable error in the Newtonian algorithm is set as 10^{-3}. In this example, the initial contour plot of the flow intensity is given in Figure 2 from which the location with high traffic flow values is shown.

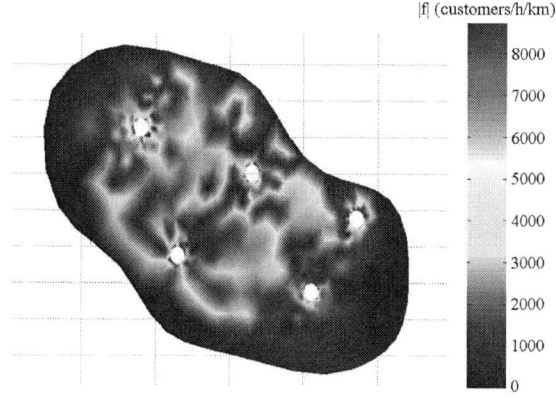

Figure 2 The initial contour plot of the flow intensity

Seting $ER = 0.1\%, RR_0 = 1\%$ and $f_{lim} = 10000$ veh/h/km [12], deleted and deformed configuration of city is shown in Figure 3, after ESO method is used to remove inefficient elements. It indicates the initial location of future artery.

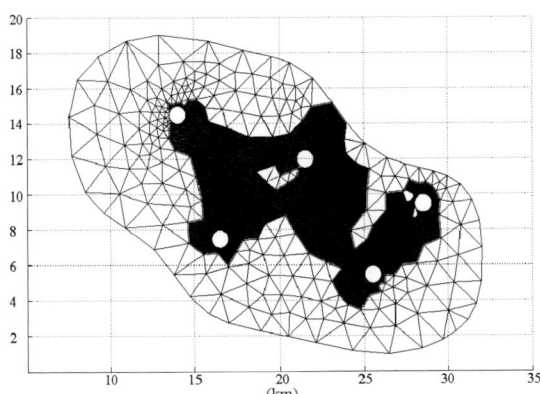

Figure 3 Topology structure on delete rate 44.53%

6. Conclusions

We have considered a city with five CBD. Within the city region, the road network is relatively dense and is considered to be a continuum firstly. We have assumed a user equilibrium condition that every user chooses a CBD and an optimal route within this continuum to the CBD, such that the total traveling and facility cost is minimized. The continuum approximation of network flow has been developed to deal with the initial location of future artery. A mathematical model has been set up to model the above problem. A finite element solution algorithm has been given to solve the resultant program. A structural optimization method has been used to get the location of future artery. Finally, a numerical example has been given to show the effectiveness of the mathematical model, the solution algorithm and the feasibility of the ESO method. We use structural optimization method to solve the land-use problem and improve the process of the user equilibrium problem solved by FEM. The satisfying location and alignment of artery can be provided though this method, if the size of the elements is appropriate. It supplies an extension for the future research.

Acknowledgement

We are grateful to Professor S. C. Wong and Dr. H. W. Ho for kindly guidance. The first author was partially supported by the funds of Chang'an University (06E08).

References

[1] H.Z. Guo and Y.J. Sun (2001). "Application of remote sensing technology in engineering geological survey of datong-Yuncheng highway (Huozhonu-Linfen part)". *ACRS/2001 22nd Asian Conference on Remote Sensing*, Singapore, pp. 639-642.

[2] M. R. Delavar, F. Samadzadegan and P. Pahlavani (2004). "A GIS-Assisted optimal urban route finding approach based on genetic algorithms". *International Archives of Photogrammetry Remote Sensing and Spatial Information Sciences*, Vol. 35, No. 2, pp. 305-308.

[3] H. Heldal (2007). "Supporting participation in planning new roads by using virtual reality systems". *Virtual Reality*, Vol 11, pp. 145-159.

[4] S. M. Easa, T. R. Strauss, Y. Hassan and R. R. Souleyrette, "Three-dimensional transportation analysis: planning and design". *Journal of Transportation Engineering*, Vol. 128, No. 3, pp. 250-258, 2002.

[5] H.W. Ho, S.C. Wong and B.P.Y. Loo (2006). "Combined distribution and assignment model for a continuum traffic equilibrium problem with multiple user classes". *Transportation Research Part B*, Vol. 40, pp. 633-650.

[6] T. Sasaki, Y. Iida and H. Yang (1990). "User equilibrium traffic assignment by continuum approximation of network flow". *Proceedings of the 11th International Symposium on Transportation and Traffic Theory*, Yokohama, pp. 233–252.

[7] O.C. Zienkiewicz, R.L. Taylor. *The Finite Element Method*. McGraw-Hill International Editions, Woburn, 2000.

[8] S.C. Wong, C. K. Lee and C. O. Tong, "Finite element solution for the continuum traffic equilibrium problems". *International Journal for Numerical Methods in Engineering*, Vol. 43, pp. 1253-1273, 1998.

[9] H. Yang and S.C. Wong (2000), "A continuous equilibrium model for estimating market areas of competitive facilities with elastic demand and market externalities". *Transportation Science*, Vol. 34 (2), pp. 216–227.

[10] H.W. Ho, S.C. Wong (2007). "Housing allocation problem in a continuum transportation system". *Transportmetrica*, Vol. 3, No. 1, pp. 21-39.

[11] Y. M. Xie and G. P. Steven (1993). "A simple evolutionary procedure for structural optimization". *Computers and Structure*, Vol. 149, No. 5, pp. 885-896.

[12] S.C. Wong (1998). "Multi-commodity traffic assignment by continuum approximation of network flow with variable demand". *Transportation Research Part B*, Vol. 32, No. 8, pp. 567–581.

Integration of an Improved Particle Swarm Optimization Algorithm and Fuzzy Neural Network for Shanghai Stock Market Prediction

Huang Fu-yuan[1,2]

[1](School of Economics and Commerce, South China University
of Technology, Guangzhou 510640, China)
[2](School of Business, Zhanjiang normal college, Zhanjiang 524048, China)
Email: hfy338@gmail.com

Abstract

Particle Swarm Optimization (PSO) algorithm and Fuzzy Neural Network (FNN) system has been widely used to solve complex decision making problems in practice. However, both of them more or less suffer from the slow convergence,"black-box" and occasionally involve in a local optimal solution. To overcome these drawbacks of PSO and FNN, in this study an improved particle swarm optimization algorithm (IPSO) is developed and then combined with fuzzy neural network to optimize the network training process. Furthermore, the new IPSO-FNN model has been applied to Shanghai stock market prediction problem, and the results indicate that the predictive accuracies obtained from IPSO-FNN are much higher than the ones obtained from neural network system(NNs). To make this clearer, an illustrative example is also demonstrated in this study. It seems that the proposed new comprehensive evolution algorithm may be an efficient forecasting system in financial time series analysis.

1. Introduction

Forecasting stock markets is a continuous effort of researchers, and attracts great attention of practitioners. The traditional technical analysis methods for stock market,such as K-line figure and moving average etc., are all based on probability theory and statistical analysis, in which some kinds of distributions are often assumed. As a matter of fact, these assumptions generally are not reasonable and non-realistic. Besides, the traditional analysis methods are more or less lack of discrimination due to the non-structured and nonlinear characteristics of stock market prediction.

In the recent years, the Particle Swarm Optimization (PSO) algorithm and the Neural Network system(NNs) have been widely used to solve financial decision-making problems due to their excellent performances in treating non-linear data with self-learning capability (Rast 1997,

Patricia and William 2000, Trond et al, 2001, and Eberhart et al, 2001). However, both of them more or less suffer from the slow convergence, black-box and occasionally involve in a local optimal solution(Arnold 2002,Selwyn 1999). On the other hand, fuzzy logic (FL) as a rule-based development in artificial intelligence can not only tolerate imprecise information, but also make a framework of approximate reasoning,this avoids the black-box. The disadvantage of fuzzy logic is the lack of effective learning capability. To overcome these drawbacks mentioned above, an improved particle swarm optimization algorithm (IPSO) is developed in this study and then combined with fuzzy logic and neural network, named the IPSO-FNN model. Also the new IPSO-FNN model has been applied to the stock market prediction problem and the results indicate that the predictive accuracies obtained from IPSO-FNN are much higher than the ones obtained from NNs. To make this clearer, an illustrative example is also demonstrated in this study.

2. Canonical particle swarm optimization algorithm (CPSO)

Particle swarm optimization (PSO), which is based on the social metaphor of bird flocking or fish schooling, is an evolutionary computation technique developed by Kennedy and Eberhart (Kennedy and Eberhart 1995, Kennedy 1997) in 1995. Although PSO has some similarities with genetic algorithm (GA), it has distinct demarcating features. Instead of using evolutionary operators to manipulate the individuals, it uses collaboration among a population of simple search agents to find optima in problem spaces. The PSO which is comparatively simple in operation and easier to understand when compared to other evolutionary computation techniques, has emerged as an important combinatorial metaheuristic technique for both continuous- time and discrete-time optimization (Eberhart and Shi 2001).

978-0-7695-3342-1/08 $25.00 © 2008 IEEE
DOI 10.1109/PEITS.2008.85

The PSO models the exploration of a problem space by a population of individuals. Each individual is treated as a volume-less particle (a point) in the D-dimensional search space. The i-th particle is represented as $X_i = (x_{i1}, x_{i2}, \cdots, x_{iD})$. The best previous position (the position giving the fitness value better than any it has found previously) of the i-th particle is recorded and represented as $P_i = (p_{i1}, p_{i2}, \cdots, p_{iD})$. The index of the best particle among all the particles in the population is represented by the symbol g. The velocity for particle i is represented as $V_i = (v_{i1}, v_{i2}, \cdots, v_{iD})$. At iteration n, the particles are manipulated according to the following equations:

$$v_{ij}(n) = w * v_{ij}(n-1) + \Delta v_{ij}(n) \qquad (1)$$

$$x_{ij}(n) = x_{ij}(n-1) + v_{ij}(n) \qquad (2)$$

$$\Delta v_{ij}(n) = c_1 * r_1(\cdot) * (p_{ij}(n-1) - x_{ij}(n-1)) + c_2 * r_2(\cdot) * (p_{gj}(n-1) - x_{ij}(n-1)) \qquad (3)$$

Where the Eq.(3) is a correction to the velocity v_{ij}, w is an *inertia weight* common in the range [0.4,0.9], $c_1 = 2.0$ and $c_2 = 2.0$ are two *positive constants*, and $r_1(\cdot)$ and $r_2(\cdot)$ are two random functions in the range [0,1].

In reference [9], the canonical form of the algorithm was:

```
Initialization
Repeat
For i=1 to population size
  CurrentEval= f(x_i)
  If CurrentEval<f(p_i) then do
    For j=1 to Dimension
      p_ij=x_ij
    Next j
    If CurrentEval<f(p_g) then g=i
  End If
  For j=1 to Dimension
    v_ij=w*v_ij+c_1*r_1(·)*(p_ij-x_ij)+
        c_2*r_2(·)*(p_gj-x_ij)
    v_ij=sign(v_ij)*min(abs(v_ij),v_max)
      x_ij=x_ij+v_ij
Next i
Until termination criterion is met
```

where v_{max} is a limit placed on the velocity to keep it from going out of bounds, and function f is evaluated, using the particle's positional coordinates as input values.

3. Improved particle sarm otimization algorithm (IPSO)

Like other global optimized algorithms such as GA and GP etc., particle swarm optimization suffers from slow convergence and oscillations in later iterations (Shi and Eberhart 1998, Eberhart and Shi 2001, Shi and Eberhart 2001). To overcome the drawbacks of PSO and speed up its convergence, an improved particle swarm optimization is proposed thustly.

The method for increasing the rate of convergence yet avoiding the danger of instability is to modify the delta rule of Eq.(3) by including a momentum term, as shown by

$$\Delta v_{ij}(n) = \alpha * \Delta v_{ij}(n-1) + \delta_{ij}(n) \qquad (4)$$

$$\delta_{ij}(n) = c_1 * r_1(\cdot) * (p_{ij}(n-1) - x_{ij}(n-1)) + c_2 * r_2(\cdot) * (p_{gj}(n-1) - x_{ij}(n-1)) \qquad (5)$$

Where α is a positive number called the *momentum constant*. Eq.(4) includes the delta rule of Eq.(3) as a special case (i.e., $\alpha=0$).

In order to see the effect of the sequence of pattern presentations on the velocities due to the momentum constant α, we rewrite Eq.(4) as a time series with index t. The index t goes from the initial time 0 to the current time n. Eq.(4) may be viewed as a first-order difference equation in the velocity correction $\Delta v_{ij}(n)$. Solving this equation for $\Delta v_{ij}(n)$ we have

$$\Delta v_{ij}(n) = \sum_{t=0}^{n} a^{n-t} \delta_{ij}(t) \qquad (6)$$

Based on this relation, we may make the following observations:

(1) The current adjustment $\Delta v_{ij}(n)$ represents the sum of an exponential velocity time series. For the time series to be convergent, the momentum constant must be restricted to the range $0 \leq |\alpha| < 1$. When α is zero, the correction of velocity operates without momentum. Also momentum constant α can be positive or negative, although it is unlikely that a negative α would be used in practice.

(2) When the $\delta_{ij}(t)$ has the same algebraic sign on consecutive iterations, the exponentially velocity sum $\Delta v_{ij}(n)$ grows in magnitude, and so the velocity $v_{ij}(n)$ is adjusted by a large amount. The inclusion of momentum in the PSO algorithm tends to accelerate convergence.

(3) When the $\delta_{ij}(t)$ has opposite sign on consecutive iterations, the exponentially velocity sum $\Delta v_{ij}(n)$ shrinks in magnitude, so the velocity $v_{ij}(n)$ is adjusted by a small amount. The inclusion of momentum in the PSO algorithm has a stabilizing effect in directions that oscillate in sign.

The following two test functions are used to demonstrate the effectiveness of the IPSO: The functions are the sphere in 30 dimensions and Rosenbrock's function in 50 dimensions. Formulas are found in Table 1. All the functions have global minimum of zero and all variables are ranged in [-30,30].

Table 1. Two test functions

Function Name	Expression
Sphere	$f_1(x) = \sum_{i=1}^{n} x_i^2$
Rosenbrock	$f_2(x) = \sum_{i=1}^{n-1}(100(x_{i+1} - x_i^2)^2 + (x_i - 1)^2)$

Figure 1 and Figure 2 show the difference of the CPSO and the modified one. From the two figures, we can draw a conclusion that the IPSO's performance is much better than the canonical one.

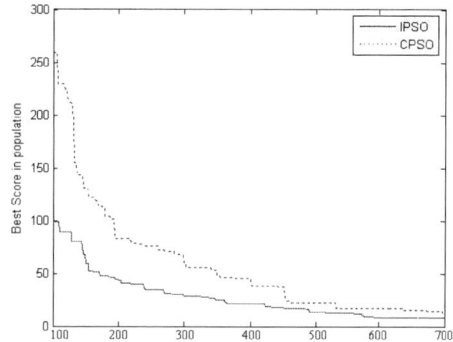

Figure 1. **The Sphere function search procedure**

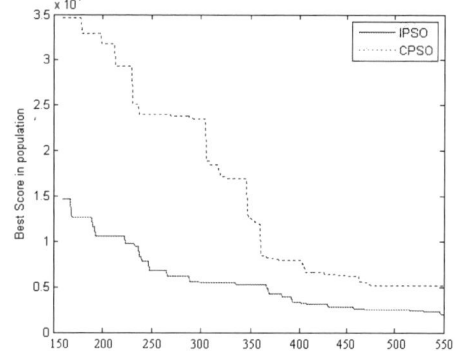

Figure 2. **The Rosenbrock function search procedure**

4. Data source

Stock data from Shanghai Stock Exchange is used as experimental data source.The data contains the daily SSE(Shanghai stock exchange) Composite Index with the corresponding date. Constituents for SSE(Shanghai stock exchange) Composite Index are all listed stocks (A shares and B shares) at Shanghai Stock Exchange. The Base Day for SSE Index is December 19, 1990. The Base period is the total market capitalization of all stocks of that day. The Base Value is 100. The index was launched on July 15, 1991.

About 240 days closing price data(from Mar 15, 2004 to Mar 15, 2005) of Shanghai Stock Exchange Composite Index(SSECI) are selected as the experimental data. The data set is divided into two subsets: one is a training sample set with 192 days data(from Mar 15, 2004 to Dec 28,2004), used to design the NNs model and the IPSO-FNN model; another is a test sample set with 48 days data to test the performance of models. Here, we use past 10 days data to forecast the price of current day.

For the convenience of data processing, the collected raw original data are normalized so that values lie between 0 and 1.

Normalization of the raw data is done using the Eq. (7) below

$$Y_t = \frac{x_t - m}{M - m} \tag{7}$$

$M = \max\{x_t\}, m = \min\{x_t\}, 1 \le t \le 238$

Where x_t is the closing price of SSECI.

5. Models for stock market prediction

5.1. Neural networks model

For comparision, A feed-forward NNs model is presented here, where the input signals are fed forward from the input layer through the networks to the processing units in output layer. The back propagation (BP) algorithm, one of the most popular learning algorithms, is employed as the NNs model's learning algorithm.

In this study, an NNs model with one hidden layer is designed and shown in Figure 3. The number of processing units can be selected by trial and error or by heuristics.

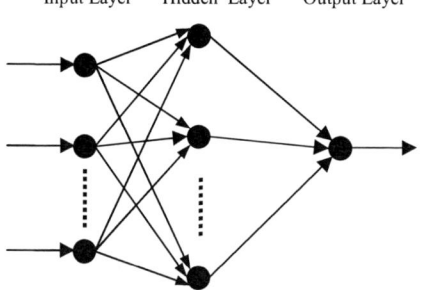

Figure 3. **The structure of NNs**

5.2. Neural networks model

According to the data source mentioned in section 4, a fuzzy neural network model is presented for stock market prediction in this section. Furthermore, to improved the model's learning capibility, an improved PSO, combined with back propagation(BP) algorithm, is employed as the model's learning algorithm.

5.2.1. Model topology structure.
According to reference [12], We can briefly describe the model and provide discussion. Consider a fuzzy system with n inputs $X_i(i=1,2,\cdots,n)$ and one output Y.

$$R_j: \text{IF } x_1 \text{ is } A_{1j} \text{ and } x_2 \text{ is } A_{2j} \text{ and ... and } x_n \text{ is } A_{nj} \quad (8)$$
$$\text{Then } Y_j=f_j(x_1,x_2,\cdots,x_n)$$

Where $R_j(j=1,2,\cdots,m)$ denotes the j-th implication, m is the number of the fuzzy implications of the fuzzy model, $x_1,x_2,\cdots,x_n$ are the premise variables, $A_{ij}(i=1,2,\cdots,n)$ is the fuzzy subset whose membership function is the Gaussian function, Y_j is the consequence of the j-th implication. Given an input $(x_1,x_2,\cdots,x_n)$, then the final output of the fuzzy model is expressed as:

$$Y= w_1*Y_1+w_2*Y_2+\cdots+w_m*Y_m \quad (9)$$

Where w_j is the weight of the fuzzy neural network.

Based on the above fuzzy system the corresponding fuzzy neural network can be constructed. Figure 4 gives a topology of fuzzy network for dealing with the stock market prediction.

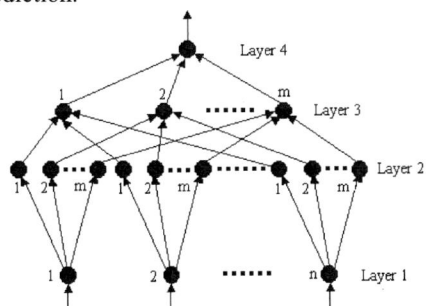

Figure 4. The fuzzy neural network for the stock market prediction

This fuzzy network consists of 4 layers. For clarity, the input-output relationship of various layers is described as:

Layer 1: Input Layer

n input neurons in the first layer are input nodes with the inputs and outputs defined by

$$I_i^{(1)} =O_i^{(1)}= x_i , i=1,2,\cdots,n \quad (10)$$

Where, x_i is the input value of the i-th node, and we have employed the notations $I_i^{(l)}$ and $O_i^{(l)}$ as the input to the i-th node of l-th layer and the output from the i-th node of the l-th layer, respectively.

Layer 2: Fuzzification Layer

$n*m$ fuzzification neurons in the second layer are fuzzy nodes which are divided into n groups, and the inputs and outputs are given by

$$I_{ij}^{(2)}=O_i^{(1)},O_{ij}^{(2)}= \mu_{ij}(x_i)=$$
$$\exp(-\left(\frac{I_{ij}^{(2)}-m_{ij}}{\sigma_{ij}}\right)^2) \quad \begin{matrix} i=1,2,\cdots,n \\ j=1,2,\cdots,m \end{matrix} \quad (11)$$

Where $\mu_{ij}(x_i)$ is Guassian function using as fuzzy membership function, m_{ij} and σ_{ij} are parameters of the Guassion functions that can be trained.

Layer 3: Defuzzification Layer

m neurons in the 3rd layer are defuzzy nodes with the inputs and outputs defined by

$$I_j^{(3)} =\prod_{i=1}^{n} O_{ij}^{(2)} ,O_j^{(3)}= I_j^{(3)}, \quad (12)$$
$$j=1,2,\cdots,m, \ i=1,2,\cdots,n$$

Layer 4:Output Layer

Single neuron in the 4th layer is the output node with the input and output defined by

$$I^{(4)} =\sum_{j=1}^{m} w_j O_j^{(3)},O^{(4)} = I^{(4)} \quad (13)$$

Where w_j is the connection weight between the j-th node of the 3rd layer and the single neuron of the 4th layer.

5.2.2. Model learning algorithm.
In this section, an improved particle swarm optimized fuzzy neural network (IPSO-FNN) is proposed by applying the IPSO to the fuzzy neural network. Here, the IPSO in section 3, is combined with the BP algorithm, has been employed to train the weights and parameters in a fuzzy neural network mentioned in section 5.2.1.

Our objective is to train the output weights and parameters of the FNN so that the fitness function is based on the square sum of error (SSE) of the FNN in each iteration. We propose a training data set as K data pairs, then the fitness function can be given as

$$f = \frac{1}{2}\sum_{i=1}^{K} (y_i - \bar{y}_i)^2 \quad (14)$$

where, $\bar{y}_i$ is the desired output and y_i is the actual output.

The major steps of the training algorithm are as follows:

Step 1: Initialization

A population of $M(M=20,30,50,\cdots$ etc.) particles are initialized with random positions $P_i=(p_{i1},p_{i2},\cdots,p_{iD})$ $p_{ij} \in [P^i_{down},P^i_{up}]$, and velocities $X_i =(x_{i1},x_{i2},\cdots,x_{iD})$ $x_{ij} \in [-V_{max},V_{max}](D=2*m*n+m$, n,m are mentioned in section 5.2.1), using the particles' positional coordinates as the FNN's weights and parameters.

Step 2: Set Parameters

In this step, all the parameters of PSO are initialized, including inertia weight w, momentum constant α, positive constants c_1 and c_2, and so on.

Step 3: Evaluate each particle's fitness value, using Eq.(14)

Step 4: Best Value Calculate and Adjust

Compare evaluation with particle's previous best value *pBest*: If current value<*pBest* then record the current position and value, and compare evaluation with group's previous best value *gBest*: If current value<*gBest* then record the group's best value and reset *gBest* index.

Step 5: Position Change

Change velocity by the Eq.(1), the delta rule use Eq.(4), and use velocity to change position of each particle.

Step 6: Loop Judgment

If a criterion is met then go to step 7 else loop to step 3.

Step 7: BP Operator

If $MSE < MSE_{max}$ then go to step 8 else do continuous BP training.

Step 8: Exit

6. Experiment

6.1. Actual models

Two actual models are constructed in this study based on the training samples. One is a 3-layer feed-forward neural network with one hidden layer. The structure of NNs is a 10-10-1 network model. On the other hand, the actual structure of IPSO-FNN is a 10-30-3-1 network model. Fuzzification layer having 30 nodes, show that each input variable has three fuzzy subsets (each variable is represented as high, mean and low).The output of models are supposed to be the predicted price of SSECI..

According to reference[13], the parameters of IPSO employed to train FNN are initialized as follows: Population size=30; Each particle of population is random initialized. Inertia weight w=0.729; Positive constant c_1=c_2=1.494; Momentum constant α =0.12; V_{up} =-10; V_{down}=10; V_{max}=3; and the end conditions of the training algorithm are: 1. the value of fitness function less than 0.001; 2. the iterations are more than 2000.

The experiments were performed on Matlab 7.0 in Dell PC version Pentium IV. After 2000 iterations, the SSE reach was 0.0010538 and that almost fits the requirements.

6.2. Result analysis

The prediction results of models are pictured in Figure 5-Figure 6. Table 2-3 summarizes the empirical results of the NNs model and the IPSO-FNN model on stock price prediction.

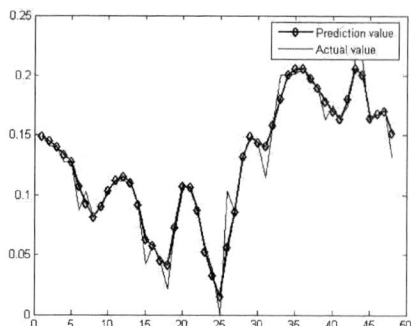

Figure 5. **Obtained result with IPSO-FNN**

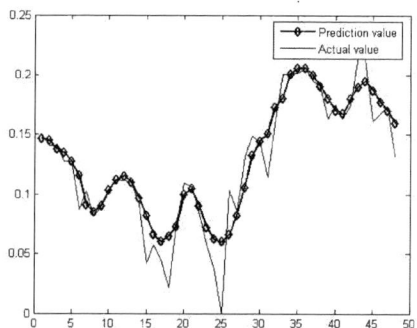

Figure 6. **Obtained result with NNs**

It is noticed that the performances of both models are almost same for the training data. However, The difference of results given by the test data is quite significant. The SSE of NNs model is 0.0083, while the SSE of IPSO-FNN model is 0.003. Specially, the percentage of correct direction(PCD) of the IPSO-FNN model is 89.6%, while the PCD of NNs is only 75%. It is clearly shown that the accuracy of stock price prediction of the IPSO-FNN model is much better than the NNs model. More importantly, the performance of the IPSO-FNN model is superior to the NNs model in increasing PCD and shows considerable promise to screen the SSECI.

Table 2. **SSE of SSECI prediction**

Samples/SSE	NNs	IPSO-FNN
Training samples	0.0010538	0.0010736
Test samples	0.0083	0.0030

Table 3. PCD of SSECI prediction

Samples/PCD	NNs	IPSO-FNN
Training samples	100%	100%
Test samples	75%	89.6%

Where PCD is presented Percentage of Correct Direction and show as follow:

$$PCD = \frac{1}{n}\sum_{i=1}^{240} pcd_i$$

$$pcd_i = \begin{cases} 1, (\overline{y}_{i+1} - \overline{y}_i)(y_{i+1} - y_i) > 0 \\ 0, \qquad\qquad\qquad otherwise \end{cases} \quad (15)$$

$\overline{y}_i$ is the desired output and y_i is the actual output.

7. Conclusion

In this study an improved particle swarm optimization is developed first, and then combined with fuzzy neural network to optimize the network training process. In addition, the new IPSO-FNN model has been applied to the stock market prediction problem, and the results indicate that the predictive accuracies obtained from the IPSO-FNN are much higher than the ones obtained from the NNs. An illustrative example is also given in this study. It seems that the proposed new comprehensive evolution algorithm may be an efficient forecasting system in financial time series analysis.

Acknowledgement

This research is supported by the Specialized Research Fund for the Doctoral Program of Higher Education of China under Grant 20060561002.

References

[1] Patricia Jackson, William Perraudin. Regulatory Implications Of Credit Risk Modeling [J]. Journal of Banking and Finance, 2000, 24, pp. 1-14.

[2] Rast M., Forecasting Financial Time Seres With Fuzzy Neural Network [J].IEEE, 1997, 1(Oct), pp. 28-31.

[3] Trond Eklund, Kai Larsen, Eivind Berhardsen. Model For Analysis Credit Risk In The Enterprise Sector[J]. Norges Bank, 2001, 72 (3), pp. 99-106.

[4] Kennedy J. Eberhart R C.Particle Swarm Optimization [A]. Proceedings of IEEE, International Conference on Neural Networks [C]. Piscataway NJ:IEEE, 1995, pp. 1942-1948.

[5] Kennedy J. The Particle Swarm: Social Adaptation Of Knowledge [A]. Preceedings of Evolutionary Computation [C] ,Indianapolis: IEEE, 1997, pp. 303-308.

[6] Eberhart R C, Shi Y H. Particle Swarm Optimization: Developments, Applications and Resources [A]. Proceedings of the IEEE Congress on Evolutionary Computation [C]. Piscataway, USA: IEEE Service Center, 2001, pp. 81-86.

[7] Arnold F. Shapiro, The merging of neural networks,fuzzy logic, and genetic algorithms, Insurance: Mathematics and Economics 2002, 31, pp. 115-131.

[8] Selwyn Piramuthn, Financial credit=risk evaluation with neural and neurofuzzy systems, European Joumal of Operational Research 1999, 112, pp. 310-321.

[9] Kennedy J. and Eberhart, R. C., with Shi Y. Swarm Intelligence [M]. San Francisco: Morgan Kaufmm Publishers. 2001.

[10] Shi Y H, Eberhart R C. Fuzzy Adaptive Particle Swarm Optimization [A]. Proceedings of the IEEE Congress on Evolutionary Computation [C]. Piscataway, USA: IEEE Service Center, 2001, pp. 101-106.

[11] Shi Y H, Eberhart R C. A Modified Particle Swarm Optimizer [A]. Proceedings of the IEEE Congress on Evolutionary Computation [C]. Piscataway, USA: IEEE Service Center, 1998, pp. 69-73.

[12] TomohiroTakagi, Michio Sugeno. Fuzzy Identification of System and its Application to Modeling and Control. Transaction on Systems, Man,and Cybernetics, 1985, 15 (1), pp. 116-132

[13] Eberhart R, Shi Y. Comparing Inertia Weights and Constriction Factors in Particle Swarm Optimization [C]. IEEECongress on Evolutionary Computation, Piscataway: IEEE Service Center, 2000, pp. 84-88.

The Collaborative Filtering Recommendation Based on Similar-priority and Fuzzy Clustering

SongJie Gong

Zhejiang Business Technology Institute, Ningbo 315012, P. R. China
zjbtigsj@163.com

Abstract

Collaborative filtering technique has been proved to be one of the most successful techniques in recommendation systems in recent years. However, most existing collaborative filtering based recommendation systems suffered from its shortage in scalability as their calculation complexity increased quickly both in time and space when the record in user database increases. So, a new collaborative filtering recommendation based on fuzzy similar-priority comparison and fuzzy clustering is presented. This method uses the fuzzy similar-priority comparison to compute user similarity and uses the fuzzy clustering technology to form nearest neighborhood, and then generates recommendations. The experimental results show that the presented algorithm can improve the performance of systems in the recommendation quality.

1. Introduction

While the rapid growth and wide application of the Internet and information technology has provided an unprecedented abundance of information resources, it has also led to the problem of information overload [1,2]. Thus, methods to help find resources of interest have attracted much attention from both researchers and vendors. To deal with the problem, the personalized recommendation systems play a more important role and collaborative filtering (CF) has proved to be one of the most effective for its simplicity in both theory and implementation.

Collaborative filtering has been successfully used in various applications. The CF method utilizes preference ratings given by various users to determine recommendations to a target user based on the opinions of other similar users [3,4]. A typical CF method employs nearest neighbors approach to derive top-N recommendations. CF algorithms are normally categorized into two general classes: memory-based and model-based algorithms. A memory-based algorithm directly calculates the similarities between the active users and other users or the similarities between the active items and other items. In contrast, a model-based algorithm first constructs a predictive model from the user database, and then uses it to make a prediction. But both the collaborative filtering make the quality of recommendation system decreased dramatically.

In this paper, we propose a collaborative filtering algorithm based on fuzzy similar-priority comparison and fuzzy clustering to improve the recommendation performance. The proposed approach addresses the issue by using the fuzzy similar-priority comparison to compute user similarity and using the fuzzy clustering technology to form nearest neighborhood, and then generates recommendations. Furthermore, the experimental results show that this method can increase the accuracy of the predicted values, resulting in improving recommendation quality of the collaborative filtering recommender system.

2. Similarity measures

A set of similarity measures are presented and a metric of relevance between two vectors. When the values of these vectors are associated with a user's model then the similarity is called user based similarity, whereas when they are associated with an item's model then it is called item-based similarity. The similarity measure can be effectively used to balance the ratings significance in a prediction algorithm and therefore to improve accuracy.

There are several similarity algorithms that have been used: Pearson correlation, cosine vector similarity, adjusted cosine vector similarity, mean-squared difference and Spearman correlation.

Pearson's correlation, as following, measures the linear correlation between two vectors of ratings.

$$sim(i,j) = \frac{\sum_{c \in I_{i,j}} (R_{i,c} - A_i)(R_{j,c} - A_j)}{\sqrt{\sum_{c \in I_{i,j}} (R_{i,c} - A_i)^2 * \sum_{c \in I_{i,j}} (R_{j,c} - A_j)^2}}$$

Where $R_{i,c}$ is the rating of the item c by user i, A_i is the average rating of user i for all the co-rated items, and $I_{i,j}$ is the items set both rating by user i and user j.

The cosine measure, as following, looks at the angle between two vectors of ratings where a smaller angle is regarded as implying greater similarity.

$$sim(i,j) = \frac{\sum_{k=1}^{n} R_{i,k} * R_{j,k}}{\sqrt{\sum_{k=1}^{n} R_{i,k}^2 * \sum_{k=1}^{n} R_{j,k}^2}}$$

Where Ri,k is the rating of the item k by user i and n is the number of items co-rated by both users.

The adjusted cosine, as following, is used in some collaborative filtering methods for similarity among users where the difference in each user's use of the rating scale is taken into account.

$$sim(i,j) = \frac{\sum_{c \in I_{i,j}} (R_{i,c} - A_c)(R_{j,c} - A_c)}{\sqrt{\sum_{c \in I_{i,j}} (R_{i,c} - A_c)^2 * \sum_{c \in I_{i,j}} (R_{j,c} - A_c)^2}}$$

Where Ri,c is the rating of the item c by user i, Ai is the average rating of user i for all the co-rated items, and Ii,j is the items set both rating by user i and user j.

3. Calculating the similarity using fuzzy similar-priority comparison

According to the user-item matrix R = (Rij) m×n , this paper calculates the similarity of users using fuzzy similar-priority comparison. Assume the user set is U = { U1, U2, …, Um }, and the item set is I = {I1, I2, …, In }. We choose a testing user Uk (1 ≤ k ≤ m) in the user-item matrix. Then, find which is similar to it and which is not similar to it. The specific method is that we compare the double users (Ui , Uj) ∈ U and the target user Uk with the item aspect Ip ∈ I(1 ≤ p ≤ n). When Ui is better similar to the Uj, we call that the Ui is the similar-priority user.

The fuzzy similar-priority comparison matrix as following, is calculated the similarity of the random (Ui , Uj) ∈ U and the Uk (i , j = 1 ,2 , …, n) with the item aspect Ip ∈ I(1 ≤ p ≤ n).

C = (Cij) m×n (Cij ∈ [0 ,1] i , j = 1 ,2 , , n)

Cij is calculated by absolute value distance method as following:

$$C_{ij} = \frac{|U_k - U_j|}{|U_k - U_i| + |U_k - U_j|} \quad C_{ji} = \frac{|U_k - U_i|}{|U_k - U_i| + |U_k - U_j|}$$

Cij + Cji = 1, and it is satisfied the fuzzy similar-priority comparison characters.

We can decide which is priority of the Ui and Uj according to the Cij, if Cij in (0.5 ,1), Ui is better than Uj; if Cij in (0 , 0.5), Uj is better than Ui; if Cij=1, Ui is absolutely better than Uj; if Cij=0.5, Ui is equivalent to Uj; if Cij=0, Uj is absolutely better than Ui.

We judge the λ level similar degree of the fuzzy similar-priority comparison matrix Cm×n, and get the similar degree orders of the other users and the target user Uk with the item aspect Ip. Set as Ip （U1, U2, …, Um） = (λ1, λ2, …, λm) , λi is the λ level similar degree of the fuzzy similar-priority comparison of the user Ui. Mi = min{ Cij }, j = 1 ～ m ; λ = max{Mi }, i= 1 ～ n. As the same, do the fuzzy similar-priority

comparison operation of the other Ij ∈ I, at last, we get the Ui user-item λ level matrix, as following.

Table 1. λ level user-item matrix

Item \ User	I_1	I_2	… …	I_n
U_1	$\lambda_{1,1}$	$\lambda_{1,2}$	… …	$\lambda_{1,n}$
U_2	$\lambda_{2,1}$	$\lambda_{2,2}$	… …	$\lambda_{2,n}$
… …	… …	… …	… …	… …
U_m	$\lambda_{m,1}$	$\lambda_{m,2}$	… …	$\lambda_{m,n}$

The similarity of the Ui and Uk is follows.

$$Sim(U_k, U_i) = \frac{\sum_{j=1}^{n} \lambda_{ij}}{n}$$

4. Using fuzzy clustering to form nearest neighbors

4.1. Standardizing the user-item matrix

When construct the fuzzy similar matrix, we have to standardize the original data and compress the original data in the closed interval as [0, 1]. We process data standardization of User$_i$ in the user vectorial fuzzy clustering. The average value and standard deviation of the n user to the j item as below:

$$\overline{R}_j = \frac{1}{n} \sum_{i=1}^{n} R_{ij} \quad s_j = \left[\frac{1}{n} \sum_{i=1}^{n} (R_{ij} - \overline{R}_j)^2 \right]^{1/2}$$

Raw data standardize for:

$$R'_{ij} = (R_{ij} - \overline{R}_j) / s_j$$

Use the extremum standardize formulae, should standardized data condense into [0,1] as:

$$\overline{R}_{ij} = \frac{R'_{ij} - R'_{min\, j}}{R'_{max\, j} - R'_{min\, j}}$$

4.2. Building the fuzzy similar matrix

the fuzzy similar matrix as following :

$$R^F = \begin{pmatrix} r_{11} & r_{12} & \cdots & r_{1n} \\ r_{21} & r_{22} & \cdots & r_{2n} \\ \vdots & \vdots & \ddots & \vdots \\ r_{n1} & r_{n2} & \cdots & r_{nn} \end{pmatrix}$$

when i = j, and rij=1 ; when i ≠ j and rij is the similarity of the two user vectors.

4.3. Calculating the fuzzy equivalence matrix

According to the above step, the fuzzy similar matrix R^F satisfy reflexivity and symmetry, but dissatisfy transfer, therefore must calculate the transitive closure. It is a complicated work and required time and space greatness to solve fuzzy matrix transitive closure. In order to set the question, both here and abroad scholar look into know clearly some approach, as directness clustering procedure, net, maximal tree and graph theory at rest [5]. In this paper, we use the graph theory. We get the similar matrix transitive closure using fuzzy similarity and the graph theory. Settings for hereon foundation upper could directness stop set λ proceed clustering, as following:

$$[x] = \{ y \mid M(x, y) \leq \lambda \}$$

4.4. Forming nearest neighbors

According to the fuzzy clustering theory, the target user's nearest neighbors mostly dispersed over some clustering centers that have the high similarity to the target user, so have not to query target user's nearest neighbors in the wholly user space. We can find the target user's great mass of neighbors in the high similarity fuzzy clustering centers

Specific algorithm as follows：
TU: target user
 result: set of the nearest neighbors
C: center of the fuzzy clustering
count: the user count of the C
λ: the similarity threshold
top-N: the nearest N neighbors
Begin
result = null;
for (int i = 0; i≤count; i++)
   if (sim (TU, Ci) ≥ λ)
     result+=Ci;}
end for
for each User in result
     SIMILARi = sim (TU, User);
end for
select top-N neighbors with the high SIMILAR;
retrun top-N neighbors;
End

4.5. Producing prediction

In order to generate prediction of a user's rating to form the dense user-item matrix, we use the item-based collaborative filtering algorithms. The producing prediction formula as following:

$$P_{u,i} = \frac{\sum_{j=1}^{n} sim(i, j) * R_{u,j}}{\sum_{j=1}^{n} sim(i, j)}$$

Ruj: the rating of the user u to the item j, sim(i, j): the similarity of the user i and the user j, n: the number of the clustering centers.

5. Experimental evaluation and results

In this section, we describe the dataset, metrics and methodology for the comparison between traditional and proposed CF algorithm, and present the results of our experiments.

5.1. Data set

We use MovieLens collaborative filtering data set to evaluate the performance of proposed algorithm. MovieLens data sets were collected by the GroupLens Research Project at the University of Minnesota [6]. The historical dataset consists of 100,000 ratings from 943 users on 1682 movies with every user having at least 20 ratings and simple demographic information for the users is included. Therefore the lowest level of sparsity for the tests is defined as $1 - 100000/943*1682=0.937$.

5.2. Performance measurement

The metrics for evaluating the accuracy of a prediction algorithm can be divided into two main categories: statistical accuracy metrics and decision-support metrics. Statistical accuracy metrics evaluate the accuracy of a predictor by comparing predicted values with user provided values. Decision-support accuracy measures how well predictions help user select high-quality items. In this paper, we use mean absolute error (MAE), a statistical accuracy metrics, to report prediction experiments for it is most commonly used and easy to understand [7].

Formally, if n is the number of actual ratings in an item set, then MAE is defined as the average absolute difference between the n pairs. Assume that p1, p2, p3, ..., pn is the prediction of users' ratings, and the corresponding real ratings data set of users is q1, q2, q3, ..., qn. See the MAE definition as following:

$$MAE = \frac{\sum_{i=1}^{n} |p_i - q_i|}{n}$$

The lower the MAE, the more accurate the predictions would be, allowing for better recommendations to be formulated. MAE has been computed for different prediction algorithms and for different levels of sparsity.

5.3. Effect of different threshold

To evaluate the performance of the fuzzy clustering threshold used in the proposed algorithm and select optimal fuzzy clustering threshold, experiments realize in three different clustering thresholds. Figure 1 illustrates the sensitivity of the algorithms in relation to the different number of neighbors applied, which compares the performance of three different fuzzy clustering thresholds.

We adopt 0.7, 0.8 and 0.9 three thresholds in the experiment. And the result is that the fuzzy clustering threshold of 0.8 has the lesser MEA. In future, we will use the fuzzy clustering threshold of 0.8 in algorithm.

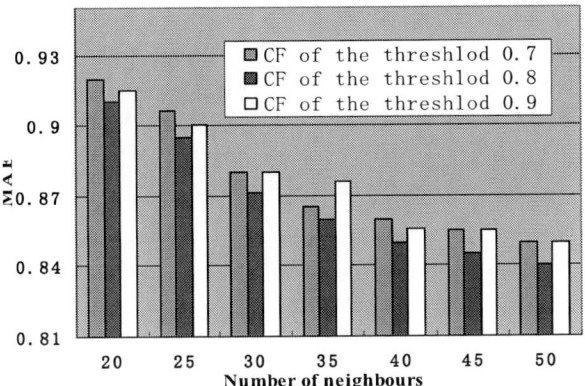

Figure 1. MAE of the different fuzzy threshold with respect to different number of neighbors

5.4. Comparing with the traditional CF

Figure 2 illustrates the sensitivity of the algorithms in relation to the different numbers of neighbors, which compares the performance of two different CF algorithms. Collaborative filtering algorithm based on fuzzy similar-priority comparison and fuzzy clustering is the proposed algorithm in this paper. The results in Figure 2 indicate that the accuracy of the proposed algorithm is better than the traditional user-based CF algorithms.

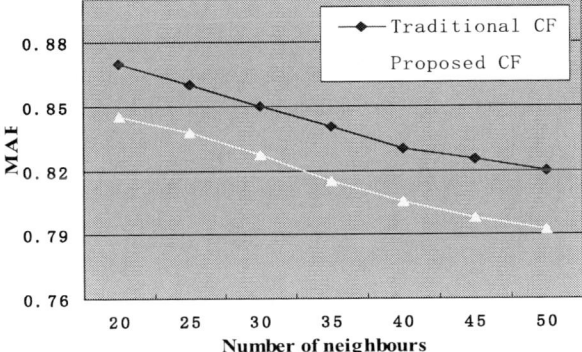

Figure 2. Comparing the proposed CF algorithm with the traditional CF algorithm

6. Conclusions

In this paper, we proposed a novel collaborative filtering recommendation algorithm based on fuzzy similar-priority comparison and fuzzy clustering to increase the scalability of a traditional collaborative filtering, the calculation complexity of which rapidly increases when the number of records in the user database increases. Experimental results show that this method can increase the accuracy of the predicted values, resulting in improving quality of the collaborative filtering recommender system.

References

[1] Bo Xie, Peng Han, Fan Yang, Rui-Min Shen, Hua-Jun Zeng, Zheng Chen, DCFLA: A distributed collaborative-filtering neighbor-locating algorithm, Information Sciences 177 (2007) 1349–1363.
[2] Yu Lia, Liu Lub, Li Xuefeng, A hybrid collaborative filtering method for multiple-interests and multiple-content recommendation in E-Commerce, Expert Systems with Applications 28 (2005) 67–77.
[3] Kwok-Wai Cheung, LilyF. Tian, Learning User Similarity and Rating Style for Collaborative Recommendation, Information Retrieval, 2004, 7, 395–410.
[4] Duen-Ren Liu, Ya-Yueh Shih, Hybrid approaches to product recommendation based on customer lifetime value and purchase preferences, The Journal of Systems and Software 77 (2005) 181–191.
[5] Jiang Lihong, Xu Boyi, Zhang Haiyan, A collaborative information filtering method and its application in information recommending system, Journal of the china society for scientific and technical information, 2005,24(6):669-673.
[6] Huang qin-hua, Ouyang wei-min, Fuzzy collaborative filtering with multiple agents, Journal of Shanghai University (English Edition), 2007,11(3):290-295.
[7] Peng Han, Bo Xie, Fan Yang, Ruimin Shen, A scalable P2P recommender system based on distributed collaborative filtering, Expert Systems with Applications 27 (2004) 203–210.

2008 Workshop on Power Electronics and Intelligent Transportation System

Support Vector Machine based Approach for State Estimation of Iraqi super Grid Network

[1]Afaneen A. Abod, [2]Abdullah H. Abdullah, [1]Mohammed K. Abd

[1]*Department of Electrical and Electronic Engineering, University of Technology, Baghdad, Iraq.*
[2]*Department of Electromechanical Engineering, University of Technology, Baghdad, Iraq.*
E-mail: waal85@yahoo.com, amaide4@hotmail.com, mka_ms2005@yahoo.com

Abstract

The correct assessment of network topology and system operating state in the presence of corrupted data is one of the most challenging problems during real-time power system monitoring, particularly when both topological and analogical errors are considered. This paper deals with Support Vector Machine method for state estimation problem in power systems including estimation and detection, which can help to improve Iraqi super grid electrical power network state estimation. The results of state estimation using the Support Vector Machine (SVM) and the conventional weighted Least Squares (WLS) State Estimator on basis of time, accuracy and robustness, particularly when both bad data and topological errors are to be considered. It has been established that the SVM based models provide results much faster, and work well even including single and multiple bad measurements, topology branches errors.

1. Introduction

Power system state estimation (PSSE) is one of the important functions executed in energy control centers in order to provide an accurate real-time database to be used by application programs, such as: economic dispatch, security analysis, etc. The system operator has to make, equitable, security related, congestion management decisions to curtail or deny power transfer rights in real time. Fast and accurate state estimation is foundation of locational marginal Pricing methodologies for transmission management costing.

Many power system state estimation (PSSE) methods are available today for the power system industry [1]. Most of the state estimation problems are formulated as over determined system of non-linear equations and solved as a weighted least squares problem. The Weighted Least Squares Estimation is by far the most popular approach in industry [2][3]. The state of the art in state estimation algorithms is presented in [4][5]. Most of the practical implementation of state estimation in electric power systems is based on the Gauss Newton methods. Singh and Alvarado [6] have formulated a topology processing similar to the state estimator algorithm and

have solved it using the least absolute value (LAV) method. Singh and Glavitsch [7] used a rule based approach. Although these methods usually perform very well under normal operating conditions (when noisy data corresponds mainly to meters inaccuracies), this is not the case when large measurement errors or topology configuration errors are to be processed. Other methods based on the application of intelligent systems for the analysis of the raw measurements have been proposed [8-10]. However, the dependency of raw measurements on the operating conditions may impose serious difficulties for obtaining representative patterns for the different types of error. Bad data identification is then a very complex and not adequately solved problem during PSSE. Alves da Silva et al. [11] used a neural network based on a multilayer perceptrons model and optimal estimate training to determine network topology.

In this work an original application of Support Vector method (SVM) for state estimation is proposed. The Support Vector method (SVM) is a general method of function estimation which does not depend explicitly on the dimensionality of the problem. The proposed estimator is studied for various cases to show its utility for state estimation in terms of accuracy and time requirements. Test results with a configuration of Iraqi super Grid Network are presented. Aspects such as efficiency, robustness, generalization capability and computational implementation of the proposed method for large-scale systems are also discussed.

2. The Iraq power system description

The transmission level in the Iraqi electrical network consists of the 400Kv network (the super grid network) and part of the 132 kV network connected to it. The aim of this work is limited to the study of only the 400Kv network with all its bus-bars and transmission lines. The network under consideration consists of 19 Bus and 30 transmission lines; the total length is 3711 Km. Fig.1 shows a configuration of this network.

978-0-7695-3342-1/08 $25.00 © 2008 IEEE
DOI 10.1109/PEITS.2008.102

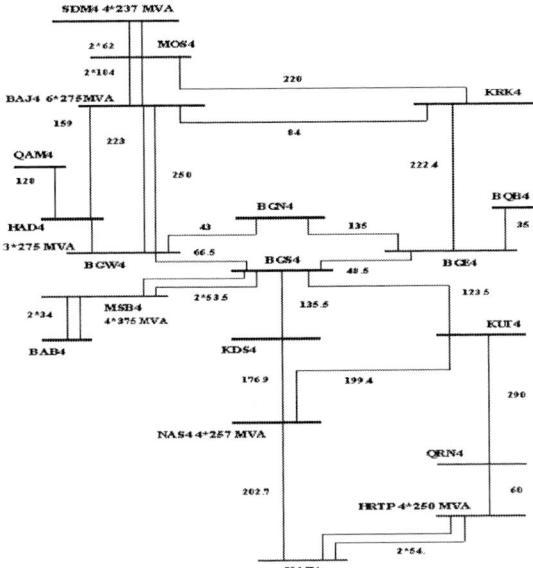

Fig.1.:Iraqi super Gird network"

3. State estimator

The state estimation provides the real time representation of the conditions in a power network. A state estimator is a data processing algorithm, which transforms meter readings and the switch status information into an estimate of the system's state (voltage magnitudes and phase angles at all the nodes). Real and reactive bus power injections, and real and reactive line flows and bus voltage magnitudes are the measurements, which are transmitted to computer control system via telemetry system. These measurements contain random noise due to instrument and phenomenon errors. The state estimation program obtains a best fit for the power system state variables by minimizing these errors. Ideally state estimation should run at the scanning rate of the telemetry system (say at every two seconds). Due to computational limitations, most practical state estimators run every few minutes or when major changes occur.

3.1. WLS State estimator

Most state estimation programs in practical use are formulated as over determined systems of nonlinear equations and solved as WLSE problem. Consider the nonlinear measurement model

$$z_j = h_j(x) + e_j \qquad (1)$$

Where

z_j is the *jth* measurement,

x is the true state vector,

$h_j(x)$ is a nonlinear scalar function relating the *jth* measurements to states, and

e_j is the measurement error, which is assumed to have zero mean and variance σ_j^2.

The WLS state estimation can be formulated mathematically as an optimization problem with a quadratic objective function and with equality and inequality constraints

$$\min\ j(x) = \frac{1}{2}\sum_{j=1}^{m}(z_j - h_i(x))/\sigma^2)$$

Subject to $g_i(x) = 0;\quad i = 1, n_g$

$$c_i(x) = 0;\quad i = 1, n_g \qquad (2)$$

is an objective function, and $g_i(x)$ and $c_i(x)$ are the functions representing power flow quantities.

The expressions for each of the above types of measurements can be expressed as follows:.

* Real and reactive power injection at bus i :

$$P_i = V_i \sum_{j \in N_i} V_j \left(G_{ij}\cos\theta_{ij} + B_{ij}\sin\theta_{ij}\right)$$

$$Q_i = V_i \sum_{j \in N_i} V_j \left(G_{ij}\sin\theta_{ij} - B_{ij}\cos_{ij}\right)$$

* Real and reactive power flow from bus i to bus j

$$P_{ij} = V_i^2\left(g_{si} + g_{ij}\right) - V_i V_j\left(g_{ij}\cos\theta_{ij} + b_{ij}\sin\theta_{ij}\right)$$

$$Q_{ij} = -V_i^2\left(b_{si} + b_{ij}\right) - V_i V_j\left(g_{ij}\sin\theta_{ij} - b_{ij}\cos\theta_{ij}\right)$$

Where

V_i, θ_i is the voltage magnitude and phase angle at bus i .

$\theta_{ij} = \theta_i - \theta_j$

$G_{ij} + jB_{ij}$ is the *ijth* elements of the complex bus admittance matrix.

3.2. Least squares support vector machine (LS-SVM)

One of the drawbacks of SVMs is tedious computation in the training phase due to the quadratic optimization problem. Suykens & al.[12] have reformulated the standard SVMs to avoid this problem and developed Least Squares Support Vector Machines (LS-SVM). The solution can be found efficiently by iterative methods such as conjugate gradient algorithm. LS-SVMs do not lead sparse solutions such as SVMs but a solution for the optimization problem is found very fast, and pruning techniques can be easily applied to enhance the sparsity.

LS-SVM algorithm is derived in a following way [12]. The estimation problem can be solved from optimization problem with equality constraints:

$$\min_{w,b,e} J(\mathbf{w}, b, e) = \frac{1}{2}\mathbf{w}^T\mathbf{w} + \gamma\frac{1}{2}\sum_{i=1}^{M}e_i^2$$

subject to $y_i[\mathbf{w}^T\varphi(\mathbf{x}_i) + b] = 1 - e_i,\ i = 1, ..., M \qquad (3)$

e_i denotes the error in the estimation of the sample x_i. One defines the Lagrangian:

$$L(\mathbf{w},b,\mathbf{e};\mathbf{\alpha}) = J(\mathbf{w},b,\mathbf{e}) - \sum_{i=1}^{M} \alpha_i \left\{ y_i \left[\mathbf{w}^T \varphi(\mathbf{x}_i) + b \right] - 1 + e_i \right\} \quad (4)$$

where α_i are Lagrange multipliers, which can be either positive or negative due to equality constraints. The conditions for optimality can be written as the solution to the following set of linear equations:

$$\begin{bmatrix} \mathbf{I} & 0 & 0 & -\mathbf{Z}^T \\ 0 & 0 & 0 & -\mathbf{Y}^T \\ 0 & 0 & \gamma\mathbf{I} & -\mathbf{I} \\ \mathbf{Z} & \mathbf{Y} & \mathbf{I} & 0 \end{bmatrix} \begin{bmatrix} \mathbf{w} \\ b \\ \mathbf{e} \\ \mathbf{\alpha} \end{bmatrix} = \begin{bmatrix} 0 \\ 0 \\ 0 \\ \vec{\mathbf{1}} \end{bmatrix} \quad (5)$$

where $\mathbf{Z} = (\varphi(\mathbf{x}_1)^T y_1, ., \varphi(\mathbf{x}M)^T y_M)^T$, $\mathbf{Y} = (y_1, ., y_M)^T$, $\vec{\mathbf{1}} = (1, ..., 1)^T$, $\mathbf{e} = (e_1, ., e_M)$ and $\mathbf{\alpha} = (\alpha_1, ., \alpha_M)^T$. The solution is also given by:

$$\begin{bmatrix} 0 & -\mathbf{Y}^T \\ \mathbf{Y} & \mathbf{Z}\mathbf{Z}^T + \gamma^{-1}\mathbf{I} \end{bmatrix} \begin{bmatrix} b \\ \mathbf{\alpha} \end{bmatrix} = \begin{bmatrix} 0 \\ \vec{\mathbf{1}} \end{bmatrix} \quad (6)$$

Hence the estimator (3) is found by solving the linear set of equations (4-6) instead of quadratic programming. The support values α_i are proportional to the errors at the data points.

4. Test result and discussions

The performance of the proposed estimator is compared with conventional WLS state estimator both the programs were coded in MATLAB for fair comparison and executed on a P4 AMD64 personal computer with 512-MB RAM.

For same data set and file operations input measurements for the Iraqi super Grid network were 98, respectively. These measurements consist of bus power measurements (real and reactive) and line measurements (real and reactive). The training patterns were generated for the 20% to 120% range of loading conditions corresponding to the base configuration. The performance comparison is made for following test cases.

4.1. All the measurements are correct (no gross errors) (case 1)

Table 1 shows the Average Absolute Error (AAE) in voltage magnitudes (p.u.) and phase angles (deg.) for the two methods, where no gross errors were present. It is observed that the accuracy of the WLS method was slightly superior when compared to the proposed method for test case 1. However, estimated states by the proposed method were quite more accurate for practical purposes. It can be seen that the CPU time required by proposed method model is much lower than the testing time of WLS.

Table1: State estimator results for Iraqi super Grid network (Case 1)

	Voltage (AAE)	Phase angle (AAE)	Time (Sec.)
LS-SVM	0.0345	0.771073	0.016
WLS	0.0788	1.131608	0.61

4.2. Gross errors in the measurement data (case 2)

Table 2 shows the estimates for voltage and bus angle for test case 2. The gross errors were introduced randomly in four measurements. The gross errors were introduces in real power on MOS4, reactive power on KRK4, real power flow in line (KRK4-BEG4) and reactive power flow in line (KRK4-BQB4). It is observed from tables 2 that the proposed estimator outperformed the WLS estimator both on account of average error for voltage magnitude and bus angles (as 0.047488< 0.07482 for V, and 0.643178<1.21107 for angles). The proposed estimator was robust when compared with WLSE state estimator. Also shows that the gross errors present in the measurement data can deteriorate the performance of the conventional state estimator and have to be removed or reweighed and are to be re-estimated by going through state re-estimation. However, the proposed state estimator is robust in such cases. The bad data can easily be detected in case of proposed method, as there is no data smearing possible unlike conventional state estimation.

4.3. Topological errors in the measurement data (case 3)

For case 3 a topological error was simulated as inclusion error of line (KRK4-BEG4). The line was actually out but the status (measurement) showed it to be in the system. WLS program was run with the line flow measurement as zero (both real and reactive), as acceptable. It is observed that the proposed estimator out performed the WLS estimator both an account of average error for voltage magnitude and bus angles (as 0.035905<0.060783 for V, and 0.771073<1.208551 for angles). The average time for the proposed neural network was 0.0550 seconds, whereas for WLS.

Table2: State estimator results for Iraqi super Grid network (Case 2)

Bus N.	True value		WLS		LSSVM		WLS		LSSVM	
	Phase angle	Voltage absolute error	Phase angle Absolute error	Voltage absolute error	Phase angle Absolute error	Phase angle	Voltage absolute error	Phase angle Absolute error	Voltage absolute error	Phase angle Absolute error
SDM4	1.0091	2.056	0.97739	2.1545	1.0154	2.1359	0.031713	0.09847	0.006315	0.079932
BAJ4	1.0199	5.9126	1.2216	6.2635	1.0131	6.0173	0.20173	0.3509	0.006813	0.10473
MOS4	1.0395	19.963	1.0672	19.728	1.133	20.971	0.027711	0.23475	0.093518	1.0078
BQB4	1.0225	7.7632	0.95437	7.6615	1.0478	7.9263	0.068134	0.10171	0.025319	0.16311
KRK4	0.9042	35.761	0.82107	38.751	0.94934	36.737	0.083129	2.9896	0.045136	0.97611
QAM4	1.05	21.28	1.2392	23.671	1.1261	21.986	0.18917	2.3911	0.076135	0.70639
BGN4	1.0116	23.762	1.0121	24.995	1.046	24.09	0.000531	1.2327	0.034428	0.32782
HAD4	0.9807	31.063	1.0393	32.848	1.0287	32.206	0.058596	1.7854	0.048002	1.1432
BAB4	1.0102	30.043	1.1724	29.667	0.98748	30.042	0.16223	0.37607	0.022725	0.00143
MSB4	1.0182	8.0142	0.98659	8.5458	1.072	8.2971	0.03161	0.53156	0.053846	0.28288
BGS4	1.0229	38.441	1.0011	42.058	1.0913	39.068	0.021829	3.617	0.068393	0.62674
BGW4	0.9021	27.083	0.82066	28.577	0.86656	26.588	0.081437	1.4939	0.035542	0.4954
KAZ4	1.0491	21.181	0.94825	22.163	1.1073	22.546	0.10085	0.98207	0.058205	1.3649
NAS4	1.0244	18.938	0.97091	20.143	1.0315	19.581	0.053495	1.2045	0.00711	0.64308
KDS4	1.0119	16.237	1.083	15.978	1.1111	17.384	0.071123	0.25939	0.099151	1.147
HRTP	0.97	19.963	0.92883	20.751	0.93795	19.878	0.041168	0.78823	0.032047	0.085484
QRN4	1.0147	49.868	1.0477	52.024	1.08	51.594	0.032953	2.1556	0.065347	1.726
KUT4	1.029	37.784	1.17	39.378	1.1139	38.728	0.14104	1.5938	0.084917	0.94414
BGE4	1.0394	28.038	1.0625	28.862	1.0001	27.644	0.023131	0.82371	0.039316	0.39424
Maximum Average error							0.07482	1.21107	0.047488	0.643178
CPU time in seconds			0.83		0.02					

Table3: State estimator results for Iraqi super Grid network (Case 3)

Bus N.	True value		WLS		LSSVM		WLS		LSSVM	
	Voltage magnitude	Phase angle	Voltage magnitude	Phase angle	Voltage magnitude	Phase angle	Voltage absolute error	Phase angle Absolute error	Voltage absolute error	Phase angle Absolute error
SDM4	1.0125	2.0560	1.016	2.0612	0.99595	2.1292	0.006944	0.073241	0.013153	0.005229
BAJ4	1.0199	5.9126	1.0553	5.8448	1.0243	5.9185	0.03539	0.005927	0.004433	0.067757
MOS4	1.0395	19.9630	1.1081	20.153	1.0644	20.677	0.068567	0.71415	0.024938	0.18995
BQB4	1.0225	7.7632	1.049	8.4858	1.0075	7.8946	0.026548	0.13137	0.014992	0.72259
KRK4	0.9042	35.7610	1.0026	36.273	0.90641	38.144	0.098415	2.383	0.002208	0.51184
QAM4	1.0520	21.2800	1.0615	21.141	1.1267	20.892	0.01155	0.38814	0.076745	0.13918
BGN4	1.0126	23.7620	1.1613	25.351	1.0523	25.307	0.14973	1.5449	0.04069	1.5894
HAD4	0.9817	31.0630	1.0579	34.118	0.98388	32.123	0.07718	1.0601	0.003182	3.0547
BAB4	1.0132	30.0430	1.1101	32.43	1.0593	31.32	0.099922	1.2766	0.049131	2.3873
MSB4	1.0182	8.0142	1.0106	8.4013	1.0566	7.9383	0.007624	0.0759	0.038397	0.38711
BGS4	1.0229	38.4410	1.096	39.362	1.1017	41.176	0.073076	2.7346	0.07881	0.92089
BGW4	0.9021	27.0830	0.86491	28.355	0.90916	29.194	0.037185	2.111	0.007064	1.2716
KAZ4	1.0491	21.1810	1.1401	21.321	1.0898	23.129	0.090876	1.9475	0.040684	0.14041
NAS4	1.0244	18.9380	0.97641	19.077	1.0567	20.761	0.047991	1.8229	0.032254	0.13947
KDS4	1.0119	16.2370	1.0357	16.008	1.0723	16.47	0.023793	0.23281	0.060399	0.22918
HRTP	0.9765	19.9630	1.0096	21.798	0.99612	21.245	0.039645	1.2821	0.02612	1.8352
QRN4	1.0147	49.8680	1.1059	50.123	1.0508	53.964	0.091163	4.0959	0.036129	0.25495
KUT4	1.0291	37.7840	1.0703	38.476	1.0697	38.716	0.041315	0.9318	0.040682	0.69203
BGE4	1.0394	28.0380	1.1674	28.15	1.1316	27.887	0.12797	0.15054	0.092191	0.11161
Maximum Average error							0.060783	1.208551	0.035905	0.771073
CPU time in seconds			0.61		0.016					

5. Summary

A proposed two-stage Support Vector Machine (SVM) scheme has been applied in this paper to solve the problem of power system state estimation for bad data detection and processing topology during real-time power system monitoring. LS-SVM is designed and tested for Iraq super Grid network. The test results presented on three systems reveal the following:

- The proposed method can be implemented for large-scale power systems.
- Both the LS-SVM estimator provide state estimation results accurately as compared to WLSE for no-gross and topological error present.
- The LSVSM state estimator is found fairly accurate even gross error and topological error present in the measurement data.
- The computation time required for the proposed LS-SVM is much better than WLSE.

6. References

[1] M. B. Do Coutto Filho, A.M. Leite da Silva and D.M. FalcZo , "Bibliography on power system state estimation (1968-1989) ", IEEE Trans. on Power System, Vol. 5, Aug. 1990, pp. 950-961.

[2] Horisberger H.P. Richard J.C and Rossier C, "Fast Decoupled Static State Estimator for Electric Power System" IEEE Trans., PAS 95, No.1, 1976, pp. 208-215.

[3] Garcia A, Monticelli A, and Abreu P, "Fast Decoupled Static State Estimation and data Processing", IEEE Trans., PAS 98, No.5, 1979, pp. 1645-1652.

[4] F.F. Wu, "Power System State Estimation: A Survey", Elect. Power Eng. System, Vol. 12, Jan. 1990, pp. 80-87.

[5] A. Monticelli, "Electric Power System State Estimation" ,Proc. IEEE, Vol. 88, No.2, Feb. 2000, pp.262-282.

[6] SINGH, H., and ALVARADO, F.L. "Network topology determination using least absolute value state estimation" IEEE/PES Summer meeting, July1994, Paper 94SM506 - 6-PWRS.

[7] SINGH, N., and GLAVITSCH, H. "Detection and identification of topological errors in on line power system analyses", IEEE Trans. Power System, Vol.6, No.l, 1991, pp. 324-331.

[8] N. Singh and H. Glavitsch, "Detection and identification of topological errors in online power system analysis", IEEE Trans Power System, Vol. 6, Feb. 1991, pp. 324-331.

[9] N. Singh, F.Oesoh, "Practical experience with Rule based on-line topology error detection", IEEE Trans on Power Systems, Vol. 9, May 1994, pp. 841-847.

[10] C.N.Lu, J.H. Teng, B.S. Chang, "Power system network topology error detection", IEE Proc.C., Vol.141, No.6, Nov.1996.

[11] DA SILVA, A.A.P., QUINTANA, V.H., and PANG, G.K.H. "Solving data acquisition and processing problems in power systems using a pattern analysis approach", IEEE Proc. C, Vol. 138, No. 4, 1991, pp. 365-367.

[12] Suykens, J.A.K., Van Gestel, T., De Brabanter, J. De Moor, B, Vandewalle, Least Squares *Support Vector Machines*, World Scientific, Singapore, 2002.

Gradual Approaching Method for Distribution Network Dynamic Reconfiguration

Yang HuPing[1*], Peng YunYan[1], Xiong Ning[2]

[1]*Information Engineering School,Nanchang University, Nanchang, 330031, P.R. China*

[2] *Department of Electrical Engineering, Shanghai Jiao Tong University, Shanghai, 200240, China*

[1]* *E-mail: yhping123@163.com，* [2]*flame1212@163.com*

Abstract

Gradual approaching method which can deal well with distribution network dynamic reconfiguration is presented in this paper. The method not only can be realized easily, but also can make use of any established static method for solving dynamic reconfiguration. At first, objective function of static reconfiguration is amended for being applied in interval optimization. And then, time enumeration method is used for determining optimal time for reconfiguration. Thus, one times of reconfiguration for interval optimization is finished. For realizing multi optimizations within specified time interval, gradual approaching method is employed. At last, correctness and effectiveness of the proposed method is validated on modified IEEE 33-bus system.

1. Introduction

Distribution network dynamic reconfiguration broadly can be classified into two categories: one is static reconfiguration based on time points, main methods include branch-exchange method [1]-[2], optimal power flow model method[3], switching groups method[4] and some artificial intelligence searching algorithm [5]-[8]. The other is dynamic reconfiguration based on the time interval (usually for one day)[9-11]. This model is aimed to determine the optimal time and network structure under a condition of load changing continuously for minimizing operation expense.

Static distribution network reconfiguration is a multi-constraints, large-scale nonlinear combinations optimization problem, which absorbs large number of scholar's focus. Some effective methods have been proposed now. Static reconfiguration can provide optimal network structure when load is constant. However, actual load in distribution power system is dynamically changeable with time. In order to have the electrical power distribution system operate safely、efficiently、economically, the structure of distribution network has to be adjusted dynamically according to the change of load, that is the dynamic reconfiguration.

Owing to the forbidding computation price, literatures of dynamic reconfiguration are not much and no good approaches for it neither at present, most of articles make use of approximate、equivalent and compromise method for reducing computation, but at the price of sub-optimal solution. In [9], variable load on the entire time series are equivalent to one or several constant one, and then optimize it by static method. The quality of optimization can not be guaranteed because it mainly depends on the effect of equivalence. In [10], the strategy of simplification and compromise is adopted to reduce the computation price. The number of switch operations is distributed according to the sequence of optimization. If the early reconfigurations take more switch operations, the latter can only choose suboptimal combination to avoid the number of switch operations being over limit. In [11], network structure is optimized based on static method for every time point at first. Then, regards each time point as a stage and takes related optimized structure as state, determines the minimal transmission loss within the entire time interval by dynamic programming method. However, the optimal structure on a time point may not be the one of the entire time interval.

In this paper, for coping with the difficult problem of distribution network dynamic reconfiguration, a gradual approaching method is presented. It can not only make use of existing static reconfiguration method, but also solve the problem with a moderate computation price. For processing time interval optimization, objective function of static model is modified firstly. Followed, in the entire time period, enumeration method is applied to determine best reconfiguration time, a dynamic reconfiguration is completed. At last, for the realization of global optimization of multi-reconfigurations, gradual approaching method is implemented.

2. Mathematical Model

In general, distribution network dynamic reconstruction take the number of switch operations as constraint condition, and minimize operating costs within specified time interval by some optimization strategy. Detailed mathematical model is shown in [11]. Owing to this model can not restrain the times of reconfiguration, which would likely result in lots of times of reconfiguration within a time interval. It is unrealistic because of low auto-level in current distribution network besides expensive worker price. In order to facilitate management and operation, times of reconfiguration is taken as restriction instead of switch operations. Hence, the times of reconfiguration can be determined according to the requirement of the actual situation, in the meantime, the numbers of switch operations are restricted implicitly.

Assuming N is the specified times of reconfiguration in a time interval, then the time period can be divided into N sub-intervals. The transmission loss of the ith sub-interval is determined by the ith optimized network structure and the ith sub-interval length. Detailed

978-0-7695-3342-1/08 $25.00 © 2008 IEEE
DOI 10.1109/PEITS.2008.104

mathematical model is shown as follow:

$$\min F = C_T \sum_{i=1}^{N} \sum_{j \in t_i} \sum_{k \in u_{i-1}} (I_{k,j}^2 \times R_k) + N \times C_R \quad (1)$$

$$s.t. \qquad N \leq N_{\max}$$

Where, F is the operation cost within the entire time period; N is the number of reconfigurations; $N_{\max}$ is the up limit of N; u_{i-1} is the network structure after the $(i-1)^{th}$ reconfiguration is executed, it corresponds to the initial network structure when i is equal to 1; t_i is the ith sub-interval; j is the time moment in sub-interval t_i; k is branch of network structure u_{i-1}; $I_{k,j}$ is the current of branch k in moment j; R_k is resistance of branch k.

In addition, network structure u_i should also satisfy radial connectivity.

3. Interval reconfiguration based on the static method

Interval reconfiguration based on static method has to conduct reconfiguration in initial moment of study period. Mathematical model is shown as follow:

$$\min F_i = \sum_{j=t_0}^{t_1} \sum_{k \in u_{t_0}} I_{j,k}^2 \times R_k \quad (2)$$

Where, t_0 and t_1 are the initial and end time of study period respectively; j is time point belong to the time interval $[t_0, t_1]$; u_{t0} is optimized network structure at initial time.

The model optimizes the network structure u_{t0} in the whole study period, but it can't optimize the time of reconfiguration. So, in essence, it still belongs to static reconfiguration and any kinds of static method can be used for solving this model.

4. Time enumeration method of dynamic reconfiguration

Static method for optimization has to be executed at the initial time of study period, but the initial time may not be the optimal time for reconfiguration. Here, time enumeration method is introduced for searching optimal reconfiguration time.

It is the main idea of time enumeration method to conduct a static reconfiguration tentatively in every time point of the whole time period. Interval transmission loss F_{int} is calculated for every time point. Assuming reconfiguration is occurred at time t_r, F_{int} is the sum of transmission loss produced by initial network structure within the time length ranging from initial time t_0 to tentative reconfiguration time point t_r and the transmission loss produced by optimized structure u_{tr} within the time length ranging from t_r to the end time of period t_1. Compare every F_{int}, the minimum is the best scheme. The expression is described as follow:

$$\min F_{int} = F_{t_r}^1 + F_{t_r}^2$$

$$= \sum_{j=t_0}^{t_r} \sum_{k \in u_{t_0}} I_{j,k}^2 \times R_k + \sum_{l=t_r}^{t_1} \sum_{m \in u_{tr}} I_{l,m}^2 \times R_m \quad (3)$$

Where, t_0 and t_1 is the initial and end time of study period respectively; t_r is the time of reconfiguration; u_{t0} is network structure at initial time; u_{tr} is network structure after reconfiguration at time t_r is executed; $F_{t_r}^1$ and $F_{t_r}^2$ are transmission loss produced by two sub-interval which is separated by t_r.

5. Gradual approaching method

Time enumeration method can be used well for one times of reconfiguration within a time period. For realizing multi-reconfigurations, mutual coordination among reconfigurations has to be taken into consideration. Hence, gradual approaching method is proposed to solve it.

Gradual approaching method is divided into two steps. At first, determining the initial time and structure for specified each reconfiguration; secondly, tuning finely the time and structure for coordination.

5.1 The initial value of time and structure determined

Due to the insertion of reconfiguration time, a time interval is divided into several adjacent sub-intervals. The best reconfiguration time should be set in the sub-interval where maximal transmission loss decline would happen. If the profit produced by one reconfiguration is more than the fee it costs, the time and structure of this reconfiguration are recorded. Otherwise, the process of initial value determination is finished.

Detailed steps on determining initial value:

1) Parameters setup: Set the initial time $t_0= 0$; end time $t_1= 24$; time interval vector $T_0= [t_0, t_1]$; times of reconfiguration $i= 0$; initial structure $u_i= u_0$; maximal times of reconfiguration is $N_{\max}$.

2) If $i \leq N_{\max}$? Yes, go to step 3); No, program terminate.

3) Running at t_0, regard two sequent points within vector T_i as one sub-interval, and then determine the length of each sub-interval $[t_i, t_{i+1}]$ and its initial structure u_i.

4) Time enumeration method is employed in each sub-interval j in order to get the optimal reconfiguration time t_{jr}、network structure u_{jr}、interval minimal transmission loss $F_{j,i+1}$ and its two sub-interval transmission loss $F_{j,i+1}^1$ and $F_{j,i+1}^2$.

5) Evaluating $\Delta F = \max_{j \in T_i} \{F_{j,i} - F_{j,i+1}\}$, if $C_T \Delta F$ $< C_R$, program terminate; Else, go to step 6).

6) Obtaining sub-interval j^* according to $\triangle F$; and then letting the $(i+1)$th reconfiguration time $t_{i+1}=t_{j^*r}$, network structure $u_{i+1}=u_{j^*r}$, substituting the sum of $F_{j^*,i+1}^1$ and $F_{j^*,i+1}^2$ for interval transmission loss $F_{j,i}$, inserting t_{i+1} into vector T_i and sorting T_i by ascending

order; at last, letting $i = i+1$, go to step 3).

5.2 The gradual approaching method of optimal time and structure

After initial time and structure are determined, gradual approaching method can is implemented to find out the global optimum. The main idea is to merge the two sequent sub-intervals as a big sub-interval. Obviously, the mergence point is the reconfiguration point at previous iteration. Time enumeration method is employed in this big sub-interval, a new reconfiguration point can be got, If the time and structure of this new point are completely consistent with the old mergence point, the procedure of adjustment is finished; otherwise take this new value as mergence point. Continue reconfiguration for the next big interval which is started by this new mergence point. Iterating repeatedly until the global optimal solution is approached.

Assuming initial value has been determined and the time vector is T_i, the approaching steps are shown as follows:

1) Set the times of iteration to 1, namely $i=1$.
2) Set reconfiguration point to be adjusted equal to 1, namely $j=1$.
3) Merge two sequent sub-interval of T_i, namely $[t_{j-1,i}, t_{j,i}]$ and $[t_{j,i}, t_{j+1,i}]$, into a big sub-interval $[t_{j-1,i}, t_{j+1,i}]$, time enumeration method is used to search the optimal reconfiguration time $t_{j,i+1}$ and structure $u_{j,i+1}$ for this big sub-interval.
4) If $t_{j,i+1}=t_{j,i}$ and $u_{j,i+1}=u_{j,i}$, program terminate; otherwise, letting $t_{j,i}=t_{j,i+1}$ and $u_{j,i}=u_{j,i+1}$, go to step 5).
5) Is t_{j+1} equal to the end time? Yes, $i= i+1$, go to step 2); no, $j=j+1$, go to step 3).

6. Study case

As for time-varying system, load increase is not always in the same way at different bus. Here, assuming study case has three kinds of loads: commerce, resident and industry. Different bus has different ratio consist of this three kinds of load, but the load/time curve of the same kind of load is equal.

Assuming k is the number of bus, i is the type of loads, j represents time point, $M_{i,k}$ is the ratio of the ith kind of load in bus k, $C_{i,j}$ is the component of the ith kind of load in time j, L_{kmax} is peak load of bus k, then the load of bus k at time j can be described as

$$L_{k.j} = \sum_{i=1}^{3} L_{k\max} M_{k,i} C_{i,j}$$

6.1 The example data of calculation

Taking modified IEEE33-bus system[12] as study case, peak value of load is promoted to 120% compared to original data; time interval interested is one day, which is divided into 24 time point, i.e., $T_0=[0, 24]$. Detailed data are shown in table 2, table 3.

Tab.2 The proportion of three types load in each bus

k	M_1	M_2	M_3	k	M_1	M_2	M_3
2	0.2	0.5	0.3	18	0.4	0.5	0.1
3	0.5	0.3	0.2	19	0.5	0.2	0.3
4	0.5	0.2	0.3	20	0.7	0.3	0
5	0.6	0.1	0.3	21	0.5	0.3	0.2
6	0.4	0.4	0.2	22	0.3	0	0.7
7	0.6	0	0.4	23	0.5	0.4	0.1
8	0.3	0.3	0.4	24	0.4	0.5	0.1
9	0.4	0.6	0	25	0.6	0.4	0
10	0.3	0	0.7	26	0.7	0.3	0
11	0.6	0.2	0.2	27	0.1	0	0.9
12	0.5	0.5	0	28	0.6	0.3	0.1
13	0.6	0.4	0	29	0.4	0.1	0.5
14	0.4	0.4	0.2	30	0.7	0.2	0.1
15	0.5	0.1	0.4	31	0.4	0.4	0.2
16	0.2	0.7	0.1	32	0.2	0.1	0.7
17	0.3	0.3	0.4	33	0.3	0.7	0

Where, the subscript 1, 2 and 3 represent business, resident and industrial load separately.

Tab.3 The load distribution in hour of three types load

j	C_1	C_2	C_3	j	C_1	C_2	C_3
0	0.1	0.1	0.3	12	0.6	0.7	0.5
1	0.1	0.1	0.3	13	0.7	0.6	0.6
2	0	0.1	0.4	14	0.8	0.5	0.8
3	0	0.1	0.2	15	1	0.6	0.8
4	0	0.4	0.2	16	1	0.7	0.7
5	0	0.3	0.3	17	0.8	0.8	0.7
6	0.1	0.4	0.3	18	0.6	0.9	0.8
7	0.2	0.4	0.4	19	0.7	1	0.9
8	0.7	0.3	0.8	20	0.6	1	1
9	0.8	0.3	1	21	0.2	0.6	0.8
10	0.8	0.5	0.9	22	0.1	0.5	0.5
11	0.7	0.6	0.9	23	0.1	0.2	0.4

In addition, the price of power is 0.6 $ kilowatt / hour, the cost of reconfiguration is 2$.

6.2 simulation results

Before reconfiguration, the number of interval is equal to 1, i.e., $T_0=[0, 24]$. Transmission loss is 2.2119MW. Using proposed method introduced in 4.1 sections determine the optimal times of reconfiguration and initial value, the result is shown in table 4.

Tab.4 the results of original values

i	j	t_i	T_i	u_i	F_i				$\triangle F$ (MW)	$C_T \triangle F$ ($)
					1	2	3	4		
1	1	0	[0 0 24]	7-8 14-15 9-10 32-33 28-29	0	1.5262	-	-	0.6857	411.42
2	2	16	[0 0 16 24]	7-8 14-15 9-10 32-33 25-29	0	0.8151	0.7077	-	0.0034	2.04
3	3	21	[0 0 16 21 24]	7-8 14-15 9-10 32-33 28-29	0	0.8151	0.6410	0.0658	0.0009	0.54

259

where, i is the times of reconfiguration, j is the sub-interval with largest transmission loss reduction, t_i is time point of reconfiguration, T_i is time interval vector, u_i is the location of tie switch after reconfiguration, F_i is transmission loss in each sub-interval, $\triangle F$ is electrical power saved by reconfiguration, $C_T \triangle F$ is the money converted from $\triangle F$.

From table 4, it is can be seen that the 3rd reconfiguration only produce 0.54$ profit, but the cost is 2.0$. Therefore, the optimal times of reconfiguration are 2 and time vector $T_2 = [0, 0, 16, 24]$.

After obtain the initial value, gradually approaching method is executed for adjusting the time and structure in order to get global optimal solution. The time of mergence point is equal to 0 which can be got by merging the first two sub-interval [0, 0] and [0, 16] of T_2 into a big sub-interval [0, 16]. The initial location of tie switch is 7-814-15 9-10 32-33 28-29. Time enumeration method is employed for getting new optimal time and structure in the big interval. The result after reconfiguration is the same with initial mergence point, namely optimal reconfiguration time at 0 and the locations of tie switch are intact too. This indicated that the initial value is global optimal solution, the algorithm is end.

7. Conclusion

Gradual approaching method is proposed in this paper for solving distribution network of dynamic reconfiguration. It makes great use of the existing static method for dynamic reconfiguration and global optimal solution can be tracked easily. The main works in this paper include:
(1) Time interval-based reconfiguration is realized by revising the static model.
(2) Single dynamic reconfiguration is fulfilled by time enumeration method.
(3) Gradual approaching method is proposed, by which multi-reconfigurations can be realized easily.

It can be seen from the study case that profit isn't further promoted by gradual approaching method. From another view, it can be concluded that the network structure has a strong compatibility with a certain range of load change. This characteristic is advantageous to proposed method for solving dynamic reconfiguration

problems because the initial value does not need to be adjusted too many times for obtaining global optimum.

References

[1] Civanlar S, Grainger JJ, Yin Lee. Et al. Distribution Feeder Reconfiguration for Loss Reduction[J]. IEEE Trans. on Power Delivery, 1988, 3(3), pp. 1127-1223.
[2] Baran M E, Wu F F. Network reconfiguration in distribution systems for loss reduction and load balancing [J]. IEEE Trans. on Power Delivery, 1989, 4(2), pp. 1401-1407.
[3] Lei Jiansheng, Deng Youman, Zhang Boming. Hybrid flow pattern and its application in network reconfiguration [J]. Proceedings of the CSEE, 2001, 21(1), pp. 57-62.
[4] M.A.Kashem, V.Ganapathy, G.B.Jasmon. A Novel Method for Loss Minimization in Distribution Networks. Internet conference on electric utility deregulation and restructuring and power technologies 2000, City University, London, pp. 251-256.
[5] Huang Y C . Enhanced genetic algorithm-based fuzzy multi-objective approach to distribution network reconfiguration[J]. IEE Proc Gener Transm Distrib, Sept, 2002, 149(5), pp. 615-620.
[6] Liu Li, Chen Xueyun. Reconfiguration of distribution networks based on fuzzy algorithm [J]. Proceedings of the CSEE, 2000, 20(2), pp. 66-69.
[7] Cheng Genjun. A Tabu search approach to distribution network reconfiguration for loss reduction [J]. Proceedings of the CSEE, 2002, 22(10), pp. 28-33.
[8] Hoyong Kim, Yunseok Ko, Kyung-Hee. Artificial neural-network based feeder reconfiguration for loss reduction in distribution systems [J]. IEEE Trans. on Power Delivery. 1993, 8(3), pp. 1356-1366.
[9] Deng Youman, Zhang Boming, Tian Tian. A fictitious load algorithm and its application to distribution network dynamic optimizations [J]. Proceedings of the CSEE, 1996, 16(4), pp. 241-244.
[10] YIN Li-yan, YU Ji-lai. Dynamic reconfiguration of distribution network with multi-time periods [J]. Proceedings of the CSEE, 2002, 22(7), pp. 44-48.
[11] Wu Jianzhong, Yu Yixin. Global optimization algorithm to time varying reconfiguration for operation cost minimization. Proceedings of the CSEE, 2003, 23(11), pp. 13-17.
[12] Comprehensive Method for Reconfiguration of Electrical Distribution Network. Power Engineering Society General Meeting,2007.IEEE June 2007, pp. 1-7.

Application of Wavelet Packet Analysis and Improved LSSVM on Rotating Machinery Fault Diagnosis

Lingling Zhao,   Kuihe Yang

College of Information, Hebei University of Science and Technology
Shijiazhuang 050018, China
zll@hebust.edu.cn

Abstract

For enhancing fault diagnosis precision, the wavelet packet analysis and least squares support vector machine are combined effectively. First, the signals are decomposed in arbitrary minute frequency bands by use of wavelet packet analysis technique. Doing energy calculation in these frequency bands to form eigenvectors is more reasonable. And then a least squares support vector machine fault diagnosis model is presented. When the least squares support vector machine is used in fault diagnosis, the Fibonacci symmetry searching algorithm is simplified and improved. It is presented to choose parameter of kernel function on dynamic, which enhances preciseness rate of diagnosis. In the model, the non-sensitive loss function is replaced by quadratic loss function and the inequality constraints are replaced by equality constraints. The simulation results show the model can effectively diagnose machinery facility faults.

Keywords: *Wavelet packet analysis, Fault diagnosis, Least squares support vector machine, KKT conditions*

1. Introduction

The support vector machine (SVM) is a new machine study method which was established by Vapnik in base of statistical learning theory (SLT) [1][2]. The SVM stresses to study statistical learning rules under small sample. Via structural risk minimization principle to enhance extensive ability, the SVM preferably solves many practical problems, such as small sample, non-linear, high dimension number and local minimum points. The SVM has been applied in pattern classification, regression forecasting, probability estimation, control theory and so on.

The least squares support vector machine (LSSVM) is an improved algorithm based on SVM [3]. The LSSVM is a kind of SVM under quadratic loss function. In LSSVM, the non-sensitive loss function is replaced by quadratic loss function and the inequality constraints are replaced by equality constraints. Via constructing loss function, the quadratic programming problem is changed as solving linear equation groups problem, which simplifies the complexity of calculation. In the paper, the wavelet packet analysis and least squares support vector

machine (LSSVM) are combined effectively. The power spectrum of fault signals are decomposed by wavelet analysis, which predigests choosing method of fault eigenvectors. The LSSVM is adopted to diagnose the diagnose machinery facility faults, which gets good diagnosis effect.

2. Wavelet packet analysis

Fourier analysis can be use to analyze frequency bands energy, and received spectrum structure eigenvectors of different faults have obtain successful application. However, Fourier analysis only considers sine wave signals. Practical diagnosis signals usually contain non- sine wave signals. Strictly speaking, these signals can not be describe using sine signals as bases. If we use sine signals as bases to describe them, the energy will be not complete.

The signals are decomposed in arbitrary minute frequency bands by use of wavelet packet analysis technique. Doing energy calculation in these frequency bands to form eigenvectors is more reasonable. The multi-distinguishing analysis of wavelet analysis map signals to child space composed of a group of positive join wavelet functions. The signals are outspreaded in different scale. The signal characters in different frequency bands are distilled, and primary signal characters in different scale are saved. However, multi-distinguishing analysis only decomposes low frequency part, and the high frequency part is held. The wavelet packet analysis is an improved approach to wavelet transform. It can offer more minute analysis method for signal. It can further analyze high frequency part which can not be decomposed in wavelet transform. The frequency distinguishing rate of whole frequency bands is enhanced. Therefore, wavelet packet decomposition can do positive join decomposition to signals in whole frequency bands. It can do decomposition in low and high frequency bands at the same time, and automatically confirms distinguishing rate of signal in different frequency bands.

In excessive distinguishing analysis, $\phi(t)$ and $\psi(t)$ meet two measure equation as follows.

978-0-7695-3342-1/08 $25.00 © 2008 IEEE
DOI 10.1109/PEITS.2008.107

$$\left.\begin{array}{l}\phi(t)=\sum_{k\in Z}h_k\phi(2t-k)\quad \{h_k\}_{k\in Z}\in l^2\\[2mm]\psi(t)=\sum_{k\in Z}g_k\phi(2t-k)\quad \{g_k\}_{k\in Z}\in l^2\end{array}\right\} \qquad(1)$$

The decomposing and re-building algorithm of Wavelet packet analysis are given as follows. Assuming $g_j^n(t)\in U_j^n$, then g_j^n can be expressed as follows.

$$g_j^n(t)=\sum_l d_l^{j,n}u_n(2^jt-l) \qquad(2)$$

We can get decomposing algorithm of wavelet packet analysis.

From $\{d_l^{j+1,n}\}$, we can get $\{d_l^{j,2n}\}$ and $\{d_l^{j,2n+1}\}$:

$$\left.\begin{array}{l}d_l^{j,2n}=\sum_k a_{k-2l}d_k^{j+1,n}\\[2mm]d_l^{j,2n+1}=\sum_k b_{k-2l}d_k^{j+1,n}\end{array}\right\} \qquad(3)$$

More, We can get re-building algorithm of wavelet packet analysis.

From $\{d_l^{j,2n}\}$, we can get $\{d_l^{j,2n+1}\}$ and $\{d_l^{j+1,n}\}$:

$$d_l^{j+1,n}=\sum_k[h_{l-2k}d_k^{j,2n}+g_{l-2k}d_k^{j,2n+1}] \qquad(3)$$

Three layers wavelet packet decomposition chart of signals is given as Figure 1 shows.

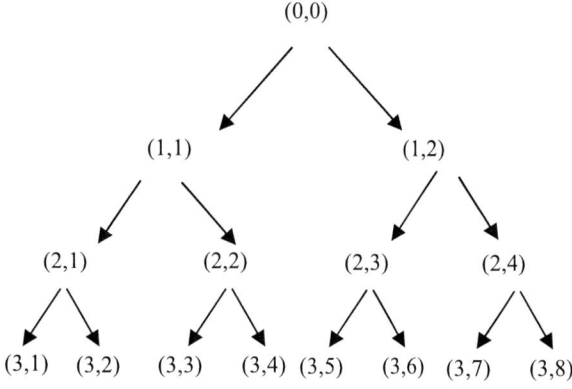

Figure 1. Structure of three layers wavelet packet decomposition

Assuming highest frequency of originality signals is f, which is normalized as number 1. Then the frequency of each layer can be expressed as follows.

(1) Originality layer: (0-1)

(2) First layer: (0-0.5) (0.5-1)

(3) Secondly layer: (0-0.25) (0.25-0.5) (0.5-0.75) (0.75-1)

(4) Thirdly layer: (0-0.125) (0.125-0.25) (0.25-0.375) (0.375-0.5) (0.5-0.625) (0.625-0.75) (0.75-0.875) (0.875-1)

3. Fault diagnosis model of LSSVM

3.1 SVM Ptinciple

The SVM is developed from hyperplane classifier under linear separable case [4]. Suppose having sample x_i and its sorts y_i, expressed as $\{(x_i,y_i)\}, x\in R^d$ $y_i\in\{1,-1\}$, $i=1,...,N$. The d is dimension number of input space. For standard SVM, its classifying margin is $2/\|w\|$. Making the classifying margin to maximum is equal to making the $\|w\|^2$ to minimum. Therefore, the optimization problem of making the classifying margin to maximum can be expressed the follow quadratic programming problem:

$$\min_{w,b,\xi}J(w,\xi)=\frac{1}{2}\|w\|^2+C\sum_{i=1}^N\xi_i \qquad(4)$$

Subject to $y_i[w^T\Phi(x_i)+b]\geq 1-\xi_i,$

$$i=1,...,N \qquad(5)$$

$$\xi_i\geq 0,\quad i=1,...,N \qquad(6)$$

where $\xi_i\geq 0$ ($i=1,...,N$) is a slack variable, which ensures classification validity under linear non-separable case, and parameter C is a positive real constant which determines penalties to estimation errors, a larger $C\psi$corresponding to assigning a higher penalty to errors. The function Φ is a non-linear mapping function, by which the non-linear problem can be mapped as linear problem of a high dimension space. In the transforming space, we can get optimal hyperplane.

The Karush-Kuhn-Tucker (KKT) Conditions play important role in both the theory and practice of constrained optimization. The KKT conditions may be stated as follows.

$$\frac{\partial}{\partial w_v}L_P=w_v-\sum_i\alpha_iy_ix_{iv}=0 \quad v=1,...,d \quad(7)$$

$$\frac{\partial}{\partial b}L_P=-\sum_i\alpha_iy_i=0 \qquad(8)$$

$$y_i(x_i\cdot w+b)-1\geq 0 \quad i=1,...,l \qquad(9)$$

$$\alpha_i\geq 0 \quad \forall i \qquad(10)$$

$$a(y_i(x_i\cdot w+b)-1)=0 \quad \forall i \qquad(11)$$

The KKT conditions are satisfied at the solution of any constrained optimization problem, with any kind of constraints, provided that the intersection of the set of feasible directions with the set of descent directions coincides with the set of descent directions. This rather technical regularity assumption holds for all support vector machines, since the constraints are always linear. Furthermore, the problem for SVMS is convex, and for convex problems, the KKT Conditions are necessary and sufficient for w, b, α to be solution. Thus solving the SVM Problem is equivalent to finding a solution to the KKT conditions.

The SVM algorithm can realize right classification to sample by solve above quadratic programming problem.

3.2 LSSVM classifying algorithm

The LSSVM is an improved algorithm based on SVM. For LSSVM, the non-sensitive loss function is replaced by quadratic loss function and the inequality constraints are replaced by equality constraints. The optimization problem has been modified as follows:

$$\min_{w,b,\xi} J_{LS}(w,\xi) = \frac{1}{2}\|w\|^2 + \frac{1}{2}\gamma\sum_{i=1}^{N}\xi_i^2 \quad (12)$$

Subject to $\quad y_i[w^T\Phi(x_i)+b] = 1-\xi_i$,

$$i = 1,...,N \quad (13)$$

where parameter γ is similar parameter C of SVM, which is used to control function $J_{LS}(w,\xi)$. In order to solve the optimization problem, the Lagrangian function is introduced as follows:

$$L_{LS}(w,b,\xi,\alpha) = J_{LS}(w,\xi) - \sum_{i=1}^{N}\alpha_i\{y_i[w^T\Phi(x_i)+b]-1+\xi_i\}$$

$$(14)$$

where α_i are Lagrangian multipliers that can either be positive or negative. In saddle points, differential coefficients for w, b, ξ and α_i are requested and let them equal to zero, and then the equation as follows can be obtained:

$$\begin{bmatrix} 0 & Y^T \\ Y & ZZ^T + \gamma^{-1}I \end{bmatrix}\begin{bmatrix} b \\ \alpha \end{bmatrix} = \begin{bmatrix} 0 \\ \vec{1} \end{bmatrix} \quad (15)$$

Where $\quad Z = \left[\Phi(x_1)^T y_1,...,\Phi(x_N)^T y_N\right]^T$,

$Y = [y_1,...,y_N]^T$, $\vec{1} = [1,...,1]^T$, $\xi = [\xi_1,...,\xi_N]^T$,

$\alpha = [\alpha_1,...,\alpha_N]^T$, and $I \in R^{(N\times N)}$ is identity matrix.

By defining $\Omega = ZZ^T = [q_{ij}]_{N\times N}$ and applying Mercer's condition to Ω matrix, each element of matrix can be expressed as follows:

$$q_{i,j} = y_iy_j\Phi(x_i)^T\Phi(x_j) = y_iy_jK(x_i,x_j) \quad (16)$$

where $K(x_i,x_j)$ is kernel function. If a function K can be found to meet condition $K(x_i,x_j) = \Phi(x_i)\cdot\Phi(x_j)$, only inner product operation are requested in high dimension space. We have kernel functions in common use as follows:

(1) Polynomial kernel function

$$K(x_i,x_j) = ((x_i \cdot x_j)+\theta)^d \quad d = 1,2,...$$

(2) RBF kernel function

$$K(x_i,x_j) = \exp(\frac{-\|x_i-x_j\|^2}{\sigma^2})$$

(3) Sigmoid kernel function

$$K(x_i,x_j) = \tanh(\beta(x_i \cdot x_j)+\theta)$$

Equation 15 can be solved by east squares method. In LSSVM, quadratic programming problem is changed as the problem of solving linear equation groups, which simplifies the complexity of calculation. After the optimization problem is solved, we can have the classifier of SVM as follows:

$$y(x) = sign\left[\sum_{i=1}^{N}\alpha_iy_i\Phi(x,x_i)+b\right] \quad (17)$$

3.3 The Improved Selecting Parameter Algorithm of LSSVM

Since RBF function calculates simply and only has a standardization parameter σ, we adopt RBF function as kernel function. In LSSVM, regularization parameter γ and standardization parameter σ of RBF function are selected as constants commonly according to experience, but for different sample set, optimal parameter values is metabolic, which affects diagnosis precise rate of faults. By examination, we find that the value of parameter γ don't affect the result obviously, so we define $\gamma=10$. On the other hand, with different σ value, the results vary obviously.

Suppose diagnosis error rate is a down single peak function in section [a, b], which has exclusive minimum in t value belonging to [a, b]. If we take freewill two points a_1 and b_1 in the section, $a_1 < b_1$, and calculate function values $f(a_1)$ and $f(b_1)$, two instances as follows will appear.

(1) $f(a_1) < f(b_1)$, then $t\in$ [a, b_1]
(2) $f(a_1) \geq f(b_1)$, then $t\in$ [a_1, b]

This shows that we can shorten section [a, b] to section [a, b_1] or [a, b_1] which still contain minimum as long as we take two different points belonging to [a, b] and compare their function values. If we want to continue shortening section [a, b_1] or [a, b_1], one point value belonging to above section should be taken to calculate its function value which is use to compare with the values of $f(a_1)$ or $f(b_1)$. The shortening rate of section is relational with the calculation times of function. In the paper, we take a=0.1, b=1, and take a_1 and b_1 in three equality part points of section [a, b].

4. Simulation results

In machinery equipment fault diagnosis technique, the study of rotating machinery fault diagnosis is most embedded and ideal, and its application is also most successful. Using excessive distinguishing analysis to character distilling of power spectrum, we can distill eigenvectors expediently and effectively. By means of wavelet packet analysis technique, we can do energy analysis to constantly appearing faults of rotating machinery [7]. Via a lot of experiments, the rotating machinery fault causes and sign corresponding table can be built. Using rotating machinery ordinary 6 faults, such as imbalance, non-middle, oil film swirling and so on, as output of neural networks, and using 9 frequency bands different spectrum apex energy values in vibration signal spectrum, the training sample set of rotating machinery faults can be formed, as table 1 shows.

Table 1 Fault causes and sign corresponding

Fault	0	1	2	3	4	5	6	7	8
frequency bands	0.01~ 0.39f	0.40~ 0.49f	0.50f	0.51~ 0.99f	f	2f	3~5f	odd number f	> 5f
Fault 1	0.00	0.00	0.00	0.00	0.90	0.05	0.05	0.00	0.00
Fault 2	0.00	0.30	0.10	0.60	0.00	0.00	0.00	0.00	0.00
Fault 3	0.00	0.00	0.00	0.00	0.40	0.50	0.10	0.00	0.00
Fault4	0.10	0.80	0.00	0.10	0.00	0.00	0.00	0.00	0.00
Fault 5	0.10	0.10	0.10	0.10	0.20	0.10	0.10	0.10	0.10
Fault6	0.00	0.00	0.00	0.00	0.20	0.15	0.40	0.00	0.25

In order to check diagnosis success rates of LSSVM model, LSSVM diagnosis models are used to diagnose simulation faults. In the paper, LSSVM fault diagnosis model uses one-against-one method to realize multiclass classification. First the LSSVM model are trained by training sample set, and then we diagnosis the simulation faults using trained models.

Suppose D_1 is simulation sample data matrix before entering noises, and D_2 is simulation sample data matrix after entering noises. The containing noises input sample data required by simulation can gain by equation below.

$$D_2(i, j) = D_1(i, j) \times (1 + \alpha \times rands(1)) \quad (17)$$

where the α is noise control coefficient, α =0, 0.2, 0.5, 0.8 respectively, and $rands(1)$ is a random function which can produce a number between -1 and 1.

Use above equation to produce 80 groups measure parameter containing noises to every fault, altogether 480 groups sample, 300 groups sample of which are used as training set and 180 groups sample of which are used test set. In not doing any pretreatment, they are used to diagnose faults by LSSVM models. The diagnosis results are showed in table 2.

Table 2 The diagnosis results

ID number	Noises control coefficient	Diagnosis preciseness rate
1	α =0.0	100%
2	α =0.2	98.6%
3	α =0.5	96.3%
4	α =0.8	93.1%

Table 2 shows the diagnosis results of LSSVM models. The diagnosis results show that the diagnosis success rates are influenced by noise control coefficient α. When there are not any noises in sample data, such as α =0.0, the diagnosis success rates of LSSVM are as high as 100%. When noises are comparatively small, such as α =0.2, the diagnosis success rate of LSSVM is 98.6%. When noises are comparatively big, such as α =0.5, 0.8, the LSSVM keeps as high as 96.3% and 93.1% diagnosis success rate. From above diagnosis success rates, although diagnosis success rates of LSSVM models appear to decline along with increasing the sample noises, the declining speed of LSSVM model is lower, and the LSSVM shows robustness.

5. Conclusion

The LSSVM has rather strongly robust diagnosis and it can be used for pattern classifying. The primary merit of SVM is that it aims at small sample, and it can gain optimal answer according to information in existence instead of sample number going to infinity. The LSSVM changes classification question into quadratic programming problem and can gain optimal answer entirely in theory, which solves the local extremum in neural networks. The LSSVM is an improved algorithm based on SVM. For LSSVM, the non-sensitive loss function is replaced by quadratic loss function and the inequality constraints are replaced by equality constraints. Consequently, quadratic programming problem is simplified as the problem of solving linear equation groups, and the SVM algorithm is realized by least squares method, which predigests the complexity of calculation. The simulation results show the LSSVM model can effectively diagnose machinery facility faults.

Acknowledgements

This work is supported by China Postdoctoral Science Foundation (2005038515) and Hebei University of Science and Technology Foundation.

References

[1] Cortes C, Vapnik V. "Support-vector network", Machine Learning, Vol. 20, pp. 273–297, 1995.

[2] Vapnik V, Golowich S, Smola A. "Support vector method for function approximation, regression estimation, and signal processing". Advances in Neural Information Processing Systems, Vol. 9 pp.281–287, 1996.

[3] Hsu C W, Lin C J. "A comparison of methods for multiclass support vector support vector machines". IEEE Trans on Neural Networks, Vol. 13, No.2, pp.415-425, May, 2002.

[4] Guo G, Li S Z, and Chan K L. "Support vector machines for face recognition". Image and Vision Computing, Vol. 19, pp. 631-638, 2001.

[5] Lendasse A, Simon G, Wertz V, et al. "Fast bootstrap for least-square support vector machines". Proceedings of European Symposium on Artificial Neural Networks. Bruges, Belgium, 28-30 April 2004, pp. 525-530.

[6] Engel Y, Mannor S, Meir R. "The kernel recursive least-squares algorithm". IEEE Trans on Signal Processing, Vol.52, No.8, pp.2275-2285, August 2004.

[7] Yu Heji, Chen Changzheng, Zhang Sheng, et al. "Intelligent diagnosis based on neural networks". Beijing, China, Metallurgy Industry Press, 2002.

[8] Ye Zhifeng, Liu Jianguo. "Probabilistic neural networks based engine fault diagnosis". Aviation Transaction, Vol.23, No.2, pp.155-157, 2002.

[9] Li Donghui, Liu Hao. "Method and application of fault diagnosis based on probabilistic neural network". Systems Engineering and Electronics, Vol.26, No.7, pp.997-999, July 2004.

[10] Chuah T C, Sharif B S, and Hinton O R. "Robust adaptive spread-spectrum receiver with neural-net preprocessing in non-gaussian noise". IEEE Transactions on Neural Networks, Vol.12, No.3. pp.546-558, August 2001.

[11] Chen S, Samingan A K, Hanzo L. "Support vector machine multiuser receiver for DS-CDMA signals in multipath channels". IEEE Transactions on Neural Networks, Vol.12, No.3, pp.604-611, August 2001.

[12] Sebald D J, Bucklew J A. "Support vector machine techniques for nonlinear equalization". IEEE Transactions on Signal Processing, Vol.48, No.11, pp.3217-3226, November 2000.

[13] ohan A K. "Support vector machines: a nonlinear modeling and control perspective". European Journal of Control, Vol. 7, pp. 311-327, July 2001.

[14] Vila J P, Wagner V, Neveu P. "Bayesian nonlinear model selection and neural networks: A Conjugate Prior Approach". IEEE Trans on Neural Networks, Vol.11, No.2, pp.265-278, May 2000.

[15] Shi Lingfeng. "A new method of infrared small weak targets detection based on wavelet analysis". System Engineering and Electronics, Vol.25, No.8, pp.1024-1027, August 2003.

[16] Wei Ruixuan, Zhang Youyun, Han Chongzhao, et al. "Study on robust modeling method for fault diagnosis of nonlinear system". Proc. of IMS'2003, National University of Deference Technology Press, 2003, pp.572-576.

An Urban Map Matching Algorithm Using Rough Sensor Data

Shu Liu[1], Zongying Shi[1], Mingguo Zhao[1], Wenli Xu[1], Kai Zhang[2]

[1]Department of Automation, Tsinghua University, Beijing 100084, China
[2]Takasago Research & Development Center,
Mistsubishi Heavy Industries, LTD, Takasago Hyogo 676-8686 Japan
alfred.liushu@gmail.com, szy@mail.tsinghua.edu.cn

Abstract

Map matching is an important task in vehicle navigation applications. In this paper, we present a map matching algorithm for complex urban environment, using only low cost sensors including GPS, speedometer and gyroscope. To deal with the complexity and inaccurate information, a novel weighting method is proposed to describe likelihoods of candidate roads, using a distance forgetting factor for history accumulation. Based on this, a Finite State Machine (FSM) matching engine and a tracing back procedure are included in the algorithm, which are designed especially for the rough data. Simulation results show that the algorithm can deal with the high uncertainties and achieve a high success rate in matching to the road.

Key words: Map matching, Forgetting factor, Finite State Machine, Urban environment

1. Introduction

Vehicle localization is an important part of transport systems. Tracking, navigation, and furthermore, behavior understanding, are always based on the knowledge of the vehicle's position. Generally speaking, there are two steps in vehicle localization. First is to determine the position of the vehicle by some sensors, such as GPS, speedometer, gyroscope. Suffering from noises, this estimated position is usually poor in precision. In complement, we have to use a technique called map matching (MM) [1], to fix the road running on and improve the accuracy of localization.

Map matching is a software technology using high precision electronic maps to reduce the errors in localization. First the vehicle is matched to a road. Then the former position measurement is projected onto the road for a better estimation. Metric adjacencies and topological connections are the main information used in map matching [2]. In the recent years, approaches including fuzzy logic [3,4], neural network [5], Kalman filter [6], and D-S reasoning [7], have been brought into the field.

So far, map matching algorithms can achieve a satisfying accuracy in highway networks and suburbs, yet not in downtown areas. In urban environment, road network is much more complex, and GPS is unreliable. Fig.1 shows the map and GPS data in our experiments. Black lines refer to roads, while black points refer to crossings and shape points on curve roads. Grey points show the position information from GPS, and grey lines connect them to a sequence. In the map, there are many short roads and curve roads. Some nearby roads have similar bearings. GPS brings even more difficulties. Suffering from dense buildings and trees, so called **urban canyons**, GPS data contain big biases, much larger than the random noises, and the biases change unpredictably. In this situation, many traditional map matching algorithms fail. New methods fitting this kind of uncertainty need to be designed [3,8].

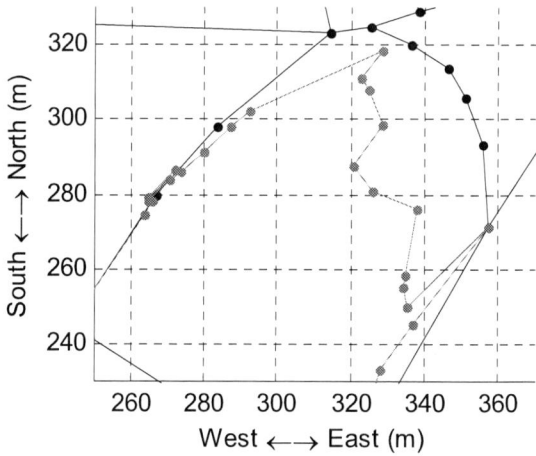

Figure 1. Map & GPS data

In this paper, even worse conditions are considered. For cost reduction, we've considered using some cheap sensors. As they give much rougher measurements, the uncertainties in the problem increase obviously. To deal with such conditions we present a map matching algorithm. We propose a weight calculation method to evaluate the likelihoods of candidate roads. A novel technique called distance forgetting factor is applied to fuse history information. Grounded on this, we present a Finite State Machine (FSM) matching engine for real time matching, and a tracing back procedure to correct history mismatches. The effectiveness of the map matching algorithm is demonstrated by simulation experiments.

978-0-7695-3342-1/08 $25.00 © 2008 IEEE
DOI 10.1109/PEITS.2008.115

2. Algorithm structure

The map matching algorithm consists of three modules: Road Weight Calculation, FSM Matching Engine, and Tracing Back. Fig.2 illustrates the whole flow of the algorithm during a sample period of GPS.

First, the method calculating weights of candidate roads is described in section 3. Then the Finite State Machine matching engine, which realizes real time matching with the weights, is introduced in section 4. The weights are also used in the tracing back procedure, which corrects history mismatches if possible. The tracing back procedure will be described in detail in section 5.

3. Road weight calculation

3.1. Stopping point integration

As poor quality speedometer and gyroscope cannot form an effective Dead Reckoning (DR) system, GPS becomes the most important reference in the algorithm. However, there are a number of errors in GPS, such as ephemeris errors, propagation errors, multi-path, and receiver noise, etc. Even when using differential GPS (DGPS), the multi-path and receiver noise cannot be eliminated. Multi-path error is caused by urban canyons. Signals have been reflected by buildings before they reach the receiver. That brings big biases to GPS.

According to experiential knowledge, such biases often change when the GPS receiver is moving. But when stopping, data from GPS point to almost the same position. They keep a same bias. During several minutes of stopping, if stopping GPS data are treated the same as running ones, an accumulated bias will be brought into matching. An example is shown by P9-P15 in Fig.3. Roads and nodes are label by R1-R7 and N1-N2, respectively. It's similar in later figures.

To avoid this kind of biases, an integration technique is designed. The GPS position sequence is integrated to a series of **geometric effective points**, which means they provide independent metric information without biases caused by stopping accumulation. When a GPS measurement comes, it is compared to the previous measurement and speedometer record. If the vehicle is stopping and the bias in GPS data has not changed obviously, the measurement is considered with the same bias as the previous one. Then it will be integrated into the latest geometric effective point. Otherwise, a new effective point will be created according to the new measurement. As in Fig.3, P9-P15, even P8-P15 will be considered one geometric effective point.

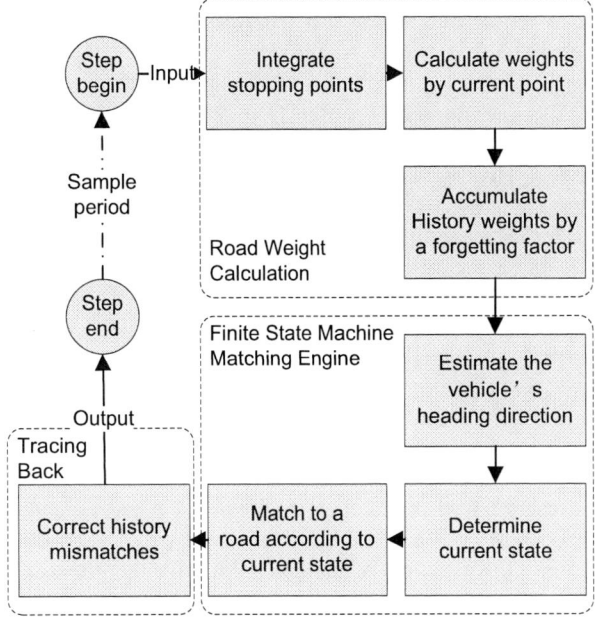

Figure 2. Algorithm flowchart

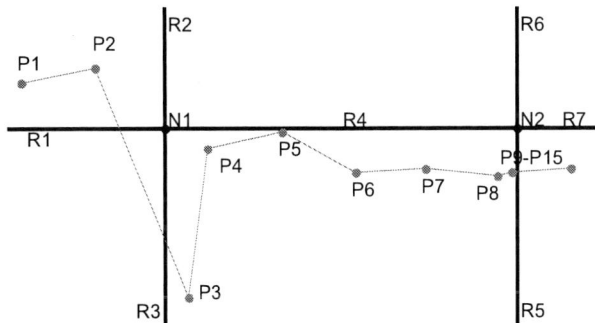

Figure 3. Illustration for weight calculation

3.2. Weight calculation according to point

Weights ranged from 0 to 1 are used to value the likelihoods of candidate roads, referring to a geometric effective point. Based on experiential knowledge, when the point is close to a road, for example, 5m or 10m, the road's weight should be almost the same. And when the point is away from the road with a long distance such as 25m or longer, the weight should decrease rapidly according to the distance's increase. In the algorithm, the weight is set as a Gaussian distribution according to the distance to a candidate road as follows.

$$w(gps, road) = \exp\left\{-\left(\min_{p \in road} d(gps, p)\right)^2 \Big/ (2\sigma^2)\right\} \quad (1)$$

Where w means weight, gps means the geometric effective point from GPS, p refers to arbitrary point on the candidate road, and σ is the deviation parameter of the Gaussian distribution..

3.3. Weight accumulation with distance forgetting factor

GPS measurements suffer from biases. Thus, using just current weights will not achieve satisfying matching results. History information has to be used in the algorithm. But GPS measurements sometimes have outliers. For example, P3 in Fig.3 is an apparent outlier. Calculate weights by an outlier and bring it into matching will cause unpredictable mismatches. An outlier elimination procedure is applied in our algorithm. When we find current GPS measurement has an unreasonable large change from the previous one, we consider either the current point or the previous one is an outlier. However, because of urban canyons, even non-outliers have big biases, and the biases are changing all the time. We cannot find out outlier from normal measurements correctly. So we adopt the following strategy: when the current GPS position is not consistent with the previous position and speedometer record, history weights are not brought into current and future weight accumulation.

When the current geometric effective point is not considered an outlier, a weight accumulation framework is suggested in the algorithm. Denote aw the accumulated weight, and $\lambda(t-k,t)$ the ratio of $w(t-k)$ in the sum.

$$aw(t) = w(t) + \sum_{k=1}^{t} \lambda(t-k,t) w(t-k) \qquad (2)$$

Considering real-time performance requirement, we assume $\lambda(t-k,t)$ can be calculated recursively

$$\lambda(t-k,t) = \lambda(t-k,t-1)\lambda(t-1,t), \forall k < t \qquad (3)$$

So that the accumulation is simplified:

$$aw(t) = w(t) + \lambda(t-1,t)aw(t-1) \qquad (4)$$

λ is in a typical form of forgetting factor. In the calculation of λ, we have some more assumptions. For history geometric effective points, the contribution to the accumulated weights should have a negative correlation with the distance to current effective point. Furthermore, λ of a too old point should be low. In detail, except for the previous point, λ of all history points should be no bigger than if it is calculated as the previous point. That equals:

$$\lambda(t-1,t) = f\big(d\big(gps(t-1), gps(t)\big)\big) \qquad (5)$$

$$\lambda(t-k,t) \le f\big(d\big(gps(t-k), gps(t)\big)\big), \forall k \ge 2 \qquad (6)$$

Here $d(p,q)$ means distance between points p and q.

To give the function an exact form, we assume when the vehicle runs in a straight line and GPS data almost follow the line, λ of non-previous points can be calculated the same as the previous one. Without loss of generality, we assume there is a time t, where from t-2 to t the vehicle runs straight. Let $g(x) = \log(f(x))$, from (3) and (5), we can deduce that

$$g\big(d\big(gps(t-1), gps(t)\big)\big) + g\big(d\big(gps(t-2), gps(t-1)\big)\big)$$
$$= g\big(d\big(gps(t-2), gps(t)\big)\big) \qquad (7)$$
$$= g\big(d\big(gps(t-1), gps(t)\big) + d\big(gps(t-2), gps(t-1)\big)\big)$$

As the two distances can be arbitrary positive real numbers, (7) is equal to

$$g(x) + g(y) = g(x+y), \forall x, y > 0 \qquad (8)$$

According to (8),

$$\forall m, n \in \mathbf{Z}^+, x \in \mathbf{R}, n \cdot g\left(\frac{m}{n}\right) = g(m) = m \cdot g(1) \qquad (9)$$

So that

$$\forall q = \frac{m}{n} \in \mathbb{Z}^+, g(q) = \frac{m}{n} \cdot g(1) = q \cdot g(1) \qquad (10)$$

Then we further assume that $g(x)$ is continuous on $(0, +\infty)$. Based on calculus knowledge, for any real number x, there exists a rational sequence r whose limit is x, so $\forall x \in (0, +\infty)$,

$$g(x) = g\left(\lim_{n \to \infty} r_n\right) = \lim_{n \to \infty} g(r_n)$$
$$= \lim_{n \to \infty} (r_n \cdot g(1)) = x \cdot g(1) = kx \qquad (11)$$

Therefore,

$$\exists a \in \mathbf{R}, \forall x \in (0, +\infty), f(x) = e^{ax} \qquad (12)$$

By an examination, we find that if and only if $a < 0$, a $f(x)$ described in (12) can satisfy (3), (5) and (6). In conclusion, let $b = e^a$,

$$\lambda(t-1,t) = b^{d(gps(t-1), gps(t))}, 0 < b < 1 \qquad (13)$$

As $\lambda(t$-1$,t)$ is a function of point-to-point distance, we call it a **distance forgetting factor**. As a result, weight accumulation equation should be

$$aw(t) = w(t) + b^{d(gps(t-1), gps(t))} \cdot aw(t-1), 0 < b < 1 \qquad (14)$$

4. Finite State Machine matching engine

4.1. Heading direction estimation

In downtown areas, GPS data suffer from enormous errors. Usually it is not reliable to locate the vehicle using position information only, even if the position estimation is improved by some kind of filtering. Information about the vehicle's heading direction need to be used in the algorithm.

We do not have an electronic compass, or other similar sensors to measure the vehicle's heading direction directly. The gyroscope measures the vehicle's angular rate. Differences in heading orientation can be accumulated by these data. However, we need an initial heading direction to generate latter ones, and should correct the result periodically as the accumulation error is divergent with time.

The heading direction estimation algorithm is used in the algorithm. At begin of the run, the estimation is not established yet. When the vehicle is first confirmed on a road, GPS data at begin and end time of the running are used to generate an approximate heading direction. The approximate result is compared to the road's bearing, to initialize the heading direction of the vehicle.

After initialization, whether the vehicle is on a long straight road is checked in each step. If no, the heading

estimation is accumulated by gyroscope data, based on previous result. Otherwise, accumulation errors will be corrected comparing to the bearing of the matched road.

This estimation benefits the algorithm greatly, especially around crossings. As sometimes the GPS data are close to a wrong road whose bearing is evidently different from the vehicle's heading. Mismatches are avoided when the heading estimation is available.

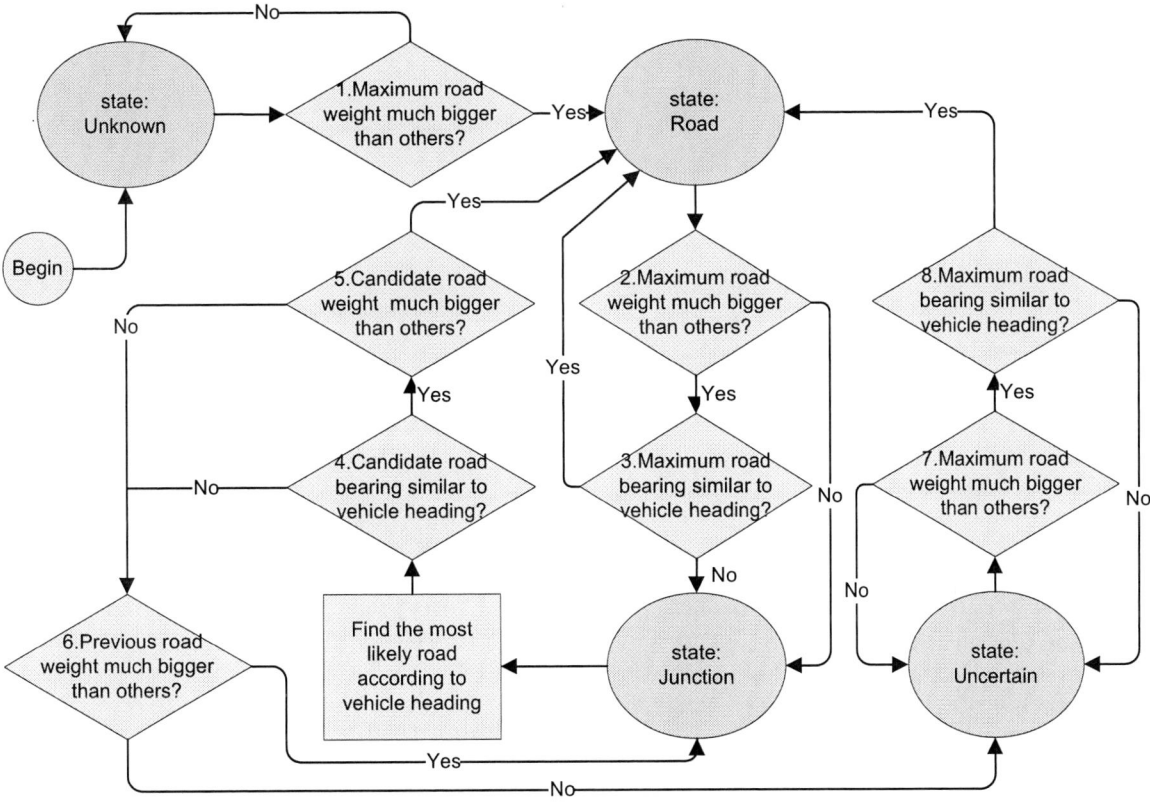

Figure 4. Illustration for Finite State Machine

4.2. Finite State Machine

Even when assisted by heading direction estimation, information for matching is not enough sometimes. We need some experiential knowledge about the vehicle's running patterns in addition. A Finite State Machine (FSM) mechanism is designed in the algorithm. There are 4 states in the FSM, named unknown, uncertain, road and junction. Each state presents for a typical running pattern.

(1)   State **unknown** refers to the situation we are not sure of the matched road, and heading estimation is not available. In this state, we can use no information other than sensor data.

(2)   State **uncertain** is the case when we are not sure of the matched road, but heading estimation is available. In this case, heading direction can be used in eliminating some unlikely roads.

(3)   State **road** is the most certain state, when we have confidence in the matched road and heading direction. It's assumed that the vehicle will run to a road next to current matched road.

(4)   State **junction** is the situation when the vehicle is about to leave a confirmed road, or take a U-turn. It's assumed that the vehicle either runs to a next-to road, or turns to the contrary direction on the confirmed road, or stays at the crossing.

The division of these 4 states brings helps remarkably in improving matching trajectory consistency.

4.3. State switches

The switches in the FSM are shown in Fig.4. Map matching begins at state unknown. When a road has a weight much bigger than others', the algorithm switches to state road and that road is set the confirmed road. As the confidence in the matched road decreases to some extent, the algorithm will change from state road to state junction. In the state of junction, if a road next to previous confirmed road gains a much higher likelihood, we match the vehicle to it and go to state road. If after a long term no new road is matched and the vehicle is far from previous crossing, we deduce that we have lost the position, and go to state uncertain.

There are two kinds of criteria used in the FSM. Firstly, weight comparisons are applied, such as criterion 1, 2, 5, 6 and 7 in Fig.5. In this case, the statement in Fig.5 equals the following inequation

$$aw_{\max} > aw_{road} \cdot threshold, \forall road \in map \qquad (15)$$

For confirmation, the maximum weight should be much higher than all others. Thresholds are selected according to criterion situations respectively, always bigger than 1.

Secondly, bearing comparisons are used in criterion 3, 4 and 8. In this case, the statement in Fig.5 equals the following inequation

$$\left|\arg\left(heading_{vehicle} - bearing_{candidate}\right)\right| < threshold \qquad (16)$$

Bearings of vehicle heading and candidate road are compared. Thresholds are selected according to criterion situations respectively. 20° is a typical value..

5. Tracing back

Sometimes matching results are not asked to be given in real time. In this situation, information received later than the time to match can be included. To further improve matching results, a tracing back procedure is added to the algorithm. Latter reliable matching results are used to correct possible mistakes in former unreliable results. Based on continuous run assumption and topological relations in the map, a set of rules are set for history correction.

A rule is applied to eliminate mismatches returning to previous roads. When the GPS biases change rapidly in a short term, mismatches may occur. A typical situation is shown in Fig.6, Part 1. From P1 to P4, the algorithm matches the vehicle to road R1->R4. Then GPS position jumps back to previous road R1. In the matching procedure, it may be matched to road R1. So P1~P8 may be matched to R1->R4->R1->R4. However, apparently the vehicle has passed the crossing for just one time if the heading keeps the same. Correct trajectory should be R1->R4.

A rule is used to eliminate a mismatched road caused by GPS biases. Mismatches often occur when GPS data has a big bias in on-track direction. An example is shown in Fig.6, Part 2. The algorithm matches P9-P14 to road R6->R8. But actually the vehicle has not ever run to road R8. This is clear after it turns to road R7. A history

correction is necessary when road R7 is confirmed at P15-P17. Matching results of P12-P14 need to be corrected.

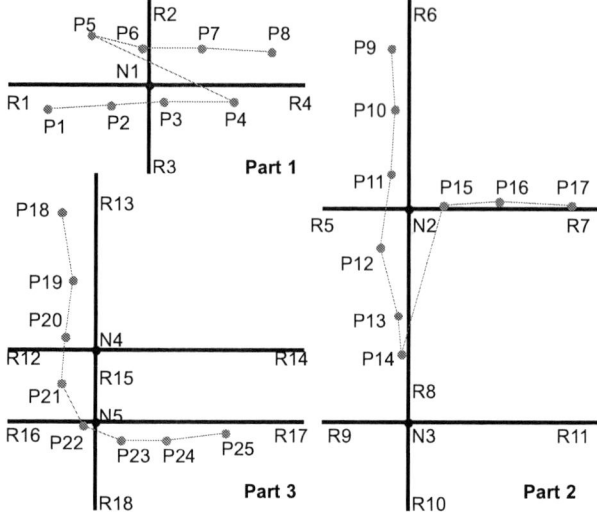

Figure 5. Illustration for tracing back rules

Another rule is set to deal with the situation shown in Fig.6, Part 3. Here road R15 is so short that the algorithm can not make sure the vehicle is on it at P21-P22. After the vehicle is matched to road R17 with confidence, R13 and R17 should be connected to form a continuous trajectory.

Some further mismatches are corrected in tracing back procedure as well. We will not describe these rare ones here. After the tracing, great improvement in the consistency of matched trajectory can be observed. This procedure is an effective addition to real time matching..

6. Simulation results

A group of simulation experiments have been taken for validation. Map of a typical urban area is used in our experiments. In the testing area, length of roads varies from 8m to 200m. As well, curve roads are contained in the map. GPS data are generated by adding noises to correct positions. Errors in GPS are often in 10~30m measured in distance, sometimes even reach up to 90m. Gyroscope and speedometer data have been weaken by artificial noises, as well. 20 sets of simulation data are tested. One of them is shown in Fig.6. The matching result of our algorithm is shown in Fig.7.

As around crossings, we cannot distinguish whether the vehicle is on the former road or the latter one exactly, a parameter **ignoring radius** is used in success rate calculation. When the vehicle has passed a crossing but the algorithm still matches it to the former road, or when the vehicle is still on the former road but the algorithm matches it to the latter road in the correct trajectory, we consider it rational and no harm to applications. When the distance between the vehicle and the intersection is shorter than this ignoring radius, such a mismatch is ignored in our statistics.

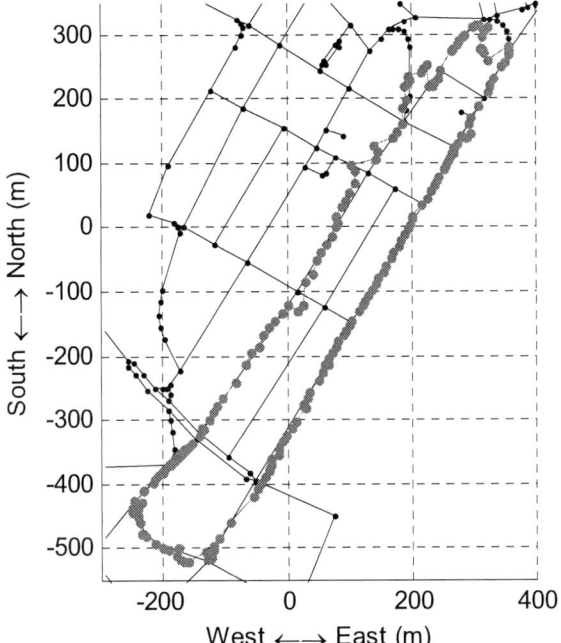

Figure 6. Map and GPS in simulation

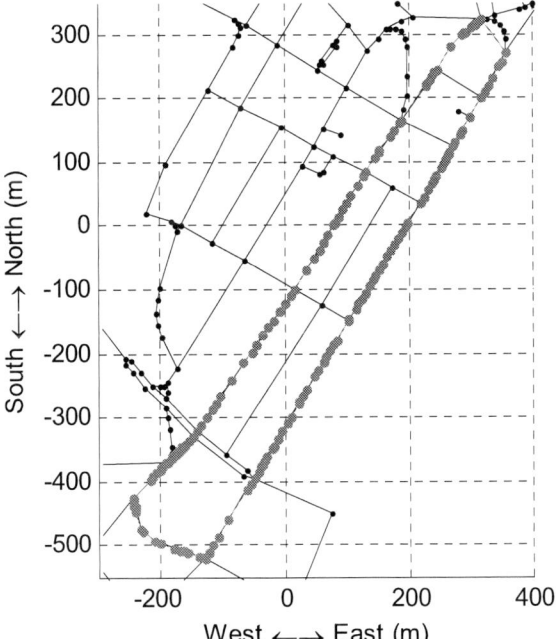

Figure 7. Matched trajectory

The average success rates of all 20 data sets are shown in Fig.8. When ignoring radius is 15m, even 95% is achieved in such environment. The effectiveness of the algorithm is proved. As well, we can see that the results after tracing have about 2% advantages than real time ones.

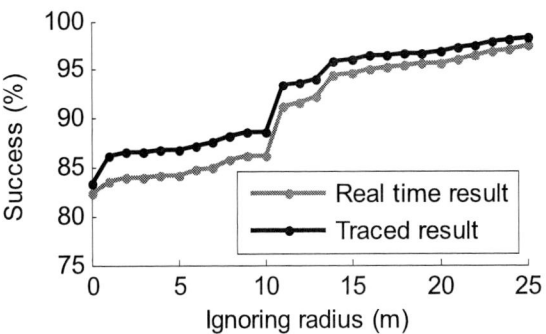

Figure 8. Success rates of the algorithm

7. Conclusion

A novel weight accumulation method is developed in this paper, called distance forgetting factor. This method makes more sufficient use of history GPS data. Based on this, a FSM matching engine and a tracing back procedure are built. All these parts compose an effective map matching algorithm in complex urban environments, using only rough sensor data. High success rate can be achieved. Simulation result shows that the algorithm can deal with the high uncertainties in such complex environments.

References

[1] R.L. French (1989), "Map Matching Origins, Approaches and Applications", *Proc. Second International Symposium on Land Vehicle Navigation.* pp.91-116

[2] M.A. Quddus (2006), "High Integrity Map Matching Algorithms for Advanced Transport Telematics Applications", *Ph.D. thesis*, Imperial College London.

[3] S. Syed, M.E. Cannon (2004), "Fuzzy Logic Based Map Matching Algorithm for Vehicle Navigation System in Urban Canyons", *Presented at ION National Technical Meeting*, San Diego, CA.

[4] Y. Zhao (1997), *Vehicle Location and Navigation Systems*, Artech House, Boston.

[5] M. Winter, G. Taylor (2003), "Modular Neural Networks for Map-Matched GPS Positioning", *IEEE Web Information Systems Engineering Conference 2003*, Wireless Geographical Information Systems, Rome.

[6] J.-S. Pyo, D.-H. Shin, and T.-K. Sung (2001), "Development of a map matching method using the multiple hypothesis technique". *In Proceedings of IEEE Intelligent Transportation Systems Conferenc*e, pages 23-27.

[7] Z.W. Chen, Y.R. Sun and X. Yuan (2002), "Development of An Algorithm for Car Navigation System Based on Dempster-Shafer Evidence Reasoning". *In Proc IEEE Intelligent Transportation Systems*, pp.534-537.

[8] W.Y. Ochieng, M. Quddus, R.B. Noland (2003), "Map-Matching in Complex Urban Road Networks", *Brazilian Journal OF Cartography*, No 55/02.

A Neighborhood-Based Trajectory Clustering Algorithm

Yunxin Tao, Dechang Pi

College of Information Science and Technology,
Nanjing University of Aeronautics and Astronautics, Nanjing, Jiangsu, 210016, China
yunxin.tao@hotmail.com, dc.pi@nuaa.edu.cn

Abstract

Existing trajectory clustering algorithm TRACLUS uses global parameters, it can not distinguish small, close, and dense trajectory clusters from large and sparse trajectory clusters. Moreover, TRACLUS needs two input parameters and is sensitive to input parameters. To avoid the shortcomings of TRACLUS, a neighborhood-based trajectory clustering algorithm named NBTC is proposed based on the improved framework. Our key insight is that neighborhood-based local density is quite different from the absolute global density used in TRACLUS. NBTC keeps the efficient of TRACLUS and needs only one input parameter. Experimental results demonstrate that NBTC can discover trajectory clusters in arbitrary shape and different densities trajectory database effectively.

1. Introduction

Clustering is the process of grouping similar data to provide a summary of data distribution patterns in a database. There are many clustering algorithms reported in the literature, density-based approach is a kind of them. Representative density-based algorithms include DBSCAN [1], OPTICS [2], NBC [3] , and TRACLUS [4]. Most of them are designed to deal with clustering of point data.

With the fast development of GPS devices, RFID sensors, and satellites, moving objects of all sizes can be easily tracked across the globe. As a result, a huge amount of trajectory data of moving objects have collected and stored in spatio-temporal database. A typical data analysis task is to find objects that have moved in a similar way [5]. The new data type arises challenge for traditional clustering algorithms.

Generally, a good clustering algorithm should be Effective, Efficient, and Easy to use (3-E criteria) [3]. Unfortunately, TRACLUS uses global parameters *Eps* and *MinLns*, it can not distinguish small, close, and dense trajectory clusters from large and sparse trajectory clusters. Parameter sensitivity is another shortcoming of TRACLUS. Thus, TRACLUS can not meet fully the 3-E criteria.

To illustrate what we state above, let us see a trajectory dataset *Synthetic* in Figure 1. We can see clearly that there are five trajectory clusters (C1~C5), in which three (C1~C3) are dense and close to each other and the other two (C4 and C5) are much sparse and locate far away. With such a dataset, no matter what parameters are taken, TRACLUS can not detect all the five trajectory clusters. If we use a larger *Eps* or a smaller *MinLns*, TRACLUS can find C4 and C5, but C1, C2 and C3 are merged into one trajectory cluster; In contrast, if we use a smaller *Eps* or a larger *MinLns*, TRACLUS can find C1, C2 and C3, but all trajectories in C4 and C5 are labeled as outlier trajectories.

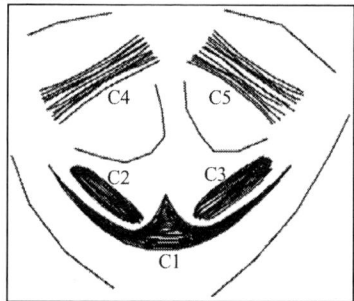

Figure 1 Trajectory dataset *Synthetic*

In this paper, we propose a neighborhood-based trajectory clustering algorithm (NBTC in abbr.). It uses the neighborhood-based line segment density factor (NLSDF in abbr.) to discover trajectory clusters. Unlike TRACLUS, NBTC can easily find all the five trajectory clusters in Figure 1. We will give the results of NBTC in experimental evaluation section. To our best knowledge, it is the first work to address this problem.

The rest of the paper is organized as follows. Section 2 presents some notions and introduces the new algorithm NBTC. In Section 3, a comprehensive experimental evaluation is made by using synthetic and real trajectory dataset. Section 4 concludes the paper.

2.Neighborhood-Based Trajectory Clustering

2.1. Notions for Neighborhood-Based Trajectory Clustering

Let D denotes a set of line segments, L, O, S, and S' be arbitrary line segments in D. In order to facilitate the readers to understand our new constructed NLSDF and NBTC algorithm, we change the definitions for points that were given in the literature [1, 3, 6] to those for line segments.

Definition 1. The *k-nearest neighbors set* of L is denoted by kNN(L). It is a set of line segments in D such that (1)$|k$NN(L)$|=k$, (2)$L \notin k$NN(L), and (3)Let S and S' be the k-th and the (k+1)-th nearest neighbors of L respectively, then $dist(L,S) \leq dist(L, S')$ holds.

Definition 2. For each L in database D, $\exists$ S $\in k$NN(L)、 $r = dist(L, S)$ such that $\forall S' \in k$NN(L), $dist(L, S') \leq r$. Then the *k-neighborhood* of L, is defined as kNB(L) = $\{S \in D \mid dist(L, S) \leq r$ and $L \neq S\}$.

Definition 3. The *reverse k neighborhood* of L is the set of line segments whose kNB contains L, which can be formally written as RkNB(L)=$\{S \in D \mid L \in k$NB(S) and $L \neq S\}$.

Definition 4. The *Neighborhood-based Line Segment Density Factor* (NLSDF) of L is defined as NLSDF(L) =$|$RkNB(L)$| / |k$NB(L)$|$.

NLSDF(L) is actually a local density of L in *relative* sense. Intuitively, the larger NLSDF is, the denser L's neighborhood is.

Definition 5. The set of *participating trajectories* of a trajectory cluster Ci is defined as PTR(Ci)=$\{$TR(Lj) $\mid \forall Lj \in Ci\}$. Here, TR(Lj) denotes the trajectory from which Lj has been extracted. Then $|$PTR(Ci)$|$ is called the *trajectory cardinality* of the trajectory cluster Ci.

2.2. New Algorithm

Figure 2 shows the improved framework of trajectory clustering based on the existing partition-and-group framework [4]. In order to avoid the shortcomings of TRACLUS and compare trajectory clustering results with it, the new algorithm named NBTC is proposed based on the improved framework.

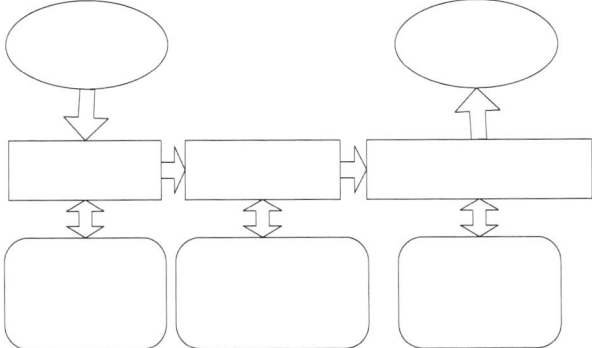

Figure 2 Improved framework of trajectory clustering

NBTC (A set of line segments D, k)
　Initialize the cluster id of each line segment L in D to UNCLASSIFIED
　Initialize clusterCount, OutlierSet, and coreLSQueue
　//Step1: calculate NLSDF for each line segment
　CalculateNlsdf(D, k);
　//Step2: scan dataset to discover trajectory clusters
　FOR each line segment L in D DO
　　IF (L.getClusterId() != UNCLASSIFIED or L.nlsdf < *thNlsdf*) THEN
　　　continue;
　　END IF
　　//label a new trajectory cluster
　　L.setClusterId(clusterCount);
　　FOR each line segment S in kNB(L) DO
　　　S.setClusterId(clusterCount);
　　　IF (S.nlsdf >= *thNlsdf*) THEN
　　　　coreLSQueue.add(S)
　　　END IF
　　END FOR
　　//expand current trajectory cluster
　　WHILE (coreLSQueue is not empty) DO
　　　L = coreLSQueue.getFirstLS();
　　　FOR each line segment S in kNB(L) DO
　　　　IF (S.getClusterId() != UNCLASSIFIED) THEN
　　　　　continue;
　　　　END IF
　　　　S.setClusterId(clusterCount);
　　　　IF (S.nlsdf >= *thNlsdf*) THEN
　　　　　coreLSQueue.add(S);
　　　　END IF
　　　END FOR
　　　coreLSQueue.remove(L);
　　END WHILE
　　clusterCount++;
　END FOR
　//Step3: check the trajectory cardinality
　FOR each trajectory cluster C DO
　　//*thTrajCard* is a threshold of trajectory cardinality
　　IF ($|$PTR(C)$| < thTrajCard$) THEN
　　　Remove C from the set of trajectory clusters;
　　END IF
　END FOR
END; //NBTC

Figure 3 NBTC algorithm

The key idea of NBTC is that: for each line segment L in a trajectory cluster, the number of line segments whose kNB contains L should be much than the number of line segments contained in kNB(L) at least. The NBTC algorithm consists of three steps. In the first step, NBTC searches kNB of each line segment, in order to reduce time cost, RkNB is created dynamically during the process of searching kNB, and then calculates NLSDF. In

Table 1 Experimental datasets

Name of datasets	Synthetic	Best Track	Deer95	Elk93
# trajectories	119	570	32	33
# points	795	17736	20065	47204
# line segments after partition	602	1607	2140	3975

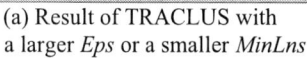
(a) Result of TRACLUS with
a larger *Eps* or a smaller *MinLns*

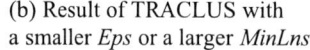
(b) Result of TRACLUS with
a smaller *Eps* or a larger *MinLns*

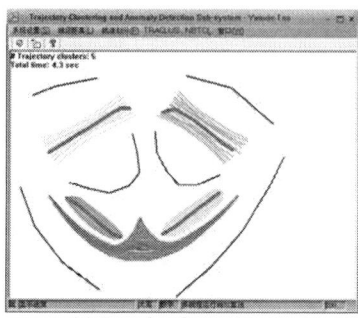
(c) Result of NBTC

Figure 4 Trajectory clustering results for different densities dataset *Synthetic*

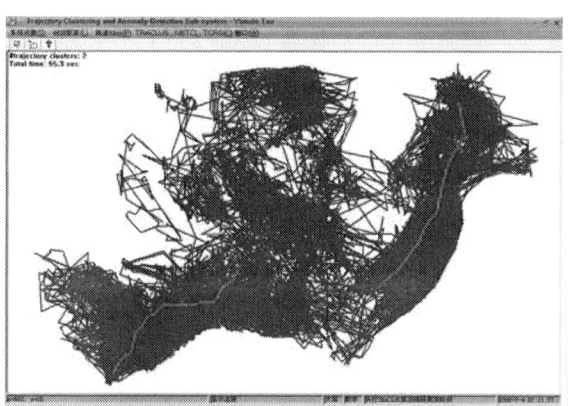
(a) Result of TRACLUS

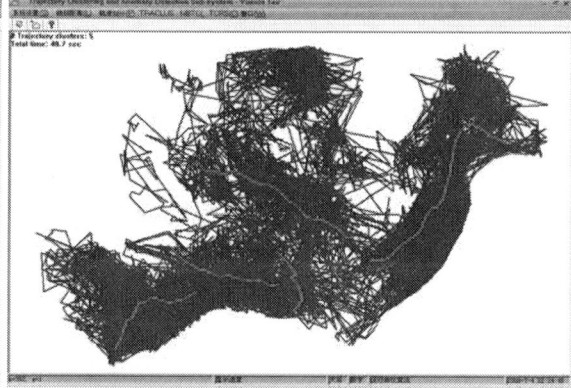
(b) Result of NBTC

Figure 5 Trajectory clustering results for real dataset *Deer95*

the second step, NBTC discovers neighborhood-based trajectory clusters. For current line segment L, if it is unclassified and NLSDF(L) ≥ *thNlsdf*, then allocate a cluster id for L, and continue to find all other line segments that are neighborhood-based density reachable from L to expand current cluster. In the third step, NBTC checks the trajectory cardinality of each cluster. If its trajectory cardinality is below the threshold, NBTC filters out the corresponding clusters. Figure 3 shows the NBTC in C++ pseudocode.

The most time-cost operation is kNB query in the first step. Let n be the total number of line segments in D, the time complexity of CalculateNlsdf is $O(n^2)$. The second and third step need scan D one time respectively, so the time complexity is $O(n)$. Therefore, the time complexity of the NBTC algorithm is $O(n^2)$. Due to the distance function of line segment is not a metric, the traditional spatial indexes can not use directly. However, the time complexity of NBTC is $O(n\log n)$ if we adopt the index technique for a non-metric distance function [7].

3. Experimental Evaluation

In this section, we evaluate the performance of our proposed algorithm by comparing with TRACLUS. We have implemented NBTC and visual inspection tool in VC6.0. All experiments have been carried out on a Pentium4 3.0GHz PC with 512MB RAM running Windows XP. Our performance study used one synthetic dataset and two real trajectory datasets, which are shown in Table 1. *Best Track* is extracted from hurricane track dataset at http://weather.unisys.com/hurricane/atlantic/, *Deer95* and *Elk93* are extracted from animal movement dataset at http://www.fs.fed.us/pnw/starkey/data/tables/. *Best Track*, *Deer95*, and *Elk93* are also the test datasets of TRACLUS.

First, we compare our approach with TRACLUS in a different densities dataset *Synthetic*. The original dataset *Synthetic* is shown in Figure 1, the trajectory clustering results of TRACLUS and NBTC are shown in Figure 4. Thick red lines are representative trajectories of each trajectory clusters. We can see that NBTC discovered all the five trajectory clusters (Figure 4(c)). In contrast, no matter what parameters we try, TRACLUS can not find all the five trajectory clusters. If we use a larger *Eps* or a smaller *MinLns*, TRACLUS can find the two sparse trajectory clusters, but the three dense trajectory clusters are merged into one trajectory cluster (Figure 4(a)); If we use a smaller *Eps* or a larger *MinLns*, TRACLUS can find the three dense clusters, but all trajectories in two sparse clusters are labeled as outlier trajectory (Figure 4(b)).

Figure 5 shows trajectory clustering results for *Deer95*. Although the upper-right region looks dense, the result of TRACLUS has no cluster in that region when the optimal parameters were taken. Moreover, several short representative trajectories are ignored by TRACLUS due to use global absolute parameters. However, NBTC can discover more meaningful trajectory clusters when we set k to be 12. Although the results for *Best Track* and *Elk93* are what we expect, we omit corresponding figures due to space limit.

Then, to test NBTC keeps the efficient of TRACLUS, two algorithms had run on each dataset in Table1 respectively. The time costs are shown in Figure 6. Since the difference between NBTC and TRACLUS lies in trajectory clustering, the time costs that shown in Figure 6 only include the run time of trajectory clustering.

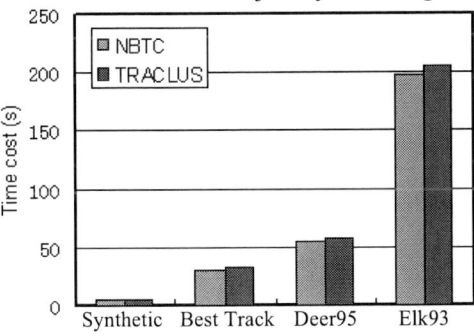

Figure 6 Time costs vary over each dataset

Finally, we study the effect of parameter k on the trajectory clustering result. To find a trajectory cluster, we must find at least a line segment L that satisfies NLSDF(L)$\geq$*thNlsdf*. Let C be the minimal cluster w.r.t. k in D, L be the first line segment found that satisfies NLSDF(L)$\geq$*thNlsdf* to expand C. Then, the minimal size of C is $|k\text{NB}(L)|+1 \geq k+1$. So k roughly determines the scale of the minimal cluster. On the other hand, in order to avoid statistical fluctuations due to k being too small, k should be at least 10 [6]. In experiments, we usually set k to a value between 10 and the scale of minimal cluster,

with which we can find most meaningful trajectory clusters. In addition, we have observed a small change of *Eps* or *MinLns* can result in different trajectory clustering result of TRACLUS. Compared with TRACLUS, NBTC is insensitive to k.

4. Conclusion

In this paper, we studied the problem of clustering the trajectory data of moving objects. After constructing neighborhood-based line segment density factor, we proposed a novel algorithm NBTC based on the improved framework. NBTC avoid the shortcomings of existing trajectory clustering algorithm TRACLUS, it can discover trajectory clusters in arbitrary shape and different densities trajectory database. Compared with TRACLUS, NBTC needs only one input parameter and is insensitive to input parameter k. The experimental results showed the effective and efficient of our approach by using real and synthetic trajectory dataset.

Acknowledgement

This work is supported by the National High Technology Research and Development Program of China under grant No. 2007AA01Z404.

References

[1] M. Ester, H.-P. Kriegel, J. Sander, and X. Xu, A Density-Based Algorithm for Discovering Clusters in Large Spatial Databases with Noise, *Proc. 2nd ACM Int'l Conf. on Knowledge and Data Mining*, pp. 226-231, 1996.

[2] M. Ankerst, M. Breunig, H.-P. Kriegel, and J. Sander, Optics: Ordering points to Identify the Clustering Structure, *Proc. 1999 ACM SIGMOD Int'l Conf. on Management of Data*, pp. 49-60, 1999.

[3] S. Zhou, Y. Zhao, J. Guan, and J. Huang, A Neighborhood-Based Clustering Algorithm, *Proc. 9th Pacific-Asia Conf. on Knowledge Discovery and Data Mining*, pp. 361-371, 2005.

[4] J.-G. Lee, J. Han, and K.-Y. Whang, Trajectory Clustering: A Partition-and-Group Framework, *Proc. 2007 ACM SIGMOD Int'l Conf. on Management of Data*, pp. 593-604, 2007.

[5] M. Vlachos, G. Kollios, and D. Gunopulos, Discovering Similar Multidimensional Trajectories, *Proc. 18th Int'l Conf. on Data Engineering*, pp. 673-684, 2002.

[6] M.M. Breunig, H.-P. Kriegel, R.T. Ng, and J. Sander, LOF: Identifying Density-Based Local Outliers, *Proc. 2000 ACM SIGMOD Int'l Conf. on Management of Data*, pp. 93-104, 2000.

[7] V. Roth, J. Laub, M. Kawanabe, and J.M. Buhmann, Optimal Cluster Preserving Embedding of Nonmetric Proximity Data, *IEEE Trans. On Pattern Analysis and Machine Intelligence*, vol. 25, no. 12, pp. 1540-1551, 2003.

A Personal Query Framework based on Personal and Role Ontology

Liying Fang , Jianzhuo Yan , Pu Wang, Bin Shi
College of Electronic Information & Control Engineering,
Beijing University of Technology, Beijing 100124, China
fangliying@bjut.edu.cn,yanjianzhuo@bjut.edu.cn,wangpu@bjut.edu.cn

Abstract

To meet requirements of personal query, a personal query framework including personal ontology and role ontology is proposed based on preference ontology. The two-tier structure of the preference semantic formalization shows the differences and commonness of the background knowledge among different users. It is possible to generate personal ontology and role ontology automatically and to maintain them by the historical query records and analysis modules in the framework. Comparing with current methods, this framework is able to implement personal semantic query and make it possible to create and maintain the two-tier ontology structure highly automatically because of the personal semantic preference formal expression.

1. Introduction

With the rapid development of information industry, personal query becomes more and more important in recent years[1][2]. Ontology has been introduced into the personal query field [3][4]. Most research focus on the user preference about vocabulary and concepts [4], however they almost pay no attention about the personal organization of concepts in user's mind. Jaime[3] notices that users may have different knowledge background, so that in INDUS user can input his own ontology with his query as well. However it needs that the user should define his ontology beforehand by himself.

In order to express personal knowledge differences with respect to a common field knowledge background, a new model is needed to describe personal knowledge background, including such as concepts which a user is interest in, the organization of these concepts, and the preference expression for each concept. Meanwhile a mechanism to maintain the model is also important for a personal query framework.

Aiming at the personal query on semantic level, this article defines preference ontology and presents a framework for personal query based on personal ontology and role ontology.

This paper is organized into 4 sections. In section 1, a brief description of personal query with ontology is introduced. Related work is analyzed in section 2 In section 3, a background of personal query is given. In the 4th section, an architecture of a personal query system is proposed at first. Followed definitions of personal ontology and role ontology, this paper focuses on the policies of automatic construction and update for these ontology The final section is our conclusion.

2. Related Work

No matter in the web search engine field[3], or in information retrieval of digital libraries field[4][5], user profile is the common key part to implement personal query. According to different forms, we divide user profiles into two types: non-ontology profile and ontology profile.

Non-ontology profile mainly includes user profiles[6] or user preference vocabulary vectors[5]. For supporting the preference vocabulary in a query, they both record these vocabularies and generalize and specialize them, and then record the results as well.

Ontology profile expresses the user profile directly or indirectly based on ontology. In Ref [4], the semantic preferences of a user are represented as vector concept weights. The concepts in a vector are all belongs to some domain ontology. However the vector can not express the preference organization of concepts. In view of different scientists (users of INDUS [3]) want to access the different view of data in INDUS information integration environment, the users of INDUS can input their own ontology when they input their quires. However INDUS needs scientists define their ontology by themselves, and the queries need to be expressed in terms of concepts in the global ontology. For applying a semantic approach to retrieving information from a blog, semblog[7] proposes the definition of personal ontology, consisting of FOAK, Contents RSS, RDFS Ontology as metadata, and constructs a user profile using the personal ontology manually. In order to avoid creating personal ontology manually [8] automatically extracts a user-interest ontology as a template ontology from a blog community, thus creating and updating ontology is easy for users.

3. Background Analysis

From the view of semantic query, only directly or indirectly based on ontology, user's profile can contain preference concepts. And directly base on ontology, user's profile can express the concepts organization in user's mind besides the concepts themselves, especially when the concepts organization in user's mind has differences with in a global ontology of an information system.

It is possible to record every user's personal ontology when the number of users is small. But after more and more users register into the information system, there will have too much personal ontology to maintain. From above , we naturally take a novel method into account, that is for a group of users who have same knowledge background, same preferences , or may be very similar, we can extract their common preference information and record only one copy in some other place other than record them in every user's personal ontology. At the same time, personal ontology only used for express the very larruping preference information of each user. So we can get a tradeoff. And then we can call the users in a group is sharing a role. One role can contain many users, and a user can have several roles.

Mining the query history and analyzing users' feedback often used to information query system to improve the degree of personalization. Meanwhile, by analyzing the relativity among the several continuous with the query records mentioned above, a system can get the implicit feedback from users and can get what are the users' real interests even more [2].

4. A framework for personal query system

4.1. Architecture

We proposed a framework (See Figure 1) which is applicable to the kind of information integration system provided with ontology expression layer. In this framework, we call the expression is global ontology, no matter if they have partial ontology or field ontology in the information system, and we suppose there already have mapping between data source and the global ontology. For expressing a user's special preference and common preference with a group users, personal ontology and role ontology make up of the user's profile in the framework. They are both used for describe a user's semantic preference. The query history database and its analysis mechanism are the parts for supporting the maintenance of personal ontology and role ontology in the framework. Figure 2 describes the query process of the framework.

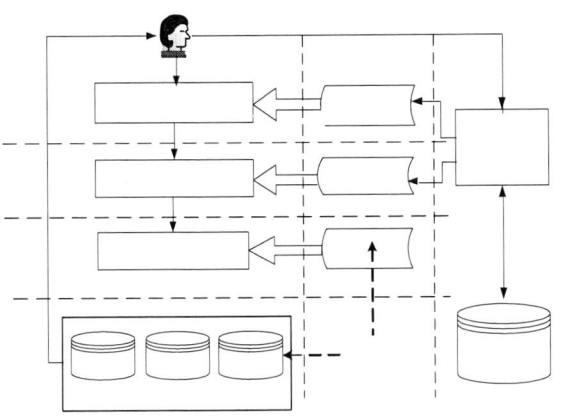

1 *user* input a *query*
2 *query* is preprocessed by user interface, and output the key vocabulary *vList* in the *query.*
3 query engine compare the preference vocabulary in *user*'s personal ontology and *vList*. To the subset of *vList* which can be found in personal ontology, store them in *vList₁*, and extract sub Preference ontology *v_subPO₁* corresponding *vList₁*. Output *vList.*, *vList₁*, and *v_subPO₁*.
4 find the *user*'s all roles *role*s. Then by analogy with step 3, query engine compare the preference vocabulary in *role*s' role ontologies and *vList₂*(=*vList.*-*vList₁*). Output sub Role Ontologies {*v_subRO*}.
5 query engine combined the *v_subPO₁* and {*v_subRO*} to a *q_userRO.*
6 according the mapping between 3 layers of ontology, rewrite the query to *rw_query* by *vList* and *q_userR*. The target is all the elements in *q_userRO* should be expressed by elements of Global Ontology.
7 information system execute the *rw_query*, and give

Figure. 2: Personal semantic query process.

Figure. 1: Personal semantic query system framework based on personal ontology and role ontology

4.2. Definitions of Personal Ontology and Role Ontology

Before the definitions of personal ontology and role ontology are given, we give another definition at first.

Definition 1: Preference Ontology=<(C, PC), (P, PP), (R, PR), (A), (I, PI)>.

In that, C/P/R/A/I denotes a concepts set, a properties set, relationships among concepts, an axiom set and a set of instances respectively.

PC is the corresponding preference values set of the C. That is, to every concept $c \in C$, there has a preference value $pc \in PC$ with it. By analogy PP is the corresponding preference values set of P. That is, $\forall p \in P$, $\ni pp \in PP$. And still we have $\forall r \in R$, $\ni pr \in PR$ and $\forall i \in I$, $\ni pi \in PI$. The range of pc, pp, pr and pi all are [0,1]: 0 for non-interesting, 1 for the highest preference.. Especially, $\forall a \in A$, there exist $a \equiv true$.

Definition 2: Personal Ontology= <userID, RList, userPO>

userID denotes some base information about a user, including a unique userId in this query system. Except userId, some other information like name, age, gender, career and so on in userID may be helpful to decide which roles the user belongs to, but these information is not necessary.

RList denotes the roles list that the user with userID belongs to.

userPO denotes a preference ontology, where the userPO is used to express the knowledge organization in a user's mind, no matter what kind of the knowledge organization in a global ontology.

Definition 3: Role Ontology= <roleID, roleName, rolePO>

roleID denotes a unique symbol to identify roles.

roleName denotes names of the roles.

rolePO denotes a preference ontology. Under each roleID, the rolePO is used to express the knowledge organization in users' mind who all are belongs to the role, and still no matter what kind of the knowledge organization in a global ontology. That is rolePO express the preference information for the role.

4.3. Automatic construction of Personal Ontology and Role Ontology

To some extend, we can consider either the personal ontology or role ontology can be view as a projection of a global ontology. Therefore by analyzing user query history, the automatic ontology generation process can be divided into two types according to different implementations: add mode and prune mode. Under the add mode, the initializations of two ontology are empty by default. And then, based on the analyzing to history query information, user's preference vocabulary and interested concepts can be found gradually. Referring to the global ontology, personal ontology and role ontology can be generated step by step. While under the prune mode, the initializations of two ontology are copies of the global ontology by default. Next, analogously depending on the

analysis of history query information, personal ontology and role ontology can be generated by pruning gradually.

4.4. Update policy of Personal Ontology and Role Ontology

Because personal ontology only reflects one user's own knowledge background, and is independent with each other, its update can be run with every feedback cycle. While a role ontology's update needs different process, because every role includes many users. When query load is relatively light, starts a special module to compare the common features in users' personal ontologies, and updates the role ontology. Rule 1 and Rule 2 are used to update rolePO, while Rule3 and Rule4 are used to update the roles themselves.

On the assumption that, there are n users: $user_1 \sim user_n$, and m roles: $role_1 \sim role_m$. Thereinto, any $role_i (i \in [1,m])$ includes the users amount $\leq n$, and any $user_j (j \in [1,n])$ belongs to $\leq m$ roles. Let $U(role_i)$ express the users in $role_i$, and $R(user_j)$ express the roles which $user_j$ has. Let subOf(•) refers to the subset of a set. Then:

Rule 1: if commonness of $subOf(userPO(U(role_i)))$ reach the remarkable value, then cut every $subOf(userPO(U(role_i)))$ from all $U(role_i)$ to $rolePO(role_i)$, and only maintain one copy of them in $rolePO(role_i)$.

Rule 2: if $\ni subOf(userPO(U(role_i)))= subOf(rolePO(role_i))$, and it's commonness less than remarkable value, then remove $subOf(rolePO(role_i))$ from role ontology.

Rule 3: if

$\ni (userPO(subOf(U(role_i))) \cap userPO(subOf(U(role_j))) \cap ... \cap userPO(subOf(U(role_t)))) \neq \varnothing$, and the commonness of it reach the remarkable value, then generate a new $role_{m+1}$, which $rolePO=(userPO(subOf(U(role_i))) \cap userPO(subOf(U(role_j))) \cap ... \cap userPO(subOf(U(role_t))))$, and add $role_{m+1}$ to the RList of $subOf(U(role_i))$ to $subOf(U(role_t))$ at the same time.

Rule 4: if $\ni rolePO(role_i) = \varnothing$, then remove $role_i$ from all RList of $U(role_i)$ first, and then remove the tuple with roleID=i from role ontology as well.

5. Conclusions

This paper defines preference ontology, and then presents a framework for personal semantic query based on personal ontology and role ontology. A user's semantic preference can be expressed well by personal ontology and role ontology. Meanwhile the two layers of ontology not only can save storage place but also can guarantee the degree of personalization. Construction and update policy make the automatic generate and maintain of these two kinds of ontology possible. Note that in the framework there exists a trade-off between personal degree of every user and common degree of a role. It depends on the setting of a remarkable value of similarity among different users.

In a view of the organization of most knowledge in human being's mind has fuzzy character, therefore if fuzzy [10] character can be introduced to the expression preference ontology, the framework maybe get better personal semantic query results. Introduce fuzzy character is our future work.

References

[1] Abiteboul S., Agrawal R., Bernstein P., Carey M., Ceri S., Croft B., et al (2005). *Proc. of the Joint Conference on Information Sciences, "The Lowell Database Research Self Assessment", Communications of the ACM,* Vol.48, pp. 111-118.

[2] Yang B., Glen J. (2006). "Retroactive Answering of Search Queries", *Proc. 15th International World Wide Web Conference.* pp. 173-183 Edinburgh, Scotland.

[3] Castillo J. A. R., Silvescu A., Caragea D., Pathak J.; Honavar V. G. (2003). *"Information Extraction and Integration from Heterogeneous, Distributed, Autonomous Information Sources--A Federated Ontology-Driven Query-Centric Approach", The 2003 IEEE International Conference on Information Reuse and Integration.* Las Vegas, USA. IEEE Press, pp. 183-191.

[4] Yan D. W., Cen Y. H., Zhang W., Mao P. (2006). "Ontology-based framework for personalized recommendation in digital libraries", *Journal of Southeast University (English Edition),* Vol.22, No.3, pp.385-388.

[5] Zhang Y., Yuan F. (2006). "A user interest model-based personalized information retrieval method"(in Chinese), *Journal of Shandong University (Natural Science)*, Vol.41, No.3, pp.120-125.

[6] Fang L. Y.,. Yan J. Z, Wang P. (2006). "A Plan of Semantic Conflict and Query Process Based on Ontology"(in Chinese), *Computer Science*, Vol.33, No.10A, pp.342-344.

[7] Ohmukai, I. Takeda H., Hamasaki M., Numa K., Adachi S. (2004). "Metadata-driven Personal Knowledge Publishing", in *International Semantic Web Conference* (ISWC 2004). Hiroshima, Japan, pp.591-604.

[8] Makoto N., Yu M., and Yoshihiro O. (2006). "Innovation Detection based on user-interest ontology of blog community", *International Semantic Web Conference* (ISWC 2006).Athens, GA, USA.

[9] Kang D. Z., Xu B.W., Lu J. J., Li Y. H. (2006). "Description logics for fuzzy ontologies on semantic web",

Southeast University (English Edition), Vol.22, No.3, pp.343-347.

The Study of the Sentence Format in Nature Language Understanding

Jianming Miao, Quan Zhang

Institute of Acoustics, Chinese Academy of Science
Beijing, 100190
mjm@mail.ioa.ac.cn

Abstract

The Sentence Format (SF) indicates the relationship among the components of the sentences, and it also can provide guiding knowledge for computer sentence processing. The sentence format is used to indicate the appearance order of the main semantics block in the sentence. The method manifested the language expression custom of the language and the synthesis utilization of each grammar category. The integrity description of the language complex sentence pattern phenomenon is described through the combination of the sentence format research and the sentence category research. The format research is advantageous to improve the analysis level of the computer sentence category processing. This paper gives the Sentence Format system in Chinese and provides the coding way. The results of test indicated that the information benefits the computer processing in NLP area.

1. Introduction

The processing to the natural sentence need not only understand the type and the number of the sentence semantics ingredient but further understand the combination relations between the ingredients. The HNC theory studies the type and number of semantic block (the sentence basic constitution ingredient defined by the HNC theory) and also thinks the semantic block arrangement in the syntax stratification and the pragmatics stratification. The HNC method puts the arrangement of semantic blocks research on the sentence format research, and makes the semantic block research separate from the language classification characteristic matter. The method rejects the interweaving difficulty of the semantic block research and the arrangement of semantic blocks research, and makes the processing difficulty reduce. The terminology of the sentence format is the continuation and development of the HNC theory to the traditional Chinese sentence pattern and the sentence expression terminology.

2. The sentence format definition

The study on the sentence format based on the sentence category expression foundation. The HNC sentence format theory called the traditional linguistics subject part and object part as the generalized object semantics block(shortly called as GBK block), and called the verb part as the characteristic semantics block(shortly called as EK block). The HNC theory divided the sentence category into the two classes according to the characteristic semantics block. The two classes are called as the basic sentence category class and the compound sentence category class. The united mathematics expression of the basic sentence category and the compound sentence category is as follows:

SCn＝GBK1+(EK)+GBKm (m=2-4)

In the foundation, we designed the four kind sentence formats based on the inspection semantics block order of each kind of semantics block changing. The formats are called as the basic sentence format, the standard sentence format, the breaching rules sentence format and the abbreviated sentence format.

2.1. The basic sentence format

The basic sentence format set up the letter "!0" on the first part, and expressed that the EK semantics block is located in the second position of the sentence, and permitted the GBK block position changing in the sentence.

2.2. The standard sentence format

The standard sentence format set up the letter "!1" on the first part, and expressed that the EK semantics block is not located in the second position of the sentence, and also must add the boundary mark between the two neighboring GBK block.

2.3. The breaching rules sentence format

The breaching rules sentence format set up the letter "!2" on the first part, and expressed that the EK semantics

block is not located in the second position of the sentence, and also didn't added the all or part boundary mark between the two neighboring GBK block.

2.4. The abbreviated sentence format

The abbreviated sentence format is defined as the sentence format of omitting the some or certain main semantics block in the sentence mathematics expression. It set up the letter "!3" on the first part. The abbreviation sentence format must be used with the other three sentence formats, and manifest the abbreviation occurred in the concrete format.

3. Code

The HNC theory divided the sentence category into the basic sentence category and the compound sentence category. The compound sentence category format code is formed on the basic sentence category code foundation. This article has only elaborated the basic sentence category format code. The reader voluntarily may combine the constitution according to the mixing requests in this foundation. The sentence format has the strict request to the code selection. The requests will be manifested in the following article.

3.1. The code of the basic sentence format

The coding rule of the basic sentence format may induce into the following three parts:

1) Because the meaning of "!01" format is same to the meaning of "!0" format, therefore we cancelled "!01" format. The "!011" format expressed that the GBK position in four main semantics block sentences has carried on the exchange, namely formed the "GBK3 + GBK2" order.

2) Besides the GBK2 block is located left side of the EK block, the "!01" format also expressed that the EK block has appeared the distortion. The "!01" format agrees the GBK1 block not appear. The format does not need restore GBK1 block. The "!021" format expressed that the GBK2 block and the GBK3 block move to left, contrasted the basic sentence format, and the GBK2 block transferred to the front of the EK block, and the GBK1 block became the GBK block of the most right flank. The "!022" format is effective to the four semantics block sentences, and indicated that the GBK block position in the right flank of the EK block had carried on the exchange. The exchange formed the "GBK1+GBK3" order.

3) The agreement of the "!03" format is similar to the "!02" format. The GBK3 block position is located in the left side of the EK block. The "!03" format expressed the EK block has appeared the distortion, simultaneously the GBK1 block does not appear. The "!031" format expressed that the GBK2 block became the GBK block in the most right flank. The "!032" format indicated that the GBK position in the right flank of the EK block had carried on the exchange. The exchange formed the "GBK2+GBK1" order.

After completing the description of the main semantics block position, we need increase the corresponding redundant code and the changing format code according to the appearance situation of the redundancy mark and the changing format situation of the GBK block.

4）The redundancy mark information of the main semantic block separately uses 7, 8, 9 letter to express that the redundant situation appear in front of GBK1, GBK2 and GBK3 block.

5）The changing format information of the main semantic block separately uses 01, 02, 03 symbol to express that the GBK1, GBK2 and GBK3 block appear the changing format situation.

6）The counter-format mark "^" can appear on the front of the mark "!" when the sentence includes the two or three blocks. It expressed that the block order of the two block sentences and the three block sentences is opposite. The four block sentence does not have the counter-format mark.

For example: The invasion by Western powers ||gradually reduced || China || to the status of semi-colonial and semi-feudal society. (The correlating sign mark can refer to the references)

!09XY02*311J=A+XY02+YB1+^YB2

The sentence format of the sentence is the "!09" format.(A: The invasion by Western powers; XY02: gradually reduced; YB1: China; YB2: to the status of semi-colonial and semi-feudal society)

3.2. The code of the standard sentence format

The "!1mnp" code was used to express the standard format. The letter "m" expresses the displacement position situation of the EK block. The displacement position number is directly wrote when the EK block shifted to the right position, and the number "4" is used to express the EK block shifted to the left position. The letter "n" expressed the position which the GBK block firstly appears. The letter "p" only appears in the four block sentences. The number "1" expressed that the second GBK block and the three GBK block is the natural order. The natural order indicated the small letter number GBK block in the front and the large letter number GBK block in the back. The number "2" expressed that the second GBK block and the three GBK block has exchanged the position.

3.3. The code of the breaching rules sentence format

The "!2mnp" code was used to express the breaching rules sentence format. The letter "m" and the letter "n" and the letter "p" are same to the three letter meaning in the standard sentence format. Moreover in the case of the breaching rules sentence format, the GBK blocks with the mark information of the semantic block are correspondingly produced according to the agreement with the redundancy and changing format coding.
For example: <What | the Chinese people | see >|| is || a prosperous socialist motherland.

$$< T1C \qquad TB1 \qquad T1 >$$
!212T1J=T1C+TB1+T1

The sentence format of the sentence is "!212" format.(T1C: What; TB1: the Chinese people; T1: see)

3.4. The code of the abbreviated sentence format

The "!3m" code was used to express the abbreviated sentence format. The letter "m" expressed that the number order of the GBK block was omitted.
For example : Marx, Engels, Lenin and Comrades [Mao Zedong ∧and Deng Xiaoping] || were all |||\ shining examples of {integrating | theory | with practice} & and of { making | theoretical innovations }/ .

$$\{ \qquad R110 \quad RB1 \qquad ^\wedge \qquad RB2 \}$$
!3109XR110*311J=(A)+R110J+RB1+^RB2
(abbreviate the action A(GBK1) block)

The sentence format is "!3109" format.(R110: integrating; RB1: theory; ^: with; RB2: practice)

4. The good and bad points of the sentence format

The English global EK block certainly is located in the second position of the sentence. The characteristic was described concisely and clearly. According to the basic sentence format, English only can appear the basic sentence format, and cannot appear the standard sentence format, simultaneously the design of sentence format had considered the language custom of the Indo-European language tense and changing format. This design of sentence format has covered many kinds of languages custom of the different language. The code design is conforms to person's memory characteristic because the design of sentence format full considered the change of the semantic block. But the design creates that the huge

number of the format complete mathematics arrangement format. It has only listed the partial common codes.

5. The result of test

The author had carried on detailed expression ability test to the sentence format expressed method. The test data originated from the standard language material in "the people's net" and "the New China net". The tests material is composed by 4,927 sentences. The result indicated the sentence format had the excellent semantic structure expression ability. After we use the sentence format to formalize the sentence corresponding information, the number that the computer can give the sentence correlation information had accounted for 95.3% of the total sentences. The result of test is satisfied with ours

6. Summaries

The sentence format is used to indicate the appearance order of the main semantics block in the sentence. The method manifested the language expression custom of the language and the synthesis utilization of each grammar category. This article showed the sentence format of the HNC theory, and elaborated the definition and code of the sentence format, and has carried on the concrete elaboration through the example. The format research is advantageous to improve the analysis level of the computer sentence category processing.

Acknowledgements

This work is supported by National Fundamental Research Project (973 Project) (Grant No. 2004CB318104).

References

[1] Huang Zeng-yang. Mathematics and physics symbol system of language in language concept space. Beijing: Ocean Press, 2004.
[2] Huang Zeng-yang. HNC (Hierarchical Network Concept) Theory [M]. Beijing: Tsinghua University Press, 1998.
[3] Zhang Ke-liang. An HNC approach to the resolution of syntactic structural ambiguity [J]. Chinese information journal, 2,004, 18 (6):43-52.
[4] Jin Yao-hong. The design and realization based on sentence category analysis system of the HNC theory. Beijing: Institute of acoustics, Chinese Academy of Science, master's degree paper, 1998.
[5] Wei Xiang-feng. The software platform for expanded sentence category analysis based on the HNC theory. Beijing: Institute of acoustics, Chinese Academy of Science , doctor's degree paper, 2005.

[6] Miao Jian-ming. The Design and Knowledge Representation of the Professional Activity Domain Sentence Category, Beijing: Institute of acoustics, Chinese Academy of Science , doctor's degree paper, 2007.

Queuing Analysis for Reconfigurable Computing

Kourosh Hassanli[1], Ali Khayatzadeh Mahani[2], Hadishahriar Shahhoseini[2], Ebrahim Teimoury[2]

[1]Department of Telecommunication Engineering,
Islamic Azad University-Jahrom Branch, Jahrom, Iran
[2]Department of Electrical and Electronic Engineering
Iran University of Science and Technology, Narmak, Tehran, 16844, Iran.
k_hassanli@iust.ac.ir, khayatzadeh@iust.ac.ir, shahhoseini@gmail.com,teimouri@iust.ac.ir

Abstract

Evaluation of reconfigurable hardware is an important part of designing the computer systems. In this paper two previously used recourse management mode for reconfigurable system, restricted and split mode described, a third mode (which is called merge mode) is introduced and all modes are analyzed. To analysis the queuing model are applied. The M/M/1/k queue is used. FPGA utilization and speedup are determined as two main metrics for system performance evaluation.

To compare the modes with each others under different circumstances the metrics is found and plotted. The performance curve show proposed mode is act better than two previous modes, and has resource management. It is more effective when the larger task must be run on the system.

1. Introduction

There are two primary methods for execution of the computational algorithms. The first is to use hardwired technology, either use application specific integrated circuit (ASIC) or a group of individual component forming a board level solution to perform the operation in hardware. The second method is to use software-programmed microprocessors that are a more flexible solution. Reconfigurable computing is intended to fill the gap between the software and hardware, achieving potentially more performance than the software method [1, 2]. FPGAs and reconfigurable computing have been shown to accelerate a variety of applications. Recent applications that have been shown to exhibit significant speedups using reconfigurable hardware include: data encryption & decryption, automatic target recognition, string pattern matching, Glomb ruler derivation, transitive closure of dynamic graphs, Boolean satisfiability, data compression and genetic algorithms for the traveling salesman problem.

In order to achieve the performance benefits, yet support a wide range of applications, reconfigurable systems are usually formed with a combination of reconfigurable logic and a Host Processor. The application program could be run by the host processor [3-7].

There are four model of reconfigurable resource which is shown in [2]. These models relate closely to the complexity of the scheduling and placement problems and contain 1D and 2D area models. (a) The most flexible area model that allows to allocate rectangular tasks anywhere on the 2D reconfigurable surface. (b) A 2D partitioned model where the reconfigurable surface is split into a statically-fixed number of allocation sites, so called blocks. (c) A 1D area model where tasks can be allocated any where along the horizontal dimension and the vertical dimension is fixed. This model is amenable to an implementation on Xilinx Virtex [2]. (d) A 1D block partitioned area model which combines the simplified placement of the model (b) with the implementation advantages of the model (c).

Currently available FPGA technology (Xilinx Virtex) is partially reconfigurable only in vertical chip-spanning columns.

This paper is concentrated on making an analytical model of reconfigurable computing. A reconfigurable system usually contains a reconfigurable resource and a host processor, which is named RR and HP respectively. To exemplify the approach in this paper the RR splits into unequal blocks and three scheduling mode are presented and discussed. Each block of RR has a queue of arrived tasks with M/M/1/K parameters. In first mode that is named restrict mode, tasks in queue q_i can only be placed into blocks that correspond to q_i (figure 1-a). In second mode each task can be allocated to the block that has larger size and its queue is empty [2]. This mode is named split mode in this paper. In third mode, which is named merge mode in this paper, two smaller and free block have been merged and a larger task which must previously assigned to a larger block could locate in the new merged block. Figure (1) shows the structure of three modes for an FPGA with three blocks.

According to these modes, an analytical model will be constructed and the most important parameters of system

978-0-7695-3342-1/08 $25.00 © 2008 IEEE
DOI 10.1109/PEITS.2008.125

(the utilization factor of RR and speedup) are determined and compared.

The rest of the paper is organized as follows: In section 2 the previous works that lead to evaluation of reconfigurable computing is presented. The analytical model and its equations are discussed in section 3. The system evaluation and performance metrics would be come in section 4, which are contained equations, important parameters and simulation results. This paper is concluded in section 5.

2. Previous works

In many applications, the systems need to be frequently reconfigured during run time to exploit the full potential of using the reconfigurable hardware. Reconfiguration overhead is therefore a major concern because the CPU often must sit idle during this process. By reducing the overall reconfiguration overhead, the performance of the system is improved. Therefore many studies have involved examining techniques to reduce the configuration overhead [8]. The techniques include configuration prefetching [9], configuration compression [8] and configuration caching [10].

Almost all previous works are concentrated on system management for reducing the configuration overhead. In this paper a technique for evaluating reconfigurable computing system is proposed. This technique is used for analysis the most important parameter. The analytical model of the system is discussed in next section.

3. Mathematical analysis

In this section the queuing analysis is used for evaluating the reconfigurable system. At first the equations for restrict mode are shown. Then it is extended for two complicated modes. The queuing structures of three modes are shown in figure 1.

3.1. Restrict mode

In this mode a private queue exist for each block. To analyze the system the following assumptions are considered:

1. All tasks enter to the appropriate queue according to their size to go to the RR (reconfigurable Resource). If the queue was full, the task will redirect to HP (Host processor) for execution.

2. The entrance process of tasks into a queue obeys from Poisson random process with mean value of λ.

3. The service process, i.e. execution time of task, is exponential with mean value of μ.

4. The queue discipline is FCFS.

5. Each queue cannot hold more than k tasks during execution of current task, and the next task during this state, redirect for execution by HP as described in assumption 1.

According to above consideration the are M/M/1/k queue is used.

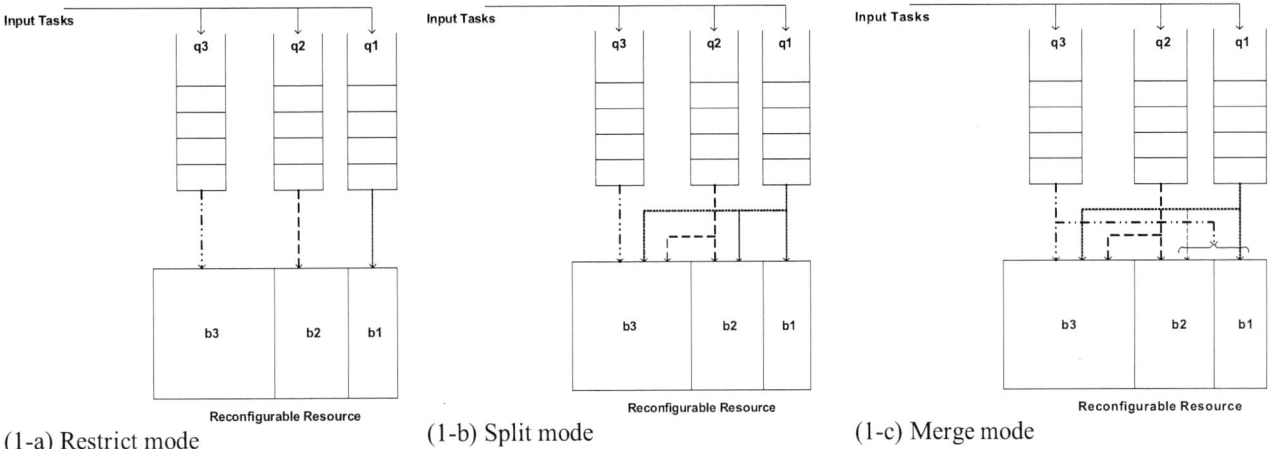

(1-a) Restrict mode   (1-b) Split mode   (1-c) Merge mode

Figure 1 Structure of three scheduling mode in reconfigurable resource

The equations for limit probabilities could be written as follows:

$$p_1 = \frac{\lambda}{\mu} p_0$$

$$p_2 = \left(\frac{\lambda}{\mu}\right)^2 p_0 \qquad (1)$$

$$\vdots$$

$$p_k = \left(\frac{\lambda}{\mu}\right)^k p_0$$

where p_i denotes the probability that i tasks exist in queue, and p_0 denotes the probability that no task exist in queue. The probability conservation can be found as follows:

$$\sum_{i=0}^{k} p_i = 1$$

$$So \qquad p_0 + \sum_{i=1}^{k}\left(\frac{\lambda}{\mu}\right)^i p_0 = 1 \qquad (2)$$

Hence p_0 and p_k can be obtained as follows:

$$p_0 = \left(1 + \sum_{i=1}^{k} r^i\right)^{-1}$$

$$p_i = \frac{r^i(1-r)}{1-r^{i+1}} \qquad i = 1,\cdots,k \qquad (3)$$

where $r = \frac{\lambda}{\mu}$. Since the behavior of all queue are the same the above equations can be used for another queues. In the rest of paper the p_i for queue j is called p_i^j.

3.2. Split mode

In this mode if the q_i, $i = 2,3,...,n$ is empty then the tasks in $q_j, j \le i$ will be allocated to q_i. So all the assumptions of restrict mode have been considered in this mode except a part of assumption 2 and 3 which is the effective entrance and service rate. So the following equations could be written for split mode:

$$\begin{cases} \lambda_i' = \lambda_i + \sum_{j=1}^{i-1} \lambda_j \prod_{j=1}^{i-1} p_0^{j+1} \\ \mu_i' = \mu_i + \mu_i \sum_{j=i+1}^{n} \prod_{k=i+1}^{j} p_0^k \end{cases} \qquad (4)$$

$$\lambda_i = \mu_i = p_0^i = 0 \qquad for \qquad i \ge n$$

where λ_1, λ_2 and λ_3 are mean value of entrance of tasks into the queues in restrict mode respectively. Also μ_1, μ_2 and μ_3 are mean values of execution time of tasks into the queues in restrict mode respectively and p_0^i is the probability that q_i was empty.

The above equations show that λ' and μ' are effective values of mean entrance tasks and mean value of execution time of tasks for split mode.

By using the new effective values the equation 3 can be written for prefer mode and the values of p_0^j and p_i^j for each queue would be computed.

3.3. Merge mode

In merge mode all the assumptions of restrict mode have been assumed except the effective entrance and service rate (that is recomputed as slit mode). Also it is assumed the block sizes would be so as all couples of lower blocks can contained first larger block. So in the merge mode smaller block can be merged if they are free, and serve as larger block. According to this assumption the input rate of smaller block is affected by lager size task.

The effective rate of λ and μ for the queues can be determined as follows:

$$\begin{cases} \lambda_i'' = \lambda_i + \lambda_{i+2} p_0^i p_0^{i-1} + \sum_{j=1}^{i-1} \lambda_j \prod_{j}^{i-1} p_0^{j+1} \\ \mu_i'' = \mu_i + \mu_i p_0^{i-1} p_0^{i-2} + \mu_i \sum_{j=i+1}^{n} \prod_{k=i+1}^{j} p_0^k \end{cases} \qquad (5)$$

$$\lambda_i = \mu_i = p_0^i = 0 \qquad for \qquad i \ge n$$

where λ'' and μ'' are the effective values of mean entrance tasks and mean execution time of tasks for merge mode.

For each of these modes the equation 3 can be rewritten, and the value of p_0^j and p_i^j would be calculated. Using above equation the system performance metrics can be calculated. The important performance metrics are found in next section.

4. System evaluation and performance metrics

Utilization of RR and system speedup, two most important metrics for system performance evaluation, are determined according to queuing equations described in section 3.

a) The Reconfigurable Resource utilization (U): U is the percentage time that the RR will be busy.

$$U = 1 - p_0 \qquad (6)$$

The utilization factor of b_1, b_2 and b_3 are named u_1, u_2 and u_3 respectively. Total utilization factor is obtained as follows:

$$U_R = \sum_{i=1}^{n} A_i U_i \qquad (7)$$

which $A_1, A_2, ... A_n$ are the weight factors that take the area of $b_1, b_2, ... b_n$ to total area.

To exemplify the analytical model, lets FPGA split into three blocks, b_1, b_2 and b_3. In which these blocks contain 20%,30% and 50% of reconfiguration area respectively. Hence total utilization factor can be written as follows:

$$U_R = 0.2U_1 + 0.3U_2 + 0.5U_3 \qquad (8)$$

The utilization factor of RR for restrict mode, prefer mode and flexible mode is computed and shows in figure 2.

b) Speedup (S):

The speedup is a runtime comparison between RR and HP. If all application programs run on reconfigurable hardware unit the maximum speedup is obtained:

$$S_{Max} = \frac{W/\Delta_H}{W/\Delta_R} = \frac{\Delta_R}{\Delta_H} \qquad (9)$$

where W is workload, Δ_H is the HP computation power and Δ_R is computation power of RR. According to assumption one of section 3, if the task queue is full the workload will run on RR. So the total workload, W, is partitioned into two parts, part one W_H that is run on HP and part two W_R run on RR. So the speedup of the system is obtained by:

$$S = \frac{W/\Delta_H}{\dfrac{W_H}{\Delta_H} + \dfrac{W_R}{\Delta_R}}$$

$$= \frac{1}{\dfrac{W_H}{W} + \left(\dfrac{\Delta_H}{\Delta_R}\right)\dfrac{W_R}{W}} \qquad (10)$$

For our example the above parameters can be written as follows:

$$\frac{W_H}{W} = \frac{\lambda_1 p_k^1 + \lambda_2 p_k^2 + \lambda_3 p_k^3}{\lambda_1 + \lambda_2 + \lambda_3}$$

$$\frac{W_R}{W} = \frac{\lambda_1(1 - p_k^1) + \lambda_2(1 - p_k^2) + \lambda_3(1 - p_k^3)}{\lambda_1 + \lambda_2 + \lambda_3} \qquad (11)$$

in which the lower script is number of task in the queue and the upper script is queue number. Mean execution time and mean arrival time of tasks for q_i are considered μ_i and λ_i respectively.

Figure 2 shows the utilization of RR for three operation modes discussed before. The figure shows the utilization factors of merge mode and split mode are greater than split mode. Figure 3 shows the speedup of three modes. It can be seen that merge mode has extensively better speedup than restrict mode. It is also better than split mode. Figure 3 shows as the execution time will be grater, which is concern to larger task; while input arrival rate of task to system is fixed, the merge mode shows more effectiveness. This is originating from usage of free capacity of system by merge mode comparing two other modes.

5. Conclusion

In this paper an analytical model of the reconfigurable computing is proposed. In the analysis a multi task FPGA is considered. The results obtained from queuing analysis have been drawn to show the behavior of performance metric according to system specifications. These results aid the designer to select the best scheduling mode for any application. The diagrams show that the restrict mode has the best speedup for all mean execution time and the prefer mode has the best utilization factor.

The future work can be concentrated on adding management overhead time to the model to have more accurate results. Applying the analysis results to other reconfigurable resources model such as most flexible 2D area model could be attended for future works.

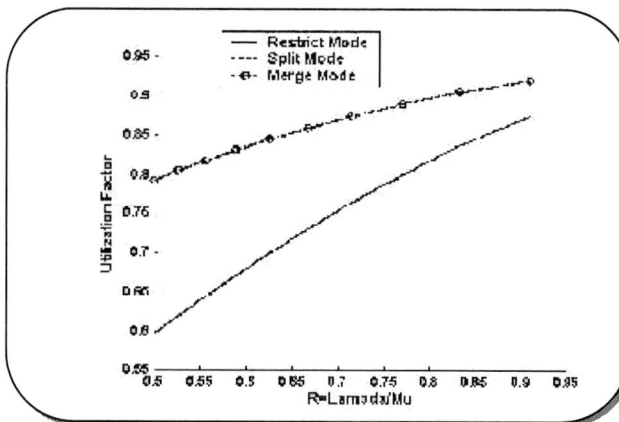

Figure 2 The utilization factor of three operation modes

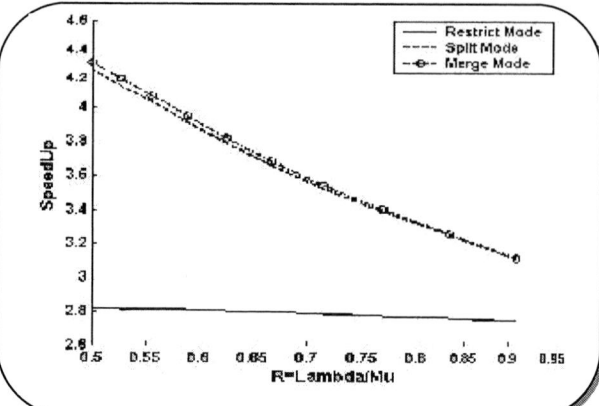

Figure 3 The speedup of three operation modes
(for $S_{Max} = 2$)

References

[1] K. Compton, S. Hauck (2002). "Reconfigurable computing: a survey of systems and software". *ACM computing surveys.* Vol. 43, No. 2, pp. 171-210.

[2] C. Steiger, H. Walder, M. Platzner, L. Thiele (2003). "On line scheduling and placement of real-time tasks to partially reconfigurable devices". *Proceeding of the 24th IEEE international real-time system symposium (RTSS 03).*

[3] Tiago Dias, Nuno Roma, Leonel Sousa and Miguel Ribeiro (2007). "Reconfigurable architectures and processors for real-time video motion estimation", *Journal of Real-Time Image Processing - Special issue on Field-Programmable Technology.* vol. 2, No. 4, pp. 191-205.

[4] Girish Venkataramani, Tobias Bjerregaard, Tiberiu Chelcea, and Seth Copen Goldstein (2006). "Hardware Compilation of Application-Specific Memory Access Interconnect". *IEEE Transactions on Computer Aided Design of Integrated Circuits and Systems.* Vol. 25, No. 5, pp.756–771.

[5] R. Tessier, V. Betz, D. Neto, A. Egier, and T. Gopalsamy (2007). "Power Efficient RAM Mapping Algorithms for FPGA Embedded Memory Blocks". *IEEE Transactions on Computer-Aided Design of Integrated Circuits and Systems.* Vol. 26, No. 2, pp. 278-290.

[6] Gokhale, Maya, Graham, Paul S. (2005). "Reconfigurable Computing: Accelerating Computation with Field-Programmable Gate Arrays". *Published by Springer.*

[7] R. Tessier (2002). "Fast Placement Approaches for FPGAs". *ACM Transactions on Design Automation of Electronic Systems.* Vol. 7, No. 2, pp 284-305.

[8] S. Trimberger, D. Carberry, A. Johnson, J. Wong (1997). "A time-multiplexed FPGA". *5th Annual IEEE Symposium on FPGAs for Custom Computing Machines.* Pp. 22 – 28.

[9] Z. Li, K. Compton, S. Hauck (2000). "Configuration caching management techniques for reconfigurable computing". *Field-Programmable Custom Computing Machines IEEE Symposiumon.* Pp. 22 – 36.

[10] S. Hauck (1998). "Configuration prefetch for single context reconfigurable coprocessors". *ACM/SIGDA international symposium on field programmable gate arrays.* pp. 65-74.

[11] S. Hauck, Z. Li, E. Schwabe (1998). "Configuration compression for the xilinxXC6200 FPGA". *IEEE symposium on FPGAs for custom computing machines.*

[12] K. Compton, J. Coley, S. Knol, S. Hauck (2000). "Configuration relocation and defragmentation for FPGAs", *IEEE symposium on field-programmable custom computing machines.*

[13] A. Aggarwal (1994). "Routing architectures for hierarchical field programmable gate arrays". *Proceedings of the IEEE International Conference on Computer Design.* pp. 475–478.

[14] V. Betz, M. Rose (1999). "FPGA routing architecture: Segmentation and buffering to optimize speed and density". *ACM/SIGDA International Symposium on FPGAs.* pp. 59–68.

Session 7

Theory, Design, and Implementation
of Circuits & Systems

Extension and Application of Space Syntax
——A Case Study of Urban Traffic Network Optimizing in Beijing

Zheng Xinqi , Zhao Lu, Fu Meichen, Wang Shuqing

School of Land science and Technology, China University of Geosciences, Beijing 100083，China

Abstract

Most of the cities all over the world have the problem of traffic jam, but Beijing is probably one of the typical cities. Researchers and government have applied the energies to finding a solution to solve this problem in Beijing, but the effect is not satisfying. Space syntax theory and its mode to solve problems can provide instructive reference for analyzing the urban traffic problem. Based on the current condition of urban traffic network in Beijing, this paper was in virtue of space syntax theory and GIS visualization technology, extended the space syntax theory, divided the traffic network according to roads' designation, and improved the axial map. We put forward the concept of Accessibility, discussed the method that how to deal with loops, and proved the optimizing measures which are supposed to revive the current traffic pressure in Beijing based on the iterative optimizing experiments. It is concluded that increasing radialized roads can help to improve the traffic status obviously of Beijing traffic network, which is in the framework of loops. Moreover, we conducted different simulating experiments, such as adding 2 more, 4 more, 6 more, 8 more, and up to 14 more radialized roads in different directions. We discovered that there is a critical of the added roads. Adding 6~8 more roads can lead to an optimal yield. In accordance with this finding, we selected the axial lines of roads on the spot according to current urban planning, and had validated the feasibility and rationality of the optimizing method. It is concluded that building some new radialized roads and rebuilding some main roads are available for improve the traffic condition and the input-output ratio is better. Consequently, the extended space syntax method can instruct road reconstructing and urban planning, and be generalized to other cities.

1. Introduction

As one of the megalopolises, the city of Beijing is the political, economic and cultural centre of China. Along with the social and economic development, the traffic demand in Beijing is soaring. There have been more than 3300 thousand vehicles by 2007. In contrast, the traffic accommodation is deficiency, which has restricted the circulating process and efficiency of goods, human and information in the city or between cities. The loss caused by poor traffic condition is about 6000 million per annum [1]. As the bottleneck, the traffic problem makes the rank of Beijing in "livable city" in China drop to the 15th [2].

In order to alleviate the traffic problem, a good many scholars have brought forward some countermeasures, such as quickening the construction of rail transit network, developing public transportation preferentially, consummating the traffic management system, and so on[3,4,5,6]. But the practical effect is not satisfying. On the other hand, the scholars at home and abroad have carried out lots of experiments and studies to solve urban traffic problem, and some have discussed the applicability of space syntax theory[7,8,9,10,11,12]. Generally, these research and study mainly analyzed the current condition of urban traffic network and studied the improvement of some local parts in the city [13,14,15], but rarely analyzed and evaluated the urban planning from the view of urban traffic.

Over the past three decades, space syntax theory has made quite great progress in some aspects, but there are also several faultinesses [16,17,18], which have emerged mostly in the phase that is from theory to specific application[12]. For example, the division of loops, the function of different road rank, the width of roads and the disposal of cloverleaf junctions haven't been solved rationally yet[9,19].

Aiming at the current condition of urban traffic network in Beijing and the practical problems in space syntax application, we extended the space syntax theory, optimized the urban road structure in Beijing, discussed the optimizing measures which are supposed to revive the current traffic pressure in Beijing based on the iterative experiments. Furthermore, we compared the optimizing rules with the current urban planning, validated the feasibility and rationality of the optimizing rules through selecting axial line of the road on the spot.

2. Space syntax theory and its extension

2.1. The keystone of space syntax theory

Each city system is composed of spatial object and free space. The former is mainly the building, while the latter is the space in which people can move freely and it is partitioned by spatial object. With a view to express free space, space syntax is a new language to describe the pattern of modern city, and is also the theory and tool to analyze the spatial structure of the city[20,21].

The basic principle of space syntax is space division, by the division methods of axial line, convex polygon and vision zone. In the condition of a city environment where the building or building colony is much denser, space syntax generally adopts axial lines to divide the space. Space syntax model mainly includes the following variables to analyze the configuration in different levels: connectivity, controlling value and integration [12].

2.2. The extension of space syntax theory

The axial map based on straight line is in common use, but it has many limitations and can not deal with loops well[22]. And in some case, the current variables may not express the entity in geographic space clearly. For example, the variables of connectivity and control can reflect the relationship between one space and the other space which intersects it directly. They express the spatial character of local structure[12]. On the other hand, in the practical research on traffic network, because of their short lengths, the connectivity values of some streets are low. But on account of the location, the connectivity value of the whole space is preferable. And therefore, they can not reflect the whole structural character well.

In our study, we put forward the division method which is based on polygonal line and the concept of Accessibility (A_i). We divide the road according to road designation and its real trend. A_i is used to reflect the accessibility of local roads in the whole network, and it can express the relationship between one space and others, so it can reflect the structural character of one space in the global level very well. Compared with connectivity, Accessibility takes the connectivity of one street and the surrounding ones into account and can incarnate the quality of urban traffic network and layout much better. It can be calculated as follows:

$$A_i = \sum_{j=1}^{k} C_j$$

where k is the amount of nodes which connect the node i directly and C_j is the connectivity of the node j.

The calculation of space syntax has been developed well[23], and Axwoman3.0 on the platform of ArcView is convenient to link up with prophase data[24]. We extended the Axwoman3.0 to calculate related variables.

3. Character of urban traffic network in Beijing and data processing

In cities, 80% of the traffic flow is concentrated on 20% of the roads[25]. The study area of this paper is the area within the fifth loop in Beijing. The urban area in Beijing accounts for only 12% in the whole city, while 1/4 of the traffic flow is in this area. Half of the vehicles in the city are within the first three loops, and most of the traffic is within the fifth loop. Our study on the character and optimization of the area within the fifth loop is of great importance to relive the traffic pressure in Beijing.

3.1. The characteristic of urban traffic network in Beijing

Loops and radialized roads act as the main framework of urban traffic network in Beijing, and they play an important role in the pattern which looks like a chessboard. The urban traffic network is centered in Tian'anmen Square, and extends along the south-north axle line and Chang'an Avenue. Over the past decade, building loops is the golden rule in developing urban traffic. For the city, the loops just like the hoops encircling the city[26]. It is investigated that most of the outlets in the 2nd, 3rd and 4th loops are the spots of traffic jam.

3.2. Data source and data processing

We attained the traffic network map of the study area in 2006 (see figure 1), based on Beijing traffic network map in 2006 and Beijing 1:50000 relief map. Then we gained the axial map according to road designation and their real trends. According to the Beijing road grading map and the distribution map of traffic flow in 2006, we derived 180 roads, including the primary roads entirely (156) and secondary road partly.

How to deal with the loops is an important consideration. After experiments, we found that the connectivity and control value of one loop are much bigger when it is treated as a whole object than it is treated after dividing. Moreover, the road extending spatially should not have the same connectivity, control or accessibility in different spatial parts. On the other hand, our objective is to discuss the optimizing measures based on the loops framework of Beijing traffic network. We need to study the spatial differentiation in related variables of loops. In this condition, dividing the loop according to road designation is much better.

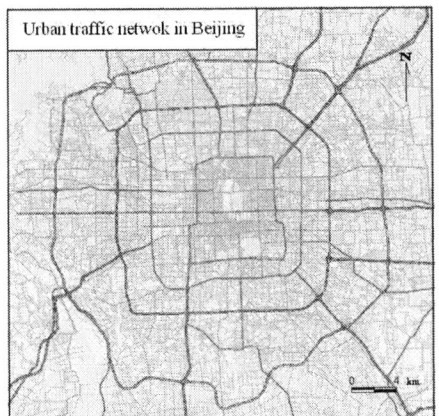

Figure 1. Traffic network of study area in 2006

4. Calculation and optimizing experiments

4.1. Calculation result and analysis

The space syntax analyzing results are showed in figure 2. The average value of integration and accessibility in Beijing urban traffic network is 1.56 and 5, separately. Generally, the integration and accessibility values in each section of loops, in the local area where the traffic network is dense and in the radialized roads connecting loops are mostly much higher. In detail, the variable value where there are fewer radialized roads is much lower. We can see that the traffic pressure in these sections, especially in the cross intersections, is much greater. There places are the sensitive points of traffic problem, and the traffic jam is prone to come into being there. It is consistent with the fact.

It is showed that the sections of road having higher accessibility and integration values are concentrated on the area within the fourth loop. Great of traffic flow is prone to pool in this area. Moreover, the radialized roads between loops and the sections having more outlets have evident split-flow function to urban traffic.

4.2. Making optimizing experiments

Nowadays, the construction of urban traffic network is following the traditional "crisscross" axis. According to above calculation results and analysis, we should building more "X" roads. On the other hand, if there are some radialized throughways, they could complement the loops and get twice the result with half the effort to improve urban traffic in Beijing.

Based on this assumption, we discussed the impact of building "X" roads on current traffic network by conducting the following 10 projects: ①add 2 roads along the east and west; ②add 2 roads along the north and south; ③add 4 roads along the northwest, southwest,

southeast and northeast; ④add 6 roads along the east, west, northwest, southwest, southeast and northeast; ⑤ add 6 roads along the south, north, northwest, southwest, southeast and northeast; ⑥, add 6 roads in arbitrary directions; ⑦add 8 roads along the east, west, south, north, northwest, southwest, southeast and northeast; ⑧ on the basis of ⑦, referring to ⑥, add 2 more according to the amount of radicalized roads in every direction; ⑨ on the basis of ⑦, referring to ⑥, add 4 more according to the amount of radicalized roads in every direction; ⑩ on the basis of ⑦, add the 6 ones in ⑥. The added radicalized roads are centered in Tian'anmen Square, and mainly in the directions that have few radicalized lines (see figure 3).

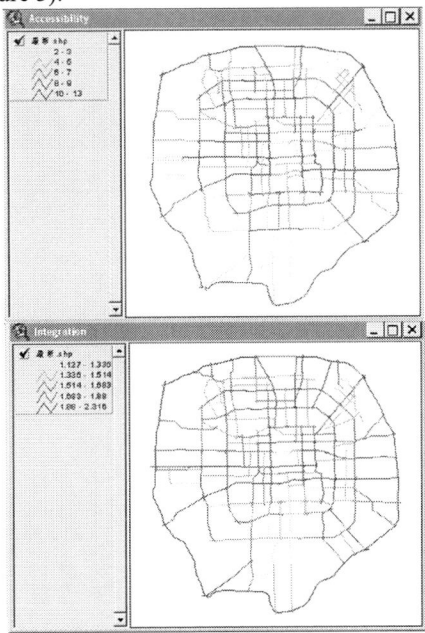

Figure 2. Accessibility and integration of axial map

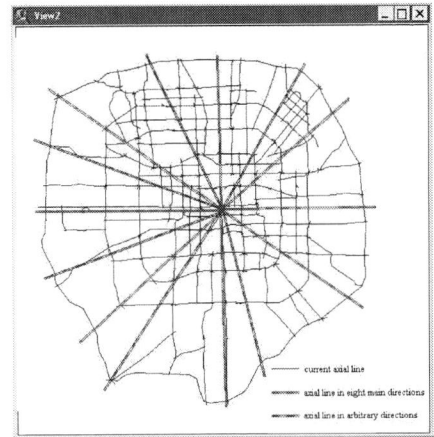

Figure 3. Distribution of added radicalized lines

4.3. Experiments results and analysis

The axial integration and accessibility under each project have risen evidently (see figure 4 and 5). It is illustrated that adding "X" radicalized roads do help to improve the traffic, and the number is more, the effect is better. Adding 2 roads can improve the integration and accessibility by about 1%~4% and 2%, separately. 4 can improve by about 10% and 5%, separately. 6 can improve by about 12% and 8%, separately. 8 can improve by about 13% and 9%, separately. 10 can improve by about 10% and 8%, separately. 12 can improve by about 15% and 13%, separately. 14 can improve by about 21% and 16%, separately.

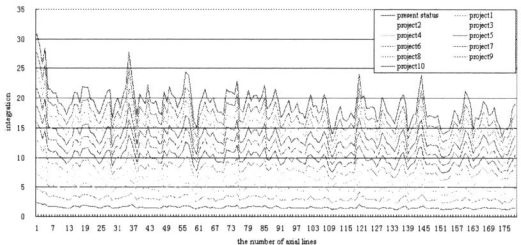

Figure 4. Contrast of integration under each project

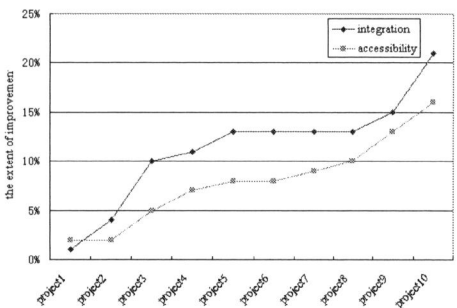

Figure 5. Improvement of integration and accessibility

In practice, we need to take the cost into consideration, so it is not true that the more the number of radicalized roads is, the better is the effect. What we pursue is the optimal input-output ratio. Moreover, we can see the 8th project is an inflexion of the curve through figure 6. In some degree, the effect of adding 6 radicalized roads on integration and accessibility is the same as that of adding 8 ones, so is that of adding 10 roads. The other variables have similar statistical characteristic. So we add 6~8 more radicalized roads is the best choice in reality.

4.4. Implementing the optimizing measures

In order to cooperate with the development of loops, there are to construct 15 more radicalized roads (see figure 6). According to current planning, we selected the axial lines of roads on the spot and had validated the feasibility

and rationality of the optimizing method. We selected Dongbeichengjiaolianluoxian, Yaojiayuan Lu, Tonghuihebei Lu, Puhuangyu Lu, Fengbei Lu, Lianhuachixi Lu, Fushi Lu and Xiwai Avenue to do simulating analysis by space syntax. The result is that the integration and accessibility of traffic network in the study area have been improved both by 12%.

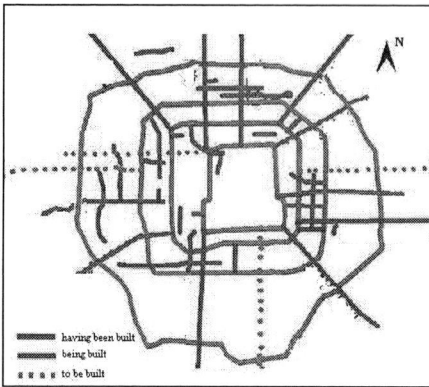

Figure 6. Urban throughways in Beijing

It is illustrated that our extension of space syntax is scientific and feasible, the optimizing measures getting from space syntax analysis are available, and strengthen radicalized roads construction in the traffic framework of loops is an effective approach to relive traffic pressure. In general, urban traffic network in Beijing can be improved by the measure of building and alteration. On one hand, build new radicalized roads according to demands. On the other hand, rebuild and broaden the main road partly.

5. Discussion

We have carried out many simulating experiments on urban traffic network in Beijing and have selected the axial lines of radicalized roads on the spot according to the current urban planning to analyze the optimizing results. It was concluded that building radicalized roads can help to relive traffic pressure efficiently. In practice, in order to exploit the whole function of traffic network and improve the serving level, we should strengthen the building of radicalized roads during the construction of loops. In order to pursue the best input-output ratio, we need to find out the optimal direction to add radicalized roads by space syntax method on the basis of studying urban planning.

Moreover, if we consider input-output ratio fully and build more radicalized roads in other pertinent directions, the effects will be more evident. However, the road building must be in the precondition of assorted development and in accordance with traffic demand. In addition, the newly-built overhead tramway and underground may also be in X cross to relive traffic pressure by adding radicalized roads in another way.

294

6. Conclusion

Although space syntax still has some limitation in studying urban traffic network which has a loop framework, it can be exploited its particular advantage after proper extension. This paper put forward the concept of accessibility and the axial map based on polygonal line based on road designation and real trend, and also discussed method that how to deal with loops. We applied the extended space syntax in the study of optimizing urban traffic network in Beijing, and have acquired profitable conclusion. After further extension, the space syntax method can instruct road reconstructing and urban planning, and be generalized to other cities.

In the further study, we will focus on the integration of space syntax and GIS based on vector data, and the 3D extension of space syntax. In addition, considering the other factors such as road width and cloverleaf junction, we will discuss the new but appropriate method to analyze urban traffic network.

Acknowledgement

The National Natural Science foundation of China, No.40571119. The National Social Science Foundation of China, No.07BZZ015. The National Key Technology R&D Program of China, No.2006BAB15803. The Talent Foundation of China University of Geosciences (Beijing), No.51900912300.

References

[1] http://58.63.114.194:86/ssds/html/2004/04/20040423120939 4054.htm

[2] http://www.ce.cn/xwzx/gnsz/gdxw/200601/02/t20060102_57 05149.shtml

[3] http://www.bjghw.gov.cn/

[4] Chen Feng, Liu Jinling, and Shi Zhongheng. "Rail transit constructing Beijing urban spatial structure", *City Planning Review*, 2006, 30(6), pp. 36-39.

[5] Zhang Jinggan. "Existing traffic jam in Beijing and Counter measures", *Urban Problems*, 2004, (5), pp. 2-5.

[6] Gao Zhonggang. "Suggestions to Beijing's traffic congestion: view from the differences between Beijing's and Shanghai's respective traffic policies", *Urban planning forum*, 2004, (4), pp. 35-39.

[7] Ben Croxford, Alan Penn, and Bill Hiller. "Spatial distribution of urban pollution: civilizing urban traffic", *The Science of the Total Environment*, 1996, 189/190(1996), pp. 3-9.

[8] Li Jiang, and Duan Jie. "Multi-sale representation of urban spatial morphology based on GIS and spatial syntax", *Journal of Central China Normal University*, 2004, 38(3), pp. 383-387.

[9] Duan Ruilan, and Zheng Xinqi. "The relation of the city road structure and the land price based on the space syntax", *Science of surveying and mapping,* 2004, 29(5), pp. 76-79.

[10] An-Seop Choi, Young-Ook Kim, Eun-Suk Oh, and Yong-Shik Kim. "Application of the space syntax theory to quantitative street lighting design", *Building and Environment*, 2006, (41), pp. 355-366.

[11] Thomson R C. "Bending the axial line: smoothly continuous road centre-line segment as a basis for road network analysis"2004-10-23. http://www.comp.rgu.ac.uk/staff/rt/Paper-with-figs.pdf

[12] Jiang Bin, Huang Bo, and Lu Feng. S*patial analysis and geovisualization in GIS*, Higher education press, Beijing, 2002.

[13] Jiang B., and Claramunt C. "Topological analysis of urban street networks", *Environment and Planning B: Planning and Design,* Pion Ltd., 2004, 31, pp. 151-162.

[14] S.Porta, P.Crucitti, and V.Latora. "The network analysis of urban streets: a dual approach", *Physica A,* 2006,369, pp. 853-866.

[15] Kim YO, and Penn A. "Linking the Spatial Syntax of Cognitive Maps to the Spatial Syntax of the Environment", *Environment and Behavior*, 2004, 36(4), pp. 483–504.

[16] Zhang Hong, Wang Xinsheng, and Yu Ruilin. "Space syntax and its research progress". *Geospatial information*, 2006, 4(4), pp. 37~39.

[17] F. Figueiredo, and L. Amorim, "Continuity lines in the axial system", in: *Proceedings of the Fifth Space Syntax International Symposium*, Delft University of Technology, Delft, 2005.

[18] Jang B., and Claramunt C. "Extending space syntax towards an alternative model of space within GIS", *Presented at the 3nd AGILE Conference on Geographic Information Science in Helsinki*, May 25-27, 2000.

[19] Batty M, and Rana S. "Reformulating Space Syntax : the Automatic Definition and Generation of Axial Lines and Axial Maps",http://www.casa.ucl.ac.uk/working_pa2pers/paper58. pdf

[20] Bill Hillier, and Chris Stutz. "New methods in space syntax". *World architecture*, 2005, (11), pp.54-55.

[21] Wu Duan. "Introduction of related theories to space syntax". *World architecture*, 2005, (11), pp.18-23.

[22] Zhao Hu, Li Lin, and Zhu Haihong. "Application of extended space syntax in urban land grading". *Journal of geomatics*, 2007, 32(2), pp. 9-11.

[23] http://en.wikipedia.org/wiki/Spatial_network_analysis_soft ware

[24] Jiang B. "Axwoman: an analytical tool for local axxesibility, venue report, centre for advanced spatial analysis".2004-10-20.
http://www.casa.ucl.ac.uk/newvenue/axwoman-introl.htm

[25] Jiang B. "A topological pattern of urban street networks: universality and peculiarity", *Physica A: Statistical Mechanics and its Applications*, 2007, 384, pp.647-655.

[26] http://www.cas.ac.cn/html/Dir/2004/04/15/9992.htm

2008 Workshop on Power Electronics and Intelligent Transportation System

The Parameter of Automatic Voltage Regulator's Effect on Steady State Stability Limit of Turbine Generator

Yong-gang Li, Jun-jie Fu, Yan-jun Zhao, Qiang Liu

School of Electrical Engineering,
North China Electric Power University, Baoding, HeBei, 071003, China
zhaoyanjun1982@163.com

Abstract

The steady state stability limit of turbine generator with automatic voltage regulator (AVR) extends steam turbine's safe range under exciting operation compared to generator without AVR. Therefore, Paper analyzes the parameter of AVR's effect on steady state stability limit of turbine generator, and verifies the conclusion with the simulation model in Matlab/Simulink environment.

1. Introduction

Nowadays with development of economy, the demand of electric energy ascends by leaps and bounds. The electric network extends larger and larger, the voltage of trunk grid becomes higher and higher. The voltage's quality of power system must be controlled much more availably. It is an effect and feasible way to control the generator's field current and absorb an excessive amount of reactive power from the power system. In recent years, there has been some renewed interest in assessing the effect of the UEL on the dynamic performance of generators. Some Methods deal with reactive power service and under-excitation limiter of the generator [1]-[2], some good design about the UEL has been obtained by the paper [3]-[4]. According to Hurwitz's criterion, Paper [5] deduces the steady state borderline of steam turbine with AVR on the map of reactive capability curves, using the mathematical model of one machine infinite bus system. This paper analyzes the parameter of AVR's effect on steady state stability limit of turbine generator, and verifies the conclusion with the simulation model in Matlab/Simulink environment.

2. Analyze the steady-state stability of generator with AVR

The electrical parameters of Generator in the leading-Phase operation is still symmetrical, and generator remains synchronous speed, so that it is a normal operation of generator that power factor has a change in operating conditions and only widens the normal operation of generators scope. Therefore, in the allow operating range, generators can run a long time into the leading-Phase operation as long as the power grid needs. When put into operation generators with AVR, its electrical characteristics show a completely different from the manual excitation and the excitation mechanical characteristics. Generators can maintain E_q' constant, and E_q will change accordance with power angle changes. The power angle limit is bigger than 90°, if generator runs into leading phase with AVR.

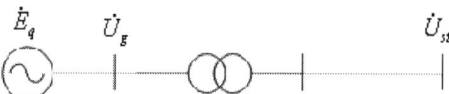

Fig.1 Diagram of a simple system

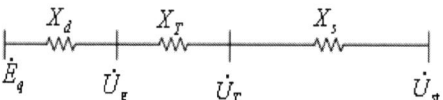

Fig.2 Resistance of a simple power system

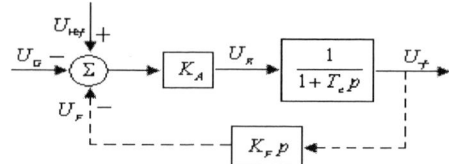

Fig.3 simplified mathematical model of excitation system

Fig.1 is diagram of a simple system, and Fig.2 is its resistance of system.Fig.3 is the exciter's simplified model. Without considering the camper winding's performance, by using small-signal analysis theory we can obtain the model, as follows:

$$-K_V \Delta U_g = \Delta E_{qe} + \left(T_e + K_F\right)\frac{d\Delta E_{qe}}{dt} \tag{1}$$

$$\Delta E_{qe} = \Delta E_q + T_{d0}'\frac{d\Delta E_q'}{dt} \tag{2}$$

$$\frac{d\Delta\delta}{dt} = \Delta\omega \tag{3}$$

978-0-7695-3342-1/08 $25.00 © 2008 IEEE
DOI 10.1109/PEITS.2008.25

$$\frac{d\Delta\omega}{dt} = -\frac{\omega_N}{T_j}\Delta P_e \qquad (4)$$

Where

K_V — equivalent gain of AVR system;

T_e — equivalent time constant of AVR system;

K_F — factor of negative feed back;

T'_{d0} — field open circuit time constant;

ω_N — rated rotational speed;

T_j — inertia time constant.

ω — per unit speed based on $2\pi f_0$

δ — Generator rotor angle

P_m — Mechanical input power

P_e — Electrical out power

E'_q — Internal voltage

E_q — Open-circuit voltage

U_{Gq} — Stator voltage, direct axis component

U_s — Infinite bus voltage

U_g — Stator voltage

X_d — Generator reactance

X'_d — Generator transient direct axis reactance

X_T — Transformer equivalent reactance

X_L — External reactance

The active and reactive power equations of single machine system with VAR can be expressed by E_q, E'_q and U_{Gq}. The equations' curves intersect at the steady running point on the map of reactive capability curves. By linearizing these equations at the steady running point and omitting quadratic and above items, we can get the matrix about the Steady-state stability of turbine-generators in consideration of AVR, considering $\Delta U_{Gq} \approx \Delta U_G$ and $\Delta P_{E_q} \approx \Delta P_{E'_q} \approx \Delta P_{U_{Gq}} \approx \Delta P_e$. As follows:

$$\begin{bmatrix} \dfrac{d\Delta E_{qe}}{dt} \\ \dfrac{d\Delta E'_q}{dt} \\ \dfrac{d\Delta\delta}{dt} \\ \dfrac{d\Delta\omega}{dt} \end{bmatrix} = \begin{bmatrix} \dfrac{-1}{T'_e} & \dfrac{-K_V R_{E'_q}}{T'_e R_{U_{Gq}}} & \dfrac{K_V\left(S_{U_{Gq}}-S_{E'_q}\right)}{T'_e R_{U_{Gq}}} & 0 \\ \dfrac{1}{T'_{d0}} & \dfrac{-R'_{E_q}}{T'_{d0} R_{E_q}} & \dfrac{S_{E'_q}-S_{E_q}}{T'_{d0} R_{E_q}} & 0 \\ 0 & 0 & 0 & 1 \\ 0 & \dfrac{-\omega_N R'_{E_q}}{T_j} & \dfrac{-\omega_N S_{E'_q}}{T_j} & 0 \end{bmatrix} \begin{bmatrix} \Delta E_{qe} \\ \Delta E'_q \\ \Delta\delta \\ \Delta\omega \end{bmatrix} \qquad (5)$$

It can be rewritten as:

$$\frac{\mathbf{d\Delta X}}{\mathbf{dt}} = \mathbf{A\Delta X} \qquad (6)$$

Where $T'_e = T_e + K_F$

$$S_{E_q} = \left.\frac{\partial P_{E_q}}{\partial\delta}\right|_{\substack{E_q=E_{q0}\\\delta=\delta_0}}, \quad S_{E'_q} = \left.\frac{\partial P_{E'_q}}{\partial\delta}\right|_{\substack{E'_q=E'_{q0}\\\delta=\delta_0}};$$

$$S_{U_{Gq}} = \left.\frac{\partial P_{U_{Gq}}}{\partial\delta}\right|_{\substack{U_{Gq}=U_{Gq0}\\\delta=\delta_0}}, \quad R_{E_q} = \left.\frac{\partial P_{E_q}}{\partial E_q}\right|_{\substack{E_q=E_{q0}\\\delta=\delta_0}};$$

$$R_{E'_q} = \left.\frac{\partial P_{E'_q}}{\partial E'_q}\right|_{\substack{E'_q=E'_{q0}\\\delta=\delta_0}}, \quad R_{U_{Gq}} = \left.\frac{\partial P_{U_{Gq}}}{\partial U_{Gq}}\right|_{\substack{U_{Gq}=U_{Gq0}\\\delta=\delta_0}};$$

According to Hurwitz's criterion, the conditions of keeping system steady are:

- All the factors of characteristic equation are greater than zero.
- Hurwitz's leading determinants are also greater than zero.

Thus, we can further draw the equation of steady-state stability limit border of turbine generator on the P-Q running capacity map.

$$P^2 = -\frac{\left(Q+\dfrac{U^2}{X_{d\Sigma}}\right)^2\left(Q+\dfrac{U^2}{X_{d\Sigma}}-\dfrac{Kp}{T'_d+T'_e}\right)}{Q+\dfrac{U^2}{X_{d\Sigma}}\dfrac{T'_{d0}+T'_e}{T'_d+T'_e}-\dfrac{Kp}{T'_d+T'_e}} \qquad (7)$$

Where:

$$Kp = \frac{T_j\left(T'_e+T'_d\right)}{\omega_n T'^2_e}\left(K_V\frac{X_d T'_e/T'_{d0}+X'_d}{X_d-X'_d}-1\right)$$

Based on the analysis of equations (7)[5], it shows that the parameter of automatic voltage regulator's affects on steady state stability limit of turbine generator. To the same Active power, the excitation system magnification is larger, the steady-state stability limit generator is smaller, and smaller magnification. Large excitation system time constant less affected the generator of steady-state stability limit.

3. The simulation model In Matlab/Simulink environment

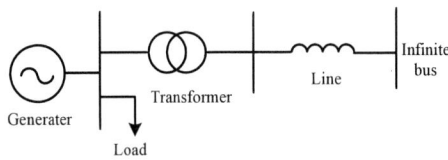

Fig.4 Diagram of simulation system

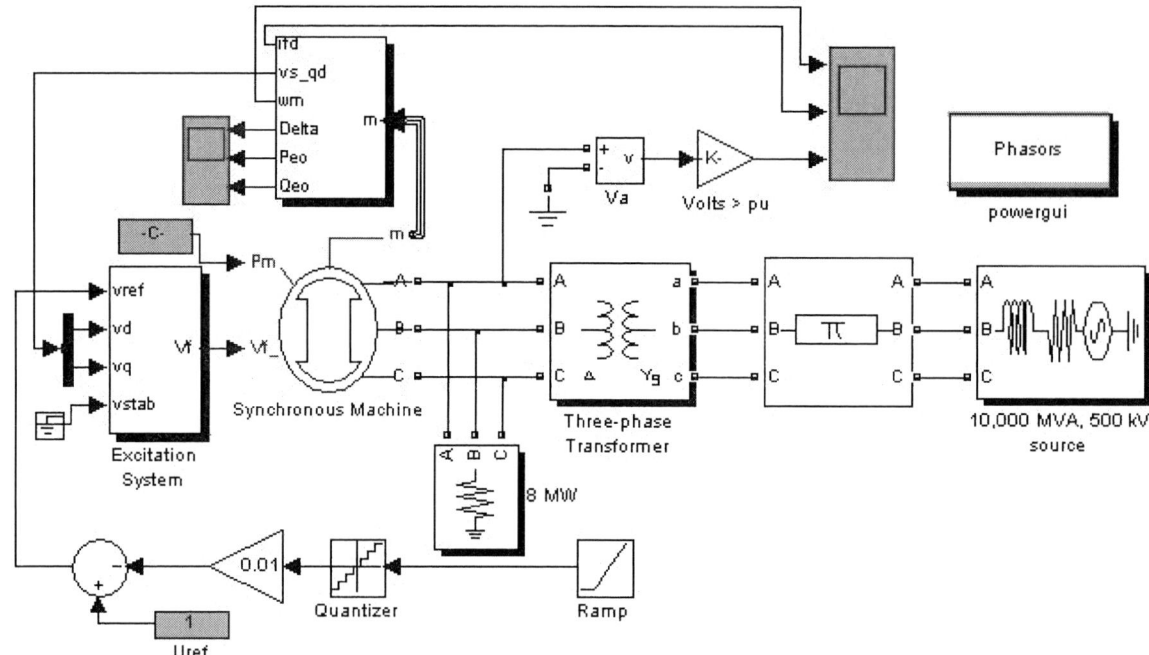

Fig.5 Model of simulation system

Fig.4 is the diagram of simulation system according to one machine infinite system; correspondingly, Fig.5 is the simulation model in Matlab/Simulink environment. The model neglects the dynamic effect of prime motor, and assumes Pm as constant. The simulation in Fig.5 is the same with the field test in the power plant, which is gradually reducing the terminal voltage setting value of the generator excitation system while maintaining active power under the condition of constant. The voltage reference value of excitation system adjusts once every 20s, after system has sufficient time to reach the steady stability. So, the generator can gradually run from the stable operating conditions into leading phase operation condition.

The parameter of generator is: $P_{GN} = 600\,\mathrm{MW}$, $f = 50\mathrm{Hz}$, $\cos\phi_n = 0.9$, $U_n = 20\mathrm{kV}$, $I_n = 19245\mathrm{A}$, $X_d = 2.270$, $X'_d = 0.267$, $X''_d = 0.204$, $X_q = 2.209$, $X'_q = 0.392$, $X''_q = 0.198$, $X_l = 0.158$, $R_a = 0.48\Omega$, $T'_d = 1.027\mathrm{s}$, $T'_{q0} = 0.173\mathrm{s}$, $T''_d = 0.035\mathrm{s}$, $T''_q = 0.035\mathrm{s}$.

The parameter of excitation system is: $T_B = 0$, $T_C = 0$, $K_a = 200$, $T_a = 0.002$, $K_e = 1$, $T_e = 0$, $K_f = 0.001$, $T_f = 0.01$.

The parameter of transformer is: k=20kV/500kV, $S_T = 705\mathrm{MVA}$, $R=0.0027$, X=0.08, R_m=500, X_m=500. Infinite system: S＝100GVA, X/R=8.

The reactance of line is 0.53, the resistance and capacitance of line is neglected.

4. Analysis of simulation data and waveform

Tab.1 test data about steady-state stability limit with 0.75p.u.(450MW) and different magnification of AVR

Ka	Qe (p.u.)	If (p.u.)	Ut (p.u.)	δ (°)
20	-0.685	2.468	0.786	118.4
50	-0.645	2.350	0.805	115.1
80	-0.580	2.200	0.828	110.0
100	-0.521	2.075	0.849	104.5
200	-0.427	1.950	0.880	95.8
300	-0.362	1.890	0.900	89.5
400	-0.328	1.870	0.910	86.2

Tab.2 test data about steady-state stability limit with 0.25p.u.(150MW) and different magnification of AVR

Ka	Qe (p.u.)	If (p.u.)	Ut (p.u.)	δ (°)
20	-0.558	0.923	0.850	132.5
50	-0.508	0.895	0.855	123.8
80	-0.482	0.748	0.875	118.5
100	-0.450	0.696	0.885	111.1
200	-0.426	0.662	0.892	104.2
300	-0.396	0.635	0.902	97.5
400	-0.390	0.637	0.901	95.8

The test data of tab.1 and tab.2 show that the larger magnification of excitation system to the same active power is, the smaller the reactive power, exciting current of generator and power angle. But the terminal voltage of generator is large.

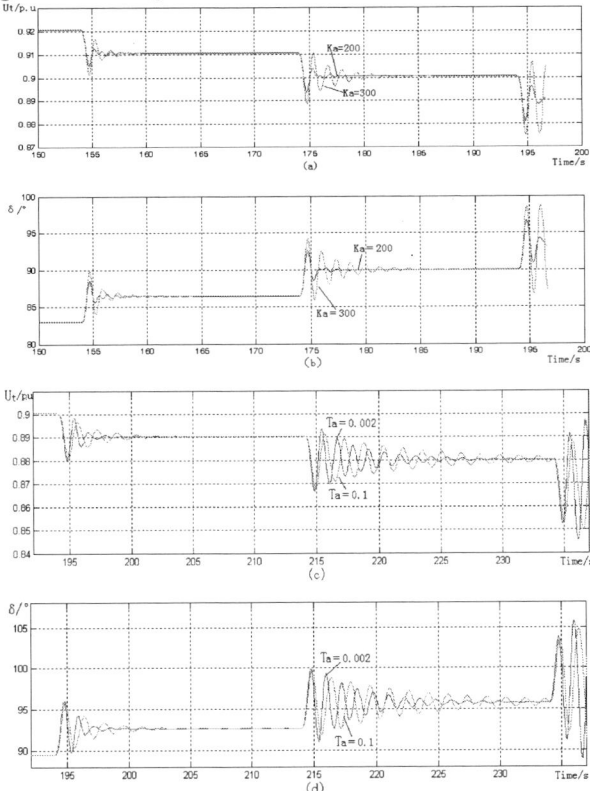

Fig.6 Simulation waveform of different Ka and Ta of the excitation system with 0.75p.u.(450MW)

In Fig.6 (a), (b) the waveform display, after t = 194s, the waveform of excitation system magnification Ka = 300 has begun concussion, loss of static stability, and the waveform of excitation system magnification Ka = 200 still shocks convergent, and maintain static stability. In Fig.6 (c), (d) the waveform display, if the time constant of excitation system Ta is smaller, the waveform fluctuations in a relatively short time, the size of Ta does not affect terminal voltage, power angle the steady-state value of generator which has the same static stability limit.

5. Conclusions

According to Hurwitz's criterion, this paper deduces the steady state borderline of steam turbine with AVR on the map of reactive capability curves, using the mathematical model of one-machine-infinite-bus system. It extends steam turbine's safe range under exciting operation compared to generator without AVR. It analyzes the parameter of automatic voltage regulator's

effect on steady state stability limit of turbine generator, and does the simulation in MatLab / Simulink environment. The simulation results show the excitation system parameters on the generator of steady-state stability limit the impact, which testifies the conclusion's correctness. To the same Active power, the excitation system magnification is larger, the steady-state stability limit generator is smaller, and the capability of generator's absorbing reactive power from the infinite system is dropping. Large excitation system time constant less affected the generator of steady-state stability limit. But the time constant of exciting system have little effect on the steady stability of generator with AVR, only change the time of system's fluctuations.

Acknowledgement

Thanks to the National Nature Science Foundation of China (NSFC):NO.565065

References

[1] Ismael EI-Samahy, Kankar Bhattacharya. A Heuristic Method to Reactive Power Service Procurement. IEEE melecon 2006, May 16-19, Benalmádena (Málaga), Spain.920-923.

[2] M.Bonanno,G.Tina. A competitive reactive power service market in a hierarchical voltage control system. IEEE melecon 2004, May 12-15, 2004, Dubrovnik, Croatia.

[3] S. S. Choi and X. M. Jia. Design of Volts per Hertz Limiter with Consideration of the Under-Excitation Limiter Control Actions. IEEE transactions on energy conversion, VOL. 16, NO. 2, June 2001.

[4] S. S. Choi and X. M. Jia. Under Excitation Limiter and Its Role in Preventing Excessive Synchronous Generator Stator End-Core Heating. lEEE transactions on power system, VOL. 15. NO. 1, pebiwary 2000. 95-101.

[5] Liu Qiang, Li Yong-gang. Steady state stability limit of turbine generator in consideration of AVR. International Conference on Electrical Machines and Systems 2007, in Korea.

[6] Johnson MA, Zarafonitis LP, Calligaris M.Prony analysis and power system stability some recent theoretical and applications research[A].Power Engineering Society Summer Meeting,2000.IEEE[C], 2000,3:1918-1923.

[7] Charles J.Mozina, Michael Reichard. Coordination of generator protection with generator excitation control and generator capability. IEEE 2007:1-17.

[8] CS Hong, S Taib MIEEE, KS Rao and I daut. Development of automatic voltage regulator for

synchronous generator. IEEE Proceedings 2004:180-184.

[9] Zbigniew L,Tadeusz L,Jacek R. Advanced spectrum estimation methods for signal analysis in power electronics[J].IEEE Trans on Industrial Electronics, 2003,50(3):514- 519.

[10] Dengjun Yan, Ping Ju,Wu,Feng, etal. On-line measurement for power angle and circuit parameter in the turbo steam generator with fem based. Proceedings of IEEE,2002:2038-2041.

2008 Workshop on Power Electronics and Intelligent Transportation System

The Integer Ambiguity Transformation Study in Mixed Navigation Based on Real-time Kinematic

Chen Yun, Wang Jijing, Wang Renying

*School of communication and transportation engineering, Changsha University of
Science and Technology, Changsha, Hunan, 410076*
Cheny1963@163.com, lovepepsi412@yahoo.com.cn, leisurely0516@126.com

Abstract

The effect of LAMBDA in RTK mixed navigation is significant in the solution of integer ambiguity. At present, there are some defects in RTK dynamic positioning, for instance, its low speed of integer ambiguity resolution and high data remand for resolution. Focus on that, this paper presents a search strategy algorithm based on LAMBDA. This algorithm can solve parts of problems which integer ambiguity resolution can not deal with, then advances the calculate speed of Real - time kinematics integer ambiguity resolution, and finally satisfy the requirement of dynamic positioning.

1. Introduction

The LAMBDA method in real-time kinematic mixed navigation is efficient in the solution of integer ambiguity. The bi-difference mode of phase observed value can clear up large parts of observation errors, and its ambiguity is integral, so we always adopt this mode to deal with GPS positioning problems [1]. But it is difficult to get correct ambiguity integer solution when we use LAMBDA method as the epoch number is small. The reason is the floating-point solution in bi-difference mode is relative strongly. To solve this problem, the transformation of relative integrals in ambiguity is necessary.

2. RTK mixed navigation and its structure

The single navigation has many defects in aspects of setting accuracy, reliability, range of use and cost; whereas mixed navigation will be the best plan to design vehicle navigation system [2]. The hardware configuration of mixed navigation is consisted of kinds of sensors, data transfer, and data processing and so on. The keystone of mixed navigation research is how to deal with the different sensor's data in the system and how to mix these data. In this paper, we combined RTK [3], GPS and

DR to be a powerful navigation system, and proposed a new form LAMBDA search strategy algorithm for conquering the shortcomings of the slow integer ambiguity resolution velocity and the high data demand of resolution [4]. The algorithm schematic diagram is showed in Fig.1. This algorithm can solve part of problems that integer ambiguity method cannot solve, and then advance resolution velocity of dynamic real time difference integer ambiguity method; finally satisfy the demand of dynamic positioning.

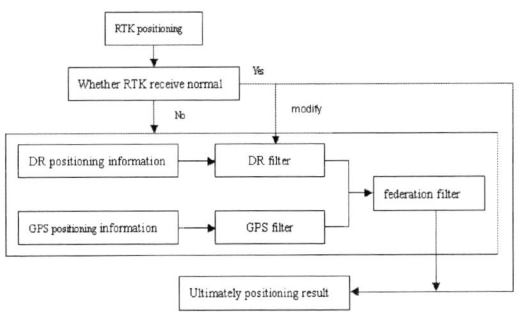

Figure 1. RTK/GPS/DR mixed navigation positioning illustrative diagram

When RTK receiver is in working order, it can take the received spot information to mixed navigation orientation information directly [5]. At the same time, it amends the spot information in DR subsystem of federation Kalman filter, and calculates precisely the spot information when DR subsystem cannot locate nicely. The benefit of this modify is advancing the setting accuracy of navigation system when it cannot receive the satellites' signals. When the satellites' signals of RTK receiver occurring cycle slip, we use new algorithm to ascertain integer ambiguity combination of moving objects' position with fast scanning, and ascertain the integer ambiguity again. When the GPS receiver signal is normal, this method is out of work. When RTK cannot receive satellites' signals or the reference station signals, the system adopt GPS/DR mixed navigation locating information to orient [6]. Because the initial spot of DR information has been amended before, the setting accuracy of mixed

978-0-7695-3342-1/08 $25.00 © 2008 IEEE
DOI 10.1109/PEITS.2008.29

navigation positioning system is much higher.

3. RTK positioning technique

RTK technique is established on the base of real time processing carrier phase of two survey station, i.e. real-time kinematic technology, its precision can reach to centimeter level at best. The carrier phase measure is discrete integrated quantity of clock variance ratio. Because of the differences of satellites' and users' clock variance ratio as well as the Doppler shift in frequency users received, the received clock variance ratio and users' clock variance ratio is different. The integer variance ratio is the difference of two clock integer ambiguity, which equals to undetermined integration constant. Doppler frequency is proportional to relative radial velocity between satellite and users; this speed is much lower than the velocity of light [7]. It is explained the accuracy demand of time measurement is depress greatly.

The RTK technology has two types: initialization method and non-initialization method. The initialization method requires the travelling carriage observe fixed a period of time, and calculate with static state relative measurement software to get phase integer ambiguity of each satellite. After that, we fixed the phase integer ambiguity and take it to be a given value in dynamic measurement in solution. The latter one which is so-called non-initialization is initialization in fact. But its initialization time is shorter, and calculates the integer ambiguity rapidly with FARA [8], OTF or some other methods. The most weakness of RTK technology is that, in despite of what kind of approach, we should do initialization to avoid satellites' loss of lock. Once loss of lock, we should capture and lock signals again. And the DFR can overcome this shortcoming effectively.

4. Integer ambiguity transformation LAMBDA method [9]

The speed of LAMBDA method(Least-squares Ambiguity Decorrelation Adjustment)is very high，as well as its efficiency。Its outstanding is the integer Gaussian transformation to variance covariance matrix of integer ambiguity, i.e. S transform, thereby reduce the search range greatly, and improve the search speed. In addition, it has its own advance in searching algorithm and co-ordinate calculating after fixed of integer ambiguity. The ambiguity of LAMBDA method is established on the base of bi-difference ambiguity, and the search range is a hyper-spheroid, and its shape is controlled by variance covariance matrix of bi-difference ambiguity. Because of the structure differences of ambiguity variance covariance matrix under different

bi-difference combinations, the search range and efficiency is varies.

To compare with the traditional LAMBDA proposed by Teunissen, there is another form of LAMBDA. The new form is base on fussy least square method and upper-triangular Cholesky decomposition under back-sequential condition.

4.1. Back-sequential conditional LS ambiguity technique

Set the expectation and dispersion is $\hat{p}_I = (p_{i+1}, \ldots, p_{s+1})^T \in R^{s+1-i}$ and $\hat{p}_i \in R$ respectively, and gave as follow:

$$E\left\{\begin{bmatrix} \hat{p}_i \\ \hat{p}_I \end{bmatrix}\right\} = \begin{bmatrix} \mu_i \\ \mu_I \end{bmatrix}, \quad D\left\{\begin{bmatrix} \hat{p}_i \\ \hat{p}_I \end{bmatrix}\right\} = \begin{bmatrix} \sigma_i^2 & D_{iI} \\ D_{Ii} & D_I \end{bmatrix}$$

LS estimate the p_i and when μ_I is limited by fixed vector p_I, we can give this form:

$$\hat{p}_{i|I} = \hat{p}_i - D_{iI} D_I^{-1} (\hat{p}_I - p_I)$$
$$(i = s, \ldots, 1) \qquad (1)$$

Here, $\hat{p}_{i|I}$ is stand for $\hat{p}_{i|(i+1),\ldots,s+1}$. The estimate is based on the ambiguity of fixed p_j, j=I+1,$\cdots$,s+1. $\hat{p}_{i|I}$ and $\hat{p}_I$ is not relevance with each other。When i=s-1，the conditional LS estimate $\hat{p}_{(s-1)|s}$, m is come from the ambiguity μ_{s+1} and μ_s of fixed p_{s+1} and p_s.

Here，$\hat{p}_{(s-1)|s}$ is certain，and m is remain constant to orderly varied p_{s+1} and p_s. Diagonalized it $D_I = diag\left\{\sigma_{s+1}^2, \sigma_{s|s+1}^2\right\}$. Then we can get:

$$\hat{p}_{(s-1)|s,s+1} = \hat{p}_{s-1} - \sigma_{(s-1),s+1} \sigma_{s+1}^{-2} (\hat{p}_{s+1} - p_{s+1})$$
$$-\sigma_{(s-1),s|s+1} \sigma_{s|s+1}^{-2} (\hat{p}_{s|s+1-p_s}) \qquad (2)$$

Here，$\hat{p}_{(s-1)|s,s+1}, \hat{p}_{s+1}$ and $\hat{p}_{s|s+1}$ is not relevance to each other。We can keep on iterating until reaching the initial ambiguity, and then the formulation of condition back-sequential conditional LS is summarized as follow:

$$\hat{p}_{i|I} = \hat{p}_i - \sum_{j=i+1}^{s+1} \sigma_{i,j|J} \sigma_{j|J}^{-2} (\hat{p}_{j|J} - p_j)$$

$$(i = s, \ldots, 1) \qquad (3)$$

$\sigma_{i,j|J}$ is stand for the covariance between $\hat{p}_i$ and $\hat{p}_{j|J}$, and the variance of $\hat{p}_{j|J}$ is $\sigma^2_{j|J}$。When $i = s-1$, $\hat{p}_{i|I}$ is equal to $\hat{p}_{s+1}$。

4.2. Upper-triangular Cholesky decomposition and new form for LAMBDA method

In equation (3), we minus p_i of both sides and can get:

$$(\hat{p}_i - p_i) = \hat{p}_{i|I} - p_i + \sum_{j=i+1}^{s+1} \sigma_{i,j|J} \sigma^{-2}_{j|J} (\hat{p}_{j|J} - p_j)$$

$$(i = s, \ldots, 1) \qquad (4)$$

In vector matrix, there is:

$$
\begin{bmatrix} \hat{p}_1 - p_1 \\ \hat{p}_2 - p_2 \\ \cdots \\ \hat{p}_{s+1} - p_{s+1} \end{bmatrix} = \begin{bmatrix} 1 & u_{12} & \cdots & u_{1s+1} \\ & 1 & \cdots & u_{2s+1} \\ & & \cdots & \cdots \\ & & & 1 \end{bmatrix} \begin{bmatrix} \hat{p}_{1|1} - p_1 \\ \hat{p}_{2|2} - p_2 \\ \cdots \\ \hat{p}_{s+1} - p_{s+1} \end{bmatrix}
$$

$$(5)$$

At the same time, there is $u_{ij} = \sigma_{i,j|J} \sigma^{-2}_{j|J} (1 \leq i < j \leq s+1)$。 Set $\hat{p} = (\hat{p}_1, \ldots, \hat{p}_{s+1})^T$, $\hat{p}_c = (\hat{p}_{1|1}, \hat{p}_{2|2}, \ldots, \hat{p}_{s+1})^T$,

and $(U)_{ij} = \begin{cases} 0 & 1 \leq j < i \leq s+1 \\ 1 & i = j \\ \sigma_{i,j|J} \sigma^{-2}_{j|J} & 1 \leq i < j \leq s+1 \end{cases}$.

At the same time, set $D_u = diag\ (\cdots,\ \sigma^2_{j|J},\ \cdots)$。 Here, back-sequential conditional LS ambiguity is not relevance with each other, their variance-covariance matrix has been diagonalized。 When the submitted error has been delivered to (5), the variance-covariance of $\hat{p}$ processed upper-triangular decomposition。 Therefore, the relevance between $\hat{p}$, $\hat{p}_c$ and their variance-covariance is as follow:

$$\hat{p} - p = U(\hat{p}_c - p) \qquad \text{and} \qquad D_{\hat{p}} = U D_u U^T$$

$$(6)$$

According to (6), the objective function of traditional LAMBDA can transfer to the minimization problem as

follow:

$$\min_p \left\| \hat{p}_c - p \right\|^2_{D_u} = \min_p \sum_{i=1}^{s+1} (\hat{p}_{i|I} - p_i) / \sigma^2_{i|I}$$

$$(7)$$

Search through the objective function, we can get the integer LS estimate of fussy vector s is $p \in I^{s+1}$.

4.3. Numerical results

The new LAMBDA method has equivalent computational efficiency with traditional LAMBDA. And these two methods have fixed base solution after the solution of integer fussy algorithm. It shows the new LAMBDA method is correct. When the epoch number is increased, the search times of ambiguity integer solution with LAMBDA will not reduce significantly, which illustrates the LAMBDA method is fit for the precision positioning in short time observation. And after the nor-correlation feasible integral number transformation, the search times of ambiguity integer solution is depressed, whereas advancing the solution velocity.

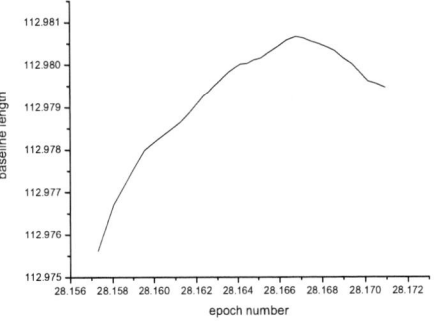

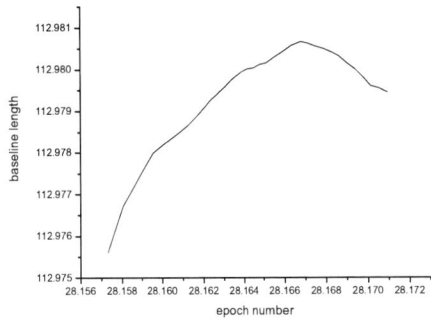

Figure 2. Comparison of the traditional LAMBDA method (up) and new LAMBDA method (down)

5. Conclusion

The RTK technology has its own predominance as well as defects in GPS navigation application. The integer ambiguity calculation in RTK technology is very important. The processing of DR and GPS is also significant in the whole mixed navigation. In this paper, we transfer the traditional LAMBDA, it search degree would not decrease when epoch number increased. And when the epoch number is less, it takes nor-correlation feasible integral number transform to original bi-difference ambiguity. It will advance the accuracy of results and decrease the search degree of ambiguity integer solution, and finally achieve the purpose of solution velocity advancing.

References

[1] Zhang Qishan, Wu Jinpei, Yang Dongkai. The Application of Intellective Vehicular Navigation Positioning System. *Science Press*, 2001.3, pp.164-185

[2] Qin Yongyuan, Zhang Hongyue, Wang Shuhua. The Theory of Kalman Filtering and Mixed Navigation. *Northwest Engineering Press*, 1998, pp.59-73

[3] Bucy R S, Renne K D. Digital Synthesis of Nonlinear Filters. *Automatics*, 1971, pp.287-289

[4] Gurtner, W. Rinex: The Receiver Independent Exchange Format. *GPS World*, 1994, pp. 49-64

[5] Teunissen PJG. The Invertible GPS Ambiguity Transformations. *Manuscript Geodesy*, 1995, pp. 325-356

[6] Teunissen PJG, De Jonge PJ, Tiberius CC. The Least-squares Ambiguity Decorrelation Adjustment: Its Performance on Short GPS Baselines and Short Observation Spans. *Journal of Geodesy*, 1997, pp.65-82

[7] Teunissen PJG. The Least-squares Ambiguity Decorrelation Adjustment: A Method for Fast GPS Integer Ambiguity Estimation. *Journal of Geodesy*, 1995, 65-86

[8] Frei E, Beutler G.Rapid Static Positioning Based on the Fast Ambiguity Resolution Approach: theory and first results. *Manuscript Geodesy*, 1990, pp.325-356

[9] Zhou Yangmei, LIU Jingnan. Another Form for LAMBDA Method. *Geo-spatial Information Science*, 2003, pp. 66-70

The Research of Car Rear-end Warning Model Based on MAS and Behavior

Liang Jun, Cheng Xian-yi and Chen Xiao-bo

School of Computer Science &Communication Engineering, JiangSu University, Zhenjiang 212013
liangjun@ujs.edu.cn

Abstract

The distance between vehicles measurement is the only factor in traditional car rear-end alarm system. An alarming model based on MAS(Multi-Agent Systems) and driver's behavior is proposed to address the above problem. It is composed of four different types of agent （ interface-agent, features-extraction-agent, recognition-agent and alarm-agent） which can either work alone or collaboration through a communication protocol based on the extended KQML. The rear-end alarming algorithm utilize Bayes decision theory to calculate the probability of collision and prevent its occurrence real-time. So autonomy and reliability was enhanced in the proposed system. The effectiveness and robustness of the model have been confirmed by the simulated experiments.

1 Introduction

The past twenty years has seen a rise of the number of traffic jams and accidents. According to the investigations and statistics, of all the traffic accidents ever took place, about 75% were caused by the misjudgment or steer miss of the drivers(see fig 1), which, to a large extent, was due to the insufficiency of the drivers' cognitive and information processing abilities when steering more speedy and complicated vehicles. As a result, the driver-and-passenger centered intelligent active safety system for automobile, along with its new driving sensors and controllers, has become a greater concern to the manufacturers and mass consumers of automobiles and affiliated products in America, Japan, Europe and countries all over the world. After attentive studying of different car rear-end warning models including mazda model on CW/CA system, honda model and an improved model by California Berklee College, the authors found out the disadvantages of most of the current alarm systems. One is that the distance between vehicles is taken as the only determinant factor of vehicle collision by current alarm systems, while the possible effects of the driver's behavior of driving is not taken into account. The other is that the technology of measurement applied by most of the current alarm systems is active measurement which entails special hardwares such as radar, infrared ray and CCD camera, etc, which are, more often than not, expensive and there could be signal interference.

As the latest development of artificial intelligence, the multi-agent system is one of the most important branches in the studies of distributed artificial intelligence. This system is widely used in automatic control, distributed information processing system, meta-synthesis of complex system, intelligent expert system and so on. Also, it offers a new exciting way for the coming out of the car rear-ends warning system.

The car rear-end warning system based on the multi-agent system is the new generation of the intelligent alarm support system. Technically based on the multi-agent system theory in the field of artificial intelligence and theoretically guided by the meta-synthesis method from qualitative to quantitative, the subject alarm system is expected to have improved performance and superior functions[1].

Hence, taking advantage of the flexibility, autonomy, interaction and intelligence of the multi-agent system and taking into account both driving behavior and traditional alarm means of distance measurement, the paper intends to propose a model of car rear-end warning based on MAS and behavior (MCRWMB), which utilizes the Bayes Decision algorithm in car rear-end warning system to the calculation of collision probability in order to prevent the automobile accidents such as tailgating or collision.

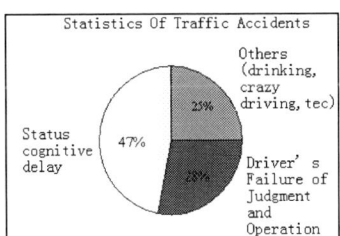

Figure 1. statistics of traffic accidents

2. MCRWMB

2.1 Behavior of driver

The history of driving behavior model can be traced back to the theory of field analysis of vehicles. The research of driving behavior didn't gain much progress in last 50 years although some work still has been done. But it became a hot research area since 2000. After the comparation and analysis of the existed theory and model, Vaa[2] pointed out that the cognition and emotion are good tools for prediction, avoidance and evaluation of dangerous situation in 2001.Then Salvucci, Krajzewicz and Delorme developed investigation on the cognitive structure of driving behavior and it show that the rapid development of cognition science and modeling methods are helpful to the research of driving behavior. Literature[3] provides micro-simulation system of city traffic based on ABM(Agent-Based Modeling), in which humanistic factor of vehicle-driver-model is added with the import of agent technique, but the driver's habits and the main body status are not embodied fully .Literature[4] provides a brief review of four major issues in this field: longitudinal driving behavior analysis and collision avoidance, lateral driving behavior analysis and lane departure warning, complex driving ability learning and driver status (fatigue, absentmindedness etc.) analysis. The likely future direction of this research field, particularly in China, is also pointed out with a special focus on the advances. Although it is affirmed that the important status of driver in driving ,but Literature lacks specific description of the driving behavior.

The model of driving behavior can be described as (E,H) abstractly, where $E=(E_1,E_2,...E_n)$ is used to express external situation such as the road, weather and so on; $H=(H_1. H_2..... H_n)$ represented the driving behavior which refers to the operation behavior on vehicles. It includes operations on steering wheel, accelerator (throttle), brake pedal, clutch pedal and related control unit like steering lamp, lighting lamp, windscreen wiper, etc. The driver usually follows his own habits or preference of (E,H). The preference model of driver can be learned based on a lot of records of his (her) driving behavior and can be used to predict his (her) future behavior H_i. If the risk evaluation function y(.) can be built simultaneously based on this, then the risk of the predicted behavior can be assessed.

2.2 Structure of MCRWMB

The multi-agent system theory provides a new way for the research of decision support system. This car rear-end warning model based on multi-agent (see Figure 2) have

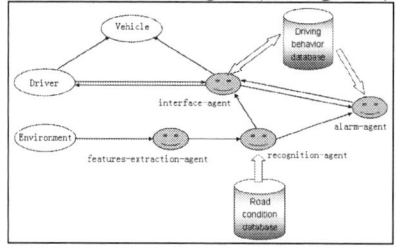

Figure 2 MCRWMB

four kinds of agents. Each agent is autonomous, and the cooperation among the agents could improve the accuracy and the real-time performance of the alarm system. For this end, it is necessary that a reasonable framework of multi-agent alarm system be designed with the ability of learning and dynamic coordination, which could give full

play to its own superiorities and change the traditional routinization of the calculation mode so as to improve the warning efficiency and reinforce the driving security.

2.3. Communications based on extended KQML

Knowledge Query and Manipulation Language(KQML) is a kind of most generally used agent communication language, which could function as a coordinator or cooperator to help realize the real-time information shared among agents through information change. Moreover, KQML has the characteristics of openness, expandability, readability, convenience and platform independence. Therefore, on the foundation of original reserved words in KQML, we define several new ones, such as <Announce>,<Precaution>,<Accept>, etc (see tab 1) so as to realize real-time communication in our multi-agent system.

Table 1 the semantics of extended KQML

3 Bayes decision with driving behavior

Name	Reserved Words	Meaning
Release	N	interface-agent send the driving message to all
Configure	Y	Alarm-agent need to be initialized before driving.
Announce	N	Interface-agent send messages about road and driving habit to other agents, and hope to get the message back
Precaution	N	alarm-agent send the result of alarm in " content" to interface-agent
Accept	N	interface-agent accept the alarm message and begin to give an alarm to driver

3.1 Bayes' formula

Bayes' formula is an extremely useful rule for combining two probability distributions. Its mathematical derivation is simple, although scientific applications of Bayes' formula can be fairly complicated. Bayes' formula follows the definition of conditional probability. P (a) is the probability that event a will occur. Then P(a|b) is the probability of event a that is conditioned by event b, i.e. the probability that event a will occur when event b has occurred. The mathematical definition of conditional probability is: $P(a|b) = P(ab) / P(b)$ (1)

Where P(ab) is the probability that events a and b will both occur. Bayes' formula is: $P(a|b) = P(b|a)P(a)/P(b)$ (2)

It follows from the definitions of the conditional probabilities P(a|b) and P(b|a). From the definition of

306

P(b|a), $P_{(ab)} = P_{(b|a)}P_{(a)}$ (3)

Substitute equation 3 into equation 1 to obtain equation 2.

And let the event of interest a happens under any of hypotheses a_i with a known (conditional) probability $P(b|a_i)$: Assume, in addition, that the probabilities of hypotheses $a_1, a_2, ..., a_n$ are known (prior probabilities). Then the conditional (posterior) probability of the hypothesis is a_i, i = 1, 2, ..., n, given that event b happened, is

$P(a_i|b) = P(b|a)P(a)/P(b)$ (4)

where $P(b) = \sum_{i=1}^{m} P(b|a_i)P(a_i)$

We call formula (4) as conditional probability. $P(a_i)$ is called as prior probability, which is gained by the analyzed data, and $P(a_i|b)$ is posterior probability, which is the renewed modificatory probability. In all, bayes' formula of conditional probability is the MCRMASB's theory basis.

3.2 MCRMASB algorithm

Combined the information of the risk assessment function of $y(H_i)$ and the grade of alarm sent by alarm-agent, we can obtain the result by bayes' formula of conditional probability.

$$P(dang|env, beh) = \frac{P(env, beh|dang)P(dang)}{\sum_{dang} P(env, beh|dang)P(dang)}$$

$$\propto P(env, beh|dang)P(dang) = P(beh, dang|env)P(env)$$

$$= P(beh|env)P(dang|env)P(env)$$ (5)

Considering the complexity of alarm environment, we introduce the fuzzy thought. In formula (5), dang is the measurement of criticality, which is a number in [0, 1], and 0 means having no danger and 1 means the severest danger. Env is also a number between 0 and 1, 0 shows that the environment is very good, such as clear whether, hollowness on road, no foot passengers and no vehicles, and 1 means that the environment is very scurviness (the complex road, many roadblocks and the bad weather). According to the linear model as follows, we can evaluate the environment.

$$env = \sum_{i=1}^{n} \alpha_i x_i$$ (6)

In formula (6), x_i is the factor to evaluate, such as weather condition, road grade and the number of roadblock in certain range; α_i is influencing factors' relative weight, which need to be appointed by the man who has great experience; beh is named as behavior tolerance, which, as mentioned earlier, is the result of risk evaluation $y(H_i)$ of driving behavior, lower behavior beh means the rear-end accident has low probability to happen. We call P(env) as prior probabilities which express occur probability of different environment, and regard it as uniform distribution when we have no prior information. P(env,beh|dang) express the occurrence probability of special environment and behavior conditioned on the danger happened, and it need to be represented by a

probability model such as GAUSS model. P(beh|env) is the probability of adopting beh behavior under the current environment. P(dang|env) is the tolerance of danger degree toward this current environment. In this paper, P(beh|env) model and P(dang|env) model are all built by GAUSS model, which can be showed by:

$$\begin{cases} beh = a * env + \varepsilon \\ dang = b * env + \varepsilon' \end{cases}$$ (7)

Where a, b is called counterchange parameter, ε and ε' are noise data.

4 Simulation experiment and analysis

Experiment 1: verifying the effectiveness of MCRWMB.

In this simulation experiment, in order to calculate easily and to predigest the grade of danger of MCRMASB, we do normalization processing toward the data of collision time. Grade of A is in [0,0.02] and B is in (0.02,0.04] and C is in (0.04,0.06] and D is in (0.06,0.08] and E in (0.08,0.10]. When we input the speed difference (Δv) and the following distance (Δs) measured previously into the model, we obtained the output, i.e., the collision time ($\Delta t = \Delta s / \Delta v$) and the change process of the risk degree, which is illustrated in Fig 3. It is shown in Fig 9 that the interface-agent gave alarm twice during the driving process. The alarm degree for the first time is relatively higher than that of the second time, which is in accordance with the experiment design. Hence the effectiveness of this model could be verified.

Comparing two curves in Figure 3, we found that MCRWMB model gave right alarm when the degree of the dangerous circumstances reached class C or above. and the above degree. whereas, at about 1000 sample point, the output of the model was class C while the collision time curve showed only a minor degree of danger at this time, which means deviation exists in the estimation of the model. However, it could just be explained that MCRWMB model adopt quite cautious "attitude" towards danger.

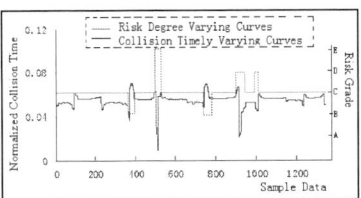

Figure 3   the change process of normal state to collision and danger of the degree

In the experiment, when the alarm-agent gave the first alarm, the following car's driver braked urgently, and then turned into another kind of normal driving state. According to the above analysis, we can see that

MCRWMB model not only can tell the driver's driving preference and give effective alarm, but also has good robustness at noise data caused by the driver's unsuitable operation in urgent conditions.

Experiment 2: use simulative brake sample data as a case to verify MCRWMB 's safety.

Here accelerator pedal motion time refers to the moment that the driver's feet begin to move from the acceleration pedal and the start point at which the driver judges whether braking is needed. It is expressed in the data that the opening of accelerate pedal change from an opposite and stable degree to 0 abruptly (See Figure 4). It is an important guideline to scale the driver's driving habits. With the neural network training and simulation, the change procedure of accelerator pedal and the result of simulation are showed in Figure 4. Obviously, we can find :(1) Sample data and simulation result's trend is very consistent;(2)Compared with sample data and simulation results, they all lie in rather safe area, so the time of applying the brake is ahead and the distance of applying the brake is increased. It is not easy to obtain a global understanding of the procedure of accelerator pedal and the result of simulation. In fact, it is observed that several different characteristics exist during the progress. However, it is enough to verify MCRWMB 's safety. It is worthy of pointing out that sample data and simulation results are fitting better in [0.91,0.94] and [0.78,0.83] section ,where is marked A, but In section B there is stated warp. This needs us to be further studied.

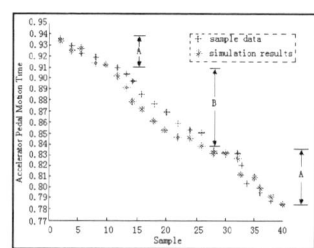

Figure 4   the result of simulation of accelerator pedal

Experiment 3: compared with other typical models

According to the settings of ANFIS model and calculation model[5], the authors did 1500 experiments to MCRWMB model, recorded the results and did the analysis and comparison adequately (see table 2 to 3).Tab 2 shows that MCRWMB model has the longest safety-critical-distance and the least probability of rear-end collision when the speed is between 50 and 120km/h . Tab 3 shows that combining driver' driving preference, MCRWMB model has longer safety distance than other model within the same length of braking time. And tab 4 shows that the forward misdiagnosis rate of MCRWMB, i.e., the ratio of the alarms which should but not be given to all the alarms that should be given, is the lowest among these models.

Table 2 the critical safety distance under different speed(km/h)

Speed	50	60	70	80	90	100	110
ANFIS	40.6	50.6	65.4	79.2	98.1	115.3	142.2
Calculation	39.6	51.1	64.2	78.8	95.5	112.8	132.8
MCRWMB	40.9	50.7	65.5	79.5	98.1	116.1	143.6

Table 3 the critical safety distance under different brake response time (s)

Brake	0.2	0.5	0.8	1.1	1.5	1.8	2.1
ANFIS	16.4	26.4	34.1	43.2	52.5	61.5	69.4
Calculation	15.6	25.8	33.2	41.3	49.9	55.8	60.5
MCRWMB	16.6	26.9	35.8	44.4	53.8	62.6	70.1

Tab 4. the rate of forward misjudgment under different early warning times(%)

Average Times	5	10	20	30	40	50	60	70
ANFIS	3.3	3.5	3.1	3.6	3.5	3.7	3.6	3.4
Calculation	4.5	4.6	4.5	4.2	4.3	4.4	4.3	4.5
MCRWMB	2.8	2.7	2.9	3.0	2.5	2.7	2.8	2.6

5 Conclusion

To sum up, the multi-agent-based car rear-end warning model has great value in reducing the number of traffic accidents and in raising the utilization ratio of roads. With the factors of driving behavior added in, this model's predominance is mainly embodied in the coordination and cooperation among agents, thus improving the reliability, correctness and autonomy of the alarm system to a large extent. Furthermore, MCRWMB model incarnate that driver has the leading position in driver-vehicle-road closed-loop system. In the meantime, the effectiveness and robustness of MCRWMB model has been verified by the simulated experiment, which is of great theoretical significance and practical value in the study of intelligent transportation system.

Acknowledgement

This work is supported by the National Natural Science Foundation of China under Grant No. 60702056.

References

[1] ChengXian-yi,Agent Computing [M] Harbin: Heilong-jiang Science and Technology Press, 2003
[2] LiuYan-fei,Wu Zhao-hui. Driver behavior modeling in ACT-R cognitive architecture [J] journal of Zhejiang University (Engineering Science) 2006 ,pp.1657-1662
[3] Li yin. Multi-Agent System And The Application In Prediction And Intelligent Transportation System [M].shanghai: East China University of Science and Technology Press,2004.11

[4] Li li, Wang Fei-yue, Zheng Nan-ning,Zhang yi. Research and Developments of Intelligent Driving Behavior Analysis [J].Acta Automatica Sinica,2007, 33(10):1014-1022

[5] GuanHsin ZhangLiCun GaoZhenHai. Research of Driver Optimal Preview Acceleration Integrated Decision Model [C]. CMESM,2006, pp144-14

2008 Workshop on Power Electronics and Intelligent Transportation System

A Voltage Fluctuation and Flicker Monitoring System Based on Wavelet Transform

Zhenmei Li   Jin Shen   Peiyu Wei   Tianze Li

School of Electric and Electronic Engineering
Shandong University of Technology, Zibo,Sandong Province, China
lzm@sdut.edu.cn

Abstract

The voltage fluctuation and flicker monitoring system based on Wavelet Transform was developed by using the virtual instrument technology. The system consists of sensors, signal conditioning circuits, data acquisitions and computers. In the system, the arithmetic model of voltage fluctuation and flicker was established. This paper proposes a method of monitoring and analyzing voltage fluctuation and flicker based on the wavelet transform. In the system, the wavelet transform was applied to virtual instrument by using the MATLAB script node in LabVIEW, in which the signals were divided into multi-frequency bands according to the voltage flicker signal character, the feature of the flicker was extracted, frequency and amplitude of voltage fluctuation signal was gained, the moment at which voltage flicker occurs was detected.The evaluation indices of short-term flicker severity Pst are obtained by using statistic sorting algorithms.

Keywords: *monitoring, flicker, voltage fluctuation, LabVIEW, wavelet transform*

1. Introduction

The application of large power and impact loads, such as large capability arc furnaces, spot welding machines and compressors, etc, had made the voltage fluctuation and flicker occurred frequently in the power system. They brought a lot of disadvantages to daily life and industry. For example, lighting equipments flicker, control unit misoperation, motor fluctuation and so on. It is very important to develop a monitoring system of voltage fluctuation and flicker in order to improve and eliminate their influence.

There are three main methods of monitoring the voltage fluctuation and flicker around the world, they are half-wave virtual value, square demodulation and full-wave rectification. But all these methods are not suitable for the time-variable voltage flickering signal detection and tine-frequency analysis, they are just suitable for analog measurement instrument. FFT also can be applied to voltage flickering detection, but because voltage flickering is low-frequency and time-variable unstable signal, while the main drawback of FFT is that the signal

is supposed to be stable. Therefore, FFT can't attain qualified accuracy in voltage flicker analysis. This paper puts forward a method for measuring voltage fluctuation and flicker based on wavelet transform, which can not only extract the feature of the voltage flicker and examine the frequency and the amplitude of the voltage fluctuation, but also detect the moment when the voltage flicker occurs. The key technologies and hardware of the system are stated below..

2. Structure and function of the system

The system consists of sensors, signal conditioning circuits, transfer boards, data acquisition (DAQ) and computers. The sensors are HYH-SB-13-U voltage sensor and HYH-SB-13-I current sensor. The signal conditioning circuit which revises error of sensor caused by individual difference, is an in-phase enlarge proportion amplifier composed of LM324. The transfer board is made by us. The data acquisition is PCI-6025E.

The software structure of the system includes data acquisition, data processing, data expression and data storage. The development platform of this system software is LabVIEW, plus Microsoft SQL Server 2005 database, with ADO technology to access the database, with datasocket technology to achieve remote data transmitting.

By analyzing and calculating value of sampling voltage and current, many results can be displayed: frequency, virtual value of voltage and current, voltage deviation, the evaluation indices of short-term flicker severity Pst. The system can also display real time waveform, such as voltage and current waveform, modulation waveform and the instantaneous flicker sensation level S(t).When some indexes exceed the national standard, the sampling data of voltage and current of that time are stored automatically.

3. The arithmetic model of monitoring voltage fluctuation and flicker

Many electrician equipments can not work properly because of the voltage fluctuation and flicker. Generally speaking, the fluorescent lamp and television are far less

978-0-7695-3342-1/08 $25.00 © 2008 IEEE
DOI 10.1109/PEITS.2008.34

sensitive to voltage fluctuation than incandescent lamp. Almost all build lighting has large number of incandescent lamp. Therefore, if the voltage fluctuation can't influence the fluorescent lamp and television, it won't cause incandescent lamp flicker. So, the operating condition of incandescent lamp can be regarded as the standard to decide whether the voltage fluctuation is acceptable or not .where the flicker can simply be comprehended as the sense to incandescent lamp.

The human brain nerve needs a limited memory time to perceive illumination fluctuation. People will not perceive illumination fluctuation if it is higher than some certain frequency. The visual feeling of human eye and brain to the illumination fluctuation are (as statistics): to the incandescent lamp with 230V, 60W, the frequency range perceivable is from 1Hz to 25Hz approximately, the flickering sensitive frequent range is from 6Hz to 12Hz approximately, the most sensitive frequency to the sine modulated wave with illumination fluctuation at 8.8Hz, the frequency range perceivable can not surpass 0.05Hz to 35Hz.

The flicker is the subjective feeling when we staring an incandescent lamp and it has some relationship with the brain. So we need to set up a mathematic model that simulates the light-eye-brain process, then count and deal with the waveform just gotten. Nowadays, the detection method recommended by IEC has been widely applied to detect the voltage flickering. But as to the square demodulation used to amplitude modulation waveform detection, its main drawback is that amplitude modulation waveform contains the multiples component of frequency of modulation volute signal, and can not detect the moment when the voltage flicker occurs. Here we construct an arithmetic model of monitoring voltage fluctuation and flicker based on the wavelet transform as shown in Figure 1.

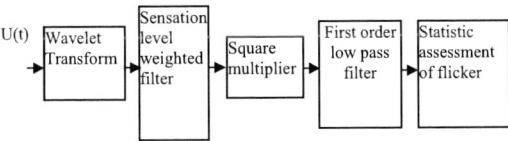

Figure 1. The arithmetic model of voltage fluctuation and flicker

By means of the wavelet transform, we can extract the amplitude modulation waveform, get the accurate time when the voltage flicker occurs and stops, and count the frequency and amplitude of voltage fluctuation. The sensation level weighted filter simulates the frequency response of human eyes to the incandescence lamps influenced by sine volute fluctuation. The **s**quare multiplier simulates the non- linearity sense process of eye –brain. The first order low pass filter uses integral function to average the smoothness, and simulates the nonlinear and memory effect of human brain nerve to

reflect vision. The output is waveform of instantaneous flicker sensation level S(t). Short-term flicker severity Pst can be attained through statistical calculation of s(t).

4. Wavelet transform and the *program designing*

4.1. Wavelet transform

Suppose $\psi(t) \in L^2(R)$ and its Fourier transform is $\overset{\wedge}{\psi}(\omega)$. Let $\overset{\wedge}{\psi}(\omega)$ be under the admissible condition, that is then

$$C_\psi = \int_R \frac{\left|\overset{\wedge}{\psi}(\omega)\right|^2}{|\omega|} d\omega < \infty \tag{1}$$

The continue wavelet transform of signal $x(t)$ is given by

$$WT_f(a,\tau) = \langle x(t), \psi_{a,\tau}(t) \rangle = \frac{1}{\sqrt{a}} \int_R x(t)\psi^*\left[\frac{t-\tau}{a}\right] dt \tag{2}$$

Where $WT_f(a,\tau)$ is wavelet transform coefficient, a is scale factor and τ is translation factor.

Reconstruction formula is as follows:

$$x(t) = \frac{1}{C_\psi} \int_0^{+\infty} \frac{da}{a^2} \int_{-\infty}^{+\infty} WT_f(a,\tau) \frac{1}{\sqrt{a}} \psi\left[\frac{t-\tau}{a}\right] dt \tag{3}$$

To minimize the redundancy of the wavelet transform coefficient, a and τ in wavelet function $\psi_{a,\tau}(t) = \frac{1}{\sqrt{a}}\psi\left[\frac{t-\tau}{a}\right]$ could be defined at some discrete value, that is

$$a = a_0^m \quad a_0 > 0, m \in Z$$

Discrete wavelet function $\psi_{a_0^j, k\tau_0}(t)$, corresponding to $\tau = ka_0^j\tau_0$, is given by

$$\psi_{a_0^j, k\tau_0}(t) = a_0^{-\frac{j}{2}} \psi\left[a_0^{-j}(t - ka_0^j\tau_0)\right] = a_0^{-\frac{j}{2}} \psi\left[a_0^{-j}t - k\tau_0\right] \tag{4}$$

Therefore, discrete wavelet transform is expressed as, then

$$WT_f(a_0^j, k\tau_0) = \int x(t)\psi^*_{a_0^j, k\tau_0}(t)dt \tag{5}$$

$$j = 0,1,2,\cdots, k \in Z$$

On the basis of multi-resolution analysis Mallat has established wavelet decomposition and reconstruction, fast algorithm which is called as Mallat algorithm. According to Mallat algorithm, the construction of wavelet function is regarded as the design of filter coefficient[1][2], thus we can obtain the fast wavelet transform algorithm.

Suppose $C_0(n)$, the discrete approximation of the transient signal $x(t)$ under the condition of the zero-scale, is known, and we can substitute $x(n)$ for the sampling sequence $C_0(n)$, that is

$$C_0(n) = x(n) \qquad n = 0,1, \ldots ,N\text{-}1$$

Where $N = 2^J$, $J \in N$.

The sequence $C_0(n)$ can be decomposed as a approximation coefficient sequence $C_1(n)$. detail coefficient sequence $d_1(n)$ and approximation one $C_1(n)$, decomposing the latter can result in a new wavelet detail coefficient sequence and approximation one, In the same way, decomposing it may result in a new wavelet detail and approximation one with the scale j, the former is rich in the higher frequent component, while the latter rich, in the lower frequent component.

The concrete expression is as follows:

$$d_j(n) = \sum_k c_{j-1}(2n-k)g(k) \qquad (6)$$

$$c_j(n) = \sum_k c_{j-1}(2n-k)h(k) \qquad (7)$$

Where h denotes low-pass filter and g high-pass filter.

By decomposing $x(t)$, the discrete approximation and detail with different resolution are available, then we can analyze the signal at will[3]. While decomposing the signal step by step, the number of the data containing the wavelet detail coefficient and approximation would reduce half each time[4], Thus signal will be decomposed with multi-scale j from 1 to J. So, the entire signal space is divided into different scale space, each corresponds to a different frequency space[5].

With the separation of the different frequency band in the signal, making use of wavelet transform coefficients with different frequency band to reconstruct the wave form belonging to different frequent band, the signal feature can be analyzed and extracted better[6][7].

4.2. The selection of wavelet functions

It is pivotal for wavelet apply to choose wavelet function, and still under research. Now, there are some different wavelet functions, such as Harr, Daubechies, Coiflets, and Symlets. For a certain signal, the analysis effect is different by using different wavelet function, because different wavelet functions have different characteristic in orthogonality compact support, smoothness and symmetry. Haar wavelet is commonly used in theory research because it's discontinuity and the defect of local property in frequency domain. Daubechies

wavelet has been applied widely since it is sensitive for non-stationary signal. Because Daubechies wavelet has characteristics of orthogonality, compact support, and fast arithmetic, so Daubechies wavelet is very suitable for detecting oddity of signal. Thus Daubechies wavelet was selected to analyze voltage fluctuation and flicker.

4.3. Detection to voltage fluctuation

Since the flicker is due to voltage fluctuation, the first step of monitoring voltage flicker is to extract component of voltage fluctuation from signal.

By analyzing the amplitude modulation waveform of single frequency to the modulation power frequency carrier wave. The instantaneous value of a voltage can be expressed as :

$$u(t) = A(1+m\cos\Omega t)\cos\omega t \qquad (8)$$

Where A is the amplitude of the power frequency carrier wave and ω its angle frequency, m the amplitude of the amplitude modulation waveform and Ω its angle frequency.

By means of the wavelet multi-resolution decomposition, the amplitude modulation waveform(weight of voltage fluctuation) is extracted from $u(t)$ and given by

$$v(t) = m\cos\Omega t \qquad (9)$$

For the amplitude modulation waveform $v(t)$ of the voltage flicker carrying about amplitude and frequency of voltage flicker, after obtaining the multi-resolution representation of the voltage flicker signal each weight of the signal is obtained in different band, since the voltage flicker signal is low-frequency, then feature of flicker is available by extracting its low-frequency band information only.

4.4. The program designing and experiment results

The program designing of monitoring voltage fluctuation and flicker based on wavelet analysis can be realized by mixed programming of MATLAB and LabVIEW. This paper discusses how to apply wavelet toolbox of MATLAB to LabVIEW so as to enhance function of signal analysis without Signal Processing Toolset, in development environment of LabVIEW, we link MATLAB script program by using MATLAB node, realize call of wavelet function, steps :

(1)Carry through handle Function →Mathematics →Formula Palette, choose right MATLAB script server, and put it to flow chart edit area. When running script node, will call MATLAB script server. Because LabVIEW uses ActiveX technology to implement MATLAB script nodes, they are available only on windows[8].

(2) Right click MATLAB script node frame edge ,in shortcut menu click Add Input and Add Output option so as to add input and output variable of MATLAB script frame ,choose proper types of variable.

(3)Input MATLAB function to MATLAB script frame

Use wavedec( ) , wavelet decomposition function and wrcoef( ), wavelet reconstruction function of Matlab toolbox to deal with the signal containing amplitude modulation waveform. The wavelet decomposition level is 6. The flow chart of monitoring voltage fluctuation and flicker is as Figure 2 shows.

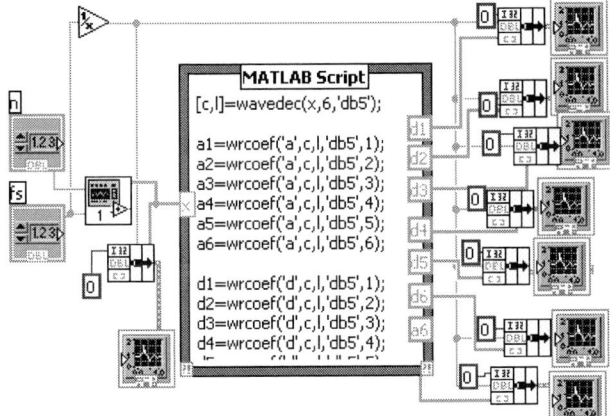

Figure 2. The flow chart of monitoring voltage fluctuation and flicker

Suppose that the input voltage signal is sine waveform with the amplitude being 220V and the frequency being 50Hz and voltage flicker with the amplitude and the frequency being 28V and 6Hz respectively occurs at t=0.1s, the sampling frequency is 3200hz. signal sequence in sub-band (0~25Hz) is available by using the wavelet DB5 decompose $u(t)$ to six scale, then, the amplitude modulation waveform of the voltage flicker is reconstructed by the signal sequence ,the frequency and amplitude of the voltage fluctuation and the moment at which the voltage flicker occurs is detected.

AS shown in Figure 3, detail signal $d1 \sim d6$ and approximation signal $a1 \sim a6$ denote signals reconstructed by wavelet coefficients in corresponding band .It is very clear from the approximation signal $d2$ that the voltage flicker occurs at $t=0.1s$ and from the test that by the wavelet analysis the moment at which the voltage flicker occurs is detected exactly. According to $a6$, the frequency and the amplitude of the flicker are 6Hz and 28V respectively and they can be extracted precisely. Table 1 shows measured results of different frequency modulation waveform.

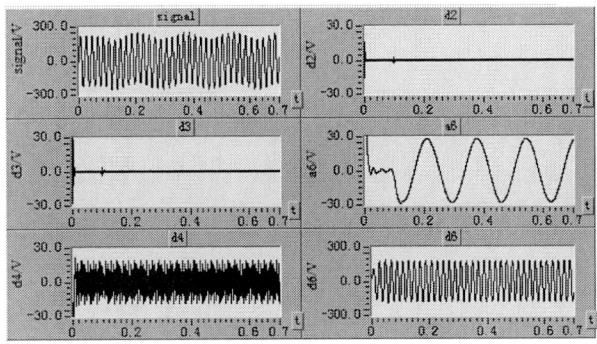

Time/s

Figure 3. The simulation results of extracting feature of flicker

Table 1. Measure results

Frequency(Hz)	Virtual value(V)		
	Theoretical value	Measured value	Error
6	19.802	19.806	0.004
8	19.802	19.809	0.007
12	19.802	19.795	0.007

5. Realization of sensation level weighted filter

The simulation of human eye's frequency choose characteristic can use the transfer function IEC recommended:

$$K(s) = \frac{K\omega_1 s}{s^2 + 2 \cdot \lambda s + \omega_1^2} \times \frac{1 + \frac{s}{\omega_2}}{(1 + \frac{s}{\omega_3})(1 + \frac{s}{\omega_4})} \quad (10)$$

K=1.74802, λ=2π×4.05981, ω_1=2π× 9.15494, ω_2=2π× 2.27979, ω_3=2π× 1.22535, ω_4=2π× 21.9.

The transfer function is a weighted filter which center frequency is 8.8Hz. Conversing expression (10) to z domain system function.

$$H(z) = \frac{\sum_{k=0}^{4} b_k z^{-k}}{1 + \sum_{k=1}^{4} a_k z^k} \quad (11)$$

Then transform it to difference equation as

$$y(n) = \sum_{k=0}^{4} b_k x(n-k) + \sum_{k=1}^{4} a_k y(n-k) \quad (12)$$

a_1=-3.548754, a_2=4.714548, a_3=-2.77601, a_4=0.610325, b_0=-0.009351, b_1=0.00329, b_2=-0.018373, b_3=-0.00032, b_4=0.009022.

6. Square multiplier and first order low pass filter

Square multiplier is Square arithmetic for output of sensation level weighting filter

The transfer function of first order low pass filter is

$$K(s) = \frac{K}{0.3s + 1} \qquad (13)$$

K=63.7864

Conversing it to z domain system function

$$H(z) = \frac{0.26471(1 + z^{-1})}{1 - 0.9917z^{-1}} \qquad (14)$$

Then transform it to difference equation as

$$y(n) = 0.26471x(n) + 0.26471x(n-1) + 0.9917y(n-1) \qquad (15)$$

We can get waveform of S(t) from above-mentioned arithmetic, then can obtain short-term flicker severity Pst .

7. Short-term flicker severity

For the voltage fluctuation caused by random changing load, such as arc furnaces, we need not only to inspect the max voltage fluctuation, but also to inspect the statistical result of voltage fluctuation in a long period (at least 10 minutes). The short-term flicker severity Pst is a statistical calculation result which shows whether the flicker is strong or weak. The method is:

Firstly, take S(t) within 10 minute, draw the curve cumulate probability (CPF) of s(t). The formulation of Pst as follows:

$$P_{st} = \sqrt{0.0314P_{0.1} + 0.0525P_1 + 0.0657P_3 + 0.28P_{10} + 0.08P_{50}} \qquad (16)$$

Where $P_{0.1}, P_1, P_3, P_{10}$ and P_{50} are detection units that the instantaneous flicker sensation level S(t) exceed 0.1% 、 1%、 3%、 10%和 50% in 10 minute.

The program of Calculating Pst was designed by using statistical sorting algorithm, because the instantaneous flicker sensation level S(t) is discrete S sequence in equality time interval, if the sum of the time period (where the S is no less than a certain S(x) is x percent of all the time period, that means S is the highest probability value within(100-x)% period of time. So the P values, $P_{0.1}$ 、 P_1、 P_3、 P_{10}、 P_{50} which correspond to 0.1%、 1%、 3%、 10% and 50% ordinate of CPF curve are the highest probability value within 99.9%、 99%、 97%、 90%、 50% of time period S(n). We just need to realign S(n)from smallness to bigness to find out corresponding probability highest value, then apply them to the formula mentioned above. This method is simple and accurate, doesn't need to draw the CPF curve.

8. Conclusion

The paper puts forward a new method which uses wavelet multi-resolution analysis to extract characteristics of voltage flicker, establishes the arithmetic model of voltage fluctuation and flicker ,which can not only detect the amplitude modulation waveform of the voltage flicker, but also can examine abrupt-change time ,frequency ,amplitude of the amplitude modulation waveform. In the system, the soft-Panel and hardware of virtual instrument and data acquisition were developed with LabVIEW, the voltage fluctuation and flicker of power network was accomplished by using the MATLAB script node in LabVIEW. The evaluation indices of short-term flicker severity Pst are obtained by using statistic sorting algorithms.

9. Acknowledgment

The work was supported by Shandong Provincial Education Department (J07YJ09).

10. References

[1] G. Beylkin, R. Coifman, and V. Rokhlin, "Fast Wavelet Transforms and Numerical Algorithms I," *Comm. Pure Applied Math.*, vol. 44, pp. 141-183, 1991.

[2] S. Mallat, A Wavelet Tour of Signal Processing. Academic Press,1998.

[3] I. Daubechies and W. Sweldens, "Factoring Wavelet Transforms into Lifting Steps," J. Fourier Analysis Applications, vol. 4, no. 3, pp. 245-267, 1998.

[4] S.G. Mallat, "A Theory for Mulitiresolution Signal Decomposition:The Wavelet Representation, " IEEE Trans. Pattern Analysis and Machine Intelligence, vol. 11, no. 7, pp. 674-693, July 1989

[5] G. Beylkin, R. Coifman, and V. Rokhlin, Wavelet in Numerical Analysis in Wavelets and Their Applications. New York: Jones and Bartlett, 1992.

[6] A.N. Akansu and R.A. Haddad, Multiresolution Signal Deomposition:Transform, Subbands and Wavelets. New York: Academic, 1992.

[7] I. Daubechies, "The Wavelet Transform, Time-Frequency Localization and Signal Analysis," IEEE Trans. Information Theory, vol. 36,no. 5, pp. 961-1005, Sept. 1990.

[8] J.H.Liu, X.Y.Zhao, "Instrument Design Based on LabVIEW, " Publishing House of Electronics Industry. pp .115-127, Jan. 2003.

2008 Workshop on Power Electronics and Intelligent Transportation System

Robust Points Tracking Method Using Euclidean Reconstruction for Augmented Reality

Chen Peng , Zhang Rui, Gao Zhang

College of Electrical Engineering & Information Technology，China Three Gorges University, 443002 Yichang, Hubei China

Abstract

Natural feature tracking is a very important research topic in computer vision field and has been used widely in Augmented Reality (AR). This paper gives a robust points tracking or transferring method based on the Euclidean reconstruction technique for AR systems. The points to be tracked include the lost natural features, and any points that are specified by the users. The proposed method distinguishes itself in following ways: Firstly, it is stable as it remains effective even when the camera is moved rapidly. Secondly, the proposed method is robust because it can operate normally as long as at least four pairs of reference point correspondences can be found during the augmentation process. Thirdly, we propose an augmented optical flow method by which the registration, annotation and video augmentation can still work even under the circumstances of large changes in illumination and viewpoint during the entire process. Several experiments have been conducted to validate the usability of the proposed approach.

1. Introduction

Registration is a crucial requirement in Augmented Reality(AR) systems. Many different methods are currently available in AR, including mechanical, magnetic, ultrasound, inertial, vision-based, etc. The registration by positioning sensors such as magnetic or gyro sensors is stable against the change of light condition and effective especially when a camera moves rapidly. However, the rotations and translations obtained only from positioning sensors are not enough accurate to achieve perfect geometrical registration. Vision based registration can solve the above problems to some degree. These kinds of methods offer a flexible and accurate tracking method for AR systems without any special cumbersome sensors. The vision-based approaches can be divided into marker-based approach and natural feature-based approach. Marker-based approach [1][2] is the most commonly used registration method in the majority of applied AR systems. In these systems, one or more man-made fiducial markers are put into the scenes, and the camera's position and orientation are computed using these markers' projections on the moving frame. However, if the markers are partially occluded or outside the field of view, the virtual contents cannot be augmented. In addition, the use of markers can be inconvenient as they require regular maintenance.

To overcome problems of the man-made markers, natural features based registration approaches have been put forward. These kinds of method take full advantage of natural features including points, planes[3-5] and so on, to achieve registration between real and virtual scenes. In some AR applications, the users may want to augment the virtual scene and annotation at a place that does not have distinct natural features. For example, the users would like to augment a virtual teapot onto a smooth table that has no distinct features. Hence, the issue is whether any virtual information can be augmented at any place. The problem is the tracking or transferring of any specified point in the real time video sequence. A lot of AR systems are based on the robust point tracking approach. Gordon and Lowe propose a markerless AR method based on Scale Invariant Feature Transform(SIFT) features [6][7]. However, SIFT features extraction is too slow to meet the real-time requirement in AR systems. Vacchetti et al. [8] proposed a 2D-3D registration method based on offline information using the TUKEY estimator. However, a 3D model is required in their method that might not be available. Simon et al. [4][5] proposed a registration method using planar structures in the scenes. They calculated the registration matrix by real-time estimation of homography between consecutive frames. Whereas, this method suffered from the problem of error accumulation, moreover, a reference plane must be specified and other planes need to be perpendicular to this plane under the condition of multiple planes. Comport et al. [9] propose a camera tracking algorithm based on visual servoing and robust M-Estimator to estimate the camera pose for markerless AR systems. However, the camera cannot be moved quickly and a lack of contrast around contours and large occlusions will lead to failure. Yuan et al. propose a point transferring method based on the projective reconstruction technique and the KLT tracker for markerless AR applications[10-12]. Although the KLT tracker is a classical natural feature tracking method, there are limitations. For example, the camera cannot move rapidly and abruptly when using the KLT tracker. If the camera moves abruptly and rapidly, all the features may be lost and the system will fail. Moreover, these methods don't consider tracking the feature points robustly and is prone to being disturbed by mismatches. Therefore, the registration may be invalidated under the

978-0-7695-3342-1/08 $25.00 © 2008 IEEE
DOI 10.1109/PEITS.2008.73

315

circumstances of large changes in illumination and viewpoint.

In this research, we propose a robust Euclidean reconstruction based point tracking and transferring method to transfer any points for registration, annotation, video augmentation in markerless AR systems. The main contributions of the research reported in this paper can be summarized as follows: Firstly, the robust RANSAC based method to estimate the registration matrix can work stably even when there are some outliers during the tracking process. Secondly, we use Euclidean reconstruction technique to compute the projections of the predefined features. Compared to the six points needed in projective reconstruction based method, our method can work normally when only four features are successfully tracked. Thirdly, we propose to augment optical flow based tracking by building a landmark representation around features. A planar patch around the feature point provides matching information that prevents drifts in feature tracking and allows establishment of correspondences across the frames with large baselines. The registration can operate normally even under the circumstances of large changes in illumination and viewpoint during the entire process. This paper is organized as follows. Section 2 illustrates the Euclidean reconstruction technique. Section 3 presents the augmented optical flow based natural feature tracking method. Section 4 introduces a robust RANSAC method to estimate the projection matrix. Section 5 gives the details of the proposed method and shows some experiments. Finally, conclusions are given in the last section.

2. Background

In this paper, both 2D and 3D points are denoted using homogeneous vectors, so that the relationship between a 3D point $\mathbf{X} = (X, Y, Z, 1)^T$ and its image projection $x = (x, y, 1)^T$ can be given as follows:

$$s\mathbf{x} = \mathbf{P}\,\mathbf{X} \quad \text{Where } P = K[R \,|\, T] \qquad (1)$$

Where s is an arbitrary scale. $\mathbf{P}$ is the projection matrix, $[R|T]$ is the extrinsic parameters of the camera. The matrix K represents the intrinsic parameters of the camera. During the tracking process, if P and X are known in advance, the image coordinate of X can be uniquely determined. The proposed tracking method is based on this principle.

When a 3D point exists on the Z=0 plane of the world coordinate, equation 1 can be rewritten as

$$s\begin{bmatrix} x & y & 1 \end{bmatrix}^T = K\begin{bmatrix} r_x, r_y, t \end{bmatrix}\begin{bmatrix} X & Y & 1 \end{bmatrix}^T = H\begin{bmatrix} X & Y & 1 \end{bmatrix}^T \qquad (2)$$

The 3×3 matrix H is the planar homography, which transforms points on the world plane to the current image. Given four or more point correspondences, we can

compute H using singular value decomposition. Then [R|t] can be recovered as follows:

$$[R\,|\,t] = [r_x, r_y, (r_x \times r_y)/\|r_x \times r_y\|, \, t] \qquad (3)$$

The epipolar geometry exists between any two camera systems. Fundamental matrix F is the mapping of image point x to its epipolar line l':

$$l' = Fx \qquad (4)$$

Also, for any corresponding point pairs $\mathbf{x}, \mathbf{x}'$

$$x'^T F x = 0 \qquad (5)$$

The fundamental matrix with the calibration matrices K,K' removed. i.e., image points are normalized by $\hat{x} = K^{-1}x, \hat{x}' = K'^{-1}x'$. Letting $E = K'^T F K$, we get

$$\hat{x}'^T E \hat{x} = 0 \qquad (6)$$

For normalized cameras, the following holds [10].

$$E = [t]_\times R \qquad (7)$$

From equation (7), we know the essential matrix E including the translation and rotation of camera only.

Given SVD of $E = U\,diag(1, 1, 0)\,V^T$, and assuming first camera is [I |0], the second camera is one of the following:

$$\begin{aligned}[R|t]_1 &= \begin{bmatrix} UWV^T \,|\, +u_3 \end{bmatrix} or \\ [R|t]_2 &= \begin{bmatrix} UWV^T \,|\, -u_3 \end{bmatrix} or \\ [R|t]_3 &= \begin{bmatrix} UW^T V^T \,|\, +u_3 \end{bmatrix} or \\ [R|t]_4 &= \begin{bmatrix} UW^T V^T \,|\, -u_3 \end{bmatrix} \end{aligned}, \quad W = \begin{bmatrix} 0 & -1 & 0 \\ 1 & 0 & 0 \\ 0 & 0 & 1 \end{bmatrix} \qquad (8)$$

Of course, a check-back step is needed to select the true position of the second camera from several image correspondences as suggested in [14]. Then the position of a 3D point in the first camera can be reconstructed using the triangulation [13] as follows:

$$\begin{bmatrix} p_3 x & - p_1 \\ p_3 y & - p_2 \\ p'_3 x' & - p'_1 \\ p'_3 y' & - p'_2 \end{bmatrix} X = 0 \qquad (9)$$

Where p_i and p'_i are the rows of the first and second camera's projection matrix P and P' respectively. x and x' are the corresponding features of the two images. We can get X by solving equation (9) using SVD.

3. Augmented optical flow tracker

For each incoming frame, we must identify which features correspond to which in the reference images. A candidate method is to track these features frame by frame using narrow baseline feature tracking techniques like Lucas-Kanade tracker [15] and so on. However these methods suffered from drift and losing features. This is especially true in the case of features going out of the field of view or occluded by users and some scene objects. Thus the valid matches will become less and less during tracking process, which will finally result in the failure of registration. A feature tracker should cope with temporal occlusion and be able to continue to track a feature, if it moves out of the image and returns back into the image. Therefore a feature must not be discarded, even if the tracking fails or the feature moves out of the image. In our work, we use affine-invariant patch and back-projection to prevent the drift and features losing problems in classical optical flow method. In man-made environments, most of the immediate surroundings of the natural features can be viewed as nearly planar. Furthermore, the scene depth is usually smaller than the distance to the scene from the camera. Therefore, it can be assumed that most of the features are affine-invariant with respect to large camera motion and moderate illumination changes. For the patch of a given feature x, the transformation from the reference image I_r to the current frame I_c can be denoted as follows:

$$\begin{bmatrix} x_c^i & y_c^i & 1 \end{bmatrix}^T = A \begin{bmatrix} x_r^i & y_r^i & 1 \end{bmatrix}^T \qquad (10)$$

Where A is the approximate affine transformation matrix given by:

$$A = \begin{pmatrix} a_{11} & a_{12} & a_{13} \\ a_{21} & a_{22} & a_{23} \\ 0 & 0 & 1 \end{pmatrix} \qquad (11)$$

It can be obtained directly from the tracked features by solving the linear system [13].For feature tracking, we search for the best matching in the immediate neighborhood of the current tracked position of the corresponding feature. The patch of the features on the reference image is first transformed using A. Then correlation is performed within ±2 pixels surrounding the tracked position since the precise location of the detected feature point is to some extent view-dependent. The correspondence is the point that has the maximal normalized cross-correlation score with the transformed patch in reference image.To recover the lost features, we first compute the candidate position on the current image using the projection matrix P. Then the operation similar to feature tracking is performed within ±3 pixels surrounding the candidate position. The correspondence is the pixel which has the maximal normalized cross-correlation score with the transformed patch. In the above computation, if the cross-correlation score is below the predefined threshold (0.6 in our case), we consider this feature is occluded and discard it simply. We also use normalized intensity values in order to take into account illumination variations. The normalization is done using:

$$I_{norm} = (I - I_{avg})/(I_{max} - I_{min}) \qquad (12)$$

Where Inorm is the normalized intensity of a pixel, I is the intensity before normalization, Iavg, Imax, and Imin are the average, maximum, and minimum intensity values of the texture patch respectively.

4. Computing   projection matrix

This section deals with the problem of compute the projection matrix P of the income frames. Since the camera intrinsic matrix is known in advance and do not change, we need only to solve the external matrix [R|t] which is also known as relative pose problems. We apply RANSAC to compute the camera pose consistent with the most matches. The three-point-pose algorithm is used to determine the candidate pose indicated by each set of matches. Because the problem is under-constrained, up to four poses for the camera are valid given the known arrangement of three points in 3D and their image projections. Several methods exist [16], of which we have selected the method of Fischler and Bolles [17] for its numerical stability. Very few RANSAC samples are needed, since the input set of matches usually contains a very small fraction of outliers. Therefore the approach is fast enough for online use.

However, their may be a large fraction of outliers for the first frame due to the unknown of the initial camera and distinct pose or viewpoint difference between current and reference images. To speed up the initialization, the $T_{d,d}$ test developed by Matas [18] is used. When the sets of three matches are chosen, a fourth feature is also randomly selected as an evaluation feature. This feature is the first to be projected to evaluate the pose. If this prediction does not find a match then that pose is abandoned. It is possible that good poses are thrown away, but because of the great speed-up this test gives, many more poses are checked. A time limit is also set for the algorithm. If the correct pose is not found within this time limit then the algorithm gives up: a new frame is taken from the camera and the algorithm is run again. This ensures that, in cases when the camera is occluded for a short period of time, new frames are periodically tried. A time limit of 100ms is chosen which is felt to be enough time to try a reasonable number of poses. If the time limit is much higher, the pose may be too far out of date when found.

5. Algorithm and experimental results

The proposed points tracking algorithm has two parts: the offline initialization stage and the online tracking stage. In the initialization stage, two reference images are obtained. The Harris corner detector is first used to pick up the most salient feature points in the two reference images. Cross-correlation and epipolar geometry are then used to obtain the point correspondences. Next, the related 3D positions of the natural features will be reconstructed from the two selected images using

Euclidean reconstruction technique. In this stage, any approximately matched points can be specified in these two reference images to define the position of the virtual objects and annotation. During the online tracking stage, if there are sufficient natural features that are tracked using the augmented optical flow tracker, the corresponding projection matrix in the image sequence can be estimated. Hence, the image coordinates of the lost natural features and any specified points can be uniquely determined using Equation (1). The virtual objects can be superimposed using the homography deduced form the correspondences of the four pairs of the specified points between the current and reference images. The complete algorithm is described as follows:

Step 1: Obtain two spatially separated images of the scene as references, detect the natural features using Harris corner detector.

Step 2: Obtain the feature correspondences and the fundamental matrix between the two reference images by the normalized cross-correlation operation and the Least Median of Squares approach (LMedS)[19]. Calculate the essential matrix by remove the calibration matrices from the obtained fundamental matrix.

Step 3: Calculate the relative pose between the two reference cameras from the estimated essential matrix. Reconstruct the 3D coordinates of the corresponding features and the specified points.

Step 4: Get the feature correspondences between the first and reference images using the normalized cross-correlation and the $T_{d,d}$ test.

Step 5: Compute the projection matrix of the current frame by the obtained feature correspondences.

Step 6: If the average reprojection error is larger than the predefined threshold (4 pixels in our case), go back to step 4, otherwise, turn to the next step.

Step 7: Transfer the specified points on to the current image, calculate the registration matrix and superimpose the virtual objects and annotation in the scene.

Step 8: Use the augmented optical flow tracker and RANSAC to obtain the corresponding natural features between the next and reference frames and turn to the step 5.

We implemented the proposed method under windows 2003(CPU 3.0 GHz, RAM 1 GB) using Visual C++ and OpenCV. The video sequences are captured using a Logitech Pro5000 camera. The image size is 320×240. The camera's intrinsic parameters are solved in advance using the method introduced in [20]. The system can run the proposed method at a speed of about 20 fps without the use of complicated virtual models. Both indoor and outdoor experiments are conducted to verify the proposed method. Readers can get the detail experiment results including some codes and augmented video sequences through the email of the authors.

In the first indoor experiment, two control images are selected with the camera placed at different positions. An initial set of feature matches is made by normalized correlation, and a classical robust approach, called the Least Median of Squares, based on the technique described in [19], is used to estimate the epipolar geometry between these two reference images. Then, more matches can be found by stereo matching with the computed epipolar geometry. Relative pose for the two views are next computed. Four pairs of noncollinear points are next specified in the two reference images to establish the world coordinate system, the four specified points are transferred during the online registration process. In this experiment, the four specified points, which are marked with the red point in Figure 1, lie on the four corners of a brown notebook. The virtual model is a teapot.

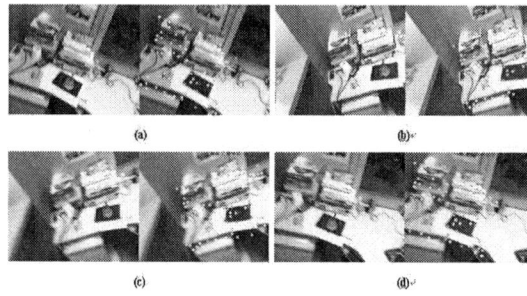

Figure 1.　Results of the first indoor experiment.

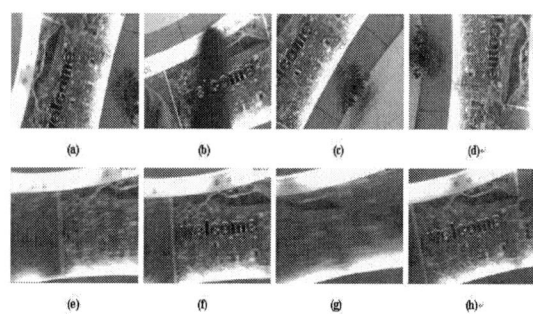

Figure 2.　Results of the second indoor experiment.

The robustness of the proposed method is tested in the second indoor experiment (Figure 2a, 2b, 2c and 2d). A virtual word "Welcome" is augmented on the wall. In the tracking process the camera has been moved through large changes of viewing angles and volumes and the scene has been occluded severely. However, due to the augmented optical flow tracker discussed in section 3, we can superimpose the virtual word successfully under the above circumstances. This experiment also proofs the usability of the proposed $T_{d,d}$ test based re-initialization method. The system can recover (Figure 2f and 2h) from registration failures (Figure 2e and 2g) quickly and does not have lag phenomenon in the tracking process.

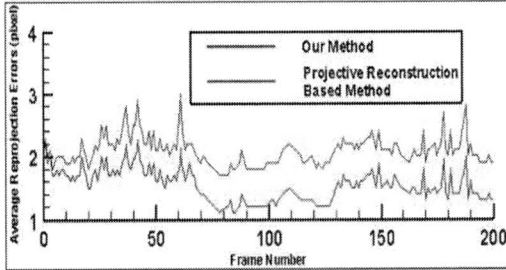

Figure 3. Errors Comparison between two methods.

An experiment is carried out to test the accuracy of the proposed method. The reprojection errors between the original specified points and their reprojections are compared. These errors are represented mathematically as follows:

$$error = \sum_{i=1}^{N} \left\| \mathbf{x}_{ki} - \widetilde{\mathbf{x}}_{ki} \right\| / N \qquad (13)$$

Here $\mathbf{x}_{ki}$ represent the projections of the original specified points in k-th frame, $\widetilde{\mathbf{x}}_{ki}$ are the tracked points obtained using equation 1, and $\left\| \bullet \right\|$ is the Euclidean distance between two points. In this experiment, the four corners of the man made marker are used as the specified points. While tracking, the corners of the marker are detected using the ARToolKit library, and their image coordinates are used as the ground truth for comparison.

We compare the proposed method with the work of Yuan et al [10-12]. Figure 3 shows the reprojection errors of the two methods. The maximum and average errors of the projective reconstruction based approach in these 200 frames are 3.1 and 2.4 pixels respectively. The maximum and average errors of our method are 2.3 and 1.7 pixels which show that our approach is more precise than the projective reconstruction based method.

6. Conclusions

In this paper, a robust point tracking method for AR systems is proposed. The points that are tracked using this method include the natural features that are lost when the augmented optical flow tracker is used and the points that are specified by the users. This method is very useful for AR applications, such as scene annotation and registration.

References

[1] A. State, G. Hirota, D. Chen, W. Garett, and M. Livingston. Superior augmented reality registration by integrating landmark tracking and magnetic tracking. Computer Graphics (Proceedings Siggraph New Orleans), pp.429–438, 1996.

[2] Y. Seo and K. S. Hong. Calibration-free augmented reality in perspective. IEEE Trans. on Visualization and Computer Graphics, vol. 6, no. 4, pp. 346–359, 2000.

[3] Y. Pang, M.L. Yuan, A.Y.C. Nee, and S.K. Ong. A Markerless Registration Method for Augmented Reality based on Affine Properties. In Proc. of AUIC, pages 25–32, 2006.

[4] G. Simon and M. O Berger. Markerless tracking using planar structures in the scene. In Proc. of ISAR, pages 120–128, 2000.

[5] G. Simon and M. Berger. Reconstructing while registering: a novel approach for markerless augmented reality. In Proc. of ISMAR, pages 285–294, 2002.

[6] D.G.Lowe. Distinctive Image Features from Scale-invariant keypoints. International Journal of Computer Vision, 60(2), 2004, 91-110

[7] I. Gordon and D.G.Lowe. What and where: 3D object recognition with accurate pose. in Toward Category-Level Object Recognition, eds. J. Ponce, M. Hebert, C. Schmid, and A. Zisserman, (Springer-Verlag, 2006), pp. 67-82.

[8] L. Vacchetti, V. Lepetit, and P. Fua, Stable Real-Time 3D Tracking Using Online and Offline Information, IEEE Transactions on Pattern Analysis and Machine Intelligence, Vol. 26, Nr. 10, pp. 1391-1391, 2004.

[9] A. Comport, É. Marchand, M. Pressigout and F. Chaumette. Real-time Markerless Tracking for Augmented Reality: the Virtual Visual Servoing Framework. IEEE Transaction on Visualization and Computer Graphics, 12(4), 2006, 615-628.

[10] M.L. Yuan, S.K. Ong, and A.Y.C. Nee. Registration Based on Projective Reconstruction Technique for Augmented Reality Systems. IEEE Trans. on Visualization and Computer Graphics, 11(3):254–264, 2005.

[11] M.L. Yuan, S.K. Ong, and A.Y.C. Nee. Registration using natural features for augmented reality systems. IEEE Trans. on Visualization and Computer Graphics, 12(4):569–580, 2006.

[12] M.L. Yuan, S.K. Ong, and A.Y.C. Nee. A generalized registration method for augmented reality systems. Computers and Graphics, 29. 980-997,2005.

[13] R. Hartley and A. Zisserman. Multiple View Geometry in Computer Vision. Cambridge University Press, 2000.

[14] J.Y.Weng and etc. Motion and structure from two perspective views: algorithms, error analysis and error estimation. IEEE Trans. PAMI, 11(5):451-476, 1989.

[15] B. Lucas, and T. Kanade. An Iterative Image Registration Technique with an Application to Stereo Vision. In Proc. of IJCAI, pages 674-679, 1981.

[16] R. M. Haralick, C. Lee, K. Ottenberg, and M. Nolle. Review and analysis of solutions of the three point perspective problem. International Journal of Computer Vision, 13(3):91–110, 1994.

[17] M. A. Fischler and R. C. Bolles. RANdom SAmple Consensus: A paradigm for model fitting with applications to image analysis and automated cartography. Communications of the ACM, 24(6):381-395, 1981.

[18] J. Matas and O. Chum. Randomized RANSAC with td,d test. Image and Vision Computing, 22(10):837–842, 2004.

[19] Z. Zhang, R. Deriche, O. Faugeras, and Q.T. Luong. A Robust Technique for Matching Two Uncalibrated Images through the Recovery of the Unknown Epipolar Geometry. Int'l J. Artificial Intelligence .78: 87–119, 1995.

[20] Z. Zhang. A flexible new technique for camera calibration. IEEE Trans. Pattern Analysis and Machine Intelligence , 2000, 22(11):1330-1334.

2008 Workshop on Power Electronics and Intelligent Transportation System

Prediction and Analyze of PCB Common-mode Radiation Based on Current-driven Mode

Zhiliang Zhang [1], Fengxun Gong [2]

[1] College of Electronics Information Engineering,
Civil Aviation University of China, Tianjin, 300300, China
[2] College of Electronics Information Engineering,
Civil Aviation University of China, Tianjin, 300300, China
zzhil01@126.com, gfxcauc@eyou.com

Abstract

Common-mode radiation commonly exists in electronic products. It is the main reason for the failure work and fall of radiation test of electronic products. The PCB ground plane inductance is the major factor of producing common-mode radiation. This paper uses the Gauss theorem and the Amp loop theorem to solute the Maxwell's equations to get the mathematical expression of the PCB ground plane inductance, and simplifies the derivation process and the complexity of the expression through the partial approximate substitutions. The result is testified by experiment. Through modular analysis of common-mode radiation, together with the PCB ground plane inductance got in this paper, the mathematical expression of current-driven common-mode radiation is got. Using the closed-form expression, calculations can be performed quickly and easily.

1. Introduction

Radiation from printed circuit boards with attached cables is of significant concern in meeting FCC regulations [1]. Among the significant problems with common-mode radiation is common-mode currents induced on cables attached to printed circuit boards. By the studying of the PCB electromagnetism characteristic, know that there are common-mode radiation and differential-mode radiation existing on PCB. The emission due to common-mode currents on peripheral conductive structures is one of the major reasons of electromagnetic interference in electronic equipment [2]. The common-mode radiation is well known as caused by the parasitic ground plane inductance, or common-mode inductance. Because of the ground plane inductance, there is a voltage drop across the inductance provides a noise source that can drive common-mode current on the external structure. All authors found up to now state that the inductance of a small PCB ground plane can not be

neglected. If the ground plane is very wide, the ground plane inductance can be negligible, but always the PCB can not be very wide. If the ground plane inductance can be computed, then the noise provided by the PCB ground plane can be predicted, then to predict the common-mode radiation.

2. Mechanisms of common-mode radiation

Two fundamental source mechanisms for driving common-mode currents on PCB have been investigated. These source mechanisms have been denoted voltage-driven and current-driven mechanisms to distinguish the differential-mode quantity that provides the driving mechanism of the common-mode current on PCB.

2.1. Voltage-driven mode mechanisms

The principle of voltage-driven mode common-mode radiation can be illustrated as Figure 1. V_{DM} is the source of DM, R is the load, C is the distributing capacitance between the signal line and the attached cable. Voltage-driven sources typically drive objects referenced to ground against other objects whose potentials are related to the signal voltage. The object typically being driven consists of the signal trace itself. If the signal traces capacitive couples to another larger object, the voltage-driven emissions can be significantly enhanced. The voltage-driven mode common-mode current can be expressed as

$$I_{CM} = j\omega \cdot C \cdot V_{DM} \tag{1}$$

978-0-7695-3342-1/08 $25.00 © 2008 IEEE
DOI 10.1109/PEITS.2008.63

320

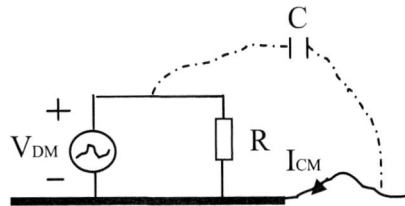

Figure 1　Voltage-driven coupling mechanisms

Differential voltage V_{DM} is signal voltage; it can be measured by come apparatus. The voltage-driven mechanism has been studied by many researchers. Several researchers have proposed methods to calculate the distributing capacitance C between the signal line and the attached cable for the purpose of calculating the induced common-mode currents by expression (1). In [3], the parameters affecting the self-capacitance of a trace are investigated, and a closed-form expression is derived. When the common-mode current is got, then the common- mode radiation can be predicted.

2.2. Current-driven mode mechanisms

In figure 2, current-driven mechanism is associated with the magnetic fields that wrap around the finite width of the signal return plane [4]. V_{DM}, R, L_S, L_G, C_G, V_{CM} denote differential-mode source, load, and inductance of the signal line, inductance of the return plane, distributing capacitance of the ground plane and common-mode voltage. The path of the differential-mode current: the differential-mode source $V_{DM} \rightarrow$ the signal line $\rightarrow$ the load $R \rightarrow$ the ground plane. For $j\omega(L_S + L_R) \ll R$, and from figure 2 the differential- mode current can be got

$$I_{DM} = \frac{V_{DM}}{R + j\omega(L_S + L_G)} \approx \frac{V_{DM}}{R}$$

（2）

Because of the inductance of the ground plane, there is a voltage drop in the ground plane. When there is a high frequency current on the ground plane, the common-mode voltage V_{DM} is the main voltage drop of the ground plane. For common-mode current is smaller than the differential-mode current, so the common-mode voltage can be expressed as

$$V_{CM} \approx I_{DM} \cdot j\omega L_G = \frac{j\omega L_G V_{DM}}{R}$$

（3）

In expression (3), V_{DM} always is the signal voltage, it can be got by measurement. The load R and frequency ω are known parameters. So there is only one unknown

parameter that is the ground plane inductance L_G. If the ground plane inductance is got, the common-mode voltage V_{DM} can be got, and then the common-mode radiation can be predicted. In the next part of this paper a closed-form expression of L_G is derived.

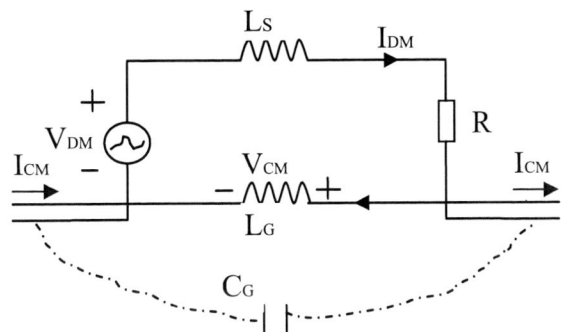

Figure 2　Current-driven coupling mechanisms

3. Inductance of the ground plane

There are many methods to calculate inductance, the normal methods are using the Newman formula, energy conservation theorem, conformal mapping, etc to solve Maxwell's equations then get the inductance. Nearly all inductance formulas found in EMI publications are based on the formulas of [5], but most referees are still using the term self-inductance for what is meant to be the effective inductance!

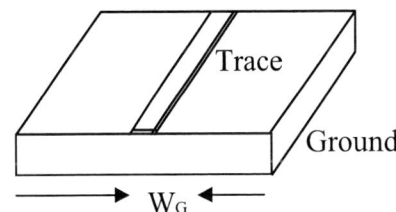

Figure 3　Mode of PCB

In this paper, the Gauss theorem and Amp loop theorem are used to get the PCB ground plane inductance [6]. Figure 3 is the mode of a PCB. W_S、W_G、H、l denote the width of signal line, width of ground plane, distance between signal line and the ground plane, length of the line, with $l << H$ and $d << W_S$. According to Amp loop theorem, the magnetic field between the signal and the ground plane can be got by

$$\oint_c Hdl = \int_s (J + \frac{\partial D}{\partial t})ds$$

（4）

321

When the high frequency signal transmitted on the signal line, there produces skin effect. Neglecting the displacement current and assuming the current flowing through a very thin surface due to the skin effect. The loop integral path is approximately as this, the two ends of return plane is a semicircle with the radius x, two straight lines with the length of W_G connect the tow semicircle.

So the strength of the magnetic field produced by the ground plane is

$$H_G \approx \frac{1}{2\pi x + 2W_G} \int_S J ds = \frac{I}{2\pi x + 2W_G}$$

(5)

Variable x denotes the vertical distance between integral curve and signal line, the scope of x is $0 < x < H$. So the magnetic flux linkage produced by unit ground plane is

$$\psi_G = \frac{1}{I_G} \int_S \mu H_G ds$$

(6)

According to Amp loop theorem, the self-inductance of conductor is

$$L = \frac{\psi}{I}$$

(7)

So the inductance of the ground plane with length l is

$$L_G = l \cdot \frac{1}{I} \int_0^H \mu \frac{I}{2\pi x + 2W_G} dx = \frac{l\mu}{2\pi} \ln(\frac{H\pi}{W_G} + 1) \quad (8)$$

In [8], several printed circuit board configurations have been measured under test. All PCBs consists of a signal track placed at a distance 1.6 mm above a wide copper ground plane of various widths. The length of the PCBs was in all cases 300 mm. The signal track has a width 3.0 mm. The dielectric was epoxy. The thickness of the copper is 35 um. PCBs with different ground plane sizes, as 5, 10, 20, 50 and 100 mm. Figure 4 shows the comparison between the data measured in [7] and the curve of expression (8).

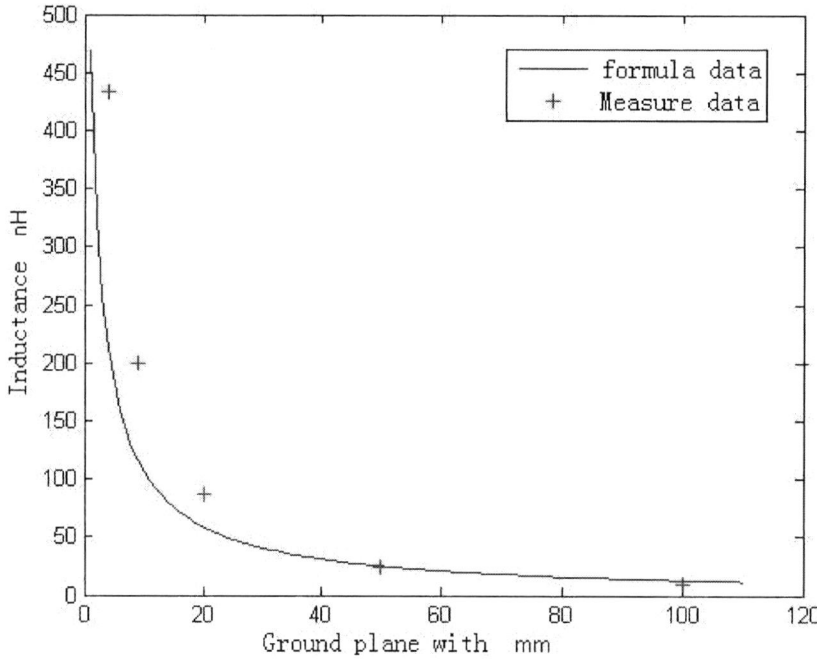

Figure 4 comparison data

Figure 4 shows that the measured data in [7] are tally with the curve of formula (8), and the error is in the allowed scope. So the expression of the PCB ground plane inductance as expression (8) can be validated by experimentation done in [7]. There may be reasons for the error: testing system error and the deviation of the approximation in the formulas derived process. But these approximations greatly simplify the derived process and the complexity of the formula.

Put the expression (8) into the expression (2) can get the common-mode voltage expression:

$$V_{CM} = \frac{j\omega V_{DM} l \mu}{2\pi R} \ln(\frac{H\pi}{W_G} + 1)$$

(9)

By expression (9) can get the information that the higher the frequency, the greater common mode radiation. If expression (9) expressed in logarithm shows that the common-mode voltage increase with frequency in the rate of 20 dB per 10 times, the ground plane inductance is the same as the frequency. The ground inductance is determined which results in much higher ground impedance than only the AC resistance does. The frequency can not be reduced, as the time goes on it will be increased. so the only way to reduce the common-mode radiation is to reduce the ground plane inductance [8]. Increase the width of the PCB ground plane can reduce the ground plane inductance.

4. Conclusion

In engineering applications, EMI always is a difficult problem to solve. In practical application, the engineer always lack of accurate mathematical expression to predict common-mode radiation. A closed-form expression for estimating the ground plane inductance of a PCB is derived in this paper. The expression can be validated by the data measured in [7]. The expression of common-mode radiation based on current-driven mode is also derived. The expression can be used to estimate the common-mode radiation on PCB. Using the closed-form expression, calculations can be performed more quickly and easily than they can be using a field modeling code. The closed-form expression also provides insight regarding the effects that various parameters have on the emissions from PCBs. The radiated emission level, can be lowered by broadening the track to a plane, or create a better coupling by reducing the distance between signal and return track.

Acknowledgement

It is a project supported by National Natural Science Foundation of China, the number is 60672172.

References

[1] Sha Fei. *EMC Technology on Electrical and Mechanical Integration Systems*, Electric Power Press , Beijing, 1999(in Chinese).

[2] D.Hockanson, J.Drewnia(1997). "Quantifying EMI resulting from finite-impedance reference planes". *IEEE Transactions on electromagnetic compatibility*, Vol, 30, No.4, pp. 286–297.

[3] Hwan-Woo Shim, Todd H.Hubing(2005). "Derivation of a Closed-Form Approximate Expression for the Self-Capacitance if a Printed Circuit Board Trace", *IEEE Transactions on electromagnetic compatibility*. Vol.47, No.4, pp.1004-1008.

[4] Thomas P. Van Doren, Fei Sha(1999). "Investigation of Fundamental EMI Source Mechanisms Driving Common-Mode Radiation from Printed Circuit Boards with Attached Cables". *IEEE Transactions on electromagnetic compatibility*,Vol.38, No.4,pp.557–566.

[5] H.W. Grover, *Inductance Calculations*. Dover publications, London, 1946.

[6] Yang Xianqing, *Electromagnetic Fields and Electromagnetic Waves*. Defense Industry Press, Peking, 2003.

[7] F.B.J.Leferink, W.C.Van Etten(2003). "Estimation of Printed Circuit Board Ground Plane Net Partial Inductance via Noise Voltage Measurements". *Electromagnetic Compatibility, 2003 IEEE International Symposium,* Vol.1, No.1, pp.242–247.

[8] Xiaoning Ye, David M. Hockanson, Min Li, etc(2001). "EMI Mitigation with Multilayer Power-Bus Stacks and Via Stitching of Reference Planes". *IEEE Transactions on electromagnetic compatibility*, Vol.43, No.4, pp.538–548.

The Experimental Study about the Influence of Methanol / Diesel Fuel Mixture on Diesel Engine Performance

Chu Weitao

Shandong Transport Vocational Collage, Weifang , 261206,China
weitaochu@sohu.com

Abstract

The influence of M0, M5, M15 methanol / diesel fuel mixture on diesel engine performance has been studied in a single-engine ZS195. Test results show that with the adding of methanol, the driving force of the engine is weaker; fuel economy has been improved; diesel smoke and CO emissions are significantly reduced; NOx emissions are more at M5, but are reduced about 8% at M15; HC emissions are increased. When the diesel engine parameters remain unchanged.

1. Introduction

Alternative fuel of internal combustion engine has been developed rapidly because of energy shortage and growing environmental pollution. Especially methanol has been regarded to develop very promisingly as a sanitary alternative fuel. Because it has an extensive source, relatively low prices; it is a highly liquid oxygen fuel; the burning speed is fast, limit of ignition point is wide, burning emissions are fewer. However, the cetane number of methanol is lower, latent heat of vaporization is larger, and there is a big difference between physical and chemical properties of methanol and those of diesel. To better understand the influence of methanol fuel on combustion and emission performance in diesel engines. This paper studied the influence of different proportion of mixed quality methanol / diesel fuel mixture on diesel power performance, economical performance and emission performance when the engine structure and technical parameters remain unchanged. So it supplies some basis for promoting methanol diesel fuel mixture.

2. The fuel characteristics methanol and diesel

Tab.1 Fuel physical and chemical characteristics of methanol and diesel

Performance indicators	Diesel	Methanol
Density(20°C)/g·L^{-1}	830-860	792
Boiling point/°C	180-370	65
Natural temperature/°C	200-220	470
Low calorific value/(MJ·kg^{-1})	42.8	19.7
Latent heat of vaporization/(kJ·kg^{-1})	250-300	1101
Viscosity(20°C)/(mPa·s)	3.9	0.611
Theory air-fuel ratio/(kg·kg^{-1})	14.6	6.4
Octane number(RON)	20-30	111
Cetane	40-55	3-5
Boundary fire volume/%	1.5-8.2	6.7-36
Mixture heat value/ (MJ·m^{-3})	3.789	3.906

There is a big difference between the physical and chemical characteristics of methanol and that of diesel, the specific parameters such as Table 1[5]. Because of simple structure of molecular methanol (CH_3OH) and high oxygen content, burning with diesel can effectively reduce the smoke and CO emissions. But methanol and diesel nature is so different that they cannot dissolve each other. Therefore in order to keep the methanol / diesel mixture fuel characteristics stable and not stratified, choosing appropriate solvent is necessary.

3. Test engine and major equipments

Test Engine main technical parameter bore / itinerary is 95 mm/115mm, emission amount is 0.8 L, power is 9.7 kw/2200rpm, compression ratio is 17. The main instruments are CWWK-6003 load speed measuring instrument,CW-260 eddy current dynamometers, FCM-05-fuel consumption, FGA-4100 exhaust analyzer and PQD-102-type smoke meters. Accordance to national standards QC/T524-1999 *Car engine performance test method*, the comparison method is applied, and meanwhile engine structure and parameters remain unchanged in the study [6]. In order to study the influence of the different proportion of methanol / diesel hybrid engine on the engine working process, the two proportions of methanol / diesel fuel mixture have been tested, the methanol quality scores are 5% (M5), 15% (M15) separately. The quality scores of co-solvent are 3%,7% successively.

4. Engine Analysis

Outside characteristics curve of diesel engines burning mixture fuel as shown in Figure 1, the use of the mixture fuelM5, the maximum output power and torque of engine dropped about 4.8%, 4.3%; With the increase of the amount of methanol, the use of M15 engine power and torque were down 8.6 %, 7.8 %. Figure 1 shows that with the lower engine speed, the decrease rate of power is smaller. With the adding of methanol and co-solvent, the low calorific value of the mixture fuel is lower; the engine power takes a downward trend. However, as a result of the low cetane number of methanol, burning period of stagnation of the mixture fuel is longer, premixed combustion increases. Because of the low boiling point of methanol, high oxygen content, the diesel is easy to spread and burning is improved, and the burning duration is shorter. So with the increase of methanol quality fraction in the mixture fuel and the change of combustion characteristics, the decline rate of engine power becomes smaller.

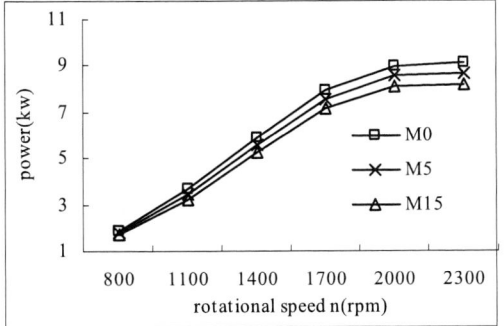

Fig.1 Full load engine power output

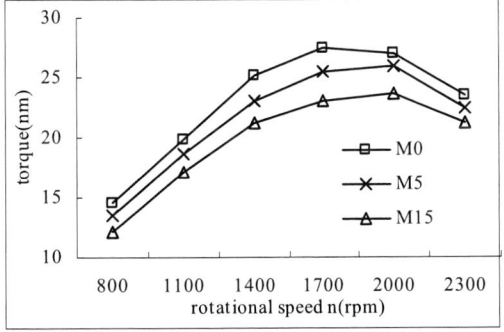

Fig.2 Full load engine torque

5.Engine of economic analysis

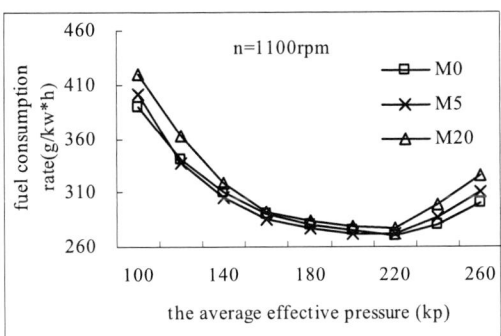

Fig.3 Effect of ratio of on economic at 1100rpm

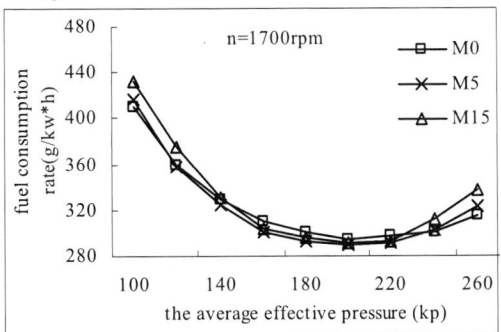

Fig.4 Effect of ratio of on economic at 1700rpm

When diesel engine is in a small load, the mixture fuel consumption rate is slightly higher or close to diesel consumption rate shown as figure 3, figure 4. The reason is that the amount of fuel spray is few and atomization is poor in a small load. Methanol is oxygenated fuel, which makes excessive air ratio greater. Too dilute mixture gas impacts flame spread, local mixture gas is too dilute to catch fire, and discharged with the waste gas emissions so that fuel consumption increases. At the same time in a low load, fewer calories give out. The cylinder wall and the residual gas temperature are low. Methanol latent heat of vaporization is high; resulting in the cylinder average temperature has decreased at the end of compression, and the cetane figure of the methanol/diesel mixture fuel is small, so that the diesel engine delay ignition period is extended too much, fuel is mainly burned in relief combustion period, the thermal efficiency decrease, the engine economy is deteriorated. With the increasing engine load, the fuel consumption rate of burning with mixture fuel has begun less than burning diesel fuel. the economy has been improved. It shows that diesel engine has enlarged its working fields that keep lower fuel consumption. The reason is that when the load is increasing, the cylinder temperature and pressure are increasing, the viscosity of the mixture fuel is small and as vaporizing micro-explosion gasification plays a major role at the effect of improving atomization, burning is more complete, thermal efficiency is improving, Thereby the economy of the diesel engine has been improved.

6.Engine emission characteristics analysis

6.1. Engine smoke emissions

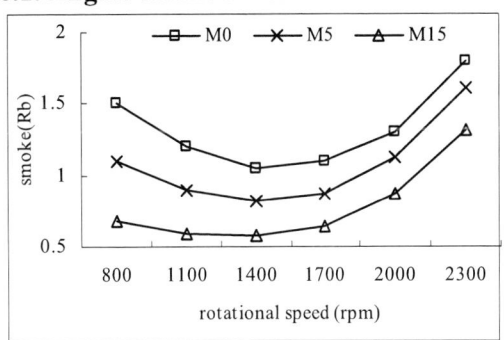

Fig.5 Smoke emissions at full load conditions

The outside characteristics curve of engine emissions smoke as shown in Figure 5, with the use of M5 and M15 engine, the smoke emissions have declined about 49%. Methanol boiling point is low, when the diesel is mixed in a certain proportion of methanol; the mixture fuel viscosity becomes smaller, in addition, "micro-burst effect". Fuel atomization of evaporation is improved; fuel-air is mixed with air well. At the same time oxygen is as high as 50%, compared with the diesel combustion, burning is more completely, so carbon particles in emissions reduce, smoke decreases significantly.

6.2 . NO$_x$ emissions characteristics of Engine

There are three conditions generating NO$_x$: high-temperature, high temperature and duration of high-temperature [7-8]. In the whole load condition, although the adding of methanol, the period of stagnation is extended, but the small proportion of methanol has little effect on the combustion temperature, and about 50% oxygen in methanol increases the amount of NO$_x$ emissions slightly. When the ratio of methanol is up to M15, because of CH3OH heat of vaporization, the temperature dropps substantially, the highest temperature combustion is inhibited; it is not conducive to the formation of NOx, so that NO$_x$ emissions decreases 8% in Figure 6 show. Overall, there is few changes in engine NO$_x$ emissions with the use of the mixture fuel.

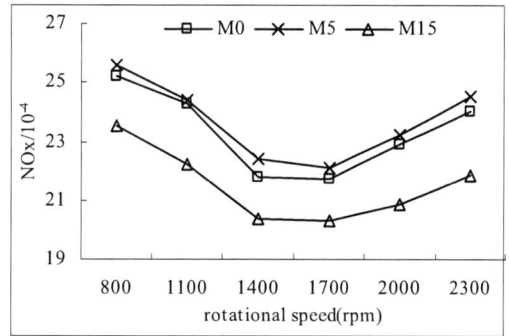

Fig.6 NO$_x$ emissions at full load conditions

6.3. Engine HC and CO emissions characteristics

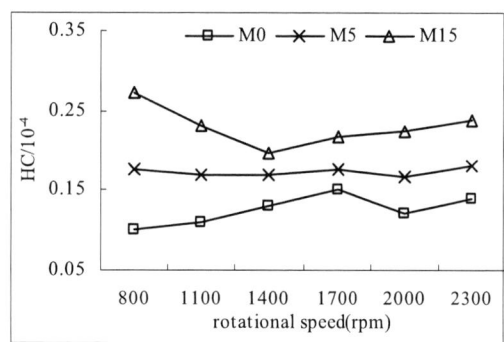

Fig.7 HC emissions at full load conditions

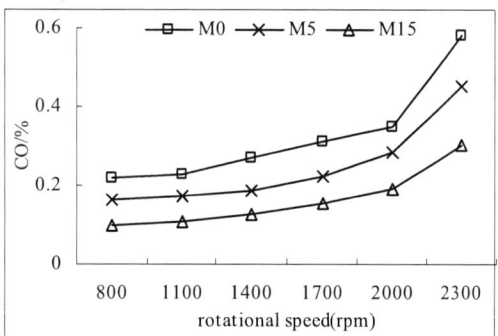

Fig.8 CO emissions at full load conditions

Unburned hydrocarbon emissions of methanol diesel mixture fuel as shown in Figure 6. Methanol is dissolved into diesel, because of easier evaporation of methanol; the fuel-air in the outlying areas may cause a dilute gas mixture increasing, thus leading to the HC emissions of engine increasing. Figure 8 is outside characteristics curve of diesel / methanol mixture fuel CO emissions. We can see that with the increase of methanol blending, CO emissions show a downward trend. It is mainly caused by the physical properties of methanol and properties of abundant oxygen. The low boiling point, low viscosity and "micro-explosion" of methanol have improved the formation and burning of the mixture gas, raised the oxygen content in the oil-rich area, and also helps to reduce the formation of CO emissions which are generated owing to being short of oxygen, particularly at the high speed range of outside characteristics curve, this trend is more evident.

7 .Conclusion

(1) Compared with the original engine, the power decreases 4.8%, 4.3% separately, when the parameters remain unchanged after the diesel engine is blended with methanol, the corresponding maximum torque decreases 8.6 %, 7.8%;

(2) With the use of M5 and M15 methanol / diesel mixture fuel in a diesel engine, the smoke emissions decrease about 49% in the whole load, CO emissions are also reduced. NOx emissions increase slightly with the use of M5, decrease around 8% with the use of M15. Compared with the original engine, HC emissions increase a little.

References:

[1] Hongbo Zou,Lijun Wang,Shenghua Liu. Effect of Pilot Diesel Quantity on the Performance and Emissions of Dual Fuel Engine Operating with Methanol and Diesel. Transactions of CSICE, 2007(5) :422-426.Shanghai Information Center (2003-06-05). "Logistics to get boost in south China city". People Daily, (1).

[2] HUANG Zuo-hua,LU Hong-bing,JIANG De-ming,ZENG Ke,LIU Bing,ZHANG Jun-qiang,WANG Xi-bin. Study on Combustion Characteristics of a DI Diesel Engine Operating on Diesel/Methanol Blends. Transactions of CSICE.2003(6):401-409.

[3] ZHANG Jun-qiang, LUHong-bing, WANG Xi-bin, JIANG De-ming. Combustion and Emission Characteristics of a DI Diesel Engine Fueled by Diesel-Methanol Blends. Journal of Combustion Science and Technology.2004(2):171-175.

[4] Gong Yanfeng Hu Tiegang Li Genbao Liu Shenghua Li Qing Li Tianwen. Effects of Methanol Diesel Blends on Performance and Emissions of a Diesel Engine. Journal of Agricultural Machinery.2007(1):49-64.

[5] HU Yi,JIANG De-ming,HUANG Zuo-hua,WEI Qi,WANG Xi-bin. Effect of Fuel Supply Advance Angle on Combustion and Emission of Diesel Engines Fuelled

with Diesel/Methanol Blend. Chinese Internal Combustion Engine Engineering.2005(4):1-4.

[6] YAO Chun-de, WANG Yan-xia, DUAN Feng, JI Qing, CHENG Chuan-hui, YU Hao-bo. Application of Diesel/Methanol Compound Combustion Mode on Turbo-Charged Diesel Engine. VEHICLE ENGINE.2005(4):6-9.

[7] WANGHui,XUGuo-qiang,CHUYi-min,et al. Study on combusting mixed methanol-diesel oil fuel in diesel engine. Journal ofHenan Agricultural University.2000(4):395-398.

[8] DUDexing, YUANGuangjie. To improve soot emission from diesel engine using dual fuel of methanol or ethanol. ACTASCIENTIAE CIRCUMSTANTIAE.2001(6):759-762.

The Development of Multifunctional Intellectual Electrical Parameters Test Instrument

Zhifang Yang

Wuhan Insititute of Technology, Wuhan 430073, China

Abstract

This article presents a test method of multifunctional electrical parameters by adopting 80C196KC single chip microcomputer as the intellectual part. The features of this design are: the hardware guarantees the measurement accuracy of forward analog channel by using data processing program in order to accomplish each electrical parameter algorithm. All the practices demonstrate this method is reliable and accurate in the on-line measurement of electrical parameters.

1. Introduction

Currently, the digital electric measuring instruments mostly adopt average value to measure the effective value of AC voltage and current. Among this AC-DC conversion method, which can measure the effective value of alternating signal, there are many faults: for example, non-true effective value measurement, unable to measure non-sine alternating periodic signals, narrow frequency-domain, switch errors, DC drift errors and so on. Moreover, the development of power electronic technology can lead the harmonic content of electric power system to increase rapidly, distort voltage and current waveform, and severely influence the safe operation of electric power system and electric devices. In order to ensure the quality of power supply, we must have real-time measurement not only with the voltage of the harmonic source and power supply point but also with current harmonic content.

The scientific research product The Multifunctional Intellectual Electrical Parameters Test Instrument, which the article presents, adopts 80C196KC as the instrument intellectual component, and the internal microsecond A/D converter of chip as data collector. Since we have point-by-point instantaneous sampling with the signals, we can overcome all the defects of the recent digital electric instrument theoretically. The instrument can realize the real-time tracking measurement of general electric parameters, by adopting 16bit single chip microcomputer, which can calculate and manage intensively.

2. The Measurement Theory of Electric Network and Quantity of Electricity

With the development of the computer, the synchronous sampling calculation method comes into being. The calculation formula of average power is:

$$P = \frac{1}{T} \int_0^T u(T) i(t) dt$$

Among this formula, u(t) and i(t) are the instantaneous value of voltage and current, T is the signal period. We can control A/D converter of ACHo and ACH1 twin channel by using 80C196KC, and at the same time, for the waveform of voltage and current, we can have point-by-point data acquisition u (n), i (n), (n=1,2,......N), change them into 10bit discrete digital sequence, and calculate the average value of power.

$$P = \frac{1}{N} \sum_{n=1}^{N} u(n) \times i(n)$$

In addition, we can infer other electrical parameters from the discrete digital sequence: true effective value of AC voltage U, true effective value of AC current I, apparent output S and power factor COS φ are:

$$U = \sqrt{\frac{1}{N} \sum_{n=1}^{N} u^2(n)} \qquad I = \sqrt{\frac{1}{N} \sum_{n=1}^{N} i^2(n)}$$

$$S = U \times I$$

$$\cos \varphi = P / S$$

Among these formulas, every transient digital input u (n) or i (n), which involves in the computing, is the recovery value that has been self-corrected and full gain self-corrected and processed by mathematical model algorithm of error correction as well.

As long as we realize synchronous sampling strictly, and satisfy the sampling frequency with $\omega_s > 2\omega_h$ (ω_h is the upper limit of the highest harmonic frequency of the measured signals), then we can defect each harmonic wave accurately and the original waveform will recur.

For any periodic function f (t), which satisfies Dirichlet condition, can be divided into a Fourier series, i.e.

$$f(t) = A_0 + \sum_{m=1}^{\infty} A_m \cos(m\omega t + \phi_m)$$

By using the synchronous sampling engineering, we can sample N times evenly during the interval 〔T1.T1+T〕and calculate:

978-0-7695-3342-1/08 $25.00 © 2008 IEEE
DOI 10.1109/PEITS.2008.129

$$a_m = \frac{2}{N}\sum_{i=1}^{N} f(T_1 + i\frac{T}{N})\cos(2\pi m \frac{i}{N})$$

$$b_m = \frac{2}{N}\sum_{i=1}^{N} f(T_1 + i\frac{T}{N})\sin(2\pi m \frac{i}{N})$$

The amplitude and initial phase of each frequency component are:

$$A_m = \sqrt{a_m^2 + b_m^2} \qquad \varphi_m = \mathrm{tg}^{-1}(b_m / a_m)$$

The occupancy of 2~31 harmonic voltages:

$$HRU_m = \frac{U_m}{U_1}\times 100\%$$

The occupancy of 2~31 harmonic current:

$$HRI_m = \frac{I_m}{I_1}\times 100\%$$

The distorting occupancy of total harmonic wave of power and that of current are:

$$THDU = \frac{\sqrt{\sum_{m=2}^{\infty} U_m^2}}{U_1} \qquad THDI = \frac{\sqrt{\sum_{m=2}^{\infty} I_m^2}}{I_1}$$

The measured voltage and current signal are changed into square-wave signals through zero comparator, and then these two square-wave signals are sent into the high speed input channel HIS.1 and HIS.2 of 80C196KC single chip microcomputer, the instrument uses the precision frequency measuring method of low-frequency signals to measure the frequency:

$$f = 10^6 (m-1) /\sum t$$

Among this formula: m is the number of square-wave signal which crosses in about two seconds, $\sum t$ is the total time that it has experienced. When measuring $\sum t$, we can set HIS-MODE as positive jump trigger, and write down the time that the first square-wave signal and the last one take during the positive jump trigger, i.e. t1 and t2, between these two the timer T1 overflows trap time m, then we have

$$\sum t = (t_1 - t_2)\times 2\,(m-1)\times 131072\ (\mu s)$$

The digital value, which is acquired through sampling, can calculate the parameters of attribute electric energy quality through FFt: voltage, current true effective value, active power, apparent output, power factor, signal frequency and harmonic analysis. Finally TM240128A graphic lattice liquid crystal indicator switches the display voltage, the column graphic chart of current signal and monocyclic and biperiodic waveform diagram through graphics.

3. The Hardware of Circuit
3.1. The Application System of Single Chip Microcomputer

The system consists of single chip microcomputer 80C196KC, program storage EPROM 27C128, data storage RAM6264, latch 74HC373, universal array logic GAL16V8 and other related circuits. 80C196KC has 16bit product instruction, which is just what we need to develop the measuring instrument. The single chip microcomputer combines with the software and then accomplish: data acquisition, digital preprocessing, FFT, calculation and display of each parameter and frame switch processing.

3.2. The Forward Analog Input Channel

The characteristics of this system are: the hardware guarantees the measurement accuracy of forward analog input channel, the block diagram of which shows as the following. This instrument, through the preset 1/100 pad and voltage comparator array, can divide the input voltage signals ranging from 1V to 1000V into eight range messages automatically. Since the bandwidth of the electrical network signal is not normally finite, the measured signals must go through the anti-mixed overlapping filter in order to prevent mixed overlapping effect and high-frequency interference. When designating the low path filter, we must take the following into consideration: guarantee signal transmission out of distortion, i.e. the phase shift and frequency between output and input become direct ratio, meanwhile, the characteristics of amplitude frequency of the filter must be flat during the domain of spectrum. The low pass filter includes Bartwase filter and resistor-capacitor unit.

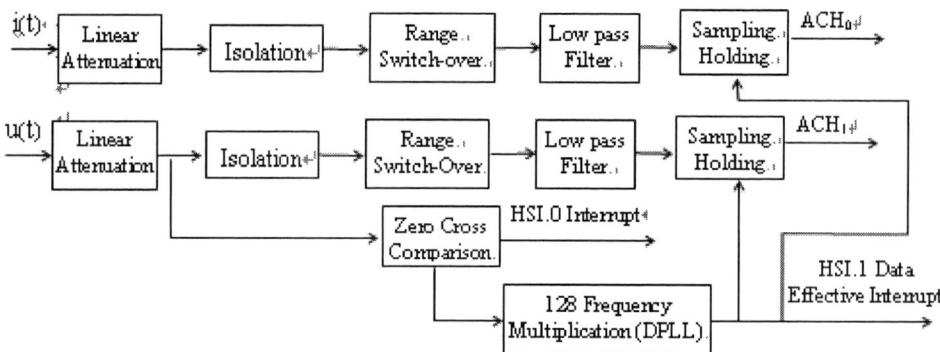

Figure 1 The Block Diagram of Forward Analog Channel

3.3. The Design of Sampling Circuit

In order to FFT the measured signal, we should deal the measured voltage and current signal with 128 point systematic sampling in one period. When power system electrical parameters that are based on microprocessing and FFT algorithm are being measured on line, the sources of their errors are: errors from the dispersion of analog value, rounding off errors during data processing computation, and quantization errors of A/D switch. For the periodic signals, they present as leakage errors that occur when the sampling frequency and signal fundamental frequency desynchronize. They are the main error sources when analyzing and processing the periodic signals. If we use the software, such as 'interpolation algorithms', which can reduce the errors and can't remove the errors thoroughly when rearranging the sequence of sampling data. In order to overcome the errors caused by the above, we designate a sampling of 128 frequency multiplication and a lock-out circuit in the instrument, its kernel is the Digital Phase-Lock Loop (DPLL) which constitutes phase-lock loop integrated circuit CD4046 and twin monostable trigger MC14538. As for its circuit working theory and accuracy analysis, the author has already stated them in the dissertation The Research on How the Tracking Fundamental Frequency Synchronous Sampling Pulse Is Produced, which was published in the 20th Volume of The Automatic Instrument. When the signal fundamental frequency begins to drift, 128 frequency multiplication circuit can track it automatically, start up an A/D switch every 20ms/128=156.25 μs, and guarantee to accomplish 128 times instantaneous sampling uniformly-spacedly in any complete period.

3.4. The Liquid Crystal Display Circuit

We can use graphic liquid crystal TM240128A as the user interface. This is a liquid crystal displayer which combines control, drive and display. It can connect with 80C196KC directly. Through programming, it can display the concrete pictures according to the users' requirement (including Chinese characters, letters, digits, tables, curves and etc.)

4. Accuracy Analysis

1. The Accuracy of u(n) and i(n)

The scope accuracy of u(n) and i(n) are decided by: A/D converter and the forward analog input channel. If we suppose the accuracy of the latter has already satisfied the demand, the A/D converter of 80C196KC adopt 10bit、±1LSB . For the sine signal, its voltage swing is $2\sqrt{2} \times U$, if we consider the waveform distortion and other factors, we can have $2\sqrt{2} \times 2$, and only distribute 1024/5.6=183 Grade to RMS; if we consider the quantization errors of A/D converter, its conversion accuracy is ±0.48%RMS. As we combine the forward analog channel, the errors can be controlled under ± 0.6%RMS.

2. The Calculation Errors

We can lead in the calculation errors because we adopt the finite length digits. As U,I and P are the product integration of instantaneous value. As we consider the data processing ability of single chip microcomputer, we use 16bit for the internal product, and 32bit for accumulation calculation, the leading-in error will be quite small, the actual leading-in error will be under ± 0.01% through emulation calculation.

3. The Errors Caused by the Computation Methods.

As it is proved: when the harmonic time of periodic signals is the definite item, and the sampling interval Ts satisfies Neiquist Theorem, and signal period happens to be the integer multiple N of the sampling period, so the integral of continuous signals, such as voltage, current and power, is equal to the sum of discrete signals, there is no measuring error theoretically.

4. The Test Results

We use standard square-wave and have a test in order to examine the property of the test instrument. By the analysis of the harmonic wave, there is some error between theoretical value and measurement value. This is because it is very difficult to have strict synchronous sampling in the current conditions. It means that this test instrument doesn't surpass the error scope that National Grade B Harmonic Wave Measurement allows.

5. Conclusion

According to the requirement to the accuracy that the multifunctional electrical parameters test instrument has on line, and the real situation of the measured signals as well, we guarantee the measuring accuracy of forward analog input channel by adopting the hardware. For 16bit single chip microcomputer calculates and processes intensively, we realize the real time tracking measurement of general electric parameters by using the software programming ensuring algorithms. This dissertation presents many technical design theories, which have been already put into practice and verified, they are generally significant for application.

References

[1] T.Grardke."Interpolation Algorithms for Discrete Fourier Transforms of Weighted Signals"IEEE Trans, Instrum. Meas,1987(6)

[2] Jiangtao Xi "A New Algorithm for Improving the Accuracy of Periodic Signal Analysis"IEEE Trans, Instrum. Meas,1996(8)

[3] Kongde Zhong,The digital signal Processing,Tsinghua University Publishing House.

[4] National Standard GB/T 14549-99, the Power Quality.

[5] The Practical Brochure of 16bit Single Chip Microcomputer of MCS-96 Series, Beijing The Electronics and Industry Publishing House,1995.

A Linear Design of Temperature Detecting Circuits with Pt Resistance

Zhifang Yang
Wuhan Insititute of Technology, Wuhan 430073, China

Abstract

A correction method for non-linear Pt resistance temperature measurement based on the principle of A/D conversion is introduced. The design principle of Pt resistance linear temperature measurement is analyzed and a new method for interfacing A/D converter with single chip computer 89c52 is provided together with the experimental data.

1. Introduction

Pt resistance is used as a standard instrument in a broad range of temperature for its accuracy and stability in measuring temperature. The relationship between the resistance and temperature can be expressed bye the following formula: $R_\theta = R_0(a\theta + b\theta^2 + 1)$ In the above formula, R_θ、R_0 are respectively the resistance when the temperature is and θ and $0\,^\circ\text{C}$. are constants and a = $3.96847 \times 10^{-3}\,^\circ\text{C}^{-1}$, b = $-5.847 \times 10^{-7}\,^\circ\text{C}^{-2}$. It is obvious that the relation between R_θ and θ is not linear. In order to correct the non-linearization of Pt resistance temperature measurement, the present paper provides a accureate correction method for the non-linearization of Pt resistance temperature measurement. The practice has proved that the method is stable in performance, simple in structure and the accuracy can reach ± 3 digits within the range of $0 \sim 200\,^\circ\text{C}$.

2. The principle of non-linear correction
2. 1 Linear Rθ/U Conversion

The relation of input and output of traditional Wheatstone bridge is not linear. To obtain the measurement with high accuracy, the linear R_θ/U conversion should be achieved. Figure 1 is a linear bridge circuit with good performance. The bridge arm is supplied with constant current.

Since $U_+ = U_-$, $U_- = R_3 U_1 /(R_\theta + R_3)$. Suppose that $R_1 + R_2 + R_3 = R_0$,

We have: $U_1 = \dfrac{2R_0 + \Delta R_\theta}{R_0} U_+$ Obviously, the output voltage U_1 is proportional to $(2R_0 + \Delta R_\theta)$, at the same time, the output voltage of the bridge is

$$U_0 = (\frac{1}{2} - \frac{R_0}{2R_0 + \Delta R_\theta})U_1 = \frac{\Delta R_\theta}{2(2R_0 + \Delta R_\theta)} U_1$$
$$= \frac{\Delta R_\theta}{2(2R_0 + \Delta R_\theta)} \cdot \frac{2R_0 + \Delta R_\theta}{R_0} U_+ = \frac{\Delta R_\theta}{2R_0} U_+$$

Hence, the linear relation between the bridges output voltage

U_0 And bridge arm resistance change ΔR_θ is obtained.

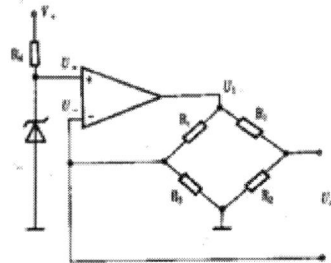

Figure1 A linear bridge circuit with good performance

2. 2 The Principle of Non-linear A/D Conversion

Since a known functional relation exists between the output voltage of the Pt resistance detected by the bridge and the temperature to be measured: $U_M = f(\theta) = A\theta + B\theta^2$. In the formula, A and B are constants. If such a functional circuit can be designed that it has the same mathematical expression as the above formula: $U_N = f(t) = At + Bt^2$, meanwhile, let $U_M = U_N$, it is easy to conclude $\theta = t$, Thus, under the condition of $U_M = U_N$, the measurement of temperature θ is conversed to the measurement of time t.

It is well known that the formula of double slope integration is:

$$U_{in} = \frac{1}{T_1} \int_0^{T_2} U_{ref}(t)\,dt \qquad (1)$$

In this expression, U_{in} is the analog input voltage, T_1 is the time of positive integration, T_2 is the time of negative integration the process of A/D conversion, and U_{ref} is the reference input voltage. When U_{ref} is a constant, a linear relation exists between U_{in} and T_2, Since T_2 can be regarded as the output of A/D conversion, under this situation, the output of A/D conversion can be thought as: $T_2 = T_1 \cdot U_{in} / U_{ref}$. It is also a most frequent method used in the linear measurement of double integration A/D conversion. The present paper takes $U_{ref}(t)$ as a variable that is the function of time t. When the functional relation of $U_{ref}(t)$ is:

$$U_{ref}(t) = M + Nt \qquad (2)$$

(M, N is undetermined constant coefficients)

The output of A/D conversion can completely compensate for the non-linearization of Pt resistance, since

$$U_{in} = A\theta + B\theta^2 \qquad (3)$$

This, together with (2), (3) into (1), we have:

$$A\theta T_1 + B\theta^2 T_1 = MT_2 + \frac{1}{2}NT_2^2$$

Suppose: $AT_1 = M$, $BT_1 = N/2$, then T_2 is equal to θ in quantity: $T_2 = \theta$。

We can see that the digital linear conversion of the Pt resistance temperature is achieved. It is easy to see that the relationship between the analog voltage input and the digital output is not linear, and their functional relation is right opposite to R ₒ /. θ .When they completely compensate for each other, the linear θ / T_2 conversion can be obtained.

Obviously, the key to the non-linearization correction by double integration A/D conversion is to satisfy the functional relation manifested in expression (3). This purpose is easily reached by the adoption of RC circuit.

3. Accurate Double Integration A/D Converter 7135

A conversion period of Icl7135 consists of three stages: the stage of automatic return to zero position, the stage of positive integration of the tested voltage and the stage of negative integration of reference voltage. Figure 2 is the oscillograms of 7135A/D output terminal (BUSY) in conversion and the integrator output terminal (INT).

3.1 Positive Integration

In the stage of positive integration, ICL7135 makes timing integration of U_{in}. The output voltage of the converter is:

$$U_0 = -\frac{1}{R_{int} C_{int}} \int_0^{T_1} U_{in}\,dt = -\frac{U_{in} T_1}{R_{int} C_{int}} \qquad (4)$$

At this stage, the reference capacitor C discharges resistor R. The external resistor R is designed here just to correct the secondary non-linear term of Pt resistance temperature. When the stage is finished, the output voltage at the two ends is:

$$U_c(t) = U_w \exp(-t/RC) \qquad (5)$$

In expression (5), U_w is the voltage of capacitor C when $t=0$. Expand the above expression when $t=T_1$ according to Maclaurin formula:

$$U_c(t) = U_w \exp(-t/RC)$$
$$= U_w - U_w \frac{T_1}{RC} + U_w \frac{(T_1/RC)^2}{2!} + \cdots + U_w \frac{(-T_1/RC)^n}{n!} + \cdots$$

If a proper parameter is chosen to satisfy $(T_1+T_2)<<TC$, then the above expression can be simplified as:

$$U_c(t) = U_w \exp(-t/RC) \approx U_w(1 - \frac{t}{RC}) \qquad (6)$$

Expression (6) is the very functional relation we seek for.

3.2 Negative Integration

At this stage, the voltage of reference capacitor C at the two ends is negatively integrated by the internal integration circuit. In the whole T_2 stage, $U_c(t)$ can be considered as linear. Integrator output returns to zero position when T_2 is finished, thus we have

$$U_0 - \frac{1}{R_{int} C_{int}} \int_{T_1}^{T_1+T_2} U_c(t)\,dt = 0 \qquad (7)$$

Sort out expression (4), (6) and (7):

$$U_{in}T_1 = U_w T_2 - \frac{U_w}{2RC}T_2^2 \qquad (8)$$

This, together with (3) gives：

$$AT_1\theta + BT_1\theta^2 = U_w T_2 - \frac{U_w}{2RC}T_2^2$$

Let the constants at the two sides correspondingly equal, that is:

$$\left.\begin{array}{c} AT_1 = U_w \\ BT_1 = -\dfrac{U_w}{2RC} \end{array}\right\}$$

We have $\theta = T_2$ in the time T_2, A/D converter automatically clock counts and outputs digital quantity through internal digital circuit, so that θ (the temperature to be tested) is expressed in digits, the digital measurement of the circuit is then fulfilled.

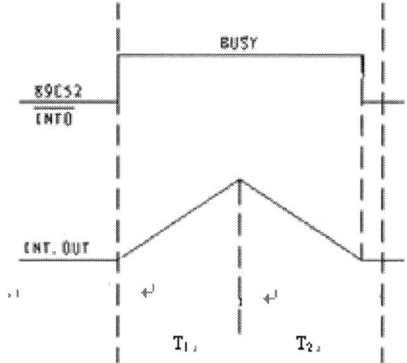

Figure2 Time T1 and time T2

At the A/D conversing stage, ICL7135 outputs BUSY. When BUSY is high level, it indicates that A/D converter is in the stage of signal positive and negative integration, that is, when signal integration of A/D converter begins, BUSY turns to high level which lasts until the negative integration is finished. We take the internal counter of the single chip computer to counter the ICL7135 clock pulses, and use BUSY as a gate-controlled signal for the counter which is made to count only when BUSY is high level, then the input signal:

A/D conversion quantity = the quantity that the counter counts when BUSY is high level-10001.

Figure3 is the schematic circuit diagram of interfacing ICL7135 with 89c52. In this figure, BUSY of 7135 is interfaced with external interrupt $\overline{INT0}$ of 89c52 and the signal of BUSY is taken as the gate-controlled signal for the counter. POL is interfaced with P1.7 of 89c52 as the signal polarity output with high level indicating positive polarity and low level indicating negative polarity.

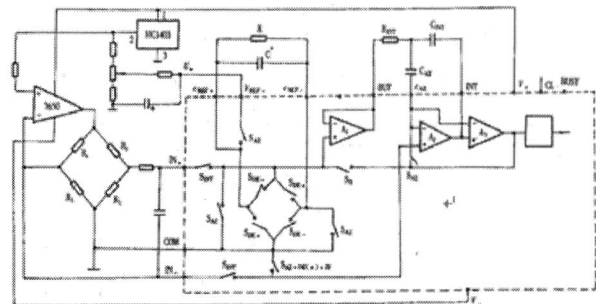

Figure 3 The schematic circuit diagram of interfacing ICL7135 with 89c52

4. Error Analysis and Experimental Results

The error of the correction method for Pt resistance non-linearization by non-linear A/D conversion proposed by the author of the present paper comes mainly from two aspects: one is the error caused by the non-linearization of the reference capacitor in the process of discharging. When the quantity of RC satisfied $RC >> (T_1 + T_2)$, the corresponding temperature error is less than $0.03\,^{\circ}\!C$.

Another aspect comes from the A/D conversion accuracy. When we choose A/D converter ICL7135, its accuracy is $\pm 0.05\%$, and $0.10\,^{\circ}\!C$ when converted into temperature. The two errors are relatively independent, so the total circuit error of temperature measurement is $\pm 0.104\,^{\circ}\!C$.

The performance of the circuit has been tested by technicians. Table 1 is the main data. (Pt resistance is Pt100.) From the test results, the biggest error of the sample is $-0.18\,^{\circ}\!C$ close to the analysis we've made.

Table 1 The main data

Standard temperature °C	Temperature read °C	Absolute error °C	Standard temperature °C	Temperature read °C	Absolute error °C
0	0.01	0.01	100	100.14	0.14
10	10.02	0.02	110	110.17	0.17
20	19.99	-0.01	120	120.09	0.09
30	29.94	-0.06	130	130.02	0.02
40	39.96	-0.04	140	140.00	0.00
50	49.99	-0.01	150	149095	-0.05
60	60.02	0.02	160	159.88	-0.12
70	70.05	0.05	170	169.84	-0.16
80	80.06	0.06	180	179.84	-0.16
90	90.11	0.11	190	189.82	-0.18

References

[1]Zhao Xuezeng , *Test and Sense Technology*, Herbing Industry University Press,1998.10.

[2]Luo Dehan,ICL7135 A/D Converter Practical Technique,*Automation and Instrument*,1993 (3).

[3] Zheng Jiangguo, An Accurate Method for Pt Resistance Temperature Measurement,*Automation and Instrument*,1997.18 (8).

Session 8

Power System

2008 Workshop on Power Electronics and Intelligent Transportation System

The Evaluation of Success Degree in Electric Power Engineering Project Based on Principal Component Analysis and Fuzzy Neural Network

Baoqian Duan[1], Yun Tang[2], Li Tian[1], Qingchao Liu[1]

1 Department of Economy and Management, North China Electric Power University, Baoding, Hebei, 071003
2 Department of international cooperation, North China Electric Power University, Baoding, Hebei, 071003
E-mail: liuqingchao2003@163.com

Abstract

Using principal component analysis (PCA) *and improved fuzzy neural network by PSO to evaluate the success degree in electric power engineering is this paper's innovative points. First we construct the algorithm model which based on PCA and BP neural network improved by PSO. Secondly using PCA to predigest the given index system and then using the relative membership degree processing the date, which as the input sample of neural network. Thirdly, use the improved BP neural network by PSO to evaluate the success degree of electric power engineering. The result denotes that it is more accuracy and speedily than BP neural network algorithm. Lastly, we give a real engineering, and get a satisfaction result.*
Keywords: evaluating degree of success, neural network, principal component analysis, PSO

1. Introduction

According the "Central Enterprises Fixed Assets Investment Project Post-Evaluation Guiding", we know that the post-evaluation is more and more important. It can summarize the project management experience and guarantee the basic construction project sustainable development. Improving the economic benefit of the project and realizing the scientific of investment decision. So this paper research is very important.

The paper use intelligent algorithm to evaluate, it is different from the traditional unintelligent algorithm. It is more accuracy. The traditional algorithm such as variable weight grey cluster[1], it do not have self-learning function, and the result is not so exactly; fuzzy comprehensive evaluation[2,3], it has more subjectivity, and the result is more experience; the traditional neural network evaluation[4], the processing input sample is not so efficiency. So we give fuzzy neural network improve by PSO, it is more accuracy and speedily than other algorithm, such as fuzzy neural network and traditional neural network.

The characteristics of BP algorithm and the PSO algorithm for global optimization capability can be considered in the neural network training, application PSO algorithm optimize the initial weights of neural network. It can be overcome the local optimum of BP neural network and shorten the calculation time. First we use principal component analysis (PCA) to predigest the index system, then we use improved fuzzy BP neural network by PSO to evaluating the project.

2. The algorithm model construction

2.1. Principal component analysis

PCA is one of the key tools in multivariate statistical analysis. It aims at constructing components, each of which contain a maximal amount of variation from the data unexplained by the other components. The user thus hopes that the information in the data can be summarized into a few principal components, which is often the case in practice. Once the principal components have been determined, all further analysis can be carried out on them instead of on the original data, as they carry the relevant information in them. PCA is thus frequently considered a first step of a statistical data analysis which aims at compression of the data: decreasing their dimensionality without losing much information. Further analysis on the principal components can consist of various methods, such as clustering, discriminant analysis, regression, etc.

Principal components contain a maximal amount of variation from the data. In mathematics this means that principal components are defined according to a maximization criterion of variance. Let $X \in R^{n \times p}$ be the data, consisting of n cases observed at p variables. Then the principal components t_i are defined as linear combinations of the data $t_i = X p_i$, where

$$p_i = \arg \max_{a} \{ \mathrm{var}(Xa) \} \quad (1)$$

Under the constraints that

$$\| p_i \| = 1 \text{ and } \mathrm{cov}(X_{p_i}, X_{p_j}) = 0 \text{ for } j < i \quad (2)$$

Exact maximization of this criterion can be done by the Lagrange multiplier method and leads to the

978-0-7695-3342-1/08 $25.00 © 2008 IEEE
DOI 10.1109/PEITS.2008.9

339

conclusion that the principal components are the eigenvectors of the variance–covariance matrix $\Sigma = n^{-1}X^TX$ (here and elsewhere, we will assume the data to be centred).

Both the variance and the variance–covariance matrix are known to be sensitive to outliers. Hence, the same conclusion holds for PCA as a whole: it is a non-robust method. A single bad outlier may cause that principal components are distorted so as to fit the outlier well, leading to bad interpretation of the results. Outliers can also cause the so-called masking effect: due to their presence, the model is distorted in such a way that based on the principal components, no outliers can be detected.

2.2. BP neural network

BP neural network is the feed-forward neural network, and it has great self-study and self-organize ability. This mathematical model comprises individual processing units called neurons that resemble neural activity. Each processing unit sums weighted inputs, and then applies a linear or nonlinear function to the resulting sum to determine the outputs. The neurons are arranged in layers and are combined through excessive connectivity. It allows specification of multiple input criteria and the generation of multiple output recommendations, without pre-assumptions regarding the form of functions related to input and output variables. The BP model eliminates the limitations of the traditional regression methods, and accurately establishes the mapping between the input and output variables. It can approximate an arbitrary nonlinear function with better precision. It is a monitored learning method, and the training course is stopped when the error function reduces to below a given tolerance.

The network learns the relationship between the input and target data by adjusting the weights to minimize the error between the input and target values. The network weights updated by formula (3) .

$$w_{ij}^m(k) = w_{ij}^m(k-1) + \Delta w_{ij} \quad (3)$$

2.3. Particle swarm optimization

Particle swarm optimization (PSO) is an evolutionary computation technique through individual improvement plus population cooperation and competition. A particle's status on the search space is characterized by two factors: its position and velocity. The position and velocity of the ith particle in the d-dimensional search space can be represented as $X_i = [x_{i,1}, x_{i,2}, \ldots, x_{i,d}]$ and $V_i = [v_{i,1}, v_{i,2}, \ldots, v_{i,d}]$, respectively. Each particle has its own best position (pbest) $Pb_i = (pb_{i,1}, pb_{i,2}, \ldots, pb_{i,d})$ corresponding to the personal best objective value obtained so far at time t. The global best particle (gbest) is denoted by Pb_g, which represents the best particle found so far at time t. The new velocity of each particle is calculated as follows:

$$v_{i,j}(t+1) = wv_{i,j}(t) + c_1r_1(pb_{i,j} - x_{i,j}(t)) + c_2r_2(pb_{g,j} - x_{i,j}(t)), j = 1,2\cdots d \quad (4)$$

where c_1 and c_2 are constants called acceleration coefficients, w is called the inertia factor, r_1 and r_2 are two independent random numbers uniformly distributed in the range of [0, 1].

Thus, the position of each particle is updated in each generation according to the following equation:

$$x_{i,j}(t+1) = x_{i,j}(t) + v_{i,j}(t+1), j = 1,2\cdots d \quad (5)$$

Generally, the value of each component in V_i by Eq. (4) can be clamped to the range $[-v_{max}, v_{max}]$ to control excessive roaming of particles outside the search space. Then the particle flies toward a new position according to Eq. (5). This process is repeated until a user-defined stopping criterion is reached.

2.4. BP neural network optimize by PSO

The characteristics of BP algorithm and the PSO algorithm for global optimization capability can be considered in the neural network training, application PSO algorithm optimize the initial weights of neural network. Using PSO algorithm for the weights of BP neural network optimization algorithm is as follows.

First using PSO algorithm repeatedly optimize ANN model weighting parameter combinations, until the fitness no longer meaningful increase. On this basis, used BP algorithm further optimize the above of network parameters which it will be more accuracy. Until searching the optimal network parameters, then we will get the optimal combination of parameters. As PSO algorithm instead of the initial optimization of neural network, network optimizes parameters based on the solution which is close to optimization, thereby effectively improving the network optimization accuracy and speed. The weight updates formula as (6).

$$w_{ij}^m(k) = w_{ij}^m(k-1) + v_i(k) \quad (6)$$

Here, $v_i(k)$ denote the variation of particle's position in time units. (Each element in the weight matrix denote one particle)

3. The evaluation of success degree in electric power engineering project

3.1 The relative membership degree matrix of index

Suggest degree of success evaluation divided into c levels, there are m evaluation factors, and then the m evaluation factors corresponding c levels evaluation structure the Standard Value Matrix of the degree of success evaluation as formula (7).

$$Y = (y_{ij})_{m \times c} \quad (7)$$

Here, y_{ij} is the jth level standard value of the ith evaluate factor.($i=1, 2,...m; j=1,2,...c$).

Suggest there are n groups' monitoring data, and the n groups' data structure the monitoring sample matrix as formula (8).

$$X = (x_{ik})_{m \times n} \quad (8)$$

Here, x_{ik} is the ith success factor value of the kth group monitoring data.($i=1, 2,...m; k=1,2,...n$).

The paper use relative membership degree in fuzzy mathematics describes the success degree. Define ith success degree index of the first level evaluation standard value for the fuzzy set of "success" in the relative membership degree $p_{i1} = 0$, and the cth level corresponding relative membership degree $p_{ic}=1$. Use linear differential equation (9) calculates the relative membership degree p_{ij} of the ith index and the jth level standard.

$$p_{ij} = (y_{ij} - y_{i1})/(y_{ic} - y_{i1}) \quad (9)$$

Also, change the various success degree indexes of monitoring data into a evaluation degree set which is the relative membership degree. For the index that the value is smaller and the success degree is higher, when $x_{ik} \leq y_{i1}$, $r_{ik}=0$; when $x_{ik} \geq y_{ic}$, $r_{ik}=1$. For the index value is bigger and degree is higher, when $x_{ik} \geq y_{i1}$, $r_{ik}=0$; when $x_{ij} \leq y_{ic}$, $r_{ij}=1$. When between y_{i1} and y_{ic}, the r_{ik} can be as formula (10).

$$r_{ik} = (x_{ik} - y_{i1})/(y_{ic} - y_{i1}) \quad (10)$$

Use formula (9) and (10), we can respectively structure the relative membership degree matrix P of the standard value index and the relative membership degree matrix R of monitoring sample index.

$$P=(p_{ij})_{m \times c} \quad (11)$$
$$R=(r_{ik})_{m \times n} \quad (12)$$

For improving the precision of training, we should do finite times interpolation for the standard index of relative membership degree matrix P. Request the index i of interpolation sample k belonging to evaluation level j relative membership degree is p_{kj}^{*}, and then the relative membership degree of the interpolation sample k belongs to evaluation level j is p_{kj}^{*}, it should meets $\sum_{j=1}^{c} p_{kj}^{*} = 1$.

Define the interpolation sample k corresponding standard level value is T_k, and then:

$$p_{kj}^{*} = p_{ij} + (p_{i(j+1)} - p_{ij}) \times q/n \quad (13)$$
$$T_k = j + q/n \quad (14)$$

Here, $i=1, 2,...,m; j=1,2,...,c; q=1,2,...,n-1$; n can be adjusted according the amount of interpolation sample. We can select partly sample and the evaluation standard sample as training sample, the corresponding level value as output sample, the others as test sample of network.

3.2. The degree of success evaluation system

According the power plant construction project, we can get the degree of success evaluation system as table 1.

Table 1. The degree of success evaluation system

The degree of success	Implementation process(A)	The accuracy of prophase decision(A_1)
		The design change cost(A_2)
		The ratio of casualty number account for total number(A_3)
		The change rate of construction period (A_4)
		The change rate of construction cost(A_5)
		The qualified rate of engineering(A_6)
		Debugging and trial production reach the standard qualified rate(A_7)
		The complete of engineering archives(A_8)
	Operation situation(B)	Power consumption rate(B_1)
		The standard coal consumption of power supply(B_2)
		The advancement of technology level(B_3)
		Utilize hour numbers each year(B_4)
	Financial benefit(C)	The total investment return rate(C_1)
		The profit rate of net worth(C_2)
		The economic internal rate(C_3)
		The financial internal rate of return(C_4)
		The economy investment recovery period(C_5)
		The financial investment recovery period(C_6)
	The effect of enviro	NO_x emission standard-reaching rate(D_1)
		SO_2 emission standard-reaching rate(D_2)

3.3 The evaluation of degree success based on PCA and improved fuzzy BP neural network by PSO optimization model

The process of applying PCA and fuzzy BP neural network for PSO optimization model to evaluate degree of success as follows:

1) Gather relevant data.

2) Use PCA model to deal with the data and get the principal component, we can get a simple index system.

3) Use the method of part 3.1, processing the related sample data to normal index relative membership degree matrix P and test sample index membership degree matrix R.

4) According the given input and output samples to identified neural network topology. This paper selects three layers BP neural network.

5) Initialize the initial position x_{mn}^{0} and speed v_{mn}^{0} of search point. They are commonly generated randomly. Selected the number of particles m, the threshold ε of fitness E, the maximum allowable number of iterative steps $T_{\max}$, acceleration factor c1, c2, inertial factor α and constraint factor β.

341

6) We can obtain Root mean square error from formula (15), and then according to formula (16) calculate every particle corresponding individual extremum ξ_i. The global extremum is the best of individual extremum. Record the serial number of particle which has best value. We should set Pb_g to the position of the best particles.

$$E_i = \sum_{k=1}^{n}(y_k - y_k')^2 \quad (15)$$

$$\xi_i = \frac{1}{\exp(E_i)} \quad (16)$$

Here, n denote the number of samples, y_k' denote the output of node k. y_k is the expect output of node k.

7) Use formula (17) to calculate and evaluate the fitness value of each particle, if it is better than individual extremum of this particle, then set the Pbi to this position of particle, and update the individual extremum. If the best individual extremum of all particles is better than current global extremum, then set the Pbg to this position of particle, record the serial number of this particle, and update global extremum.

$$fitness = \frac{1}{1 + \frac{1}{2}\sum_{k=1}^{N}(y_k - y_k')^2} \quad (17)$$

8) Using formula (4) and (5) to update the speed and position of each particle.

9) Checking it if accord with termination condition. If current position meet expected requirement, then stop iterative, output the optimization weight and threshold of network, else return to step 7.

10) Regard optimization of the network weights and thresholds as the initial parameters of neural network.

11) Regarding the normalization date as the learning sample of neural network, and call optimized neural network learning.

12) Keeping the result of network learning, for calling.

13) Input the date of needing evaluation, calling the network which had been training.

14) Output the evaluation result.

From above we can see the success degree evaluation based on PCA and improved fuzzy BP neural network by PSO optimization model as Figure 1.

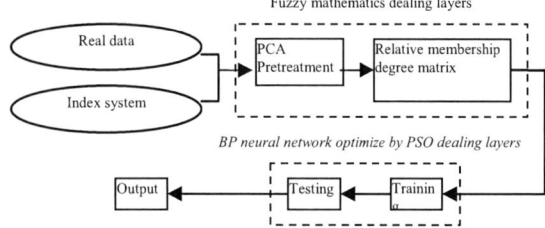

Figure 1. The success degree evaluation flow figure

3.4 The evaluation result and analyses

We use 20 power plant data for the principal component evaluating. Limited the length, here can't list the data. Through the principal component analysis, we can get the each index contribution rate. We adapt the principle that accumulative more than 80%, and get the 12 indexes which are A_2, A_3, A_5, A_6, A_8, B_1, B_3, B_4, C_1, C_3, C_6, D_1.

According to "Sanmenxia power plant feasibility report" in 2003 and other literature, we can get the evaluation standards of quantitative index and use fuzzy method can get the qualitative index evaluation standards as shown in table 2. We also get the monitoring data of Sanmenxia power plant as measured sample as show in table 3.

Table 2. The evaluation criteria index

Index	A standard	B standard	C standard	D standard
A_2	30	100	300	500
A_3	0	0.3	0.5	0.8
A_5	0	0.05	0.1	0.2
A_6	1	0.85	0.75	0.6
A_8	0.85	0.75	0.6	0.45
B_1	6.46	6.66	6.96	7.26
B_3	0.85	0.75	0.6	0.45
B_4	5500	4800	3727	3700
C_1	15	8	4.18	4.1
C_3	50	20	8	7.7
C_6	8	10	14.03	14.09
D_1	100	95	90	85

Use the method of part 3.1, processing the above data. We can get the standard index relative membership degree matrix P and measured sample index relative membership degree matrix R.

Then use formula (11) and (12) to interpolate in the standard index relative membership matrix P and then we can get more sample. Here, $n=7$, then there can get 22 learning samples. Use sample 1, 2, 3, 5, 6, 7, 8, 9, 10, 12, 13, 14, 15, 17, 18, 19, 21, 22 as training samples as show in table 4(x denotes the standard level value and y denotes the evaluation value level), the others as test samples. Then training the measured sample, we can get the accurate evaluation.

Table 3. Monitoring data

Index	A_2	A_3	A_5	A_6	A_8	B_1
Value	300	0.2	0	1	0.81	6.46
Index	B_3	B_4	C_1	C_3	C_6	D_1
Value	0.81	5147	12.9	31.19	9.92	100

Table 4 Learning samples

	A_2	A_3	A_5	A_6	A_8	B_1	B_3	B_4	C_1	C_3	C_6	D_1	x	y
1	0	0	0	0	0	0	0	0	0	0	0	0	0.1	1
2	0.0214	0.0543	0.0357	0.0543	0.0357	0.0357	0.0357	0.0557	0.0914	0.1014	0.0471	0.0471	0.11429	1
3	0.0429	0.1086	0.0714	0.1086	0.0714	0.0714	0.0714	0.1114	0.1829	0.2029	0.0943	0.0943	0.12857	1
4	0.0643	0.1629	0.1071	0.1629	0.1071	0.1071	0.1071	0.1671	0.2743	0.3043	0.1415	0.1414	0.14286	1
5	0.0857	0.2171	0.1429	0.2171	0.1429	0.1429	0.1429	0.2229	0.3657	0.4057	0.1886	0.1886	0.15714	2
6	0.1071	0.2714	0.1786	0.2714	0.1786	0.1786	0.1786	0.2786	0.4571	0.5071	0.2357	0.2357	0.17143	2
7	0.1286	0.3257	0.21429	0.3257	0.2142	0.2142	0.2142	0.3342	0.5486	0.6086	0.2829	0.2829	0.18571	2
8	0.15	0.38	0.25	0.38	0.25	0.25	0.25	0.39	0.64	0.71	0.33	0.33	0.2	2
9	0.21	0.4157	0.3043	0.4157	0.3043	0.3043	0.3043	0.4757	0.69	0.75	0.4243	0.3771	0.21429	2
10	0.27	0.4514	0.3586	0.4514	0.3586	0.3586	0.3586	0.5614	0.74	0.79	0.5186	0.4243	0.22857	2
11	0.33	0.4871	0.4129	0.4871	0.4129	0.4129	0.4129	0.6471	0.79	0.83	0.6129	0.4714	0.24286	2
12	0.39	0.5229	0.4671	0.5229	0.4671	0.4671	0.4671	0.7329	0.84	0.87	0.7071	0.5186	0.25714	3
13	0.45	0.5586	0.5214	0.5586	0.5214	0.5214	0.5214	0.8186	0.89	0.91	0.8014	0.5657	0.27143	3
14	0.51	0.5943	0.5757	0.5943	0.5757	0.5757	0.5757	0.9043	0.94	0.95	0.8957	0.6129	0.28571	3
15	0.57	0.63	0.63	0.63	0.63	0.63	0.63	0.99	0.99	0.99	0.99	0.66	0.3	3
16	0.6314	0.6829	0.6829	0.6829	0.6829	0.6829	0.6829	0.9914	0.9914	0.9914	0.9914	0.7086	0.31429	3
17	0.6929	0.7357	0.7357	0.7357	0.7357	0.7357	0.7357	0.9929	0.9929	0.9929	0.9929	0.7571	0.32857	3
18	0.7543	0.7886	0.7886	0.7886	0.7886	0.7886	0.7886	0.9943	0.9943	0.9943	0.9943	0.8057	0.34286	3
19	0.8157	0.8414	0.8414	0.8414	0.8414	0.8414	0.8414	0.9957	0.9957	0.9957	0.9957	0.8543	0.35714	4
20	0.8771	0.8943	0.8943	0.8943	0.8943	0.8943	0.8943	0.9971	0.9971	0.9971	0.9971	0.9029	0.37143	4
21	0.9386	0.9471	0.9471	0.9471	0.9471	0.9471	0.9471	0.9986	0.9986	0.9986	0.9986	0.9514	0.38571	4
22	1	1	1	1	1	1	1	1	1	1	1	1	0.4	4

$$
P=\begin{bmatrix}
0 & 0.15 & 0.57 & 1 \\
0 & 0.38 & 0.63 & 1 \\
0 & 0.25 & 0.50 & 1 \\
0 & 0.38 & 0.63 & 1 \\
0 & 0.25 & 0.63 & 1 \\
0 & 0.25 & 0.63 & 1 \\
0 & 0.25 & 0.63 & 1 \\
0 & 0.40 & 0.99 & 1 \\
0 & 0.65 & 0.99 & 1 \\
0 & 0.71 & 0.99 & 1 \\
0 & 0.33 & 0.99 & 1 \\
0 & 0.33 & 0.67 & 1
\end{bmatrix}
\qquad
R=\begin{bmatrix}
0.57 \\
0.25 \\
0 \\
0 \\
0.1 \\
0 \\
0.1 \\
0.12 \\
0.19 \\
0.44 \\
0.32 \\
0
\end{bmatrix}
$$

The paper uses fuzzy BP neural network for PSO optimization model to evaluate degree of success which based on the evaluating accuracy, so we will respectively use BP neural network and fuzzy BP Neural Network for PSO optimization model to evaluate, and then compare the result. This need adapt the same value as the input parameter of neural network. The node point number of input layer is 12; the node point number of output layer is 1. Network training selects strong adaptability BP neural network which has three layers, hidden layer use sigmoid function as excitation function. Input layer use linear neuron. Hidden layer has 25 nodes. The network learning rate $\eta = 0.69$, momentum factor $\mu = 0.81$. The population size of PSO algorithm training weight of the neural network $n=10$, initial inertia weight $w(0) =0.86$, with the number of iterations linear decline to 0.5, $c_1=c_2=2$, connection weight is interval variable at [-1, 1]. First, we check the accuracy of the method. We use training sample and testing sample as table 3. Then we can obtain the BP neural network and fuzzy BP neural network improved by PSO calculate accuracy comparing as table5, and speed comparing as figure 2 and figure 3.

Table 5 The data accuracy comparing

The sequence of sample	4	11	16	20
Practical data	0.1428	0.2428	0.3142	0.3714
BP network algorithm	0.1424	0.2413	0.3115	0.3670
Fuzzy BP network for PSO algorithm	0.1416	0..2425	0.3140	0.3709

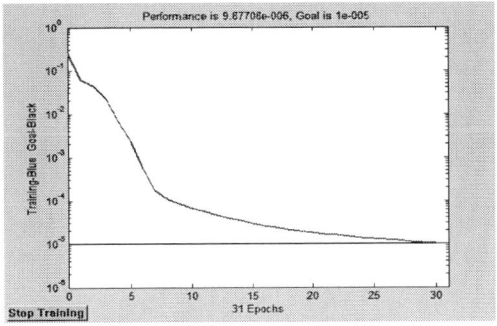

Figure 2 BP network algorithm

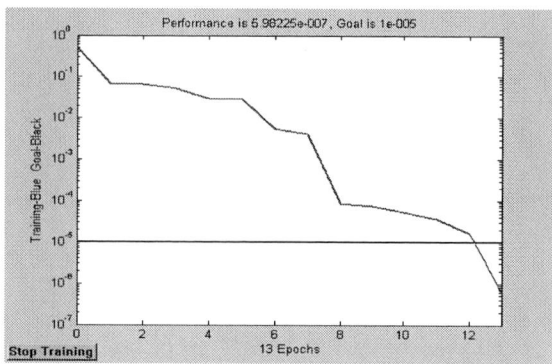

Figure 3 Fuzzy BP neural network improved by PSO algorithm

From table 5 we can see that fuzzy BP neural network for PSO algorithm is more accuracy than BP network algorithm, more closely the practical data. So we use fuzzy BP neural network for PSO algorithm to evaluate the measure sample as table 3 will be more accuracy. The evaluation result is 0.2108. From the data we can see that the degree of success is the second level.

4 Conclusions

The paper establish PCA and fuzzy BP neural network for PSO optimization model for the degree of success evaluation, it has more accuracy and speedier. The method is more efficiently than other traditional algorithm. It provides a reliable and credible ways to the power plant construction project in degree of success evaluation and enhances its work efficiency. It is also provides a reliable method for post-evaluation and give the operator more experience. This evaluation method can be also for the other project if we change some index.

This paper need more research about exploring information system software. It will be more efficiently for the operator to control the project.

Reference

[1] Zhang Huiying. "Evaluating Degree of Success in Highway Construction Projects Based on VariableWeight Grey Cluster", *Journal*, Highway, Beijing,2006.08, pp.241-246.

[2] Chen Hui and Li Jun. An Exploration into the Model of Success-degree Evaluation of Technical, *Journal*, JOURNAL OF WUYI UNIVERSITY (Natural Science Edition), Jiangmen, 2003.01 ,pp.44-47.

[3]Li Zuoyong, Ding Jing and Peng Lihong, *The principle and method of environmental quality evaluation*, Chemical Industry Press, Beijing, 2004.

[4]Zheng-miao,LI Jing and WANG Bi-ling, Evaluation on Environmental Quality of Heavy Metals in Soils and Vegetables Based on Geostatistics and GISXIE, *Journal*,

Chinese Journal of Environmental Science, Beijing, 2006.10, pp.2110-2116.

[5]ZHANG Wen-ge,GUAN Xin-jian and XU qing-shan, Water environment quality evaluation based on fuzzy nearness method, *Journal*, Water Resources Protection, Beijing, 2006.2, pp.19-22.

[6]WANG Xiao-peng,ZENG Yong-nian and DING Sheng-xi, Quality Assessment Model for Complex Environment System Based on Refinements to Principal Component Analysis, *Journal*, Systems Engineering-Theory & Practice,Beijing, 2005.11, pp.112-118.

[7]Lei Yingjie and Zhang Shanwen. *Application of genetic algorithm toolbox based on matlab*, Xi An, XIDIAN UNIVERSITY PRESS, 2005.

[8] J. Kennedy, R.C. Eberhart and Y. Shi, *Swarm intelligence*, Morgan Kaufmann Publishers, San Francisco 2001.

A Canceling Noise Research for Underground Mine Powerline Carrier Communication Based on Adaptive Theory

Shaoliang Wei, Jialin Cao and Yimin Chen
Mechanical & Electronic Engineering and
Automation Shanghai University, China
Wsl_ify@163.com

Shaoliang Wei, Fengyu Cheng, Deming Nie and
Hengwen Li
Shandong Science and Technology University,
China

Abstract

Because communication underground coal mine is difficulty, this paper analyzed the underground power line noise, and indicated that we can design the adaptive filter based on the adaptive theory and deal with the channel noise. Experiments show that, through the adoption of improved Adaptive Noise Canceling technology and DSP system we can achieve to remove the noisy and gain pure signal.

1. Introduction

The geological conditions underground coal mine are very complex and the working environments are poor because the mine roadway up dozens of kilometers, operating points spreading, mining machinery and electrical equipment and staff moving greatly, roadway excavation face narrow space. In addition, the large number of risks of explosion because of carbon monoxide, gas and coal dust and other air mixture are the great potential to cause a great accident. Therefore, the establishment of a better communication system for the coal mine underground is a very important significance to modernize the mine management, increase the labor productivity, and strengthen the security to protect the lives and property of the mines. At present, cable telephone communication, power line carrier communication, sensor communication, telecommunication, leakage communication, and radio communication are the main forms of communication in China's coal mine. Both at home and abroad, researchers have done c-certain researches on coal mine underground communication-on in recent years, but because of the special nature of un-reground coal mine environment, the research is smaller. Existing research is generally limited to the method of m-mobile communication. The study results show that, a variety of mobile communication methods have their limitary application under coal mine, and it is difficult to form a mine-wide communication network system under coal mine. As the power lines underground coal mine cover a broad area, if we can

make full use of power lines to communicate, that we can completely solve the problem underground communication. But there are lots of complex interference noise in power line communication under coal mine, so how to reduce and eliminate noise interference are the most important part of our research. The carrier of voice communication is canceling noise in this paper.

2. Power line noise analysis

The noise in power line channel is mainly generated from the work of the electrical equipments and environment noise. Different locations, different times, the work of the electricity network equipment are different, so power line channel noise is inevitable with alterable time and location. But because of a 24-hour work system under coal mine, large-scale operations equipment's switching is not frequently, so underground mine power line network brings advantageous communications. The surrounding environment noise is changed by different communication locations.

Power line channel noise under cola mine can be divided into five categories:

- Colored background noise
Generated by many small power source of the noise s-superposition, its power spectral density is decreasing with the increasing of frequency, and changes very slowly over time.
- Surrounding Environment Noise
The noise is produced by the surrounding environment which input to the receiver randomly when the voice is communicating.
- Periodic pulse noise (not synchronous with the frequency of power grid)
This noise is mainly produced by the switching power supply under coal mine, and most of them repeated by 50-200 kHz frequency. It is some discrete spectrum in the frequency domain.
- Periodic pulse noise (synchronous with the frequency of power grids)
It's produced by large DC equipments which exist under cola mine. The noise frequency synchronizes with t-he frequency of power grid, its repeat cycle is 50 Hz or

100 Hz, and its duration is short (a few ms), and the power spectral density decreasing with the increasing of lower frequency.

- Asynchronous impulse noise

It's produced by network equipment on the moment of a switch, its duration from several microseconds to milliseconds, and the time of arrival is random. Sometimes the power spectral density of it is 50 dB higher than the back-ground noise.

Frequency shift, line filter, and shielding technique are used in traditional carrier telephone under cola mine, these methods play a very good role in restraining the e-electromagnetic interference from network, but there still exists noise signal with the same frequency of voice signal, because of the channel has a strong time-varying characteristics, the traditional anti-jamming measures can not extract useful weak voice signal from the strong noise. With the development of adaptive theory and DSP hardware technology, it is possible to reduce steady noise which has the same frequency as the carrier telephone, and to extract weak voice signal.

3. The improvement of adaptive noise canceling

The core of the adaptive noise canceling technology is established on an adaptive filter of real-time, obviously tracking results, and good convergence. It adjusts the adaptive filter algorithm parameters and tracks noise changes in the process of the importation of statistical characteristic which is unknown or changeable to reach the best filtering effect. There are many adaptive filter algorithms, and the common use is the least squares algorithm (LMS), recursive least squares (RLS), improvement in the LMS algorithm, and so on. With the characteristics of simple structure, easy to implement, and real-time, LMS algorithm cooperates with DSP hardware circuit can eliminate the noise in voice at real-time.

Although the effect of other algorithm in noise canceling is better, its complex structure is not suitable for using in real-time canceling environment Figure 1 is a cross LMS filter noise canceling structure, the mainly input sampling signal is $d(t)$, which is a mixed-signal superimposed by useful signal $s(t)$ and noise signal $n_0(t)$. $x(t)$ is the sampling input signal, including the noise signal associated with $n_0(t)$. The output signal $y(t)$ comes from the input signal $x(t)$ superimposed by adaptive filter, and the compared error between $d(t)$ and $y(t)$ is

$$e(t) = d(t) - y(t) = s(t) + n_0(t) - y(t).$$

According to LMS adaptive theory, the error signal $e(t)$ should be the least to reach $y(t) \approx n_0(t)$ to eliminate the noise in the main input.

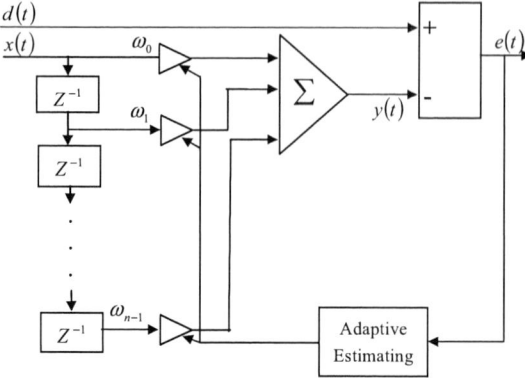

Figure1. Structure of adaptive noise canceling

Adaptive filter matrix requires to be solved under the principle of the least square error, through calculating the sampling point data $x(t) \sim x(t-N+1)$ and adjusting filter coefficients $\omega_0 \sim \omega_{n-1}$ to make the estimation error between the output signal $y(t)$ and the main sampling input signal to be the least. The relation between $d(t)$ and the estimation error $e(t)$ as follows:

$$
\begin{aligned}
e(t) &= d(t) - y(t) \\
&= d(t) - [\omega_0, \omega_1, \cdots, \omega_{n-1}]^T [x(t), x(t-1), \cdots, x(t-n+1)] \\
&= d(t) - WX(t)
\end{aligned}
\tag{1}
$$

$$W = [\omega_0, \omega_1, \cdots, \omega_{n-1}]^T ,$$
$$X(t) = [x(t), x(t-1), \cdots, x(t-n+1)]$$

The mean square value of $e(t)$ is:

$$E[e^2(t)] = E[[d(t)-WX(t)]^2] = E[d^2(t)] - 2WE[d(t)X(t)] + WE[X(t)X^T(t)]W^T$$

$E[e^2(t)]$ is a quadratic function and the filter coefficients is W, it has only a minimum value point which can be used to obtain the best filter coefficients W. The gradient of the mean square error and the filter coefficient vector is:

$$
\begin{aligned}
\nabla W\{E[e^2(t)]\} &= 2E\{e(t)\nabla W[e(t)]\} \\
&= 2E\{e(t)\nabla W[d(t)-WX(t)]\} \\
&= -2E[e(t)X(t)]
\end{aligned}
\tag{2}
$$

According to the fall theory of the most optimal control, first from a group of filter coefficients and choose a point in N-dimensional space to calculate the gradient of the point $E[]$, an then the gradient vector reverse at the minimum mean square error in the direction, so the filter coefficients recursive formula is derived as follows:

$$
\begin{aligned}
W_{j+1} &= W_j - (\beta/2)\nabla W\{E[e^2(t)]\} \\
&= W_j + \beta E[e(t)X(t)]
\end{aligned}
\tag{3}
$$

β is the convergence factor, used to control the speed of convergence. Actual calculation of a stochastic gradient $e(t)X(t)$ can be used to approximate calculate instead of $E[e(t)X(t)]$, that is $W_{j+1} = W_j + \beta e(t)X(t)$.

Because the $X(t) = \nabla W\{E[e(t)]\}$ is the same gradient direction as $\nabla W\{E[e(t)]\}$, the first recursive algorithm is only estimate the filter coefficient, and ultimately converges with the calculation of the filter coefficient tracking study.

Figure 1 shows the structure of the Adaptive Noise Cancellation, and the need for adaptive estimation closely related to the noise reference signal $X(t)$, but in the actual underground communication process, it is difficult to be associated with the noise reference signal. Therefore, in Figure 2 we have adjusted the traditional adaptive filter structure. $d(t)$ is used only for system input, as the frequency band width of voice signal is inconsistent with noise signal, and the scene strong interference signal is general for broadband signal, we can use this structure in Figure 2 to enhance the carrier signal drown-ed in noise. Figure 2 is the same adaptive filter structure as Figure 1. Z^{-M} is the delay unit. If the delay unit output is $d(t-M) = x'(t)$, the adaptive filter characteristic-s is the same as Figure 1, and then the following equation can be given:

$$e(t) = d(t) - y(t) \tag{4}$$
$$= d(t) - [W_0, W_1, \cdots, W_{n-1}]^T[x'(t), x'(t-1), \cdots, x'(t-n+1)]$$

$$W_{j+1} = W_j + \beta e(t)X'(t) \tag{5}$$
$$= W_j + 2\mu e(t)X'(t)$$

μ is the convergence factor in this equation. $e(t)$ is the best approximation in the minimum mean square error criterion when $E[]$ reached to the minimum. We ca-n extract $y(t)$ when the output of $y(t)$ is the carrier signal after the random noise was filtered out

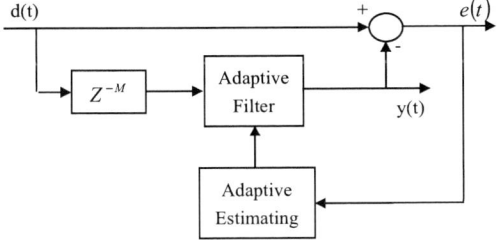

Figure 2. Development structure of adaptive noise canceling

4. Experimental research findings

By using DSP technology in laboratory, we have designed a carrier telephone (in Figure 3) by using adaptive filtering technology. The core of the adaptive filter is included in the DSP part in this figure. We have chosen the T-MS320C30 made by the TI Company of United States as the central processing chip in our hardware circuit. The filter structure is in accordance with the design shown in Figure 2. The data of Adaptive Filter is processing respectively in the sending end and the receiving end, the AD converter samples the voice signal and noise signal and sends to the communications channel in the sending end. On the receiving end, DA converter directly samples the carrier IF. In the filter structure shown in Figure 2, the n-noise repeated by the carrier frequency signal in the process of enhancing and demodulated by audio power amplifier to drive speaker.

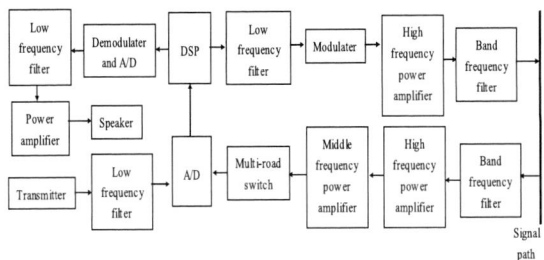

Figure 3. Structure of carrier telephone with DSP noise canceling

In experimental course of the study, the convergence condition of μ in （5）is: $0 < \mu < \dfrac{1}{\lambda_{max}}$, λ_{max} is the largest eigenvalue of reference input auto-correlation matrix, we value $0 < \mu < 0.04$ in accordance with the input power to meet $0 < \mu < \dfrac{1}{P_{max}}$ in this system when computing. Test results show that the maximum value of μ is the convergence critical value and the value impacts the algorithm imbalance greatly. In the way of reducing μ may improve the degree of imbalance, but it may also slow down the convergence. In order to resolve the contra-diction between convergence speed and the imbalance, w-e have processed the voice signal by sub-processing and have designed a fast convergence method, we value $\mu_{max} = 0.16$ in order to fast the adaptive filter to converge in a short period of time at the beginning, then after the convergence, we convert $\mu = 0.01$ smoothly to reduce the level of imbalance in the later calculation. The reasonable choice of N is the important indicator to reduce the noise signal, increasing N can increase the use of information and iterative calculation more accurately when N is small, but when N reach to a certain value, the increasing N will dramatically increase the computing capacity, and then the offset and noise output will increase, so we value N=31 in this experimental study to achieve the best result of filtering. In Fig.4a, we gain a sound recording signal. In Fig.4b, the sound recording signal is translated t-through carrier telephone without DSP noise canceling. In Figure4c, the calculation of DSP accords to the adaptive filter structure of Fig.2, and the carrier frequency signal from the noise superimposed carrier signal whose frequency is 50kHz has been strengthened.

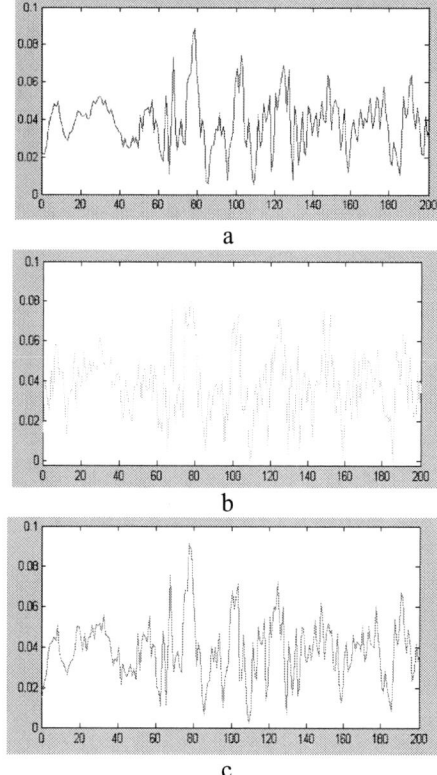

Figure 4. Transmitting and receiving signal
(a. A sound recording signal. b. Signal without DSP noise canceling. c. Signal with DSP noise canceling)

Experiments show that, Adaptive Noise Cancellation technology in the telecommunications carrier to eliminate noise and extract useful signal is a very effective method. But how to use the technology better to the high-speed digital communication under coal mine is worth of further studying.

References

[1] Shao-liang Wei, "A canceling noise research on carrier telephone of electric locomotive underground," 1997.
[2] GERALD SCHICKHUBER, OLIVER MCCARTHY. "Control Using Power Liner-A European View," IEEE Transactions On EMC. Vol. 16, No. 2, 1997, pp. 16-19

[3] "Summary of an IEEE Guide for Power-Line Carrier Applications,"IEEE Trans. Power Apparatus and Systems, vol. PAS-99. No. 6, Nov.-dec. 1980, pp. 2334-2337
[4] J. Newbury, "Communications requirements and standards and for low votlage signaling,"IEEE Trans. Power Delivery, vol. 13,1998,pp. 46-52
[5] Amitava D.R. "Networks for Homes," IEEE Spectrum, vol. 12 ,No.36, pp. 42-49
[6] John N. William M. "Polential Communication Services Using Power Line Carriers and Broadband Intergrated Services Digital Network, " IEEE Transaction on Power Dslivery, vol. 4, 1999,pp. 1197-1201.

Research on Circulation of Parallel Three-Phase Converters in MW Wind Power System

Chunxue Wen [1,2], Jianlin Li [1], Xiaoguang Zhu [1], Honghua Xu [1]

1 Institute of Electrical Engineering Chinese Academic of Sciences, Beijing, China
2 Graduate University of Chinese Academy of Sciences, Beijing, China
E-mail: wcx@mail.iee.ac.cn

Abstract

In this paper, modeling of parallel three-phase converters is first presented. Then definition of the circulating currents of paralleled converters and the circulating-current-generating mechanism are clearly explained. Thus, based on this definition, an averaged model of the circulating current is proposed. It is seen from this model that the circulating current consists of mainly the zero-sequence components. The governing differential equation also shows explicitly the relation between the circulating currents and the affecting factors such as different pulse width-modulation strategies, different circuit parameters etc. With this understanding, a simple coordinate control is then presented to reduce the circulating current. The phenomenon of the intrinsic circulating current is also explained. Furthermore, a parallel converters system used for wind power application is constructed, and the proposed control is implemented. Both simulation and experimental results verify the validity of the proposed theory and control.

1. Introduction

With wind power generation system in the stand-alone capacity expansion, including double-fed and direct-drive wind-power generation system, the converter has the capacity to MW-class level. Because of power level in existing devices and switching frequency and many other constraining factors, the converter directly parallel way with simple, modular design and easy to extend the benefits is an ideal solution to raise power levels of converters.

Parallel converters can increase power levels, system reliability and efficiency and reduce switching losses and the current & voltage ripples [9]. In addition, parallel structure is suitable for the modular system, easy re-configuration system, and improves the flexibility of the system [10]. However, the parallel converters, regardless of the converters switches in how to conduct the closed-loop control, circulating current will be automatically generated. The circulating current between the converters will lead to current distortion, the unbalanced load and the overall system performance decline.

In order to avoid circulation, the traditional method is to use an independent AC or DC power supply, or in AC side of parallel system using a transformer. However, due to an increase of isolation transformers or additional power supply, the system has become huge in size and the higher cost. In order to reduce the size of parallel system and the construction costs, the AC or DC parallel side should use phase reactors to provide high-zero sequence impedance [1][2]. However, the reactors only have higher impedance in some high frequency, and low-frequency circulation can not be inhibited. In view of this shortcoming, now is studying three-phase PWM Rectifier parallel system of staggered intermittent voltage space vector (SVM) modulation to improve it [3]. This SVM modulation method without zero vectors eliminates the impact of the pure zero-sequence current circulation in parallel circuit. On the other hand, the method based on the literature [4] added a new variable to adjust and control the vector of time rather than completely removing zero vectors. However, when the system is running at saturation, zero vectors role must be canceled [5]. In addition, the literature [5][6] also introduced dq axis nonlinear control strategy for each independent converter to limit net zero sequence current. However, almost no literature on general circulation models of parallel converters in MW wind power system is proposed. More about the general integrity circulation model and optimizing control study are needed.

In this paper, the three-phase parallel PWM Boost Rectifier which applies to the MW class wind power generation system is taken for example, a precise definition of circulation is given, and parallel system of the average circulation model is also proposed. The mechanism of circulation can be learned through this model, and the various reasons such as PWM modulation strategy that cause circulation can have a better understanding. According to circulation models, a coordinate control approach to reduce the circulation is given in this paper. The model study found that, as long as the various parallel converters is not fully synchronized control, the circulation current which is called inherent circulation will exist, even in the coordinate control.

2. Derivation of Circulating Current Model
2.1. Definition of Circulating Current

To illustrate the definition of circulating current, first of all the single-phase system is shown in Figure 1. Two single-phase converters are parallel. When the two converters in the same load are under the synchronized control, $I_1=I_2$, that is without circulating current between

parallel circulation links. Otherwise, I_1 will be not equal to I_2. Circulating current C1 will loop in converter 1 and is given in figure 1(a), $C_1 \triangleq (I_1 - I_2)/2$. On the other hand, when considering converter 2 in figure 1 (b) below, the circulating current is defined as C_2, $C_2 = (I_2 - I_1)/2$.

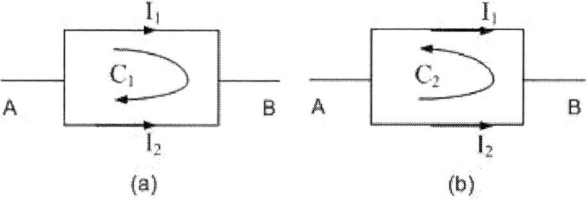

(a)   (b)

Fig. 1 Definitions of circulating currents in two converters
(a) C_1 for converter 1 and (b) C_2 for converter 2

As shown in the three-phase Boost Rectifier parallel system in Figure 2, circumstances of the circulation are more complicated in comparison with the aforementioned single-phase system. For the sake of simplicity, a-phase analysis is taken as an example. Assuming that in a certain period of time, $i_{ak}>0$, $i_{bk}<0$, $i_{ck}<0$, $k \in \{1,2\}$ set up. In addition, assuming that the two converters switch state is same, and the two converters fully synchronize operation. Figure 2 indicates that the loops connecting voltage source e_a and e_b have four:

Loop 1: A—A_1—S_{11}—P—N—S_{51}—B_1—B—O—A
Loop 2: A—A_1—S_{11}—P—N—S_{52}—B_2—B—O—A
Loop 3: A—A_2—S_{12}—P—N—S_{51}—B_1—B—O—A
Loop 4: A—A_2—S_{12}—P—N—S_{52}—B_2—B—O—A

For loop 2 and 3 all loop current flows through another converter to form a circulating current. In addition, for the loop A—A_1—S_{11}—P—N—S_{12}—A_2—A, although no voltage source, the circulation may be arising from the current line redistribution. If the two converters do not operate fully synchronized, in the following loop 5: A—A_1—S_1—P—N—S_{42}—A_2—A will appear another kind of circulation. In the loop 5, the EMF is the only DC output voltage. Circulation wave is decided by PWM switching function. Similarly, this circulation also exists in the b and c phase.

In the understanding of the mechanism of circulation the definition of circulating current can be extended to a three-phase n Boost Rectifier parallel system. The k-phase circulation of j-converter consists of the circulation component C_{kjm}.

$$C_{kj} = C_{kj1} + C_{kj2} + \cdots + C_{kjn} \triangleq \sum_{m=1}^{n} C_{kjm} \qquad (1)$$

$$C_{kjm} \triangleq \frac{i_{kj} - i_{km}}{n}, k \in \{a,b,c\}; j,m \in \{1,2,\ldots,n\}$$

Where, i_{kj} is the k-phase current of j-converter, $C_{kjj}=0$. According to the definition of circulating current, we can

see that the circulation is in connection with the oscillation of switching process.

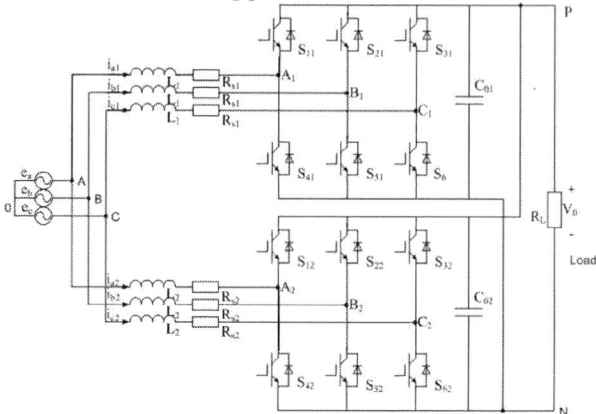

Fig. 2 three-phase Boost converter parallel system

2.2. Modeling of Circulating Current

For the analysis of the circulation model, the average state space can be used, and assuming that all the switches are ideal switch. Then a-phase equation of parallel converter can be given as following through figure 2.

$$\begin{cases} L_1 \dfrac{di_{a1}}{dt} + i_{a1} R_{s1} = e_a - d_{11} v_0 - v_{NO} \\[2mm] L_2 \dfrac{di_{a2}}{dt} + i_{a2} R_{s2} = e_a - d_{12} v_0 - v_{NO} \qquad (2) \\[2mm] d_{qj} + d_{q+3,j} = 1, j \in \{1,2\}, q \in \{1,2,3\} \end{cases}$$

Where, constants L_1, L_2, R_{s1}, R_{s2} are two converters equivalent series inductance and resistance. Variables d_{11}, d_{12} are the duty cycle function of the first converter switch (for connecting a-phase). The equation $d_{qj} + d_{q+3,j} = 1$ indicates that the switches of each converter under the same bridge are complementary.

Then the circulation matrix expression corresponding to a-phase in different converters can be drawn as following.

$$\begin{bmatrix} C_{a1} \\ C_{a2} \end{bmatrix} = \frac{1}{2} \begin{bmatrix} -\dfrac{1}{Z_1} & \dfrac{1}{Z_2} \\[2mm] \dfrac{1}{Z_1} & -\dfrac{1}{Z_2} \end{bmatrix} \begin{bmatrix} d_{11} v_0 \\ d_{12} v_0 \end{bmatrix} + \frac{1}{2} \begin{bmatrix} \dfrac{1}{Z_1} - \dfrac{1}{Z_2} \\[2mm] -\dfrac{1}{Z_1} + \dfrac{1}{Z_2} \end{bmatrix} (e_a - v_{NO})$$

$$(3)$$

Where, $Z_j = L_j \dfrac{d}{dt} + R_{sj}, (j=1,2)$. Similarly, the expressions of b and c phase are also available to obtain.

According to the model above, we can see that DC output voltage v_0 and all the AC voltage e_a, e_b, e_c are major voltage sources causing circulation. Moreover, the converter switching function will have different switch states that can also produce closed current flow paths between two converters when the different parallel control strategy or PWM modulation methods are used. Inductance and equivalent impedance Z_k which connected each converter in each closed path will also affect magnitude of circulating current. Formula (3) shows that AC voltage source will not affect the circulation when $Z_1=Z_2=\cdots=Z_n$. If all converters have the same structure and use the synchronous control methods, $d_{1k}=d_{1j}$, $d_{2k}=d_{2j}$, $d_{3k}=d_{3j}$, $k,j \in \{1,2,\ldots,n\}, k \neq j$, the circulation will not exist in the system. Certainly, this is only a perfect condition, it is impossible to have this kind of situation actually. It is worth mentioning that although the converter structure may be the same and the control may also be synchronous, the circulation will still produce in loops because of the difference of the inductance parameters in the lines. In addition, this circulation is inherent and unable to eliminate through controllers.

2.3. Model Simulations

In order to explain the model significance in formula (3), the following simulations are given. Firstly, for the simplified analysis, assuming that $Z_1=Z_2 \triangleq Lp + R$ (p is differential operator), then according to formula (3) the circulation model of one of converters (such as No.1) can be obtained to be as follows.

$$(Lp + R)\begin{bmatrix} C_{a1} \\ C_{b1} \\ C_{c1} \end{bmatrix} = \frac{1}{2}\left(\begin{bmatrix} d_{11} \\ d_{21} \\ d_{31} \end{bmatrix} - \begin{bmatrix} d_{12} \\ d_{22} \\ d_{32} \end{bmatrix} \right)v_0 \qquad (4)$$

Formula (4) has given three first-order linear differential equations and does not include e_a-v_{NO} item compared to formula (3) because of $Z_1=Z_2$. After the switching duty cycle function and v0 in right side are decided, the circulation can be obtained directly. The simulations are executed by PSIM and MATLAB software in the circuit as shown in figure 2, and the switching duty cycle function and v_0 as input are connected to another circuit module. Figure 3(a) shows the connection in detail, where n=2, $Z_1=Z_2$. In addition, Figure 3(b) gives internal structure of $1/Z_1$ and $1/Z_2$ modules. Simulation parameters of the model in figure 3 are given as follows.

$$L_1 = L_2 = 1.6mH, R_{s1} = R_{s2} = 0.2\Omega, \omega = 100\pi rad / s$$
$$C_{01} = C_{02} = 2200\mu F, R_L = 10\Omega, v_0 = 250V$$
$$\begin{cases} e_a(t) = 100\cos(\omega t) \\ e_b(t) = 100\cos(\omega t - 120^0) \\ e_b(t) = 100\cos(\omega t + 120^0) \end{cases}$$

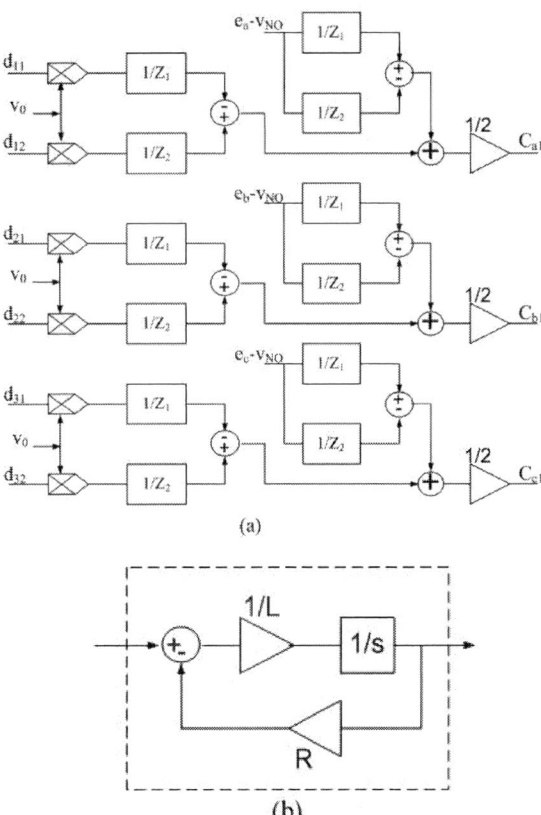

(a)

(b)

Fig. 3 the circulation model of the first converter in parallel system

(a) Simulation block diagram; (b) 1◁Zψmodule diagram

The simulation results for the model in figure 3 are introduced. Owing to contents constraint in this paper, only three examples are given to illustrate the validity of the proposed circulation model.

Example1 — equal switching frequency and load sharing but with 180° phase shift: To simplify the analysis, simulation use SPWM modulation method, switching frequency is 5kHz, and t is from 0 to 100ms under the zero initial condition. Three phase circulation waveforms C_{a1}, C_{b1}, C_{c1} in the first converter are given in figure 4.

351

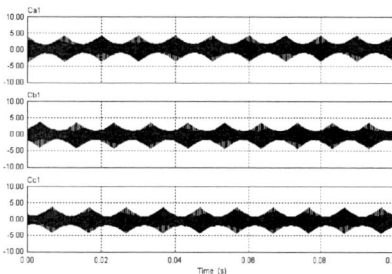

Fig. 4 Three phase circulation waveforms C_{a1}, C_{b1}, C_{c1}

Example 2—Unbalanced load sharing: Under the same simulation conditions as example 1, the two converters have different load, and the load ratio corresponding to two converters is 1.2:1.0. The simulation results are shown in figure 5, and the simulation calculation method is also same as example 1.

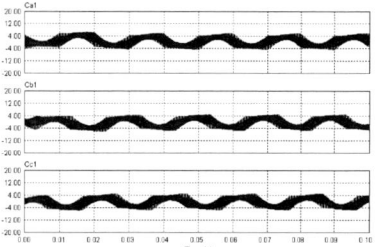

Fig. 5 Three phase circulation waveforms C_{a1}, C_{b1}, C_{c1}

Example 3 — Unequal switching frequency: The simulation conditions are similar to example 1, but the switching frequency of second converter is 10kHz. Corresponding simulation circulation waveforms are shown in figure 6.

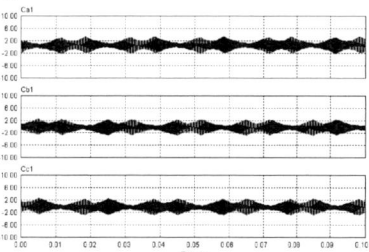

Fig. 6 Three phase circulation waveforms C_{a1}, C_{b1}, C_{c1}

To sum up, it is seen that the proposed circulation model can indeed properly predict each phase circulating current in each converter.

3. Coordinate Control Strategy

From the circulation model and some simulation results it can be seen that a large circulating current always exists although each converter uses closed loop control. The current can cause line current waveforms distortion, lower power quality and system efficiency, more power loss and overload of converters. Therefore, when the converters are parallel it is needed to reduce the circulation by coordinate control. In order to get a useful model for coordinate control, the circulation model of the nth converter needs to be decoupled to positive-sequence,

negative-sequence and zero-sequence components by Fourier series as follows.

$$
\begin{bmatrix} \vec{C}_{Zn} \\ \vec{C}_{Pn} \\ \vec{C}_{Nn} \end{bmatrix} = \frac{1}{3} \begin{bmatrix} 1 & 1 & 1 \\ 1 & e^{j2\pi/3} & e^{-j2\pi/3} \\ 1 & e^{-j2\pi/3} & e^{j2\pi/3} \end{bmatrix} \begin{bmatrix} \vec{C}_{an} \\ \vec{C}_{bn} \\ \vec{C}_{cn} \end{bmatrix} \quad (5)
$$

Where, $\vec{C}_{Zn}$, $\vec{C}_{Pn}$, $\vec{C}_{Nn}$ are behalf of zero-sequence (ZSCC), positive-sequence (PSCC) and negative-sequence (NSCC) circulation vectors of the nth converter. According to the formula (8) (10), the matrix model is as follows:

$$
\begin{bmatrix} \vec{C}_{Zn} \\ \vec{C}_{Pn} \\ \vec{C}_{Nn} \end{bmatrix} = \frac{1}{3n} \begin{bmatrix} \vec{D}_{Z1} & \vec{D}_{Z2} & \cdots \vec{D}_{Zn} \\ \vec{D}_{P1} & \vec{D}_{P2} & \cdots \vec{D}_{Pn} \\ \vec{D}_{N1} & \vec{D}_{N2} & \cdots \vec{D}_{Nn} \end{bmatrix} \begin{bmatrix} 1/Z_1 \\ 1/Z_2 \\ \vdots \\ -(n-1)/Z_1 \end{bmatrix} v_0 + \frac{1}{3n} \begin{bmatrix} X_Z \\ X_P \\ X_N \end{bmatrix} (\frac{-1}{Z_1} + \frac{-1}{Z_2} + \cdots + \frac{n-1}{Z_n})
$$

$$(6)$$

Where, $\vec{D}_{Zk}$, $\vec{D}_{Pk}$, $\vec{D}_{Nk}$ are ZSCC, PSCC, NSCC vectors of the nth converter, $\vec{E}_a$, $\vec{E}_b$, $\vec{E}_c$ and $\vec{V}_{NO}$ are vectors corresponding to e_a, e_b, e_c and v_{NO}, For harmonic components $\vec{E}_a$, $\vec{E}_b$, $\vec{E}_c$ are zero.

The reduction analysis of formula (6) indicates that in order to make ZSCC and NSCC circulation eliminated the zero-sequence component of each converter line current should be controlled to zero. Control methods can be used simple line currents feedback control. For positive-sequence component if a-phase current of each converter meets $i_{a1}(t) = i_{a2}(t) = \cdots = i_{an}(t) \neq 0$, PSCC does not exist.

On the basis of aforementioned theory analysis we can see that coordinate control should be to eliminate ZSCC and NSCC, and PSCC should be controlled by load balance. The coordinate control block diagram using simple PI control is given as follows.

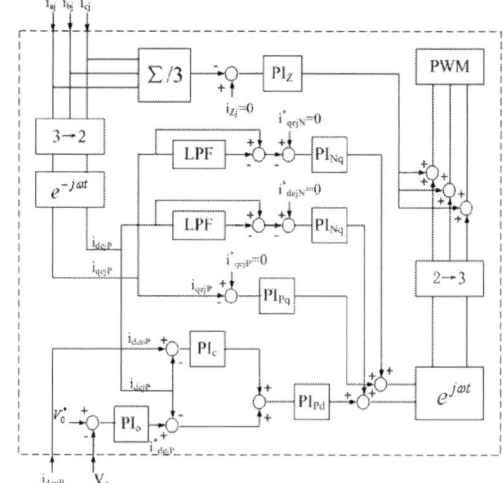

Fig.7 Control block diagram

The figure tells us that one converter is the reference to another, and PI_P, PI_Z, PI_N controllers restrain PSCC, ZSCC and NSCC alternatively. The simulation results to parallel converters systems shown in Fig. 2 are given as follows, and the simulation conditions are shown in table 1.

Table 1 simulation parameters

parameters	Converter 1	Converter 2
frequency	5kHz	10kHz
R_s	0.2 Ω	0.3 Ω
L	1.6mH	2mH
C_0	2200uF	2200uF
AC input voltage	$e_a(t) = 100\cos\omega t V$ $e_b(t) = 100\cos(\omega t - 120^0)V$ $e_c(t) = 100\cos(\omega t + 120^0)V$	
R_L	10 Ω	
DC output v_0	250V	

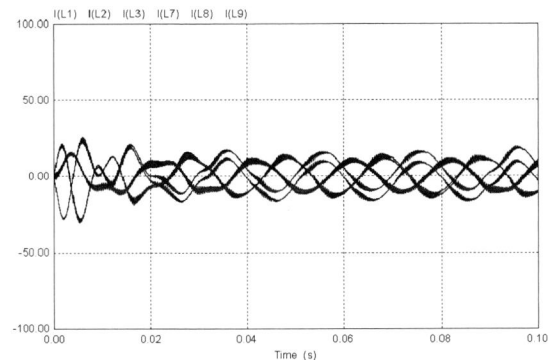

Fig. 7 three-phase input current of parallel system without coordinate control

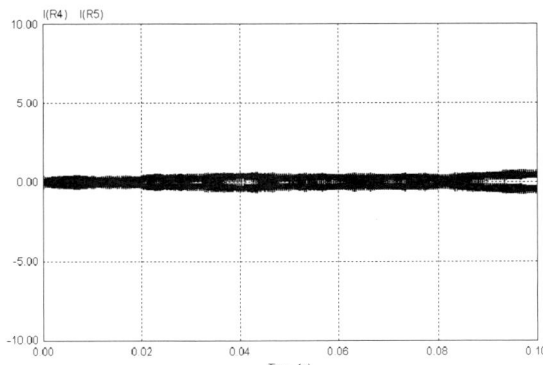

Fig. 8 zero-sequence circulating current without coordinate control

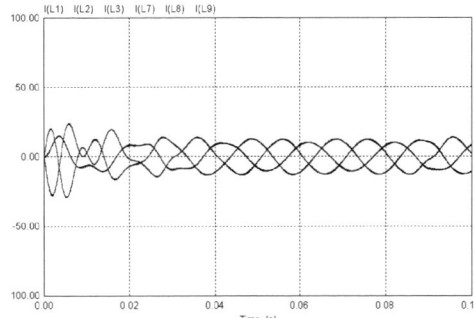

Fig. 9 three-phase input current of parallel system under coordinate control

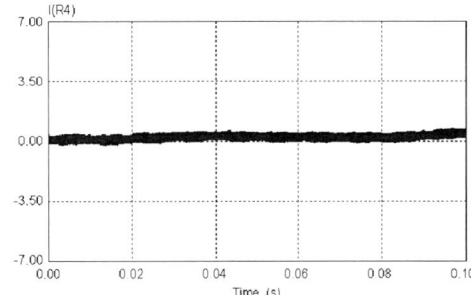

Fig. 10 zero-sequence circulating current under coordinate control

The simulation figures above describe the impacts of circulation before and after using coordinate control and indicate that this control method can reduce the circulation greatly. Moreover, we can see that an inherent circulation exists because of different parameters of parallel converters, the actual switching speed limitation, and limited regulation bandwidth.

4. Experimental Results

In order to verify these theories, the parallel experimental system is given as follows. IPM (PM100DSA60) and DSP (TMS320F2407) are used. Experimental conditions are similar as simulation conditions.

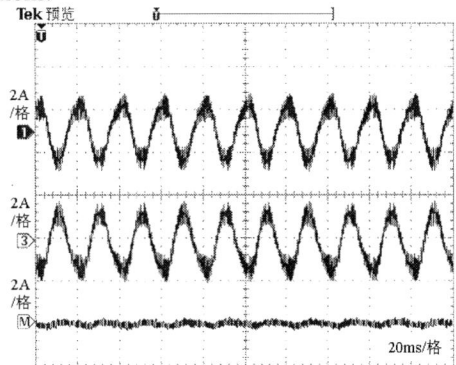

Fig. 11 two phase current of one converter and ZSCC without coordinate control

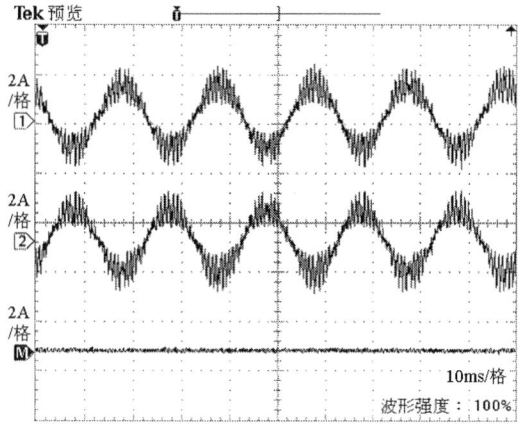

Fig. 12 two phase current of one converter and ZSCC under coordinate control

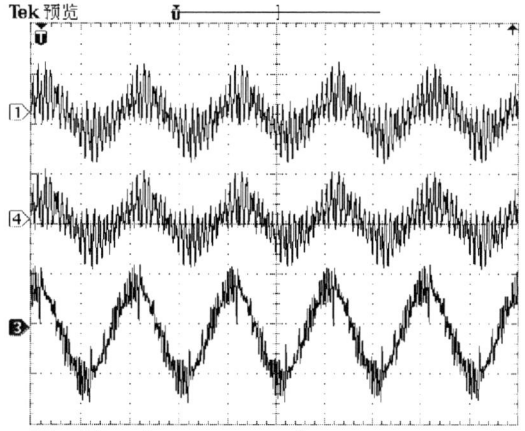

Fig. 13 a-phase currents of two converters and synthesizing current of parallel system

Figure 11 and 12 show the line currents and ZSCC of two converters without and under coordinate control. These results indicate that ZSCC is obvious without coordinate control and attenuate greatly as well as two converters achieve load current sharing under the control. Moreover figure 13 gives a-phase currents of two converters and their synthesizing output current. Compared to line currents of two converters alternatively, the distortion of the synthesizing current is reduced significantly, and this experimental result makes clear that parallel system can enhance power quality.

5. Conclusion

A precise circulating current definition of parallel converters is given and the circulation models including ZSCC, PSCC and NSCC are established by definition in this paper. Moreover, this paper also specifies the mechanism of the circulation and proposes the coordinate control method correspondingly. Finally, two Boost rectifiers parallel topology suitable for MW wind power system is given. Meanwhile, simulations and experiments based on this topology are also given to verify the validity of theory in this paper.

References

[1]. Y. Sato and T. Kataoka, "Simplified control strategy to improve Ac-input current waveform of parallel-connected current-type PWM rectifiers," *Proc. Inst. Electr. Eng.-Electr. Power Appl.*, vol. 142, no. 4, pp. 246–254, Jul. 1995.

[2]. K. Matsui, Y. Murai, M. Watanabe, M. Kaneko, and F. Ueda, "A pulse width-modulated inverter with parallel-connected transistors using current-sharing reactors," *IEEE Trans. Power Electron.*, vol. 8, no. 2, pp. 186–191, Apr. 1993

[3]. K. Xing, F. C. Lee, D. Boroyevich, Z. Ye, and S. K. Mazumder, "Interleaved PWM with discontinuous space-vector modulation," *IEEE Trans. Power Electron.*, vol. 14, no. 5, pp. 906–917, Sep. 1999

[4]. Z. Ye, D. Boroyevich, J. Y. Choi, and F. C. Lee, "Control of circulating current in two parallel three-phase boost rectifiers," *IEEE Trans. Power Electron.*, vol. 17, no. 5, pp. 609–615, Sep. 2002

[5]. S. K. Mazumder, "A novel discrete control strategy for independent stabilization of parallel three-phase boost converters by combining space vector modulation with variable-structure control," *IEEE Trans. Power Electron.*, vol. 18, no. 4, pp. 1070–1083, Jul. 2003

[6]. Mazumder, S.K, "Continuous and discrete variable-structure controls for parallel three-phase boost rectifier," *IEEE Trans. Ind. Electron.*, vol. 52, no. 2, pp. 340–354, Apr. 2005

[7]. J. M. Guerrero, J. Matas, L. G. De Vicunagarcia De Vicuna, M. Castilla, and J. Miret, "Wireless-control strategy for parallel operation of distributed-generation inverters," IEEE Trans. Ind. Electron., vol. 53, no. 5, pp. 1461–1470, Oct. 2006.

[8]. Z. Ye, K. Xing, S. Mazumder, D. Borojevic and F. C. Lee. "Modeling and Control of Parallel Three-phase PWM Boost Rectifiers in PEBB-Based DC Distributed Power Systems". IEEE APEC. Vol. 2, pp. 1126-1132,. Feb. 1998

[9]. Ching-Tsai Pan,Yi-Hung Liao, "Modeling and Coordinate Control of Circulating Currents in Parallel Three-Phase Boost Rectifiers". IEEE TRANSACTIONS ON INDUSTRIAL ELECTRONICS, VOL. 54, NO. 2, pp. 825-838. APRIL 2007

[10]. T.-P. Chen, "Circulating zero-sequence current control of parallel three-phase inverters". IEEE Proceedings. Vol. 153, pp. 282-288. Mar. 2006

SOA Based Electric Power Real-time Data Warehouse

Cuiru WANG

Dept. of Computer, North China Electric Power University, Baoding, 071003, China
hdwcr@126.com

Shuangxi LIU

Dept. of Computer, North China Electric Power University, Baoding, 071003, China
hdlshx@126.com

Abstract

At present, traditional data warehouse can no longer satisfy the electric power enterprise's decision support, real-time data warehouse can provide the electric power enterprise with daily tactical operation technical support, and the article discussed the real-time data warehouse implement method, proposed feasible real-time data warehouse architecture based on SOA. This SOA based real-time data warehouse architecture uses the web service to pack the various source electric power database systems, and the various changed data was captured by the data capture web service. When it comes to the data transformation and flow, the update strategy based on message queue and xml are used for the real-time updating. Also the multi-level real-time data cache is used for real-time data storage. Using this real-time data warehouse architecture, we can implement the electric power real-time data warehouse easily.

1. Introduction

The changeable market makes the information and knowledge becomes the most valuable resources of electric power companies. Using advanced information technology to integrate existing software and hardware resources, lets the personnel and equipment in best running status, accurately understanding current production situation and the forecast market demand, has become a important way for the electric power enterprise to enhance the industry competition [1]. Therefore the real-time data warehouse not only act as the electrical network movement recorder, but also provides formidable data foundation for the real-time analysis and the high-level application[2].However, because of the information construction's imbalance and independence between various areas and the department caused the present electric power enterprise informationization can not to be able to construct the effective real-time data warehouse system using the traditional data warehouse plan, and provided the enterprise with decision supporting.

Real-time data warehouse is the combination of the real-time behaviors and the data warehouse [2]. Currently, the dominant method of replenishing data warehouses and data marts is to use extraction, transformation and load (ETL) tools that "pull" data from source systems periodically at the end of a day, week, or month and provide a "snapshot" of your business data at a given moment in time. A real-time data warehousing solution and framework can commonly be divided into three fundamental tiers with data flows between them. The three layers are Presentation Layer, Architecture Layer, and Middleware Layer. These tiers or layers must be seamlessly integrated and function as one to ensure the immediate success and long-term benefits of a real-time data warehouse. The most important technology is the real-time data update of the data warehouse. But the traditional ETL process can not support the exact real time data updating. After the SOA technology appeared, we can do the real-time acquisition based on SOA and web service. Most researchers proposed the CTF (Capture-Transformation-Flow) process to update date in real-time data warehouse. CTF is a simple and effective data transfer technology in the heterogeneous systems [3]. This is the real-time date updating process. Data can be transferred to data warehouse phase table from the OLTP system in a very low-delay when changed. This article proposed an electrical power real-time data warehouse based on SOA (service-oriented architecture), it can provide the enterprise with real-time data service, and moreover the SOA technology can develop the real-time data warehouse based on the existing system investment.

2. Real-time data warehouse architecture

Figure 1 describes the real-time data warehouse architecture, components and functions as follows:

1) OLTP systems: application databases needed to be integrated, including generate electricity information database, user database, electric power information database, contract information database and so on.

2) Change data capture (CDC) and ETL: the real-time data of the OLTP systems was captured by the CDC service, non-real time data was loaded to the data warehouse by the ETL service.

3) Real-time data cache: all real time data from the OLTP systems was stored in the real-time data cache.

4) Data warehouse: All static data was stored in the data warehouse.

5) Real-time data integrate (RDI): data in each real-time data cache will be integrated by RDI service.

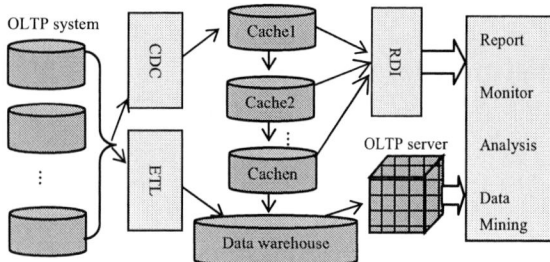

Figure 1. The architecture of real-time data warehouse

6) OLAP server: Use for support complicate analyze operations, and can do celerity, neatly query process according to the analyzer's need.

7) BI tools: include real-time data query, monitor, analyze application, report forms and the OLAP analyze.

3. Data capture based on web service

Today, more and more businesses using a data warehouse are beginning to realize they cannot achieve point-in-time consistency without continuous, real-time change data capture. There are several techniques used by data integration/replenishment software to move data. Essentially, integration tools either push or pull data on an event driven or polling basis.

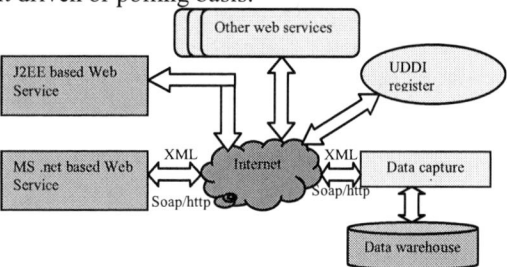

Figure 2. Data capture base on web service

In this article, we use Web Services as the base data capture structure. Figure 2 described the data capture system including many distributed isomerous subsystems (various OLTP systems), a data gather service and the UDDI registration organization. Each isomerous subsystem has a Web Services server, Web Services shield the isomerous subsystem's internal detail, announces its service connection outward, can respond the gather service's request, and provide the service. UDDI (Universal Description Discover and Integration) is the register center on the internet, and provide the register service. Web Services provides the registration service on Interact the organization. Users can look up for the available Web Services through UDDI registration center. The function of the data gather service is capturing the changed data from each Web the Services. The data gather service exchange the XML data with Web service through SOAP/http (Simple object access protocol/Hypertext Transfer Protocol).

3.1. Define Web Services interface by WSDL

The WSDL provides a kind of pattern and format to describe Web Services. WSDL separate the Web Services' abstract functional description with the detail of the concrete realization. WSDL first describes the message format of the service provider and service user. The message itself is described abstractively, and then ties up to the concrete network protocol and the message format. The message is composed of certain type's data item. The multiple exchange messages between the service provider and user is described as operations. The operation set is called the port type. The port set is called service type, in which each port carries out a port type, but service type contains all details of the service interact.

3.2. The realization and release of the Web Services interface function

Because the function provided has already existed in the system, the current duty is the compilation of few codes, installs a shell for the original system, and encapsuleds the original system. Mainly realizes two functions in the procedure. Firstly, analyze the request message of the Web Services client; secondly, encapsulated the result of the system's component function by XML grammar (WSDL); finally transmits for the Web Services client side. After the Web Services function realized, we need to release the service. Register the function need to be released in the UDDI registration center, and then the user may discover the service needed in the UDDI registration center. In this article, WSDL is sent to the data gather service, is also so-called static binding. After the Web Service function realized and released, users can access it through internet.

3.3. Data capture

The data gather service module captures the changed data by Web Service. Web Services receiving the asynchronous information in XML form. In this paper, the data gather service is complete by Web Services, uses the XML documents to carry on the asynchronous exchange. When gather service needs to gather the data, it translates the content catalog into the XML format query message. The Web Services will return response messages after accepts the request. The response message is encapsulated a SOAP envelope.

The main title (on the first page) should begin 1-3/8 inches (3.49 cm) from the top edge of the page, centered, and in Times 14-point, boldface type. Capitalize the first letter of nouns, pronouns, verbs, adjectives, and adverbs; do not capitalize articles, coordinate conjunctions, or prepositions (unless the title begins with such a word). Leave two blank lines after the title.

4. Data transmit based on XML and message queue mechanism

Message queue service is a kind of loose coupling distribute application integration mode. In this mode, the sending and receiving are asynchronous. That is to say the sender and receiver can execute the other codes without wait for the succeed message from the other side. This method greatly increased the capability of affair process. The message send mechanism has the resume ability while breakdown, which make it possible for the integrating of the sender and receiver which were built on different physical platforms.

In the data transmission process, the XML form is used as the intermediate data format, which provides a unified data accessing, transmission format. And it shields the difference of the data format stored in the different relational databases; facilitates the integration of the heterogeneous data source.

In this paper, the mmessage queue manager (MQM) is used for the message management. It is the "heart" of the message middleware. It provides a message queue interface for the queue and message management of the procedure, in order to facilitate the procedure communication.

The message queue manager uses the network equipment existed (e.g. TCP/IP, SNA or SPX) to transmit the message to other queue manager through the channel. The main function of the message manager is shown as following:

- Manage the message queue for the application procedure;
- Provide the program interface for application procedure (Message Queue Interface- MQI);
- Transmit the message to the other queue manager based on the existed network equipment.
- Update the database and queue simultaneously, so the PUT/GET can run simultaneously;
- Do the message partition and the reorganization if necessary, the manager can combine some messages to one physical message and transmit it to the destination, then the destination carries on automatic split and reorganization (considering the capability);

The data transmit mode based on massage queue and XML is shown in Figure 3. Each cache maintains a massage queue, and managed by the message queue manager. The date format transmitted in the queue is XML. The final data loads to the true data warehouse similarly passes through a message queue maintains by the data warehouse. After passed through all caches, the data was loaded to the static data warehouse. The system uses the multistage real-time data caches to store the real-time data, simultaneously different caches store data with different freshness, i.e. the data was generated in different time. Along with the time increased, the preliminary cache data will flow to high-level cache through message queue Q, i.e. cache will be defers to the respective update cycle to carry on the data the refresh, for example: In the system, Cache-0, Cache-1, Cache-2, the Cache-3 update cycle respectively is 5 minute, 10 minute, 30 minute, 60 minutes, they deposit data in 5 minute, 10 minute, 30 minute, 60 minute separately.

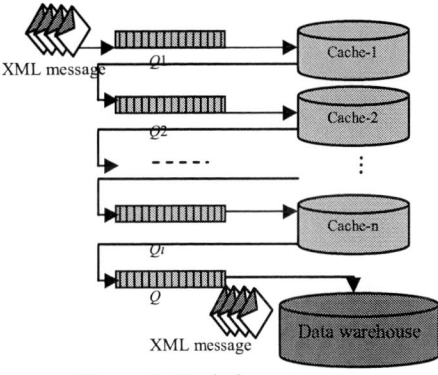

Figure 3. Updating process

The algorithm for the cache-i and Q_i updating:

```
cache_updating{
while (T_i begins)
//T_i is the updating cycle of the cache-i
    while (Q_i is not empty){
// Q_i is the data queue maintained by the cache-i
        d = Q_i.dequque();
//Q is the data queue maintained by the data
        put d into dataset M;
        foreach( m in M)
          integrate m into C_i;
        select the dataset N which older than T_{i+1}-T_i in
C_i;

          foreach( m in N)
            Q_{i+1}.inquque(n);
    }
return C_i, Q_{i+1};
}
```

The algorithm for the data warehouse updating:

```
Datawarehouse_Updating{
while (Q is not empty ){
//Q is the data queue maintained by the data warehouse
      d = Q.dequque();
//d is the xml message in the updating data flow
      put d into dataset M;
      foreach( m in M)
      integrate m into C;
      //C is the data set in data warehouse
      }
return C;
}
```

5. Real-time data query

The multi-level cache we used in this real-time data warehouse not only facilitates the real-time data updating, but also help the manager to reduce the real-time data query load of the real-time data warehouse. Using this technology, the real-time data query load is equalized on the caches, and the system's stability will be increased. For example: Cache-0, Cache-1, Cache-2, the Cache-3 updating cycle are 0 minute, 10 minute, 30 minute, 60 minutes respectively, the data freshness degree are 0, 1, 2, 3 respectively, if the data freshness degree which inquires Q needs is 20 minutes, then it can be satisfied on Cache-2, and the query service will combine the data in Cache-2-- Cache-n and the data warehouse together to complete this query.

When a query Q arrives, the system does not need to know where the real-time data locates and how to access it. In this real-time data warehouse, the Real-time data integrate (RDI) service will automatically get the real-time part of the required data from the multi-level caches and combine it together with the historical part to satisfy the query's requirement. What the query Q needs to do is to declare its freshness requirement of the data, according to which the system may decide which cache to be used to best serve the query.

The algorithm for answering queries with multi-level caches:

AnswerQuery{
  $S = \varphi$; //S is the maximum dataset for Q,
  $k = 1$; // the number of caches is n,
  while $(F((D_i) > F(Q))$
  // D_i is the dataset in cache-i;
  //F(D) is the freshness level of D;
    $k = k + 1$;
  for $(i = k; i \leqslant n; i++)$
    $S = S \cup D_i$;
  if (Cache-i is overloaded){
    $T_i = T_i - (T_i - T_{i-1})/10$;
    Updating cache-i;
  }
  return S for Q;
}

This method assigns the query in each data cache, lightens the data warehouse's burden, and reduces the conflict occurrence. In some situations, the frequent query may concentrate on the data in a period, and makes massive inquiries, increases the query load of the relative caches. In order to solve this problem, we can adjust the update cycle of the cache while implements the data warehouse. If some buffer's load is too heavy, reduces this buffer's update cycle, the data quantity of this cache will be decreased, and the query load will be decreased.

6. Electric power data storage structure

Data warehouse constructing mainly include two aspects, the design of interface of data warehouse and OLTP system and data warehouse [4]. The 2, 3 sector of this article introduced the interface design method and the data extract, transformation and loading method. This section will introduce the data warehouse storage structure design. The electrical power data warehouse is a huge system, data relations are intriguing, but the relation database modeling has certain difficulty to this kind of complex data relations, therefore we use the object-oriented method to carry on the data modeling. Simultaneously considering the data warehouse system's openness, but the object-oriented database management system does not support the SOL query language, and is difficult to integrate with the existing system and share information, therefore we use the relation object database (RODBMS) to implement the electric power data warehouse designing.

Now we will illustrate the real-time data warehouse data modeling by the electric quantity management. The data set needed for electric management query organizes 4 dimensions (time, region, industry and user) and a measure attribute (electricity consumption).Time dimension includes period, day, month and year; region includes circuitry, transformer substation and the terminal; user dimension includes user characteristics. The preceding text mentioned that the electrical power system real-time data warehouse's data storage part uses star connection model shown in Figure 4:

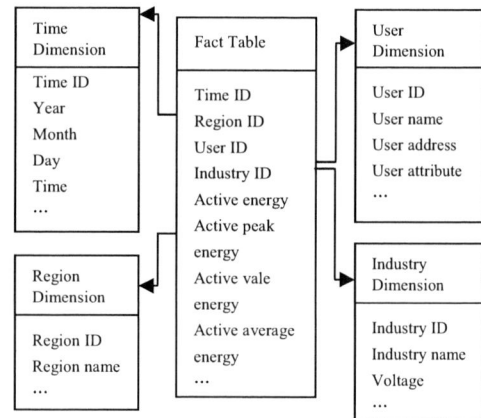

Figure 4. Data model for electric quantity management

Fact table contained external keys links to the dimension table, might connect with the dimension table rapidly, simultaneously the fact table also contained the non-external key information which will be inquired frequently. Fact table also includes the main data which accessed frequently. So we can carry on the query by scanning the fact table, thus enhanced the query speed greatly. Data pre-connection and found selective data redundancy simplifies the electric power system real-time data warehouse access and analyzes process design method.

7. Conclusions

Using the SOA technology, we can construct the loose coupling electrical power real-time data warehouse, not only uses the existing system, but also facilitates the system's expansion. This system can provide reliable data transmission, satisfies the electric power enterprise's request, and provides the formidable support for its decision-making. The real-time data warehouse application provide a good data environment for electric power enterprise's information system, further more, the on-line analysis processing and the data mining technology provides enormous information support for its strategy and tactical decision.

8. References

[1] LIU Jizhen, FANG Fang, NIU Yuguang. Real-time database in electric power enterprise [J].*Electric power*.2004.37(2).73-77.

[2] Shi Xiaohui. SOA based real-time data warehouse study and application[D].Beijing: Beijing University of Posts and Telecommunications,2003.10.

[3] Xu Li, Ma Ruixin. Research and Implementation of SOA-Based Real-Time ETL[J]. *Applications of the Computer Systems*.2007.4.

[4] Zhu Yijun, Wang Cheng, Zhang Feng. Data Warehouse and Its Design in Power System [J]. *Computer Simulation*.2004.7.

[5] Zhang Jun, Zhang Zhongneng, Reserch of Real-Time Data Warehouse Archtecture[J]. *Computer engineering*.2004.12.

[6] Jiang Zhen, Huang Xia. Research of Real-time Data W arehouse Technology[J]. *Applications of the Computer Systems*.2007.7.

[7] Zhou Long, Yao Yaowen. Discussion about ETL based on banking application [J]. *Computer Applications*. 2004,24(10):147-150.

[8] Ziyu Lin,Dongqing Yang, Guojie Song,Tengjiao Wang. Dealing with Query Contention Issue in Real-time DataWarehouses by Dynamic Multi-level Caches [C]. *Seventh International Conference on Computer and Information Technology*. 2007.60.

Unified Control of Photovoltaic Grid-connection and Power Quality managements

Xiaogao Chen, Qing Fu, Shijie Yu, Longhua Zhou

Institute for Solar Energy Systems

Sun Yat-sen University, Guangzhou, 510275, China

Chenxiaogao99@163.com, fuqing@mail.sysu.edu.cn, yushijie_zsu@163.com, zlhmaster@126.com

Abstract

This paper presents a control system that combines photovoltaic (PV) grid-connected generation and power quality managements. With the proposed structure, the system can not only realize photovoltaic generation, but also suppress current harmonics, compensate reactive power, eliminate voltage sags or swells and mitigate instantly power interruption and other power quality problems. An adaptive predictive maximum power point tracking (MPPT) algorithm is employed to improve the efficiency of PV generation. To simplify the control, the synchronous reference frame method is applied in modeling the inverter. The structure of a faster internal current control loop and a slower external voltage control loop is used in order to obtain a good dynamic response. The simulation results verify the good performance of the proposed controller.

1. Introduction

Nowadays the solar energy has become one of the most promising renewable energy due to its inexhaustible and environmental advantages. The photovoltaic technology has been deeply researched and popularized[1]. Grid-connected PV generation is one of the major development trends of photovoltaic applications. Meanwhile, with the development of the power electronics industrialization process, a large number of nonlinear loads have appeared. Harmonics and reactive current from nonlinear load induce that the power quality problems become more and more serious.

PV grid-connected generation operates during the day and has to stop at night. This affects the stabilization of power system and the utilization of equipment. Therefore, in order to increase the utilization, the PV system can be designed to also provide the function of power quality managements [2]-[3].

This paper combines PV grid-connected generation device and shunt active power filter. The unified system can supply active power as well as current harmonics and reactive power compensation when sunshine is available. At weak irradiation, it has the all function of shunt active

power filter. Besides, the system is equipped with large capacity storage batteries so that it can not only realize PV generation, harmonics suppression and reactive power compensation, but also compensate voltage sags or swells and instantly power interruption. The system is modeled in the rotating dq frame with a decoupled control strategy for reducing the control complexity. The simulation results show that the proposed controller yields good performances.

2. The proposed grid-connected PV System

The unified system (as shown in fig.1) is composed of PV generation and energy storage system and the power quality comprehensive compensator.

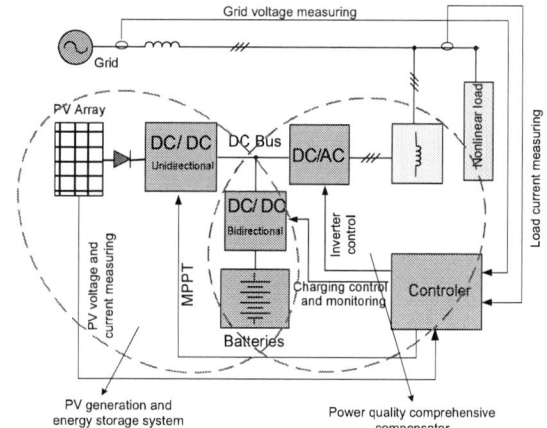

Fig.1 Grid-connected PV system with power quality comprehensive compensator

Solar energy is converted into electricity through PV array. Then the voltage across PV array is amplified with a unidirectional DC/DC boost converter so as to track the maximum power point of PV array conveniently[4]. After that, the DC/AC inverter converts DC power into AC power which is injected into the utility grid or directly supplies to local loads. At the same time the system can provide a strongly voltage compensation when the grid voltage sags or swells because it is equipped with large

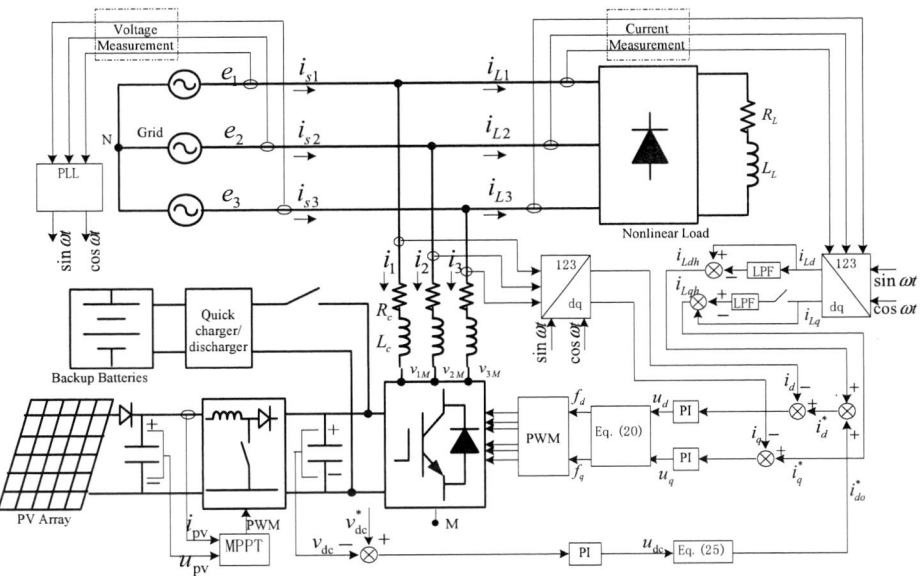

Fig.2 The control block diagram of the proposed system

capacity storage devices. PV array also enhances the ability of its power compensation. When the grid voltage rises suddenly, the batteries are charged quickly, absorbing energy from the grid, thereby the grid voltage swells is restrained. In the same way, when the grid voltage drops suddenly, the batteries release energy quickly to grid, suppressing the voltage sags. Compared to the conventional photovoltaic generation devices[5], it can not only inject active power into grid, also has the functions of reactive power compensation and harmonic suppression, eliminating grid voltage sags and swells. In addition, the system can act as shunt active power filter and UPS in the absence of sunshine or the weak sunshine. Fig.2 shows the control block diagram of the proposed unified control system.

3. MPPT Control

MPPT control technology is one of the important problems and one of the key technologies in photovoltaic system. An adaptive predictive MPPT control algorithm is put forward which is based on the Perturbation and Observation (P&Q) idea. The preliminary principle of the adaptive predictive control is shown in fig.3.

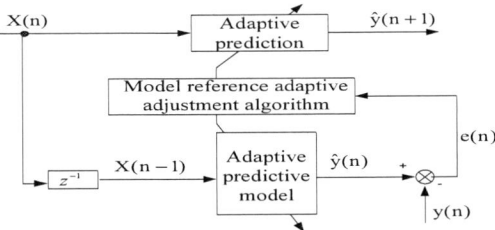

Fig.3 Adaptive predictive control of output power of PV array

In the figure 3,

$$\begin{cases} X(n-1) = [x(n-1), x(n-2), \cdots, x(n-N)] \\ X(n) = [x(n), x(n-1), \cdots, x(n-N+1)] \end{cases} \quad (1)$$

where the $x(n)$ is the current voltage of PV array and $x(n-1), \cdots, x(n-N+1)$ are the historical voltages.

$y(n)$ is the actual output power at the current sampling moment. $\hat{y}(n)$ and $\hat{y}(n+1)$ are the predictive output powers at the current sampling moment and next moment respectively. The error is

$$e(n) = y(n) - \hat{y}(n) \quad (2)$$

The adaptive prediction [6]-[7] is based on the Finite Impulse Response (FIR) model.

$$\hat{y}(n+1) = \sum_{k=0}^{N-1} h_k x(n-k) = H'X(n) \quad (3)$$

where, $H' = [h_0, h_1, \cdots, h_{N-1}]$ is adaptive predictive coefficient matrix.

The adaptive adjustment algorithm is based Least Mean Square (LMS) criterion, the aim of which is to solve the optimum h_k $(k = 0,1,\cdots,N-1)$ by minimizing the mean square error of $e(n)$.

The difference from the conventional P&Q MPPT control method[8] is that the PV output power of next sampling time is observed by prediction mechanism instead of voltage perturbation, so that it eliminates the loss power suffered from voltage perturbation. The MPPT control is implemented in the boost DC/DC independently. Fig.4 shows the control flow of the proposed MPPT.

361

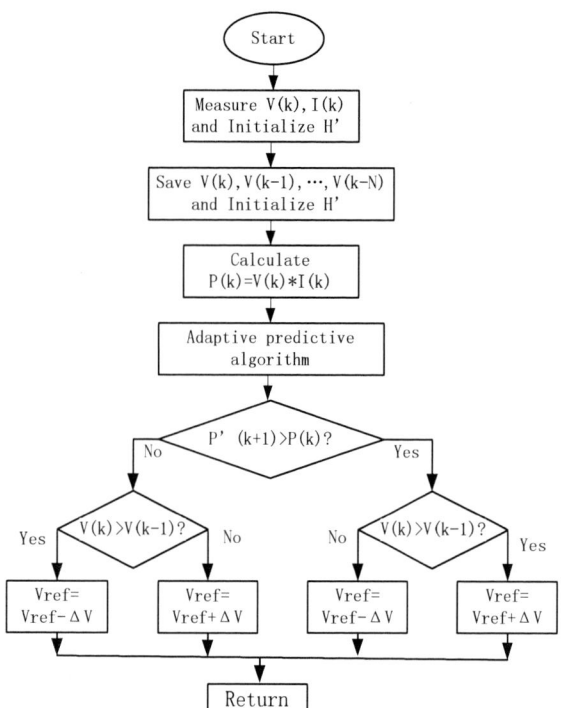

Fig.4 Flow of the proposed MPPT control

4. Inverter modeling

According to Kirchoff's rules for voltages and currents, differential equations (4) in the stationary abc frame can be obtained at the connection point of the inverter. See in the fig2.

$$e_k = L_c \frac{di_k}{dt} + R_c i_k + v_{kM} + v_{MN} \qquad (k=1,2,3) \quad (4)$$

With the assumption that the zero-sequence is absence and the ac voltages are balanced, one can write the relation:

$$v_{MN} = -\frac{1}{3}\sum_{m=1}^{3} v_{mM} \quad (5)$$

Define the switch function s_k of the kth leg of the inverter as in (6),

$$s_k = \begin{cases} 1, & \text{if the upper device is On and the lower one is Off} \\ 0, & \text{if the upper device is Off and the lower one is On} \end{cases} \quad (6)$$

Hence, from Eq. (4), (5), (6) one can derive:

$$\frac{di_k}{dt} = -\frac{R_c}{L_c} i_k - \frac{1}{L_c}\left(s_k - \frac{1}{3}\sum_{m=1}^{3} s_m\right) v_{dc} + \frac{e_k}{L_c} \quad (7)$$

Further, a switching state function f_k is defined as follows.

$$f_k = \left(s_k - \frac{1}{3}\sum_{m=1}^{3} s_m\right) \quad (8)$$

On the other hand, on the dc side the following equation holds:

$$\frac{dv_{dc}}{dt} = \frac{1}{C} i_{dc} = \frac{1}{C}\sum_{m=1}^{3} s_m i_m \quad (9)$$

and it can be verified that:

$$\sum_{m=1}^{3} f_m i_m = \sum_{m=1}^{3} s_m i_m \quad (10)$$

which allows to obtain the equation (11),

$$\frac{dv_{dc}}{dt} = \frac{1}{C}\sum_{m=1}^{3} f_m i_m = \frac{1}{C}\begin{bmatrix} f_1 \\ f_2 \\ f_3 \end{bmatrix}^T \begin{bmatrix} i_1 \\ i_2 \\ i_3 \end{bmatrix} \quad (11)$$

After the coordinate transformation to the synchronous reference frame:

$$\frac{dv_{dc}}{dt} = \frac{1}{C}\left(C_{123}^{dq}\begin{bmatrix} f_d \\ f_q \end{bmatrix}\right)^T \left(C_{123}^{dq}\begin{bmatrix} i_d \\ i_q \end{bmatrix}\right) \quad (12)$$

where the coordinate transformation matrix is:

$$C_{123}^{dq} = \sqrt{2/3}\begin{bmatrix} \sin \omega t & -\cos \omega t \\ \sin(\omega t - 2\pi/3) & -\cos(\omega t - 2\pi/3) \\ \sin(\omega t + 2\pi/3) & -\cos(\omega t + 2\pi/3) \end{bmatrix}$$

the following is obtained:

$$\frac{dv_{dc}}{dt} = \frac{f_d i_d}{C} + \frac{f_q i_q}{C} \quad (13)$$

According to equation (7) and (8), one can obtain:

$$\frac{d}{dt}\begin{bmatrix} i_1 \\ i_2 \end{bmatrix} = -\frac{R_c}{L_c}\begin{bmatrix} 1 & 0 \\ 0 & 1 \end{bmatrix}\begin{bmatrix} i_1 \\ i_2 \end{bmatrix} - \frac{1}{L_c}\begin{bmatrix} f_1 \\ f_2 \end{bmatrix} v_{dc} + \frac{1}{L_c}\begin{bmatrix} e_1 \\ e_2 \end{bmatrix} \quad (14)$$

Applied the coordinate transformation [9], equation (14) can be written in the dq frame as the following equation,

$$\frac{d}{dt}\begin{bmatrix} i_d \\ i_q \end{bmatrix} = -\begin{bmatrix} R_c/L_c & -\omega \\ -\omega & R_c/L_c \end{bmatrix}\begin{bmatrix} i_d \\ i_q \end{bmatrix} - \frac{1}{L_c}\begin{bmatrix} f_d \\ f_q \end{bmatrix} v_{dc} + \frac{1}{L_c}\begin{bmatrix} e_d \\ e_q \end{bmatrix} \quad (15)$$

Finally, the whole dynamic model in the dq frame is obtained from (13) and (15),

$$\frac{d}{dt}\begin{bmatrix} i_d \\ i_q \\ v_{dc} \end{bmatrix} = \begin{bmatrix} -R_c/L_c & \omega & -f_d/L_c \\ -\omega & -R_c/L_c & -f_q/L_c \\ f_d/C & f_q/C & 0 \end{bmatrix}\begin{bmatrix} i_d \\ i_q \\ v_{dc} \end{bmatrix} + \frac{1}{L_c}\begin{bmatrix} e_d \\ e_q \\ 0 \end{bmatrix} \quad (16)$$

5. Inverter control

In order to achieve better performance, a faster inner current tracking loop and a slower outer dc voltage regulation loop are adopted. The current and voltage control loops are designed using PI compensators. The phase-locked-loop (PLL) detects the amplitude and the position of the grid voltage vector.

5.1. Current tracking control

From the model (16), one can write:

$$\begin{cases} L_c \dfrac{di_d}{dt} + R_c i_d = L_c \omega i_q - v_{dc} f_d + e_d \\ L_c \dfrac{di_q}{dt} + R_c i_q = L_c \omega i_d - v_{dc} f_q + e_q \end{cases} \quad (17)$$

We define the equivalent inputs as following:

$$\begin{cases} u_d = L_c \omega i_q - v_{dc} f_d + e_d \\ u_q = L_c \omega i_d - v_{dc} f_q + e_q \end{cases} \quad (18)$$

Here, u_d and u_q can be get by using PI compensators for achieving a fast dynamic response.

$$\begin{cases} u_d = k_p \Delta i_d + k_i \int \Delta i_d dt \\ u_q = k_p \Delta i_q + k_i \int \Delta i_q dt \end{cases} \quad (19)$$

Where, $\Delta i_d = i_d^* - i_d$, $\Delta i_q = i_q^* - i_q$, i_d^*, i_q^* are the reference of i_d and i_q respectively.

Then control inputs (20) can be derived from (18):

$$\begin{cases} f_d = \left(e_d + L_c \omega i_q - u_d\right)/v_{dc} \\ f_q = \left(e_q + L_c \omega i_d - u_q\right)/v_{dc} \end{cases} \quad (20)$$

Thus, through the input transformation (20), the coupled dynamics of the current tracking problem have been transformed into decoupled dynamics[9]. Hence, the currents i_d and i_q can be controlled independently by acting upon inputs u_d and u_q respectively.

5.2 DC bus voltage control

The inverter dc link voltage controller calculates the current to maintain the dc link voltage by passing the dc link voltage error through a PI compensator.

An equivalent input u_{dc} is defined:

$$u_{dc} = f_d i_d + f_q i_q \quad (21)$$

The error $\Delta v_{dc} = v_{dc}^* - v_{dc}$ is passed through a PI compensator, so the voltage controller output is:

$$u_{dc} = k_{p1} \Delta v_{dc} + k_{i1} \int \Delta v_{dc} dt \quad (22)$$

Assuming the grid voltage are given by (23),

$$\begin{cases} e_1 = E \sin \omega t \\ e_2 = E \sin(\omega t - 2\pi/3) \\ e_3 = E \sin(\omega t + 2\pi/3) \end{cases} \quad (23)$$

So we can get:

$$\left[e_d, e_q\right]^T = C_{123}^{dq}\left[e_1, e_2, e_3\right]^T = \left[\sqrt{3/2}E, 0\right]^T \quad (24)$$

Given that the current loop is ideal, the following approximations hold:

$$f_d v_{dc} \approx e_d, \quad f_q v_{dc} \approx e_q.$$

So the control effort is derived according to (21) and (24),

$$i_{do}^* = \dfrac{u_{dc} - f_q i_q}{f_d} = \dfrac{u_{dc} v_{dc} - f_q v_{dc} i_q}{f_d v_{dc}} \approx \sqrt{\dfrac{2}{3}} \dfrac{v_{dc}}{E} u_{dc} \quad (25)$$

The reference current i_{do}^* is added to the reference current i_d^* of the current loop as shown in fig.2 in order to inject active power from PV array to the grid.

6. Simulation results

According to the control structure shown in fig.2, the proposed system is simulated in Matlab/Simulink with the parameters listed in Table 1.

Table 1 Simulation parameters

$E = 220\sqrt{2}$V, $f = \omega/2\pi = 50$Hz
$R_L = 100\Omega$, $L_L = 1$mH; $R_c = 0.1\Omega$, $L_c = 5.5$mH
PV : $I_{sc} = 12.75$A, $V_{oc} = 93.5$V $I_m = 10.7$A, $V_m = 74.5$V
$C_{dc} = 1000\mu$F, $C_{pv} = 470\mu$F
$k_p = 89$, $k_i = 3000$; $k_{p1} = 0.056$, $k_{i1} = 4.68$

With the adaptive predictive algorithm mentioned above, the predictive power of PV array is almost the same as the measured power as shown in fig.5. It demonstrates that the adaptive predictive algorithm is of high accuracy. From the fig.6, the maximum power can be maintained at about 766W under the irradiation of 800W/m² although has small oscillations. During the simulating, PV generation starts at 0.1s. Seen from the fig.7, only the current harmonics is compensated before 0.1s, and the grid current is greatly improved. After 0.1s the current harmonics suppression, reactive power compensation and PV generation start simultaneously. The grid current and the load current are displaced 180 degrees because the power generated by the PV array is greater than the load consumption. Fig.8 shows the current control loop (d-axis and q-axis) has the good tracking performances before and after the PV generation works.

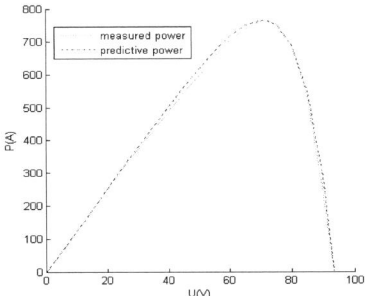

Fig.5 The actual and the predictive power curves of PV array

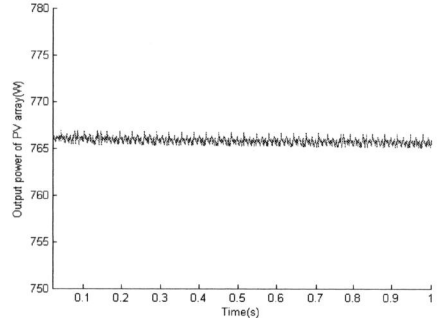

Fig.6 The output power of the PV array

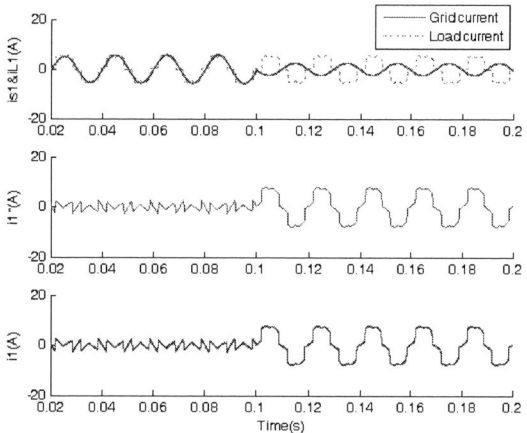

Fig.7 PV generation and improving current quality

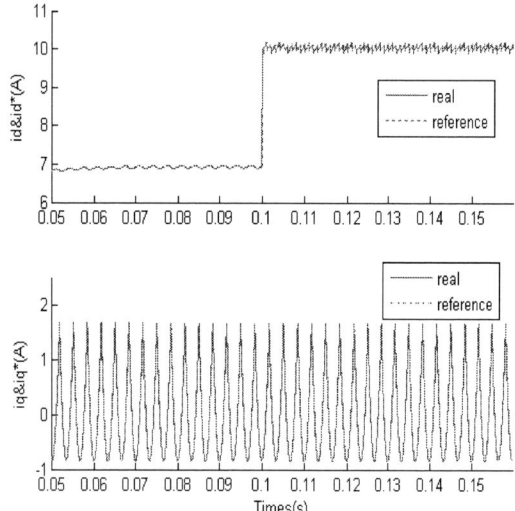

Fig.8 Tracking performances of the current loop: d-axis and q-axis

When the nonlinear load has 30% drop and 30% step at 0.1s respectively, the current harmonics compensation waveforms are shown in fig.9. From the fig.9 (a) and (b), when there are variations in nonlinear load, current harmonics can be detected quickly and the compensation currents can track the reference currents efficiently.

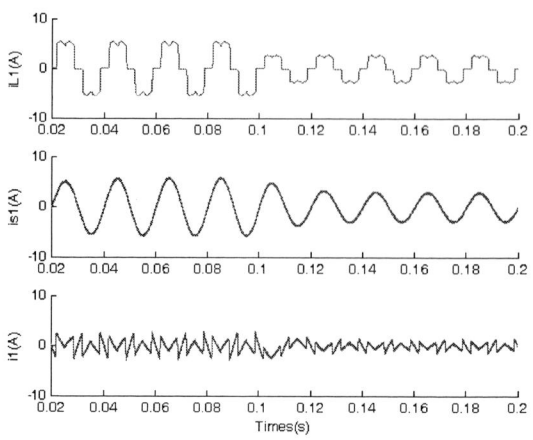

(a)   When a drop change in the nonlinear load

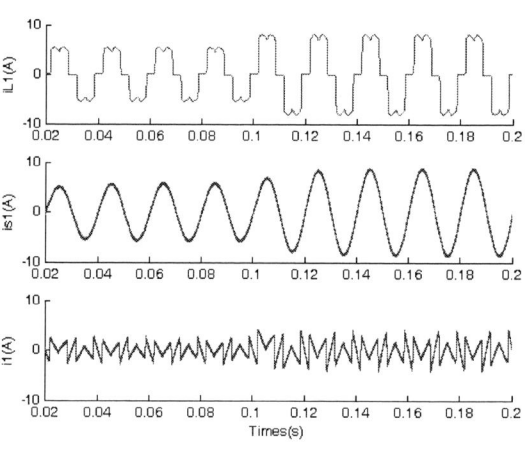

(b)   When a step change in the nonlinear load

Fig.9 Compensation characteristics under the variations of the nonlinear load

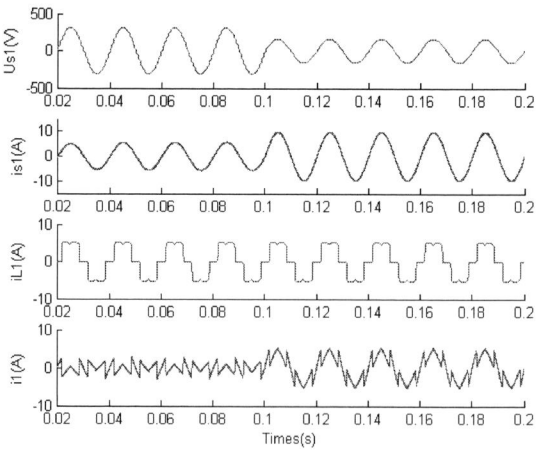

(a)   Eliminating grid voltage sags

364

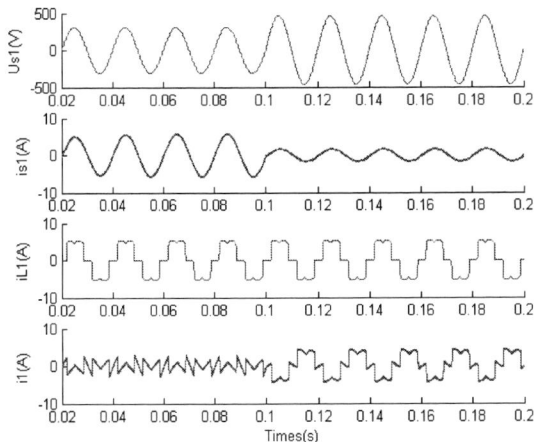

(b) Eliminating grid voltage swells

Fig.10 Compensation characteristics when the grid voltage sags or swells

When the grid voltage is suffered from 150V sags and swells at 0.1s respectively, the power compensation characteristics are shown in fig.10. When the grid voltage sags, active power from dc bus link is needed to be injected into the grid in order to maintain the load current. Likewise, when the grid voltage swells, the dc bus link absorbs active power from the grid.

The prototype of the system is under construction and the experiment results will be shown in the later paper.

7. Conclusion

The proposed PV system in this paper has the functions of PV grid-connected generation and power quality compensation. An adaptive predictive algorithm based on P&Q control idea is employed and achieved good results in MPPT control of PV generation. Voltage vector oriented reference frame models are used to control the inverter. Theoretical analysis and MATLAB simulations show that the control structure of double loop has good dynamic performances and the proposed system can realize PV generation, harmonics suppression and reactive power compensation, as well as eliminating voltage sags or swells.

8. References

[1]. BARKER, P.P, BING, J.M. "Advances in solar photovoltaic technology: an applications perspective", *Proceedings of IEEE Power Engineering Society General Meeting*, vol.2, 2005, pp.1955-1960.

[2]. M. C. Cavalcanti, G. M. S. Azevedo, B. A. Amaral, K. etc, "Efficiency Evaluation in Grid Connected Photovoltaic Energy Conversion Systems", *Proceedings of the2005*

IEEE Power Electronics Specialists Conference, 2005, pp. 269-275.

[3]. M.C. Cavalcanti, G.M.S. Azevedo, B.A. Amaral, etc. "A photovoltaic generation system with unified power quality conditioner function", *Industrial Electronics Society, 2005.31st Annual Conference of IEEE*, 2005, pp.750-755.

[4]. Y. C. Kuo, T. J. Liang, and J. F. Chen, "Novel maximum power point tracking controller for photovoltaic energy conversion system", *IEEE Trans. Industrial Electronics*, vol. 48, 2001, pp. 594-601.

[5]. F. Antunes, A.M. Torres, "A three-phase grid-connected PV system", *Industrial Electronics Society, 2000. 26th Annual Conference of the IEEE*, vol.1, 2000, pp.723 – 728.

[6]. Ficarra, M., Rodellar, J., Bordonau, J., etc; "Adaptive predictive control algorithm for compensation of parameters of a power electronics system", *the IEEE Electronics Letters*, Vol.31, 1995, pp.329-330.

[7]. Y. Xue, Y. Wu, H. Zhang, "An Adaptive Predictive Current-Controlled PWM Strategy for Single-phase Grid-connected Inverters", *Industrial Electronics Society, 33rd Annual Conference of the IEEE*, 2007, pp.1548-1552.

[8]. C. Hua and C. Shen, "Comparative Study of Peak Power Tracking Techniques for Solar Storage System", *Proceedings of the 1998 IEEE Applied Power Electronics Conference (APEC'YB)*, vol.2, 1998, pp. 679-685.

[9]. N. Mendalek and K. Al-Haddad, "Modeling and nonlinear control of shunt active power filter in the synchronous reference frame", *IEEE Harmonics and Quality of Power International Conference*, 2000, pp.30-35.

Research on Testing Technology of High-power Photovoltaic Arrays

Yinglei He , Qing Fu , Xiangfeng Li , Xiaogao Chen
Institute for Solar Energy Systems
Sun Yat-sen University, Guangzhou, 510275, China
lei_behappy@163.com , fuqing@mail.sysu.edu.cn, lxfcy83@tom.com, Chenxiaogao99@163.com

Abstract

With the test on the characteristics of photovoltaic arrays, we can assess the performance of photovoltaic arrays effectively. It's help to the research and development of photovoltaic system. The paper introduces the testing technology of photovoltaic arrays and proposes an algorithm to predict the characteristics of photovoltaic arrays. Using this algorithm method the curve which shows the characteristics of photovoltaic arrays can be established according to the field-testing data under arbitrary temperature and irradiation. The testing technology helps to reflect the actual condition of the photovoltaic arrays.

Key words: *Photovoltaic (PV) Arrays, Testing predict algorithm, Simulation of I-V curve*

1. Introduction

Accompanying the rapid development of the economy, the global energy problems have become increasingly severe. No delay to take the sustainable development path to utilize new green energy sources. People pay more and more attention to the high-quality and renewable solar energy, therefore, testing and predicting PV arrays in order to put solar energy into full use become a focus [1].

PV arrays vary with environment remarkably, the output power of PV arrays change with different illumination and temperature [2]. The parameters of PV arrays provided by manufacturers are given in the standard testing conditions (S_{ref} = 1000W/m^2, T_{ref} = 25 ℃). But in applications conditions such as temperature and illumination around PV arrays change, these parameters do not actually reflect the characteristics of PV arrays. In addition, PV system assembles the cells according to the demand of output power. PV arrays have different performances in different connections (serial or parallel), changes with the connections will also cause connection losses. Considering the years use of PV arrays and the changing working condition, the parameters of PV arrays can't hold the line. Therefore relying solely on the standard parameters of PV arrays, PV system design will always be difficult to achieve the desired effect.

This paper presents the testing technology of high-power PV arrays and a predictive algorithm, which identifies the mathematical model of PV arrays based on field test data and thus obtain PV arrays' predictive characteristic curve under arbitrary conditions. PV arrays field test generally use the traditional method just like resistance testing and capacitance rapid charging dynamic testing method, the paper use the later method since it has a greater advantage in testing PV arrays.

2. Mathematical-physical model

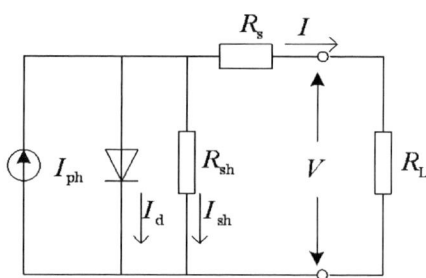

Fig.1 Equivalent circuit of PV cell

To test the characteristics of PV arrays, firstly, we should establish the mathematical-physics model [3,4], which reflects the influence of temperature, illumination and other factors on the characteristics of PV arrays. Generally PV cells can be equivalent to the circuit shown in Fig. 1 [4]. Based on the analog circuit analysis, Eq. (1) is obtained:

$$I = I_{Ph} - I_d - I_{sc}$$
$$= I_{Ph} - I_o[\exp\{q(V + R_s I) / AKT\} - 1] - (V + R_s I) / R_{sh} \quad (1)$$

where I is the output current of PV arrays, I_{ph} is PV current, I_o is the saturated current of the diode. V is the output voltage of PV arrays, R_s is the serial resistance of PV arrays. A is the characteristic factor of the diode. K is the Boerziman constant. T is the temperature of PV arrays. R_{sh} is the parallel resistance of PV arrays.

Since the expression $(V + R_s I) / R_{sh} \ll I_{ph}$ and R_s is far smaller than the forward continuity resistance, I_{ph} can be approximately equal to I_{sc} [5]. Introduce undetermined coefficient C_1 and C_2 in Eq.(1) and substitute AKT / q for $C_1 V_{oc}$, I_o for $C_2 I_{sc}$, it follows that,

978-0-7695-3342-1/08 $25.00 © 2008 IEEE
DOI 10.1109/PEITS.2008.81

$$I = I_{sc} - (1 - C_1\{\exp[V/(C_2 V_{oc})] - 1\}) \tag{2}$$

Among

$$C_1 = (1 - I_m/I_{sc})\exp[-V_m/(C_2 V_{oc})] \tag{3}$$

$$C_2 = (V_m/V_{oc} - 1)\left[\ln(1 - I_m/I_{sc})\right]^{-1} \tag{4}$$

A practical simplified model is established. It should be noted from the above expression that we can get the characteristics of PV arrays just with four major parameter which including short-circuit current I_{sc}, open-circuit voltage V_{oc}, the circuit I_m and voltage V_m at the max power point. Through the model we can we can assess the performance of PV arrays easily and effectively.

3. Prediction Algorithm

Using a appropriate mathematics-physical model to set up the relation between PV arrays and environmental factors[6].

$$\Delta T = T - T_{ref} \tag{5}$$

$$\Delta S = \frac{S}{S_{ref}} - 1 \tag{6}$$

$$I'_{sc} = I_{sc}\frac{S}{S_{ref}}(1 + a\Delta T) \tag{7}$$

$$V'_{oc} = V_{oc}(1 - c\Delta T)\ln(e + b\Delta S) \tag{8}$$

$$I'_m = I_m\frac{S}{S_{ref}}(1 + a\Delta T) \tag{9}$$

$$V'_m = V_m(1 - c\Delta T)\ln(e + b\Delta S) \tag{10}$$

where the coefficient a is $0.0025/^0C$, b is 0.5, c is $0.00288/^0C$.

In order to distinguish the four major parameters in different conditions we definite them respectively as followed:

In the standard condition: I_{sc_ref}, U_{oc_ref}, I_{m_ref}, U_{m_ref}, S_{ref}, T_{ref}. While in actual testing condition: I_{sc}, V_{oc}, I_m, V_m, S, T. and predict new condition: I'_{sc}, V'_{oc}, I'_m, V'_m, S', T'.

Take the following steps we can get the characteristic of PV arrays.

First, put the actual testing data into the Eq.(7)~(10), calculate out the parameter in the standard condition

$$I_{sc\_ref} = \frac{I_{sc}S_{ref}}{S(1 + a\Delta T)} \tag{11}$$

$$V_{oc\_ref} = \frac{V_{oc}}{(1 - c\Delta T)\ln(e + b\Delta S)} \tag{12}$$

$$I_{m\_ref} = \frac{I_m S_{ref}}{S(1 + a\Delta T)} \tag{13}$$

$$V_{m\_ref} = \frac{V_{oc}}{(1 - c\Delta T)\ln(e + b\Delta S)} \tag{14}$$

Secondly, substitute the parameter above into Eq.(7) to (10), in this way, four predictive major parameter in new environment is obtained.

$$\Delta T' = T' - T_{ref} \tag{15}$$

$$\Delta S' = \frac{S'}{S_{ref}} - 1 \tag{16}$$

$$I'_{sc} = \frac{I_{sc}S'(1 + a\Delta T')}{S(1 + a\Delta T)} \tag{17}$$

$$V'_{oc} = \frac{V_{oc}(1 - c\Delta T')\ln(e + b\Delta S')}{(1 - c\Delta T)\ln(e + b\Delta S)} \tag{18}$$

$$I'_m = \frac{I_m S'(1 + a\Delta T')}{S(1 + a\Delta T)} \tag{19}$$

$$V'_m = \frac{V_m(1 - c\Delta T')\ln(e + b\Delta S')}{(1 - c\Delta T)\ln(e + b\Delta S)} \tag{20}$$

After getting the key parameters in new condition, we can use the practical mathematical-physics model obtain the characteristic curve. The method not only has complex computation, but also brings in large calculation error by employing the fixed-point DSP. Taking into account that the characteristics of PV arrays is linear similarity, referring to the current environment the characteristic curve of PV arrays in the predict condition can be imitated. The regional characteristics of PV arrays will be shown.

4. Experimental results

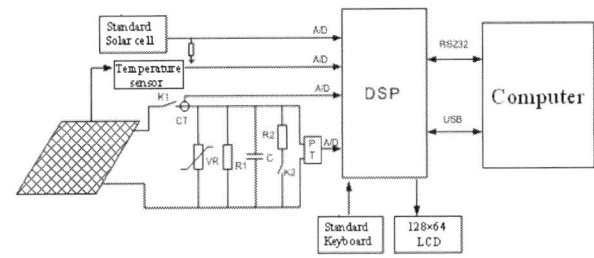

Fig.2 Structure of PV arrays characteristics meter

The general scheme of large-power PV arrays characteristics meter is shown in Fig.2. The standard PV cells whose function is to test the illumination, combined with the temperature sensor(AD590) to acquire the standard parameter under the standard condition in which illumination is $1000W/m^2$ and temperature is $25\,^\circ\!C$. As is seen in Fig.2, VR is a varistor that prevent over-voltage, resistance R_1 plays a role in slow discharging whereas resistance R_2 fast discharging. In the situation that K_1 off and K_2 on, R_2 is help to release the energy from the capability in favor of the next testing .As previously described, the testing system adopt the capacitance rapid charging dynamic testing method. Through testing the changing voltage and current during the charging process, the characteristics curve can be drawn. The moment the K_1 close, the capacitance start charging, when the circuit current considered to be the short-circuit current. At the time that the charging current turn into zero when the voltage of capacitance is considered to be the open-circuit voltage. The testing system sample the working point through the whole charging process, the sampled point form PV arrays characteristics curve in the current condition. What's the advanced technology in the system is that it employs the DSP (TMS320F2812) a high-speed and large-capacity data processor to collect PV arrays characteristic data. The system uses the C language to set up a programme aim at saving, querying real-time monitoring the illumination, statistic and such other function.

Compare the predict data using the predictive algorithm with the actual data, it can find out if the predictive algorithm works. To get a reasonable assessment we pick out three set of data from different illumination. Among the comparison V , I belong to the actual data, however, I' is for the predict data using the predictive algorithm. δ is the relative error which is equal to $(I-I')I_{sc}\times100\%$. From the Tab.1 list the characteristics compared data in detail, the datum are no better than each other with little error range from 1% to 3%.The result show up vividly in Error analysis Fig.3. So we verify that the system is possibly to predict the characteristics of PV arrays.

5. Conclusion

The predictive algorithm presented in this paper is adopted to give a more accurate assessment to PV arrays .Thanks to the prediction the PV system will work in the best condition putting solar energy in full use. With PV arrays testing technology further improvement, it will plays increasingly vital role in the PV applications.

Tab.1 Tested data of PV arrays characteristics

V	I	I'	δ
$S = 522\text{W/m}^2$			
0.00	8.24	8.24	0.00%
2.00	8.19	8.24	-0.60%
4.00	8.08	8.23	-1.82%
6.00	8.02	8.22	-2.42%
8.00	7.94	8.18	-2.91%
10.00	7.75	8.07	-3.88%
12.00	7.59	7.80	-2.54%
14.00	7.10	7.15	-0.61%
16.00	5.28	5.56	-3.40%
17.00	3.97	4.03	-0.73%
$S = 651\text{W/m}^2$			
0.00	10.20	10.20	0.00%
2.00	10.12	10.20	-0.78%
4.00	10.04	10.19	-1.47%
6.00	9.96	10.17	-2.06%
8.00	9.82	10.12	-2.94%
10.00	9.68	9.99	-3.04%
12.00	9.43	9.68	-2.45%
14.00	8.72	8.89	-1.67%
16.00	7.12	7.00	1.18%
18.00	2.51	2.40	1.08%
$S = 825\text{W/m}^2$			
0.00	12.74	12.74	0.00%
2.00	12.67	12.73	-0.47%
4.00	12.62	12.72	-0.78%
6.00	12.44	12.68	-1.88%
8.00	12.26	12.61	-2.75%
10.00	12.05	12.41	-2.83%
12.00	11.68	11.96	-2.20%
14.00	10.87	10.94	-0.55%
16.00	8.59	8.60	-0.08%
18.00	2.68	3.20	-4.08%

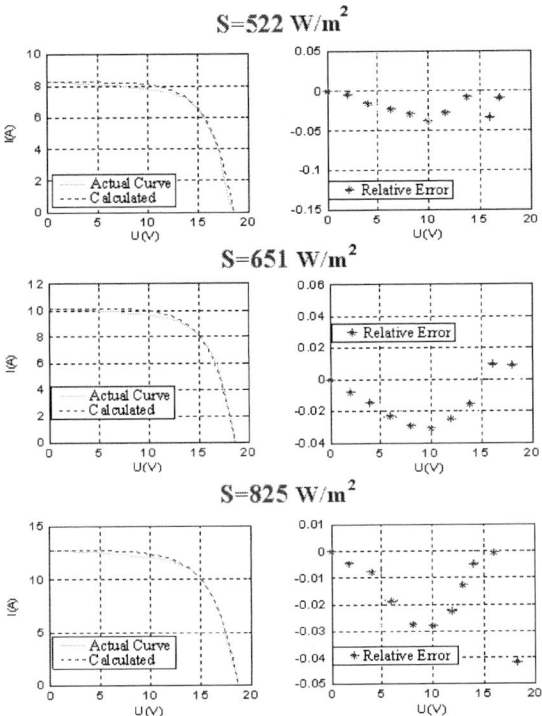

Fig.3 Error analysis of tested data

References

[1] Luo Xuelian (2005). "Solar Energy cells and its application". *Wuhan Scientific college academic Journal*. Vol. 18, No. 10, pp.36-37.

[2] Jae-Hyun Yoo (2001). "Analysis and Control of PWM Converter with V-I Ouput Characteristics of Solar Cell". *IEEE*, Vol. 2, pp.1049-1054.

[3] Gnter.Ryot (1991). "PV use of solar energy". *Yu Shijie、He Huiruo translate，Hefei University of Technology*

[4] Su Jianghui,Yu Shijie and Zhao Wei (2001). "An investigation on engineering analytical model of silicon cells". *Acta Energiae Solaris Sinica*, Vol. 22

[5] MehmetAkbaba，IsaQmaber and Adel Kemal (1998). "Matching of Separately Excited DC Motors to Photovoltaic Generators for Maximum Power Output". *Solar Energy*, Vol. 63, No. 6, pp.375-385.

[6] Han Yu (2006). "Research and design for solar energy arrays simulation". *Zhejiang University*.

[7] Dong Mi (2007). "Optimal Design and Research on Control Strategy of Grid-Connected PV System", *South China University*

[8] Altas (2007). "PV Arrays Simulation Model for Matlab-Simulink GUI Environment". *Clean Electrical Powe*r, pp.341-345.

2008 Workshop on Power Electronics and Intelligent Transportation System

Direct Power Control Based on Three-Phase Shunt Hybrid Power Filter Case Study

Wenjin Dai, Ming Chen

Department of Information Engineering,Nanchang University,Nanchang,China
E-mail: dwj480620@yahoo.com.cn, dtjcccm@sina.com.cn

Abstract

The research based on shunt hybrid power filter in this paper is of better performance. Due to the limited switching frequency(typically below 1 kHz) of high-power solid-state devices (GTO/IGCT), current control is simplified by merging the use of synchronous reference frame (SRF) and instantaneous reactive power(IRP), and the vector algorithm without transformation matrix IRP is used in this paper. The reliability of this method has been verified by the simulation results based on MATLAB.

1. Introduction

Active power filters are basically power electronic devices used to compensate the current or voltage harmonics and the reactive power flowing in the power grid. The reduction of the harmonic and reactive currents becomes an increasingly required issue owing to the wide use of power electronic equipment. Traditionally, passive LC filters have been used to remove line current harmonics and to improve the power factor. However, when implemented, these passive filters present many drawbacks such as tuning problems, series and parallel resonance. Other industrial applications use power factor correction (PFC) devices for reactive power and current harmonics compensation. In these circuits, switched capacitors banks are typically connected in parallel to current–source-type loads. Seen from the load side, the capacitance of the PFC and the source inductor create a parallel resonant circuit. Looking from the source side, the PFC capacitors and the line inductor represent a series resonant circuit. To prevent resonance due to current harmonics in power systems with PFC equipment, typical shunt or series active topologies have been proposed [1, 2, 3]. However, these topologies suffer from a high kVA-rating of the filter power stage [4, 5]. The boost-type rectifier that constitutes the shunt active filter requires generally a high DC-link voltage [3, 6, 7] in order to compensate the high order harmonics effectively. On the other hand, a series active filter needs a transformer capable to withstand full load current in order to compensate the

voltage distortion. Therefore, a hybrid filter topology has been developed achieving the desired damping performance with a significant reduction of the KVA-rating required by the power active filter[3,6,8,9].

The TSHPF (three-phase shunt hybrid power filter) uses three power factor correction capacitors connected in series with the secondary winding of the transformer. The primary winding of the transformer is connected to a three-phase voltage-source inverter. The compensation capacitor needed for PFC is used as part of the filter to create a single harmonic trap. Three relatively low-rated active filters compensate the residual harmonics and prevent the excitation of the resonant circuit [10]. The direct current control technique is based on the classic demodulation method that allows extracting easily the source current reference from the distorted waveform of the load current. Multiple techniques are studied to increase the system bandwidth at low switching frequency:

First, among most current controller designed for active filters, multiple synchronous reference frame based current controller is proven to have the highest bandwidth with modest feedforward gain.

Second, an equivalent stationary frame based current controller is easily derived based on the classic vector theory. The results are similar as that derived from convolution[11] or physical concept[12] in recent literature.

Thirdly, instantaneous reactive power (IRP) theory[13] and synchronous reference frame (SRF)[14] based methods are briefly compared. The connections between them are explained. Based on these, the power reference is created by the IRP and a linear direct power controller is designed accordingly. It is similar to the multiple reference frame current controllers, which are suitable for high power/low switching frequency applications.

Finally, a new approach for three-phase loads compensation based on the instantaneous reactive power theory without transfer matrix also uses in the power reference. Simulations are provided and the proposed theory is verified on a three-phrase shunt hybrid power filter.

978-0-7695-3342-1/08 $25.00 © 2008 IEEE
DOI 10.1109/PEITS.2008.86

370

2. Modular structure

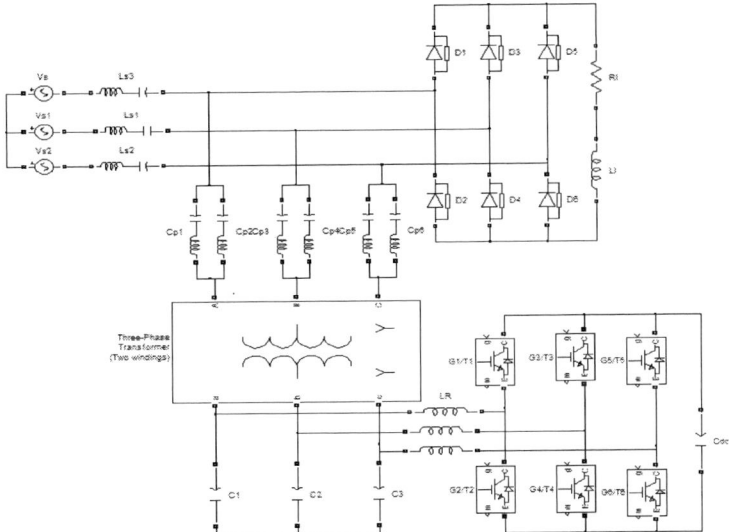

Fig. 1. Topology of Main Circuit for Three-phase Shunt Hybrid Power Filter

Fig. 1 shows the configuration of the TSHPF. It consists of an active power filter, three output filters, the coupling transformer and the passive filter. The active power filter is a voltage-source inverter which solves the dynamic compensation of the harmonic. There is a super capacitor in the DC terminal of the inverter. The AC terminal connects to the output filter(L_R, C_R) in order to filter the burr of the high frequency due to the on-load of the switching device. The coupling transformer carries out the electrical isolation between the active filter and the passive filter, and choices the appropriate ratio of voltages according to the grade of the voltage and the current. The passive filter shunted in the power grid consists of the 5th and 7th monotone filter groups.

Passive filters in shunt hybrid and shunt active power filters act as four roles: providing the harmonic current with low resonace channel; compensating the reactive power; isolating net voltage in order to decrease the withstanding voltage for active power filter; the medium for active power filter harmonic compensating voltage converting into harmonic current.

The output voltage of VSI consists of the switch frequency of the inverter and the switch harmonic wave near integral frequency. If the output voltage of the inverter is directly added into the circuit, it is obvious for the net to bring new high frequency harmonic wave pollution. Thus the output filter must be used to sieve high frequency burr caused for the inverter's on and off.

Coupling transformers acts as the rule of electric isolation and balancing the primary and the secondary voltage and current in shunt hybrid and shunt active power filters.

The expected voltage waveform through voltage source inverter is achieved by PWM control. The proposed PWM control signal uses a comparative method[15] of triangular carrier based on improving regular sample in this paper. Considered the basic principle of PWM control, the voltage of Dc side can't be too low, its minimum value relates with triangular carrier whose amplitude relates to modulation wave's.

3. Current control

Current controller design is very critical for the current controlled voltage source inverter system, especially for non-sinusoidal low switching frequency application such as medium voltage active power filter design. Existing current control techniques[16] can be classified into two main groups, linear and non-linear controllers. Non-linear controller is not suitable for low switching frequency applications and will not be discussed here.

In the linear group, there are four controllers that are generally studied for active filtering application: P stationary, PI stationary, PI synchronous and predictive (deadbeat) with constant switching frequency. As shown in Fig. 2, these conventional used current controllers of active power filters have been extensively studied[17,18], but most methods are not suitable for low switching frequency applications.

The conventional current controller can be summarized as:

When the reference current is a dc signal, zero steady-state error can be secured by using a conventional proportional integral (PI) controller.

When the reference current is a sinusoidal signal, straightforward use of the conventional PI controller would lead to steady-state error due to finite gain at the operating frequency. A synchronous-frame PI controller is popularly used which guarantees zero steady-state error. When the reference current is a non-sinusoidal signal, predictive/deadbeat controller is often used as a viable solution. Performance of the predictive controller, on the other hand, is subject to accuracy of the plant model as well as accuracy of the reference current prediction. This generally requires high switching frequency due to the nature of predictive controller.

Actually, the requirements of designing a non-sinusoidal current controller for active power filter can be considered as designing a multiple frequency current controller, which is compared with generally studied single frequency current controller

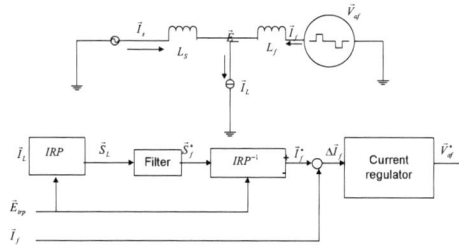

Fig. 2. Typical Current Controlled Voltage Source Inverter Based Parallel Active Power Filter

Generally, the dominant harmonics in medium voltage line are the 5th, 7th, 11th, 13th harmonics. A multiple synchronous reference frame based current control method [19] can be designed to reach zero steady state error for multiple harmonics, therefore it has superior benefits in low switching frequency application where feed forward gain has to be kept modest to limit the switching frequency at the carrier frequency. To simplify the complex rotation frame transformation, an equivalent stationary frame current controller can be developed by convolution derivation[11] or the direct physical concept[12]. However, the well-known cross-coupling term in rotation frame is not considered and the relationship between rotation frame and stationary frame is not straightforward to understand.

4. Direct power control based on SRF and IRP

Instantaneous reactive power [5] and synchronous reference frame [6] based active filtering theories have been popularly studied. To investigate the difference and connection between instantaneous reactive power theory and synchronous reference frame based control, mathematical models are derived separately. With complex vector, Eqs. (1)-(8) give the mathematical equations of synchronous reference frame based control; Eqs. (9)-(19) give the

mathematical equations of instantaneous reactive power based control.

The derivation of synchronous reference frame based control is:

$$\vec{V} = V_\alpha + jV_\beta \quad ; \quad \vec{I}_f = I_{f\alpha} + jI_{f\beta} \quad ; \quad \vec{E} = E_\alpha + jE_\beta \tag{1}$$

$$\vec{V} - \vec{E} = L_f \frac{d\vec{I}_f}{dt} \tag{2}$$

$$\vec{V} \times e^{-j\omega t} - \vec{E} \times e^{-j\omega t} = L_f \frac{d\vec{I}_f}{dt} \times e^{-j\omega t} \tag{3}$$

$$L_f \frac{d(\vec{I}_f \times e^{-j\omega t})}{dt} = L_f \frac{d\vec{I}_f}{dt} \times e^{-j\omega t} + L_f \vec{I}_f \frac{de^{-j\omega t}}{dt}$$

$$= L_f \frac{d\vec{I}_f}{dt} \times e^{-j\omega t} + L_f \vec{I}_f \times (-j\omega e^{-j\omega t})$$

$$= L_f \frac{d\vec{I}_f}{dt} \times E^{-j\omega t} - j\omega L_f \vec{I}_f \times e^{-j\omega t} \tag{4}$$

$$\Rightarrow L_f \frac{d\vec{I}_f}{dt} \times e^{-j\omega t} = L_f \frac{d(\vec{I}_f \times e^{-j\omega t})}{dt} + j\omega L_f \vec{I}_f \times e^{-j\omega t} \tag{5}$$

$$\vec{V} \times e^{-j\omega t} - \vec{E} \times e^{-j\omega t} = L_f \frac{d(\vec{I}_f \times e^{-j\omega t})}{dt} + j\omega L_f (\vec{I}_f \times e^{-j\omega t}) \tag{6}$$

$$\vec{V}_\omega = \vec{V} \times e^{-j\omega t} ; \quad \vec{E}_\omega = \vec{E} \times e^{-j\omega t} ; \quad \vec{I}_{f\omega} = \vec{I}_f \times e^{-j\omega t} \tag{7}$$

$$\vec{V}_\omega - \vec{E}_\omega = L_f \frac{d\vec{I}_{f\omega}}{dt} + j\omega L_f \vec{I}_{f\omega} \tag{8}$$

The derivation of instantaneous reactive power theory is:

$$\vec{V} = V_\alpha + jV_\beta ; \quad \vec{I}_f = I_{f\alpha} + jI_{f\beta} ; \quad \vec{E} = E_\alpha + jE_\beta \tag{9}$$

$$\vec{E}_{irp} = E_\alpha - jE_\beta \tag{10}$$

$$\vec{V} - \vec{E} = L_f \frac{d\vec{I}_f}{dt} \tag{11}$$

$$\vec{V} \times \vec{E}_{irp} - \vec{E} \times \vec{E}_{irp} = L_f \frac{d\vec{I}_f}{dt} \times \vec{E}_{irp} \tag{12}$$

$$L_f \frac{d(\vec{I}_f \times \vec{E}_{irp})}{dt} = L_f \frac{d\vec{I}_f}{dt} \times \vec{E}_{irp} + L_f \vec{I}_f \frac{d\vec{E}_{irp}}{dt} \tag{13}$$

$$\Rightarrow L_f \frac{d\vec{I}_f}{dt} \times \vec{E}_{irp} = L_f \frac{d(\vec{I}_f \times \vec{E}_{irp})}{dt} - L_f \vec{I}_f \frac{d\vec{E}_{irp}}{dt} \tag{14}$$

$$\vec{V} \times \vec{E}_{irp} - \vec{E} \times \vec{E}_{irp} = L_f \frac{d(\vec{I}_f \times \vec{E}_{irp})}{dt} - L_f \vec{I}_f \frac{d\vec{E}_{irp}}{dt} \tag{15}$$

$$\vec{V}_{irp}^2 = \vec{V} \times \vec{E}_{irp} \quad ; \quad \vec{E}_{irp}^2 = \vec{E} \times \vec{E}_{irp} \quad ; \quad \vec{S}_f = \vec{I}_f \times \vec{E}_{irp} \tag{16}$$

$$\vec{V}_{irp}^2 - \vec{E}_{irp}^2 = L_f \frac{d\vec{S}_f}{dt} - L_f \vec{I}_f \frac{d\vec{E}_{irp}}{dt} \tag{17}$$

Where $\vec{E}_{irp} = \sqrt{E_\alpha^2 + E_\beta^2} e^{-j\omega t}$, IRP based equation will become SRF based equation.

If

$$\vec{E}_{irp} = \sum_{k=1}^{n} E_k e^{-j\omega t} \tag{18}$$

$$\Rightarrow -L_f \vec{I}_f \frac{d\vec{E}_{irp}}{dt} = L_f \sum_{k=1}^{n} [(jk\omega E_k) \times (\vec{I}_f e^{-j\omega t})] \tag{19}$$

Eqs. (1) and (2) are the system model in stationary frame.

Eq. (3) is transforming the system model from stationary frame to the synchronous frame.

Eqs. (4) and (5) is showing the origin of the cross-coupling term in current differential term.

Eqs. (6)–(8) lead the derivation to the system

372

model in synchronous frame.

Eqs. (9)–(11) are the system model in current differential format.

Eq. (12) is transforming the system model from current differential format to power differential format.

Eqs. (13) and (14) is showing the origin of the cross-coupling term in power differential term.

Eqs. (15)–(17) lead the derivation to the system model in power differential format.

Eqs. (18) and (19) shows a general cross-coupling term if Eirp is non-sinusoidal term.

Based on the correlation between Eqs. $-L_f \vec{I}_f (d\vec{E}_{irp}/dt)$ (1)–(8) and (9)–(19), the similarities between them show that IRP is equivalent as SRF when E_{irp} only has one frequency component.

The cross-coupling term in IRP (shown in Eqs. (17) and (19)) reveals that the general nature of the cross coupling term in SRF is a cross-differential phenomenon, which can be explained as:

When a time domain signal is shifted in the frequency domain, the shifted frequency will be reflected back to the time domain with one more cross-differential term.

If the shifting is linear (one frequency or multiple frequency with the same size), the size of the cross-differential term is proportional to the distance (or frequency), which is how much the signal is shifted in the frequency domain.

If the shifting is non-linear (multiple frequency with different size), the size of the cross-differential term is not proportional to the distance (or frequency), which is how much the signal is shifted in frequency domain, but is the result of the sum of shifted distance (or frequency) multiplying their own size.

In the non-linear situation, the portion of the different shifted frequency is not observable in time domain, so cross decoupling in this case cannot be realized for multiple frequency.

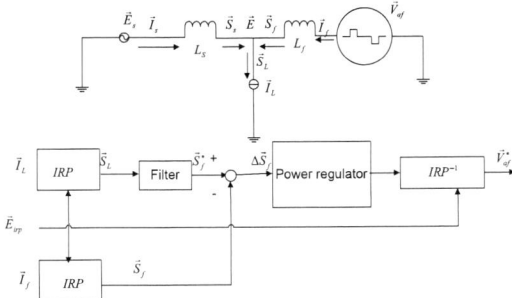

Fig. 3. Proposed Direct Power Controlled Voltage Source Inverter Based Active Power Filter

Similar to Fig.2, direct power control (Fig. 3) generates a power reference and feedback (controlled) power based on the IRP theory. The error is fed into a power regulator and the output of the power regulator is then transformed back by the IRP to be the voltage reference of the inverter. This new concept combines the IRP and current control into one controller called

direct power controller. The power reference is created by IRP and a linear direct power regulator is designed accordingly.

5. Simulation results

In order to validate the direct power control theory, a typical three-phase shunt hybrid power filter is considered in simulation, which is shown in Fig. 1. The non-linear load is simplified as a diode rectifier with inductance and resistance load. Simulations results of the TSHPF are demonstrated in Fig.

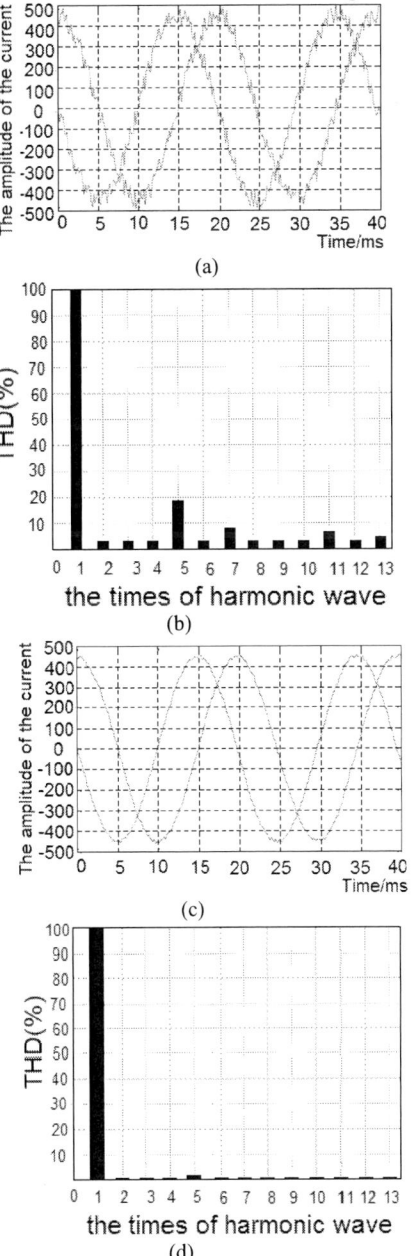

Fig. 6. Simulation Results of Single Phase Current Wave Form and Spectrum

In the Fig.6, (a) shows the amplitude of the current before the compensation of the current. (b) Shows the total harmonic distortion before the compensation of the current. (c) Shows the amplitude of the current after the compensation of the current. (d) Shows the total harmonic distortion after the compensation of the current. The above current wave forms describe that current wave forms in network improve distortion wave into proximity sine-wave and harmonic parameters are decreased greatly, when the application of shunt hybrid and shunt active power filters is developed.

6. Conclusion

A new control method on three-phase shunt hybrid active power filter is proposed in this paper. The state of theory derives from the comparative about the calculation between instantaneous reactive power (IRP) theory and synchronous reference frame (SRF) transformation. The control schemes are improved based on the typical voltage source active power filter. The results of the analysis show that the control method modified the original theory are put into use in power system. The Matlab-Simulink results show that the performance of shunt hybrid power active filter is improved and reduce the harmonic current after the method is used.

References

[1] H. Fujita, H. Akagi, "The unified power quality conditioner: the integration of series and shunt-active filters", IEEE Trans. Power Electron. 13 (2),1998, pp. 315–322.

[2] B. Singh, K. Al-Haddad, A. Chandra, "Performance comparison of two current control techniques applied to an active filter", in: Eighth International Conference on Harmonics and Quality of Power, vol. I, Athens, Greece, October 14–16, 1998, pp. 133–138.

[3] B. Singh, K. Al-Haddad, A. Chandra, "A review of active filters for power quality improvement", IEEE Trans. Ind. Electron. 46, 1999,pp. 960–971.

[4] S. Bhattacharya, D. Divan, "Synchronous frame based controller implementation for a hybrid series active filter system", in: Conference on Record IEEE-IAS Annual Meeting, 1995, pp. 2531–2540.

[5] F.Z. Peng, H. Akagi, A. Nabae, "A new approach to harmonic compensation in power systems—a combined system of shunt passive, series active filters", IEEE Trans. Ind. Appl. 26 (6) ,1990,pp. 983–990.

[6] J. Jacobs, D. Detjen, R. De Doncker, "A new hybrid filter versus a shunt power filter", in: European Conference on Power Electronics and Applications (EPE 2001), Graz, Austria, 2001.

[7] S.T. Senini, P.J.Wolfs, "Systematic identification and review of hybrid active filter topologies", in: Proceedings of the IEEE PESC'02, 2002, pp. 394–399.

[8] H. Fujita, H. Akagi, "The unified power quality conditioner: the integration of series and shunt-active filters", IEEE Trans. Power Electron. 13 (2),1998,pp. 315–322.

[9] S. Rahmani, K. Al-Haddad, F. Fnaiech, "A series hybrid power filter to compensate harmonic currents and voltages",in: IEEE Industrial Electronics Conference IECON02, Seville, Spain, 2002.

[10] D. Detjen, J. Jacobs, R. De Doncker, "A New Hybrid Filter to Damped Resonances and Compensate Harmonic Currents in Industrial Power Systems with Power Factor Correction Equipment", APEC, Anaheim, CA, 2001.

[11] D.N. Zmood, D.G. Holmes, G.H. Bode, Frequency-domain analysis of three-phase linear current regulators, IEEE Trans. Ind. Appl. 37 (March/April (2)) (2001).

[12] X.Yuan,W. Merk, H. Stemmler, J. Allmeling, Stationary-frame generalized integrators for current control of active power filters with zero steady-state error for current harmonics of concern under unbalanced and distorted operating conditions, IEEE Trans. Ind. Appl. 38 (March/April (2)) (2002).

[13] H. Akagi, A. Nabae, Control strategy of active power filters using multiple voltage source pwm converters, IEEE Trans. Ind. Appl. IA-22 (May/June) (1986) 460–465.

[14] S. Bhattacharya, T.M. Frank, D.M. Divan, B. Banerjee, Active filter system implementation, IEEE Ind. Appl. Mag. 4 (September/October) (1998) 47–63.

[15] Luo An, Tu Chunming. Design of harmonic analysis and elimination system in power grid[J]. J. Cent. South Univ. Technol., 2001, 32(6): 631-634.

[16] M.P. Kazmierkowski, L. Malesani, Current control techniques for three-phase voltage-source PWM converters: a survey, Ind. Electron. IEEE Trans. 45 (October (5)) (1998) 691–703.

[17] H. Akagi, New trends in active filters for power conditioning, IEEE Trans. Ind. Appl. 32 (6) (1996) 1312–1322.

[18] S. Buso, M. Luigi, M. Paolo, Comparison of current control techniques for active filter applications, IEEE Trans. Ind. Electron. 45 (5) (1998) 722–729.

[19] J. Allmeling, A control structure for fast harmonics compensation in active filters, IEEE Trans. Power Electron. 19 (March (2)) (2004).

A Novel Three-phase Active Power Filter Based on Instantaneous Reactive Power Theory

Wenjin Dai[1], Yu Wang[2]

Department of Information Engineering, Nanchang University
Nanchang, JiangXi, 330031, China
dwj480620@yahoo.com.cn, yuebook@yahoo.cn

Abstract

This paper presents a novel three-phase active power filter (APF). The control for this active power filter is based on the Instantaneous Reactive Power Theory (IRPT), which also knows as the p-q theory. APF is connected to non-linear and unbalanced load by shunt and the p-q theory application allows a compensation strategy-constant power. When APF is connected, the instantaneous power supplied by the source is constant. In this paper, the p-q theory is reformulated without using mapping matrices, which makes to obtain the compensation currents easier. Compared with conventional p-q and p-q-r theory, the different cases are considered and then simulated in order to show validity of the instantaneous reactive power theory. The practical cases at MATLAB simulation has been given to verify the theoretical analysis.

1. Introduction

Nowadays, with the widespread increase of power electronic loads in industry, the significant non-linear loads appear in a considerable amount of harmonic injection and low power factor in power systems. They tend to introduce harmonics in voltage and current waveforms at the point of common coupling. The existence of current and voltage harmonics in power systems increases losses in the lines, decreases the power factor and can cause timing errors in sensitive electronic equipments [1].

Conventionally, the passive filters have been used to eliminate current harmonics and to increase the power factor, which are simple and low cost. However, the use of passive filter has many disadvantages, such as large size, tuning and risk of resonance problems. Recently, because of the rapid progress in modern power electronic technology, the presented work was oriented mostly on the active filters instead of passive filters. The basic difference between the passive and active filters is that the active filters have the capability to compensate random varying currents [2, 3]. One of the most popular active filters is the Shunt Active Power Filter (SAPF). SAPF have been researched and developed, that they have gradually been recognized as a feasible solution to the problems created by non-linear loads. They are used to eliminate the unwanted harmonics and compensate fundamental reactive power consumed by non-linear loads through injecting the compensation currents into the AC lines. In addition to eliminating harmonic currents and improving the

power factor, SAPF can keep the power system balance under the condition of non-linear and unbalanced loads.

This SAPF is based on instantaneous reactive power theory (IRP) or p-q theory. With the theory, the more ambitious objective is achieved when the instantaneous power supplied by the source is constant and with the same value as the average power consumed by the load, which is named constant power compensation. If the source voltage is balanced and sinusoidal, the current will be balanced and sinusoidal after compensation. But when the source voltage is non-sinusoidal and unbalanced, the current will not be balanced and sinusoidal. This paper demonstrates a reformulation of p-q theory and all the compensation objectives can be obtained, such as unity power factor compensation or balanced and sinusoidal source current strategy. It avoids the mapping matrices (It is hard work in mathematical analysis), and it makes use of the victoria representation, which simplifies to obtain the compensation currents. The different compensation strategies have been applied to a practical case. Then the simulation results are given to confirm the validity of the instantaneous reactive power algorithms.

2. Active power filter

Fig.1 shows the basic compensation principle of SAPF. A SAPF is designed to be connected in parallel with the load, to detect its harmonic current and to inject into the system a compensating current. The resulting total current drawn from the AC main is sinusoidal. Ideally, the SAPF needs to generate just enough reactive and harmonic current to compensate the non-linear loads in the line [4].

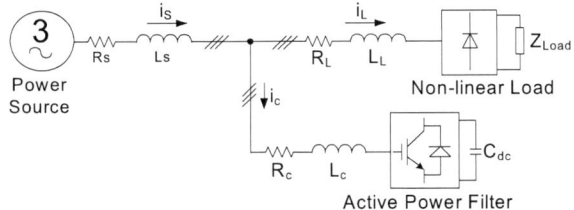

Fig.1 Basic Compensating Principle of SAPF

A current controlled voltage source inverter is used to generate the compensating current and is injected into the utility power source grid. Thus it can cancel the harmonic components drawn by the non-linear load and keeps the utility line current sinusoidal. The voltage source inverter with IGBT switches and an energy storage capacitor on DC bus is implemented as a SAPF. The main aim of the three-phase SAPF is to compensate harmonics and reactive power.

3. p-q theory

The instantaneous reactive power theory is the formulation with the largest diffusion along these years. For this reason, it is the most used as control strategy in SAPF. The p-q theory was developed for 3-phase 4-wire systems with balanced and sinusoidal source voltages. The compensation objective assumed was the obtainment of a constant instantaneous source power. This theory is based on a translation from the phase reference system (123) to the $0\,\alpha\,\beta$ system. The transformation matrix associated is as follows:

$$
\begin{bmatrix} e_0 \\ e_\alpha \\ e_\beta \end{bmatrix} = \sqrt{\frac{2}{3}} \begin{bmatrix} \frac{1}{\sqrt{2}} & \frac{1}{\sqrt{2}} & \frac{1}{\sqrt{2}} \\ 1 & -\frac{1}{2} & -\frac{1}{2} \\ 0 & \frac{\sqrt{3}}{2} & -\frac{\sqrt{3}}{2} \end{bmatrix} \begin{bmatrix} u_1 \\ u_2 \\ u_3 \end{bmatrix} \tag{1}
$$

$$
\begin{bmatrix} i_0 \\ i_\alpha \\ i_\beta \end{bmatrix} = \sqrt{\frac{2}{3}} \begin{bmatrix} \frac{1}{\sqrt{2}} & \frac{1}{\sqrt{2}} & \frac{1}{\sqrt{2}} \\ 1 & -\frac{1}{2} & -\frac{1}{2} \\ 0 & \frac{\sqrt{3}}{2} & -\frac{\sqrt{3}}{2} \end{bmatrix} \begin{bmatrix} i_1 \\ i_2 \\ i_3 \end{bmatrix} \tag{2}
$$

From Eqs. (1) and (2), it can be deduced that

$$
i_{N} = i_1 + i_2 + i_3 = i_0\sqrt{3} \tag{3}
$$

The different power terms are defined as follows:

$$
\begin{bmatrix} p_0 \\ p_{\alpha\beta} \\ q_{\alpha\beta} \end{bmatrix} = \begin{bmatrix} e_0 & 0 & 0 \\ 0 & e_\alpha & e_\beta \\ 0 & -e_\beta & e_\alpha \end{bmatrix} \begin{bmatrix} i_0 \\ i_\alpha \\ i_\beta \end{bmatrix} = [T] \begin{bmatrix} i_0 \\ i_\alpha \\ i_\beta \end{bmatrix} \tag{4}
$$

P0 is the zero sequence (real) instantaneous power, $p_{\alpha\beta}$ is the real instantaneous power and $q_{\alpha\beta}$ is the imaginary instantaneous power. Considering the [T] inverse matrix, the calculation of the current components from the different power terms is possible. The expression is shown as that:

$$
\begin{bmatrix} i_0 \\ i_\alpha \\ i_\beta \end{bmatrix} = \frac{1}{e_0 e^2{}_{\alpha\beta}} \begin{bmatrix} e^2{}_{\alpha\beta} & 0 & 0 \\ 0 & e_0 e_\alpha & -e_0 e_\beta \\ 0 & e_0 e_\beta & e_0 e_\alpha \end{bmatrix} \begin{bmatrix} p_0 \\ p_{\alpha\beta} \\ q_{\alpha\beta} \end{bmatrix} \tag{5}
$$

$$
e^2{}_{\alpha\beta} = e^2{}_\alpha + e^2{}_\beta
$$

Although original p-q theory is developed from the mapping matrices, an alternative develop is possible. This model does not use mapping matrices. Therefore, it makes easier the theory treatment and its application to the active filters control. The

vectored frame allows obtaining any compensation strategy in a simple way, while the mapping matrices have been always limited to the constant power strategy [5, 6]. In fact, the vectored frame is a conceptual approach. Three new voltage vectors will be defined as Fig. 2.

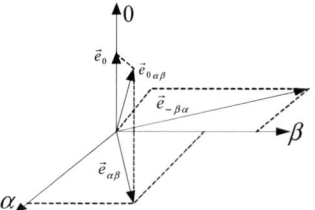

Fig.2 New Voltage Vectors in $0\alpha\beta$ Coordinates System

The voltage and current space vectors are defined as follows:

$$
\vec{e}_{\alpha\beta} = \begin{bmatrix} 0 \\ e_\alpha \\ e_\beta \end{bmatrix}; \vec{e}_{-\beta\alpha} = \begin{bmatrix} 0 \\ -e_\beta \\ e_\alpha \end{bmatrix}; \vec{v} = \vec{e}_0 = \begin{bmatrix} e_0 \\ 0 \\ 0 \end{bmatrix}; \vec{i} = \begin{bmatrix} i_0 \\ i_\alpha \\ i_\beta \end{bmatrix} \tag{6}
$$

$\vec{e}_{-\beta\alpha}$ is the orthogonal voltage vector and $\vec{e}_0$ is the zero sequence voltage vector. It is always verified as that :

$$
\vec{e}_{0\alpha\beta} = \begin{bmatrix} e_0 & e_\alpha & e_\beta \end{bmatrix}^t = \vec{e}_{\alpha\beta} + \vec{e}_0 \text{ and } \vec{e}_{-\beta\alpha}.\vec{e}_{0\alpha\beta} = 0 \tag{7}
$$

In the $0\alpha\beta$ reference frame, the current vector $\vec{i}$ will be expressed as the sum of its projections over the vectors $\vec{e}_0$, $\vec{e}_{\alpha\beta}$ and $\vec{e}_{-\beta\alpha}$, this can be expressed as follows:

$$
\begin{aligned}
\vec{i} &= \frac{p_{\alpha\beta}(t)}{\vec{e}_{\alpha\beta}.\vec{e}_{\alpha\beta}} \vec{e}_{\alpha\beta} + \frac{q_{\alpha\beta}(t)}{\vec{e}_{-\beta\alpha}.\vec{e}_{-\beta\alpha}} \vec{e}_{-\beta\alpha} + \frac{p_0(t)}{\vec{e}_0.\vec{e}_0} \vec{e}_0 \\
&= \frac{p_{\alpha\beta}(t)}{\vec{e}_{\alpha\beta}.\vec{e}_{\alpha\beta}} \vec{e}_{\alpha\beta} + \frac{q_{\alpha\beta}(t)}{\vec{e}_{\alpha\beta}.\vec{e}_{\alpha\beta}} \vec{e}_{-\beta\alpha} + \frac{p_0(t)}{\vec{e}_0.\vec{e}_0} \vec{e}_0
\end{aligned} \tag{8}
$$

Where $p_{\alpha\beta} = \vec{e}_{\alpha\beta}.\vec{i}$ is the instantaneous real power in α - β components, $q_{\alpha\beta} = \vec{e}_{-\beta\alpha}.\vec{i}$ is the instantaneous imaginary power and p0(t) is the zero sequence instantaneous real power. They are identical to those power perms defined in Esq. (4). The fact that the orthogonal voltage vector norm and the voltage vector without zero-sequence component norm are the same has been considered in Esq. (8). Since now the compensation currents will be obtained from the proposed strategy within the model presented in Esq. (8), without using mapping matrices.

4. Constant power compensation

The strategy assumed by the p-q theory has been the obtainment of a constant instantaneous power in the source side with the only restriction of getting a null average instantaneous power exchanged by the compensator p_c. Lower case represents instantaneous values. In fact, total power required by the load can be expressed as:

$$p_L(t) = p_{L\alpha\beta}(t) + p_{L0}(t) = p_{L\alpha\beta} + \widetilde{p}_{L\alpha\beta}(t) + p_{L0} + \widetilde{p}_{L0}(t) \quad (9)$$

where the upper case are referred to the average values and the terms with the character $\sim$ over it are referred to the power oscillatory component. To calculate the compensator current, and according to Fig. 3:

$$p_C(t) = p_L(t) - p_S(t) = p_L(t) - p_L \quad (10)$$

p_L is the total active power incoming to the load and $p_S(t) = p_L$ after compensation. Taking into account Esq. (8), and the independence of $0\alpha\beta$ coordinates, the Esq. (9) can be expressed as the follows:

$$p_{C\alpha\beta}(t) = p_{L\alpha\beta}(t) - p_{L\alpha\beta} = \widetilde{p}_{L\alpha\beta}(t) \quad (11)$$

$$p_{C0}(t) = p_{L0}(t) - p_{L0} = \widetilde{p}_{L0}(t) \quad (12)$$

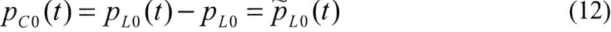

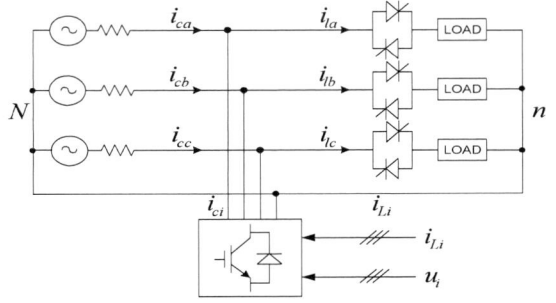

Fig.3 Three-phase Four-wire Compensated System

These equations means that the average value $\langle pc(t) \rangle = p_C = 0$, where $\widetilde{p}_L(t)$ represents the AC or oscillatory part of $p_L(t)$. On the other hand, the instantaneous imaginary power exchanged by the compensator must be the same as the instantaneous imaginary power required by the load $q_C(t) = q_{L\alpha\beta}(t)$. Therefore, from Esq. (5), the compensation current is shown as the follows:

$$\begin{bmatrix} i_{C0} \\ i_{C\alpha} \\ i_{C\beta} \end{bmatrix} = \frac{1}{e_0 e^2_{\alpha\beta}} \begin{bmatrix} e^2_{\alpha\beta} & 0 & 0 \\ 0 & e_0 e_\alpha & -e_0 e_\beta \\ 0 & e_0 e_\beta & e_0 e_\alpha \end{bmatrix} \begin{bmatrix} \widetilde{p}_{L0}(t) \\ \widetilde{p}_{L\alpha\beta}(t) \\ q_{L\alpha\beta}(t) \end{bmatrix} \quad (13)$$

This strategy achieves a constant instantaneous power supplied by the source. But, it does not eliminate the neutral current. After compensation, the source supplied a constant instantaneous power with a value identical to the load active power with a neutral current not null.

The results presented in Esq. (13) have been got by mean of the mapping matrices. In this paper, with the model of the p-q theory without mapping matrices, the constant power compensation strategy can be developed. Besides, the elimination of the neutral is possible. The corresponding procedure is presented as follows: After compensation, the source current $\vec{i}_S(t) = K\vec{e}_{\alpha\beta}$. Taking into account that $\vec{e}_0 . \vec{e}_{\alpha\beta} = 0$, from Esq. (6), the source total instantaneous real

power must be:

$$p_S(t) = \vec{e}_{0\alpha\beta} \cdot \vec{i}_S = K\vec{e}_{0\alpha\beta} \cdot \vec{e}_{\alpha\beta} = K\vec{e}_{\alpha\beta} \cdot \vec{e}_{\alpha\beta} = p_L \quad (14)$$

If $p_S(t) = p_L$, the proportionality factor can be obtained and it gets $K = p_L/e^2_{\alpha\beta}$. Therefore, from this and Esq. (7), the compensation currents may be expressed as a vector:

$$\begin{aligned} \vec{i}_C(t) &= \vec{i}_L(t) - i_S(t) \\ &= \frac{p_{\alpha\beta}(t)}{e^2_{\alpha\beta}} \vec{e}_{\alpha\beta} - \frac{p_L}{e^2_{\alpha\beta}} \vec{e}_{\alpha\beta} + \frac{p_{L0}(t)}{e^2_0} \vec{e}_0 + \frac{q_{L\alpha\beta}}{e^2_{\alpha\beta}} \vec{e}_{-\beta\alpha} \\ &= \frac{\widetilde{p}_{\alpha\beta}(t) - p_{L0}}{e^2_{\alpha\beta}} \vec{e}_{\alpha\beta} + \frac{p_{L0}(t)}{e_0} \vec{e}_0 + \frac{q_{L\alpha\beta}}{e^2_{\alpha\beta}} \vec{e}_{-\beta\alpha} \end{aligned} \quad (15)$$

$\widetilde{p}_{\alpha\beta}(t)$ represents the oscillatory part of the instantaneous real power in the 0-α-β system. From Esq. (15), such component of the source current is as the follows:

$$i_{C0} = \frac{p_{L0}(t)}{e^2_0} e_0, \quad i_{C\alpha} = \frac{\widetilde{p}_{L\alpha\beta}(t) - p_{L0}}{e^2_{\alpha\beta}} e_\alpha - \frac{q_{L\alpha\beta}}{e^2_{\alpha\beta}} e_\beta$$

$$i_{C\beta} = \frac{\widetilde{p}_{L\alpha\beta}(t) - p_{L0}}{e^2_{\alpha\beta}} e_\beta + \frac{q_{L\alpha\beta}}{e^2_{\alpha\beta}} e_\alpha \quad (16)$$

5. Proposed control algorithm

The p-q theory is suitable for ideal three-phase systems but is inadequate under non-ideal mains voltage cases. Under non-ideal mains voltage conditions, the sum of components $(v^2\alpha + v^2\beta)$ will not be constant and the alternating values of the instantaneous real and imaginer power have current harmonics and voltage harmonics. Consequently, the SAPF does not generate compensation current equal to current harmonics and gives to mains more than load harmonics than required. To overcome these limitations, the p-q theory based a new control algorithm to decrease total harmonic distortion for desired level is proposed. The instantaneous reactive and active powers have to calculate after filtering of mains voltages. In this paper, the control theory is designed for three-phase four-wire inverter system. The proposed method has a simple algorithm, which allows compensating harmonics, reactive power, neutral current and imbalance load currents under unbalanced non-linear load and non-ideal mains voltage cases. The proposed p-q theory based method block diagram for the four-leg SAPF is shown in Fig.4. Since the mains voltages applied to control algorithm of SAPF is to be balanced and sinusoidal, the proposed voltage harmonics filter block diagram is show in Fig.5. In the proposed method, the instantaneous voltages are first converted to synchronous d-q coordinates as the follows:

$$\begin{bmatrix} v_d \\ v_q \end{bmatrix} = \sqrt{\frac{2}{3}} \begin{bmatrix} \sin(wt) & \sin(wt-120°) & \sin(wt+120°) \\ \cos(wt) & \cos(wt-120°) & \cos(wt+120°) \end{bmatrix} \times \begin{bmatrix} v_a \\ v_b \\ v_c \end{bmatrix}$$

(17)

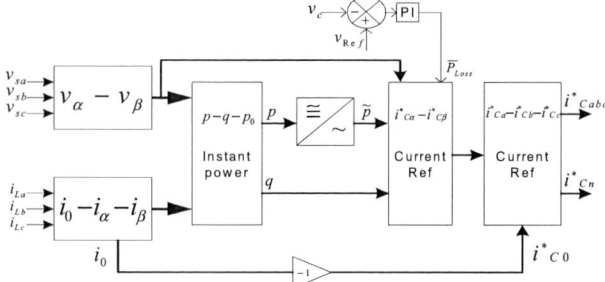

Fig.4 Block Diagram for Proposed Method Based on p-q Theory

The produced d-q components of voltages are filtered by using the 5th order low-pass filters (LPF) with a cut-off frequency at 50 Hz. These filtered d-q components of voltages are reverse converted α - β coordinates as expressed in Esq.(17). These α - β components of voltages are used in conventional IRP theory. Hence, the non-ideal main voltages are converted to ideal sinusoidal shape by using LPF in d-q coordinate.

$$\begin{bmatrix} v_\alpha \\ v_\beta \end{bmatrix} = \begin{bmatrix} \sin(wt) & \cos(wt) \\ \sin(wt) & -\cos(wt) \end{bmatrix} \begin{bmatrix} \bar{v}_d \\ \bar{v}_q \end{bmatrix} \quad (18)$$

Thus, the mains voltages are assumed to be an ideal source in the calculation process. Since the APF input voltages have no zero-sequence components, zero-sequence power is (p0) to be always zero. These reference currents calculated by the control algorithm equations should be supplied to the power system by switching of the IGBT of the inverter. The method for generation of the switching pattern is achieved by the instantaneous current control of the four-leg APF line currents. The actual four-leg APF line currents are monitored instantaneously, and then compared to the reference currents generated by the control algorithm. A hysteresis-band PWM current control is implemented to generate the switching pattern of the VSI. The hysteresis-band PWM current control is the fastest control method with minimum hardware and software but variable switching frequency is its main shortcoming.

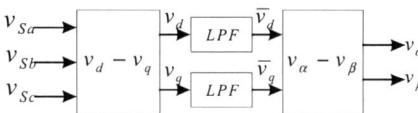

Fig. 5 Voltage Harmonics Filtering Block Diagram

6. Simulation results

For the harmonic current filtering, reactive power compensation, load current balancing and neutral current elimination performance of the four-leg APF with the proposed method, the p-q theory have been examined under unbalanced voltage case. The purpose of the designed case studies is to show the validity and performance of the proposed APF control strategy, even if the mains voltages are highly distorted and unbalanced. The presented simulation results were obtained by using Matlab-Simulink Power System Toolbox for

a three-phase four-wire power distribution system with a shunt APF.

Three-phase thyristor rectifier and single-phase diode rectifier non-linear loads are connected to the power system, in order to produce an unbalance, harmonic and reactive current in the phase currents and zero-sequence harmonics in the neutral current. The four-leg APF is switched on 0.15s later. After 0.2s, a single-phase diode bridge rectifier load is connected to evaluate the dynamic performance of the APF. Firing angle of three-phase thyristor rectifier is α =30o and RL load is connected on the DC side. The DC side of single-phase diode rectifiers is connected RC filtered ohmic load. Since the reactive power compensation performance of the APF is showed clearly, load and source current are enlarge to two times in phase c. The comprehensive simulation results are discussed as follows:

Unbalanced loads or single-phase are not evenly distribution between the phase of three-phase system will cause voltage unbalance. Excessive voltage unbalance can cause motor overheating and failure of power conversion components and increases the stresses of power electronics. When three-phase power system is not balanced, effective values of phase voltages is not equal and there will be fundamental negative-sequence voltage component in the unbalanced voltage. Harmonic current suppression and load current balancing simulation results with the instantaneous reactive power theory and the proposed method for this APF under unbalanced voltage are shown in Fig. (6). Detailed summary of A、B、C-phase and neutral currents and their total harmonic distortion (THD) levels are shown in Table 1. Harmonic spectra of load and source current under ideal voltage case is shown in Fig. (7).

Since negative-sequence component of unbalanced voltage with the proposed method is eliminated, after compensation three-phase source currents are balanced and sinusoidal. The unbalanced voltage in three-phase four-wire power system will not affect the four-leg APF performance with proposed algorithm.

Table 1 Source Currents and Their THD% under Unbalanced Voltage

	Without APF	With APF
A-phase	22.10	3.23
B-phase	27.31	3.31
C-phase	24.17	3.54
Neutral	39.70	-

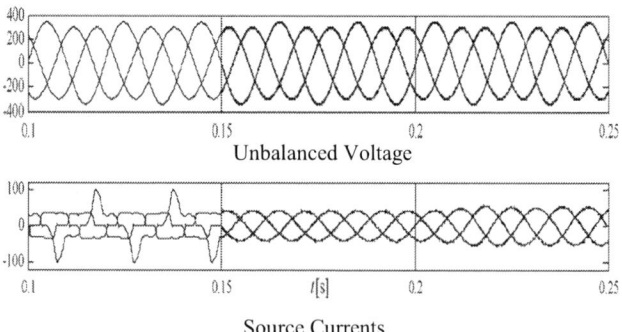

Unbalanced Voltage

Source Currents

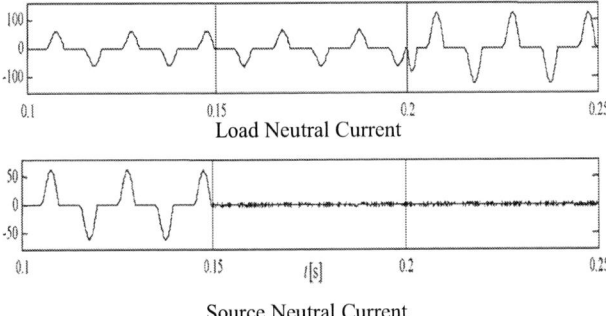

Load Neutral Current

Source Neutral Current

Fig. 6 Source Currents and Source Neutral Current under Unbalanced Voltage

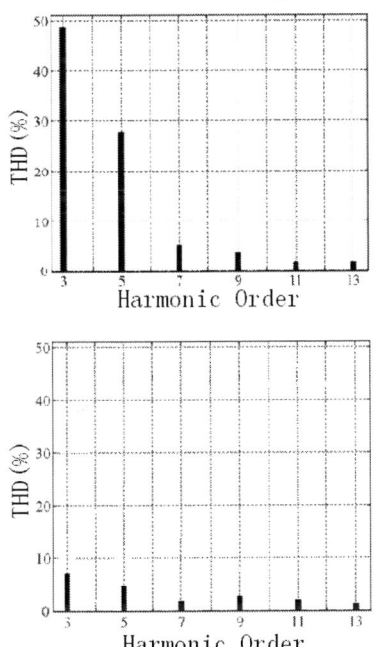

Fig. 7 Harmonic Spectra of Lload and Source Current under Unbalanced Voltage

7. Conclusion

This paper has developed an analysis of the p-q instantaneous reactive power theory based on a proposed method. This three-phase four-wire shunt APF control algorithm has been proposed to improve the performance of the APF under unbalanced voltage cases. The new control theory has been presented which is suitable for the shunt APF design under unbalanced voltage. The computer simulation has verified the effectiveness of the proposed control scheme. The simulation results are found quite satisfactory to eliminate harmonics and reactive power components under balanced and unbalanced voltage.

8. References

[1] Yuce E, et al. Stability problems in universal current-mode filters, Int. J. Electron. Common. (AEU) (2006), doi: 10.1016/j.aeue.2006.09.007

[2] H. Kim, H. Akagi, The instantaneous power theory on the rotating p-q-r reference frames, in: IEEE Proceeding on Power Electronic and Drive System, 1999.

[3] A.E.Emanuel, Summary of IEEE Standard 1459: definitions for the measurement of electric power quantities under sinusoidal, nonsinusoidal, balanced and unbalanced conditions, IEEE Trans. Ind. Appl. 40(3) (2004) 869-876.

[4] Saetieo S, Devaraj R, Tomey D. The design and implementation of a three-phase active power filter based on sliding mode control. IEEE T Ind Appl 1995;31:993 – 1000.

[5] A. Hom, L.A. Pittorrino, J.H.R. Enslin, Evaluation of active power filter control altorithms under non-sinusoidal and unbalanced conditions, in: Proceedings of the Seventh International Conference on Harmonics and Qulity of Power, pp. 217-224.

[6] M. Depenbrock, V. Staudt, H. Wrede, A theoretical investigation of original and modified instantaneous power theory applied to four-wire systems, IEEE Trans. Ind. Appl. 39 (4) (2003) 1160-1167.

Study and Design of Three-Phase Voltage-Source PWM Rectifier in Movable Power Station

Xinggui Wang[1], Xiaoying Li [2]

[1]Department of Electrical and Information Engineering,
Lanzhou University of Science and Technology,
Lanzhou，Gansu，730050，China
[2]Department of Electrical and Information Engineering,
Lanzhou University of Science and Technology,
Lanzhou，Gansu，730050，China
wangxg82@tom.com, linda_800909@163.com

Abstract

This paper presents the research and design of a modified PWM rectifier in movable power station drive system.The topological structure is analyzed.A novel SVPWM algorithm is proposed and implemented based on TI's DSP TMS320LF2407.The experimental results show that this PWM rectifier reduces harmonic pollution and advances voltage utility ratio.The power quality of movable power station is improved effectively.

1. Introduction

Movable power station is widely used in fields such as military command post, oil-drilling well as long-term power supply or in hospital,school and other fields where emergency power supply is needed. In movable power station system, there is harmonic distortion mainly caused by nonlinear loads especially phase-controlled rectifier[1]. The distortion of voltage or circuit will not only do harm to generator set,transformer and other equipment but also threat the stablility of the whole system[2].

Movable power station belongs to small capacity electric network. It is apt to be influenced by harmonic distortion. So harmonic elimination is especially important.There are many methods of harmonic suppression,such as reactive harmonic suppressors and active filter[3].These installations have disadvantages of complex structure and high maintainance cost.So more active and effective way to solve harmonic pollution is to eliminate harmonic source.The actual method is adopting PWM rectifier which has capabilities of bi-directional power flowand high power factor to take place of phase-controlled one[3].

This paper adopted three-phase voltage-source PWM rectifier, analyzed fixed switching frequency SVPWM control algorithm and designed a small capacity system based on TI's DSP TMS320LF2407.

2. D-Q model of PWM rectifier

PWM rectifier model circuit is shown as Figure 1:

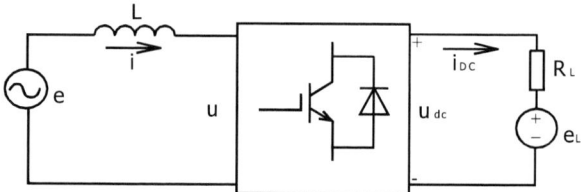

Figure 1 PWM rectifier model circuit

This circuit is composed of AC loop,DC loop and power switching bridge.When power loss is ignored,expession (1) is deduced according to power balance relation:

$$ui = u_{DC}i_{DC} \qquad (1)$$

In expression (1):
  u,i—the voltage and current of AC side;

u_{DC}, i_{DC}—the voltage and current of DC side.

It can be concluded by rxpression (1) that we can control the DC side through the AC side,vice versa.

Among various PWM rectifiers,voltage-source(VSR in short) PWM rectifier is more common in use as its simple constructure and lower power loss[4].So in movable power station drive system a three-phase voltage PWM rectifier which is shown in Figure 2 is choosed.

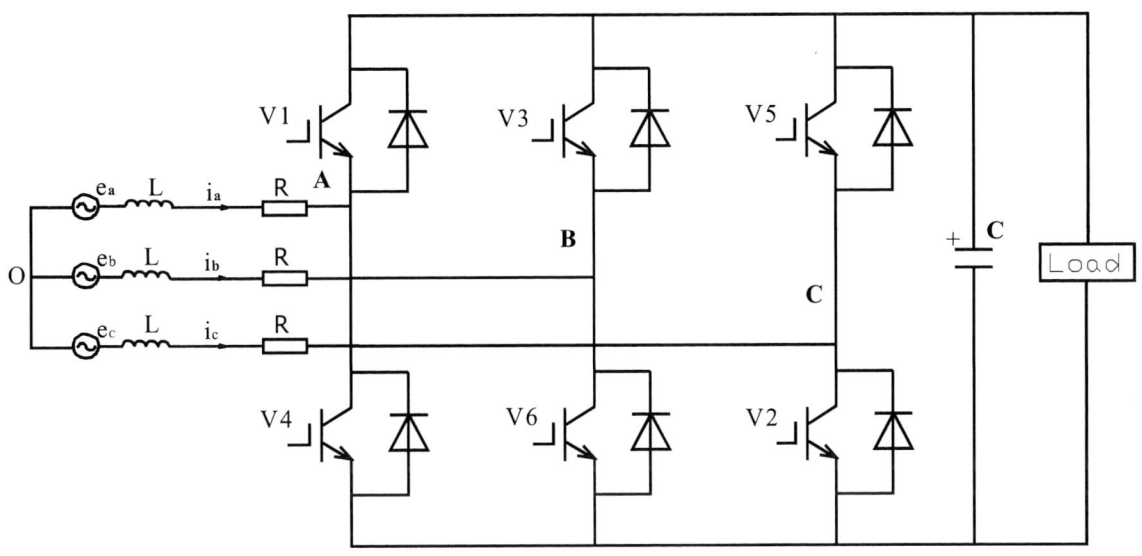

Figure 2 Three-phased voltage-source PWM rectifier

The mathematical models of VSR PWM rectifier in the d-q synchronous rotation frame of axes are as following:

$$\frac{di_d}{dt} = \frac{e_d}{L} + \omega i_q - \frac{R}{L} i_d - \frac{v_{dc}}{L} \mu_d$$

$$\frac{di_q}{dt} = \frac{e_q}{L} - \frac{R}{L} i_q - \omega i_d - \frac{v_{dc}}{L} \mu_q \qquad (2)$$

$$\frac{dv_{dc}}{dt} = \frac{3}{2C} (i_d \mu_d + i_q \mu_q) - \frac{v_{dc}}{R_L C}$$

In equation (2):

i_d, i_q —ac-side current components of d-axes and

q-axes;

v_{dc} —voltage of dc-side;

e_d、e_q —supply electromotive force components of

d-axes and q-axes;

L、R—ac-side inductance and resisitance;

C—dc-side capacitance ;

ω —ac-side fundamental angular frequency;

μ_d、μ_q —control input.

From equation(2),the current of d-axis and q-axis are decoupled[5]. According to above d-q model, i_d and i_q can

be regulated by μ_d and μ_q.

3. Control Strategy

3.1. Principle of voltage space vector PWM control

There are several control strategies of three-phase VSR PWM rectifier[4].When applied in practical system, people incline to control strategy which is simple in algorithm and convenient in realization. Research has suggested that SVPWM control scheme promotes voltage utlization ratio comparing to SPWM control[6].In movable power station drive system,fixed switching frequency direct-current control SVPWM scheme is selected to reduce harmonic content and switching loss.

In Figure 2,the drive signal of each power switch in one bridge arm is complementary.If Vdc denotes the voltage of dc-side and Va0,Vb0,Vc0 stands for ac-side phase-voltage respectively. When V1 is on and V4 is off,va0=vdc;when V4 is on and V1 is off,Va0=0.Define Sa，Sb，Sc as switching variable of A,B,C bridge arm. Output voltage of each arm can be denoted by production of SaSbSc and Vdc. As Sa,Sb,Sc has two possible values: 0 or 1.So three phase VSR PWM rectifier has 8 switching states,state0 ～ state7.These 8 states and corresponding voltage of each state are shown in Table1.

Table 1 Switching state and output voltage

Swiching state	Sa Sb Sc	Vab/Vdc	Vbc/Vdc	Vca/Vdc	Va0/Vdc	Vb0/Vdc	Vc0/Vdc
0	000	0	0	0	0	0	0
1	001	0	-1	1	-1/3	-1/3	2/3
2	010	-1	1	0	-1/3	2/3	-1/3
3	011	-1	0	1	-2/3	1/3	1/3
4	100	1	0	-1	2/3	-1/3	-1/3
5	101	1	-1	0	1/3	-2/3	1/3
6	110	0	1	-1	1/3	1/3	-2/3
7	111	0	0	0	0	0	0

In above 8 switching states, state0 and state1 are zero state while other 6 states are nonzero states.According to the definition of switching state,equation (3) can be written as follows:

$$\begin{cases} u_{AO} = S_a \cdot U_D \\ u_{BO} = S_b \cdot U_D \\ u_{CO} = S_c \cdot U_D \end{cases} \quad (3)$$

And equation (4) can be deduced as :

$$\begin{cases} u_{AB} = u_{AO} - u_{BO} = (S_a - S_b) \cdot U_D \\ u_{BC} = u_{BO} - u_{CO} = (S_b - S_c) \cdot U_D \\ u_{CA} = u_{CO} - u_{AO} = (S_c - S_a) \cdot U_D \end{cases} \quad (4)$$

The matrix form of equation(4) is as equation(5):

$$\begin{bmatrix} u_{AB} \\ u_{BC} \\ u_{CA} \end{bmatrix} = U_D \cdot \begin{bmatrix} 1 & -1 & 0 \\ 0 & 1 & -1 \\ -1 & 0 & 1 \end{bmatrix} \cdot \begin{bmatrix} S_a \\ S_b \\ S_c \end{bmatrix} \quad (5)$$

When switching state is state4(100),that is Sa=1,Sb=Sc=0.The equivalent space voltage vector $\vec{U}$ is on the a-phase axis($\omega t = 0$).Other nonzero states also are equivalent to space voltage vectors of space position $\omega t = 60°,120°,180°,240°$ and $300°$ respectivy.Zero state 000 and 111 is equivalent to zero vecor $\vec{U}_Z$ .All is shown in Figure 3.

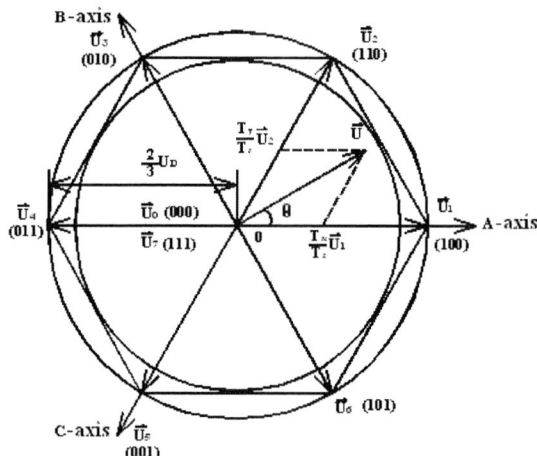

Figure3 Space voltage vector of three phase VSR rectifier

3.2. Control strategy and its realization based on DSP

(1) Control scheme of VSR PWM rectifier

Combining the d-q model of VSR PWM rectifier,the inverter's current can be controlled by the swicthing of voltage space vector.The PI controller is adopted as current regulator in the d-q synchronous rotation frame of axes to generate instruction signal of voltage space vector.And voltage space vector of VSR will follow the instruction voltage vector by SVPWM control .

(2) Control system based on TMS320LF2407

The framework of control system is shown in Figure 4

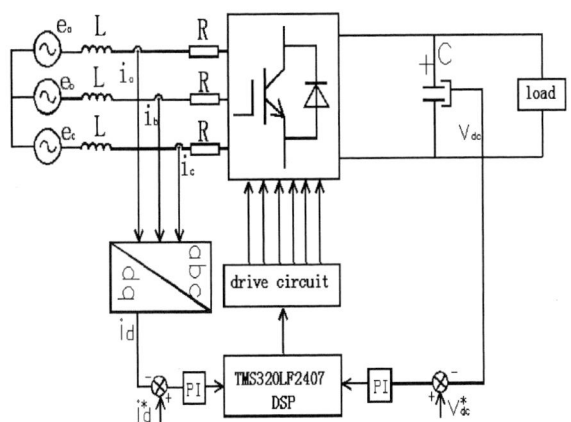

Figure 4 Framework chart of control system

Hall sensors are used to detect output dc voltage and input current.There are one voltage loop and one current loop in this control system. The space voltage vector PWM algorithm is completed in DSP.Protection and drive circuit is designed too,but is not included in this paper.

4. Prototype comprision and experimental results

The movable power station supply can be equivalent to three-phase balanced electromotive force as Figure1 shown. A prototype based on Figure 1 has been implemented and tested. The experimental parameters are set as following:input voltage is 220V/50Hz,L is10mH, IGBT is 1000V/25A,C is 2200µF/450V.

To verify the performance of this appratus,experiment is completed.Partial experimental results are given as follows.

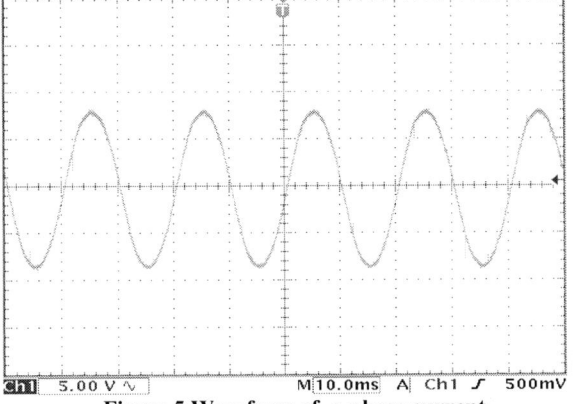

Figure 5 Waveform of a-phase current

5. Conclusions

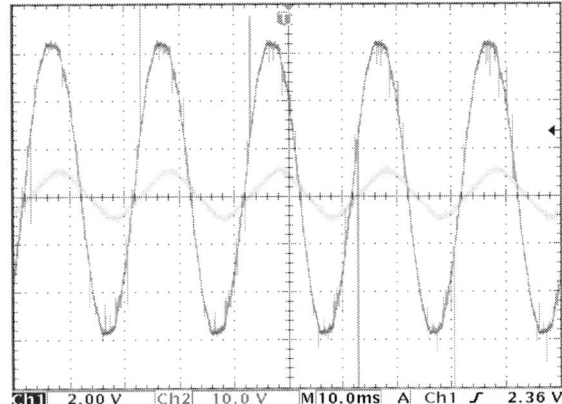

Figure 6 Waveform of a-phase voltage and current

The design and implementation of three-phase voltage PWM rectifier in movable power station drive system has been presented in this paper.Fixed switching frequency SVPWM direct current control strategy is studied and realized based on DSP.The validity and feasibility is proved by experiments.The experimental results also demonstrate that this PWM rectifier has low harmonic distortion and unity power factorThis PWM inverter can be used in drive system to provide variable DC power supply besides improving the power factor of movable power station.It will have good application prospect in oil field, military command post and other places where movable power station is used.

References

[1] Yingmin Yan."The analysis of endangerous harmonic existing in military generating set".*Movable Power Station & Vehicle.* 2001,No.2,pp.27-28 (in Chinese).

[2] Hongwen Liu."Study on harmonic analysis of movable power station".*Movable Power Station & Vehicle.* 2005,No.2,pp.24-26 (in Chinese).

[3] L.S.Czarnecki. "Harmonics and power phenomena".*Wiley Encyclopedia of Electrical and Electronics Engineering, John Wiley & Sons, Inc.,* Supplement 1.2000, pp. 195-218.

[4] Xing Zhang, Chongwei Zhang, PWM Rectifier And Control. Beijing, China: Mechanical Industry Publishing Company, 2003.

[5] Zhifeng Lu, Bo Zhang, Weihua Deng(2005). "The research on nonlinear decoupling control of three-phase voltage PWM rectifier". *Power Electronics,.*Vol. 39, pp. 40–44.

[6] Jian Xiong(1999). "Comparison Study of Voltage Space Vector PWM and Conventional SPWM".*Power Electronics*, No.1,pp.25-28.

Modeling and Simulation of Monitor-Control Network in Ship Power Station

Dandan Chen [1], Li Xia [1], Haifeng Wang [2]

[1] College of Electric and Information Engineering,
Naval University of Engineering, Wuhan, HuBei, 430033 China
[2] Equipment Section of Quality Controlling,
Naval Unit No.91663, Qingdao, ShanDong, 266011, China
chendandanwhf@163.com

Abstract

In order to improve the whole design level of ship power system and reduce unnecessary expense, according to the actual situation of the ship power system, the transmission performance of monitor control network in ship power station is researched before prototype. Firstly, according to the actual power station, monitor-control network framework is given. The data incoming rule within power station is analyzed thoroughly and mathematical models of different kinds of data flows are developed. Secondly, a set of CAN bus models are designed and MAC layer process model is introduced in detail. Finally, the transmission delays in the traditional Ethernet, switched Ethernet and CAN are obtained by simulation. The result shows the real-time performance of these networks in the ship power station. The switched Ethernet is the best. The delay of critical messages in the CAN bus are small enough to meet the real-time request of the monitor control in ship power station. The delay of the traditional Ethernet is so long that it could not meet the real-time request.

1. Introduction

As the development of control, communication and network, the fields of data exchange cover device level, control level and management level. Fieldbus and industrial Ethernet have a fast progress these years. For the performance of these communication networks, many experts have done lots of researches and obtained great harvest [1-2]. In the monitor control network of ship power station, there are many kinds of messages, large traffic and strict request of real-time. However, the bandwidth of the fieldbus is limited and the delay of Ethernet is not determinate, which becomes a problem that whether fieldbus or Ethernet could support these traffic and meet the real-time request. If we can study the transmission performance of all kinds of messages based on the actual monitor in ship power station before prototype, it will not only improve the whole design level, but also reduce unnecessary expense.

Recently Ethernet networks are being introduced even at the Field level. CAN bus is the only fieldbus that enters into ISO standard and is used widely in the ship. Thus in this paper, a monitor network built by traditional Ethernet, switched Ethernet and CAN bus in ship power station is simulated and compared. The paper is organized as follows: in section 2, based on an actual power station the monitor-control network framework is given and the data incoming rule within power station is analyzed thoroughly and mathematical models of different kinds of data flow are developed. Section 3 designs the simulation models. A set of CAN bus models including the CAN MAC layer model, node model are built. Finally, section 4 compares the real-time performance of the monitor network built by traditional Ethernet, switched Ethernet and CAN bus in ship power station.

2. Monitor-control network framework and data flows

The monitor network in ship power station is mainly composed by area controller, simulate measure modules, manipulation boards, simulate screen of power supply and gateway. The manipulation boards and simulate screen of power supply locate in the main switchboard room, central control room and back-up control room. The topology of the monitor network is usually bus in traditional Ethernet and fieldbus or star in the switched Ethernet. Based on the above conditions, the network framework is built as figure 1.

After building the network model, the data flows in the network should be analyzed. An accurate and proper modeling of data flow is necessary for getting a close-to-reality simulation result. Based on the function of power station, data flows could be classified as below:

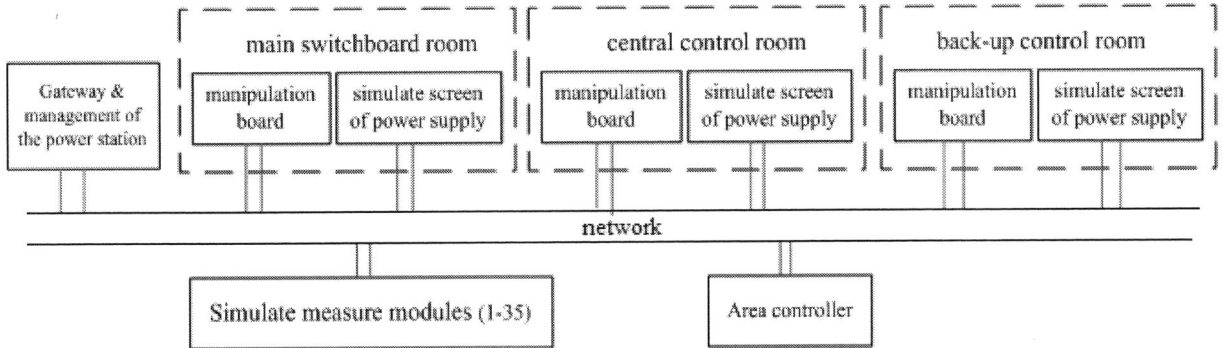

Figure 1 Monitor control network framework in ship power station

2.1 Periodic real-time refreshing data

This kind of data includes analog sampling signals such as voltage, current, frequency and power, binary signals such as device status, manipulate part and control way. These data are comparatively stable, usually are relatively fixed end-to-end communication. They are sent on scheduled time. The length of the data is constant and occupies fixed bandwidth. They are the main factors that affect the stable performance of the monitor network. This type of data could be simulated by periodic messages with fixed interval time and length. In this paper, the interval time is 0.5s. Taking a power station as an example, area controller will send 114 binary signals and 205 analog signals in a period. The binary signals are denote by 1bit and the analog signals are denote by 2bytes. In the Ethernet, the length of this message is 425 Bytes and the others are 46Bytes. In CAN bus, the data will be sent in multi-frames. The data length of the device status message is 5Bytes and the others are 8Bytes.

2.2 Stochastic command, response and request data

This kind of data includes manipulation commands that the manipulation boards and gateway send the startup, stop, merge, etc., response for the execute status after the area controller received the command and the request data. The character of this kind of data is that the sending time is not fixed, the amount of the data is few, but the timeliness is higher than periodic data. Because the poisson distribution is always used to simulate stochastic process in probability theory, in this paper this kind of data is modeled as poisson distribution with $1/\lambda = 100\text{ms}$. In the Ethernet, the length of the message is 46 Bytes. In the CAN bus, the length of the data is 8Bytes and the priority is higher than periodic data.

2.3 Burst data [3]

This kind of data is mainly the protection information sent by area controller, switch position changed information and SOE (sequence of event) information. The character is that the length is short, transmits in concentrate time, timeliness is high and obvious sudden. This kind of data does not simply appear with probability of P, but is affect by the arriving status of the former data. It has the character of the time aftereffect. These data will come suddenly in a moment and will continue a moment without sending. This kind of data should be modeled by ON/OFF model that the continue time of ON obeys Pareto distribution with $k = 512\,\mu s$, $\alpha = 1.1$ and continue time of OFF obeys negative exponent distribution with $1/\lambda = 20\text{ms}$. In the Ethernet, the length of the message is 46 Bytes. In the CAN bus, the length is 8Bytes and the priority is the highest.

3. Simulation model

OPNET is an excellent simulator which provides a comprehensive development environment supporting the modeling of communication networks and distributed systems. Both behavior and performance of modeled systems can be analyzed by performing discrete event simulations. The OPNET environment incorporates tools for all phases of a study, including model design, simulation, data collection, and data analysis. Because it includes a set of Ethernet models, many experts simulate the performance of the Ethernet based on OPNET [4-5]. Simulating monitor network built by the Ethernet and switched Ethernet are so easy that we just give the result. However, without fieldbus models in OPNET, it must design a set of CAN models in the first. OPNET simulator system is formed by network model, node model and process model. The link model and node model of the CAN bus are similar with Ethernet models. The process model of MAC layer in CAN is the important and difficult part. It will be introduced in detail.

3.1 Node model
Based on the node model called ethcoast_station_adv,

node model of CAN could be built easily. As figure 2, delete defer module and exchange mac module for mac_can module which will be introduced later. Then modify busty_gen which is data source module by adding priority attribute and destination attribute. In order to produce various kinds of message by one node, the number of the source modules should be adjusted according to the kinds of message actually needed. According to the actual power station, in this paper, there are three bursty_gen modules in CAN node model.

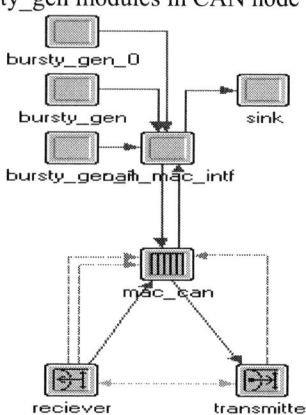

Figure 2　Node model of CAN

3.2 Process model

Process model is composed of state transfer graph, macro define block, function block, the define blocks of state variables and temporary variable and so on. The designed process model of MAC layer in CAN is in figure 3. There are ten states which include six unforced states and four forced states. The functions of states are as follow:

init state: sets the simulate parameters and initializes the statistic variables.

idle state: waits the arrival of the message. If there is low_arrival which means message from physical layer arrived, goes to rx. If the message is from the higher layer, it will be put into preparing queue which is arrayed by priority. Note that the identifier and the rest part of the message should be put into two queues. When there is message in the queue preparing to be sent, namely, !queue_empty then go to frm_out state.

frm_out state: a forced state that dequeues the first identifier in the queue, then go to tx_prepare.

tx_prepare state: checks whether the bus is free. If bus is free, begins to send the message and goes to tx_send. If there is low_arrival, goes to rx. If the message is from the higher layer, it will be put into preparing queue.

tx_send state: a forced state that sends the identifier of the message.

tx_wait state: enables physical layer to listen, when the sending finish, goes back to idle state. If there is low_arrival, go to rx. If the message is from the higher layer, it will be put into preparing queue.

rx state: a forced state that deals with the message. If it is a data packet, sends it to higher layer and goes back to idle. If it is identifier, goes to arbitration state.

arbitration state: compares the identifiers and chooses the packet with the highest priority. If message from physical layer arrived, goes to rx state. If exceeds a bit time, goes to rec state.

rec state: a forced state which judges whether the packet with highest priority comes from this node. If it is from this node, sends the rest part of the message and goes to data_send. If not, judges whether this node has joined into competition. If has, it means this competition

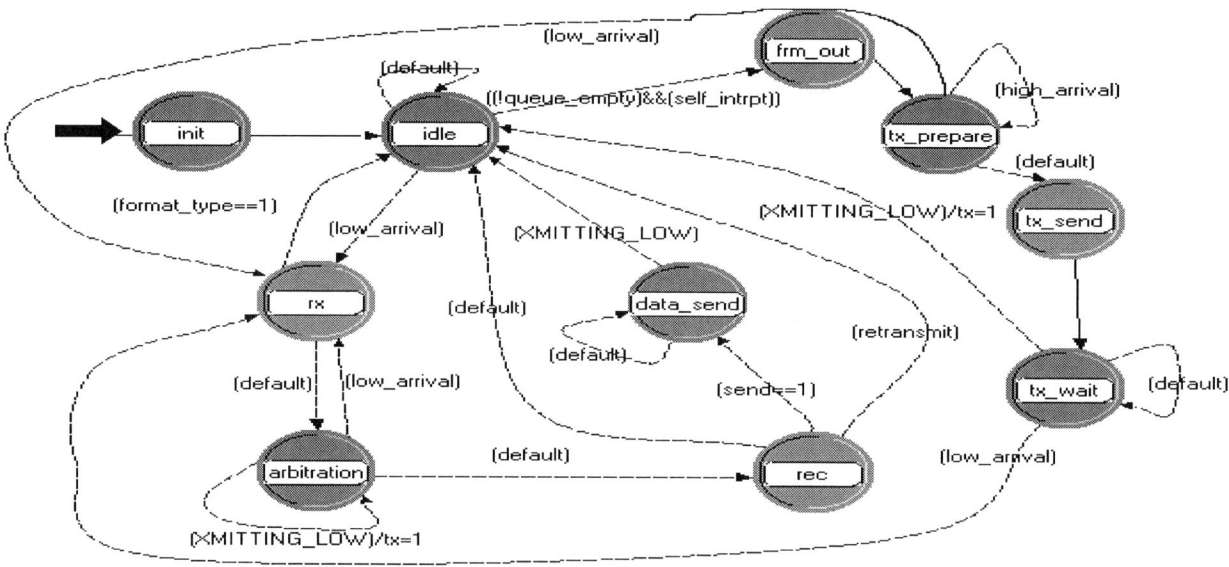

Figure 3　MAC layer process model of CAN

386

fails. Puts the identifier into preparing queue again and goes to idle state to wait next sending. If has not joined into competition, still goes to idle state. While judges whether goes on sending, judges the destination of the packet with highest priority. If the destination is this node, prepares for receiving.

data_send state: waits for the finish of message sending, destroys the message and goes back to idle.

Thus the whole process of message transition completes.

The importance of process model of MAC layer in CAN is dividing a CAN message into two parts which are identifier and the rest including data. When the bus is free, send the identifier. Then compare the received identifiers and choose the identifier with highest priority. If it is from this node, then send the other part of the message. If it is not from this node and this node has sent the identifier, stop sending and rejoin in the competition until the bus is free. The length of the data field in CAN message is not a constant, but the length of the identifier is a constant. Sending the identifier firstly could ensure that all the nodes receive the identifiers at the same time. It could make the arbitration easy. Dividing a CAN message into two parts and send the identifier firstly avoid the confine that the length of the data field should be same. In this paper, the CAN bus models could simulate the transmissions of messages with different length.

4. Result of simulation

In the simulation, the bandwidth of the Ethernet is 10M, the bandwidth of CAN is 125k, and data flows begin to send at 5s. The mean delay in traditional Ethernet and switched Ethernet are in figure 4, 5. The mean delay in traditional Ethernet is up to 8ms which could not meet the request in IEC 61850 that the max delay is no more than 4ms when control command transmits in the network. However, in the switched Ethernet the mean delay is less than 0.4ms. If the monitor network is built by CAN, we can get the delay of three kinds of data in figure 6, 7, 8. The mean delay of burst data is about 1ms, which is means it has high timeliness. The delay of stochastic data is less than 3.5ms which could not meet the real-time request. The delay of periodic data is about 0.1s, which could realize the real-time refreshing.

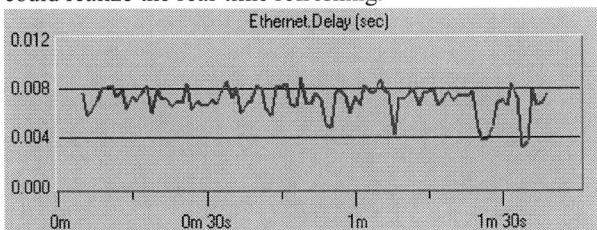

Figure 4 Delay in traditional Ethernet

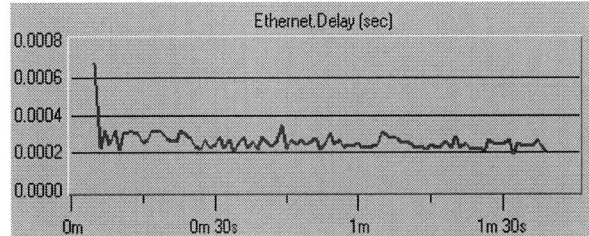

Figure 5 Delay in switched Ethernet

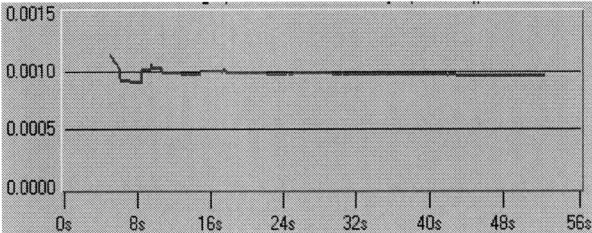

Figure 6 Mean delay of the outburst data in CAN

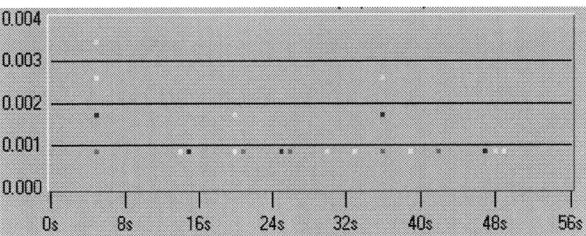

Figure 7 Delay of the stochastic data in CAN

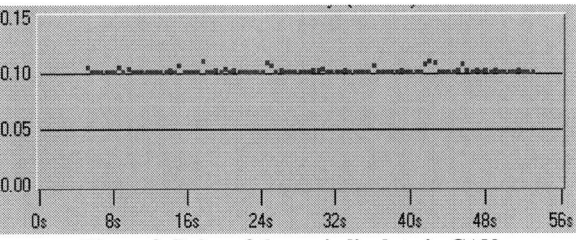

Figure 8 Delay of the periodic data in CAN

From the simulation, we could find that the real-time performance of switched Ethernet is the best. Though the bandwidth of the CAN is small, due to its priority arbitration mechanism, it could meat the real-time of the monitor in the power station. For the traditional Ethernet, the bandwidth is 10 times larger than the largest bandwidth of the CAN (1M), however, the timeliness is poor because of its collision avoidance mechanism.

5. Conclusion

This paper simulated the transmitting of three kinds of data flows in ship power station by analyzing characters of the data flow and incoming rule within power station thoroughly, designed a set of CAN bus models by OPNET simulator and obtained transmission delay of different data flows in different networks. The result showed that

CAN bus and switched Ethernet could meet the real-time request of the monitor-control in the ship power station. Power station is a part of power system. For the whole power system in the ship, the monitor network will include more devices and data flows. It will be more complexity and expect to research later.

6. References

[1] Lian F L, Moyne J R, Tilbury D M (2001). Performance evaluation of control networks: Ethernet, Controlnet and Devicenet. IEEE Control Systems. Vol.21, No.1, pp.66-83.

[2] Lee K C, Lee S (2002). Performance evaluation of switched Ethernet for real-time industrial communications. Computer Standards & Interfaces. Vol.24, No.5, pp.411-423.

[3] Wu Zaijun, Du Yansen (2005). Analysis of realtime performance of communication network in substations. Automation of Electric Power Systems. Vol.29, No.8, pp.45-49.

[4]]Skeie T, Johannessen S, Brunner, C (2002). Ethernet in substation automation. IEEE Control Systems Magazine. Vol. 22, No.3, pp.43-51.

[5] Shen Hongtao, Li Ruifang (2007). Simulative research on substation automation network communication system. Electric Power Automation Equipment. Vol.27, No.6, pp.114-117.

2008 Workshop on Power Electronics and Intelligent Transportation System

A New Harmonic Analysis Method for AC/DC/AC Converters

LV Zhaorui, XIA Li, WU Zhengguo
Naval University of Engineering Wuhan 430033 China
navylvzhaorui@163.com

Abstract

A new direct harmonic analysis approach for AC/DC/AC conversion system is presented and discussed. It uses the characteristic of the vectors and their sequence characteristic to obtain the voltages and currents transfer rules of the converters in conversion system, and then get the direct harmonic analysis method of the whole AC/DC/AC. The method can deduce the expression of ac harmonic current from ac voltage, and does not require iteration between AC and DC side. The results obtained by direct harmonic method were verified using time domain simulations.

1. Introduction

AC/DC/AC converter is becoming a significant load component for modern power grids. It can employ power electric devices to generate the variable frequency power supply for AC motor speed control[1]. The current distortion generated by AC/DC/AC has been a matter of concern for many years. Assessment and mitigation of the harmonic current have become an important aspect of power quality management[2].

At present, a plentiful research productions have already obtained in the harmonic analysis for AC/DC/AC converters. Paper [3] deals the PWM inverter with a dc current source, and realizes the harmonic calculation for adjustable speed drives(ASD), but it is too simple to regard a converter as a dc source. Paper [4] presents an analytical approach for studying harmonics and interharmonics current generated by VSI-fed ASD, the method has important value to investigate the harmonic production mechanism, but it's not systemic enough. Rosario Carbone etc make a contribution to the AC/DC/AC converters' harmonic analyzing, they have done thorough research on the iterative harmonic analysis, and adopt Paralleled Compensation Technique(PCT) to improve the iterative method convergence characteristic[5][6][7]. But the approach has the shortcoming of complicated calculation, especially in the circs of multi-converters.

The direct harmonic analysis for converters is not only algorithmic simply, but also has no convergence problem relative to the iterative method. The article [8] presents

direct harmonic analysis method to calculate the AC side harmonics and interharmonics in an HVDC system. This paper aims to use the direct harmonic method on AC/DC/AC converters. The authors make the best of each component's sequence characteristic of voltage, current and switching function to study the transfer rules of PWM inverter, and realize direct harmonic calculation of the whole AC/DC/AC converter finally. The latter time simulations validate the feasibility of this method. Figure 1 is sketch map of the AC/DC/AC conversion system.

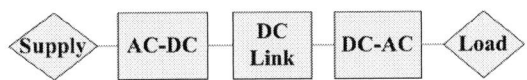

Figure 1. AC-DC-AC conversion

2. Direct harmonic method for AC/DC converters

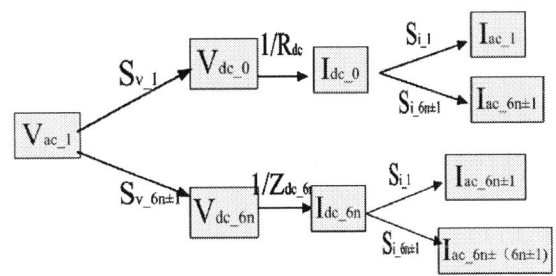

**Figure 2. Direct harmonic analysis
approach of AC-DC**

The main thinking of direct harmonic analysis method for 12-pulse HVDC systems paper [8] presents is shown in figure 2. For a 6-pulse converter, the calculation of ac fundamental current I_{ac_1} as a example. V_{ac-1} is the ac fundamental voltage of rectifier, V_{ac-1} modulated with fundamental vector of voltage switching function S_{v-1} is DC voltage V_{dc-0} , and then get the DC current I_{dc-0}, I_{dc-0} multiply by current switching function S_{i-1} is I_{ac-1}, the expression of a phase fundamental current is as formula 1:

$$I_{ac\_a\_1} = \frac{3}{2R_{dc}} \cdot S_{i\_1} \cdot \overset{*}{S}_{v\_1} \cdot V_{ac\_a} \qquad (1)$$

978-0-7695-3342-1/08 $25.00 © 2008 IEEE
DOI 10.1109/PEITS.2008.98

Where, R_{dc} is DC resistance value, $\overset{*}{S}$ is transpose of vectors.

And the $6n \pm 1$ harmonic current produced by DC current I_{dc_0} can be expressed:

$$I_{ac\_a\_6n\pm1} = \frac{I_{ac\_a\_1}}{S_{i\_1}} S_{i\_6n\pm1} \quad (2)$$

$S_{i\_6n\pm1}$ is $6n \pm 1$ components of current switching function. The meanings of the other subscripts are the same.

Formula 3, 4 is $6n \pm 1$ ac harmonic current produced by $6n$ ripple of DC current, and formula 5, 6 present the $6n \pm (6n \pm 1)$ ac harmonic current.

$$I_{ac\_a\_6n+1} = \frac{3}{2 Z_{dc\_6n}} \cdot S_{i\_1} \cdot \quad (3)$$

$$(\overset{*}{S}_{v\_6n+1} \cdot V_{ac\_a} + S_{v\_6n-1} \cdot V_{ac\_a})$$

$$I_{ac\_a\_6n-1} = \frac{3}{2 Z_{dc\_6n}} \cdot \overset{*}{S}_{i\_1} \cdot \quad (4)$$

$$(\overset{*}{S}_{v\_6n+1} \cdot V_{ac\_a} + S_{v\_6n-1} \cdot V_{ac\_a})$$

$$I_{ac\_a\_6n\pm(6n\pm1)} = \frac{I_{ac\_a\_6n+1}}{S_{i\_1}} S_{i\_6n\pm1} \quad (5)$$

$$I_{ac\_a\_6n\pm(6n\pm1)} = \frac{I_{ac\_a\_6n-1}}{\overset{*}{S}_{i\_1}} \overset{*}{S}_{i\_6n\pm1} \quad (6)$$

From the six formulas, the ac harmonic current can be deduced directly from fundamental voltage when the load is linear. But the load which rectifier supply in AC/DC/AC converter is inverter, we can't use the method introduced above. Direct harmonic method from DC voltage to DC current must be researched to realize the whole AC/DC/AC converter's harmonic analysis.

3. Direct harmonic method for AC/DC/AC

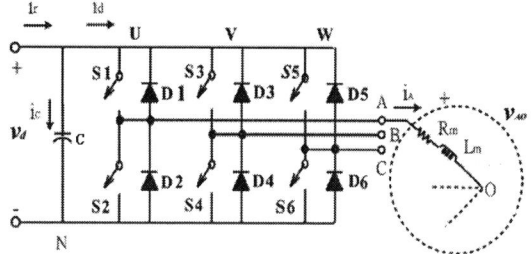

Figure 3. Schematic circuit of the PWM inverter

Figure 3 is the Schematic circuit of three phase voltage converter, the load is linear for simplicity's sake. The switching control strategy is PWM, and the switching functions of three phase bridges are: S_{AN}, S_{BN}, S_{CN}.

3.1 Voltage transfer rules from DC to AC

$$V_{AN} = S_{AN} \cdot V_d$$
$$V_{BN} = S_{BN} \cdot V_d \quad (7)$$
$$V_{CN} = S_{CN} \cdot V_d$$

$$\begin{bmatrix} V_{A0} \\ V_{B0} \\ V_{C0} \end{bmatrix} = \begin{bmatrix} 2/3 & -1/3 & -1/3 \\ -1/3 & 2/3 & -1/3 \\ -1/3 & -1/3 & 2/3 \end{bmatrix} \begin{bmatrix} V_{AN} \\ V_{BN} \\ V_{CN} \end{bmatrix} =$$

$$-\frac{1}{3}\begin{bmatrix} 1 & 1 & 1 \\ 1 & 1 & 1 \\ 1 & 1 & 1 \end{bmatrix}\begin{bmatrix} V_{AN} \\ V_{BN} \\ V_{CN} \end{bmatrix} + \begin{bmatrix} 1 & 0 & 0 \\ 0 & 1 & 0 \\ 0 & 0 & 1 \end{bmatrix}\begin{bmatrix} V_{AN} \\ V_{BN} \\ V_{CN} \end{bmatrix} \quad (8)$$

When V_{AN}, V_{BN}, V_{CN} are positive or negative sequence, $V_{AN} + V_{BN} + V_{CN} = 0$, then:

$$\begin{bmatrix} V_{A0} \\ V_{B0} \\ V_{C0} \end{bmatrix} = \begin{bmatrix} V_{AN} \\ V_{BN} \\ V_{CN} \end{bmatrix} = \begin{bmatrix} S_{AN} \\ S_{BN} \\ S_{CN} \end{bmatrix} \cdot V_d \quad (9)$$

And when V_{AN}, V_{BN}, V_{CN} are zero sequence, $V_{AN} = V_{BN} = V_{CN}$,

$$\begin{bmatrix} V_{A0} \\ V_{B0} \\ V_{C0} \end{bmatrix} = 0 \quad (10)$$

Thus, at the load side there have no zero sequence voltages, so there are no zero sequence currents too for symmetrical systems.

Suppose the frequency of DC voltage vector V_{dc_n} is $n\omega_1$, ω_1 is fundamental frequency of AC/DC/AC supply side. Inverter switching function component is S_{mk}, the frequency is $m\omega_c + k\omega_2$, ω_c is carrier frequency, and ω_2 is load side fundamental frequency. The frequency, phase sequence and expression of phase 'a' at load side produced by the two vectors are shown in table 1.

Table 1. Inverter voltage transfer rules from dc to ac

	$k = 3p+1$	$k = 3p-1$
$m\omega_c + k\omega_2$ $+ n\omega_1$	$V_{dc\_n} \cdot S_{mk}$ positive	$V_{dc\_n} \cdot S_{mk}$ negative
$m\omega_c + k\omega_2$ $- n\omega_1$ or $n\omega_1 - m\omega_c$ $- k\omega_2$	$\overset{*}{V}_{dc\_n} \cdot S_{mk}$ positive $V_{dc\_n} \cdot \overset{*}{S}_{mk}$ negative	$\overset{*}{V}_{dc\_n} \cdot S_{mk}$ negative $V_{dc\_n} \cdot \overset{*}{S}_{mk}$ positive

In the table, p is integer, the frequency of phase voltage is:

$$m\omega_c + k\omega_2 + n\omega_1, \text{ and}$$

$$m\omega_c + k\omega_2 - n\omega_1 \text{ (or } n\omega_1 - m\omega_c - k\omega_2)$$

the frequency of phase voltage is ($m\omega_c + k\omega_2 - n\omega_1$ or $n\omega_1 - m\omega_c - k\omega_2$) depended on the magnitude of $m\omega_c + k\omega_2$ and $n\omega_1$.

when n=0, phase 'a' voltage frequency is $m\omega_c + k\omega_2$, it satisfy the first row's expression, and the phase sequence is the same as S_{mk}.

3.2 Current transfer rules from DC to AC

Because the inverter neither consumes nor produces instantaneous power with ideal switching operations, so:

$$v_d i_d = v_{A0} i_A + v_{B0} i_B + v_{C0} i_C \quad (11)$$

Current transmitting can be denoted approximately [2]:

$$I_d = S'_{AN} I_A + S'_{BN} I_B + S'_{CN} I_C$$

Where,

$$\begin{bmatrix} S'_{AN} \\ S'_{BN} \\ S'_{CN} \end{bmatrix} = \begin{bmatrix} 2/3 & -1/3 & -1/3 \\ -1/3 & 2/3 & -1/3 \\ -1/3 & -1/3 & 2/3 \end{bmatrix} \begin{bmatrix} S_{AN} \\ S_{BN} \\ S_{CN} \end{bmatrix} \quad (12)$$

$$I_d = \begin{bmatrix} S_{AN} & S_{BN} & S_{CN} \end{bmatrix} \begin{bmatrix} 2/3 & -1/3 & -1/3 \\ -1/3 & 2/3 & -1/3 \\ -1/3 & -1/3 & 2/3 \end{bmatrix} \begin{bmatrix} I_A \\ I_B \\ I_C \end{bmatrix} \quad (13)$$

There have no zero sequence current in i_A、i_B、i_C which have introduced above, so:

$$I_d = \begin{bmatrix} S_{AN} & S_{BN} & S_{CN} \end{bmatrix} \begin{bmatrix} I_A \\ I_B \\ I_C \end{bmatrix} \quad (14)$$

If the frequency of AC current I_{ac} is ω, the frequency of switching function harmonic component $S_{m'k'}$ is $m'\omega_c + k'\omega_2$, the expression of DC current vector is obtained according to the two vectors' sequence characteristic and magnitude of the frequency.

$$I_d = \begin{cases} \dfrac{3}{2} I_{ac} \cdot S_{2\_m'k'} & m'\omega c + k'\omega_2 + \omega, different \\ \dfrac{3}{2} I_{ac} \cdot S^*_{2\_m'k'} & \omega - m'\omega c - k'\omega_2, same \\ \dfrac{3}{2} I^*_{ac} \cdot S_{2\_m'k'} & m'\omega c + k'\omega_2 - \omega, same \end{cases}$$

In the formula above, 'different' represents 'the phase sequences are different', 'same' represents 'the phase sequences are same'.

From the voltage and current transfer rules above, direct expression from DC voltage to current can be obtained. For example, when the DC current V_{dc_n}, voltage function S_{mk} and current function $S_{m'k'}$ satisfy the hereinafter conditions:

S_{mk} is positive, $S_{m'k'}$ is positive too,

$$n\omega_1 - m\omega_c - k\omega_2 > 0, \text{ and}$$

$$n\omega_1 + m\omega_c + k\omega_2 - m'\omega_c - k'\omega_2 > 0$$

The two frequencies of DC current vectors are:

$$I_d = \begin{cases} V_{dc\_n} \left(\dfrac{1}{R_m + j\omega_{L1} L_m} \cdot S_{2\_mk} \cdot S^*_{2\_m'k'} \right) \\ V_{dc\_n} \left(\dfrac{1}{R_m + j\omega_{L2} L_m} \cdot S_{2\_mk} \cdot S^*_{2\_m'k'} \right) \end{cases} \quad (15)$$

$\omega_{L1} = n\omega_1 + m\omega_c + k\omega_2$, $\omega_{L2} = n\omega_1 - (m\omega_c + k\omega_2)$

The two frequency of DC current are:

$$n\omega_1 + m\omega_c + k\omega_2 - m'\omega_c - k'\omega_2$$

and $n\omega_1 - m\omega_c - k\omega_2 + m'\omega_c + k'\omega_2$。

Especially, when $m = m', k = k'$, two frequencies of current vectors are all $n\omega_1$.

The harmonic current in the capacitance of DC side is:

$$I_C = -\frac{1}{j\omega_C C} V_{dc\_n}, \quad \omega_C = n\omega_1 \quad (16)$$

Then, the DC side current after rectifier:

$$I_r = I_d + I_c \quad (17)$$

From equations (15), (16) and (17), each harmonic current can be calculated directly by DC voltage. Integrated equations (15)-(17) with formulas (1)-(6), the direct harmonic analysis method can be got for the whole AC/DC/AC converter.

3.3 Feasibility explains of direct harmonic analysis method

The method above can calculate the harmonic currents produced by fundamental voltage in the supply side of AC/DC/AC converter, and did not consider the transfer of harmonic voltage. This method is simple on calculation, and has no problem of convergence, it is at the cost of precision. But the arithmetic has feasibility all the same:

1. the converters are currently connected on the main in power grid, and the voltage distortion on main is very small all the time, the effects is little without considering harmonic voltages.

2. the current frequencies produced by harmonic voltages are high, and the high frequency information will be weakened greatly when they get across the capacitance and inductance in DC side.

4. Harmonics results and simulation validated

To validate the direct harmonic analysis method discussing above, harmonic calculation and time simulation on two AC/DC/AC conversion systems are processed. The main parameters of the system 1 are shown in table 2.

Table 2. Main parameters of AC-DC-AC conversion system 1.

Supply	amplitude(r.m.s): 587V; frequency(ω_1): 50Hz。
rectifier	6-pulse uncontrolled rectifier
DC	C: 5.0mH。
PWM inverter	frequency (ω_2): 20Hz; modulation ratio (m_a): 0.8; frequency ratio (m_f): 21。

In order to minish the frequency range this paper studied, the output frequency is set as 20Hz, and the m_f is 21.

The article start from the AC voltage to calculate the harmonics and interharmonics currents imported the systems by direct harmonic analysis method. A time simulation is carried out on the same conversion system. FFT is processed on the AC current information at supply side. The fundamental frequency of FFT is the greatest common divisor of supple side frequency 50Hz and load side frequency 20Hz, that is 10Hz. The antitheses table of simulation and results of calculation is presented as table 3.

In the table, 'harm' is short for 'harmonics', 'interharm' is short for 'interharmonics'.

Table 3. Harmonic components antitheses table of supply side ac current for system 1.

Harm (Hz)	Interharm (Hz)	calculation	simulation
50		52.97	53.08
250		14.30	15.67
	310	0.14	0.15
350		4.22	4.38
	410	0.23	0.25
	430	0.16	0.11
	530	0.12	0.074
550		3.34	3.22
650		1.52	1.77
	790	0.03	0.044
850		1.31	1.16
	890	0.063	0.089
950		0.92	0.99

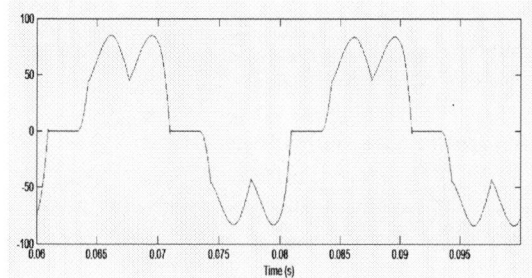

Figure 4. Waveform of ac current in the supply side for system 1.

As the PWM is in operation, the AC harmonic components in load side mainly are: 340Hz (negative sequence), 380Hz (positive sequence), 460Hz (negative sequence), 500Hz (positive sequence), 820 Hz (negative sequence) and 860Hz (positive sequence). According to the sequence characteristic of each component, they will mainly produce ripple currents with frequency of 360 Hz、480 Hz and 840 Hz . And then, the frequency of interharmonics current in supply side mainly include 310 Hz,410 Hz,430 Hz,530 Hz,790 Hz and 890 Hz, analysis and results of calculation are inosculated with one another. The supply side AC current waveform is shown as figure 4 for system 1.

The different parameters of system 2 to system 1 mainly include:Supply voltage V=380v, the output frequency ω_2 of PWM converter is 30Hz, and the frequency ratio (mf) is 16.

Table 4. Harmonic components antitheses table of supply side ac current for system 2.

Harm (Hz)	Interharm (Hz)	calculation	simulation
50		105.7	106.4
250		26.48	27.56
	340	0.64	0.76
	440	0.62	0.71
350		7.03	7.79
	520	0.37	0.41
550		4.92	5.58

	620	0.42	0.35
650		3.23	3.60
850		1.41	1.77
	910	0.10	0.12
950		1.26	1.59

The frequency of harmonics and interharmonics components of current for system 2 under 1000 Hz are presented in table 4. the analysis and current waveform are not introduced for the limit of length.

5. Conclusion

A new approach called direct harmonic analysis method is applied to AC/DC/AC converter. Voltages and currents transfer rules were deduced using the harmonic vectors and their sequence characteristic. The method which started from supply side AC fundamental voltage can calculate the harmonics and interharmonics currents at the same side, and does not need the iterative process between AC and DC side. Direct harmonic analysis method has the virtue of simple formulas and explicit approaches, and is more fit for the situations of little voltage distortion and multi-converters.

Reference

[1] EPRI TR-101140, Adjustable Speed Drives: Application Guide, Research Projects 2951-12 and 2951-04, December 1992.

[2] Mohammed Bashir Rifai, Thomas H. Ortmeyer, William J. McQuillan. Evaluation of Current Interharmonics from AC Drives. IEEE Transactions on Power Delivery, Vol. 15, NO. 3, July 2000.1094-1098

[3]Wilsun Xu,Hermann W.Dommel,M.Bret Hughes,etc. Modeling of Adjustable Speed Drive for Power System Harmonic Analysis. IEEE Transactions on Power Delivery,vol.14,No.2,April 1999.595-601.

[4] Gary W. Chang, Shin-Kuan Chen. An Analytical Approach for Characterizing Harmonic and Interharmonic Currents Generated by VSI-Fed Adjustable Speed Drives. IEEE Transactions on Power Delivery,vol.20,No.4,October 2005.2585-2593.

[5] Rosario Carbone, Francesco De Rosa, Roberto Langella,etc. Modelling of AC/DC/AC Conversion Systems with PWM Inverter. Power Engineering Society Summer Meeting, 2002 IEEE. Vol 2,July 1004-1009.

[6] F. De Rosa, R. Langella, A. Testa. Evaluation of harmonics and interharmonics produced by AC/DC/AC conversion systems. in Proc. Int. Conf. Harmonics Quality Power, New York, Sep. 12–15, 2004.495-500.

[7] Rosario Carbone, Francesco De Rosa, Roberto Langella, etc. A New Approach for the Computation of Harmonics and Interharmonics Produced by Line-Commutated AC/DC/AC Converters. IEEE Transactions on Power Delivery,vol.20,No.3, July 2005. 227-2234.

[8]L Hu, L Ran. Direct Method for Calculation of AC Side Harmonics and Interharmonics in an HVDC System. IEE Proc-Gener. Transm.Distrib.,Vol.147, No.6, November 2000.329-335.

Simulation and Optimization of the Power Station Coal-Fired Logistics System Based on Witness Simulation Software

Yabin Li [1], Rong Li [2]

[1]School of Mechanical Engineering,
North China Electric Power University, Baoding , HeBei, 071003, China
[2]Information and Network Management Center,
North China Electric Power University, Baoding , HeBei, 071003, China
lybhnjz@163.com

Abstract

Based on analyzing characteristics and relations of the power station coal-fired logistics system, including the choice of coal suppliers, the railroad transport system, the power station stockpile system, the power station coal transfer system and human resources, by Witness software the thesis sets up a simulation and optimization model of fuel coal logistics system and detailedly analyzes one example. The model results not only macroscopically image bottleneck points and resources availability, but define optima separate stage ordering goods batches and stockpile quantities. The thesis is powerful and significant for thermal power station to establish scientific reasonable logistics system strategies, reduce the cost and increase competition ability.

1. Introduction

According to statistics, the velocity of national electric power increase is very quick in the last few years. In the total output of electrical energy, the thermal power occupies about 80%. Among them, the cost of a thermal power station coal-fired logistics system in China, including the purchase of coal, transport and storage et al.., is often accounted for 60% to 70% of the cost of power generation. Under the condition of guarantying coal-fired supply, how to cause the least coal-fired logistics system expense, is the primary mission and goal of the coal-fired logistics system administration[1].

In recent years there are also some literatures on transport and stockpile models of coal-fired thermal power stations. Literature [2] has theoretically merely studied the power station coal-fired logistics system's characteristic, and not given the substantive solution. Literature [3] has studied coal purchase and transportation questions on Taiwan Electricity companies, and established the transportation plan of many suppliers selection and the stockpile control mix integer project model. Literature [4] has studied the thermal power station coal-fired stockpile optimization strategy, and established the dynamic project model of order time and order quantity. Literature [5] has studied an integrated coal transportation and stockpile model under the condition of rail direct transportation.

On the one hand, about the above-mentioned documents conditions are limitedly considered, and the study of human resources is also lack in thermal power station coal-fired transportation and stockpile problems; on the other hand, it is crucial not to realize the simulation of the power station coal-fired logistics system, so it is very difficult to find what time and what position bottlenecks occur in the power station coal-fired logistics system and thus it is absent to promptly and effectively improve the characteristics of the power station coal-fired logistics system.

Therefore, this paper, according to the premise of coal boiler requirements, on the basis of the comprehensive analysis of coal-fired rail transportation and stockpile that are two important links of logistics system, major studies, sets up and simulates by Witness2006 software the multi-cycle logistics system of coal-fired thermal power stations. And by demonstrating the entire process of its logistics systems, the model makes users clearly understand its logistics problems in the process and adjust the model parameters in a timely. And the model makes the total costs of the logistics system as the optimal goal, which includes coal purchase cost, transportation cost, stockpile cost, capital cost, and applies Witness software optimization modules for optimization, so as to implement the best value of the power station coal-fired logistics system.

2. Analysis of the power station coal-fired logistics system

The power station coal-fired logistics system has three major components, namely, suppliers' selection, railroad transport systems and power stations.

2.1. Suppliers' selection

2.1.1. Guarantying coal supplies. When coal production capacity or supply bottlenecks encounters blocked, it is easy for the supply chain to trigger coal-shortage crisis. Therefore, power enterprises can unite with coal enterprises to handle an ore for reaching the purpose participating in competition on the headstream, and also can sign the middle-long supply and demand contract with coal enterprises under strategic partnership, which can cut down incomplete coal risk, coal purchase cost, and build healthy developed economic supply chaining [6].

2.1.2. Ensuring coal qualities. Because the fired coal has the strict request to the indexes, when power stations select suppliers they must consider qualities and quantities of coals in the lowest purchase cost, as well as after mixed, to be able to achieve the coal-fired requests.

2.2. Railway transport systems

More than 80 % of power station coals need railway transports, and the vast majority of coal rail transports are used to rail direct transport. Of course, this way has certain requirements on the number of direct cargo trains, the consignor loading capacities, the rail carrier capacities, economic freight weights, plant unloading capacities and stockpile system conditions.

2.3. Power stations

2.3.1. stockpile systems. In the coal-fired logistics system, the plant stockpile system is essential aspect, which is closely related to security, stability and economic operation. It is decided by the factors that are power station demanding fuel quantities per day, security stockpile, ordering cycle, delayed arrival time, the ability unloading capacities and so on. [7].

2.3.2. Factory coal-fired conveying systems. It is mainly composed of dropping coal machines, belt conveyers, screening machines, measurement machines, sampling machines, transportation machinery and so on. This system mainly researches unloading coals machinery related to unloading ability and transport machinery related to the boiler load.

2.3.3. Human resources. In this study of the logistics system, the human resources mainly are for checking and unloading coals, machinery service and conveying coals. And, it is very important for quality testers to check coals qualities and quantities, which not only relates the round turn economic interest, but also relates the accuracy of coal-fired consumption computation used for electricity generation. And the magnitudes of moisture contents of coals arriving to factories are affective to coal-fired qualities and quantities [8].

3. Modeling principle by Witness

Based on the thermal power station coal-fired logistics system, the modeling principle is: ①power stations may select suppliers according to suppliers' quote, freight, coal composition and delayed payment, et al.; ② the quantities and the time of supplying fired coals depend on power stations' demands and the minimum cost principle; ③ coals transport systems should be consistent with the requirements of railway system and ensure the economy transport weigh and transport efficiency; ④ power station stockpile system should ensure that the optimal stockpile costs, and the use of human resources, as well as the scheduling of electrical equipments should meet maximize efficiency.

4. Constructing models

4.1. Parameters definition

The model input parameters: $Q(k)$ as the demand for power station at stage k, t; E as the proportion of a supplier' supplies occupying total demands (top limit),%; $C_i(k)$ as order cost of coal i at stage k ,yuan/time; $B_i(k)$ as unit railway direct transport costs of coal i at stage k, yuan/t; $P_i(k)$ as unit purchase price of coal i at stage k, yuan/t; $G(k)$ as power station unit storage costs at stage k, yuan/t; S as stockpile capacity of a power station, t; L_i as lower limit of railway direct transport capacity, t; $S(k)$ as safety stockpile quantities of a power station at stage k, t ; U_i as the arriving rate of supplier i providing coals ,%; U as the minimum acceptable arriving rate of a power station,% ;R as the discount rate,%; T_i as the delayed period of a power station paying for coal i, d; X_{ij} as index j value of coal i,%; Y_j as a boiler requesting coal index j value,%; $W(k)$ as the wages of power station workers at stage k, yuan/month.

The model output parameters: $Z_i(k)$ as the quantities of ordering coals from suppliers i at stage k, t, $V(k)$ as the stockpile of power station fired coals at initial stage k, t; $Cost_i(k)$ as power station coals purchase total costs and transport total costs from supplier i at stage k; $Stock(k)$ as power station stockpile costs at stage k; $Cost$ as the total cost of thermal power station coal-fired logistics system.

4.2. Assumption of constraint conditions and determination of objective functions

Construction of the model based on the following assumptions:

(1) In order to maintain consistent with the ordering

fired coal plan, assume the research cycle of the model for one year, namely, designing a one-year coal-fired subscription plan. And the research phases of the model are monthly divided into 12 stages, and the specific research object is one day;

(2) In order to ensure the security and stability of coal-fired supplies, the level of a power station depending on each supplier is not more than 50%, and each supplier provide only one type of coal, namely:

$$Z_i(k) \leq 0.5 * Q(k), i=1,2,\dots,n, \quad k=1,2,\dots,12 \quad (1)$$

(3) Assuming that applying the rail direct transport mode, trains have minimum and maximum capacity constraints;

(4) In order to the actual situation of unequal paths from suppliers to power stations, and to ensure the accessibility of the model, the model utilize different transport costs per km to express unequal paths, namely, under the conditions of different fired coals with unequal transport costs per km and the coal-fired invariable unit transport cost per km at every stage, the paths in the model show the equal distance from different suppliers to the same power station;

(5) In general, the coal enterprises pay for the freight in advance, and after coals arrive to the plant the power station together pays for purchase costs and the freight. Therefore, in order to reflect capital time value on the impact of the coal-fired cost, the model introduce a discount rate and a power station delayed payment period for coal i to accurately reflect the true value cost, and the delayed period is different with different suppliers;

(6) Ordered coals arrive in the same stage [9];

(7) The stockpile is the quantities of the beginning of each stage (month, day), and unit storage cost is related to storage quantities and stages, and there are the largest stockpile and safety stock restrictions, namely:

$$\sum_{i=1}^{n} Z_i(k) + V(k) \leq S, \quad \forall i, k \quad (2)$$

$$V(k+1) = V(k) + \sum_{i=1}^{n} Z_i(k) - Q(k) \geq S(k), \forall i,k \quad (3)$$

(8) The coals-mixed quality requirements of the boiler mainly reflect coal-fired caloric power, the volatile content, moisture content, sulfur content and ash content, and so on. These indexes must be controlled to set limits. When the raw coal arrives at the power station, there is the lowest request regarding the coal arriving percentage. After quality testers checkout indexes and quantities, it can be arranged by the dropping coal personnel to unload the coal to the coal field, namely:

$$\sum_{i=1}^{n} X_{ij} * Z_i(k) \leq (\geq) Y_j * Q(k), \quad \forall i,j,k \quad (4)$$

$$\sum_{i=1}^{n} U_i * Z_i(k) \geq U * Q(k), \quad \forall i,k \quad (5)$$

The objective function is set to the minimum total cost of the logistics system, namely:

$$Cost_i(k) = C_i(k) + Z_i(k) * (P_i(k) + B_i(k))/(1+R)^{T_i} \quad (6)$$

$$Stock(k) = G(k) * (\sum_{i=1}^{n} Z_i(k) + V(k+1) + V(k))/2 \quad (7)$$

$$Cost = \sum_{k=1}^{12} [\sum_{i=1}^{n} Cost_i(k) + Stock(k) + labors * W(k)] \\ \forall i,k \quad (8)$$

Total cost of the logistics system consists of three parts, namely, coal-fired purchase and transportation costs, stockpile costs and labor costs.

4.3. Construction of simulation and optimization models

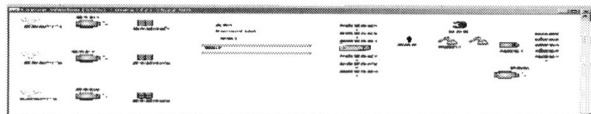

Figure 1. Simulation and optimization models of the power station coal-fired logistics system

Based on assumptions, applying Witness software constructs the model as shown in Figure 1. In the Model, the suppliers may have more than one, and there is a coal-loading machine and a loading buffer zone corresponding to one supplier. Under assumptions (4), corresponding to the power station there are a number of trains and one circular direct railway (track 1 and track 2). The model also orderly sets up the power station buffer to unload coals, three quality testers, two coal-whippers, the power station stock, one dig and transport coal machine and nine operators, as well as one boiler and so on.

The coal supplies in the model depend on the needs of power stations, with ton as coal units and hours as simulation time; the adjusted cycle of coal-unloading and coal-mining machines are one time per 100,000 time work, and the adjusted time is 24-hour long. And maintenance cycle meets NEGEXP (20,1) distribution, and repairing time meets POISSON (12,2) distribution; a worker works eight hours under the condition of three shifts of work system, and so on.

The model applies optimizer 4.3 module for optimization. The model makes assumption conditions as restrictions, 'cost' as the objective function and Simulated Annealing method as algorithms.

5. Case study

5.1. Parameters assignment

Aimed directly at South China thermal power station A coal-fired logistics condition, this paper applies the model of simulation and optimization. The plant purchase coals from three coal suppliers, with the procurement period of 12 months and one month as a stage (k =12).

Initially stockpile $V(0)$ = 105000 t, E=0.5, $C_i(k)$ = 900 Yuan, $G(k)$ = 0.4 Yuan, S = 420000 t, L_i =1200t, $S(k)$=80000 t, R=0.02, T_i=10 days, U_1=0.98, U_2=0.97, U_3=0.97, U=0.97, $W(k)$=5,000 Yuan, 3 quantity testers, 9 operators and other variables as shown in table 1.

5.2 Analysis

After the system operation, the simulation results can macroscopically show the bottlenecks of the logistics system operation and the information of resource utilization efficiency at all aspects, which can be used for targetedly repeatedly adjusting logistics system resource allocation to ensure the final maximize resource utilization, and thereby minimize logistics costs. After the optimization, optimization results of the procurement strategies, the stock quantities and the total cost as shown in table 2.

Table 1. Parameters of the power station coal-fired logistics system

k	Time	B1	B2	B3	P1	P2	P3	Q
1	0-720	91.35	96.6	94.5	388.5	393.75	393.75	202860
2	720-1440	91.35	96.6	94.5	367.5	362.25	367.5	210168
3	1440-2160	89.25	94.5	92.4	367.5	372.75	378	273168
4	2160-2880	89.25	94.5	92.4	378	378	383.25	275436
5	2880-3600	87.15	92.4	90.3	388.5	393.75	393.75	277200
6	3600-4320	89.25	94.5	92.4	399	399	399	293832
7	4320-5040	89.25	94.5	92.4	399	399	393.75	315000
8	5040-5760	89.25	94.5	92.4	388.5	393.75	388.5	357084
9	5760-6480	86.1	90.3	89.25	378	378	383.25	319284
10	6480-7200	86.1	90.3	89.25	367.5	372.75	372.75	268884
11	7200-7920	90.3	95.55	93.45	378	378	378	252000
12	7920-8640	91.35	96.6	94.5	388.5	388.5	388.5	231084

j	X_1	X_2	X_3	Y				
1*	0.25	0.35	0.3	0.3				
2*	0.02	0.03	0.04	0.03				

Note:1*-volatilize content ; 2*-sulfur content

Table 2. Optimized results of the power station coal-fired logistics system

k	Z_1 (k)	Z_2 (k)	Z_3 (k)	V (k)	Cost(Million)
1	101430	0	80430	84000	
2	105084	105084	105084	189084	
3	136584	136584	136584	325668	
4	137718	137718	94332	420000	
5	138600	0	138600	420000	1200.293
6	0	0	0	126168	
7	115332	0	157500	84000	
8	178542	0	178542	84000	
9	159642	159642	0	84000	
10	134442	134442	134442	218442	
11	126000	96642	126000	315084	
12	0	0	0	84000	

6. Conclusions

Based on the analysis of characteristics of the thermal power station coal-fired logistics system, applying witness simulation software builds the thermal power station coal-fired multi-cycle logistics system model under the condition of a rail direct transport. Objectives is the minimum total cost of the logistics system , including coal purchase costs, transportation costs, stockpile costs, capital costs. Logistics system take full account of the various constraints relying on the needs of a coal-fired power station, including suppliers selection, railway

transport capacity constraints, transport quality constraints, stockpile capacity constraints, security stock constraints, boiler requirements for coal qualities, as well as blending ratio, et al.. Through simulation and optimization of the model, optima separate stage ordering goods batches and stockpile quantities can be defined, which provide fundamental strategies to formulate scientific and rational logistics system. The analysis of the practical sample through the model further explains the construction process of the model and verifies the validity and the feasibility of the model.

Acknowledgement

It is a project supported by North China Electric Power University scientific research fund.

References

[1] Hyde M(1998). "Optimizing plant and system coal inventories". *IEEE Transaction on Power Systems*, Vol. 32, No. 1, pp. 337-342

[2] Ying Li, and Mengjun Li(2007). "Study on logistics cost control of thermal power plant". *Enterprise Logistics,* Vol. 26, No. 5, pp. 117-118

[3] Shih Li-Hsing(1997). "Planning of fuel coal imports using a mixed integer programming method". *Int. J. Production Eeonernics*, Vol. 51, No. 6, pp. 243-249

[4] LF Li, and PQ Huang(2003). "Research of fuel inventory management in power plant". *Industrial Engineering and Management*， Vol. 32, No. 6, pp. 68-71

[5] Xueming Cao, and Boliang Lin(2006). "An integrated coal transportati0n and inventory model under condition of rail direct transportation". *Journal of Beijing Jiaotong University*, Vol. 30, No. 6, pp. 27-31

[6] Haixian Yang, and Xinwei Li(2004). "Improvement electricity coal supply chain". *China Power Enterprise Management*， Vol. 21, No. 7, pp. 30-31

[7] Lixia Chen(2006). "Sate storage management of coal in power plant". *Industrial Safety and Dust Control*, Vol. 32, No. 2, pp. 61-63

[8] Xichun Fan, and Xiaohua Wang(2004). "Effect of coal-acceptance on stock coal weight in plant coal-fired". *Jilin Electric Power*, Vol. 170, No. 9, pp. 51-52

[9] Lingfeng Li, and Peiqing Huang(2003). "Research of fuel inventory management in power plant". *Industrial Engineering and Management*, Vol. 25, No. 6, pp. 67-71

Study on Voltage Stability of Distribution Networks

Wu Xiaomeng[1,3] , LIU Jian[2] , Yan Suli[3]
1. Xi'an University of Technology, Shaanxi, Xi'an, 710048,CHINA
2.Xi'an University of Science & Technology, Xi'an, 710054,CHINA
3.Xi'an Shiyou University, Xi'an, Shaanxi, 710065, China
Jessica93@163.com

Abstract

A new criterion of voltage stability for distribution network is deduced based on the existence of solution of power flow, i.e., the first class criterion. The profile of load is included in the proposed criterion. The weakness of the existing criterion based on load-voltage property, i.e., the second class criterion, is described. It is pointed that the first class criterion is the necessary condition of voltage stability but not the sufficient condition. To guarantee voltage stability, both the first and the second class criterion must be satisfied. The indexes of voltage stability margin and safe distance are defined. The proposed method is demonstrated by an example showing its feasibility. It is also shown that the proposed indexes can evaluate the voltage stability and the tolerance of disturbance.

1. Introduction

The IEEE give the definition of the voltage stability ,voltage collapse and the voltage safety at the earliest[1-2].The phenomenon of voltage lose steady occurred many times in the United States ,Japan, Europe and so many countries[3-4].Therefore research the voltage unsteady and voltage collapse problems ,have very significant society meanings. In recent years, with this phenomenon, many countries opened an exhibition of researchers and got many achievements [5-17]. However, the exist research of voltage stability mostly take transmission power network as object and the less involve in distribution network. Toward at a in motion distribution network, the main reason about voltage instability is because load exceed the control capacity and cause voltage reduction and can't control.

In consideration complexity of variety load, during the high load peak period, the short duration of high level may threat to voltage stability of the system. For take the machine which take out of oil and moves in quasi-periodic as the lord for the oil-field distribution network, when such many machines in high level motion state, the fugacious load would consumedly exceed average level, probably a threat to the voltage stability. In addition, some oil-field distribution networks with longer transmission wire, particularly in the check to fix under the way, the larger of the power supply wire's radius, the more easy voltage instability to occur. So study to voltage stability of distribution networks is advantage to safety more for the networks.

G.B.Jasmon researched on voltage stability of distribution network, and derived the voltage stability index of the network which only contains two nodes, and also expansion the application in multistage point of distribution network by equivalent resistance [18]. In some extent, the indexes reflect the relation between total load and voltage stability, but can't reflect the load distribution influence on voltage stability. And for the distribution network which existence branches, is unsuitable to analysis its voltage stability problem. Ranjan R and Das D etc, put forward a index of aim to load distribution reflection in distribution network, but it not strict and it is only an essential condition for voltage stability. Aim at above problem, in this paper we put forward a new kind of voltage stability index of distribution networks.

The judgment method of voltage stability can mostly divided into types, namely basis the existence of power flow solution judgment method and basis load-voltage characteristic judgment method. Because of both judgment method's criterion are very not homology, so we get the relation of two types in this paper, and account for the meaning, the voltage stability indexes are expandable and perfectible of the distribution networks.

2. The voltage stability criterion of distribution networks based on the existence of power flow solution

For a distribution networks which consists of N nodes, it also include branch lines and compensation capacitor. Suppose any branch line is b_{ij}, i and j are respectively two nodes of the line, power flow direction is from node i and j, the load flow from node j is P_j+jQ_j, and the impedance of the branch line *bij* is $R_{ij}+jX_{ij}$.

Therefore,

$$U_j = U_i - (R_{ij} + jX_{ij})(P_j - jQ_j)/U_j^* \quad (1)$$

Where, U^i and U^j are respectively the voltage vector of right angle coordinate formalization for node i and j. Derive from equation (1)

$$U_{R,j}^2 + U_{I,j}^2 = U_{R,i}U_{R,j} + U_{I,i}U_{I,j} - (P_jR_{ij} + Q_jX_{ij}) \quad (2)$$

$$U_{I,j} = [U_{R,j}U_{I,i} - (P_jX_{ij} - Q_jR_{ij})]/U_{R,i} \quad (3)$$

Where, $U_{R,i}$ and $U_{I,i}$ are respectively real and negative of U_i; Similarly, $U_{I,i}$ and $U_{I,j}$ of U_j. By rearranging equations (3) and (2) we can obtain:

$$U_{R,j}^2(U_{R,i}^2 + U_{I,i}^2) - U_{R,j}[2U_{I,i}(P_jX_{ij} - Q_jR_{ij}) + U_{R,i}^3 + U_{R,i}U_{I,i}^2] +$$
$$(P_jX_{ij} - Q_jR_{ij})^2 + U_{R,i}U_{I,i}(P_jX_{ij} - Q_jR_{ij}) + U_{R,i}^2(P_jR_j + Q_jX_{ij}) = 0$$
$$(4)$$

$$U_{I,j}^2(U_{R,i}^2 + U_{I,i}^2) + U_{I,j}[2U_{R,i}(P_jX_{ij} - Q_jR_{ij}) - U_{I,i}^3 - U_{I,i}U_{R,i}^2] +$$
$$(P_jX_{ij} - Q_jR_{ij})^2 - U_{R,i}U_{I,i}(P_jX_{ij} - Q_jR_{ij}) + U_{I,i}^2(P_jR_{ij} + Q_jX_{ij}) = 0$$
$$(5)$$

If we want to pledge the voltage stability of the distribution networks, equations (4) and (5) must have real roots, which are quadratic in form and the determinant B^2-4ac>0.Hence, we can obtain:

$$4[(P_jX_{ij} - Q_jR_{ij})^2 + (P_jR_{ij} + Q_jX_{ij})U_i^2] \leq U_i^4 \quad (6)$$

Where, U_i is the mold of the U^i. In this paper we say form (6) is the first class of voltage stability condition for distribution networks.

Define first class voltage stability index of branch line b_{ij} is L_{ij}. Consequently,

$$L_{ij} = \frac{4}{U_i^4}[(P_jX_{ij} - Q_jR_{ij})^2 + (P_jR_{ij} + Q_jX_{ij})U_i^2] \quad (7)$$

and form (6) can be written as,

$$L_{ij} \leq 1 \quad (8)$$

For a whole distribution network, the first class of voltage stability index is determine of the maximum value of the L_{ij} in the branch lines. Therefore the first class criterion voltage stability index for a whole network is

$$L = Max\{L_b\} \quad (9)$$

Where, L_b is the collection for first class voltage stability indexes of all branch lines in the distribution network; L can evaluate the condition of voltage stability ,the less value of the L, the stronger voltage stability of networks. We can define the first class voltage stability margin of branch line b_{ij} as follows:

$$B_{ij}=1-L_{ij} \quad (10)$$

Therefore, the whole distribution network, s first class voltage stability margin is,

$$B=1-L \quad (11)$$

Clearly, the larger value of the voltage stability margin, the higher stability of the networks.

For only two nodes ($i = 0, j = 1$) of the single line distribution networks, as the node of power for voltage reference point, that is the voltage $V = 1.0$. And the equation (6) is show as,

$$4[(P_1X_{01} - Q_1R_{01})^2 + (P_1R_{01} + Q_1X_{01})] \leq 1 \quad (12)$$

We can know from analysis of above all, L_b is clearly reflect the relation between load distribution and voltage stability is proposed given in this paper.

3. The voltage stability criterion of distribution networks based on load-voltage characteristic

In reference [1], according to the b_{ij} branch line's load-voltage characteristic of send and receive side. We derived a voltage stability criterion. Which receive side's load-voltage characteristic formula is

$$I_{ij} = S_j/U_j \quad (13)$$

Where, I_{ij} is the mold value of the flow current; U_j is the mold of U^j; $S_j = \sqrt{P_j^2 + Q_j^2}$. The formula is actually reflection the load characteristic of distribution power system. The send side's load-voltage characteristic equation is as follow:

$$I_{ij} = \sqrt{(U_i \cos\delta - U_j)^2 + (U_i \sin\delta)^2}/Z_{ij} \quad (14)$$

Where, δ is the deviation of the voltage angle of send and receive side; $Z_{ij} = \sqrt{R_{ij}^2 + X_{ij}^2}$ equation (14) is actually reflect the characteristic of power supply in distribution system.

Equation (13) and (14) are respectively correspond to curve 1and 2 show in Fig.1. Generally, two curves have two nodes A and B, among them A is stability of work point, the B is instability of work point, the slope of the two curves are minus. Voltage stability condition is defined in reference [17]:

$$|K_{1,ij}| \leq |K_{2,ij}| \quad (15)$$

Where $|k_{1,ij}|$ and $|k_{2,ij}|$ are the absolute value of slope correspond to curve 1 and 2.

If real system dissatisfaction equation (15), when occur voltage perturbation, variety current provide by send side will can not satisfy variety current flow in load side, and cause voltage perturbation even more voltage lose steady. Hence, we call (15) to the second class voltage stability condition. By equation (13), we can obtained,

$$|K_1| = \frac{S_j}{U_j^2} \quad (16)$$

Due to changes of U_j will cause changes of U_i, therefore processing of its compute is complex and often draw math method support from reference [17], that is,

$$K_2 = \frac{dI_i}{dU_i} = \frac{I_{R,i}\dfrac{dI_{R,i}}{dU_i} + I_{I,i}\dfrac{dI_{I,i}}{dU_i}}{\sqrt{I_{R,i}^2 + I_{I,i}^2}} \quad (17)$$

Where, $I_{R,i}$ and $I_{I,i}$ are respectively the current's real and negative which flows from node i (take U_i as reference voltage vector).

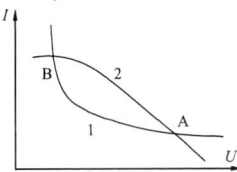

Fig.1 V-I characteristic of load side (curve-1) and source side (curve-2) of an electric power system

$$\frac{dI_{R,i}}{dU_i} \approx (G_{ij}\cos\delta_{ij} + B_{ij}\sin\delta_{ij})\frac{\Delta U_j}{\Delta U_i} \quad (18)$$

$$\frac{dI_{I,i}}{dU_i} \approx (B_{ij}\cos\delta_{ij} - G_{ij}\sin\delta_{ij})\frac{\Delta U_j}{\Delta U_i} \quad (19)$$

Where, ΔU_i (ΔU_j) is node $i(j)$ gradually altered its reactive power so the voltage deviation obtained from the adjoining power flow compute.

4. Relation of two voltage stability criterions for distribution networks

The concept of voltage stability criterion is clear in reference [17], but not strict. Because it held water only when curve 1 and 2 in diagram 1 had nodes. But more heavy load or under the line with bigger resistance, the two curves may have no nodes, for this reason, thus the voltage lose steady.

The load-voltage characteristic in send and receive side mean into the form of vector and have:

$$\boldsymbol{I}_{ij} = \boldsymbol{S}_j^* / \boldsymbol{U}_j^* \quad (20)$$

$$\boldsymbol{I}_{ij}\boldsymbol{Z}_{ij} = \boldsymbol{U}_i - \boldsymbol{U}_j \quad (21)$$

Where, Iij is current vector flow from branch line b_{ij} flow from node j.

Unite and rearranging equations (20) and (21), the result derived as same as the quadratic equation from equation (1). If curve 1 and 2 have nodes, thus equations (20) and (21) have real roots, equation (6) is the first class voltage stability condition.

Generally distribution networks have small δ, thus equation (14) can reduce as:

$$I_{ij} = \frac{U_i - U_j}{Z_{ij}} \quad (22)$$

Unite equation (13) and (22), a quadratic equation in terms of U_j is obtained:

$$U_j^2 - U_i U_j + S_j Z_{ij} = 0 \quad (23)$$

Where, the condition can pledge equation (13) have real roots is the approximate first class voltage stability condition. That is,

$$S_j \le \frac{U_i^2}{4Z_{ij}} \quad (24)$$

The approximate first class voltage stability is,

$$\hat{L}_{ij} = \frac{4Z_{ij}S_j}{U_i^2} \quad (25)$$

Equation (6) is also show as,

$$[U_i^2 - 2(P_j R_{ij} + Q_j X_{ij})]^2 \ge 4S_j^2 Z_{ij}^2 \quad (26)$$

That is

$$U_i^2 \ge 2(P_j R_{ij} + Q_j X_{ij}) + 2S_j Z_{ij} \quad (27)$$

Consider,

$$P_j^2 X_{ij}^2 + Q_j^2 R_{ij}^2 \ge 2P_j Q_j R_{ij} X_{ij} \quad (28)$$

Thus,

$$\begin{aligned}S_j^2 Z_{ij}^2 &= P_j^2 R_{ij}^2 + Q_j^2 X_{ij}^2 + P_j^2 X_{ij}^2 + Q_j^2 R_{ij}^2 \\ &> \left(P_j R_{ij} + Q_j X_{ij}\right)^2 = P_j^2 R_{ij}^2 + Q_j^2 X_{ij}^2 + 2P_j Q_j R_{ij} X_{ij}\end{aligned} \quad (29)$$

That is,

$$4S_j Z_{ij} \ge 2(P_j R_{ij} + Q_j X_{ij}) + 2S_j Z_{ij} \quad (30)$$

Therefore, as long as equation (24) established, the equation (6) will also established. That is when established the approximate first class voltage stability condition, and the first class voltage stability condition will established.

The first class voltage stability is an essential condition of distribution networks. For fixed power, it is voltage stability will must not maintenance if disaffection the condition.

The branch lines which satisfy the first class voltage stability condition can work in steady point A in Fig.1, under no disturbance situations. But considered the disturbance existing in real system, distribution networks disaffection the second class voltage stability condition will lose steady. Thus we derived voltage stability condition for distribution network as follow: not only satisfy the first class voltage stability condition but also satisfy the second class voltage stability condition.

Under the circumstance that given the load and branch line resistance, by equation (15), we can compute the second class critical voltage $U_{T,ij}$ of the line.

The second class voltage stability index is defined H_{ij}, it is

$$H_{ij} = |K_1|/|K_2| \quad (31)$$

Thus equation (15) is description as follow,

$$H_{ij} \leq 1 \tag{32}$$

The second class voltage stability index of the whole network is

$$H = Max\{H_b\} \tag{33}$$

Where, H_b is assemble of the second class voltage stability indexes of all branch lines in the whole networks; H reflect the voltage counter-disturbance capability and the less of H, the stronger capability of voltage counter-disturbance. The second class voltage stability margin of branch line b_{ij} is defined as follow:

$$D_{ij} = 1 - H_{ij} \tag{34}$$

The whole distribution network's voltage stability margin is defined as follow:

$$D = 1 - H \tag{35}$$

Clearly, the larger of the second class voltage stability margin, the stronger voltage counter-disturbance capability of the distribution networks. Voltage counter-disturbance safety distance of the line b_{ij} is defined:

$$DU_{ij} = U_j - U_{T,ij} \tag{36}$$

The voltage counter-disturbance safety distance of the whole networks is defined

$$DU = Min\{D_b\} \tag{37}$$

Where, D_b is the assemble of the voltage counter-disturbance safety distance in the whole networks; the larger of the D, the better of the counter-disturbance capability.

5. Analysis Procedure on Voltage Stability of the Distribution Networks

The steps of the voltage stability analysis as follows:

(I) Compute power flow of the distribution networks. If convergence of the power flow solution, that is illustrate the any branch line of the network all satisfy the first class voltage stability condition, then go to step(2); otherwise go to step(4).

(II) According to the above results and equations (7), (9), (10) and (11) separately compute every branch lines and the whole network, s first class voltage stability index and voltage stability margin.

(III) According to the results of power flow and equation (31) and (33)~(37) separately compute every branch lines and the whole network, s second voltage stability index, voltage margin and counter-disturbance safety distance, then go to step (5).

(IV) Hypothecate the system, s voltage fixed, neglect each branch line, s reactive power loss, according to the KCL and KVL and from the power supply to the receive side order, calculate the voltage of each node one by one. And judgment each line, s voltage stability use of equation (6) and (24), make use of equation (7), (9) and equation (10), (11) respectively calculate each line and

the whole network, s first class voltage stability index and margin, find out the weak link of the distribution system.

(V) Obtain the analysis result, and then be over.

6. Analysis of example

In this paper, the distribution network with 9 nodes application in simulate analysis show in Fig.2, where $P_{Li} + jQ_{Li}$ indicates the load of node-i, other network's data show in appendix. For the sake of study in different conditions voltage stability exponent of the load. Used data in reference [18] is a foundation load, multiply a coefficient k on this basic, such computing obtain the real calculation load. The larger value of k, show the more heavy of load. Adopt on the method in this paper, while the value of k is different, we can obtain the voltage stability analysis results show in Tab.1 and Tab.2.

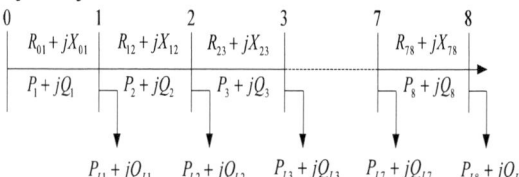

Fig.2 A distribution network with 9 nodes

Table 1 Results of voltage stability analysis (k=0.8)

Branch	L	B	H	D	Stability Result (I and II)
1	0.025	0.975	0.217	0.783	Y, Y
2	0.013	0.987	0.215	0.785	Y, Y
3	0.017	0.983	0.384	0.616	Y, Y
4	0.022	0.978	0.416	0.584	Y, Y
5	0.031	0.969	0.493	0.507	Y, Y
6	0.031	0.969	0.493	0.507	Y, Y
7	0.038	0.962	0.756	0.244	Y, Y
8	0.035	0.965	0.731	0.269	Y, Y
General	0.038	0.962	0.756	0.244	Y

Table 2 Results of voltage stability analysis (k=4.0)

Branch	L	B	H	D	Stability Result (I and II)
1	0.198	0.802	0.431	0.569	Y, Y
2	0.244	0.756	0.375	0.625	Y, Y
3	0.256	0.744	0.578	0.422	Y, Y
4	0.273	0.727	0.734	0.266	Y, Y
5	0.284	0.716	0.876	0.124	Y,Y
6	0.276	0.724	1.213	-0.213	Y, N
7	0.291	0.709	3.472	-2.472	Y, N
8	0.287	0.713	2.987	-1.987	Y, N
General	0.291	0.709	3.472	-2.472	N

By Tab.1 and Tab.2, we know, the voltage stability margin and voltage counter-disturbance safety distance put forward in this paper can measure the voltage stability and counter-disturbance capability indeed.

7. Conclusion

(1) Based on the existence of power flow solution and based on load-voltage characteristic's judgment method, we obtained the same first class voltage stability condition, and point out this stability condition is the essential condition in distribution networks.

(2) Consider the fluencies about disturbance of the networks, only satisfy the first class voltage stability condition of the distribution networks also occurs voltage instability. In this paper, we definition the voltage stability margin and voltage counter-disturbance safety distance index etc, analyzed the existing of second voltage stability criterion problem based on load- voltage characteristic, indicated voltage stability of distribution networks must satisfy the first class and second class voltage stability condition meanwhile.

Acknowledgement

It is supported by the novel project of China National Petroleum Corporation(Z06086).

References

[1] C. W. Taylor. Power systems voltage stability [M], EPRI Power System Engineering Series. McGraw Hill, 1994

[2] T. V. Custem, C. Vournas. Voltage stability of electric power systems [M], Kluwer Academic Publishers, 1998

[3] Prada, R.B.; Souza, L.J.; Voltage stability and thermal limit: constraints on the maximum loading of electrical energy distribution feeders [J]. IEE Proceedings-Generation, Transmission and Distribution, Volume 145, Issue 5, Sept. 1998 Page(s) : 573 - 577

[4] Lu Weixing, Shu Yinbiao, Shi Lianjun. WSCC disturbance on August 10,1996 in the United States [J]. Power System Technology, 1996, 20(9): 40-42.in Chinese

[5] Duan Xianzhong, He Yangzan, Cheng Deshu. Review of voltage stability in Electric power system [J]. Power System Technology, 1995, 19(4): 20-24.in Chinese

[6] Wei Hua, Ding Xiaoying. An algorithm for determining voltage stability critical point based on interior pPoint theory [J]. Proceedings of the CSEE, 2002, 22(3): 27-31.in Chinese

[7] Wang Qinghong, Zhou Shuangxi, Hu Guogeng. Power system Voltage stability analysis using expanded power flow models [J]. Power System Technology, 2002, 26(10): 25-29.in Chinese

[8] Liu Jian, Bi Pengxiang, Dong Haipeng. Simplified Analysis & Optimization of Complicated Distribution Networks [M], China Electric Power Press, Beijing, 2002.in Chinese

[9] Guimaraes, M.A.N.; Lorenzeti, J.F.C.; Castro, C.A. Reconfiguration of distribution systems for stability margin enhancement using tabu search [C]. Power System Technology, 2004. PowerCon 2004. Volume 2, 1556-1561

[10] Liu Daowei, Xie Xiaorong, Mu Gang. An on-line voltage stability index of power system based on synchronized phasor measurement [J]. Proceedings of the CSEE, 2005, 25(1): 13-17.in Chinese

[11] Bao Lixin, Duan Xianzhong, He Yangzan. Dynamical analysis of voltage stability in state space [J]. Proceedings of the CSEE, 2001, 21(5): 17-22 in Chinese

[12] Zhu Zhenqing, Yang Xiaoguan, Zhou Ge. Practical algorithm of voltage stability in power system [J]. Power System Technology, 1998, 22(4): 13-16. in Chinese

[13] Qiu Xiaoyan, Li Xiyuan, Lin Wei. Methods for contingency screening and banking for on-line voltage stability assessment [J]. Proceedings of the CSEE, 2004, 24(9): 50-55.in Chinese

[14] Zhou Shuangxi, Jiang Yong, Zhu Lingzhi. Review on steady state voltage stability indices of power systems [J]. Power System Technology, 2001, 25(1): 1-7. in Chinese

[15] Li Xinran, He Renkui, Zhang Jian. Effect of load characteristics on power system stead-state voltage stability and the practical criterion of voltage stability [J]. Proceedings of the CSEE, 1999, 19 (4) : 23-30.in Chinese

[16] Zhang Ming, Bi Pengxiang, Liu Jian. Study on Voltage Stability of Distribution System [J]. Electric Power Construction, 2002, 23(10):41-43.in Chinese

[17] Li Baoguo, Lu Baochun, Ba Jinxiang. A practical method to judge static voltage stability of load nodes based on V-I characteristic [J]. Relay, 2004, 32(11):1-4.in Chinese

[18] G. B. Jasmon, L . H. C. C. LEE. Distribution network reduction for voltage stability analysis and loadflow calculations. International Journal of Electrical Power& Energy Systems, 1991,13(1): 9-13.

[19] R. Ranjan, B. Venkatesh, D.Das. Voltage stability analysis of radial distribution networks [J]. International Journal of Electric Power Components and Systems, 2003,31:501-551.

A New Modeling Method for the Switching Power Converter

Weiping Wang [1], Yujie Shi [2], Dianbing Yang [3]

[1,2,3]*Zhengzhou Information Science and Technology Institute, Zhengzhou,*
Henan, 450002, China
weeping1109@163.com, syj_zz@126.com

Abstract

A new modeling method for the complicated switching power converter, which is called parameters of differential equations fitting algorithm, is proposed in the paper. The models of the Buck converter and Boost converter are built using this method, which prove that the method is feasible.

1. Introduction

Switching power supply(SPS) is a closed-loop automatic voltage-regulating system, whose design depends on mathematical model of each part. Switching power converter is the core of SPS and its model is really hard to make. For switching power converters with simple timing control, some methods, such as state-space averaging method[1] and symbol analysis method[2],can be used to establish small-signal models to analyze their dynamic characteristics. However, for switching power converters with complex timing control, it's very difficult to model by these methods. Besides, these methods brings much approximation in modeling process,which leads to imprecise models. Literature [3] presents a method called step-response fitting algorithm, which is applicable for complex converters. It can establish precise models for typical second-order system, but it makes imprecise model for high-order system. At the same time, it requires rich modeling experience ,repeated testing and heavy workload. The paper presents a new modeling method with the help of PSPICE and MATLAB software, which is called parameters of difference equations fitting algorithm. The method is based not on the circuit model, but on input and output data of the switching converter. Therefore, it is applicable to complex switching converters.

2. Principle of Parameters of Differential Equations Fitting Algorithm

2.1.Modeling for linear time-invariant continuous system

Linear time-invariant continuous system can be described with linear constant coefficient differential equation or impulse transfer function. If the continuous system discretizes with zero-order maintaining law[4], discrete system can be described using the linear constant coefficient differential equation (1) or the impulse transfer function (2).

$$y(kT)+a_1y[(k-1)\ T]+a_2y[(k-2)\ T]+\cdots+a_ny[(k-n)\ T]$$
$$=b_0x(kT)+b_1x[(k-1)\ T]+\cdots+b_mx[(k-m)\ T]$$

(1)

$$H(z)=\frac{b_0+b_1z^{-1}+b_2z^{-2}+......+b_mz^{-m}}{1+a_1z^{-1}+a_2z^{-2}+......+a_nz^{-n}}$$

(2)

Equation (1) can also be expressed with equation (3)

$$y(kT)=-\sum_{i=1}^{n}a_iy[(k-i)\ T]+\sum_{j=0}^{m}b_jx[(k-j)\ T]$$

(3)

In equation (3), a_i and b_j are constants，m and n are non-negative integers, $m \le n$ .When the sampling frequency is sufficiently high, the response of the discrete system is equal to that of the original continuous system with a given input signal ,except impulse input signal[4]. That is, these samples from the original continuous system will meet differential equations (1). What we have to do is to find the parameters $a_1......a_n,b_0b_1......b_m$, n and m according to the sampling sequences $x(kT)$ and $y(kT)$ and their relationship (3).

Assume that the output signal $y(t)$ of the continuous system with the input signal $x(t)$ is known. We sample them with the synchronous sampling cycle T , and get sequences $x(kT)$ and $y(kT)$. Define n and m are constants, so that the system can be described with equation (3). N groups of samples are taken respectively from $x(kT)$ and $y(kT)$ (N equals to $n+m+1$,i.e.the number of coefficients in equation (3),).Each group contains $m+1$ input samples and $n+1$ output samples. By substituting them into equation (3),we obtain

$$
\begin{cases}
y(k_1T) = -\sum_{i=1}^{n} a_i y[(k_1-i)\ T] + \sum_{j=0}^{m} b_j x[(k_1-j)\ T] \\[2mm]
y(k_2T) = -\sum_{i=1}^{n} a_i y[(k_2-i)\ T] + \sum_{j=0}^{m} b_j x[(k_2-j)\ T] \\[2mm]
\qquad\qquad\qquad \cdot \\
\qquad\qquad\qquad \cdot \\
\qquad\qquad\qquad \cdot \\
y(k_NT) = -\sum_{i=1}^{n} a_i y[(k_N-i)\ T] + \sum_{j=0}^{m} b_j x[(k_N-j)\ T]
\end{cases}
$$

$$(4)$$

Define $\mathbf{b} = \begin{bmatrix} y(k_1T) & y(k_2T) & \ldots & y(k_NT) \end{bmatrix}^T$

$\mathbf{X} = \begin{bmatrix} b_0 & . & . & b_m & -a_1 & . & . & -a_n \end{bmatrix}^T$

$$
\mathbf{A} = \begin{pmatrix}
x(k_1T) & x[(k_1-1)T] & \ldots & y((k_1-n)T) \\
x(k_2T) & x[(k_2-1)T] & \ldots & y((k_2-n)T) \\
\cdot & \cdot & \cdot & \cdot \\
\cdot & \cdot & \cdot & \cdot \\
\cdot & \cdot & \cdot & \cdot \\
x(k_NT) & x[(k_N-1)T] & \ldots & y((k_N-n)T)
\end{pmatrix}_{N\times N}
$$

Therefore, the equations can be expressed as

$$\mathbf{b} = \mathbf{A} \times \mathbf{X} \qquad (5)$$

According to linear algebra theory, n-variant systems of nonhomogeneous linear equations have solutions if and only if the rank of coefficient matrix $\mathbf{A}$ is equivalent to that of its augmented matrix $\mathbf{B} = (\mathbf{A}, \mathbf{b})$.

When $rank(\mathbf{A}) = rank(\mathbf{B}) = N$, the equation (5) has only one solution, that is, $\mathbf{X} = \mathbf{A}^{-1} \times \mathbf{b}$ .When $rank(\mathbf{A}) = rank(\mathbf{B}) = r < N$, the equation (5) has an infinite number of solutions. Therefore, the necessity condition of equation (5) having an only solution is that $rank(\mathbf{A}) = N$. For the step input signal, the rank of $\mathbf{A}$ isn't full because of $x(k_iT) = x[(k_i-1)T] = \ldots\ldots x[(k_i-m)T](i=1,2\ldots N)$ .So the only solution doesn't exist. However, the problem doesn't exist for the slope input signal. When we can only obtain the step-response curve, the following data processing method can be used.

Suppose $h(kT)$ corresponds to the unit-impulse response of the system. The unit-step response is gotten by:

$g(kT) = h(kT) * \varepsilon(kT)$.

Then, with the following deductions, we obtain the unit-slope sequence.

$$
\begin{aligned}
g(kT) * \varepsilon(kT) &= h(kT) * \varepsilon(kT) * \varepsilon(kT) \\
&= h(kT) * (k+1)\varepsilon(kT) \\
&= h(kT) * k\varepsilon(kT) + h(kT) * \varepsilon(kT) \\
&= h(kT) * k\varepsilon(kT) + g(kT) \\
y(kT) &= h(kT) * (kT)\varepsilon(kT) \\
&= T \times [h(kT) * k\varepsilon(kT)] \\
&= T \times [g(kT) * \varepsilon(kT) - g(kT)]
\end{aligned}
$$

That is, we can get $y(kT)$ just by the operation with $g(kT)$. And then the solution of equation (5) can be gotten using N groups of sequences $y(kT)$ and $(kT)\varepsilon(kT)$.

According to linear algebra theory, there are three cases.

A，$rank(\mathbf{A}) \neq rank(\mathbf{B})$, no solution.

B，$rank(\mathbf{A}) = rank(\mathbf{B}) = r < N$, equatio(5) has an infinite number of solutions. This case can't happen to a system, so it doesn't conform to the reality.

The two cases mentioned above don't conform to the reality. It means that the n , m and model form chosen are inconsistent with the actual model. We need to reselect them.

C，$rank(\mathbf{A}) = rank(\mathbf{B}) = N$, equation (5) has only one solution $\mathbf{X}$ .But we can't assure that it is the actual parameters. Because it exists that the sample points of response curves of two systems overlap. To ensure that the solution is the actual parameters for the system, we need to substitute another group of input and output data into the equation (1) with the parameters $\mathbf{X}$ .If they meet the equation (1), the only solution $\mathbf{X}$ is basically regarded as the actual parameters(For further confirmation, substitute more groups of data into equation (1)). Otherwise, it means that the n, m and model form chosen are inconsistent with the actual model. The only solution $\mathbf{X}$ isn't the actual parameters for the system. We need to reselect them.

About the model form mentioned above, there are various forms of n-order system. For example, second-order system has the following forms.

$$
\begin{aligned}
y(kT) + a_1 y[(k-1)\ T] + a_2 y[(k-2)\ T] \\
= b_0 x(kT) \\
y(kT) + a_1 y[(k-1)\ T] + a_2 y[(k-2)\ T] \\
= b_0 x(kT) + b_1 x[(k-1)\ T] \\
y(kT) + a_1 y[(k-1)\ T] + a_2 y[(k-2)\ T] \\
= b_0 x(kT) + b_1 x[(k-1)\ T] + b_2 x[(k-2)\ T]
\end{aligned}
$$

etc.At first, the order and model form of a system are unknown, therefore, the modeling process is actually a searching process. We can write a program, then this process can be completed by computers.

To sum up, modeling steps for linear time-invariant continuous system are summarized as follows

A, First of all, sample input and output signals of the system synchronously with the same cycle ,then obtain input and output sequences (If we just obtain the step-response sequences, handle them into slope-response sequences).

B, Assuming a differential equation model for the system (that is, choose a group of b_i and a_j), substitute corresponding samples into the equation (4). when it

405

doesn't satisfy $rank(\mathbf{A}) = rank(\mathbf{B}) = N$, reselect the system model. Otherwise, obtain the only solution $\mathbf{X}$.

C, Substitute another (or more) group of input and output data into the equation (1) with the parameters $\mathbf{X}$ to test. If these meet the equation (1), the solution is the actual parameters. Otherwise, reselect the system model and calculate until getting the actual parameters of the system..

D, After taking the actual parameters $\mathbf{X}$ into equation (2), continue the equation using zero-order retainer law, then obtain impulse transfer function of the linear time-invariant continuous system. Thus modeling finish.

At last, we can use MATLAB to simulate the step-response curve of the continuous system (namely fitting curve). The results show that the fitting curve coincides well with the actual curve.

2.2. Modeling for Switching Power Converters

PWM switching converter is a controlled object of the switching automatic voltage-regulating system. Its input signal is duty cycle function $D(s)$ (produced by a calibration device) and output signal is the output voltage $V(s)$ of PWM switching converter, which is also the output voltage of the automatically voltage-regulating system. Therefore, the dynamic model of the switching converter can be described with impulse transfer function $G(s) = \dfrac{V(s)}{D(s)}$. Unluckily, the converter is a nonlinear system, that is, the relationship between system parameters and working point (duty cycle D) is nonlinear. When SPS is working, the dynamic parameters of the system change nonlinearly with D. Therefore, only small signal linear model nearby a given working point D can be established. that is, the relationship between system parameters and working point (duty cycle D) is nonlinear. When SPS is working, the dynamic parameters of the system change nonlinearly with D. Therefore, only small signal linear model nearby a given working point D can be established.

According to First Method of Lyapunov[5], assuming that a system is described by a nonlinear state equation

$$\dot{\mathbf{x}} = \mathbf{f}(\mathbf{x}) \qquad (6)$$

Nearby the working point $\mathbf{x_e}$ (working point is the quilibrium point), linearization of equation (6) is

$$\dot{\mathbf{y}} = \mathbf{Ay} \qquad (7)$$

Here, $\mathbf{y}$ is the increment function nearby the working point $\mathbf{x_e}$, $\mathbf{A}$ is the Jacobian matrix.

$$\mathbf{A} = \left.\frac{\partial \mathbf{f}}{\partial \mathbf{x}}\right|_{\mathbf{x}=\mathbf{x_e}} = \begin{bmatrix} \dfrac{\partial f_1}{\partial x_1} & \dfrac{\partial f_1}{\partial x_2} & \cdots & \dfrac{\partial f_1}{\partial x_n} \\ \dfrac{\partial f_2}{\partial x_1} & \dfrac{\partial f_2}{\partial x_2} & \cdots & \dfrac{\partial f_2}{\partial x_n} \\ \cdot & \cdot & \cdot & \cdot \\ \cdot & \cdot & \cdot & \cdot \\ \cdot & \cdot & \cdot & \cdot \\ \dfrac{\partial f_n}{\partial x_1} & \dfrac{\partial f_n}{\partial x_2} & \cdots & \dfrac{\partial f_n}{\partial x_n} \end{bmatrix}_{\mathbf{x}=\mathbf{x_e}} \qquad (8)$$

As equation (7) has a corresponding linear constant coefficient differential equation, we can use parameters of difference equations fitting algorithm to model it. Just keep in mind that the model is a small-signal model nearby a given working point. With the nonlinear relationship between system parameters and working point D, the models are different at different working points. So it's necessary to model at different working points within the working scope, then compare those models and choose the one with the worst dynamic characteristics for designing the closed-loop system. This is because if the closed-loop is designed to meet the worst model, it can meet all models at other working points within the scope of work.

Therefore, once the working point of the PWM switching power converter is given, such as $D = 0.3$, the system can be described by a linear constant coefficient differential equation. When we give a burst step signal $D = 0.3$ to switching converter, the response curve shows dynamic characteristic at the working point. Then, it can be used to build the dynamic small-signal model for the converter.

3. Examples of modeling for Switching converters

3.1. Modeling for BUCK converter

Figure 1 shows the step-response curve $g(t)$ simulated by PSPICE at $D = 1/3$. Sample $D\varepsilon(t)$ and $g(t)$ with the synchronous sampling cycle $T = 0.02ms$, then model for BUCK converter using the proposed method. Finally, the differential equation model is obtained as following

$$y(kT) + a_1 y[(k-1)\ T] + a_2 y[(k-2)\ T]$$
$$+ a_3 y[(k-3)\ T] = b_1 x[(k-1)\ T]$$

$a_1 = -1.9686, a_2 = 1.2081, a_3 = -0.1836, b_1 = 0.3929$.

Using the command d2c of MATLAB with zero-order retainer law, we can transform it into impulse transfer function of the actual continuous system. The result is as follow

$$H(s) = \frac{9126s^2 + 1.82e009s + 1.052e014}{s^3 + 8.476e004s^2 + 9.602e008s + 1.495e013}$$

Simulate the step-response curve $g'(t)$ of $H(s)$ using MATLAB,then show $g(t)$ and $g'(t)$ in Figure 2.As we see, $g'(t)$ fits very well with the actual curve $g(t)$.

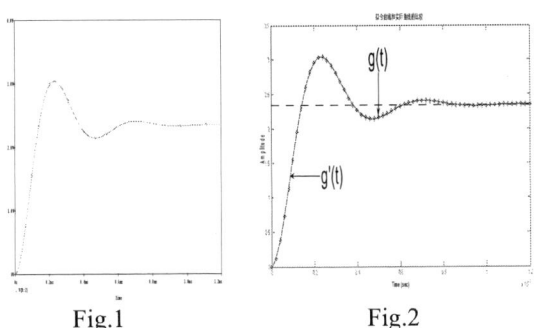

Fig.1

Fig.2

Fig.1 the step-response curve of BUCK at $D=1/3$

Fig.2 $g'(t)$ the fitting curve —

     $g(t)$ the actual curve ***

3.2 Modeling for BOOST converter

Figure 3 shows the full-response curve simulated by PSPICE at $D=1/6$.Remove the zero-input response, then the step-response $g(t)$ can be obtained. Sample $D\varepsilon(t)$ and $g(t)$ with the synchronous sampling cycle $T = 0.05ms$, then model BOOST converter using the proposed method. Finally, the differential equation model is obtained as follow

$$y(kT) + a_1 y[(k-1)\ T] + a_2 y[(k-2)\ T]$$
$$= b_1 x[(k-1)\ T] + b_2 x[(k-2)\ T]$$

$a_1 = -1.8339, a_2 = 0.8982, b_1 = -1.6440, b_2 = 3.7978.$

Using the command d2c of MATLAB with zero-order retainer law, we can transform it into impulse transfer function of the actual continuous system. The result is as following

$$H(s) = \frac{-5.846e004\ s + 9.138e008}{s^2 + 2148s + 2.726e007}$$

Simulate the step-response curve $g'(t)$ of $H(s)$ using MATLAB,and show $g(t)$ and $g'(t)$ in Figure 4.As we see, $g'(t)$ fits very well with the actual curve $g(t)$.

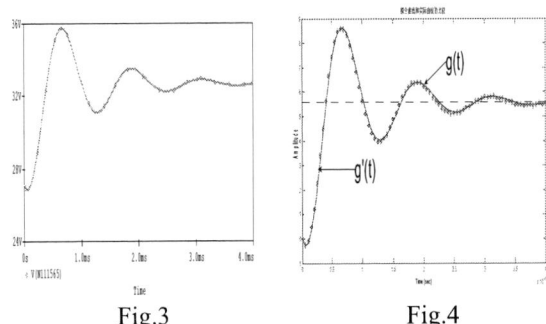

Fig.3

Fig.4

Fig.3 the full-response curve of Boost at $D=1/6$

Fig.4 $g'(t)$ the fitting curve —

     $g(t)$ the actual curve ***

4. Conclusion

It's very difficult to establish dynamic models for complex switching converters using analytical methods .However, using the new method with the help of PSPICE and MATLAB software ,it becomes really easy. The process of model matching can be realized on computers,which reduces workload greatly.Finally,we can see that the fitting models are close to the actual models.

References

[1] R.D. Middlebrooke, Slobodan Cuk (1976). "A General Unified Approach to Modeling Switching-converter Power Stage".*IEEE PESC*.Vol. 36, No.6, pp.251-256 .

[2] S.S.Qiu，L. M. Filarmsky ,and BT.lin (1999). "A new method of analysis for PWM switching power converters".*Int.J. of Electronics*. Vol. 86, No.11, pp.1395-1410.

[3] Liu Xiaobao,Shi Yujie,and Liang Ningning (2007). "Dynamic System Modeling and Closed-Loop Control of Switching Power Converter". *Journal of Information Engineering University*.Vol. 8, No.2, pp.238-241.

[4] Shi Yujie (2008). Automatic Control Theory (volumeⅡ). *National Defense Industry Press* , Beijing

[5] Wu Qi (2004).Automatic Control Theory (volumeⅡ). *Tsinghua University Press*, Beijing

Research on Wind Power Systems

Luo Qi [1,2]

[1] Information Engineering School, Zhongnan Branch of Wuhan University of Science and Technology, Wuhan, 430223, Hubei, China

[2] School of Computer Science and Engineering, Wuhan Institute of Technology, Wuhan, 430073, China

Abstract

In order to meet increasing need of energy and the press of environment protection, in the world many country increase the develop degree of new energy. Today the clean, pollute-free, green and renewable energy has become the focus of world's energy development. Wind energy is the most representational energy in clean reproducible energy, It has important signification of the environment protection and ecosystems balance, it can reduce the depend on the conventional source of energy and improve the energy structure. In the paper, the distributions of wind speed and capacity factor are proposed in the paper. And then, the Utilization ratio of wind power in many countries is also introduced. This is a belief abstract for an invited talk at the workshop on power Electronics and Intelligent transportation system.

1. Introduction

Wind power is the conversion of wind energy into a useful form, such as electricity, using wind turbines. At the end of 2007, worldwide capacity of wind-powered generators was 94.1 gigawatts.[1] Although wind currently produces about 1% of world-wide electricity use,[2] it accounts for approximately 19% of electricity production in Denmark, 9% in Spain and Portugal, and 6% in Germany and the Republic of Ireland (2007 data). Globally, wind power generation increased more than fivefold between 2000 and 2007.

Most wind power is generated in the form of electricity. Large scale wind farms are connected to electrical grids. Individual turbines can provide electricity to isolated locations. In windmills, wind energy is used directly as mechanical energy for pumping water or grinding grain.

Wind energy is plentiful, renewable, widely distributed, clean, and reduces greenhouse gas emissions when it displaces fossil-fuel-derived electricity. The intermittency of wind seldom creates problems when using wind power to supply a low proportion of total demand, but it presents extra costs when wind is to be used for a large fraction of demand. However these costs even for quite large percentage penetrations are considered to be modest.

2. Wind energy
2.1 Distribution of wind speed

Windiness varies, and an average value for a given location does not alone indicate the amount of energy a wind turbine could produce there. To assess the frequency of wind speeds at a particular location, a probability distribution function is often fit to the observed data. Different locations will have different wind speed distributions. The Rayleigh model closely mirrors the actual distribution of hourly wind speeds at many locations.

Because so much power is generated by higher wind speed, much of the energy comes in short bursts. The 2002 Lee Ranch sample is telling; half of the energy available arrived in just 15% of the operating time. The consequence is that wind energy does not have as consistent an output as fuel-fired power plants; utilities that use wind power must provide backup generation for times that the wind is weak. Making wind power more consistent requires that storage technologies must be used to retain the large amount of power generated in the bursts for later use.

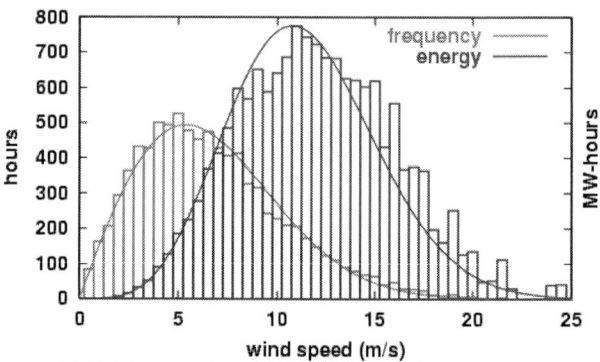

Figure 1 Distribution of wind speed (red) and energy (blue) for all of 2002 at the Lee Ranch facility in Colorado.

2.2 Capacity factor

Since wind speed is not constant, a wind farm's annual energy production is never as much as the sum of the generator nameplate ratings multiplied by the total hours in a year. The ratio of actual productivity in a year to this theoretical maximum is called the capacity factor. Typical capacity factors are 20-40%, with values at the upper end of the range in particularly favourable sites. For example, a 1 megawatt turbine with a capacity factor of 35% will not produce 8,760 megawatt-hours in a year (1x24x365), but only 0.35x24x365 = 3,066 MWh, averaging to 0.35 MW. Online data is available for some locations and the capacity factor can be calculated from the yearly output.

Unlike fueled generating plants, the capacity factor is limited by the inherent properties of wind. Capacity factors of other types of power plant are based mostly on fuel cost, with a small amount of downtime for maintenance. Nuclear plants have low incremental fuel cost, and so are run at full output and achieve a 90% capacity factor. Plants with higher fuel cost are throttled back to follow load. Gas turbine plants using natural gas as fuel may be very expensive to operate and may be run only to meet peak power demand. A gas turbine plant may have an annual capacity factor of 5-25% due to relatively high energy production cost.

According to a 2007 Stanford University study published in the Journal of Applied Meteorology and Climatology, interconnecting ten or more wind farms allows 33 to 47% of the total energy produced to be used as reliable, baseload electric power, as long as minimum criteria are met for wind speed and turbine height .

3. Utilization of wind power

The modern wind power industry began in 1979 with the serial production of wind turbines by Danish manufacturers Kuriant, Vestas, Nordtank, and Bonus. These early turbines were small by today's standards, with capacities of 20 to 30 kW each. Since then, they have increased greatly in size, while wind turbine production has expanded to many countries all over the world.

There are now many thousands of wind turbines operating, with a total capacity of 73,904 MW of which wind power in Europe accounts for 65% (2006). Wind power was the most fastest growing energy source at the end of 2004.[citation needed] World wind generation capacity more than quadrupled between 2000 and 2006. 81% of wind power installations are in the US and Europe, but the share of the top five countries in terms of new installations fell from 71% in 2004 to 62% in 2006.

In 2007, the countries with the highest total installed capacity were Germany, the United States, Spain, India, and China (see chart).

By 2010, the World Wind Energy Association expects 160GW of capacity to be installed worldwide, up from 73.9 GW at the end of 2006, implying an anticipated net growth rate of more than 21% per year.

Denmark generates nearly one-fifth of its electricity with wind turbines -- the highest percentage of any country -- and is fifth in the world in total wind power generation. Denmark is prominent in the manufacturing and use of wind turbines, with a commitment made in the 1970s to eventually produce half of the country's power by wind

In recent years, the United States has added more wind energy to its grid than any other country; U.S. wind power capacity grew by 45% to 16.8 gigawatts in 2007. Texas has become the largest wind energy producing state, surpassing California. In 2007, the state expects to add 2 gigawatts to its existing capacity of approximately 4.5 gigawatts. Iowa and Minnesota are expected to each produce 1 gigawatt by late-2007. Wind power generation in the U.S. was up 31.8% in February, 2007 from February, 2006. The average output of one megawatt of wind power is equivalent to the average electricity consumption of about 250 American households. According to the American Wind Energy Association, wind will generate enough electricity in 2008 to power just over 1% (4.5 million households) of total electricity in U.S., up from less than 0.1% in 1999. U.S. Department of Energy studies have concluded wind harvested in just three of the fifty U.S. states could provide enough electricity to power the entire nation, and that offshore wind farms could do the same job.

India ranks 4th in the world with a total wind power capacity of 6,270 MW in 2006, or 3% of all electricity produced in India. The World Wind Energy Conference in New Delhi in November 2006 has given additional impetus to the Indian wind industry. The windfarm near Muppandal, Tamil Nadu, India, provides an impoverished village with energy. India-based Suzlon Energy is one of the world's largest wind turbine manufacturers.

In December 2003, General Electric installed the world's largest offshore wind turbines in Ireland, and plans are being made for more such installations on the west coast, including the possible use of floating turbines.

In 2005, China announced it would build a 1000-megawatt wind farm in Hebei for completion in 2020. China reportedly has set a generating target of 20,000 MW by 2020 from renewable energy sources — it says indigenous wind power could generate up to 253,000

MW. Following the World Wind Energy Conference in November 2004, organised by the Chinese and the World Wind Energy Association, a Chinese renewable energy law was adopted. In late 2005, the Chinese government increased the official wind energy target for the year 2020 from 20 GW to 30 GW.

Mexico recently opened La Venta II wind power project as an important step in reducing Mexico's consumption of fossil fuels. The 88 MW project is the first of its kind in Mexico, and will provide 13 percent of the electricity needs of the state of Oaxaca. By 2012 the project will have a capacity of 3500 MW.

Another growing market is Brazil, with a wind potential of 143 GW. The federal government has created an incentive program, called Proinfa, to build production capacity of 3300 MW of renewable energy for 2008, of which 1422 MW through wind energy. The program seeks to produce 10% of Brazilian electricity through renewable sources.

South Africa has a proposed station situated on the West Coast north of the Olifants River mouth near the town of Koekenaap, east of Vredendal in the Western Cape Province. The station is proposed to have a total output of 100MW although there are negotiations to double this capacity. The plant could be operational by 2010.

France has announced a target of 12,500 MW installed by 2010.

Canada experienced rapid growth of wind capacity between 2000 and 2006, with total installed capacity increasing from 137 MW to 1,451 MW, and showing an annual growth rate of 38%. Particularly rapid growth was seen in 2006, with total capacity doubling from the 684 MW at end of 2005. This growth was fed by measures including installation targets, economic incentives and political support. For example, the Ontario government announced that it will introduce a feed-in tariff for wind power, referred to as 'Standard Offer Contracts', which may boost the wind industry across the province. In Quebec, the provincially-owned electric utility plans to purchase an additional 2000 MW by 2013[4] [5] [6] [7].

References

[1] Robert Zavadil et al, Making Connections: Wind Generation Challenges and Progress, *IEEE Power and Energy Magazine*, Nov/Dec. 2005, pgs. 27-37

[2] Edgar A. DeMoe et al, Wind Plant Integration: Cost, Status and Issues, *IEEE Power and Energy Magazine*, Nov/Dec. 2005, pgs. 39-46

[3] Wind Energy: Rapid Growth. *Canadian Wind Energy Association*. Retrieved on 2006.04

[4] Canada's Current Installed Capacity. Canadian Wind Energy Association. Retrieved on 2006.12

[5] Standard Offer Contracts Arrive In Ontario. Ontario Sustainable Energy Association (2006). Retrieved on 2006

[6] International Energy Outlook, 2006, Energy Information Administration.

Simulation Study of Fuzzy Based Unified Power Flow Controller on Power Flow Controlling

Lütfü Sarıbulut
E-mail: lsaribulut@cu.edu.tr

Mehmet Tümay
E-mail: mtumay@cu.edu.tr

Çukurova University, Department of Electrical & Electronics Engineering,
Balcalı, Adana, Turkey

Abstract

Unified Power Flow Controller (UPFC) is the most widely used FACTS device to control the power flow and to optimize the system stability in the transmission line. The controller used in the control mechanism has an important effect on the performance of UPFC. According to this, the performance of UPFC for several case studies was observed by using different control mechanisms based on Proportional-Integral (PI) and fuzzy controllers in this study. Fuzzy Logic Controller (FLC) was developed by using Takagi-Sugeno inference system and used in UPFC. The case studies are applied to prove the capability of UPFC on power flow and the effectiveness of controllers on the performance of UPFC at the different operating conditions. PSCAD/EMTDC program is used to model UPFC and to verify the performance of UPFC with different controllers in the transmission line.

1. Introduction

In recent years, great electric power demands have been imposed upon high voltage transmission networks in the worldwide. Also, the construction of new generating units and transmission circuits becomes more difficult because of economic and environmental reasons [1]. One of the important issues of transmission lines is the controlling of the power flow and the other is reactive power compensation. A new technology concept known as Flexible Alternating Current Transmission Systems (FACTS) technology was presented in the late of 1980s [2]. FACTS technology consists of devices depended on using the reliable and high speed power electronic devices instead of mechanical controllers. Thus, the utilization of the existing power system comes into optimal condition and the controllability of the system is increased.

Unified Power Flow Controller (UPFC) is the member of FACTS device that has emerged for the control and optimization of power flow in electrical power transmission systems [3]. It consists of a series converter namely Static Synchronous Series Compensator (SSSC) and a shunt converter namely Static Synchronous Compensator (STATCOM) connected by a common DC

link capacitor. It can simultaneously perform the function of transmission line real/reactive power flow control in addition to UPFC bus voltage/shunt reactive power control [4].

The control mechanism and the controller have important effect on the performance of UPFC. In literature, several control mechanisms are used in the UPFC model. A novel fuzzy inference system described in matrix form is proposed and used to improve the dynamic control of real and reactive power [5]. Two fuzzy logic controllers based on Mamdani type fuzzy logic are used. One of the controllers is a proportional fuzzy logic controller (PF-UPFC) and the other is a Hybrid fuzzy logic UPFC (HF-UPFC) [3]. The selection of suitable location for UPFC is studied and composite-criteria-based fuzzy logic is used to evaluate the network contingency ranking [6]. The power-feedback control scheme is used in the control mechanism of UPFC [7]. The power fluctuation is damped readily and the value of reactive power is minimized as possible by using several time constants. In addition, there is no value changed in the real power. The control method of variable interval-fuzzy-mutual is used in the control mechanism of UPFC [8]. In the simulation results, there is a high overshoot values occurred both real power and bus voltage during the three phase faults applied. However, the real power value is increased but there is no value changed in the reactive power. The performance of UPFC is observed by using three different controllers [9]. In the simulation results, the variation of the real power direction can be observed easily. However, the value of reactive power is kept at zero because of there is no reactive power flow in the system. The performance of Pulse Width Modulation (PWM) based UPFC is observed [10]. According to results, the values of real and reactive power are changed in large values with UPFC because of the low values of bus voltage.

This paper presents the performance evaluation of UPFC in several case studies by using different control mechanisms based on PI and fuzzy controllers. *"Takagi-Sugeno Inference System"* is used in the decision making of fuzzy model and *"Weighted Average"* method which is the special case of *"Mamdani model"* is used in the

defuzzification process in the created fuzzy model. PSCAD/EMTDC program is used to create UPFC model and to obtain the results of case studies [11].

2. Modeling of UPFC System

UPFC is a combination of two back to back three-phase converters. One of the converters is connected in shunt and the other is in series with the transmission line by coupling transformers. The shunt converter is named as STATCOM and the series converter is named as SSSC. The converters are connected by a common DC-link where the capacitor is coupled and it allows a bi-directional real power flow between the shunt output terminals of the STATCOM and the series input terminals of the SSSC. The basic system configuration of UPFC is shown in Figure 1.

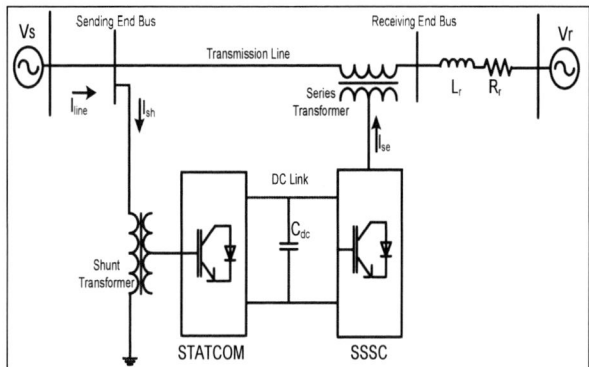

Figure 1 Basic scheme of UPFC system

2.1. The control mechanism of shunt converter

The shunt converter has an important function in the operation of UPFC system and it can be rolled as a STATCOM [12]. It operates to draw a controlled current from the transmission line. The shunt converter can provide the real power drawn by the series branch and the losses, which maintain DC voltage, and can provide reactive compensation without an external electric energy source to the system independently [13]. PI controller based and fuzzy controller based control mechanisms of shunt converter are shown separately in Figure 2 and Figure 3, respectively.

2.2. The control mechanism of series converter

The series converter actualizes the main function of UPFC. The magnitude and the phase angle of series-injected voltage are controlled by series converter to provide the desired real and reactive power flow in the transmission line. The current of the line flows through the series converter resulting in an active and reactive power exchange with the ac system [14]. The control mechanism of series converter is shown in Figure 4.

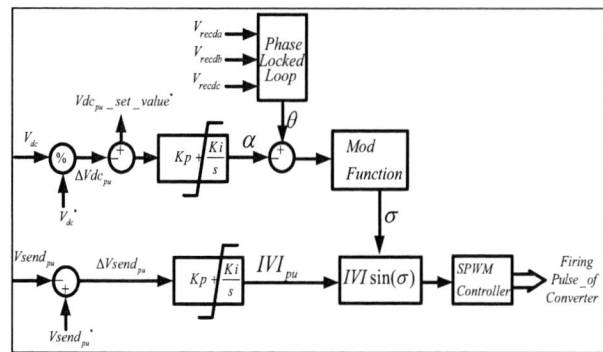

Figure 2 PI controller based control mechanism

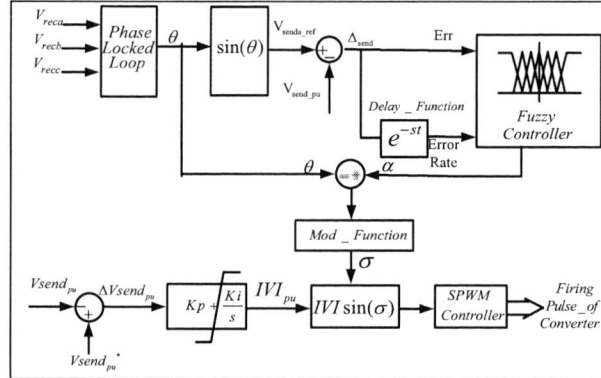

Figure 3 Fuzzy controller based control mechanism

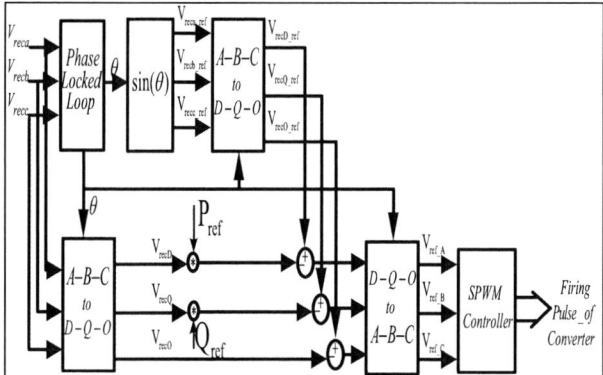

Figure 4 Control mechanism of series converter

3. Implementation of Fuzzy Controller

FLC are formed by simple rule based on "*if x and y then z*". These rules are defined by taking help from person's experience and knowledge about the plant behavior. The performance of the system is improved by the correct combinations of these rules. Each of the rules defines one membership which is the function of FLC. More sensitivity is provided in the control mechanism of FLC by increasing the numbers of membership function [15-17]. In this study, the inputs of the fuzzy system are assigned by using 7 membership functions and the fuzzy system to be formed in 49 rules. Hence, the sensitivity in the control mechanism is increased. The steps of proposed fuzzy control system are given in Figure 5.

412

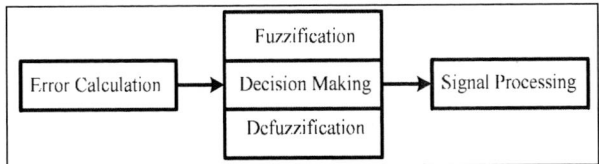

Figure 5 General steps of fuzzy logic control system

Error Calculation: The signals of supply voltage for each phase are measured and converted into per unit (pu.) value. The error (err_A) is calculated from the difference between the source voltage value and the reference value obtained from Phase Locked Loop (PLL). The difference between variation of error at current sampling and its previous sampling is called error rate (Δerr_A). For phase A, the error and error rate are defined as:

$$err_A = V_{\sin(\theta)_{pu}} - V_{send_{pu}} \tag{1}$$

$$\Delta err_A = err_A(n) - err_A(n-1) \tag{2}$$

where $V_{\sin(\theta)_{pu}}$ is a PLL voltage possessed the same phase with the sending end voltage, $V_{send_{pu}}$ is the phase of sending end voltage and n is the sampling time.

Fuzzification: The numeric input variable measurements are transformed by fuzzification part into the fuzzy set variable, which is clearly defined boundary, without a crisp. These linguistic variables of error/error rate are shown in Figure 6.

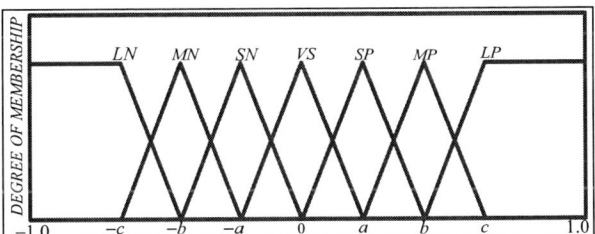

Figure 6 Error /error rate of fuzzy membership functions

Decision Making: The fuzzy models are created by using Sugeno system. The I_{th} rule is given in the following form [17]:

$$L^{(1)} : If \ x_1 \ is \ F_1^l \ and....and \ x_n \ is \ F_n^l, then \tag{3}$$

$$y^l = c_0^l + c_1^l x_1 + c_2^l x_2 + ... + c_n^l x_n \tag{4}$$

where F_i^l denotes fuzzy set, c_i^l is the real coefficients, y^l is the output set and $x_1 ..., x_n$ are the inputs.

The error and error rate are described as linguistic variables such as large negative (LN), medium negative (MN), small negative (SN), very small (VS), small positive (SP), medium positive (MP) and large positive

(LP). The input values of the fuzzy controller are connected to the output values by the if-then rules. The relationship between the input and the output values can be achieved easily by using "*Takagi-Sugeno inference*" method. The output values are characterized by memberships and named as linguistic variables such as negative big (NB), negative medium (NM), negative small (NS), zero (Z), positive small (PS), positive medium (PM) and positive big (PB). The membership functions of output variables and the decision tables for FLC rules are seen in Table 1.

Table 1. Table of fuzzy decision

Error rate /Error	LP	MP	SP	VS	SN	MN	LN
LP	PB [1]	PB [2]	PB [3]	PM [4]	PM [5]	PS [6]	Z [7]
MP	PB [8]	PB [9]	PM [10]	PM [11]	PS [12]	Z [13]	NS [14]
SP	PB [15]	PM [16]	PM [17]	PS [18]	Z [19]	NS [20]	NM [21]
VS	PM [22]	PM [23]	PS [24]	Z [25]	NS [26]	NM [27]	NM [28]
SN	PM [29]	PS [30]	Z [31]	NS [32]	NM [33]	NM [34]	NB [35]
MN	PS [36]	Z [37]	NS [38]	NM [39]	NM [40]	NB [41]	NB [42]
LN	Z [43]	NS [44]	NM [45]	NM [46]	NB [47]	NB [48]	NB [49]

Defuzzification: In the defuzzification process, the controller outputs represented as linguistic labels by a fuzzy set are converted to the real control signals. "*Sugeno's Weighted Average*" method which is the special case of "*Mamdani Model*" is selected for the defuzzification process. The defuzzification is achieved by using following equations:

$$y = \frac{\sum_{l=1}^{M} w^l y^l}{\sum_{l=1}^{M} w^l} \tag{5}$$

$$w^l = \prod_{i=1}^{n} M_{F_i^l}(x_i) \tag{6}$$

where w^l is the overall truth value of the rule $L^{(1)}$, $M_{F_i^l}(x_i)$ is the membership function described the meaning of the linguistic variable F_i^l.

Signal Processing: The control signals are produced from the output of FLC process. They are used in the generation of switching signals for converter by comparing a carrier signal as shown in Figure 2 and Figure 3.

4. Case Studies

PSCAD/EMTDC program is used to simulate the modeling of UPFC and the test system. The parameters of simulated system are selected low ratings to be enabled the implementation of system in the laboratory environment. The test system is shown in Figure 7 and the parameters are given in the Appendix. Two generators are used and named as sending end and receiving end generators, respectively. UPFC is constructed at the sending end bus before the line impedance. Case studies were carried out to test the performance of UPFC with PI based and FLC based control mechanisms.

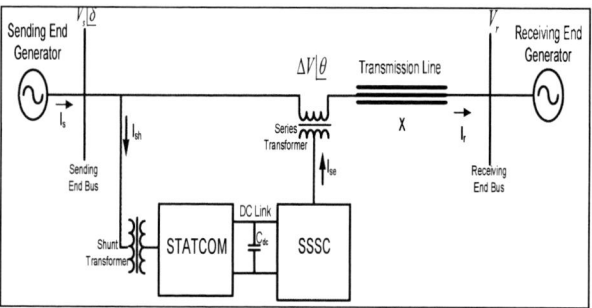

Figure 7 Test system for case study 1

In the first case study, the receiving end generator is delayed from the sending end generator according to several phase angles and the results of real and reactive power values in the line are taken to compare as with and without UPFC. The results are given in Table 2.

Table 2 Power flow results for phase variations in transmission line

Phase angle of Receiving end Generator (^{0})		60^0	40^0	30^0	15^0
without UPFC	P(KW)	3.44	2.67	2.11	1.11
	Q(KVAR)	-1.4	-0.97	-0.72	-0.34
PI Controller with UPFC	P(KW)	4.14	3.09	2.43	1.37
	Q(KVAR)	-0.42	-0.15	-0.09	0.04
FUZZY Controller with UPFC	P(KW)	4.33	3.14	3.14	1.41
	Q(KVAR)	-0.07	-0.08	-0.08	0.07

In the second case, the three phase fault is applied to the sending end bus. It is started at the 1.4 sec and it is continued 0.2 sec. The control mechanism of series converter senses fault, it immediately activates the electronic bypass to protect the series converter. After the fault, the electronic bypass is removed by the control mechanism automatically. The test system is shown in Figure 8.

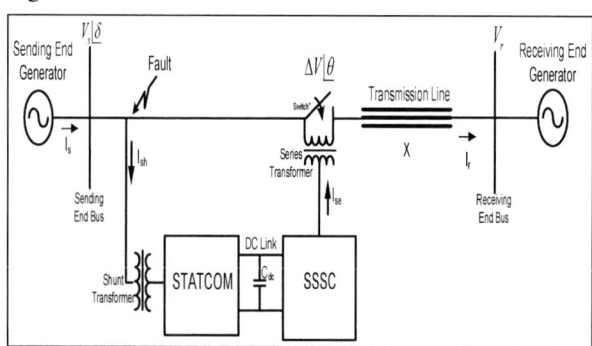

Figure 8 Test system for case study 2

The variations on the receiving end reactive power (Q_{re}) and the receiving end voltage in pu. ($Vpu._{re}$) are illustrated as a graphically in Figure 9 and Figure 10.

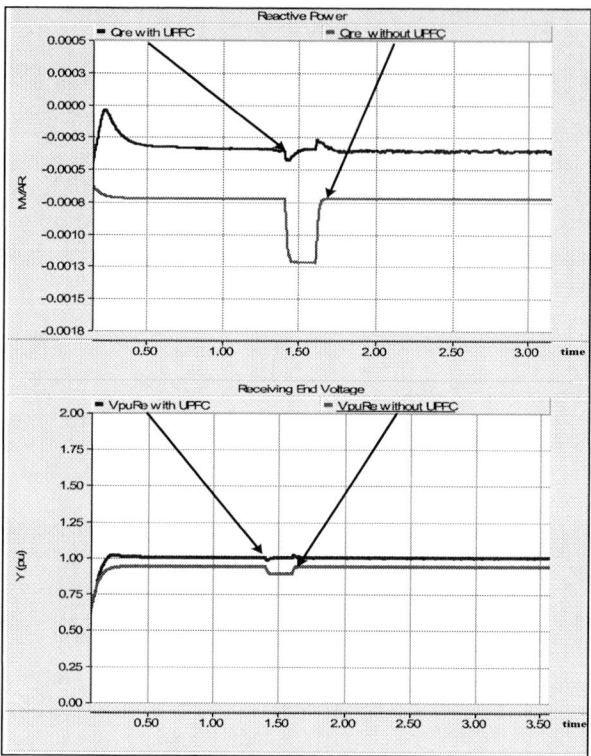

Figure 9 PI controller results at 3-phase balanced fault

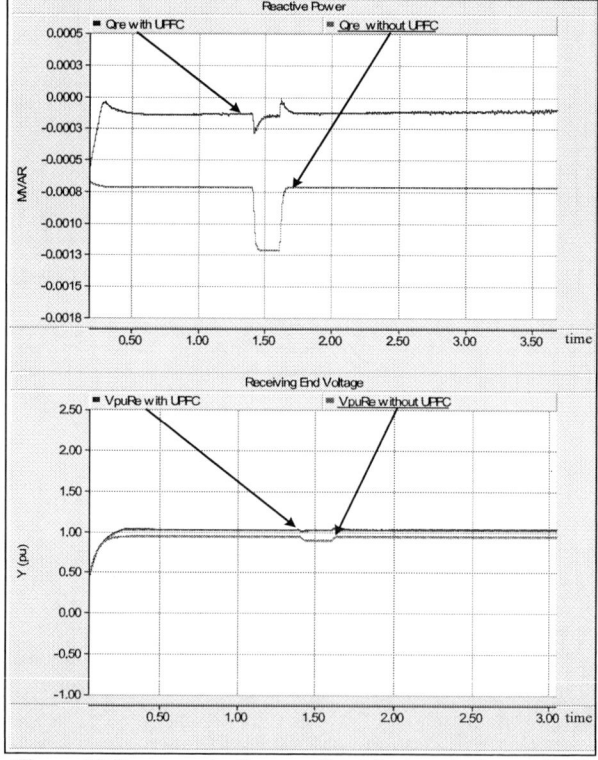

Figure 10 Fuzzy controller results at 3-phase balanced fault

5. Conclusion

In this paper, the performance of UPFC on real/reactive power flow is examined with different control mechanisms based on PI controller and fuzzy controller. It was shown that UPFC improves the system performance under the transient and the normal conditions. Two cases were carried out to prove the effect of UPFC on real/reactive power flow. The case studies demonstrated that UPFC control the power flow in the transmission line, effectively. Beside, FLC based control mechanism showed better performance then PI controller based control mechanism.

PSCAD/EMTDC program is used to model UPFC and to take simulation results from the test system. FLC are added in main library of PSCAD/EMTDC as a new component by using *Takagi-Sugeno inference system* in decision making and *Weighted Average* method in the defuzzification process. Thus, the innovation is supplied into the main library of PSCAD/EMTDC.

Appendix

The technical details of sending / receiving end generators:

Based MVA (3-phase)	: 0.01 [MVA]
Base Voltage (L-L)	: 0.380 [kV]
Base Frequency	: 50.0 [Hz]
Phase	: 0.0 [°]

The technical details of shunt / series converter transformers:

Based MVA (3-phase)	: 0.003 / 0.003 [MVA]
Winding #1/ #2 Voltage (L-L)	: 0.380 / 0.040 [kV]
Winding #1 / #2 Voltage (L-L)	: 0.110 / 0.110 [kV]

The system parameters:

V_{dc}	: 0.042 [kV]
C_{dc}	: 20 mF
R_r	: 4 Ω
L_r	: 10 mH

Vdc voltage controller PI parameters:

K_p / K_i	: 4.6 / 0.0001

Bus voltage controller PI parameters:

K_p / K_i	: 4.05 / 0.005

References

[1] X. Y. Zhou, H. F. Wang, R. K. Aggarwal, (2004), "Detailed Modeling and Simulation of UPFC Using EMT", *39th International Universities Power Engineering Conference*, UPEC, Vol. 1, pp. 256 – 260.

[2] Y. H. Song, A. T. Jons, (1999), "Flexible AC Transmission Systems (FACTS)", *IEE Power and Energy Series 30*.

[3] A. A. Eldamaty, S. O. Faried, S. Aboreshaid, (1999), "Damping Power System Oscillations Using a Fuzzy Logic Based Unified Power Flow Controller", *Electrical and Computer Engineering*, 2005, pp. 1950 – 1953.

[4] S. Kannan, S. Jayaram, M. M. A. Salama, (2004), "Real and Reactive Power Coordination for a Unified Power Flow Controller", *IEEE Transactions on Power Systems, Volume 19*, Issue 3, pp. 1454 – 1461.

[5] R. Orizondo, R. Alves, (2006) "UPFC Simulation and Control Using the ATP/EMTP and MATLAB/Simulink Programs", *Transmission & Distribution Conference and Exposition*, IEEE/PES, pp. 1 – 7.

[6] D. Thukaram, L. Jenkins, K. Visakha, (2005), "Improvement of System Security with Unified Power Flow Controller at Suitable Locations under Network Contingencies of Interconnected Systems", *IEE Proceedings Generation, Transmission and Distribution*, Vol. 152, Issue 5, pp. 682 – 690.

[7] H. Fujita, Y. Watanabe, H. Akagi, (1999), "Control and Analysis of a Unified Power Flow Controller", *IEEE Transactions on Power Electronics*, Volume 14, Issue 6, pp. 1021 – 1027.

[8] B. Lu, L. Hou, B. Li, Y. Liu, (2007), "A New Unified Power Flow Fuzzy Control Method", *Innovative Computing, Information and Control*, Second International Conference, pp. 479 – 479.

[9] Y. Qing, L. Norum, T. Undeland, S. Round, (1996), "Investigation of Dynamic Controllers for a Unified Power Flow Controller", *Industrial Electronics, Control, and Instrumentation*, Vol. 3, pp. 1764-1769.

[10] M. W. Mustafa, A. A. Zin, A.F. Kadir, (2002), "Steady State Analysis of Power Transmission Using Unified Power Flow Controller", *Transmission and Distribution Conference and Exhibition*, Vol. 3, pp. 2049 – 2053..

[11] Visual Power System Simulation, Web site available at http://www.pscad.com .

[12] Y. Zhang, L. Liu, P. Zhu, X. Liu, Y. Kang, Y. Gao, (2005), "Double Closed Loop Control and Analysis for Shunt Inverter of UPFC", *Electrical Machines and Systems*, Vol. 2, pp. 1118 – 1123.

[13] L. Liu, P. Zhu, Y. Kang, J. Chen, (2005), "Design and Dynamic Performance Analysis of a Unified Power Flow Controller", *IEEE Industrial Electronics Society*, pp.6.

[14] I. Papic, P. Zunko, D. Povh, M. Weinhold, (1997), "Basic Control of Unified Power Flow Controller", *IEEE Transactions on Power Systems*, Vol. 12, Issue 4, pp. 1734-1739.

[15] T. S. Chung, Y. Xiaodong, D. Z. Fang, C. Y. Chung, (2004), "Development of Adaptive UPFC Supplementary Fuzzy Controller for Power System Stability Enhancement", *Electric Utility Deregulation, Restructuring and Power Technologies*, Vol. 1, pp. 216 – 221.

[16] G. L. Sheng, W. Y. Lin, L. Sheng, "Research on Flexible Power Supply System for Arc Furnace Based on UPFC", *Industrial Elect. and Applications*, 2007, pp. 227-230.

[17] T. Takagi, M. Sugeno, (1985), "Fuzzy Identification of Systems and Its Applications to Modeling and Control". *IEEE Transaction on Systems, Man and Sybern*, Vol. pp. 1: 116-132.

Session 9

Intelligent Transportation System

2008 Workshop on Power Electronics and Intelligent Transportation System

Framework on Hierarchical Optimization of Traffic Count Location for City Traffic System

Heng Wang, Ke_Ping Li, Jian Sun, Ying Liu
Tongji University, Shanghai, China
whmmx9802@gmail.com

Abstract

This paper develops a framework about how to solve the traffic count location problem. According to target scope, Traffic count location problem can be divided into 3 categories: network count location problem(NCLP), corridor count location problem(CCLP), and link count location problem(LCLP). Programming method and micro-simulation tool are used to solve the related count location problem.

1. Introduction

Intelligent Transportation System (ITS), incorporate a smoothing or rationalizing process that is intended to make the existing supply of transportation infrastructure better suited to meet current and future transportation demand. By employing advanced technologies, the infrastructure is made more efficient, effective and safer while precluding increased land use and environmental pollution.

A prerequisite for ITS services is the collection of timely and accurate information about traffic and road condition. For many years, traffic surveillance has been achieved by inductive loop detectors, which can sense the presence of a vehicle.

One of the important services of ITS is the support of transportation planning and OD trip table is the essential data to transportation planning. Conventionally, an OD trip table is estimated from a large scale survey which is costly, time-consuming, and labor-intensive. Therefore, OD trip table estimation from traffic counts is regarded as a convenient and practical way to obtain up-to-date information about travel demand patterns. The traffic counting location problem can be considered as a preprocess of the OD estimation problem. It has been overlooked, but it is a practically important problem. The identification of informative links for a given number of stations in the road network, is referred to as the network count location problem (NCLP).

Another important service of ITS is the traffic management. Traffic characteristic (traffic rate, occupancy and speed) should be surveyed correctly.

Certainly, count location on the link of road network will influence the precision of traffic characteristic survey. Locating detector on a single link is referred to as link count location problem (LCLP).

Expressway system is the artery of city, special surveillance is to be needed for travel time prediction and traffic affair identification. Identifying traffic detector density on expressway is referred as corridor count location problem (CCLP).

Figure1. presents research scope .

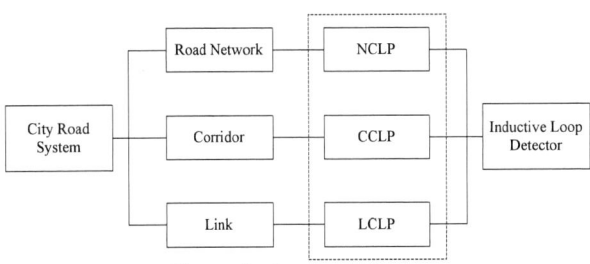

Figure 1. Research Scope

2. Network Count Location Problem (NCLP)

Yang and Zhou (1998) conducted a comprehensive investigation on the traffic counting locations for effective estimation of OD trip tables from traffic counts. Based on the theory of Maximal Possible Relative Error (MPRE) in OD trip table estimation (Yang et al., 1991), they derived four rules to locate traffic counting points: the OD covering rule, the maximal flow fraction rule, the maximal flow intercepting rule and the link independence rule. NCLP is formulated as an integer mathematical program, where the O-D covering rule and the link independence rule are incorporated as constraints, and the total net traffic flows observed are taken as the objective function to be maximized.

2.1. Notation

Following Yang and Zhou(1998)

$G(\lambda)$ Measure of estimation error of OD matrix

t_w^* True trips between OD pair w

t_w Estimated trips between OD pair w

978-0-7695-3342-1/08 $25.00 © 2008 IEEE
DOI 10.1109/PEITS.2008.15

λ_w Relative deviation of the estimated trips and true one for OD pair w

p_w^a Proportion of trips using link a for OD pair w

2.2. Modified Location Rules

According to Yang and Zhou (1998), we have

$$G(\lambda) = \sqrt{\frac{\sum\limits_{w \in W} \lambda_w^2}{w_n}} \qquad (1)$$

$G(\lambda)$ is used as the evaluation index of OD trip table estimation. Apparently, the less of $G(\lambda)$, the more accurate of OD trip table estimation.

If the OD trip table estimation is accurate enough, t_w^* is equal to t_w. So we can get

$$\lambda_w = \frac{(t_w^* - t_w)}{t_w} = 0 \quad (t_w^* \geq 0, t_w > 0, w \in W) \qquad (2)$$

According to concept of statistics, we have $E(\lambda) = 0$, and $S(\lambda) = \sqrt{\sum\limits_{w \in W} \lambda_w^2 / w_n}$. So $G(\lambda) = S(\lambda)$. Namely, $G(\lambda)$ represents discrete degree between relative error λ and average value.

Let $u_w^a = p_w^a t_w / v_a$ denote the fraction of flow between OD pair $w \in W$ in link flow v_a. From u_w^a, we have

$$\sum_{w \in W} u_w^a = \sum_{w \in W} p_w^a t_w / v_a = 1 \qquad (3)$$

Let $\overline{\lambda_w^a}$ be the upper bound of relative error in link flow v_a. We have

$$\overline{\lambda_w^a} = 1/\mu_w^a - 1 = p_w^a t_w / v_a - 1 \qquad (4)$$

Let k_a denote the number of routes from OD pair $w \in W$ in link a, from (3), we have

$$\sum_{k=1}^{ka} u_k^a = 1 \qquad (5)$$

As for link a , we have the deviation:

$$S_a^2(\lambda_a) = \frac{\sum\limits_{k=1}^{ka} \lambda_k^2}{k_a} \leq \frac{\sum\limits_{k=1}^{ka} \overline{\lambda_k^2}}{k_a} \qquad (6)$$

and

$$\sum_{k=1}^{ka} \overline{\lambda_k^2} = \sum_{k=1}^{ka} (\frac{1}{u_k^a} - 1)^2 \qquad (7)$$

According to *Cauchy Inequality* and (7), we have

$$\sum_{k=1}^{ka} \overline{\lambda_k^2} = \sum_{k=1}^{ka} (\frac{1}{u_k^a} - 1)^2 \geq \frac{\left[\sum\limits_{k=1}^{ka} (\frac{1}{u_k^a} - 1)\right]^2}{k_a}$$

$$= \frac{(\sum\limits_{k=1}^{ka} \frac{1}{u_k^a} - k_a)^2}{k_a} \qquad (8)$$

According to *Mean Value Inequality* and (8), we have

$$\frac{(\sum\limits_{k=1}^{ka} \frac{1}{u_k^a} - k_a)^2}{k_a} \geq \frac{\left(\dfrac{k_a^2}{\sum\limits_{k=1}^{ka} u_k^a} - ka\right)^2}{k_a} = k_a(k_a - 1)^2 \qquad (9)$$

From (7), (8), (9), we have

$$\sum_{k=1}^{ka} \overline{\lambda_k^2} \geq k_a(k_a - 1)^2 \qquad (10)$$

If and only if $u_1^a = \cdots = u_k^a = \cdots u_{ka}^a$, we have

$$\sum_{k=1}^{ka} \overline{\lambda_k^2} = k_a(k_a - 1)^2 \qquad (11)$$

then

$$S_a^2(\lambda_a) \leq \frac{k_a(k_a - 1)^2}{k_a} = (k_a - 1)^2 \qquad (12)$$

From (11) and (12), we have the conclusion: for $S_a^2(\lambda_a) \in (0, (k_a - 1)^2)$, when k_a decreases, $S_a^2(\lambda_a)$ decreases. So decreasing the number of routes through one link can promote the estimation precision.

So we can get the modified location rules:

(1) choose the links that have few enough routes;

(2) choose the links that path flows from all OD pairs are as close as possible

2.3. Formulate NCLP Model

According to Yang and Zhou (1998) and rules discussed above , we can get the NCLP programming model as follows:

$$\min \ f(z) = \sum_{w \in W} y_w \overline{\lambda_w^2} = \sum_{w \in W} y_w (\frac{1}{u_w} - 1)^2 \qquad (13)$$

$$s.t. \ u_w = \underset{rw \in R_W}{Max}(\underset{rw \in R_W}{Max}(\delta_w^a z_a \mu_w^a)) \qquad (14)$$

$$\sum_{a \in L} z_a = \tilde{l} \qquad (15)$$

$$\sum_{rw \in R_W} \sum_{a \in w} z_a \geq y_w (w \in W) \qquad (16)$$

$$\sum_{a \in L} \delta_w^a z_a \geq 1 (w \in W) \qquad (17)$$

$$y_w = 0,1; z_a = 0,1; \delta_w^a = 0,1 (a \in L, w \in W) \qquad (18)$$

We use the Root Mean Square Error (RMSE) and Mean Relative Error (MRE) to compare modified location rules with Yang's OD covering model.

$$RMSE = \sqrt{\frac{\sum\limits_{k=1}^{n} (q_k - v_k)^2}{n - 1}} \qquad (19)$$

420

$$MRE = \frac{\sum_{k=1}^{n} |q_k - v_k|}{\sum_{k=1}^{n} v_k} \tag{20}$$

The results are represented in table 1. From that, we can get that the results of modified model is slightly better than OD covering model.

Table 1. Results comparison

Programming model	RMSE	MRE
OD covering model	47	0.066
Modified model	44	0.060

3. Corridor Count Location Problem (CCLP)

The problem studied in this paper can be stated as: given a freeway segment and a given number of fixed location counts (such as loop detectors), where these counts should be placed so that their deployment is 'optimal' in terms of providing travel time estimates.

3.1. Instantaneous Travel Time
The calculation of instantaneous travel times can be illustrated in Figure 2. In this figure, solid lines represent trajectories of individual vehicles in the 'space-time' diagram. Mathematically, we have:

$$\hat{T}^{m,i}(t) = \sum_{a=1:A} \frac{L_a}{v_a(t)} \tag{21}$$

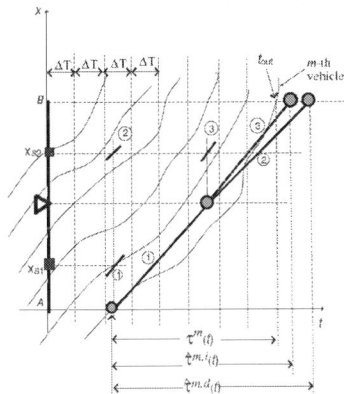

Figure 2. Travel Time from Vehicle Trajectory

Here $\hat{T}^{m,i}(t)$ is the instantaneous travel time of m-th vehicle, La is the length of a-th link, A is the total number of links of the route.

3.2. Objective Function
If we know the actual and estimated travel times of the m-th vehicle, the travel time estimation error for this vehicle can be expressed as:

$$e^m = \hat{T}^m - T^m \tag{22}$$

Assume we have in total K counts, the route can thus be divided into K links. Therefore, we have:

$$e^m = \sum_{k=1}^{k} (\hat{T}_k^m - T_k^m) \tag{23}$$

Assume that there are M such vehicles and the optimal count locations should minimize the Mean Squared Error (MSE) of the travel time estimation of all these vehicles.

$$E = \sum_{m=1}^{M} (e^m)^2 / M$$
$$= \sum_{m=1}^{M} \left\{ \left[\sum_{k=1}^{K} (e_k^m)^2 \right] + 2 \sum_{m=1}^{M} \sum_{k=1}^{K} (e_i^m e_j^m) \right\} \tag{24}$$

Here e_i^m and e_j^m are the travel time estimation errors of the m-th vehicle on link i and link j ($i \neq j$), respectively.

If we assume both e_i^m and e_j^m are random variables and more importantly they are independent to each other, the second term in equation (24) is zero.

$$\hat{E} = \sum_{k=1}^{k} (\hat{E}_k) \tag{25}$$

$$\hat{E}_k = \sum_{m=1}^{M} (e_k^m)^2 / M \tag{26}$$

3.3. Discretization for the Link

We first discretize the given route r into small segments, called sections. The premise is that if the length of a section is small enough, it does not matter where to place a sensor within the given section.

Denote $T_{sk,yk}^m$ the instantaneous travel times of the m-th vehicle traversing Link k,. we thus have:

$$T_{sk,yk}^m = t_{yk}^m \Delta x - t_{sk-1}^m \Delta x \tag{27}$$

$$T_{sk,yk}^m = \frac{(y_k - s_k + 1)\Delta x}{v_{nk}, h_1^m} \tag{28}$$

3.4. Programming Formulation

Since we assume a count is in the middle of its associated link, the optimal count location problem can be solved via finding the optimal starting and ending locations of the links.

$$\hat{E}_k = \frac{\sum_{m=1}^{M} \left(\frac{(y_k - s_k + 1)\Delta x}{v_{nk}, h_1^m} - t_{yk}^m \Delta x + t_{sk-1}^m \Delta x \right)}{M} \tag{29}$$

So we can get the our programming model:

$$\min_{sk,yk,\forall k=1,\dots,K} \sum_{k=1}^{K} \left\{ \hat{E}_k(s_k, y_k) \right\} \tag{30}$$

The above optimization model is a linear integer problem since $\hat{E}_k(s_k, y_k)$ can be computable for any given $(k, sk, yk), \forall k = 1, \dots, K$.

To determine the optimal number of counts needed by the studied segment, Figure 3 shows, how the MSEs change as we vary K from 3 to 60

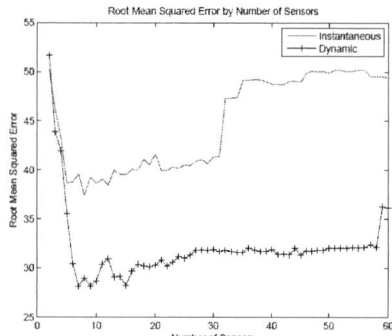

Figure 3. the Objective Value VS K

4. Link Count Location Problem (LCLP)

The main goal of the research is to find the best count location on the link with various signal controls in urban areas, and to validate and apply the outputs to provide an average link travel speed to drivers. Because the speed from one count is a point-speed on an arterial, the location selected must be a representative speed to obtain an average arterial travel speed/time. There are three types of signal controls, pre-timed signal control, actuated signal control, and coordinated signal control in the consideration.

4.1. Objective Function

The optimal count location for each case is selected by the following equation.

$$A^* = \frac{1}{n}\sum_i^n \left| AS - DS_i \right| \qquad (31)$$

Where, n= number of simulations
AS= Average link travel speed
DS=Speeds from each count

4.2. Influencing Factors Consideration

According to the Traffic Delay Theory, influencing factors on signalized intersection is dependent on several major factors which are:
  1) Signal cycle length
  2) Green time
  3) Traffic volumes
  4) Road capacity
  5) Speed limit
Traffic volumes, however, are not fixed in the real world, so the range of traffic volumes has to be decided. Traffic volumes should cover both under saturation and over saturation flow condition. The selected lowest traffic volume has to have LOS A for any signal coordination,

maximum traffic flow rate is defined for when the queue from the downstream intersection reaches the end of the link after 5 minutes.

4.3. Solution

The objective function can't be solved using a mathematical process, because the factors can't provide the travel speed and the speed from detectors directly. So, lab-based micro-simulation tools should be taken to solve these problems. Such as VISSIM, PARAMICS, CORSIM et al.

5. Conclusion

We use the programming method and simulation method to solve macro, mesco, and micro-count location problem. In this paper, the author is intended to provide the framework and popular method of systematically solving related traffic count location.

References

[1] Yang, H. and Zhou, J. (1998) Optimal traffic counting locations for origin-destination matrix estimation. Transportation Research Part B, 32, 109-126.
[2] Yang, H., Iida, Y. and Sasaki, T. (1991) An analysis of the reliability of an origin destination trip matrix estimated from traffic counts. Transportation Research Part B, 25, 351-363.
[3] Sunho oh, Modeling loop detector location for data collection on urban arterial for ATIS application, PHD dissertation, Uath university, 2003
[4] Jeff Ban, Ryan Herring, JD Margulici, Optimal sensor placement for freeway travel time estimation, California center for innovative transportation
[5] Anett Ehlert, Michael G.H. Bell, Sergio Grosso, The optimisation of traffic count locations in road networks, Transportation Research Part B: Methodological, Volume 40, Issue 6, July 2006, Pages 460-479

2008 Workshop on Power Electronics and Intelligent Transportation System

Game Theory Choice Model for Public Transportation Priority –Take Nanjing for Example

Chen Wang, Jun Chen
Southeast University
atalandisi@163.com

Abstract

Public transportation priority is the most effective way to improve the efficiency of the public transportation which is the most important part of the urban transportation. For the choice of the Public transportation priority, the methods usually used now in China are not very imprecise for consideration of the indices' weights. The paper introduces game theory and sets up the choice model for the public transportation priority by identifying the function of payoff and the aggregate of participants and stratagems. Then it analyses the existence of NASH equilibrium and puts forward the arithmetic of Iterated Elimination of Strictly Dominated Strategies in the general condition and the Mixed Strategies in the special condition .At last, it verifies the feasibility of the model with the example in Nanjing.

1. Introduction

The most important thing now for urban transportation in china is to improve condition of the public transportation. Public transportation priority has been taken care of more and more widely as a very important both concept and act in this country. However, the research now for the public transportation priority in China is mostly concerned about the implementation of the public transportation priority technology while a moderate selective model or system has not been taken enough attention. How to choose a moderate technology, regarding the different factors and need of the development, to improve the traffic condition and the maximum benefit of the whole society is the most important task at present in China.

There are so many methods for multi-objective selection .The most used methods are Analytic Hierarchy Process(AHP) 、 fussy logic 、 expert decision system and so on. The methods mentioned above all need to take the weights of indices into account. For they mostly depends on mans' decisions, especially the AHP which has too many bottom schemes, the model could be very complicated and the calculation enormous, outcome imprecise[1].

Game theory is a method which disregards the weight of indices. It is impervious to the amount of the factors, schemes and the participants. In this passage, a game theory model has been set up which transforms the multi-objective selection to a precise mathematical question. Game theory model which seeks for the maximum of the benefit under the given restrictions has a great applicability and is easy to settle[2].

According to the instance illuminated above, in the passage, the knowledge of the game theory is applied to the multi-objective selection in order to set up a game theory choice model for the public transportation priority technologies.

2. Model

In this part, a game theory choice model for the public transportation priority technologies will be established step by step.

Defining or establishing a game theory model needs at least three facets: the participants of the game theory, the stratagem aggregate of the game theory and the benefit function of the participants. Therefore, establishing a game theory choice model for the public transportation priority technology also needs the three facets[3].

2.1. The participants of the game theory

The factors involved in the public transportation are so many and complex. To ensure the maximum benefit of the technologies, the main factors are defined to be the participants of the game theory. They are put together to be a aggregate N.

When choosing the members of the aggregate N, lots of aspects should be considered such as traffic, society, economic, technology and so on. Then, the main factors of the every aspect are chosen using the Analytic Hierarchy Process. The factors chosen out are the factors which will be considered for the choice model of for the public transportation priority technology, that is, the members of the aggregate N. The hierarchy scheme of the factors which affect the public transportation priority technologies are shown in Fig.1.

978-0-7695-3342-1/08 $25.00 © 2008 IEEE
DOI 10.1109/PEITS.2008.18

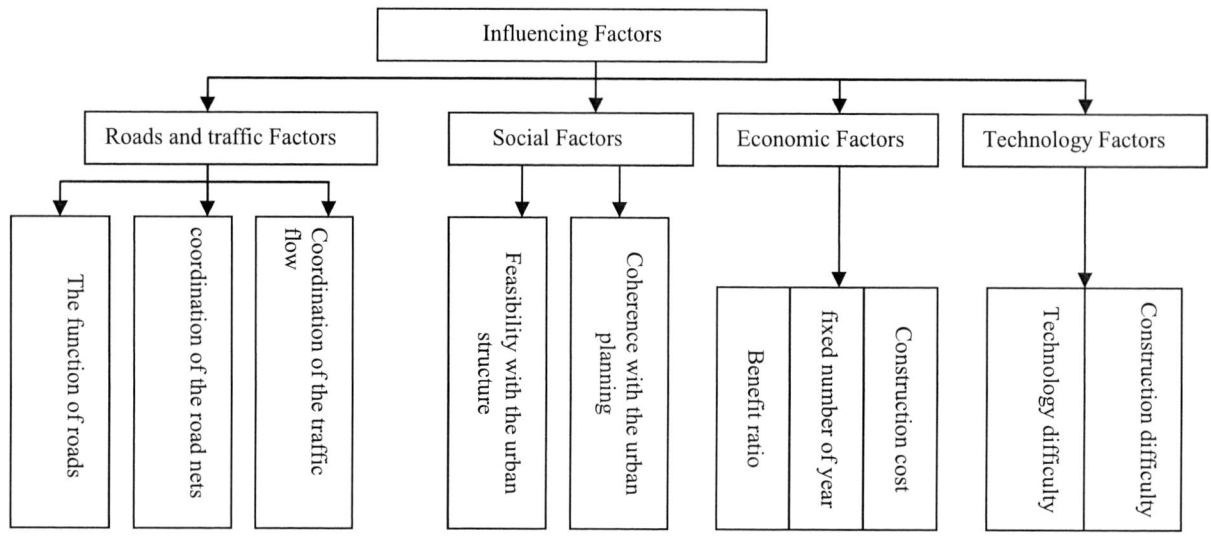

Fig.1 Analytic hierarchy Scheme of the factors which affect the public transportation priority technologies

2.2. The stratagem aggregate of the game theory

The purpose of the choice model for the public transportation priority technologies is to choose some technologies as the final technologies for the implementation of the public transportation priority. Therefore, the public transportation priority technologies are defined the stratagem aggregate of the game theory.

As for the public transportation priority technologies, there are many different choices. As a whole, there are two parts: priority technology for space, priority technology for time.

Technologies of priority for spacing[4]: 1.dedicated import lane 2.hackle import lane 3.feedback line 4.interlace segment

Technologies of priority for time[5]: 1.expand priority 2.transfer priority 3.dedicated phase 4.transportation weight 5. prefab time priority 6.green light reset priority

Let S be the stratagem aggregate of the game theory, that is ,every participant i has the same choice of technology Si which consists of the S.

Suppose there are n main factors chosen and m public transportation priority technologies to be chosen , that is , S i=A = { a1 , a2 ,   , an},.The stratagem aggregate of the game theory can be defined as $S = \{a | (a_i) i \in N\}_{n*m}$

。

2.3 The benefit function of the participants

When the benefit function of the participants is got, the influencing degree of the various technologies based

on various factors should be calculated at first[6]. Let C be the influencing degree.

To quantify the benefit function of the participants, some methods, such as questionnaire dispensing, interviewing, statistical analysis, cluster analysis, structure equation model and importance-satisfaction analysis, could be introduced to get the influencing degrees of some qualitative factors. Some influencing degrees of the factors which can not be quantified could be got by qualitative analysis, expert-estimating system, fuzzy analysis and so on.

Let the benefit function of the participants be u. It depends on the rational decision and the preference of the participants. To the participants I ,when facing the different technologies a and b ,if $C_a \geq C_b$,then $u(C_a) \geq u(C_b)$ .The aggregate of the benefit function of the participants for the different public transportation priority technology based on the different factors is defined as $U = \{u | u_i(c), i \in N\}_{n*m}$ .And the matrix of the benefit function is shown in table 1.

Tab.1 matrix of the benefit function

Factors	Technologies				
	A1	A2	A3	...	Am
N1	u11	u12	u13	...	u1m
N2	U21	u22	u23	...	u2m
...	...	...	...	...	...
Nn	un1	un2	un3	...	unm

2.4 Game theory choice model for the public transportation priority

Suppose there are m public transportation priority technologies , n factors , the game theory can be defined as :

$$G=\{S,U\}; \qquad S = \{a \mid (a_i) i \in N\}_{n*m} \qquad ;$$

$$U = \{u \mid u_i(c), i \in N\}_{n*m}$$

Where S is the stratagem aggregate of the game theory for the public transportation priority technologies , U is the benefit function of the participants.

3. The solution of the model

For the information of participants in the model is well known and the choosing behaviors are accomplished once, the model which is established above is a static game theory model of full information. The key point to get the answer of the model is to find the NASH equilibrium.

3.1 The existence of the NASH equilibrium

The NASH theorem defines: In a standard game theory where the finite numbers of the participants are, if the stratagem aggregate to every participants are finite, there will be at least one NASH equilibrium [7].

As for the choice model for the public transportation priority technologies, the aim is to the find the best technology. To get the answer of the game theory model, the NASH equilibrium based on some condition should be found. In the game theory choice model for the public transportation priority technologies, the number of the participants and the technologies are all finite. Based on the NASH theorem, there is at least one NASH equilibrium of the model.

3.2 The general condition

Though there is at least one NASH equilibrium $S^* = (a_{ij}^*)_n$ (a_{ij}^* presents the technology chosen based on the factor i) of the model, the only best technology for the public transportation priority is uncertain. In the NASH equilibrium $S^* = (a_{ij}^*)_n$,only when the same technology will be chosen under the various factors ,that is , $a_j^* = a_k^*$ in S^* ,could the best

technology be chosen. However, for the different participants may have different preference, the technologies chosen based on the different factors may not be the same.

To get the best technology, iterated elimination of strictly dominated strategies [8]is introduced to get the dominant-strategy equilibrium to find the technology which all the factors are inclined to. The difference of the NASH equilibrium and the dominant-strategy equilibrium is all dominant-strategy equilibrium is NASH equilibrium while not all NASH equilibrium is dominant-strategy equilibrium.

The step of iterated elimination of strictly dominated strategies is shown below:

(1) Check NASH equilibrium $S^* = (a_{ij}^*)_n$ got from the game theory model G .If $a_j^* = a_k^*$, $S^* = (a_{ij}^*)_n$ will be the best answer and q^* will be the best technology If not ,switching to the next step;

(2) Establish the new game theory model $G = \{S = \{a_j \mid a_j = a_{ij}^* (a_{ij}^* \in S^*)\}, U = u_j(a_j)\};$

(3) Introduce the standard λ_i for the elimination to iterated elimination of strictly dominated strategies. If $u_i(a_j^*) \le u_i(\lambda_j)$,the technology will be kept and $S = \{a_j^*\}$ while other technologies will be eliminated.

Minish λ_i with some rule and do the elimination to iterated elimination of strictly dominated strategies again until $\lambda \le \delta$ if S is vacant aggregate. δ is a standard for verifying ,it can be enacted by the influencing degree and some other rules. But it can not be too small to eliminated all the technologies. As long as the λ_i and δ are enacted in a proper range , a best technology could always be found under circulation. If not, establish a new game theory model $G = \{S = \{a_j^*\}, U = u_i(a_j^*)\}$ and find the NASH equilibrium and come back to step 1.

3.3 The special condition

Besides the condition mentioned above, a condition when there are not only one NASH equilibrium caused by the inconspicuous differences between the different stratagems based on the same factors could cause the iterated elimination of strictly dominated strategies disabled[9]. In this condition, it shows the difference between the public transportation priority technologies based on the same factor is not obvious. Therefore, only single technology could hardly be chosen and some combination of the technologies could be taken into account. For example, the combination of the space and signal priority methods could be used in a intersection and several priority technologies could be taken in one area at one time.

Based on the analysis, the mixed strategy is introduced to solve the problem in the special condition, that is, when the single strategy is difficult to decide, the combination will be taken into account to achieve the best benefit.

Suppose there are k strategies based on factor i: $S_i = \{S_{i1}, \cdots S_{ik}\}$, then the distribution of probability for the mixed strategy for factor i is $(x_{i1}, \cdots, x_{iK})$,where p_{ik} presents the probability of strategy s_{ik} .For x_{ik} is a degree of probability , $0 \leq x_{ik} \leq 1$ and $x_{i1} + \cdots + x_{iK} = 1$ are in existence. Let x_i be the one mixed strategy of S_i .The strategy for factor I is:

$$S_i^* = \{x^i = (x_1^i, x_2^i, ..., x_{m_i}^i) \,\Big|\, \sum_{j=1}^{m_i} x_j^i = 1, x_j^i \geq 0, j = 1, 2, ..., m_i\}$$

For the matrix $\Gamma = \{S_1, S_2, A\}$,if (x^*, y^*) is mixed equilibrium :

$$\sum_{j=1}^{n} a_{ij} y_j^* \leq \sum_{i=1}^{m} \sum_{j=1}^{n} a_{ij} x_i^* y_j^* \leq \sum_{i=1}^{m} a_{ij} x_i^* ; i = 1, 2, ..., m, j = 1, 2, ..., n$$

Change this to

$$\sum_{j=1}^{n} a_{ij} y_j^* \leq U \leq \sum_{i=1}^{m} a_{ij} x_i^* ; i = 1, 2, ..., m, j = 1, 2, ..., n$$

Establish the equations and transform:

$$\begin{cases} \sum_{j=1}^{n} a_{ij} y_j \leq U; \ \sum_{j=1}^{n} y_j = 1; i = 1, 2, ..., m \\ \sum_{i=1}^{m} a_{ij} x_i \geq U; \ \sum_{i=1}^{m} x_i = 1; j = 1, 2, ..., n \\ x_i \geq 0, y_j \geq 0; i = 1, 2, ..., m, j = 1, 2, ..., n \end{cases}$$

$$\Rightarrow s.t. \begin{cases} \sum_{i=1}^{m} a_{ij} x_i \geq U \\ \sum_{i=1}^{m} x_i = 1; j = 1, 2, ..., n \\ x_i \geq 0; i = 1, 2, ..., m \end{cases} \text{ and } \begin{cases} \sum_{j=1}^{n} a_{ij} y_j \leq U; \\ \sum_{j=1}^{n} y_j = 1; i = 1, 2, ..., m \\ y_j \geq 0; j = 1, 2, ..., n \end{cases}$$

$$\Rightarrow s.t. \begin{cases} \sum_{i=1}^{m} a_{ij}(x_i / U) \geq 1 \\ \sum_{i=1}^{m} (x_i / U) = 1 / U; j = 1, 2, ..., n \\ x_i / U \geq 0; i = 1, 2, ..., m \end{cases}$$

$$\min U$$

$$\begin{cases} \sum_{j=1}^{n} a_{ij}(y_j / U) \leq 1; \\ \sum_{j=1}^{n} (y_j / U) = 1 / U; i = 1, 2, ..., m \\ y_j / U \geq 0; j = 1, 2, ..., n \end{cases}$$

Then it transforms to linear programming question:

$$X = \min \sum_{i=1}^{m} x_i$$

$$s.t. \begin{cases} \sum_{i=1}^{m} a_{ij} x_i \geq 1 \\ x_i \geq 0; i = 1, 2, ..., m \end{cases}$$

$$Y = \max \sum_{j=1}^{n} y_j$$

$$s.t. \begin{cases} \sum_{j=1}^{n} a_{ij} y_j \leq 1; \\ y_j / U \geq 0; j = 1, 2, ..., n \end{cases}$$

There are many methods to do the linear programming question. In the next chapter, a software named Lindo[10] will be introduced to do it.

4. Case study

To validate the feasibility of the model, the public transportation priority technology for Nanjing is taken into account. The model is used to make some decision for several parts of the area in this district (Fig.2)

Fig.2 Roads and areas of Nanjing district

Tab.2 Matrix of the part H2 benefits of different factors for different priority technologies

Factors	Technologies				
	A1	A2	A3	A4	A5
N1	31	-12	25	44	56
N2	17	66	2	40	33
N3	3	27	-12	14	37
N4	22	14	15	5	1
N5	-12	2	17	43	-25

Standardize the data to decimal system, new matrix is obtained:

Tab.3 Standard matrix the benefits of different factors for different priority technologies

Factors	Technologies				
	A1	A2	A3	A4	A5
N1	3.1	-1.2	2.5	4.4	5.6
N2	1.7	6.6	0.2	4.0	3.3
N3	0.3	2.7	-1.2	1.4	3.7
N4	2.2	1.4	1.5	0.5	1.0
N5	-1.2	2.0	1.7	4.3	-2.5

All the streets and areas in the Fig.2 are labeled.

Five technologies are put forward and five factors are chosen to be the main factors which will play the most important role. Five factors include coherence with the urban planning (N1), constriction cost (N2), coordination of the road nets (N3), construction difficulty (N4) and fixed number of year (N5). Five technologies include hackle import lane (A1), feedback line (A2), interlace segment (A3), dedicated phase (A4) and green light reset priority (A5).

The questionnaire and expert decision system are applied to obtain the corresponding influencing degree.100 questionnaires had been hand out to the experts, officials, professors and citizens. There choices (good , hard to say , bad) are enacted for every technologies based on the different factors to get the influencing degree. Let the choice for good be 1 points, for hard to say 0 points , for bad -1 points. $C = \sum_{t=1}^{100} p^t_{ij}$,where p^t_{ij} is the tth people's points for the jth technologies based on the ith factors. Because the benefit is only depends on the influencing degree ,let $u(C)=C$ After the investigation , the matrix of the benefits of different factors for different priority technologies is obtained. The matrix of the first part of H2 is shown below:

Find the NASH equilibrium is the first step, $S^* = \{a_{15}^*, a_{22}^*, a_{35}^*, a_{41}^*, a_{54}^*\}$ is got. Obviously, $a_j^* \neq a_k^*$ .So the second step is to establish

$$G = \{S = \{a_j | a_j = a_{ij}^* (a_{ij} \in S^*)\}, U = u_j(a_j)\};$$

then new $S^* = \{a_{41}^*, a_{22}^*, a_{54}^*, a_{15}^*\}$ is obtained. Then it comes to the third step to eliminate scenarios .Enact the proper λ_i which depends on the real condition in the limit of $u_i(a_j^*) \leq u_i(\lambda_j)$ and circulate the step. In this condition , $\lambda_1 = 4$, $\lambda_2 = 5$, $\lambda_4 = 3$, $\lambda_5 = 3$ is enacted. At last, $u_4(a_1^*) \leq u_4(\lambda_4)$ is obtained and other scenarios are eliminated, that is ,A1 will be adopted.

The size of λ_i is determined by the experts and the officials who make a decision for the strategic transportation planning of the city. Based on the different λ_i ,choice will be different.

When considering the part H3 , the situation is changed .The matrix of the H3 is shown below(Tab.4):

Tab.4 Matrix of the part H3 benefits of different factors for different priority technologies

Factors	Technologies				
	A1	A2	A3	A4	A5
N1	31	31	25	-22	51
N2	13	11	40	12	11
N3	1	32	-17	12	33
N4	25	10	15	16	24
N5	-17	11	13	40	-23

Standardize the data to decimal system, new matrix is obtained:

Tab.5 Standard matrix of the part H3 benefits of different factors for different priority technologies

Factors	Technologies				
	A1	A2	A3	A4	A5
N1	3.1	3.1	2.5	-2.2	5.1
N2	1.3	1.1	0.4	1.2	1.1
N3	0.1	3.2	-1.7	1.2	3.3
N4	2.5	1.0	1.5	1.6	2.4
N5	-1.7	1.1	1.3	4.0	-2.3

In this condition, some degrees have small differences .It causes the invalidation of iterated elimination of strictly dominated strategies. The mixed strategy is introduced to solve the problem. According to the data matrix, linear programming is set up:

$$\min x_1 + x_2 + x_3 + x_4 + x_5$$

$$s.t. \begin{cases} 3.1x_1 + 3.1x_2 + 2.5x_3 - 2.2x_4 + 5.1x_5 \geq 1 \\ 1.3x_1 + 1.1x_2 + 0.4x_3 + 1.2x_4 + 1.1x_5 \geq 1 \\ 0.1x_1 + 3.2x_2 - 1.7x_3 + 1.2x_4 + 3.3x_5 \geq 1 \\ 2.5x_1 + 1.0x_2 + 1.5x_3 + 1.6x_4 + 2.4x_5 \geq 1 \\ -1.7x_1 + 1.1x_2 + 1.3x_3 + 4.0x_4 - 2.3x_5 \geq 1 \\ x_1, x_2, x_3, x_4, x_5 \geq 0 \end{cases}$$

Use the software Lindo to get the answer(Tab.6):

Tab.6 The answer of the linear programming

	N1	N2	N3	N4	N5	X_i^*
A1	3.1	1.3	0.1	2.5	-1.7	0.334129
A2	3.1	1.1	3.2	1.0	1.1	0.307876
A3	2.5	0.4	-1.7	1.5	1.3	0
A4	-2.2	1.2	1.2	1.6	4.0	0.357995
A5	5.1	1.1	3.3	2.4	-2.3	0
min	0.831514					

For X3, X5 is or near zero, eliminate the impacts of the scenarios A3 and A5.Take the mixed strategy with A1, A4 and A5. From the data, three scenarios as near as make no differences for the probability. The expert and the officials can make a plan with this principle.

And the final decision based on the game theory model is shown below (Tab.7):

Tab.7 Final decision based on the game theory model

	A1	A2	A3	A4	A5
H1	0	0	0	1	0
H2	0	1	0	0	0
H3	0.33	0.31	0	0.36	0
H4	0.30	0	0.25	0.02	0.43
H5	0.11	0	0.42	0.47	0
Z1	0.23	0.11	0.26	0.24	0.16
Z2	1	0	0	0	0
Z3	0	0	1	0	0
Z4	0	0	0	0	1
Z5	0.01	0.22	0	0.51	0.26

5. Conclusions

In this passage, a game theory choice model is introduced to the public transportation to select the proper public transportation priority technologies. It is better than some other methods mostly used in China because there is no need to consider the weights of indices which involve too many human decisions. What have done in the passage is shown below:

(1) Game theory is introduced and established to solve the multi-objective problem in public transportation priority.

(2) Two different conditions have been separated to be discussed and two different solutions have been put forward.

(3) Use the choice model to the case in Nanjing .With the implementation in practice, the feasibility of the model is validated.

6. Acknowledgement

This research is supported by National Basic Research Program of China (Project 2006CB705500) and National High-tech R&D Program (2008AA11Z201). Thanks for Professors Chen who gave lots of help both for the suggestions and the statistics of Nanjing city.

7. Reference

[1] Zhu Jianmei, "Game Model of Selection of Competitive Transportation Corridors", Journal of Southwest Jiaotong University, 2003,38(3): 336-340

[2] M,Pusillo,"L.Equilibria and Well posedness " International Game Theory Review ,2006 ,Vol. 8 ,No. 1 :33 -44.

[3] Wang Zeke, Li Jie. *Introduction to Game theory.* Beijing: Renming University Press, 2004

[4] JI Yanjie，Deng Wei, "A Review of the Development and Current Situation on Bus Priority at Intersections", Journal of Transportation Systems Engineering and Information Technology, 2004，4(1): 30-34

[5] Zhao Yue, Chen Fanghong, "An approach of public transport priority", Traffic and Transportation, 2005,(7): 59-62

[6] HU Guangming, LEI Hongyao, "Research of Application of Game Theory in the Appraisal of the Planning Project of Layout of Public Transport Junction", Central South Highway Engineering, 2006,31(6):116-121

[7] RobertGibbons. *A Primer In Game Theory.* Beijing: China Society Science Press, 1999：2-11

[8] Kuhn Harold W. *Game Theory Classic.* Beijing: Renming University Press, 2005

[9] C.D. Aliprantis and S.K. Chakrabarti , *Games and Decision Making*, New York&Oxford, Oxford University Press, 2000

[10] LINDO Reference. www.lindo.com

Research on the Establishment of Xi'an Advanced Public Transport System

Feng Zhongxiang[1], Liu jing[2], Dong xianyuan[1], Gao kunpeng[1]

[1] School of Automobile, Chang'an University, Xi'an, Shaanxi,China

[2] Construction Machinery School, Chang'an University, Xi'an, Shaanxi, China

E-mail: 21179191@qq.com

Abstract

In order to resolve the problems about public transport in Xi'an, this paper pointed out an effective way to establish an Advanced Public Transport System(APTS) which adapt to the development of Xi'an urban transportation based on the theory of APTS. Then a framework about Xi'an APTS was put forward. It particularly analyzed the functions of the various subsystems of the APTS in Xi'an in a systematic way. According to the status in Xi'an traffic, it proposed a development strategy and implementation plan about Xi'an APTS.

Key words: Advanced Public Transport System; Public Transport; Intelligence Scheduling System; Urban traffic

1. Public transport status of Xi'an

Urban public transport is not only the important infrastructure of city, but also is the fundamental way to resolve the urban traffic problems. Since the 1990s, Xi'an public transport has greatly improved. The public transport's lines have increased from 54 in 1990 to more than 200 in 2006. Now, Xi'an has been formed a public transport system which its main body is bus, taxi as complementary part. There are more than 6000 buses in Xi'an, which have covered the road fully that the bus can pass. There are more than 10000 bus stops and 2500 bus station groups, and the bus bay mainly distribute in the Dacaishi, the South Gate, Changlepu, Banpu, Puzi Village, the West Gate and so on. The main bus junction stations are at railway station, the Bell Tower, the South Gate, Xiaozhai, the Tu gate, Inter-city bus station and so on. However, due to economic, historical, traffic conditions and other reasons, the present public transport in Xi'an is still out of date. Mainly as follows:

(1)The lack of bus transfer junction station, the less land used of bus stations and the less than 10 bus bays in entire Xi'an, this is not convenient for passengers waiting and their security can not be ensured. It is also not convenient for the passengers because lack of transfer facilities in the bus transfer junction station.

(2)It is a ubiquity problem that many buses stop in one same station, that makes buses very hard to stop and the order is much jumbled. According to the random investigate in the Bell Tower bus stop; as many as 50 people were stayed at a station which has 3 public transport lines and waiting time is more than 8 minutes.

(3)The line layout is irrational, and waiting time is too long. According to a investigate in Xi'an, there are 20% bus stops must wait for bus for more than 10minutes, and the proportion of transfer is more than 1/3.The irrational line of buses, the buses overcrowding in the urban area ,and it is too empty in the suburb. The bus stop is irrational too, the distance between some stops is too long, and another is too short.

(4)The speed is too low, the most buses which starting from the railway station have an average velocity of only 10.2 km / h, the maximum velocity is only 17.9 km / h, it is lower than the urban road traffic reasonable velocity of 16-25 km / h. The proportion of people, whose riding time is more than 1 hour, is high and through transport proportion is only 64.3%.

(5)There are many dilapidated, poor condition vehicles and a low level of intelligent vehicles, and only a few people using the IC card.

Besides the conditions have been told above, there are many other problems such as the bad serving attitude ,the poor sanitation in some vehicles ,the low level of vehicles' safety grade and so on.

2. The necessity for establishment of Xi'an APTS

In order to solve the traditional public transport problems and promote economic development at the same time, therefore, in this case the APTS came into being. The reasons for APTS as Figure 1:

Under the premise of the public transport network distribution, scheduling and other pivotal basic theoretical research, the so-called advanced public transport system applies modern communication,

978-0-7695-3342-1/08 $25.00 © 2008 IEEE

DOI 10.1109/PEITS.2008.19

information, electronic control, computer, network, GPS, GIS to public transport with systems engineering theory and methods. The APTS contains intelligent scheduling system, information services, electronic payment system to achieve informationization, modernization and intelligence of public transport scheduling, operations and management. It can provide more secure, comfortable and convenient public transport services. So the system attracts public to choose public transport, ease urban traffic congestion, solve urban traffic problems effectively and create greater social and economic benefits.[1]

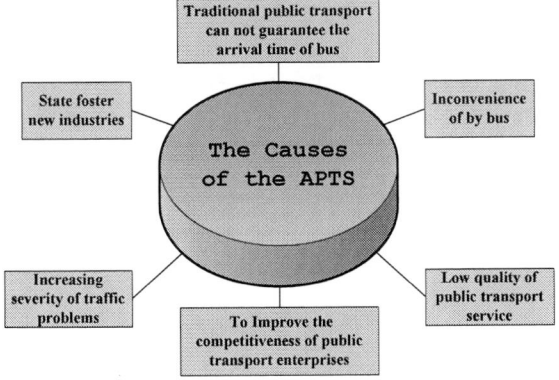

Fig.1 The causes of APTS

In the 1970s and 1980s of the last century, some developed countries have already begun to implement APTS, such as Japan's CTCS, the United States ITS MATA96, PAPT project of Paris, France. At the same time, some domestic major cities have already made a trial of advanced scheduling system such as BeiJing ,ShangHai and HangZhou.

There are many advantages of APTS in Xi'an:

(1)It is conducive to increasing economic efficiency. The APTS can not only greatly improve transport efficiency to save a lot of fuel and time but also reduce traffic accidents to decrease economic losses.

(2)It is conducive to improving social benefits. The travel time saved from the use of APTS can help to improve people's quality of life. Decrease of traffic accidents will make people's lives more comfortable and security.

(3)It is conducive to improving environment. Improving the efficiency of public transport will reduce emissions caused by traffic jam and reduce the number of small cars for convenient travel, which improve air quality.

(4)It is conducive to urban construction. Convenient public transport may prompt the urban population to the suburbs of Xi'an, and so people will live far and far from cities and their work places. This can also promote local economic development and integration of urban and rural, which has great significance for the increasing attractiveness of cities, expanding influence of Xi'an.

(5)It is conducive to high-tech development. This huge market of APTS which contains a large number of high technologies can help to promote the development of these areas in turn.

Data shows that APTS can reduce 15% to 18% of travel time, improve 12% to 23% of the reliability of services on time, and get 35% of the annual investment return. Advanced public transport system provides a higher security through the long-range surveillance to state of public transport vehicles and passengers activities.

3. The Establishment of Xi'an APTS

The overall objective of Xi'an Advanced Public Transport System is to achieve modernization, digital and intelligent of public transport in Xi'an, so the Xi'an APTS will improve the degree of crowdedness, improve the utilization of the road, improve the public transport service standards and increase the convenience of traffic.

APTS is a complex information system, it has many aspects. The whole system contains internal system and external system, internal system is compose of four subsystems, external system is compose of two subsystems. Fig. 2 shows the architecture of APTS

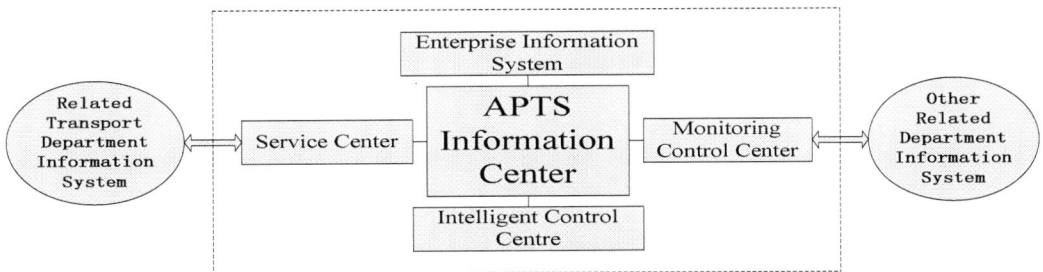

Fig.2 The Architecture of APTS

The APTS Information Center as a bridge to contact other subsystems in the whole system. The APTS Information Center controls the whole system macroscopically, processes and saves date, makes all subsystems to carry out data sharing.

3.1. Intelligent control centre

Intelligent control centre is the core of the Advanced Public Transport Information System, the traditional scheduling is only in origin and destination, and the vehicle can not be monitored when it on the road, waybill and the dates are recorded by man, this method not only waste the human and material resources but also make a mistake easily.[2] Intelligent control centre can resolve these problems well. Fig.3 shows the intelligent control system.

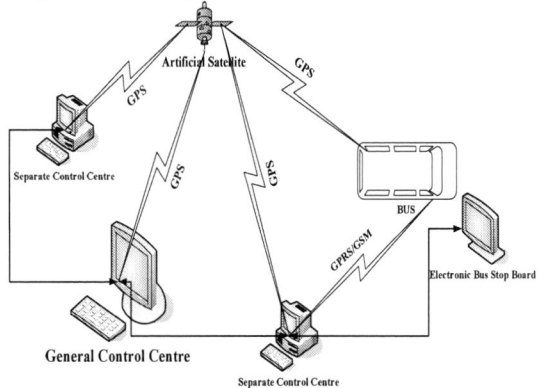

Fig.3 Intelligent Control System

The system is composed of General control centre, Separate control centre, Vehicle digital platform and Electronic bus stop board. The main principle of the work and functions are as follows:

3.1.1. General control centre.
General control centre take charge of dispatch all buses in Xi'an, and it sum up, integrate, processing the traffic information data which collected by Separate control centre, then according to the historic traffic environment, highway condition, traffic network records and the strategy which provided by the system administrator, it uses proper arithmetic to switch these dates become useful dates and can be transmitted by GIS software. At last, General control centre transmit these information to Separate control centre by a standardized format.

3.1.2. Separate control centre.
Separate control centre can monitor bus, obtain the real-time date of the running vehicle, and sent the collected information to general control centre with GPS (global positioning system).another function of the Separate control centre is to transmit the standardized date from the General control centre to electronic bus stop board by special telephone line of ISDN、DDN,and transmit it to the vehicle screen by GPS/GSM technology. The deferent information contains necessary traffic information, weather forecast, advertisement and time,etc.The third function of the Separate control centre is to dispatch the vehicle in real-time according to the vehicle condition and its request.

3.1.3. Vehicle digital platform.
Vehicle digital platform is the core platform between the vehicle and the control centre, it has many functions. The main functions of vehicle digital platform are as follows:

(1)It can take count of passengers automatically with intelligent footplate or infrared sensor.

(2)It can carry out electronic toll collection with IC card and magcard.

(3) It can report the name of bus stop with people's voice automatically.

(4)It can send the vehicle information to Separate control centre by GPRS/GSM.

3.1.4. Electronic bus stop board.
The electronic bus stop board communicates with Separate control system by special telephone line. Its main function is to issue information as bus stop information, the arrival time of bus, line information, and advertisement and so on, so the passengers can choose bus easily.

3.2. Monitoring control center

The main function of Monitoring control center is to monitor the highway condition, vehicle condition and traffic accident condition[3].fig.4 shows the content of Monitoring control center.

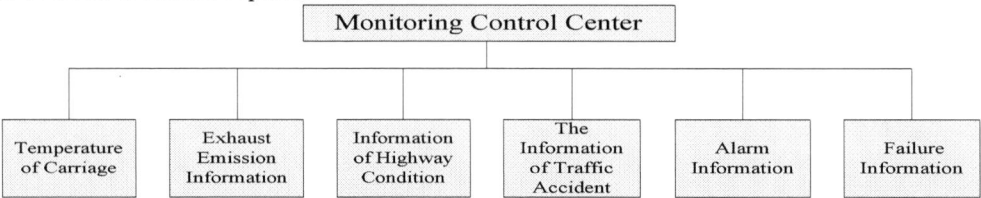

Fig.4 Monitoring Control Center

The temperature of carriage and exhaust emission are monitored and controlled by drive recorder. The purpose of monitoring the temperature inside is to monitor the drivers whether to use air-conditioning in regulations, ensure that the inside temperature up to standard. The purpose of monitoring the exhaust emission is to ensure that the exhaust emission of the running vehicles up to standard.[4]

The equipments on the roadside can get the highway condition and traffic accident condition information. On the important roadside and crossing must install the video signal data acquisition equipment, laser sensor equipment to monitor the highway condition, vehicle condition and send the information to the Monitoring control center. The useful information will be transmitted to the Vehicle digital platform with GPRS/GSM.

Alarm information and failure information is sent to the Monitoring control center by Vehicle digital platform. when the information is transmitted to the Monitoring control center, it will dispatch the people to go to the accident spot[5].

3.3. Service Center

3.3.1. Public transport information inquiry system.
There are three inquiry modes in the public transport Information Inquiry System, the first mode is to set up many teuchscreens on the junction of park and shift where has many passengers. They can inquire about the public transport information they want. The second mode is to use phone and SMS(Short Message Service) inquire about the public transport information at any moment and in everywhere. The last mode is to use internet to inquire about the public transport information.

The provided information of public transport Information Inquiry System contains bus information, ticket information and transfer information. Bus information contains bus origin stop, departure frequency, departure time, stop station and so on. Ticket information contains fare, the way of buy ticket. Bus transfer information contains transfer stop, transfer times, transfer distance and so on.

3.3.2. Guidance system for blindman or amblyopia person.
Although, the trip frequency of these groups is low, The Advanced Public Transport system provide Guidance System for blindman or amblyopia person in every bus stops specially.

3.3.3. The other information inquiry system.
This Information Inquiry System can inquire about Xi'an weather condition ,Xi'an road information and so on.

3.4. Enterprise information system

To establish APTS, the public transport company must attain digital management. The goal of the public transport company information system is to make the work of all departments, post, employee attain digital management. It is a huge database. The process of date input, processing and output are as fig.5

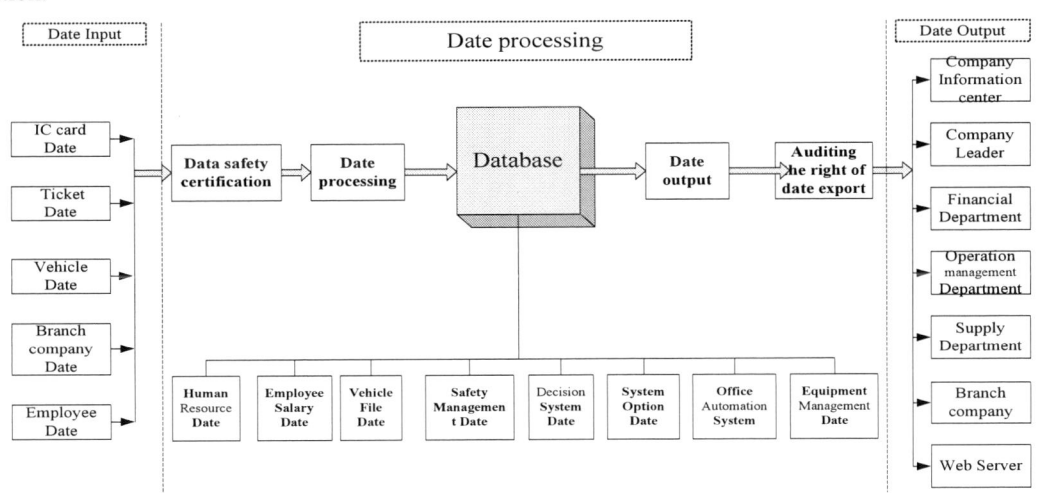

Fig.5 Date processing

According to working time, working content and production of the employee, the enterprise information system organizes production activity and assesses the employee. The enterprise information system is touch with the Vehicle digital platform, and exports the date of passengers by credit card to database.

3.5. The external information system

There are two aspects in the external system; one is related transport department information system, it contains Transportation Bureau, The municipal government, construction bureau and so on. Public transport Company can achieve OA of the Transportation Bureau and it can know the policies and documents about transport of the municipal government and construction bureau by advanced pubic transport system. The other information system contains Train Information System, passenger transport Information System, parking Information System, taxi Information System, city Information System and so on. And we can use all-in-one card for passenger transport,taxi,parking and public transportation to increase the join capability of the various transportation mode and to save time.

4. The development strategy and implementation plan about Xi'an APTS

4.1. The development strategy

4.1.1. Policy setting. Government should formulate some policy and guide some technology about the APTS vigorous development. For example: the advantage of software industry, electronic information and talented people in Xi' an。

To speed up training talents, Chang'an University is the authority in transport field. Xi'an should depend on the Chang'an University to train the technical talents. Then Xi'an will become the training base of advanced public transport talents.

To increase capital investment, it should use the way as joint-venture, issuance of bonds, bank lending to raise fund besides the government's subsidy to establish the Xi'an advanced public transport system.

To train related industries, it should pass a series of policies. In this way, it can encourage the industry development

4.1.2. Establishing the Research Team of Xi'an APTS. This team will research and formulate the master plan of Xi'an APTS, instituting the frame of Xi'an APTS, organizing some expert to design the key technology and the security system of the Xi'an APTS, and applying for the establishment of the National Advanced Transport System's frame.

4.1.3. To spread propaganda. It should spread the propaganda of Xi'an Advanced Transport System in leaders, corporations and citizens.

4.2. The implementation schemes

There are many difficulties to establish Xi'an APTS

Firstly, Xi'an doesn't have good external information sources. The Developed countries can obtain data from the database of the city intelligent transportation system to establish the APTS, but the informationization level of Xi'an advanced public transport is lower than them.

Secondly, to establish advanced public transport must cost much money and need high request of

exterior environment. To most public transport companies, the conditions to carry the program are not ripe. So, Xi'an should start at the local and eventually develop into an APTS network. To achieve this, it must in a planned way and with focuses. The course of implementation should pay attention to two points：

(1)It should take full advantage of the existed transport infrastructure, for example, broadcasting stations, some alterable information board, and increase the using efficiency of transport infrastructure and the efficiency of transportation, rebuild and upgrade the equipments. At the same time, it must consider compatibility and expansibility of the system.

(2)The establishment of the APTS should keep pace with the rebuilding of road net and the adjustment of public transportation. Data accumulation in the pilot area is the basis for the establishment of the future system.

5. Conclusion

The Xi'an Advanced Public Transport System is not a simple concept; It is complex, huge system engineering. The purpose of the advanced public transport is to serve the traveller.It emphasizes people foremost. The Xi'an APTS increases the trip efficiency of Xi'an citizen. It also develops a new industry for Xi'an. The advanced public transport is the inevitable tendency of the public transport. The advanced transport system has attracted many public transport departments, and has implemented successfully in Beijing,Hangzhou.so the departments of Xi'an public transport should pay attention to it and develop it vigorously.

6. References

[1] Teng,Jing, "Real-time operational control of singular bus line under APTS", *Journal of Tongji University*, v 34, n 6, June, 2006, pp.744-747

[2] Leonard, Elizabeth I, "Program synthesis from formal requirements specifications using APTS", *Higher-Order and Symbolic Computation*, v 16, n 1-2, March/June, 2003, pp.63-92

[3] Liu,chuang. "Research on the framework of urban transit intelligent management system",*Urban Public Transport,*v2,Feb,2003,pp..28-29.

[4] Shen,hui, "Developing intelligent public traffic system to improve efficiency of Wuhan's public traffics". *Engineering Journal of Wuhan University*,v6,June,2001,pp.22-26

[5] Li,qingli. Intelligent Public Transportation Control System Based On ITS,*Electronic Technology*,v8,Aug, 2003,pp.36-37

A Forecast Model of Urban Passenger Flow Containing New Railway Project

XIE Hui   YAN Kefei   WEN Ya

The school of transportation engineering, Tongji University, Shanghai, 201804, PR China
Xiehui110@126.com   yankf@mail.tongji.edu.cn   wenyachina@sina.com

Abstract

The urban passenger flow forecast is an important part of urban traffic planning. The urban passenger flow forecast containing the new railway project is much different from the common passenger flow forecast. The difficult is ascertaining the share rate of the railway passenger flow. Traditionally, the share rate is a fixed constant based on the experience of the cities which have already run the railway project. However, this way is not enough sure. In the paper, a split model on the railway has been built, which has based on the analysis of the influence scope of railway and the influence on urban passenger traffic volume of railway (e.g. the traffic model choice). Additionally, in the model, a variable like the cosine function (coming from SP investigation of wuhu) denotes the railway passenger flow share rate, which reflects that the railway passenger traffic volume varies with distance. Finally, the model is used to forecast the urban passenger volume involving the new railway project of Wuhu urban comprehensive transportation planning (WUCTP), and attains practical opportunity well.

1. Introduction

With the city scale's extension, many big cities are actively building or planning to build railway facilities. Railway has gradually become one of the most important parts of the urban passenger traffic system. Railway, as a kind of rapid and mass urban passenger transportation, has changed the trip accessibility of passengers alone its line. At the same time, it impacts the urban space layout for land use. For example, quickening the process of suburb's urbanization and improving the land development intensity alone the line will influence the urban passenger traffic's generation and distribution. Additionally, railway has the characteristic of using less land (e.g. subway and light rail). So, railway passenger flow volume should be eliminated when dealing with the urban passenger flow volume prediction of land traffic system planning (excluded railway). Thus, the prediction of railway passenger share volume will always precede the prediction of urban passenger flow volume when the city contains the new railway projection.

There are various achievements in the prediction methods of urban railway passenger flow volume. For example, Guo Xiucheng et al. (2000) have presented a URT's passenger flow forecast with the joint mode split\model on cooperant and competitive OD matrix Based on analyzing disadvantage of URT's passenger flow forecast by the joint mode split\assignment model on all modes OD matrix. It concludes three steps: firstly, all modes OD matrix can be gained by trip production and distribution forecast. Secondly, cooperant and competitive OD matrix can be achieved by layered and politic mode split method. Lastly, URT forecast can be done by joint mode split\assignment model. Li Chunyan et al. (2006) have brought forward disaggregated activity-based forecasting model of railway passenger. Activity-based forecasting method researches people's behavior. A chain of activities describes the order of different activities during a person's run of the day. Wang Haiqiang et al. (2005), who aims at the correlative factor of trip spatial distribution, have built a computing model for generalization cost correlating the zone's progression of importance degree based on the analysis of common method of passenger volume forecast about intercity railway and applied generalization cost as resistance of traffic to the bi-restriction gravity model, which makes the trip distribution reflect fully the social economy and land utilization among zones. This model increases the veracity of forecast on a degree. Ye Zhongping et al. (2006) have brought forward a kind of traffic demand prediction method that bases on the reminder traffic volume. It considers that the difference between traffic demand and traffic supply (namely reminder traffic demand) is railway demand. Additionally, there are others, such as Lu Huapu, Ma Heling and so on, have further researched on the prediction methods of railway passenger flow volume. These methods base on "four-step" method, in which the analysis is not well enough at the aspect of the reasons of railway passenger traffic generation, and they consider that railway passenger flow volume is only the normal trend traffic volume due to omitting the railway's inducted traffic volume, so the result is lower. Additionally, the data requires too much and the maneuverability is not better. Besides, Extension Proportion (EP) and Conversion Probability (CP) are applied to conform railway passenger volume, namely, railway passenger volume is ascertained according to the different distance away from the railway station. For example, the share rate of railway passenger volume of the place 500m away from the station is more than 60%, 500~800m is between 50%~30%, and the place beyond 800m isn't considered. This way is easy and convenient; however, there are some disappointments: the rate used in recent, mid, and long period will not be changed, but it is often got by experts through analogy other country's experiences, so this is an inaccurate way and lack of the basis of the theory. In the paper, a split model on the railway has been built, which has based on the analysis of the influence scope of railway and the influence on urban passenger traffic volume of railway (e.g. the traffic model choice). Additionally, in the model, a variable like the cosine function denotes the railway passenger flow sharing rate, which reflects that the railway passenger traffic volume varies with distance.

The rest of paper is organized as follow: section 2 analyses the new railway project's effect on urban passenger flow volume. Section 3 researches the scope influenced by the railway station. Section 4 contains the numerical example. Finally section 5 achieves conclusions and brings forward some other researches.

2. Effect analysis of new railway project on urban passenger flow volume

To urban transit system, not only the growth of residents' income and population will affect the urban passenger flow volume, but also the traffic system's promotion will inspire the growth of residents' trip. New railway project's influence on urban passenger flow volume is mainly reflected in the accessibility of road network and residents' travel time. It can improve the traffic service condition (as seen in the figure 1, S1→S2), so as to induce the growth of urban passenger flow volume (as seen in the figure 1 (Q2-Q1) (Enjian Yao et al (2005)).

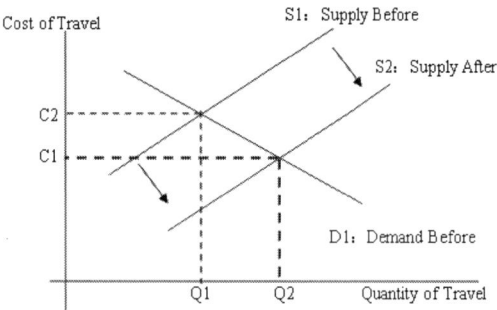

Fig1. Analysis of supply-demand (Enjian Yao et al (2005))

2.1 Accessibility

Normally, traffic accessibility means whether is convenient to arrive at another site from one site in a city, sometimes, it means whether is convenient to get one site

from other site too. To the city built new railway project, its railway network has not been constructed. For getting the sharing rate of railway, the traffic accessibility here could be considered as that the convenience between the traffic zone and the zone affected by railway, which can be expressed with the distance between traffic zone and railway stations. New railway project means new railway stations in traffic zone, and the railway becomes the new manner for residents' trip. It's the mathematical expression is:

Suppose P as point-set, it reflects the set of resident trip OD. And here, it is simplified into gravity center of the traffic zone. Suppose Q as the set of zone, so P and Q is one-to-one correspondence. l is supposed as the set of traffic zone contained railway station. It can suppose that railway is the composition of the railway stations to passengers, and can be expressed with countable point-set $S_i \in IR^2$, $i = 1, 2, \cdots, k$;

$$\ell = \left\{ x \in IR^2; x \in S_i, \forall i = 1, 2, \cdots, k \right\} \quad (1)$$

The solution set of X means the point-set on the plane, so $X \subseteq \ell$, namely $\forall x \in X$, and then $x \in \ell$.

There are more railway stations in the traffic districts, which promotes the accessibility of the areas. Consequently, the population and job opportunities will grower accordingly. Accessibility is proportional to the population and the job while in the inverse distance between zones. If the shortest distance is $D(p, x)$ from p to x, accordingly to Gravitation traffic accessibility model (Hansen, W.G. (1959)), between p and l including x railway station, the accessibility is as follow:

$$A = g \frac{m_p m_\ell}{D(p, \ell)^a} \quad (2)$$

Where m_p, m_ℓ is respectively the population and job of traffic zone p and l; g is the harmonize modulus so as to make A between 0 to 1; a is the decay coefficient, generally, its numerical value is between 1 to 3. So the service scope of railway can be got from the formula as follow:

$$D(p, x) \leq r \quad (3)$$

Namely, the traffic zone $p \in P$ can use the railway station $x \in \ell$, when $D(p, x) \leq r$. $D(p, x)$ shows the shortest distance between p and x; r is a preset of service radius, it is varied with the access modes of transport.

2.2 The trip time of resident

The trip time of resident means the time from the origin to the destination. This paper is about the changing of the trip time of resident under the influence of railway.

So the investigation area can be shrunk into the districts in the scope of railway service.

The distance to the nearest station maybe is shortened due to railway, and its rapid, part of passengers will save trip time. Follow the above, the distance to the nearest station is $D(p_i, x_i)$, after the railway is put into operation; and the distance away from the station is $D(p_j, x_j)$ too. The distance between two stations is $D(p_j, x_j)$.

After all the railway operation, the time spent from the origin p_i to destination p_j by the passenger is

$$t_b = \frac{D(p_i, x_i)}{v_{it}} + \frac{D(x_i, x_j)}{v_m} + \frac{D(p_j, x_j)}{v_{jt}} \quad (4)$$

Where v_{it} is the average speed of the access modes of transport from origin i to station; v_m is the average speed of the railway; v_{jt} is that the average speed of the access modes of transport from the station to destination j.

Suppose t_{ij} is the time from the origin p_i to the destination p_j before the railway is put into operation. And after railway is put into operation, the saving time through railway is

$$t_a = t_{ij} - t_b \quad (5)$$

3. The share rate of railway passenger flow modal

3.1 Forecast model

It is the most important that Ascertaining the sharing rate of the passenger flow of the railway in the research of forecast model of the urban passenger flow volume which involves new railway projection. According to the above, the EP is a easiest way, and the most possible to cause the inaccuracy of the prediction is the problem of the railway sharing rate. The common way is to use other city's experience. In the paper, a split model on the railway has been built, which has based on the analysis of the influence scope of railway and the influence on urban passenger traffic volume of railway (e.g. the traffic model choice). Additionally, in the model, a variable like the cosine function denotes the railway passenger flow sharing rate, which reflects that the railway passenger traffic volume varies with distance. The model is as follow:

$$w_i = \frac{\exp(v_i)}{\sum_j \exp(v_j)} \quad (6)$$

Where w_i is the probability of choosing this traffic modes; i, j is the kinds of traffic modes; v_i, v_j is the utility function of taking the traffic modes i, j;

In the paper, the main task is to distinguish the passenger flow of choosing the railway and other modes. Whether selecting the railway is related to the accessibility of the railway station and the time of resident's trip. So its utility function is as follow:

$$v_i = a_1 A_i + a_2 t_i + \varepsilon \quad (7)$$

Where A_i, t_i is respectively the accessibility and the trip time of traffic modal i; its formula is as above; a_1, a_2 is the variable of the utility function; ε submits to normal distribution.

This model takes the accessibility and trip time & distance as the variable of the utility function, and reflects the sharing rate of railway passenger flow volume will change with the layout of the station and the distance of railway station. What's more, the sectional drawing shined upon the accessibility and share rate flat surface is similar to the diagram of the cosine function (coming from SP investigation of Wuhu, as seem the figure 2). The distance to the station is shorter; the accessibility is better; the sharing rate is bigger. This meets the characteristic of railway passenger flow distribution.

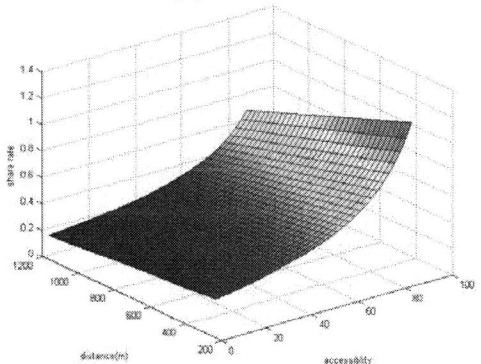

Fig2. The relation among with accessibility, trip distance, share rate of railway (a part of SP results)

Before railway operation, the OD matrix is $[q_{ij}]$. And the rate z_i taken by zone i is

$$z_i = \frac{\sum_j q_{ij} \cdot w_j}{\sum_j q_{ij}} \quad (8)$$

So the traffic generation volume of zone i which has railway is

$$p_i' = p_i \cdot z_i = p_i \cdot \frac{\sum_j q_{ij} \cdot w_j}{\sum_j q_{ij}} \quad (9)$$

Additionally, the traffic attraction volume of the zone i is

$$a_j' = a_j \cdot z_j = a_j \cdot \frac{\sum_i q_{ij} \cdot w_i}{\sum_i q_{ij}} \qquad (10)$$

The urban passenger flow generation volume which contains new urban railway projection is

$$\begin{cases} p_i'' = p_i - p_i' \\ a_j'' = a_j - a_j' \end{cases} \qquad (11)$$

And then, the passenger flow volume distribution OD can be got by making traffic distribution forecast of the urban road through using of generation and attraction volume.

3.2 Parameter estimation

Maximum likelihood estimate is a kind of common mathematics statistical method which is often used of calibrating the parameter of the function. Considering the characteristic of high praescriptio of urban traffic system, we suppose that resident will choose the most valuable modal to travel. So we can calibrate the parameter of the model with Maximum likelihood estimate method.

First of all, we give a definition.

Definition 1: suppose $f(x, \theta)$ as the probability function of the variable x, θ as the parameter of the variable x, ($X_1, X_2, \cdots, X_n$) as the sample value of the variable x, thus the probability function of the point ($x_1, x_2, \cdots, x_n$) or likelihood function is the function of θ, it is defined as follow:

$$L(\theta) = L(\theta, \ x_1, \cdots, x_n) = \prod_{k=1}^{n} f(x_k, \theta), \theta \in \Theta \qquad (12)$$

Where Θ is the set of parameters, such as $\exists \hat{\theta} = \hat{\theta}(x_1, \cdots, x_n) \in \Theta$, and $L(\hat{\theta}) \geq L(\theta), \theta \in \Theta$ is count amounts, thus $\hat{\theta}(x_1, \cdots, x_n)$ is the maximum likelihood value of parameter θ, $\hat{\theta}(X_1, \cdots, X_n)$ is the maximum likelihood estimate value of parameter θ.

Suppose the definition of the choosing variable parameter is as follow:

$$s_{ij} = \begin{cases} 1 & \text{if } i \text{ modes choiced} \\ 0 & \text{otherwise} \end{cases} ; \qquad (13)$$

And its probability function is

$$L(\theta) = \prod_i (w_i)^{s_{ij}} \qquad (14)$$

Both sides' logarithmic, it can be turn into the formula as follow:

$$\ln L(\theta) = \ln(\prod_i (w_i)^{s_{ij}}) = \sum_i s_{ij} \cdot \ln(w_i) = \sum_i s_{ij} \cdot \ln(\frac{\exp(v_i)}{\sum_j \exp(v_j)}) \qquad (15)$$

If there are solutions of the equation above, so

$$\frac{\partial \ln L(\theta)}{\partial \theta} = 0 \qquad (16)$$

The solution set of the equation （16） is the maximum likelihood estimate value of the model.

3. Case analyses

Now Wuhu is taken as a case example to illustrating the modal above. Wuhu is a city of Anhui province, which possesses the most powerful urban function, economic capability and population in Wan river area. The prediction suggests that Wuhu will achieve the GDP 350 billion, 200 million people and the 6.5~7 million passenger flow per day in future. To solve the problem of resident trip demand, two railway line projections are planned in future (2020) in WUCTP. Concrete conditions of railway trend and stations layout is showed in the figure 3.

According to WUCTP, there will be 59 traffic districts in totally, and the peak hour OD at the year of 2019, the time impedance of districts and the traffic production volume in the year of railway operation are known. According to the models and formulas above, the peak hour passenger volume of attraction and generation can be got easily (as seem the figure 4), as well as the peak hour passenger volume of attraction and generation excluded the passenger flow volume of the railway (as seem the figure5) and the assignment picture of the traffic volume on the road net (as seem the figure 6).

In the process of evaluation, the forecast method of the urban passenger flow volume containing the new railway project has been favorable commented and attended by the experts, as well as its good simulated result.

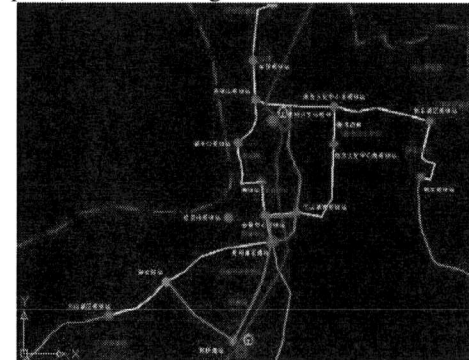

Fig3. The Design of Rrailway Network in Wu Hu

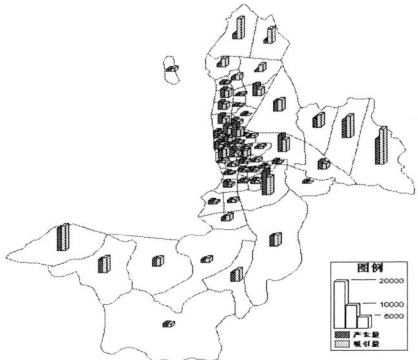

Fig4.The Peak Passenger Flow of Railways

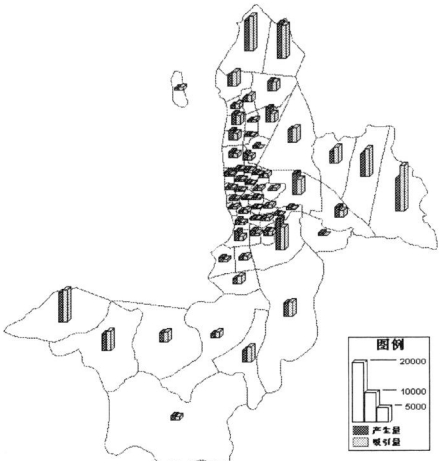

Fig5. The Peak Flow of Road Network.

Fig6. The Flow Assignment of Road Network

4. Conclusion

The forecast model of the urban passenger flow volume involving the new railway project is put forward based on the influence of the new railway project on urban traffic volume. It is a kind of quantify prediction method of urban passenger volume. Considered the variable of making the accessibility and trip time as the utilizable function, the passenger flow sharing rate model of urban railway is put forward, which reflects the sharing rate of railway passenger flow volume will change with the layout of the station and the distance of railway station. Thus this model is more accurate than the common methods.

Because the model considers the influence of the railway and makes indirect prediction of urban passenger flow volume, it can not be put into the prediction of new other projects. Therefore, in the future, a firmly important task is to find a direct and easy model which can be put into urban passenger flow volume prediction containing various new projects.

References

[1] Guo xiucheng, Lv zhen. "Study of URT's Joint Modal Split Assignment Model on Cooperative and Competitive OD Matrix". *China Journal of Highway and Transport*, 2000, 13(4) pp. 91-94.

[2] Li chunyan, Chen jinchuan et al. "Research of Activity-Based Passenger Forecast Used in Beijing Metro Line 5". *Journal of Transportation System Engineering and Information Technology*, 2006, 6(6), pp.143-149.

[3] Wang haiqiang, Ma dezhong et al. "Study on Trip Spatial Distribution of Intercity Railway Based on Importance Degree". *Communications Standardization*, 2005(2/3), pp. 93-96.

[4] Ye zhongping, Wu ruilin. "Forecast about Urban Rail Transit Demand on Remainder of Traffic Demand". *Urban Mass Transit*, 2006 (3), pp. 18-21.

[5] Lu huapu, Wang jianwei et al. "Study on The Methods to Forecast the Passenger Volume of Intercity Rapid Railway". *China Civil Engineering Journal*, 2003, 36(1), pp. 41-45.

[6] Ma heling, Guo zhiyong. "Discussion on the Disaggregate Model of the Flux Forecasting of Urban Rail Transportation". *Journal of Huazhong University of Science and Technology (Urban Science Edition)*, 2002, 19(1), pp. 65-67.

[7] Enjian Yao, Takayuki Morikawa. "A Study of an Integrated Intercity Travel Demand Model". *Transportation Research PartA*.39, 2005, pp.367-381.

[8] Hansen, W.G. "How Accessibility Shapes Land Use". *Journal American Institute of Planners*, Vol.25, 1959, pp. 73-6.

[9] Wuhu city government. "Wuhu urban comprehensive transportation planning" (WUCTP), *The College of Architecture and Urban Planning of Tongji University, the college of transportation engineering of Tongji University*, Shanghai, 2008.

Distributing Stations in a Given Urban Rail Transit Line

Wei Jin-li [1], Zhang Meng-meng [2]

[1] Collage of Automobile and Traffic,
Qingdao Technological University, Qingdao, Shandong, 266520, China
[2] Collage of Traffic,
Shandong Jiaotong University, Jinan, Shandong,250031,China
Wjl827025@163.com, 573275197@qq.com

Abstract

Urban rail transit brings the benefits of various aspects of society has become a worldwide consensus. When emplacing the stations on network, taking into account the high cost of rail transportation, we not only consider actualizing the optimized network as a whole, but also consider improving the operating efficiency of every line. For a given rail transit line, emplacing the station reasonably can increase the attraction of lines, expand the coverage of the stations and improve the operating efficiency. Therefore, a new Method of emplacing stations was found to improve the line coverage of trips : Firstly, searching the distributing center of passenger flow on the line to produce various stations-distribution projects; then, emplacing the stations by considering the constraints of the space between each ones. For the projects with the same number of the sites, the project of which the network can cover the largest trips is the optimal one; otherwise, for the projects which have different numbers of sites, the best one can be chosen by integrative evaluation. Finally, an example was taken to certificate the operation of the method and take reference for researchers.

1. Introduction

Urban rail transit was a vital and lasting project of a city, once a line completed, it was very difficult to change. So taking into account the high cost of rail transit, and with the purpose of using the lowest number of stations to cover the greatest number of passengers, distributing station in given lines should not only to achieve overall optimization of the network, but also to achieve relative optimization on operation of each lines. Generally, after the initial network completed, laying the specific stations in given lines only depended on experience, it included a great lot of qualitative analysis, and made many stations distribution unreasonable, limited effective exertion of the role of rail transit. Therefore, according to the theory of

traffic areas [1], this paper combined distribution patterns of passengers flow in given lines closely, laid stations one by one by meeting the station spacing constraint to form a variety of stations-distribution projects, then chose the best one by constraints of line coverage of trips and integrative evaluation.

2. Forming projects of stations distribution

2.1. Finding station-choices

1. Searching for the main distributing center of passenger flow by processing the OD matrix of Long-term forecasts.

Set the diagonal elements to 0 to eliminate inner trips of traffic area, summed the data of every row and line to get the producing volume and attracting volume outside traffic area, and sorted them. According to the size of population and land, users could select the top 10 by 20 or 10% ~ 15% of the numbers of total area [2] as the main distributing center of passenger flow, then marked them in the given line in city layout.

2. Checking points that neglected.

According to the land planning, non-main distributing center of passenger flow were analyzed to determine whether potential choice sites are missed. If existed, integrate them into sets, and adjust relevantly:

(1) If the given line has points of intersection with the built lines, put them the first choices of transfer stations.

(2) If several points concentrated in a smaller area or less than a certain distance, combined them into one point.

2.2. Generating dummy stations-distribution

The principles were shown as follows:
1. Overlapping: If collection and distribution points near the built station, superposed them;
2. Approach: the points should be close to main source of distribution center of passenger flow;

3. Convenience: they should be located in the road intersections or Plaza, to facilitate passengers transferring in different traffic modes;

4. Maneuverability: they should be constructed easily, to prevent having difficulties in construction after the dummy points became into real stations;

5. Favorableness: it should be easy to pave tracks between dummy points.

2.3. Forming stations-distribution projects

Assumption: there were W choice stations in a given line; the specific method was shown in Figure 1.

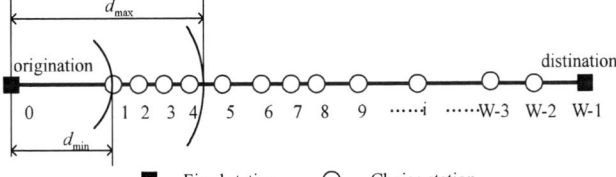

Figure 1　Sketch map of laying stations in given line

First, taking origination of the line as centre, drawing arc respectively with radius of d_{min} (the minimum station spacing) and d_{max} (the largest station spacing), to define the location interval of the first station. If only one choice station locates in this interval, set its location as a theoretical position; If several choice stations locate in this interval (as shown in Figure 1:point1、2、3、4 are all the choice stations of the first station), they would take respectively as starting point to continue to define the viable stations in the next interval with d_{min} and d_{max}, then continue defining until the destination of the line, it could form a variety of stations distribution projects, as shown in Figure 2.

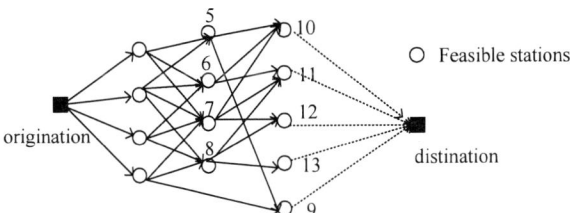

Figure 2　Sketch map of forming stations distribution projects

d_{min} and d_{max} could be ascertained from city development, experience and the technical factors. According to the survey analysis, d_{min} could be getting smallest to 0.4 km, d_{max} could be getting largest to 3 km. Generally, the larger span between d_{min} and d_{max}, the more projects of stations distribution formed in accordance with the method above. In particularly, when d_{min} gets the minimum in all the choice stations, d_{max} was made an right choose within a certain range, that is $d_a \leq d_{max} \leq L_a$ (d_a was the optimal average station spacing, L_a was

average trip distance)[3], it could be get all available stations distribution projects according to the method above.

3. Selecting the optimal station distribution project

Suppose that using the above method could form M projects, and each project had Nm stations, $m=1,2,…,$ M, and $N_m \leq W$. Taking into account the constraint of the optimal average station spacing and the goals of laying station (using as little as possible stations to cover as much as possible the passengers flow), only the projects that had the number of stations $N_m \leq N+2$ (N could be get by the length of line L and d_a: $N=int(L/d_a)+1$) needed to be selected.

For the projects of same numbers of stations, the optimal one had the greatest ability to attract passengers, so it could be gotten by calculating the line coverage of trips; for the projects of different numbers of stations, it needs to establish Integrative Evaluation System [4] to choose the best one.

3.1 Calculating the line coverage of trips

1. Model assumptions

To analyze expediently, this paper made the following assumptions: in given rail transit line, the origination or destination of passengers trips were in the attraction scope of the rail transit station (especially to one trip attraction range, and supposing that passengers arrived station by foot), for the transfer passengers, the transfer station was in the attraction range, that is just considering the walk scope of passengers. At the same time, supposing that the number of passengers by bus just had relation with land use condition of traffic area, and the trip ratio didn't change with the change of station's location.

2. Method of calculating station coverage volume of trips

The volume of station coverage of trips primarily depends on the coverage scope and the trip patterns of surrounding people; it had closely relationship with the structure of surrounding land use.

(1) Ascertaining the coverage scope of station

The coverage scope of station primarily depends on the reasonable walking distance from the station. According to the survey [5], the appropriate distance for Chinese residents walk to rail transit station is 500 to 600 m for the centre, and is 800 to 1,000 m for the suburb.

(2) Choice of passenger trip pattern

Taking into account that use of discrete distribution could generate conflicting conclusion [3], this paper used continuous distribution pattern. Supposing that C is the impact area of rail transit line, it is divided into several areas c_j by population $_j$, that is:

441

$$C = \sum_{j=1}^{J} c_j \qquad j = 1, 2, \cdots J \qquad (1)$$

Each area is polygon, its population density is ρ_j (ten thousand people / km^2), and if the employment data of each area is known, ρ_j could be the weight sum of population density and employment density.

(3) Establishment of station coverage model

According to Lesley assumption [6], the station attract range was circular, as shown in Figure 3, supposing that $B(X_i, r)$ was an stations aggregate that the distance between the station to the X_i ($i = 1, 2, ... M$) was not more than r, known as the attract circle of attraction radius was r, for each station X_i, there were K different attraction intensity r_k ($k = 1, 2, ... K$), corresponding to concentric zone around the site, then $B(X_i, r_k) = B_{ik}$. And r_k could be getting through the distance between passengers walk points to rail transit station.

Taking into account the mobility of passenger trip, it is more convenient to make station coverage discrete, so set the adjacent attract strength r_k, r_{k-1} as constant. And the volumes of station attraction with the population density, attraction intensity and coverage range meet the gravity model. The $R_m(X_i)$ of the attraction of station X_i, $i = 1, 2, ...N_m$ in the Mth project is:

$$R_m(X_i) = \sum_{j=1}^{J}\sum_{k=1}^{K} \frac{\alpha\rho_j}{\left((r_{k-1}+r_k)/2\right)^2} S\left((B_{ik} \cap B_{i(k-1)}) \cap c_j\right) \quad (2)$$

In formula: α—calibration constants, it has the same meaning as constant K of the gravity model;

ρ_j—the weight sum of population density and employment density (ten thousand people / km^2) ;

$S\left((B_{ik} \cap B_{i(k-1)}) \cap c_j\right)$—The cross area between traffic survey area and attraction ring (km^2), which could be obtained through geometric calculation or a software with function of GIS (km^2).

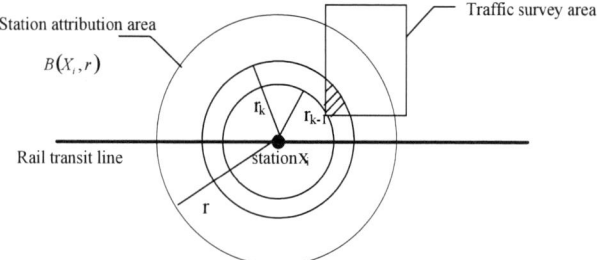

Figure 3   Sketch map of Calculating station coverage of trips

3. Getting the line coverage of trips

The line coverage of the mth project $R_m(l)$ was:

$$R_m(l) = \sum_{i=1}^{N_m} R_m(X_i) \qquad (3)$$

3.2. Getting the optimal project

Supposing that there were U projects had the same number of stations in M projects. So the line coverage of relative best project was:

$$R = max\{R_u(l)\} \qquad u = 1, 2, ..., U \qquad (4)$$

Then put this relative best project and the other $(M-U)$ projects together, through integrative evaluation to choose the optimal project.

4. Case Analysis

Xi'an Metro line 2 prongs the main north-south corridor of Xi'an. 24 stations could be calculated according to constraint of the optimal average station spacing. There were 26 choice stations in Line 2, the minimum distance between stations was 603 m, the longest was 2,500 m, so d_{min} is chosen 600 m, d_{max} is chosen 2,500 m, and formed four projects by the method in this paper. Project 1 had 24 stations; project 3 and 4 had 25 stations, project 2 had 26 stations.

According to forecast results on passenger flow of Metro Line 2[7], using the model of station coverage to calculate the coverage of each choice station. And attraction radius in central area chose 600 m, 800 m to the suburb, α is a constant, and may not be considered in comparison. The results of station coverage of Long-term planning were shown in table 1.

Table 1   station coverage of Long-term planning （Unit: pcu）

No.	coverage	No.	coverage	No.	coverage	No.	coverage
1	38848	8	440075	15	554338	22	230233
2	60326	9	452105	16	542004	23	200431
3	87246	10	464671	17	492753	24	172176
4	361918	11	478929	18	467311	25	117895
5	249287	12	509627	19	458837	26	77689
6	297731	13	550694	20	489654		
7	374420	14	571494	21	274924		

Calculating line coverage of project 3 and 4 according to Table 1 to get the result that abandoned project 3, then using integrative evaluation in project 1, 2, 4[4], and got the optimal project is project 1.

5. Conclusion

Laying stations in given rail transit line should consider many factors, and station spacing is an important one, this paper took into account the station spacing constraint fully, chose the optimal project by combining the line coverage with the integrative evaluation, to avoid irrationality of subjective assumptions in deciding the location of station. It formed a new method on station distribution, had strong maneuverability. Due to the time and the limited capacity of author, some details still need to be improved, for example: considering the mobility of passenger trip and other relevant factors, some parameters in station attraction model may become common variables; the method to ascertain them had to be further studied.

References

[1] Wang Wei, Yang Xin-miao, Chen Xue-wu. Planning method and management technology of urban public transits system [M]. *Press of Science*. 2002

[2] Qin Yu. Study on planning and designing rail transit hubs [J]. Shanghai: traffic engineering college, *Tongji University*. 2002

[3] Sun Zhuang-zhi. Study of a number of theoretical issues on urban rail transit network planning [J] Beijing: *Northern Jiaotong University Doctoral Thesis*. 2002.

[4] Wei Jin-li. Study on Distributing Station in Urban Rail Transit Line [J]. *Xi'an: Dissertation Submitted to Chang'an University for the Master Degree*. 2006.

[5] Hu Gang. Research for optimization practical technology of urban public transit network and stop [J]. Nanjing: *Southeast University Doctoral Thesis*. 2003

[6] Julian Ross. Railway stations: planning, design and management [M]. Oxford, Boston, *Architectural Press*. 2000

[7] Study on forecasting passenger flow of Xi'an Metro Line 2 [Z]. Xi'an: *Chang'an University*. 2005

2008 Workshop on Power Electronics and Intelligent Transportation System

Intelligence Toll Management System of Highway Traffic

Bo Yan, Yehua Huang

School of economics and commerce, South China University of Technology, Guangzhou 510006, Guangdong, China
E-mail: yanbo@scut.edu.cn

Abstract

With the development of new technologies such as computer network and communication system, the highway traffic net is to be more consummate, traditional charge cannot meet the demand that require quickly, accurately, and efficiently. According to the advanced technologies of highway traffic communication network and the data transmission requirement for toll management system, this paper put forward the project for toll management system of highway traffic based on computer network, used IC card management system as the passing medium, and gives a sufficient discussion on the goal and structure of toll management system.

1. Introduction

Intelligent toll management system of highway traffic is designed to resolve the problem of different charging unit on the road, and the charging unit set their own closed facilities for charging independently. Traditional toll management is lack of uniform technical criterion and supervisory mechanism [1-2]. This paper research toll management system of highway traffic is easy to upgrade to meet the centralized management. System and other related systems to achieve the sharing of resources, and it provide communications and information services to transport system and community.

As toll management system of highway traffic used advanced, reliable technology and mature equipment, therefore it can reduce the staff on duty in the toll of the control room, lower operating cost management. And the use of networking system for manage department is easy centralized supervision and management. This will change the traditional "one company for one way, construction integrating with manages" management structure and operation of the charges shortcomings. It will be an inevitable trend in the future development of a modern payment system.

2. System Analysis and Description

As the defect of traditional toll management system leads to every highway section carry out an independent of highway charge. And the adjacent sections are connected with the main line toll stations, which resulting in the main line of toll Station is intensive, so it restricted the road traffic capacity into full play. Facilities of charge

and technology are also incompatible with each other, which would give rise to an increase of building and operating costs. Toll systems can not effective interconnection will resulting in Lack of effective integration of resources and so on [3-5].

This system uses artificial judgment vehicle type in the entrance, release traffic-ticket of non-contact IC card, exit artificial recheck vehicle type, recovery of non-contact IC card and collect tolls. When the vehicle to pass the vehicle detector, it will check the number of vehicles. The computers is semi-automated statistical management charge standard. Traffic-ticket uses the repeated non-contact IC cards. Charge business is mainly to receive cash, but also have credit cards and others payment methods prepare for highway. Each section by section from the jurisdiction of the management system can be designed for two-level computer administration system, that is the computer system and toll charges center computer system, and set aside the interface with the parent body.

The entire highway system is consists of various management centers that owned the charges center computer system. And all sub-toll stations manage themselves computer systems, as well as sub toll stations of exit/ entrance lane toll system. Charging system has the following basic requirements: full closed the charge structure, traffic-ticket of non-contact IC card, toll stations and the sub center is two-level administration system, advanced communications equipment: Fast Ethernet LAN and WAN SDH fiber.

3. Architecture of Application Systems

3.1. Toll System

Intelligent toll management system of highway traffic has two-level administration system, their mutual relations: charges sub center is responsible for the supervision and management of toll stations, toll station manages each toll lane. Each one is a complete computer management system, sub-centers with toll stations through SDH synchronous digital network rapidly transfer data of charges, toll stations through LAN to implement monitoring and acquire related charges information. Toll station and sub center relations is shown in Figure 1.

978-0-7695-3342-1/08 $25.00 © 2008 IEEE
DOI 10.1109/PEITS.2008.39

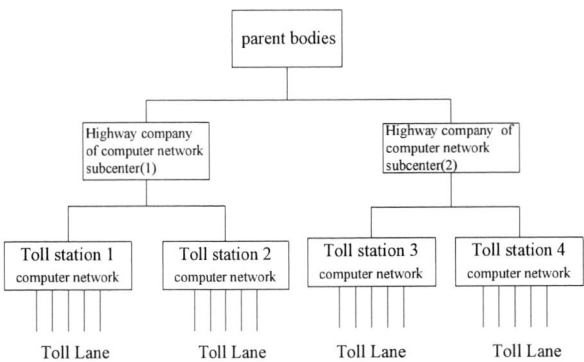

Figure 1. Relationship diagram of toll station and sub center

3.1.1. The computer system of charges sub center. The computer system of charges sub center is set in the central control room used for the collection of the changes data that from each station's computer system so as to complete the combined audit. The computer system of charges sub center through communication systems

connected with all the toll station of the computer system. Charges sub center is constituted by a computer system that link several small computers LAN and the corresponding software components, and its main function is to provide statistics and road data of each jurisdiction charges collected by the station computer system, to manage the IC card, to scheduling and real-time enquiry, to establish the network model of demolition clearing accounts. And according to data of traffic charges established analysis and decision-making module, the computer system through the toll stations will transmitted charge standard and system parameters that from the Center to the lane controllers. The computer systems of charge sub center, computer of tall stations and the lane controller work independently. Toll station and the computer of lane controller can not be affected to operate independently when the computer system of charges sub center fault occurs. Figure 2 shows network structure relationship diagram of toll sub center.

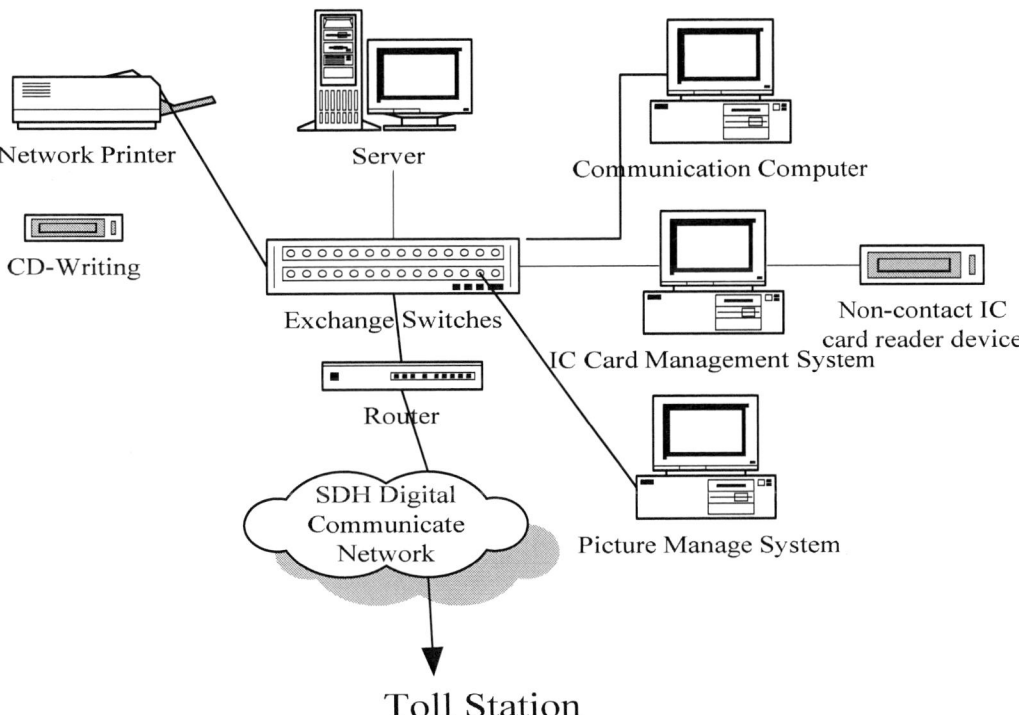

Figure 2. Network structure relationship diagram of toll sub center

lane charges and staff charges, as well as the management of this information through the computer network connections spread to other toll stations within functional departments. And toll station regular or irregular report

exit\entrance vehicles information through routers to charge sub center. Their mutual relation is shown in Figure 3.

445

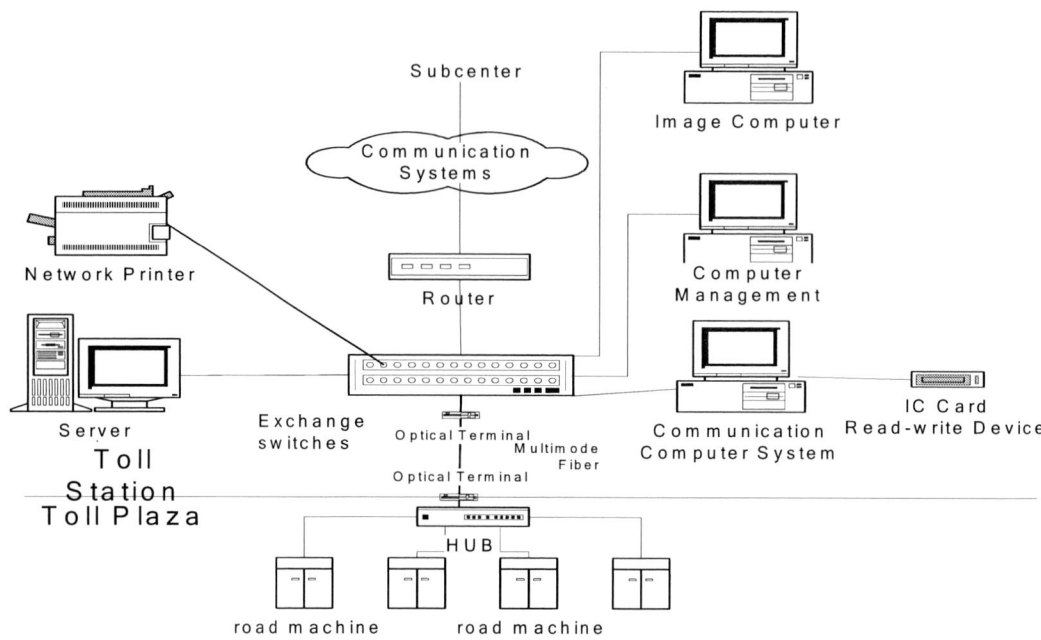

Figure 3. The computer system structure diagram of toll system

Switched hub controller complete the road network within the communications equipment of the exchange of information request. Server network hub with the ramp with the exchange of 10 M-hub port and other ports used 100 M. HUB lane and toll stations use multi-mode optical fiber, optical terminal complete the sigh photoelectric conversion.

(1)Entrance lane equipment. Entrance lane equipment is constituted by the entrance lane controllers, toll collector terminal, vehicle detector, manual railings, automatic railings, rain lights, yellow flash alarm device, multi-channel digital control unit and other equipment,

including the corresponding power cable and signal cable. Through computer networks to establish communication between toll station and charges lanes, it transmitted the state charges lane, equipment operation or fee-based information to toll station. Charges lanes and toll stations have independent Telephone Network and CCTV systems, toll stations supervision can directly order, they can also write the date, time and other information into non-contact IC card and transferred them to the server computer of toll station. Figure 4 shows the formation diagram of entrance lane equipment.

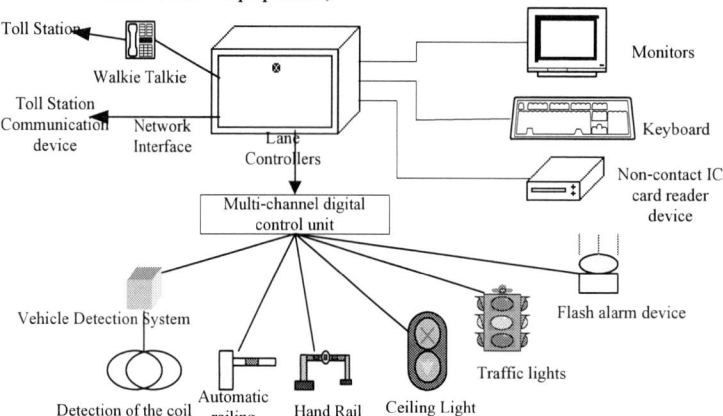

Figure 4. Formation diagram of entrance lane equipment

(2)Exit lane equipment .Exit lane equipment is the same equipment as entrance lane equipment, just increase

of printer paper, the charges for display, image acquisition equipment, which is shown in Figure 5.

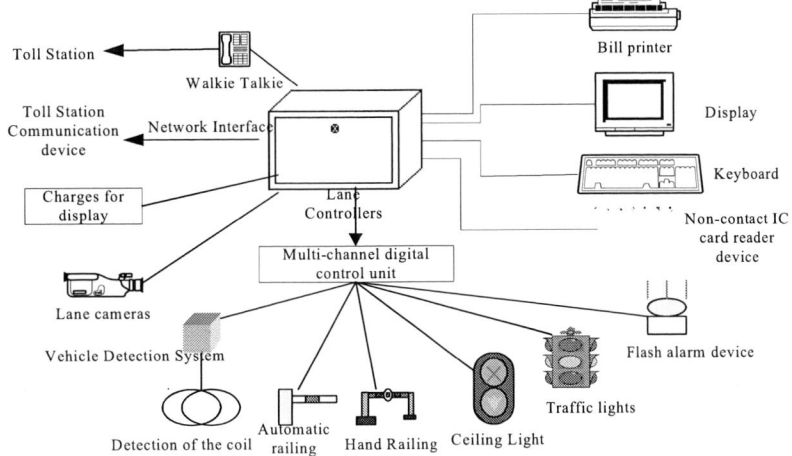

Figure 5. Formation diagram of exports lane equipment

3.1.3. IC card system of online charging system.
According to the current highway toll collection facilities, combine the highway network integration requirements, the entire road network charge settlement system use the three-level administration system. Toll Station System is a real-time management centers in the entrance and exit charge lane operation, the completion of the charges data transfer. According to interconnection charge required for solving the problems, in the Highway Authority must design IC card management centre, it manage the access cards, stored-value cards and bill business [6-7].

Figure 6 illustrates the major components of the system, card-issuing centers and settlement center represent the functions of the operating management agencies. Card-issuing centers undertake the card file management, initialization, the distribution of access cards, stored-value card issuing and management of bills. Settlement Centre bears all the collection of transaction data (card issuing stored-value of trading data and vehicular traffic data of transactions), classification of transaction data and account settlement.

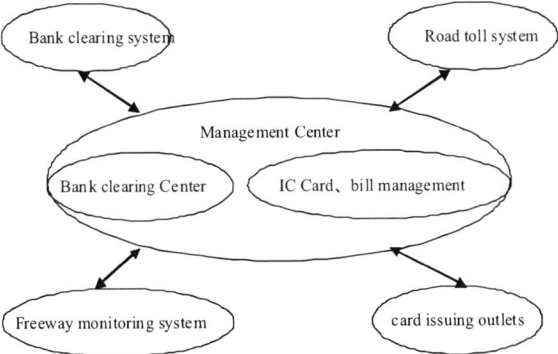

Figure 6. Formation diagram of IC card Online Charging System

(1)IC card issuance system

Toll system used IC card includes secret key card, access card, stored value cards, the staff cards and other types of cards. According to the card issuing system function and physical structure, which is divided into the secret card key system and CPU card issuance system. Secret key card issuance system in charge of the generation system by the needs of various key card system and CPU card issuance system mainly generated by the system in need of all kinds of card users, including access cards, stored value cards, the staff cards, and other types of cards . The structure of card issuing system is shown in Figure 7.

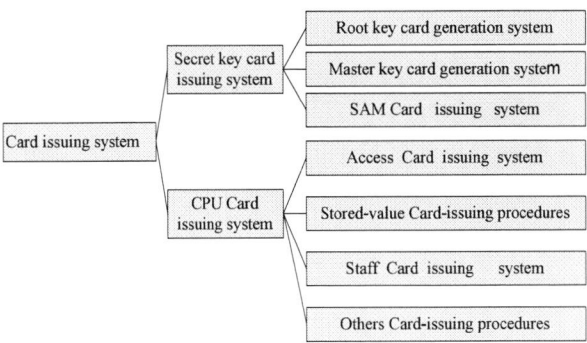

Figure 7. The structure of card issuing system

① Secret key card issuance system Secret key card issuance system provides various secret key generation mechanism systems, such as encryption algorithms. Secret key card issuance system is illustrated in Figure 8.

② Access card issuance system. All of the access card of the system to initialize, and in accordance with system requirements, it establish related data files and secret key documents. Secret key documents provide the verification of the legality of the card.

447

③Stored value card issuance systems. It is used for electronic payment cards for personal. When personal of card holders who write the basic information and account information, establish the corresponding wallet documents and secret key file documents.

(2)IC card recharge system

This part is mainly designed for electronic payment. Under the control of recharge secret key cards in the successful test the legality of the cards, stored value cards are recharged.

(3)IC Card pass systems

Current system is mainly reflected in the IC card access function. It completes access billing and charging function.

(4)IC Card Management System

IC card management refers to macroscopic management of the various types of IC cards, mainly in the course of the use of management to achieve reasonability, efficiency and security.

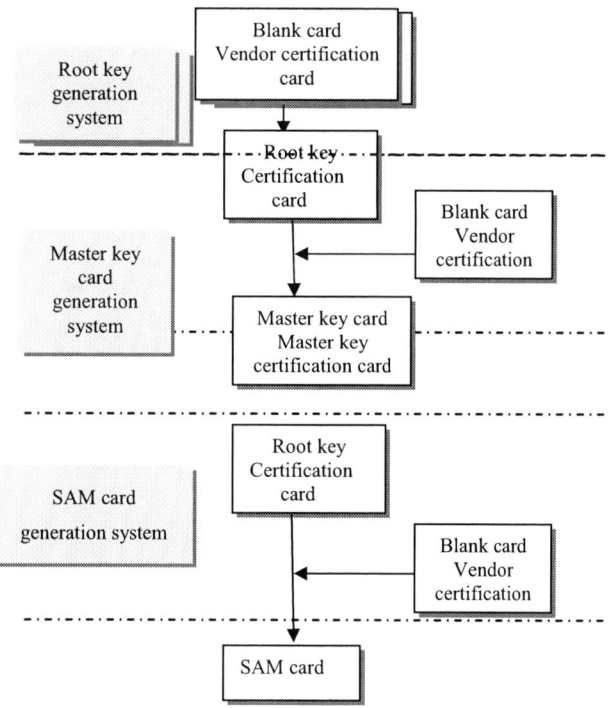

Figure 8. The structure of Secret key card system

4. Conclusion

Highway interconnection charge system as an important component part of intelligent transport system is constantly developing. It is a complex systematic project that involved traffic engineering, computer network technology and high-tech communicate and monitoring. In short, this paper put forward the solution that is in order to ensure the entire network is built a high automation, economical and practical, secure and reliable network, with the traffic and future development of adaptation. The solution also has an important reference value and forward-looking for the majority of our traffic at the current interconnection charges system.

References

[1]Herzel Yamin,,Michael Shlepakov,Chen Menachem, "Miscellaneous applications. II. Tracking Systems, Toll Collection, Oil Drilling, Car Accessories, Oceanography", *Industrial Applications of Batteries,* 2007, pp. 617-647.

[2] Huang zheng quan, "Design and Implementation of Networking Toll Collection Systems for Highways", *Beijing,Beihang University,*2003.

[3] Mike Slinn ,Paul Matthews and Peter Guest, "Intelligent Transport Systems," *Traffic Engineering Design (Second Edition),* 2005, pp 209-217.

[4] Tim Horberry, Tore J Larsson. "Forklift safety, traffic engineering and intelligent transport systems", *Applied Ergonomics, Volume 35, Issue 6,* 2004, pp 575-581.

[5]Luo guiliang, "Highway network monitoring and information management IC card system",*Jilin,Jilin University College of Computer Science,* 2006.

[6] K. W. Ogden," Privacy issues in electronic toll collection," *Transportation Research Part C: Emerging Technologies, Volume 9, Issue 2,* April 2001, pp.123-134.

[7]David Levinson,Elva Chang,"A model for optimizing electronic toll collection systems", *Transportation Research Part A: Policy and Practice, Volume 37, Issue 4,* May 2003, pp 293-314.

Congestion Pricing and Sustainable Development of Urban Transportation system

Jianhu Zheng

(Department of Automobile Engineering, Minjiang University, Fuzhou, Fujian, 350108, China)
zjianhu1028@163.com

Abstract

The rapid growth in urbanization and motorization generally contributes to an urban transportation system that is economically, environmentally and socially unsustainable. The result has been a relentless increase in traffic congestion. Road congestion pricing has been proposed many times as an economic measure to fight congestion in urban traffic, but has not seen widespread use in practice because of Some potential impacts of road pricing remain unknown. The paper first reviews the concept of sustainable transportation system, which should meet the goals of economic development, environmental protection and social justice collectively. And then, based on the characteristics of sustainable transportation system, how congestion pricing can contribute to economic growth, environmental protection and social justice is examined. Examination result shows that congestion pricing is a powerful way to promote the sustainable development of urban transportation system.

1. Introduction

Urban transportation is a pressing concern in mega cities around the world. Along with China's rapid development of urbanization and motorization, traffic jams has become a more and more serious problem, resulting in greater time delay, increase of energy consumption and air pollution, decrease of reliability of road network. In many cities traffic congestion is seen as a hindrance to economic development. Numerous methods can be used to address congestion and reduce transport density, including building new infrastructure, improving maintenance and operation of infrastructure, and using the existing infrastructure more efficiently through demand management strategies, including pricing mechanisms.

Congestion pricing has long been proposed as an effective measure to combat traffic congestion. The principle objective of congestion pricing is to alleviate congestion by implementing surcharge for the use selected congested facilities during peak time periods. By shifting some trips to off-peak periods, to routes away

from congested facilities, or to higher-occupancy vehicles, or by discouraging some trips altogether, congestion pricing schemes would result in savings in time and operating costs, improvements in air quality, reductions in energy consumption and improvements in transit productivity. There are lots of successful applications in some countries and regions in the rest of the world. Following Singapore in the early 1970s and Norwegian toll rings in the mid-1980s, the city of London introduced its area toll in February 2003; up till now, it is the most well-known example of a large metropolitan area that has implemented congestion pricing.

However, congestion pricing has not seen widespread use in practice due to theoretical and political reasons. Some potential impacts of road pricing remain unknown, and the sustainability of congestion pricing for urban development requires further study.

Sustainability is normally taken as basic objectives in the assessment of transportation policy. The idea of sustainable transportation emerges from the concept of sustainable development in the transport sector and can be defined as follows [1], "sustainable transportation infrastructure and travel policies that serve multiple goals of economic development, environment stewardship and social equity, have the objective to optimize the use of transportation systems to achieve economic and related social and environment goals, without sacrificing the ability of future generations to achieve the same goals". Sustainable transportation systems require a dynamic balance between the main pillars of sustainable development, i.e. economic development, environmental protection and social justice for current and future generations.

In the context of sustainable transport systems, how congestion pricing can contribute to economic growth, environmental protection and social justice is examined in this paper. The rest of the paper is structured as follows. In section 2, the impacts of congestion pricing on economic development are described. Section 3 presents the contribution to environment protection, and the relationship between social justice and congestion pricing is analyzed. Some conclusions are given in section 5.

2. Economic development

Traffic congestion, resulting in the increase of travel time, traffic accident, energy consumption and environment deterioration, has produced numerous economic losses. It was reported that economic loss caused by traffic congestion amounted to 40 million Yuan per year in Beijing and 1/3 of GDP in Shanghai in 2003 [2]. In many cites, traffic congestion has seen as a hindrance to economic development.

Any sustainable transportation management polices should meet the goals of improving the effects of transport on economic development, and without adversely impacting the environment and the potential for further economic growth. How effective would congestion pricing strategy be in reducing congestion, lowering pollutants and greenhouse gas emissions, cutting fuel use, and reducing other adverse impacts of current transportation system?

The basic economic principles of congestion pricing can be illustrated in Figure 1. Theoretically, individual users decide whether or not to use a particular road by weighing the costs they will to bear against the benefits to themselves. Total social benefits can be measure by the area under D in Figure 1. The user costs indicated on the MPC curve reflect only the costs borne by each user as new users (i.e. "marginal" users) are added. However, the marginal user occasions additional social costs, such as air pollution and delay to other users, which he does not bear. The total costs borne by each marginal user and the social costs occasioned by him are the marginal social costs of each trip. Marginal social costs are indicated by the MSC curve in figure 1.

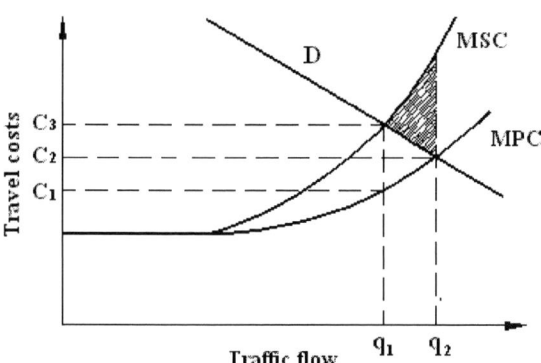

Figure 1 Effect of congestion pricing

If there are n vehicles in the transport system, and mean user cost is represented by MPC, one marginal user added will increase mean user cost to MPC+$\triangle$MPC. Thus the marginal social costs can be formulated as [3]:

$$MSC = (n+1)(MPC + \Delta MPC) - nMPC$$
$$= MPC + \Delta MPC + n\Delta MPC$$

The unregulated 'no-toll' equilibrium occurs at the intersection of MPC and D, resulting in an equilibrium

flow of q_2 and an equilibrium price of C_2. However, the social optimal is found in Figure 1 at the intersection of MSC and D, where the marginal willingness to pay for trip is C_3, and the flow is q_1, which is less than in the unregulated equilibrium. Congestion pricing makes the system optimal by charging marginal congestion cost imposed by a traveler on others [4].

Congestion pricing is an important means of transportation demand management, and initially only affects transportation decisions. Practices prove that congestion pricing can effectively regulate traffic travel time and space distribution, promote effective utilization of road resources, and enhance the efficiency of transportation operations. Congestion pricing implementation in Singapore has showed that traffic volumes decreased by 17% in peak time periods [5], and London's experience also indicated that pricing schemes was successful [6]. As we convert all the savings in travel time resulting from decreased congestion to monetary units, we can conclude that the reduction of congestion will promote sustainable economic development.

Toll will affect travelers' budget constraints and will result not only in mode switching but also in broader changes in the economy that will be accompanied by the geographic redistribution of trips. Some concerns that congestion pricing may have negative effects on the economy of the central area, particularly on retail [7]. A counterargument, however, states that the reduced congestion is supposed to lower the costs of the downtown businesses, making them more competitive. The location of retail activity, on the other hand, is also driven by individuals' preferred shopping locations. As well known, individuals tend to shop near their place of residence, so retail firms cannot easily move out of the core area, because so many people live in center area. This dependence on customer convenience explains why retail production decreases less than the output of other primary industries in the core area.

Congestion pricing reduces congestion during the peak time and increases congestion during other periods. By shifting toward public transit and high occupancy vehicles, congestion pricing reduced the number of trips to the pricing areas across all time periods. Except reduced congestion, if improvement of air quality and decrease of fuel consumption are taken into account, the economic efficiency resulting from congestion pricing would be considerable.

3. Environmental protection

In China, environmental problem become increasingly serious. It was reported that china is the second emitter of CO_2, and 7 cities are in the list of seriously-polluted cities in the world.

The environmental effects of transportation cover a wide range of different impacts, including for example air pollution, noise, and climate change. Motor vehicles are the dominant producers of urban air and noise pollution, including carbon monoxide, oxides of nitrogen, and airborne particulates. These pollutants are key factors in many respiratory ailments such as asthma as well as "a range of other human health effects, from headaches and eye irritation to cancer"[8].

The World Bank estimates that 0.5 million people in developing countries die each year from transport-related air emissions, with a similar death toll from traffic accidents[9].

Transportation cannot be replaced because it is the part of the production chain. For this reason, transportation systems must be developed and standardized, the effectiveness of transportation service must be increased, while the environmental pollution must be decreased or prevented.

Emission from road traffic is a complex system with an output that cannot be completely measured. It is natural to analyze the emissions from a sample of vehicles under different driving conditions. The California Air Resources Board pointes out that congestion-stop and go traffic significantly increases emissions. As an example, one report estimates that a 10-mile trip, using an average 1987 automobile, results in running exhaust HC emission of 1grams at a speed of 55 mph but that HC emissions would be 7 grams at an average speed of 20 mph, typical of stop-and-go conditions[10]. The relationship between emissions and speed can be seen in Table 1.

Table 1 Emissions level in different speed [11,12]

Speed (km/h)	CO (g/km)	CO_2 (g/km)	Fuel consumption (g/km)
10	4.85	310	130
20	2.80	240	96
30	2.05	200	79
40	1.70	170	65
50	1.25	150	55
60	1.15	145	48
70	1.05	138	43
80	0.95	130	40
100	0.8	110	50

With respect to air quality impacts, congestion pricing can be expected to improve air quality in two ways. The first impact comes from the reduced level of congestion on the facility, compared with the level of congestion on the un-priced road. The second impact on emissions is due to the reduced number of vehicles on the road facilities. To the extent that higher prices succeed in reducing vehicle miles traveled, there will be fewer vehicles on the road.

It is clear that reduced travel demand in peak periods reflects fewer and shorter trips being made and results in higher travel speeds during peak periods, with consequent reductions in vehicles emissions during peak periods. However, the bulk of daily travel occurring in off-peak periods may be negatively affected, as some peak travelers shift their time of travel to off-peak periods to avoid tolls. Overall beneficial impacts on air quality may thus be smaller than that suggested simply by peak period travel demand reductions. This reduction cannot be measured directly, but will have to be calculated from the information on traffic diversion, congestion reduction, and ride-sharing increase.

In the long term, congestion pricing could lead to increases in use of alternative travel modes for all daily trips, thus reducing total daily emissions.

4. Social justice

Sustainable development has three widely agreed meta-goals: sustainable economic development, environmental protection and social justice. All three goals must be addressed together if development is to be sustainable. Much attention has been paid to economic development and environmental protection, but less attention has been paid to social justice. Social justice (equity) contains economic equity and environmental equity.

Economic equity issues resulting from congestion pricing are difficult to address completely. Some people have argued that road pricing is regressive, in that it will bear more heavily on poorer car users, short distance journeys, and on those living adjacent to the cordons; more generally concern has been expressed at the impact on those who, for whatever reason, have no choice but to travel by car. These arguments can be countered to some extent by devising more flexible charging regimes. It has to be accepted that any form of road pricing will introduce some inequities. The key is to keep these to minimum, and to find ways of compensating those who do not benefit from congestion pricing. In practice, the lowest income travelers, who typically travel by public transport or on foot, are most likely to benefit.

For the conventional transportation system, the tax rates on gasoline, which are the same regardless of whether transport users are traveling during congested or un-congested periods, provide no incentive for users to use infrastructure more efficiently. Congestion pricing is expected to reduce this unfair by implementing surcharge for the use selected congested facilities during peak traffic periods.

Under congestion pricing, residents inside the cordon should pay extra charge in peak periods, and seemly have an increase in travel costs. In fact, the reduction in congestion brought by the cordons including time savings, resulting in lower travel costs to cordon residents. If congestion pricing scheme is well designed, and revenues collected through charges should be directed to local authorities and earmarked for further improvements in the transport situation and mitigation of the congestion effects. It is no doubt that congestion pricing should enhance economic equity among all infrastructure users.

Equity issues addressed in transport have been largely concerned with economic equity, including the relationship between public and private transport, the impacts of congestion pricing on peripheral areas and underprivileged population groups. However, environmental-equity issues have been little within a transport context. Transport produces direct effects such as atmospheric emissions and noise, and also indirect effects, through its influence on the location of polluting facilities and affected people.

Environmental equity refers to the social distribution of environmental quality (and specifically the distribution of NO_2 by deprivation status). That is equal access to a clean environment and equal protection from possible environmental harm irrespective of income, class or other differentiating feature of socio-economics status [13]. Transportation planners should point out that the reduction of congestion levels and increased trip speed on an entire facility will benefit users of public transit.

It will important to explain to the public that congestion pricing represents a step towards a more equitable method of paying for transportation systems. Existing transportation improvement programs are paid for by gasoline tax. Congestion pricing requires those choosing travel in peak time period to pay significantly more than those choosing travel in any other time periods. Congestion pricing also reduces inequity in exposure to NO2 with the extent of the reduction congestion areas according to the charge option. In other words, the NO_2 inequity reduces linearly with improvements in city-wide air quality and with reduction in total distance traveled on the road network.

From an environmental equity perspective, the effectiveness of congestion pricing is sensitive to the spatial distribution of socio-economic characteristics; hence the best scheme design may be different for each application city.

As can be seen, social justice is most difficult to address, and practices indicate that the revenue from road pricing plays a crucial role. If the revenue is recycle to permit investment in public transport, or maintenance of higher service levels, it helps to provide a clear alternative, which both reinforces the impact on mode choice and reduces the concern of those on whom the charge is imposed. If it is used to improve the road infrastructure, it can be seen that traffic diverted from the controlled area is being catered for. If it is used to pay for environmental improvements, it will help to reinforce the wider benefits of charging. How to use the revenue and improve the efficiency of urban transport system is play a key role in sustainable development.

5. Conclusions

With the continuing growth in traffic demand and decreasing scope for expansion of road infrastructure, traffic congestion is not a problem that will go away soon. Congestion pricing is proposed as an important demand-management strategy for tackling traffic, but still receives strong social and political opposition because of the uncertain impacts on economic development, environmental protection and social justice.

This paper reviews the relationship between congestion pricing and three goals of sustainable transportation system, and the results show that congestion pricing is a promising traffic management strategy, which can promote urban economic development, improve environmental quality and reduce inequity. Recent advances in electronic toll collection (ETC) technologies have made congestion pricing technologically feasible. It is safe to predict that coming decades will witness an increasing number of implement or attempts to implement congestion pricing. The design and evaluation of congestion pricing will require a deeper understanding of the direct and indirect impacts on sustainable development.

Acknowledgement

This work was supported by the Young Talents Project of Fujian province (2007F3078). Appreciation is also extended to reviewers for their helpful comments and suggestions on improving this paper.

References

[1] Spaethling D (1996). Sustainable transportation: the American experience. *Proceedings of seminar C, Planning for sustainability of the 24th European Transport Forum*, PTRC Education and Research Services Limited, London, September.

[2] Leilei Liu (2006). Promoting Public Transport Development for China Cities. *China Academy of Transportation Sciences*, Ministry of Communications, December (in Chinese).

[3] Li Feng, Shen Jiadong (1999). Economic analyses of urban traffic congestion. *Journal of Shanghai Tiedao University*, Vol.20, No.2, pp.1-5 (in Chinese).

[4] C. Robin Lindsey, Erik T (2000). Verhoef. Traffic congestion and congestion pricing. *Tinbergen Institute Discussion paper*.

[5] Mark Goh (2002). Congestion management and electronic road pricing in Singapore. *Journal of Transport Geography*, No.10, pp.29-38.

[6] Vickerman R (2005). Evaluation the wider economic impacts of congestion charging schemes: The limitations of conventional modeling approaches. *The 45th ERSA Congress*, Amsterdam, Netherlands.

[7] Santos G. B., Shaffer (2004). Preliminary results of the London congestion charging scheme. *Public Works, Management and Policy*, Vol.9, No.2, pp.164-181.

[8] Chertok M., Voukelatos A., Sheppeard V., etc. (2004). comparison of air pollution exposure for five commuting

modes in Sydney-car, train, bus, bicycle and walking. *Health Promotion Journal of Australia*, Vol.15, pp.63-75.

[9] World Bank (2002). City on the move: a World Bank urban transport strategy review. *Washington DC*.

[10] Air Resources Board (1989). The air pollution-transportation linkage, *state of California. Sacramento*, pp.4-5.

[11] Joumard, R (1999). Estimation of Pollutant Emissions from Transportation. *Transport Research*, Office for Special Publications of the European Communities, Luxembourg.

[12] Leon S., Diaz M. I., Mandoza B (1997). Traffic, noise and air pollution in Las Palmas de Gran Canaria: An evaluation of the effects of the ring road. *Urban Transport and the Environment for the 21st century III*, WIT Press, UK, pp.353-360.

[13] Mitchell G (2005). Forecasting environmental equity: air quality responses to road user charging in Leeds, UK. *Journal of Environmental Management*, Vol.77, No.3, pp.212-226.

Solving Traveling Salesman Problems by Genetic Differential Evolution with Local Search

Li Jian[1], Chen Peng[2], Liu Zhiming[1]

1. Department of Computer Science and Engineering, Hubei University of Education, Wuhan, 430074, China

2. College of Electrical Engineering & Information Technology, China Three Gorges University, Yichang, 43002, China

Leejan4ever@gmail.com

Abstract

To solve traveling salesman problems (TSP), a genetic differential evolution (GDE) was introduced, which was derived from the differential evolution (DE) and incorporated with the genetic reproduction mechanisms, namely crossover and mutation. The Greedy Subtour Crossover (GSX) was employed to generate an offspring to denote the difference of the parents. A modified ordered crossover (MOX) was employed to perform mutation to generate trial vector with a user defined parameter, the parameter were used to control the rates of the target vector components and the mutated vector components in the trial vector. Moreover, a 2-opt local search was implemented to enhance local search performance. GDE was implemented to the well-known TSP with 52, 100 and 200 cities with variable parameters. Based on analysis and discussion on the results, typical values of the parameters were given, with which GDE provided effective and robust performance.

1. Introduction

The simple description of TSP could be: The salesman must visit a list of cities, all the cost between every two cities are given. The salesman's task is to find the cheapest tour connecting them all, visiting each city only once, and return to the city of origin. Cost here could be distance, time, money, etc. Various problems in science, engineering, etc. can be formulated as TSP, such as vehicle routing, scheduling problems, integrated circuits designs and physical mapping problems [1]. Traveling salesman problem (TSP) is a NP-hard problem [2] whose computational complexity rises exponentially by increasing the number of nodes.

Many researches have been done for solving TSP. Tsai proposed an evolutionary algorithm called the heterogeneous selection evolutionary algorithm [1]; Nguyen described a hybrid genetic algorithm [3]; Baraglia combined genetic and local search heuristics as a hybrid heuristic [4]. There are also researches on PSO algorithm for solving TSP. Clerc proposed discrete particle swarm optimization (DPSO) algorithm, but it is

poor in performance than other algorithms, even in some simple TSP [5].

The traditional DE is based on the real valued operators and doesn't suit TSP, a combination optimization problem. An approach for TSP with differential evolution (DE) with a position-order encoding method (POEM) was introduced, where the vectors were calculated with DE and then sorted to provide position order to present the tour [6]. The method was implemented to the cases with 16 and 48 cities, and failed to find the optimal tour for the larger case in some runs.

To solve TSP, a genetic differential evolution (GDE) was introduced in the paper, which was derived from DE and incorporated with the genetic reproduction mechanisms, namely crossover and mutation. The Greedy Subtour Crossover (GSX) was employed to generate an offspring to denote the difference of the parents. And a modified ordered crossover (MOX) was employed to implement the mutation and crossover based on three mutually different individuals. Moreover, a 2-opt local search was implemented to enhance local search performance. The approach was implemented to the case with 52, 100 and 200 cities and the simulation results have shown its feasibility and effectiveness.

2. Traditional Differential Evolution

The tradition differential evolution (DE) is a simple population based stochastic parallel search evolutionary algorithm for global optimization [6]. In DE, the population consists of real valued vectors with dimension D that equals the number of control variables. The size of the population is adjusted by the parameter Np. The population of a DE is randomly initialized within the initial parameter bounds. The optimization process is conducted by means of the three main operations: mutation, crossover and selection. In each generation, each individual of the current population becomes a target vector. For each target vector, the mutation operation produces a mutant vector, by adding the weighted difference between two randomly chosen vectors to a third vector. The crossover operation generates a new vector, called trial vector, by mixing the parameters of the mutant vector with those of the target vector. If the trial vector obtains a better fitness value than the target vector, then trial vector replaces the target vector in the next generation. The evolutionary operators are described below.

1) Mutation

DE relies upon the population itself to perturb the vector parameter. For each $i \in [1,...,N_p]$, the weighted difference of two randomly chosen population vectors, X_{r2} and X_{r3}, is added to another randomly selected population member X_{r1} to build a mutated vector V_i.

$$V_i = X_{r1}^t + F(X_{r2}^t - X_{r3}^t) \tag{1}$$

In Eq.(1), i, $r1$, $r2$ and $r3$ are mutually different indices from the current generation t. F is the user defined parameter, called step size, which is typically chosen from the range [0, 2]. If V_i is outside variable limit, it will then be fixed to the upper or lower limit.

2) Crossover

A trial vector u_i is created incorporating the mutated vector V_i and the target vector X_i.

$$u_i = {}_{i,k}^{t+1} = \begin{cases} v_{i,k}, if(rand_{k,i} \le CR \parallel k = I_{rand}) \\ x_{i,k}^t, if(rand_{k,i} > CR \&\&k \ne I_{rand}) \end{cases} \tag{2}$$

Where $rand_{k,i} \in [0,1]$ and I_{rand} is chosen randomly from the interval $[1,...,D]$ once for each vector to ensure that at least one vector component originates from the mutated vector V_i. Eq.(2) is applied for each vector component. CR is the DE control parameter, called the crossover rate, and is a user defined parameter within range [0, 1]. Trial parameter with randomly chosen index, I_{rand}, is taken from mutant vector to ensure that the trial vector does not duplicate X_i.

3) Selection

If the trial vector u_i has an equal or better objective function value than the target vector X_i, it replaces the target vector in the next generation.

$$X_i^{t+1} = \begin{cases} u_i, if(f(u_i) \le f(X_i^t)) \\ X_i^t, otherwise \end{cases} \tag{3}$$

3. The genetic Differential Evolution for TSP

DE is based on the real valued operators, so it does not suit combination optimization problems. To solve TSP, this paper introduced a genetic differential evolution (GDE), which was derived from DE and incorporated with the genetic reproduction mechanisms, namely crossover and mutation.

1) Coding

For the vector $X_i^t = (x_{i,1}^t, x_{i,2}^t,...,x_{i,D}^t)$, d is the number of all cities $x_{i,d}^t$ denotes the d_{th} city.

2) Modified ordered crossover

A modified ordered crossover (MOX) was employed to cross two vectors, which was derived from the ordered crossover (OX) [7]. In MOX, a sequential segment of a source tour with user defined length *Len* is added to a target vector, and then the duplicate cities in the target vector will be removed, which is the crossover operator in GDE. The MOX is described as follows.

Given two vectors r1 and r2, which are target tour and source tour, respectively, and r1 = [1, 2, 3, 4, 5|, 6, 7, 8, 9, 10]; r2 = [1, 3, 7, 10, 9, 6, 2, 8, 5, 4]; *Len* is set to 4.

Firstly, a city in r1 is selected randomly, for example city 5, and then the sequential 4 cities behind city 5 in r2 are selected as [4, 1, 3, 7] (if it meets the end of the tour, keep on selecting from the head of the tour). Secondly, r1 takes the city behind city 5 as its head and then r1 is rebuilt as [6, 7, 8, 9, 10, 1, 2, 3, 4, 5]. Thirdly, the duplicate cities both in the segment and r1 are removed from r1 as [6, 8, 9, 10, 2, 5]. At last the segment is added to r1 as [4, 1, 3, 7, 6, 8, 9, 10, 2, 5]. And then in MOX, the parameter *Len* defined the crossover rate of the two tours, which is similar to *CR* in traditional DE.

3) Mutation

In GDE, the mutated vector is built with a target vector and the crossover of the two vectors with the Greedy Sub tour Crossover (GSX) [8]. In GSX, a greedy heuristic rule is employed, where from an arbitrary city, unselected cities in both parents will be inserted to the offspring on each side of the city one by one, respectively, hence the edges of the parents are segregated each other. Similar to Eq.(1), i, $r1$, $r2$ and $r3$ are mutually different indices from the current generation t. And an offspring vector p_i is obtained as $p_i = GSX(X_{r2}^t, X_{r3}^t)$, which denotes the *difference* of the two tours. After that, the mutated vector is obtained as $V_i = MOX(X_{r1}^t, p_i, Len_1)$, where X_{r1}^t is the target tour and p_i is the source tour in MOX. Len_1 works as F in Eq.(2), which defines the rates of X_{r1} and the *difference* in V_i.

4) Crossover

After the mutation, a crossover based on MOX is implemented as $u_i = MOX(X_i^t, V_i, Len_2)$. Len_2 works as CR in Eq.(3) to control the rates of vector components from target vector and mutated vector. Len_2 is a user defined parameter within the range $[1,...,D]$.

5) Local *Search*

To enhance local search performance, for each u_i, a 2-opt local search is performed. The 2-opt algorithm is a special case of the k-opt algorithm [9], where in each step k links of the current tour are replaced by k links in such a way that a shorter tour is achieved. In other words, in each step a shorter tour is obtained by deleting k links and putting the resulting paths together in a new way, possibly reversing one ore more of them.

6) Selection

The selection of GDE is the same with DE, if the tour obtained after local search has an equal or short distance than the target vector X_i, it replaces the target vector in the next generation.

In GDE, the definitions of the real valued mutation and crossover operators are modified for TSP, while the concept of GDE is the same with DE. Firstly, its mutation does not use a predefined probability density function to generate perturbing fluctuations. Secondly, it relies upon the population itself to perturb the vector parameter. Len_2 works as CR to select components from the target vector and the mutated vector. At last, its selection operator is the same with DE.

4. Numerical Simulations

To validate the feasibility and effectiveness of the proposed approach, GDE with N_p=30 and 100 generations have been implemented to TSP cases with 52, 100 and 200 cities [10],for 30 independent runs where Len_1 and Len_2 were set to variable percents of the dimensions of the cases. The simulation results are listed in table 1 to table 3, where Opt is the known optimal value of the case, Best and Worst are the best and worst values provided by GDE, Std is the standard deviation, EG is the mean effective generation, Rate is the success rate of the algorithm find the optimum, Err is the mean error of the runs.

It can be seen from Table 1, with various values of Len_1 and Len_2, GDE has consistently found the optimum for berlin52. And with each value of Len_1, the effective generation with $Len_2 = 25$ is the minimum. And with the same Len_2, the effective generation with $Len_1 = 25$ is also the minimum.

Table 1. Results of berlin52

Len1 (%)	Len2 (%)	Opt	Best	Err (%)	Worst	Std	EG	Rate (%)
25	25	7542	7542	0	7542	0	**16.0**	100
	50	7542	7542	0	7542	0	21.3	100
	75	7542	7542	0	7542	0	16.3	100
50	25	7542	7542	0	7542	0	**21.2**	100
	50	7542	7542	0	7542	0	29.1	100
	75	7542	7542	0	7542	0	26.1	100
75	25	7542	7542	0	7542	0	**21.3**	100
	50	7542	7542	0	7542	0	27.2	100
	75	7542	7542	0	7542	0	23.9	100

Table 2. Results of kroA100

Len1 (%)	Len2 (%)	Opt	Best	Err (%)	Worst	Std	EG	Rate (%)
25	25	21282	21282	0	**21282**	**0**	**59.4**	100
	50	21282	21282	0.036	21356	20.2	64.4	80
	75	21282	21282	0.13	21431	43.5	73.2	60
50	25	21282	21282	**0.088**	**21433**	**37.1**	**79.4**	63.3
	50	21282	21282	0.32	21516	64.9	86.5	13.3
	75	21282	21282	0.32	21577	81.6	80.2	26.7
75	25	21282	21282	**0.081**	**21368**	**26.3**	**77.1**	46.7
	50	21282	21282	0.19	21542	53.3	78.3	40.0
	75	21282	21282	0.41	21694	113.7	80.5	23.3

Table. 3. Results of kroA200

Len1 (%)	Len2 (%)	Opt	Best	Err (%)	Worst	Std	EG	Rate (%)
25	25	29368	29662	**1.5**	**30199**	**137.7**	89.4	0
	50	29368	29733	2.2	30629	198.1	**84.9**	0
	75	29368	**29634**	2.5	30661	260.2	91.1	0
50	25	29368	**29725**	**2.4**	30454	193.3	89.7	0
	50	29368	29889	3.7	30950	241.8	82.6	0
	75	29368	29748	3.1	31106	336.6	**82.5**	0
75	25	29368	29867	**2.5**	**30454**	**146.7**	82.7	0
	50	29368	**29643**	2.9	30976	282.9	**82.5**	0
	75	29368	29644	2.7	30661	275.1	89.3	0

Table 4. Results of kroA100 with new values of the parameters

Len1 (%)	Len2 (%)	Opt	Best	Err (%)	Worst	Std	EG	Rate (%)
5	5	21282	21282	0	21282	0	61.4	100
10	10	21282	21282	0	21282	0	50.0	100
15	15	21282	21282	0	21282	0	**47.3**	100
20	20	21282	21282	0	21282	0	55.0	100

Table 5. Results of kroA200 with new values of the parameters

Len1 (%)	Len2 (%)	Opt	Best	Err (%)	Worst	Std	EG	Rate (%)
5	5	29368	**29420**	**1.0**	**29878**	111.8	88.7	0
10	10	29368	29532	1.2	29913	**100.2**	89.4	0
15	15	29368	29553	1.2	30034	108.2	**88.1**	0
20	20	29368	29469	1.5	30000	130.8	88.2	0

And the similar phenomena can be observed in Table 2 for kroA100, where with the same value of Len_1 or Len_2, the less the values of Len_2 or Len_1 the better the results were obtained in all the terms. And with the minimum values of Len_1 and Len_2, all the runs converged to the optimum, while with the bigger values some runs failed.

In Table 3, all the runs failed to found the optimum for kroA100, while the mean errors are no more than 4%, which proved the feasibility and effectiveness of GDE. Similar to the results in table I and table II, for each value of Len_1, GDE provided the best result in Err, Worst and EG term with Len_2=25 which proved that with a less Len_2 the algorithm is more effective and robust. And on the other hand, for each Len_2, with the lowest Len_1, GDE provided the best or comparative results for the terms.

It can be observed from Table 4, for kroA100, with Len_1 and Len_2 being set to lower values, GDE has consistently found the optimum. And by comparisons with the results in Table 2 with Len_1 and Len_2 equal to 25, the new values of the parameters provided better results, except those of Len_1 and Len_2 equal to 5. For kroA200, with the new values of parameters, the results were improved remarkably, and with Len_1 and Len_2 equal to 5, GDE provided the best results in the terms of Best, Err and Worst, which are also better than those in Table 3. And for the Std term, the results in Table 5 are all better than those in Table 3, which proved that with the new parameter values, the approach was more robust and consistent. Moreover, the best tour obtained by GDE is very close to the optimum.

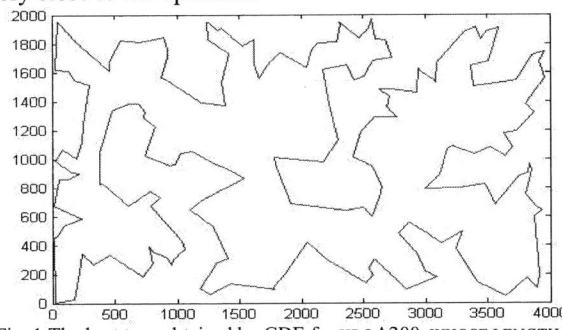

Fig. 1 The best tour obtained by GDE for KROA200, WHOSE LENGTH IS 29420

With the analysis and discussion above, GDE was proved to be feasible and effective for TSP. With proper parameter values, GDE presented robust and consistent performance. But different from DE, where the values of F and CR can be set to a very wide range, in GDE the value of Len_1 and Len_2 should be set to a narrower range as [0.05, 0.25] of the dimensions of the cases. And the values are objective dependency. For those low dimensions cases, the parameters can be set to greater, while for the cases of high dimensions the values should be set to less and about 5%, because GDE is a genetic coding based algorithm, for TSP, if too many sequential segments of a tour are broken, it may cause the loss of high quality genetics.

5. Conclusion

This paper introduced a novel genetic differential evolution (GDE) to solve traveling salesman problems (TSP), which was derived from the differential evolution (DE) and incorporated with the genetic reproduction mechanisms, namely crossover and mutation. The Greedy Subtour Crossover (GSX) was employed to generate an offspring to denote the difference of the parents. A modified ordered crossover (MOX) was employed to perform mutation to generate trial vector with a user defined parameter, the parameter were used to control the rates of the target vector components and the mutated vector components in the trial vector. Moreover, a 2-opt local search was implemented to enhance local search performance. GDE was implemented to the well-known TSP cases with 52, 100 and 200 cities with variable parameters. Based on analysis and discussion of the results, typical values of the parameters were given, with which GDE provided effective and robust performance.

References

[1] H. K. Tsai, J. M. Yang, Y. F. Tsai, and C. Y. Kao, "An Evolutionary Algorithm for Large Traveling Salesman Problems," IEEE Transactions on SMC -Part B, Vol. 34, No. 4, pp 1718-1729, Aug. 2004

[2] P. Crescenzi, and V. Kann, (1998, Aug) "A Compendium of NP Optimization Problems," [Online]. Available : http://www.nada.kth.se/theory/compendium/

[3] H. D. Nguyen, I. Yoshihara, K. Yamamori, and M. Yasunaga, "Implementation of an Effective Hybrid GA for large-Scale Traveling Salesman Problems," IEEE Transactions on SMC -Part B, Vol. 37, No. 1, pp 92-99, Feb. 2002

[4] R. Baraglia, J. I. Hidalgo, and R. Perego, "A Hybrid Heuristic for the Traveling Salesman Problem," IEEE Transactions on Evolutionary Computation, Vol. 5, No. 6, pp 613-622, Dec. 2001

[5] M. Clerc, "Discrete particle swarm optimization," New Optimization Techniques in Engineering, Springer-Verlag, 2004, pp. 219-240

[6] Storn Rainer, Price. "Differential evolution- a simple and efficient adaptive scheme for global optimization over continuous spaces ," Technical Report TR-95-012, ICSI, 1995.

[7] Soak, Sang-Moom. "New genetic crossover operator for the TSP," Lecture Notes in Artificial Intelligence, 2004, Vol. 3070, pp 480-485.

[8] Hiroaki Sengoku, Ikuo Yoshihara. "A fast TSP solver using GA on Java,"[EB/OL]. http://www.gcd.org/sengoku/docs/arob98.pdf

[9] Muyldermans, Luc. "Exploring variants of 2-opt and 3-opt for the general routing problem," Operations Research, 2005, 53(6), pp 982-992.

[10] http://elib.zib.de/Pub/Packages/mptestdata/tsp/tsplib/tsp/index.html /[OL]

A Vehicle Scheduling Model and Efficient Algorithm for Single Bus Line

Jingqi XU[1], Haode LIU[2], Jing TENG[1]

[1]*School of Transportation Engineering,*
Tongji University, Shanghai, 200092, P.R. China
[2]*Department of Surveying and Geo-informatics,*
Tongji University, Shanghai, 200092, P.R. China
njxxjq@163.com, beuture@163.com, tengjing@263.net

Abstract

Based on analyzing the past research on vehicle scheduling in transit scheduling, a new optimization model to minimize the total vehicle operation cost is proposed. This optimization model is a nonlinear integer programming and can be solved rapidly with the global solver in LINGO, as long as the appropriate value of fleet size. It is the advantage of the optimization model and solution methodology in searching the global optimum solution quickly. Finally, the scientific and reasonableness of them is proved by experimental data in the paper.

1. Introduction

Scheduling is one of the most important programs in transit operation. And it is classified as headway-based and schedule-based. Headway-based method means only the departure time of first-cycle vehicle should be predetermined. Following trips will be arranged following the return order and the headway. Generally, this method is applied in complex and heavy traffic condition. And schedule-based method means that each vehicle runs according to the predetermined schedule. [1] The schedule-based vehicle scheduling in transit planning can offer higher-level services under the condition of bus priority measures and will be discussed in this paper. An optimization model of vehicle scheduling is established to minimize the operation cost of the fleet. According to investigation, fewer transit operations are using optimization theory and computer aided system to schedule vehicles in developing countries. The manual method of vehicle scheduling is time-consuming and error-prone. And little changes will bring about wasteful duplication of effort. Therefore, bus operation departments need an optimization model and practical algorithm of vehicle scheduling deeply, which not only take account of transit operation cost, but also calculate accurately and rapidly as well.

2. Model formulation

2.1. Review

The problem of vehicle scheduling in a single-depot for single bus line is complex, and considerable effort is devoted to solve it in an exact way. Ozdemir pointed out that scheduling vehicles can also be defined as set covering problem (SCP) or set partitioning problem (SPP).[2] Haghani modeled single depot vehicle scheduling with route time constraints (SDVSRTC) based on continuous trips and SCP model. In order to improve efficiency, the multi-depot was converted into single-depot by means of trips processing. [3] Valouxis and Housos established column generation and quick shift combination (CGQS) to solve the model.[4] Jan-Ming Su and Wen-Sung Yu apply a genetic algorithm with a new method which can execute without trips over-cover or uncover in the process of crossover and mutation on the problem. In this algorithm, they use a weighted Hölder norm in the fitness function to combine different objectives. [1] Most studies solved the model by heuristic algorithms, especially the genetic algorithm. However, this paper will try to establish the appropriate model and solve it in the global solver of LINGO.

2.2. Model formulation

This paper focuses on putting forward the vehicle scheduling optimization model which can be solved in the global optimum solution. For illustrating easily, only the timetable in a morning peak is used here.

2.2.1. Hypotheses of model:

(1) The time unit of the model is minute and the model is accurate to minute.

(2) All the bus vehicles are the same type.

(3) There is only one terminal on the route. It means that bus vehicle runs in the cycle of the route.

(4) The timetable of the route is given.

(5) The travel time and speed of single journey (up or down) are constant in the certain period of time.

(6) The minimum dwell time at the depot is constant.

(7) The fleet size in the model takes no account of vehicle repair and maintenance.

(8) The vehicle operation cost is constituted by three parts. Firstly, travel cost (direct ratio to travel distance); secondly, cost of the dwell time at the depot (direct ratio to dwell time); last, fixed cost of each vehicle.

2.2.2. Model proposing:

(1) Objective function:

The objective is to minimize the fleet operation cost. When the timetable of the route is given, travel cost is constant. Thus, the objective function regardless of the travel cost is

$$\min Z = \sum_{i=1}^{I} \sum_{j=1}^{J} \sum_{k=1}^{K} C_{ijk} y_{ijk} + K C_0$$

Where: Z = value of the objective function; i = the event of ending a trip at time a_i; j = the event of the start of a trip at time b_j; C_{ijk} = the cost of the dwell time at the depot; C_0 = the fixed cost of each vehicle; $I = J$ = the number of the trips; and K = the fleet size.

(2) Independent variable [5]:

$$y_{ijk} = \begin{cases} 1, & \text{Ending } a_i \text{ is connecting to Start } b_j \text{ by Vehicle } k. \\ 0, & \text{otherwise.} \end{cases}$$

(3) Constrains:

▲Vehicles arrangement: Each vehicle should be arranged no more than once at any time.

$$\sum_{i=1}^{I} \sum_{j=\alpha}^{<\alpha+\Omega+\Delta>} y_{ijk} \leq 1 \quad \left(k=1,2,\cdots,K\right)$$

Where: α = any time; Ω = travel time of single trip; and Δ = minimum dwell time at the depot.

▲Trips implementation: No trip is over-cover or uncover.

$$\sum_{i}^{I} \sum_{k=1}^{K} y_{ijk} = 1 \quad \left(j=1,2,\cdots,J\right)$$

(4) Coefficient calibration:

▲Cost of the dwell time at the depot (C_{ijk}):

$$C_{ijk} = \begin{cases} c \cdot \left(b_j - a_i\right) & b_j - a_i > 0 \\ +\infty & b_j - a_i < 0 \end{cases}$$

Where: c = coefficient of dimension change.

▲Fixed cost of each vehicle (C_0):

$$C_0 \gg C_{ijk}$$

The exact value of C_0 won't affect the solution of the optimization model, other than the value of objective function.

▲Fleet size (K):

The value of K will directly determine that the optimization model can be solved or not. And the following expression is recommended to calculate K.

$$K = \left\lceil \frac{\Omega+\Delta}{T/J} \times 1.2 \right\rceil$$

Where: T = length of the required period of time; and $\lceil \; \rceil$ means the top integral function.

(5) Other input: the timetable of the single route during a required period of time.

3. Solution methodology

The optimization model above is 0-1 nonlinear integer programming with single objective. Therefore, the global optimum solution must exist, as long as feasible domain of this model exists.

In view of the above-mentioned facts, the global solver in LINGO is highly recommended in this condition.

LINGO is a comprehensive tool designed to make building and solving linear, nonlinear and integer optimization models faster, easier and more efficient. LINGO provides a completely integrated package that includes a powerful language for expressing optimization models, a full featured environment for building and editing problems, and a set of fast built-in solvers. [6]

Nonlinear solvers employing methods like successive linear programming (SLP) or generalized reduced gradient (GRG) return a local optimal solution to an NLP problem. However, many practical nonlinear models are non-convex and have more than one local optimal solution, such as this optimization model. The global solver in LINGO guarantees globally optimal solutions to general nonlinear problems with continuous and/or discrete variables.

The LINGO global optimization procedure (GOP) employs branch-and-cut methods to break an NLP model down into a list of subproblems. Each subproblem is analyzed and either a) is shown to not have a feasible or optimal solution, or b) an optimal solution to the subproblem is found, e.g., because the subproblem is shown to be convex, or c) the subproblem is further split into two or more subproblems which are then placed on the list. Given appropriate tolerances, after a finite, though possibly large number of steps a solution provably global optimal to tolerances is returned. Traditional nonlinear solvers can get stuck at suboptimal, local solutions. This is no longer the case when using the global solver.

4. Numerical experiment

The experiment data came from Bus No.210 (from West Beijing Rd. to Yonghe New Village in Shanghai). The cycle running time of the route is 70min during morning peak (7 a.m. – 9 a.m.) and the minimum dwell time at the depot is 6 min. And Table 1 shows the timetable of the route during one morning peak.

Used the optimization model and solution methodology above, model parameters should be demarcated here.

$I = J = 31$, $\Omega = 70 \min$, $\Delta = 6 \min$, $T = 180 \min$,

$$K = \left\lceil \frac{\Omega + \Delta}{T/J} \right\rceil \times 1.2 = 16, \ C_0 = 10000, \ c = 1.$$

Table 1. Input data - Timetable

7:02	8:16	9:28
7:06	8:22	9:36
7:14	8:24	9:40
7:24	8:28	9:42
7:34	8:32	9:44
7:38	8:34	9:48
7:44	8:44	9:54
7:46	8:52	9:58
7:56	8:56	10:06
8:02	9:06	
8:10	9:16	

Space lacks for a detailed description of it. Only the main part of the program in LINGO is shown as following:

```
----------------------------------------------
MIN=@SUM(TRIPS:COST*Y)+FIXEDCOST*@SU
M(NUMBER:VEHICLE);
@FOR(TRIPS:@BIN(Y));
@FOR(NUMBER:@BIN(VEHICLE));
@FOR(TIMES(J):@SUM(TRIPS(I,J,K):Y)=1);
@FOR(NUMBER(K):VEHICLE(K)=@IF(@SUM(T
RIPS(I,J,K):Y) #EQ# 0,0,1));
@FOR(NUMBER(K):@FOR(TIMES(L):@SUM(TRI
PS(I,J,K)|J #GE# ENDSTART(L) #AND# J #LE#
INTERVAL(L):Y)<=1));
----------------------------------------------
```

The computer with a Pentium 4 CPU 2.93GHz and 521M Ram is used to complete process.

The calculation time is about half an hour, and the Solver Status box shows that the model is PINLP (pure integer nonlinear programming) and shown as Figure 1.

The state of the solution is global optimum and the value of objective function （General Cost) is 1.30141×10^7. No constraints in the model are violated. The number of iterations completed thus far by LINGO's solver is 5295.

To validate the optimization model and the solution, $K=20$, 24 is substituted into the model. The results are the same. But compared with the solver ($K=16$) above, it is found that the elapsed time is two times or more long. This means that the value of K should be close to the minimum fleet size, the more the better. The empirical value is also feasible, but the random value is forbidden, because it may result in non-solution.

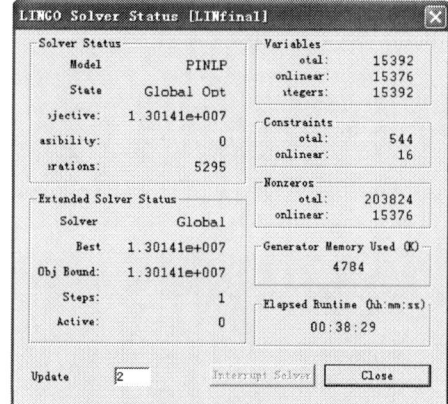

Figure 1. Solver Status (K = 16)

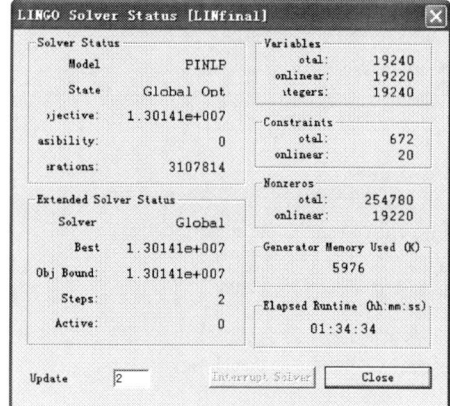

Figure 2. Solver Status (K = 20)

Table 2. Result - Running table

Veh. No.	1	2	3	4	5	6	7	8	9	10	11	12	13	14
Departure Time	07:02	07:06	07:14	07:24	07:34	07:38	07:44	07:46	07:56	08:02	08:10	08:16	08:24	08:28
	08:22	08:34	08:32	08:44	08:52	08:56	09:06	09:28	09:16	09:36	09:42	09:48	09:40	10:06
	09:44	09:58	09:54	N/A	N/A	N/A	N/A	N/A	N/A	N/A	N/A	N/A	N/A	N/A

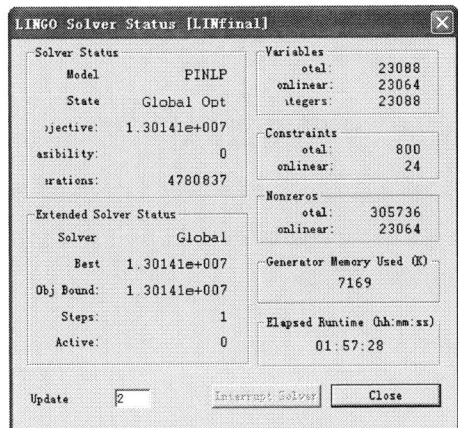

Figure 3. Solver Status (K = 24)

According to the results, the vehicle running scheme can be finished as Table 2, from which it can be seen that the foremost three vehicles need to finish three trips, and the others finish two trips.

5. Conclusion

In this paper, an optimization model of vehicle scheduling is established, which is to minimize the total cost of a single line's operation. And the global solver in LINGO is used to solve the model. It is shown that this problem is a pure integer nonlinear programming and can be solved to get the optimal solution. After comparing the calculation results, we find that choosing appropriate fleet size (the value of K) is favorable to improve algorithm efficiency of the model. In a word, the model of vehicle scheduling in a single-depot for single bus line and the global solver in LINGO are extremely effective in the engineering application of vehicle scheduling. However, some problems are not solved in my research and need be studied in the future. Firstly, as an important objective, optimization with different vehicle dispatching mode (cycle running in this paper) should be considered and corresponding constraints need be complemented. Secondly, how to ensure the application of the model and solution methodology need be studied in detail, such as translating them into optimization module in application software and so on.

Acknowledgments

The research was supported by the National Natural Science Foundation of China "Study on Theory and Methodology of Transit Coordination Dispatching for Transfer Hub" (Grant No. 70601022) and Program for Young Excellent Talents in Tongji University. And the experimental equipment used in Numerical Experiment was provided by Key Laboratory of Road and Traffic Engineering of ministry of education, Tongji University.

References

[1] Jan-Ming Su, Wen-Sung Yu, "Single-depot bus drivers and vehicles scheduling problem", *Transportation Planning Journal*, Taipei, Vol.35 No.2, 2006, pp.131-158.

[2] H.T. Ozdemir, "Graph Based Evolutionary Algorithms for Transportation Problems", Ph.D. Dissertation, Department of EECS, Syracuse University, 2001.

[3] A. Haghani, M. Banihashemi, and K.H. Chiang, "A Comparative Analysis of Bus Transit Vehicle Scheduling Models", *Transportation Research Part B*, Vol.37, 2003, pp. 301-322.

[4] C. Valouxis, and E. Housos, "Combined Bus and Driver Scheduling", *Computers and Operations Research*, Vol.29, 2002, pp. 243-259.

[5] A. Ceder, "Urban transit scheduling: framework, review and examples." *Journal of Urban Planning and Development*, ASCE, Vol.128 No.4, 2002, pp.225-244.

[6] LINOO SYSTEMS INC., "*LINGO User's Guide*". http://www.lindo.com

[7] Xin-sheng Yuan, Da-hong Shao , and Shi-lian Yu, "*Application of LINGO and EXCEL in Mathematic Modeling*", Science Press, Beijing, 2007.

Stress Test of Road Network Capacity with Cube Software

Li Li

Transportation Research Center of Beijing University of Technology, Beijing, China
bluesabrina@emails.bjut.edu.cn

Yu Quan

Transportation Research Center of Beijing University of Technology, Beijing, China
yuquan@bjut.edu.cn

Bian yang

Transportation Research Center of Beijing University of Technology, Beijing, China
bianyang@bjut.edu.cn

Yi Ping

The University of Akron, USA
pyi@uakron.edu

Abstract

As one of important issues in transportation planning, capacity of road network offers theoretical support to the quality evaluation of urban network in traffic control and management. This paper investigates the capacity of road network from a new perspective in which the experimental traffic engineering concept is adopted through macro traffic simulation and planning with the Cube software. By considering the impact factors on roadway capacity and evaluation index system of road network operational quality, we carried out stress tests over the roadway network and developed a process for conducting such tests. The Cube based process was then used in a field test in the Huairou District, Beijing, where the capacity of the roadway network has been estimated.

1. Background

Urban network is an essential carrier for transportation; this important infrastructure affects traffic system operation and the social economic development. The rapid economic development leads to accelerated urbanization and improved transportation mobility. For example, in Beijing, the total number of motor vehicles in recent years increased dramatically. Up to March 2008, the total number of motor vehicles had reached 3.2 million and still grows with the rate of 1000 vehicles per day. Because the increase in roadway capacity is far less than in the travel demand, this problem puts a tremendous pressure on the roadway network, resulting in increased traffic congestion, frequent accidents, level of service reduction, and increased travel inconvenience for citizens.

Needless to say, it is critical to plan urban network and determine roadway capacities to balance traffic demand and supply.

The stress test method, which is based on macro traffic simulation and planning software-Cube, uses the idea of experimental traffic engineering to study network capacity. Unlike the traditional modeling approach, Cube based approach relies on experiments to analyze the actual traffic problems. In this way, the open-loop and static problem analysis process used in the past is replaced by a closed-loop cycle that studies the dynamics of the system and improves the reliability of the findings.

2. Experimental traffic engineering

Experimental traffic engineering is defined as using traffic simulation technology to test traffic project designs in the computer platform and momentarily evaluate and optimize them through extraction and analysis of real-time data [1]. This new concept is presented in recent years by Professor Sasaki, the University of Kyoto in Japan, in an academic lecture in China [2].

3. Stress test of roadway network capacity using Cube

In order to find a research method which is practical, accurate and innovative in studying capacity of the roadway network, this research applies the experimental traffic engineering concept to establish the stress test method on the Cube platform.

3.1 Conception of Stress Test

978-0-7695-3342-1/08 $25.00 © 2008 IEEE
DOI 10.1109/PEITS.2008.62

Stress test is defined as building the test platform similar to the actual traffic environment and constantly increasing traffic flow load (pressure) to the road network until it reaches the paralysis state. The test exams the roadway network operating conditions under different levels of stress and the capacity combinations.

Stress test is a special form of performance test, like destructive test in other engineering areas, and its purpose is to find the bottleneck affecting the system capacity by continuously increasing load until reaching resources saturation to degrade the performance of the tested roadway network or to induce fault [3].

3.2 Introduction of Macro Traffic Simulation and Planning Software-Cube

Stress test in the field of finance or software development often requires use of an appropriate tool to complete the usually large-scale and complex test task. Its application in the field of transportation, however, is just starting to emerge. At present, there is not any professionally developed software tool designed specifically for stress test of the roadway network. However, by following the experimental traffic engineering approach to set up the purpose and analyze the results of stress test, we can develop a Cube based macro traffic simulation and planning tool to obtain the needed comprehensive functions, good stability, and wide applicability of stress test.

The developer of Trips, MVA, cooperated with the Urban Analysis Group in California and established Citilabs, the largest center of transportation software system development and application in the world [4]. The lab has developed a macro traffic simulation and planning software package-Cube, which integrated a number of traffic analysis software models, such as Trips, TP+, Voyager, Cargo, etc.

Cube is the latest multi-purpose transportation simulation and planning software, in which, Cube Base, the core component, is similar to Microsoft Windows® interface by integrating all other parts in Cube System into a complete, systemic modeling and analysis tool [5]. Cube includes network building model, four-stage process model and output result model, and can independently accomplish all the tasks of stress test to achieve the objective and requirements of testing roadway network capacity.

3.3 Process of Stress Test

As a testing technology, stress test should follow the specifications and requirements of the project during the implementation course [6] to ensure that the test on capacity of roadway network yields practical and reliable results. Experiences in stress test in software and financial systems help us establish a stress test process based on Cube shown in Figure 1.

3.3.1 Investigation of Roadway Network and Traffic Conditions After identifying the roadway network to be analyzed, the first step of stress test is collecting various roadway and traffic data according to key factors affecting capacity of road network, laying the foundation for testing capacity. These data include roadway conditions, traffic conditions, and effectiveness of network operations.

3.3.2 Selection of Evaluation Index of Network Operating Quality Using stress test to study capacity of roadway network is an experimental analysis process. In this process, selection of the evaluation index is particularly important because it can determine whether a traffic flow has reached capacity level during the test. According to the characteristics of Cube, we selected the average degree of saturation of traffic flow and the average speed to set up the evaluation index of road network operating quality as a standard for network capacity stress test.

3.3.3 Building Stress Test Roadway Network Before performing the test, the roadway network has to be built in Cube to closely represent the actual roadway and traffic environment, which is significant and basic for stress test. This includes constructing the network by drawing the nodes and links or by using an existing network map obtained directly from a GIS file. Traffic zones are then formed and delineated in accordance with land use, natural obstacle, transit hub and station. At last, the attributes of the roadway are set, including the road type (LINKTYPE), the district number (JURJSDICTION), the capacity index (CAPINDEX), the time or speed signs (TIME_SPDFLAG), time or speed (TIM_SPD), capacity (CAPACITY).

3.3.4 Creating Operating Modules According to the application procedure of Cube, we can write programs or use existing modules such as MVNET、AVROAD、MVGRAM、MVHWAY to construct the four-stage operation model step by step for stress test, including trip production, traffic distribution, traffic modal split and traffic assignment. At each stage, we can choose and change different parameters and functions to use.

3.3.5 Testing and Monitoring Road Network Performance After the establishment of operational modules, stress test over the roadway network can be started. Cube applies MVGRAF or Graphics model to graphically display the result of traffic distribution and, at the same time, researcher can monitor the operation status of roadway network and record the results of each test. If the test result does not satisfy to the preset evaluation index over network operating quality, traffic flow load (pressure) will be increased to retest. Once the road network reaches maximum level of saturation or the operation of roadway network breaks down, fulfilling with evaluation index of road network operating quality, stress test will be stopped.

3.3.6 Analyzing Results of Stress Test After the stress test, all data will be recorded and collected for analysis. By using figures and tables to depict and compare the test results, we can obtain the capacity of roadway network which has satisfied the evaluation index of road network operating quality. In the meanwhile, it is necessary to summarize other data, such as stress increased value, the number of test times, reason of failure in the test, and the changing status of network performance during the test.

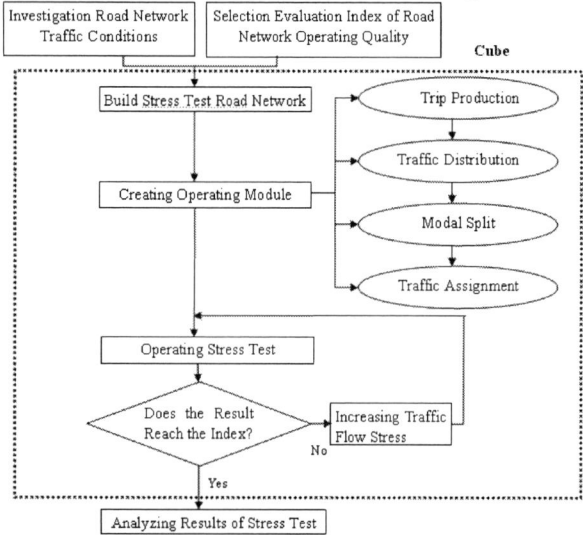

Figure 1 Stress test process of road network capacity

4. Field Test Case Study

In order to verify the feasibility and practicality of the stress test method based on Cube, we applied stress test on the Huairou District roadway network and estimated its roadway network capacity.

4.1 Analysis of Huairou District Roadway Network

Huairou District, located in the eastern part of Beijing, is a major town for tourism. Its roadway network includes three arterial roads and eleven sub roads in the city zone. According to detailed investigation on roadway conditions, traffic impact factors and network performance over 12 main roads along with a 24-hour OD survey (by visiting and filling out questionnaire), we obtained a great deal of basic data about road and traffic,

and OD data as exemplified in Table 1. At present, the traffic flow on this network is low, where of the roadway capacity (supply) exceeds the demand. However, with fast economic development of Huairou District, traffic flow will grow rapidly in the near future. Therefore, it is essential to study capacity of this road network so that effective strategies can be developed for traffic control and management.

Table 1 Trip production of road network in Huairou District

Time	Trip Production （pcu/h）
1	4802
2	6517
3	10633
4	11319
5	13034
6	13034
7	13377
8	13720
9	14063
10	19208
11	20923
12	25725
13	26411
14	26754
15	29155
16	31556
17	32242
18	35329
19	36015
20	37730
21	38416
22	39788
23	45619
24	46305

Based on the objective of this stress test, after extensive analysis and survey we identified and chose the evaluation index system of network operating quality to be as follows: network average level of saturation is 0.6 and the average speed drops to 28km/h.

4.2 Network capacity estimate by stress test

Using Cube, we built the test environment in accordance with actual roadway net work in Huairou District. Then we divided the network into 12 traffic zones and set attribute for all roads. The virtual roadway network is displayed in Figure 2.

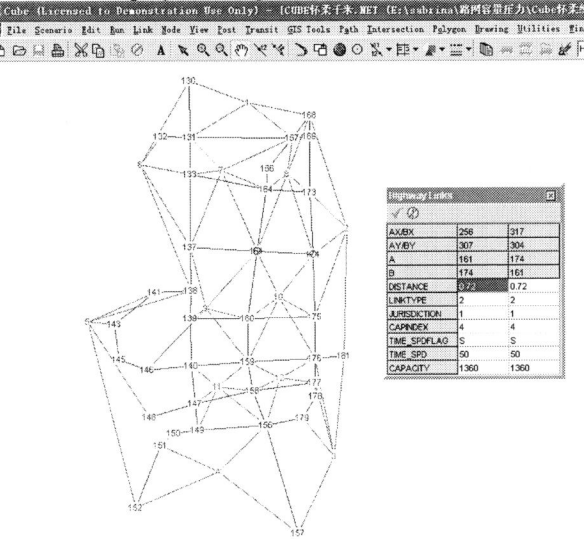

Figure 2 Test environment of road network in Huairou District

Next, we created the operation module for the stress test, as shown in Figure 3, including four stages as trip production, traffic distribution, traffic modal split and traffic assignment.

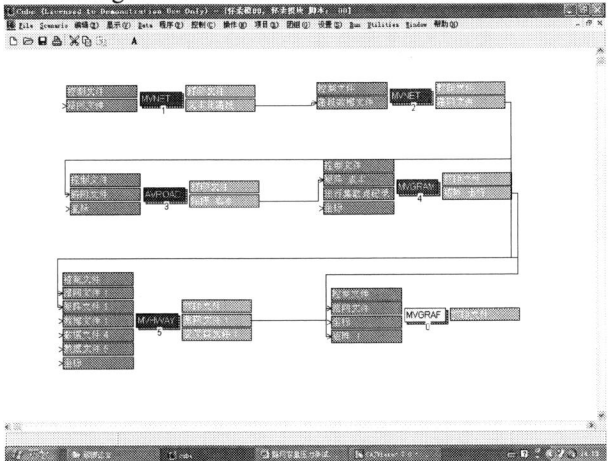

Figure 3 Flow chart of operation module

After creating the operation module, stress test was conducted. In order to determine the increased value of traffic, surveyed OD data were analyzed through regression and the result is displayed in Figure 4. The regression equation is

$$y = 1765x + 2600 \qquad (1)$$

Where, y is the whole traffic flow of road network in Huairou District（pcu/h）；

x is the time.

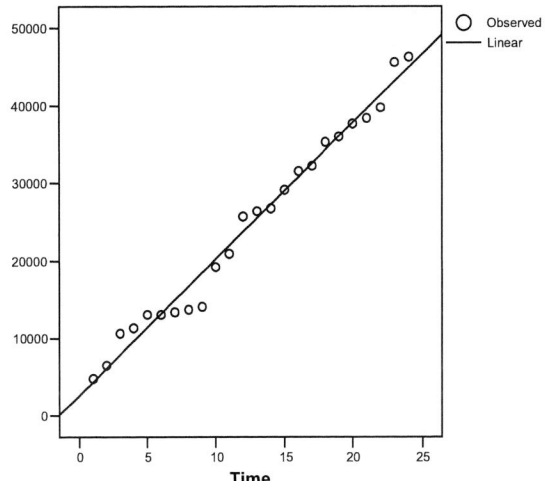

Figure 4 Regression line of OD data

Stress of traffic flow is changed with the rising trend of the regression equation, and after a number of tests, roadway network capacity were reached gradually. Tests results are collected as shown in Table 2.

Table 2 Results of stress test

Time of Test	Traffic Flow of Road Network（pcu/h）	Average Saturation of Road Network	Average Speed of Road Network（km/h）
1	46297	0.28	49.58
5	52020	0.31	47.06
10	60845	0.34	45.29
15	69670	0.41	45.97
20	78495	0.46	39.05
25	87320	0.50	35.67
30	96145	0.55	30.08
35	104923	0.60	27.41

After 35th test, the average saturation level of roadway network reached 0.60 and the average speed of road network was 27.41 km / h, reaching evaluation index when road network still operates normally. At this time, the traffic flow is stable with acceptable delay, indicating a secondary service level which is between expedite and low-grade congestion. Therefore, the traffic flow of roadway network, 104923 pcu / h, is considered as its capacity. Related parameters of road network are shown in Figures 5 to 6.

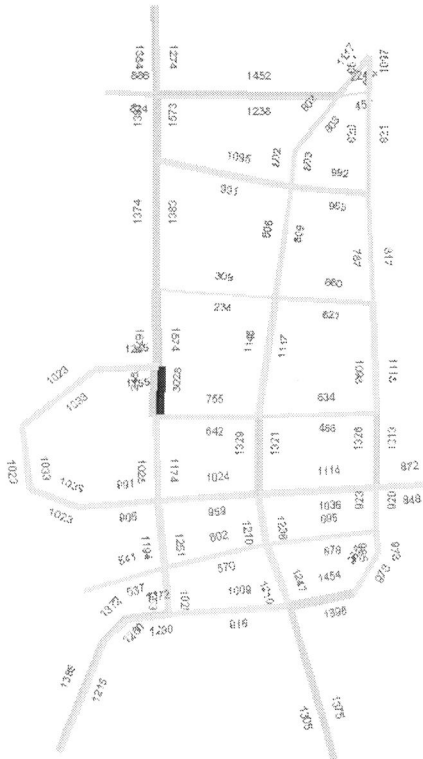

Figure 5 Traffic flow of stress test in 35th time

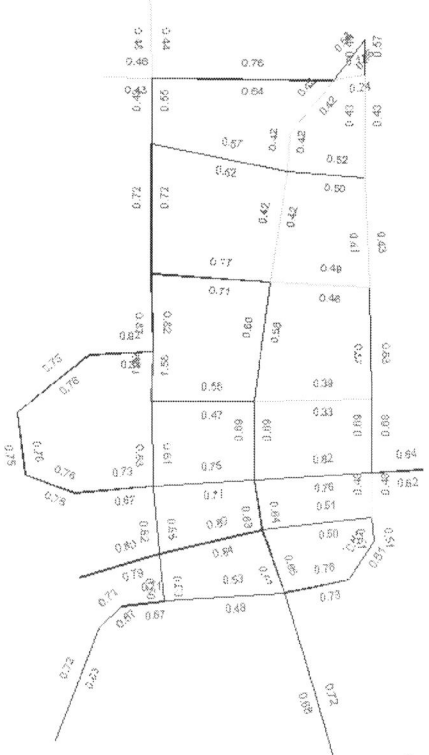

Figure 6 Average saturation of stress test in 35th time

5. Conclusion

This study established a new stress test method for roadway network capacity based on Cube and applied it in Huairou District to obtain the local network capacity. Stress test method does not require complex theoretical model, but only needs to follow a standardized test process. Using this method, traffic analysts can evaluate the capacity of roadway network, find potential problems, and propose to improve traffic operation feasibly, plan network reasonably and manage traffic system scientifically.

Acknowledgements

The project was sponsored by Ministry of Housing and Urban-Rural Development of the People's Republic of China (MOUHURD) and China Construction Department and Beijing Science and Technology Commission, and Intelligent Traffic Management System Demonstration Project of Huairou Sponsored by Beijing Municipal Science and Technology Commission. Many researchers of Transportation Research Center devote their efforts to the projects.

References

[1] GAO Yong, YU Lei, CHEN Xu-mei, CHU Qin. *Theory and Practical Application of Experimental Traffic Engineering Method* [J]. Journal of Transportation Systems Engineering and Information Technology. 2005.8

[2]Sasaki. *Future Prospects of Experimental Traffic Engineering* [Z]. Academic Report of Beijing Jiaotong University.2004.3

[3]YE Xin-ming, FENG Xiao-li. *The Flow of Stress Test in Software* [J]. Journal of Neimenggu University. 2002.1

[4]YANG Tao, TANG Guo-xi. *Application of CUBE/TRIPS Software in Traffic Impact Analysis*[J]. Communications Standardization Issue. 2006.12

[5]User Manual of Cube, http://www.ftc.net.cn

[6]Yogesh Deshpande and Steve Hansen. Web Engineering: Creating a Discipline among Disciplines. IEEE Software,2001(2):82-87

[7] JIANG Chang-hua, ZHU Min, CHEN You-guang. *Web-based Application Stress Testing*. [J] Computer Application. 2003.10

Traffic Congestion Information Promulgating System Based on GIS-T

Bing Zhang
*College of Transportation,
Southeast University, Nanjing,
China, 21009*
zbing1981@sohu.com

Wei Deng
*College of Transportation,
Southeast University, Nanjing,
China, 21009*
dengwei@seu.edu.cn

Ling Mao
*College of Transportation,
Southeast University, Nanjing,
China, 21009*
maolin19821126@163.com

Abstract

Through the traffic congestion detecting and analyzing, this paper puts forward the framework and primary functions of the traffic congestion information promulgating system (TCIPS) based on Geographic Information Systems (GIS) and develops the system by using VB6.0 and MapX5.0. TCIPS can intuitively and visualization provides real-time traffic congestion information for traffic managers and users, e.g. the queue lengths at intersections. And this paper provides technical guidelines for applying the dynamic segmentation technology, which can make the dynamical display of traffic congestion roadway more convenient, and supply the effective implement for the analysis and inquiry. Finally, a prototype platform of TCIPS is introduced according to the actual traffic situation in Changchun city.

1. Introduction

Nowadays the traffic congest has become a serious problem that the whole society is confronted with both at home and abroad. Several indicators confirm this trend. For example, over the years, the percentage of the peak-hour vehicle-miles traveled that occurs under congested conditions has steadily increased, although at a slower pace in recent years. In 1996, that percentage was 54% for the urban interstate system and 45% for the urban national highway system[1]. The associated cost was also huge, For example, in 1997 congestion cost travelers in 68 urban areas in the United States 4.3 billion hours of delay, 6.6 billion gallons of wasted fuel consumed, and $72 billion of time and fuel cost[2]. In a word, congestion usually results in time delays, increased fuel consumption, pollution, stress, and added traffic incidents. Faced with this reality, many urban areas have opted for implementing alternative management measures. Therefore, we develop a Traffic Congestion Information Promulgating System (TCIPS),

which can intuitively and visualization provide real-time traffic congestion information for traffic managers and users and identify alternative actions to alleviate congested roadway conditions[3].

2. System Overall Design

2.1. System Design Goal and Principle

The goal of this system is to study, design, and establish a system which can supply traffic congestion information to traffic managers and users. It is an important component of traffic management system. Traffic congestion information data can be collected, processed, and analyzed by this system. GIS is applied as information distributing platform in this system, and congestion information can be distributed by an intuitive and understandable method.

2.2. System Framework and Function Analysis

This system is mainly consisted of traffic data collection module, communication module, database module and information display module[4]. The functions of these components are respectively introduced as follows:

(1)Traffic Data Collection Module

GIS and traffic data from traffic and road management branches are inputted in this module, e.g. data collected by detector, survey data, image data, video data, and map vector data are processed digitally. The results are classified and stored in database; thereby data source applied by the whole system is generated. The module mentioned here can apply data collection module of traffic management system.

(2) Communication Module

System data collection, transmission, and distribution all rely on communication module. Under the condition of existed communication network, which to ensure reliable and real-time data transmission, communication methods with excellent anti-jamming performance, broad bandwidth and high transmission rate must be applied. Current cable transmission, wireless network transmission and so on can already meet demand of the system.

(3) Database Module

This work is supported in part by National 863 Foundation of P.R.China (2007AA11Z202)

978-0-7695-3342-1/08 $25.00 © 2008 IEEE
DOI 10.1109/PEITS.2008.64

All the data in the system and processed information should be stored and transmitted by database. Because the traffic data formats are not uniform, data are stored in system database in various forms. Running rate and stability of system depend on inter-organization and connection between data, which requires that storage format design, data correlation, and proper traffic data transfer should be executed according to data characteristics during database design.

(4) Traffic Information Analysis Module

Collected traffic data is analyzed by various traffic models and algorithms in this module, and results are transmitted to distribution module. The functions of this module can process different traffic congestion information, also can judge and predict congestion areas and degrees, and so on.

(5) Traffic Information Display Module

Through dynamic segmentation, congestion areas and degrees are displayed dynamically on GIS platform. Also, queue length of vehicles at signal intersection is displayed dynamically on the electronic map. Then traffic congestion information is promulgated in an intuitive and vivid form.

Different modules are connected by communication system. The relationship between them is shown in figure 1. Database design and dynamic congestion information display module design are the focuses of system design in this paper, which are discussed respectively as follows.

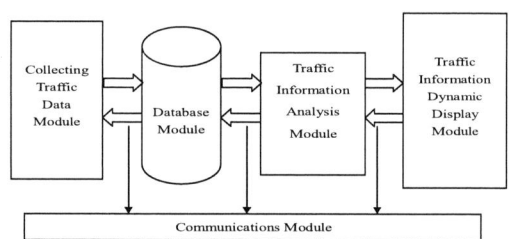

Figure 1. TCIPS logic structure sketch

2.3. System Physical Structure

TCIPS are different according to distributed information of public transportation and daily information needed by traffic management sections. Traffic congestion information promulgation includes two parts, one part is inner information promulgation used by traffic manager, which mainly supplies services such as management decision, control, emergency treatment, faced with technical personnel and decision-making group in traffic management section, within traffic management intranet to commanding center at different levels. The other part supplies traffic participants with services such as indicator, variable sign, VMS, telephone, cell phone, TV, traffic message broadcast, in-vehicle guidance system, internet, in order to acquire the effects of predicting and controlling

traffic states, and make road network in the best operation conditions, i.e.traffic congestion information promulgation. Physical structure sketch is shown in figure 2 below.

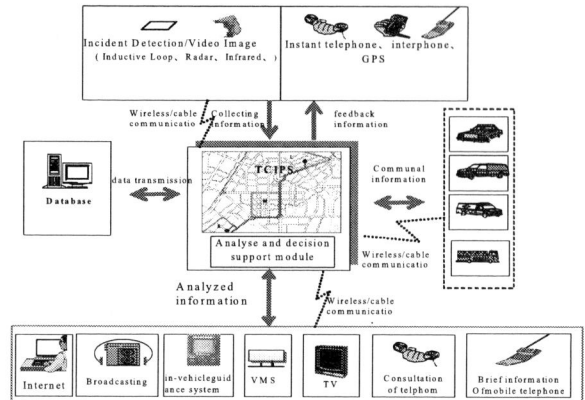

Figure 2. TCIPS physical structure sketch

3. System Database Design

Designing a reasonable structure is helpful to query, transfer, and extend system database, which can furthermore ensure function realization, security, and stability of traffic information system. Traffic congestion information database includes spatial database and attribute database. In traffic information promulgation module, to realize dynamic segmentation, the relations between spatial and attribute database should be designed effectively.

3.1. Spatial Database Establishment

In the traffic congestion information promulgating system based on GIS, various spatial information and attribute data of graphic element are stored and managed in spatial database. Geographic entities are organized by layers in spatial database, mainly including point, line, and area entity. Each geographic entity has its own attribute and is stored in attribute list in the form of one record. Entities with similar attributes or functions are stored in the same layer when needed in practical application. Each layer corresponds with one attribute list, to realize operation and inquiry of attributes of graphic element. Actually, it is a map file of MapInfo in the system, including spatial database of non-related data structure with spatial information and attribute database with related data structure[5].

Spatial data mainly include point, line, area entity such as urban road network, urban terrain, traffic facility map, administrative regionalization, road transect and vertical section. Spatial data on the road map are collected and input by electronic map or engineering drawing of

AutoCAD format. Then the data are edited and processed by MapInfo. At last, basic layers are integrated by Geoset Manager, in order to link and operate. Electronic map applied by TCIPS includes nine layers: traffic incident layer, traffic detector layer, segment layer, intersection layer, traffic facility layer, river layer, railway layer, and construction layer. Among them, traffic layer is upon road layer. Conditions of vehicle flow at each intersection are displayed on this layer. Traffic conditions at intersections can be classified by several degrees when needed. Traffic layer mainly includes intersection number, name, vehicle flow, average velocity, and average travel-time. Added points or lines needed later (e.g. graphic element at traffic accident point, graphic element of safety facility) can be generated by linking attribute data.

3.2. Attribute Database Establishment

In TCIPS, first of all, attribute database of important intersections is established, based on spatial database of urban roads. Secondly, related segment database is established according to attribute database of road segments. Finally, traffic congestion attribute database is established.

System attribute database includes spatial attribute database and extended attribute database. Spatial attribute database includes traffic flow information, attribute information of urban roads, and attribute information related to other traffic facilities. Abundant data needed by parameter information of traffic flow and traffic congestion judgment are stored in extended attribute database, where parameter information of traffic flow is strongly real-time, data formats are various, and traffic congestion information includes vast media files which need frequent record and short maintenance cycle. Suppose all the attribute information is bound to map entities, and computer memory is taken up at the same time, it will influence system map refreshing and displaying, and lead to reduce system running rate. Therefore, attribute data include entity data related to GIS, e.g. GIS attribute data, and parameter data of traffic flow and traffic accident data over the years extended.

Traffic attribute data in GIS attribute list are applied to inquire and analyze traffic congestion information by users, not including parameters of traffic flow and all the information of traffic-reference point, while they are stored in extended database and defined according to data list field. Entity attribute list of traffic flow and accident is formed by part of extended data list field. Traffic reference point data are from writing and electronic data related to urban construction, which are processed by manual inputting, scanning, and format changing and so on. At last, the data are input into SQL Server 2000 database.

3.3. Association with Spatial Database and Attribute Database

Spatial data are the basic loaders to indicate, query, analyze space and construct road network. Since the range of urban road network is large, traffic information is refreshed at any time. The spatial data of different traffic information in road network are on adjusting, which need to maintain and refresh spatial data frequently, in order to ensure integrity, accuracy, and validity of spatial data.

Attribute data are stored in a list in the form of database, usually extended database. There are many alternatives here and relation database is often chosen, such as SQL Server, Oracle, Sybase, Access, dBase, Paradox, and so on. Generally, SQL Server and Oracle are used to develop large-scale database, Access and dBase are used to develop medium and small-scale database.

The attribute and spatial data are linked by data index mechanism in GIS. Graphs and characters can be checked quickly with each other too. Data index mechanism is a method associating spatial objects with attribute data.

History data records of traffic information are stored in extended database, which should also be connected with processing module of extended detectors. Besides, there should be index linking attribute data of graphic elements. Related database can be realized on this point. Database Structure is shown in figure 3 below.

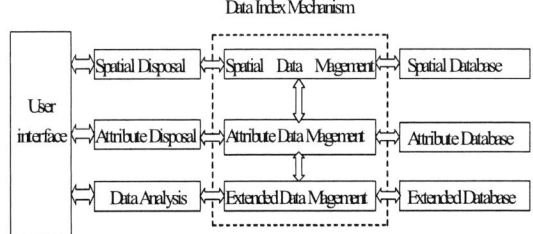

Figure 3. Database structural sketch

4. The Traffic Information Dynamical Displaying Module Design

The module detects the space position information of the congestion and the degree congestion information which are promulgated by the GIS flat. Now the traffic information is displayed and promulgated by traditional segmentation method in usual promulgating systems. Traditional segmentation is divided into some fixation road segment, which must be segmented little enough to reflect the truth of the attribute data[6]. Every segment corresponds to a correlation attribute data register. When displaying some road congestion, the attribute value is amended by analysis and account, and the display effects will be changed. But the traditional method has a disadvantage that it can not display dynamically and do not locate exactly. So the dynamic segmentation technology is applied in the module, which can't change the spatial data automated segmentation according to the degree of congestion, and display dynamically the congestion with different color bright. The queue lengths at intersections are displayed in the system real time. The twinkling segmentation objects mean that more serious congestion. For less serious congestion, the systems use three kinds of color segments (red /orange and green) to refer to different kinds of traffic congestion respectively. If traffic congestion has been disappeared completely and traffic operation is restored to normal state, the corresponding segments can be changed with the color in GIS-T interface, which can enhance the veracity and visually of traffic information promulgating.

Dynamic Segmentation is a technology of analyzing and displaying the linear object's attributes dynamically. With this technique, we can segment the route dynamically, and storage the attributes of all the roads in an unattached table (event table) [7]. Linear Reference System is the one dimension system and it can join directly the multiattribute and linear objects.

The attribute information of incident (point incident and Line incident) is stored in the corresponding incident attribute tables in Dynamic segmentation. But the real roads (spatial information) are not change. Attribute information and spatial information is separated in the system, which is convenient for database maintenance and management, and the data redundancies are reduced. The results of segmentation aren't virtual entities changed. In fact, it is just logical segmentation. The results are stored in the relevant incident lists by GIS software, and the related segmentation is generated when needed. The topology relations data model is described in figure 4 below.

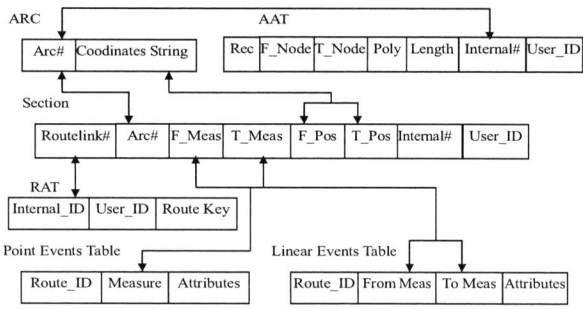

Figure 4. "Node-Arc" topology relations data model

In this paper, the steps of the dynamic segmentation are discussed by the example of dynamic display of the queue lengths on approaches at intersections. First, stopping line is applied as the origination point, and the linear reference system is established according to the traffic signs in artery as reference points. There are two methods of the congestion displaying according to the characteristics of road and the distance between two artery intersections next to each other. ① When the queue length exceed the distance of the farthest intersections of inferior road, it is considered that there is no circumambulated road between intersections of the two arteries , and the color is changed by the spatial objects of the entire road. For example, the red is denoted as the seriously congested road. ② When the queue length doesn't exceed the distance of the farthest intersections of inferior road and there are the intersections of circumambulated inferior road, the dynamic segmentation technique is applied to display the length of congestion road dynamically. The line objects can be highlighted in GIS-T interface as the length of congested road by the module in which the queue lengths of intersections and the end point coordinates of the queue by the reference points are calculated. The detailed steps of the algorithm are described in figure 5 below.

5. Prototype Platform Design

There are many kinds of developing tools of GIS-T. MapInfo7.0 is selected as system platform in this paper. MapInfo has many functions such as strong graph handing, statistic analysis, query and option, and can visit various databases such as SQL, Oracle, Sybase and Informix. A further developing tool of MapX5.0 and VisauBasic6.0, which have many advanced characteristics, is applied to develop the function module. The Map of city is stored in MapInfo grid tables, and the spatial property information is stored in MapInfo data tables. Other affiliated information about traffic congestion is stored in SQL 2000 database. This system is programmed by Visual Basic language, and OLE is used to establish the communication between MapInfo and Visual Basic. The interface of prototype platform according to the actual traffic situation in Changchun city is shown in figure 6 below.

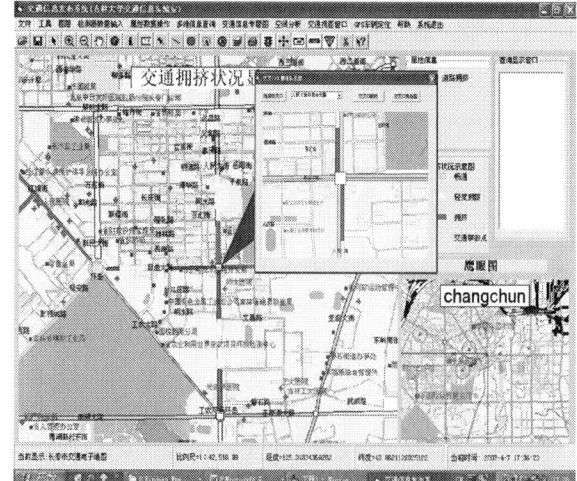

Figure 6. Interface of TCIPS

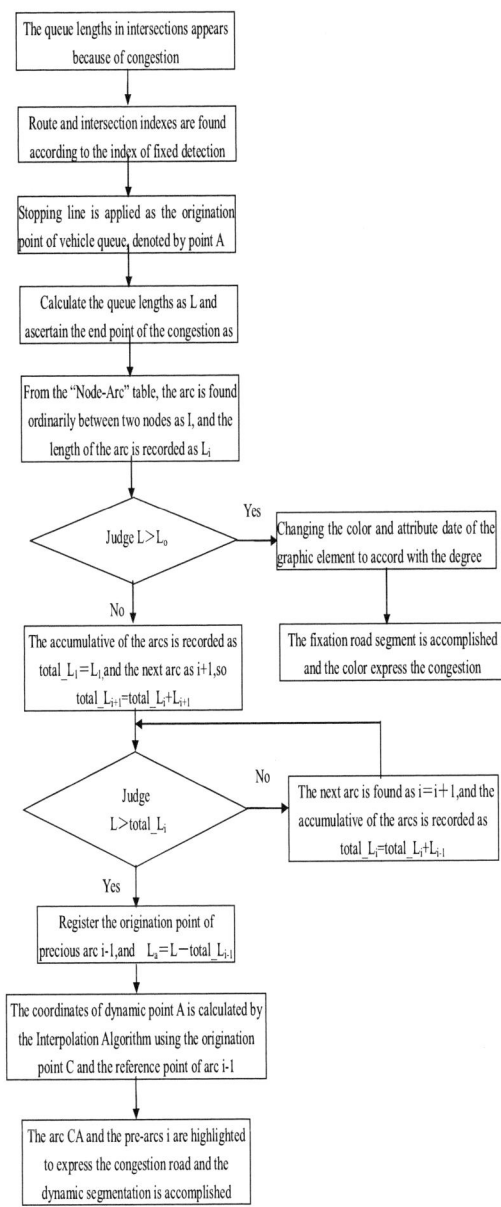

Figure 5. Dynamic segmentation algorithm

6. Conclusions

This system based on GIS-T can realize the functions of dynamic management and analyze of the traffic congestion information, which can display the spatial places on the map vividly. It is useful to reduce the traffic congestion as well as to analyze the factors of traffic jam and help traffic managers to make decisions according to traffic congestion conditions and spatial characteristics of the around roads. The core of this system is to promulgate the traffic congestion information dynamically based in GIS-T. TCIPS is valuable to improve the efficiency of urban traffic management and utilize the traffic resource

fully. The next developing way of this system is to perfect its function gradually, e.g. further optimize traffic control and management.

Reference

[1]Cesar A. Quiroga, "Performance measures and data requirements for congestion management systems," *Transportation Research Part C*, 8 (2000), pp. 287-306.

[2]Schrank, D.L., Lomax, T.J, "The 1999 Annual Mobility Report," Texas Transportation Institute.

[3] Lindquist, E,"Assessing effectiveness measures in the ISTEA management systems," Southwest Region University Transportation Center, Texas Transportation Institute, College Station, TX, 1999, p.103.

[4]Yi Fu, Zhiheng LI, Kezhu SONG, "Integrated Traffic Management Platform Design Based on GIS-T," *6th International Conference of ITS Telecommunications Proceedings*, 2006 IEEE, pp.29-32.

[5]Gong Jianya, "Technology and Some Theory of Present GIS," Wuhan: Surveys and Science Technology University Press, 1999.

[6]Shen Jie, Lu Guo-nian, "Dynamic Segmentation and Its Application in GIS,"*Journal of Nanjing Normal University (Natural Science)*, vol.25, no.4, pp.106-108, 2002.(In Chinese)

[7] DUEDER KJ, VRANA R, "Dynamic Segmentation Revisited: A Mile Point Linear Data Model," *Urisa Journal*, vol.4, no.2, pp.92-105, 1992.

Optimization for Urban traffic assignment by Congestion toll levied on Reginal Expressway

Meng-Memg Zhang[1], Jinli Wei[2]

1 Department of Communications Engineering, Shandong Jiaotong University, Jinan 250023, China
2 Qingdao Technological University, Qindao,China
E-MAIL: mengmeng8169@126.com

Abstract

The objective of this paper is to study the toll standard of regional expressway network so as to optimizing traffic assignment and consequently reducing the holistic cost of the network. Based on Wardrop's principles and expense analyses, a function called generalized expense model is brought forward to calculate the optimal toll rate. Genetic algorithm (GA) is applied to solving the model according to its characteristics. The paper takes a network planning as an example for analysis. The traffic distribution without toll controlling is depicted by VISUM traffic simulation. Contrast analysis, which is made between the model solution and the simulation results, demonstrates that the model is valid and practicable.

1. Introduction

With the vehicles increasing, the regional expressway network becomes more and more congested. In order to make full use of urban street network and induce vehicles choosing reasonable route, the paper considers levy congested toll on vehicles running on the expressways [1]. Therefore，constituting a set of new toll standard and toll policy is severe significant for obtaining reasonable traffic assignment in the network and increasing the social benefit.

Traffic impedance [2] is the criterion when drivers select route, and travel time, running cost, toll and so on should be considered as a whole in depicting traffic impedance.

In this paper, transportation planning theory and system optimization principle are applied to establish the model for optimizing network volume. Toll standard is introduced into the model as a component of traffic impedance. In view of the characteristics of the model, GA is applied to solving the model. The model can calculate the optimal traffic volume and the corresponding toll rate of the regional expressway network.

2. Function analyses

2.1. The basic form of objective function

The whole regional expressway network is considered as a system in this paper, through regulating the toll rate of expressway network, traffic volume could get the optimal assignment in the network. Two principles should be considered when assigning traffic volume in expressway network, which are Wardrop's first principle (user equilibrium) and Wardrop's second principle (system equilibrium) [3]. The objective function should be the minimum of users' total running cost in the whole expressway networks [4]. This function is described as follows:

$$\min S(T_{ijr}) = \sum_a v_a S_a(v_a) \tag{1}$$

T_{ijr} traffic volume in route r heading form traffic zone i to traffic zone j;

S total expense of the whole expressway network

v_a traffic volume in expressway section a

$$v_a = \sum_{ijr} T_{ijr} \delta_{ijra} \tag{2}$$

δ_{ijra} a dummy，which is a binary variable. When traffic volume in route r passing expressway section a heading from i to j, $\delta_{ijra}=1$; else, $\delta_{ijra}=0$

$S_a(v_a)$ the expense function of section a when the traffic volume on which is v_a

2.2. The basic form of generalized expense function

$S_a(v_a)$, the expense function of expressway section, is usually called traffic impedance of road section in transportation engineering. Traffic impedance is customarily expressed by travel time. Federal Highway Administration of American recommends the most commonly used traffic impedance function and its form is shown in formula (3):

$$t = t_0(1 + \alpha(V/C)^\beta) \tag{3}$$

But travel time may not always be the main factor when drivers select travel routes, especially for expressway networks, toll may be an important factor to be considered in selecting travel route [5]. Therefore, only travel time is inadequate to be taken as traffic impedance, and certain comprehensive function should be constituted to depict the travel expense that influence route selection.

In order to meet the above requirement, a kind generalized expense function is put forward in this paper, which covers travel time, running cost and toll. Its basic form is as follows:

$$S_a(v_a) = CP_a(v_a) + Ct_a(v_a) + Toll_a(v_a) \qquad (4)$$

S_a generalized travel expense in expressway section a. Unit: Yuan

CP_a running cost in expressway section a. Unit: Yuan

Ct_a travel time cost in expressway section a. Unit: Yuan

$Toll_a$ toll of expressway section a. Unit: Yuan

v_a traffic volume in expressway section a

2.3. Conversion of generalized expense function

2.3.1. Conversion of CP_a and Ct_a
The running cost (CP_a) is composed by fuel consumption and various depreciations. The depreciations have loose relationship with travel velocity and hence only fuel consumption needs to be considered when studying the effect of traffic disequilibrium to vehicles' operational cost.

Travel time cost (Ct_a) is the product of unit time value multiplying travel time. Unit time values differ from each other because of diversity of individual property and travel motive. Therefore, only the expectation of the total could be calculated, as is shown in formula (5)

$$Ct_a = E(VOT) \cdot t_a \cdot v_a \qquad (5)$$

E(VOT) the expectation of unit time values

$$E(VOT) = \sum_{i=1}^{m} p_i E_i(VOT) \qquad (6)$$

pi the percentage of vehicle type i in all traffic volume in the regional expressway network；

Ei(VOT) the expectation of time value of vehicle type i.

2.3.2. Conversion of $Toll_a$
Based on the consumer surplus theory in economics, the consumer surplus is the difference between the prices that consumers willing to pay and the price that consumers actually pays. Road users are accustomed to choose route that has the maximum surplus value. This relation can be abstracted as formula (7):

$$\max \quad \int_0^{v_a} [E(VOT)t'_a(v) + CP'_a(v) - Toll_a(v)]dv \qquad (7)$$

$t'_a(v)$ travel time variation caused by traffic volume change

$CP'_a(v)$ fuel consumption expense variation caused by traffic volume change

Based on the above analysis, the relationship between toll standard and traffic volume can be deduced as formula (8):

$$Toll(v) = E(VOT)t'(v) + CP'(v) \qquad (8)$$

3. Generalized expense model and its solution by GA

3.1. Generalized expense model

Base on the previous analysis, the objective function in formula (1) can be translated to formula (9) and it is the generalized expense model.

$$\min \sum_a v_a \{ \{ a[U_a(0) + \alpha U_a(0)(v_a/C_a)^{-\beta}]^2 \\ - b[U_a(0) + \alpha U_a(0)(v_a/C_a)^{-\beta}] + c \}(v_a+1) \qquad (9) \\ + E(VOT)t_a(0)[1 + \alpha(v_a/C_a)^\beta + \alpha\beta v_a^{\beta-1}/C_a^\beta] \\ + \alpha\beta U_a(0)(v_a/C_a)^{-\beta}[b - 2aU_a(0) - 2a\alpha U_a(0)(v_a/C_a)^{-\beta}] \}$$

$$s.t. \quad \begin{cases} \sum_a v_a Toll_a \geq OC \\ \sum_r T_{ijr} = T_{ij} \\ T_{ijr} \geq 0 \end{cases}$$

α, β, a, b and c are constants，and $\alpha = 0.0171$, $\beta = 1.395$, $a = 54.618$, $b = 0.15$, $c = 4$.

$U_a(0)$, $t_a(0)$ and C_a are all known when the expressway network is predetermined

OC the expense of road construction

3.2. Model solution based on GA

The model is a nonlinear programming model, involving many variables and constrains, and hence it is hard for traditional methods to solve the model. In another aspect, the optimum should be obtained from numerous feasible solutions, and this phenomenon is similar to the survival of the fittest. In this paper, the chromosome, which has the most vital forces, corresponds to the optimum of traffic assignment; therefore, GA is applied to optimize the traffic assignment in the regional expressway network.

3.2.1. Encoding. Float encoding is adopted in order to insure the precision. The variables are s_f $(j = 1, 2, \cdots, a)$ and each variable denotes a value of gene. Its physical meaning is the traffic saturation of each expressway section.

3.2.2. Fitness function. The length of bit strings of chromosome denotes the number of road sections in the expressway network.

$$F(X) = \begin{cases} C_{\max} - f(X), & if \quad f(X) < C_{\max} \\ 0, & if \quad f(X) \geq C_{\max} \end{cases} \quad (10)$$

C_{max} a biggish number that is predetermined

3.2.3 Genetic operator Elite selection, arithmetic crossover operator, random partnership and uniform mutation operator are adopted in this paper.

3.2.4. Operational parameters The length of encoding strings equal to a, the size of colony is 80The probability of crossover is 0.7, the probability of mutation is 0.01Generation number for termination is 1000.

4. Example analysis

The perspective expressway network planning is adopted in this paper as an example to be analyzed [6].

4.1. The results calculated by the model and GA

Based on the model and GA adopted in this paper we can get the traffic assignment results, traffic saturation of expressway section and toll rates of the regional expressway network, as is shown in table1.

Tab.1. Traffic assignment results by GA

Sequence number of expressway sections	Traffic volume	Traffic saturation of expressway section	Toll rate
1	4896	0.51	0.00
2	13344	1.39	0.30
3	18816	1.96	0.20
4	18720	1.95	0.30
5	6912	0.72	0.00
6	15120	1.8	0.30
7	7644	0.91	0.00
8	15876	1.89	0.90
9	2912	0.56	0.00
10	8632	1.66	0.40
11	3978	0.51	0.00
12	11844	1.41	0.20
13	12432	1.48	0.30
14	7392	0.88	0.00
15	5568	0.58	0.00
16	16032	1.67	0.30
17	11648	1.82	0.50
18	7056	0.84	0.00
19	6552	0.78	0.00
20	4602	0.59	0.00
21	12096	1.44	0.50
22	17280	1.35	0.20
23	20352	1.59	0.10
24	8064	0.63	1.00

4.2. Traffic assignment without optimizing traffic volume

VISUM traffic simulation is adopted to depict the traffic assignment after stopping tolling without optimizing traffic volume so as to compare with the results obtained by the model and GA in this paper.

In the expressway network, if some expressway sections (sequence number are 5,7,8,10,14,15,18,21 and 24) come to the toll deadlines and tolling stop without taking any control measures, then the traffic assignment situation will be the same as table 2.

Tab. 2. Traffic assignment results after stop tolling

Sequence number of expressway sections	Traffic volume	Traffic saturation of expressway section
1	15168	1.58
2	2016	0.21
3	6192	0.64
4	25440	2.65
5	8976	0.93
6	1806	0.21
7	5040	0.60
8	15582	1.85
9	16042	3.08
10	3380	0.65
11	8697	1.11
12	0	0%
13	28518	3.39
14	798	0.09
15	31728	3.30
16	31632	3.29
17	992	0.15
18	27174	3.23

19	9240	1.10
20	25311	3.24
21	1764	0.21
22	57920	4.52
23	11648	0.91
24	30464	2.38

4.3. Contrast analysis

Some indexes are listed in Table 2 and Table 3 so as quantitative contrast analysis can be made to illustrate the difference between toll control and without control.

We can see from table 3, after stop tolling without control, there are eight sections on which traffic load is less than 0.6, that is to say the using rate of expressway is very low; while there are eight sections on which traffic load is greater than 2.0 that means the traffic situation on these sections become supersaturating. In a word, without toll controlling, traffic volume will imbalanced distribute in the expressway network.

With toll controlling, there is only one section on which traffic load is less than 0.6 and we can not find any section on which traffic load is greater than 2.0. Traffic volume in the whole network distribute soundly.

5. Conclusions

Starting with the general form of objective function, we go from abstract to concrete, analyzing the expense function step by step and making it more reasonable than traffic impedance measured only by time. We not only put forward generalized expense function, which covers three main components and especially suits for Chinese expressway situation, but also manage to make the model solving feasible and accurate by applying GA. Actual planning example is used to illustrate the validity of the model and contrast analysis is carried out between the model calculating results and VISUM traffic simulation results.

The following are some aspects need to be further studied.

In generalized expense model, it is hard to introduce security and easiness as components to depict their relationship with traffic volume. But these relations indeed exist and exert unassailable function in affecting traffic distribution.

The object of this study is confined in regional expressway network. Traffic volume in low-grade highways could not be regulated by the study in this paper. The regional highway system should be brought into further study to make the results more valid and applicable.

6. References

[1] Levine, Jonathan; Inam, Aseem, The market for transportation-land use integration: Do developers want smarter growth than regulations allow? Kluwer Academic Publishers, Dordrecht, Netherlands, Vol.11, No.4, pp.409-427, Nov. 2004

[2] Rousseau, Guy; Clymer, Tracy, Travel demand modeling and conformity determination: Atlanta regional commission case study, Transportation Research Record, No.1871, pp. 172-176, Feb.2002

[3] Chiou, S.-W.Suh-Wen . Joint optimization for area traffic control and network flow. Computers and Operations Research, Volume 32, Issue 11, pp. 2821-2841, Nov. 2005.

[4] Yildirim, Mehmet Bayram; Duman, Ekrem, Computational methods for toll pricing problems, IIE Annual Conference and Exhibition 2004, Houston, TX, United States, pp. 169-170, May. 2004.

[5] Yan, Jia; Small, Kenneth A.; Sullivan, Edward C. Choice models of route, occupancy, and time of day with value-priced tolls, No. 1812, pp. 69-77, Feb. 2002.

[6] Mengmeng Zhang. STUDY on method of calculating the toll of the expressway network based on optimizing traffic flow. Harbin: Dissertation Submitted to Harbin Institute of Technology for the Master Degree. 2005.12

Active Safety in Autonomous Land Vehicle

Daxue Liu	Xiangjing An	Zhenping Sun	Hangen He
National University of Defense Technology, Changsha 410073, P.R. China	*National University of Defense Technology, Changsha 410073, P.R. China*	*National University of Defense Technology, Changsha 410073, P.R. China*	*National University of Defense Technology, Changsha 410073, P.R. China*
daxue_l@yahoo.com.cn	anxj@ie165.com	sunzp1976@163.com	hehangen@yahoo.com

Abstract

In this paper, active safety is divided into four stages, which are perception enhancement, driving warning, assistant driving, and autonomous driving. The autonomous driving stage which synthesized all the other stages is emphasized. The relationship of the autonomous driving with Intelligent Transport Systems (ITS) program as well as its development in China is introduced. As an example of autonomous driving system, the prototype HongQi Autonomous Land Vehicle is analyzed systemically. The control subsystem, the environments recognition subsystem and the experiments on the highway of the vehicle are presented in detail. It performs excellently through the thousands kilometers of autonomous driving. The vehicle can be a test-bed for various active safety techniques and parts of the achievements can be commercialized.

1. Introduction

It has been over 100 years since the first vehicle came into the world. With the increase of speed and the number of vehicles driving on the road, the issue of safety driving has become more and more important. In recent years, many motor companies in the world have focused on activity safety.

Active safety is relative to passive safety which is to help drivers and passengers stay alive and minimize injury in an event of crash, such as air bags and adjustable steering wheel; on the other hand, the feature of active safety is to help drivers avoid crashes, such as brake assistance, traction control and etc. Active safety has become an important component in Intelligent Transport Systems (ITS) program.

The rest of the paper is organized as following. Section 2 analyses the active safety and divides it into four stages. The relationship of autonomous driving with the ITS program is also introduced in this section. Section 3 describes the prototype of HongQi Autonomous Land Vehicle. In this section, the control subsystem, the environments recognition subsystem and the experiments

in the highway of the vehicle are presented in succession. Conclusion is deduced at the end of the paper.

2. The analysis of the active safety

According to the effect of the active safety system to the driver's behaviors, active safety can be divided into four stages, which are perception enhancement, driving warning, assistant driving, and autonomous driving. The four stages are connected tightly with each other and are all targeted for increasing the safety of driving.

Perception enhancement system aims to increase the capacity of driver's perception by sensors. It enlarges the field of vision or enhances the driver's reflection to the environments. Night vision enhancement and attention control are examples of this stage.

Driving warning system monitors the behaviors of driver/vehicle and the environments around the vehicle during the course of driving. When it detects that potential hazard might occur under the current driving command, then warning will be given. This system includes collision warning (CW), lane departure warning (LDW), and other warnings.

Assistant driving is an extension of the driving warning system which starts to work when the driver has no reaction to the warning given by the warning system. It includes the collision avoidance (CA), lane keeping system, adaptive cruise control (ACC) and so on.

The autonomous driving aims to take the place of the drivers and to drive without participation of human. It is the combination of the three stages, and is the final goal of the active safety.

In 1972, Stanford University designed the first autonomous mobile robot-Shakey[1]. This is the first successful attempt to make machine move autonomously. It accumulated techniques for the subsequent autonomous vehicle and active safety research.

Japan is one of the first countries in the world engaging in the R&D for the ITS. In 1996, National Police Agency and four other related government bodies compiled a "Comprehensive Plan for ITS in Japan" which was a long-term vision of ITS's goals[2]. In August 1999, these government bodies released an article named "System Architecture for ITS". The third development area in this architecture is assistance for safe driving. The related user's service includes provision of driving and road conditions information, danger warning, assistance for driving and automated highway systems.

The autonomous driving research is mainly centralized in institutes and universities in China.

The National University of Defense Technology started to do autonomous driving research since 1980's. CITAVT-I, CITAVT-II, CITAVT-III, CITAVT-IV prototypical autonomous driving system came into being successively. In 2000 CITAVT-IV reached the autonomous speeds of 75.6km/h which was the highest domestic speed at that time[3]. In 2003, the prototype of HongQi Autonomous Land Vehicle achieved the peak autonomous speed of 170km/h.

THMR-V was developed by Tsinghua University. The vehicle can autonomously move in a complex campus road environment[4]. In 2003, THMR-V achieved the peek autonomous speed of 150km/h in structured road environments[5]. JLUIV-V is an electric automatic

Table 1.Vehicle related items in the "System Architecture for ITS in China"

Service area	service
4. vehicle safety and driving assistance, AVCSS	(13) Vision enhancement
	(14) Longitudinal collision avoidance
	(15) Lateral collision avoidance
	(16) Intersection collision avoidance
	(17) Safety condition(inspection)
	(18) Pre-crash restraint development
	(19) Automatic vehicle drive

intelligent vehicle developed by Jilin University. In 2006, the vehicle was tested in Jilin University and the Culture Center of Jilin Province[6].

From 1999, China started drafting the system architecture for ITS. In 2001, the "System Architecture for ITS in China (version 1)" was released. The revised version released latter[7]. In the official version of the system architecture, items related to active safety are in the fourth service area. Seven services are included in this area [8].

3. The Prototype HongQi Autonomous Land Vehicle

In order to demonstrate active safety techniques in China, the National University of Defense Technology and FAW Group Corporation of China launched the HongQi autonomous driving project. The first phase of the project started in 2001, and had been finished in 2003.

The prototype HongQi Autonomous Land Vehicle is based on a CA7460 HongQi Grand passenger saloon

manufactured by FAW Group Corporation. It is equipped with two forward-looking cameras and two onboard industry computers. The steering wheel, the fuel pedal and the brake pedal have been modified, and can be controlled by computer. The autonomous driving system consists of three subsystems, which are environment recognition subsystems, control subsystems and Human Machine Interface (HMI).

The environment recognition subsystem includes two digital cameras, one for lane detection, and the other for

Figure 1.The Prototype HongQi Autonomous Land Vehicle

vehicle detection. The cameras are fixed on each side of the cab behind the windscreen. The control subsystem consists of steering controller, throttle controller and brake controller. The two subsystems communicate with each other through VME bus. The system architecture is show in figure 2. The computers of the above two subsystems located in the same cabinet put in the luggage of the saloon. The HMI subsystem includes a TV monitor, a LED-based control panel. The driver sets the functions of the system through control panel.

All control components of the vehicle are miniaturized and inosculated with the original driving system. No additional controllers can be seen in the carriage, and thus the original style is kept intact.

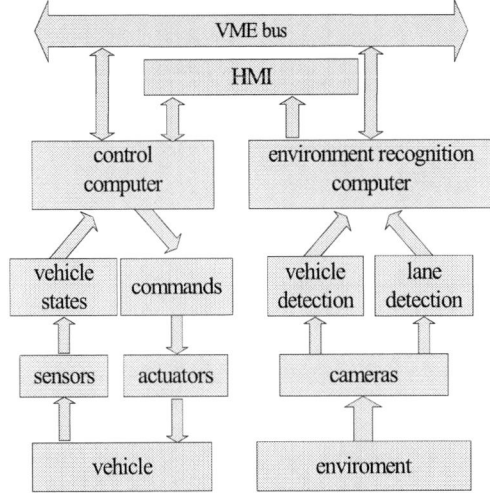

Figure 2.The System Architecture of the Prototype HongQi Autonomous Land Vehicle

3.1. Environment Recognition Subsystem

HongQi uses the left camera for lane detection. Thanks to DMA device, acquisition can reach a high frequency to meet the needs of real-time control, which is over 25Hz. Considering the powerful processing capacity and the plenty of development tools of the standard processing system, a processor with AltiVex instruction is chosen for image processing. These AltiVex instructions accelerate the performance of application based on computationally intensive algorithms that localized recurring operations on small native data.

In order to remove the perspective effect from the acquired image, the Inverse Perspective Mapping (IPM) method is employed[9, 10]. It remaps the original image into a new 2-dimention domain, in which the information content is homogenously distributed among all pixels, which allowing the efficient implementation of the SIMP structure.

The original image and the processed result used IPM is shown in figure 3. By means of IPM, the perspective effect can be reduced, and the parallel character about two lanes can be used easily. Considering the lane change/merge maneuver of HongQi, three lanes should be extracted simultaneously.

HongQi uses the right camera for vehicle detection. The geometrical attributes as well as the lanes given by the left camera is used to locate the vehicles in front of HongQi. The distance and the speed of the detected vehicles can be calculated.

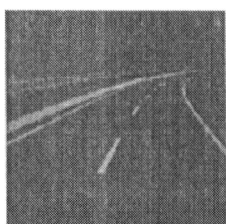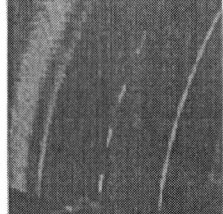

Figure 3.Original Image (left) and the IPM Result (right)

3.2 Control Subsystem

The lanes are described with a set of points given by the environment recognition subsystem. Task plan, behavior decision, behavior plan and operation generate are performed by the control subsystem. The framework of this subsystem is shown in figure 4.

The framework of this control subsystem is divided into four layers[11, 12]. The four layers are different from each other in temporal and spatial resolution. They focus on different environment information. The task plan layer disassembles the tasks that come from human into subtasks, and it has the coarsest temporal and spatial resolution. The behavior decision layer converts the subtask into behaviors, such as acceleration, decelerate, lane change and lane merge. The behavior plan layer is an interface between the behavior decision layer and the operation generate layer, and it translates the result of the upper layer to trajectory that can be understood by the sub layer. The operation generate layer sends the command to actuators, which make the vehicle under control.

The internal parts in the four layers have the similar architecture. Each layer has five modules, which are environment modeling, learning, decision, supervision and knowledge base[13] The environment-modeling module fetches the information concerned from the environment recognition subsystem and forms a local map. The decision module gives the result of decision based on the knowledge base and the local map. The knowledge base is a database developed by human off board. The learning module can improve the performance of the system. The supervising module is a human machine interface to monitor and modify the work of the other modules.

3.3 Experiments in Highway

In order to test the control framework and the image process program, we have made a lot of experiments on Changsha Loop3 highway and the JingZhu highway. The Changsha Loop3 highway is located in mountain area and there are many bridges over it. The JingZhu highway is one of the busiest highways in China. It is very representative for autonomous driving experiment made on them. There are varying road scenarios that include changing illumination conditions, territory sloping and a generally high amount of traffic.

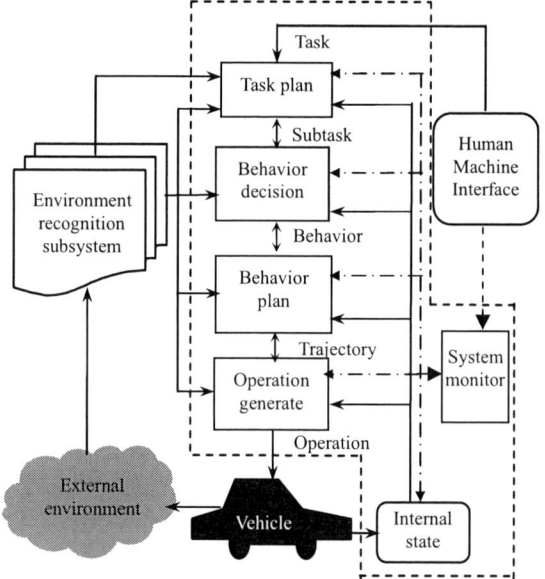

Figure 4.Four Layers Control Framework

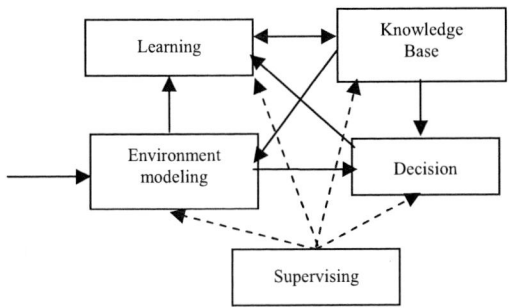

Figure 5.General Structure of Control Unit

The experiments lasted from April to July. The system endures the test of high temperature over 35 degrees centigrade. HongQi works well while passing through the shadows of the bridges for the shot time control algorithm.

The image-processing algorithm has robust performs in different light conditions, such as the sun in front of the cameras, with the sun high or low on the horizon, during the evening or the day, and even in little rain. The control mechanics endures the high temperature. In different conditions including upgrade, downgrade, surface with water, etc., the control error is always small.

In the end of the first phase the project, the prototype HongQi Autonomous Land Vehicle achieves the following achievement: On highway in normal traffic, the average driving speed is about 130km/h, and the peak speed is 170km/h. Overtaking function is also included.

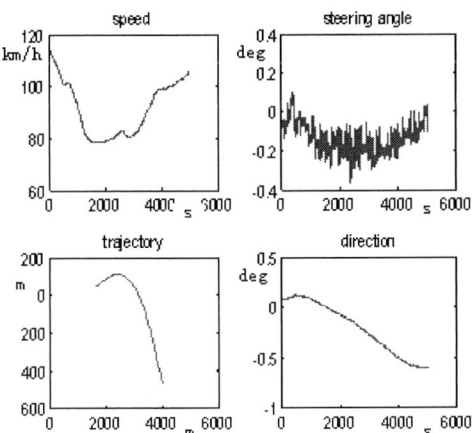

Figure 6.An experiment on Changsha Loop3 Highway

4. Conclusions

Active safety is the future of vehicle safety. Great efforts have been devoted in this area and great benefits have gained from it. Autonomous vehicle, as the final goal of the vehicle and the synthesis of all the active safety techniques, has achieved great attention word widely. It has become part of ITS research and many successful prototypes have developed in China.

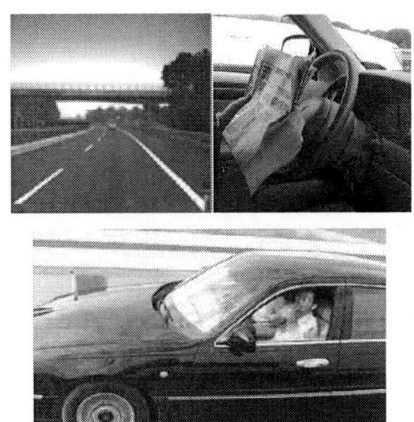

Figure 7.Autonomous Driving on Highway

The prototype HongQi Autonomous Land Vehicle exhibits excellent performance during the thousands kilometers of autonomous driving. It endures the high temperature, the different light conditions and the various terrain conditions. Many active safety technologies can be tested on the vehicle, such as Lane Departure Warning, Adaptive Cruise Control and so on. Parts of the achievement can be commercialized now.

5. References

[1] N.J. Nilsson, "Shakey the robot", SRI A.I. Center Technical Note 323, 1984.

[2] "System Architecture for ITS in Japan", ITS Japan, 2007. [Online]. Available: http://www.its-jp.org/english/arch_e/index.htm.

[3] S. Zhenping, A. Xiangjing, and H. Hangen, "CITAVT-IV--An Autonomous Land Vehicle Navigated by Machine Vision", Robot, Vol. 24, pp. 115-121, 2002.

[4] Z. Pengfei, H. Kezhong, O. Zhengzhu, and Z. Junyu, "Multifunctional Intelligent Outdoor Mobile Robot Testbed-THMR-V", Robot, Vol. 24, pp. 97-101, 2002.

[5] N. Kai and H. Kezhong, "THMR-V: An Effective and Robust High-Speed System in Structured Road " in 2003 IEEE International Conference on Systems, Man and Cybernetics, Washington, USA 2003. pp. 4370-4374.

[6] L. Jin, L. Guo, R. Wang, R. Zhang, and L. Li, "System Design and Navigation Control for Vision based CyberCar " in 2006 IEEE International Conference on Vehicular Electronics and Safety Shanghai, China 2006. pp. 459-464.

[7] Z. Ke, Q. Tong-yan, L. Dong-mei, W. Chun-yan, H. Rui-hua, and L. Hao, "The Latest Achievements of Chinese National ITS Architecture," journal of transportation systems engineering and information technology, Vol. 5, pp. 6-11. 2005.

[8] "Service Areas and Service System in the Architecture of ITS in China", Ministry of communications of the people's republic of China. 2005. [Online]. Available: http://www.moc.gov.cn/2006/06jiaotongzs/200606/t20060618_42057.html.

[9] A. B. M. Bertozzi, "GOLD: A Parallel Real-time Stereo Vision System for Generic Obstacle and Lane Detection," IEEE Transaction on Image Processing, Vol. 7, pp. 62-81, 1998.

[10] A. B. M. Bertozzi, and A. Fascioli, "Stereo Inverse Perspective Mapping: Theory and Application," Image and Vision Computing, Vol. 16, pp. 585-590, 1998.

[11] C. Ünsal, "Intelligent Navigation of Autonomous Vehicles in an Automated Highway System: Learning Methods and Interacting Vehicles Approach," Ph.D. dissertation, Virginia Polytechnic Institute and State University, Blacksburg, Virginia, 1997.

[12] M. I. J. Miura, and Y. Shirai, "A Three-level Control Architecture for Autonomous Vehicle Driving in a Dynamic and Uncertain Traffic Environment," in 1997 IEEE Conf. on Intelligent Transportation Systems, Boston, MA, 1997. pp. 706-711.

[13] M. T. Rosenstein and A. G. Barto, "Supervised Learning Combined with an Actor-Critic Architecture," CMPSCI Tech. Rep. 02-41, 2002.

2008 Workshop on Power Electronics and Intelligent Transportation System

Context Sensitive Design — Route Selection Design of Changsha-Jishou Freeway

Yingxue Zhang[1], Qisen Zhang[1], Chunhua Han[2] and Zhaohui Liu[1]

[1]School of Traffic and Transportation Engineering,
Changsha University of Science and Technology, Changsha, Hunan, 410076, China
[2]Braun Intertec Corporation 11001 Hampshire Avenue S Minneapolis, MN 55438, U.S.A
cscuzyx@126.com, chan@braunintertec.com

Abstract

Context Sensitive Design (CSD) means to take the project development and geometric design into consideration together, reduce the influence of highway project to its context, such as environment, economy, historical locations, communities, landscape and so on. Changsha to Jishou (Chang-Ji) freeway is one section of Changsha to Chongqing freeway, which is one of the eight big west channels as planned in China. It passes through two cities of Changde and Huaihua and one autonomous region of Xianxi in western Hunan Province, along which the cultural heritage and the natural resource are rich, and the landscape environment is exquisite. This paper carries on a basis of massive investigations and analysis on cultural heritage and natural environment along the route of Chang-Ji freeway. By using the CSD method, considering safety, availability and sensitive context, the proposal optimization of route design is put forward. At last, the paper introduces the practice effect of one section of the freeway.

1. Introduction

Changde to Jishou(Chang-Ji) freeway is one of the eight big channels to west in national plan — a section of Changsha to Chongqing freeway, the overall route extends from northeast to the west, which begins at Doumu lake in Changde City of Hunan Province in the east (meets to Xujia Bridge ,which connects Chang-Zhang freeway), which passes through Taoyuan County and Yuanling County in the west, then enters Luxi autonomous County of western Hunan, and finally ends in Jishou City with a total route distance of 230km. The cultural heritage is rich and the natural environment is exquisite along the route, if the route design is unreasonable, it will certainly influence safety and comfort, destroy the cultural heritage and the natural environment, increases the pollution of atmosphere, the water source and the noise.

Context Sensitive Design(CSD) [1][2][3] takes the road project development and the geometric design into

consideration, which meets the needs of minimum affect and maximum benefit for the project construction to its context (natural environment, economy, historical locations, communities, landscape and so on). The key of CSD is in each stage of project development, the design not only considers safety and availability, but also considers all essential environment factors except itself in road context at the same degree, which can realize sustainable and coordinated development of the highway construction and the environment. CSD method is used to select the route of Chang-Ji freeway in this paper, generalizes analysis on region context, the cultural heritage and the natural environment characteristic along the route, selects the related evaluation indexes, and evaluates the route design.

2. Investigation and Analysis of Cultural Heritage and Natural Environment along the Route

2.1. Cultural heritage

The cultural relic archaeology, investigation and exploration along Chang-Ji freeway are strictly according to the "Field Archaeology Service Regulations" of National Article Commodity Bureau, which uses work methods of prospect in those areas. The investigation along the route has discovered 26 ruins, the kiln sites, the grave groups and ancient construction vestige altogether, in which including 17 ancient ruins, 2 ancient kiln sites, 6 ancient graves and 1 ancient construction vestige. Historical age is from Old Stone Age to Ming and Qing Dynasty of China, the distributing area of historic sites is from 200 square meters to 80000 square meters.

There have four historical and culture ancient towns along the route which preserved well, and there have many old trees, some have over four hundred years old as sour Chinese jujubes, some have over two hundred years old as maples, and so on. All these cultural relics or historical heritages should be protected well with appropriate and scientific methods [4] [5] when construct highway because of their beyond retrieve.

978-0-7695-3342-1/08 $25.00 © 2008 IEEE
DOI 10.1109/PEITS.2008.76

2.2. Natural environment

After having carried on the investigation and analysis about the natural and cultural landscapes along the route of Chang-Ji freeway, we show the main landscapes in table 1.

Table 1 Main landscapes along Chang-Ji freeway

Serial No.	City/County Name	Main Landscapes
1	Changde City	The region is mainly paddy field, the terrain is plain, the landscape is mainly about farming culture
2	Taoyuan County	Primarily hill region, the terrain is relatively more flat, the relative difference in elevation is small, the landscape is mainly about hill and mountainous region, and the countryside village, as the close neighbor of Taohuayuan scenic area, it is also the historical culture landscape
3	Yuanling County	Foothill and low mountain region, the landscape is primarily hill and mountainous region, the countryside village landscape, with the Yuanjiang river landscape, the historical culture, the religious culture landscape, the subtropics green rich forest landscape
4	Chenxi County	Foothill and low mountain region, the landscape is primarily hill and mountainous region, the countryside village landscape, with the Yuanjiang river landscape, the religious culture landscape and the subtropics green rich forest landscape
5	Luxi County	Mountain areas with high and steep hillside, elevation difference is large, the situation is complex. The landscape is mainly about peak forest, the canyon landscape, the knoll and mountainous region and the Yuanjiang river bank landscape, with the subtropics green foliage forest landscape and the countryside village landscape
6	Jishou City	Mountain areas with high and steep hillside, elevation difference is large, the situation is complex. The landscape is mainly about peak forest, the canyon landscape, the knoll and mountainous region landscape, with the countryside village landscape

The rivers along the route belong to Yuanjiang river system, which is mainly Yuan water, the first branches of Yuan water and Wu water and other branches of Yuan water and Wu water. The surface water of Yuanjiang and its branches are little polluted with a better water quality. A well perfect water conservation irrigation system is constituted along the planned road, the irrigation and water conservation condition is better. Along the route there have not the large-scale wild rare and precious animals and without wild beasts and birds habitat of the national protection. At present, the vegetation coverage situation is good along the route, which is primarily about natural vegetation. We found that there had not the national protection plant species according to the Specification [6] in the scope of appraisal after investigation. Generally speaking, the structure of the land utilization is insufficiently reasonable along the route at present: few varieties, excessive utilization with light protection, the degraded soil fertility and the backward reclamation way.

3. CSD-Route Design Practice in Chang-Ji Freeway

(1) Route on both banks of Wushui and Wanrongjiang rivers

The route enters the coastal region of Wu Shui and Wan Rongjiang, considering traffic safety, cost, geology, environment, landscape, and so on, the route design comes from two compared plans, one is plan main route A, the other is plan route S.

Route A: After the route crossing Wu Shui in Long Zhiping, it passes Pingshangtian and crosses Wu Shui twice in Jiheping and Hutoudi; and it passes Gaoyixing and crosses Wu Shui twice again; then it enters along the river bank after passing north Zhongqieji village; it passes Shangqieji and Ousixi and crosses Wushui in Kanxizhai and leaves away along the river route; it passes Wuchangping and crosses Wushui near Hexi Hydrometrical Station; sets Hexi interchange on the opposite bank of Tonghe in Hexi Town, which connects 319 national highway; after passing Daodixi and Hongyanshan to Jiabianhe, and crossing Wanrongjiang twice, designs route on both banks of Wanrongjiang, the right route crosses Wanrongjiang once and the left side crosses Wanrongjiang twice; it passes Qifangqiao and arrives at K218+469.478.The total length is 24.789 km.

Route plan S: The route starts separating with route A in Pingshangtian, without crossing Wushui, flows from the north of Xiaopiliu power plant, passes the northern side of Qiepengxi and Niudaxi and the southern side of Paizhaiping Village, crossing Simahe and Zhangmuxi at the southern side of Langanpo to Anuoa, suppose Anuo interchange connecting with 319 national highway, after passing through the interchange, crosses Tonghe and passes Niujiaotuo and southern side of Bailatuo to Qifangqiao and recloses with main rout A. The total length is 21.909 km.

After analysis the two design plans, it concludes that plan route S is better than plan main route A when considering traffic safety, sensitive context and design criterions[7][8]of China.

·Horizontal indicators of plan route S are more advantageous for traffic safety than plan main route A .

·The total length of bridges and tunnels of route S is shorter than route A. So it can save project cost.

·The geologic condition along route S is better than route A.

·Route S is away from the rivers of Wu Shui and Wan Rongjiang, G319 national highway, and residential area is less along route S with small disturbance to inhabitants from the environment protection.

·Route S is reasonable for the region road net layout as it is away from the rivers and G319 national highway.

·Traffic condition of the area along route A is better

than route S. So the using value of the soil of route A is higher. However, the volume of earth and rock of route S is more than route A. And the transportation distance of material is further in some areas of route S.

The map of route layout of Wu Shui and Wan Rongjiang and the picture of natural environment protection are showed in Figure 1.

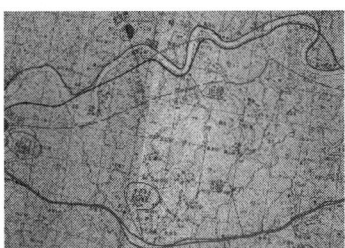 

Figure 1 Route layout of Wu Shui and Wan Rongjiang rivers to protect natural environment

(2) Design selection between high embankment and viaduct

The route of Yingtaowan tunnel in Huaihua section starts at Douziping of Liangshuijing Town, along Lian Xi and Mengxi mountainside, passes through Yaojia and Yingtao tunnels to Longdongxi reservoir in Liangshuijing Town. The route passes through a ravine terrain, which is deep and narrow, and also curved with steep hill and high difference altitude; landscape environment is natural and beautiful, natural slope of terrain is a little big, and the geological condition is complex; there have many mineral resources in route corridor, the shortcoming of the influence of the project to the natural environment is bigger than the advantage. In order to take the active control over investing and pay great attention to the mineral resources and the natural environment protection, the viaduct and the high embankment plans are proposed to the tunnel connecting route in the preliminary design stage, and the natural environment influence appraisal is carried on to two plans[9]. At last, the viaduct design is adopted to substitute for the high embankment and realized the union plan of bridge and tunnel. Figure 2 is the picture of Yingtao Yaojia viaduct construction.

Figure2 Picture of Yingtao Yaojia viaduct construction

(3) Historical Sites Protection

Viaduct bridge site in south Rinmin Road is located in Qianzhou of Jishou city, an autonomous area in western Hunan Province. The southeast of bridge site is

1.5km.length from Qianzhou. In the west, it is closely next to the general electrified plant. The bridge site first crosses south Rinmin Road, passes through Jiao-Liu railway and then crosses over a stream with a viaduct from southeast to northwest in the city of Jishou. When selecting the viaduct site, it considers the protection of Taixu temple- a historical site, which sits nearby the route, avoids more negative influence to the environment simultaneously and housebreaking in the inhabited area. Before selecting the bridge site design, many route plans are design, compared and evaluated, finally the plan of the bridge site with the least influence to CSD elements is selected [10]. The way of crossing south Rinmin Road in Jishou city and Jiao-Liu railway with a viaduct bridge is adopted, which protects Taixu Temple successfully as showed in figure 3.

Figure 3 Selection viaduct site to protect historical heritage

(4) Keep the primitive style of the ancient town

There has an ancient town architectural complex and ancient town vista with a history of nearly 200 years dating from Ming and Qing dynasties in the village of Madiyi in the town of Madiyi, whose preservation is rather complete. When the route arrives near the village, it goes nearby the ancient town and crosses the farmland area behind it after the analysis and contrast of the plan in order to protect the historical culture and its integrality. The route is adaptive to its context well, without any sight conflict and negative influence. The style of the ancient town and village countryside landscape shows in figure 4.

Figure4 The primitive style of an ancient town

(5)Protection the old and famous trees

Old and famous tree refers to the old big trees, commemoration trees, rare and precious trees in severe danger as well as the trees with unique landscape value; it is one kind of non-renewable natural resource and cultural heritage, which has important scientific, historical and sight value. Some trees are also carrier and representation of local conditions and customs and folk culture. Chang-Ji freeway route design should try to go by pass the old trees as far as possible; for example, in K87+720, design retaining walls for old trees protection; in K144+680, change the geometric design with 500,000yuan to protect an old tree, as shows in figure 5.

Figure5 Protect an old tree with retaining wall

4. Conclusions

In this paper, 5 segments of design route of Chang-Ji freeway in west China to protect context along its corridor are analyzed and selected by using the CSD thought. This not only can promote natural and man made landscape, but also promote highway itself and other resources and community values along the route. In the project, there has a lot of effort needs to do in order to protect natural environment furthest.

Acknowledgement

It is a project supported by the West Transportation Construction Science and Technology of China (2003 318 798 05) and the project supported by the Natural Sciences Foundation of Hunan Province of China (07JJ3111).

References

[1] Flexibility in Highway Design. *U.S. Department of Transportation Federal Highway Administration.* (FWHA Pub. No. FWHA-PD-97-062).

[2] Understanding Flexibility in Transportation Design-Washington. *Washington State Department of Transportation (WSDOT) and the Safety and Aesthetics Interdisciplinary Group (IDG).*April, 2005.

[3] Environmental Research Needs in Transportation. *Report of a conference Transportation Research Board Washington, D.C.* 2002.

[4] Zhang Wenrui. "The relation between cultural relics protection and economy construction". *http://www.cqvip.com.*

[5] Tang Yuyang. "Unmoving Cultural Heritage: Views of Trends in Science Protection". *CHINA Cultural Heritage.*2004, pp84-85.

[6] Industry Standard of the People's Republic of China. *Design Specification of Highway Environmental Protection (JTJ/T 006 — 98) [S].*The People's Communications Publishing, Beijing: 1998(in Chinese).

[7] Industry Standard of the People's Republic of China. *Highway Engineering Technique Standard (JTG/B01 — 2003)/[S].*The People's Communications Publishing House,Beijing,2003(in Chinese).

[8] *Chinese Ministry of Communications Specifications for environment impact assessment of highway (JTG B03-2006)/[S].* The People's Communications Publishing House, Beijing: 2006(in Chinese).

[9] Zhang Chonglu, Zhang Yingxue and Ning xiangxiang. "Analysis the Designs about Viaduct and High Fill Embankment of Chang-Ji Freeway based on CSD". *Journal of Highway of Automotive Application,* Vol. 3, pp102-104(in Chinese).

[10] Zhang Chonglu, Zhang Yingxue and Ning Xiangxiang. "Protection the Culture Heritage of Chang-Ji Freeway When Design Route". *Journal of Highway of Automotive Application,* Vol. 117, pp.78-80(in Chinese).

2008 Workshop on Power Electronics and Intelligent Transportation System

Modeling of Urban Traffic System based on Dynamic Stochastic Fluid Petri Net

Li Jingyu[1], Li Qiqiang[2]

[1]College of Information and Electrical Engineering,
Shandong Jianzhu University, Jinan, Shandong 250101, China
[2]School of Control Science and Engineering,
Shandong University, Jinan, Shandong, 250061, China
lijingyu@sdjzu.edu.cn, qqli@sdu.edu.cn,

Abstract

Urban traffic system, which is characterized by dynamics, concurrence and synchronization, is a hybrid dynamic stochastic system involving discrete and continuous behaviors. The modeling and analyzing of urban traffic system is the key of the deep understanding of the whole traffic and better guiding and controlling of urban traffic system, especially under the current traffic status, which is more and more congested and complicated. So we proposed the Dynamic Stochastic Fluid Petri Net (DSFPN) of well mathematical analysis ability, which can give a good description of the dynamic and stochastic character of the urban traffic system. The dynamic attributes and the stochastic actions of urban traffic system can be then modeled by Dynamic Stochastic Fluid Petri Net. Then an example of an intersection are modeled and presented in this paper. The simulation and performance analysis of urban traffic system model is introduced to indicate the validity of the DSFPN.

1. Introduction

Urban traffic net [1], whose traffic flow has stochastic character, is a large complex system affected by many factors. Especially under the acting of the human beings, the whole system becomes more complicated because of the existence of more uncertain factors. Now the modeling of the urban traffic flow system mainly focuses on the macro-model [2, 3] and the micro-model [4] based on the theory of fluid dynamics and car-following respectively. Due to the veracity and practicability of these two models, they have played an important role in the development of the traffic flow modeling [5]. However, the macro traffic flow model can only describe the dense and uniform flow, but can not explain the phenomenon of congestion, stat-stop and jam. The micro traffic flow model can only describe the high-density traffic flow about the local parts of the system, but can not give the whole understanding of the traffic flow [6]. Considering all of these reasons, we propose the method of modeling the urban traffic flow system with the Dynamic Stochastic Flow Petri net.

Petri net [7] as a graphic and mathematic modeling tool has not only the strict math definition and theory to support it but also the strongpoint of easy looking,

understanding and using. Petri net can well model the systems with characters of concurrence, synchronization, conflict and sequence. It has been extended in many aspects and applied to different fields [8-14] from it being. Now, the Dynamic Stochastic Fluid Petri net is given basing on the original Petri net and under the environment of the urban traffic flow system. Compared with the basic Petri net, the DSFPN takes the dynamic and stochastic property into account, and can model the complex system which consists of discrete and continuous variables. The urban traffic flow system is the system that agrees with the description of the above complex system, so we can model it with DSFPN. By this method, we can not only describe the dynamic and stochastic property of the whole system, but also can analyze the phenomenon of the congestion, jam and so on.

The rest of paper is organized as follows. Section 2 introduces the considered DSFPN formalisms. In Section 3, we introduce the property of DSDPN. In section 4 we introduce a crossing in traffic flow system as an application example, model and analyze the traffic flow system with DSFPN. Section 5 presents some conclusions.

2. The DSFPN formalisms

The basic Petri net is a four-tuple $N=(P,T,A,m_0)$, where. P is a finite set of places, T is a finite set of transitions, A is a set of oriented arcs with $A=(P \times T) \cup (T \times P)$, and m_0 is the initial marking of P.

The proposed Dynamic Stochastic Fluid Petri Net considering the character of dynamics, stochastic and flow is based on Petri net and increases the capacity of dealing with the optimal problem for mono-objective and multi-objective system. The definition is as follows.

2.1 The definition of the DSFPN

The DSFPN $N=(P,T,A,m_0,\lambda,W,F,C,P_r,G,t)$

978-0-7695-3342-1/08 $25.00 © 2008 IEEE
DOI 10.1109/PEITS.2008.82

485

is a 12-tuple, Where:

- P is a finite set of places with $P = P_d \cup P_c$, $P_d = \{p_{d1}, p_{d2}, \cdots, p_{d|P_d|}\}$ is the finite set of discrete places, $P_c = \{p_{c1}, p_{c2}, \cdots, p_{c|P_c|}\}$ is the finite set of flow places. P_d and P_c satisfies $P_d \cap P_c = \phi$. $|P_d| \geq 1$ is the dimension of the set of discrete places, and $|P_c| \geq 1$ is the dimension of the set of flow places.

- T is the finite set of transitions with $T = T_E \cup T_I$, where $T_E = \{t_{E1}, t_{E2}, \cdots, t_{E|T_E|}\}$ is the finite set of time transitions, $T_I = \{t_{I1}, t_{I2}, \cdots, t_{I|T_I|}\}$ is the finite set of immediate transitions. T_E and T_I satisfies $T_E \cap T_I = \phi$. $|T_E| \geq 1$ is the dimension of the finite set of time transitions, and $|T_I| \geq 1$ is the dimension of the finite set of immediate transitions.

- A is the finite set of oriented arcs with $A = A_d \cup A_c$, where $A_d = (P_d \times T_I) \cup (T_d \times P_I) \to N$ represents the set of discrete arcs, $A_c = (P_c \times T_E) \cup (T_E \times P_c) \to R_0$ represents the set of flow arcs.

- $m_d = (\# p_k, p_k \in P_d)$ stands for the marking vector of the discrete places, and $m_c = (\# c_k, c_k \in P_c)$ stands for the flow level vector of the flow places. The whole status of the DSFPN is represented with $m = (m_d, m_c)$, using M to denote the entire marking sets, $|M| \geq 1$ stands for the dimension of the marking vector. The initial marking is $m_0 = (m_{d0}, m_{c0})$. At time t, the whole status marking vector of DSFPN is $m(t) = (m_d(t), m_c(t))$, and the initial marking vector is $m_0(t) = (m_{d0}(t), m_{c0}(t))$.

- λ is the firing rate function of the time transition T_E, $\lambda : T_E \times M \to IR^+$. If a time transition t_{Ej} is enabled at the marking m, the firing rate is denoted $\lambda(t_{Ej}, m)$. At time t, it is represented $\lambda(t_{Ej}, m(t))$.

- W is the weight function of the immediate transition T_I, $W : T_I \times M \to IR^+$. At time t, it is represented as $W(t)$.

- F is the flow rate function defined on the arc connecting the flow place and the time transition, $F : A_c \times M \to IR^+ \cup \{0\}$. At time t, it is denoted $F(t)$.

- C stands for the property set of arcs. If it is multi-objective, it is denoted $C = \{c^1, c^2, \ldots, c^g\}$, where $G = \{1, 2, \ldots, g\}$ stands for multi-objection set. For the arc $(i, j) \in A_c, k \in G$, $C^k = \{c_{ij}^{kz_k}(t)\}_{z_k = 1,2,\ldots,D}$ stands for D possible arc sets leaving at time t for the kth objective. For each z_k, $c_{ij}^{kz_k}(t)$ is non-negative real number.

- P_r is the happening probability set of the multi-objective, $P_r = \{p_r^1, p_r^2, \ldots, p_r^g\}$. For each $c_{ij}^{kz_k}(t)$, the relevant probability is $\rho_{ij}^{kz_k}(t) \in P_r^k$.

2.2 The transition rules of DSFPN

Rule 1: The flow independence

The enabling of transition in DSFPN is irrespective to the flow marking, and totally depends on the discrete marking, represent as follows:

$$\varepsilon(m) = \varepsilon(m') = \varepsilon(m_d)$$

Where $\varepsilon(m)$ stands for the enabling marking set at $m = (m_d, m_c)$, $\varepsilon(m')$ stands for the enabling marking set at $m' = (m_d, m_c')$. Form above, we can know that the enabling is only relevant to discrete marking m_d and irrespective to the flow marking.

Rule 2: The enabling and firing effect on discrete and flow marking

At the marking $m = (m_d, m_c)$, the transition $t \in \varepsilon(m)$ is enabled, resulting the new marking $m' = (m_d', m_c')$, that is $m \xrightarrow{t} m'$. For discrete marking there is:

$$\forall p_k \in P_D, \quad m_k' = m_k - A_d((p_k, t), m) + A_d((t, p_k), m)$$

At time τ, the changing rate of flow marking $m_c(\tau)$ is:

$$\forall c_k \in P_C, \quad r_k(m(\tau)) = \sum_{t \in \varepsilon(m(\tau))} A_C((t, c_k), m(\tau)) - A_C((c_k, t), m(\tau))$$

2.3 The graphical representation of DSFPN

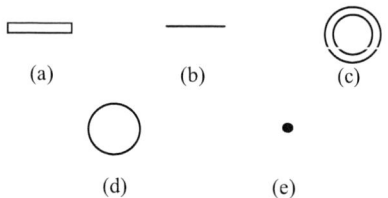

Figure 1　The elements of DSFPN

a——time transition, b——immediate transition, c—— flow place, d——discrete place, e——a discrete marking

3. The property of DSFPN

Property 1: The Dynamic property

The whole DSFPN has obvious dynamic property: the status marking $m(t)$, the firing rate $\lambda(t_{Ej}, m(t))$ of time transition T_E, the weight $W(t)$ of immediate transition, the flow rate function $F(t)$ of arc, the arc property $C^k = \left\{ c_{ij}^{kz_k}(t) \right\}_{z_k=1,2,...,D}$ and the relative probability $\rho_{ij}^{kz_k}(t)$ are all time dependent. At different time, DSFPN takes on different states and properties. So the whole traffic net exhibits dynamic characters as time goes by. For a special time interval S, we can divide it into several small time intervals, denoted as $S = \left\{ t_0 + s \triangle t \right\}_{s=\{0,1,2,...,I\}}$, which takes on different dynamic properties.

Property 2: Stochastic switching property

If an immediate transition t_{Ej} is enabled in marking m, it fires with probability

$$\frac{W(t_{Ej}, m)}{\left(\sum\limits_{\theta \in T_I, 且在m下使能} W(\theta, m) \right)}$$

Property 3: Mono-objective stochastic property

For an arbitrary arc $(i, j) \in A_c$, $c_{ij}^1(\bullet)$ stands for the arbitrary possible mono-objective cost function, so there is: $\Pr\{ x_1 + c_{ij}^1(x_1) \le x_2 + c_{ij}^1(x_2) \} = 1 \, \forall x_1, x_2 \in X \text{ and } x_1 \le x_2$ If taking the time as the mono-objective, there is:

$$\Pr\{ \tau + c_{ij}^1(\tau) \le t + c_{ij}^1(t) \} = 1 \, \forall \tau, t \in S \text{ and } \tau \le t$$

Property 4: Multi-objective stochastic property

For an arbitrary arc $(i, j) \in A_c$, $c_{ij}^k(\bullet)$ stands for the arbitrary possible cost function at the kth objective in the multi-objective net, $\omega^k(\bullet)$ stands for the possible objective value difference function, so there is:

$$\Pr\{ c_{ij}^k(x_1) \le \omega^k(x_2 - x_1) + c_{ij}^k(x_2) \}$$
$$\forall x_1, x_2 \in X, x_1 \le x_2 \text{ and } k \in G$$

Property 5: Flow rate conservative property

At time t, the marking of DSFPN is $m(t) \in M(t)$, fluid can leave place $c_k \in P_c$ along the arc $(c_k, t_E) \in A_c$ at rate $R[(c_k, t_E), m(t)]$. For each $t_E \in T_E$ enabled in $m(t)$, the instantaneous rate at which fluid builds at a place $c_k \in P_c$ at time t is then given by $r_{c_k}[m(t)] =$

$$\sum\limits_{t_E \in T_E, 且在m下使能} R[(t_E, c_k), m(t)] - \sum\limits_{t_E \in T_E, 且在m下使能} R[(c_k, t_E), m(t)]$$

Property 6: Liveness

we call the system modeled by DSFPN live, if there

always exists a firing sequence that make the marking transfer from m to m' for two arbitrary valid markings $m(t) = (m_d(t), m_c(t))$ and $m'(t) = (m_d'(t), m_c'(t))$, otherwise we call it deadlock.

Property 7: Boundness and security

A DSFPN is said to be bounded-safe if the fluid level of the flow place dose not exceed a upper-bound and the number of tokens in the discrete place also do not exceed, whatever its states are. For the flow place, we have

$$\frac{dc_k(t)}{dt} = \begin{cases} [r_k(m_t)]^+, & c_k(t) = 0 \\ [r_k(m_t)]^-, & c_k(t) = B_k \\ r_k(m_t), & 0 < c_k(t) < B_k 且 r_k(m_{t-})r_k(m_{t+}) \ge 0 \\ 0 & 0 < c_k(t) < B_k 且 r_k(m_{t-})r_k(m_{t+}) < 0 \end{cases}$$

We let $c_k(t)$ be the fluid level at time t in a flow place $c_k \in P_c$. We assume that there is an upper bound on the fluid content, that is, $c_k(t) \le B_k$ for all $t \ge 0$.

Proposition 1:

We call the net Discrete Free Choice Petri Net(DFCPN).

if the discrete place in the net satisfies $\forall p_d \in P_d, |p_d^\cdot| > 1 \Rightarrow {}^\cdot(p_d^\cdot) = \{p_d\}$.

Proposition 2:

$N_d = (P_d, T, F)$ is a discrete subnet of PN, $D, S \subseteq P_d, D \ne \phi, S \ne \phi$. we call D a non-empty siphon(deadlock) if and only if ${}^\cdot D \subseteq D^\cdot$. We call S a non-empty trap if and only if $S^\cdot \subseteq {}^\cdot S$. A siphon(trap) is said the minimal if and only if the non-empty proper subset of D (S) is not a non-empty siphon(trap).

Proposition 3:

The state machine component N_1 of the net N is defined as a subnet generated by places having the following two properties: (1) Each transition in N_1 at most has one incoming arc and one outgoing arc; (2) A subnet consists of places, all of their input and output transitions, and their connecting arcs.

In lecture [15], it gives the proof of the liveness and security theorem, due to it is not the main intention of this lecture, we just extend it to DFCPN.

Theorem 1:

A DFCPN is live iff every siphon in N contains a marked trap.

Theorem 2:

A live DFCPN is safe if and only if N is covered by strongly-connected State Machine(SM) components each of which has exactly one token in M_0.

4. Example

4.1 The crossing of urban traffic flow system

We take the crossing of urban traffic flow system as an example. There are many types of crossing, such as t-shaped crossing, crossroad, multi-crossed road and so on. We take the four-phase crossroad as an example.

We first introduce the traffic flow directions of the four-phase crossroad:

The first phase is east-west (west-east) straightly running and right turning; the second phase is east-west (west-east) left turning and right turning; the third phase is south-north (north-south) straightly running and right turning; the fourth phase is south-north (north-south) left turning and right turning.

We denote west, south, east and north respectively as 1, 2, 3 and 4, so we have:

(1) The east-west (west-east) straightly running traffic flow is 13、31; the east-west (west-east) right turning traffic flow is 12、34; the east-west (west-east) left turning traffic flow is 14、32;

(2) The south-north (north-south) straightly running traffic flow is 42、24; the south-north (north-south) right turning traffic flow is 41、23; the south-north (north-south) left turning traffic flow is 43、21.

The figure of the four-phase traffic flow crossroad is presented as follows:

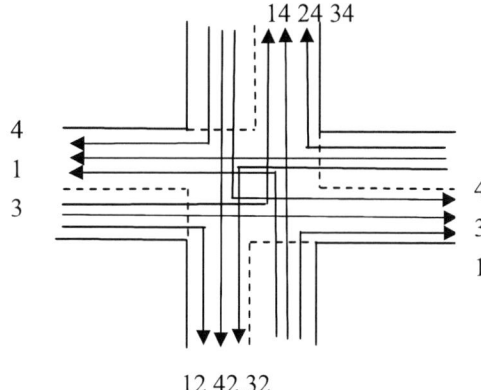

Figure 2 The four-phase traffic flow crossroad

4.2 The DSFPN model of the urban traffic flow system crossing

Assume that the crossroad is the general one, which is not limited in the direction of turning right, that is, it is always permitted to turn right. The modeling of the traffic flow crossroad with DSDPN is shown if figure 3.

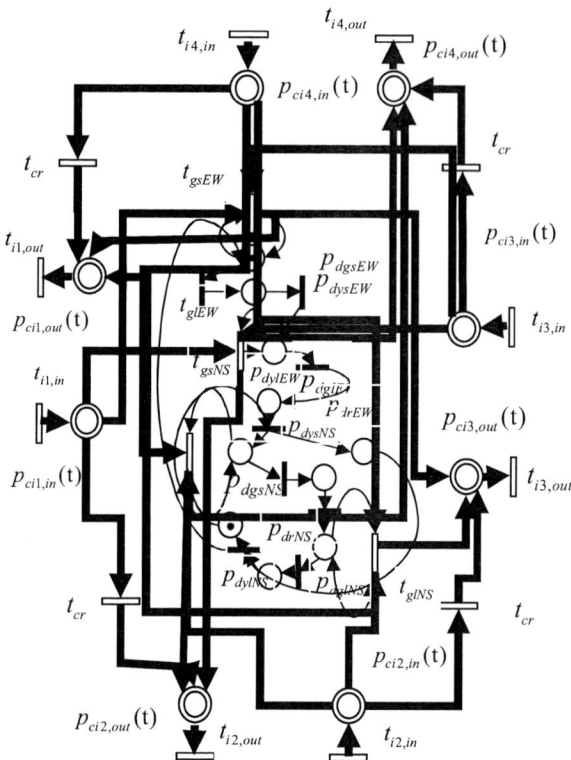

Figure3 The traffic flow crossroad DSFPN

In figure 3, $p_{ci1,in}(t)$, $p_{ci2,in}(t)$, $p_{ci3,in}(t)$ and $p_{ci4,in}(t)$ stand respectively for the crossroad traffic flow incoming place of west, south, east and north at time t; $p_{ci1,out}(t)$, $p_{ci2,out}(t)$, $p_{ci3,out}(t)$ and $p_{ci4,out}(t)$ stand respectively for the crossroad traffic flow outgoing place of west, south, east and north at time t; $t_{i1,in}$, $t_{i2,in}$, $t_{i3,in}$ and $t_{i4,in}$ stand for the incoming transitions; $t_{i1,out}$, $t_{i2,out}$, $t_{i3,out}$ and $t_{i4,out}$ stand for the outgoing transitions; t_{gsEW} and t_{gsNS} stand respectively for the east-west and north-south traffic flow running straightly across the crossroad; t_{glEW} and t_{glNS} stand respectively for the east-west and north-south traffic flow turning left at the crossroad; t_{c1} and t_{c2} stand for the immediate transition that the traffic light transfer from east-west (south-north) to south-north (east-west); p_{dgsEW} and p_{dgsNS} are discrete places, representing that the green light of running straight is on in the direction of east-west and south-north respectively; p_{dysEW} and p_{dysNS} are discrete places, representing that the yellow light of running straight is on in the direction of east-west and south-north respectively; p_{dglEW} and

p_{dglNS} are discrete places, representing that the green light of turning left is on in the direction of east-west and south-north respectively; p_{dylEW} and p_{dylNS} are discrete places, representing that the yellow light of turning left is on in the direction of east-west and south-north respectively; p_{drEW} and p_{drNS} are discrete places, representing that the red light is on in the direction of east-west and south-north respectively.

4.3 The traffic running analysis of the model

The running analysis of crossroad DSFPN model under the control of traffic lights is as follows:

Step 1: The initial marking of p_{dgsEW} and p_{drNS} is 1, which represents that the signal light of the south-north direction is red, and the signal light of the east-west direction is green, and now the transition t_{gsEW} is enabled. The traffic flow runs straight from $t_{i1,in}$ into the crossroad at the flow rate f_{gsEW} in the east-west direction. The firing rate of transition t_{gsEW} is $\lambda(t_{gsEW}, m(t))$, and satisfies some probability distribution, that is, the distribution of the crossroad traffic flow satisfies some stochastic time-delayed property. Once t_{gsEW} is fired, the running straight green light in the east-west direction and the red light in the south-north direction will continue until the turning left yellow light in the east-west direction ends in the step 4.

Step 2: The marking in the place p_{dgsEW} moves into place p_{dysEW} after the green light of running straight in the east-west direction, at this time in the east-west direction, it is in the state of yellow light with a yellow light time delay.

Step 3: The marking in the place p_{dysEW} transfers into the place p_{dglEW}, which stands for the east-west direction is in the state of turning left green light, the transition t_{glEW} is enabled. The traffic flow in the east-west direction turns left at the rate f_{glEW}, The firing rate of transition t_{glEW} is $\lambda(t_{glEW}, m(t))$, and satisfies some probability distribution with a time delay of the duration of green light for turning left.

Step 4: The marking in the place p_{dglEW} transfers into place p_{dylEW}, the light in the east-west direction is yellow, lasting a duration of one yellow light on.

Step 5: The marking of p_{dylEW} and p_{drNS} is 1, t_{c1} is

enabled, then the marking of the places p_{drEW} and p_{dgsNS} will be 1, the traffic flow transfer from the east-west to the south-north, whose analysis is the same as the east-west direction above.

5. The performance analysis of the DSFPN model

5.1 Liveness analysis

The liveness of DSFPN is mainly depends on the control part of the system, that is the signal control part

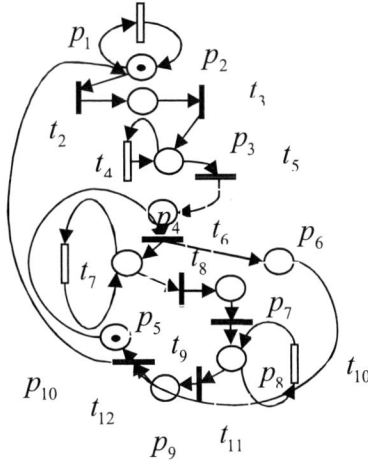

Figure 4   The traffic control

in figure 4. The following liveness analysis aims at the control part of the model.

From the structure analysis of the model in figure 4, we can see the model satisfies the **Proposition 1**. According to the **Proposition 1**, we can conclude that the model in figure is a Discrete Free Choice Net. The following is the analysis of the siphons and traps of the net.

First, we give a definition:

$D_1 = \{p_1, p_2, p_3, p_4, p_6\}$ and
$D_2 = \{p_5, p_7, p_8, p_9, p_{10}\}$

From the model figure, we can know:

(1) ${}^{\cdot}D_1 = \{t_1, t_2, t_3, t_4, t_5, t_6, t_{12}\}$ and
${}^{\cdot}D_1 = \{t_1, t_2, t_3, t_4, t_5, t_6, t_{12}\}$, so we can get
${}^{\cdot}D_1 = D_1{}^{\cdot}$;

(2) ${}^{\cdot}D_2 = \{t_6, t_7, t_8, t_9, t_{10}, t_{11}, t_{12}\}$ and
$D_2{}^{\cdot} = \{t_6, t_7, t_8, t_9, t_{10}, t_{11}, t_{12}\}$, so we also can get
${}^{\cdot}D_2 = D_2{}^{\cdot}$;

According to the Proposition 1, we know that:

(1) ${}^{\cdot}D_1 \subseteq D_1{}^{\cdot}$, ${}^{\cdot}D_2 \subseteq D_2{}^{\cdot}$, D_1 and D_2 are both siphons, from the traffic net analysis above, we can

know that the system only have the two siphons;

(2) $D_1' \subseteq {}^{\cdot}D_1$, $D_2' \subseteq {}^{\cdot}D_2$, so D_1 and D_2 are both traps;

We can know that all siphon in the net only has one trap, according to the **Theorem 1,** we can conclude that the traffic control net above is live.

5.2 Security analysis

In the figure 4, t_1, t_4, t_7, t_{10} are the continuous transitions which have no effect on judging security of the system, so we can have a predigestion of the above net, shown in figure 5.

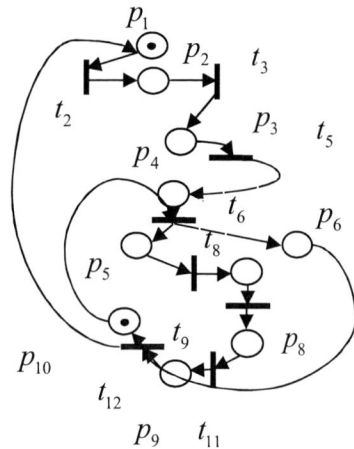

Figure 5   The simple traffic control model

In the above net, two strong-connected State Machine (SM) component is generated by the set of places $\{p_1, p_2, p_3, p_4, p_6\}$ and $\{p_5, p_7, p_8, p_9, p_{10}\}$, which is shown in the figure 6.

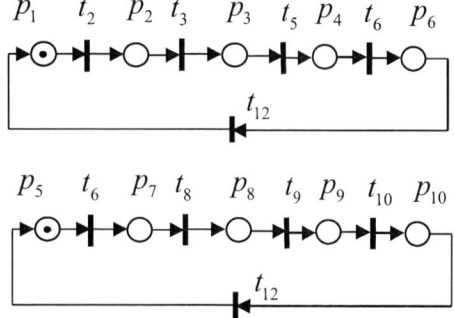

Figure 6   The SM component Model figure of the traffic control model

In the figure 6, we can know that the two SM component satisfies the **Proposition 3**, and the initial marking of the SM component only has one token, According to the **Theorem 2**, we can conclude that the simple model of traffic control is safe. Because the transitions t_1, t_4, t_7, t_{10} have no effect on the security, we can conclude the whole model of traffic control is safe.

6. Conclusion

The Dynamic Stochastic Flow Petri Net is the extension of the basic Petri net. It can model systems with dynamic, stochastic and fluid properties. The system modeling with DSFPN has not only the character of easy looking, but also good mathematic analysis ability, which make us have a deeper understanding of the whole complex dynamic stochastic flow system. Through the modeling and running analyzing of the crossroad in the urban traffic flow system, we can have a more visual knowledge about the inner dynamic and stochastic property of the system. According to the analysis of the liveness and security, we demonstrate the feasibility of the net.

References

[1] Lu Huapu. Intelligent traffic system [M]. Beijing: China Communications Press, 2003.

[2] Aw A, Rascle M. Resurrection of "second order" models of traffic flow [J]. SIAM J. Appl. Math, 2000, 60: 916-938.

[3] Helbing D, Hennecke A, Shvetsov V, Treiber M. Micro- and macro-simulation of freeway traffic [J]. Mathematical and Computer Modeling, 2002, Vol.35, 517-547.

[4] Wang Xiaoyuan, Juan Zhicai, Ja Hongfei. Micro-simulation models of traffic flow of developing and evaluating ITS [J]. Journal of Traffic and Transportation Engineering, 2002, 2(1): 64-66.

[5] Gong Xiaoyan, Tong Shuming, Wang Zhixue, Chen Dewang. Survey on freeway traffic flow modeling [J]. Journal of Traffic and Transportation Engineering, 2002, 2 (1):74-79.

[6] C. Tolba, D. Lefebvre, P. Thomas and A. El, Moudni.

Continuous and timed Petri nets for the macroscopic and microscopic traffic flow modeling. Simulation Modeling Practice and Theory, 13(2005): 407-436.

[7] Yuan Chongyi. Petri Net [M]. Nanjing: Southeast University，1989.

[8] Lin Chuang. Stochastic Petri net and system performance evaluation [M]. Beijing: TSingHua University Press, 2001.1.

[9] Trivedi K S, Kulkarni V G. FSPNs: fluid stochastic Petri nets[C]. In: Proceedings of Petri Nets'93.Chicago, 1993, 24-31.

[10] Horton G, Kulkarni V G, Nicol D Metal. Fluid stochastic Petri nets: theory, applications and solution [J]. European Journal of Operation Research, 1998,105 (1):184-201.

[11] Wolter K. Second order fluid stochastic Petri nets: an extension of GSPNs for approximate and continuous modeling [C]. In: Proceedings of WCSS'97, Singapore, Sept.1997, 328-332.

[12] Bouyekhf. M, Abbas-Turki. A, Grunder. O, EI. Moudni. A, Fluid stochastic Petri net for control of an isolated two-phase intersection. Proceedings of the IEEE multiconference on computational engineering in system applications, France, 2003。

[13] Mariagrazia Dotoli, Maria Pia Fanti. An urban traffic network model via colored timed Petri nets. Control Engineering Practice, 2006, 14: 1213-1229.

[14] M. Gribaudo, M. Senero, A. Horvath, A. Bobbio, Fluid stochastic Petri nets augumented with flash-out arc: modeling and analysis. Discrete event dynamic systems: theory and applications, 2001, 11, 97-117.

[15] Murata, T. Petri Nets: Properties, analysis and applications [J]. Proc. of the IEEE, 77(4):541--580, 1989.

2008 Workshop on Power Electronics and Intelligent Transportation System

Study on Driving Behavior and Traffic Conflict at Highway Intersection

Xing Ge[1], Jian Lu[1], Qiao-jun Xiang[1], Peng-ying Wang[1]
[1]*School of Transportation, Southeast University ,*
Nanjing, Jiangsu, 210096 , China
e-mail: gx7980@yahoo.com.cn

Abstract

Traffic conflict studies have been accepted because the data and the effectiveness of a treatment of safety problems can be collected and evaluated in a short period of time. Driving behavior is the reflection of the vehicles operating characteristic, and it has close relation with traffic safety. In the paper, the process of vehicle coming and going at intersection, and driving behavior when conflict occurring were analyzed by field observation and video record observation. Based on the above analysis, driving behavior was classified into fiver types. Because each driving behavior has influence of different degree on traffic safety, driving behavior needs to be weighted. The paper used expert consultation method, and obtained multiple expert' scoring for each kind of driving behavior. Then the more reasonable weights were given by statistical method determining weights. Comprehensive value of driving behavior during statistical time is gotten according to the survey data. Finally, the relation model between driving behavior and conflict was established.

1. Introduction

Intersections are areas of high-ways that produce conflicts among vehicles and pedestrians because of entering and crossing movements. The driver decides when to complete the required maneuver, based on elements of the decision context including distances, velocities, and car performance. Driver decisions in these situations and their impact on safety and capacity considerations are typically reflected in probabilistic approaches that consider the vehicles operation and traffic condition at highway intersections. The changes of those driving behaviors are strongly associated with traffic conflicts and traffic safety [4][5].

In an overview of previous studies, Verschuur, W.L.G and Hurts, K established safe and unsafe driving behaviour model. In the model, the following risky behaviours and characteristics related to driving

were measured through self-report[6]. Xuedong Yan analyzed Effect of restricted sight distances on driver behaviors during unprotected left-turn phase at signalized intersections by identifying the changes of those driving behaviors and traffic conflict associated with drivers' restricted sight distances. The analyses in the paper provide a better understanding of the relationship between highway visibility and traffic safety and operation[7]. Meanwhile, driving behaviors has been used as an important measurement to analyze and predict traffic conflicts and accident rates at intersections [8][9]. However, those study focus on gap acceptance behaviors generally. It is inadequate to reflect driving behavior by gap acceptance decision because the movement of vehicles passing the intersection is quite complex especially at unsignalized intersections as conflict happening.

This paper gives some analysis about the process of traffic conflict happening at unsignalized intersections and the definition of driving behavior based on field observation and video record observation. Driving behavior is classified into five kinds, i. e. alone, pushing, waiting1, waiting2 and following. Before a relation model between driving behavior and traffic conflict is established by the survey data, each driving behavior is weighted for influence of different degree on traffic safety.

2. Driving behavior characteristics analysis

The essence of traffic conflict is the manifestation of traffic behavior unsafe factors, which can induce traffic accident, or avoid it for appropriate taking evasive action [10]. The process of traffic conflict is presented as Figure 1.

978-0-7695-3342-1/08 $25.00 © 2008 IEEE
DOI 10.1109/PEITS.2008.83

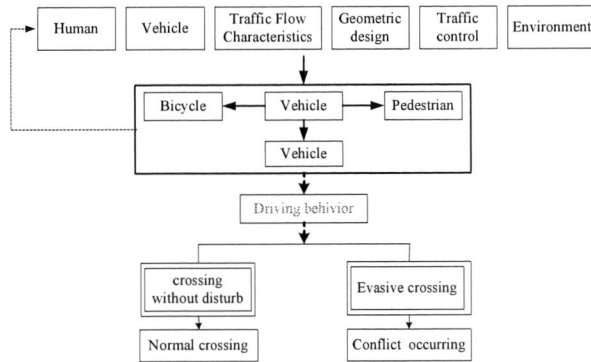

Figure 1 The process of traffic conflict happening

From Figure 1, Driving behavior plays an important role in the process of traffic conflict occurring. The driver behavior of vehicle crossing intersection is a reflection of vehicle operation characteristic, and affected by various factors; meanwhile it is consequence of different factors influencing. As the different characteristics and temperament of drivers, and the different environment in which they are driving a vehicle, these make vehicles have different running acts. This concretely manifests as different cognitive performance to the same things, such as for a heavy traffic flow at major road, the vehicles at minor road is to choose to decelerate and wait for passing carefully, or do not stop and continue to push and force vehicles at major road slow down or stop.

At unsignalized intersection, vehicles have different operation action, owing to the behavior based on the field video observation, the direct relationship between driving behavior and traffic conflict is discovered, when existing more pushing vehicles, more traffic conflict can be occurred with traffic chaos at the intersection; while having more following vehicles, which represents that vehicles is more heavy in the direction, and usually not easy to be disturbed by other directions, the traffic conflicts is less, usually are same direction conflicts.

To analyze the process of vehicles crossing intersection and driver behavior when traffic conflict occurring, it is pre-requisite to classify driving behavior. The paper classify it into four types: alone, following, waiting and pushing. The definitions are described as bellow:

Alone: vehicle comes and goes cross conflict spots at intersection with slight block, not being disturbed by other vehicles, so the rate of conflicts happening is less.

Following: vehicle comes and goes cross conflict spots at intersection following others without block;

the rate of conflicts happening is also less, mainly conflicts between vehicles in same direction.

Waiting: vehicle is disturbed by other vehicles and waits crossing intersection, the driving behavior is discontinuous, including two situations:

(1) When vehicle coming and going unsignalized intersection, considering safety, the prudent drivers usually decelerate or stop, looking around until the gap between opposite vehicles is enough to cross. Or at stop or yield and the bad sight distance intersections, drivers who abide by traffic regulations choose to stop and wait a gap. The conflict commonly occurs between vehicles on minor road that grades of cross road and traffic volumes have obvious differences.

(2) To avoid conflicts with other vehicles or disturbance by other factors, drivers choose to cross intersection with evident deceleration and wait to march forward.

Pushing: in the process of crossing intersection, vehicle is disturbed, but driver chooses to squeeze to cross. All are pushing. The type keeps closest relationship with conflict, and the degree of conflict is highest.

According to above definitions of several driving behaviors, the influence of each one on conflict is different, and each one can induce conflict, so it is not possible to reflect vehicles' operation using any one behavior, and must use the comprehensive value of driving behaviors obtained by analyzing the relation between them and conflicts.

3. Weight determination

Based on influences of various driving behaviors on conflict, weight of different behaviors must be determined. There are lots of methods to determine weight, mainly subjective and objective methods. The paper adopts statistics method to determine weight of driving behaviors. Statistics method is easier to work and overcomes objective liberty of expert inquiry method to a certain extent [11].

The processing of statistical method is described as follows: Based on independent scoring of selected experts, the score of each driving behavior is collected, and adopting mathematical statistics, these scores are analyzed, and then the logical weight will be obtained.

According to the statistical method, forty experts were consulted and give their own weight vector $U=\{u_1, u_2, \cdots u_m\}$ on each elements contained factors (driving behavior):

$$A_i = \left(a_{i1}, a_{i2}, \cdots, a_{im}\right)\left(\sum_{j=1}^{m} a_{ij} = 1, i = 1, 2, \cdots, n, \quad n = 40\right) \quad (2)$$

Then each factor u_j (j=1,2,···, m, m=5) is treated by adopting one-way factor method:

(1) The maximum M_j and minimum m_j must be collected in all its weights a_{ij} (i=1,2,···,n) for u_j, namely:

$$M_j = \max_{1 \le i \le n} a_{ij}, \quad m_j = \min_{1 \le i \le n} a_{ij} \qquad (3)$$

(2) Choose proper positive integer K and calculate the group distances of K groups divided by n weights from small to big.

(3) Calculate the numbers and frequency of weight in each group of weights.

(4) According to the distribution of numbers and frequency, determine the weight a_j of u_j, then obtain the weight vector A=($a_1,a_2,···,a_m$).

Based on above methods, the frequency of each weight is rather concentrated by judging various driving behaviors like alone u_1, following u_2, waiting1 u_3, waiting1 u_4, pushing u_5, and both maximum number and frequency are in intervals as (0.4-0.8), (0.7-1.0), (0.3-0.7), (0.3-0.45), (0.37-0.42). According as the real observation, value 0.06, 0.08, 0.05, 0.41 and 0.40 are chose respectively as the corresponding weight of above five driving behaviors.

4. Relationship analysis

To analyze relationship between driving behavior and conflict, a video investigate for 6-hour was made at intersection of S123 and Zhenfang West Road in April, 23rd, 2007, and the correlative dates were obtained.

The intersection observed has no any control measures with traffic chaos, and S123 is arterial highway, Zhenfang West Road is minor arterial highway. Usually the vehicles on Zhenfang West Road yield to ones on S123. The ratios of various driving behaviors will be shown Figure 2.

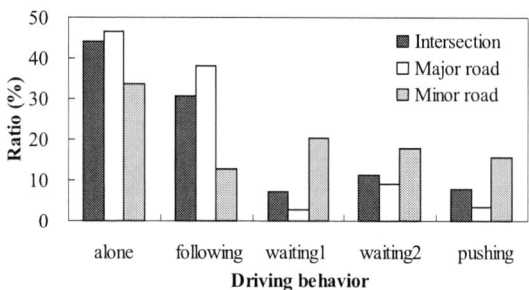

Figure 2 Ratio of driving behavior

Through statistics, the ratios of various driving behaviors are obtained respectively: 44%, 30%, 7%, 11%, 8%, and approximately 26% (waiting, pushing) of vehicles were be blocked when crossing intersection. The ratio 53% of waiting and pushing behaviors on Zhenfang West Road is far more than the one 15.5% on S123. And the vehicles were blocked more on minor road than major one. From the videos observation, the probability of conflicts on minor road is much bigger than one on major road. The different ratio of various driving behaviors can reflect not only traffic conditions but also safety and conflict happening at intersection.

To explore the relationship between driving behavior and conflict, two indices: conflict rate and driving behavior aggregative ratio, Will be introduced. The former is the ratio of conflict numbers per hour and mixed passenger car unit (TC/MPCU), and its value reflects a certain safety level. It is relatively valid and reliable that using conflict rate to evaluate the safety level of an intersection[12]. The latter is the ratio of driving behavior comprehensive value (the sum of various driving behaviors numbers and their corresponding weight in statistic time period) and vehicles numbers, and it reflects the weight of various driving behaviors.

In this paper the statistic time interval is 10min, by recording and processing the investigation dates, the conflict rate and driving behavior comprehensive ratio were obtained. Considering that the dates' discrete degree can be greater because of variance of different statisticians on conflict and driving behavior, the writer handles dates smoothly and eliminates some abnormal ones. Based on the handled dates, adopting regression analysis method, the regression equation between conflict rate and driving behavior comprehensive ratio are presented in Figure 3.

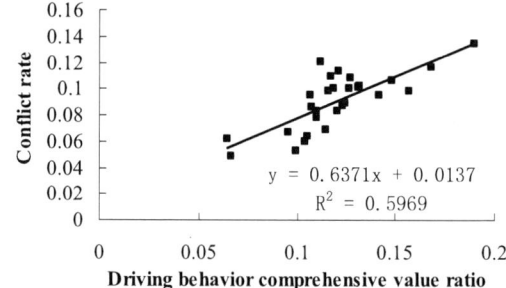

Figure 3 Relation between driving behavior and conflict

According to Figure 3, the relationship of driving behavior and conflict is approximately linear (R^2=0.6), the conflict rate increases at intersection with the increasing of driving behavior comprehensive ratio. The above conclusion reveals that drivers' different behavior coming and going intersection keeps close relationship with traffic conflict. Thereby, it is feasible to further open out the conflict forming process through analyzing drivers' driving behavior, and judge the safety condition of intersection.

5. Conclusions

The paper presents a relation model between driving behavior and conflict at unsignalized intersection based on the impact on traffic operating and safety. The definition of each kind of driving behavior was derived from analysis on traffic conflict happening and observation of real behaviour, and various driving behavior was used with different weights by expert scoring and statistical method. The study indicated that driving behavior that incorporate characteristic of vehicles crossing intersection, stream interaction and traffic safety different from gap acceptance behavior are helpful in explaining traffic performance at intersections with incomplete enforcement of priorities. These may also contribute to further study unsafe driving behavior and mechanism of producing conflict in different circumstances.

In this study, the linear regression model was used to analyze the change of proportion of different driving behavior and conflict rate. The results show the larger conflict rate increases, the larger comprehensive value rate of driving behavior. Meanwhile, it was found that proportion of waiting and pushing behavior was higher, vehicles movement was blocked more seriously and traffic operating is more unsafe at intersection from statistical data. The model served as a predict of conflict rate of "uncontrolled" intersections is possible and driving behavior is a critical component. Moreover, results of the study support evaluation for traffic safety on highway intersection and can be theoretical basis for making safety management strategy.

Acknowledgement

The work that is described here was supported by grants from the National Natural Science Foundation of China (50522207).

References

[1] Parker, M. R., and Zegeer, C. V(1989). "Traffic conflict technique for safety and operations: Observers Manual", Report FHWA-IP-88-027, U.S. Department of Transportation, Washington, DC

[2] N. M. Katamine (2000). "Various volume definitions with conflicts at unsignalized intersections", *Journal of Transportation Engineering*, Vol. 126, No. 1, pp.27-35

[3] Su Zhang. *Traffic conflicts technique in China*, Southwest Jiao Tong University Press, Chengdu, 1998

[4] Isam Kaysi, Ghassan Alam (2000). "Driving behavior and traffic stream interactions at unsignalized intersections", *Journal of Transportation Engineering* , Vol. 126, No. 6, pp.498-506

[5] Zhang en-liang, Xiao gui-ping, Nie lei (2006). Influencing analysis and strategy study of traffic environment on driver psychology. *Journal of Highway and Transportation Research and Development (application technique edition)* , Vol.11, pp.164-165

[6] William L.G. Verschuur and Karel Hurts(2007). Modeling safe and unsafe driving behaviour, *Accident Analysis and Prevention*, Vol.34, No.6, pp.779–792

[7] Xuedong Yan, Essam Radwan(2007). Effect of restricted sight distances on driver behaviors during unprotected left-turn phase at signalized intersections. *Transportation Research Part F*, Vol.10, pp.330 ﹣ 344

[8] Alexander, J., Barham, P. and Black, I (2002). "Factors influencing the probability of an incident at a junction: results from an interactive driving simulator", *Accident Analysis and Prevention* , Vol.34, No.6, pp.779–792

[9] Spek, A. C. E, Wieringa, P. A, Janssen, W. H (2006). "Intersection approach speed and accident probability", *Transportation Research Part F: Traffic Psychology and Behavior*, Vol. 9, pp.155-171.

[10] Cheng wei. *Theory and Application of Traffic Conflict Technique on urban traffic*, Science Press, Beijing, 2006

[11] Yin-sheng Yang, Kui-yuan Zhang. *Application fuzzy mathematics method*, Jilin University Press, Changchun, 2001

[12] Cheng Wei, Ding Tong-qiang, Li Jiang (2004). "Evaluation of Traffic Conflict Based on Gray Theory at Intersection", *Journal of highway and transportation research and development*, Vol.21, No.6, pp.97-100

Study on Information Sharing Platform of Distributed Transportation Monitoring Systems Based on Interoperability and Internet

Jian LEE, Yuan-hua JIA, Gu-chang AO

College of Traffic and Transportation，
Beijing Jiaotong University, Beijing, 100044, China
E-mail: njleegend@yahoo.com.cn

Abstract

Integrated traffic information sharing and uniform management has achieved more and more attention in China, for most of transportation monitoring systems belongs to different management departments, so information of them can not be shared by common, even of one road. Each monitoring system can be viewed as an isolated information island, and this phenomenon cannot be changed by gigantic investment. In this paper, by applying ideas and methods of system analysis and integration, we put forward a platform-designing scheme based on internet as data transmission carrier and interoperability as database operation method, and with this, we can realize information sharing of separated monitoring systems through internet.

1. Introduction

Thousands of monitoring systems that existed in nearly 50,000 km expressway network were constructed by isolated model in China because of the management, investment and other reasons, and they are not synchronized and coordinated. According to the viewpoint of system engineering, this will inevitably undermine the overall strength of system, such as traffic control, emergency secure etc.

On the other hand, increasing demands for mobility of goods and people and urban development all puts pressure for a more practical, efficient and sustainable transport system. With the development of Intelligent Trans. System (ITS), integration became a main trend of ITS development. With technical means of ITS to optimize and improve the regional road traffic control and management is an effective countermeasure, and to achieve this, we must gain the support of information-sharing platform and advanced control technology.

2. Information sharing platform research condition

Traffic information sharing platform is a database or data warehouse system that combines services, technology, management, control and planning [1-2].

The importance of integrated traffic monitoring systems in providing sharable information is being recognized by researchers and transportation authorities in recent years, and the development of transportation monitoring systems information sharing platform has been approached differently by researches and traffic engineering on the basis of different subjects and technologies such as data communications, networking, web database, network engineering and system engineering, etc.

In this field, the recent works include ITS data management system development and ITS data fusion (the United States) [3-6], ITS platform construction for Smart-way (Japan) [7], 5-T system (Italy), TRIDENT and UTMC (UK)[8], and I-transportation platform (Singapore)[9].

Traffic information-sharing platform technology study in China is still in its infancy. The representative works include the comprehensive platform architecture based on various data processing techniques[2], the architecture use of multi-agent technology[10], description of platform structure and the types of service of city road network[11], data management module of inter cities traffic[12], and metadata based platform module with the techniques of multi database and data warehouse[13-14].

3. Existent problems and shortages

At present, about 10 cities in China has identified traffic information sharing as key development content during the progress of ITS demonstration construction. In the past and at present, those platforms that has been built or to be build adopted multi-level structure mostly, the construction of hardware attaches great importance and cost a lot in the construction of regional control center and communication network.

The traditional mode of centralized control system must overcome the existing obstacles of information heterogeneity of various systems. Moreover, this not only need to increase the substantial investment, but also massive data in the upper system will cause data redundancy multiplied and this will reduce the effect of real-time traffic control. Meanwhile, data mining and processing means to massive original data is still lack and the sharing model is available in the study.

4. Problem solving method and scheme

Aimed at solving the technological problem, we use system design principles for reference, the first one is, maintain former subsystems' character and structure; second, take full advantages of former subsystems and gain the support of them.

4.1. Related Techniques — distribution and interoperability

According to the concept of web GIS, in system designing, distribution means that data, information and processing programs can be located in several servers of different sites, and distribution scheme and computing method is propitious to data processing's assignment and optimizing over computer system[15]. Adopt this scheme not only can overcome the bottleneck of responding that exist in centralized system, but also can solve problems such as data isomer, data sharing and operation complexity of web GIS[16].

In addition, the other technique we need in the platform study and design is interoperability. In traditional meaning, interoperability is the ability to communicate and share data across programming languages and platforms [17].

According to the reference above, interoperability among information systems can be defined as the ability to communicate and share data across different information systems and the ability to make decisions or take actions based on shared information.

In generally meaning, interoperability is a mutual behavior, and it can be divided into design-time interoperability and run-time interoperability.

The former means interoperability among systems be designed early in the period of system construction and it is suitable to close, well-rounded and concentrated system; and the later means interoperability will not work only when mutual communication need produced among different structural systems and it is suitable to open system more. Both kinds of interoperability need to base on plentiful standards, and besides some static standards and criterions such as data structure, format, grammar, and communication protocol that common to them, run-time interoperability need more system criterions such as registration, combination and service progress.

Therefore, the design of this platform should obey the following demands and concepts.

First, this platform or system should base on distributed topology and only by this means can we assure the combination of multi-resources' data that located in different monitoring systems.

Second, the platform must be accessed in real time and this requires that the platform's data transmission and information issuance base on Internet. Further, in system designing, we should keep to the mechanism of 'thin client terminal' that means the requirement to users' soft and hard ware as low as possible.

Third, the platform should be designed as an open system with strong expansibility, that means the platform can be expanded and mend more functions according to users' need. During virtual application period, this platform will allow more roads' monitoring systems join into, and at the same time, it can shield out some of function modules.

4.2. Integrated solutions — information sharing platform

According to the demands and concepts of design above, and characters of distribution and interoperability, the platform we put forward in this article is an internet-based interactive collaboration virtual regional integration control system. In this platform, several kinds of agents for implementation of different functions are included.

The platform includes several monitoring centers of different roads, and each center includes basic data collection and monitoring system data layer, the middle layer and application layer. The middle layer includes transport layer features databases, model base, knowledge base, information transmission and decision making intelligent bodies.

Monitoring systems that included in this platform only need to visit each other's characteristics database, without having to visit underlying database, such sharing model not only ensure the security of the internet platform, but also conducive to resolving current difficult problems of data sharing because of divided administrative.

This platform exchanging information with the outside world based on service-oriented architecture (SOA). Mainly refers to transportation and statistical information exchange among various monitoring systems and related traffic management departments, and provide the community with the regional state of road traffic information.

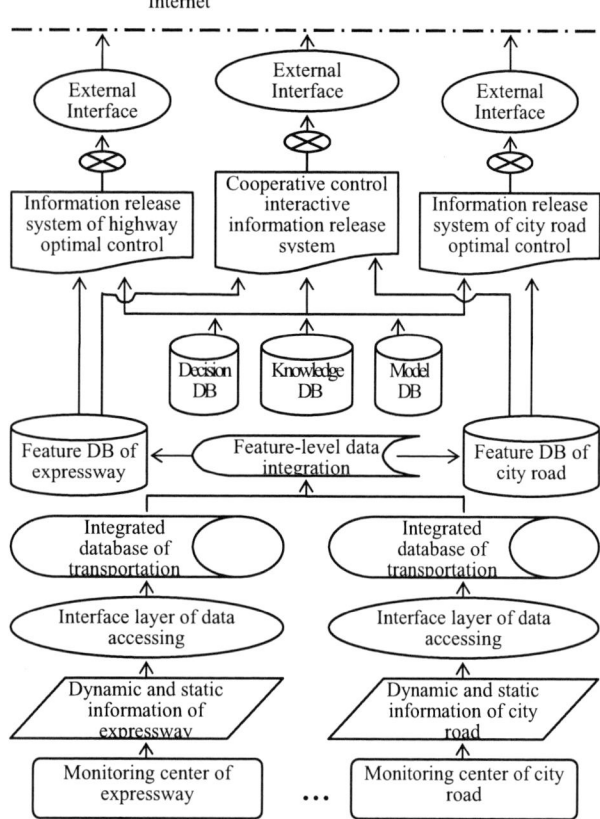

Figure 1. Design process of platform

4.3. Topology of information sharing platform

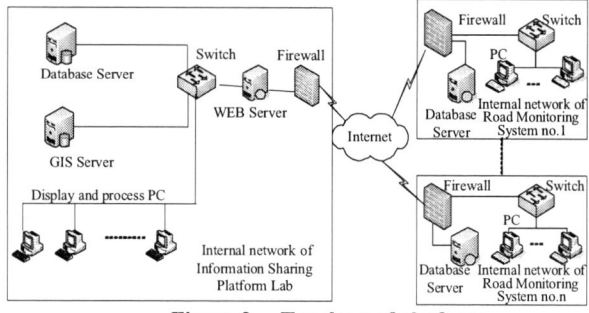

Figure 2. Topology of platform

5. Conclusion and discussion

Each kind of hardware equipment has its physical bottleneck, no matter how wide the cable is or how vastness the database is, so there lays a series of problems in the platform to be solved for the characters of hardware the platform adopts. Nowadays, information quantity increasing and data dealing demand bring forward high request on the platform. We can anticipate that the platform we design here will face information congestion

and time delay because of gigantic visiting volume with more and more roads joining into this platform in the coming future. Therefore, broadcast mode can be adopted in the platform's information release instead of user's requirement by order.

Acknowledgement

The work is supported by National High-Tech Research and Development Plan (863 Program), No. 2007AA11Z213.

Reference

[1] YANG Dong-yuan, XU Shi-wei, and JIA Jun-gang, "Shared Information Platform for Expressway Management Information System"[J]. *Journal of Tongji University*, Tongji University Press, Shanghai, 2000, 28 (6): 664-669.

[2] LI Rui-min, and LU Hua-pu, "Study on Intelligent Transportation Integrated Information Platform" [J]. *Central South Highway Engineering*, Central South University Press, Changsha, 2005, 30(2): 30-33.

[3] LI Rui-min, LUHua-pu, and SHI Qi-xin, "Research of Development and Trend of Integrated Transportation Information Platform"[J], *Journal of Highway and Trans.Research and Development*, China Communication Press, Beijing, 2005, 22(4): 90-94.

[4] Department of Transportation,USA, *The National ITS Architecture version 6.0*[R],Washington, 2006.

[5] Florida Department of Transportation District. *Surface Trans.Security and Reliability Information System Model Deployment Final Deployment Plan*[R], Tallahassee, 2004.

[6] GUAN Ji-zhen, "System Framework and Integration of ITS Common Information Platforms"[J], *Journal of Trans. Systems Engineering and Information Technology*, China Science Press, Beijing , 2002, 2(4): 11 - 16.

[7] Road Bureau, Ministry of Land. *Infrastructure and Transport. ITS Handbook 2003-2004*[R], Tokyo, 2003.

[8] Glasgow City Council and Birmingham City Council. *UTMC 10 User Handbook*[R], Birmingham , 2000.

[9] QIAO Rui-long. "Intelligent Transportation System Development Survey of Singapore"[J]. *Intelligent Transportation Systems*, Qinghua University Press, Beijing, 2003, 5(1): 5-9.

[10] LI Rui-min, ZHEN Ying-jie, and LU Hua-pu, "A Multi-agent basedIntegrated Transport Information Platform"[J]. *Journal of Transportation Systems Engineering and Information Technology*, China Science Press, Beijing , 2006,6(1): 28−32.

[11] XU Hong-feng, YANG Xiao-guang, and PENG Chun-lu, "Study on the Planning of the Highway Transportation Information Platform" [J], *Intelligent Trans. Systems*, Qinghua University Press , Beijing, 2005, 7(3):13-16.

[12] HUANG Shan, LU Hua-pu, and QIAN Zhen., "Data Management Modules Research of Intercity Traffic Information Platform Software"[J], *Computer and Communications*, Wuhan Institute of Technology Press, Wuhan, 2005, 23(6): 1-4.

[13] GU Wei-hao, CHEN Guang-wei., "Information Sharing Platform Model"[J]. *Chinese Railway Science*, China Railway Press, Beijing, 2003, 24(4): 49-53.

[14] ZHANG Feng, LI Bao-luo, "Metadata based Information Sharing Platform"[J]. *Database Thch.and Application*, China Telecom. Press, Beijing, 2007, 3(2): 54-57.

[15] BIAN Fu-ling, TAN Xi-cheng, "P2P Network Model Conformed to the Service of Distributed Virtual Geographic Environment"[J], *Geomatics and Information Science of Wuhan University*, Wuhan University Press, Wuhan, 2007, 32 (11): 1028-1034.

[16] AI Hong, "Distributed Acces Control"[J], *Computer Engineering and Design*, China Aerospace Industry Press, Beijing, 2007, 28(21):5110-5112.

[17] ZHANG Shu-liang, YIN Li-li, and SHI Miao-miao. "The Summary of Spatial Interoperability Architecture Integrated Mode"[J]. *Geo-InformationScience*, China Science Press, Beijing, 2006, 2(4): 88-95.

2008 Workshop on Power Electronics and Intelligent Transportation System

Quantitative Measure for Transportation Network Efficiency

QIN Jin[1,2]   SHI Feng[1]   Miao Li-xin[2]   CHEN Qun[1]

1 School of Traffic and Transportation Engineering, CentralSouth University, Changsha, 410075
2 Research Center for Logistics, Graduate School at Shenzhen, Tsinghua University, Shenzhen, 518055
csu_qinjin@hotmail.com

Abstract

A new quantitative measure for transportation network efficiency is proposed in this paper. The measure can be used to assess the performance of a transportation network, because it can reflect the influence on the network efficiency induced by flows, travel cost and travel behavior under the equilibrium of the transportation network. Numerical application denotes the method can measure the efficiency of the transportation network rationally, and the results of the measure accords to the practical situation.

1. Introduction

Networks and, especially, complex networks, have been the subject of intense research activity in recent years. Indeed, this subject of networks has been studied in operational research, transportation engineering, economics, physics, and sociology. In this field the reliability or vulnerability of the network is the study emphasis all through. However, in order to be able to evaluate the vulnerability or the reliability of a network, a measure that can quantifiably capture the efficiency or performance of a network must be developed. In this paper, the quantitative efficiency measure for transportation network is studied.

JIAO Peng-peng and LU hua-pu (2005) regarded the efficiency of transportation network as construction cost, pollution cost and energy consumption. YU Li-jun and JIN Wen-zhou(2006) used the accessibility, mobility, productivity and utility to analysis the transportation efficiency. Latora, Marchiori(2001) defined the efficiency by the shortest path lengths between the nodes, and this measure had been applied by the above authors to a variety of networks, but only the shortest path was considered in the measure, and other important factors could effect the network efficiency were ignored. For example, the network configuration obviously has an impact on network performance or efficiency and this impact can be found in many network design problems. In addition, the traffic flow and the behaviors of users in transportation network should impose on the efficiency too, because the flow is an important indicator of the network travel cost, and the travel choice on the time or mode of the users can work on the travel cost too.

Therefore, a new quantitative measure for transportation network efficiency that captures flows, the costs associated with "travel," and user behavior, along with the network configuration, is more appropriate in evaluating transportation networks.

In this paper, a new transportation network efficiency measure was proposed and the new measure was related to the Latora and Marchiori (2001) measure used in the "complex" network literature. We compared the resulting derived from our measure to it and also provide illustrative examples.

2. Traffic network equilibrium model

We now recall the general traffic network equilibrium model, which is widely used and applied in practice. Given a transportation network $G = (N, A)$, in which N is set of nodes, A is set of roads, and $|A| = n_A$. The set of O/D pairs of nodes is denoted by W. The demand for O/D pair $(r,s) \in W$ is denoted by q_{rs}. The set of paths joining an O/D pair $(r,s) \in W$ by P_{rs} and $\cup P_{rs} = P$. We denote the nonnegative flow on path $p \in P_{rs}$ by f_{rs}^p and the flow on road a by x_a. We define a binary variable δ_a^p, if road a is contained in path p, $\delta_a^p = 1$, otherwise $\delta_a^p = 0$. The cost on road a by $t_a = t_a(x_a), a \in A$.

A network equilibrium can be defined as follows:

$$Min \ Z(\mathbf{x}) = \sum_{a \in A} \int_0^{x_a} t_a(y) d_y \qquad (1)$$

$$\text{s.t.} \qquad \sum_{p \in P_w} f_w^{rs} = q_{rs}, \qquad \forall (r,s) \in W \qquad (2)$$

$$x_a = \sum_{(r,s) \in W} \sum_{p \in P_{rs}} f_{rs}^p \delta_a^p, \quad \forall a \in A \qquad (3)$$

$$f_{rs}^p \geq 0, \forall r, s, p \qquad (4)$$

In the model, the objective function (1) is to minimize the total travel cost in the network, constraint (2) is the conservation on flows. The road flows are related to the path flows through the conservation of flow equations (3), expression (4) is the constraint of nonnegative on flows.

There many algorithms for the solutions of the above

978-0-7695-3342-1/08 $25.00 © 2008 IEEE
DOI 10.1109/PEITS.2008.92

model (1)-(4), such as Frank-Wolfe Method, Gradient Projection Method etc. And the classical Frank-Wolfe Method is selected in here to solve the model. The detailed steps of the Frank-Wolfe Method can be found in HUANG hai-jun(1994).

3. Transportation Network Efficiency Measure

Up to now, there were few quantitative transportation network efficiency measures in the literatures.

We could recall that the network efficiency measure of Latora and Marchiori (the L-M measure) (2001), which was proposed to measure the efficiency of networks was defined as follows:

$$E = E(G) = \frac{1}{n(n-1)} \sum_{i \neq j \in G} \frac{1}{d_{ij}} \qquad (5)$$

Where n is the number of the nodes in network G, d_{ij} is the shortest path length between nodes i and j, if no path between nodes i and j, then $d_{ij} = +\infty$.

Obviously going with the rise of the traffic flow, the shortest path length between nodes i and j will increase, then the network efficiency E computed by expression (5) will descend linearly at the same time, and E is a strict monotone decreasing function on traffic flow, so we could infer that the efficiency will reach its maximum when the traffic flow in the network is equal to 0. This result of the L-M measure impossibly is rationally.

The authors considered an reasonable transportation network efficiency measure must have some properties as follows: ① In a given transportation network, for fixed traffic demand, the bigger the total travel time, the smaller the efficiency of the network, namely, the network efficiency is in inverse proportion to the total travel time; ② In a given transportation network, if the total travel time is fixed, the bigger the traffic flows in network, the bigger the efficiency of the network too, namely, the network efficiency is in proportion to the traffic flows.

Based on the above assumptions, a new network efficiency measure for transportation networks is proposed now. The measure is defined in the context of network equilibrium, and it captures demands and costs, and the underlying behavior of "users" of the network. The formal definition is as follows:

For a given transportation network G and OD demand vector $\mathbf{q}$, then the network efficiency $\varepsilon(G, \mathbf{q})$ can be defined as:

$$\varepsilon = \varepsilon(G, \mathbf{q}) = \frac{1}{n_A} \sum_{a \in A} \bar{x}_a \Big/ \bar{t}_a \qquad (6)$$

Where n_A is the number of roads in the network, and $\bar{x}_a$ and $\bar{t}_a$, respectively, the flow and travel time on rods a in the equilibrium of the network. Furthermore, we could extend the travel time in expression (6) into generalized cost, namely, the cost could include the environmental cost etc. In addition, because some roads must be treated individually in practice, we still could add weight coefficient $\lambda_a \geq 0$ for any roads $a \in A$, which can stand out or reduce the impact of the specifically roads on the network efficiency. The settings of these factors have no influence on the following research, so we don't analysis these factors here.

Note that the transportation network efficiency measure given in (6) has a meaningful economic interpretation which is that it measures the average performance by cost in terms of roads, with the performance being measured by the flows and the cost on the road. It could be regarded as the average traveler quantity that can be operated when a unit travel cost was consumed in the equilibrium transportation network. The higher the traffic that can be handled at a given price (which also reflects the cost and, from an engineering perspective, the travel time), the higher the efficiency or performance of the transportation network.

For example, in the context of transportation network only has one road, and $\bar{x}_a = 50$ vehicles, $\bar{t}_a = 0.5$ hour, then according to expression (6), $\varepsilon = 100$ (vehicles/hour), so the network can process 100 vehicles in an hour, If $\bar{t}_a = 1$ hour, then we can get $\varepsilon = 50$ (vehicles/hour), and this network would be half as efficient as the former network. In addition, we must notice that depending upon the transportation network under consideration, the unit of measurement would correspond to the type of flow on the network.

4 Numerical example

Consider the transportation network in Figure 1 in which there are four nodes 1, 2, 3 and 4, five roads a, b, c, d and e. Assume that the roads cost functions are given by BPR functions:

$$t_a(x_a) = 10\left(1 + 0.15\left(\frac{x_a}{4}\right)^4\right),$$

$$t_b(x_b) = 15\left(1 + 0.15\left(\frac{x_b}{6}\right)^4\right),$$

$$t_c(x_c) = 12\left(1 + 0.15\left(\frac{x_c}{3}\right)^4\right),$$

$$t_d(x_d) = 15\left(1 + 0.15\left(\frac{x_d}{10}\right)^4\right)$$

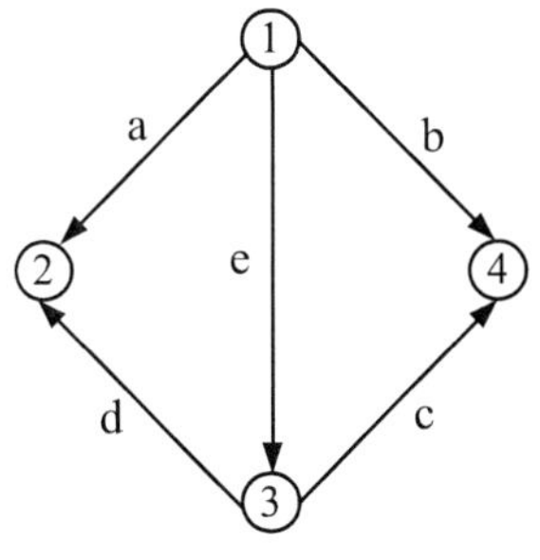

Fig.1 Transportation network

$t_a(x_e) = 20\left(1 + 0.15\left(\dfrac{x_e}{8}\right)^4\right)$, There are five OD pairs: $(1,2)$, $(1,4)$, $(1,3)$, $(3,2)$ and $(3,4)$ in the network, with demands given, respectively, by $q_{12} = 11$, $q_{13} = 2$, $q_{14} = 6$, $q_{32} = 3$, $q_{34} = 1$。The equilibrium flow can be easily found using Frank-Wolfe Method as follows: $\overline{\mathbf{x}} = \{\overline{x}_a, \overline{x}_b, \overline{x}_c, \overline{x}_d, \overline{x}_e\} = \{8.23, 4.93, 2.07, 5.77, 5.84\}$.

Our network efficiency computed by (6) is then $\varepsilon = 0.249$, and efficiency computed by (5) is $E = 0.020$, the total travel time in the network is 114.3。The compare between these absolute value can't demonstrate the rationality of the efficiency ε, but we can use the impact on the network efficiency which come from the variety of the factors to prove it.

4.1. Impact of the travel behavior

We now analysis the impact of different behaviors or configurations of users in the network on network efficiency, namely, when the traffic demands in the network are fixed, the relation between the network efficiency and the travel behaviors.

For example, if all demands in the network are transferred to OD pairs (1,2), as $q_{12} = 23$, $q_{13} = 0$, $q_{14} = 0$, $q_{32} = 0$, $q_{34} = 0$. So in the equilibrium network the roads b and c will not be used, accord with the practice the network efficiency should decrease obviously. And we could get that $E' = 0.0176$, $\varepsilon' = 0.191$, and compared with the original efficiency, respectively the new efficiency decrease 12% and 23.3%, at the same time the total travel time in the network is 158.2. Similarly, if all demands in the network are transferred to OD pairs (1,4), as $q_{14} = 23$, $q_{12} = 0$, $q_{13} = 0$, $q_{32} = 0$, $q_{34} = 0$, then we could get $E'' = 0.0171$, $\varepsilon'' = 0.102$, and respectively the new efficiency decrease 14.5% and 59.1%, the total travel time is 277.1. According to the change of the total travel time, we could know that the measure proposed in this paper is more scientific and more rational than the L-M measure, because its results accord with the fact much more.

Furthermore, we analysis the variation trend of efficiency in the process which the demand q_{12} transfers to demand q_{14} gradually.. The Δ_1 is defined as the transferred number from q_{12} to q_{14}, so $0 \le \Delta_1 \le q_{12}$. In this process, we could know the trend of the network efficiency that is illustrated with fig 2. We could also define Δ_1 as the transferred number from q_{14} to q_{12}, $0 \le \Delta_2 \le q_{14}$, then the trend of the network efficiency is illustrated with fig 3.

From the Fig.2 and Fig.3, we know that when the

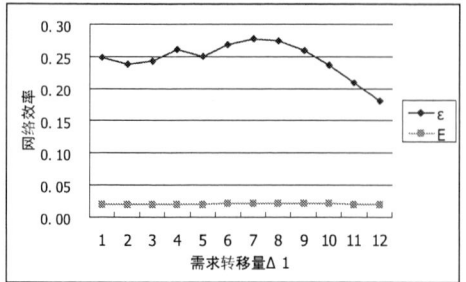

Fig.2 q_{12} transforms to q_{14}

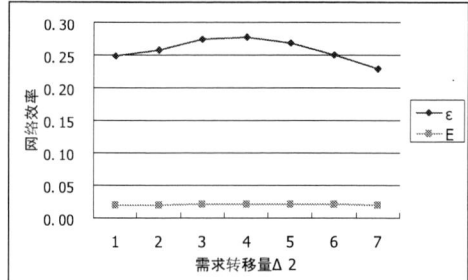

Fig.3 q_{14} transforms to q_{12}

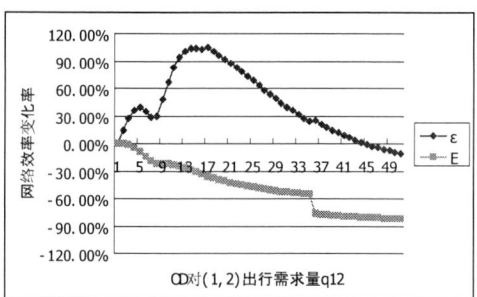

Fig.4 q_{12} vs. network efficiency

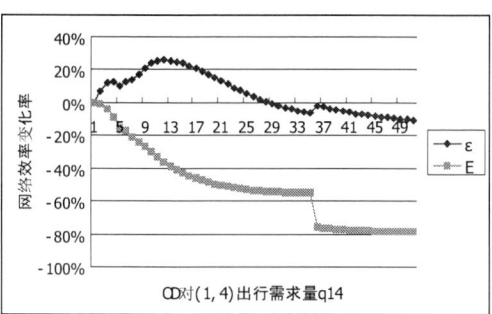

Fig.5 q_{14} vs. network efficiency

travel behaviors of the users in the network change, the network efficiency E computed by the L-M Method is nearly fixed, but at the same time, the network efficiency ε computed by the method proposed in this paper fluctuates according to the change of the travel behavior in the network. So we could conclude that the new method is more rational than the L-M Method on the reflection of the travel behavior.

4.2. Impact of travel demand

The Fig.4-Fig.5 describe the change rate of network efficiency relative to its start value when the q_{12}, q_{14} increase from 1 to 50 respectively. In these figures, we know that the efficiency E is the monotone decreasing function with respect to travel demand. But the efficiency ε also fluctuates according to the change of demand, and they all increased before decrease, moreover, they all have their maximums. So we should know the efficiency ε is more scientific than the efficiency E.

From the Fig.4 – Fig.5, we also know that for a given transportation network, there must exist an optimal traffic flow which could maximize the network efficiency. So when the traffic flow changes, we could use some methods to change the transportation network configuration, such as add new roads, or we also could use other methods to change the traffic flow, such as route guidance, so the network efficiency could be increased.

4.3. Impact of travel network configuration

The following Fig.6 and Fig.7 described that the sensitivity of the efficiency E and ε to the demand after delete road a and b respectively. From these figures, we know that the efficiency E is also the monotone decreasing function with respect to travel demand under this condition. And the values of E before and after delete the roads are all closely, so it could not reflect the impaction of the transportation network configuration on the transportation network efficiency. But we can see that the efficiency ε reflect this impaction rationally. And in the change of ε, the famous Braess Paradox phenomenon can be explained.

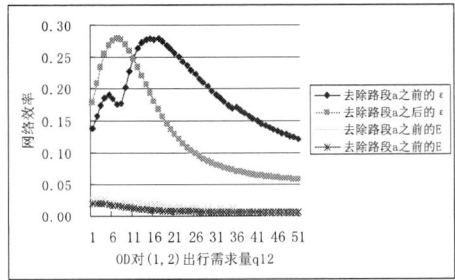

Fig.6 Network efficiency before and after remove road a

If the transportation network efficiency before delete some roads in smaller than which after delete the roads, then we could think that the Braess Paradox occurs in the transportation network. For example, in the Fig.9, when $q_{14} = 2$ and other OD demand have not change, the efficiency ε is equal to 0.237 and 0.285 before and after delete road b respectively, and the total travel time is

equal to 95.58 and 88.76 respectively. So it show that we remove the road b, the efficiency ε increases and total travel time reduces on the contrary, which is just the Braess Paradox.

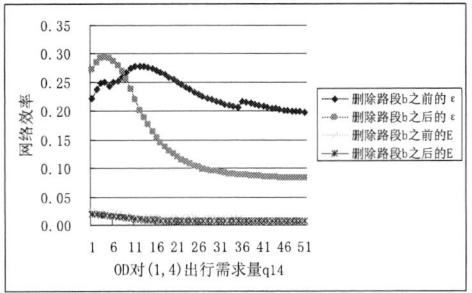

Fig.7 Network efficiency before and after remove road b

Furthermore, we could find in the Fig.6 that when the value of q_{14} is in [1，8], then the efficiency ε will increases after delete road b, but when value of q_{14} is greater than 8, then the efficiency ε will reduce after delete road b, and the efficiency changes analogously according with road a.

The Braess paradox occurs in a certain part of demand range accords to the result in Dietrich Braess and Anna Nagurney(2005). And the efficiency ε reflects this rule perfectly.

Moreover, in the Fig.4-5, and Fig.6-7, we could find that the efficiency ε has two extremum point before deleting the road a or b, this is attribute to two paths link the OD pair (1, 3). And after the road b is removed, there is only one path links OD pair (1, 3), so only one extremum point exist in these figures. But in all the figures we could only find one extremum point all the time. So we think that the efficiency ε accords with the fact much more than E.

5. Conclusions

In this paper a new quantitative measure for transportation network efficiency is proposed. The method computes by the flows and travel times on the roads in the equilibrium transportation network, and it captures flows, costs, and behavior of travelers, in addition to network. The network measure is well-defined, even in the case of disconnected networks. The numerical example denotes that the measure provides more scientific decision information than the old methods. And the results of the example also reflect the traffic demand and network configuration have impaction on the network efficiency. So in the practical traffic management, we could improve the transportation network operation efficiency by controlling(adding or reducing) the traffic flow or change the network configuration.

6. Reference

[1]. JIAO Pengpeng, LU Huapu, WANG Jianwei. "Optimization of urban road network based on transport efficiency". *Journal of Tsinghua University(Science and Technology),* 2005, March, pp. 297-300

[2]. YU Li-jun; JIN Wen-zhou. "A Study on Measures of Transportation Efficiency". *Highway,* 2006, Oct, pp. 102-106

[3]. Latora, Marchiori. "Efficient behavior of small-world Networks". *Physical Review letters,* 2001, Oct, pp. 1-4

[4]. Latora, Marchiori. "Is the Boston subway a small-world network?". *Physica A*, 2002, May, pp. 109-113

[5]. Latora, Marchiori. "How the science of complex networks can help developing strategies against terrorism". *Chaos, Solitons and Fractals,* 2004, Oct, pp. 69-75

[6]. HUANG Hai-jun. Urban transportation network eguilibrium analysis:theory and practice. China Communication Press, Beijing, 1994

[7]. Dietrich Braess, Anna Nagurney, Tina Wakolbinger. "On a Paradox of Traffic Planning". *Transportation Science,* 2005, June, pp. 446-450

[8]. PEI Yu-long~1, GAI Chun-ying. "Study on Operation Reliability of Highway 墩 network". *Journal of Highway and Transportation Research and Development.* 2005, May, pp. 119-124

2008 Workshop on Power Electronics and Intelligent Transportation System

Combined Land-use and Transportation Demand Modeling Based on Equitableness

Ming Yang, Xiucheng Guo, Lan Wu

Transportation College, Southeast University, Nanjing, 210008, China

yangming45@hotmail.com, seuguo@163.com, wl_nfu@sina.com

Abstract

Land use changes inevitably have impacts on travel costs and trip production. The impacts may be positive or negative to transportation users. Importantly, equity issues arise as the positive benefits and the negative ones may be unevenly distributed. This paper will propose an equity-based Land Use-Transportation Model, which examines the benefit distribution among transportation users as well as investigates equity concerns associated with the impact of land development on the equilibrium O-D travel cost. The study uses a bi-level programming model integrating equity issues into the travel demand model. The upper level sub-model maximizes traffic production with equity constraints, while the lower level sub-model addresses a combined trip distribution/assignment user equilibrium problem. Genetic algorithm based method is applied to solve the bi-level model.

1. Introduction

In the field of regional travel demand modeling and transportation planning, it has been widely accepted that land-use decisions and residential/ commercial development patterns contribute to local traffic generation, choice of travel mode, and congestion [DOT 2001]. For planning problems related to land use and transportation, the maximum total zonal trip generation in a traffic zone can be used as a evaluation metric, and can help back-calculate the maximum amount of new development that generate pre-defined traffic impacts on different users of the network. This paper is intended to investigate the interrelationship between land use and transportation from this aspect.

It can be seen, for example, that the existing vacant land is developed to accommodate economic growth, which lead to more trips and generally higher vehicle trip generation. Intuitively, higher travel cost from an origin to a destination may occur because of increases in traffic volume. Although Fisk [1] verified that on some route an increase in one O-D pair's demand may result in a decrease in another O-D pair's travel costs, it is true that network users' overall travel cost in the network will increase. This problem should be considered as paradox since the conditions of a transportation network

generally will be increasingly severe with incremental trip production. The percentages of the increase in travel cost for different groups of network users may fluctuate sharply. To some extent, this raises an equity issue in that changes in O-D travel cost for network users are not evenly distributed among users. The definition of equity may vary in economics and in transportation research, though the fairness is the common character. There exists unequal distribution of costs/benefits among different groups of households such the increasingly growing inequality between rich and poor. This inequality may affect spatial distribution. For this kind of equity issue, many researchers used the Gini coefficient [2] to show the degree of income inequality among various groups of households. However, this paper is not investigating how the inequality affects trip generation, while it is investigating how increases in traffic lead to different equilibrium travel cost to different zones and may further result in inequality. A travel cost ratio is developed as a proxy measure of equity to shed light on benefit distribution. It should be noted that the proposed proxy equity measure cannot capture the loss of consumer surpluses from the suppressed demand and the unequal distribution of the cost-benefit. Meng and Yang [3] examined the equity issue in terms of the change of equilibrium O-D travel cost in the continuous road network design problem. They also verified that to different network users the benefit distribution is different in terms of the change of travel cost. If such concerns about inequity are neglected, it may be difficult to gather public support but rather arouse public opposition to the implementation of a zonal development plan or a road construction project. In fact, it may be unfair to the community. Therefore, it is essential to develop an evaluation method to understand how additional trip generation associated with new development may impact on social equity. In short, this equity issue of land-use transportation problems is classified as the Land-Use Transportation Problem based on Equity (LUTPE). LUTPE is intended to measure the relationship by estimating the potential trip generation of zonal development based on the equity consideration. Planners and real estate developers have made a large number of explicit recommendations regarding the land use and transportation connection. Nonetheless, equity management of land use and

978-0-7695-3342-1/08 $25.00 © 2008 IEEE
DOI 10.1109/PEITS.2008.100

505

transportation has in many cases been inconsistent and insufficient. Although equity issues in continuous network design problem have been examined theoretically and empirically [3], such models have not been applied to land-use and transportation problems.

From the user perspective, every user tries to minimize his or her travel cost when traveling from origin to destination. New land development attracts and produces additional trips, which may cause congestion and increase travel costs. Such a connection requires the consideration of land use changes in travelers' route choice behavior. However, the dynamic and unsteady relationship between land use and transportation are not considered in static four-step transportation planning models. In addition, conventional travel demand models have overwhelmingly focused on the traffic demand from origin to destination while overlooking the attraction of destination.. Kim [4] pointed out that once the transportation network is determined, zonal travel demand is endogenously determined together with link congestion costs, optimal amounts of production and resulting factors of destinations. In this study, we adopt the model of Equilibrium Trip Distribution/Assignment with Variable Destination Costs (ETDA-VDC) to integrate land use and transportation and to capture the activity characteristics of individual trip-producing zones.

Based upon game theory, the LUTPE is a typical Stackerlberg game in which the leader is planner and the follower is network user who complies with the deterministic user equilibrium (DUE) principle in route choice. Thus, a bi-level programming approach can perfectly model the LUTPE. In the bi-level programming model for the LUTPE, the upper level sub-problem maximizes the production of each residential zone subject to some physical constraints as well as the equity constraint; while the lower level sub-problem characterizes the network users' decisions with regard to routes, origins, and destinations, in response to the traffic planners' decisions. It is well-known that the bi-level programming models from transportation network optimization problems with DUE constraints are difficult to solve due to the non-differentiable property of these models, implicit relations between DUE link flows and the decision variables of problems, and the large scales of actual transportation networks [5]. Besides various heuristics and gradient-based methods including the sensitivity analysis based algorithms [6] [7], the random search techniques can be applied to solve bi-level programming models generated from transportation systems. For example, Friesz et al. [8] applied the simulated annealing algorithm for the continuous network problems with DUE constraints. Given that sensitivity analysis based method may not always be the right approach for bi-level problems, we adopt one of the random search methods — the Genetic Algorithms (GA) based method — to solve the bi-level model. The GA is a derivative-free optimization method and is one of the evolutionary algorithms which simulating the natural evolutionary process of human

beings results in stochastic optimization techniques [9]. Owing to their simplicity, minimal problem restrictions, global perspective, and implicit parallelism, the GA is fit to solve the bi-level programming model in this paper. However, a caveat is that the accuracy of GA based method is depending much on the computing time and population size.

2. Illustrative inequity example

In this section, we employ an artificial example to show the existence of inequity generated by land-use development.

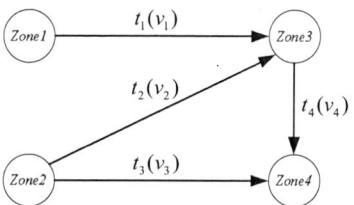

FIGURE 1 A Simple Network

Consider the simple network shown in Fig. 1, consisting of two O-D pairs, from Zone 1 to Zone 4 and from Zone 2 to Zone 4 respectively. Let $q_{14} = 50$ and $q_{24} = 450$ be the demands from Zone 1 to Zone 4 and Zone 2 to Zone 4 respectively. The travel cost functions are:

$$t_1(v_1) = 0.5 + \frac{v_1}{450}, t_2(v_2) = 3.0 + \frac{v_2}{200}$$

$$t_3(v_3) = 0.5 + \frac{v_3}{50}, t_4(v_4) = 3.0 + \frac{v_4}{50}$$

Assuming that the behavior of the network users' route choice follows the DUE, it is a simple matter to obtain the equilibrium travel cost for each O-D pair as follows:

$$\bar{u}_{14} = 4.90 \text{ and } \bar{u}_{24} = 9.21$$

Assume that Zone 2 has an increased projection due to the land use development and its traffic production q'_{24} increased to 600. After implementation of this project, the equilibrium travel cost for these two O-D pairs will be:

$$u_{14} = 7.46 \text{ and } u_{24} = 12.64 .$$

In this case, the corresponding ratios of the O-D travel costs both after and before the land use development project is,

$$\frac{u_{14}}{\bar{u}_{14}} = 1.52 \text{ and } \frac{u_{24}}{\bar{u}_{24}} = 1.37 .$$

It then follows that the equilibrium travel cost ratio for the users traveling from Origin 1 to Destination 4 is greater than for those traveling from Origin 2 to Destination 4, after Zone 2 was developed. This means that people in Zone 1 cannot gain benefits from the development of Zone 2 and that the travel cost ratio for users in Zone 1 are even higher than those of Zone 2.

This then poses the question of whether the development of Zone 2 is appropriate as well as how much additional production is appropriate.

3. Combined trip distribution/assignment problem

A cost-effective solution to transport problems should consist of land use pattern, transport system and set of road pricing policies that together bring demand and supply into balance in an efficient and equitable way. It can be seen that appropriate equilibrium traffic assignment in urban transportation networks combined with land-use and transportation as an integrated problem is an increasing trend. The network equilibrium problem was originally formulated by Beckmann, McGuire and Winston [10]. In a simpler version of this problem, the number of trips is fixed and specified. The ETDA-VDC model can incorporate an attractiveness measure of destination reflecting the activity opportunities available there, and determine the travelers' choice of destination and routes simultaneously, for any given number of trips originating from each origin by solving an equivalent convex programming problem. The model is therefore particularly suitable for our present analysis of zonal development potential with equity constraints.

Let $G = (N, A)$ be a directed and connected network, where N is the set nodes and A is the set of direct links. Let I be the set of origins, J be the set of destinations, where $I, J \subseteq N$ and $\langle i, j \rangle$ be the O-D pair from the origin $i \in I$ to destination $j \in J$. Each link $a \in A$ has a flow-dependent travel cost $t_a(v_a)$, which describes per unit flow or average cost on that link, and we assume that $t_a(v_a)$ is continuous and monotonically increasing with the amount of flow v_a. It is also associated that each link $a \in A$ a capacity C_a that is fixed and can be specified either as the physical economical or the environment capacity of that link. It is assumed that the distribution of the current O-D demand $\bar{d}_{ij}$, $i \in I, j \in J$ is given and remains stable subsequently. This existing demand will be assigned to the network in the conventional DUE manner. On the other hand, the total traffic growth at each residential zone due to new land-use development is distributed among various destination zones by a multinomial logit model, depending on the path and destination costs. In our model the destination cost or attractiveness c_j, $j \in J$, is assumed to be not fixed, but an increasing function, $c_j = c_j(d_j)$, of the number of trips, d_j the newly attracted flow to destination j,

where $d_j = \sum_{i \in I} d_{ij}$. In reality, the destination cost function could be either a strictly increasing or decreasing function in its argument of d_j, reflecting either negative or positive externalities. This postulation could be justified to stem from the consequences of various physical or economic processes such as congestion effects. Hence, the ETDA-VDC model can be formulated as the following strictly convex minimization problem

$$\min_{v,d} z(v,d) = \sum_{a \in A} \int_0^{v_a} t_a(\omega)d\omega +$$

$$\frac{1}{\theta} \sum_{i \in I} \sum_{j \in J} d_{ij}(\ln d_{ij} - 1) + \sum_{j \in J} \int_0^{\sum_{i \in I}(d_{ij} + \bar{d}_{ij})} c_j(\omega)d\omega \quad (1)$$

Subject to

$$\sum_{j \in J} d_{ij} = o_i, \ i \in I, \quad (2)$$

$$\sum_{r \in R_{ij}} h_r^{ij} = \bar{d}_{ij}, \ i \in I, \ j \in J, \quad (3)$$

$$\sum_{r \in R_{ij}} f_r^{ij} = d_{ij}, \ i \in I, \ j \in J, \quad (4)$$

$$v_a = \sum_{i \in I} \sum_{j \in J} \sum_{r \in R_{ij}} \left(f_r^{ij} + h_r^{ij} \right) \delta_{ar}^{ij}, \ a \in A, \quad (5)$$

$$f_r^{ij} \geq 0, \ i \in I, \ j \in J, \ r \in R_{ij}, \quad (6)$$

$$h_r^{ij} \geq 0, \ i \in I, \ j \in J, \ r \in R_{ij}, \quad (7)$$

$$d_{ij} \geq 0, \ i \in I, \ j \in J. \quad (8)$$

where o_i, $i \in I$ is the total growth in automobile trip production at zone i generated from land-use development; and d_{ij}, $i \in I, j \in J$ the number of trips starting from i and ending at chosen destination j; and f_r^{ij} and h_r^{ij} are the path flows associated with d_{ij} and $\bar{d}_{ij}$, respectively; θ is the positive parameter in multinomial logit model, $\mathbf{v} = (\cdots, v_a, \cdots)^T$ and $\mathbf{d} = (\cdots, d_{ij}, \cdots)^T$.

There have been a number of efficient algorithms solving this sort of strictly convex minimization models, [11] [12] [13] [14] [15]. This paper utilizes the conventional convex combination algorithm to solve the problem defined in eqns. (1)-(8), a detailed breakdown analysis of the algorithm can be referred in Yang et al. [16].

4. The Bi-level programming model for LUTPE

As can be seen in the illustrative example, the inequity problem is prevalent, and can lead to severe results. Given that the distribution of the travel cost ratios is inconsistent, it is straightforward to assume that by setting these ratios within a range of α to β. α and β are given equity measurements. For any given growth of trip productions at various residential zones, ETDA-VDC model can predict trip distribution among alternative destination zones and relevant DUE O-D travel cost. Then we can maximize the number of trips generated at each origin zone subject to some physical constraints and the vital equity constraints which force the values of the ratios of equilibrium O-D travel cost for all O-D pairs yielded by the ETDA-VDC model to be in the interval $[\alpha, \beta]$. Namely, the equity constraints can be mathematically characterized by the following equations.

$$\alpha \le \frac{u_{ij}(\mathbf{o})}{\bar{u}_{ij}} \le \beta, i \in I, j \in J \tag{9}$$

where $\mathbf{o} = (...,o_i,...)^T$, $u_{ij}(\mathbf{o})$ denotes the DUE travel cost between the O-D pair $\langle i, j \rangle$ and it is an implicit function with respect to the vector of all trip generation $\mathbf{o}$, $\bar{u}_{ij}$ and u_{ij} represent the travel cost from i to j before and after implementing the incremental production into the network. Clearly, the physical constraints such as the number of trips generated at each residential zone should be less than a present upper bound, o_i^{max}, $i \in I$, thus translating the maximum amount of vacant land or development potential that is suitable and available for residential use:

$$\sum_{j \in J} d_{ij}(o) \le o_i^{max} - \bar{o}_i, \ i \in I \tag{10}$$

Where $o_i = \sum_{j \in J} d_{ij}$, $\bar{o}_i = \sum_{j \in J} \bar{d}_{ij}$. The total number of trips attracted to each destination zone should also be within an upper bound, $d_j^{max}, j \in J$, translating the maximum number of job opportunities, parking capacity, etc:

$$\sum_{i \in I} d_{ij}(o) \le d_j^{max} - \bar{d}_j, \ j \in J \tag{11}$$

Where $d_j = \sum_{i \in I} d_{ij}$, $\bar{d}_j = \sum_{i \in I} \bar{d}_{ij}$, being the existing trip attraction. Any other factors that influence zonal development potential could be accommodated in appropriate forms of constraints such as the road capacity constraints. The objective is to maximize the trip generation o_i, $i \in I$ of each zone.

$$\max_{\mathbf{o}} \mathbf{o} = (...,o_i,...)^T \tag{12}$$

This objective function is a multi-objective function. The weighted approach is one of the most popular methods to solve this kind of multi-objective functions. This paper adopts a weighted summation of the trip production to transfer the multi-objective function (12)

into a single objective function by weighting $F(\mathbf{o}) = \sum_{i \in I} w_i o_i$. The weights reflect the costs and priorities of land development of the various origin zones and given by the planner.

Synthesizing the above statements, we have the following bi-level optimization model for maximizing the sum of total trip generations:

$$\max_{\mathbf{o}} F(\mathbf{o}) = \sum_{i \in I} w_i o_i \tag{13}$$

Subject to

$$\alpha \le \frac{u_{ij}(\mathbf{o})}{\bar{u}_{ij}} \le \beta, i \in I, j \in J, \mathbf{o} = (...,o_i,...)^T \tag{14}$$

$$v_a(\mathbf{o}) \le C_a, \ a \in A \tag{15}$$

$$\sum_{j \in J} d_{ij}(o) \le o_i^{max} - \bar{o}_i, \ i \in I \tag{16}$$

$$\sum_{i \in I} d_{ij}(o) \le d_j^{max} - \bar{d}_j, \ j \in J \tag{17}$$

$$o_i \ge 0, \ i \in I \tag{18}$$

Where $\{u_{ij}(\mathbf{o}), i \in I, j \in J\}$ can be obtained by solving the ETDA-VDC model defined in (1)-(8) for any given $\mathbf{o}$.

$d(\mathbf{o})$ is obtained by solving the lower level traffic equilibrium programs (1)-(8). In this model, the upper level sub-problem is a multi-objective function and maximizes the production of each residential zone subject to some physical constraints and the equity constraint; while the lower level sub-problem characterizes the network users' decisions with regard to route, origin and destination.

Note that the imposition of the equity and link capacity constraint (14) and (15) at different (lower or upper) levels will result in different behavioral interpretations, and yield different solutions. The lower-level network equilibrium sub-problem acts as a constraint. However, in our model, the equity constraint and capacity constraint are imposed in the upper-level sub-problem only, and thus the lower-level sub-problem represents a normal distribution and assignment model that admits no queues. Hence, the sum of the trip generations at origin zones is maximized without the occurrence of queuing on any link of the network. For the explanation of putting the constraints to the lower-level sub-problem, please refer to Yang et al. (2000).

In our bi-level model, we assume that the travel cost ratio of each path ranges from α to β. This assumption could be extended according to different types of paths. In other words, different path $r, r \in R_{ij}$ can has different $\alpha_r, r \in R_{ij}$ and $\beta_r, r \in R_{ij}$, demonstrated as vectors $\boldsymbol{\alpha}$ and $\boldsymbol{\beta}$. Let $\mathbf{r}(o)$ denotes the vector of the travel cost ratio of different path, and then the formula (15) could be simply replaced by:

508

$$\boldsymbol{\alpha} \le \mathbf{r}(o) \le \boldsymbol{\beta} \tag{19}$$

The above equation illustrates that the model has a flexible character in the calculation of maximum zonal production, and hence has broad potential applications as a planning tool.

In this research, set α equals to 1 and do not consider the paradox problem. The other equity parameter β should not be excessive high otherwise the constraint will be invalidated and the desired production will be close to the maximum limit of o_i^{max}.

5. GA-Based method for the Bi-Level programming model

The proposed bi-level programming model, like any other form of bi-level programming problems, is intrinsically nonconvex, and hence might be difficult to solve with a global optimum [17]. The difficulty is due to the fact that the equilibrium link flow $v_a(\mathbf{o})$, $a \in A$, and the O-D travel time $u_{ij}(\mathbf{o})$ generally are nonconvex, continuous but non-differentiable functions associated with $\mathbf{o}$. In view of the difficulty in applying the standard algorithmic approaches in search of the global optimum, this study adopted the GA method, which is particularly suitable for the models proposed here.

The basic idea of the GA approach is to code the decision variables of the upper level problem to finite strings and calculate the fitness of each string by solving the lower level problem. After the selection of the samples, followed by the crossover and mutation operations of GA, the optimal string may be achieved. In this research, the real vector $\mathbf{v}(o_1, o_2, \ldots, o_i)$ was used as the chromosome to present a solution.

GA-Based Method:

Step 0. Initialization: Code the decision variable $\mathbf{o}$ of the upper level problem to finite strings $(o_1, o_2, \ldots, o_i)$; determine the transform equation to map the objective function of the upper level problems to a fitness function.

Step 1. Select at random the initial population $\mathbf{v}(1)$. Set $k = 1$.

Step 2. Calculate the fitness function for individuals $v_j(k)$, $k = 1, 2, \ldots, N$ by solving the lower level user equilibrium traffic assignment problem, and reproduce the population $\mathbf{v}(k)$ according to the distribution of the fitness function values.

Step 3. Carry out the crossover operation through a random choice with probability P_c.

Step 4. Carry out the mutation operation through a random choice with probability P_m. This yields a new population, $\mathbf{v}(k+1)$.

Step 5. If k = maximum number of generations, the individual sample with the highest fitness is adopted as the optimal solution of the problem. Alternatively, set $k = k + 1$ and return to Step 2.

It can be noted that the constraints of the upper level model are inequality constraints. Thus, the fitness function should incorporate the constraint violations by means of a penalty method. The fitness function with the penalty factor ρ is described as follows:

$$F(x) = \sum_i o_i - \rho \Big[\max(0, v_a - C_a) + $$
$$\max\left(0, \sum_{j \in J} d_{ij}(o) - o_i^{max} + \bar{o}_i\right) + $$
$$\max\left(0, \sum_{i \in I} d_{ij}(o) - d_j^{max} - \bar{d}_j\right) $$
$$\sum_{i \in I} \sum_{j \in J} \max\left(0, u_{ij} / \bar{u}_{ij} - \beta, \alpha - u_{ij} / \bar{u}_{ij}\right) \Big] \tag{20}$$

It was found that implementing the GA approach was a simple procedure. The optimization process involved solving the lower level problem by conventional optimization techniques while the rest of the work was left to the "black box" of the GA. Fortunately, the lower level problems were not difficult to solve based on the successful studies of researchers.

6. Conclusion

This study proposed and solved the bi-level optimization programming models that explicitly address social equity issues associated with changes in the O-D travel cost. The study is also of great interest to planners and policy decision-makers as the proposed travel cost ratio in this paper can be conveniently and efficiently used for equity evaluation purposes. Our study found that this travel cost ratio enables the prediction of the amount of additional demand accommodated by the road network, and hence establishes a useful policy tool to address the impact of land use changes on traffic growth. In the bi-level model solution, a penalty function method with a genetic algorithm approach was applied to solve the bi-level programming models with equity constraints. As demonstrated in the Nguyen-Dupuis network example, the proposed model and algorithms proved to be meaningful tools for the development of equity-based transportation and land-use, and at the same time attempting not to cause excessive negative impacts on certain groups of users.

Further research can compare the result by activating equity constraint and activating capacity constraint. Another area for relevant research is how the proposed model could be applied to estimate optimal land price and optimal density to accommodate traffic growth.

Reference

[1] Fisk, C.S. 1979. "More Paradoxes in the Equilibrium Assignment Problem". *Transportation Research*, 13B, pp. 305-309.

[2] Dixon, P. M., Weiner, J., Mitchell-Olds, T., and Woodley, R. 1987. "Bootstrapping the Gini Coefficient of Inequality". *Ecology*, 68, pp.1548-1551.

[3] Meng, Q. and H. Yang. 2002. "Benefit Distribution and Equity in Road Network Design". *Transportation Research*, 36B, pp.19-35.

[4] Kim, T.J. 1983. "A Combined land Use-Transportation Model When Zonal Travel Demand Is Endogenously Determined", *Transportation Research,* 17B, pp.449-462.

[5] Yang, H., and M.G.H. Bell. 2001. "Transport Bi-level Programming Problems: Recent Methodological Advances". *Transportation research*, 35B, pp.1-4.

[6] Abdulaal, M., L.J. LeBlanc. 1979. "Continuous Equilibrium Network Design Models". *Transportation research*, 13B, pp.19-32.

[7] Fisk, C.S. 1984. "Game Theory and Transportation Systems Modeling". *Transportation Research* 18B, pp.301-313.

[8] Friesz, T.L., H.J. Cho, N.J. Mehta, R.L. Tobin, and G. Anandalingham. 1992. "A Simulated Annealing Approach to the Network Design Problem with Variational Inequality Constraints". *Transportation Science*, 26, pp.181-197.

[9] Goldberg, D. 1989. "Genetic Algorithms in Search, Optimization and Machine Learning". Addison-Wesley, Reading, Ma.

[10] Beckmann, M. C., B. McGuire and C. Winston. 1956. "Studies in the Economics of Transportation", *Yale University Press*, New Haven.

[11] LeBlanc, L. J., E. K. Morlok and W. P. Pieskalla. 1975. "An Efficient Approach to Solving the Road Network Equilibrium Traffic Assignment Problem". *Transportation Research* , 9B, pp.309-319.

[12] Nguyen, S. 1974. "An Algorithm for the Traffic Assignment Problem", *Transportation Science*, Vol. 8, pp.203-216.

[13] Bar-Gera, H. and D. Boyce. 2003. "Origin-based Algorithm for Combined Travel Forecasting Models". *Transportation Research*, 37B, pp.405-422.

[14] Erlander, S. 1977. "Accessibility, Entropy and the Distribution and Assignment of Traffic". *Transportation Research*, 11B, pp.149-153.

[15] Evans, S.P.1976. "Derivation and Analysis of Some Models for Combining Trip Distribution and Assignment". *Transportation Research* , 10B, pp.37-57.

[16] Yang, H., M.G.H. Bell, and Q. Meng. 2000. "Modeling the Capacity and Level of Service of Urban Transportation Networks". *Transportation Research* , 34B, pp.255-275.

[17] Friesz, T.L., R.L. Tobin, H.J. Cho, and N.J. Mehta. 1990. "Sensitivity Analysis Based Heuristic Algorithms for Mathematical Programs with Variational Inequality Constraints". *Mathematical Programming* , 48, pp.265-284.

2008 Workshop on Power Electronics and Intelligent Transportation System

Safety Evaluation for First Class Highway Design Based on Unascertained Measure Model

TU Sheng-wen[1,2], GUO Xiu-cheng[1], He Jian-ming[3]

[1]School of Traffic & Transportation Engineering, Changsha University of Science and Technology, Changsha, Hunan, 410076,China

[2]School of Communication, Southeast University, Nanjing, Jiangsu, 210096,China

[3]China Highway Engineering Consulting Group LTD., Beijing,100097, China

tusw2004@163.com

Abstract

Regarding the characteristics of uncertainty on the relationship between design factors of first class highways and traffic safety, the unascertained measure model of the safety evaluation for the design of first class highways is established on the basis of unascertained measure space. Through unascertained measure of single indicator, calculation of comprehensive measure for multiple indicators and the weight determination process of indicators, the comprehensive evaluation for the design of first class highways is realized. The example research indicates that it's theoretically feasible to use unascertained measure model to study the safety evaluation problem for the design of first class highways and the evaluation results are credible and this provides a new method for the safety evaluation of highway design under unascertained conditions.

1. Introduction

The implementation stages of highway safety evaluation are mainly classified into design stage and operation stage, for the safety evaluation of design stage, the domestic and foreign research achievements are mainly focused on three aspects: the method based on highway safety examination system[1], the method based on microcosmic highway traffic accidents forecast model [2-3], the method based on highway safety design decision-making support system[4]; in recently years, some scholar also brought forward the highway safety design method based on vehicle dynamics simulation[5]. Quite a lot of research achievements had reached systematic and practical degree and this had active acceleration function for enhancing road design quality and improving road operation safety. However, the current research has a remarkable defect: Most of them established the interrelation between traffic accidents and road parameters based on the statistics and the two do not have necessary or always substantial "causal relationship". The design factors affecting highway safety mainly include: alignment quality, roadside danger grade, intersection safety level, access density, traffic security facility configuration and so on. For the safety evaluation personnel, these factors have remarkable uncertainty on the actual security response

after the highway is put into operation. The traditional regression analysis method can not comprehensively deal with these types of uncertain information and so we need seek for new solution.

"Unascertained mathematics theory" is a new type of science on research of incomplete information. It was brought forward by academician Wang Guangyuan in 1990s and its frame system was initially established after the further research of scholars like Liu Kaidi. Currently, it's applied in many fields, such as preliminary evaluation on ship safety [6], preliminary evaluation of coal mine safety [7], levee safety assessment [8]; if compared with such evaluation methods as fuzzy comprehensive evaluation, grey clustering method, matter element analysis, BP artificial neural network, the unascertained evaluation model is more precise, the evaluation results are reasonable, precise with high resolution and is more suitable for safety pre-evaluation. The relationship between the highway design factors and the traffic safety is the information that currently can not be ascertained due to the complexity of traffic accidents occurrence mechanism, so it belongs to the scope of unascertained information and is suitable to be dealt with unascertained mathematics theory. This article takes first class highway as the research object and face to the design stage to establish the unascertained measure model of the safety evaluation for first class highway and provide decision-making basis for improving design quality.

2. Unascertained measure model [6-8]

Assumed that certain evaluation object has m pieces of evaluation factor and if certain evaluation factor has n pieces of evaluation indicators, which are represented with $x_1,x_2,...,x_n$, then the indicator space $X=\{x_1,x_2,...,x_n\}$,x_i represents the ith measure value of the evaluation factor. There are p pieces of evaluation grade of $c_1,c_2,...,c_p$ for x_i, the evaluation space is marked as $U,U=\{c_1,c_2,...,c_p\}$. Setting the grade k is "stronger" than grade k+1, we record them as $c_k>c_{k+1}$.And if $c_1>c_2>c_3>...>c_p$, we call $\{c_1,c_2,...,c_p\}$ is a sequential division category of the evaluation space.

1.1. Unascertained measure of single indicator

978-0-7695-3342-1/08 $25.00 © 2008 IEEE

DOI 10.1109/PEITS.2008.106

Assumed that $\mu_{ik}=\mu(x_i \in c_k)$ indicates the degree that measure value x_i belongs to the kth evaluation grade c_k, and μ satisfies:

$$0 \leq \mu(x_i \in c_k) \leq 1 \qquad (1)$$

$$\mu(x_i \in U)=1 \qquad (2)$$

$$\mu\left|x_i \in \bigcup_{l=1}^{k} c_l\right|=\sum_{l=1}^{k}\mu(x_i \in c_k) \qquad (3)$$

Where, $i=1,2,...,n$; $k=1,2,...,p$. We call the μ that satisfies (1)~(3) to be unascertained measure. Formula (2) indicates that μ satisfies polarity for the evaluation space and formula (3) indicates that μ satisfies additivity for the evaluation space. If μ dissatisfies polarity and additivity, then it is considered to be un-credible theoretically.

1.2. Comprehensive measure evaluation vector for multiple indicators

If $\mu_k=\mu$ ($M \in c_k$) indicates the degree that the evaluation factor M belongs to the kth grade, then:

$$\mu_k = \sum_{i=1}^{n}\omega_i\mu_{ik} \quad (k=1,2,...p) \qquad (4)$$

Where, ω_i indicates the significance degree of the evaluation indicator x_i compared with other evaluation indicator, namely, the weight ω_i satisfies:

$$0 \leq \omega_i \leq 1, \sum_{i=1}^{n}\omega_i = 1 \qquad (5)$$

Apparently, $0 \leq \mu_k \leq 1, \sum_{i=1}^{n}\mu_k = 1$, so the u_k determined by formula (5) is the unascertained measure. We call vector $u_k=(u_1,u_2,...u_k)$ to be comprehensive measure evaluation vector for multiple indicators.1.3 Recognition rule for confidence of degree

If $c_1>c_2>c_3>...>c_p$, then we call $\{c_1,c_2,...,c_p\}$ to be

a sequential division category of the evaluation space and we can introduce the "degree of confidence" evaluation rule as follows:

Let λ to be degree of confidence ($\lambda>0.5$, which is often set $\lambda=0.6$ or 0.7), and set

$$k_0 = \min\left|k : \sum_{l=1}^{k}\mu_l > \lambda, k=1,2,...,p\right| \qquad (6)$$

Then evaluation factor M can be judged to belong to the k_0^{th} evaluation grade c_{k_0}.

3. Safety evaluation indicators system for first class highway design

The first class highway in China is equivalent to multilane with median in the foreign highway classification system. Up to now, the domestic and foreign research on traffic safety mainly focuses on freeway and two-lane highway and there is much less research on first class highway. As the first class highway has both the characteristics of high alignment standard, fast driving speed of freeways and the openness of two-lane highway, we can refer to the related research achievements of freeway and two-lane highway to determine the safety evaluation indicators system of the first class highway. In this paper, the alignment quality, roadside danger grade, intersection safety level, access density, configuration level of traffic security facility is adopted to set up the safety evaluation indicators system for first class highway design. All the indicators are classified into three grades: grade I indicates secure, grade II indicates generally secure and grade III indicates insecure. The classification of all the indicators is based on literature [9]-[13] and it can be referred to table 1 for details.

Table 1 The classification of safety evaluation indicators system for first class highway design

Indicators	grade I ,c_1	grade II ,c_2	grade III,c_3
Measure for alignment quality,x_1	≥10	1~10	0~1
Measure for roadside danger grade,x_2	0~3.5	3.5~10.5	≥10.5
Intersection safety level,x_3	0~70	70~220	≥220
Access density(#/km),x_4	0~0.5	0.5~2	≥2
Configuration level of traffic security facility,x_5	80~100	60~80	≤60

Table 2 The measuring data of safety evaluation indicators for first class highway design

Indicators	Value
Measure for alignment quality,x_1	2.7
Measure for roadside danger grade,x_2	3.7
Intersection safety level,x_3	230
Access density(#/km),x_4	0.4
Configuration level of traffic security facility,x_5	76

4.1. Establishment of single indicator measure function

Firstly, the single indicator measure function $\mu(x_i \in c_k)$ should be established to calculate the measure values μ_{ik} of all the evaluation indicators, so as to calculate the measure space $(\mu_{ik})_{5\times5}$ of all the evaluation

indicators of the first class highway; where $i = 1,2,3,4,5,k = 1,2,3$. According to the classification standard in table 1, by taking the stricter classification standard, the indicator characteristic value of c_1 grade regards the lower limit value of the interval number, and the indicator characteristic value of c_3 grade takes the upper limit value of interval numbers. Meanwhile, c_2 grade takes the median of interval numbers as the grade classification standard.

According to the definition of unascertained measures, the unascertained measure function of all the evaluation indicators can be constructed as shown in Fig 1-5.

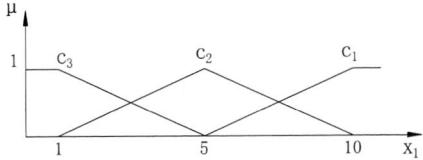

Fig.1 Measure function of single indicator for alignment quality

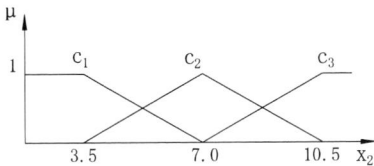

Fig.2 Measure function of single indicator for roadside danger grade

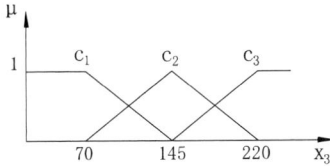

Fig.3 Measure function of single indicator for intersection safety level

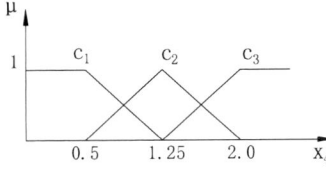

Fig.4 Measure function of single indicator for access density

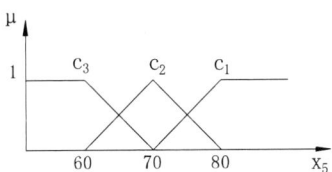

Fig.5 Measure function of single indicator for configuration level of traffic security facility

It's easy to validate that the constructed measure functions $\mu(x_i \in c_k)$ satisfy $0 \le \mu_{ik} \le 1$, and $\sum_{k=1}^{p} \mu_{ik} = 1$ (namely, satisfy the polarity and additivity conditions). Therefore it is the unascertained measure function.

From the data and the measure function in table 2, the evaluation matrix for single indicator measure can be calculated as follows:

$$
(\mu_{ik})_{5 \times 3} =
\begin{array}{ccc}
c_1 & c_2 & c_3
\end{array}
\begin{bmatrix}
0 & 0.378 & 0.622 \\
0.943 & 0.057 & 0 \\
0 & 0 & 1 \\
1 & 0 & 0 \\
0.6 & 0.4 & 0
\end{bmatrix}
\begin{matrix}
x_1 \\ x_2 \\ x_3 \\ x_4 \\ x_5
\end{matrix}
\quad (7)
$$

4.2. Calculation of evaluation vector for multiple indicators measure

In order to calculate evaluation vector for multiple indicators measure, the weight of all the indicators should be determined firstly. When the property value μ_{ik} and the evaluation indicator are determined, the property weight has been correspondingly determined and is hidden in the ascertained measures of the property measurement value, so it is not generally specified subjectively and should be determined according to the actual measured value of the property. In this paper we refer to the ideology of literature [14] and borrow the concept of information entropy to define the peak value of x_i:

$$
v_i = 1 + \frac{1}{\log p} \sum_{k=1}^{p} \mu_{ik} \cdot \log \mu_{ik} \quad (8)
$$

Where, p is the dimension of evaluation space and the value of v_i reflects the importance of the indicator x_i. Define the weight of property x_i to be:

$$
\omega_i = v_i / \sum_{i=1}^{n} v_i \quad (9)
$$

Then from evaluation matrix of single indicator measure and formula (7), (8) and (9), the weight of all the evaluation indicators can be determined to be: $(\omega_1, \omega_2, \omega_3, \omega_4, \omega_5) = (0.111, 0.223, 0.279, 0.279, 0.108)$. According to formula (4) we can get the measure vector for multiple indicators to be:

$$
\mu_k = (0.554, 0.098, 0.348) \quad (10)
$$

4.3. Recognition for degree of confidence

Take degree of confidence $\lambda = 0.6$ and we can judge the safety evaluation grade for Juning around-city highway to be grade II according to comprehensive evaluation vector formula (10) and degree of confidence evaluation rule formula (6), namely, the safety grade is general security. Actually, through the method of safety checklist, the first class road design has something to improve regarding the alignment quality, roadside safety and traffic security facilities, so the evaluation conclusion of the unascertained measure model is credible.

5. Conclusions

The factors that affect the safety of first class highway design mainly include: alignment quality, roadside danger grade, intersection safety service level, access density, service level of traffic security facility system and so on. The traffic safety effect of these factors has apparent uncertainty. Regarding this type of characteristic, the unascertained measure model is introduced in this paper for the safety evaluation of first class highway design. The example research shows that the application of unascertained measure model in the safety evaluation of highway design is theoretically reasonable and the calculation results are credible. This provides a new method for safety evaluation of highway design. Meanwhile, the unascertained measure model has definite significance and the calculation process is quite simple and has certain popularization and utility value.

References

[1]Guo ZHong-yin, Fang Sou-en, *Highway Safety Engineering*, Beijing: China Communication press, 2003(in Chinese)

[2]Andrew Vogt, Joe Bared,*Accident Models for Two-lane Rural Segments and Intersections,*Paper No.98-0294 , Transportation Research Record 1635.

[3]Fitzpatrick K.,L. Elefteriadou, D. Harwood, J. Collins ,etc, *Speed Prediction for Two-Lane Rural Highways*, Report FHWA-RD-99-171, FHWA U.S. Department of Transportation, 2000.

[4]Justin Bansen, Karl Passetti,*Application of the IHSDM: a case study,*3rd international symposium on highway geometric design,2005: 1221

[5] XU Jin , PENG Qi-yuan , SHAO Yi-ming, *Simulation System for Safety Estimation of Alignment and Road Surface Condition in Designing Phase,* China Journal of Highway and Transport,2007, Vol . 20 No. 6,pp.36-42(in Chinese)

[6]DUAN Aiyuan, ZHAO Yao, *Preliminary evaluation on ship safety based on unascertained measure,* Journal of Shanghai Maritime University,2007, Vol. 28 No. 2,pp.21-23(in Chinese)

[7]YAN Le-lin, WANG Guo-qi, XU Man-gui ,etc, *The Unascertained Measure Model and Application for Preliminary Evaluation of Coal Mine Safety,* JOURNAL OF CATASTROPHOLOGY, 2004, Vol. 19 No. 2,pp.18-22(in Chinese)

[8] DING Li, etc, *Application of Unascertained Mathematics in Levee Safety Assessment,* Water Resources and Power, Vol. 23 No. 4,pp.29-32(in Chinese)

[9] WU De-hua．FANG Shou-en. *Quantity Evaluation Model of Line Design Quality of Freeway[J].* Journal of Tongji University(Natural Science), 2005,33(11): 1469-1473(in Chinese)

[10] CHEN Le-sheng, etc, *Research on Classified Method on Danger Degree of Roadside of Ordinary Highways in Mountainous Areas,* Highway, 2005, No11,pp.159-163(in Chinese)

[11] Pan Fu-quan , Lu Jian,etc, *Computation model of safety level of service for non-signalized intersection,* Journal of Traffic and Transportation Engineering, 2007, Vol. 7 No. 4,pp.104-111(in Chinese)

[12] Kyriacos C. Mouskos ,etc, *Impact of Access Driveways on Accident Rates at Multilane Highways,* A Final Report for NJ-99-008-NCTIP5, U. S. Department of Transportation,1999

[13] WANG Jian-jun ,etc, *Evaluation of Traffic Safety Facilities of Expressway,* Journal of Transportation Systems Engineering and Information Technology, 2007, Vol. 7 No. 4,pp.66-70(in Chinese)

[14] CAO Qing-kui , etc, *Calculation method of objective index weight by entropy,* Journal of Hebei Institute of Architectural Science and Technology, 2000, Vol. 17 No. 3,pp.40-42(in Chinese)

2008 Workshop on Power Electronics and Intelligent Transportation System

Some Remarks on ON/OFF Network Traffic

Chu Chen
School of Electronic and
Information Engineering,
South China University of
Technology,
Guangzhou, Guangdong,
510641, China
chuch@scut.edu.cn

Yong Xu
School of Computer Science
and Engineering,
South China University of
Technology,
Guangzhou, Guangdong,
510641, China
yxu@scut.edu.cn

Ling Zhang
School of Computer Science
and Engineering,
South China University of
Technology,
Guangzhou, Guangdong,
510641, China
ling@scut.edu.cn

Abstract

One of the most attractive self-similar traffic generating methods is ON/OFF model due to its clear physical explanation. However, the simulated traffic may show very strange characteristics, e.g. the estimated Hurst parameters for simulated traffics are larger than 1 and are almost equal despite of their different ON period parameters because of wrong sampling interval. In this paper, we point out that we should pay attention to two aspects when we use ON/OFF model to simulate network traffic. First, the first part of the simulated traffic is non-stationary and should be cut off. Second, the simulated traffic might be extremely over-sampled. The simulated traffics with appropriate sampling interval and non-stationary part cut off show satisfying result.

1. Introduction

Since the seminal study of Leland etal [1], self-similar network traffic model has been commonly accepted. Many self-similar traffic simulating methods have been put forward including FGN and FARIMA [2]. The ON/OFF method developed by Taqqu et al [3] is successful due to its plausible physical explanation. Recently, Ricciato constructed two simulated data sets X_1 and X_2 with ON/OFF method [4]. The only difference between the two data sets was their Pareto tail index parameter of ON periods: $\alpha_1 = 1.25$ and $\alpha_2 = 1.6$. The wavelet estimator [5, 11] reported that the Hurst parameter for both data sets were almost the same and equaled to 1.48, far from their theoretical values: $H_1 = 0.875$ and $H_2 = 0.7$. Ricciato could not give a clear explanation for this result. In [6], the authors obtained the same result as Ricciato's, but they

thought Ricciato got the slope α reported by the LDestimator code instead of Hurst parameter and the Hurst parameter H was related to the slope α by the formula $H = (3-\alpha)/2$, which would correspond to $(3-1.48)/2 = 0.76$. However, Matthew and Darryl[7] pointed out that the linear behaviour is absent beyond a certain octave due to under-sampling and the LDestimator can not reported the right Hurst value because of the wrong sampling interval. In this paper, we point out that the simulated ON/OFF model traffics are non-stationary and extremely over-sampled.

2. Non-stationary part of the simulated data set

The ON/OFF method simulate network traffic as follow [4]: Consider a flow arrival process where flow arrivals are Poisson with intensity λ and flow duration are Pareto with parameters k (shape factor) and α (tail index). Let $X(k)$ denote the number of parallel flows by a fixed sampling interval t_s, according to [3] such process is self-similar with Hurst parameter $H = (3-\alpha)/2$.

The first parts of the simulated data sets are undoubtedly non-stationary. As we know, the number of flow can not be larger than 1 at the very beginning and can not be larger than 2 at the second sampling point either. The reason is that we ignore the flows which arrive before the simulation starts. These flows might be still active. The number of these flows depends on the ON period Pareto tail index parameter. We constructed four data sets by using the similar simulating conditions as Ricciato's. All the simulated data sets have the same mean flow duration which equals to 26.67. Figure 1 shows the non-stationary characteristic of the simulated traffics.

The non-stationary effect can not be completely get rid off because some of the flow durations can be infinite. The flow duration T is a Pareto random variable with pmf $f(t)$

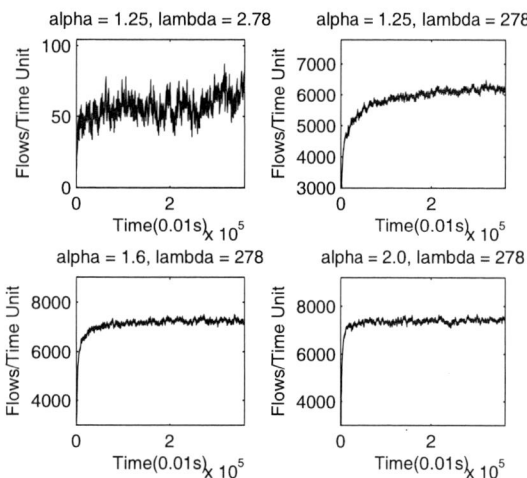

Figure 1. Simulated traffic with different α and λ. The beginning parts of these traffics are non-stationary.

and cdf $F(t)$ as below:

$$f(t) = \alpha k^{\alpha} t^{-\alpha-1} \qquad (1a)$$

$$F(t) = P[T \leq t] = 1 - (k/t)^{\alpha} \qquad (1b)$$

We can diminish non-stationarity by cut off the first part of the sequence. In particular, T has finite mean and infinite variance if $1 < \alpha < 2$. The mean flow duration T_{mean} is:

$$T_{mean} = \frac{k\alpha}{\alpha - 1} \qquad (2)$$

Let us consider a simulated network traffic with N flows. The theoretical mean number of flows is:

$$m_x = \frac{NT_{mean}}{T_{fin}} \qquad (3)$$

Where T_{fin} is the duration of the simulation. Although the number of flows increases when simulation starts, the increasing rate declines. In this paper, we define the stationary part begins from T_1, where the mean number of flows is larger than $95\% m_x$. Thus the number of flows whose durations is larger than T_1 can not be more than $5\% m_x$:

$$N \times P\{T \geq T_1\} \leq \frac{5m_x}{100} \qquad (4)$$

According to (2) and (3), we have two equivalent formulas as follow:

$$T_1 \geq (1 - \frac{1}{\alpha}) \times (20T_{fin})^{\frac{1}{\alpha}} \times T_{mean}^{1-\frac{1}{\alpha}} \qquad (5a)$$

$$T_1 \geq (20 - \frac{20}{\alpha}) \times k^{(1-\frac{1}{\alpha})} \times T_{mean}^{\frac{1}{\alpha}} \qquad (5b)$$

If we want to get a stationary process, then the T_1 must be less than T_{fin}. So we get the following relationship:

$$\frac{T_{mean}}{T_{fin}} \leq 20^{\frac{1}{1-\alpha}} (\frac{\alpha}{\alpha - 1})^{\frac{\alpha}{\alpha-1}} \qquad (6)$$

In our experiments, the mean flow duration T_{mean} is 26.67s and the simulation duration T_{fin} is 3600s. The formula (2) is satisfied with $\alpha_1 = 1.25$, $\alpha_2 = 1.6$. But the value of T_1 changes according to different α value. The results are showed in table 1.

Table 1. Different stationary beginning time with mean flow duration 26.67s and simulation duration 3600s

α	1.2	1.4	1.6	1.8	2.0
$T_1(s)$	3215.8	2152.2	1395.1	955.2	692.8

Figure 1 shows the non-stationary characteristics of the simulated traffics. The simulated traffic can be considered as a stationary process after a certain time T_1 and we can calculate its value according to the parameter T_{mean}, T_{fin} and α. Although the wavelet method can correctly report the right Hurst parameter by setting the number of zero moments M of the wavelet function ψ for long-range dependent sequence with trend [5], we should remove the non-stationary effect by cutting the first part of the time series the network traffics are considered as stationary generally.

3. The sampling effect

The Nyquist Sampling Theorem states that if a continuous bandwidth-limited signal contains no frequency components higher than half the frequency at which it is sampled, then the original signal can be recovered without distortion. On the other side, a under-sampled sequence will lose some information and cause wrong Hurst parameter estimation for self-similar process [7]. In order to show more details, we simulated FGN using Paxson's FFT method[8] and then under-sampled the original sequence by 10 times. The results of different Hurst estimating method for the two kinds of sequences are showed in table 2.

We used three other common Hurst estimators [9, 12] besides the wavelet method to estimated the Hurst parameter for the original FGN and their under-sampling sequences. The three other methods are aggregate variance method, R/S method and periodogram method. From table 2, we can see that all the Hurst values of the sampled sequences decrease significantly especially for the periodogram method which belong to the frequency domain estimator. As for the

Table 2. Estimation results for H using 10 independent FGN sequences (1,000,000 long) and their under-sampling sequences (100,000 long)

	Aggvar	R/S	Per	Wave
fgn060	0.599	0.606	0.602	0.605
unsam_fgn060	0.526	0.543	0.527	-
fgn070	0.698	0.699	0.703	0.705
unsam_fgn070	0.623	0.641	0.597	-
fgn080	0.791	0.795	0.804	0.815
unsam_fgn080	0.745	0.762	0.706	-

Table 3. Estimation results for H using 10 independent FGN sequences (100,000 long) and their over-sampling sequences (1,000,000 long)

	Aggvar	R/S	Per	Wave
fgn060	0.599	0.609	0.602	0.603
ovsam_fgn060	0.648	0.627	1.167	1.481 (0.624)
fgn070	0.688	0.710	0.704	0.701
ovsam_fgn070	0.732	0.709	1.211	1.487 (0.725)
fgn080	0.773	0.784	0.806	0.804
ovsam_fgn080	0.812	0.779	1.256	1.492 (0.826)

wavelet method, the plot shows that there is not linear part exit.

Since under-sampling influences the value of Hurst parameter a lot. One might ask weather the over-sampling also has some influence on Hurst parameter estimation. The answer is definite. Note here we define the over-sampling by duplicating each value over n time indices. The over-sampled FGN is not FGN in the finest scale and it is still FGN in large scales ($> n$). However, most of the estimators will report these sampled sequence with larger Hurst parameter than the original FGN. The results are more complicated than that of under-sampling situations as we showed in table 3.

Generally speaking, the over-sampling results in overestimation of the Hurst parameter except for R/S method. The over-sampling has little effect on R/S method and we do not know whether the estimated Hurst parameter will increase or decrease. As for wavelet method, there are two approximately linear parts for each case. One linear part is between octave 2 to octave 5 with the estimated Hurst parameter about 1.48 for all cases, the other linear part is between octave 6 to octave 9 and the estimated H is a little larger than the Hurst parameter of the original FGN. The large scales correspond to low frequencies and small scales to high frequencies [10]. So the wavelet estimator also shows the over-sampled FGN is self-similar only with

Table 4. Hurst estimation for simulated ON/OFF traffic with 10 runs

	Aggvar	R/S	Per	Wave
$X_1(0.875)$	0.992	0.850	1.496	1.486
$X_2(0.7)$	0.993	0.889	1.495	1.485

the low frequencies (large scales).

The reason that over-sampling results in much overestimation of Hurst estimation in frequency domain estimator results from the time-frequency transfer. The basic rule of time-frequency transfer is that the time domain has inverse ratio to the frequency domain. To over-sample a time series means broadening the time domain and narrowing the frequency domain. The Periodogram estimates Hurst parameter by a regression of the logarithm of the periodogram on the logarithm of the frequency. Therefore, to narrow the frequency will overestimate the Hurst parameter.

4. Analysis of simulated ON/OFF model

Both the estimated Hurst parameters for the simulated data sets in [5] are approximate 1.48, which is very close to the estimated Hurst for over-sampling FGN sequences during octave 2 to octave 5. But the wavelet method plots have only one linear part for simulated ON/OFF traffics as showed in Ricciato's paper. We believed the simulated traffics are over-sampled. First of all, let us compare the simulated ON/OFF traffic X_1 to the well known BC-pAug traffic. The theoretical Hurst parameter of X_1 is 0.875. We calculated the number of active flows from 600s in order to eliminate the non-stationary effect. The mean of generated time series is 8.18 and the standard deviation is 2.43. The Hurst parameter estimated value of the BC-pAug traffic (in packets) is about 0.82. This packet series also has a sample mean of 3.18 and standard deviation of 2.61.

Figure 2 shows that the simulated data set X_1 is much different from the real Ethernet traffic. The traffic number of X_1 remains 5 during the first 300 time unit in the finest time scale, which unlikely happens in real network. Figure 2 also shows that the simulated data X_1 does have some kind of self-similarity within the rough scales. These are the exact characteristics of over-sampled time series. We also calculated the mean of Hurst parameter for 50 different simulated X_1 and X_2 with non-stationary part cut off. Table 4 shows the results.

From table 4, we can see almost all of the reported values are larger than the theoretical values, especially for the Periodogram and wavelet methods. These results are similar to the over-sampling sequence estimations except for two differences. First, none of the estimators can discriminate X_1 from X_2. Second, the wavelet estimator has only one linear

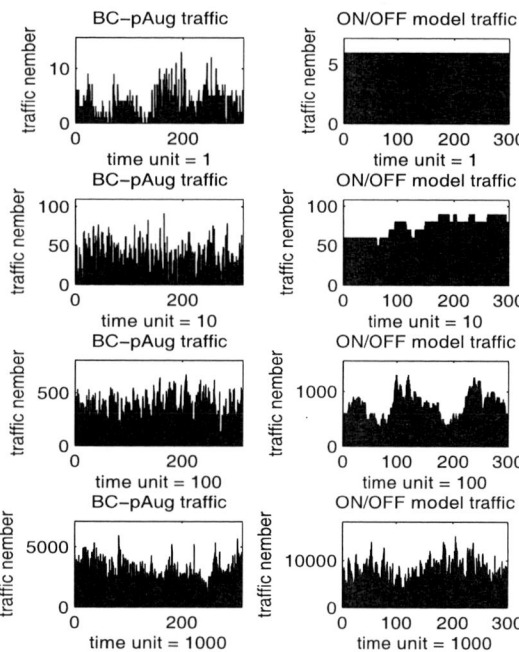

Figure 2. Comparison of BC-pAug traffic and simulated ON/OFF model traffic. The later can be consider extremely over-sampled.

part instead of two linear parts for each data set. However, the simulated traffic maybe over-sampled much more times than 10. As we can see from Figure 2, the 10 aggregate level of simulated ON/OFF traffic can also be considered over-sampled.

The algorithm of ON/OFF traffic model will cause over-sampling. Theoretically, we need infinite number of ON/OFF flows to generate a self-similar process, which is impractical. To sum finite number of flows can not correctly reflect the changes of a simulated traffic. We can either increase the number of flows greatly or decrease the sample frequency to obtain correct simulated traffic. And we should notice that the under-sampling also affects the estimation of Hurst parameter. If the Periodogram estimator reports Hurst parameter larger than 1 for a sample sequence, the over-sampling much likely exists and we should decrease the sequence's sample frequency. Table 3 and Table 4 show that the R/S method is the most insensitive estimator to sample frequency while the Periodogram method is the most sensitive one. Thus, the Periodogram method can roughly test weather a sequence has appropriate sample frequency and the R/S method can roughly report its Hurst parameter. In our experiment, we generated different data

Table 5. Hurst estimation for different sample frequent ON/OFF traffic with 10 runs

	Aggvar	R/S	Per	Wave
Y_1	-	-	-	1.478
Y_2	0.937	0.814	1.455	-
Y_3	0.844	0.802	1.210	-
Y_4	0.808	0.795	0.846	0.814

sets with different sampling interval form 0.01s to 10s. The simulated data sets Y_1 with $\alpha_1 = 1.4$, $k_1 = 7.62$. The total number of flows is 360 and the arrival times are uniformly distributed in half an hour, which means that the flow arrival is Poisson with intensity $\lambda = 0.2 arrivals/sec$. From Y_2 to Y_4, we increased the sampling interval each by 10 times and also increased the time duration each by 10 times in order to keep the same sequence length. We also cut off the non-stationary part for each data set. Table 5 shows the average Hurst parameter estimation for different situations with 10 runs.

Table 5 shows the parameter estimation for different sampling interval ON/OFF traffic. According to 6, the data set Y1 is non-stationary. The Periodogram method reports value less than 1 only for the simulated traffic with 10s sampling interval. Note for wavelet method, there in no approximately linear part for data sets Y_2 and Y_3.

As we can see from figure 2, the simulated traffic with 0.1s sampling interval is much different from the real Ethernet traffic for they were over-sampled. We can infer the approximately correct sampling interval is 10s for ON/OFF simulated traffic with parameter $\alpha = 1.4$ from table 5. So we estimated the Hurst parameter for self-similar sequences generated by ON/OFF method with sampling interval equals 10s.

We generated 10 independent data sets for each different α value of 1.2, 1.4, 1.6, 1.8, 2.0. The location parameters k are 4.44, 7.62, 10, 11.85 and 13.33 respectively in order to keep the same mean of flow duration. The flow density $\lambda = 0.2 arrivals/sec$ and the sample interval $t_s = 10s$. The total run time is 50 hours and the non-stationary part is cut off.

Table 6. Hurst estimation for simulated ON/OFF traffic with 10 runs

α(H)	1.2(0.9)	1.4(0.8)	1.6(0.7)	1.8(0.6)	2(0.5)
R/S	0.839	0.802	0.740	0.680	0.636
Per	0.899	0.849	0.782	0.738	0.682

From table 6, we can see that the R/S method can almost report the right value when H is 0.9 or 0.8. For smaller H, it tends to overestimate the value. The Periodogram method

518

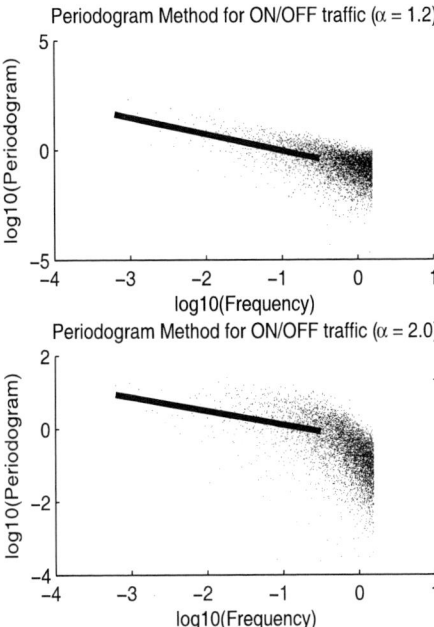

Figure 3. Periodogram plots for same sampling interval ON/OFF traffic with different parameter α. The bottom plot obviously does not correctly report Hurst parameter and this simulated traffic can not be modeled by self-similar process.

still can not report the correct value and the difference becomes more obvious as the theoretical H value decreases. However, it can roughly tell the extent of self-similarity of the data sets. Are the lower H value simulated ON/OFF traffics still over-sampled? Figure 3 can give us some clues.

The upper and bottom plots of Figure 3 are the Periodogram plots of ON/OFF traffic with $\alpha_1 = 1.2$ and $\alpha_2 = 2.0$ respectively. The sampling intervals both are 10s. The bottom plot obviously does not correctly report Hurst parameter and this simulated traffic can not be modeled by self-similar process. In other words, we need different sampling interval for different parameter α ON/OFF simulated traffics.

5. Conclusion

The characteristics of ON/OFF simulated traffic are much more complicated than we expect. We concluded in this paper that the wavelet method does report Hurst parameter 1.48 for the data sets simulated in [4] because of wrong sampling interval. We also pointed that the sampling interval have a close relationship with the ON time period parameter α. In addition, the simulated traffic is

non-stationary. We can greatly diminish the non-stationary effect according the value of parameter α too. Therefore, we need pay much more attention when we use ON/OFF method to simulate self-similar network traffic.

References

[1] Leland W, et al (1994). "On the Self-similar Nature of Ethernet Traffic (Extended Version)". *IEEE/ACM Transactions on Networking*, Vol. 2, No. 1, pp.1–15.

[2] K. Park and W. Willinger. *Self-Similar Network Traffic and Performance Evaluation*, Wiley, NewYork, 2000.

[3] Murad S. Taqqu, Walter Willinger and Robert Sherman (1997). "Proof of a Fundamental Result in Self-Similar Traffic Modeling". *ACM Sigcomm Computer Communication Review*, Vol. 27, No. 2, pp.5–23.

[4] F.Ricciato (2006). "Some Remarks to Recent Papers on Traffic Analysis". *ACM Sigcomm Computer Communications Review*, Vol. 36, No. 3, pp.99–102.

[5] P.Abry and D.Veitch(1998). "Wavelet Analysis of Long-range-dependent Traffic". *IEEE Transactions on Information Theory*, Vol. 44, No. 1, pp.2–15.

[6] D. Malone, K.Duffy, and C. King (2007). "Some Remarks on LD Plots for Heavy-tailed Traffic". *ACM Sigcomm Computer Communications Review*, Vol. 37, No. 1, pp.41–42.

[7] Matthew Roughan, Darryl Veitch (2007). "Some Remarks on Unexpected Scaling Exponents". *ACM Sigcomm Computer Communications Review*, Vol. 37, No. 5, pp.71–74.

[8] Vern Paxson (1997). "Fast, Approximate Synthesis of Fractional Gaussian Noise for Generating Self-Similar Network traffic". *ACM Sigcomm Computer Communications Review*, Vol. 27, No. 5, pp.5–18.

[9] Murad S. Taqqu, Vadim Teverovsky and Walter Willinger (1995). "Estimators for Long-range Dependence: an empirical study". *Fractals*, Vol. 3, No. 4, pp.785–798.

[10] Stilian Stoev, Murad S. Taqqu, Cheolwoo Park, J.S. Marron (2005). "On the wavelet spectrum diagnostic for Hurst parameter estimation in the analysis of Internet traffic". *Computer Networks*, Vol. 48, No. 3, pp.423-445.

[11] D. Veitch, P. Abry, *Matlab code for the wavelet based analysis of scaling processes*, Available from: http://www.cubinlab.ee.unimelb.edu.au/~darryl /secondorder_code.html

[12] Chu Chen, *Hurst parameter estimate*, Available from: http://www.mathworks.com/matlabcentral/fileexchange /loadCategory.do?objectType=category&objectId33 &objectName=Fractals

2008 Workshop on Power Electronics and Intelligent Transportation System

A Target Classification Algorithm Based on Transportation Sensing Network

CUI Xun-xue, QIU Guo-xin, ZENG Jian-qin, XING Li-jun, LIU Qi

New Star Research Institute of Applied Technology, Hefei, 230031, P.R.China

xxcui@tsinghua.org.cn

Abstract

Target classification is an important enabling technology for the monitoring task in transportation sensing networks. In the paper the magnetic signal and seismic acceleration signal are collected, analyzed and transmitted for a mobile road target. A classification algorithm based on the sensor network is proposed, which adopts a peak and valley pattern of hybrid detection signals from different sensors. The hardware equipment system of sensor node is devised. The terminal nodes are small with low power consumption so that the transportation sensing network is easy to deploy. The advantage of the algorithm lies on its sorting exactness of several kinds of targets. The results from many field tests have been shown that it is capable of identifying mobile targets on some roads in intelligent transportation system.

1. Introduction

The transportation sensing network is recognized as an important component of the intelligent transportation systems. More and academic researchers and people from industry are engaged in developing them due to the good promise and potential with various applications. Wireless sensor networks offer the potential to significantly improve the efficiency of existing transportation systems [1, 2, 3]. Currently, collecting traffic data for traffic planning and management is achieved mostly through wired sensors. The equipment and maintenance cost and time-consuming installations of existing sensing systems prevent large-scale deployment of real-time traffic monitoring and control. Small wireless sensors with integrated sensing, computing, and wireless communication capabilities offer tremendous advantages in low cost and easy installation.

Target classification is an important enabling technology for the monitoring task in transportation sensing networks. Some applications of sensor network technologies in intelligent transportation systems (ITS) include parking lot monitoring, traffic monitoring, and traffic control et al. Vehicle classification information is one of the important measurements that we need to obtain in practice, which is invaluable for various aspects of transportation including engineering and planning. If a mobile target on some key roads can be recognized, it would provide helpful traffic information [4, 5].

In order to improve recognition performance, we adopt the concept of peak and valley pattern which is used to deal with the hybrid detection signals from different sensors. In the paper, the magnetic signal and seismic acceleration signal from two micromation sensors are collected, analyzed and transmitted for a mobile road target. On the basis of the two feature vectors, we propose a novel algorithm model to classify target signals. We obtained experimental results of the performance of classification of mobile target signals. These results are shown that classification accuracy is high, and the algorithm is capable of identifying mobile targets on some roads in intelligent transportation system.

The paper develops as follows. The next section describes the basic problem background which includes the classification principle of mobile target and type selection of the used sensors. Section 3 gives a brief overview of the related work. Section 4 proposes the classification algorithm model for transportation targets. Section 5 describes the experiment performance evaluation of classification algorithm and section 6 concludes the paper.

2. Problem background

2.1. Classification principle of mobile target

In a transportation system mobile targets can be classified according to different classification standards. In the paper we divide them into six kinds, including free personnel, personnel with mental object, small vehicle, passenger car, heavy vehicle and tracklayer vehicle. This classification method can be used either in most of applications or in military field.

The detailed recognition description is given as follows. Free personnel are the persons who walk without any mental object, while personnel with mental object can be detected by the sensing signals from magnetic sensors nearby. They would create a tiny seismic signal.

As many sorts of vehicles exist, they can be divided by different standards in various application fields. Usually they can be divided according to its engine model, dimensions, the total mass and wheelbase. After the GB/T3730.1-1988 standard in China [6] is consulted, we divide the usual vehicles into three types, which include heavy vehicle, small vehicle and passenger car. A heavy vehicle has a total mass exceeding 4500kg, otherwise it should be belong to the type of small vehicle. The length of passenger car should exceed six meters taking no

978-0-7695-3342-1/08 $25.00 © 2008 IEEE
DOI 10.1109/PEITS.2008.109

520

account of its weight. A tracklayer vehicle is made up of track-style structure. As this type of target has a high strength of seismic signal in the ground that is obviously different from other types of vehicle, thus we take it as a special type of target. The category of tracklayers exists in most of battlefields, e.g. armored cars and tanks.

2.2. Type selection of sensors and node system

With the MEMS technology is developed rapidly, the trend of sensor micromation becomes common in many applications. Usually a single sensor is used to provide the information of road targets. In the paper two kinds of tiny sensors are adopted to collect the target data. They are the magnetic Honeywell 1052 sensor and the seismic acceleration ADXL202E sensor. A networked node with small size is easy to be deployed and needs less power energy.

The HMC 1052 is a Honeywell chip [7], whose magneto-resistive sensors convert the magnetic field to a differential output voltage, capable of sensing magnetic fields as low as 30 μgauss 0. Ferromagnetic material, such as iron, with a large permeability, changes the earth's magnetic field.

The ADXL202E is a low-cost, low-power, complete 2-axis accelerometer with a digital output. The ADXL202E will measure accelerations with a full-scale range of ± 2 g. The ADXL202E can measure both dynamic acceleration and static acceleration [8].

Two sensors are integrated in the terminal node of a transportation sensing network. The node should be exactly placed to match the road direction so that an intense indication by a mobile target can be obtained. As the collected data from the two sensors is plentiful, the sample frequency is set to 20Hz in our node system. The node MCU is Atmelgal128L, and the communication chip is CC2420 in network node.

3. Related work

From the related research work about vehicle classification in sensor network, the California PATH is an important project [9, 10]. In the project a single magnetic sensor is used for classification. The researchers believe that wireless magnetic sensor networks offer a very attractive, low-cost alternative to inductive loops for traffic measurement in freeways and at intersections. Sensor data from passing vehicles at the same site are processed and classified into 6 types. A simple algorithm is proposed to use this magnetic information to classify the vehicle into six types: passenger vehicles, SUV, Van, bus, mini-truck, and truck. It is reported that sixty percent of the vehicles are classified correctly, when length is not used as a feature. They believe that if two sensors are used, vehicles can be classified with accuracy better than

80 percent. In the paper we are base on this idea to develop magnetic and seismic sensors to detect targets.

In ref. [11] a vehicle classification algorithm is proposed to use inductive signatures obtained from a prototype innovative loop sensor, known as a 'blade'. A probabilistic neural network, a neural network implementation of multivariate Bayesian classification scheme, and a heuristic classification algorithm are employed to classify vehicle types. Vehicle feature vectors representing the vehicle shapes are extracted from blade signatures, and then utilized as inputs of the proposed algorithm. The classification performances are investigated with four different types of vehicles including passenger car, pick-up truck, sports utility vehicle, and van. The shortcoming of this algorithm is that it cannot be executed quickly with a high computational complexity.

Usually existing target classification algorithms based on sensing data appear to suffer from three shortcomings [12]: (1) recognition model only adopts sensing data from single sensor; (2) low computational efficiency results in high wireless communication cost; (3) implementation of algorithm with hardware is not easy.

In the paper two kinds of micro-sensors are combined in a single network node. By using the hybrid sensing signals a classification algorithm is proposed to classify the transportation targets. The recognition systems based on transportation sensing networks are attractive because of their low cost, ease of installation, flexibility of deployment and high classification precision.

4. Classification algorithm model for transportation targets

4.1. Basic principle of classification algorithm

Magnetic and seismic sensor can detect two kinds of signal about a target. When the target moves by a magnetic-resistance, all dipoles of the target is sensed. For a vehicle target, a variation of earth magnetic field reveals the type feature of a vehicle. Thus it is easy to distinguish vehicles with magnetic disturbance signals.

An acceleration sensor collects the seismic data from the ground produced by a mobile target. Using the information the type of free personnel, wheel vehicles and tracklayers can be distinctly distinguished. If the magnetic field signal is combined to be analyzed, the target classification would be accomplished.

The hybrid sensing signal provides plenty of feature information which is superior to the data from a single sensor, especially for the two compensatory sensors. Many samples are used to train for all types of targets, and then a passing target is recognized in a real time manner in a networked node. The target classification result is transmitted to a gateway node, and finally the statistical data are sent to a traffic monitoring center to provide the managers for analysis.

According to the classification principle and the hybrid sensing data above-mentioned, the basic principle of a classification algorithm proposed in the paper is given as follows:

The variation ratio of continue sample data is compared to a threshold. When the comparison result is plus and above the threshold, the sample feature is characterized as +1; if it is negative and its magnitude is above the threshold, the sample feature is determined as -1; as the magnitude of variation ratio is not bigger the threshold, the featured signal is decided by 0. The magnetic and seismic signals of a target are characterized as peak and valley pattern, which exhibits respectively a mountain and a valley skeleton in the two axial directions.

4.2. Algorithm description

The classification model is designed as follows.
4.2.1. Threshold design

The threshold of detection signal is determined by the sensing output voltage from the corresponding sensor. A detailed threshold is decided by plenty of experiments. Here the threshold design method is set by:

$$threshold = \left| \frac{wavy\ value - base\ value}{the\ number\ of\ int\,erval\ po\,int\,s} \right|$$

Where the wavy value is an output voltage collected by a sensor node with interference as no target pass by. The base value is the detection data in a situation with neither mobile target nor interference. The base value would be varied with different roads and places. The number of interval points is the number of data between wavy value and base value. It is often bigger than or equals 2. In TABLE I an example of signal data collected by magnetic sensor is shown within a period of time. Here the base value is 64, the wavy value is 63, and the number of interval points is determined as 2. Thus the threshold of magnetic signal is computed as 05.

TABLE I
AN EXAMPLE OF SIGNAL DATA COLLECTED BY MAGNETIC SENSOR

Output signal (mv)	64	64	64	64	63	63	64	64	64
Sampling time	15:26:48	15:26:48	15:26:49	15:26:49	15:26:49	15:26:49	15:26:49	15:26:49	15:26:49

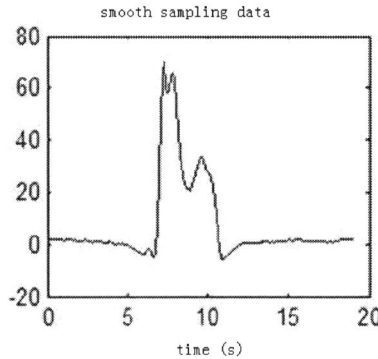 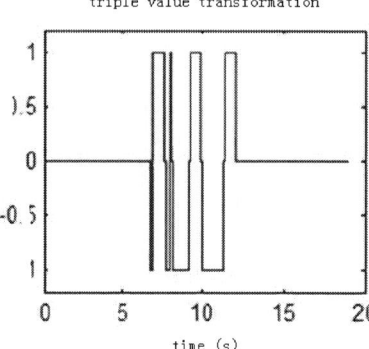 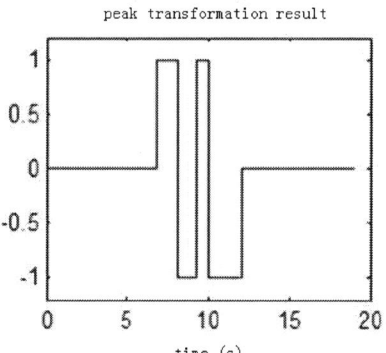

Fig. 1 An example of peak transformation process

4.2.2. Transform method of peak value

The peak transformation pattern is a process that feature signals produced by a target are changed into triple symbols {+1, 0, -1}, where the key character for recognition is obtained. This process has a very high compress ratio such that the real-time classification has a less computation cost and low energy consumptions. Its detail is described as follows.

(a) A character signal of target is changed into a triple symbol {+1, 0, -1}:

$$c(k) = \frac{\Delta a(k)}{\Delta k} \quad (1)$$

$$\overline{c(k)} = \begin{cases} +1, & c(k) > C_{threshold} \\ -1, & c(k) < -C_{threshold} \\ 0, & other \end{cases} \quad (2)$$

Where $c(k)$ is the variation rate of sampling data, i.e. slope; $\overline{c(k)}$ is the triple symbol transformed by the peak pattern; $C_{threshold}$ is the threshold.

522

(b) To avoid frequent peaks and valleys resulted from small changes of feature signal, the self-adaptive state machine is adopted to filtrate these changes. If the value scope and continual time reach a pre-assign value, those peak and valley data are treated as valid statistics. A sample of transformation process is illustrated in Fig.1.

(c) According to the property of the type recognition of target, in the paper the dimension of vector transformed from peak and valley is uniformly set as 10. For example, a vector of peak symbol {+1,-1,0,-1,+1} can be expressed as {+1,-1,0,-1,+1,0,0,0,0,0}. That is to say, if the element of a vector is less than 10, the rest site is filled with 0.

4.2.3. Construction of sample base

By using peak transformation pattern, all smooth signals collected by a sensor node are changed into peak and valley style. A sample base can be built for recognition algorithm through many experiments. We finished the analysis of plenty of test data. For the two-axis magnetic sensor and seismic sensor, the constructed character set of classification sample is listed in TABLE II.

TABLE II
MAGNETIC AND SEISMIC SAMPLE CHARACTER SET FOR TARGET CLASSIFICATION

Target type	Magnetic symbol vector	Seismic symbol vector
Free personnel	{0,0,0,0,0,0,0,0,0,0}	{+1,-1,0,+1,-1,0,+1,-1,0,+1} or {-1,+1,0,-1,+1,0,-1,+1,0,-1}
Personnel with mental object	{+1,0,-1,+1,-1,0,0,0,0,0}	
Small vehicle	{+1,-1,0,0,0,0,0,0,0,0}	{+1,-1,+1,-1,+1,0,0,0,0,0} or {-1,+1,-1,+1,-1,0,0,0,0,0}
Passenger car	{-1,+1,-1,+1,-1,0,0,0,0,0}	
Heavy vehicle	{+1,-1,+1,-1,0,-1,0,0,0,0}	
Tracklayer vehicle	{-1,+1,-1,0,0,0,0,0,0,0}	{+1,-1,+1,-1,+1,-1,+1,-1,+1,-1} or {-1,+1,-1,+1,-1,+1,-1,+1,-1,+1}

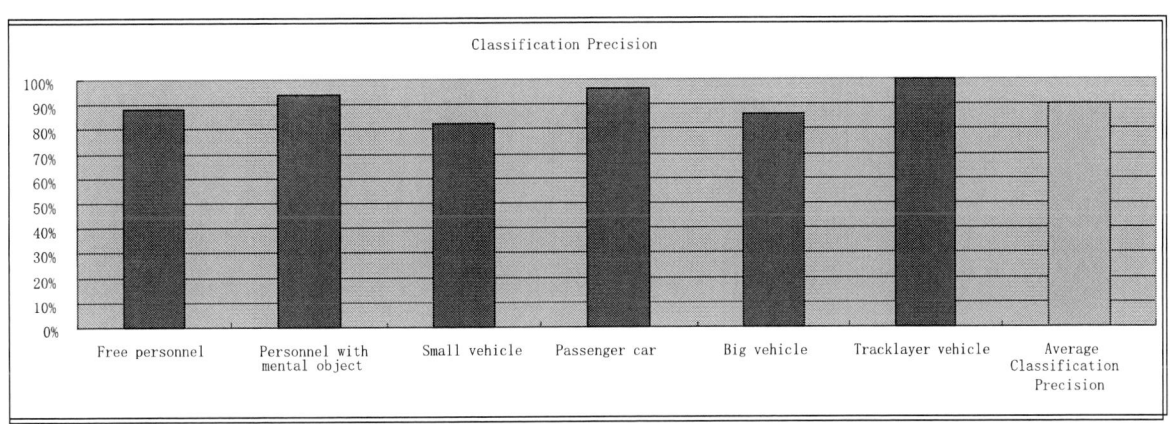

Fig. 2 The statistical result of classification precision about six kinds of targets

4.2.4. Target pattern recognition method

As a target is tested for recognition with its sampling signals, the character set is compared with the detection signals. It is calculated as:

$$R = \left| c_1 - c_{b1} \right| + \left| c_2 - c_{b2} \right| + \cdots + \left| c_{10} - c_{b10} \right| \quad (3)$$

Where $\{c_1, c_2 \cdots c_{10}\}$ is a vector of peak values, and $\{c_{b1}, c_{b2} \cdots c_{b10}\}$ is the produced character set of sample.

After a sensor node has collected the combinational characters of a target, these signals are treated as recognition features to compare with the corresponding regular feature vector in the sample character set. The Euclidean distance is calculated between them. From the six distances, the smallest one is selected, and thus the corresponding target type of this character is taken as a classification result. For example, suppose that {+1,-1,+1,-1,0,-1,+1,0,0,0} is the sampling symbol vector of a magnetic sensor. Using equation 3 it is compared with the regular magnetic character of each type of vehicles. Then we can obtain that $R_{\text{small vehicle}}=4$, $R_{\text{passenger car}}=11$, $R_{\text{heavy vehicle}}=1$. Here the value of $R_{\text{heavy vehicle}}$ is the smallest one. Thus this target is determined as a big vehicle from the above process.

Similarly, the seismic signals can be treated in this manner. Through the seismic character of a target, it is easy to distinguish personnel, wheel vehicles, and tracklayers. Then the magnetic signals are combined to decide the match extent. The detailed sort of a target is finally classed from the whole match procedure.

5. Performance experimental evaluation

We have built a prototype of the sensor node for target classification, which consists of a processor, a radio, a magnetometer, an acceleration sensor, a battery and a cover for protection from the vehicles. In our node system RS232 serial port and 11520Baud/s is adopted to transmit data. The sampling data is saved in an Excel database file. In the experiments a sensor node was stuck in the center of a single lane section in local traffic. The sampling frequency is 20Hz. It is possible that a higher sampling rate might reveal the 'missing' peak.

The sensor nodes can be placed on bridges and overpasses. Sensor data from 230 passing targets at the same site are processed and classified into 6 types. A summary of the experimental results is shown in Fig. 2. It is revealed that the hybrid sensing system can distinguish mobile targets at approaching 90% average accuracy for six kinds of them.

The classification algorithm achieves a correct classification rate of 90 percent in real time. The algorithm can be implemented online by the sensor node itself, in contrast to other methods based on high scan-rate inductive loop signals, which require extensive offline computation. Moreover, it is superior to the approach by using a single sensor which achieves a correct classification rate of 60 percent [9].

6. Conclusion

In the paper the magnetic signal and seismic acceleration signal are collected, analyzed and transmitted for a mobile road target. A classification algorithm based on the sensor network is proposed, which adopts a peak and valley pattern of hybrid detection signals from different sensors. The hardware equipment system of sensor node is devised. The terminal nodes are small with low power consumption so that the transportation sensing network is easy to deploy. The advantage of the algorithm lies on its sorting exactness of several kinds of targets. The results from many field tests have been shown that it is capable of identifying mobile targets on some roads in intelligent transportation system.

Acknowledgment

This work was partially supported by the National Natural Science Foundation of China under grant No. 60773129; the Excellent Youth Science and Technology Foundation of Anhui Province of China under grant No.08040106808.

References

[1] Tubaishat M, Zhuang P, Qi Q, And Shang Y. Wireless sensor networks in intelligent transportation systems[J]. Wireless Communications and Mobile Computing, (to appear), 2008.

[2] Coleri S, Cheung S Y, Varaiya P. Sensor networks for monitoring traffic[C]. Forty-Second Annual Allerton Conference on Communication, Control, and Computing, U. of Illinois, Sep 2004.

[3] Ding J G, Cheung S Y, Tan C W et al. Signal processing of sensor node data for vehicle detection[C]. 7th International IEEE Conference on Intelligent Transportation Systems (ITSC), Oct 2004.

[4] Brooks R R, Ramanathan P, Sayeed A M. Distributed target classification and tracking in sensor networks[J]. Proceedings of the IEEE, 2003, 91(8): 1163-1171.

[5] Aroraa A, Duttaa P, Bapata S et al. A line in the sand: a wireless sensor network for target detection, classification, and tracking[J]. Computer Networks, 2004, 46(5): 605-634

[6] http://www.21chinaits.com/

[7] Honeywell. Vehicle Detection Using AMR Sensors. Application Note-AN128:1-2. Http://www.magneticsensors.com.

[8] Analog Devices, Inc. http://www.analog.com

[9] Cheung S Y, Coleri S, Dundar B, et al. Traffic measurement and vehicle classification with a single magnetic sensor[R]. California Partners for Advanced Transit and Highways (PATH) Project Working Papers: Paper UCB-ITS-PWP-2004-7, September, 2004.

[10] Cheung S Y, Coleri S, Dundar B et al. Traffic measurement and vehicle classification with a single magnetic sensor[C]. 84th Annual Meeting, Transportation Research Board, 2005.

[11] Cheol Oh, Stephen G R. Recognizing vehicle classification information from blade sensor signature[J]. Pattern Recognition Letters. 2007, 28(9): 1041-1049.

[12] Ding J. Vehicle detection by sensor network nodes[D]. MS thesis, Department of Electrical Engineering and Computer Science, University of California, Berkeley, CA, Fall 2003.

2008 Workshop on Power Electronics and Intelligent Transportation System

The Optimal Route Guidance Modeling for ITS Influenced by Dynamic Factors

Wensheng ZHANG[1], Mingsheng WANG[2]

[1] College of Surveying Engineering,
Shijiazhuang Railway Institute, Shijiazhuang,HeBei, 050043,China
[2] College of Traffic Engineering,
Shijiazhuang Railway Institute, Shijiazhuang, HeBei,050043,China
xyns@hotmail.com, vms_wang@163.com

Abstract

The optimal route guidance, is an important part of ITS. The factors influencing route guidance, such as the position of vehicles, the information on road conditions, the free parking spaces, etc., are dynamically variable. These dynamically factors are constrained conditions.How to construct the optimal model of vehicle route guidance under such constrained conditions is a difficulty that puzzles ITS. This paper studies how to select optimal driving route based on dynamic conditions for ITS. Firstly, find factors influencing guidance route and find which are significant and which are subordinate, so as to determine their weights relative to all factors. Then by the help of Fuzzy Comprehensive Evaluation (FCE), the guidance model has been built, which elicits the optimal guidance route influenced by dynamic factors.

1 Introduction

Among the present studies on route guidance, the recognized classic arithmetic on shortest route is presented by the Netherlander mathematician Dijkstra (Ahuja R K, et al., 1990, Cherkassky B V, et al., 1996) in 1959. Much improved arithmetic, which is based on Dijkstra Arithmetic, has its advantages on facets of space complexity, time complexity, reliability and application range (Cherkassky B V, et al., 1993; Zhan F B, Noon C E, 1998). Farther，the heuristic search arithmetic,(Jagadesh G R, et al., 2003)，fluid neural network algorith(Wen Huimin,Yang Zhaosheng, 1999) and genetic algorithm(Chang W A，Ramakrishna R S, 2002), were used to analyse the shorest route. In the 1990s, the Italian scholars Dorigo and Colorini used Ant Colony Algorithm (ACA) to solve the problem of vehicle traffic flow guidance (Zhou P, et al., 2004). But all of the model and arithmetic is only applicable to have solutions of guidance route at static conditions. In reality, there are plenty of factors influencing guidance, for example, the traffic, the road conditions, the parking spaces, the psychology of driver, etc. Some of the

factors are dynamically variable, How to construct the optimal model of vehicle route guidance under such constrained conditions is a difficulty that puzzles ITS.

The paper presents a dynamic guidance modeling means based on analysis of complexity and uncertainty of transportation network. First, dynamic and static factors which influence vehicle guidance have been found and sorted according to their essentiality with the aid of AHP (Analytic Hierarchy Process). Then the driving velocity at one moment which passes all available roads under varied influencing factors will be calculated by applying FCE (HE Zhongxiong, 1983). Because varied influencing factors, such as real time information on traffic and road condition and psychological tendency of drivers have been considered in the calculation of guided route, among all the calculated routes from the position of vehicle to the destination, the fasted route is the optimal guided route at that moment. By calculation in a certain time section, for example 30s, the real time optimal route guidance under dynamic influencing factors will be calculated.

2 The Modeling of Optimal Route for ITS Influenced by Dynamic Factors

Before modeling of optimal route, the influencing factors should be analysis and ranking. The influential factors are classified into dynamic ones and static ones. The dynamic are factors influencing route guidance with time-variable values, including the traffic flux, the position of driving vehicle, the parking spaces, driver's psychological dependence on roads etc. The static are constant factors, such as the road grade, the road length, the delay time of traffic light, etc. It is essential to find influential factors as many as possible and determine their essentiality, in order to acquire Factors Set, because it is vital to the validity and reliability of calculated results. Herein, we utilize AHP to classify the numerous influential factors and determine their weights (the grade of essentiality).

2.1 Constructing the Grading System of Influential Factors Based on AHP

There are four aspects influencing guidance route, namely, road condition, time, fee and psychology of drivers. Every aspect includes several factors. The sum of all of the factors constitutes constrained conditions of optimal guidance route. We set the optimal guidance route under constrained conditions as aim layer, select road, time, fee and psychology as standard layer, and make several influential factors of every standard as index layer, so that the grading system of intelligent guidance under constrained conditions have been constructed based on AHP. Comparing by twos is the fundamental of AHP. To construct the priorities for 4 standards, two standards should be compared and the essentiality of one standard to another will be given. That is to say, the four standards (road, time, fee and psychology) should be compared by twos, so the more important standard can be known. Then the priority of one standard to another will be given. The priorities are values between 1 and 10. Larger value means more important. These priorities constitute a matrix. The priorities of every standard have been achieved by calculating it.

The evaluation consistency should be tested when comparison by twos is carrying out, which is reasonable if the consistency ratio is lower than 0.10. In the same way, priorities of factors in Index Layer under the standards of road, time, fee and psychology, have been calculated, as shown in Table 1. The details of calculation will not be elaborated here.

To determine the weight of every index and how much it accounts for all influential factors, it's necessary to build up an overall sequence of the aforementioned priorities. The formulation is as follow:

By assuming $\omega_A(X_j)$ as the weight of X in A and the weights of no-influence-indexes as zero, the overall sequence of all factors has been calculated by the following equation with the results shown in Table 2.

$$\omega_A(X_j) = \sum_{i=1}^{n} \omega_A(B_i)\omega_{B_i}(X_j), (j=1,2,3...,m) \tag{1}$$

Hereinto, A is the overall aim; B is the priorities of standards in standard layer; X is the priorities of indexes of standards.

Table1. The priorities of all standards

Road		Time		Fee		Psychology	
Index	Priority	Index	Priority	Index	Priority	Index	Priority
Crowd extent of roads	0.634	Delay in crossing	0.128	Type of garage	0.320	Reservation of garage	0.800
Road length	0.260	Route depth	0.083	Parking fee	0.557	Psychological dependence on road	0.200
Road grade	0.106	Walking time	0.257	Driving fee	0.123		
		Garage space	0.532				

Table 2. The overall sequence of influential factors for optimal route at dynamic conditions

Standard B $\omega_A(X_j)$ Index X	Road 0.536	Time 0.229	Fee 0.157	Psychology 0.077	$\omega_A(X_j)$
Crowd extent of roads	0.634	0	0	0	0.340
Road length	0.260	0	0	0	0.139
Road grade	0.106	0	0	0	0.057
Delay in crossing	0	0.128	0	0	0.029
Route depth	0	0.083	0	0	0.019
Walking time	0	0.257	0	0	0.059
Garage space	0	0.532	0	0	0.122
Type of garage	0	0	0.320	0	0.050
Parking fee	0	0	0.557	0	0.087
Driving fee	0	0	0.123	0	0.019
Reservation of garage	0	0	0	0.800	0.062
Psychological dependence on roads	0	0	0	0.200	0.015

The weights in Table 2 have experienced consistency analysis by AHP to affirm that they are accordant with the consistency requirement.

2.2 The Optimal Route Guidance modeling

The paper presents the route guidance modeling based on FCE. That is to say, first the parameters influencing driving velocity of vehicles which constitute the Factors Domain U are found; the driving velocity of vehicles is set as Evaluation Domain V. Then the driving velocity influenced by both dynamic factors and static factors are resolved by FCE. The driving time of vehicles at varied time, varied route will be predicted. The route on which vehicles drive fastest is the optimal guidance route. Approaches of guidance route arithmetic based on FCE will be presented as follows:

(1)To define Factors Domain $U = \{u_1, u_2, \cdots, u_n\}$ and Evaluation Domain $V = \{v_1, v_2, \cdots, v_m\}$;

The factors which influence driving velocity of vehicle have been defined as Factors Domain, which comprises of 12 performances, namely: crowd extent of roads, road length, road grade, delay in crossing, road depth, walking time, garage vacancy, garage type, parking fee, driving fee, the reservation of garage and driver's psychological dependence on road. The speed grades have been defined as Estimation Domain V, which has 5 grades, $V = \{v1 \quad v2 \quad v3 \quad v4 \quad v5\}$

(2)To estimate single factor $u_i (i=1,2,\cdots,n)$. Then the Fuzzy Set $(r_{i1}, r_{i2}, \cdots r_{im})$ on V will be achieved;

(3) The development of fuzzy relationship matrix R

according to the Fuzzy Set of single factor:

$$R = \begin{bmatrix} r_{11} & r_{12} & \cdots & r_{1m} \\ r_{21} & r_{22} & \cdots & r_{2m} \\ \vdots & \vdots & \cdots & \vdots \\ r_{n1} & r_{n2} & \cdots & r_{nm} \end{bmatrix}$$

Because the statuses of all indexes of single factor aren't equal, it's necessary to weigh every index. $A = (a_1, a_2, \cdots, a_{12})$ represents the distribution of weights of a factor. The integrated fuzzy estimation considering road grade and crowd extent of roads will be gained by combining A and Fuzzy Relation R with an adaptable operator.

$$B = A \circ R = (b_1 \quad b_2 \quad b_3 \quad b_4 \quad b_5) = (a_1 \quad a_2 \cdots a_{12}) \circ \begin{bmatrix} r_{11} & r_{12} & r_{13} & r_{14} & r_{15} \\ \vdots & & \ddots & & \vdots \\ r_{121} & r_{122} & r_{123} & r_{124} & r_{125} \end{bmatrix}$$

During the aforementioned process, the speed grade have been defined as Estimation Domain V, which has 5 grades

$$V = \{v_1 \quad v_2 \quad v_3 \quad v_4 \quad v_5\}$$

Hereinto, v_1, v_2, v_3, v_4 and v_5 denote the speeds respectively very fast, fast, normal, slow and very slow. The corresponding velocities are: c_1, c_2, c_3, c_4, c_5.

The final fuzzy comprehensive evaluation results reflect the driving velocity on every available road to destination acted by all influential factors. If the evaluation results predict that the driving velocity on certain road reaches or approaches v_1, then the road will be selected preferentially, which means that the fastest route is the optimal guidance route. Because it is calculated by considering all influential factors, such as traffic, road grade and garage spaces etc; and the

essentialities of these factors have been reflected by their weights. The optimal guidance influenced by static and dynamic factors at certain moment is gained. Then, the transient guidance route of running vehicles will be given by calculating repeatedly as mentioned before. The methodology of fuzzy comprehensive evaluation is written in books about FCE, so it will not be explained here.

3 Analyses and Discussion

It is known from the knowledge of fuzzy estimation that $b_j = \max_i \min[a_i, r_{ij}] = \vee[a_i \wedge r_{ij}]$, if Zadeh Fuzzy Operator $M(\wedge, \vee)$ is adopted in a composite calculation. Only the most outstanding influencial factors affect the results. Whereas composite calculation with operator M ($\bullet, +$), $b_j = \sum_i^n a_i r_{ij}$ is a comprehensive estimation which considers all of the factors according to their weights, which means that the calculation results exhibit the influence from all factors. Therefore, among the familiar four operators $M(\wedge, \vee)$, $M(\bullet, \vee)$, $M((\wedge, \oplus)$ and $M(\bullet, +)$, the largest difference of calculated results exists between the above mentioned two operators.

Therefore, prediction of velocity from 150 groups of simulated data (information on road condition and road grade) has been carried out adopting M ($\bullet, +$) and $M(\wedge, \vee)$. The comparison of results is shown in Fig. 1. It can be seen that the accordance between predicted driving velocities based on two operators exists.

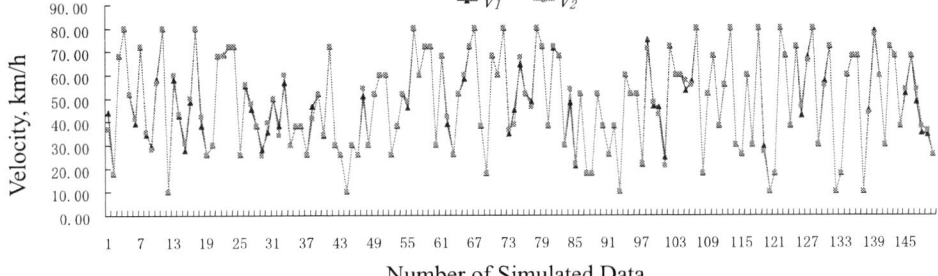

Fig.1 Comparison of estimated velocities based on two operators

Hereinto, v1 and v2 are forecasted velocities based on operators M ($\bullet, +$) and $M(\wedge, \vee)$ respectively. The parameters have been set as:

$$c_1 = 80km/h, c_2 = 60km/h, c_3 = 50km/h, c_4 = 30km/h$$

The absolute error between two predicted results of

150 groups are calculated by the following formula,

$$\overline{\Delta v} = \frac{\sum_{i=1}^{150} |v_2(i) - v_1(i)|}{150} = \frac{123.85}{150} \approx 0.83km/h$$

If the average velocity is $\overline{v} = 50km/h$, when travel time is 1h, the difference of travel times predicted by two methods is less than 1 minute:

$$\overline{\Delta t} = \frac{\overline{\Delta v}}{\overline{v}} \times 60 = \frac{0.83}{50} \times 60 \approx 0.99\text{min}$$

Therefore, it is believable that the arithmetic presented here is of high precision and practicability.

4 Conclusions

The paper presents optimal route guidance modeling of vehicles influenced by dynamic factors, which has been tested by simulated information on road condition and road grade. The test shows that the modeling is of high simplicity, applicability and precision.

Acknowledgment

It is an international cooperation project supported by Science & Technology Department of HeBei Province. The authors would like to gratefully acknowledge Professor Stefano Bocchi and Dr. Stefano P. Desperati of Milan University for providing research materials and Professor Wu Lixin of China University of Mining & Technology for instruction.

References

[1] Ahuja R K, Mehlhorn K, (1990), "Faster Algorithms for the Shortest Path Problem", Journal of the Association for Computing Machinery, 37(2): 213-223.

[2] Chang W A, Ramakrishna R S, (2002), "A genetic algorithm for shortest path routing problem and the sizing of populations", IEEE Transactions on Evolutionary Computation, 2002,6(6): 566-579.

[3] Cherkassky, B.V., Goldberg, A.V, Radzik T, (1993), "Shortest paths algorithms: theory and experimental evaluation", Technical Report 93-1480, Computer Science Department, Stanford University.

[4] Cherkassky B V, Goldberg A V, (1996), "Shortest Paths Algorithms: Theory and Experimental Evaluation", Mathematical Programming, 73: 129-174.

[5] F Benjamin Zhan, Noon C E, (1998), "Shortest path algorithms: An Evaluation using real road networks", Transportation Science, 32(1): 65-73.

[6] He Zhongxiong, (1983), Fuzzy mathematics and its application, Tianjin Press of Science and Technology, Tianjin.

[7] Jagadesh G R, Shrinkthan T, Quek K H, (2003),Heuristic techniques for accelerating hierarchical routing on road networks, IEEE Transactions on Intelligent Transportation System, 2003, 3(4):301-309.

[8] Wen Huimin,Yang Zhaosheng(1999). "Study on the shortest path algorithm based on fluid neural network of in vehicle traffic flow guidance system. Proceeding of the IEEE International Conference on Vehicle Electronics. Changchun: IEEE, 1999: 110- 113.

[9] Zhou P, Li X P, Zhang H F, (2004), An ant colony algorithm for job shop scheduling problem, Proceedings of the 5th World Congress on Intelligent Control and Automation, 2004: 2889-2903.

Research on Intelligent Transportation System Technologies and Applications

Luo Qi [1,2]

[1] School of Computer Science and Engineering, Wuhan Institute of Technology,
Wuhan, 430073, China

[2] Information Engineering School, Zhongnan Branch of Wuhan University of Science and
Technology, Wuhan, 430223, Hubei, China

Abstract

ITS (Intelligent Transportation System) has been developed since the beginning of 1970s, which makes human, vehicles, roads united and harmonic and establishes a wider range, fully efficient, real-time and accurate information manage system. In the paper, intelligent transportation technologies such as Wireless communications, Computational technologies, and Sensing technologies have been proposed. Intelligent transportation applications are also introduced. This is a belief abstract for an invited talk at the workshop on power Electronics and Intelligent transportation system.

1. Introduction

Intelligent Transportation Systems (ITS) is a broad range of diverse technologies applied to transportation to make systems safer, more efficient, more reliable and more environmentally friendly, without necessarily having to physically alter existing infrastructure [1] . The range of technologies involved includes sensor and control technologies, communications, and computer informatics and cuts across disciplines such as transportation, engineering, telecommunications, computer science, finance, electronic commerce and automobile manufacturing [2]. ITS is an emerging global phenomenon benefiting public and private sectors alike. For example, ITS makes it possible to implement a number of government regulations and processes (e.g., customs and immigration clearance, transportation safety compliance, road/bridge toll collection) more economically, and to improve corporate productivity through time savings, reduced operating costs and energy consumption, and enhanced reliability and safety [3][4] .

2. Intelligent transportation technologies

Intelligent transportation systems vary in technologies applied, from basic management systems such as car navigation; traffic signal control systems; container

management systems; variable message signs; automatic number plate recognition or speed cameras to monitoring applications, such as security CCTV systems; and to more advanced applications that integrate live data and feedback from a number of other sources, such as parking guidance and information systems; weather information; bridge deicing systems; and the like. Additionally, predictive techniques are being developed in order to allow advanced modeling and comparison with historical baseline data. Some of the constituent technologies typically implemented in ITS are described in the following sections.

2. 1 Wireless communications

Various forms of wireless communications technologies have been proposed for intelligent transportation systems. Short-range communications (less than 500 yards) can be accomplished using IEEE 802.11 protocols, specifically WAVE or the Dedicated Short Range Communications standard being promoted by the Intelligent Transportation Society of America and the United States Department of Transportation. Theoretically, the range of these protocols can be extended using Mobile ad-hoc networks or Mesh networking.

Longer range communications have been proposed using infrastructure networks such as WiMAX (IEEE 802.16), Global System for Mobile Communications (GSM), or 3G. Long-range communications using these methods are well established, but, unlike the short-range protocols, these methods require extensive and very expensive infrastructure deployment. There is lack of consensus as to what business model should support this infrastructure.

2. 2 Computational technologies

Recent advances in vehicle electronics have led to a move toward fewer, more capable computer processors on a vehicle. A typical vehicle in the early 2000s would have between 20 and 100 individual networked

---

978-0-7695-3342-1/08 $25.00 © 2008 IEEE
DOI 10.1109/PEITS.2008.124

microcontroller/Programmable logic controller modules with non-real-time operating systems. The current trend is toward fewer, more costly microprocessor modules with hardware memory management and Real-Time Operating Systems. The new embedded system platforms allow for more sophisticated software applications to be implemented, including model-based process control, artificial intelligence, and ubiquitous computing. Perhaps the most important of these for Intelligent Transportation Systems is artificial intelligence.

2. 3 Sensing technologies

Technological advances in telecommunications and information technology coupled with state-of-the-art microchip, RFID, and inexpensive intelligent beacon sensing technologies have enhanced the technical capabilities that will facilitate motorist safety benefits for intelligent transportation systems globally. Sensing systems for ITS are vehicle and infrastructure based networked systems, e.g., Intelligent vehicle technologies. Infrastructure sensors are indestructible (such as in-road reflectors) devices that are installed or embedded on the road, or surrounding the road (buildings, posts, and signs for example) as required and may be manually disseminated during preventive road construction maintenance or by sensor injection machinery for rapid deployment of the embedded radio frequency powered (or RFID) in-ground road sensors. Vehicle-sensing systems include deployment of infrastructure-to-vehicle and vehicle-to-infrastructure electronic beacons for identification communications and may also employ the benefits of CCTV automatic number plate recognition technology at desired intervals in order to increase sustained monitoring of suspect vehicles operating in critical zones.

3. Intelligent transportation applications

3. 1 Electronic toll collection

Electronic toll collection (ETC) makes it possible for vehicles to drive through toll gates at traffic speed, reducing congestion at toll plazas and automating toll collection. Originally ETC systems were used to automate toll collection, but more recent innovations have used ETC to enforce congestion pricing through cordon zones in city centers and ETC lanes.

Until recent years, most ETC systems were based on using radio devices in vehicles that would use proprietary protocols to identify a vehicle as it passed under a gantry over the roadway. More recently there has been a move to standardize ETC protocols around the Dedicated Short Range Communications protocol that has been promoted

for vehicle safety by the Intelligent Transportation Society of America, ERTICO and ITS Japan.

While communication frequencies and standards do differ around the world, there has been a broad push toward vehicle infrastructure integration around the 5.9 GHz frequency (802.11.x WAVE).

Via its National Electronic Tolling Committee representing all jurisdictions and toll road operators, ITS Australia also facilitated interoperability of toll tags in Australia for the multi-lane free flow tolls roads.

Other systems that have been used include barcode stickers, license plate recognition, infrared communication systems, and Radio Frequency Identification Tags (see M6 Toll tag). Figure 1 is Electronic toll collection at Costanera Norte Freeway.

Figure 1 Electronic toll collection at Costanera Norte Freeway

3. 2 Automatic road enforcement

A traffic enforcement camera system, consisting of a camera and a vehicle-monitoring device, is used to detect and identify vehicles disobeying a speed limit or some other road legal requirement and automatically ticket offenders based on the license plate number. Traffic tickets are sent by mail. Applications include:

Speed cameras that identify vehicles traveling over the legal speed limit. Many such devices use radar to detect a vehicle's speed or electromagnetic loops buried in each lane of the road.

Red light cameras that detect vehicles that cross a stop line or designated stopping place while a red traffic light is showing.

Bus lane cameras that identify vehicles traveling in lanes reserved for buses. In some jurisdictions, bus lanes can also be used by taxis or vehicles engaged in car pooling.

Level crossing cameras that identify vehicles crossing railways at grade illegally.

Double white line cameras that identify vehicles crossing these lines.

High-occupancy vehicle lane cameras for that identify vehicles violating HOV requirements.

Turn cameras at intersections where specific turns are prohibited on red. This type of camera is mostly used in cities or heavy populated areas. Figure 2 is Automatic speed surveillance and enforcement equipment

Figure 2 Automatic speed surveillance and enforcement equipment

4. Conclusion

Intelligent transportation system (ITS) is a foreland research task which applied the electronics to the traffic field in the world.

References

[1] Y. Zhao, Vehicle Location and Navigation Systems. Norwood, MA: Artech House, 1997.

[2] FCC, "FCC acts to promote competition and public safety in enhanced wireless 911 services," Washington, DC: WT Rep. 99-27, Sept. 15, 1999.

[3] Y. Zhao, "Vehicle navigation and information systems," in Encyclopedia of Electrical and Electronics Engineering, J. G. Webster, Ed. New York: Wiley, 1999, vol. 23, pp. 106–118.

[4] C. Drane and C. Rizons, Positioning Systems in Intelligent Transportation Systems. Norwood, MA: Artech House, 1998.

Session 10

Other Topics

Study on the Turnover Intention of Knowledge Employees Influenced by Person-Organization Fit (POF)

Pei-lan GUAN[1], Xiao-jun WU [2]

[1] School of Economics and Management
Wuhan University, Wuhan, Hubei, P.R.China, 430072
[2] School of Economics and Management
Wuhan University, Wuhan, Hubei, P.R.China, 430072
Xiaojun9669@163.com   gplwhu@126.com

Abstract

This article abstracts the essential organization influence variable from the variable analysis on the cause influencing the turnover intention of knowledge employees, i.e. the influence of enterprise value on knowledge employees. Meanwhile, basis on the closely relationship between value and POF, it makes a conclusion that the POF can decrease the turnover intention of knowledge employees and designs theoretic hypothesis model. It also proposes outlooks to the further study in future.

1 Introduction

With the increasingly deepened economy globalization and integration, the demand for knowledge employees of enterprises expands so that the knowledge employees became the objectives for which enterprises are scrambling. Knowledge employees' frequently voluntary turnover is a typical feature of the social talent flow in today, in particular, after China's entry into WTO, those domestic enterprises will be confronting to the challenge from powerful transnational enterprises, focusing on struggling for talents. Therefore, our study on the cause variable of knowledge employees' turnover and predicting, recognizing and controlling the turnover behaviors of knowledge employees to reduce their voluntary turnover behaviors and minimize ventures to the enterprise due to the knowledge employee drains that becomes a crucial task, which needed in enterprise managerial theory as well as managerial practices. Moreover, our introduction of China's practical factors by regarding knowledge employee as research objective to study the causes for and characteristics of Chinese knowledge employees' turnover and to propose the preserve strategy for knowledge employees that is of guidance significance for maintaining and promoting the enterprises' core competitive power.

Thanks to knowledge employees' relatively higher professional techniques and more proficient professional knowledge, they are with fairly strong achievement motivation. Consequently, on the one hand, they shall require their own ability to adequate with the resources provided by organization, on the other hand, they pay much more attention to whether their own core value are harmonious with the one of enterprises. The theory of POF, an organization theory specialized on the values of enterprise employees and organization, from whose visual angle we perhaps find a new thought in solving the current problem of the knowledge employees' turnover in Chinese enterprises.

2 Retrospect of Related Main Content

2.1 Knowledge Employee

2.1.1 The Definition and Scope.

The concept of "knowledge employee" has been proposed by Peter • Drucker, the master in management science, in 1956, who deemed a knowledge employee as the one makes his life by knowledge acquired in school more than by physical strength or manual workmanship. The Canadian well-known scholar, the chief umpire officer of Canadian Outstanding Funds Appraisal activities Frances.Horibe (1999) said: In brief, knowledge employees are those people who create wealth by their brains more than their hands. They bring additional values to products via their creative ideas, analysis, judgments, collection and designs. Afterward, the foregoing 2 definitions have generally been adopted in studies on knowledge employee[1].

The international consultant enterprise-Anderson Consulting Company, during the researches on new economy for several decades, proposed knowledge employees are mainly including the following members: (1) Professional; (2) those assistant professional personnel with senior professional techniques; (3) Middle and senior managers. They are usually working in the following fields: research & development, design, marketing, advertising, planning, sales, asset management,

978-0-7695-3342-1/08 $25.00 © 2008 IEEE
DOI 10.1109/PEITS.2008.11

accounting program, law affairs and advising of financing and management.

2.1.2 Features.

As for the features of knowledge workers, many scholars and specialists home and abroad have proposed their own views. The representatives are including Peter.Drucker, the master in management science and Frances.Horibe(1999) [2] , the Canadian knowledge management specialist. The latter one regarded that the essential feature of the knowledge employees distinguished from those average employees is the knowledge employees are holding knowledge capital as the mean of production. Therefore, they have following 7 major features comparing with average employees: (1) independence and self-determination; (2) creative laboring; (3) it is hard to monitor their working process directly; (4) it is hard to measure their laboring fruit; (5) comparatively strong achievement motivation; (6) contemn authority; (7) strong flow intention.

2.2 Person-Organization Fit (POF)

The study on Fit springs from psychology. The organizational behavior scientists have introduced the concept of POF into management science in 1950s. Since then, aiming at professions (Holland, 1985), job selection (Hackman and Oldham, 1980) and organization environment (Joyce and Slocum, 1984), people have proposed theories on Fit, which absorb the content of interactive psychology to take into account how to combine the features of individual with the features of situation to influent together the response of a specific individual to an existed environment [3]. All definitions about POF of today can be dated back to the ASA (attraction-selection-attrition) theoretic frame that proposed by Schneider(1987) [4]. This frame indicates that the individuals and organizations sharing the same value and goal subject to select counterpart interactively.

Kristof (1996) [5] proposed an Integrated Model for Fit after he had integrated the concept of person-organization fit on basis of the previous research conclusions. This Model integrated the viewpoints of identical fit and complementary fit, demand-supply and requirement-capability. Kristof reckoned the person-organization fit exists to some extent when one of the following conditions fulfilled: (a) person and organization, at least either of them is able to provide the counterpart with required resources; (b) person and organization hold similarities in certain characteristics; (c) or the above both 2 conditions fulfilled. In despite of the definition of Kristof wins agreement of majority scholars because of its comprehensive nature, in practical research, most people shall adopt the other definition of POF proposed by Chatman (1991), who considered that POF refers to the identity between the individual value model

and organization value model. The individual value hereinto refers to some person's value in an organization. Although in many aspects, the organization would influence the employees' behaviors and attributes, POF remains as an effective approach in assessing the interaction between individual and organization for value is elementary factor, relatively enduring, additionally, the individual values can be directly compared with organization values.

2.3 Turnover Intention

The turnover of employee refers to the process of an individual who gets the substantial benefits from the organization terminating the membership of his organization. It has been divided into two types: the voluntary turnover and the passive turnover. The turnover intention, belonging to the category of voluntary turnover, refers to the thought that the working staff deliberately made a determination to leave the organization where he had been working for a period after careful consideration (Mobley，1977) [6]. The turnover intention is different from the turnover behaviors. The former emphasizes an attitude trend while the latter focuses on the actual behaviors. They are different yet because the turnover intention will transform into different ways and will not always lead to turnover behaviors. However, it is generally considered to be the last phase after the employee gets the idea of turnover for attempting to seek for a new post as well as the best way of predicting turnover behaviors（Cai Kunhong，2000）[7].

3 Study on the Turnover Intention of Knowledge Employees Influenced by POF

3.1 Analysis on the Cause Variable Influencing the Turnover Intention of Knowledge Employees

There are a great number of studies on the influenced factors of knowledge employees' turnover with so many turnover models overseas. Within those models, the 1st edition Price-Mueller Model [8] issued by Price is being widely used and this Model has been instituted on basis of deep analysis on existed turnover achievements in some disciplines (sociology, psychology and economy). After then, the Price-Mueller Model had been revised importantly for many times, each one of revision based on the previous model introducing new decisive variable discovered by empirical researches. There are 4 categories variable related to turnover in latest model variable: environment variable, individual variable, structure variable and process variable. All of them have been distinguished from the former abundant empirical researches. There are 2 environment variables: family responsibility and opportunity. There are 3 individual variables: universal training, job involvement and positive/negative emotion. 7 exogenous variables are structure variables: independence, result justice, work pressure, salary and reward, promotion opportunity, job monotones and social support. Model presumes that all those structure variables indirectly influence turnover behaviors via its influence on job satisfaction and organizational commitment. There 4 process variables: job satisfaction, organizational commitment, behaviors of hunting for job and turnover intention. For a long time, the job satisfaction has been regarded as a key decisive variable.

Comparing with the overseas mature research on turnover factors, the study on the turnover of employee in China is a relatively weak domain because it began later than other western countries and now most of studies rest on the level of qualitative research. As for the research on knowledge employees, majority of which focuses on how to motivate knowledge employees instead of the problems of turnover that are yet to be paid more attention. While the established researches mostly, carry out qualitative analysis from the following 3 aspects: individual factor, enterprise factor and social environment factor, in lack of quantitative research. What's more, the variables in application are also following the models proposed by overseas scholars so that the models are in lack of the account of environmental factors of Chinese culture. A conclusion of the researches on the influenced factors of turnover intention of knowledge employee by Chinese scholars is shown hereinafter (see Table 1):

Table 1: Concourse for the research on the influenced factors of turnover intention of knowledge employee

Influenced Variables of Turnover Intention of Knowledge Employee	Representative Scholar
Factors of Economics, Politic Movement and Organizational Culture	Ding Xiuling (2002) [9]
Imbalance between Human Capital Value and Return and Disaccorded Goals of Individual and Organization	Jin Juan (2003)[10]
Challengeable Jobs, Independent Right and Salary (subordinated position)	Jiang Chunyan (2001)[11]
Salary, Administration and Individual Factors	Zhang Mian (2003)[12]
Organizational Commitment, Job Satisfaction, Behaviors of Hunting for Jobs, Opportunities, Job Involvement, Expectation Matching Degree, Positive Emotion, Occupation Maturity, Promotion Opportunity and Job Monotones	Zhang Mian ect.(2003) [13] [14]
Job Independence, Achievement Motivation, Study, Career Development, Job Monotones, Job Involvement, Job Independent Right, Level of Salary, Opportunities, Promotion, Results, Fairness, Job Satisfaction, Organizational Commitment, Conducts of Hunting for Job, Occupation Maturity, Switching Cost, Turnover Trend, Career Matching Degree, Relations.	Wu Yaqun (2005)

No matter in the researches at home or overseas, different scholars have acquired almost same conclusion, i.e. a high salary had knowledge employees increased their organizational commitment and job satisfaction but not had them decreased their turnover intentions. There are such relatively significant factors in knowledge employees' values as job independence, achievement motivation, studies, career development and job monotones. The turnover intention has been influenced by adjustment function of value. Those value factors in traditional models such as job involvement, job independent right, salary level, opportunities, promotion, justice results, job satisfaction, organizational commitment and behaviors of hunting for jobs ect., have strong relations with the turnover intentions of knowledge employees.

Therefore, based on this conclusion, from the perspective of POF, we shall propose a hypothesis about the turnover intention of knowledge employees: as long as the individual value of a knowledge employee agreed with his organizational value permitted by some enterprise, the turnover intention of this knowledge employee would remarkably decrease. In the following discourse, the writer is going to make a further demonstration on this hypothesis.

3.2 Establishment of the Model about POF and the Turnover Intention of Knowledge Employee

3.2.1 POF and Job Satisfaction of Knowledge Employee and Organizational Commitment.

A couple of researches prove that POF is with active effective of how an individual treats his job. According to the viewpoint of Identical Fit, it has been proved that value fit can lead to a better job satisfaction and higher organizational commitment. This relation has been demonstrated from the practical researches on the following different knowledge employees: junior accountant（Chatman, 1991[15]）, leaders working in governmental department（Boxx et,1991）[16], MBA students, medium level accountants and middle-level managers（O'Reilly et,1991）[17].

According to the viewpoint of Matching Fit, the virtual fit-to-purpose has strong positive relations with job satisfaction and organizational commitment of employees. Voucouver and Schmitt（1991）[18] have researched on teachers and presidents in more than 350 high schools, from which they discovered the fit-to-purpose of superior-subordinate(president and teacher) and subordinate - subordinate strongly and evidently shown in positive relations with the satisfaction and loyalty of employees.

As the causal model in respect of turnover behaviors, the organizational commitment and employee satisfaction have often been as the mediator variable in individual characteristic research and the influence of working experiences on turnover process. Doughterty, Bluedor and Keon (1985), Lachmanh and Aranga (1986) [19] proposed the description of organizational commitment and job satisfaction is the forecast for employees' turnover intention. Williams and Hazer (1986) proposed the following viewpoints: (1) the employee individual characteristics and job characteristics will not directly influence the turnover trend but directly influences the turnover trend via the influence on satisfaction and commitment. (2) The satisfaction is the key decisive factor of commitment, and the employee individual and job characteristics will not influence commitment directly but indirectly influence commitment via the influence on satisfaction. (3) The satisfaction will not directly influence turnover intention, but indirectly influence turnover intention via commitment. The research by Steer (1997) showed the job satisfaction directly act on turnover intention.

3.2.2 POF and Employee's Turnover Intention.

Addition to its indirect influence on the employee's turnover intention by its effect to employee satisfaction and organizational commitment, POF has positive relations with employee's turnover intention as a variable

The analysis on survival ratio made by O'Reilly（1991）[17] and other scholars indicates that the value fit is a major factor to influence the flow of employees; according to the analysis on value fit level, the employee's turnover trend can be predicted effectively within 2 years. Meanwhile, it was also proved by Chatman（1991）[15] that at the beginning of the new employee's entry to an enterprise or that the new employee had undergone interviews and socialized process for one year, the employee's turnover intention can be predicted effectively according to the fit level between the enterprise and the employee. The research of Voucouver and Schmitt（1991，1994）shows that: between superior and subordinate or, between colleagues, the high-level fit for goal has negative relations with turnover intention; whereas the employees with low-level fit show relatively high turnover intention. The Chinese scholars Zhang Mian, Zhang De and Yu Dan had also take 742 enterprise employees from Xi' an as samples, discovering that the expected fit has evident positive influence on job satisfaction as well as organizational commitment while negative influence on turnover intention[20].

3.2.3 Establishment of Model.

According to the foregoing analysis, we can see that the establishment of POF not only has the most direct influence on individual job satisfaction, organizational commitment and turnover intention, but job satisfaction and organizational commitment would further influence turnover intention. In a word, the enterprise implements management from the angle of POF that can predict on turnover behaviors of knowledge employees and take corresponding measures. Certainly, the establishment of POF shall not be accomplished in an action. For individual, there is a process for choices between his understanding of organizational value and determination to enter into this organization. On the other hand, for organization, not every employee is sure to be of satisfactory. Thus, in the process of the establishment of POF, it is necessary to make a choice and carry out socialization (like training and motivation, etc.). Then we can set up the corresponding model on the effect of POF to the turnover intention of intellectual staff. (See chart 1)

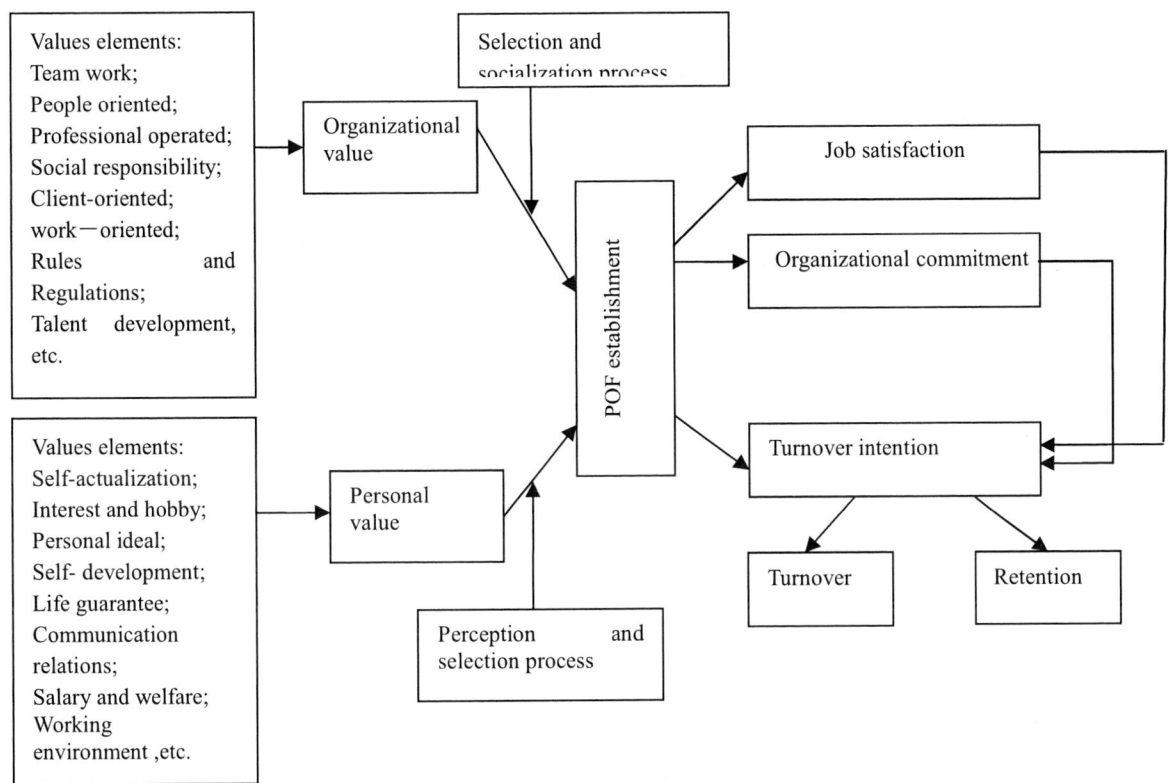

Chart 1: Model for the influence of POF on knowledge employees

4. Conclusion and Outlook

This article, under the puzzledom of Chinese enterprises facing knowledge employee high drains ratio and after demonstrating positive effects of POF to knowledge employee, proposes certain corresponding management countermeasures from the point of view of POF, which may have some guidance significance to the management practice of Chinese enterprises. However, at the same time, we should notice that there are some problems to be solved urgently in the study on POF and the turnover intention of knowledge employees.

First of all, in the cause analysis on the turnover intention of knowledge employee, we had drawn the conclusion that there is strong relation between value factors and the turnover intention at the cost of inadequate examination on specific variable of the value factors. In addition, most variables are derived from overseas models of the turnover being lack of profound analyzing on the variable of based on Chinese practical situations. As a result, it leads to applicability problems for this conclusion in China.

Secondly, in the detailed reasoning of POF to the turnover intention of knowledge employee, this article is mainly using some research materials for demonstration of foreign experts and scholars. It is difficult to find out the demonstration research by Chinese scholars in this field when searching relevant documents. It indicates that

in those researches on the turnover intention of knowledge employees influenced by POF, something yet to do for us is to pay more attention to the management research values. For instance, what are the influenced range and intension of different enterprises with different ownerships? What kinds of organizational behaviors did they take before individual and organization achieved the authentic fit?

Finally, although the fit has evident positive correlation with personal performance, the research indicated that at the organizational level, the high-level fit sometimes creates negative result. Because one important result caused by the POF can form organization of a relative homogeneity. This homogeneity can create active action in the early period of organization development, but if the homogeneity of organization is formed by "inbreeding", then this homogeneity caused ossification, conservative and a lack of innovation, accordingly reduced adaptability of organization to the varying outside environment（Walsh, 1987）[21]. Therefore, after realizing the active influence of POF on the turnover intention of knowledge employee, a cautious attitude should be taken towards specific organizational application. We cannot only highlight in positive aspects but ignore the negative aspects. We must combine it with the practical situation of the enterprises.

References

[1] Qi Jianguo, *Knowledge Economy and Management*, social science academic press : 2001，2-4(in Chinese)

[2] Frances • Horibe: *The Management of Knowledge Employees* , China Machine Press 2000(in Chinese)

[3] Chatman J. Matching people and organization: Selection and socialization in public accounting firms . *Administrative Science Quarterly*, 1991, 36: 459~484

[4]Schneider B, Goldstein HW. The ASA framework: An update . *Personnel Psychology*, 1995, 48: 747-773.

[5] Kristof AL. Person-organization fit: An integrative review of its conceptualizations, measurement, and implications . *Personnel Psychology*, 1996, 49:1-49.

[6]Mobley W H. Intermediate Linkages in the Relationship Between job Satisfaction And Employee Turnover .*Journal of Applied Psychology*，1977，62(2):237-239

[7] Cai Kunhong. Researches on the Satisfaction, Job Involvement, Organizational Commitment and Turnover Intention of Employees in Huajin Group . *A dissertation for Master Degree of Institute of Human Resource Management in National Sun Yat-sen University*(in Chinese)

[8]Price J L. Reflections on the determinate of voluntary turnover. *Journal of International Manpower* .2001

[9] Ding Xiuling. On Flow and Retention of Knowledge Employees. *Journal of Nanjing University of Economics* .2002，5:21-23. (in Chinese)

[10] Jin Juan. An Analysis of the Drain of Professional Managers in China. *Journal of Beijing University of Posts and Telecommunications* (Social Sciences Edition), 2003,7:35-38(in Chinese)

[11] Jiang Chunyan, Zhao Shuming. The Characteristics, Causes of and Countermeasures for Knowledge Employees Flow. *China Soft Science Magazine* .2001，2:85-88. (in Chinese)

[12] Zhang Mian, Zhang De. A Study on the Factors Influencing Voluntary Turnover of Employees. *China Soft Science Magazine* .2001，2:85-88. (in Chinese)

[13] Zhang Mian, Zhang De. An Empirical Study on a Path Model of Turnover Intention among Technical Staff in Enterprises. *Nankai Business Review* ，2003，4: 12-20. (in Chinese)

[14] Zhang Mian, Li Shuzhuo. A Review of Psychological Casual Models of Employee Voluntary Turnover. *Advances in Psychological Science*，2002，3: 330-341. (in Chinese)

[15] Vancouver J B, Schmitt N W. An exploratory examination of person-organization fit: Organization goal congruency. *Personnel Psychology,* 1991, 44:333－352

[16] O'Reilly CA III, Chatman J, Caldwell D F. People and organizational culture: A profile comparison approach to assessing person-organization fit. *Academy of Management Journal*, 1 991, 34:487－516

[17]Chatman J. Matching people and organization: Selection and socialization in public accounting firms . *Administrative Science Quarterly,* 1991, 36: 459~484

[18]Boxx WR, Odom RY, Dunn MG, Organizational values and value congruency and their impact on satisfaction, commitment, and cohesion. *Public Personal Management*, 1991, 20, 195-205

[19]Williams L J,Hazer J T Antecedent and Consequences of Satisfaction and Commitment in Turnover Models: A Reanalysis Using Latent Variables Structural Equation Methods.*Journal of Applied Psychology*, 1986, 71(2):219-231

[20] Zhang Mian, Zhang De, Yu Dan. An Empirical Study on the Effects between Met Expectation and Individual Outcomes. *Forecasting,* 2003, (4). (in Chinese)

[21] Walsh WB, Person-environment congruence: A response to the Moos perspective. *Journal of Vocational Behavior*, 1987, 31: 347-352

2008 Workshop on Power Electronics and Intelligent Transportation System

An Approach for Multi-Dimensional Separation Concerns at Architecture level

Lin-lin Zhang, Shi Ying, You-cong Ni, Jing Wen, Kai Zhao

The State Key Lab of Software Engineering，Wuhan University，Wuhan, Hubei, China
zllnadasha@yahoo.com.cn

Abstract

Traditional architecture design approaches suffer from the crosscutting features and behaviors, which scattered and tangled in the components and connectors, result in the final architecture solutions uneasy to evolve and reuse. In this paper, we propose an approach for multi-dimensional separating crosscutting concerns at architecture level，including: (1) an architectural meta concern space can be reused for deriving more concrete system-specific concerns; (2)a "1+X" model can be used for representing crosscutting and non-crosscutting architectural concerns and relationships; (3) a process is used to address those in a multi-dimensional way. This provides a rigorous analysis of architecture-level concerns as well as important insights into various architecture design solutions. Besides, this makes it possible to, not only eliminate the crosscutting features and behaviors at early architecture design stage, but also encapsulate the crosscutting concerns with concepts and techniques related to aspect in later architecture design activities.

1. Introduction

Nowadays, software architecture becomes the hot topic in the fields of software engineering. Yet tradition architecture design approaches suffer from the crosscutting features and behaviors, which scattered and tangled in the components and connectors, not only contaminates both components and connectors, also blurs the bounder of them, finally result in the last architecture solutions uneasy to evolved and reused. Therefore, how to effectively handle the crosscutting elements in the architecture design activities by existing approaches and generate a pure architecture solution become urgent.

Aspect-Oriented Software Development (AOSD) [1] is emerging as a promising technique to promote enhanced modularization and composition of crosscutting concerns throughout the software lifecycle. Both at requirement and architecture level, aspects provide a new abstraction to represent concerns that naturally cut across modularity units in either a requirement or an architecture description, such as viewpoint and concerns requirements [2], or aspectual components [3]. Yet mapping a concern at requirement level to a counterpart at architecture level is largely unsupported. Consequently, this leads to the gap between the two life stages.

In this paper, we focus on the concerns at architecture level and propose an approach of MDSoC[4], for architects to separate architectural concerns in multi-dimensional way. This makes it possible to effectively cope with the crosscutting concerns at architecture level, then encapsulate those using aspects or design decisions in the later architecture design activities. The key characteristics of our approach include:

- the definition of an architectural meta concern space from where concrete system-specific concerns can be derived based on the specific features of the problem domain at architecture level;
- a "1+X" model of MDSoC, which divides the architecture concern into 1 main dimension and X secondary dimensions, vividly describing concerns and relationships .
- a process of MDSoC in architectural meta concern space, which can guide the architects to address all of architectural concerns.

Section 2 discusses motivation for our approach. Section 3 shows our architectural meta concern space, and the "1+X" model of MDSoC will be described in section 4. Section 5 explores a process of multi-dimensions separating the architectural concerns, combining a running example—e-bank system. Finally, section 6 discusses some related work, as well as the conclusions and future work in section 7.

2. Motivation

Here we start with unifying the definition of the term concern in this paper according to [5]:"an architectural concern is an interest, which pertains to the system's architecture design, its representation or any other matters that are critical to one or more stakeholders." An architectural crosscutting concern is a concern at architecture level, which cannot be modularly represented within the selected decomposition, such as components or connectors. Consequently, it scatters and tangles within other components and connectors, which blurs the bounder of the components and connectors.

During the whole architecture process, from designing, analyzing, and evolving to reusing, various kinds of concerns are involved, but the main two can not be ignored—functional and non-functional concerns. Functional concerns presents the business logical requirements at architecture level, while non-functional

978-0-7695-3342-1/08 $25.00 © 2008 IEEE
DOI 10.1109/PEITS.2008.30

concerns points the requirements about the system quality attributes and constraints. Contrasted with functional concerns, non-functional concerns are concerned with much more stakeholders, such as architects, maintenance men, and quality checkers, etc. Besides, non-functional concerns are related to many facets of problem space, including business requirements, system requirements, quality requirements, techniques constraints, and so on. As mentioned above, every architectural concern denotes a specific systematic goal and should be separately identified and specified at architectural level in order to be traced in the later life stages.

The traditional methods for architecture design almost take regard both business processes and use cases scripts as main concerns, then devise the sketch of architecture comprised of components and connectors, finally add the non-functional requirements into both components and connectors casually. Lack of systematically handling these crosscutting concerns leads to scattering and tangling in the architectural elements, which contaminates the components and connectors and makes the final architecture solution complex and uneasy to understand.

So it is even more crucial to separate the crosscutting concerns at the architecture level that would form suitable approach for identifying, representing and composing all the concerns. The multi-dimensional approach proposed in this paper solves the above issues by eliminating the crosscutting features in the traditional architecture solutions. It starts with the definition of architectural meta concern space.

3. Architecture meta concern space

Based on analysis and observation on architectural concerns mentioned above, in this paper we separate the architectural space into two different spaces in Fig. 1.

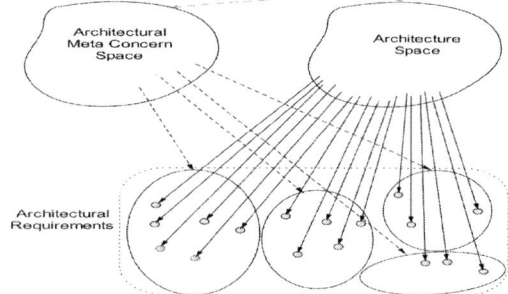

Figure 1. Architectural concern space

- Architectural meta concern space, which comprises abstract set of typical concerns, both functional and non-functional, which appears many times in various system.
- Architectural space, which is made up of the various solutions that architects want to carry out and comprises various domain- or system-specific concerns.

Every architecture solution in architecture space has a number of ideal features, which leads to construction of the architectural requirements (gray dots in Fig. 1), through interviews, analysis of business logic, or architectural requirements templates [6], etc., shown by the solid arrows in Fig. 1. Then we group these requirements into bigger-grained ones, that are specific and concrete concerns (ovals in Fig. 1), according to the architectural meta concern space (represented by dashed lines). So we can get a conclusion that there is a conceptual binding between the abstract representations of concerns in architectural meta concern space with their concrete description in architectural space, denoted by the gray dashed arrow in Fig. 1. Note that not all abstract concerns can be used in a system architecture design, in most cases only a subset of those can be used in a specific system.

4. "1+X" model at architecture level

Our "1+X" model for MDSoC in architectural meta concern space is shown in Fig. 2. Here all architectural concerns are divided into two groups: one named main or primary dimensional concern represents the logic function of a system, X named secondary dimensional concerns represent the crosscutting concerns, including the functional and non-functional architectural requirements. In meta concern space there is only one horizontal flat referring to the main concerns, orthographically with X flats denote secondary dimensional concerns $D_1,...,D_X$, which vividly show the secondary crosscuts the main concerns. In D_1 dimension, C_1 are refined into $C_{11},..., C_{1n}$, which only belongs to domain- or system-specific space and will be explain in section 5.4 in detail.

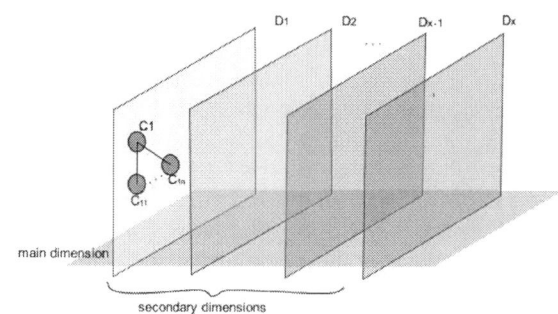

Figure 2. "1+X" model of MDSoC

4.1. Running example

The case study we have chosen is an e-bank system, having the following key characteristics:

A new account is created for a new customer, or added to an existing customer in a bank, while an existing account can be cancelled for some reasons. A customer can modify the passwords of his own accounts. Transfer must be provided in local, including inside-bank and outside-bank referring to transfer occurs between different branches in the same kind of bank and between different banks, respectively. Users can query their account balance in details, such as when and where did the users deposit and withdraw how much cashes. Besides, system must record the all operations for every account, and all the transactions must be performed in a rapid and safe way. Lastly, system should be always in active state in order to provide services for customers at any time.

5. A process of MDSoC at architecture level

Here we propose a process for multi-dimensions separating of architectural concerns which comprises four steps as following: identify architectural concerns, specify architectural concerns, refine architectural concerns, and compose architectural concerns.

5.1. Identify architectural concerns

Rational employs FURPS+ [6] to capture architectural requirements through providing architectural requirement questionnaire templates for stakeholders in order to collect the requirements as comprehensive as possible. So we can easily capture the functional, usability, reliability, etc., by using analysis mechanisms and others templates. Through FURPS+, we have identified the critical architectural concerns in our e-bank system.

In e-bank system, the concerns on the main dimension are listed as follows: (1) account management, including: ① create a new account, ② cancel an account, ③modify the password; (2) account query, including: ④query the balance, ⑤review the detail; (3) customers transactions management, including: ⑥ inside-bank transactions; ⑦ outside-bank transactions.

Throughout the paper we focus on four secondary dimensional concerns for the e-bank system, they are, response time, security, availability and log. Response time requests the system services should be finished within a given time limit, such as 15 seconds. Security is fully described in Tab.4. Availability indicates that customer services should be available 999/1000 requests. Log means that all operations involved in business of e-bank system should be recorded.

5.2. Specify architectural concerns

Here we employ concern templates to specify the main and secondary dimensional concerns in e-bank system. For each concern whether it belongs to crosscutting or not, two forms are provided, the architectural meta concern space and architecture space—our e-bank system, respectively. Therefore, there are four forms provided by this paper, two for main dimensional concerns (cf. Tab.1 & 2), and the other two for secondary dimensional concerns (cf. Tab.3 &4).

Table 1 "Add" in architectural meta concern space

Meta Concern Name	Add
Dimension	Main dimension
Description	Add a new record into database.
Examples	Add a user; Add a department; Add a book, etc.

Table 2 "Add" in architecture space

Concern Name	Add An Account
Dimension	Main
Description	Add an account to an existing customer in a bank, or create an account for a non-existing customer in a bank.

Table 3 "Security" in architectural meta concern space

Meta Concern Name	Security
Dimension	Secondary dimension
Description	Ensure the system always be in safe.
Examples	Confidentiality; Integration; Authorization; Authentication.

Table 4 "Security" in architecture space

Concern Name	Security
Dimension	Secondary dimension
Description	E-bank system must be available always, so that user can transact anywhere and anytime; authorize customers in the access to the bank; authenticate the customers that interact with the system; ensure the information security during transaction.
Concerns involved	Availability; Authorization; Authentication; Confidentiality.

5.3. Refine architectural concerns

Refinement of concerns includes refining the main and secondary dimensional concerns, the main are listed above from ① to⑦, so our attentions will be paid more to the secondary dimensions. Refining those means to determine their influence on later development stages and identify their mapping into a function, a decision or an aspect [2]. The result of refinement of four secondary concerns is shown in Tab.5.

Table 5 Secondary concerns refinement

Secondary concerns	Influence	Mapping
Response time	Design, implement	Function, Aspect
Security	Design, implement	Functions, Aspect
Availability	No	Decision
Log	Design, implement	Function, Aspect

Security concern in e-bank system can be further divided into four concerns (cf. Tab. IV), each of them will influence the design of the classes realizing the requirements constrained by themselves, and then will be further mapped into a specific function as well as an aspect. *Availability* concern influences architecture design only, so it will map into a design decision during the refinement. Others secondary concerns in our case and their influence and mappings are shown in Tab. V. After acquiring aspects from secondary concerns, the architects can take some special manners to address them at the later architecture design stage.

5.4. Compose architectural concerns

In this step the main dimensional concerns in e-bank system should be composed with secondary dimensional concerns, showing the initial version of whole system. Here we introduce a new conception—composite points, where the composition of main and the secondary dimensional concerns will be taken place possibly. In our e-bank system, we state those points in Tab. 6.

Table 6 Composition points in e-bank system

SDC MDC	Response Time	Confidentiality	Availability	Log
①~⑦	①~⑦	⑥~⑦	①~⑦	①~⑦

SDC is secondary dimensional concern, while MDC is main dimensional concern.

After this, the architects must resolve the conflict by negotiating with the stakeholders and setting the priority on one of them, when two concerns have opposite effects on the same points. This will be solved in future work.

6. Relate work

Hyperspaces [4] employ hyperslices and hypermodule to decompose and compose the overlapped concerns in a multi-dimensional way. Cosmos [7] models concern-spaces through concerns, the relationships between them, and predicates. Concerns are classified as logical and physical representing concepts and elements of software system respectively. Yet both hyperspaces and Cosmos only propose basic models, not providing explicit multi-dimensional separating methods with respect to some specific life stages.

Some similarity to our work about applying MDSoC into early life stages can be found in [2, 8]. However, [2] aimed at requirement level, while our paper at architecture level. [8] also proposed MDSoC at architecture level, but focused on the existing ADLs failed to separate concerns, so it is quite different from our work.

There are quite numbers of research work have made in the area connecting requirement engineering with architecture smoothly, but only a few touch the aspect-oriented software development. [9] outlines a traceability schema that will provide support for recording the lost information generated during mapping requirement engineering to architecture. [10]

7. Conclusions and future work

Traditional architecture design approaches fail to cope with the crosscutting concerns, so we propose a new approach based on multi-dimensional separating of architectural concerns. The architecture concern space comprised by meta concern space and architecture space, a multi-dimensional separating model in architectural meta concern space, and a modeling process are also provided to illustration it.

The definition of an architectural meta concern space provide an effective way to reuse, we can derive domain-specific concerns from there when design other system architecture. Multi-dimensional separation in this paper may be not similar with other existing ways. We designate one main dimension representing non-crosscutting concerns, and a number of secondary dimensions representing crosscutting concerns. In our model, flats denoting main dimension and secondary dimensions are vertical each other. Lastly, the process of multi-dimensional separating architectural concerns can guide architects for handle crosscutting concerns and hand over the output for following activities at architecture design stage.

Our future work will focus on handle the conflicting concerns and developing case studies to further validate the proposed approach. In the near future we aim to design and develop modeling tool which will provide support the MDSoC at architecture level and carry out mapping from model of MDSoC into ADL.

Acknowledgements

This work is funded by National Natural Science Foundation of China (No. 60773006) and Research Fund for the Doctoral Program of Higher Education of China (No. 20060486045).

References

[1] Aspect-Oriented Software Development Web Site, http://www.aosd.net.

[2] A. Moreira, J. Araujo, and A. Raishd, "A concern-oriented requirements engineering model", Proc. CAiSE 2005, LNCS, vol.3520, pp.293-308.

[3] M. Pinto, L. Fuentes, and J.M. Troya, "DAOP-ADL: an architecture description language for dynamic component and aspect-based development". Proc. GPCE 2003, LNCS, vol. 2830, pp. 118–137.

[4] P. L. Tarr, H. Ossher, W. H. Harrison, and S. M. Sutton, "N degrees of separation: multi-dimensional separation of concerns", Proc. ICSE, 1999, ACM, pp. 107-119.

[5] K. van den Berg, J. M. Conejero, and R. Chitchyan, "AOSD ontology 1.0–public ontology of aspect-

orientation", AOSD-Europe Report, Deliverable D9, 27 May 2005.

[6] P. Eeles, "Capturing architectural requirements". Rational Technique Report. 2004.

[7] S. Sutton, I. Rouvellou, "Modeling of software concerns in Cosmos". Proc. AOSD 2002, pp. 127-133.

[8] M. M. Kandé, A.Strohmeier, "On the role of multi-dimensional separation of concerns in software architecture". Proc. OOPSLA 2000, pp. 1-6.

[9] R. Chitchyan, M. Pinto, L. Fuentes, and A. Rashid, "Relating AO requirements to AO architecture" [EB/OL].http://www.aosd-europe.net/allPublications /AnalysisDesign/Architecture/chitchyan_pinto_fuentes_ra shid_ea05.pdf.

[10] J. Liu, R. R. Lutz, and J.M. Thompson, "Mapping concern space to software architecture: a connector-based approach". Proc. MACS 2005, pp.1-5.

[11] R. Chitchyan, M. Pinto, A. Rashid, and L. Fuentes. "COMPASS: composition-centric mappiing of aspectural requirements to architecture", Transactions on AOSD IV, LNCS 4640, 2007, pp.3-53.

GADAM: an authorization model based on attribute delegation in grid

Wang Rongbin[1,3] Chen Shuyu[2] Li Jing[4]

1.College of Computer Science, Chongqing University, 400044,Chongqing, China
2. College of Software Engineering, Chongqing University, 400044,Chongqing, China
3. Chongqing expressway development ,ltd. 400042, Chongqing, China
4.Department of Computer and Modern Education Technology, Chongqing Education College,
400045,Chongqing,China
Email:wangrongbin@vip.163.com Tel:1399638610

Abstract

Inherited the flexibility merit of the access control model based on attribute(ABAC), the authorization model based on attribute delegation in grid(GADAM) is put forward, which Expressed the attribute set with attribute expression in this model, and adopted the method based on attribute delegation directly to assign and transmit the permission. Complied with the constraint mechanism of atomic authorization and the simplest authorization attributes in delegation strictly, the resource search along opposite direction of the authorization chain for the requester, if trusted authorization source could be found, the final attribute set possessed by requester is trusted. According to the analysis, the model has the characters of better security and flexibility, which was suitable for gird.

Keywords: grid; attribute; delegation; trusted; authorization

1   Introduction

Most of the research on grid security are central on role-based access control model (RBAC) [1,2] currently. As the attribute-based access control model (ABAC) [3] has the characters that are more flexible, fine-grained and dynamically suitable to authorization in distribute environment than RBAC, it is more suited for grid. In grid, how can make it trusted to the resource that the permission or attributes in ABAC after assigned and delegated much times is a difficult problem, whereas there are few systematical research on how the attributes in ABAC expressed, transmitted and verified .The authorization model based on attribute delegation (GADAM) presented in this paper researches on the authorization in ABAC. The remainder of this paper is as follows: Section 2 describes the related works on this topic and section 3 states the fundamentals of GADAM , and the formal definition of GADAM is described, and how to verified the authorization is also presented. In section 4, the GADAM's

characters and security are analyzed. Finally, section 5 underlines some conclusions and future research lines.

2   Related works

The authorization in grid will be delegated across many times, and the delegation has two types, the one is among the authorities, the other is among user. [4] expresses the authorization polices with Datalog, construct five model: $RT_0, RT_1, RT_2, RT^T, RT^D$, these models for authorization are expressed clearly. But the model has other questions: there are no constraint in permission transmission, and after many steps the permission or role delegated, the intrinsic hierarchy of roles may be broken, so the hidden security trouble is exist.

It's important to constrain the multi-stage authorization in grid. [5] draws the time property into RBAC authorization , which must comply with the time restriction for the role activation and session. [6] represents the delegation based on restrict attribute, which is suitable for distributed environment. The CAS[7] designates the permission to the users with capability certificate directly, the authorization method is not fit for giant and dynamically changeable users in grid. All these models are based on RBAC, which aren't appropriate for grid.

VOMS[8]、Akenti[9] 、PERMIS[10] are aim to express and transmit the users' attribute with attribute certificate, and don't take advantage of the attribute to construct access control model based on user' attributes, neither have the attribute delegation between users. How to utilize the character of ABAC which is fine-grained and flexible to deploy in grid, and research on the attribute expression, delegation, assign and verification of ABAC in grid systematically is an important direction.In this paper, the GADAM research on attribute expression, attribute delegation and constraint to resovle these above question.

3   The authorization model   based on attributes delegation (GADAM)

3.1 ABAC Model

In ABAC, the subject S represents users, process, at al. P represent permission. the others elements definition as follow:.

Definition 1 Basic Elements

（1） atomic permission P_{atom}

$P=\{p_i\}, i \in [1,n]$, p_i is the leaf of permission tree, which couldn't be split, namely the most fine-grained permission, so we define it as p_{atom}.

（2） attribute express ats

$ats=<attr><op><value>$. $attr$ is attribute type; op is operator, $op=\{\geq,\leq,>,<,=,\neq,IN,\ NOT\ IN\}$;value is attribute value. value_type $=\{$Number, Character, Date,Time, Other$\}$..

$AT = \bigcup_{i=1}^{n} ats_i$, n=$|AT|$, AT is the. constraint attribute set.

$S(AT)$ represent the subject S has the attribute set AT

（3） grid resources RES

$RES = \{r_i \mid i \in [1,n]\}$ .r_i is the subject which can provide the services.

（4） access request REQ

$REQ =< s(at), res, p))$

REQ is a triple tuple, subject s with attributes set at request the permission p on resource res.

（5） access control polices POC

$POC = \bigcup_{i}^{n} RULE_i$ $n =| POC |$ · $RULE =< at, res, p, c >$

POC composed of rules as $RULE$, the $RULE$ is tetrad, which denote that the subject with attribute set at can execute p on resources res, and must comply with the constraint c.

Definition 2 attribute-based access control model ABAC

ABAC=$\{AUTH, S, RES, AT,, P, C\}$,it denotes that the subject s must has the attribute at which authorized by authority $AUTH$ can execute p on resources res, and satisfy the constraint c. ABAC has such relation:

$ASA \subseteq AT \times S$:the relation of assignation AT to S.

$PAA \subseteq AT \times P$:the relation of assignation Pto AT.

$Attributes : P \rightarrow 2^{AT}$:the reflection of p to AT,

$Attributes(p) = \{(p_i, at) \mid (p_i, at) \in PAA\}$.

$PSA \subseteq P \times RES$:the relation of P and RES.

Definition 3 atomic attribute set AT_{atom}

In ABAC, if there is attribute set AT,

$AT= \{at | \exists at \subseteq AT \land at_i \in at, (at, P_{atom})$
$\in PAA \land (at - at_i, P_{atom}) \notin PAA\}$

we definition AT_{atom} is corresponding to the atomic P_{atom}.

$AT = \{AT_{atom\ i}\}, i = 1, ..., n$

3.2 Attribute Delegation

In ABAC for grid, the attribute set represent permission, so the assignation and delegation of permission can implemented by assigning and delegating attribute set.

Definition 4 delegation between authority DGTA_A

In the grid VO, there are several authorities who assign the permission to subjects. In GADAM, the authority as broker possesses many attribute set which get from resource. The resource delegates those attribute set to authority according to the polices in local domain, or the VO delegation the attribute set to the other authority according to the VO global polices directly, and the global polices in VO have been composed from many autonomic domains. So the relation of delegation between the resource and authority is :

$(res, auth, at) \in DGTA\_A \subseteq RES \times AUTH \times AT$

$Delegete\_a : AUTH \rightarrow 2^{AT}$ is the reflection relation from the resource to authority. When the authority possess attribute set, they can delegate to subject.

Definition 5 delegation between subject DGTA_S

In grid. when requesting a task, the request may across many autonomic domains, so the subject must delegate their permission or attribute to other subject, such as the proxy in Globus. The delegation from subject to subject is formally expressed as:

$(s_i, s_j, at) \in DGTA\_S \subseteq S \times S \times AT$,i,j=1,...,n,i $\neq$ j,

$Delegete\_s : S \rightarrow 2^{AT \times S}$ is the reflection relation from the subject to other subject.

We can perorate: $DGTA \subseteq AT \times S$ .and. $DGTA = DGTA\_A \cup DGTA\_S$.

Definition 6. authorization chain Auth_chain

After the attribute set delegated from the resource to the terminate subject (the requester), the delegation form a path, whose starting point is resource, and destination is the requester. The ordered sequence of the node in the delegation path is called authorization chain. The formal expression is:

$Auth\_chain = (d_1, d_2, .., d_n)$ $d_i, d_j \in Auth\_chain$, when n=0, it's null sequence. d_1 is the start point , which is authority or resource . d_n is the requester.

3.3 GADAM model

Definition 7. GADAM

GADAM=$\{VO, ABAC, DGTA, AAC, SAC, AV\}$. Namely, the GADAM has these elements:. Grid virtual organization

VO, access control model ABAC, attribute delegation DGTA, atomic authorization constrain AAC, simplest authorization constraint SAC, authorization verification AV. Our model is based on ABAC, so ABAC is an important element.

In GADAM, the permission represented as attribute is expressed and transmitted, so it's very important to constrain the process of the attribute delegation strictly, otherwise the integrity and reliability will be destroyed.

As fig.1, firstly, the resource assign the permission to specific attribute set in polices which represent as PAA. then the local resource RES or global resource RES in VO delegates the attribute set AT in the polices to other authority (broker) AUTH according to the resource, letting AUTH be trusted by authorization authorities, which represented as DGTA_A. After that, the AUTH delegates certain AT to subject S corresponding to the resource requested by S. When S requests one domain resource ,he may delegate his AT to other subject in local domain or other domains which represent as DGTA_S. In the process of delegation DGTA_A and DGTA_S., we must constrain the AT delegation strictly.

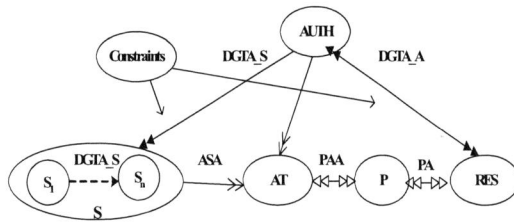

Fig.1 ADAM model scheme

Definition 8 Atomic Authorization Constrain   AAC

In GADAM, $\forall (A,B,AT) \in DGTA$, $task \infty P$ denotes that the permission P must needed by the task $task$, we know: $P = \bigcup_{i=1}^{n} p_{atom\ i}$,then, when A delegate AT to B the follow rules must be satisfied:

(1) $(at'_{atom\ i}, p'_{atom\ i}) \in PAA$, $AT' = \bigcup_{i=1}^{n} at'_{atom\ i}$

(2) $\forall at \in at'_{atom\ i} \wedge at \in at'_{atom\ j} \wedge (i,j)$ $\in [1,n]$, $AT_{atom} = AT' - at$,则 $AT = AT_{atom}$

So we call it atomic authorization constrain AAC ,denoted that the least attributes .needed for the P_{atom} delegated to the delegatee B by A.

Definition 9. The Simplest Authorization Constrain SAC

$\forall (A,B,AT) \in DGTA$, when A delegate AT to B the follow rules must be satisfied:

(1) $(AT_{atom}, p_{atom}) \in AAC$, $(A,B,AT_{atom}) \in DGTA$

(2) $\exists at \in AT_{atom} \wedge at \in B(AT)$, then $AT = AT_{atom} - at$

So we call it the simplest authorization constraint as SAC, namely, when B has some attributes at of AT delegated by A, then it's no need for A to delegates these attributes at to B any more, so the delegation is most concise and simplest

3.4 Authorization verification

(1) the authorization rules in GADAM

In GADAM, we use Datalog and log-program expression to define the basic authorization rules syntax as follow:

（1）$H \leftarrow B$

（2）$B = B_1; B_2, B_3 : B_4$

H is the head of the expression, and is also the conclusion. B is body of the expression, which is the precondition. B is the union of B_1, B_2, B_3, B_4, semicolon ";" denotes the selection relation "or", comma "," denotes "and". colon ":"denotes the exclusion relation "except".

To express the authorization rules conveniently, we define several functions as follow:

Has(s, at): subject s has attribute set at.

Actives(s(at),at_0,ss): subject s actives the attribute set at_0 of his attribute set at in session ss.

Delegates(s_1,s_2,at,c): subject s_1 delegates at to subject s_2,the delegation must be complied with the constraint c, in GADAM, $c(c = AAC \cup SAC)$, and other constraint as duty separation.

Authorize(a,s,at,c): authority a authorize (delegate in GADAM) at to subject s, he delegation must be complied with the constraint c, c may be alive time or other precondition

Originates(e,auth_chain): the source of the authorization chain $auth\_chain$ is entity e:

Then we define the authorization rules as follow:

Definition 10. authorization rules in GADAM

R_1: $Has(s,at) \leftarrow Delegates(s_1,s,at,c);$
$\qquad\qquad\qquad authorizes(a,s,at,c)$

R_2: $Delegates(s_1,s,at,c) \leftarrow$
$\qquad\qquad\qquad (s_1,s,at) \in auth\_chain(p)$

R_3: $(s_1,s,at) \in auth\_chain(p) \leftarrow$
$\qquad\qquad authorizes(a,s_1,at',c), at \subseteq at'$

548

R_4: $authorizes(a, s_1, at', c) \leftarrow$

$\quad Delegates(a, res, at'', c), at' \subseteq at''$

R_5: $Delegates(a, res, at'', c) \leftarrow res = Originates.e$

The above rule R_1 represent if subject s has attribute set at, he must get at through delegation from other subject s_1 or from authority a directly. Rule R_2 represent s_1 and s must be neighbor in $auth\_chain$ when s_1 delegate the at to s, s_1 must has the attribute at' from trusted authority a, and at is the subset of $at'(R_3)$. We know that the authority a get these attribute set at'' from res, the res is the trusted source of $auth\_chain(R_4 \setminus R_5)$. So the subject's attribute set at can be found his trusted source res search along with the $auth\_chain$.

(2) authorization verification algorithm

Definition 11. Trusted Attributes TA

When the resource receives the request REQ from subject S, the resource will verify whether the identity and attribute set AT of the requester is trusted. To verify the attribute set is our focal point, firstly, the resource searches if the trusted source could de found along with the authorization chain, meanwhile, the resource verify if the attribute set satisfy POC in vo and local domain. All above condition are satisfied, then the attribute set AT of subject S is trusted. The formal express as follow:

$TA(REQ.at) = Has(s, at) \wedge (at \geq POC.at')'$

$\wedge task \infty p \wedge Assign(p', at') \wedge REQ.p \leq p$

As Fig.2, we construct the authorization directed graph by connecting the entity which have authorization(delegation) relation each other, the start point is delegator, the end point is delegatee, the authorization is the side, the tier is corresponding to the hierarchy of the authorization relation. s(at) is end conclusion, so it is the end point in the directed graph.

the authorization verification as follow:

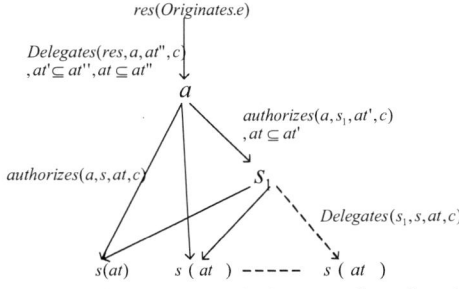

Fig.2 Trust authorization source directed graph

Algorithm 1 auth_very_algorithm:

Input: request REQ of subject S, POC;

Output: the result of verification:

yes/no/ undefined

Method:

step1: analyze the REQ, get the key element（s, res, at, p）.

step2: if Originates(e, auth_chain)= REQ.res   then step3, otherwise step 7.

step3: allocate the polices global POC in VO and local domain POC, analyze the POC, get the key element（at, res, p,c）.

step4: if REQ. re $\subseteq$ POC.res, REQ.at $\supseteq$ POC.at, REQ.p $\subseteq$ POC,,then step 6, otherwise step 7.

step5: return result(undifined).

step6: return result(yes).

step7: return result(no).

4   Analysis of GADAM

4.1 The security of GADAM

In GADAM, we can make the follow conclusion:

Theorem 1. GADAM satisfies the least privilege constraint

Proof: we provide $task \infty p$, $P = \{p_{atom1}, p_{atom2}, \ldots, p_{atom\,n}\}$, $\because \forall P_{atomi} \rightarrow 2^{at_{atomi}}, i \in [1, n]$, and compose the atomic attributes according to the atomic permission p_{atom}, $AT' = \bigcup_{i=1}^{n} at'_{atom\,i}$, delete the redundant attribute: $AT_{atom} = AT' - at$, $\because p = \bigcup_{i=1}^{n} p_{atom\,i}$ $\therefore task \infty P \rightarrow 2^{MIN(AT)}$, $i \in [1, n]$. **Proof is over.**

Theorem 2. GADAM satisfies the separation of duty(SoD)

We will not discuss about SoD in GADAM, but GADAM supports any SoD constraint. We can adopt ":" to add any exclusion expression in GADAM authorization rules. For example: the duty of subject A and B are excluded each other, they can't have the same permission(attribute).so the authority can't authorize the same attribute to them at the same time, we can express as follow:

$Has(A, at) \leftarrow delegates(a, A, at) : delegates(a, B, at)$.

Theorem 3. the S(AT) satisfied the AAC and the authorization rules is trusted.

proof:（1）trusted authority could be found: In GADAM, $\forall s(at) \rightarrow Has(s, at)$,the at of s is from: : the resource res delegate to subject s directly : $Delegates(res, s, at)$,obviously, the trusted source is the res;   the other subject or authority in authorization chain $auth\_chain$ delegate to subject s .As fig.2, $\because s(at)$ is the end point in the orient authorization graph,

$\therefore \quad \exists (s_i, s, at) \in auth\_chain \lor Delegates(s_i, s, at, c)$
$\land Delegates(s_{i+1}, s_i, at^i, c) \forall at^{i-1} \subseteq at^i, i = [1, n]$, if s_i isn't the start point in the orient authorization graph, when $i=n, \exists (s_{n+1}, s_n, at^n) \in Delegates(s_{n+1}, s_n, at^n, c) \land s_{n+1} = res$, so the trusted authorization source is found.

（2）**the access control polices could be satisfied:** We know $POC=$（$RES.task, p, at$）,the trusted authority $AUTH$ or RES delegate at^i to s_i, then s_i delegates at^{i-1} to next subject s_{i-1}, then there is a authorization chain $auth\_chain= \{res, auth, s_i, s_{i-1}, ..., s_1\}, i = [1, n]$ .and there is the delegation relation $Delegaqtes(s_i, s_{i-1}, at^{i-1}, c)$, $\because at \subseteq at^i, at \subseteq at^{i-1}, ..., at \subseteq at^1$, and the end point in the authorization chain $auth\_chian$ is $S(at^1)$, $\therefore s(at^1) = REQ.at \supseteq POC.at$,namely, the access control polices are satisfied.

According to the definition of trusted attributes , the S(AT) is trusted. **Poof is over.**

4.2   The characters of GADAM

the GADAM has such character as :

■ the model express the attribute structure, transmission, delegation and verification systematically.

■ The permission assignation is flexible: Because the GADAM has inherited the merit of ABAC, and adopt the method of attribute delegation to solve the permission assignation , which could be satisfied to the dynamic need in grid environment.

■ The constraint is strict: the model strictly constrain the delegation of permission (attribute) to ensure the security of authorization, which satisfies the least privilege and SoD.

■ The control is fine-grained: Corresponding to RBAC, the attribute in attribute set could be composed many atomic subset, so could reflect the atomic permission, which could be the most fine-grained and could be satisfied for grid environment.

5   Conclusions

We have put forward the GADAM in this paper, define many important element such as attribute set ,expression and so on, and represent the authorization based on attribute delegation, and strictly constrain the process of attribute delegation, so can ensure the security and character suitable for grid. After analyze the model characters ,we can make the conclusion. That the GADAM suitable for grid environment. but GADAM hasn't research on the activation, revocation and update the attribute, it will be the next work for us.

Acknowledgement

It is a project supported by New Century Educational Talents Plan of Chinese Education Ministry (under contract No.NCET-04-0843), and the Natural Science Fund of Chongqing Science and Technology Committee (under contract No. 2005BB2192).

References

[1] Hai Jin, Weizhong Qiang, Xuanhua Shi,at. RB-GACA: A RBAC based grid access control architecture. Int. J. Grid and Utility Computing, 2005. 1(1):61-69.

[2] Weizhong Qiang, Hai Jin, Xuanhua Shi. Joint Management of Authorization for Dynamic Virtual Organization. Proceedings of the The Fifth International Conference on Computer and Information Technology(CIT'05). Washington, DC.2005.375 - 381

[3] Shen Haibo, Hong Fan. An Attribute-Based Access Control Model for Web Services, In: Proceedings of the Seventh International Conference on Parallel and Distributed Computing, Applications and Technologies. Washington. IEEE Computer Society Press. 2006 74-79 .

[4]Li N,Mitchell J c . RT:a role-based trust—management framework[C] Seamonst K E,ed. Proceedings of the DARPA Information Survivability Conference and Exposition (DISCEX'03).   Washington:IEEE Press, 2003:214-226.

[5]Dong Guang-yu, Qing Si-han, Liu Ke-long. Role-Based Authorization Constraint with Time Character. Journal of Software (in chinese). 2002.13(08):1521-1527.

[6] Ye Chunxiao ,Wu Zhongfu, Fu Yunqing, et al. An Attribute-Based Extended Delegation Model, Journal of Computer Research and Development (in Chinese), 2006, 43(6): 1050～1057.

[7]Pearlman L,welch V,Foster I,et al . A community authorization service for group collaboration[M]. In:Wemer B ed. Proceedings of IEEE Workshop on Policies for Distributed Systems and Networks.Monterey,Califomia,2002. Alamitos:IEEE Computer Sodety.2002,5—59.

[8]Alfieri R,Cecchini R,Ciaschini V,et al. From gridmap-file to VOMS:authorization in a grid environment [J] . Future Generation Computer Systems,2005,21(4):549—558.

[9] R.Mary, Thompson, Abdell Ah Essial. Certificate-based Authorization Policy in a PKI Environment. ACM Transactions on Information and System Security (TISSEC).ACM Press 2003, 566－588.

[10]D. W. Chadwick and A. Otenko. The PERMISX.509 Role Based Privilege Management Infrastructure. 7th ACM Symposium on Access Control Models and Technologies （ SACMAT2002). Monterey, California: ACM Press. 2002. 135-140.

Research on the Application of Probationary System in Tacit Knowledge Sharing

Zhihong Li, Jun Li

School of Business Administration,
South China University of Technology, Guangzhou, Guangdong, 510640, China
bmzhhli@scut.edu.cn, lijunfrank@163.com

Abstract

Knowledge sharing, especially tacit knowledge sharing plays a vital role in the formation of corporate competitive advantages. To begin with, this paper analyzes the characteristics of probationary system that contributes to effectively cultivating the successors in the process of tacit knowledge sharing. Then, it explores the main obstacles on practical application of this system in enterprises. Finally, the paper provides some feasible countermeasures, which can greatly improve the efficiency of tacit knowledge sharing in enterprises.

1. Introduction

As the development of economy and the adjustment of industrial structure in China, it demands for senior technicians increasingly. The senior technicians in a special field can deal well with the key technical issues, so they play an important part in the development of enterprises. At present, however, we are facing a serious problem that many of the senior technicians have retired or are about to retire. The young technicians without practical experience do not have the ability to solve the problems well and effectively in the crucial situation. As Michael Polanyi (1967) wrote in The Tacit Dimension, a significant challenge is the fact that "we can know more than we can tell". While people record key information in written specifications and a number of other transfer deliverables, there remains a good deal of knowledge in the minds of individuals that will never find its way to the printed paper or online knowledge repository. Tacit knowledge is the accumulation of experience and not in documents or other explicit forms[1].

Probationary system linking theory with practice plays a crucial part in tacit knowledge transfer in an enterprise. Probationers will face many changes in the work environment and methods, and will have more opportunities to communicate with senior technicians, which facilitating the transfer and sharing of tacit knowledge among individuals in an organization.

2. Probationary system and its role in tacit knowledge sharing

2.1. Probationary system

Probationary system has a very long history. In ancient China, the prince was directly entitled to manage some national affairs, which would contribute to the accumulation of management experience. British Navy, 300 years ago, was in a similar situation that in wartime, if warship captain killed or injured, there was a person could immediately implement the corresponding duties-- probationer. In modern enterprises, in order to guarantee long-term development, they should execute the probationary system that each management position has one or several probationers with the conditions of the position. Generally speaking, a general manager or manager should have a probationer to make sure that a qualified people to fill the vacancy when they retire or transfer to other departments.

In summary, from the perspective of tacit knowledge sharing, probationary system is that, in order to ensure that tacit knowledge transfers from knowledge provider to recipient smoothly and completely, recipient works in the provider's position and communicates with him to capture key knowledge and technologies, which will contribute to the continuity of critical tacit knowledge in the organization.

2.2. Probationary system's role in the promotion of tacit knowledge sharing

Tacit knowledge is the most important source of corporate wealth, and its efficient transfer is the key to business success[2]. Probationary system's roles in the promotion of tacit knowledge sharing mainly are as follows:

(1) To provide probationers with opportunities to capture tacit knowledge

Tacit knowledge is the gradually accumulation of individual practical experience in the long-term work period[3]. Probationary system provides opportunity to contact and solve problems involved in the key

technology, which as well as an important way to cultivate senior technicians through capturing the practical experience. After the collection of operational experience, probationers will be no longer confused by these problems, can inspect the equipments independently and apply the basic knowledge and the skills to the practical work, which will greatly improve the interest of probationer to get tacit knowledge.

(2) To enhance the degree of inter-trust between knowledge provider and recipient

According to the research, the higher the degree of inter-trust between staff, the more positive behaviors happen[4]. Generally, individuals often share their knowledge with a reliable person; therefore, trust is the prerequisite of choosing a recipient. Probationers and senior technicians work together to study and solve some technical problems, which indeed promotes mutual understand and enhance the degree of inter-trust between them. This will be conducive to the tacit knowledge transfer smoothly and improve the quality of tacit knowledge sharing.

(3) To enhance the senior technician's ability to capture tacit knowledge

Compared with the traditional mode of teacher-apprentice, probationers bring senior technicians with pressure that if they do not fulfill their duties well or have competitive skills, the probationers can replace them. Fear of being replaced, the senior technicians will not be reluctant to solve the problems faced during the work. Even the organization do not give material or spiritual incentives in solving key technical difficulties, the senior technicians will abandon the idea of pursuing incentives and then positively deal with the technical problems. Because of their fear of being replaced by the probationers, the senior technicians will try to master the core skills relating to the business's future, which will turn into their tacit knowledge.

(4) To improve probationer's initiative to capture tacit knowledge

In the traditional management methods, staff found it hard to transcend the positional limitations and lost the initiative and creativity when they worked. Probationers know that they have been affirmed by enterprise and may get promotion. Because of the inherently need for promotion, they would assume responsibility for the most critical work in order to obtain further affirmation and attention. This has greater efficiency value than external stimulation, and improves the performance of probationers.

3. Obstacles to probationary system's application in tacit knowledge sharing

Tacit knowledge is a kind of understanding and the core of knowledge management in an enterprise. The capture of tacit knowledge often depends on working together with senior technicians, communication networks and working groups. However, there are some questions in the tacit knowledge sharing due to the knowledge-sharing environment, providers and other factors. The problems of probationary system in tacit knowledge sharing are as follows:

3.1. Threat of taking responsibility

In reality, probationers always work under the guidance of the senior technicians, so they should take full responsibility of the errors probationers have done. As fearing of improper operation or even accident, senior technicians seldom allow probationers to independently operate machines, which, to a certain extent, curbed their enthusiasm about creation.

3.2. Fear of cultivating competitors

Considering the ownership and competitive status, tacit knowledge owners often are reluctant to share that with other, or control the transfer of knowledge. Renzal (2006) survey found that concerns over losing their own advantages have a negative impact on knowledge sharing[5]. Worrying that share tacit knowledge will lose their competitiveness, the senior technicians will control the spread of valuable ideas, technologies or other useful information.

3.3. Lack of effective incentive mechanism

It is not easy for the senior technicians to get the tacit knowledge, therefore, they don not share it with other without any awards or honors. Reasonable incentive mechanism has great influence on tacit knowledge sharing in enterprises. Traditional incentive systems such as time-based are no longer effective for knowledge-based staff, especially the senior technicians who know core knowledge and technology[2]. At present, most China's enterprises lack of a good atmosphere in which can speed up knowledge flows, transfers and re-creates. Without effective incentive mechanism, organizations are likely to be in a hard situation and do not have the competence to transfer tacit knowledge among individuals.

3.4. Low-efficiency tacit knowledge transfer mechanism

In the probationary system, the tacit knowledge transfer is strictly accordance with the hierarchy-chain. Tacit knowledge transfer in double-way (Kx), as shown

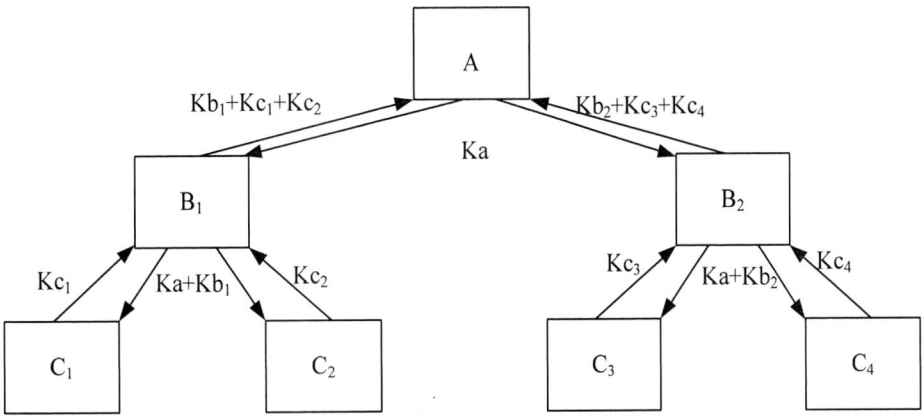

Figure 1 Way of tacit knowledge sharing in hierarchy model

in figure 1[6], beyond the traditional one-way flow. But tacit knowledge can not be transferred in the same level in this model, increasing the possibility of distortion and loss and making relatively low-efficiency knowledge sharing. In China, employees in a higher level often meet psychological obstacles on the exchange of information with their staff. However, staff in the same level exchange and share information or useful tacit knowledge efficiently and smoothly.

4. Research on the measures to improve tacit knowledge sharing in enterprises

On the basis of the characteristics and some obstacles to probationary system's application in tacit knowledge sharing in the enterprises, measures and approaches to improve China's enterprises tacit knowledge sharing are mainly showed as follows:

4.1. To select probationers who inherit tacit knowledge from the senior technicians

In enterprises, there are many young technicians who generally lack the practical experience of working in the key positions, operating senior machines and dealing with unexpected problems. These young people are the targets in selecting program, and the key is that how to define the criterion for selecting the probationers?

The author takes two variables: desire and capacity as the criterion. Desire refers to the degree that the staff is willing to positively work. Even in a same enterprise, employees always differ from performance due to their enthusiasm about job. Capacity reflects the ability of the staff engaged in that work to solve the problems. Human Resources Department should record the performance on these two criterions of young workers(Table 1).

Human Resources Department's role is to collect information about all young technicians and divide them into four categories in accordance with the relevant

standards. Enterprises should strive to foster and provide heavy responsibility for them who are strong both in capacity and desire.

Table 1 Staff performance evaluation framework

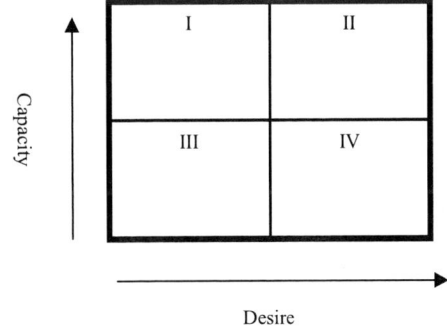

Desire

I   strong capacity, weak desire
II   strong capacity, strong desire
III weak capacity, weak desire
IV   weak capacity, strong desire

4.2. To establish a standardized database of operating process

The enterprise should make strict regulations for senior technicians express clearly the complicated process in detail. Because each part of the whole work has been recorded, everyone involved in that work can look up these procedures written down. Therefore, a large number of tacit knowledge will turn into explicit knowledge, raising the possibility of tacit knowledge sharing.

Knowledge management departments which classify and code the tacit knowledge initiate the establishment of the operational process database. Probationers who meet the similar problems can look up the specifications recorded in the explicit form, what is more, can know the factors in a specific issue taken into account logically. These paper-based records facilitating the process of capturing and sharing knowledge assets speed up the information transfer. The process of turning individual tacit knowledge into corporate knowledge can help the enterprise systematically manage the tacit knowledge and improve the efficiency of the tacit knowledge sharing.

4.3. To establish a knowledge-based incentive mechanism

The establishment of a knowledge-based incentive mechanism contributing to exchanging knowledge is a key measure to speed up enterprise tacit knowledge sharing. Enterprises need to give knowledge providers some benefits or honors, or they will be reluctant to let other share their knowledge[7]. The knowledge and experience senior technicians owned are not born. Therefore, if they share tacit knowledge with others without any pay, they will largely reduce the momentum for more knowledge. Moreover, this does not conform to the market economy. Professor Zhang Wuchang said that the transaction costs include negotiation cost, the formulation and implementation cost, information cost and supervision cost[8]. In the market economy, it is necessary to pay certain cost for getting scarce resources.

Enterprises should give deliberate consideration for such issues as the staff's fair treatment and self-satisfaction. To make the staff share knowledge on their own initiative rather than external requirements, enterprises need coordinate the incentives from both material and spiritual aspects. Material incentives should be linked with the tacit knowledge's amount and quality they provided.

Incentive mechanism can be material or spiritual. A "contribution-based" material incentive mechanism which always be used in different organizations lets providers realize that they can get awards if they share knowledge with others. Spiritual rewards include honors or indirect benefits from the contribution. Spiritual incentive satisfying desire to be respected and admired gives the staff more power and responsibilities. An effective incentive mechanism will facilitate tacit knowledge transfer from individual to enterprises designated groups, and increase their knowledge capital. That helps the individual advantages turn into enterprise competitive advantages and reduces the dependence of individual knowledge.

4.4. To create the "round table" communication model which can improve the efficiency of tacit knowledge sharing

Double-way communication model feedbacks timely. And it not only ensures the accuracy of information but also help shorten the distance between the two sides, establishing enterprise competitive advantage[9]. Probationers exchange freely, and have a new perspective through conversations. Such talks may give rise to debates, but these debates will lead to probationers present new improvements to previous solution. This kind of dynamic exchange between individuals has positive influence on tacit knowledge transfer.

Under sincere and open knowledge exchange situation, the probationers can promote mutual trust, and trigger the desire to share tacit knowledge. The "Round Table" communication mechanism will make everyone has equal opportunities to capture information and knowledge. In that horizontal linkage and multi-interaction mechanism, each individual may be as a recipient or provider of tacit knowledge, reducing the possibility of knowledge loss. Therefore, in some extnet, it enhances the speed and efficiency of knowledge sharing. As shown in figure 2[6], where A, B, C, D, E are in the same level in an enterprise.

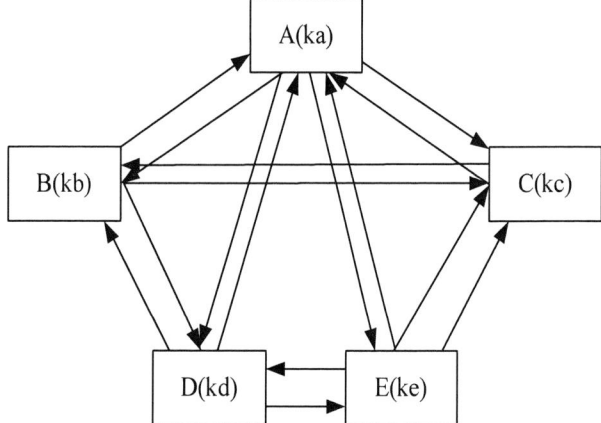

Figure 2 "Round table" tacit knowledge sharing model

5. Conclusions

Probationary system, which has been used to train successor in the world for several hundred years. Today, as knowledge has become the key factor for enterprises to gain competitive advantage, probationary system still has many advantages that can not be overlooked in the tacit knowledge sharing. However, in some business, there are some obstacles which reduce or hinder the tacit knowledge sharing. This paper has put forward several measures on how to solve these obstacles, but they are not systematic and in-depth. We look forward to more effective measures to improve tacit knowledge sharing.

References

[1] Eucker (2007). "Understanding the impact of tacit knowledge loss". *Knowledge Management Review*, Vol. 10, No. 1, pp.10-12.

[2] Zhang Qingpu, Li Zhichao (2003). "Research on Flows and Transformation of Enterprise Tacit Knowledge". *China Soft Science*, Vol. 1, No. 1, pp.88-91 (in Chinese).

[3] Shi Qinfen, Cui Zhiming (2004). "Analysis of Tacit Knowledge Transfer Characteristics and Mode". *Dialectics of Nature Study*, Vol. 20, pp.62-63 (in Chinese).

[4] Yang Yuhao, Cao Keyan (2007). "Explore the Motivate Constraints on Staff Knowledge Sharing". *Scientific Intelligence Development and Economy*, Vol. 17, pp.238-239 (in Chinese).

[5] Renzal B (2006). "Trust in management sharing: the mediating effects fear and knowledge document". *The International Journal of Management Science*, Vol. 6, No. 5, pp.1-15.

[6] Liu Liping (2006). "Cost, Benefits and Incentive Mechanisms on Internal Knowledge Sharing". *Business Research*, Vol. 340, pp.99-101 (in Chinese).

[7] Zhang Haiying (2002). "Development and Use of Tacit Knowledge Management". *Intelligence Science*, Vol. 20, No. 6, pp.655-657 (in Chinese).

[8] Zhang Wuchang. *Experience Research on System Changes*. Economic Science Publishing House, Beijing, 2003 (in Chinese).

[9] Rui Mingjie, Chen Xiaojing (2007). "Economic Analysis of Enterprises Tacit Knowledge Sharing". *Management Journals*, Vol. 4, pp. 272 (in Chinese).

Research on Factors Influencing Knowledge Transfer and Managerial Mechanisms in the Community of Practice

Zhihong Li, Jun Li, Minxia Li
School of Business Administration,
South China University of Technology, Guangzhou, Guangdong, 510640, China
bmzhhli@scut.edu.cn, lijunfrank@163.com

Abstract

Increasingly, the ability to transfer knowledge in an organization is viewed as basis for competitive advantage. One promising method used to facilitate knowledge transfer is the creation of Community of Practice (CoP) in an organization. The purpose of this paper is to explore the key factors that have significant influences on the ability to transfer knowledge in a community of practice. To begin with, it describes the characteristics of a CoP and discusses these influential factors respectively, then integrates into a conceptual framework. Finally, it concludes some managerial mechanisms to instruct the organizations to transfer knowledge well.

1. Introduction

With the development of knowledge economy era, a consensus is emerging that intellectual capital is displacing land, physical labor and financial capital as the primary source of economic and social value. The organizations that are able to transfer knowledge effectively from an individual to others will gain a sustained competitive advantage.

The community of practice which acts as the supplement to formal organizations has already attracted scholars' attention due to its role in the knowledge creation and spread[1]. At present, the domestic academic circles concentrate more on knowledge transfer and sharing in the formal social network in the organizations. In the enterprises, the community of practice has not obtained expected understanding from management and lacks their support and approval. This paper intends to explore the factors influencing knowledge transfer in the community of practice and provide some managerial mechanisms for organizations to manage knowledge effectively.

2. Swee's framework of factors influencing knowledge transfer in an enterprise

Wang Zhongtuo proposed that knowledge transfer referred to the exchange of information and knowledge through different methods, and it was the transition from knowledge production to consumption behavior, which was the process of knowledge spread and application in organizations[2]. The goal of knowledge transfer is to let people recognize and utilize the knowledge as much as possible.

Some overseas scholars thought that organizational culture, motivational mechanism and communication willingness were important attributes in effective knowledge transfer[3, 4]. And Swee's research result is representative of that, he explored some primary factors influencing effective knowledge transfer in an enterprise, and then integrated the key factors into a comprehensive framework (Figure 1).

Swee elaborated that leader played a role in building the culture of seeking problem, solving problem, collaboration and providing the supportive conditions which satisfied the knowledge transfer. The supportive structure which broke down the hierarchical chain in knowledge transfer contributed to horizontal communication in the enterprise. The nature of the relationship between knowledge recipients and providers could be a barrier to effective knowledge transfer. Explicit knowledge would be transferred through more technology-driven, structured processes such as information systems, Lotus Notes. Tacit knowledge might be best transferred through more interpersonal means and using processes that are less structured.

3. A conceptual framework of factors influencing knowledge transfer in the community of practice

3.1. The characteristics of the community of practice

Many corporations have promoted and benefited from cultivating CoPs, albeit under different names[5] such as "learning communities" at Hewlett-Packard, "family

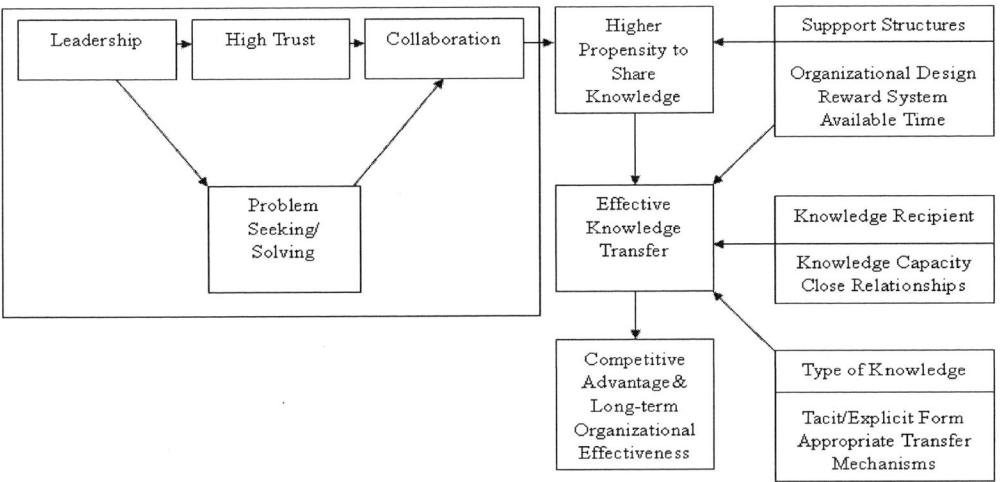

Figure 1 A framework of factors influencing knowledge transfer in an enterprise

groups" at Xerox, "thematic groups" at the World Bank, "peer groups" at British Petroleum, and "knowledge networks" at IBM Global Services. As Brown and Duguid (1996) said, significant learning and innovation arose from informal "communities of practice", where people worked. This informal approach to learn knowledge not only offers a way for members to consult learning materials from the organization, but also combines information with learning materials they create and share through ideas exchange in the informal network where peers, mentors, and subject matter experts solve problems. There are some differences between CoP and formal organizations (such as work groups and project teams)[6]. The biggest one is that work groups and project teams are characterized by being more organized than CoP. Compared with the formal organizations, the chief characteristics of community of practice are as follows:

(1) Spontaneity

The formal organizations usually formed by the appointment of higher authority or through related stipulations, but the CoP is a spontaneous organization in which members are contextually bound by shared interest in learning and applying a common practice. In the community of practice, members who participate in transferring knowledge are voluntary, and these activities occur naturally when they discuss a specific field which all of them care about.

(2) High trust

Trust is an important dimension of effective CoP functioning. Members in the CoP exchange their knowledge and discuss a topic with others is based on high trust among them. High trust leads to greater openness between members. So long as it exists, people will maintain the informal social network. Transferring knowledge will enhance mutual trust between both sides and urge them to participate in the knowledge exchange

and cooperation, which is also the foundation of long-term co-operative relationship in the community of practice.

(3) Innovation

Members of the CoP come from different departments, which can promote profound understanding on their work due to wide-range knowledge and experience exchange. Generally speaking, the knowledge innovation frequently occurs when different disciplines meet. Therefore, communicating with experts or staff who are from different fields is advantageous to the creation of new ideas and further innovation[7]. It is common that new practice methods, service and product are developed when they cooperate.

(4) Well-communication

Information and knowledge can flow smoothly and fast among members in the CoP, because they have shared interests and use common language to exchange ideas[8]. When the members participate in the collective discussion, they will have common understanding on the topic and its significance. As a result, they will show their ability and wisdom by transferring specific knowledge. Generally speaking, collective discussion is a process of organizing, managing and adjusting the communication network in the community of practice.

3.2. Analysis of factors influencing knowledge transfer in the community of practice

There are many common influential factors in knowledge transfer such as supportive structure, trust, knowledge recipient, type of knowledge between CoP and an enterprise mentioned in Swee's model. Because knowledge transfer in the CoP is based on common interests, mutual trust as well as well-communication among members, so the process of transferring knowledge in a CoP will be also affected by individual and interpersonal factors.

(1) Liking

Liking is an affection based on admiration and common interests. The key factors affecting knowledge transfer is member's value orientation. Nonaka (1994) proposed that "without shared knowledge is difficult to share others' thinking process". So, the closer between both sides, the more knowledge transfer. In the community of practice, members possessing common experience and liking are willing to share their cherish knowledge with each other. When the topics differ from personal liking, the member tends to quit and choose another community where they have the same subject to discuss.

(2) Affiliation

Affiliation refers to value derived from the connection to the community, which could cause the people to realize that he has already become a member. This kind of feeling urges the member to develop sense of responsibility. Members who transfer knowledge do not request rewards immediately, because they expect also obtain help from others when they meet problems. That is, one's tendency to contribute knowledge to other people in the community originates his trust laid on peers as well as the responsibility for maintaining informal network.

(3) Reputation

Reputation refers to the overall quality or character as seen or judged by the community. Davenport and Prusak(1998) thought that the owners who transferred knowledge might wish to be "a gentleman of wisdom", and they were willing to share the precious knowledge with other people. Transferring knowledge not only enables the recipient to solve problems, but deepens the provider's understanding on his field. In the community of practice, member will obtain respect and admiration if he helps one deal with problems by sharing his knowledge and experience, which directly promotes the knowledge creation.

(4) Altruism

Altruism is the behavior of someone that, although not beneficial or perhaps even harmful to himself, benefits others (Hoffman, 1981). Some person are glad to help others naturally, but do not strive for the reward. In the community where social network is based on the high trust among members, albeit without any reward, the members will not be reluctant to transfer their own knowledge and experience to help others solve work difficulties, because they believe that if they meet difficulties someday, the other people will support his work without demanding any payment and hesitation. The existence of altruism may extend the motivational mechanism in the community.

(5) Personal needs

Many people are reluctant to transfer knowledge, because they are not always willing to share their practical experience which has been accumulated for long term with others. Therefore, member's decision on whether transfer knowledge or not depends on his willingness, and the only restraint is his sense of responsibility for community. Davenport and Prusak (1998) believed that a man would not share limited knowledge until he profited from the process. Similarly, members wish that they can obtain material compensation through his contribution. Except material needs, the members usually want to get spiritual rewards, such as promotion, honors and so on. Personal needs play an important role in effective knowledge transfer in the community.

3.3. A conceptual framework of factors influencing knowledge transfer in the community of practice

The informal social network not only satisfies personal liking, but establishes high trust among the members, and the feeling of being approved can bring strong reputation and altruism for them when they disseminates their own specialized knowledge to others, promoting members' further cooperation. Practical cooperation as well as supportive structure that breaks down the hierarchical chain in knowledge transfer urge the members to have the higher propensity to share knowledge, which is also the prerequisite of effective knowledge transfer. Well-communication and close relationship between both sides facilitate the knowledge's flowing from the owner to recipient, and this strengthens the formation of competitive advantage and long-term organizational effectiveness.

On the basis of analysis of factors influencing knowledge transfer in a CoP, this paper proposes a conceptual framework (Figure 2), which shows the interactive mechanism of influential factors and explains how to manage knowledge effectively in the community of practice.

4. Managerial mechanisms in the community of practice

In the community, members who come from different departments establish high trust through knowledge dissemination. Moreover, it is advantageous to create new knowledge and raise individual work efficiency. Through knowledge exchange and sharing, the community will increase its knowledge storage and develop organizational superiority. In order to facilitate knowledge transfer in the community of practice, the organizations should establish some managerial mechanisms.

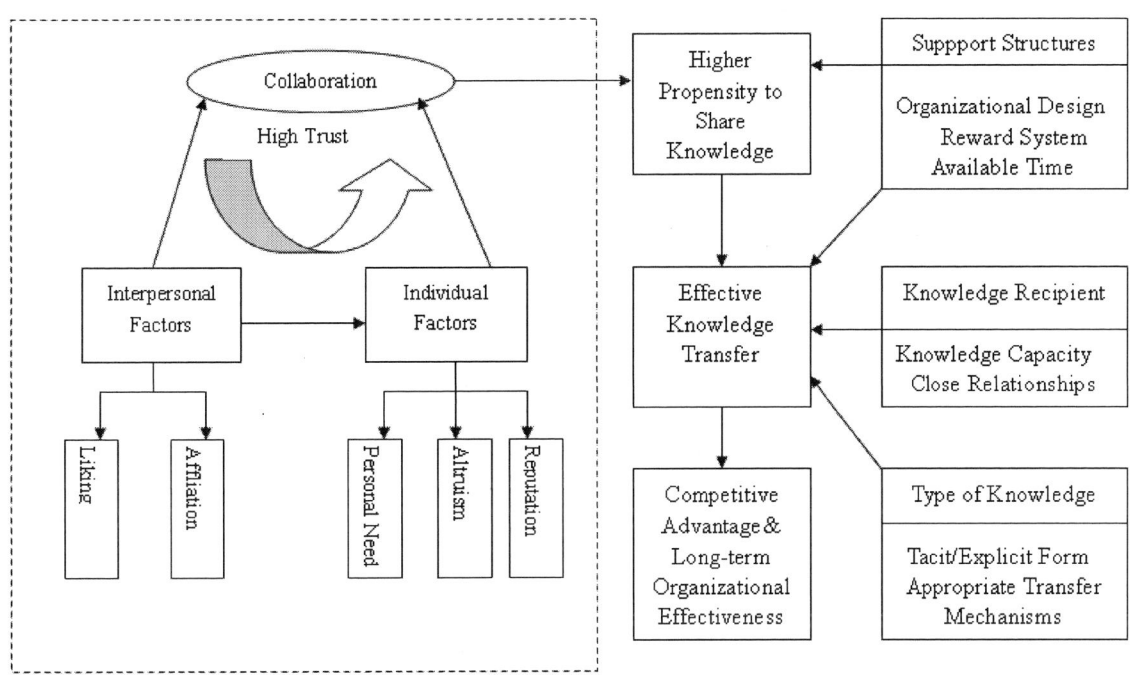

Figure 2 A conceptual framework of factors influencing knowledge transfer in community of practice

4.1. Build the trustful and co-operative culture

A fundamental variable in co-operation between individuals is level of trust. A high level of trust is an essential condition for a willingness to cooperate. Exchange experience openly in the community will promote the level of trust and desire of cooperation. If more people participate in the activity of exchange ideas, the community's knowledge will present the "Scale-up Effect", which accelerates the speed of forming a knowledge chain of "trust, transfer, study and share".

The organization needs to pay more attention to build the organizational culture of high trust in order to encourage members' cooperation. It could operate discussions on how to span the barriers to further cooperative practices and express their new ideas to get rid of obsolete thinking model. If members are allowed to communicate freely with the outside which breaks the boundary of project teams or groups, it provides opportunity to exchange experience and knowledge at different levels. And invite experts to participate in the community can attract more people to learn and transfer knowledge.

4.2. Cultivate the CoP manager

The CoP is a new organizational form, and it does not have any standards or regulations. However, there is an increased awareness that it needs to be managed[9]. Seemingly, the idea of having a manager to coordinate the work in a CoP conflicts with our traditional understanding of CoP. In a sense, the manager encourages people to participate in the activity, seek to new excited topic, and maintain the stability and vitality of community. The manager will bring drive, vision and enthusiasm to the people. Each community is composed of members who come from the different fields and have different cultural context, so the manager is responsible for confirming and clarifying the scope and the making disciplines. On the foundation of "organizational strategy, community subject, knowledge innovation", the manager should care about organizational strategy and keep informal and formal networks balance.

4.3. Construct a knowledge-tracking module

Conceptually, the role of the knowledge-tracking module is to monitor the level of knowledge contribution in the community. All members are encouraged to record the valuable knowledge, then the knowledge track tool which acts as "repeater" will confirm the new records. After the track tool records the information about that such as how many visitors and when they download this information as well as the feedback regarding this, it will transfer the date to the knowledge officials who are responsible for certain topics evaluation. The evaluation standards include that integrity which means the elaborated conclusion needs reasonable logical reasoning process, applicability which refers to the value that be utilized on some occasions in the whole organization, and acceptability that decides the level of the recorded knowledge can be captured by others.

4.4. Establish the contribution-based motivational mechanism

Linking knowledge contribution with individual rewards, to some extent, may enhance the member's responsibility for transferring and sharing knowledge, which will promote the development of community. Rewards are given to the members in proportion to the extent to which they are involved with the community. That means the level of the reward should be proportional to the quantity of questions and responses posted in the community, the frequency to communicate with others, and the number of uploading and downloading useful files. Motivational mechanisms can be material or spiritual. Example of material form is monetary reward which is common incentive used in the community. Spiritual rewards include honors or indirect benefits from the contribution. This sort of motivational mechanism can be important to one who believes that contributions could get reputation from others.

5. Conclusions

Effective knowledge transfer among members is an important condition to keep CoP prosperous. The contribution of this paper is to elaborate and integrate some of the key factors that can influence the effectiveness of the knowledge transfer process, and explore some managerial mechanisms to transfer knowledge which may instruct the organizations to manage knowledge well. But, it can not solve all problems in the community of practice, and it has not carried out quantitative analysis on the influential factors. We look forward to more insights on how to transfer knowledge effectively in a community of practice.

References

[1] Seely Brown, Paul Duguid (2000). "Balancing act: how to capture knowledge without killing it". *Harvard Business Review*, Vol. May-June, pp.73-80.

[2] Wang Zhongtuo. *Knowledge Systems Engineering*. Scientific Publishing House, Beijing, 2004 (in Chinese).

[3] Syed Omar Sharifuddin Syed-Ikhsan, Fytton Rowland (2000). "Knowledge management in a public organization: a study on the relationship between organizational elements and the performance of knowledge transfer". *Journal of Knowledge Management*, Vol. 8, No. 2, pp.95-111.

[4] Swee C. Goh (2002). "Managing effective knowledge transfer: an integrative frame work and some practice implications". *Journal of Knowledge Management*, Vol. 6, No. 1, pp.23-30.

[5] P. Gongla, C. R. Rizzuto (2001). "Evolving communities of practice: IBM Global Services experience". *IBM Systems Journa*, pp.40.

[6] Wenger, E., Snyder, W (2000). "Communities of Practice: the organizational frontier". *Harvard Business Review*, Vol. 78, No. 1, pp. 39-45.

[7] Kanter, R. M (1988). "When a thousand flowers bloom: structural, collective and social conditions for innovation in organizations". *Research in Organizational Behavior*, Vol. 6, No. 10, pp. 169-211.

[8] Eric H. Kessler, Paul E. Bierlv, Shanthi Gopalakrishnan (2000). "Internal vs. external learning in new produced development: effects on speed, cost and competitive advantage". *R&D Management*, Vol. 30, No. 3, pp.213-223.

[9] Thomas N. Garavan, Ronan Carbery, Eamonn Murphy (2007). "Managing intentionally created community of practice for knowledge sourcing across organizational boundaries: insights on the role of the community of practice manager". *The Learning Organization*, Vol. 14, No. 1, pp.34-36.

BA Extended Model Based on the Competition Factors

Shaohua Tao[1], Chaofeng Guo[1], Jingju Gao[2], and Song Hang[1]

[1]Department of Computer Science and Technology, Xu Chang University, Xu Chang, 461000, P.R. China; [2] Department of Computer Science, Central China Normal University, Wu Han, 430079, P.R.China.
Email: shtao_2008@ 126.com

Abstract

The discovery of scale free network captured new sights for our understanding on complex system. Considering the growth and preferential attachment, BA model provided a lot of enlightenment for us. However, the two basic assumptions in BA model are still too simple for explaining many real phenomena, there are quite far distance to real networks. In this paper, we try to expand BA model, namely in the basis reality network, increases certain influence factors. We proposed the competitiveness factors in networks which affected the numbers of link of the nodes. According to competitive relations between nodes, advanced a new competition BA extended network model, the network model growth algorithm is given. This model algorithm is achieved by computer programming and gotten the experimental data and simulation graphics.

1. Introduction

Recently, the researchers discovered the scale-free networks in many fields, such as ecosystem, social relationship, and food chain and so on[1-4]. After further studying ER model, some researchers found that ER is a static network model, but in real word many networks have dynamic properties, especially, the node entered the network early and the probability of link bigger[5]. The old nodes have a great deal connection easily when the network increased the certain scale, which is network growth. On the other hand, each node has the same connection probability when connected to other nodes in ER network, that is, all of node in network are equal, but this fall short of the real world[6]. For instance, the new creating web site connected to other web site always choosing the famous web. Hence, the web site having a great deal connection would have further connection, which is preferential attachment. The growth and preferential attachment mechanism give a reasonable explanation to hubs[7].

The BA model was proposed by Barabási A L, and Albert R, and given the theory explanation to scale-free network. While the small-world and scale-free network models capture the basic properties, it is the results that simplify the real world networks and the preferential attachment of BA is also assumed. Further studies revealed that captured microcosmic mechanism more details can better reveal the law of complex networks evolving, hence, it is necessary that further studies the complex networks modeling to unanimous with real world. This paper advanced the competition mechanism existing between nodes and analyzed the node links number effecting by competition. Hence, a new BA model with competition bring forward according to the competition relationship between nodes. The network model is created simply and the arithmetic of model is simulated by computer, finally the experiment data and simulation gotten.

2. A new network extended model

The new nodes have preferential attachment to certain old nodes, and a few old nodes have a great deal link numbers, which is "richer and richer"[2]. But to some network, such as energy sourced network, traffic network and enterprise network would have some real limits. For example, in traffic network the new road lead to some short cut or free road far great than old roads always have traffic peak. In some enterprise network the new parts would have more relation to some parts close linked itself not to core part, which is competition relationship between nodes. Hence, we introduced a new parameter to extended model, which is competition of nodes. The new extended model considered the network structure and competition to study the dynamic evolving of network.

2.1. The network model with competition

978-0-7695-3342-1/08 $25.00 © 2008 IEEE
DOI 10.1109/PEITS.2008.58

To study the scale-free network dynamic evolution and considered structure and the competitiveness of node, the model evolving divided into the following three steps:

(1) Initial: suppose have node v_0, they form a full-coupled networks, each node is given a competitive strength parameters s_i。

(2) Growth: each time add a new node, and let the new node connected with v_0, so each has new edges v. Because added new node network search costs and selecting connection costs may not, so here we take a parameter M_i, use it denote the edges growth of the number of node i, such as $M_i = v$. Connections of node in accordance with the competitive strength parameters to prefer choose, and the probability of the old node was selected as following:

$$\prod v \to i = s_i \Big/ \sum_j s_j$$

j is the total number of nodes, that is, the greater the competitive strength of the nodes may have been chosen

(3) Node competitiveness of the strength of the dynamic evolution: a certain period of time, when an increase of nodes n, the total number of nodes N, of the competitiveness of the strength parameters of the node i adjusted to:

$$s_i = s_i + \Delta s_i$$

The connection probability of node is

$$\prod v \to i = s_i + \Delta s_i \Big/ \sum_j (s_j + \Delta s_j) \qquad (1)$$

where, $\Delta s_i = \Delta M_i \Big/ \sum_{i=1}^{N} M_i$, $\Delta s_j = \dfrac{\Delta M_j}{\sum_{j=1}^{N} M_j}$

In this period increase the total number of edges ΔM_i of node i, the other nodes during this period increase the total number ΔM_j.

2.2. The distribution of the competition model

If the new nodes connected with node i, which increases the degree s_i probability by equation (1), then

$$\frac{\partial s_i}{\partial t} = v \frac{s_i + \Delta s_i}{\sum_j (s_j + \Delta s_j)} \qquad (2)$$

The initiation condition of equation (2) is: node i occurs system at t_i, which degree is $s_i(t_i) = v$。

For $s_i \geq 0$, $t \geq 0$, we can change the probability $s_i(t_i)$ into

$$p(s_i(t_i) < s) = p\left(t < \frac{2vt_i + \sum \Delta s_j}{2v(v + \Delta s_i)^2}(s + \Delta s_i)^2 - \frac{\sum s_j}{2v} \right)$$

$$= p\left(t_i > \left(t + \frac{\sum \Delta s_j}{2v} \right)\left(\frac{v + \Delta s_i}{s + \Delta s_i} \right)^2 - \frac{\sum \Delta s_j}{2v} \right)$$

$$= 1 - p\left(t_i \leq \left(t + \frac{\sum \Delta s_j}{2v} \right)\left(\frac{v + \Delta s_i}{s + \Delta s_i} \right)^2 - \frac{\sum \Delta s_j}{2v} \right) \qquad (3)$$

and t obeys uniform distribution

$$p(t_i) = \frac{1}{v_0 + t}$$

Take it into equation (3), and then get the probability distribution

$$p(s_i(t_i) < s) = 1 - \frac{1}{v_0 + t}\left[\left(t + \frac{\sum \Delta s_j}{2v} \right)\left(\frac{v + \Delta s_i}{s + \Delta s_i} \right)^2 - \frac{\sum \Delta s_j}{2v} \right]$$

So the degree distribution of vertices is

$$p(s) = \frac{\partial p(s_i < s)}{\partial s}$$

$$= -\frac{1}{v_0 + t}\left(t + \frac{\sum \Delta s_j}{2v} \right)(v + \Delta s_i)^2 \left(-2(s + \Delta s_i)^{-3} \right)$$

$$= \frac{2}{v_0 + t}\left(t + \frac{\sum \Delta s_j}{2v} \right)(v + \Delta s_i)^2 (s + \Delta s_i)^{-3} \qquad (4)$$

From equation (4), when $t \to +\infty$, the degree distribution of vertices is

$$p(s) \approx 2(v + \Delta s_i)^2 (s + \Delta s_i)^{-3}$$

Where, $\Delta s_{i} = \Delta M_{i} \Big/ \sum_{i=1}^{N} M_{i}$, we can get

$$p(s) \approx 2\left(v + \frac{\Delta M_{i}}{\sum_{i=1}^{N} M_{i}}\right)^{2}\left(s + \frac{\Delta M_{i}}{\sum_{i=1}^{N} M_{i}}\right)^{-3}$$

3. Simulation and analysis

3.1. The impact of competitive

The initial assumption that there are four state nodes, they form a full-coupled network, each node is given competitive strength parameters, the node $n1$ the competitiveness of strength parameters for the 0.2, $n2$ for 0.5, $n3$ for 0.1and $n4$ for the 0.8. Simulate real network, in a certain time period, assuming growth of four new nodes. Because each new node network search costs and selecting may not like, so here each new nodes under the different conditions and set their respective growth of edge are: 4,8,5,2. Connections of node in accordance with the competitive strength parameters to prefer choose, that is, the greater the competitive strength of the nodes may have been chosen. The connection is completed, we come to the competitive strength of each node to new network model. The node $n1$ the competitiveness of strength parameters for the 0.466667, $n2$ for the 0.766667, $n3$ for the 0.333333, $n4$ for the 1.033333, $n5$ for the 0.133333, $n6$ for the 0.133333, $n7$ for the 0.133333 and $n8$ for the 0.066667.

Based on the above data from the experiment, eight nodes for the competitiveness of the strength parameters is gotten, and the simulation of extended model was done by matlab, shown as figure.1.

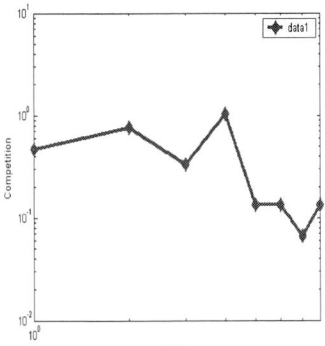

Fig 1. The competition strength

In this experiment, not only follows the model of BA algorithm thinking, and the biggest difference is

considered the competitiveness of the nodes in real network. The initial network settings full-coupled network, in accordance with the competitive strength for choosing the best connection when add new nodes. In order to more closely real-life phenomena, This paper also takes into account each new node network search costs and selecting the best contact connection costs may not like, so we allow the edges different respectively.

As shown, is the assumption that when we consider these factors, according to the competitiveness strength of nodes got the simulation graph. The results show that scale-free networks may indeed existence the competition between the nodes, this competitive strength will be affected the scale-free network growth to a certain extent.

3.2. The parameters of network model

We general research the static statistics of complex network to obtain their properties. These statistics typically include the distribution, the average shortest path and the concentration coefficient, as follows.

If a network has node N , the distribution k_{i} is for all the edge degree of node i , The distribution for the arbitrary choice of a node of the edge is the probability k . In competitive network model, the total of 5,000 nodes, the distribution network shown in Figure 2.

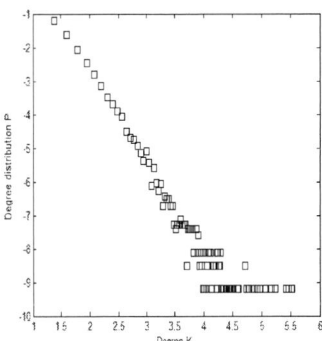

Fig. 2. The degree and the degree distribution

The average shortest path L for the network, the definition of any two nodes n_{i} , n_{j} , in the network, which the distance $d(n_{i}, n_{j})$ between their edges are the minimum. The average shortest path L for the network

$$L = \frac{\sum_{i,j=1}^{N} d(n_{i}, n_{j})}{C_{N}^{2}}$$

The average shortest path of network is as shown in Figure 3.

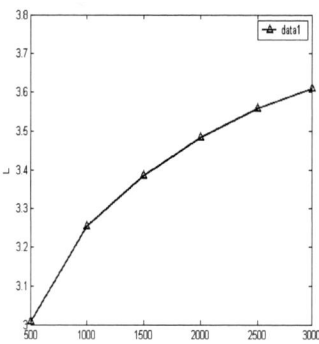

Fig 3. The average shortest path

The clustering coefficients of networks, The nodes i, it may be connected to the nodes on the $N-1$. For the greatest possible number of edges $m(i) = \dfrac{(N-1)*(N-2)}{2}$, the clustering of node n defined as $C(n) = \dfrac{e(i)}{m(i)}$, $e(i)$ are the real edges. The clustering coefficient of the entire network is defined as:

$$C = \sum_{i=1}^{N} C(n)$$

The network clustering coefficients are shown in Figure 4.

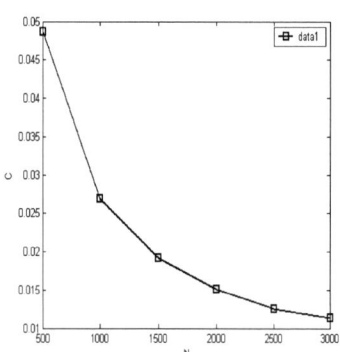

Fig 4. The clustering coefficient

4. Conclusion

This paper has been widely familiar with the BA scale-free networks on the basis of closer to the actual network and its algorithm is modified. We proposed the competition between the nodes of the network impact the connection quantity, and in accordance with the competition relationship of nodes advanced a new of the scale-free network extended model, and presents a simple model of growth algorithm. This model algorithm is achieved by computer programming and gotten the experimental data and simulation graphics.

The upper limits of node degree is not fixed of the scale-free networks, but also its dynamic evolution mechanism is very complicated and therefore a reasonable and realistic significance of the evolutionary model is very valuable, this model can hope for this area provides some reference.

Acknowledgment

The work was supported by the Natural Science Foundation of Henan Province (Grant No. 0611052800 and Grant No.2008A520023).

References

[1]Strogatz S H, Exploring complex Networks, Nature, 2001, pp.268-276.

[2] Albert R, and Barabási A L, Statistical Mechanics of Complex Networks. Rev. Mod. Phys, 2002, pp. 47-97,.

[3] Dorogovtsev S N, and Mendes J F F, Evolution of Networks, Adv. Phys, 2002, pp.1079-1187.

[4] Newman M E J, The Sturcture and Function of Complex Networks, SIAM Review, 2003, pp.167~256.

[5] Wang X F, and Chen G R, Complex Networks: small-world, Scale-free and Beyond, IEEE Circuits & Systems Magazine, 2003, pp.6-20.

[6] Amaral L A N, and Ottino J M, Complex Networks: Augmenting the Framework for the Study of Complex Systems, Eur. Phys.J.B, 2004, pp.147-162.

[7] Barabási A L, and Albert R, Emergence of scaling in Random Networks, Science, 1999, pp.509-512.

A Reflective NetGAP Logic Framework Design

Yu Songsen[1,2,3], Zhan Yiju[2], Cai Qing-lin[2] and Wang Yong-hua[2]

[1] School of Computer, Nanchang University, Nanchang 330031, China

[2] School of Technology, SUN YAT-SEN University, Guangzhou 510275, China

[3] Department of Software Engineering, Shenzhen Institute of Information Technology,

Shenzhen 518029, China

yss8109@163.com, zhanyiju@mail.sysu.edu.cn

Abstract

Adopting expert hardware to physical isolate the Trusted network from the Non-Trusted network can prevent all kinds of attack basing on network layer and operating system layer etc. This paper comes up with network isolation system model based on Reflective architecture. The Reflective GAP is hardware-based and its purpose is to rapidly mirror buffers between the Non-Trusted memory and the Trusted memory. We realize it by LVDS bus and high speed double switch technique. Its activities are achieved using store & forward of memory blocks. Its software system comprises with seven main basic modules. The Reflective GAP can keep good security and implement high Real-time data transmission.

1. Preface

Although the technique of traditional Network Isolation card ensures network security, it forms an "isolated island" owing to lack of information exchange mechanism and at the same time limits the development of its application [1]. In the recent years, the rapidly-developed NetGAP technique is on the basis of physical isolation. While NetGAP technique ensures the security, it not only solves the difficulty of information exchange between networks, but also breaks through the application difficulty caused by security [2,3].

NetGAP can physically isolate the internal and external networks. Its hardware includes PCI interface control circuits, network interface circuits and logic control circuits. At present NetGAP mostly bases on SCSI switch technique or bus switch technique. Its system always exists data exchange lowness and real-time bad. In order to overcome these problems, we come up with Reflective GAP design, making use of LVDS bus and high speed double switch architecture to automatically adapt its operation speed and mode to the network and securely isolate the internal and external networks.

2. Constitution

The NetGAP appliance features 2 built-in Single Board Computers (SBC):

• Non-trusted, connected to the non-trusted network

• Trusted, connected to the trusted network

Inside the NetGAP device, the Trusted and Non-Trusted are connected using specialized hardware. This hardware consists of proprietary Security Boards connected by a high-speed LVDS bus. The Security Boards themselves are plugged into each SBC via a PCI bus.

The Security Boards together with the LVDS bus implement Reflective GAP™ Technology. Each Security Board contains duplicate double switch architecture, with 4 switches in total. One double switch set is used for incoming traffic and the other is used for outgoing traffic. The double switch architecture ensures a complete physical disconnection between the trusted and Non-Trusted without impairing the bus throughput.

The NetGAP implements Reflective GAP technology using two embedded RISC processors separated by Reflective GAP. The entire assembly is implemented as shown by the diagram below [4,5].

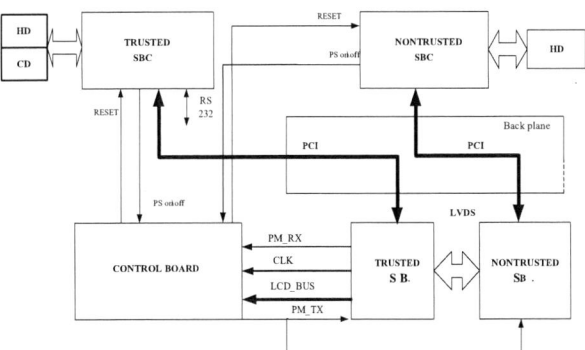

Fig1. Constitution diagram of Reflective NetGAP

3. Work flow

Reflective GAP is implemented in both hardware and software. When the GAP process is divided into two high-level tasks, one for incoming traffic and one for outgoing traffic, it can be describes as follows:

Incoming Data:

The software runs on the NON-TRUSTED side terminates the TCP/UDP session and retrieves the data at the application layer. After receiving known amounts of data, the software signs the data and stores it on a known location in the Non-Trusted SDRAM memory for the TRUSTED to retrieve it.

When data passes through the hardware, it goes through two stages:

Check by the hardware to verify that the software signed it correctly, and then it is marked "valid" or "invalid" by the hardware. The data is encoded by the hardware to prevent hostile data to be executed on the TRUSTED.

Software on the TRUSTED side inspects the hardware flags for each data pack and retrieves only the valid data packages. After retrieving the data, it is checked against the rule-base to determine if it is "legal".

The rule base inspects the data for: Source and destination addresses, Special words and Protocol violations.

If the data is authorized the software retransmits it to the trusted network.

Outgoing Data:

Software on the TRUSTED receives data from the trusted network aimed to the network using TCP/UDP protocols. The software terminates the TCP session and retrieves the data at the application layer.

The data is screen by the rule base, and if it is authorized by the security policy to leave the trusted network it is then placed on known memory location.

The NON-TRUSTED GAP hardware retrieves the data and passes it on to its application. The application on the NON-TRUSTED retransmits the data with TCP/UDP protocol to the non-trusted network. The underlying assumption in this scenario is that the data originates in the trusted side, and therefore there is no coding and no electronic signature on outbound traffic.

4. Reflective gap
4.1. Role

The Reflective GAP interacts with the two processors through their local bus in a DMA transfer means. The Reflective GAP is the Trusted; it is responsible to create DMA session in both sides.

The Reflective GAP is hardware-based and its purpose is to rapidly mirror buffers from the Non-Trusted memory to the Trusted memory, and vice versa; Hence the term "reflection". The Reflective GAP is not based on any communication protocol and all its activities are achieved using store & forward of memory blocks.

The Reflective GAP consists of two separate data paths, one for transmission and the other for receiving. As the data path from Non-Trusted to Trusted is shown in the direction from left to right, the data path from Trusted to Non-Trusted is the opposite direction from right to the left. Each double switch mechanism is implemented in hardware inside the Reflective GAP logic, for traffic incoming into it.

4.2. Responsibilities

All the Non-Trusted (and the entire non-trusted network behind it) can do is send data blocks to the Non-Trusted Reflective GAP. The Non-Trusted Reflective GAP transfers them to the Trusted, which places passive data blocks, garbled, into memory buffers designated by the software on the Trusted computer. It is now up to the Trusted to handle the data as it pleases. This is similar to Sneaker net, where the operator puts a diskette with the data in the drive, and the computer on the trusted network can read the data and manipulate it as it pleases. All logic in the Reflective GAP is controlled by hardware and cannot be modified, in the same manner that a floppy drive cannot be modified to write to a write-protected diskette. A Reflective GAP implements a GAP between the Trusted and the Non-Trusted, and this GAP cannot be breached even if the Non-Trusted computer is completely taken over by hackers.

A similar data path takes place for data sent from the Trusted to the Non-Trusted, utilizing the remaining components of the Non-Trusted and Trusted.

5. Main hardware module
5.1. Trusted SBC

The Trusted is the Computer inside the GAP device that is connected to the trusted network. On the Trusted runs proprietary software that delivers to the trusted network traffic from the Non-Trusted. In short, the Trusted receives garbled data, checks it against the security policy that is stored on the Trusted and makes all security decisions in a safe environment that is isolated from the non-trusted network. If the security policy dictates it, the data is dropped. If all is well, the Trusted generates a completely new session with the computer on the trusted network to which the information was addressed, decodes the data again (knowing that it is safe because it has been inspected) and sends it to its destination.

5.2. Non-Trusted SBC

The Non-Trusted is the only Computer inside the GAP that is connected to the non-trusted network. On the Non-

Trusted runs proprietary software that listens to incoming traffic from the non-trusted network. The Non-Trusted Software sends data blocks across the GAP by delivering it to the Bus Interface. Before doing so, the Non-Trusted Software digitally signs each data block with a key that is regenerated in every boot of the system, and is passed to the Non-Trusted Software by the Security Board. The Non-Trusted Software uses a key transformation mechanism after each use of the key. This renders eavesdropping to previous transmissions useless because the key is transformed into another in between transmissions. The purpose of the digital signature is to make it difficult for other software to impersonate the Non-Trusted Software. Although even if that takes place, the unique properties of the GAP ensure that the security of the GAP chip is not compromised.

5.3. Control Board

The control board was designed in order to replace the old LCD card and to add some more crucial features related to the power management of the NetGAP.

The control board response to commands from the hardware push buttons and software requests initiated by the NG ADMIN or the NG terminal consol in the following way:

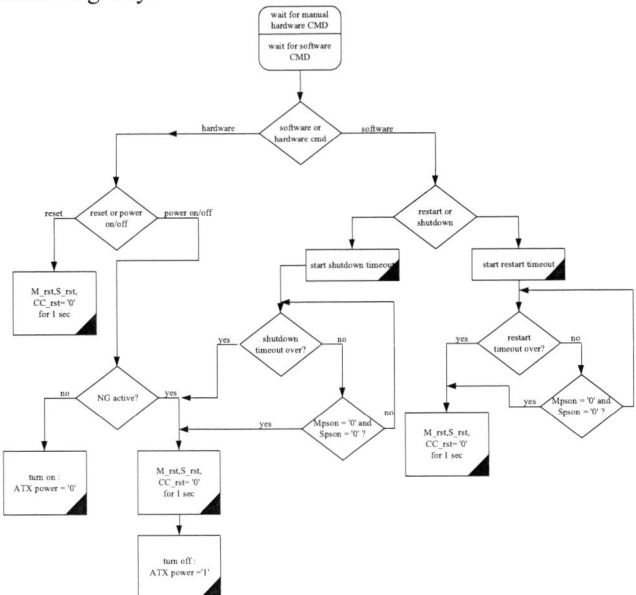

Fig2. Command flow of control board

Reset will be executed 1 sec before the ATX shuts down every time that soft shut down is requested, in order to prevent damage to the HD while abnormal shutdown is being executed. Hardware commands have priority over the software request and also cancel every software request.

6. Basic software module

There are seven main software modules that comprise the "GAP" system. Six out of the seven modules are actually twin modules that reside on both sides of the "GAP", the trusted and non-trusted sides. The trusted side ("Trusted") is the side that is connected to the trusted network, while the non-trusted side ("Non-Trusted") is the side that is connected to the non-trusted network. The remained module (the one that does not have a twin module) is the security module that handles the protocol security decisions in the system. For security reasons, it resides, on the trusted side of the "GAP".

A single datagram goes through five modules in its life cycle in the system. The Fig3. shows the basic seven modules and the flow of a single datagram.

A datagram can come from both sides of the network (the trusted and non-trusted sides). In both cases, it passes the same route of modules. The only difference between one route to the other, is where the datagram crosses the "GAP" chip and moves to from one CPU to the second CPU.

The route of a datagram starts at the "TCP module" which is a proxy module. It stops the datagram, packs the data buffer to an application message and passes the message to the "Parser Modules". The "Parser Module" does the protocol parsing analysis, creates a proprietary "parsed message" and sends it to the "Trusted Module". The "Trusted Module" does most of the security decisions and transfers the "approved" parsed messages to the "Regenerator Module". Here the "Parsed" data is regenerated to a similar but not equal to the original "buffered form" message. Finally, it is handled by the seconds "TCP Module" proxy, which mirrors the original session.

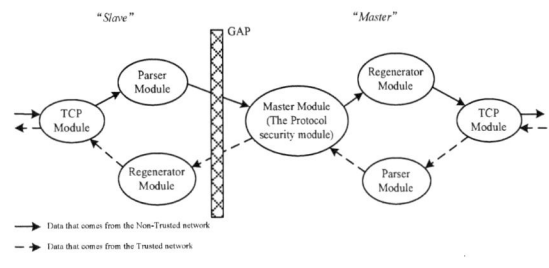

Fig3. Basic software module

6.1. TCP Module

The "TCP Module" is a proxy module that handles all TCP/UDP sessions in the system. It relies on the OS TCP stack (or some other TCP stack). Additional IP protocols like DHCP and ICMP will require additional proprietary secured modules. The "GAP" system does not reflect IP packets to the trusted side.

The "TCP Module" receives its input from both the "Regenerator Module" and the network. The "Trusted Module" controls the "TCP Module" (via the "Regenerator Module") using command messages. These command messages are translated by the "TCP Module" to perform network tasks. Other messages that the "TCP Module" is handling are the "data messages". In this case, the "TCP Module" sends the buffers that reside in those messages. When data arrives from the network, the "TCP Module", which is a proxy application, strips the data from the TCP packet and creates an inner application data message.

The "TCP Module" maintains a table of protocol sessions. Each entry in the table ties together all of the sockets and other resources that associated with the session. A session can comprise more than one socket (e.g. FTP with its command and data sockets). Because of performance matters, we also map between socket id and an entry in the table.

6.2. Parser Module

The "Parser Module" requires protocol data messages. It performs syntax checks and constructs the same message in a proprietary compact representation. This is message is called the "parsed message".

The "Parser Module" holds "Parser Units" for each protocol session that it is currently handling. Each data message that is received from the "TCP Module" is delegated to the appropriate "Parser Unit" for parsing. After the parsing process is finished a parsed message is sent to the "Trusted Module

6.3. Trusted Module – The Main Security Module

The "Trusted Module" is the singular point where the entire protocol session messages are processed from both the trusted and non-trusted sides. It is the only module in the system that receives messages from the two sides of the network. The reason for this uniqueness relies on its assignment to handle the protocol state-machine and to make security decisions. The "Trusted Module" makes security decisions based on a global configurable rule-base. According to the protocol state machine and the global rule base, this unit can either generate command messages to the two "TCP Modules" (e.g. "Open new connection", "close session" etc...), or pass the "parsed message" to the next module; the "Regenerator Module".

6.4. Regenerator Module

The "Regenerator Module" is the "reverse image" of the "Parser Module". The term "reverse image" refers to the task that the "Regenerator Module" performs.

"Regenerator Module" takes parsed messages and regenerates them to application data messages in the standard protocol format. These messages are similar but not equal to original application data messages that the "TCP Module" has generated. The "Regenerator Module" passes its messages to the last module in the chain, the second "TCP Module".

7. Conclusion

In above sections, we introduced our proposed network isolation system model based on Reflective architecture. The Reflective GAP is hardware-based and its purpose is to rapidly mirror buffers from the Non-Trusted memory to the trusted memory, and vice versa. Its activities are achieved using store & forward of memory blocks. And, we researched the implementation of this model by LVDS bus and high speed double switch technique.

The "GAP" system comprises with seven main software modules. Six out of the seven modules are actually twin modules that reside on both sides of the "GAP", the trusted and non-trusted sides. The remained module is the security module that handles the protocol security decisions in the system.

In conclusion, the Reflective GAP implements a GAP between the Trusted and the Non-Trusted, and cannot be breached even if the Non-Trusted computer is completely taken over by hackers.

Reference

[1]Peddireddy,Taraka D.,Vidal,Jose M. Multiagent network security system using FIPA-OS, *IEEE SoutheastCon 2002*, Apr 5-7 2002, p229-233

[2]Yin Yong, Research on Theory and Technology in Physical Isolation System and Its Application in Netwroked Manufacturing Environment, *Huazhong University of Science and Technology Doctor thesis*, Oct,2004

[3]Zheng,Wei Xu,Wenbo, Design and Realization of Netgap Based on Physical Isolation, *tiny computer information*, vol 12-2,2005

[4] NetGAP High Level Design, *www.Spearheadsecurity.com*, Apr 9, 2003

[5] ViGap technique white book, *Weisi combine company*. Oct 3, 2003

Research on the Real Option Decision Model about Venture Capital Exit

YANG Qing[1], PAN Ane[2], LI Jue[3]

Management School, Wuhan University of Technology, Wuhan, 430070, P. R. China
[1]yangq@whut.edu.cn, [2]panane@126.com, [3]lij93@163.com

Abstract

The exit mechanism from a venture investment is the primary way the venture capitalist can realize a positive return on the investment. The main exit routes are Initial Public Offerings (IPO), Trade Sale, Buyback and Write-off. It is an important issue for the venture capitalists to choose the most appropriate exit route. This paper sets up a real option decision model based on the comprehensive analysis of exit costs and benefits, the venture capital funds life, the venture capitalists cooperation willing, and other influential factors, in order to provide a scientific base for the exit decision-making.

1. Introduction

The venture capital exit mechanism is the systems by which the venture capital funds liquidates the venture investment when the venture-funded firms are mature or no prospects. To exit the capital may be the most critical step in the process of the venture capitalists' investment (Bygrave and Timmons，1992)[1].

According to the analysis of Black and Gilson (1998)[2], the main exit routes of venture capital are Initial Public Offerings (IPO), Trade Sale, Secondary Sale, Buyback and Write-off. As the third party purchases the shares from the venture capitalists in both Trade Sale and Secondary Sale, Trade Sale can be combined with Secondary Sale into Merger and Acquisition. In view of the real option, the venture capitalists have the exit decision control at any time from the beginning to the end of the investment [3]. Thus it is possible for us to adopt the real option theory to explore the venture capital exit.

2. Primary models about venture capital exit

2.1 The assumptions

Assume the cash flow from the venture project is $x_t - C$, in which C is constant and above zero, x_t is subject to geometrical Brownian Motion as follows:

$$dx_t = \mu x_t dt + \sigma x_t dz \qquad (1)$$

Presume the value of the venture project is $V(x_t)$, then

$$V(x_t) = \int_t^{+\infty} (x_h - C)e^{-\rho(h-t)} dh \bigg|_{x_t} = \frac{x_t}{\rho - \mu} - \frac{C}{\rho} \qquad (2)$$

where ρ is the discounted rate.

At the time of the venture capital exit, the venture firm is nearly mature and its cash flow is stable. Thus suppose that different exit time and routes will not influence cash flow or intrinsic value of the venture firm and meanwhile that the venture capitalist completely exit venture investment without consideration of various exit route combinations.

Essentially the profit and loss of the venture capital exit results from the difference of the venture firm valuation. That is to say different risk-preference individuals adopt different discounted rates in the valuation of the venture project. From the profit analysis of the venture project exit, we can obtain that $\rho_1 > \rho_2 > \rho_3 > \rho_4 > \mu > \gamma$, where ρ_1、ρ_2、ρ_3、ρ_4 are respectively the discounted rates adopted by the venture capitalist, the entrepreneurship, strategic investor, and the share purchaser in the public market, and γ is the safe interested rate (Zhang Jingyu, 2004)[4]. Suppose that $\delta_i = \rho_i - \mu$, and $i = 1,2,3,4$.

2.2 Exit profit and loss function

The main exit routes of venture capital are IPO, Trade Sale, Buyback and Write-off. Let us denote the profit and loss function at the time of t in every exit route by $\Omega_I(x_t)$, $\Omega_M(x_t)$, $\Omega_B(x_t)$ and $\Omega_W(x_t)$ respectively.

In IPO,

$$\Omega_I(x_t) = s\left[(1-\theta)\left(\frac{x_t}{\delta_4} - \frac{C}{\rho_4}\right) - C_I\right], \text{where } s \text{ is}$$

the percentage of share capital owned by the venture capitalist, θ is the percentage of the sale expense appropriated to the amount of share capital issued, and C_I is the fixed expense for IPO.

In Trade Sale, $\Omega_T(x_t) = s\left(\dfrac{x_t}{\delta_3} - \dfrac{C}{\rho_3}\right) - C_T$, where

C_T is the transaction cost for Trade Sale.

There are two types of Buyback which have different impacts on the exit profit and loss function. In the active

Buyback, $\Omega_{B1}(x_t) = s\left(\dfrac{x_t}{\delta_2} - \dfrac{C}{\rho_2}\right) - C_B$, where C_B

is the coordination cost. In the passive Buyback, both the payment price and form are definite, so that $\Omega_{B2}(x_t) = G_0$, where G_0 is constant.

In the Write-off, $\Omega_W(x_t) = H_0$, where H_0 is usually constant.

2.3 Exit timing

With the above assumption that the venture capital exit is complete without consideration various exit route combinations, venture capitalists have to make an exclusive choice of exit route. Thus it is necessary to calculate the value of various exit routes firstly, and then to make a comparison and analysis. At the same time since the benefits from the passive Buyback and Write-off are fixed, this paper firstly calculates the option values of IPO, Trade Sale and the active Buyback. And then compare with the benefits from the passive Buyback and Write-off; lastly make an optional exit decision.

From the analysis of Dixit and Pindyck（1994）[5], when choosing the exit time, the venture capitalist should take the Bellman equation which meets the optional stop issue into account:

$$F(x_t) = \max\left\{\Omega(x_t), s(x_t - C)dt + (1 + \rho_1 dt)^{-1} \times E[F(x_t + dx_t + dt)|x_t]\right\}$$

(3)

Within the continuous part (that is to keep the venture investment in the venture project), equation (3) can be changed into the following:

$$\rho_1 F(x_t) = s(x_t - C) + \frac{E[dF(x_t)|x_t]}{dt}$$

(4)

According to Ito Theory，equation (4) can be changed into the following:

$$\frac{1}{2}\sigma^2 x_t^2 F''(x_t) + \mu x_t F'(x_t) - \rho_1 F(x_t) + s x_t - C = 0$$

(5)

Its marginal condition can be obtained as follows:

$$F(x^*) = \Omega(x^*) \tag{6}$$

$$F'(x^*) = \Omega'(x^*) \tag{7}$$

The equation (6) and (7) are the value matching condition and smooth binding condition respectively. Consolidating equations from (5) to (7), we can get the optional exit timing or threshold of venture capital in the different exit routes.

(1) The IPO exit route

In this exit route, the real option value at the time of t is:

$$F_I(x_t) = A_1 x_t^{\lambda_1} + \frac{s x_t}{\delta_1} - \frac{sC}{\rho_1} \tag{8}$$

where

$$\lambda_1 = \frac{-(\mu - \sigma^2/2) + \sqrt{(\mu - \sigma^2/2)^2 + 2\sigma^2 r}}{\sigma^2}$$

The results can be calculated as follows:

$$x_I^* = \left[\lambda_1 C_I + \lambda_1 C\left(\frac{1-\theta}{\rho_4} - \frac{1}{\rho_1}\right)\right] \Big/ \left[(\lambda_1 - 1)\left(\frac{1-\theta}{\delta_4} - \frac{1}{\delta_1}\right)\right] \tag{9}$$

$$A_1 = s\left[\left(\frac{1-\theta}{\delta_4} - \frac{1}{\delta_1}\right)x_I^* - \left(\frac{1-\theta}{\rho_4} - \frac{1}{\rho_1}\right)C - C_I\right]\frac{1}{x_I^{*\lambda_1}} \tag{10}$$

Rewrite them as:

$$F_I(x_t) = \begin{cases} \Omega_I(x_t) & x_t \geq x_I^* \\ A_1 x_t^{\lambda_1} + \dfrac{s x_t}{\delta_1} - \dfrac{sC}{\rho_1} & x_t < x_I^* \end{cases} \tag{11}$$

Then the expected exit time is:

$$\begin{cases} T_I = t + \tau_I & \tau_I \geq 0 \\ T_I = t & \tau_I < 0 \end{cases} \tag{12}$$

where τ_I can be obtained from the following equation:

$$x_t \times \exp(\mu \tau_I) = x_I^*$$

(2) The trade sale exit route

In this exit route, the real option value at the time of t is:

$$F_T(x_t) = A_2 x_t^{\lambda_1} + \frac{s x_t}{\delta_1} - \frac{sC}{\rho_1} \tag{13}$$

The result can be calculated as follows:

$$x_T^* = \lambda_1\left[sC\left(\frac{1}{\rho_3} - \frac{1}{\rho_1}\right) + C_T\right] \Big/ \left[s(\lambda_1 - 1)\left(\frac{1}{\delta_3} - \frac{1}{\delta_1}\right)\right] \tag{14}$$

$$A_2 = \left[\left(\frac{1}{\delta_3} - \frac{1}{\delta_1}\right)s x_T^* - \left(\frac{1}{\rho_3} - \frac{1}{\rho_1}\right)sC - C_T\right]\frac{1}{x_T^{*\lambda_1}} \tag{15}$$

$F_T(x_t)$ can be rewritten as:

$$F_T(x_t) = \begin{cases} \Omega_T(x_t) & x_t \geq x_T^* \\ A_2 x_t^{\lambda_1} + \dfrac{s x_t}{\delta_1} - \dfrac{sC}{\rho_1} & x_t < x_T^* \end{cases} \tag{16}$$

Then the expected exit time is:

$$\begin{cases} T_T = t + \tau_T & \tau_T \geq 0 \\ T_T = t & \tau_T < 0 \end{cases} \tag{17}$$

where τ_I can be obtained from the following equation:

$$x_t \times \exp(\mu \tau_T) = x_T^*$$

(3) The active buyback exit route

In this exit route, the real option value at the time of t is:

$$F_{B1}(x_t) = A_3 x_t^{\lambda_1} + \frac{s x_t}{\delta_1} \tag{18}$$

The result can be calculated as follows:

$$x_{B1}^* = \lambda_1 \left[sC \left(\frac{1}{\rho_2} - \frac{1}{\rho_1} \right) + C_B \right] \Big/ \left[s(\lambda_1 - 1) \left(\frac{1}{\delta_2} - \frac{1}{\delta_1} \right) \right] \tag{19}$$

$$A_3 = \left[\left(\frac{1}{\delta_2} - \frac{1}{\delta_1} \right) s x_{B1}^* - \left(\frac{1}{\rho_2} - \frac{1}{\rho_1} \right) sC - C_B \right] \frac{1}{x_{B1}^{*\,\lambda_1}} \tag{20}$$

$F_{B1}(x_t)$ can be rewritten as:

$$F_{B1}(x_t) = \begin{cases} \Omega_{B1}(x_t) & x_t \geq x_{B1}^* \\ A_2 x_t^{\lambda_1} + \frac{s x_t}{\delta_1} - \frac{sC}{\rho_1} & x_t < x_{B1}^* \end{cases} \tag{21}$$

Then the expected exit time is:

$$\begin{cases} T_{B1} = t + \tau_{B1} & \tau_{B1} \geq 0 \\ T_{B1} = t & \tau_{B1} < 0 \end{cases} \tag{22}$$

where τ_{B1} can be obtained from the following equation:

$$x_t \times \exp(\mu \tau_{B1}) = x_{B1}^*$$

2.4 Selection of the exit timing

When there is a passive Buyback term in venture capital, assume it is possible for the venture firm to exit by the passive Buyback way after T_{B2}, then the expected gain at the time of t is:

$$F_{B2}(x_t) = \begin{cases} \Omega_{B2}(x_t) & T_{B2} < t \\ e^{-\rho^1(T_{B2} - t)} \Omega_{B2}(x_t) & T_{B2} \geq t \end{cases} \tag{23}$$

And the optional exit route is determined by the following equation:

$$F(x_t) = \max \{ F_I(x_t), F_T(x_t), F_{B1}(x_t), F_{B2}(x_t), \Omega_W(x_t) \} \tag{24}$$

Without the passive Buyback term, the optional exit route is determined by the following equation:

$$F(x_t) = \max \{ F_I(x_t), F_T(x_t), F_{B1}(x_t), \Omega_W(x_t) \} \tag{25}$$

3. Extensive model about venture capital exit

In the primary model about venture capital exit, only costs and gains are concerned, with no consideration the life of venture funds, the various exit conditions, the cooperation willing of venture capitalists, and etc.

The cooperation willing of venture capitalists can be understood that when venture capitalists are not willing to cooperation completely, they would increase the negotiation cost or compensate the entrepreneur. This means the exit cost is increasing [6]. Denote the venture capitalist cooperation willing index in the exit route of IPO, Trade Sale, and active Buyback respectively by q_I、 q_T and q_{B1}, and $q_I, q_T, q_{B1} \in [0,1]$. And assume the matching exit costs are $\dfrac{C_I}{q_I}$, $\dfrac{C_T}{q_T}$ and $\dfrac{C_{B1}}{q_{B1}}$ respectively. When venture capitalists will not cooperate at all, the exit cost is nearly infinite.

Suppose that in the venture capital life T, the condition matching level of the exit route of IPO, Trade Sale, and active Buyback is p_I, p_T, p_{B1} and p_{B2}, where $p_I, p_T, p_{B1}, p_{B2} \in [0,1]$, and

$$\frac{p_I(1-\theta)}{\delta_4} > \frac{1}{\delta_1} , \frac{p_T}{\delta_3} > \frac{1}{\delta_1} \text{ and } \frac{p_{B1}}{\delta_2} > \frac{1}{\delta_1} .$$ When do not correspond with the above conditions at all, it means the condition matching level is too low to exit venture investment with a certain route. Then the expected exit gains are as follows respectively:

In the IPO exit route,

$$\Omega_I(x_t) = p_I s \left[(1-\theta) \left(\frac{x_t}{\delta_4} - \frac{C}{\rho_4} \right) - \frac{C_I}{q_I} \right].$$

In the trade sale exit route,

$$\Omega_T(x_t) = p_T \left[s \left(\frac{x_t}{\delta_3} - \frac{C}{\rho_3} \right) - \frac{C_T}{q_T} \right].$$

In the active Buyback exit route,

$$\Omega_{B1}(x_t) = p_{B1} \left[s \left(\frac{x_t}{\delta_2} - \frac{C}{\rho_2} \right) - \frac{C_B}{q_{B1}} \right].$$

In the passive Buyback exit route,

$$\Omega_{B2}(x_t) = p_{B2} G_0.$$

In the liquidation exit route, $\Omega_W(x_t) = H_0.$

Similarly to part 2, consolidating equations from (5) to

(7), we can get the decision threshold, the option value, and the expected exit timing of venture capital in the exit routes of IPO, Trade Sale and the active Buyback.

We can obtain the decision rules as follows: Rule 1, do not consider the exit routes of IPO, Trade Sale and the active Buyback when the expected exit timing is beyond the venture funds life T. Rule 2, make a selection from the residual exit routes according to the above equation (24) or (25).

4. Case

Venture capitalist has a venture project exit assessed. The relative parameters are as follows: $\mu = 0.15$, $\sigma = 0.4$, $\delta_1 = 0.08$, $\delta_2 = 0.06$, $\delta_3 = 0.04$, $\delta_4 = 0.02$, $C = 5$, $s = 0.6$, $C_I = 350$, $C_T = 60$, $C_B = 20$, $T = 2$, $\theta = 0.07$, $G_0 = 200$, $H_0 = 60$. And the time of passive Buyback is mature. The condition matching levels are $p_I = 0.5$, $p_T = 1$, and $p_{B1} = p_{B2} = 1$. The cooperation willing indexes of venture capitalists are $q_I = 1$, $q_T = 0.8$ and $q_{B1} = 1$.

With the consideration of exit costs and benefits, the results are as follows (see in figure 1):

$$F_I(x_t) = \begin{cases} 4.7464 x_t^{1.3136} + \dfrac{0.6x_t}{0.08} - \dfrac{3}{0.23} & x_t < 43811 \\ 0.6 \times \left[0.97 \times \left(\dfrac{x_t}{0.02} - \dfrac{5}{0.17} \right) - 350 \right] & x_t \ge 43811 \end{cases}$$

(26)

$$F_T(x_t) = \begin{cases} 1.8716 x_t^{1.3136} + \dfrac{0.6x_t}{0.08} - \dfrac{3}{0.23} & x_t < 35.0439 \\ 0.6 \times \left(\dfrac{x_t}{0.04} - \dfrac{5}{0.19} \right) - 60 & x_t \ge 35.0439 \end{cases}$$

(27)

$$F_{B1}(x_t) = \begin{cases} 0.6208 x_t^{1.3136} + \dfrac{0.6x_t}{0.08} - \dfrac{3}{0.23} & x_t < 35.5916 \\ 0.6 \times \left(\dfrac{x_t}{0.06} - \dfrac{5}{0.21} \right) - 20 & x_t \ge 35.5916 \end{cases}$$

(28)

From figure 1, when $x_t = 40$, $F_I(x_t)$ is maximum, and the expected exit timing in IPO is $\tau_I = 0.6080$, which means the venture capital will exit in IPO route about half a year later.

With the comprehensive consideration of exit costs, gains, timing, and cooperation willing with the entrepreneurs, the results are as follows (see in figure 2):

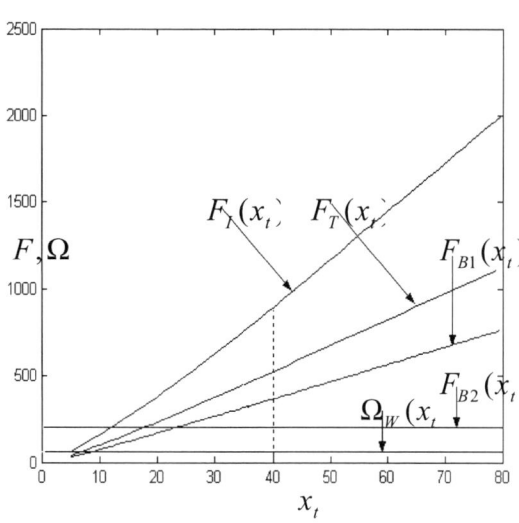

Figure 1 Exit decision analysis with the consideration of costs and benefits

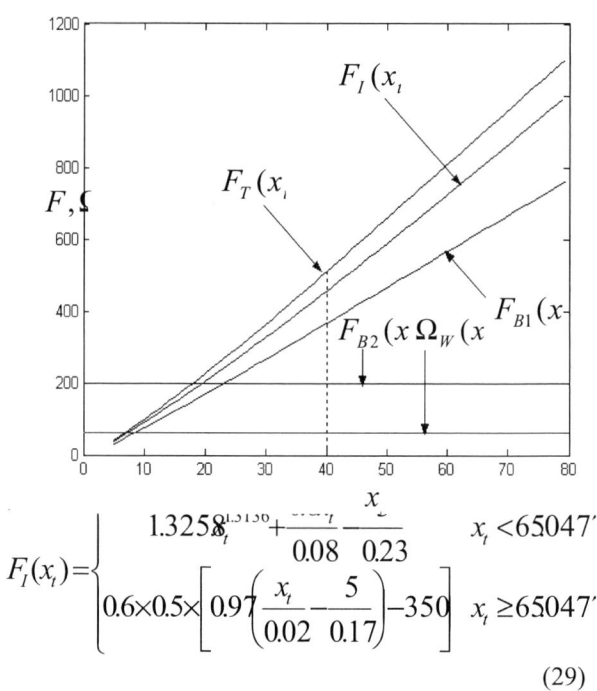

$$F_I(x_t) = \begin{cases} 1.3258 x_t^{1.3136} + \dfrac{0.6x_t}{0.08} - \dfrac{3}{0.23} & x_t < 65047 \\ 0.6 \times 0.5 \times \left[0.97 \left(\dfrac{x_t}{0.02} - \dfrac{5}{0.17} \right) - 350 \right] & x_t \ge 65047 \end{cases}$$

(29)

$$F_T(x_t)=\begin{cases}1.7499x_t^{1.3136}+\dfrac{0.6x_t}{0.08}-\dfrac{3}{0.23} & x_t<43.4214\\[2mm]0.6\times\left(\dfrac{x_t}{0.04}-\dfrac{5}{0.19}\right)-\dfrac{60}{0.8} & x_t\geq43.4214\end{cases}$$

(30)

$F_{B1}(x_t)$ is the same as the equation (28).

Similarly assume $x_t=40$, the expected exit timing in IPO is $\tau_I=3.2416$, which is beyond the venture funds life, therefore, IPO is excluded. From figure 2, $F_T(x_t)$ is maximum, the expected exit timing is $\tau_M=0.5472$, which means the venture capital will exit in Trade Sale route about half an year later.

With the comparison of equation (26) and (29), when the exit condition cannot be matched completely, the exit option value will decrease, and meanwhile the exit decision is delayed. With the comparison of equation (27) and (30), when the cooperation willing of venture capitalist is lowering, the same situation exists as well.

5. Conclusions

The main exit routes of venture capital are Initial Public Offerings (IPO), Trade Sale, Buyback and Write-off. How to choose the exit route is an important issue venture capitalist highly concerned about. With the comprehensive consideration exit costs and benefits, venture capital funds life, venture capitalist cooperation willing, and other influential factors, this paper sets up a real option decision model about venture capital exit in order to provide a scientific base for the venture capital exit decision.

Reference

[1] Bygrave, W. D. and Timmons, J. A.. Venture Capital at the Crossroads. Harvard Business School Press, Cambridge, MA, 1992

[2] Black, B. S. and Gilson, R. J.. Venture Capital and the Structure of Capital Markets: Banks Versus Stock Markets. Journal of Financial Economics, 1998, 47: 243-277

[3] Bernardo, A., and B. Chowdhry. 2002. Resources, Real Options, and Corporate Strategy. Journal of Financial Economics 63:211-234

[4] Zhang Chenyu, Ran Lun, Li Jinlin. Exit Decision and Control in Venture Capital. Journal of Beijing University of Technology, 2004(9):841-844 (in Chinese)

[5] Dixit, A. K. and Pindyck, R. S.. Investment under Uncertainty. Princeton University Press，1994

[6] Anderson, S. P., and M. Engers. 1994. Strategic

Investment and Timing of Entry. International Economic Review 35: 833-853

Study on Optimal Model and Algorithm of Sorting Sequence of Stage Plan on Marshalling Station

Li Lei [1], Cui Bing mou [2]

[1] College of Transportation,
Zhejiang Normal University, Jinhua, ZheJiang, 321004, China
2 College of Traffic and Transportation,
Lanzhou Jiaotong University, Lanzhou, Gansu, 730070, China
lilei@zjnu.cn,cuibm@163.com

Abstract

Arrangement of sorting sequence is an important part for formulation of stage plan on marshalling station. In this paper, we establish a model to minimize the waiting time of wagon as objective function ,and we present simulated annealing algorithm to optimize the problem of sorting sequence, through searching feasible solution in solution space, for each initial solution, we use distribution car flow algorithm pretreatment and calculate the fitness value, then according to the size of fitness value, we estimate the merits of sorting sequence, in light of the actual situation we describe implementation of the correlation method of simulated annealing algorithm. When applied to the computational experiments, the results indicate this method is feasible and effective.

1. Introduction

The central task of framing stage plan on railway marshalling station gets departure train full-axis and departure train punctuality. The reasonable arrangement of sorting sequence plays a key role to the aim of departure train full-axis and departure train punctuality on marshalling station.

The determining of sorting sequence should consider some problems such as the situation of changing shift in the hump operating district, delay time of workers eating and so on, because of this problem related to the sorting sequence, considering these problems can make the sorting sequence have the effective application in the marshalling station, but how to identify the sorting sequence is NPC problem, and there is no good precise algorithm.

Liang Chuncai[1] proposed the heuristic algorithm by analyzing the characteristics of the thinking process of dispatchers scene points for thinking process. Cao Jiaming[2], Cui Bingmou[3]proposed the integer linear programming model for this problem. He Shiwei[4] presented the mixed 0-1 programming model. Wang Zhengbin[5] proposed hybrid genetic algorithm.

Based on the work of the predecessors, for the problem of sorting sequence this paper proposes a general model and an effective algorithm-- simulated annealing algorithm to solve the model and identify the sorting sequence.

2. Model formulation

2.1. Determine the coding of solution

First according to railway bureau issued stage plan, we calculate the earliest sorting time (te_i) of train stock in this stage, then according to the train departure plan we calculate the latest marshalling time (tl_i) of departure trains [3].

In the application of simulated annealing algorithm to solve the problem of train sorting sequence, because of problems inherent in the unique, if adopting traditional way implements solution space, dealing with the unfeasible solution will be complex, so the most convenient way is the natural coding, that is, directly using sorting sequence of train stock as solution code.

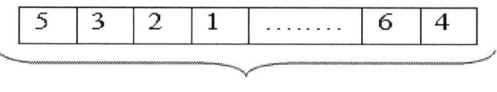

Figure 1 Coding of solution

The position in the code represents of the sorting sequence, each figure in the code represents arrival sorting train of this stage (every stage includes 3h).The sorting time of the each train stock must be not less than the earliest sorting time(te_i), otherwise, the solution of coding is not feasible.

Through the prediction &accurate message, information of each train can be known such as arrival time, marshalling number, the direction number and car number of every direction.

2.2. Generating rules of random code

If generating sorting sequence by simply using random number, then there will be a lot of infeasible solutions, which will also affect the convergence of the solution.

In order to obtain an initial feasible solution, the initial process that is as follows:

Step1: According to the earliest sorting time of train stock to sort the trains of this stage, generate a new collection of J;

Step2: Identify the elements of the code: when the hump locomotive idles and hump is not taken up, from the collection J elements before the next start busy time of the hump locomotive are selected to construct collection j_1 from which can be used the first elements of the code, calculate the number(*count*) of elements from collection j_1, randomly generates a number between [0, *count*) ,then selects a element A from collection J, add the element to the first location of the code, element A will be moved out from the collection J;

Step3: Repeat Step2, determine other elements of the code;

Step4: When the collection of elements in J is empty, the end.

2.3. Object function

The criterion which is measured the advantage and disadvantage of the sorting sequence is that the waiting time of wagon is the minimum.

$$\min F = \sum_{i=1}^{m}\sum_{j=1}^{n_i}\sum_{k=1}^{k}[(x_{xj}^{ft} - x_{xjk}^{dt})\times s_{ijk}]$$

x_{ij}^{ft} --the arrival time of the direction of the j train set in the train i;

x_{ijk}^{dt} --the departure time of the j train set in the train i which is distributed to the k departure train;

s_{ijk}--the car number of the j train set in the train i which is distributed to the k departure train;

k--the number of departure trains;

m--the number of sorting train stock;

n_i--the train set number of the train i.

3. Determine fitness function of sorting sequence

3.1. Fitness function

This model can be used directly as a fitness function of simulated annealing algorithm.

$$f = \sum_{i=1}^{m}\sum_{j=1}^{n_i}\sum_{k=1}^{k}[(x_{ij}^{ft} - x_{ijk}^{dt})\times s_{ijk}]$$

The determining process of the sorting sequence has a main content that is so called calculation of car flow. Through given sorting sequence that is code of solution, and pretreating the solution by using algorithm of distribution car flow, we get car staying time that is a fitness value.

3.2. Distribution car flow principle and algorithm

3.2.1. Distribution car flow principle:

Distribution car flow principle:

(1) When the type of train is not same, according to grade from high to low distributing car flow;

(2) When the type of train is same, according to the direction number of the train from small to large distributing car flow;

(3) When the type of train and the direction number of train are same, according to the departure order of train distributing, that is, first departing first distributing car flow.

3.2.2 Distribution car flow algorithm

Step1: According to the condition that car flow must meet the following time that is the latest marshalling time of train (tl_i). get set A, each element in set A include some information such as train number of arrival train, car direction, the opportunity number[6] of cars, the number of the car flow, and the train number of departure train of the meeting following time;

Step2: According to the size of the opportunity number of car flow sorting by ascending construct a set J, each element in set J include some information such as train number of arrival train, car direction, the opportunity number of car, the number of the car flow;

Step3: According to the grade of departure train from high to low sorting by descending construct a set D, each element in set D include some information such as the grade of departure train, train number of departure train;

Step4: According to the train departure plan, marshalling plan, marshalling train; the car flow whose opportunity number is less first be distributed the departure train d^i, then modify the number of the car flow of the same opportunity number in the set J. When the departure train d^i is full axis, modify the set D, D/d^i is equal D;

Step5: If the car flow whose number of opportunity is the least equals zero, finding the less number of opportunity; goto Step2;

Step6: The car flow is first distributed the departure trains which can be full axis, then distributed the departure trains which can be not full axis, at the same time modifying the set D, D/d' is equal D;

Step7: If D is empty set goto Step9;

Step8: If car flow of departure train is lack, adjust, goto Step9, otherwise, goto Step10;

Step9: For some pick-up trains, transship trains which are not full axis can depart, in the condition of meeting the latest marshalling time of the departure train, supplying the high-grade departure train, in order to make the departure train full-axis, goto Step10;

Step10: The end, saving results of departure trains, calculating the function value f.

3.2. Initial temperature and update scheme [7-8]

3.2.1. Selection of initial temperature

If the initial temperature is too low, the algorithm can't escape local optimum. If the initial temperature is too high, the possibility of moving to a worse state is proved. So it is necessary to select a proper temperature. This paper adopted through randomly generate a group state, determine the most biggest difference $|\Delta_{max}|$ between two goal value, then determine t_0 through formulation $t_0 = -\Delta_{max}/\ln n_c$, where n_c is accepted probability whose value is nearly 1, for example 0.8, 0.9.

3.2.2. Method of temperature dropping

The Method of temperature dropping is used to control the algorithm whether end, this paper uses the function $t_{k+1} = \lambda t_k$, where t_k is the former temperature state, λ is a random value uniformly distributed between 0 and 1 and it is constantly changing with the operational process, because if the λ is too small, the searching time will be long, reversely, if λ is too great, that the temperature will drop too fast get a local optimal solution.

3.3. Algorithm Iteration

3.3.1. Algorithm iteration rule

Present a accepted ratio R (ratio of the accepted times and refused time), iteration step upper limit U and lower limit L, at every same temperature iteration steps must be not less than L times, iteration total steps and accepted times should be recorded, when iteration steps exceed L, if the ratio of the accepted times and total iteration times is greater than R, stop iteration at this temperature and drop temperature, otherwise, iteration continues until the upper limit steps U.

3.3.2 Algorithm termination condition

The termination temperature of algorithm is zero, so we give a very small positive number ε, when the iteration temperature t_k is less than the positive number ε, iteration ends.

3.4. Realization process of algorithm

Step1: First determine the initial temperature, give a accepted ratio R, step upper limit U and lower limit L, by the generating rule of random state generates a solution, pretreating the solution through the distributed car flow algorithm, getting the fitness $f(s)$ of the solution. and finding the most optimal fitness $f(s^*)$, the most optimal solution $s^* = s$;

Step2: Supposed $u=0$, $l=0$, according to the Metropolic rule, $\Delta f = f(s^*) - f(s)$, $l=l+1$;
If $\Delta f \geq 0$, then $f(s^*) = f(s)$, $s^* = s$, $u=u+1$;
If $\Delta f \leq 0$, then $\min\{1, \exp(-(f(j) - f(i))/ t_k)\} \geq \text{random}[0,1]$, $u=u+1$;
If $l \geq L$ and $l/u \geq R$ goto Step4; otherwise, goto Step6.
If $l=U$ goto Step4; otherwise, goto Step3, where u represents the iteration number and l represents the accepted number.

Step3: Generating a solution s by the generating rule of random state，goto Step2;

Step4: $t_{k+1} = \lambda t_k$, $k=k+1$, generating a solution s by the generating rule of random state, goto Step2;

Step5: If $t < \varepsilon$ meeting the condition of termination, goto Step6;

Step6: Output the most optimal results.

4. Experimental results

The simulation tests the wagon data of 20:00-00:00 on marshalling station, wagon data are presented in Table11, Table2. The value of every parameter in algorithm: $R=0.5$, $U=500$，$L=100$，$\varepsilon=2\times10^{-6}$，$n_c=0.8$.

The algorithm has been implemented in Java and the algorithm was executed on a Pentium IV Celeron 2400MHz (256Mb RAM) computer. The satisfied solution can be solved when the program run time within 35s. The optimal result of sorting sequence in Table3.

Table 1 Train timetable between 20:00-00:00

Arrival Train No.	Arrival Time	Arrival Train No.	Arrival Time
1	20:02	9	22:00
2	20:15	10	22:15
3	20:30	11	22:22
4	20:58	12	22:55
5	21:02	13	23:08
6	21:26	14	23:25
7	21:33	15	23:48
8	21:50		

Table 2 Train departure plan

Departure Train No.	Departure Time	Departure Train No.	Departure Time
1	21:40	6	22:42
2	22:00	7	23:16
3	21:43	8	22:00
4	21:50	9	21:40
5	22:20		

Table 3 Result of sorting sequence

Arrival Train No.	Arrival Time	Sorting sequence	Arrival Train No.	Arrival Time	Sorting sequence
1	20:02	1	9	22:00	11
2	20:15	2	10	22:15	9
3	20:30	3	11	22:22	10
4	20:58	4	12	22:55	13
6	21:02	5	13	23:08	12
5	21:26	7	14	23:48	15
7	21:33	6	15	23:25	14
8	21:50	8			

5. Conclusions

In this paper we researched the problem of the sorting sequence on marshalling station, and we built model and proposed algorithm. Through the method of analysis and simulation results show that the algorithm is effective, and the running time is acceptable. This study for the information construction of marshalling station has practical significance.

6. References

[1] Liang Chuncai (1980). "Determination of Car Sorting Sequence on Marshalling Station". *Railway Transport and Economy*. Vol. 02, No. 04, pp.141-146.

[2] Cao Jiaming, Fan Zheng, Mao Jieming (1993). "Decision Support System for Optimizing Marshalling Station Operation---Sorting Subsystem". *Journal of the China Railway Society*, Vol. 15, No. 4, pp.67-68.

[3] Cui Bingmou (1996). "A Study on Some Problems of System for Working Out a Stage Plan on Marshalling Station". *Journal of Lan Zhou Railway College*, Vol. 15, No. 2, pp.64-68.

[4] He Shiwei, Song Rui, Zhu Songnian (1997). "Optimal Model and Algorithm on Stage Plan of Sorting and Marshalling Operation for Marshalling Station". *Journal of the China Railway Society*, Vol. 19, No. 3, pp.1-8.

[5] Wang Zhengbin, Du Wen, Wu Baiqing (2008). "Model and Algoritnm for Estimation of Wagon Flow of Stage Plan Based on Break-up and Make-up Sequences". *Journal of Southwest Jiaotong University*, Vol. 43, No. 1, pp.91-95.

[6] Hu gang. *Research on Automatic Programming and Adjustment of Stage Plan on Marshalling Station*. Southwest Jiaotong University, Chengdu, 2000.

[7] Xing Wenxun, Xie Jinxing. *Modern Optimization Methods*. TsingHua University Press, Beijing, 1999.

[8] Wang Ling. *Intelligent Optimization Algorithms with Applications*. TsingHua University Press, Beijing, 2001.

2008 Workshop on Power Electronics and Intelligent Transportation System

A New Class of Highly Fault Tolerant Erasure Code for the Disk Array

Dan Tang, Xiaojing Wang, Sheng Cao, Zheng Chen
Chengdu Institute of Computer Applications Chinese Academy of Sciences
Chengdu, China
tangdan99@gmail.com

Abstract

We present a new class of erasure codes of size n×n (n is a prime number) called T-code, a new family of simple, highly fault tolerant XOR-based erasure codes for storage systems (with fault tolerance up to 15). T-code is not maximum distance separable (MDS), but has many other advantages, such as high fault tolerance, simple computability, and high efficiency of coding and decoding. Because of its superior quantity over many other erasure codes for the storage system, this new coding technology is more suited in RAID or dRAID systems.

1. Introduction

Along with the rapid development of information technology, we need better storage systems. Larger capacity, quicker speed and more reliability are in urgent demand for information storage in practical application, and the reliability has become one of the most important performance metrics in storage systems [1]. Redundant Array of Inexpensive Disks (RAID) is a popular technique used to improve the reliability and performance of secondary storage. To meet requirements such as utilization, performance and data protection, there are several types of RAID implementation including RAID0, RAID3, RAID5 and RAID6 [2]. RAID0 provides the highest performance but lacks redundancy; so it cannot tolerate any fault. Except for RAID6, other types of RAID can tolerate only one fault in a disk array; so we cannot reconstruct information if two disk errors appear simultaneously. At present, new technologies have led to the availability of inexpensive, high capacity disk drives in SATA (Serial ATA) formats, allowing storage managers to build low cost, high capacity arrays to protect their data. However, these affordable options also pose their own risks as SATA drives fail more often than Fiber Channel (FC) or SCSI drives. Take the implementation of RAID5 as an example, if one drive fails, another will take over. But failure of two drives at the same time causes loss of data and system downtime. Therefore, when these less reliable drives are used, advanced protection is needed to guard against multiple drive failures and provide fault tolerance and high availability of data. Furthermore, increasing the number of hard drives in a disk array effectively raises the expected failure rate of the first hard drive. During system recovery using a spare drive, the failure rate of the second hard drive is also multiplied. A subsystem composed of multiple drives needs added protection to guarantee data availability in case two or more disks fail simultaneously.

In recent years, RS (Reed-Solomon) code [4] has been used to provide error correction for multiple failures in storage systems. Some RAID6 systems use RS codes to correct two errors in a disk array. RS code is a highly fault tolerant MDS code in the family of BCH codes, but the coding and decoding of RS codes need finite field operations, which require plenty of resources in the storage system, especially in large-scale systems. Complex calculations limit the progress of RS codes employed in a storage system. Generally speaking, what is needed is a coding technology which not only has the ability of correcting multiple errors, but also has the quality of simple calculation. The array code is a kind of linear code, but which stores its information in a two-dimension array. A common character of array codes is that the encoding and decoding procedures need only XOR operation which is much simpler than the calculation in the finite field. Thus if there is a kind of array code which can correct more errors in the disk array, it would be very suitable for large-scale storage systems.

The paper is organized as follows. In section II, we describe the method of constructing and decoding the T-code. In section III we will analyze the basic qualities of T-code, some other erasure codes will be discussed in section IV, and we conclude with a short summary in the end.

2. T-code Description

In T-code, information symbols are placed in an array of size m×n, with n being a prime number and m <n. Parity symbols are generated by XOR operations of some information symbols , which are placed in an array of size t×n, t<m. Parameters m, t and n should satisfy an equation: m + t = n.

2.1. Disk Array Layout of T-code

Supposing there are n disks in our disk array and each disks has n blocks, we have a disk array of size n×n

---

978-0-7695-3342-1/08 $25.00 © 2008 IEEE
DOI 10.1109/PEITS.2008.79

578

and label it A=[a(i,j)], $0 \leq i < n-1, 0 \leq j < n-1$, with a(i, j) being the symbol at the i*th* row and j*th* column. The information symbols are stored in the front m rows of A, so we name these rows the data array, and denote which as D=[d(i,j)], $0 \leq i < m-1, 0 \leq j < n-1$. While the parity symbols are placed in the last t rows of A, so we call the last t rows the parity array, and label it P=[p(i,j)], $0 \leq i < t-1, 0 \leq j < n-1$.

Figure 1 is the disk array layout of T-code when m=4 and n=7.

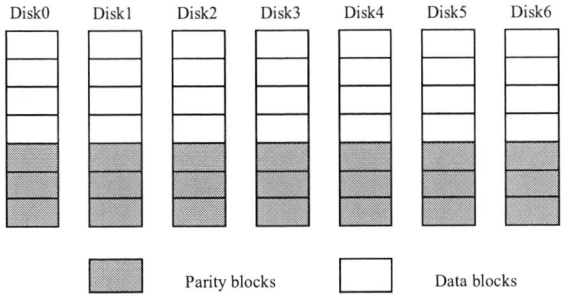

Figure.1 The disk array layout

2.2. Encoding Procedure

As described in section 2.1, we refer to the array in which information symbols are stored as label D, and that in which parity symbols are stored as label P. The parity symbols of T-code can be constructed according to the following rule:

$$p(i,j) = \sum_{k=0}^{m-1} d(k, < (i+1) \bullet (k+1) + j >_n)$$

$$i \geq 0, j \geq 0 \qquad (1)$$

Where Σ means the XOR operation of multiple symbols.

$<m>_n$ means the operation: m modulo n.

p(i, j) is the symbol at the i*th* row and j*th* column in the array P.

d(i, j) is the symbol at the i*th* row and j*th* column in the array D.

2.3. Efficient Decoding Method

If errors appear in the disk array, we can correct them as follows:

1) Randomly select a useful parity symbol and check the information symbols which need to be verified. If there are no erased information symbols, mark the parity symbol as useless; if there is only one information symbol erased, mark the parity symbol as useless and recover the information symbol according to the following algorithm:

$$d(i',j')=a(i,j) \oplus (\sum_{s=0}^{n-k-1} d_{i-(n-k),s})$$

$$i-(n-k) \neq i', s \neq j$$

2). Repeat step1 until there is no useful parity symbol or until every useful parity symbol contain more than one erasure errors. If all the erased information symbols have been recovered, the decoding procedure is successful.

Note: In step2, once the circumstance of every useful parity symbol contains more than one erasure errors have been found, it indicates that the algorithm above cannot correct all errors, and it is necessary to exit the decoding procedure.

2.4. Fault Tolerance

T-code is a kind of highly fault tolerant erasure codes, using it as the coding method in disk arrays; one can correct multiple errors simultaneously. Unfortunately, we have not yet been able to find a mathematical function to get the concrete error number of T-code can tolerant to different n size disk arrays. However, we have much experimental computer data that is valuable in practice for storage systems.

Table I lists some data concerning the fault tolerance of T-code.

TABLE I . DATA CONCERNING THE FAULT TOLERANCE OF T-CODE

The Number of A's Columns	The Number of D's Rows	The Number of P's Rows	The Number of Fault Tolerance
7	4	3	2
7	3	4	3
7	2	5	4
11	7	4	3
11	4	7	5
11	3	8	6
13	10	3	2
13	5	8	5
13	3	10	8
17	12	5	3
17	9	8	4
17	4	13	9
19	11	8	4
19	9	10	5
19	4	15	11
23	8	15	7
23	6	17	10
23	3	20	15

3. Features of T-code
3.1. Analysis of T-code

In storage systems, encoding and decoding complexity is an important value criterion, and it constitute the main difference between array codes and RS codes. Code rate is another important property of a coding method. We analyze the complexity and code rate of T-code as follow:

Encoding complexity: in a disk array, each disk divides into blocks, and each block has the same size.

According to the equation (1) of section 2.2, one needs to execute $m \times n \times t = m \times n \times (m-n)$ XOR operations.

Decoding complexity: according to the decoding algorithm, the decoding complexity of T-code is unsure. If there are c errors in the disk array, the XOR operations to be executed lie between $c \times m^2$ and $(c \times m^2)!$.

Code rate: Different from other array codes, the code rate of T-code can be determined by the requirement of different applications. For example, if m=7 and n=11, the code rate of T-code is approximately 0.64.

3.2. Other Characteristics

T codes have instances of exceptionally high fault tolerance (we gave the parameters which can be used to constructed 15 fault tolerance T-code and conjecture that the construction of more fault tolerance T-code should be possible). There are few codes in the storage system literature that meet these fault tolerance levels. The only viable options for high fault tolerance to date seem to be Reed Solomon codes, with their high computational costs, and N-way mirroring with its very low code rate. Moreover, T-code needs only the XOR operation which can be executed rapidly by regardless of software or hardware. Thus the coding progress of T-code is much faster than that of the RS code which needs the finite field computation. According to the section 2.2, each parity unit is the XOR operation result of m information units. In other words, every single parity unit is decided by the number m, not by the number n, so that the coding rate dose not slow down rapidly when the storage system has more disks.

4. Related Work

The T-code can be compared to any other erasure code suitable for storage systems. We have mentioned some already such as Reed-Solomon. In this section, we will discuss some representational array codes, and compare them with T-code.

The EVENODD [5] code has been devised by M.Blaum in 1993, which can tolerate up to two disk failures in RAID architectures. EVENODD is the first known scheme for tolerating double disk failures that is optimal with regard to both storage and performance. It consists of simple XOR computations and only requires parity hardware, which is typically present in standard RAID5 controllers. Hence, EVENODD can be implemented in standard RAID5 controllers without any hardware changes. Furthermore, the number of XOR operations involved in implementing EVENODD in a disk array with 15 disks is about 50% of the number required when using RS scheme. The update complexity of EVENODD codes approaches 2 as the number of the columns of the code increases. But it was proven in [5] that for any linear array codes with only parity columns, the update complexity is strictly larger than 2(the obvious lower bound).

The X-code [6] is another class of MDS array codes. X codes are of minimum column distance 3, namely, they can correct either one column error or two column erasures. The key novelty in X-code is that it has a simple geometrical construction which achieves encoding/update optimal complexity, i.e., a change of any single information bit affects exactly two parity bits. Compared to EVENODD, the construction of X codes is that all parity symbols are placed in rows rather than columns.

STAR [7] is the further extension of EVENODD, which can correct three erasure errors in a disk array simultaneously. It is a kind of MDS codes with the minimum column distance 4. In 2005, James Lee Hafner devised a new coding method called WEAVER. It is a kind of highly fault tolerant XOR-based erasure codes for storage systems (with fault tolerance up to 12). WEAVER codes are in general not maximum distance separable (MDS), but which has many other advantages, such as host I/O size uniformity, simple computability and so on. However, WEAVER has a fatal flaw that its code rate is very low(less than 50%), and which seriously limits its application in storage systems.

From what have been discussed above, we can sum up as follows. EVENODD codes and X codes can correct no more than two disk errors in a disk array, which cannot satisfy the modern storage systems obviously. STAR codes can tolerate up to three erasure errors, but it cannot set our hearts at rest if we employ STAR codes in dRAID systems which may appear much more errors than in RAID systems. The fault tolerance ability of WEAVER is strong enough to adapt to RAID and dRAID systems. However, once it was employed in storage systems, we had to use much more additional disk space to store parity info. T-code can correct multiple erasure errors in a disk array; and compared with WEAVER, T-code's code rate is much higher when they correct the same number of errors, in other words, T-code can save lots of disk space when the two coding methods correct the same number of errors. in table II ,we compare the code rate between WEAVER and T-code.

TABLE II. CODE RATE COMPARISON

The Number of Fault Tolerance	The Code Rate of WEAVER	The Code Rate of T-code
1	0.50	0.50
2	0.33	0.89
3	0.25	0.79
4	0.20	0.62
5	0.17	0.55
6	0.14	0.43
12	0.07	0.27

Author names and affiliations are to be centered beneath the title and printed in Times 12-point, non-boldface type. Multiple authos may be shown in a two- or three-column format, with their affiliations below their respective names. Affiliations are centered below each author name, italicized, not bold. Include e-mail addresses if possible. Follow the author information by two blank lines before main text.

5. Conclusion

We have presented T-code, a new class of array code. The significant difference between T-code and other known array codes is the highly fault tolerance, although T-code is not a MDS code and its fault tolerance ability is uncertain in terms of mathematics. T-code has many other features such as computational simplicity and high coding efficiency. All these features make the T codes suitable for any storage system with high fault tolerance and performance requirements; they are perhaps best suited to dRAID systems which may contain more errors because of the instability of the network. One further research problem is to find the mathematical function so that we can get the precise error number of T codes can tolerant for different number n. If we can solve that problem, T-code could be an integrated coding system.

ACKNOWLEDGMENT

This paper is supported by Auto-reasoning Lab of CICA-CAS and National Basic Research Program of China, No.2004CB318003. We want to thank the Auto-reasoning Lab of CICA-CAS for the opportunity to run many of the larger experiments on their system (and their assistance). Testing n sizes in 5-41 ranges. The author also wants to extend his thanks and appreciation to Professor Daniel Kister for his support.

References

[1] Xin Q, Miller E.L, Schwarz T.J. Reliability mechanism for very large storage systems. In Proceeding of the 20th IEEE/11th NASA Goddard Conference on Mass Storage systems and Technologies, 2003, 146-156.

[2] D. A. Patterson, Garth Gibson, and R. H. Katz. A Case for Redundant Arrays of Inexpensive Disks (RAID). In International Conference on Management of Data (SIGMOD), 109-116, June 1988..

[3] Park C. Efficient placement of parity and data to tolerate two disk failures in disk array systems. IEEE Transaction on Parallel and Distribute Systems, 1995, 6(11): 1177-1184.

[4] Plank J.S. A tutorial on Reed-Solomon Coding for fault-tolerance in RAID-Like systems. Software

Practice and Experience (SPE) 1997, 27(9): 995-1012.

[5] M. Blaum, J. Brady, J. Bruck, and J. Menon.EVENODD: An Efficient Scheme for Tolerating Double Disk Failures in RAID Architectures. IEEE Transaction on Computer, vol. 45, 192-202, 1995.

[6] L. Xu, and J. Bruck.X-code: MDS Array Codes with Optimal Encoding.IEEE Transaction on Information theory, 45(1): 272-276, Jan 1999.

[7] C. Huang and L. Xu. STAR: An Effcient Coding Scheme for Correcting Triple Storage Node Failures. FAST-2005: 4th Usenix Conference on File and Storage Technologies, December, 2005.

[8] J. L. Hafner: WEAVER Codes: Highly Fault Tolerant Erasure Codes for Storage Systems., FAST-2005: 4th Usenix Conference on File and Storage Technologies, December, 2005

2008 Workshop on Power Electronics and Intelligent Transportation System

A hybrid Active Compensation Method for Current Balance Based on Y,d11 connection traction transformer

Wang guo[1], Ren enen[1], Tian mingxing[1]
[1] School of Automation and Electrical Engineering,
Lanzhou Jiatong University, lanzhou, GanSu, 730070, China
wangguo2005@eyou.com

Abstract

Electrified railway traction loads present large single phase loads to the supply system. Most of traction transformer use Y,d11 connection in china. Typically zero sequence components are removed using a phase to phase transformer connection, however a large negative sequence component remains. With the developing of high speed and heavy load of the train, low power factor, high contents of harmonic currents, and negative-sequence current in traction power systems become more critical. This paper presents a novel hybrid compensation method to compensate reactive power, harmonic, and negative-sequence currents in two feeders of a traction substation. The system structure of current balance compensator method is constituted by an AC-DC-AC single-phase inverter, base frequency and high-frequency harmonic passive filter to suppress harmonic current, negative-sequence current and reactive current. It detects the active current difference of two phase traction loads, and then makes active power flows from the light load side to the heavy side through the compensator. The result is two phase traction loads of the transformer reach to balance state to insure the current symmetry of three-phase side of the y, d11 connection traction transformer.

1. Introduction

Single-phase supply system is widely used in the electrified railway. It has the advantages such as low cost, easy implementation and maintenance over three-phase system. Electrical locomotives are treated as single-phase loads, the speed and load condition of the train always changes frequently, which can cause the low power factor, high contents of harmonic currents, and negative-sequence currents in traction power systems. As a single phase sub-system of the three-phase power delivery system, we must solve the problem of load matching[1-3].

Conventional passive power quality compensators, such as reactive power compensation capacitors, passive filters, phase sequence rotation and so on, are single-phase equipment and are installed in each feeder of a traction substation separately[4]. These projects are only alleviating three phases unbalance in a certain extent but it can not solve the problem of load matching radically.

This paper proposes an electrified railway supply system topology, using a single-phase active filter, base frequency and high-frequency harmonic passive filter to compensate the negative sequence current, reactive power and harmonics produced by locomotives. View from the power supply system side, the electrical traction system will act as a balanced resistance load.

2. Condition of eliminate the negative-sequence current

The use of Y,d11 connection transformer is mostly widely used in china electrified railway. We deduce a balance qualification from Y,d11 connection transformer.

The relation of the primary and the secondary current is the following equation:

$$\begin{bmatrix} \dot{i}_A \\ \dot{i}_B \\ \dot{i}_C \end{bmatrix} = \frac{\sqrt{3}}{K} \begin{bmatrix} -2 & -1 \\ 1 & -1 \\ 1 & 2 \end{bmatrix} \begin{bmatrix} \dot{i}_\alpha \\ \dot{i}_\beta \end{bmatrix} \qquad (1)$$

Where

$\dot{I}_A, \dot{I}_B, \dot{I}_C$ the three phases system current in phase A, phase B and phase C,

$\dot{I}_\alpha, \dot{I}_\beta$ the load current in phase α and phase β,
$K=N_1/N_2$ the turns ratio of Y,d11 transformer.

Each sequence of current is presented as:

$$\begin{bmatrix} \dot{i}_1 \\ \dot{i}_2 \\ \dot{i}_0 \end{bmatrix} = \frac{\sqrt{3}}{3K} \begin{bmatrix} -1 & a^2 \\ -1 & a \\ 0 & 0 \end{bmatrix} \begin{bmatrix} \dot{i}_\alpha \\ \dot{i}_\beta \end{bmatrix} \qquad (2)$$

Where

I_1 the primary positive sequence current component of the traction transformer,

I_2 the primary negative sequence current component of the traction transformer,

978-0-7695-3342-1/08 $25.00 © 2008 IEEE
DOI 10.1109/PEITS.2008.88

I_0 the primary zero sequence current component of the traction transformer.

Three phases currents are asymmetry when I_α is not equals $e^{-j60°}I_\beta$ [5].

3. Topology selection of the current balance compensator

3.1. Load-compensator interaction

It is instructive to consider the interaction between the topology of compensator and the load. This allows a topology to match to a load type. Topology may be examined in terms of the following general criteria[6]:

• Is the topology capable of sinking load harmonic currents and compensating reactive power?

• Can the compensator control the base harmonic power flows from traction α side into traction β side, and insure two traction loads currents are equal?

• Can the topology prevent resonance, hold load voltage and restrain terminal voltage distortion?

• Are there redundant components?

3.2. Configuration of the compensator

Passive compensator had widely used in electrified railways. Active compensator with small capacity and low voltage is introduced to improve the effect of passive compensation. The hybrid active compensation is considered the best of topology. The appropriate topology for two element system is shown in fig.1. Fig.2 shows the efficacious connection topology for three element system.

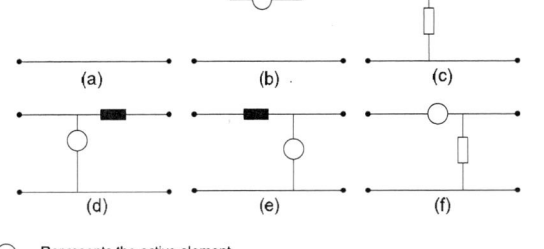

—○— Represents the active element

—▢— Represents a passive element which is low impedance at the harmonic frequencies

—■— Represents a passive element which is low impedance at the fundamental frequency

Fig.1 Hybrid compensation topologies with one active and one passive element

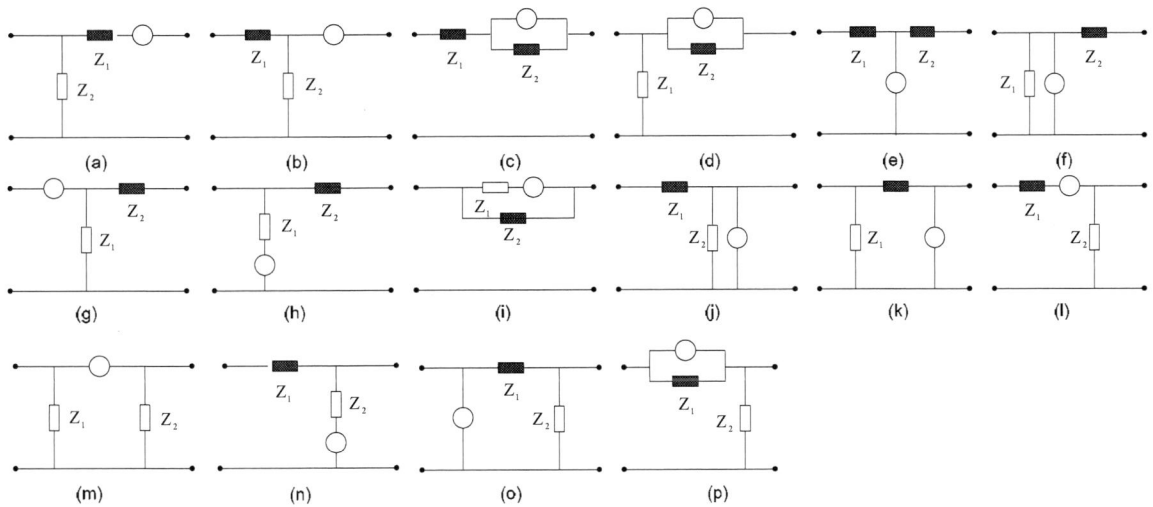

Fig.2 Hybrid compensation topologies with one active and two passive element

These topologies are proposed in reference[6]. Of configurations in Figure 2 and 3, we select the suited hybrid topology for the electrified railway comprehensive compensation. Hybrid topology is connected between two traction loads, but the voltages of two traction loads are unequal in phase. Configuration 3(i) offers a unique harmonic isolator solution which does not act through converting harmonic voltage to harmonic currents. Impedance Z_2 should have a high harmonic impedance to isolate the supply and load buses. The Z_1 should present a low harmonic. Harmonic isolating voltages are generated

by small circulating harmonic currents in the filter. A large portion of fundamental current may be diverted to Z_2. A high fundamental impedance in Z_1 assists this diversion and the active element then carries only small amounts of fundamental current and voltage. This configuration in reduction fundamental voltage and current offers strong potential for reduced ratings.[6,10]

In view of the original fixed capacitors compensation and reducing the active component ratings , we select figure 3(i) as the hybrid compensation topology of Y,d11 connection traction transformer[6].

583

4. Novel current balance compensator

The configuration of novel comprehensive compensator is shown in fig.4.

In the traction power supply system shown in figure 4, the traction transformer is a Y,d11 connection transformer, the three-phase 110kV power supply is converted to 27.5kV and the phase contrast is 1200. The current balance compensator is composed of single-phase converter, base harmonic passive filter, harmonic passive filter and shunt capacitors. The proposed novel current balance compensator has the following advantages[4,9]:

- Through a Y,d11 connected transformer, a base harmonic passive filter(BHPF) and a single phase converter are connected in parallel, and are connected in series between two traction loads. The harmonic passive filter (HPF) and shunt capacitors are connected to traction load in parallel. The performance of passive filter and reactive compensation are improved greatly. At the same time, most of the fundamental current is diverted to the BHPF branch, the reactive power is compensated by the shunt capacitors, and most of the high harmonic is reduced by HPF. This results in a great reduction of the capacity rating of the single phase converter.

- The harmonic passive filter consists of 3rd, 5rd, 7rd. Most of the harmonic currents are compensated.

- In order to reduce the capacity further, the fixed reactive power compensation capacitors banks are connected in parallel with single phase converter .the shunt capacitors compensate most of the reactive power.

- The single phase converter takes the whole traction loads, passive filters and shunt capacitors as the compensation object.

- Use a single phase converter to compensate the current balance, and the reactive power, harmonic currents in transformer secondary transformer side.

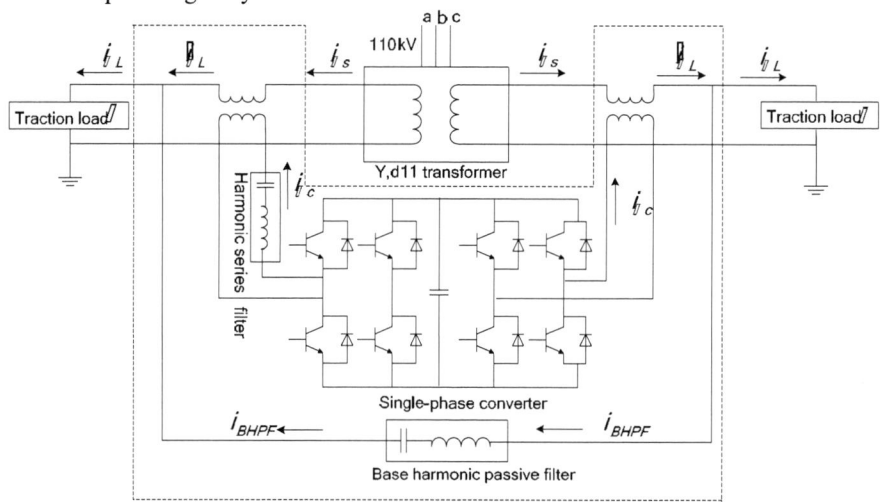

Fig.4 System configuration of the current balance compensator method on Y,d11 connection transformer

4.1. Compensation principle of the hybrid active power balance current compensator

Electric locomotives are treated as single phase load, the speed and the load condition power factor always changes frequently, which can cause the uncontrolled load current. In order to insure two traction loads symmetry, the single phase converter takes the whole traction loads, passive filters and shunt capacitors as the compensation object to compensate current balance with the reactive power and harmonic currents compensation. The single phase converter can not produce active power, but transfer active power between two traction loads using converting function. The expression of and phase currents are:

$$
\begin{cases}
i_{\alpha L} = i_{\alpha s} + i_{\alpha c} + i_{BHPF} \\
i_{\beta L} = i_{\beta s} + i_{\beta c} - i_{HPF} - i_{BHPF}
\end{cases}
\tag{3}
$$

It is independent of Active power, reactive power and harmonic currents, because of the α and β traction loads are not controlled each other. We suppose the voltage of α phase is:

$$ u_\alpha (t) = U_\alpha \sin \omega t $$

The analysis of the current of a phase is:

$$i_{\alpha L}(t) = \sum_{n=1}^{\infty} I_n \sin(n\omega t + \phi_n)$$

$$= I_{\alpha p}\sin\omega t + I_{\alpha q}\cos\omega t + \sum_{n=2}^{\infty} I_n \sin(n\omega t + \phi_n) \qquad (4)$$

$$= i_{\alpha p}(t) + i_{\alpha q}(t) + i_n(t)$$

Where:

$I_{\alpha p}(t)$ the active power current component of base harmonic current,

$I_{\alpha q}(t)$ the reactive power current component of base harmonic current,

$I_n(t)$ harmonic current.

The current of $i_{\alpha q}(t)$ and $i_n(t)$ is the component of the α phase current to be reduced, and the active power of $i_{\alpha p}(t)$ need to be transferred. The current expression of compensation is:

$$i_{\alpha c} + i_{BHPF} = i_{\alpha q}(t) + i_n(t) + \Delta i_{\alpha p}(t)$$

$$= I_{\alpha q}\cos\omega t + i_n(t) + \Delta I_{\alpha p}\sin\omega t \qquad (5)$$

So $i_{\alpha s}(t) = (i_{\alpha p} - \Delta i_{\alpha p})\sin\omega t = I_{\alpha s}\sin\omega t$

The instantaneous power of traction load α and power supply are expressed in equation (6).

$$P_{\alpha L}(t) = u_\alpha(t)i_{\alpha L}(t) = \frac{U_\alpha I_{\alpha p}}{2} + P_{\alpha h}(t)$$

$$P_{\alpha s}(t) = u_\alpha(t)i_{\alpha s}(t) = \frac{U_\alpha I_{\alpha s}}{2}(1 - \cos 2\omega t) \qquad (6)$$

We also calculate the each expressions of traction load β.

The output power from transformer is equal to the power that is consumed by load in a cycle base on the law of conservation of energy. A definite integral to instantaneous power of two traction loads and power supply is given in equation(7).

$$\int_T \left[P_{\alpha L}(t) + P_{\beta L}(t) \right] dt = \int_T \left[P_{\alpha s}(t) + P_{\beta s}(t) \right] dt$$

$$\Rightarrow \left(\frac{I_{\alpha p}}{2} + \frac{I_{\beta p}}{2} \right) U_\alpha T = \left(\frac{I_{\alpha s}}{2} + \frac{I_{\beta s}}{2} \right) U_\alpha T \qquad (7)$$

The power currents that is

$$I_{\alpha s} = I_{\beta s} = \frac{I_{\alpha p} + I_{\beta p}}{2} \qquad (8)$$

So the power current in phase α and β is equal to the average of the two traction loads of base harmonic current.

The currents of the single phase converter that hybrid active power balance current compensator (HAPBCC) transfers between two traction phases are:

$$i_{\alpha cp} = -i_{\beta cp} = \frac{1}{2}(i'_{\alpha L} - i'_{\beta L}) \qquad (9)$$

So the compensation current is only to detect the $i'_{\alpha L}$ and $i'_{\beta L}$ in figure 4 with this balance compensation method. Harmonics produced by HSF, BHPF and traction loads are blocked by single-phase converter.

4.2. single-phase equivalent of the hybrid active power balance current compensator

Each of HSF and BHPF is equivalent to an ideal impedance, and single-phase converter is equivalent to an ideal controlled current source $i_{\alpha c}$ and $i_{\beta c}$. The hybrid active power balance current compensator is illustrated in figure 5.

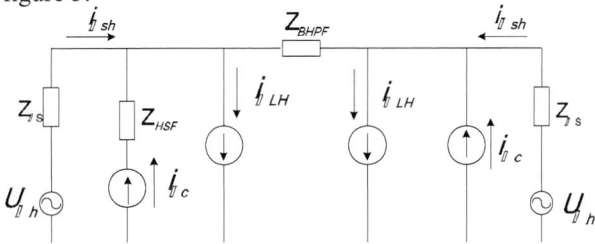

Fig.5 single phase equivalent circuit of hybrid active power balance current compensator

In figure 5:

$U_{\alpha h}$, $U_{\beta h}$ the virtual value of phase α and β power supply voltage,

$Z_{\alpha s}$, $Z_{\beta s}$, Z_{BHPF}, Z_{HSF} the equivalent impedance of α power supply, power supply, BHPF and HSF,

$i_{\alpha LH}$, $i_{\beta LH}$ the harmonic current component of α and β traction load,

$i_{\alpha sh}$, $i_{\beta sh}$ the harmonic current component of α and β power supply,

$i_{\alpha c}$, $i_{\beta c}$ the equivalent harmonic current of single-phase converter.

The source current can be expressed as:

$$\begin{cases} i_{\alpha sh} = \dfrac{1}{Z_\Sigma}U_{\alpha h} - \dfrac{1}{Z_\Sigma}U_{\beta h} + (1-k_\alpha)\dfrac{Z_{\beta S}+Z_{BHPF}}{Z_\Sigma}i_{\alpha LH} + (1-k_\beta)\dfrac{Z_{\beta S}}{Z_\Sigma}i_{\beta LH} \\[4mm] i_{\beta sh} = -\dfrac{1}{Z_\Sigma}U_{\alpha h} + \dfrac{1}{Z_\Sigma}U_{\beta h} - (1-k_\alpha)\dfrac{Z_{BHPF}Z_{\alpha S}}{Z_\Sigma Z_{\beta S}}i_{\alpha LH} + (1-k_\beta)\dfrac{Z_{\alpha S}+Z_{BHPF}}{Z_\Sigma}i_{\beta LH} \end{cases} \qquad (9)$$

Where:

$Z_{\Sigma}=Z_{\alpha s}+Z_{\beta s}+Z_{BHPF}$, $i_{ac}=k_{\alpha}i_{Alh}$, $i_{\beta c}=k_{\beta}i_{Blh}$,

k_{α} is the equivalent plus of harmonic detection, control circuit, current compensation generator of α phase,
k_{β} is the equivalent plus of harmonic detection, control circuit, current compensation generator of β phase.
If $U_{\alpha h}\approx U_{\beta h}$, then from (9), (10) we deduced that

$$
\begin{cases}
i_{\alpha sh} \approx (1-k_{\alpha})\dfrac{Z_{\beta S}+Z_{BHPF}}{Z_{\Sigma}}i_{\alpha LH} + (1-k_{\beta})\dfrac{Z_{\beta S}}{Z_{\Sigma}}i_{\beta LH} \\[3mm]
i_{\beta sh} \approx -(1-k_{\alpha})\dfrac{Z_{BHPF}Z_{\alpha S}}{Z_{\Sigma}Z_{\beta S}}i_{\alpha LH} + (1-k_{\beta})\dfrac{Z_{\alpha S}+Z_{BHPF}}{Z_{\Sigma}}i_{\beta LH}
\end{cases}
\tag{10}
$$

From equation (9), we get the following views.

• If we ensure $k_{\alpha}=k_{\beta}=1$, the harmonic component produced by electrical locomotives nonlinear loads can be eliminated.

• If we ensure $U_{\alpha h}\approx U_{\beta h}$, the harmonic component in power supply system can be eliminated.

5. Conclusions

The novel hybrid active power balance current compensator proposed in this paper is composed of a single phase converter, one base harmonic passive filter and one harmonic series filter. This hybrid active power balance current compensator is proposed based on Y,d11 connection traction transformer. The features are gained:

• The compensator can control the base harmonic power flows from traction α side into traction β side, and eliminate the negative sequence in primary side of traction transformer.

• The single phase converter takes the whole traction loads, base harmonic passive filters and original compensation of fixed shunt capacitors as the whole compensation object.

• The unbalance of traction currents, harmonic currents, reactive power, in two feeders of a traction substation can be compensated together.

The feasibility of the novel hybrid active compensator is needed by simulation in the subsequent time.

Acknowledgement

It is a project supported by natural foundation in GanSu Province(3ZS042-B25-032).

References

[1] Lin haixue, Three Phases Unbalance of power system, Beijing: China electric power publication, 1998.
[2] Lu runyu. (1998). High-frequency harmonic of power system, China electric power publication,1998.
[3] Zeng Guohong, Hao Rongtai, "A novel three-phase balanced traction supply system based on active power filter", Journal of the china railway society,25(1),2003,pp.48-53.
[4] Sun zhuo, jiang xinjian, "A novel of active power quality compensator topology for electrified railway", IEEE transactions on power electronics,19(4),2004,pp.1036-1042.
[5] Ouyang fan, zhou youqing, Guo ziyong, "A compensation method for current balance based on balance transformer", Transactions of china electro technical society, 22(5), 2007, pp.53-57.
[6] S.T. Senini, P.J. Wolfs, "Systematic identification and review of hybrid active filter topologies",Power Electronics Specialists Conference, pesc, IEEE 33rd Annual, 2002, pp.394-399.
[7] Lin qionglin, Liu huijin, Sun jianjun etc, "Topology analysis of the active power filter with large capacity", High Voltage Engineering 32(2),2006, pp. 70~77.
[8] Wu mingli, Li qunzhan, "Three-Phase Harmonic Modelling for Power System and Electric Traction System", Jounal of the china railway society,21(1), 1999, pp.43-47.
[9] Sun zhuo, Jiang xinjian, "Study of novel traction substation hybrid power quality compensator", International Conference on Power System Technology, 2002, pp. 480-484.
[10] S.senini, P.Wolfs, "An approach to harmonic flow control using hybrid series active/passive topologies", Journal of electrical and electronics engineering, austrilia,2001, vol 21,no 2, pp109-118

Research on Virtual Disassembly Simulation Based on Constraint Matrix

LI Shi-ting ZHU Bo CAI Qi

(Naval University of Engineering，WuHan，HuBei，P.R.China)
E-mail:lishiting0620@163.com

Abstract

Disassembly is one of the important processes of complex equipments maintenance. On the basis of analyzing the maintenance characteristic of complexity equipments, a method to generate disassembly sequence is presented. Using the knowledge of Graph Theory, the method realizes to generate automatically the optimal disassembly sequence through establishing the constraint matrix. And the method is applied to virtual maintenance simulation platform (VMSP)developed by our work group. On the platform，the whole process of maintenance is simulated.

1.Introduction

With the development of equipments direction to high speed, high power and high integration, the new methods and technology for the equipments maintenance are requested. The conventional maintenance methods using physics prototype to study maintenance and maintainability have many shortcomings. For example, operator's error may cause the expensive and complex equipments fault during maintenance training, which leads to severe economic loss. However, it is necessary to disassemble equipments for quickly repairing, and the main reason is that disassembly is one of the important processes of maintenance and maintaining. To repair the fault component quickly, amount of disassembly training for the technician are requested. To resolve the contradiction between trying to increase the training time and trying to decrease disassembly frequency of the physics prototype for training, the virtual reality (VR) technology is adopted. VR is a synthetic, three-dimensional, interactive environment typically created by a computer. It provides a unique avenue to enhance the visualization of complex three-dimensional objects and environment with its real time, more interactive and spatial ability. Using the virtual prototype and virtual environment created by computer, whenever and wherever that the technician can perform the maintenance training. Disassembly sequence is key factor to maintenance, which may be various combinations for equipment. It should be noted that the constraint relation among components is miscellaneous for the complex equipment. According to the above reasons, the virtual

disassembly simulation system is developed, which can be applied for the training of maintenance technicians.

The remainder of this paper is organized as follows. Section 2 describes the relate work about disassembly sequence and methods. Section 3 presents the disassembly sequence model and components of program structure. Section 4 presents a case to verify the validity of model. Finally, conclusions will be provided at the end of the paper.

2.Related work

Over the years, researchers have proposed several approaches to disassembly sequence planning. These include the AND/OR graph based approach such as the AND/OR graph and the disassembly Petri net [1,2], and the disassembly tree approach [3]. A detailed review and a discussion of these approaches has been presented elsewhere [4,5]. It has been established that the disassembly sequence planning for maintenance is more complicated than those for assembly sequence planning and recycle purposes as it is target driven, that is, dependent upon the components to be maintained, and may require either partial or complete disassembly. Thus, a novel representation scheme for disassembly sequence planning known as disassembly constraint graph (DCG) had been established by the authors [6]. Basically, the DCG is able to generate all the possible disassembly sequences for a target component, which is the system component to be maintained. A sequence-based optimization approach such as genetic algorithm (GA) is deployed to generate the near optimal sequence-based on such criteria as minimum disassembly time or cost. These criteria take into account factors such as changes of disassembly tool and disassembly orientations for the disassembly operations.

3.Disassembly sequence simulation

3.1.Modeling constraint matrix

The structure model is showed in Fig.1, the model can distinctly express the constraint relation among the components. The structure model is able to generate all the possible disassembly sequences for a target component. In structure model, the direction of arrow represents

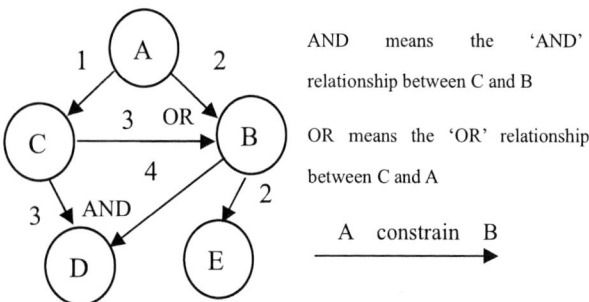

AND means the 'AND' relationship between C and B

OR means the 'OR' relationship between C and A

A —— constrain —— B

Fig.1. structure model of directed graph

constraint relation between two components, the value on the arc represents weight value, the node represents the component, 'AND' and 'OR' represents logic relation of among constraints. For example, let us assume node "E" as a target component. So, before disassembling E, the constraint of E must release, that is, B should be firstly disassemble. According to the same principle, then the constraint of B, "A" and "C", is disassembled. Finally, two disassembly sequences are generated: $1. A \rightarrow B \rightarrow E$, $2. A \rightarrow C \rightarrow B \rightarrow E$. If the value of arc means time that the disassembly operation take from node to node, the near optimal sequence is obtained based-on such criteria as minimum disassembly time. The constraint matrix corresponding to constraint relation model is showed as follows.

$$\begin{array}{ccccc} A & B & C & D & E \end{array}$$
$$\begin{bmatrix} 0 & 1 & 1 & 0 & 0 \\ 0 & 0 & 0 & 1 & 1 \\ 0 & 1 & 0 & 1 & 0 \\ 0 & 0 & 0 & 0 & 0 \\ 0 & 0 & 0 & 0 & 0 \end{bmatrix} \begin{array}{c} A \\ B \\ C \\ D \\ E \end{array}$$

3.2. Algorithms

Basically, a disassembly constraint graph is a hybrid graph representation for the equipments with nodes representing the minimal maintenance unit and edges encoding the disassembly constraints for maintenance.

One data body of graph records the corresponding information, including of the name of component, the disassembly code and constraint relation matrix etc. Data body of append information storages the information of disassembly tools and component code etc. So, the algorithms are realized in compile procedure by defining the data structure body, parts of definition in C++ is showed as follows:

typedef struct
{ String vex_name// name of component is key field
  String assemble_name; // name of assemble
  int assemble_attribute; // attribute of assemble
  int vex_no;
} Vex_Info;   //node name

4. A case study

4.1. Generating disassembly sequences of the valve body

A case study on the maintenance of valve is used here to demonstrate the process of generating disassembly sequence. The faulty components that need to be disassembled are assumed to be casing in the valve, namely the valve body. As mentioned above, firstly, the disassembly constraint graph is built，the constraint relation is showed in Fig2. Apparently, if the valve body E will be disassembled successfully, the constraint components for it, shelf D and shelf bolts B, are disassembled firstly. According to the similar principle, along the negative direction of arrow, the constraints are

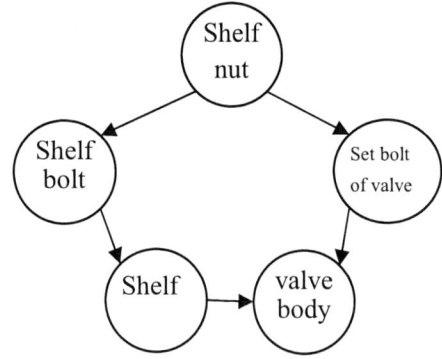

Fig.2. Constraint relation of valve body

gradually released. Finally, the disassembly sequence of valve body is obtained (Fig.3). The processes of releasing

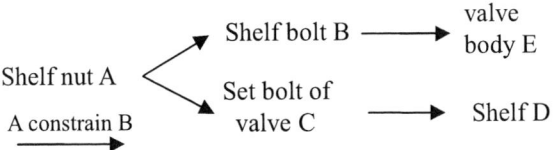

Fig.3.Disassembly sequence of valve body

constraints are implemented by the computer to run the programme. When the constraints among the components are very complex, the benefit of algorithm becomes oblivious. The disassembly sequence is denoted with adjacent matrix in computer, and the adjacent matrix of valve body can be showed as follows.

$$
\begin{array}{ccccc}
A & B & C & D & E \\
\begin{bmatrix}
000 & 000 & 000 & 000 & 000 \\
001 & 000 & 000 & 000 & 000 \\
001 & 000 & 000 & 000 & 000 \\
000 & 000 & 001 & 000 & 000 \\
000 & 001 & 000 & 001 & 000
\end{bmatrix}
& & & &
\begin{array}{c}
A \\ B \\ C \\ D \\ E
\end{array}
\end{array}
$$

4.2. Improved model

From the adjacent matrix, it is oblivious that the matrix has been improved comparing with the matrix in section 2. The main reason is that former matrix is not sufficient to express constraint information, while the constraint structure is very complex. The improved adjacent matrix expresses the constraint relation by using three decimal numbers, such as "000","001". Its implication is that the first number means the constraint relation, such as "0" represents no constraint, "1" represents "AND" constraint relation and "2" represents "OR" constraint relation; the second and third numbers mean the evaluation index of disassembly, such as maintenance time, cost etc. For example, "204" represents the constraint relation is "OR" between two nodes, and when the index is disassembly time, it will take four unit interval to complete the disassembly step. So, the method provides the maintenance engineers with near optimal disassembly sequences in relation to the target component

for virtual maintenance in virtual environment.

4.3. Virtual disassembly

To verify the suitability of having generating disassembly sequences of valve, virtual maintenance simulation platform (VMSP) developed by our work group has been used to realize the process of virtual maintenance in the virtual environment. The principle graph of interactive control is showed in Fig.4.The main

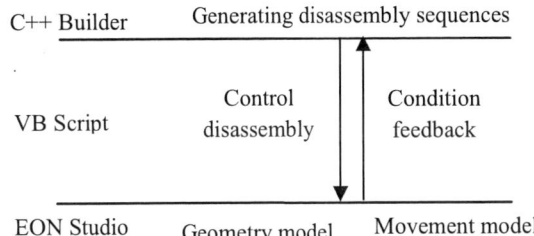

Fig.4. Principle graph of interactive control

interface of VMSP is showed in Fig.5, and the list of components is showed in Fig.6. The core part of VMSP adopts EON Professional[6] software with a user-friendly graphic user interface (GUI), which omits the burden to write source code. EON Professional supports the second development with java script and VB Script.

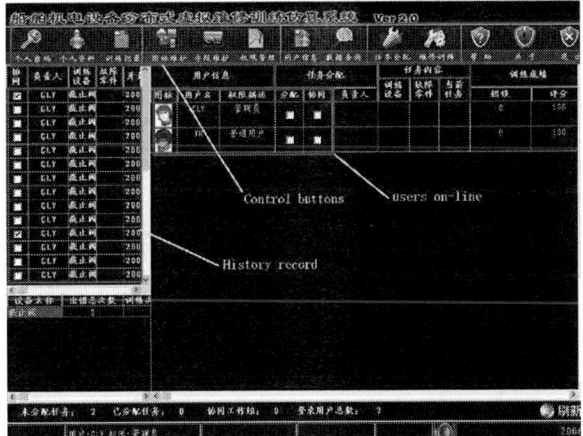

Fig.5. The main interface of VMSP

As depicted in Fig.7, the window of virtual disassembly, the design of the left-hand information window provides the users with an overview of the maintenance process and a fixed information window to track the information of different maintenance phases. This is used to facilitate the learning and training process. The virtual disassembly process is depicted as follows. Firstly, the user sets up the target component of fault facility, and analyses the fault cause. Secondly, the users can not only choose the disassembly sequences that generated automatically by computer, but set up disassembly sequences by themselves. Subsequently, disassembly operations can be classified into two kinds: automatic disassembly and manual disassembly, which are carried out according to the sequence.

Fig.6. The window of generating disassembly sequence

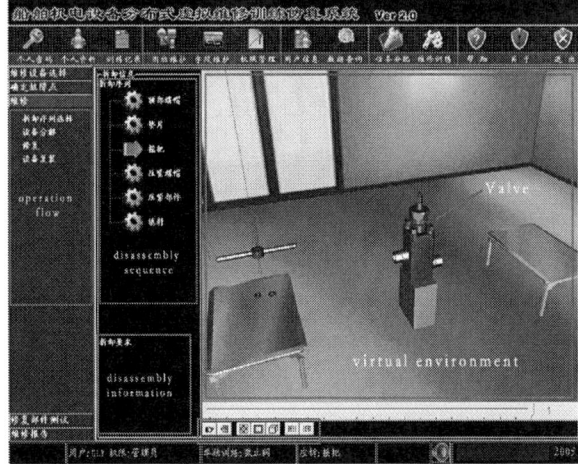

Fig.7. The windows of virtual disassembly

The near optimal disassembly sequence that the system follows is output in the disassembly sequences window, where the 'current target' for target component is denoted by green arrow. On the interface (Fig. 7), the bottom disassembly information window shows the information on the current disassembly process, such as disassembly tools and planning. Thus, the users can visualize the disassembly process, and get the knowledge of the 3D geometric relationship among the components with VR display. The virtual environment for maintenance is displayed in the VR display window. This window allows the users to navigate and visualize the environment. Moreover, to enhance interactivity, toolbars such as navigation toolbar and system standard toolbar have also been incorporated.

5.Conclusions

Maintenance involves disassembly sequence planning and disassembly process, which are both knowledge and experience intensive. It requires the maintenance staff, new staff in particular, to undergo rigorous training so as to improve their knowledge and skills for maintenance tasks. This paper presents a disassembly architecture model on the base of constraint relation graph of components. The model is used in VMSP to practice the disassembly for certain component. A case study has been performed to illustrate the suitability of the model. Furthermore, VMSP can apply to virtual disassembly simulation of other mechanical and electrical products. The study result has been used in some units, and received good evaluations. According to their feedback information, next work is considered for components stability in virtual disassembly.

References

[1] L.S. Homem de Mello, A.C. Sanderson, A correct and complete algorithm for the generation of mechanical assembly sequences, IEEE Transactions on Robotics and Automation 7 (2) (1991) 228–240.

[2] E. Zussman, M.C. Zhou, A methodology for modeling and adaptive planning of disassembly processes, IEEE Transactions on Robotics and Automation 15 (1) (1999) 190–194.

[3] H.C. Zhang, T.C. Kuo, A graph-based disassembly sequence planning for EOL product recycling, in: Proceedings of the 21st IEEE/CPMT International

Electronics Manufacturing Technology Symposium, Austin, USA, 1997, pp. 140–151.

[4] J.R. Li, L.P. Khoo, S.B. Tor, A novel representation scheme for disassembly sequence planning, International Journal of Advanced Manufacturing Technology 20 (2002a) 621–630.

[5] J.R. Li, S.B. Tor, L.P. Khoo, A hybrid disassembly sequence planning approach for maintenance, ASME Transactions, Journal of Computing and Information Science in Engineering (JCISE) 2 (2002b) 28–37.

[6] EON studio Reference handbook. http:\\ www. eonreality.com.

The study on e-commerce standardization in China

He Shaohua

Female, born in August 1951, professor of Guangdong Electronic Commerce Market Application Technology Key Laboratory, professor and doctoral tutor of SIM of Wuhan University.

Email: wdshua@gmail.com

Yang Fan

Male, born in April 1982, Ph.D. candidate in Sim of Wuhan University, research area is e-commerce and knowledge management

Email: cavalieryf@163.com

Abstract

Standardization is the only way for the development of e-commerce, but whether domestic or international e-commerce standards have not satisfactory. This article analyzes the e-commerce standardization problems and obstacles from 3 parts: position and classification, formulation, monitoring and implementation. This article accounts that the source of the problems is the contradiction between e-commerce technical standards and service standards, thus leading to the centralized management of e-commerce standard is not clear. In the formulation of standards, this article proposes to solve the problem of slow e-commerce standardization progress in China by the ways from top to bottom and bottom to the top. Finally, this article puts forward three main points of monitoring and implementation on e-commerce standardization.

At present, e-commerce is blooming dramatically in the world; it has become an important component of the world economic structure. Researching, designing and implanting international and national standards series of e-commerce, strengthening the regular management of e-commerce market behavior, ensuring the fair and just trade activities of the network, have become premier issues of pushing forward the development of network economy. Such topic will be discussed as below.

1. Introduction

There is a point of view that the world's first "internet business standard" or "e-commerce standard" is "The Standard for Internet Commerce, Version 1.0-1999", which was published in the December 14, 1999 in San Francisco. Leading by Ziff-Davis Magazine, which organized 301 famous Internet and IT enterprises, related news, non-governmental organizations and academics, and through a half-year and two rounds of voting to 7 parts, 47 clauses, the standard was finally confirmed.

Another point of view is that e-commerce standard can be traced back to 1977 U.S. data encryption standard DES, and subsequently a series of cryptographic techniques related Federal Information Processing Standards (FIPS) developed by the U.S. National Institute of Standards and Technology (NTST).

These two kinds of viewpoint actually reflect two different understanding on e-commerce: the core of e-commerce standards are technical standards or service standards. Although there are no contradiction between service standards and technical standards, but the uncertainty and incoordination of them lead to a slow development status quo of e-commerce standard system construction.

Position and classification of E-commerce standard system

The first problem is the unclearness of position and classification of E-commerce standard system.

International organizations for standardization of e-commerce are so many, such as ISO / IEC, ITU, UN / ECE / CEFACT, ISOC, Rosettanet, OASIS, W3C/IETF, IFIP, OFX, FSTC and BIPS, it can be said that all the standardization organizations have involved in construction of e-commerce standardization.

In ISO/IEC, technical committees related with e-commerce standardization are: ① ISO/IEC/JTC1--Information Technology Standardization Committee; ② ISO/TC154-- Processes, data elements and documents in commerce, industry and administration, which is mainly responsible for international standardization and registration of business, and administration processes and supporting data used for information interchange between and within individual organizations and support for standardization activities in the field of industrial data. By now, TC154 has developed 19 international standards, mainly involving EDI and grammar rules in the electronic commerce, trade data elements, the basic semantics of data dictionaries and paper format; ③ ISO/TC46-- information and documentation; ④ ISO/TC68--

banking, securities and other financial services; ⑤ ISO/TC184--Automation systems and integration; ⑥ IEC/TC3—Information Structures, Documentation and Graphical Symbols; ⑦ IEC/TC56—Dependability; ⑧ IEC/TC93--Design Automation. ITU subordinates: SG3--Tariff and accounting principles, SG4--Telecommunication Management, SG7--data communications network and open system communication, SG13--Next Generation Networks, SG16--Multimedia terminals, systems and applications.

Take Rosettanet for example, which is an open, non-profit consortium of industry leaders, it is focusing on collaborative development and rapid dispersion of open e-commerce standards to integrate business processes of world-wide high-tech trade network. At present, there are more than 400 enterprises of information technology, electronic components and semiconductor manufacturing participate in the Rosettanet's standards development, strategy and action plan. Rosettanet emphasizes the standardization of framework and processes, on other business units, data elements and other specific business model, it basically adopt existing open e-commerce international standards, which are exclusive in such field. The core of Rosettanet standards system is the Partner Interface Processes (PIPs), it provides business process standard between partners of trade. On the PIPs, Rosettanet also developed a data dictionary regulation and implementation framework regulation.

From the above examples, we can find at least three problems:

Firstly, despite many concerns are on e-commerce standardization, most attention has focused on the technical standards of e-commerce and ignored the other relevant aspects. In addition, many e-commerce standards is developed under the sub-Committee of technical standards of Standardization Committee or lower organization, which leads to the absence of co-ordination and systematicness. Take ISO for instance, although there are several standardization committees are engaged in the formulation of standards related to e-commerce, but has not been a professional committee to research and develop e-commerce services standards and e-commerce standard system. June 1997 set up the ISO/IEC/JTC1 "e-commerce business working group" (BT-EC), set up on June, 1997, also only work on the area of user interface, basic functions, data and the object code definition.

Even in the field of e-commerce standard, because of organizations scattering, different concerns, distribution of benefits and other issues, makes the incompatibility among technical standards and overlapping, duplication and disordered competition among organizations.

Secondly, the research on entire international and national e-commerce service standards system need to strengthen. In many cases, research of e-commerce service standards is completed within the industrial organization; it is difficult to be a international standards accepted by ISO or other international standardization organizations. Due to reasons of different politics, economies, culture and lifestyle in different countries and regions, makes e-commerce service standard system construction more complex and more difficult. But the establishment of international business system, the improvement of appropriate laws and regulations, and the creation of certain technical frameworks, has provided a good basis for the establishment of e-commerce service system. We should seize the opportunity and strive to create a standard system for e-commerce services.

Thirdly, e-commerce services standard system is out of step with its relevant technical standards system. The absence of e-commerce services standard system can easily lead to the disorder of e-commerce technology standard system; correct application and dispersion of a series of technical standards need guidance and feedback, which rooted in the e-commerce services. For instance, there are dozens of e-commerce payment agreements in using, and the result in dozens of payment means and channels in the actual shopping process, which increase user costs and system maintenance costs. If we can regulate the payment service standards in advance, even in different channels of different platforms to purchase products, seamless conversion between platforms and agreements the will be much easier.

Comparatively speaking, the problems of China's position and classification of e-commerce service standards system will be more sophisticated. As China's institutional reasons, the standardization work is always an important work of relevant administrative departments but not professional institutions. Whether national standards or industrial standards, they are all presided, planned, researched, developed, disseminated, implanted and supervised the government's functional departments. The same phenomena exist in the area of development of e-commerce standards.

Government-led operation mode, on one hand is to some extent in order to ensure the authority of content and scope of the standards, provide a powerful guarantee for the implementation; on the other hand, this government-led model of research may result in unclearness of position and centralized location more easily.

First, the current China's e-commerce standard work is still in the stage of following international level, substantive formulation and implementation still progress slowly. Many ills of foreign e-commerce standard development also exist in China. At the same time, tracking research is lack of timeliness and forward-looking. Introduction advanced foreign technology to China, further develop to the standard, the standards may be out of date. Moreover, there are a number of international standards coexist in the world; the choice of standard is also a problem we should consider.

Second, the establishment of China's e-commerce standard system still targeted management issues. Government-led mode avoid duplicate construction and some other problems, but bring disadvantages to the development of e-commerce standard system, one of which is the add a administrative location on the academic location of research of e-commerce standard system. Standard research unit's e-commerce standard positioning more or less. The unit attribute to which department, which experts take part in the study, which content should be studied, which organization participate in the formulation, such issues will be the impacted by the administration. Furthermore, the understanding and research focus on the e-commerce standardization are different because of different agencies, which will lead to a situation that replacing entire with some parts of it. Especially in current days, international e-commerce standard positioning are not clear yet, this will be more complicated in China.

The appearance of position and classification problems of e-commerce standard system is related to the characteristics of their own. Broadly speaking, all the relevant IT technical standards can become part of e-commerce standard system, and with the other regulations related with e-commerce activities, e-commerce standard system will be inevitably cumbersome. However, this should not become the barriers of we develop e-commerce standard system. For example, should we establish a new standard committee to develop e-commerce standard system? Generally speaking, e-commerce standards belong to information service standards. Can we stand on a higher degree of information service standards to solve the problems of e-commerce standards position and classification? Such issues are worthy of our study.

If we can solve the problem of position and classification, we can more easily integrate several standardization research teams. For example, the integration of ISO and IEC, SET, Rosettanet, China's domestic IGRS (Intelligent Grouping and Resource Sharing), and other organizations are the best examples of integrating their respective advantages of standardization.

2. The formulation of e-commerce standard system

Standards in the development of e-commerce have been considerable attention in the world. In 1998, the U.S. government released the outline of the U.S. e-commerce, clearly proposed to establish some common standards, "to ensure the equal rights between internet-consumers and store-consumers." South Korea has also developed e-commerce standard scheme in 2000.

China's State Council issued "General Office of the State Council: Several Suggestions on Quickening the Development of E-commerce" in 2005, it declared that: "(China will) establish and improve national e-commerce standard system, raise awareness of standardization, fully mobilize the activity of all its bearings, step up improving the national e-commerce standard system; encourage the enterprises as the mainstay, joint with universities and research institutions, study the formulation of e-commerce key technical standards and norms, participate in the development of formulation and amendments of international standards, actively promote e-commerce standardization process.

2006 China CPC Central Committee General Office and the Office of the State Council issued the "2006--2020 national information technology development strategy," said: "strengthen the government guide, relying on major information technology application projects, to companies and trade associations as the main body, speed up industrial technology standards system.

Improve the technology systems of information application and technology norms and standards of industries and products; promote interoperability network, open operating systems and information sharing. Accelerate the basic information standards of population, legal entities and geographical space, coding and etc. Strengthen protection of intellectual property rights. Strengthen international cooperation, actively participated in international standard-setting. "

January 2007, the overall planning and coordinating organization of Chinese e-commerce standards work - the total group of national e-commerce standardization was set up in Beijing. Its purpose is to give better supports to enterprises and experts in related fields to play a role in the establishment and improvement of e-commerce standard system, systematic coordinate and scientific develop China's national e-commerce standards. Its main tasks are: to research on China's overall demand of standards for the development of e-commerce, to put forward technology policy of national e-commerce standardization to the State Administration of Standardization Committee and the relevant administrative departments; to formulate national e-commerce standard system, to put forward general guidelines and annual work plans of the national standardize work; to put forward formulation and amendments programs of national e-commerce standards, to coordinate and organize national professional standardizations technical committees to draft national e-commerce standards; to supervise and inspect the research progress of national e-commerce standards, to coordinate important technology problems of e-commerce standards; to promote large-scale and backbone enterprises to actively participate in e-commerce standardization work, to push advocacy, implementation and promotion of national standards forward; to help Standardization Administration of China to get strategic decision consultation of e-commerce standardization policies and standardization project.

However, we can see that the intensity is not reflected in actual effect in China's e-commerce standard system. There are three issues:

First, the standard-setting process issues.

China's standards system construction is government-led. Nowadays, there are two important ways to formulate e-commerce standards in China. One is the introduction of international standards, demonstrated by the experts and then become a national standard directly.

However, whether international standards and domestic e-commerce status quo can link up, whether China's domestic enterprises organizations can endorse the international standards; these are difficult "localization" problems. Once the lack of a standard application of the soil, it has no vitality.

The other way is to use the outstanding achievements and successful experience of pilot enterprises in various industries, to sum up the experience, to rose to a high degree of standard solutions, and then these present enterprise standards would expand to trade standards, countries would rise these standards to the national standards by legal procedures at last. This process for formulating trade standards and national standards which is common in foreign countries is relatively more difficult in China. First, the ultimate realization of the process takes time. Chinese e-commerce standards construction model is still in the stage of constructing their own, which is the initial stage. Second, the standardization awareness of Chinese enterprises and the actual support from government are limited. It makes many companies have very good abilities to research and product, but they can not afford their products and services upgraded to a standard application.

Second, the development of e-commerce standard system is not scientific and prospective enough. E-commerce standard system is an organic whole fused by many subsystems, the infiltration among each subsystems are logic, scientificity, unity, diversification and extensibility. E-commerce standard system should be synchronized with the rapid development of e-commerce.

China's e-commerce standard system needs improvement, in terms of both system and content. For example, a XML-based e-commerce standards technology system is not constructed; business process standard and online payment standard are very few; the compatibility and integrity of electronic documents format standard are not enough; there are serious cross-repeat in some standards caused by the lack of overall plans and the serious cross-repeat among several standards committees.

Third, e-commerce standard has very low adaptability in market. According to Chinese National Standards Institute study shows that about half of the above standards are almost not in use, in particular the electronic document format. This was mainly due to the blindness in the formulation of international standards, without localization.

The crux of these problems above is the uncertainty of requirement and demand guide of standard-setting. "Demand-oriented, user-led" is the successful experience of standardization in advanced Western countries. In recent years, China has also taken steps to change this situation, as in China's "the development of e-commerce standardization '11th Five-Year' plan" that pointed in, should "adhere to the enterprises as the mainstay, market-oriented principles and enhance the standards applicable in market ", "relying on domestic e-commerce enterprises, give full play to the association, intermediary organizations and the role of research institutions to formulate a group of independent innovation of China's e-commerce standards. " These requires into practical work further.

There are several points should pay attention to in the operation of national e-commerce standard system:

First, top-bottom unified planning for e-commerce standard system. A national-level standards system should include national standards, trade standards, local standards, enterprise standards and other several standards. National e-commerce standard is followed by the e-commerce standards at all levels under/below it, which is the common ground and basic of such standards. The standard system following the national standard system should reflect their applicability, they are inseparable community, and therefore, all standards levels are the basis of national e-commerce standards. All localities and organizations should develop self-adaptive standards system under the guidance of national e-commerce standard system; at the same time, enterprises and units can also apply for trade standards or even national standards for their excellent products by effective channels. In concerns about e-commerce standard system construction, such a top-bottom pyramid-shaped basis should be considered in the first place. The application-oriented, market-oriented, unified planning, so that we can transform the standards into productive forces, we can scientific develop.

Second, bottom-top construction for e-commerce standard system. Enterprise standards system is the basis of national standards system, all standards levels must implement through enterprise. So, the only way for formulating and developing e-commerce standards is taking enterprises as the mainstay and basing on the enterprise scientific research efforts, it is a bottom-top way. The administrative departments and governments should provide funding, technology, channels, and actively help enterprises to transform outstanding achievements into standards, because standardization is an effective way to promote innovation.

In addition, the e-commerce standard system framework research needs multi-side participants.

Standards system covers wide and complicated in China's National E-commerce standard system Framework (Fig.1: China's National E-commerce standard system Framework),but the system has not been fully established, even in the basis part of national e-commerce standard system framework. At the same time, it should include e-commerce standards subsystem framework. Therefore, all localities can also participate in localized

construction of e-commerce standard system includes national e-commerce standard system through the application.

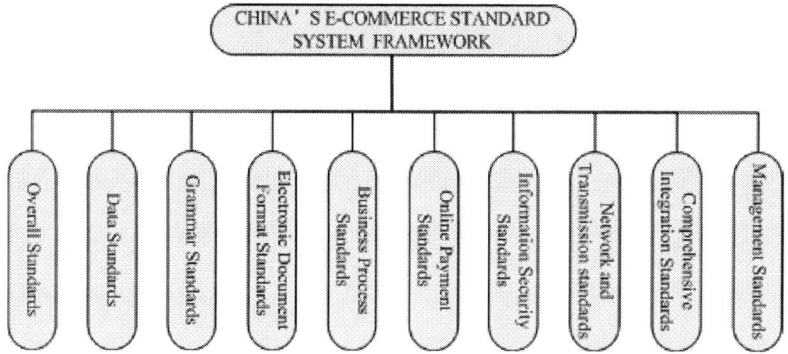

Figure 1. China's e-commerce standard system framework

E.g.: Concentrate on the key step of the e-commerce, the government of Guangdong province , China , researches to build the e-commerce standard system to encouraging the development of the province. They encourage the agency corporate with the college and commercialize R & D achievements to get the e-

commerce standard system (Such as Coding items, electronic documents, information exchange, business processes etc.) and then join the R&D of the whole China.

Fig.2 , Fig.3 is the research system and system standard of the e-commerce standards platform:

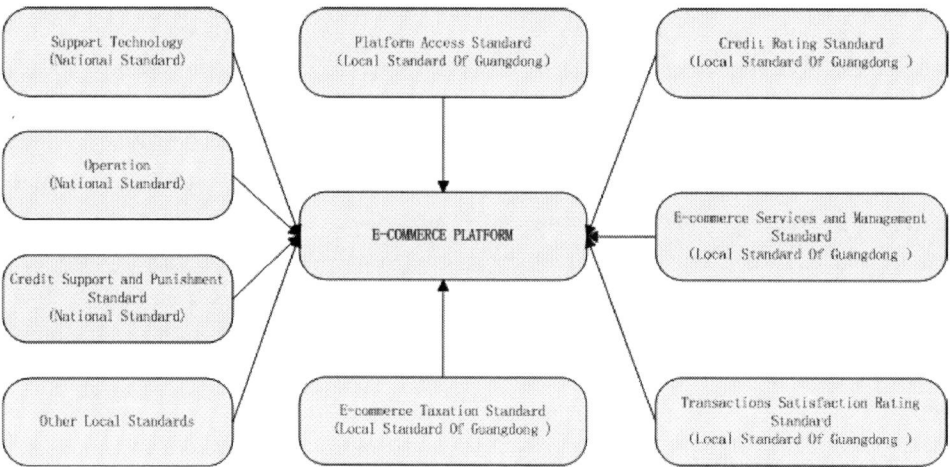

Figure 2. Guangdong E-Commerce Standard Platform Research System

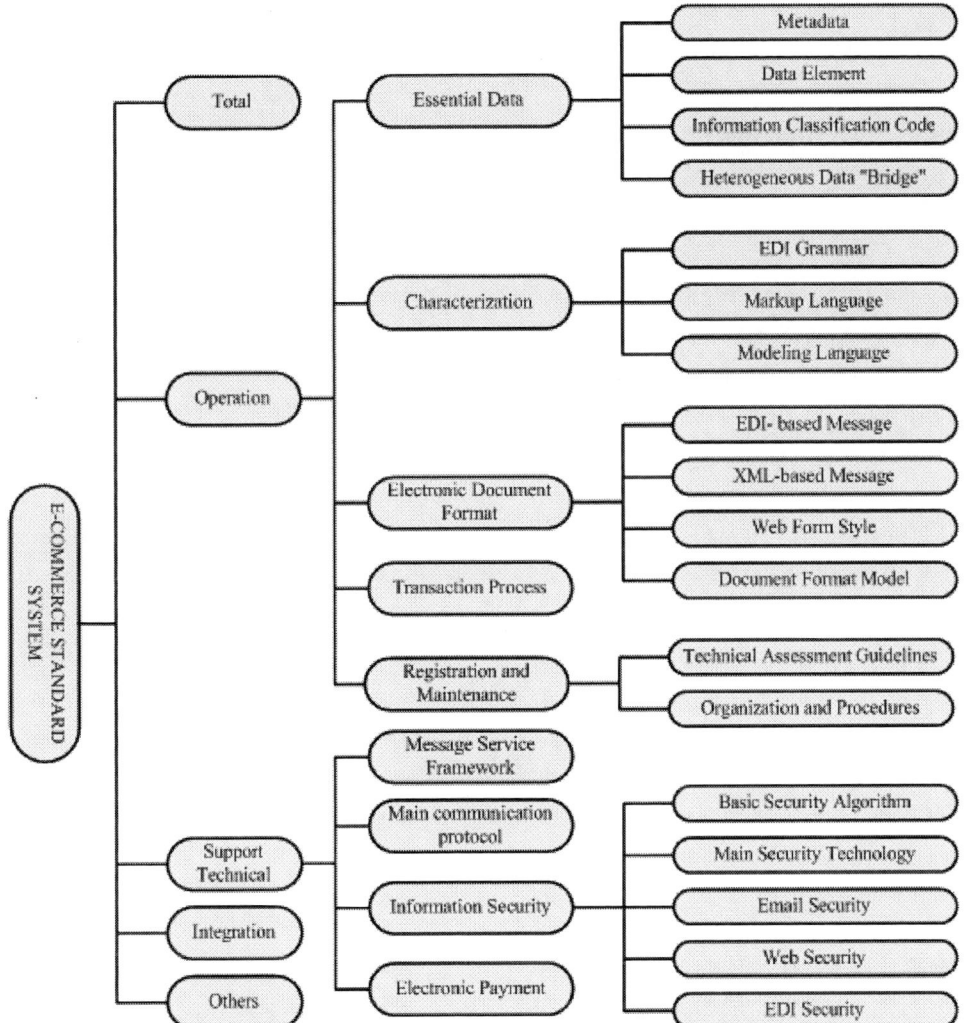

Figure 3. Guangdong E-commerce standard system

During making the e-commerce standard system of Guangdong China, it is asked to use the country standard as much as possible to keep the same step with the development of the China e-commerce standard system. The province governments should make the local standards if there are not concerned country standards and industry standards. The province government of Guangdong could make the local e-commerce standards in some area because of their advantage in the development of the e-commerce. The local government should share the information with the concerned department of the central government, which can help finding the need of the agency and the society. And then they

could be involved in the new e-commerce standard system.

E.g.: To the e-commerce credit standards, it needs industry associations, governments, enterprises to finding the essence of the e-commerce credit system to helping making the standard to build a great e-commerce credit environment, before the building of society credit system is not finished. CASME (China Association of Small and Medium Enterprises), CECA (China E-Commerce Association)and CAI(China Association of Information) do much job (such as "Credit Star Plan" to develop the e-commerce credit management system of small and medium enterprises)to standardize the e-

599

commerce ,by leading of China's Credit Standardization Technical Committee.

3. Issues on implementation and supervision of e-commerce standard

Firstly, unified planning of implementation should combine with local conditions.

The healthy development and cycle of e-commerce depends on the harmonized standard. Thus, standard with broad applicability, strong authority and finer versatility should be chosen in the implementation of e-commerce standard. However, local condition should be considered in the operation, i.e. different standards should be implemented in accordance with actual situation.

For example, e-commerce service and management standards should be developed accordingly in the more economically developed areas of China; index of logistics standards, response time, etc. of Guangdong, Shanghai, Beijing and other major cities should be higher than the national standard.

Secondly, combine the key area of the e-commerce standards & the key area of the e-commerce industry.

The China's 11th Five-Year E-commerce Development Plan has pointed out several key development areas, including Electronic authentication system, online payment system, modern logistics system, e-commerce credit system, electronic documents, information exchange, business processes, etc., which basically focused on the supportive tech and professional operation. In the process of implicating and practicing standards, resources, caption & intelligent support should be effectively concentrated on these areas.

E.g.: online payment has 10 year history in China, has made up a complete industry system of banks, Third-party payment platform &

enterprises, composing a perfect base of standardization.

E.g.:, modern logistics system, the base of E-commerce, has been one of the fastest developing areas in China. With lots of international enterprises joining this competition, relevant standards has been in urgent need to standardize the competition and optimize the industry structure.

Thirdly, combine the market competition & government acts of standards implication.

In the process of standards implication, the statement which government acts playing a main role in China must be turned into a better one which puts the market competition to the most important position instead of government acts. In the old times, when implicating E-commerce standards, China was used to making government acts, which had promoting the development of E-commerce in some way, while ignoring the effect of marketing. In this highly market-oriented E-commerce area, market mechanism should earn fully respect and utilization.

References:

[1]. General Office of the State Council: Several Suggestions on Quickening the Development of E-commerce, http://www.gov.cn/zwgk/2005-08/15/content_21825.htm

[2]. 2006--2020 National Information Technology Development Strategy,

https://news.xinhuanet.com/newscenter/2006-05/08/content_4522878.htm

[3]. The E-commerce Standardization of "11th Five-Year Plan", Yearbook of China's E-commerce (2004-2008), pp207

[4].The E-commerce of "11th Five-Year Plan", http://www.sdpc.gov.cn/zcfb/zcfbtz/2007tongzhi/W02 0070620595393012331.pdf

The application of the analytical hierarchy process in Performance evaluation system in commercial bank's IT department

Wang Jiang-tao[1, 2] Zhou Hong[1]

1 School of Economics & Management, Beihang University
Beijing 100083, P.R China
2Bank of China Xinjiang branch, Xinjiang, 830002, P.R China

Abstract

The balanced scorecard is a series of financial evaluation indices and non-financial evaluation system; it provides prompt, overall, and objective support for managements to understand the operation situation. The balanced scorecard serves the informationalization of the commercial bank. IT balanced scored starts from the business strategy, focusing on the information system construction, contributing to the enterprise value. The right evaluation of the informationalization level relies on the scientific, reasonable, and feasible evaluation indices system. The construction of the comprehensive evaluation system for the commercial bank's IT department, using AHP, will reflect the situation of the information system, single out the major factors, and ensure the credibility of the evaluation. And the evaluation indices might be operable.

1. Introduction

The Analytical Hierarchy Process (AHP) is a decision-making methodology that breaks decision-related elements down into such layers as goals, criteria, and plans, based on which the qualitative and quantitative analysis is carried out. Originally proposed by Professor T. L. Satty, an American operations researcher at University of Pittsburgh, during the 1970's, this methodology is an effective way of converting semi-qualitative and semi-quantitative problems into quantitative calculations. This process would first of all hierarchize complex decision-making systems, then establish the modeling judgment matrix by comparing layer-by-layer the levels of importance of various correlated factors, and then provide the decision-making process with the basis through a set of quantitative calculation methods [1].

As a set of corporate performance evaluation system originally proposed by Professor Robert S. Kaplan of Harvard University and Mr. David P. Norton, President of Palladium Consulting Firm, in 1992, Balanced Scorecard (BSC) has sparked widespread interest from the management and accounting research sector. BSC represents an integration of a series of accounting performance measurement parameters and non-accounting performance measurement parameters; it is more of a kind of management process that focuses primarily on the realization of organizational and strategic goals of businesses [2]. It covers four areas: accounting, customers, internal business, as well as learning and growth [3].

Balanced Scorecard creates a very good underlying framework for strategic management activities of businesses, and helps them to successfully execute their business strategies [4].

The IT Balanced Scorecard (IT BSC) is an IT performance evaluation system evolved on the basis of the Balanced Scorecard. The underlying framework of the IT Balanced Scorecard is intended to carry out IT performance evaluations in four areas, i.e. IT value contribution, IT user satisfaction, IT internal processes, and IT learning and reforms; the purpose is to realize the convergence between the strategic goal of informationalization efforts and the strategic goal of corporate business. The IT Balanced Scorecard can be seen as the tool to execute the corporate informationalization strategy[4] It can bring together the vision, the mission, and the development strategy of informationalization with the performance evaluation of informationalization. It transforms the mission and strategy of informationalization into specific goals and evaluation indices to provide organic combination between strategy and performance, thereby helping commercial banks to successfully execute their business strategies.

Performance management in terms of execution of strategies of commercial banks is, in a sense, also a kind of decision making. In many cases, the performance measurement parameters are very hard to be measured directly and objectively, therefore some subjective judgment is needed. As far as these characteristics are concerned, performance management is also a multi-objective decision making problem that has a quite complicated structure, that has many decision criteria, and

978-0-7695-3342-1/08 $25.00 © 2008 IEEE
DOI 10.1109/PEITS.2008.114

that is not easy to be quantified. Therefore, it is quite appropriate to use the Analytical Hierarchy Process, on the basis of the IT Balanced Scorecard as the underlying framework, to determine the weights of various factors and address the evaluation issue in performance management at commercial banks' information departments, and it is also advantageous over other techniques.

2. The performance evaluation hierarchical structure of the information department in the commercial bank

This study has established the performance evaluation hierarchical structure for the information department in the commercial bank based on the characteristics of informationalization construction in the commercial bank, and has built a comprehensive evaluation model for the information department in the commercial bank by using the Analytical Hierarchy Process (AHP) and specific cases at the author's school. The premise of correctly evaluating the performance of the information department in the commercial bank lies in the establishment of a scientific, reasonable, and feasible evaluation index system. However, given the fact that each commercial bank has its own particular situations, there is not a

unified model for the establishment of such an evaluation index system, and the evaluation index may vary depending on the differences in the characteristics of products, time of evaluation, and purpose of evaluation, etc, of the commercial bank. In general, the evaluation index system should be able to effectively reflect the basic conditions of informationalization in the commercial bank and focus on the main factors to ensure the completeness and credibility of the evaluation work. While at the same time, the evaluation indices have to be easy to operate.

（1）This study has, based on a summarization of the specific performance situations in the information department in the commercial bank, carried out evaluations in such criteria areas as IT value contribution (B1), IT user satisfaction (B2), IT internal processes (B3), and IT learning and reforms (B4). Specific evaluation indices are proposed for each area, leading to the establishment of an valuation index system with the hierarchical structure.

This hierarchical analytic model primarily includes four layers. The top is the objective layer A, i.e. performance evaluation for the information department in the commercial bank; the middle is the criteria layer B, which is followed beneath by the actions layer C. The detailed descriptions of the framework are listed below, see Figure 1.

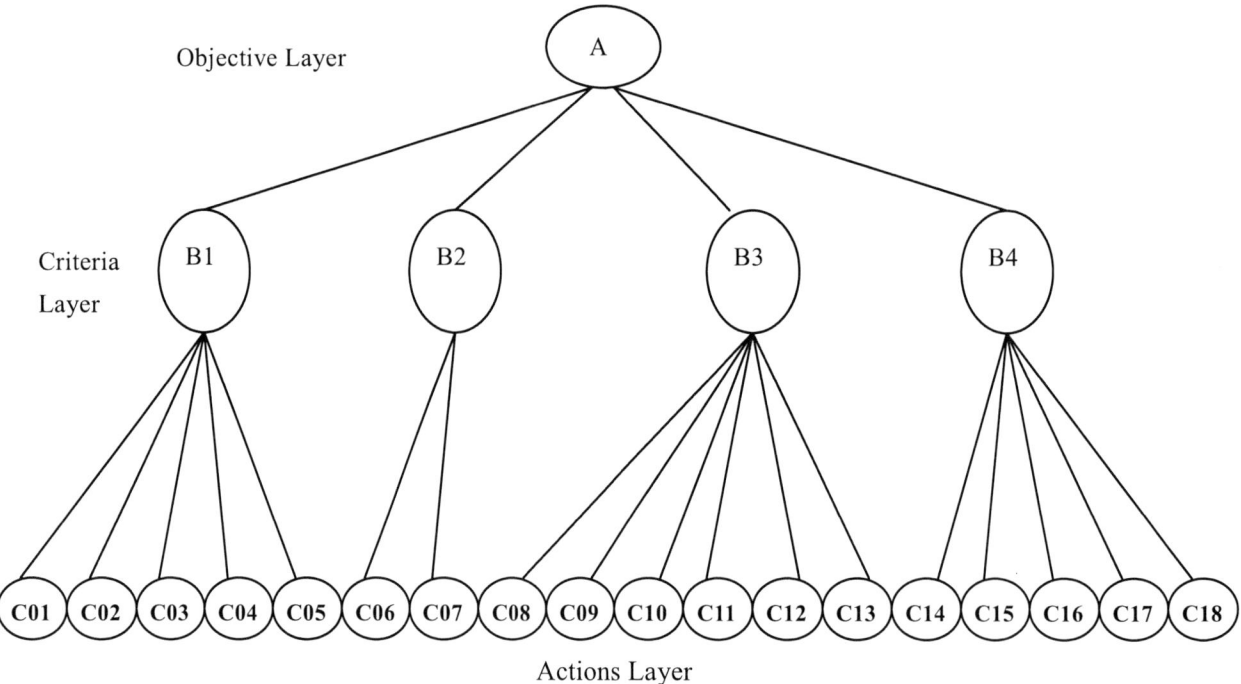

Figure 1 Performance Hierarchical Analytic Model for Information Department in Commercial Bank

(2) The evaluation index system for the performance level of the information department in the commercial

bank represents the performance of the information department in the commercial bank, and includes a total of

4 criteria layers, i.e. A = { B1,B2,B3,B4} ={ IT value contribution, IT user satisfaction, IT internal processes, IT learning and reforms}.

1. IT value contribution (B1): this parameter represents the indices of the information department in the commercial bank in the financial area. It is not isolated, but rather needs to be not only tightly connected to the informationalization development strategy of the commercial bank to centrally reflect the informationalization strategic goals and how they are fulfilled in the commercial bank, but also serve as the final objective and measurement standard for the other three non-financial areas to form a vertical cause-and-effect chain between informationalization financial and non-financial performance motives in the commercial bank that should cover all four areas of the balanced scorecard.

IT value contribution includes a total of 5 sub-indices, i.e. B1 = {C01,C02,C03,C04,C05} ={net profit of the entire bank, controllable expenses of the department, controllable expenses-business reception expenses of the department, controllable expenses-travel expenses of the department, and electronic equipment maintenance and consumable supplies expenses}, where, C01 (net profit of the entire bank) is associated with the operational indicators of current-year net profit of the entire bank of the commercial bank, the net profit of the commercial bank refers to the remaining amount of the total income earned by the bank during its business operation process after deduction of various costs and taxes, and only such amount is the net profit, i.e. net earning, of the commercial bank. The most essential key of strengthening banking profit management lies in the ability to expand the scale of profitable assets in the bank, to improve the utilization rate of capital and the rate of return of profitable assets, to reduce costs and asset-risk-induced losses, to increase productivity, thereby achieving the goal of increasing profits. It is the key performance index that all departments in the commercial bank must undertake, and also demonstrates that the strategic goal of informationalization construction in the commercial bank is to promote business growth.

2. IT user satisfaction (B2): this parameter primarily represents the customer area, meaning that the commercial bank should evaluate its operational results from the customer's stand point. In the fierce market competition, the survival or death of the commercial bank is dependent on the currency votes by customers. The commercial bank can continue to survive and grow only if and when, during

informationalization construction, it can really understand its external customers, continue to satisfy and create demands among its internal and external customers, make its financial information products to adapt to market requirements, and realize transformation of product value.

Combining the customer area into the Balanced Scorecard performance management system can help the commercial bank to initiate informationalization construction on the basis of the needs of major customers and develop accurate marketing strategies and tactics, thereby generating satisfactory financial performance. IT user satisfaction includes a total of 2 sub-indices, i.e. B2 = { C06,C07 } ={external customer satisfaction, internal customer satisfaction}.

3. IT internal processes (B3): this parameter primarily represents the condition of informationalization governance in the commercial bank by mainly examining 6 sub-indices, i.e. B3 = { C08 , C09,C10,C11,C12,C13} ={information technology development planning, software management, hardware management, system management, telecom services, network operation}.

4. IT learning and reforms (B4): this parameter primarily represents such information in the commercial bank as informationalization construction, arrangement of training tasks planned by the human resources department, learning and innovation of employees, and career planning and development, etc, by mainly examining 5 sub-indices, i.e. B4 = {C14,C15,C16,C17,C18} ={operational guidance and team building in the entire bank, performance coaching and communication, external job training, internal job training, career planning and innovation}.

(3) The architecture for informationalize performance evaluation in the commercial bank.

Index Dimensions	Performance Indices	Index Dimensions	Performance Indices
IT Value Contribution (B1)	Net profit of the entire bank (C01)	IT Internal Processes (B3)	Information technology development planning

			(C08)
	Controllable expenses of the department (C02)		Software management (C09)
	Controllable expenses of the department (C03)		Hardware management (C10)
	Controllable expenses of the department (C04)		System management (C11)
			Telecom services (C12)
	Allocated expenses of the department (C05)		Network operations (C13)
IT User Satisfaction (B2)	External customer satisfaction (C06)	IT Learning and Reforms (B4)	Operational guidance and team building in the entire bank (C14)
			Performance coaching and communication (C15)
			External job training (C16)
	Internal customer satisfaction (C07)		Internal job training (C17)
			Career planning and innovation (C18)

3. Principle of Analytical Hierarchy Process[5]

Originally proposed by American operations researcher T. L. Satty in the 1970's, the Analytical Hierarchy Process is a kind of multi-objective decision analysis methodology that combines both qualitative and quantitative analyses. It absorbs and utilizes the characteristics of behavioral science to quantify the experience and judgment of the decision maker. This method is particularly suitable in the circumstances where the structure of the target (factor) is complex and there is absence of necessary data. As a commonly used systematic analysis method in the systems science, it has become one of the teaching tools for systematic analysis.

(1) Determining the weights of various evaluation factors in the index system.

The Analytical Hierarchy Process offers quantification of decision maker's evaluation, decision-making, and thinking process against complex systems. The basic idea is to carry out two-by-two comparisons, judgment, and calculations on various elements at the same layer on the basis of elements of the upper layer as the criteria so as to clearly identify the levels of importance of various elements in the index system. In order to obtain authoritative judgment values, the author has distributed "Expert Judgment" questionnaires to related managers of operation sectors and technology management departments. The final assessment values are acquired by using judgment measures to convert summarized expert opinions.

(2)Establishing the judgment matrix.

The judgment matrix is intended to carry out two-by-two comparisons of lower layer elements on the basis of an element at the upper layer as the criteria in order to determine the values of elements in the matrix; its judgment measures are listed in the table below (with A being elements):

Table 1 Table of Judgment Measures

Scale a_{ij}	Definition
1	Factor i is as important as factor j
3	Factor i is slightly more important than factor j
5	Factor i is more important than factor j
7	Factor i is much more important than factor j
9	Factor i is absolutely more important than factor j
2, 4, 6, 8,	The ratio of the importance levels of factor i and factor j lies in between any two adjacent levels mentioned above
Inverse of 1, $\dfrac{1}{2}, \dfrac{1}{3}, \dfrac{1}{4}, \dfrac{1}{5}, \dfrac{1}{6}, \dfrac{1}{7}, \dfrac{1}{8}, \dfrac{1}{9}$	The judgment value obtained by comparing factor j and factor i is the reciprocal inverse of a_{ij}, $a_{ji} = \dfrac{1}{a_{ij}} \qquad a_{ii} = 1$

Using this method against the second layer in the model, doing two-by-two comparisons among IT value contribution, IT user satisfaction, IT internal processes, and IT learning and reforms fives the following judgment matrix A:

$$A = \begin{bmatrix} B01/B01 & B01/B02 & B01/B03 & B01/B04 \\ B02/B01 & B02/B02 & B02/B03 & B02/B04 \\ B03/B01 & B03/B02 & B03/B03 & B03/B04 \\ B04/B01 & B04/B02 & B04/B03 & B04/B04 \end{bmatrix}$$

$$= \begin{bmatrix} 1 & 2 & 2 & 2 \\ 1/2 & 1 & 2 & 3 \\ 1/2 & 1/2 & 1 & 3 \\ 1/2 & 1/3 & 1/3 & 1 \end{bmatrix}$$

(3) Using the Square Root Technique to obtain the solutions of the maximum characteristic root and the corresponding eigenvector of the judgment matrix [6].

1. Calculating the geometric mean of all elements in each row of the judgment matrix $w_i = \sqrt[n]{\prod_{j=1}^{n} a_{ij}}$, $i = 1, 2 \ldots, n$;

thus giving $\mathbf{w} = (w_1, w_2 \ldots, w_n)^T$ where w_i indicates an element of the eigenvector;

2. Normalizing it $\overline{w}_i = w_i \Big/ \sum_{j=1}^{n} w_j$, $i = 1, 2 \ldots, n$; thus giving $\overline{\mathbf{w}} = (\overline{w}_1, \overline{w}_2 \ldots, \overline{w}_n)^T$ as the approximate value of the eigenvector, which is also the relative weight of a single factor;

3. Calculating the maximum eigenvalue λ_{Max} of the judgment matrix, since $\mathbf{A}\overline{\mathbf{w}} = n\overline{\mathbf{w}}$; thus giving $\lambda_{Max} = \frac{1}{n} \sum_{i=1}^{n} \frac{(\mathbf{A}\overline{\mathbf{w}})_i}{\overline{w}_i}$; where A indicates any one judgment matrix, therefore it is possible to calculate the eigenvector $\overline{\mathbf{w}}^{(1)}$ of the judgment matrix P; eigenvalue λ_{Max} ;

4. Conducting consistency check. The test coefficient $CR = \dfrac{CI}{RI}$, where $CI = \dfrac{\lambda_{Max} - n}{n - 1}$ is the consistency index, RI is the average consistency index, whose values are listed in the table below; when $CR \leq 0.1$, it can be deemed that the judgment matrix offers satisfactory consistency, otherwise, it would be necessary to re-adjust the judgment matrix; doing

top-to-bottom calculations layer by layer in such a way, until the weights of all elements in the bottom layer are identified.

Table 2 Judgment Matrix Value Table

Dimensionality of $\overline{A}$	1	2	3	4	5	6	7	8
					9			
$R \cdot I$	0.00	0.00	0.58	0.96	1.12	1.24	1.32	1.41	1.45

By calculating the CR of the criteria layer in B, CR=0.08<0.1, it is then possible to calculate the weight of the next layer. Through calculating the weights of the four factors in that layer, we can identify the individual weight of each of the four factors in the IT Balanced Scorecard to be: the index for IT value contribution is 38.1%, the index for IT user satisfaction is 29.8%, the index for IT internal processes is 21.1%, and the index for IT learning and reforms is 11%.

(4) Constructing the judgment matrix for the second layer.

1. The judgment matrix of actions layer C relative to the B1 elements in the criteria layer B is Q:

Thus, the eigenvector $\mathbf{w}_1^{(2)}$ of the judgment matrix Q is calculated; i.e. $\overline{w}_1^{(2)} = (w_1^{(2)}, 0, 0, 0, 0, 0, 0, 0, 0, 0, 0, 0, 0, 0, 0, 0, 0, 0)^T$, where the eigenvalue is 5.01, and the consistency test coefficient RI=0.002<0.1.

From the judgment matrix Q, the five parameters of IT value contribution are calculated: net profit of the entire bank (C01), controllable expenses of the department (C02), controllable expenses of the department (C03), controllable expenses of the department (C04), controllable expenses of the department (C05), whose weights are39.4%, 23.4%, 12.4%, 12.4% and 12.4%, respectively.

2. Constructing the judgment matrix for the second layer, where the judgment matrix of actions layer C relative to the B2 elements in criteria layer B is R:

Thus, the eigenvector $\mathbf{w}_2^{(2)}$ of the judgment matrix R is calculated; i.e. $\overline{w}_2^{(2)} = (0, 0, 0, 0, 0, w_2^{(2)}, 0, 0, 0, 0, 0, 0, 0, 0, 0, 0, 0, 0)^T$, where the eigenvalue is 2, and consistency check is not needed for the first and second orders.

From the judgment matrix R, the two indices in IT user satisfaction are calculated, i.e. external customer satisfaction (C06) and internal customer satisfaction (C07) are 25% and 75%, respectively.

3. Constructing the judgment matrix for the second layer, where the judgment matrix of actions layer C relative to the B3 elements in criteria layer B is S:

Thus, the eigenvector $\mathbf{w}_3^{(2)}$; of the judgment matrix S is calculated; i.e. $\overline{w}_3^{(2)} = (0, 0, 0, 0, 0, 0, 0, w_3^{(2)}, 0, 0, 0, 0, 0)^T$, where the eigenvalue is 6.43, and the consistency test coefficient is RI=0.07<0.1.

From the judgment matrix R, the six indices of IT internal processes are calculated: information technology development planning (C08), software management (C09), hardware management (C10), system management (C11), telecom services (C12), and network operation (C13): whose weights are 38.7%, 8.1%, 10.6%, 17.1%, 8.8% and 16.8%, respectively.

4. Constructing the judgment matrix for the second layer, where the judgment matrix of actions layer C relative to the B4 elements in criteria layer B is T:

Thus, the eigenvector $\mathbf{w}_4^{(2)}$ of the judgment matrix T is calculated; i.e. $\overline{w}_4^{(2)} = (0, 0, 0, 0, 0, 0, 0, 0, 0, 0, 0, 0, 0, w_4^{(2)})^T$, where the eigenvalue is 5.04, and the consistency test coefficient is RI=0.008<0.1.

From the judgment matrix T, the five indices of IT learning and reforms are calculated: operational guidance and team building in the entire bank (C14), performance coaching and communication (C15), external job training (C16), internal job training (C17), and career planning and development (C18), whose weights are 7%, 9.3%, 18%, 32.8% and 32.8%, respectively.

(5) Calculating combined weights. The combined weight of action layer C relative to the target layer A:

$$\overline{\mathbf{W}} = \overline{\mathbf{w}}^{(1)} \left[\overline{\mathbf{w}}_1^{(2)} \overline{\mathbf{w}}_2^{(2)} \overline{\mathbf{w}}_3^{(2)} \overline{\mathbf{w}}_4^{(2)} \right]^T$$; The combined weights are calculated below:

C01	C02	C03	C04	C05	C06	C07	C08	C09
0.150	0.089	0.047	0.047	0.047	0.075	0.224	0.082	0.017
C10	C11	C12	C13	C14	C15	C16	C17	C18
0.022	0.036	0.019	0.035	0.008	0.010	0.020	0.036	0.036

(6)It is seen from the above data that net profit of entire bank (C01) has a big weight in IT value contribution (B1) and internal customer satisfaction (C07) has a big weight in IT user satisfaction (B2). This means that experts believe the financial parameters of the entire bank are very important, which would require the information technology department in the commercial bank to strengthen its collaboration with operational departments to work hard on innovations and increase the core competitiveness of financial products; while at the same time would also require the information technology department in the commercial bank to strengthen its collaboration with operational departments and work hard to improve internal customer satisfaction. The experts, however, showed less interest in such areas as work progression in information technology management, learning and growth of employees, and safe production of computers, etc.

4. Conclusions

It is a quite complicated issues to carry out comprehensive evaluation on the information department in the commercial bank, because it is related to many aspects. The 4 major index systems and 18 specific evaluation parameters are established here for comprehensive evaluation on the information department in the commercial bank through analyzing the characteristics of the information department in the commercial bank based on the Balanced Scorecard as the underlying framework, and an AHP structural model is built for comprehensive evaluation on the information department in the commercial bank. In actual practices, related evaluation indices and weights may be subject to some adjustments as needed. Anyway, the right people should be chosen to evaluate the individual indices of evaluation on the information department in the commercial bank, and the utmost effort should be made to ensure the evaluation be objective so as to give highly scientific conclusions.

References

[1]LIN Qi-Ning, *"Decision Analysi"* , Beijing: Beijing University of Post and Telecom Press, pp.102-105,2003.
[2]HAO Xiao-Ling, SUN Qiang. *"Information-based Performance Evaluation Frameworks, Implementations and Case Studie"* . Beijing: Tsinghua University Press, pp.36-55, 2005..
[3]ZHAO Guo-Jie, ZHAO Hong-Mei. *"Balanced Scorecard-based Creation of Performance Evaluation Systems for Commercial Banks"*. Modern Finance and Economics, pp.4-10, (5)2004.
[4]HAO Xiao-Ling, SUN Qiang. *"Informationalized Performance Evaluation Frameworks, Implementations and Case Studies"*. Beijing: Tsinghua University Press, pp.36-55, 172-177.2005..
[5]YUE Chao-Yuan. Decision Making *"Theory and Methodology"*. Beijing: The Science Press, pp. 206-209,2003..
[6]XIE Mei. *"The Application of Analytical Hierarchy Process in Balanced Scorecard [J]"*. Financial & Accounting Update (Cai Kuai Tong Xun), 2004, (2):39..

Transform Government Functions Effectively, And Innovate New Rural Industrial Development Model——Based on the Theory and Empirical Study of Deqing

Li Bing Ning
College of Economic Management, Guangdong Institute of Science and Technology, Guangzhou, Guangdong Province, 510640, China

Abstract

New socialist countryside construction is an important historical task in Chinese modernization process; economic development is the necessary material basis of new socialist countryside construction. Deqing county party committee and government have transformed government functions timely and effectively, identified development as top priority, and innovated the model of new rural industrial development in view of the actual underdevelopment of the mountains economic, making economy and society develop rapidly and healthily and raise the people's production and living standard.

I. The theory of the governmental and construction of new rural areas

1. The theory of the government and its administration

According to the principles of political economy, the government is a kind of significant system of the economic and social development in a country or an area. The theory of the government and its administration arise with the separate of the state from society and develop through the process of the state extending to society.

The basic theory of government functions burgeon during the period from 1500s to late 1700s, when mercantilists began to pay great attention to the function issues of the government in socioeconomic life and claimed the construction of a new-type market order and the start of world market via the government's power.

From early 1800s to 1920s, liberalists, like Smith and Loch, discussed the necessity and possibility of the government function that could only serve as "night watchman" from different angles, putting forward "the theory of limited government functions" and arguing that government functions and limits of authority only lay in providing with security of the citizens freedom and property and that the government should not interfere economic activities.

In 1930s, to solve the common phenomenon of the market ineffectiveness in western world, Keynes proposed "the theory of government interference", thinking that the market itself owned drawbacks like the blindness of the source distribution, which could not be overcome from within and could only be solved by government interference. He also advocated that the government should attach itself to economic field, encourage investment and consumption and undertake deficit financing.

In 1970s, the side effects of "the theory of government interference" emerge, i.e. the government ineffectiveness. In order to solve the problem, James Buchanan, Black Duncan, Arrow, Anthony Downs put forward "the theory of public choosing", which applies the assumption of "economic person" in economics to the government theory to present the speciality of political and official "rational economic person". The theory thinks it is the government's over-interference that caused the being unmarketable encountered by the western government, and the government should be the limited one, whose basic function should go to the maintenance of the market order and normal operation and which should perform free and non-interfering policy.

Since 1990s, a new type of governmental theory, such as "new public administration", "new public management" and "government reconstruction", has gradually arose in face of "the legalization crisis of public administration". Under this circumstance, most countries in the world urge their government to undertake revolution in governmental functions. Under the influence of thoughts like abandoning bureaucratism, surpassing bureaucratism, creating high efficient governmental organization, individualizing public departments and being market-oriented, people claim that the government should make public affairs, especially public sources and service, individually managed so as to promote communitism and set up an ideal civil society based on the government, the market and the community. The government is also asked to enforce the cooperation between the public departments and private business and develop non-governmental organizations actively. At the same time, the government should try to reconstruct itself with enterprise spirit and improve its competitiveness by applying the organizing culture in business management to the governmental departments.

In October, 2000 Robert B.Denhardt and Janet V.Denhardt published the article The new public service: Serving rather than steering in the US journal Public Administration Review. Since then the theory of the new public service represented by Denhardt has brought discussions and researches in the academic community, who try to make up for the drawbacks of the new public service theory.

Scholars in China translated the article into Chinese, and since then the term of the new public service gradually has become popular among government departments and the academic community. It is widely discussed how the new public service theory can be utilized to direct the governmental reformation and how it can be realized in practical.

The political proposition of constructing service-oriented government was firstly put forward at the Third Plenary Session of the Sixteenth Central Committee of the Chinese Communist Party and approved at the Six Plenary Session of the Sixteenth Central Committee of the Chinese Communist Party, "the resolution about major issues on constructing socialist harmonious community by The Central Committee of the Chinese Communist Party and the State Council", clearly claims that we should construct service-oriented government, strengthen its functions of social management and public service. As to the public service oriented government, it has several meanings in different aspects. In terms of economy, the existence of the government is to correct the failure of the market, which means providing society with public goods and services that can not be effectively offered by market, making fair regulations to enhance its supervisions, ensuring the effectiveness of the market competition and its basic functions in the allocation of resource. In terms of politics, the power of the government is bestowed by the people. So the government should make sure that it has provided all levels of society with

2. The essence of new rural area construction and its major participants

New rural area construction is a significant decision made by the central committee of the Party to coordinate the national overall layout of the modernization construction, to construct a better-off society integratively and to construct a harmonious socialist society. The new rural area construction covers the five following aspects: actively advancing the urban and rural plans as a whole; developing modern agriculture; deepening rural reforms; facilitating the development of rural public service and increasing peasants' income. *The notions on promoting the building of a new socialist countryside by the central committee and the state counsil* put forward the general aims from a macro view: advancing production, ample

living materials, civilized rural customs, tidy village appearance and democratic management. The essence of building a new socialist countryside is to transform the old countryside based on natural economy and with heavy traditional mental load as well as low work division and coordination into a new type of countryside based on modern commodity economy with high level of specialization and peasants armed with modern notions.

The new rural construction mainly involve the four kinds of participants: the government, mediate organizations, enterprises and individuals (mainly peasants), all of whom are playing an important role in the development of social economy. What the government is to do is make laws and regulations, provide public service as infrastructure facilities, maintain the market order and so on. The mediate organizations will function as a bridge helping with better communication between the government and enterprises and between enterprises. Enterprises will provide more job opportunities for the peasants to promote the development of rural economy and improve people's living standard by their advancing production. Individuals are the core force to build a new countryside, so the quality of the individuals determines the development of social economy. At present, the rural economy has fallen behind, so an appropriate and even dominant function is needed for the great-leap-forward development of the rural economy. As Rose stated, "The government bears the final responsibility for the development, recession or stagnation " Thus, how to innovate the government service items, service forms and service methods to adjust the government to satisfy the aim of "advancing production, ample living materials, civilized rural customs, tidy village appearance and democratic management" has become a brand new question governments at all levels should consider carefully. With most of the countryside governed by the county level, the governments at the county level have to shoulder the hard but glorious task.

II. Deqing County: Transform government functions effectively and innovate new industrial development model

Located in the central west of Guangdong Province, under the Zhaoqing City, Deqing County is a typical mountainous area with a combination of "80% of mountains, 10% of water and 10% of farmland " . It presides over 13 towns (neighborhood), 175 village committees, with a land area of 2,258 square kilometers, and a population of 360,000, including the agricultural population 287,000, or about 80 percent. In 2002, the 16th CPC National Congress report made it clear that "coordinating urban and rural economic and social development," building modern agriculture, developing the rural economy and increasing the income of the farmers comprehensively, building a well-off society as a major task to effectively solving the "three rural" issue are to be put in a prominent position on the Party and government agenda. The Fifth Plenum of the 16th CPC National Congress has set the goals of building a new socialist countryside in China's modernization process as a major historical task, and clearly proposed the target of "production development, well-off life, civilized rural customs, clean village capacity and democratic management," Deqing county Committee of CPC and the county government, responding positively to this call, made timely and effective transformation of government functions. Considering the actual economic underdevelopment of the mountain area, they identified its development as the most important task. Adhering to the scientific concept of development and using it as a guide to the overall economic and social development, they vigorously implement the development strategy of "strengthening the county by means of industry, enriching the people by means of agriculture and making money by means of tourism ". At the same time they adopted vigorously the strategy of developing competitive industries, consolidating and improving traditional industries and testing and promoting new-tech industries. They made utmost effort to create a county with three titles of national powerfuls, namely, China's modern powerful forestry chemical county, China's modern powerful agricultural efficiency county and the China's powerful tourism county ".

They tried every effort to achieve a sound coordination of speed, quality and efficiency, a harmonious combination of regions, cities and rural areas, a sustainable concordance of population, resources and environment and an agreeable development of industry, agriculture and tertiary industry, thus achieving rapid, healthy economic and social development and raising the people's living standard together with an advanced production standard.

1. Promoting the citrus growing as a leading industry and creating a "government- help-farmers" agricultural development model

They launched a revolution for farmers from undergoing an extensive citrus growing to shaping regional brand to enhance competitive advantage of the leading industry。 The development and expansion of the scale of the leading industry, according to traditional practice, was set by administratives of different levels and the farmers just did what had been planned. The leaders of Deqing committee of CPC and the county government disposed of this practice normally adopted in planned economy times. According to Tan Pei'an, Secretary of the CPC committee, with the transformation of government functions and the operational autonomy to the farmers of agricultural products, the government main task is to offer

assistance to the farmers and solve the urgent problems beyond their reach and abilities. Based on this thinking, TAN Pei'an ,the secretary ,led a group of officials to Great Hall of the People in Beijing at the end of 2003, marketing Gonggan , the citrus of tribute to ancient Chinese royal family, and officially launched the brand project of "King of Chinese Citruses. Ever since then, the Committee and the Government leaders have organized farmers every year to hold news press in municipalities like Beijing and Shanghai and other cities in Northwest, Northeast, Southwest, Northwest and Central China, presenting and advertising Deqing Gonggan, which eventually gave birth to the "China-Citrus" winner in December2005 at the "China's First Brand Fruit Contest"。

2. Transforming the single service for the industry to the construction of leading industry protection system.

The leaders of Deqing county Party committee and government have deeply realized that the scattered planting and operation could not guarantee the product quality and the brand. Therefore, it is necessary to enhance the agricultural organization, establish the standardization planting of the citrus, and construct a perfect quality guarantee system with a series of guarantee services for the sustainable development of the whole citrus industry in Deqing.

Firstly, to establish the high production standardization planting system and agriculture loan system. A series of steps should be taken to realize the objective, including accelerating the construction of the female park of Gonggan and Sha Tangju, establishing the long cooperation with the agriculture colleges, improving the breeds of Gonggan and Sha Tangju, setting up the first Gonggan Standardization Production Technology Regulations nationwide, and providing the farmers with standard and regulation which will guide them to develop the high value, high production, and high quality agriculture industry. Deqing has established 1126 Three High agricultural bases, and 122 agriculture leading enterprises, which has promoted the agriculture increase and raised the income of farmers. Currently, the average output of Gonggan and Sha Tangju per acre has increased to five to six thousand kilogram, and the highest output reaches eight thousand kilogram. Especially, the highest output of Sha Tangju per acre could reaches fifty thousand kilogram, which brings fifty to sixty thousand Yuan income. Deqing government has established a series of policies to support the agriculture loan, and recruited the village Party secretary as the agricultural credit coordinator. In 2006, the total amount of agriculture loan from all the rural credit union in the county reached three hundred and eighty million Yuan. The following will make detailed introduction on the finance system.

Secondly, to construct the diseases and insect pest prevention system and food safety protection system. The government will hold training classes and special topic lectures by the related experts and professors for the farmers, and cooperate with the high colleges to control the spreading of diseases and insect pest of citrus including improving the supervision of the agricultural material market, forbidding the sales, usage of the high poison and high left agricultural drugs and fertilizers, and developing the green and organic citrus. Since 2003, the government has held 42 lectures and more than 560 training classes and large scale science and technology exchanges which benefit two hundred and fifty one thousand people.

Thirdly，create the distribution system of agricultural resources. Set up the distribution companies of agricultural resources and purchase fertilizers and chemical pesticides directly from reputable and reliable manufacturers. In this way, not only the prices are low, but also the products which are fake, poisonous and high vestige can be stamped out and the cost of production can also be reduced. The companies supply the technical services to the farmers and guide the farmers to spreading chemical fertilizer and pesticides scientifically, which can save the cost and increase the yield of farmland.

Fourthly, further developing and improving the marketing system. Government betters the marketing system by creating brand, promoting, and setting up network. Deqing tribute oranges and sugar oranges gain the title of Famous Chinese Fruite and King Orange of China. Hundreds of sales networks have been set up in over 100 big cities in China, and 20 fruit companies are also built up and a large number of lavish consumers have come into being to purchase, process, and sell the oranges. It owns more than large orange purchasing and processing places and 18 product lines of orange cleaning, rating, waxing, packing. The number of purchasing and processing reaches 1,000 tons per day.

3．Forming industrialization mode combined of "external-aid economy leading, agricultural-industrialization spurring and third industry drawing".

Deqing possesses unique advantages of developing planting and wooding owing to its special pleasant natural environment with the balancing proportion of mountains, waters and fields. To make good use of it, Deqing must extend its industry chains, realizing holistic development mode, increasing interest space of the processed products, to convert the resource advantages into competition merits. Through agricultural industrialization, Deqing has created an industrialized mode with special Chinese mountain counties' characteristics, which includes external-aid economy leading, agricultural-industrialization spurring and third industry drawing, by setting up pole-industry

system such as wooding chemical industry, wood deep-processing, cementing construction materials and eco-stainless steel.

The characteristics of this mode is: to establish industry development zone and appeal the investment from developed countries or areas including Hong Kong and Macao to create the "develop pole"; to realize and develop modern efficiency agriculture, to promote agriculture resource processing, and to establish the industry chains of "agriculture- industry-trade" to realize the collaborative development of the three fields; and to increase farmers' income by developing tour industry which meanwhile will bring benefit to various economy fields.

Presently, the whole county has set up 102,000 hectares of turpentine base, 12,000 hectares of fast growing forests, industry raw material forests and industry recourse forests base. The living and on-standing wood accounts for 6.389 million square meters. About 40,000 peasants have come to work in the fruit fields and fruit-processing factories, with average daily salary of 30 yuan RMB. Direct and indirect tourism related workforce has reached more than 7,600 people and more than 5,600 poverty-stricken people benefited from tourism development. In 2006, the three backbone industries--wooding chemical industry, wood deep-processing, and wind-machine manufacturing- created whole production value of 1.478 billion yuan RMB, account for 53.12% of the whole county's TPV (total production value), among which, wooding chemical industry realized 680 million of VTP, increased by 32.1 percent, and realized 20.71 million yuan of tax revenue. In the first four months of this year, the four backbone industries of wooding chemical industry, wood deep-processing, wind-machine manufacturing and cementing construction materials achieved 542 million yuan RMB, increased by 55%, and realized tax revenue of 19.44 million yuan RMB.

4. Promoting the fast development of the third industry through tourism

During the 10th Five-Year Plan period, Deqing county have successively constructed the following Tourist Sites: Panglong Gorge Ecological Tourist Spots, Song Street, Jinlin Region of Rivers and Lakes and Flower World, and developed a very popular travel route in Guangdong Province: the Travel of Dragons—— Travel to Mother Dragon's Hometown, Deqing, as a result of which, tourism has developed into a leading burgeoning industry, with an average annual growth rate of 20 percent. The fast growth of tourism also promotes the rapid development of tertiary industries such as restaurant, accommodation, entertainment, commerce and transport, with an average annual growth rate of 15.4 percent.

In 2006, some new tourist spots including Panlong

Heaven and Warm Mineral Springs in the Flower World were open to the tourists. Nine tourist hotels with more than 1000 beds were added. Seven no less than three-star hotels were under construction. Deqing cooperated with the South Lake National Travel Agency of Guangdong Province by assigning the management right of the tickets of Mother Dragon's Ancestral Temple, Sanyuan Tower and Study Palace to it and established a new market-based travel management system, which promoted the rapid development of the County's tourism. As a result, the number of the tourists to the county increased by 40 percent and the occupancy rate of the hotel room increased by 35 percent. In the year, Deqing County received 2.23 million tourists, with the tourism incomes of 550 million yuan, up by 21.7% and 23.7% over the same period last year respectively. And the tax revenue of the accommodation and catering went up by 55.6 percent.

During the travel golden week of the Spring Festival in 2007, Deqing County received 261 thousand tourists, with the tourism incomes of 59.55 million yuan, up by 8.7 percent and 11 percent respectively compared with the travel golden week of the Spring Festival in 2006. From January to June, the whole county received 1392 thousand tourists, with the tourism incomes of 310 million yuan, up by 18% and 13% over the same period last year respectively. Eight tourist hotels with more than 500 beds were newly added, as a result of which, the total number of hotel beds in the county reached four thousand. Two no less than three-star hotels are under way and 2 Sales Department of the travel agency have been added. The rapid growth of tourism not only promoted the fast development of tertiary industries but also greatly increased the local farmers' income. Some farmers living in the scenic spots or nearby obtained extra income by engaging in agriculture and service trades at the same time, which effectively combined primary industry and tertiary industry.

Acknowledgement

It is a project supported by Guangdong Provincial Science and Technology Agency, Guangdong soft science researches, No. 2007B070900059 and references provided by Deqing county committee in Guangdong Province. The paper is also assisted by Ms Huang Pei Hong when completing.

References

[1] Douglass C. North and Robles Thomas, "The Rise of the Western World", translated by Li Yi Ning [M], Huaxia Publishing Co., Ltd., 1999

[2]Yi Xian Rong, "The Introduction to Modern Agreement Economics"[M], *China Social Sciences Press*, 1997

[3]Lu Xian Xiang, "The Western New Institutional

Economics"[M], *Development Press of China*, 1996

[4] Douglass C. North, "The Construction and Changes of the Economic History", translated by Chen Yu and Luo Wei Ping [M], *SDX Joint Publishing Company, Shanghai People's Press*, 1994

[5] Yang Rui Long, "Facing the Institution" [M], *Development Press of China*, 2000

[6] Li Ping, "Analysis of the Changes in Means of Economic Growth" [M], *Press of Southwest University of Finance and Economic*, 2001

[7] Guo Feng, "Analysis of the Innovation of China's Rural Economic Institution" [M], *The Commercial Press*, 2000

[8] Elinor Ostrom and Larry Schroeder, "Institutional Incentives and Sustainable Development" [M], 2000

[9] References of Deqing County, "The Innovation of Governmental Service, the Improvement of Governing Capacity and the Construction of New Wealthy and Harmonious Rural Areas" [R], Mar, 2007

[10] References of Deqing County, "Innovating upon Modes of Rural Information Service, Promoting the Construction of New Socialist Rural Areas" [R], Jan, 2007

A Dietary Investigation and Analysis on Students Majored in P.E. and Sports

Yanxia Peng

P.E. Department, Branch School of Huazhong Normal University, Wuhan, 430212, China

Abstract

An one week dietary investigation towards the students majored in P.E. and sports showed that most of the objects being question aired either lacked of nutrition kownlege,normal ways of nutrition or had an unreasonable dietary structure,therefore,the dietary most objects in took contained almost no potato. All these habits resulted in an unbalanced nutrition in taken with more energy produced by supper as well as midnight tea and less energy produced by breakfast so that there were 10.08% objects being question aired bad BMI more than over25 .It is suggested that the nutrition education to the university students be strengthened and food intake habit be adjusted.

1. Introduction

With the development of the society economics' and medicine science in our country, the health care levels of our residents are promoted. The nutrition is becoming one of the most important sciences for people taking care of their healths.It is highly recognized by most people that nutrition is the essence of the life. The nutrition makes an important effect on the human body's growth and development. In order to understand the dietary questions being in the students in the institute of physical education, the dietary investigation and the nutrition knowledge and the habit of dietary questionnaires had been made in the college students who were major in physical education.

2. Research objects and methods

2.1 Research objects

There are 119 students who are major in physical education in Wuhan Institute of Physical Educaton, and we can see the details in the below table.

Table1 the basic conditions of the research objects

Number	Age	Height (cm)	Weight (kg)
119	22.53=1.07	176.10 ±4.11	71.09 ±7.31
	BMI 22.92 ±2.11	BMI>2.5 12	

2.2 Research objects

These 119 students need receive a questionnaires investigation and one week dietary investigation, and 50 testing samples are extracted in random. The main content of the questionnaire is containing the concept of proper nutrition, knowledge and the habit of dietary. And there are 119 questionnaires and all the all were withdrawed, the ratio of the withdraw is 100%.Both the questionnaire reliability (0.81) and availability (0.91) are all met the need of statistics. The specific investigation contents are including the amount of the average intake food per day and all the intake nutrition elements, the ratio of energy nutrition elements, the original ratio of the protein and fat, the energy distribution among breakfast, lunch and dinner, and so on. The analysis are done using some data and index [1][2][3].

3. Results and Analysis

There are many aspects analysis below though the dietary investigation.

3.1 The structure of food

Table2 is the results of the investigation. From this table, it can be seen that the food intake amount is occupied 86% of all the provided, which can basically meet the need of these students; cereal is occupied 91% of all the provied, and which can also approach to the meet. It is obviously showed us that the meat, egg and milk are more surplus for intake, which are occupied 221%,192%and 147%of all the provided.Fish,beans and vegetables are much lack, and which are occupied 48%,50% and 41% of all the provided.

Table2 the analysis between the food intake and provide

The kinds of food	one		two			total
	cereals	potatos	meat	eggs	fish	
Intake	364	.0	166	73	24	1155
Provide	400	100	75	38	50	1338
Ratio	91	0	221	192	48	86
The kinds of food	three		four		five	
	milk	beans	vegetable	fruit	fat	
Intake	147	25	186	150	20	1155
Provide	100	50	450	150	25	1338
Ratio	147	50	41	100	80	86

3.2 The origin of the energy

The cord question in nutrition is the balance of the energy. It is generally believed that an adult person should intake 90% DRIs energy, and lower 80% is seemed as the lack of energy. And other nutrition can reach the 80% DRIs above, which can prevent most people from lacking nutrition [4]. From the investigation results, it can be seen that the energies intake being in investigate objects are much lower, and which can be only occupied 82.07% DRIs (seen from table 3).The proteins intake are obviously surplus, which can be occupied 128.26% DRIs.From table 4,which is the results of the energy-nutrition investigates, it can be seen that the ration that protein occupied the total energy is 3.52% higher than the reference numher,which is 18.52%,however,the ratio of fat and carbohydrates are much lower, which are only 23.00% and 58.47%.Table5 is the analysis on the origin ratio of protein and fat, from this table, it can be seen that the animal fats intake are 38.67% higher than the reference number, which can be occupied the 88.67% total fats, and the same situation is being in the quality proteins.

Table 3 Each nutrition analysis between intakes and provide

Nutrition	Intake	DRI	Ratio(%)
Protein(g)	102.61	80	128.26
Energy(kcal)	2215.96	2700	82.07

Table 4 The ratio of energy-nutrition

	Protein	Fat	Carbohydrates	Total
Intake(g)	102.61	56.64	323.94	1455.01
Energy(kcal)	410.44	509.76	1295.76	2215.96
Ratio(%)	18.52	23.00	58.47	100
Reference(%)	15	25	60	100

Table 5 the origin ratio of protein and fat

	protein		total
	quality	common	
Intake(g)	51.79	50.64	102.61
%	50.65	49.35	100
Reference(%)	>33.33[4]		

	fat		total
	animal	vegetable	
Intake(g)	50.22	6.42	56.64
%	88.67	11.33	100
Reference(%)	50	50	

3.3 Inorganic salt and Vitamin

Table 6 is the inorganic salt and Vitamin nutrition analysis between intakes and provide. From this table, it can be seen that the question being in inorganic salt and vitamin is that the most prominent VC and calcium intake is obviously insufficient; the DRIs only 61.47% and 60.58 %.VA intake was also low, representing only 75% of the DRIs. Zinc, Magnesium and VB1 intake accounts for 81.37% and 90% of the DRIs, which can meet the basic needs. Phosphorus, Iron, Selenium, VB2 intake can both more than 100% DRIs, and which also meet the needs.

To sum up, leading to the targets imbalance of nutrient intake is the main reason caused by the too much animal food intake, or plant food intake and calcium intake too little. More information that caused such a dietary fat (especially animal fat), cholesterol and animal protein intake too much will enable the cardiovascular and cerebrovascular diseases, chronic diseases and tumors more modern. And fruits,vegetables,potato products contained in high-quality protein, fatty acids, rich in calcium and vitamin both prevention of chronic diseases and tumors of the modern role.

Table 6 Inorganic salt and Vitamin nutrition analysis between intakes and provide

Nutrition	Intake	DRI	Ratio(%)
Ca(mg)	491.77	800	61.47
P(mg)	1193.53	700	170.50
Fe(mg)	20.0	15	133.30
Zn(mg)	12.64	15.5	81.55
Se(ug)	53.99	50	107.98
Mg(mg)	284.78	350	81.37
VB1(mg)	1.26	1.4	90.00
VB2(mg)	1.42	1.4	101.41
Vpp(mg)	23.82	14	170.14
VC(mg)	60.58	100	60.58
VE(mg)	20.89	14	149.21
VA0(ugRE)	600.50	800	75.06

4. Conclusion and Suggestion

4.1 Dietary investigation summary

To sum up, the students under investigation have a weak and lack of concept of nutrition; even some students have bad eating habits. Their dietary status are as follows: total food intake to meet basic needs, cereal intake is adequate, but less species (only rice and flour), animal food (except fish tired) was excessive. Fish, beans and vegetables serious shortage of potato intake are no, the intake of fruits can basically meet the need. The structure of food led to energy intake to meet basic needs, but a higher proportion of protein, carbohydrate and low fat ratio. The high proportion of animal fat is not so well. And the proportion of high quality protein, Calcium and VC are obviously inadequate, however VA is still adequate, and the remaining salts and vitamins can also meet the need [5][6].

4.2 Suggestion

It is very important for students to reduce the pork intake, eating more fish, beans and its products, the daily they should insist drinking milk, eating food with high levels of calcium. This will improve and adjust the quality of protein and fat, and increase calcium intake and prevention of chronic diseases.

With attention to grain size, eat more vegetables, fruit and potato. In this way, it can not only adjust the structure of meals, food diversification, but also to correct the lack the salts and vitamins of meals (especially VC). The food is also available in the dietary fiber and fruit in organic acids, acid treatment of constipation also. So these can make a excellent effects on human health.

The students should increase the energy intake for their breakfast and improve its quality. To rectify the bad habit of skipping breakfast, breakfast stresses the energy ratio up to 30%, and we should pay attention to their nutritional balance. Science breakfast should be one with little protein, fat-based food, but supplemented by vitamins, eggs, milk, snacks, vegetables, while they may reduce the food rich in protein and fat food.

References

[1]. Proposals of China Nutrition Association. *Chinese residents dietary guidelines*[R],1997

[2]. Proposals of China Nutrition Association. Suggestions and application of a balanced diet[R],1998

[3]. Proposals of China Nutrition Association. The reference of Chinese residents dietary nutrients intake[R],2000

[4]. Yancheng Gao. Nutiology [M]. Beijing:the press of Beijng University of Physical Education,1992

[5]. Chen JS, Chen Xs. Changes of the dire composition of Chinese people and a proper diet guidance.Chiese J Preventive Mde 1993,27:266-269

[6]. Editiorial Boar of China,s Health Year Book.(1993)Natural changes and death causes of Chinese populatio.China Health Year Book, People,s Health Pudlishing House, Beijing PP 421-422

Short Term Traffic Flow Prediction Based on Online Learning SVR

Dehuai Zeng[1,2], Jianmin Xu[1], Jianwei Gu[3], Liyan Liu[3], Gang Xu[2],

[1] School of Civil Engineering and Transportation,
South China University of China, Guangzhou, 510640, China
2 Institute of Intelligent Technology, Shenzhen University, Shenzhen, 518060, China
3 Guangzhou Post and Telecom Equipment Co.LTD., Guangzhou, 510663, China

Abstract

A number of different forecasting methods have been proposed for traffic flow forecasting including historic method, real-time method, time series analysis, and artificial neural networks (ANN), but accuracy and time efficiency in prediction are a couple of contradictions to be hard to resolve for real-time traffic information prediction. In order to improve time efficiency of prediction, a new short-term traffic flow prediction model and method based on accurate online support vector regression (AOSVR) is proposed in this paper, which can update the prediction function in real time via incremental learning way. A comparison of the performance of AOSVR with ANN, real time, and historic approach is carried out. Experiments results demonstrate that the AOSVR predictor can reduce significantly both relative mean errors and root mean squared errors of predicted travel times. Therefore, AOSVR based traffic flow prediction is applicable and performs well for traffic data analysis.

Index Terms—**support vector regression, traffic flow prediction, intelligent transportation systems**

1. Introduction

In recent years, the Intelligent Transportation System (ITS) has been developed very rapidly. The traffic control system, incident detection system and transport guidance system are the main research fields of the ITS. But they have one important thing in common: the real-time traffic flow prediction. The prediction results are strongly related with the effectiveness of these systems. Therefore, traffic flow forecasting has attracted much attention because of its importance in both the theoretical and empirical aspects of ITS deployment.

The traffic researchers have built many short-term traffic flow prediction models, most of which oversimplify the complex temporal correlation presenting in the traffic flow. By now there are approximately 30 prediction methods[1~5]: the dynamic traffic flow distribution methods, the historical-mean methods, the regression analysis methods, the time series methods, the Kalman filter methods, the neural network methods, the fuzzy neural network method, the fuzzy-neural method, the nonparametric regression methods, etc. Although

these methods have alleviated difficulties in traffic modeling and prediction to some extent, from a careful review we can still find some problems. the historical mean methods [2], [3] and regression analysis methods [4] have common drawbacks, which suppose that traffic flow and travel time are both strictly periodic and ignore the uncertainty and nonlinearity of traffic flow. The Kalman filter methods [2]–[6] are unsuitable for predicting the traffic flow which sample interval is less than 5 min. The nonparametric regression methods [1] need a huge historical database which occupancies many memory and takes much time to predict the traffic flow. The neural network methods suffer from problems like the existence of local minima and the limited generalization ability. In order to improve prediction performance, the hybrid model combined with wavelet analysis and neural network is proposed, but accompanied with low efficiency due to the inherent theory flaw from neural networks.

SVM is a new machine learning method that is put forward by V. Vapnik et al. and based on SLT (Statistics Learning Theory) and SRM (structural risk minimization). The application of SVM to time-series forecasting has shown many breakthroughs and plausible performance. Moreover, the rapid development of support vector machines (SVM) in statistical learning theory encourages researchers actively focus on applying SVM to various research fields like document classifications and pattern recognitions. The time-varying properties of SVR applications resemble the time-dependency of traffic forecasting, combined with many successful results of SVR predictions encourage our research in using SVR for travel-time modeling.

SVM possess great potential and superior performance as is appeared in many previous researches [7]. This is largely due to the structural risk minimization (SRM) principle in SVM that has greater generalization ability and is superior to the empirical risk minimization (ERM) principle as adopted in neural networks [8]. In SVM, the results guarantee global minima whereas ERM can only locate local minima. SVM can solve some flaws of the neural networks, and has many unique advantages in the fields of small samples and high-dimensional nonlinear manifested.

Unfortunately, classical SVM learning algorithm [9] does not support online learning. In order to adapt to new

978-0-7695-3342-1/08 $25.00 © 2008 IEEE
DOI 10.1109/PEITS.2008.134

samples, the traditional SVM has to discard all the previous training results and re-train the new classifier on the whole data set. In this paper, an SVM-based online learning algorithm is presented and applied to web prediction problem. This online learning algorithm is based on incremental chunk for LS-SVM (Least Square Support Vector Machines) classifier [10]. The training of the LS-SVM can be placed in a way of incremental chunk avoiding computing large-scale matrix inverse but maintaining the precision when training and testing data. This online algorithm is especially useful for the large data set and practical applications where the data come in sequentially.

The remainder of this paper is organized as follows. In Section 2, the standard LS-SVM is reviewed briefly and the accurate online LS-SVM based on accurate online learning is derived. And in Section 3 the construction of the prediction model is given. The experimental results of applying the proposed algorithm to the traffic flow prediction arc presented in Section 4. Finally, conclusion is drawn in section 5.

2. Accurate online Support Vector Regression

21. Standard Support Vector Machine

Least Square Support Vector Machine (LS-SVM) is a new technique for regression. When LS-SVM is used to model urban traffic, the input and output variables should be chosen firstly. Given a training data set $\{(x_1,y_1),\ldots,(x_l,y_l)\}$ with input data $x_i \in R^n$ and output data $y_i \in R$. In order to get the function dependence relation, SVM map the input space into a high-dimension feature space and construct a linear regression in it. The regression function is expressed with

$$y = f(x) = w^T \varphi(x) + b \qquad (1)$$

with $\varphi(\cdot): R^n \to R^{n_\varphi}$, a function which maps the input space into a so-called higher dimensional (possibly infinite dimensional) feature space, w and b are the regression parameters to be solved.

LS-SVM regression estimation involves primal and dual model formulations. Given the training data set $\{(x_1,y_1),\ldots,(x_l,y_l)\}$, the goal is to estimate the model (1), where f is parameterized as in (4), we can formulate the following optimization scheme to infer our parameters

$$\min_{w,b,e} L_P(w,e) = \frac{1}{2}\|w\|^2 + \frac{\gamma}{2}\sum_{i=1}^{l} e_i \qquad (2)$$

$$s.t. \quad y_i = w \cdot \varphi(x_i) + b + e_i, \quad i = 1,2,\ldots,l$$

Where error variables $e = (e_1, e_2,\ldots,e_l)^T, e_i \in R$, the regularization constant $\gamma > 0$ is included to control the bias-variance trade-off. The above statement is in fact the same formulation as is used in case of ridge regression [12] in the feature space defined by $\varphi(\cdot)$. Note that in some

cases w becomes infinite dimension, and the above problem formulation cannot be used to solve the problem. Therefore, we perform the computations in another space, called the dual space of Lagrangian multipliers after applying Mercer's theorem. Consider the Lagrangian of (2) given by

$$L_D(w,b,e_i,\alpha) = \frac{1}{2}w^T w + \frac{\gamma}{2}\sum_{i=1}^{l} e_i \qquad (3)$$
$$- \sum_{i=1}^{l} \alpha_i(w^T \varphi(x_i) + b + e_i - y_i)$$

Here $\alpha = (\alpha_1, \alpha_2,\ldots,\alpha_l)^T, \alpha_i \in R$ are Lagrangian multipliers. The first order conditions for optimality are given by:

$$\begin{cases} \dfrac{\partial L_D}{\partial w} = 0 \to w = \sum_{i=1}^{l}\alpha_i\varphi(x_i), \\[2mm] \dfrac{\partial L_D}{\partial b} = 0 \to 0 = \sum_{i=1}^{l}\alpha_i, \\[2mm] \dfrac{\partial L_D}{\partial e_i} = 0 \to \alpha_i = \gamma e_i, i=1,2,\ldots,l \\[2mm] \dfrac{\partial L_D}{\partial \alpha_i} = 0 \to y_t = w^T\varphi(x_t) + b + e_i, i=1,2,\ldots,l \end{cases} \qquad (4)$$

That is:

$$\begin{bmatrix} I & 0 & 0 & -Z \\ 0 & 0 & 0 & -\vec{1} \\ 0 & 0 & \gamma & -I \\ Z & \vec{1} & I & 0 \end{bmatrix}\begin{bmatrix} w \\ b \\ e \\ \alpha \end{bmatrix} = \begin{bmatrix} 0 \\ 0 \\ 0 \\ y \end{bmatrix} \qquad (5)$$

here

$$Z = (\varphi(x_1), \varphi(x_2),\ldots,\varphi(x_l))^T, y = (y_1, y_2,\cdots,y_l)^T \ I = (1,1,\ldots,1)^T,$$
$$e = (e_1, e_2,\ldots,e_l)^T, \ \alpha = (\alpha_1, \alpha_2,\ldots,\alpha_l)^T$$

From formula (4), combining the first and the last condition yields

$$y_i = \sum_{i=1}^{l}\alpha_i\varphi(x_i)^T\varphi(x_i) + b + e_i \qquad (6)$$

Replacing the third expression from (4) into (6) gives

$$y_i = \sum_{i=1}^{l}\alpha_i k(x_i, x_j) + b + \frac{\alpha_i}{\gamma}, \ i=1,2,\ldots,l \qquad (7)$$

$$0 = \sum_{i=1}^{l}\alpha_i,$$

that is:

$$\begin{bmatrix} \vec{1} & ZZ^T + \gamma^{-1}I \\ 0 & \vec{1}^T \end{bmatrix}\begin{bmatrix} \alpha \\ b \end{bmatrix} = \begin{bmatrix} y \\ 0 \end{bmatrix} \qquad (8)$$

Now, using Mercer's condition:

$$\varphi(x_i) \cdot \varphi(x_j) = k(x_i, x_j) \equiv \Omega_{ij}, \ i,j = 1,2,\ldots,l \qquad (9)$$

to replace the dot product $\varphi(x_i)^T\varphi(x_i)$, let $C = \Omega + \gamma^{-1}I$, then the formula (8) can be written as:

$$\begin{bmatrix} \vec{1} & C \\ 0 & \vec{1}^T \end{bmatrix}\begin{bmatrix} \alpha \\ b \end{bmatrix} = \begin{bmatrix} y \\ 0 \end{bmatrix} \qquad (10)$$

From formula (8), the regression parameters α and b can be solved as follows:

$$b = \frac{\vec{1}^T C^{-1} y}{\vec{1}^T C^{-1} \vec{1}} \tag{11}$$

$$\alpha = C^{-1}(y - b\vec{1}) \tag{12}$$

Thus, the regression function is expressed in dual form

$$y = w^T \varphi(x) + b = \sum_{i=1}^{l} \alpha_i k(x_i, x_j) + b \tag{13}$$

2.2 Online learning LS-SVM

Considering that LS-SVM model based on the first N pairs of data has been constructed, and the new data pair $\{(x_l,y_l),\ldots,(x_{l+K},y_{l+K})\}$ is fed, Let

$$\begin{bmatrix} \vec{1} & C \\ 0 & \vec{1}^T \end{bmatrix} = A_l, \quad \begin{bmatrix} \alpha \\ b \end{bmatrix} = \alpha_l, \quad \begin{bmatrix} y \\ 0 \end{bmatrix} = Y_l$$

Then formula (10) changes into

$$A_l \alpha_l = Y_l \Rightarrow \alpha_l = A_l^{-1} Y_l \tag{14}$$

The subscript l means that the current model is based on the first l pairs of data. For $l+K$ pairs of data, one has

$$\alpha_{l+K} = A_{l+K}^{-1} Y_{l+K} \tag{15}$$

Where

$$A_{l+K} = \begin{bmatrix} A_l & B^T \\ B & C \end{bmatrix}$$

$$B = \begin{bmatrix} y_{l+1} & \Omega_{l+1,1} & \Omega_{l+1,2} & \cdots & \Omega_{l+1,l} \\ \vdots & \vdots & \vdots & \vdots & \vdots \\ y_{l+1} & \Omega_{l+K,1} & \Omega_{l+K,2} & \cdots & \Omega_{l+K,l} \end{bmatrix}$$

$$C = \begin{bmatrix} \Omega_{l+1,l+1} & \Omega_{l+1,l+2} & \Omega_{l+1,l+3} & \cdots & \Omega_{l+1,l+K} \\ \vdots & \vdots & \vdots & \vdots & \vdots \\ \Omega_{l+K,l+1} & \Omega_{l+K,l+2} & \Omega_{l+K,l+3} & \cdots & \Omega_{l+K,l+K} \end{bmatrix} + \gamma^{-1}I$$

$$Y_{l+K} = [y_{l+1}, y_{l+2}, \cdots, y_{l+K}]^T$$

If A_l^{-1} can be used to obtain A_{l+K}^{-1} without totally recalculating, then the incremental SVM learning task is done. According to Ref. [9-11], the following two lemmas hold:

Lemmas1: *For a matrix* $A = \begin{bmatrix} A_{11} & A_{12} \\ A_{21} & A_{22} \end{bmatrix}$, *where* A_{11}^{-1} *and* A_{22}^{-1}

exist, the following formula is true

$$A^{-1} = \begin{bmatrix} [A_{11} - A_{12}A_{22}^{-1}A_{21}]^{-1} & A_{11}^{-1}A_{12}[A_{21}A_{22}^{-1}A_{12} - A_{22}]^{-1} \\ [A_{21}A_{22}^{-1}A_{12} - A_{22}]^{-1}A_{21}A_{11}^{-1} & [A_{22} - A_{21}A_{11}^{-1}A_{12}]^{-1} \end{bmatrix} \tag{16}$$

Lemmas 2: *For matrics A, B, C and D, where* A^{-1} *and* C^{-1} *exist, the following equation is true*

$$(A + BCD)^{-1} = A^{-1} - A^{-1}B(C + DA^{-1}B)^{-1}DA^{-1} \tag{17}$$

From Lemmas 1 & 2, Theroem 1 can be inferred as follows:

Theroem 1: *the matrics* A_{l+K}^{-1} *in equation (15) can be obtained form* A_l^{-1} *without computing the matrix inverse*

from Eq.(16), one can obtain

$$A_{l+K}^{-1} = \begin{bmatrix} A_l & B^T \\ B & C \end{bmatrix}^{-1} \tag{18}$$

$$= \begin{bmatrix} [A_l - C^{-1}B^T B]^{-1} & A_l^{-1}B^T[BA_l^{-1}B^T - C]^{-1} \\ [BA_l^{-1}B^T - C]^{-1}BA_l^{-1} & [C - BA_l^{-1}B^T]^{-1} \end{bmatrix}$$

Applying Eq.(17) to the top left submartix in Eq.(18) yields

$$[A_l - C^{-1}B^T B]^{-1} = A_l^{-1} - A_l^{-1}B^T[-C + BA_l^{-1}B^T]^{-1}BA_l^{-1} \tag{19}$$

It is clear that A_{l+K}^{-1} can be computed from A_l^{-1} without totally recalculating, which avoids expensive inversion operation. Therefore the corresponding coefficients and bias $\alpha_{l+K} = [\alpha, b]^T$ can be calculated from formula (11~12). Based on the original results, the prediction function should update as following, according to the new samples added:

$$y = \sum_{i=1}^{l+K} \alpha_i k(x_i, x_j) + b \tag{20}$$

3 Traffic flow predictor model based AOSVR

3.1 Data preparation

Traffic prediction can be differentiated by the data source used in the predicting process. The key issue in predicting is how to make use of these sources of information. In a real-time case, data from real-time collection are used to make continuous and simultaneous predictions for several minutes in future. We focused on the traffic flow prediction in common work days because the traffic flow in work days was much larger than that in weekend. Since traffic data may be missed or corrupted, we select a better portion of the dataset that covers a 25-km stretch of a busy section of the Guangyuan Highway, from Guangzhou to Dongguan. The traffic flow data is available from a Web server of Guangzhou traffic monitoring center. Thus the previous traffic flow sequence data can be used to predict the future traffic flow.

The traffic flow data were sampled per 5 mins from 7:00 a.m. to 7:00 p.m. in every day. Since traffic flow data were always sampled with noise, the data preprocessing need to be implemented to correction errors, of which the threshold test and traffic flow theory-based check are two commonly used methods. At last, the data should be normalized treatment to improve the efficiency of computation.

3.2 predictor design

In this paper, corrected AOSVR with Guass kernel is made use to build a real-time prediction model for traffic information. The algorithm in details is:

Step1: Given a prediction origin *0,* construct a set of training samples and a set of testing samples.

618

Step2: Construct a predictor, which means model construction and parameters selection. Here corrected SVR with Guass kernel is made use to build a forecast model. As discussed before, error parameter γ determines the trade-off between margin maximization and training error minimization. A large γ assigns higher penalties to errors and lower generalization. We focused on the choice of an RBF kernel for traffic flow prediction.

Step 3: Initialize the LS-SVM algorithm, train the proposed predictor from traffic flow samples $\{(x_1,y_1),\ldots,(x_l,y_l)\}$ as training data set, the AOSVR parameters α, b are adapted by formula (1)~(13), Thus, the prediction value of traffic flow y is:

$$y=\sum_{i=1}^{l}\alpha_i k(x_i,x_j)+b$$

Step 4: Update the prediction regression function along with the continuous dynamically acquired data. Supposing K new samples are chosen to added, the new training set is $\{(x_1,y_1),\ldots,(x_{l+K}, y_{l+K})\}$, the AOSVR parameters α, b can be updated in time, the prediction model can maintain high accuracy and getting practical more and more.

The forecast mechanism can be seen in Fig.1.

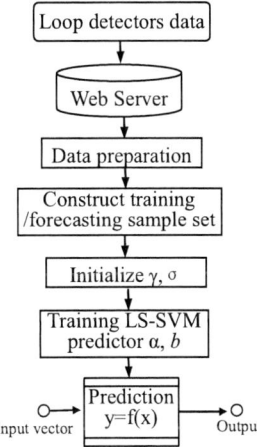

Fig.1 Mechanism of AOSVR prediction algorithm

4 Experiments and Results

To evaluate the prediction applicability of traffic flow with AOSVR, some common baseline traffic flow prediction methods are exploited for performance comparison. We used 3 weeks traffic data as training set and 1 week traffic data as testing set in order to find the similarities and the instant dynamics from the daily traffic data. Since there were large amount of raw data in this database, we simply randomly selected these samples in a short period in each day. The MLANN-based short term traffic flow prediction model was programmed by using Matlab 2008 neural network toolbox, the AOSVR based forecast model was programmed by using mySVM software kit. The operating environment is: CPU Pentium

2.4 GHz, Memory 1G, MS WinXP operation system.

4.1 Traffic flow predictor and error measurement

As discussed previously, there are many parameters that must be set for travel-time prediction with AOSVR. We have tried several combinations, and finally chose Radial Basis Function (RBF) kernel with σ =16, error parameter γ =118. The SVR experiments were done by running mySVM software kit.

In order to illustrate the effectiveness of the proposed method, the MLANN is also used to forecast the traffic flow. The transfer function between the input layer and the hidden layer is *Sigmoid* function. The input layer, the hidden layer and the output layer are composed of 5, 8 and 1 neurons respectively.

To evaluate its adaptability to different predict methods, three error indicators: Relative Mean Errors (RME), Mean Absolute Relative Error (MARE) and Root Mean Squared Errors (RMSE) are applied as performance indices.

$$rmerr=\frac{1}{N}\sum_{i=1}^{N}\frac{y_i-y_i^*}{y_i}$$

$$marerr=\frac{1}{N}\sum_{i=1}^{N}\frac{|y_i-y_i^*|}{y_i}$$

$$rmserr=\sqrt{\frac{1}{N}\sum_{i=1}^{N}\left(\frac{y_i-y_i^*}{y_i}\right)^2}$$

Where y_i is the actual value and y_i^* is the predicted value.

4.2 Experiment results and discussion

The experiment results are shown in Fig. 2, we can see that the whole traffic flow always changes sharply in two different periods of time including in the morning when people come to work, and in the evening when people go home. As expected, the MLANN predictor is usually slow to reflect the changes of traffic patterns. Since SVR can converge rapidly and avoid local minimum, the SVR predictor performs very well in our experiments. From the figure, AOSVR predictor can catch the characteristic of whole traffic flow and accurately forecast the traffic flow. But the forecasting errors of few points are comparatively big. The reason is that there are not so many traffic flow points that the found nearest neighbors of some phase points are far away from those phase points. Along with the increase of time series, the selection of near neighbor will be much reasonable, which will be helpful to improve the forecasting precision.

Three evaluation indices of MLANN and AOSVR predictors are listed in Table 1, we can see that the AOSVR predictor gave higher accuracy than the MLANN predictor. Besides the OSVM updated forecast function via online learning algorithm, which was more suited for the practical application. In generalized performance, the AOSVR predictor is also superior to the MLANN predictor. This means that SVRPM is the most excellent predictor. Faced with variable traffic condition, support vector machine regression shows strong powerful robustness.

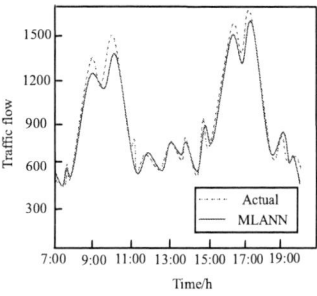

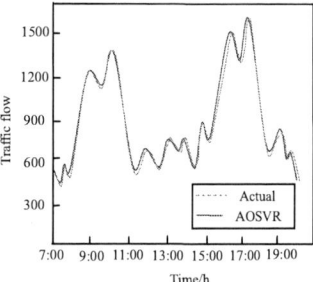

Fig.2. Prediction comparison using MLANN and AOSVR

Tab. 1 Evaluation Indices between MLANN and AOSVR

Predictor	Evaluation Indices		
	rmerr%	marerr%	rmsrerr%
MLANN	0.82	4.26	12.44
AOSVR	0.49	1.84	5.82

However, the average training time of AOSVR ranges from 30 seconds to 50 seconds, the training of neural network costs almost 6 seconds. According the theory of support vector regression, it is necessary to cost a large number of mathematical operations for searching the optimal hyper plane. Moreover, the training of AOSVR in fact is a convex quadratic programming problem, which can obtain the global optimal solution with hybrid GA optimized parameters. Contrarily, gradient descent algorithm was used for neural network, which often result in local optimal solution.

5 Conclusions

Traffic flow prediction models are expected to play an important role in ITS. Specifically, they will support advanced traffic control systems and traveler routing aids. In this paper, we have presented a new method for predicting the traffic flow based on online support vector machine. This method was based on least square support vector machine, and used online learning strategy to dynamic update forecast function, which was more suited

for the practical application. Experimental results demonstrate that our proposed method gains high prediction accuracy and satisfied the demands of real time traffic flow prediction. However, there are still some problems need to be solved. Methods optimizing the parameters of online learning algorithm using GA should be considered in the future to improve adaptability for variable traffic flow.

Acknowledgement

This work is supported by National High-Tech Plan (863) with grand No.2006AA11Z211, Natural Science Foundation of China with grand No. 50578064, and China Postdoctor Foundation with grand No. 20070410827.

References

[1] Davis, G. A. and Nihan, N. L. "Nonparametric regression and short term freeway traffic forecasting." Journal of Transportation Engineering, pp.178-188, 1991.

[2] Liang Zhao Fei-Yue Wang "Short-term traffic flow prediction based on ratio-median lengths of intervals two-factors high-order fuzzy time series", Proc. of Conf. on Vehicular Electronics and Safety, Dec. 2007, pp.1~7

[3] Smith. B. L. and Demetsky. M. J., "Traffic flow forecasting: comparison of modeling approaches." Journal of Transportation Engineering, vol. 123, no. 4, p.261-266, 1997.

[4] Kaysi. I., Ben-Akiva. M., and Koutsopoulos. H.,"An integrated approach to vehicle routing and congestion prediction for real-time driver guidance." Transportation Research Record 1408, TRB, Washington, D.C., pp. 66-74,1993.

[5] Hobeika A.G,Chang kyun Kim,"Traffic-flow-prediction system based on upstream traffic [A]". Vehical Navigation and Information System Conference [C],1994.

[6] Ben-Akiva. M., Cascetta. E. and Whittaker. J., "Recent progress in short-range traffic prediction." Compendium of Technical Papers, Institute of Transportation Engineers (ITE), pp. 262-265,1993.

[7] D. C. Sansom, T. Downs and T. K. Saha, "Evaluation of Support Vector Machine Based Forecasting Tool in Electricity Price Forecasting For Australian National Electricity Market Participants", in Proceedings of Australasian Universities Power Engineering Conference, 2002.

[8] J.W.C. van Lint, S.P. Hoogendoorn and H.J. van Zuylen, "Robust and adaptive travel time prediction with neural networks," proceedings of the 6th annual TRAIL Congress (part 2), December 2000.

[9] Mangasarian, O.L. and Musicant, D.R., "Lagrangian Support Vector Machines", Journal of Machine Learnlng Research, Jan.2000, pp. 161-177.

[10] Suykens, J.A.K. and Vandewalle, J.,"Least Squares Support Vector Machine Classifiers", Neural Process Letter. Sept. 1999, pp.293-300.

[11] Junshui Ma, "Accurate On-line Support Vector Regression", Neural Computation, Vol.15, 2003, pp2683~2703

[12] G.H. Golub, C.F. Van Loan, Matrix Computations, Baltimore MD: Johns Hopkins University Press, 3th edition, 1996.

Short Term Traffic Flow Prediction Using Hybrid ARIMA and ANN Models

Dehuai Zeng[1,2], Jianmin Xu[1], Jianwei Gu[3] , Liyan Liu[3] , Gang Xu[2]

1 School of Civil Engineering and Transportation,
South China University of China, Guangzhou, 510640, China
2 Institute of Intelligent Technology, Shenzhen University, Shenzhen, 518060, China
3 Guangzhou Post and Telecom Equipment Co. LTD., Guangzhou, 510663, China

Abstract

According to the complexity of the traffic historical data and the randomness of a lot of uncertain factors influence, a hybrid predicting model that combines both Autoregressive Integrated Moving Average (ARIMA) and Multilayer Artificial Neural Network (MLANN) is proposed in this paper. ARIMA is suitable for linear prediction and MLFNN is suitable for nonlinear prediction. This paper also investigates the issue on how to effectively model short term traffic flow time series with a new algorithm, which estimates the weights of the MLFNN and the parameters of ARMA model. Experimental results with real data sets indicate that the combined model can be an effective way to improve forecasting accuracy achieved by either of the models used separately.

*Keywords: **Traffic flow prediction, MLFNN, time series, ARIMA model, hybrid model***

1. Introduction

In recent years, Intelligent Transportation Systems (ITS) have achieved great developments. There are three kinds of data in transportation systems, which are historical data, real-time data and short-term forecasting data. The ability to predict traffic variables such as speed, travel time or flow, based on real time data and historic data, collected by various systems in transportation networks, is vital to the ITS components such as in-vehicle route guidance systems (RGS), advanced traveler information systems (ATIS), and advanced traffic management systems (ATMS). So short-term traffic flow forecasting, which is to determine the traffic flow in the next time interval usually in the range of ten minutes to half an hour, is one of the important problems in the research area of ITS.

A considerable amount of effort has been expended on short-term traffic flow forecasting and some models are proposed, such as random walk, historical average, time series models (including ARIMA, seasonal ARIMA), Kalman filter theory, neural network approaches, non-parametric methods, simulation models, local regression models and layered models known as the ATHENA model and the KARIMA model [1]-[6]. At present, the research on dynamic traffic prediction is still in development, not constructing a mature theories system.

ARIMA plays an important role in system identification, signal restoration and anomaly detection, which is required to solve such systems in real time [7]. Unfortunately, ARIMA methods are not capable of accurately forecasting the traffic flow time series as it is based on the theory of stationary stochastic processes and follow normality assumption. The universal approximation capability of a neural network is one of the most exciting propertied and has potentials for applications to problems such as system identification, signal processing, prediction, control, and pattern recognition [8]. A multilayer artificial neural network can approximate any nonlinear continuous function to an arbitrary accuracy. Neural Network needs very few assumptions to learn patterns of input variables for prediction purpose. But Zhang et al. [14] pointed out that the artificial neural network model really had advantages while dealing with a large amount of historical load data with non-linear characteristic, but he neglected the linear relations including the data. Therefore, one promising approach to solve such problems is to employ hybrid ARIMA-ANN model. ARIMA can deal with the linear part in the historical load data, the nonlinear part of historical load data is treated with ANN model.

The main purpose of this paper is to review known techniques involving the use of ARMA and MLFNN, and to propose an algorithm which requires less processing time and offer improved computational performance.

2. ARIMA Time Series Prediction Model

An Autogreesive Moving Average ARMA(*p*,*q*) is defined by [9,10]

$$y_t = \sum_{i=1}^{p} \phi_i y_{t-i} + \sum_{j=0}^{q} \theta_j \varepsilon_{t-j} \qquad (1)$$

Where y_t is the time seriers, ε_t is a purely random process with mean zero and variance σ^2 ,

978-0-7695-3342-1/08 $25.00 © 2008 IEEE
DOI 10.1109/PEITS.2008.135

$y_{t-i}, i = 1, \cdots, p$ are time lagged values of the time series and $\phi_1 \cdots \phi_p, \theta_1 \cdots \theta_q$ are the model parameters.

One extension to the ARMA (p,q) class of processed, which greatly enhance their value as empirical descriptors of non-stationary time series, is the class of autoregressive-integrated-moving average. Non stationary time-series process can be transformed by differencing the series one or more than one times, to make them stationary. The number of times d that the integrated process must be differenced to be stationary is series y_t is called an autoregressive-integrated-moving average process of order (p,d,q) and denoted by ARIMA (p,d,q).

ARIMA processes were popularized [11] in the early 1970s; as a result, ARIMA processes are mostly known as Box-Jenkins models. The ARIMA approach to forecasting is based on the following ideas: 1) The forecasts are based on linear functions of the sample observations; 2) The aim is to find the simplest models that provide an adequate description of the observed data. The ARIMA (p,d,q) for a stationary time series process predict values of a dependent time series with a linear combination of its own past values and past errors described by the following equation.

$$\phi(B)\nabla^d y_t = \theta(B)\varepsilon_t \qquad (2)$$

Where $\phi(B)$ and $\theta(B)$ are the polynomials of degree p and q.

$$\phi(B) = 1 - \phi_1 B - \phi_2 B^2 - \cdots - \phi_p B^p$$

$$\theta(B) = 1 - \theta_1 B - \theta_2 B^2 - \cdots - \theta_q B^q$$

Where p is the autoregressive order, d is the degree of differencing involved, q is the moving average order, B is the lag operator which gives the previous value of the series when placed in front of any variable with a time subscript:

$$B(y_t) = y_{t-1}$$

$$(1-B)^d y_t = y_t - y_{t-d} = \nabla^d y_t$$

Where   is known as the backward difference operator.

Fitting an ARIMA process to an observed time series proceeds in the following four stages: a)Model Identification, b)Estimation of parameters, c)Model diagnostic checking, d)Forecast verifications.

a) Model Identification

To the data of time series, y_t model identified is through sample relevant function (ACF), lean relevant function (PACF) and application to ARMA. The concrete steps are:

Step1: if ACF $\{\hat{\rho}_s, 1 \leq s \leq m\}$ fluctuates slightly from head to foot in zero after some operation step, it can judge y_t promptly in obedience to MA (q) model and roughly confirm the orders of q in MA (q).

Step2: Because PACF $\{\hat{\rho}_s, s \geq 1\}$ of AR(p) model dose not cut end, when $s \geq q, \hat{\phi}_s$ is obeying asymptotic normal distribution, N(0,1/N) is similar to MA(q) model test, so AR(p) model decides PACF. Formula of PACF is:

$$PACF(n+1) = ACF(n+1)$$
$$-[ACF(n+1)]^{n+1}/(1-[ACF(n)]^{n+1}) \qquad (3)$$

Step3: to hybrid model ARMA (p, q), both ACF and PACF can't confirm p, q value of ARMA model alone. Commonly, model is fitted from low order to high order. To time series ARMA(p, q), ACF and PACF can't assure the orders, there are three kinds of the methods mainly: ACF and PACF, FPE criterion; AIC criterion. AIC criterion is used in this paper:

$$AIC(n,m) = In\hat{\delta}_a^2 + 2(m+n+1)/N \qquad (4)$$

Where $AIC(n,m) = \min_{1 \leq n, m \leq L} AIC(n,m)$

From low-order to high-order the different value of p, q is set up the model respectively and parameter estimate. Comparing with AIC value of each model, it makes minimum optimization.

b) Estimation parameters

After identifying the ARIMA model parameters of the process are estimated. ARIMA process parameter estimation determines model coefficients through software application of least squares and maximum likelihood methods.

c) Model diagnostic checking

The third step in Box-Jenkins model building is to check the model accuracy using diagnostics tests. After chosen a particular ARIMA model and having estimated its parameters, we next see whether the chosen model fits the data reasonably well. For diagnostic checking of white noise of a time series model we used Q-statistic.

d)Forecast verifications

Using graphs, simple statistics and confidence intervals determine the validity of forecasts and track model performance to detect out of control situations.

3 ARIMA-MLANN hybrid traffic flow prediction model

To quantities of historical load data, none of them is a universal model that is suitable for all circumstances. ARIMA model to complex nonlinear problems may not be adequate. On the other hand, using *ANN* model to linear problems has yielded mixed results. For example, Markham and Rakes (1998) found that the performance of *ANN* for linear regression problems depends on the sample size and noise level. If ARMA and *ANN* are combined, different parts of data are treated with advantage of difficult models, it can improve forecast accuracy. A method is taken into account, namely ARMA linear and neural networks nonlinear. The ARMA-BPNN hybrid model consists of MLANN in series with an ARIMA model as shown in Fig. 1[12,13].

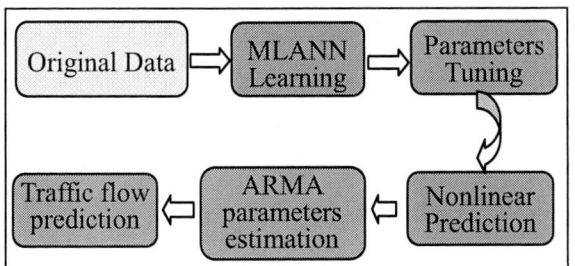

Fig.1 Hybrid traffic flow prediction model

In this paper the focus was feed-forward neural network in which input nodes are connected forward to each and every node in hidden layers, until they reach the output layer. Consider a feed forward network with one hidden layer. Fig. 2 shows the structure of the MLANN which is a fully interconnected layered feed forward network and more than one hidden layer can be constructed

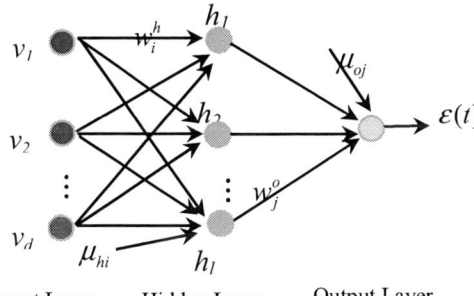

Input Layer   Hidden Layer   Output Layer

Fig.2 Structure of Multilayer feed forward network

The input layer is represented by $v(t)$, the hidden layer is represented by a vector $l = (h_1, h_2, \cdots, h_l)$ and output layer is represented by $\varepsilon(t)$

The output of the jth hidden unit is obtained by first forming a weighted linear combination of the 'd' input values and adding a bias. The activation of hidden unit 'j' can be obtained by transforming the linear sum using a logistic activation function $g(\cdot)$

$$h_j(t) = g(w_i^h v(t) + \mu_{hi}) \quad (5)$$

Where w_i^h are the weights on connections from the input node and μ_i^h are the weights on connections from the bias unit. For the output layer, the node is defined as

$$a_k(t) = \sum_{j=1}^{L} w_j^o h_j(t) + \mu_{oj}) \quad (6)$$

$$\varepsilon(t) = f(\sum_{j=1}^{L} w_j^o g(w_i^h v(t) + \mu_{hi}) + \mu_{oj}) \quad (7)$$

Where L is the number of the neurons at the hidden layer, w_j^o are the weights on the connection from the jth hidden node to the output, μ_{oj} is the weight on the connection from the bias to the output node. $f(\cdot)$ is a nonlinear activation function, if it is linear then formula(7) can be rewritten as:

$$\varepsilon(t) = \sum_{j=1}^{L} w_j^o g(w_i^h v(t) + \mu_{hi}) + \mu_{oj} \quad (8)$$

If the connection from input layer to the output layer is directed and the activation function of the output layer is linear, then the network becomes

$$\varepsilon(t) = \sum_{i=1 \to c}^{i} \alpha_{ic} x_i + \sum_{j=1}^{L} w_j^o g(w_i^h v(t) + \mu_{hi}) + \mu_{oj} \quad (9)$$

Considering Fig. 1, the objective is to develop a recursive algorithm by which one can adjust the weights of MLANN and the parameters of the ARMA model so that the application of a set of inputs produces a desired set of outputs.

Trying to use the BP to adjust the weights of BPNN, it is a gradient procedure that first computes the gradient E of the error function with respect to each weight of the network. It tell us how a small change in the weight will affect the over all error E. One knows what the desired output $y(t)$ of the system is, but the desired output $\varepsilon(t)$ of MLANN is unknown. Intuitively, the error at the output of MLANN must be related to the error at the output of the ARMA model. BP is the most commonly used learning algorithm. This idea is used to derive the updating equations for the weights and biases of BPNN which can be accomplished by defining the following error measure at the output of the ARMA model:

$$E(t) = \frac{1}{2}(y_d(t) - y(t))^2 \quad (10)$$

Where $y_d(t)$ is the desired output of the system, and $y(t)$ is the output of the hybrid model. Replacing $y(t)$ in formula (10) by formula (1), one gets

$$E(t) = \frac{1}{2}(y_d(t) - \sum_{i=1}^{p} \phi_i y_{t-i} - \sum_{j=0}^{q} \theta_j \varepsilon_{t-j})^2 \quad (11)$$

$$= \frac{1}{2}(y_d(t) - \phi_1 y_{t-1} - \cdots - \phi_p y_{t-p} - \theta_0 \varepsilon_t - \cdots - \theta_q \varepsilon_{t-q})^2$$

The input of the ARMA model at time t is ε_t, which is the output of BPNN, ε_t is replaced by formula (7), Thus, $E(t)$ can be written as

$$E(t) = \frac{1}{2}(y_d(t) - \phi_1 y_{t-1} - \cdots - \phi_p y_{t-p}$$
$$- \theta_0 f(\sum_{j=1}^{L} w_j^o h_j(t) + \mu_{oj}) - \cdots - \theta_q \varepsilon_{t-q})^2 \quad (12)$$

To minimize the error $E(t)$, the weights of the output layer w_j^o and the bias μ_{oj} should be updated in the negative direction of the gradient of $E(t)$, $\nabla E(t)$. If activation function $f(\cdot)$ is a linear, the gradient of $E(t)$ with respect to the weight w_j^o and the bias μ_{oj}

$$\nabla E(t) = \frac{\partial E(t)}{\partial w_j^o} = -(y_d(t) - y(t)) \cdot \theta_0 \frac{\partial}{\partial w_j^o} (\sum_{j=1}^{L} w_j^o h_j(t) + \mu_{oj}) \quad (13)$$
$$= -\theta_0 (y_d(t) - y(t)) h_j(t)$$

Thus, the weight and bias of the output layer of MLANN are updated according to the following rules:

$$\frac{\partial E(t)}{\partial \mu_{oj}} = -\theta_0 (y_d(t) - y(t)) \quad (14)$$

$$w_j^o(t+1) = w_j^o(t) + \lambda \theta_0 (y_d(t) - y(t)) h_j(t) \quad (15)$$

$$\mu_{oj}(t+1) = \mu_{oj}(t) + \lambda \theta_0 (y_d(t) - y(t)) \quad (16)$$

Where λ is the learning rate. Since the error at the output of MLANN has been determined, this error can be propagated backward to the hidden layer and the BP algorithm can be used to update the weights and the biases on the hidden layer which are given by

$$w_i^h(t+1) = w_i^h(t) + \lambda \delta_i^h(t) v(t) \quad (17)$$

$$\mu_{hi}(t+1) = \mu_{hi}(t) + \lambda \delta_i^h(t) \quad (18)$$

Where

$$\delta_i^h(t) = \hat{g}(w_i^h v(t) + \mu_{hi}) \theta_0 (y_d(t) - y(t)) w_i^h \quad (19)$$

Where $g(\cdot)$ is the derivation of the activation function of the hidden neurons.

In summary, the proposed methodology of the hybrid system consists of two steps. In the first step, a MLANN model is used to analyze the nonlinear part of traffic flow time series. In the second step, an ARIMA model is developed to model the residuals from the ANN model. Since the BPNN model cannot capture the linear structure of the data, the residuals of nonlinear model will contain information about the linearity. The results from the neural network can be used as predictions of the error terms for the ARIMA model. The hybrid model exploits the unique feature and strength of ARIMA model as well as MLANN model in determining different patterns. Thus, it could be advantageous to model linear and nonlinear patterns separately by using different models and then combine the predictions to improve the overall modeling and predicting performance.

4 Experiment results and discussion

4.1 Evaluation index of prediction result

To evaluate its adaptability to different predict methods, three error indicators: Relative Mean Errors (RME), Mean Absolute Relative Error (MARE) and Root Mean Squared Errors (RMSE) are applied as performance indices.

$$rmerr = \frac{1}{N} \sum_{i=1}^{N} \frac{y_i - y_i^*}{y_i}$$

$$marerr = \frac{1}{N} \sum_{i=1}^{N} \frac{|y_i - y_i^*|}{y_i}$$

$$rmserr = \sqrt{\frac{1}{N} \sum_{i=1}^{N} \left(\frac{y_i - y_i^*}{y_i} \right)^2}$$

Where y_i is the actual value and y_i^* is the predicted value

4.2 Data set preparation

Traffic prediction can be differentiated by the data source used in the predicting process. The key issue in predicting is how to make use of these sources of information. In a real-time case, data from real-time collection are used to make continuous and simultaneous predictions for several minutes in future. We focused on the traffic flow prediction in common work days because the traffic flow in work days was much larger than that in weekend. Since traffic data may be missed or corrupted, we select a better portion of the dataset that covers a 45-km stretch of a busy section of the Guangyuan Highway in Guangzhou. The traffic flow data is available from a Web server of Guangzhou traffic monitoring center. Thus the previous traffic flow sequence data can be used to predict the future traffic flow.

The traffic flow data were sampled per 8 mins from 7:00 a.m. to 7:00 p.m. in every day. To assess the forecasting performance of different models, different model is adopted to predict the same datum respectively using matlab simulation in this paper. Each data set is divided into two samples of training and testing. The training data set is used exclusively for model development and then the test sample is used to evaluate the established model.

4.3 Results

Tab.1 Performance comparison of different predictor

Predictor	Evaluation Indices		
	rmerr%	marerr%	rmsrerr%
ARIMA	0.92	4.26	12.44
BPNN	0.89	3.94	11.64
Hybrid	0.58	2.34	5.68

According to ARIMA definitely traffic data is fitted, a subset autoregressive model ARIMA(2,3,2) has been found to be the most parsimonious among all ARIMA models that are also found adequate judged by the residual analysis. The neural model used is a $3 \times 4 \times 1$ network. Table 1 gives the forecasting results for the sunspot data. Results show that while applying neural networks alone can improve the forecasting accuracy over the ARIMA model in the 39-period horizon, the performance of MLANN is getting worse as time horizon extends to 63 periods. This may suggest that neither the neural network nor the ARIMA model captures all of the patterns in the data. The results of the hybrid model show that by combining two models together, the overall forecasting errors can be significantly reduced except for the 39-period forecasting. Compared different predictor RMSR, we can see predicting accuracy of hybrid model increases by 46% than that of ARIMA model.

The comparison between the actual value and the forecast value for the 63 points out-of-sample is given in Fig. 3. Although at some data points, the hybrid model gives worse predictions than either ARIMA or ANN forecasts, its overall forecasting capability is improved.

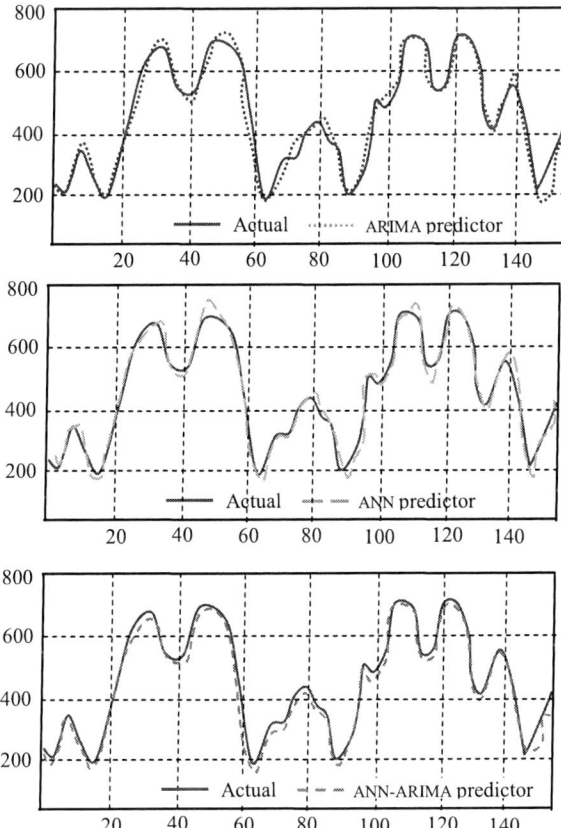

Fig. 3. Comparisons of predicted travel flow using different model.

5 Conclusions

In this paper, we propose to take a combining approach to traffic flow forecasting. The hybrid model takes advantage of the unique strength of ARIMA and ANN in linear and nonlinear modeling. ARIMA and ANN have different roles for a large amount of historical load data. The linear ARIMA model and the nonlinear ANN model are used jointly, aiming to capture different forms of relationship in the traffic flow time series data. For complex problems that have both linear and nonlinear

correlation structures, the combination method can be an effective way to improve forecasting performance. The experiment results demonstrate that the proposed hybrid model is able to outperform each component model used in isolation.

Acknowledgement

This work is supported by National High-Tech Plan (863) with grand No.2006AA11Z211, Natural Science Foundation of China with grand No. 50578064, and China Postdoctor Foundation with grand No. 20070410827.

References

[1] B. M. William, "Modeling and Forecasting Vehicular Traffic Flow as a Seasonal Stochastic Time Series Process," Doctoral Dissertation, University of Virginia, Charlottesville, 1999.

[2] Changshui Zhang, Shiliang Sun, " A Bayesian Network Approach to Time Series Forecasting of Short-Term Traffic Flows," The 7th Int. IEEE Conf. on Intelligent Transportation Systems, 3-6 Oct., 2004, pp. 216-221.

[3] Guoqiang Yu, Changshui Zhang, "Switching ARIMA Model Based Forecasting for Traffic Flow," IEEE Int. Conf. on Acoustics, Speech, and Signal Processing, vol. 2, 17-21 May, 2004, pp. 429-432.

[4] I. Okutani and Y. J. Stephanedes, " Dynamic Prediction of Traffic Volume through Kalman Filer Theory," Transportation Rearch, Part B, vol. 18B, 1-11, 1984.

[5] J. Hall and P. Mars, "The Limitations of Artificial Neural Networks for Traffic Prediction," 3rd IEEE Symp. on Computers and Communications, Proceedings, 1993, pp. 8-12.

[6] B. L. Smith and M. J. Demetsky, "Traffic flow forecasting: comparison of modeling approaches," ACSE Journal of Transportation Engineering, vol. 123, no. 4, 1997, pp. 261-266.

[7] Chising Leung, Kwokwo Wong, Puifai Sum, A Pruning Method for the Recursive Least Squared Algorithm, Neural Networks, 14, 2001: pp. 147-154.

[8] M. M. Gupta, L. Jin, and N. Homma, Static and Dynamic Neural Networks, New York: Wiley, 2003.

[9] J. Johnston and J. DiNardo, *Econometric methods*, 4th ed. Singapore: McGraw-Hill, 1997.

[10] Baibing Li, Bart De Moor, Identification of Influential Observations on Total Least Squares Estimate, Linear Algebra and Its Applications, 348, 2002: pp. 23-39.

[11] Box, G.E.P and Jenkins C.M.(1970). Time series analysis, forecasting and control, Holdenday, San Francisco.

[12] Fatima, S. Hussain, G., Statistical models of KSE100 index using Hybrid Financial Systems, Proc. of IEEE Int. Conf. on Engineering of Intelligent Systems,2006, pp:1-6

[13] Tianqi Yang, A Time Series Data Mining Based on ARMA and MLFNN Model for Intrusion Detection, Journal of Communication and Computer, Vol.3, No.7, 2006, pp16-22

[14] ZHANG G P.Time Series Forecasting Using a Hybrid ARIMA and Neural Network Model[J].Neurocomputing,2003,50(1):159-175.

Author Index

Abd, Mohammed K.	252	Dong, Xianyuan	430
Abdullah, Abdullah H.	252	Du, Rong-Hua	167
Abod, Afaneen A.	252	Duan, Baoqian	339
An, Xiangjing	476	Enen, Ren	582
Ao, Gu-Chang	496	Fan, Liguo	227
Cai, Qi	587	Fang, Liying	276
Cai, Qing-Lin	565	Feng, Xiao-Rong	195
Cao, BingGang	131, 136	Feng, Xing-Jie	195
Cao, Jialin	345	Feng, Zhongxiang	430
Cao, Sheng	578, 188	Fu, Jun-Jie	296
Chen, Baichao	110	Fu, Meichen	291
Chen, Chu	515	Fu, Qing	360, 366
Chen, Dandan	384	Gao, Jingju	561
Chen, Jun	423	Gao, Kunpeng	430
Chen, Ming	370	Gao, Zhang	315
Chen, Peng	454, 315	Ge, Xing	492
Chen, Qun	500	Geng, Jianyan	33
Chen, Shuyu	546	Gong, Fengxun	320
Chen, Xiao-Bo	305	Gong, SongJie	248
Chen, Xiaogao	360, 366	Gu, Jianwei	616, 621
Chen, Yimin	345	Gu, Wanyi	77
Chen, Yun	301	Guan, Pei-Lan	535
Chen, Zheng	578, 188	Guo, Chaofeng	561
Cheng, Fengyu	345	Guo, Jinxu	161
Cheng, Xian-Yi	305	Guo, Wang	582
Ci, Tiejun	394	Guo, Xiucheng	505
Cui, BingMou	574	Guo, Xiu-Cheng	511
Cui, Xun-Xue	520	Han, Chunhua	481
Dai, Wenjin	370, 375	Hang, Song	561
Deng, Jun	199	Hassanli, Kourosh	284
Deng, Qing-Hua	56	He, Hangen	476
Deng, Wei	467	He, Jian-Ming	511
Ding, Lei	127	He, Shaohua	592

Author Index

He, Yang 18
He, Yanxiang 181
He, Yinglei 366
He, Yuyao 237
Hou, Zhongsheng 122
Hua, Jiwei 8
Huang, Chunfeng 38
Huang, Fu-Yuan 242
Huang, Yehua 444
Huang, Zhenyue 113
Jia, Yuan-Hua 496
Jiang, Guoqiang 233
Jiang, Nan 72
Jin, Cong 48, 52, 56
Jin, Shangtai 122
Jin, Shu-Wei 48, 52
Jing, Teng 203
Kou, Hongjuan 157
Lee, Jian 496
Lei, Zhaoming 8
Leng, Zhaoxia 88
Li, Chao 18
Li, Di 83
Li, Guo 153
Li, Hengwen 345
Li, Jian 454
Li, Jianlin 349
Li, Jing 546
Li, Jingyu 485
Li, Jue 569
Li, Jun 551, 556
Li, KePing 419
Li, Lei 574

Li, Li 462
Li, Meixia 105
Li, Minxia 556
Li, Qiqiang 485
Li, Shi-Ting 587
Li, Tianze 310
Li, Xiangbin 153
Li, Xiangfeng 366
Li, Xiaoying 380
Li, Xueren 153
Li, Yabin 394
Li, Ying-Feng 175
Li, Yong 42
Li, Yong-Gang 296
Li, Zhenmei 310
Li, Zhihong 551, 556
Li, Zhizhong 100
Liang, Jun 305
Liang, Tao 8
Liao, Zhiling 113
Lin, Shuyong 143
Liu, Daxue 476
Liu, Fuqiang 3
Liu, Haode 458
Liu, Jian 88, 100, 399
Liu, Jing 430
Liu, Jun 56
Liu, Liyan 616, 621
Liu, Qi 520
Liu, Qiang 296
Liu, Qingchao 339
Liu, Qingfeng 88
Liu, Shu 266

Author Index

Liu, Shuangxi	355	Ren, Wuling	223
Liu, Xu	23	Rong, Jian	214
Liu, Ying	419	Saribulut, Lütfü	411
Liu, Yongqiang	12	Shahhoseini, Hadishahriar	284
Liu, Zhaohui	481	Shang, Junna	171
Liu, Zhaowei	209	Shangguan, Lixian	149
Liu, Zhiming	454	Shen, Jin	310
Long, Bo	131, 136	Shi, Bin	276
Lu, Bo	161	Shi, Feng	500
Lu, Jian	492	Shi, Yujie	404
Lv, Zhaorui	389	Shi, Zhong-Ke	175
Mahani, Ali Khayatzadeh	284	Shi, Zongying	266
Mao, Ling	467	Sun, Hexu	8
Miao, Jianming	280	Sun, Jian	419
Miao, Li-Xin	500	Sun, Jifeng	149
Mingxing, Tian	582	Sun, Xuekang	77
Mo, Yikui	199	Sun, Zhenping	476
Ni, You-Cong	541	Tan, Jingxing	67
Ni, Yunfeng	100	Tang, Dan	578, 188
Nie, Deming	345	Tang, Shoupeng	72
Ning, Li Bing	608	Tang, Yun	339
Ouyang, Yuanxin	18	Tao, Shaohua	561
Pan, Ane	569	Tao, Yunxin	272
Peng, Xiaoming	181	Teimoury, Ebrahim	284
Peng, Yanxia	613	Teng, Jing	458
Peng, YunYan	257	Tian, Li	339, 181
Pi, Dechang	272	Tu, Sheng-Wen	511
Ping, Yi	462	Tümay, Mehmet	411
Qi, Luo	529, 408	Wang, Bo	12
Qin, Jin	500	Wang, Chen	423
Qiu, Guo-Xin	520	Wang, Chuanqi	42
Qu, Zhaoyang	38	Wang, Cuiru	355
Quan, Yu	462	Wang, Dan	42

Author Index

Wang, Haifeng 384
Wang, Heng 419
Wang, Huamin 88
Wang, Jiang-Tao 601
Wang, Jie 233
Wang, Jijing 301
Wang, Jingyuan 199
Wang, Jiuhe 117
Wang, Mingsheng 525
Wang, Peng-Ying 492
Wang, Pu 276
Wang, Renying 301
Wang, Rongbin 546
Wang, Sen 100
Wang, Shuqing 291
Wang, Weiping 404
Wang, Xiaojing 578, 188
Wang, Xinggui 380, 127
Wang, Xinhong 3
Wang, Xiu 214
Wang, Yong-Hua 565
Wang, Yu 375
Wei, Jinli 472
Wei, Jin-Li 440
Wei, Lianyu 209
Wei, Peiyu 310
Wei, Shaoliang 345
Weitao, Chu 324
Wen, Chunxue 349
Wen, Hang 27
Wen, Jing 541
Wen, Ya 435
Wu, Lan 505

Wu, Qi 218
Wu, William 12
Wu, Xiao-Jun 535
Wu, Xiaomeng 399
Wu, Zhengguo 389
Xia, Li 384, 389
Xia, Peirong 117
Xiang, Qiao-Jun 492
Xie, Hui 435
Xing, Li-Jun 520
Xiong, Ning 257
Xiong, Zhang 18
Xu, Bing 127
Xu, Gang 616, 621
Xu, Hong 94
Xu, Honghua 349
Xu, Jianmin 616, 621
Xu, Jingqi 458
Xu, Shiyan 237
Xu, Shiyu 171
Xu, Wenli 266
Xu, Xiangzheng 110
Xu, Yong 515
Xu, Zhujun 223
Yan, Bo 444
Yan, Hong-Sen 218
Yan, Jianzhuo 276
Yan, Kefei 435
Yan, Suli 399
Yang, Bian 462
Yang, Dianbing 404
Yang, Fan 592
Yang, Hong-Bing 218

Author Index

Yang, HuPing	257	Zhang, Qisen	481
Yang, Kuihe	261	Zhang, Quan	280
Yang, Ming	505	Zhang, Rui	315
Yang, Qing	569	Zhang, Shuguang	157
Yang, Xiaoguang	72	Zhang, Wensheng	525
Yang, Xiaoling	94	Zhang, Xinhua	113
Yang, Zhifang	332, 328	Zhang, Yilin	3
Yao, Yubin	42	Zhang, Yingxue	481
Yin, Liang	67	Zhang, Zhiliang	320
Yin, Xiaofeng	67	Zhao, Kai	541
Ying, Shi	541	Zhao, Lingling	261
Yu, Changjun	23	Zhao, Lu	291
Yu, Shijie	360	Zhao, Mingguo	266
Yu, Songsen	565	Zhao, Weiquan	83
Zeng, Dehuai	616, 621	Zhao, Yan-Jun	296
Zeng, Jian-Qin	520	Zhao, Zhijin	171
Zhai, Yikui	143	Zheng, Jianbin	161
Zhan, Yiju	565	Zheng, Jianhu	449
Zhang, Bing	467	Zheng, Xinqi	291
Zhang, Dening	157	Zhong, Xiaochum	214
Zhang, Haitao	214	Zhou, HaoBin	131, 136
Zhang, Jinlong	117	Zhou, Hong	601
Zhang, Jiuhua	61	Zhou, Jinchuan	105
Zhang, Kai	266	Zhou, Longhua	360
Zhang, Lijiang	33	Zhou, Zhi-Na	175
Zhang, Ling	515	Zhu, Bo	587
Zhang, Lin-Lin	541	Zhu, Qunxiong	94
Zhang, Meng-Meng	440, 472	Zhu, Xiaoguang	349
Zhang, Qiong	143	Zuo, Feng	227

IEEE Computer Society Conference Publications Operations Committee

CPOC Chair
Chita R. Das
Professor, Penn State University

Board Members
Mike Hinchey, *Director, Software Engineering Lab, NASA Goddard*
Paolo Montuschi, *Professor, Politecnico di Torino*
Jeffrey Voas, *Director, Systems Assurance Technologies, SAIC*
Suzanne A. Wagner, *Manager, Conference Business Operations*
Wenping Wang, *Associate Professor, University of Hong Kong*

IEEE Computer Society Executive Staff
Angela Burgess, *Executive Director*
Alicia Stickley, *Senior Manager, Publishing Services*
Thomas Baldwin, *Senior Manager, Meetings & Conferences*

IEEE Computer Society Publications
The world-renowned IEEE Computer Society publishes, promotes, and distributes a wide variety of authoritative computer science and engineering texts. These books are available from most retail outlets. Visit the CS Store at *http://www.computer.org/portal/site/store/index.jsp* for a list of products.

IEEE Computer Society *Conference Publishing Services* (CPS)
The IEEE Computer Society produces conference publications for more than 250 acclaimed international conferences each year in a variety of formats, including books, CD-ROMs, USB Drives, and on-line publications. For information about the IEEE Computer Society's *Conference Publishing Services* (CPS), please e-mail: cps@computer.org or telephone +1-714-821-8380. Fax +1-714-761-1784. Additional information about *Conference Publishing Services* (CPS) can be accessed from our web site at: *http://www.computer.org/cps*

IEEE Computer Society / Wiley Partnership
The IEEE Computer Society and Wiley partnership allows the CS Press *Authored Book* program to produce a number of exciting new titles in areas of computer science and engineering with a special focus on software engineering. IEEE Computer Society members continue to receive a 15% discount on these titles when purchased through Wiley or at: *http://wiley.com/ieeecs*. To submit questions about the program or send proposals, please e-mail jwilson@computer.org or telephone +1-714-816-2112. Additional information regarding the Computer Society's authored book program can also be accessed from our web site at: *http://www.computer.org/portal/pages/ieeecs/publications/books/about.html*

Revised: 21 January 2008

CPS Online is our innovative online collaborative conference publishing system designed to speed the delivery of price quotations and provide conferences with real-time access to all of a project's publication materials during production, including the final papers. The **CPS Online** workspace gives a conference the opportunity to upload files through any Web browser, check status and scheduling on their project, make changes to the Table of Contents and Front Matter, approve editorial changes and proofs, and communicate with their CPS editor through discussion forums, chat tools, commenting tools and e-mail.

The following is the URL link to the **CPS Online** Publishing Inquiry Form:
http://www.ieeeconfpublishing.org/cpir/inquiry/cps_inquiry.html

IEEE
445 Hoes Lane
Piscataway, NJ 08854-4141

ISBN 978-0-7695-3342-1